C. Truesdell · W. Noll

The Non-Linear Field Theories of Mechanics

Second Edition with 28 Figures

Springer-Verlag
Berlin Heidelberg New York
London Paris Tokyo
Hong Kong Barcelona Budapest

Professor Clifford Truesdell
4007 Greenway, Baltimore, Maryland 21218, USA

Professor Walter Noll
Carnegie-Mellon University, Pittsburgh, Pennsylvania 15213, USA

ISBN 3-540-55098-4 Springer-Verlag Berlin Heidelberg NewYork
ISBN 0-378-55098-4 Springer-Verlag NewYork Berlin Heidelberg

Library of Congress Cataloging-in-Publication Data
Truesdell, C. (Clifford), 1919 – The non-linear field theories of mechanics / Clifford Truesdell, W. Noll. p. cm.
Includes bibliographical references and indexes.
ISBN 3-540-55098-4 (alk. paper) -- ISBN 0-387-55098-4 (alk paper)
1. Continuum mechanics. 2. Nonlinear mechanics. I. Noll, W. (Walter), 1925 – . II. Title. QA808.2.T69 1992 531--dc20
92-3947

Printing: Mercedes, Berlin;
Binding: Lüderitz & Bauer GmbH, Berlin
55/3020-5 4 3 2 1 0 Printed on acid-free paper.

Preface

This volume is a second, corrected edition of The Non-Linear Field Theories of Mechanics, which first appeared as Volume III/3 of the Encyclopaedia of Physics, 1965. Its principal aims were to replace the conceptual, terminological, and notational chaos that existed in the literature of the field by at least a modicum of order and coherence, and second, to describe, or at least to summarize, everything that was both known and worth knowing in the field at the time.

Inspecting the literature that has appeared since then, we conclude that the first aim was achieved to some degree. Many of the concepts, terms, and notations we introduced have become more or less standard, and thus communication among researchers in the field has been eased. On the other hand, some ill-chosen terms are still current. Examples are the use of "configuration" and "deformation" for what we should have called, and now call, "placement" and "transplacement", respectively. (To classify translations and rotations as deformations clashes too severely with the dictionary meaning of the latter.)

We believe that the second aim was largely achieved also. We have found little published before 1965 that should have been included in the treatise but was not. However, a large amount of relevant literature has appeared since 1965, some of it important. As a result, were the treatise to be written today, it should be very different.

On p. 12 of the Introduction we stated "... we have subordinated detail to importance and, above all, clarity and finality". We believe now that finality is much more elusive than it seemed at the time. The "General theory of material behavior" presented in Chapter C, although still useful, can no longer be regarded as the final word. The "Principle of Determinism for the Stress" stated on p. 56 has only limited scope. It should be replaced by a more inclusive principle, using the concept of "state" rather than a history of infinite duration, as a basic ingredient. In fact, forcing the theory of "materials of the rate type" into the general framework of the treatise as is done on p. 95 must now be regarded as artificial at best, and unworkable in general. This difficulty was alluded to in footnote 1 on p. 98 and in the discussion of B. Bernstein's concept of a "material" on p. 405. This major conceptual issue was first resolved in 1972 ["A New Mathematical Theory of Simple Materials" by W. Noll, Archive for Rational Mechanics and Analysis, Vol. 48, pp. 1–50], and then only for simple materials. The new concept of material makes it possible, also, to include theories of "plasticity" in the general framework, and one can now do much more than "refer the reader to the standard treatises", as we suggested on p. 11 of the Introduction.

Here is a list of topics that pertain to the field and have seen major development – since 1965:

The theory of large elastic deformations treated by the tools of modern non-linear analysis, especially those created to solve problems posed by elasticity, a pilot example.

Solutions of problems arising in the mechanics of fluids displaying non-linear response.

Application of modern analysis to the theory of materials with memory.

Mathematical theories intended to describe the behavior of liquid crystals.

Theories of materials with microstructure.

Theories of thermomechanics.

Here is a list of books appearing after 1965 on subjects cognate with NLFT. We regret having overlooked any that should have been included.

DAY, W. A., The Thermodynamics of Simple Materials with Fading Memory, Springer Tracts in Natural Philosophy, Vol. 22, 1972.

WANG, C.-C., & TRUESDELL, C., Introduction to Rational Elasticity, Leyden, Nordhoff, 1977.

WANG, C.-C., Mathematical Principles of Mechanics and Electromagnetism, Vol. 1. N. Y. & London, Plenum, 1979.

GURTIN, M. E., An Introduction to Continuum Mechanics, N. Y. etc., Academic Press, 1981.

OWEN, D. R., A First Course in the Mathematical Foundations of Thermodynamics, N. Y. etc., Springer, 1983.

TRUESDELL, C., Rational Thermodynamics, N. Y. etc., Springer, second edition, 1984.

KRAVIETZ, A., Materialtheorie, mathematische Beschreibung des phänomenologischen thermo-dynamischen Verhaltens, Heidelberg etc., Springer, 1986.

RENARDY, M., HRUSA, J., & NOHEL, J., Mathematical Problems in Viscoelasticity, Pittman Monographs and Surveys, Vol. 35, Longman, 1987.

CIARLET, P. G., Mathematical elasticity, Vol. 1, Amsterdam etc., North Holland, 1988.

CAPRIZ, G., Continua with Microstructure, Springer Tracts in Natural Philosophy, Vol. 35, 1989.

ERICKSEN, J. L., Introduction to the Thermodynamics of Solids, Chapman & Hall, 1991.

VIRGA, E. G., Book on Liquid Crystals, to appear in 1992.

Note on the corrigenda and addenda. Typographical errors and minor infelicities of style have been silently emended. The few inserted sentences in square brackets remark false conjectures, omitted conditions, and possibly confusing statements; some are also cancelled by rules and emphasized by new footnotes, marked with an asterisk.

We have not tried to bring the book up to date.

C. T. & W. N. 1992

Contents.

The Non-Linear Field Theories of Mechanics.

By

C. TRUESDELL and W. NOLL.

A. Introduction[1].

1. Purpose of the non-linear theories. Matter is commonly found in the form of materials. Analytical mechanics turned its back upon this fact, creating the centrally useful but abstract concepts of the mass point and the rigid body, in which matter manifests itself only through its inertia, independent of its constitution; "modern" physics likewise turns its back, since it concerns solely the small particles of matter, declining to face the problem of how a specimen made up of such particles will behave in the typical circumstances in which we meet it. Materials, however, continue to furnish the masses of matter we see and use from day to day: air, water, earth, flesh, wood, stone, steel, concrete, glass, rubber, ... All are *deformable*. A theory aiming to describe their mechanical behavior must take heed of their deformability and represent the definite principles it obeys.

The rational mechanics of materials was begun by JAMES BERNOULLI, illustrated with brilliant examples by EULER, and lifted to generality by CAUCHY. The work of these mathematicians divided the subject into two parts. First, there are the *general principles*, common to all media. A mathematical structure is necessary for describing deformation and flow. Within this structure, certain physical laws governing the motion of all finite masses are stated. These laws, expressed nowadays as integral equations of balance, or "conservation laws", are equivalent either to *field equations* or to *jump conditions*, depending on whether smooth or discontinuous circumstances are relevant. Specifically, the axioms of

[1] *Acknowledgment.* This treatise, while it covers the entire domain indicated by its title, emphasizes the reorganization of classical mechanics by NOLL and his associates. He laid down the outline followed here and wrote the first drafts of most sections in Chaps. B, C, and E and of a few in Chap. D. Among the places where he has given new results not published elsewhere, shorter proofs, or major simplifications of older ideas may be mentioned Sects. 22, 27, 30, 34, 44, 46, 52, 63, 64, 65, 68b, 83, 99, 100, 101, 106, 107, 112, 117, 119, 121, and 122. The larger part of the text was written by TRUESDELL, who also took the major share in searching the literature. While NOLL revised many of the sections drafted by TRUESDELL, it is the latter who prepared the final text and must take responsibility for such oversights, crudities, and errors as may remain.

Both authors express their gratitude to the U.S. National Science Foundation for partial support through research grants to The Johns Hopkins University and The Carnegie Institute of Technology. In addition, TRUESDELL was a fellow of that Foundation in 1960—1961, and his work was supported in part also by the U.S. National Bureau of Standards from 1959 to 1963. Several colleagues and friends, chief among them being B. D. COLEMAN, J. L. ERICKSEN, M. E. GURTIN, and R. A. TOUPIN, generously answered questions and criticized various parts of the manuscript in the successive revisions circulated privately since 1958. To Prof. Dr. K. ZOLLER and to Dr. C.-C. WANG we express our deep gratitude for their careful check of most of the calculations and for help in revision of the style, both in manuscript and on the proof sheets.

We record here also our heartfelt thanks to Springer-Verlag, which to the usual peerless quality of its work has added willingness to co-operate in our every wish, even so far as to tolerate an unprecedented measure of alteration and addition after the text had been set in type.

continuum physics assert the balance or conservation of *mass*, *linear momentum*, *moment of momentum*, *energy*, *electric charge*, and *magnetic flux*. There is a seventh law, a principle of *irreversibility*, expressed in terms of the *entropy*, but the true form of this law, in the generality we keep here, is not yet known[1]. The reader of this treatise is presumed to be familiar with these piers of continuum mechanics and also to have some competence in the classical linear or infinitesimal theories; such a reader will be able to follow our analysis, which we have attempted to keep self-contained. However, a detailed modern exposition of the general principles, *The Classical Field Theories*, by C. TRUESDELL and R. TOUPIN, with an Appendix on *Invariants* by J. L. ERICKSEN, has been published in Vol. III/1 of this Encyclopedia. Frequent references to sections and equations of that work, indicated by the prefix "CFT", are given so as to provide a helping hand at need; they do not imply that the reader of this treatise is expected to have read CFT or to keep it by him.

The general physical laws in themselves do not suffice to determine the deformation or motion of a body subject to given loading. Before a determinate problem can be formulated, it is usually necessary to specify the *material* of which the body is made. In the program of continuum mechanics, such specification is stated by *constitutive equations*, which relate the stress tensor and the heat-flux vector to the motion. For example, the classical theory of elasticity rests upon the assumption that the stress tensor at a point depends linearly on the changes of length and mutual angle suffered by elements at that point, reckoned from their configurations in a state where the external and internal forces vanish, while the classical theory of viscosity is based on the assumption that the stress tensor depends linearly on the instantaneous rates of change of length and mutual angle. These statements cannot be universal laws of nature, since they contradict one another. Rather, they are *definitions of ideal materials*. The former expresses in words the constitutive equation that defines a *linearly and infinitesimally elastic material*; the latter, a *linearly viscous fluid*. Each is consistent, at least to within certain restrictions, with the general principles of continuum mechanics, but in no way a consequence of them. There is no reason *a priori* why either should ever be physically valid, but it is an empirical fact, established by more than a century of test and comparison, that each does indeed represent much of the mechanical behavior of many natural substances of the most various origin, distribution, touch, color, sound, taste, smell, and molecular constitution. Neither represents all the attributes, or suffices even to predict all the mechanical behavior, of any one natural material. No natural body is perfectly elastic, or perfectly fluid, any more than any is perfectly rigid or perfectly incompressible. These trite observations do not lessen the worth of the two particular constitutive equations just mentioned. That worth is twofold: First, each represents in ideal form an *aspect*, and a different one, of the mechanical behavior of nearly all natural materials, and, second, each does predict with considerable though sometimes not sufficient accuracy the observed response of *many different natural materials in certain restricted situations*.

Pedantry and sectarianism aside, the aim of theoretical physics is to construct mathematical models such as to enable us, from use of knowledge gathered in a few observations, to predict by logical processes the outcomes in many other circumstances. Any logically sound theory satisfying this condition is a good theory, whether or not it be derived from "ultimate" or "fundamental" truth. It is as ridiculous to deride continuum physics because it is not obtained from

[1] Note added in proof: Major progress toward finding this law has been made by COLEMAN in work described in Sect. 96bis.

nuclear physics as it would be to reproach it with lack of foundation in the Bible. The conceptual success of the classical linear or infinitesimal field theories is perhaps the broadest we know in science: In terms of them we face, "explain", and in varying amount control, our daily environment: winds and tides, earthquakes and sounds, structures and mechanisms, sailing and flying, heat and light.

There remain, however, simple mechanical phenomena that are clearly outside the ranges of the infinitesimal theory of elasticity and of the linear theory of viscosity. For example, a rod of steel or rubber if twisted severely will lengthen in proportion to the square of the twist, and a paint or polymer in a rotating cup will climb up an axial rod. Moreover, the finite but discrete memory of the elastic material and the infinitesimal memory of the viscous fluid are obviously idealized limiting cases of the various kinds of cumulative memories that natural materials show in fast or slow or repeated loading or unloading, leading to the phenomena of creep, plastic flow, strain hardening, stress relaxation, fatigue, and failure.

2. Method and program of the non-linear theories. The *non-linear field theories* also rest upon constitutive equations defining ideal materials, but ideal materials more elaborate and various in their possible responses. Of course the aim is to represent and predict more accurately the behavior of natural materials, and in particular to bring within the range of theory the effects mentioned above, typical in nature but altogether wanting in the classical linear or infinitesimal theories.

Insofar as a constitutive equation, relating the stress tensor to the present and past motion, is laid down as defining an ideal material and is made the starting point for precise mathematical treatment, the methods of the linear and non-linear theories are the same, both in general terms and in respect to particular solutions yielding predictions to be compared with the results of experiment in certain definite tests, but in other ways they differ.

α) *Physical range.* When two different natural materials are brought out of the range in which their responses are approximately linearly elastic or linearly viscous, there is no reason to expect their mechanical behaviors to persist in being similar. Rubber, glass, and steel are all linearly elastic in small strain, but their several responses to large strain or to repeated strain differ from one another. It is easy to see mathematically that infinitely many non-linear constitutive equations, differing not only in quantity but also in quality, may have a common linear first approximation. Thus, both from theory and from physical experience, there is no reason to expect any one non-linear theory to apply properly to so large a variety of natural substances as do the classical linear or infinitesimal theories. Rather, each non-linear theory is designed *to predict more completely the behavior of a narrower class of natural materials.*

β) *Mathematical generality.* Because of the physical diversity just mentioned, it becomes wasteful to deal with special non-linear theories unless unavoidably necessary. To the extent that several theories may be treated simultaneously, they surely ought to be. The *maximum mathematical generality* consistent with concrete, definite physical interpretation is sought. The place held by material constants in the classical theories is taken over by material functions or functionals. It often turns out that simplicity follows also when a situation is stripped of the incidentals due to specialization. For example, the general theory of waves in elastic materials is not less definite but is physically easier to understand as well as mathematically easier to derive than is the second-order approximation to it, or any theory resulting from quadratic stress-strain relations.

γ) *Experiential basis*[1]. While laymen and philosophers of science often believe, contend, or at least hope, that physical theories are directly inferred from experiments, anyone who has faced the problem of discovering a good constitutive equation or anyone who has sought and found the historical origin of the successful field theories knows how childish is such a prejudice. The task of the theorist is to bring order into the chaos of the phenomena of nature, to invent a language by which a class of these phenomena can be described efficiently and simply. Here is the place for "intuition", and here the old preconception, common among natural philosophers, that nature is simple and elegant, has led to many great successes. Of course physical theory must be based on experience, but experiment comes after, rather than before, theory. Without theoretical concepts one would neither know what experiments to perform nor be able to interpret their outcome.

δ) *Mathematical method.* The structure of space and time appropriate to classical mechanics requires that certain *principles of invariance* be laid down. Alongside principles of invariance must be set up *principles of determinism,* asserting which phenomena are to be interconnected, and to what extent. In more popular but somewhat misleading terms, "causes" are to be related to "effects". Principles of these two kinds form the basis for the construction of constitutive equations. General properties of materials such as isotropy and fluidity are related to certain properties of invariance of the defining constitutive equations.

ε) *Product.* After suitably invariant principles of determinism are established, we are in a position to specialize intelligently if need be, but in some cases no further assumptions are wanted to get *definite solutions* corresponding to physically important circumstances. In addition to such solutions, absolutely necessary for connecting theory with experience and experiment, we often seek also *general theorems* giving a picture of the kind of physical response that is represented and serving also to interconnect various theories.

The physical phenomena these theories attempt to describe, while in part newly discovered, are mainly familiar. The reader who thinks that one has only to do experiments in order to know how materials behave and what is the correct theory to describe them would do well to consult a paper by BARUS[2], published in 1888. Most of the effects BARUS considered had been known for fifty to a hundred years, and he showed himself familiar with an already abundant growth of mathematical theories. That he interpreted his own sequences of experiments as confirming MAXWELL's theory of visco-elasticity has not put an end either to further experiments reaching different conclusions or to the creation of other theories, even for the restricted circumstances he considered. If the basic problem were essentially experimental, surely two hundred years of experiment could have been expected to bring better understanding of the mechanics of materials than in fact is had today.

This and many other examples have caused us to write the following treatise with an intent different from that customary in works on plasticity, rheology, strength of materials, etc. We do not attempt to fit theory to data, or to apply the results of experiment so as to confirm one theory and controvert another. Rather, just as the geometrical figure, the rigid body, and the perfect fluid afford simple, natural, and immediate *mathematical models* for some aspects of everyday experience, models whose relevance or application to each particular physical

---

[1] For further remarks on methods of formulating constitutive equations, see Sects. 293 and 3 of CFT.

[2] BARUS [1888, *1*].

situation must be determined by the user, we strive to find a rational ingress to more complex mechanical phenomena by setting up clear and plausible theories of material behavior, embodying various aspects of long experience with natural materials.

3. Structure theories and continuum theories[1]. Widespread is the misconception that those who formulate continuum theories believe matter "really is" continuous, denying the existence of molecules. This is not so. Continuum physics presumes *nothing* regarding the structure of matter. It confines itself to relations among gross[2] phenomena, neglecting the structure of the material on a smaller scale. Whether the continuum approach is justified, in any particular case, is a matter, not for the philosophy or methodology of science, but for ***experimental test.*** In order to test a theory intelligently, one must first find out what it predicts. Few of the current critics of continuum mechanics have taken so much trouble[3].

Continuum physics stands in no contradiction with structural theories, since the equations expressing its general principles may be identified with equations of exactly the same form in sufficiently general statistical mechanics[4]. If this identification is just, the variables that are basic in continuum mechanics may be regarded as averages or expected values of molecular actions.

It would be wrong, however, to infer that quantities occurring in continuum mechanics *must* be interpreted as certain particular averages. Long experience with molecular theories shows that quantities such as stress and heat flux are quite insensitive to molecular structure: Very different, apparently almost contradictory hypotheses of structure and definitions of gross variables based upon them, lead to the same equations for continua[5]. Over half a century ago, when molecular theories were simpler than they are today, POINCARÉ[6] wrote, "In most questions the analyst assumes, at the beginning of his calculations,

---

[1] Other remarks on this subject are given in Sect. 1 of CFT.

[2] The word "macro*scopic*" is often used but is misleading because the scale of the phenomena has nothing to do with whether or not they can be seen (σκοπεῖν). "Molar", the old antithesis to "molecular", is also a fit term to the extent that only massy materials are considered.

[3] What it is surely to be hoped is the high-water mark of logical confusion and bastard language has been reached in recent studies of the aerodynamics of rarefied gases, where the term "non-continuum flow" often refers to anything asserted to be incompatible with the Navier-Stokes equations. Even the better-informed authors in this field usually decide by *ex cathedra* pronouncement based on particular molecular concepts, rather than by experimental test, when continuum mechanics is to be regarded as applicable and when it is not.

[4] Cf. the recent work of DAHLER and SCRIVEN [1963, *19*] on the statistical mechanics of systems with structure: "Both approaches, continuum and statistical, yield the same macroscopic behaviour, regardless of the nature of the molecules and submolecular particles of which the physical system is composed."

[5] The structural theories of NAVIER are no longer considered correct by physicists, but the equations of linear viscosity and linear elasticity he derived from them have been confirmed by experiment and experience for a vast range of substances and circumstances. MAXWELL derived the Navier-Stokes equations from his kinetic theory by using, along with certain hypotheses, a definition of stress as being entirely an effect of transfer of molecular momentum, but experience shows the Navier-Stokes equations to be valid for many flows of many liquids, in which no one considers transfer of momentum as the main molecular explanation for stress. In recent work on the general theory of ensembles in phase space, different definitions of stress and heat flux as phase averages lead to identical field equations for them. Examples could be multiplied indefinitely.

Cf. TRUESDELL [1950, *16*, § 1]: "History teaches us that the conjectures of natural philosophers, though often positively proclaimed as natural laws, are subject to unforeseen revisions. Molecular hypotheses have come and gone, but the phenomenological equations of D'ALEMBERT, EULER, and CAUCHY remain exact as at the day of their discovery, exempt from fashion."

[6] POINCARÉ [1905, *2*, Ch. IX].

either that matter is continuous, or the reverse, that it is formed of atoms. In either case, his results would have been the same. On the atomic supposition he has a little more difficulty in obtaining them — that is all. If, then, experiment confirms his conclusions, will he suppose that he has proved, for example, the real existence of atoms?" While the logical basis of POINCARÉ'S statements remains firm, the evidence has changed. The reader of this treatise is not asked to question the "real" existence of atoms or subatomic particles. His attention is directed to phenomena where differences among such particles, as well as the details of their behavior, are unimportant. However, we cannot give him assurance that quantum mechanics or other theories of modern physics yield the same results. Any claim of this kind must await such time as physicists turn back to gross phenomena and demonstrate that their theories do in fact predict them, not merely "in principle" but also in terms accessible to calculation and experiment.

The relative position of statistical theories, engineering experiment, and the rational mechanics of continua was surveyed as follows by v. MISES[1] in 1930:

„Lassen Sie mich diesen kurzen Andeutungen zwei Schlußbemerkungen anfügen. Die eine wird nahegelegt durch die Weiterbildung, die die Grundlagen der Mechanik in jüngster Zeit auf Seiten der Physiker gefunden haben. Man wird nicht vermuten, daß die Modifikationen, zu denen die Relativitätstheorie oder die Wellenmechanik führen, für die hier von mir behandelten Probleme von Bedeutung sein könnten. Aber es steht anders mit der *Statistik*. Es ist denkbar, daß eine einigermaßen befriedigende Darstellung der typischen Erscheinungen an festen Körpern im Rahmen der Differentialgleichungs-Physik gar nicht möglich ist, daß es keine Ansätze gibt, die in Erweiterung oder Zusammenfassung der bisherigen das Charakteristische der bleibenden Formänderungsvorgänge richtig wiedergeben. In der Hydromechanik der turbulenten Bewegungen scheint es ja schon festzustehen, daß der statistische Grundzug der Erscheinung schon beim ersten Ansatz einer brauchbaren Theorie berücksichtigt werden muß. Fragt man aber, ob wir von der „statistischen Mechanik" her Hilfe für unsere Aufgaben erwarten dürfen, so sieht es damit wohl schlecht aus. Es zeigt sich ja umgekehrt, daß dort, wo man ganz unzweifelhaft mit statistischem Material zu tun hat, etwa in der Mechanik der Kolloide, das Beste, was überhaupt erreichbar ist, durch Herübernahme von Ansätzen aus der Mechanik der Kontinua gewonnen wird. — Die zweite Bemerkung kehrt an den Ausgangspunkt meines Berichtes zurück, das Verhältnis der Technik zu den Bestrebungen der rationellen Mechanik nach Aufklärung des mechanischen Verhaltens der wirklich beobachtbaren Körper. Kein Zweifel: Die Aufgaben der Materialprüfung drängen nach wenigstens vorläufigen praktischen Lösungen, und sie sucht sie in Ansätzen der hier beschriebenen Art, aber ohne rationelle Grundlage, meist ohne Kenntnis des Vorhandenen, unter ständig zunehmender Verwirrung der Begriffe und Bezeichnungen. Zahllose Aufsätze, die an neue experimentelle Feststellungen anknüpfen, suchen immer von neuem die Grundbegriffe zu definieren, Meßverfahren, ja Maßeinheiten für die Stoffeigenschaften einzuführen — es gibt wenigstens ein halbes Dutzend verschiedener Plastizitätsmesser — ohne jede vernünftige theoretische Grundlegung. Von amerikanischer Seite ist schon der Vorschlag gemacht worden, zur Klärung der Verhältnisse einen Ausschuß von Fachleuten einzusetzen. Ich glaube, daß wir den Fortschritt zunächst auf andere Weise suchen müssen: durch sorgfältige Beachtung der logischen Grundlagen der Theorie und der bisherigen mathematischen Ansätze, deren Ausgestaltung *allein dazu führen kann, die experimentelle Forschung in geordnete und fruchtbare Bahnen zu leiten.*"

Since 1930, the data on which v. MISES based this summary have been replaced by other, more compelling facts. While intensive and fruitful work has been carried out both in statistical theories of transport processes and in experiment on materials, on a scale overshadowing all past efforts, the reader of this treatise will see that the rational mechanics of continua has grown in even greater measure.

It should not be thought that the results of the continuum approach are necessarily either less or more accurate than those from a structural approach. The two approaches are *different*, and they have different uses.

First, a structure theory implies more information about *a given material*, and hence less information about *a class of materials*. The dependence of viscosity on

---

[1] v. MISES [1930, *2*].

temperature in a gas, for example, is predicted by the classical kinetic theory of moderately dense gases, while in a continuum theory it is left arbitrary. For each different law of intermolecular force, the result is different, and for more complicated models it is not yet known. A continuum theory, less definite in this regard, may apply more broadly. The added information of the structural theory may be unnecessary and even irrelevant. To take an extreme example, a full structural specification implies, in principle, all physical properties. From the structure it ought to be possible to derive, among all the rest, the smell and color of the material. A specification so minute will obviously carry with it extreme mathematical complexity, irrelevant to mechanical questions regarding finite bodies.

Second, structural specification necessarily presents *all the attributes* of a material simultaneously, while in continuum physics we may easily separate for special study *an aspect* of natural behavior. For example, the classical kinetic theory of monatomic gases implies a special constitutive equation of extremely elaborate type, allowing all sorts of thermo-mechanical interactions, with definite numerical coefficients depending on the molecular model. For a natural gas really believed to correspond to this theory, these complexities are sometimes relevant, and the theory is of course a good one. On the other hand, it is a highly special one, offering no possibility of accounting for many simple phenomena daily observed in fluids. For example, it does not allow for a shear viscosity dependent on density as well as temperature, or for a non-zero bulk viscosity, both of which are easily handled in classical fluid dynamics. Doubtless it is true that natural fluids for which such viscosities are significant have a complicated molecular structure, but this does not lessen the need for theories that enable us to predict their response in mechanical situations, perhaps long before their structure is determined.

Third, a continuum theory may obtain by *a more efficient process* results shown to be true also according to certain molecular theories. For example, a simple continuum argument suggests the plausibility of the Mooney theory of rubber, which was later shown to follow also from a sufficiently accurate and general theory of long-chain molecules. A more subtle but more important possibility comes from the general principles of physics. For example, certain requirements of invariance and laws of conservation may be applied directly to the continuum, rendering unnecessary the repeated consideration of consequences of these same principles in divers special molecular models, so that the continuum method may enable us to derive directly, once and for all, results common to many different structural theories. In this way we may separate properties that are truly sensitive to a particular molecular structure from those that are necessary consequences of more general laws of nature or more general principles of division of natural phenomena.

Fourth, the information needed to apply a continuum theory in an experimental context is *accessible to direct measurement*, while that for a structural theory usually is not. For example, in the classical infinitesimal theory of isotropic elasticity it is shown that data measured in simple shear and simple extension are sufficient to determine all the mechanical response of the material. Both the taking of the data and the test of the assertion are put in terms of the kind of measurement for which the theory is intended. The non-linear theories show this same accessibility, though in more complex form.

In summary, then, continuum physics serves to *correlate the results of measurements* on materials and to isolate *aspects* of their response. It neither conflicts with structural theories nor is rendered unnecessary by them.

The foregoing observations refer to those structural theories in which the presumed structure is intended to represent the molecules or smaller particles of natural materials. In regard to the mechanismomorphic structures imagined by the rheologists, we can do no better than quote some remarks of COLEMAN and NOLL[1] in a more special context:

"It is often claimed that the theory of infinitesimal viscoelasticity can be derived from an assumption that on a microscopic level matter can be regarded as composed of 'linear viscous elements' (also called 'dashpots') and 'linear elastic elements' (called 'springs') connected together in intricate 'networks'....

"We feel that the physicist's confidence in the usefulness of the theory of infinitesimal viscoelasticity does not stem from a belief that the materials to which the theory is applied are really composed of microscopic networks of springs and dashpots, but comes rather from other considerations. First, there is the observation that the theory works for many real materials. But second, and perhaps more important ..., is the fact that the theory looks plausible because it seems to be a mathematization of little more than certain intuitive prejudices about smoothness in macroscopic phenomena."

4. General lines of past research on the field theories of mechanics. While, reflecting the stature of the researchers themselves, the early researches on the foundations of continuum mechanics did not show any preference for linear theories, with the rise of science as a numerous profession in the nineteenth century it was quickly seen that linearity lends itself to volume of publication. The linear theories of heat conduction, attraction, elasticity, and viscosity, along with the linear mathematical techniques that could be applied in them, were developed so intensively and exclusively that in the minds of many scientists down to the present day they are synonymous with the mechanics of continuous media. It would be no great exaggeration to say that in the community of physicists, mathematicians, and engineers, less was known about the true principles of continuum mechanics in 1945 than in 1895.

Blame for this neglect of more fundamental study may be laid to two contradictory misconceptions: First, that the classical linear or infinitesimal theories account for everything known about natural materials, and, second, that these two theories are merely crude "empirical" fits to data. The second is still common among physicists, many of whom believe that only a molecular-statistical theory of the structure of materials can lead to understanding of their behavior. The prevalence of the former among engineers seems to have grown rather from a rigid training which deliberately confined itself to linearly biased experimental tests and deliberately described every phenomenon in nature, no matter how ineptly, in terms of the concepts of the linear theories.

Of course, at all times there have been a few scientists who thought more deeply or at least more broadly in regard to theories of materials. Various doctrines of *plasticity* arose in the latter part of the last century and have been cultivated diffusely in this. These theories have always been closely bound in motive, if often not in outcome, to engineering needs and have proliferated at once in detailed approximate solutions of boundary-value problems. Their mechanical foundation is insecure to the present day, and they do not furnish representative examples in the program of continuum physics. Similarly, the group of older studies called *rheology* is atypical in its nearly exclusive limitation to one-dimensional response, to a particular cycle of material tests, and to models suggested by networks of springs and dashpots.

While only very few scientists between 1845 and 1945 studied the foundations of continuum mechanics, among them were some of the most distinguished savants of the period: ST. VENANT, STOKES, KIRCHHOFF, KELVIN, BOUSSINESQ, GIBBS, DUHEM, and HADAMARD. Although phenomena of viscosity and plasticity were not altogether neglected, the main effort and main success came in the theory of

[1] COLEMAN and NOLL [1961, 7, § 1].

finite elastic strain. The success, however, was but small. When the brothers COSSERAT published their definitive exposition[1] in 1896, its 116 pages contained little more than a derivation of various forms of the general equations. Beyond the laws of wave propagation and the great theorem on elastic stability obtained shortly afterward by HADAMARD, no concrete progress was made in the finite theory for the following fifty years. Not only did the want of concepts such as to suggest a simple notation lay a burden of page-long formulae on the dragging steps of writer or reader, but also there was no evidence of a program of research. Linear thinking, leading to easy solutions for whole classes of boundary-value problems, obviously would not do, but nothing was suggested to take its place, except, perhaps, the dismaying prospect of creeping from stage to stage in a perturbation process.

In that period, however, many papers on the subject were published. When not essentially repetitions of earlier studies, these concerned special theories or approximations, most of which have later turned out to be unnecessary in the cases when they are justified. Knowledge of the true principles of the general theory seems to have diminished except in Italy, where it was kept alive by the teaching and writing of SIGNORINI.

A new period was opened by papers of REINER[2] and RIVLIN[3]. The former was the first to suggest any *general approach* or *unifying principle* for non-linear constitutive equations[4]; the latter was the first to obtain *concrete, exact solutions* to specific problems of physical interest in non-linear theories where the response is specified in terms of *arbitrary functions* of the deformation. Both considered not only finitely strained elastic materials but also non-linearly viscous fluids. RIVLIN was the first to see the far-reaching simplification effected in a non-linear theory by assuming the material to be incompressible.

In 1952 was published a detailed exposition, *The Mechanical Foundations of Elasticity and Fluid Dynamics*[5], in which both the old and the new trends were summarized. On the one hand, the numerous special or approximate theories were set in place upon a general frame and related to each other insofar as possible, especially so as to make clear the arbitrary and unsupported physical assumptions and the insufficient if not faulty mathematical processes by which they had been inferred. On the other, the concrete and trenchant gains won by the new approaches were presented in full and with emphasis.

A summary[6] of the researches of 1945—1952, referring especially to problems of flow, has stated:

"By 1949 it could be said fairly that all work on the foundations of rheology done before 1945 had been rendered obsolete. The phenomenon of normal stresses had been shown to be of second order, while departures from the classically assumed linear relation between shearing tractions and rates of shearing are of third order in the rates. The old viscometers, designed without a thought of normal stresses, had fixed opaque walls to help the experimenter overlook the most interesting effect in the apparatus or to prevent his measuring the forces supplied so as to negate it. By theory, the phenomenon of normal stresses was straight-

---

[1] E. and F. COSSERAT [1896, *1*]. Essentially the same material, but expressed in tensor notation, is contained in the widely read paper of MURNAGHAN [1937, *2*].

[2] REINER [1945, *3*] [1948, *9*].

[3] RIVLIN [1948, *12*, *14*] [1949, *15*, *16*, *17*, *18*].

[4] Considerations of invariance had occurred earlier, notably in the work of POISSON and CAUCHY, but always in rather special contexts. Cf. Sect. 19A.

[5] TRUESDELL [1952, *20*] [1953, *25*]. Corrected reprint, *Continuum Mechanics* I, International Science Review Series, Gordon & Breach, N.Y., 1966.

[6] TRUESDELL [1960, *58*, pp. 13, 15].

away seen to be a universal one, to be expected according to all but very special kinds of non-linear theories. Of course a result so universally to be expected must have been occurring for a long time in nature, and it was quickly seen that many familiar effects, such as the tendency of paints to agglomerate upon stirring mechanisms, as well as some carefully concealed mysteries of the artificial fiber industry, are examples of it, though a century of linear thinking in physics had blinded theorists to the possibility that simple mechanics, rather than chemistry, is all that is needed in explanation. ...

"While ... [this research] gained a number of theoretical predictions of remarkable completeness, these are the least of what it gave us. Next is the fact that, with little exaggeration, *there are no one-dimensional problems*: A situation which is one-dimensional in a linear theory is automatically two-dimensional or three-dimensional in any reasonable non-linear theory. More important is the *independence in theory* which resulted from the realization that any sort of admissible non-linearity would yield the correct general kind of behavior, and that to account for the phenomena, far from being difficult, was all too easy[1]. Of a theory, we learned that both less and more had to be expected. To calculate the creep in a buckled elliptical column with a square hole in it is too much until the response of materials shall be better understood than it is today; to be satisfied with a normal stress of the right sign and order, with an adjustable coefficient, is too little until the response of the *same* material in a *variety* of situations is determined and correlated, with no material constants or functions altered in the process. What is needed is a theory of theories."

Since 1952, it cannot be said that the older type of work has ceased; rather, in the common exuberance of modern publication, an easy place is found not only for continued search of avenues known to be sterile, but also for frequent rediscovery of special theories included and criticized in *The Mechanical Foundations*, and of special cases of solutions presented there in explicit generality. Beyond this, and heedless of it, a small school of younger scientists, of backgrounds and trainings as various as mathematics, physics, chemistry, and engineering, has developed the newer approaches. Not only have major results been obtained in the classical general theory of finite elastic strain, to the point that there is now a technology of the subject, but also success beyond any fair expectation has been met in a very general theory of non-linear viscosity and relaxation. A great range of the mechanical behavior of materials previously considered intractable if not mysterious has been brought within the control of simple, precise, and explicit mathematical theory. Just a little earlier, relevant experiments had begun on a material which lends itself particularly well to measurements of the effects of large deformation and flow: polyisobutylene. It should not be thought, however, that the theories apply only to high polymers. The non-linear effects are typical of mechanics, and there is reason to think they occur in nearly all materials — for example, in air and in metals — but generally their variety is so great that it is difficult to separate one from another. High polymers are distinguished not so much for the existence as for the simplicity of the non-linear effects they exhibit. The new researches on the general theories, preceded by the classical foundation established in the last century, form the subject of the present treatise.

Of the several kinds of attack to which the new continuum mechanics has been subject, only two deserve notice, because only these have some basis in truth. First, some scientists of the "practical" kind presume that pages full of tensors and arbitrary functions or functionals

---

[1] Detailed substantiation is given by our analysis of the Poynting effect in Sect. 54 and our presentation of normal-stress effects in Sects. 106—115.

can never yield results specific enough to apply to the real world. Second, the analyst who has been taught that everything begins with existence and uniqueness theorems may reject as being only "physics" or "engineering" anything that does not consist solely of convergence proofs and estimates. We hope that critics of the former kind will notice in our text the multitude of exact or approximate solutions of specific problems for elastic materials and for simple fluids as well as certain explicit calculations for more general materials, with results fit for comparison with measurements; while this treatise is purely mathematical in content, we have included by way of an existence proof some tables and graphs of data on experiments done expressly in response to the analyses here summarized. We hope that critics of the latter kind will notice page after page of definite theorems and strict proofs and will allow that mathematics is not confined to any rigid pattern; in particular, we hope that this treatise will be admitted in evidence that mathematics enables us to correlate information available on various aspects of a class of physical theories even when that information is too imperfect to lay down a "well set problem" in the style of the common theories of the last century. Finally, we trust that those who regard as essential to modern science the expense and labor of numerical computation on large machines will easily find for themselves a thousand points in our subject where such a taste can be gratified at any time.

5. The nature of this treatise. In 1955 it was planned to contribute to this Encyclopedia two articles that would in effect bring *The Mechanical Foundations* up to date and complete it by a correspondingly detailed presentation of aspects of the foundations of continuum mechanics omitted from it. The former part of the project, concerning the general principles of continuum physics, has been finished and printed as *The Classical Field Theories* (CFT) in Vol. III/1. The latter part has had to be modified[1].

In the first place, the flow of important publication on the basic principles of non-linear theories and on experiment in connection with them has increased tenfold: Scarcely a month passes unmarked by a major paper. What follows here has been not only rewritten but also several times reorganized so as to incorporate researches published after we had begun — in some cases, researches growing straight from the difficulties we ourselves encountered in the writing. Second, the special or approximate theories of elasticity or viscosity, to explaining and interrelating which a considerable part of *The Mechanical Foundations* was devoted, have lost their value because of the greater efficiency and enlightenment the more general methods have since been shown to offer. Third, the theories usually named "plasticity" remain in essentially the same state as they were in 1952, when they were intentionally omitted from *The Mechanical Foundations*[2].

For these reasons, the present treatise is of lesser scope than was originally planned. First, although we have taken pains to include a new and general foundation for the continuum theory of dislocations, we have not felt able to do more in regard to the usual theories of "plasticity" than to refer the reader to the standard treatises, e.g. to the article by FREUDENTHAL and GEIRINGER in Vol. VI of this Encyclopedia. Second, we have omitted most of the special theories of elasticity and viscosity, for them referring the reader to *The Mechanical Foundations*[3].

Work in this field is often criticized for opaque formalism. Some of those not expert in the subject have implied that the specialists attempt to make it seem more difficult than it is. In the original development of any science, the easiest way is often missed, and the lack of a pre-organized common experience and vocabulary, often called "intuition" by those whose concern is paedagogy or professional amity rather than discovery, makes the path of the creator hard to follow. In writing the treatise we present here, earnest and conscious effort has been put out to render the subject simple, easy, and beautiful, which we believe it is, increasingly with the repeated reconsideration of the groundwork and the major results which have appeared in the last decade. On the other hand, we have not followed the lead of some experts in other fields who have lightly entered

[1] In the mean time a general exposition of the field has been published by ERINGEN [1962, *18*] and reviewed by PIPKIN [1964, *67*].

[2] Recently GREEN and NAGHDI [1965, *18*] have proposed a rational theory of finitely deformed plastic materials, but they adopt a yield condition as in the older literature.

[3] TRUESDELL [1952, *20*, §§ 48–54, 60, 81–82] [1953, *25*].

this with too hasty expositions that by their slips and gaps prosper in making the subject appear to their unwary readers as being simpler and easier (though less beautiful) than in fact the physical behavior of materials in large and rapid deformation can be.

Instead of completeness, we have attempted to achieve *permanence*. As the main subjects of this treatise we have selected those researches that formulate and solve *once and for all* certain clear, definite, and broad conceptual and mathematical problems of non-linear continuum mechanics. We not only hope but also believe that the major part of the contents is not controversial or conjectural, representing instead unquestionable conquests that will become and remain standard in the subject. After the classic researches done before 1902, nearly everything in this treatise was first published, at least in the form here given to it, after 1952. We do not pretend, however, to be exhaustive[1] even for the most recent work or for citation of it, since we have subordinated detail to importance, and, above all, to *clarity and finality*.

Our citations refer either to the original sources or to works containing related developments not given in this treatise. Thus, since scant service would be done any reader by directing him to the numerous textbooks and paedagogic "introductions", we follow the precedent of *The Classical Field Theories*, criticized by one reviewer for preferring very old or very new references.

Properly, our title should have indicated restriction to classical mechanics, for relativistic field theories lie outside our scope. Since, however, the term "classical" suggests to many a domain long mastered — indeed, one reviewer criticized *The Classical Field Theories* for including material he did not already know — that word seems inappropriate in the title of a treatise devoted mainly to very recent work. Specifically, we consider *the mechanical response of materials in three-dimensional Euclidean space*. While often the dimension 3 can be replaced effortlessly by n, the main conceptual structure is closely bound to Euclidean geometry. Relativistic generalization has required major changes in views and details which were not yet known when this treatise was planned[2].

This treatise is written, not for the beginner, but for the specialist in mechanics who wishes to gain quickly and efficiently the solid and complete foundation necessary to do theoretical research, either in applications or in further study of the groundwork, in non-linear continuum mechanics. We use the term *non-linear* in the sense of material response, not of mathematical analysis[3].

Accordingly, after an introductory chapter fixing notations and listing a number of mathematical theorems for use in the sequel, this treatise is divided into three major parts, as follows.

Chapter C presents a *general approach*, based upon principles of *determinism, local action*, and *material frame-indifference*, to the mechanical properties of materials. For the special case of a *simple material*, in which the stress at a particle is determined by the cumulative history of the deformation gradient at that particle,

---

[1] It seems necessary to add, however, that when we merely cite and describe a work, without presenting its contents in detail, we do so merely to help the reader find his way in the literature, implying *neither endorsement nor criticism*.

[2] A bibliography of older work on relativistic theories of materials, mainly fluids, is given at the end of CFT. BRESSAN [1963, *13, 14, 16*] and BRAGG [1965, *3*] are the first authors to consider correctly finite deformations and accumulative effects in relativity.

[3] The classical theory of viscous incompressible fluids, for example, is governed by non-linear partial differential equations, but we do not include it here since its defining constitutive equation is a linear one. In fact, since the acceleration is a non-linear function of velocity and velocity-gradient, all theories of the motion of continua are non-linear in the spatial description, so the analytical distinction is an empty one except in regard to methods of approximation.

all three fundamental principles may be expressed in a final and explicit mathematical form. Qualities distinguishing one kind of material from another are then defined by invariant properties of the response functionals; the terms "materially uniform", "homogeneous", "solid", "fluid", and "isotropic" are made precise in terms of mathematical systems constructed from the functionals. Finally, it is shown that if the response functional of a simple material is sufficiently smooth in a certain sense, then BOLTZMANN'S equations of linear visco-elasticity result as an approximation in motions whose histories are nearly constant. Thus the general theory of simple materials is seen to furnish a properly invariant generalization of classical visco-elasticity to arbitrary states of deformation and flow; likewise it includes not only as special cases but also in suitable senses of approximation the classical theories of finite elasticity and linear viscosity.

In statics, the stress in any simple material reduces to a function of the finite strain. Materials having this property also in time-dependent deformations are said to be *elastic*, and most of Chapter D is devoted to them. Here we present the theory of *finite elastic strain*, not only its principles but also its general theorems and the known special exact solutions or approximate methods, in generality and completeness not before attempted. When, as proposed by GREEN, the work done in elastic deformation is stored as internal energy, so that the stresses are derivable from a stored-energy function as a potential, the material is called *hyperelastic*. Nearly all previous studies concerned this case exclusively. While we develop its distinguishing properties and general theorems, our emphasis lies on the more embracing concept, due to CAUCHY. Generalizations of hyperelasticity to allow for thermal conduction, polarization, and couple stresses are then sketched. The last sections of the chapter concern the partly more general and partly exclusive concept of *hypo-elasticity*, according to which the time-rate-of-change of stress is an explicit function of the stretchings, shearings, and spin at a material element, along with the present stress. The behavior of a hypo-elastic material depends essentially upon the initial stress.

Chapter E concerns *fluidity*. Most of its contents is given over to an exhaustive survey of what is known about *simple fluids*. These are distinguished from other simple materials by having the maximum possible isotropy group; all are in fact isotropic. While they are capable of exhibiting complicated effects of stress-relaxation and long-range memory, these are proved to have no influence on certain special kinds of flow, which turn out to include all those customarily used in viscometric tests. For these special flows, the response functional is shown to manifest itself only through three *viscometric functions*. One of these may be interpreted as a non-linear shear viscosity; the other two, as differences of normal stresses. The exact solutions of the dynamical equations are developed for these flows, as well as for some others of similar kind. The chapter closes with consideration of materials embodying various other concepts of fluidity.

By including general effects of rates and of relaxation, we cover a broader range of physical phenomena than did *The Mechanical Foundations*, although we narrow the topic by omitting most special or approximate theories. The main difference, however, is one of depth. In the present treatise the method of inquiry and formulation, far less formal than the approaches known in 1952, goes straight to the physics of each situation. We have sought, and we believe we have often succeeded in finding, *simple and clear* mathematical expression for the physical principles or hypotheses.

6. Terminology and general scheme of notation. We employ at will the notations of RICCI'S tensor calculus[1], of linear vector and matrix algebra, and of GIBBS'

[1] Short introductions to tensor analysis have been given by LICHNÉROWICZ [1950, *10*] and REICHHARDT [1957, *17*]. A large body of definitions, identities, and theorems especially useful in continuum physics is presented in the Appendix to CFT.

vector analysis. While our viewpoint emphasizes the direct concepts and methods of vector spaces[1], the analysis will be followed easily by a reader familiar with any of the standard approaches, especially since all the major results are stated in component form as well as in abstract notation.

The space under consideration in this treatise is always an n-dimensional Euclidean space of points or vectors. In many cases the dimension of this space is immaterial, but for all results whose form depends upon n we shall take n as 3 without reminding the reader that we have done so.

Bold-face type is reserved for *points, vectors, tensors,* or functions whose values are quantities of one of these kinds. The *components* of such quantities are denoted by attaching indices to a kernel index that is the same letter in light-face. For example, $\ddot{\boldsymbol{x}}$ is the acceleration, while $\ddot{x}^k$ and $\ddot{x}_k$ are its contravariant and covariant components, respectively[2]; similarly, $\boldsymbol{T}_{\mathrm{R}}$ is the Piola-Kirchhoff stress tensor, having components $T_{\mathrm{R}k}{}^{\alpha}$, $T_{\mathrm{R}}^{k\alpha}$, $T_{\mathrm{R}}{}^{k}{}_{\alpha}$, and $T_{\mathrm{R}k\alpha}$, as may be relevant in particular cases. The index R, it will be noted, is printed in Roman type; this fact shows that it is a descriptive mark, not a tensor index, for tensorial indices are always printed in italics or in Greek minuscules, never in any other style. There will be some specified exceptions to the rule on kernel indices; for example, the components of the unit or Euclidean metric tensor $\mathbf{1}$ are g^{km}, δ^k_m, and g_{km}, as in general usage. The physical components of a tensor $\boldsymbol{A}$ are denoted by $A\langle k\ldots m\rangle$. We shall use physical components only for orthogonal co-ordinate systems. When a particular co-ordinate system is used, number indices are often replaced by symbols for co-ordinates. For example, $T\langle r\theta\rangle = T\langle 12\rangle$ is a physical component of the stress tensor in cylindrical co-ordinates: $x^1 = r$, $x^2 = \theta$, $x^3 = z$.

Two points $\boldsymbol{y}$ and $\boldsymbol{z}$ uniquely determine a vector $\boldsymbol{v}$, which is represented by the arrow going from $\boldsymbol{z}$ to $\boldsymbol{y}$. We denote this vector by[3]

$$\boldsymbol{v} = \boldsymbol{y} - \boldsymbol{z}. \tag{6.1}$$

Using this concept of *difference of points*, one can define the derivative of a point-valued function $\boldsymbol{z} = \boldsymbol{z}(s)$ of a real variable s in the customary manner:

$$\frac{d}{ds}\boldsymbol{z}(s) = \lim_{\Delta s \to 0} \frac{\boldsymbol{z}(s+\Delta s) - \boldsymbol{z}(s)}{\Delta s}. \tag{6.2}$$

Such a derivative is a vector-valued function of the real variable s.

A vector field $\boldsymbol{v}(\boldsymbol{x})$ is a vector-valued function of the point variable $\boldsymbol{x}$. Let $\boldsymbol{e}_k(\boldsymbol{x})$ be a set of n linearly independent vector fields, so that, for each $\boldsymbol{x}$, $\boldsymbol{e}_k(\boldsymbol{x})$ forms a basis. The *dual basis* $\boldsymbol{e}^k$ of $\boldsymbol{e}_k$ is uniquely defined by the relations

$$\boldsymbol{e}^k \cdot \boldsymbol{e}_m = \delta^k_m. \tag{6.3}$$

Components of a field $\boldsymbol{v}(\boldsymbol{x})$ with respect to $\boldsymbol{e}_k(\boldsymbol{x})$ and $\boldsymbol{e}^k(\boldsymbol{x})$ are defined as being the scalar products of $\boldsymbol{v}$ with the basis vectors:

$$v_k = \boldsymbol{v} \cdot \boldsymbol{e}_k, \qquad v^k = \boldsymbol{v} \cdot \boldsymbol{e}^k. \tag{6.4}$$

In general, the basis $\boldsymbol{e}_k(\boldsymbol{x})$ may vary in any way from point to point. If $\boldsymbol{x} = \boldsymbol{x}(x^k)$ is the point whose co-ordinates are x^k in some general co-ordinate system, the derivatives

$$\boldsymbol{e}_k = \frac{\partial \boldsymbol{x}}{\partial x^k}, \tag{6.5}$$

taken in the sense of (6.2), define a vector basis, which is called the *natural basis* of the co-ordinate system. In this case the components v_k and v^k defined by (6.4) are called *covariant*

[1] Detailed expositions have been written by HALMOS [1958, *22*] and GREUB [1963, *28*].

[2] Since the space is Euclidean, readers who so prefer may always employ a Cartesian co-ordinate system. In such a system corresponding covariant, contravariant, and physical components are equal, so that the placement of indices is immaterial.

[3] Another often-used notation is $\boldsymbol{v} = \overrightarrow{\boldsymbol{z}\boldsymbol{y}}$.

and *contravariant*, respectively, and may be shown to satisfy the equations of transformation usually associated with those terms. In this treatise we shall occasionally employ a variable basis that is not the natural basis of any co-ordinate system. The components (6.4) are then called *anholonomic components* of the vector field $\boldsymbol{v}$ (cf. Sect. 10 of the Appendix to CFT). If the variable basis $\boldsymbol{i}_k$ is everywhere orthonormal, i.e., if $\boldsymbol{i}_k \cdot \boldsymbol{i}_m = \delta_{km}$, we use the notation

$$v\langle k\rangle = \boldsymbol{v} \cdot \boldsymbol{i}_k. \tag{6.6}$$

In the special case when $\boldsymbol{i}_k = \sqrt{g^{kk}}\,\boldsymbol{e}_k = \sqrt{g_{kk}}\,\boldsymbol{e}^k$, where $\boldsymbol{e}_k$ is the natural basis of an orthogonal co-ordinate system, the formula (6.6) gives the physical components. Similar remarks apply to tensor fields.

We shall customarily view the second-order tensor $\boldsymbol{A}$ as the *linear transformation* that assigns to the vector $\boldsymbol{u}$ a certain other vector,

$$\boldsymbol{v} = \boldsymbol{A}\boldsymbol{u}, \tag{6.7}$$

or, in contravariant and covariant components,

$$v^k = A^k{}_m u^m, \qquad v_k = A_k{}^m u_m. \tag{6.8}$$

Alternatively, the second-order tensor $\boldsymbol{A}$ may be regarded as the *bilinear form* that assigns to the pair of vectors $\boldsymbol{u}, \boldsymbol{v}$ the scalar

$$\boldsymbol{u} \cdot \boldsymbol{A}\boldsymbol{v} = u^k A_{km} v^m = u^k A_k{}^m v_m = u_k A^{km} v_m = u_k A^k{}_m v^m. \tag{6.9}$$

Since most tensors occurring in this treatise are of second order, the adjective "second-order" is often omitted. Finally, if a single Cartesian co-ordinate system is used throughout, it is permissible to identify the tensor $\boldsymbol{A}$ with the matrix $\|A_{km}\|$ of its components. However, if curvilinear co-ordinates are used, the tensor $\boldsymbol{A}$ should be carefully distinguished from the matrices of its several types of components.

The notations $\boldsymbol{A}^T$, $\operatorname{tr}\boldsymbol{A}$, and $\det\boldsymbol{A}$ denote the *transpose*, the *trace*, and the *determinant* of the second-order tensor $\boldsymbol{A}$. If A^{km} are the contravariant components of $\boldsymbol{A}$, then $A^{mk}u_m$ are the contravariant components of $\boldsymbol{A}^T\boldsymbol{u}$. Also[1]

$$\operatorname{tr}\boldsymbol{A} = A^k{}_k = A_k{}^k = g^{km} A_{km} = g_{km} A^{km}, \tag{6.10}$$

$$\det\boldsymbol{A} = \det\|A^k{}_m\| = \det\|A_m{}^k\|. \tag{6.11}$$

The *product* $\boldsymbol{A}\boldsymbol{B}$ of the tensors $\boldsymbol{A}$ and $\boldsymbol{B}$ is defined by the composition of the transformations $\boldsymbol{B}$ and $\boldsymbol{A}$. The covariant components of $\boldsymbol{A}\boldsymbol{B}$ are given by

$$(\boldsymbol{A}\boldsymbol{B})_{km} = A_{kp} B^p{}_m = A_k{}^p B_{pm}. \tag{6.12}$$

The tensor $\boldsymbol{Q}$ is *orthogonal* if any of the following equivalent conditions holds:

$$\boldsymbol{Q}\boldsymbol{Q}^T = \boldsymbol{1}, \qquad Q_k{}^m Q^p{}_m = \delta^p_k, \qquad g_{mq} Q_k{}^m Q_p{}^q = g_{kp}. \tag{6.13}$$

The *magnitude* $|\mathsf{A}|$ of a tensor A of arbitrary order is defined by

$$|\mathsf{A}| = \sqrt{\mathsf{A}_{k\ldots m} \mathsf{A}^{k\ldots m}}. \tag{6.14}$$

In particular, for a vector $\boldsymbol{v}$ and a second-order tensor $\boldsymbol{A}$ we have

$$v = |\boldsymbol{v}| = \sqrt{\boldsymbol{v}^2}, \qquad |\boldsymbol{A}| = \sqrt{\operatorname{tr}(\boldsymbol{A}\boldsymbol{A}^T)}. \tag{6.15}$$

The *tensor product* (dyadic product) of the two vectors $\boldsymbol{u}$ and $\boldsymbol{v}$ is denoted by $\boldsymbol{u}\otimes\boldsymbol{v}$. It may be identified with the linear transformation that assigns to a vector $\boldsymbol{a}$ the vector $\boldsymbol{u}(\boldsymbol{v}\cdot\boldsymbol{a})$. The contravariant components of $\boldsymbol{u}\otimes\boldsymbol{v}$ are given by

$$(\boldsymbol{u}\otimes\boldsymbol{v})^{km} = u^k v^m. \tag{6.16}$$

The *gradient* of a vector field $\boldsymbol{v} = \boldsymbol{v}(\boldsymbol{x})$ is the second-order tensor field, which will be denoted by $\nabla\boldsymbol{v} = \nabla\boldsymbol{v}(\boldsymbol{x})$, defined as the linear transformation that assigns to a vector $\boldsymbol{a}$ the vector given by the following rule:

$$(\nabla\boldsymbol{v})\,\boldsymbol{a} = \lim_{s\to 0} \frac{\boldsymbol{v}(\boldsymbol{x}+s\boldsymbol{a}) - \boldsymbol{v}(\boldsymbol{x})}{s}. \tag{6.17}$$

---

[1] Note that $\det\boldsymbol{A} \neq \det\|A_{km}\| \neq \det\|A^{km}\|$ unless $\det\|g_{km}\| = g = 1$.

Under suitable assumptions of smoothness it can be proved that the operation $\nabla \boldsymbol{v}$ is a linear transformation, as presumed in applying the term "tensor". The various components of $\nabla \boldsymbol{v}$ are denoted by $v^k{}_{,m}$, $v_{k,m}$, $v_k{}^{,m}$, $v^{k,m}$; they are the *covariant and contravariant derivatives* of the components of the vector field $\boldsymbol{v}$. The *partial derivative* operators, for any function of the co-ordinates x^k, are denoted by ∂_m. In particular, $\partial_m v^k$ and $\partial_m v_k$ stand for the partial derivatives of the components of the vector field $\boldsymbol{v}$.

It is customary in much of analysis not to distinguish between the values of a function and the function itself. We follow this procedure whenever no confusion is likely to result from it. However, in some cases, as for example in the theory of response functionals, it is necessary to observe the distinction.

General scheme of notation

We attempt to employ notations that reflect as clearly as possible differences of kind among mathematical entities, but in some cases hindrance rather than help would follow from holding too close to the scheme below, laid down here as a general guide but not a rigid code.

Full-sized characters:

Italic light-face letters A, a, ... (except X, Z): Scalars or the kernel indices of vectors or second-order tensors.

X, Z: Particles.

$\boldsymbol{X}$, $\boldsymbol{Y}$, $\boldsymbol{Z}$, $\boldsymbol{x}$, $\boldsymbol{y}$, $\boldsymbol{z}$: Points.

Italic bold-face majuscules $\boldsymbol{A}$, $\boldsymbol{B}$, ... (except $\boldsymbol{X}$, $\boldsymbol{Y}$, $\boldsymbol{Z}$): Second-order tensors and tensor fields.

Italic bold-face minuscules $\boldsymbol{a}$, $\boldsymbol{b}$, $\boldsymbol{u}$, $\boldsymbol{v}$, ...: Vectors and vector fields.

Sans-serif light-face majuscules A, B, ...: Kernel indices of tensors of order 3 or more, physical dimensions.

Sans-serif bold-face majuscules $\boldsymbol{\mathsf{A}}$, $\boldsymbol{\mathsf{B}}$, ...: Tensors of order 3 or more, or functions whose values are such tensors.

Greek light-face letters α, β, γ, ..., Γ, Δ, ...: Scalars, scalar-valued functions of tensors, kernel indices of mappings, thermodynamic parameters, scalar material constants.

Greek bold-face minuscules $\boldsymbol{\xi}$, $\boldsymbol{\varkappa}$, $\boldsymbol{\chi}$, ...: Configurations, mappings, and other point-valued functions.

German bold-face majuscules $\boldsymbol{\mathfrak{F}}$, $\boldsymbol{\mathfrak{G}}$, ...: Functionals (vector-valued or tensor-valued).

German light-face majuscules $\mathfrak{F}$, $\mathfrak{G}$, ...: Kernel indices of functionals.

German bold-face minuscules $\boldsymbol{\mathfrak{f}}$, $\boldsymbol{\mathfrak{g}}$, ...: Tensor-valued functions of tensor variables.

German light-face minuscules $\mathfrak{f}$, $\mathfrak{g}$, ...: Kernel indices of tensor-valued functions, scalar-valued functionals.

Script majuscules $\mathscr{A}$, $\mathscr{B}$, ...: Sets, bodies, regions, etc.

Script minuscules g, u, ...: Groups.

Hebrew letters $\aleph$, $\beth$, ...: Response coefficients for isotropic materials.

Indices:

Italic and Greek light-face indices k, m, α, β, ...: Tensorial indices.

Roman indices A, a, ...: Descriptive marks.

German minuscule and Greek majuscule indices $\mathfrak{a}$, $\mathfrak{b}$, ..., Γ, Δ, ...: Enumerative indices. The corresponding numbers are printed in italics: *1*, *2*, ...

Bold-face indices $\boldsymbol{A}$, $\boldsymbol{B}$, ...: Tensor argument of a special scalar invariant, gradients with respect to tensors.

Index of frequently used symbols.

(Only symbols used in more than one section are listed. If a different notation is used in CFT for the same quantity, it is given within square brackets at the left; also, it has not been possible to adhere rigidly to these notations, so that sometimes within a single section these same letters are used for quantities other than those listed below.)

Symbol	Name	Place of definition or first occurrence
$\boldsymbol{A}_{\mathfrak{n}}$, $A_{\mathfrak{n}km}$	$\mathfrak{n}$'th Rivlin-Ericksen tensor	(24.14)
A, $\mathsf{A}_k{}^\alpha{}_m{}^\beta$	Elasticity tensor	(44.3)
$[\boldsymbol{c}^{-1}]$ $\boldsymbol{B}$, B^{km}	Left Cauchy-Green tensor	(23.5)
$\boldsymbol{B}_{(t)}$, $B^{km}_{(t)}$	Relative left Cauchy-Green tensor	Sect. 23
B, B^{kmpq}	Elasticity tensor	(45.4)
$\mathscr{B}$	Body	Sect. 15
$\boldsymbol{C}$, $C_{\alpha\beta}$	Right Cauchy-Green tensor	(23.4)
$\boldsymbol{C}_{(t)}$, $C_{(t)km}$	Relative right Cauchy-Green tensor	(23.10)
$\boldsymbol{C}^{(t)}$	History of right Cauchy-Green tensor	(29.5)
C, C^{kmpq}	Elasticity tensor	(45.2)
$[\boldsymbol{d}]$ $\boldsymbol{D}$, D_{km}	Stretching tensor	(24.10)
$\boldsymbol{D}_{\mathfrak{n}}$, $D_{\mathfrak{n}km}$	$\mathfrak{n}$'th stretching tensor	(24.12)
$\boldsymbol{E}$	Green-St.Venant strain tensor	(63.8)
$\tilde{\boldsymbol{E}}$, $\tilde{E}_{km}$	Infinitesimal strain tensor	(41.12)
$\boldsymbol{F}$, $F^k{}_\alpha = x^k{}_{,\alpha}$	Deformation gradient	(21.9)
$\boldsymbol{F}_{(t)}$, $F^k{}_{(t)\alpha}$	Relative deformation gradient	(21.13)
$\boldsymbol{F}^{(t)}$	History of the deformation gradient	Sect. 28
F, $\mathsf{F}^{km}{}_{pq}$	Elasticity tensor	(48.13)
$\boldsymbol{G}(s)$	History of a certain relative strain measure	(31.10)
$\boldsymbol{G}^*(s)$	History of a certain rotated relative strain measure	(29.27)
$\mathfrak{G}$, $\mathfrak{G}_\varkappa$, $\mathfrak{G}_{\boldsymbol{K}}$	Response functional of a simple material	(28.3), (28.6), (28.8)
$\boldsymbol{H}$, $H^k{}_\alpha = u^k{}_{,\alpha}$	Displacement gradient	(41.9), (44.23)
H, H^{kmpq}	Hypo-elastic response function	(99.9)
J	Jacobian of deformation	(21.17)
$\boldsymbol{K}$	Local configuration	Sect. 22
$\boldsymbol{L}$, $L_{km} = \dot{x}_{k,m}$	Velocity gradient	(24.1), (24.2)
$\boldsymbol{L}_{\mathfrak{n}}$, $L_{\mathfrak{n}km} = \overset{(\mathfrak{n})}{x}_{k,m}$	$\mathfrak{n}$'th acceleration gradient	(24.5)
L	Dimension of length	
L, L^{kmpq}	Infinitesimal elasticity tensor	(41.17), (41.26), (50.11)
M	Dimension of mass	
$\boldsymbol{N}$		(106.7), (106.8)
$\boldsymbol{P}$	Local deformation	(21.21), (22.15)
$\mathscr{P}$	Part of a body	Sect. 15
$\boldsymbol{Q}$, $\boldsymbol{Q}(t)$	Arbitrary orthogonal tensor	(6.13)
$\boldsymbol{Q}(\boldsymbol{n})$, Q_{km}	Acoustical tensor	(71.4)
$\boldsymbol{R}$	Local rotation tensor	(23.1)
$\boldsymbol{R}_{(t)}$	Relative local rotation tensor	Sect. 23
$\tilde{\boldsymbol{R}}$, $\tilde{R}_{km}$	Infinitesimal rotation tensor	(41.15)
$[\boldsymbol{t}]$ $\boldsymbol{T}$, T^{km}	CAUCHY's stress tensor	(16.5)
$\overline{\boldsymbol{T}}$, $\overline{T}_{\alpha\beta}$	Convected stress tensor	(29.7)
$\tilde{\boldsymbol{T}}$, $\tilde{T}^{\alpha\beta}$	Second Piola-Kirchhoff stress tensor	(43A.9)
$\boldsymbol{T}_{\mathrm{E}}$, T^{km}_{E}	Extra stress	(30.11)
$[\boldsymbol{T}]$ $\boldsymbol{T}_{\mathrm{R}}$, $T_{\mathrm{R}k}{}^\alpha$	First Piola-Kirchhoff stress tensor	(43A.3)

Symbol	Name	Place of definition or first occurrence
T_a	Principal force	(48.7)
T	Dimension of time	
U	Speed of propagation	(71.1)
U_{ab}	Speed of propagation of a principal wave	Sect. 74
$\boldsymbol{U}$	Right stretch tensor	(23.1)
$\boldsymbol{U}_{(t)}$	Relative right stretch tensor	Sect. 23
$\boldsymbol{U}^{(t)}$	History of the right stretch tensor	(29.1)
$\boldsymbol{V}$	Left stretch tensor	(23.1)
$[\boldsymbol{w}]$ $\boldsymbol{W}$, W_{km}	Spin tensor	(24.9)
$\boldsymbol{W}_{\mathfrak{n}}$, $W_{\mathfrak{n}km}$	$\mathfrak{n}$'th spin	(24.11)
X, Y, Z	Particles (also, in special problems, Cartesian co-ordinates of $\boldsymbol{X}$)	Sect. 15
$\boldsymbol{X}$, $\boldsymbol{Z}$, X^α, etc.	Points in the reference configuration (material co-ordinates)	(21.1), (21.2)
$\boldsymbol{a}$, a^k	Amplitude of a wave	(71.1)
$[\boldsymbol{f}]$ $\boldsymbol{b}$	External body force density	(16.1)
$\boldsymbol{d}$	Director of an anisotropic fluid	Sect. 127
$\boldsymbol{d}_\sigma$	Natural basis of co-ordinate system in the reference configuration	(21.3)
g_{km}, $g_{\alpha\beta}$	Components of the metric or unit tensor	Sect. 6
$\mathscr{g}$	Isotropy group	Sect. 31
$\mathfrak{g}$	Response function of an elastic material	(43.1)
$\boldsymbol{h}$	Heat flux vector	(72.4), (79.2)
$\mathfrak{h}$	Response function of an elastic material	(43.8)
$\boldsymbol{i}, \boldsymbol{j}, \boldsymbol{k}$	Orthonormal triad	
$\boldsymbol{n}$	Unit normal vector	
$\boldsymbol{n}_{\mathrm{R}}$	Unit normal vector, used in the reference configuration	
$\mathscr{O}$	Orthogonal group	
p	Hydrostatic pressure	
$\boldsymbol{p}$	Position vector of the place $\boldsymbol{x}$	
q	Heat absorption	(79.2)
s	Time lapse	(26.10), (28.4)
s_0	Natural time-lapse	(28.15), (108.18)
t	Time	(15.8)
t_a	Principal stress	(48.1)
$\boldsymbol{t}$	Stress or traction vector	(16.2)
$\boldsymbol{t}_{\mathrm{R}}$	Traction vector in the reference configuration	(43A.5)
$\mathfrak{t}$	Response function of an elastic material	(43.10)
$\boldsymbol{u}$, u^k	Displacement vector	(44.22)
$\mathscr{U}$	Unimodular group	
v_a	Principal stretch	(48.1)
$\boldsymbol{x}$, $\boldsymbol{y}$, $\boldsymbol{z}$, x^k, etc.	Points in space (spatial co-ordinates)	Sect. 6, (21.5), (21.6)
x, y, z	Cartesian co-ordinates of $\boldsymbol{x}$	
$x^k{}_{,\alpha}$	Deformation gradient	(21.9)
$\dot{\boldsymbol{x}}$, $\dot{x}^k$	Velocity	(15.9)
$\ddot{\boldsymbol{x}}$, $\ddot{x}^k$	Acceleration	(15.10)
$\boldsymbol{\Gamma}$, $\Gamma_{\mu\lambda}{}^\gamma$	Material connection	(34.4)
$\alpha_1, \alpha_2, \ldots, \alpha_\Gamma$	Elasticities	(66.3)

Symbol	Name	Place of definition or first occurrence
$\alpha_1, \alpha_2, \ldots, \alpha_\Gamma$	Viscosities	(119.15)
$\gamma, \gamma_{\mathrm{loc}}, \gamma_{\mathrm{con}}$	Specific productions of entropy	(79.7), (79.9), (79.10)
$\delta^k_m, \delta^\alpha_\beta$	Kronecker delta	
ε	Internal energy density	(79.1)
ζ	Acceleration-potential or shearing function (to be distinguished by the context)	(30.28), Sect. 108
η	Specific entropy	Sect. 45, (79.4)
θ	Temperature (also used as polar co-ordinate)	Sect. 45, 79
$\varkappa$	Shearing	(108.1)
$\boldsymbol{\varkappa}$	Reference configuration	(21.1)
$\boldsymbol{\varkappa}_X$	Localization of $\boldsymbol{\varkappa}$ at X	(22.2)
λ	Lamé coefficient of linear viscosity or linear elasticity (to be distinguished by the context)	(41.5), (41.27), (119.2)
$\hat{\lambda}$	Thermo-mechanical potential	(81.11)
μ	Lamé coefficient of linear viscosity or linear elasticity (to be distinguished by the context)	(41.5), (41.27), (119.2)
μ_0	Natural viscosity	(108.18), $(108.22)_1$
$\hat{\mu}$	Shear modulus function	(54.10)
$\tilde{\mu}$	Shear viscosity function	(108.10)
$\boldsymbol{\xi}, \xi^k$	Place at time τ	(22.3)
ϱ	Mass density	(15.6)
ϱ_{R}	Mass density in the reference configuration	(43A.4)
$\sigma, \bar{\sigma}$	Strain-energy functions	(82.9), (84.5)
σ_1, σ_2	Normal-stress functions	(108.5), (108.6)
τ	Shear-stress function	(108.4)
υ	Potential of the external body force	(30.25)
ψ	Specific free energy	(82.4)
$\boldsymbol{\chi}, \chi^k$	Configuration, motion	(15.3), (15.8)
$\aleph_0, \aleph_1, \aleph_2, \aleph_\Gamma$	Elastic response coefficients (through Sect. 103 but not afterward)	(47.8)
$\beth_0, \beth_1, \beth_{-1}, \beth_\Gamma$	Elastic response coefficients (through Sect. 103 but not afterward)	(47.9)
∂	Boundary operator	
∂_x, ∂_t, etc.	Partial derivative operator	
$\mathbf{1}$	Unit tensor	Sect. 6
$I_a(\boldsymbol{A}), I_{\boldsymbol{A}}, II_{\boldsymbol{A}}, III_{\boldsymbol{A}}$	Principal invariants of $\boldsymbol{A}$	(8.2)
$\bar{I}_a(\boldsymbol{A}), \overline{II}_{\boldsymbol{A}}, \overline{III}_{\boldsymbol{A}}$	Moments of $\boldsymbol{A}$	(8.4)

6A. Appendix. Cylindrical and spherical co-ordinates. The following formulae are frequently used in this article and are likely to be useful also in future work on special solutions.

α) *Cylindrical co-ordinates r, θ, z:*

Components of the unit tensor:

$$g_{rr} = g^{rr} = g_{zz} = g^{zz} = 1, \qquad g_{\theta\theta} = r^2, \qquad g^{\theta\theta} = \frac{1}{r^2}. \tag{6A.1}$$

Non-zero Christoffel symbols:

$$\begin{Bmatrix} \theta \\ r\,\theta \end{Bmatrix} = \begin{Bmatrix} \theta \\ \theta\, r \end{Bmatrix} = \frac{1}{r}, \qquad \begin{Bmatrix} r \\ \theta\,\theta \end{Bmatrix} = -r. \tag{6A.2}$$

Contravariant components of the acceleration $\ddot{\boldsymbol{x}}$ of a velocity field $\dot{\boldsymbol{x}} = \dot{\boldsymbol{x}}(\boldsymbol{x}, t)$:

$$\ddot{x}^r = \partial_t \dot{x}^r + \dot{x}^k \partial_k \dot{x}^r - r(\dot{x}^\theta)^2, \quad \ddot{x}^\theta = \partial_t \dot{x}^\theta + \dot{x}^k \partial_k \dot{x}^\theta + \frac{2}{r} \dot{x}^r \dot{x}^\theta, \quad \ddot{x}^z = \partial_t \dot{x}^z + \dot{x}^k \partial_k \dot{x}^z. \tag{6A.3}$$

Physical components of the divergence of a symmetric second-order tensor field $\boldsymbol{T}$:

$$\left.\begin{aligned}
(\operatorname{div} \boldsymbol{T})\langle r\rangle &= \partial_r T\langle rr\rangle + \frac{1}{r}\partial_\theta T\langle r\theta\rangle + \partial_z T\langle rz\rangle + \frac{T\langle rr\rangle - T\langle\theta\theta\rangle}{r},\\
(\operatorname{div} \boldsymbol{T})\langle\theta\rangle &= \partial_r T\langle r\theta\rangle + \frac{1}{r}\partial_\theta T\langle\theta\theta\rangle + \partial_z T\langle\theta z\rangle + \frac{2}{r} T\langle r\theta\rangle,\\
(\operatorname{div} \boldsymbol{T})\langle z\rangle &= \partial_. T\langle rz\rangle + \frac{1}{r}\partial_\theta T\langle\theta z\rangle + \partial_z T\langle zz\rangle + \frac{1}{r} T\langle rz\rangle.
\end{aligned}\right\} \tag{6A.4}$$

β) Spherical co-ordinates r, θ, φ:

Components of the unit tensor:

$$g_{rr} = g^{rr} = 1, \quad g_{\theta\theta} = r^2, \quad g^{\theta\theta} = \frac{1}{r^2}, \quad g_{\varphi\varphi} = (r\sin\theta)^2, \quad g^{\varphi\varphi} = \frac{1}{(r\sin\theta)^2}. \tag{6A.5}$$

Non-zero Christoffel symbols:

$$\left.\begin{aligned}
&\left\{{\theta \atop \theta\, r}\right\} = \left\{{\theta \atop r\,\theta}\right\} = \left\{{\varphi \atop \varphi\, r}\right\} = \left\{{\varphi \atop r\,\varphi}\right\} = \frac{1}{r}, \quad \left\{{r \atop \theta\,\theta}\right\} = -r, \quad \left\{{r \atop \varphi\,\varphi}\right\} = -r\sin^2\theta,\\
&\left\{{\varphi \atop \varphi\,\theta}\right\} = \left\{{\varphi \atop \theta\,\varphi}\right\} = \cot\theta, \quad \left\{{\theta \atop \varphi\,\varphi}\right\} = -\sin\theta\cos\theta.
\end{aligned}\right\} \tag{6A.6}$$

Contravariant components of the acceleration $\ddot{\boldsymbol{x}}$ of a velocity field $\dot{\boldsymbol{x}} = \dot{\boldsymbol{x}}(\boldsymbol{x}, t)$:

$$\left.\begin{aligned}
\ddot{x}^r &= \partial_t \dot{x}^r + \dot{x}^k \partial_k \dot{x}^r - r[(\dot{x}^\theta)^2 + \sin^2\theta\,(\dot{x}^\varphi)^2],\\
\ddot{x}^\theta &= \partial_t \dot{x}^\theta + \dot{x}^k \partial_k \dot{x}^\theta + \frac{2}{r}\dot{x}^\theta \dot{x}^r - \sin\theta\cos\theta\,(\dot{x}^\varphi)^2,\\
\ddot{x}^\varphi &= \partial_t \dot{x}^\varphi + \dot{x}^k \partial_k \dot{x}^\varphi + \frac{2}{r}\dot{x}^r \dot{x}^\varphi + 2\cot\theta\,\dot{x}^\theta \dot{x}^\varphi.
\end{aligned}\right\} \tag{6A.7}$$

Physical components of the divergence of a symmetric second-order tensor field $\boldsymbol{T}$:

$$\left.\begin{aligned}
(\operatorname{div} \boldsymbol{T})\langle r\rangle &= \partial_r T\langle rr\rangle + \frac{1}{r}\partial_\theta T\langle r\theta\rangle + \frac{1}{r\sin\theta}\partial_\varphi T\langle r\varphi\rangle +\\
&\quad + \frac{1}{r}[2T\langle rr\rangle - T\langle\theta\theta\rangle - T\langle\varphi\varphi\rangle + \cot\theta\, T\langle r\theta\rangle],\\
(\operatorname{div} \boldsymbol{T})\langle\theta\rangle &= \partial_r T\langle r\theta\rangle + \frac{1}{r}\partial_\theta T\langle\theta\theta\rangle + \frac{1}{r\sin\theta}\partial_\varphi T\langle\theta\varphi\rangle +\\
&\quad + \frac{1}{r}[3T\langle r\theta\rangle + \cot\theta\,(T\langle\theta\theta\rangle - T\langle\varphi\varphi\rangle)],\\
(\operatorname{div} \boldsymbol{T})\langle\varphi\rangle &= \partial_r T\langle r\varphi\rangle + \frac{1}{r}\partial_\theta T\langle\theta\varphi\rangle + \frac{1}{r\sin\theta}\partial_\varphi T\langle\varphi\varphi\rangle +\\
&\quad + \frac{1}{r}[3T\langle r\varphi\rangle + 2\cot\theta\, T\langle\theta\varphi\rangle].
\end{aligned}\right\} \tag{6A.8}$$

B. Tensor functions.

This chapter, after explaining the concepts of tensor functions as we use them in this treatise, lists and in some cases proves theorems of invariant theory frequently useful in modern continuum mechanics. Some readers may prefer to pass directly to Sect. 14, where the proper subject of the treatise begins, leaving the present chapter for reference as needed.

I. Basic concepts.

7. Definitions. In this chapter we restrict attention mostly to second-order tensors. These form a space $\mathscr{L}$ of dimension n^2. Often we confine our considerations to symmetric tensors, which form a subspace $\mathscr{S}$ of $\mathscr{L}$ of dimension $\frac{1}{2}n(n+1)$.

Any function whose arguments are tensors in $\mathscr{L}$ or $\mathscr{S}$ and whose values are scalars or tensors is called a *tensor function*.

Let

$$\varepsilon = \varepsilon(\boldsymbol{A}) \tag{7.1}$$

be a scalar-valued tensor function of one tensor variable $\boldsymbol{A}$ in $\mathscr{L}$ or $\mathscr{S}$. More precisely, the domain of the function ε is a suitable subset of $\mathscr{L}$ or of $\mathscr{S}$, and the range of ε is a set of real numbers. Let A_{km} be the covariant components of $\boldsymbol{A}$ relative to some vector basis $\boldsymbol{e}_k$. Then the function (7.1) corresponds to a real-valued function of the n^2 real variables A_{km} if $\boldsymbol{A} \in \mathscr{L}$ or of the $\frac{1}{2}n(n+1)$ real variables A_{km}, $k \leqq m$, if $\boldsymbol{A} \in \mathscr{S}$. The form of this function depends, of course, on the choice of the basis $\boldsymbol{e}_k$. The component form of (7.1) will thus be written

$$\varepsilon = \varepsilon(A_{pq};\, \boldsymbol{e}_r) \quad \text{or} \quad \varepsilon = \varepsilon(A_{pq}) \tag{7.2}$$

depending on whether we do or do not wish to stress the dependence on the basis $\boldsymbol{e}_r$. This dependence must be observed in particular when the components of a tensor field referred to curvilinear co-ordinates are substituted for the A_{km}. In this case the basis vectors and hence the functional form of $(7.2)_2$ will in general vary from point to point. Similar remarks apply to scalar-valued functions of several tensor variables. Examples of scalar-valued tensor functions are

$$\varepsilon(\boldsymbol{A}) = \operatorname{tr} \boldsymbol{A}^m, \qquad \varepsilon(\boldsymbol{A}) = \det \boldsymbol{A}, \qquad \varepsilon(\boldsymbol{A}, \boldsymbol{B}) = \operatorname{tr}(\boldsymbol{ABAB}). \tag{7.3}$$

Let

$$\boldsymbol{B} = \mathfrak{f}(\boldsymbol{A}) \tag{7.4}$$

be a tensor-valued function of a tensor variable $\boldsymbol{A}$. The component form of (7.4) involves n^2 functions $\mathfrak{f}_{km}$ of n^2 real variables A_{km} if $\boldsymbol{A}, \boldsymbol{B} \in \mathscr{L}$, or $\frac{1}{2}n(n+1)$ functions $\mathfrak{f}_{km}$, $k \leqq m$, of $\frac{1}{2}n(n+1)$ real variables A_{km}, $k \leqq m$, if $\boldsymbol{A}, \boldsymbol{B} \in \mathscr{S}$. As in (7.2) we may or may not stress the dependence on the basis $\boldsymbol{e}_k$, as the context requires:

$$B_{km} = \mathfrak{f}_{km}(A_{pq};\, \boldsymbol{e}_r), \qquad B_{km} = \mathfrak{f}_{km}(A_{pq}). \tag{7.5}$$

For tensor functions of several variables, the situation is analogous.

Examples of tensor-valued tensor functions are *tensor polynomials*. These are linear combinations, with constant coefficients, of products of the variables $\boldsymbol{A}_1, \boldsymbol{A}_2, \ldots, \boldsymbol{A}_{\mathfrak{l}}$, taken in any order. Each term of a tensor polynomial is of the form

$$c\, \boldsymbol{A}_{\mathfrak{k}_1}^{p_1} \boldsymbol{A}_{\mathfrak{k}_2}^{p_2} \ldots \boldsymbol{A}_{\mathfrak{k}_r}^{p_r}, \tag{7.6}$$

where c is a constant, each $\mathfrak{k}_j$ is one of the indices $1, 2, \ldots, \mathfrak{l}$, and p_j is a non-negative integer. An infinite sum of terms of the form (7.6) is called a *tensor power series* provided that it converges, at least for suitably restricted arguments.

A tensor power series in one variable has the form

$$\boldsymbol{B} = \mathfrak{p}(\boldsymbol{A}) = \sum_{k=0}^{\infty} c_k \boldsymbol{A}^k. \tag{7.7}$$

If all but a finite number of the constant coefficients c_k reduce to zero, $\mathfrak{p}(\boldsymbol{A})$ is a *tensor polynomial in one variable*. If the argument $\boldsymbol{A}$ is in $\mathscr{S}$, then the series (7.7) will have a sum $\boldsymbol{B} \in \mathscr{S}$.

The tensor power series constitute only a very special class of tensor functions. Even if all components $\mathfrak{f}_{km}(A_{pq})$ of a tensor function have convergent power series expansions or are polynomials, the tensor function need not be a tensor power series or tensor polynomial, respectively, as can be seen from the following examples:

$$\mathfrak{f}(\boldsymbol{A}) = \boldsymbol{A}^T, \qquad \mathfrak{f}(\boldsymbol{A}) = (\operatorname{tr} \boldsymbol{A})\, \boldsymbol{1}, \qquad \mathfrak{f}(\boldsymbol{A}) = \sqrt{\boldsymbol{A} - \boldsymbol{1}}. \tag{7.8}$$

In the last example, $\boldsymbol{A}$ is to be restricted to such tensors as render $\boldsymbol{A} - \boldsymbol{1}$ positive semi-definite and symmetric.

Generalizations of tensor power series of one variable are known in the literature as matrix functions[1]. However, the emphasis in most studies of this subject is on complex matrices, while we are concerned exclusively with real matrices. For real matrices most of the definitions of matrix functions proposed in the literature reduce to special cases of isotropic tensor functions, which will be discussed in the following section.

A tensor function of one tensor variable $\boldsymbol{A}$ is said to be *linear* if it is additive and homogeneous in $\boldsymbol{A}$. Every scalar-valued linear tensor function $\lambda(\boldsymbol{A})$ has the form

$$\lambda(\boldsymbol{A}) = \operatorname{tr}(\boldsymbol{L}^T \boldsymbol{A}) = L^{km} A_{km}, \tag{7.9}$$

where $\boldsymbol{L}$ is a tensor in $\mathscr{L}$ or $\mathscr{S}$, depending on whether the domain of λ is in $\mathscr{L}$ or $\mathscr{S}$. We note that $\operatorname{tr}(\boldsymbol{A}^T \boldsymbol{B}) = A^{km} B_{km}$ satisfies the axioms for an inner product in the space $\mathscr{L}$ or $\mathscr{S}$. Hence, (7.9) is simply a special case of the general theorem which states that, in an inner-product space, every linear form has a representation as an inner product. We use the following bracket notation:

$$\boldsymbol{B} = \mathsf{L}[\boldsymbol{A}], \tag{7.10}$$

when L is a tensor-valued linear tensor function. In this case L may be regarded as a linear transformation of the space $\mathscr{L}$ or of the space $\mathscr{S}$. A component form of (7.10) is

$$B_{km} = \mathsf{L}_{kmpq} A^{pq}, \tag{7.11}$$

and L may be identified with the fourth-order tensor whose covariant components are L_{kmpq}. If $\boldsymbol{A}$ and $\boldsymbol{B}$ are restricted to $\mathscr{S}$ in (7.10), then the components of L satisfy the relations

$$\mathsf{L}_{kmpq} = \mathsf{L}_{mkpq} = \mathsf{L}_{kmqp} = \mathsf{L}_{mkqp}. \tag{7.12}$$

L is a symmetric linear transformation of the space $\mathscr{L}$ or $\mathscr{S}$ if and only if the components obey the relations

$$\mathsf{L}_{kmpq} = \mathsf{L}_{pqkm}. \tag{7.13}$$

A tensor function of several tensor variables is called *multilinear* if it is linear in each of its variables. Multilinear tensor functions may be identified with tensors of higher order. For example, a scalar-valued bilinear tensor function has the form

$$\lambda[\boldsymbol{A}, \boldsymbol{B}] = \mathsf{L}_{kmpq} A^{km} B^{pq}. \tag{7.14}$$

The relations (7.12) hold when $\boldsymbol{A}$ and $\boldsymbol{B}$ are restricted to $\mathscr{S}$, and the relations (7.13) are valid if and only if

$$\lambda[\boldsymbol{A}, \boldsymbol{B}] = \lambda[\boldsymbol{B}, \boldsymbol{A}] \tag{7.15}$$

for all $\boldsymbol{A}$ and $\boldsymbol{B}$.

8. Invariants and isotropic tensor functions. A scalar-valued tensor function $\varepsilon(\boldsymbol{A}_1, \boldsymbol{A}_2, \ldots, \boldsymbol{A}_{\mathfrak{l}})$ is said to be *isotropic* if the relation

$$\varepsilon(\boldsymbol{A}_1, \ldots, \boldsymbol{A}_{\mathfrak{l}}) = \varepsilon(\boldsymbol{Q}\boldsymbol{A}_1\boldsymbol{Q}^T, \ldots, \boldsymbol{Q}\boldsymbol{A}_{\mathfrak{l}}\boldsymbol{Q}^T) \tag{8.1}$$

holds for all orthogonal tensors $\boldsymbol{Q}$ and all $\boldsymbol{A}_{\mathfrak{k}}$ in the domain of definition of ε. We say that ε is *isotropic relative to* $\mathscr{g}$, where $\mathscr{g}$ is a subgroup of the group $\mathscr{o}$ of all orthogonal tensors, if (8.1) holds for all $\boldsymbol{Q}$ in $\mathscr{g}$. A scalar-valued isotropic tensor function $\varepsilon(\boldsymbol{A})$ of one variable is also called an *orthogonal invariant*, or briefly an *invariant* of $\boldsymbol{A}$. In the case of several variables, the term *simultaneous invariant* is used. The examples (7.3) are all invariants.

[1] For a brief discussion of this topic, see Sect. 42 of the Appendix to CFT.

Of particular importance are the *principal invariants* $I_k(\boldsymbol{A})$, $k=1, 2, \ldots, n$, defined as the coefficients of the following polynomial in λ:

$$\det(\lambda \mathbf{1}+\boldsymbol{A})=\lambda^n+I_1(\boldsymbol{A})\,\lambda^{n-1}+\cdots+I_{n-1}(\boldsymbol{A})\,\lambda+I_n(\boldsymbol{A}). \tag{8.2}$$

We have, in particular,

$$I_1(\boldsymbol{A})=\operatorname{tr}\boldsymbol{A}, \qquad I_n(\boldsymbol{A})=\det\boldsymbol{A}. \tag{8.3}$$

Other important invariants are the *moments* $\bar{I}_k(\boldsymbol{A})$, defined by

$$\bar{I}_k(\boldsymbol{A})=\operatorname{tr}\boldsymbol{A}^k. \tag{8.4}$$

In three dimensions ($n=3$) we use the notation

$$I_1(\boldsymbol{A})=I_{\boldsymbol{A}}, \quad I_2(\boldsymbol{A})=II_{\boldsymbol{A}}, \quad I_3(\boldsymbol{A})=III_{\boldsymbol{A}} \tag{8.5}$$

for the principal invariants and a similar notation for the moments: $\bar{I}_{\boldsymbol{A}}$, $\overline{II}_{\boldsymbol{A}}$, $\overline{III}_{\boldsymbol{A}}$. When no confusion should result, the subscript $\boldsymbol{A}$ is omitted. For a detailed discussion of these and other special invariants, the reader is referred to Sects. 38 and 39 of the Appendix to CFT.

Important examples of simultaneous invariants are the traces of products of the argument tensors. They are of the form

$$\varepsilon(\boldsymbol{A}_1, \ldots, \boldsymbol{A}_\mathfrak{l})=\operatorname{tr}\boldsymbol{A}_{\mathfrak{k}_1}^{p_1}\ldots\boldsymbol{A}_{\mathfrak{k}_r}^{p_r}, \tag{8.6}$$

where the p_j are positive integers.

A tensor-valued function $\mathfrak{f}(\boldsymbol{A}_1, \ldots, \boldsymbol{A}_\mathfrak{l})$ is said to be *isotropic* if the relation

$$\boldsymbol{Q}\,\mathfrak{f}(\boldsymbol{A}_1, \ldots, \boldsymbol{A}_\mathfrak{l})\,\boldsymbol{Q}^T=\mathfrak{f}(\boldsymbol{Q}\boldsymbol{A}_1\boldsymbol{Q}^T, \ldots, \boldsymbol{Q}\boldsymbol{A}_\mathfrak{l}\boldsymbol{Q}^T) \tag{8.7}$$

is satisfied for all orthogonal tensors $\boldsymbol{Q}$ and all $\boldsymbol{A}_\mathfrak{k}$ in the domain of definition of $\mathfrak{f}$. We say that $\mathfrak{f}$ is *isotropic relative to* $\mathscr{g}$ if (8.7) holds for all $\boldsymbol{Q}$ in the subgroup $\mathscr{g}$ of orthogonal tensors. Examples of isotropic tensor functions are the tensor polynomials, the tensor power series, and the functions (7.8) mentioned in the previous section.

Isotropic tensor functions can also be characterized as follows: *A tensor function is isotropic if and only if the forms of its component functions depend on the basis* $\boldsymbol{e}_k$ *only through the components* $g_{km}=\boldsymbol{e}_k\cdot\boldsymbol{e}_m$ *of the unit tensor. In particular, a tensor function is isotropic if and only if the forms of its component functions are the same for all orthonormal bases.* The latter characterization has a useful consequence when tensor fields are substituted for the arguments of an isotropic tensor function and when these tensor fields are represented by their physical components in an orthogonal co-ordinate system. The forms of the component functions are then independent of the point in space.

We prove the above theorem for an invariant $\varepsilon=\varepsilon(\boldsymbol{A})$ with the component form

$$\varepsilon=\varepsilon(\boldsymbol{A})=\varepsilon(A^{pq}\boldsymbol{e}_p\otimes\boldsymbol{e}_q)\equiv\varepsilon(A^{pq};\,\boldsymbol{e}_r), \tag{8.8}$$

say. An orthogonal transformation $\boldsymbol{Q}$ maps the basis $\boldsymbol{e}_k$ onto another basis $\boldsymbol{e}_k^*$:

$$\boldsymbol{e}_k^*=\boldsymbol{Q}\boldsymbol{e}_k. \tag{8.9}$$

Since $\varepsilon(\boldsymbol{A})$ is isotropic,

$$\varepsilon(\boldsymbol{A})=\varepsilon(\boldsymbol{Q}\boldsymbol{A}\boldsymbol{Q}^T) \tag{8.10}$$

for all orthogonal $\boldsymbol{Q}$, and hence by (8.8) and (8.9)

$$\left.\begin{aligned}\varepsilon(A^{pq};\boldsymbol{e}_r)&=\varepsilon(\boldsymbol{Q}A^{pq}\boldsymbol{e}_p\otimes\boldsymbol{e}_q\boldsymbol{Q}^T),\\&=\varepsilon(A^{pq}\boldsymbol{Q}\boldsymbol{e}_p\otimes\boldsymbol{Q}\boldsymbol{e}_q),\\&=\varepsilon(A^{pq}\boldsymbol{e}_p^*\otimes\boldsymbol{e}_q^*);\end{aligned}\right\} \tag{8.11}$$

i.e.

$$\varepsilon(A^{pq};\,\boldsymbol{e}_r)=\varepsilon(A^{pq};\,\boldsymbol{Q}\boldsymbol{e}_r). \tag{8.12}$$

Therefore, by the theorem of CAUCHY we shall prove in Sect. 11, the values of ε depend on the $\boldsymbol{e}_k$ only through the orthogonal invariants $g_{km}=\boldsymbol{e}_k\cdot\boldsymbol{e}_m$ of the vectors $\boldsymbol{e}_k$. The method of proof is the same for general isotropic tensor functions.

The proof above shows also that a tensor function is isotropic relative to a subgroup $\mathscr{g}$ of the orthogonal group if and only if the forms of its component functions remain unchanged under changes of basis of the form (8.8) with $\boldsymbol{Q}\in\mathscr{g}$.

The concept of isotropy and isotropy relative to a subgroup $\mathscr{g}$ of the orthogonal group can be extended to functions whose arguments and values consist of any combination of scalars, vectors, and tensors of any order. Isotropy relative to $\mathscr{g}$ means that the functional relation remains unchanged if every tensor A occurring as an argument or value is replaced by the tensor $\bar{\mathsf{A}}$ whose components $\bar{\mathsf{A}}_{k_1\ldots k_\nu}$ are related to the components $\mathsf{A}_{k_1\ldots k_\nu}$ of A by the following transformation rule:

$$\bar{\mathsf{A}}_{k_1\ldots k_\nu}=Q_{k_1}{}^{m_1}\cdots Q_{k_\nu}{}^{m_\nu}\mathsf{A}_{m_1\ldots m_\nu}, \tag{8.13}$$

where $\boldsymbol{Q}$ is any member of $\mathscr{g}$. In the special cases when A is a vector $\boldsymbol{v}$ or a second-order tensor $\boldsymbol{A}$, (8.13) reduces to

$$\bar{\boldsymbol{v}}=\boldsymbol{Q}\boldsymbol{v}, \qquad \bar{\boldsymbol{A}}=\boldsymbol{Q}\boldsymbol{A}\boldsymbol{Q}^T, \tag{8.14}$$

respectively. For example, a vector-valued function $\mathfrak{h}(\boldsymbol{A},\boldsymbol{v})$ of a symmetric tensor $\boldsymbol{A}\in\mathscr{S}$ and a vector $\boldsymbol{v}$ is isotropic relative to $\mathscr{g}$ if

$$\boldsymbol{Q}\mathfrak{h}(\boldsymbol{A},\boldsymbol{v})=\mathfrak{h}(\boldsymbol{Q}\boldsymbol{A}\boldsymbol{Q}^T,\boldsymbol{Q}\boldsymbol{v}) \tag{8.15}$$

for all $\boldsymbol{A}\in\mathscr{S}$, all vectors $\boldsymbol{v}$, and all $\boldsymbol{Q}\in\mathscr{g}$.

If $\mathscr{g}$ is the full orthogonal group $\mathscr{o}$, functions isotropic relative to $\mathscr{o}$ are called simply *isotropic*. If $\mathscr{g}$ is the proper orthogonal group, functions isotropic relative to $\mathscr{g}$ are called *hemitropic*.

When the dimension n is 3 or any odd number and when only tensors of even order occur as arguments and values, there is no distinction between isotropic and hemitropic functions. This is true because, for an improper orthogonal $\boldsymbol{Q}$ ($\det\boldsymbol{Q}=-1$), the transform $\bar{\mathsf{A}}$ of a tensor A of even order remains unchanged when $\boldsymbol{Q}$ is replaced by $-\boldsymbol{Q}$, which is proper-orthogonal since $\det(-\boldsymbol{Q})=-\det\boldsymbol{Q}=+1$. The distinction between isotropic and hemitropic functions disappears, in particular, for the tensor functions considered in Sect. 7. It does not disappear in general. For example, since for any two vectors $\boldsymbol{v}$ and $\boldsymbol{w}$ and any orthogonal $\boldsymbol{Q}$,

$$\boldsymbol{Q}\boldsymbol{v}\times\boldsymbol{Q}\boldsymbol{w}=(\det\boldsymbol{Q})\,\boldsymbol{Q}(\boldsymbol{v}\times\boldsymbol{w}) \tag{8.16}$$

the cross product $\boldsymbol{v}\times\boldsymbol{w}$ is a hemitropic vector-valued function of the two vectors $\boldsymbol{v}$ and $\boldsymbol{w}$ that is not isotropic. Likewise, the scalar-valued function

$$\delta(\boldsymbol{A},\boldsymbol{v})=e^{kmp}A_{km}v_p, \tag{8.17}$$

in which e^{kmp} are the components of the determinant tensor[1], is hemitropic but not isotropic.

The term *simultaneous invariant* is used not only for scalar-valued isotropic functions of several second-order tensor variables, but also for scalar-valued isotropic or hemitropic functions of any set of vectors and tensors of any order.

9. Gradients of tensor functions. We consider a scalar-valued tensor function $\varepsilon(\boldsymbol{A})$ and assume that the corresponding component function $\varepsilon(A_{km})$ has con-

[1] The components e^{kmp} are $1/\sqrt{g}$, $-1/\sqrt{g}$, or 0, according as kmp is an even permutation of 123, an odd permutation, or neither.

tinuous partial derivatives with respect to all its variables A_{km}. The functions $\mathfrak{f}^{km}$ given by the definition $\mathfrak{f}^{km} = \frac{\partial \varepsilon}{\partial A_{km}}$ are then the contravariant components of a tensor-valued tensor function, which will be denoted by

$$\frac{\partial \varepsilon}{\partial \boldsymbol{A}} = \varepsilon_{\boldsymbol{A}}(\boldsymbol{A}) = \varepsilon_{\boldsymbol{A}} \tag{9.1}$$

and which will be called the *gradient* of ε. The gradient $\varepsilon_{\boldsymbol{A}}$ can also be defined in an intrinsic way by

$$\operatorname{tr}\{\varepsilon_{\boldsymbol{A}}^T \boldsymbol{C}\} = \operatorname{tr}\{\varepsilon_{\boldsymbol{A}} \boldsymbol{C}^T\} = \frac{d}{ds}\,\varepsilon(\boldsymbol{A} + s\boldsymbol{C})\Big|_{s=0} = \frac{\partial \varepsilon}{\partial A_{km}} C_{km} \tag{9.2}$$

where $\boldsymbol{C}$ is an arbitrary tensor. In the case when $\varepsilon(\boldsymbol{A})$ is defined only for symmetric tensors $\boldsymbol{A} \in \mathscr{S}$, then also $\boldsymbol{C}$ must be restricted to $\mathscr{S}$ in $(9.2)_{2,3}$, and $\varepsilon_{\boldsymbol{A}} = \varepsilon_{\boldsymbol{A}}^T$. In order to obtain the components $\frac{\partial \varepsilon}{\partial A_{km}}$ of $\varepsilon_{\boldsymbol{A}}$, one must first extend the domain of definition of ε to the space $\mathscr{L}$ by putting $\bar{\varepsilon}(\boldsymbol{A}) \equiv \varepsilon\left(\tfrac{1}{2}(\boldsymbol{A} + \boldsymbol{A}^T)\right)$ and then take the derivatives of $\bar{\varepsilon}(\boldsymbol{A})$ with respect to the n^2 components A_{km} of $\boldsymbol{A} \in \mathscr{L}$. After these derivatives have been taken, $\boldsymbol{A}$ must be restricted again to $\mathscr{S}$.[1]

The *partial gradients* of a function of several tensor variables are defined in an obvious way. If $\varepsilon(\boldsymbol{A})$ is an invariant, then its gradient $\varepsilon_{\boldsymbol{A}}(\boldsymbol{A})$ is an isotropic tensor function, as can be seen easily. The partial gradients of a simultaneous invariant are also isotropic.

We now compute the gradients of the principal invariants $I_k(\boldsymbol{A})$ mentioned in the previous section. Consider first the invariant $\varepsilon(\boldsymbol{A}) = I_n(\boldsymbol{A}) = \det \boldsymbol{A}$, and assume that $\boldsymbol{A}$ is invertible. We have

$$\det(\boldsymbol{A} + s\boldsymbol{C}) = s^n (\det \boldsymbol{A}) \det\left(\frac{1}{s}\,\mathbf{1} + \boldsymbol{A}^{-1}\boldsymbol{C}\right);$$

hence, using (8.2) when $\lambda = 1/s$ and when $\boldsymbol{A}$ is replaced by $\boldsymbol{A}^{-1}\boldsymbol{C}$, we see that

$$\det(\boldsymbol{A} + s\boldsymbol{C}) = (\det \boldsymbol{A})\{1 + I_1(\boldsymbol{A}^{-1}\boldsymbol{C})\,s + \cdots + I_n(\boldsymbol{A}^{-1}\boldsymbol{C})\,s^n\}.$$

Differentiation with respect to s and then putting $s=0$ gives

$$\frac{d}{ds}\det(\boldsymbol{A} + s\boldsymbol{C})\Big|_{s=0} = (\det \boldsymbol{A})\, I_1(\boldsymbol{A}^{-1}\boldsymbol{C}) = \operatorname{tr}\{(\det \boldsymbol{A})\,\boldsymbol{A}^{-1}\boldsymbol{C}\}.$$

Since $\boldsymbol{C}$ is arbitrary, it follows from (9.2) that

$$\frac{\partial I_n(\boldsymbol{A})}{\partial \boldsymbol{A}} = \frac{\partial \det \boldsymbol{A}}{\partial \boldsymbol{A}} = \det{}_{\boldsymbol{A}} \boldsymbol{A} = (\det \boldsymbol{A})\,(\boldsymbol{A}^{-1})^T. \tag{9.3}$$

If an orthonormal basis is used, this formula corresponds to the relation

$$\frac{\partial}{\partial A_{km}}\{\det \|A_{pq}\|\} = \bar{A}_{km} \tag{9.4}$$

where the $\bar{A}_{km}$ are the cofactors of the matrix $\|A_{km}\|$. The gradient $\det_{\boldsymbol{A}} \boldsymbol{A}$ is another example of an isotropic tensor function that is not a tensor polynomial even though its components are polynomials.

In order to determine the gradients of the other principal invariants $I_k(\boldsymbol{A})$, we take the gradient, with respect to $\boldsymbol{A}$, of the equation (8.2), which may be written in the form

$$\det(\lambda\,\mathbf{1} + \boldsymbol{A}) = \sum_{k=0}^{n} \lambda^{n-k} I_k(\boldsymbol{A}), \tag{9.5}$$

[1] For example, if ε is defined by $\varepsilon(\boldsymbol{A}) = A_{12}$ for $\boldsymbol{A} \in \mathscr{S}$, then $\frac{\partial \varepsilon}{\partial A_{12}} = \frac{\partial \varepsilon}{\partial A_{21}} = \frac{1}{2}$.

where $I_0(\boldsymbol{A})=1$. By (9.3) we have

$$\frac{\partial}{\partial \boldsymbol{A}} \det(\lambda \mathbf{1}+\boldsymbol{A})=[\det(\lambda\mathbf{1}+\boldsymbol{A})][(\lambda\mathbf{1}+\boldsymbol{A})^{-1}]^T; \tag{9.6}$$

hence, by (9.5),

$$\left.\begin{aligned} \mathbf{1}\det(\lambda\mathbf{1}+\boldsymbol{A}) &= (\lambda\mathbf{1}+\boldsymbol{A})^T \sum_{k=0}^{n} \lambda^{n-k}\frac{\partial I_k}{\partial \boldsymbol{A}},\\ \mathbf{1}\sum_{k=0}^{n}\lambda^{n-k} I_k &= \mathbf{1}\sum_{k=0}^{n}\lambda^{n-k+1}\frac{\partial I_k}{\partial \boldsymbol{A}} + \boldsymbol{A}^T\sum_{k=0}^{n}\lambda^{n-k}\frac{\partial I_k}{\partial \boldsymbol{A}}. \end{aligned}\right\} \tag{9.7}$$

Comparing the coefficients of the powers of λ, we get the recursion formula

$$\frac{\partial I_{k+1}}{\partial \boldsymbol{A}} = I_k \mathbf{1} - \boldsymbol{A}^T\frac{\partial I_k}{\partial \boldsymbol{A}}, \qquad k=0,1,\ldots,n \tag{9.8}$$

where $I_0 \equiv 1$, $I_{n+1}\equiv 0$. Using induction, we arrive at the following expressions for the *gradients of the principal invariants*:

$$\frac{\partial I_k(\boldsymbol{A})}{\partial \boldsymbol{A}} = \left[\sum_{j=0}^{k-1}(-1)^j I_{k-j-1}(\boldsymbol{A})\,\boldsymbol{A}^j\right]^T. \tag{9.9}$$

This relation follows also in the case when $k=n+1$, with $I_{n+1}=0$, and gives the theorem of HAMILTON and CAYLEY:

$$\boldsymbol{A}^n - I_1(\boldsymbol{A})\boldsymbol{A}^{n-1} + I_2(\boldsymbol{A})\boldsymbol{A}^{n-2} - \cdots + \cdots + (-1)^n I_n(\boldsymbol{A})\mathbf{1} = \mathbf{0}. \tag{9.10}$$

For three dimensions ($n=3$), (9.9) reduces to

$$\frac{\partial I}{\partial \boldsymbol{A}}=\mathbf{1}, \quad \frac{\partial II}{\partial \boldsymbol{A}} = I\mathbf{1}-\boldsymbol{A}^T, \quad \frac{\partial III}{\partial \boldsymbol{A}} = III(\boldsymbol{A}^{-1})^T = [\boldsymbol{A}^2 - I\boldsymbol{A} + II\mathbf{1}]^T, \tag{9.11}$$

which formulae correspond to the component relations[1]

$$\left.\begin{aligned} \frac{\partial I}{\partial A_{km}} &= g^{km}, \qquad \frac{\partial II}{\partial A_{km}} = I g^{km} - A^{mk},\\ \frac{\partial III}{\partial A_{km}} &= III(A^{-1})^{mk} = A^m{}_p A^{pk} - I A^{mk} + II g^{mk}. \end{aligned}\right\} \tag{9.12}$$

Let $\mathfrak{f}(\boldsymbol{A})$ be a tensor-valued tensor function with components $\mathfrak{f}_{km}(A_{pq})$, and assume that these component functions have continuous partial derivatives,

$$(\mathfrak{f}_{\boldsymbol{A}})_{km}{}^{pq} = \frac{\partial \mathfrak{f}_{km}}{\partial A_{pq}}. \tag{9.13}$$

These partial derivatives are the components of a tensor of order four, which is called the *gradient* of $\mathfrak{f}$ and is denoted by

$$\frac{\partial \mathfrak{f}}{\partial \boldsymbol{A}} = \mathfrak{f}_{\boldsymbol{A}} = \mathfrak{f}_{\boldsymbol{A}}(\boldsymbol{A}). \tag{9.14}$$

The gradient may also be defined directly by

$$\frac{\partial \mathfrak{f}}{\partial \boldsymbol{A}}[\boldsymbol{C}] = \mathfrak{f}_{\boldsymbol{A}}(\boldsymbol{A})[\boldsymbol{C}] = \mathfrak{f}_{\boldsymbol{A}}[\boldsymbol{C}] = \frac{d}{ds}\mathfrak{f}(\boldsymbol{A}+s\boldsymbol{C})\Big|_{s=0}. \tag{9.15}$$

For each fixed $\boldsymbol{A}$, under suitable conditions of smoothness, $\mathfrak{f}_{\boldsymbol{A}}[\boldsymbol{C}]$ can be shown to be a linear tensor function of $\boldsymbol{C}$, so that the notation conforms with (7.10). In accord with the considerations at the end of Sect. 7, this linear tensor function

[1] See also (38.16) of the Appendix to CFT.

corresponds to the fourth-order tensor with components (9.13), and $\mathfrak{f}_{A}[C]$ has the components

$$(\mathfrak{f}_{A}[C])_{km}=\frac{\partial \mathfrak{f}_{km}}{\partial A_{pq}}C_{pq}. \tag{9.16}$$

When $\mathfrak{f}(A)$ is defined only for symmetric tensors $A\in\mathscr{S}$, before computing the derivatives $\frac{\partial \mathfrak{f}_{km}}{\partial A_{pq}}$ one must first extend $\mathfrak{f}$ by putting $\bar{\mathfrak{f}}(A)=\mathfrak{f}(\frac{1}{2}(A+A^T))$ for $A\in\mathscr{L}$ and then differentiate $\bar{\mathfrak{f}}$, as in the case of scalar-valued functions.

The tensor function $\mathfrak{f}_{A}(A)[C]$ of the two tensor variables A and C is linear in the second variable C but not necessarily in the first variable A. If $\mathfrak{f}(A)$ is an isotropic function, then also $\mathfrak{f}_{A}(A)[C]$ is an isotropic function.

Partial gradients of tensor functions of several variables and higher gradients can be defined in an obvious way. The usual rules of differentiation are valid for the gradients defined here, in particular, the rule for the differentiation of composite functions (chain rule). Attention must be paid to the order of multiplication when differentiating products of tensor functions.

The following formulae are easy to establish:

$$\frac{\partial A}{\partial A}[C]=C, \qquad \frac{\partial A^T}{\partial A}[C]=C^T, \tag{9.17}$$

$$\frac{\partial A^m}{\partial A}[C]=\sum_{k=0}^{m-1}A^k C A^{m-1-k}, \qquad m>0, \tag{9.18}$$

$$\frac{\partial A^{-1}}{\partial A}[C]=-A^{-1}CA^{-1}. \tag{9.19}$$

We determine now the gradients of the moments (8.4): $\bar{I}_k(A)=\operatorname{tr}A^k$. Since $\bar{I}_1(A)=I_1(A)=\operatorname{tr}A$, we have by (9.9)

$$\frac{\partial \operatorname{tr} A}{\partial A}=\operatorname{tr}_A A=1. \tag{9.20}$$

By the chain rule, by (9.2), and by (9.18), we get

$$\operatorname{tr}\left(\left[\frac{\partial \bar{I}_k(A)}{\partial A}\right]^T C\right)=\operatorname{tr}\left(\sum_{j=0}^{k-1}A^j C A^{k-1-j}\right)=k\operatorname{tr}(A^{k-1}C). \tag{9.21}$$

Since C is arbitrary, we can infer the following formula for the *gradients of the moments*:

$$\frac{\partial \bar{I}_k(A)}{\partial A}=k(A^{k-1})^T. \tag{9.22}$$

Component forms of (9.22) in three dimensions are[1]

$$\frac{\partial \bar{I}_A}{\partial A_{km}}=g^{mk}, \qquad \frac{\partial \overline{II}_A}{\partial A_{km}}=2A^{mk}, \qquad \frac{\partial \overline{III}_A}{\partial A_{km}}=3A^m{}_p A^{pk}. \tag{9.23}$$

II. Representation theorems.

10. Representation of invariants of one symmetric tensor. In this subchapter we assume that all tensors are *symmetric*, i.e., we confine our attention to the space $\mathscr{S}$ of symmetric tensors. Assume that A and B are two symmetric tensors whose principal invariants coincide:

$$I_j(A)=I_j(B), \qquad j=1, 2, \ldots, n. \tag{10.1}$$

[1] See also (38.16) of the Appendix to CFT.

The proper numbers $a_1, a_2, \ldots, a_n$ of $\boldsymbol{A}$, with appropriate multiplicity, are the solutions of the equation

$$x^n - I_1(\boldsymbol{A})\, x^{n-1} + \cdots + (-1)^n I_n(\boldsymbol{A}) = 0. \tag{10.2}$$

By (10.1), the corresponding equation for $\boldsymbol{B}$ is the same. Therefore the proper numbers a_k of $\boldsymbol{A}$ coincide with the proper numbers b_k of $\boldsymbol{B}$. We may order the b_k so that

$$a_k = b_k, \quad k = 1, 2, \ldots, n. \tag{10.3}$$

Since $\boldsymbol{A}$ is symmetric, there is an orthonormal basis $\boldsymbol{e}_k$ of proper vectors of $\boldsymbol{A}$ such that $\boldsymbol{A}\boldsymbol{e}_k = a_k \boldsymbol{e}_k$. Similarly, we can find an orthonormal basis $\boldsymbol{f}_k$ of proper vectors of $\boldsymbol{B}$ such that $\boldsymbol{B}\boldsymbol{f}_k = b_k \boldsymbol{f}_k = a_k \boldsymbol{f}_k$. There is a unique orthogonal transformation $\boldsymbol{Q}$ that maps the basis $\boldsymbol{e}_k$ into the basis $\boldsymbol{f}_k$, so that $\boldsymbol{Q}\boldsymbol{e}_k = \boldsymbol{f}_k$. It follows that

$$\boldsymbol{Q}\boldsymbol{A}\boldsymbol{e}_k = a_k \boldsymbol{Q}\boldsymbol{e}_k = a_k \boldsymbol{f}_k = b_k \boldsymbol{f}_k = \boldsymbol{B}\boldsymbol{f}_k = \boldsymbol{B}\boldsymbol{Q}\boldsymbol{e}_k; \tag{10.4}$$

hence $\boldsymbol{Q}\boldsymbol{A} = \boldsymbol{B}\boldsymbol{Q}$, or $\boldsymbol{B} = \boldsymbol{Q}\boldsymbol{A}\boldsymbol{Q}^T$. Conversely, if $\boldsymbol{B} = \boldsymbol{Q}\boldsymbol{A}\boldsymbol{Q}^T$ for some orthogonal $\boldsymbol{Q}$, then (10.1) holds.

We have shown that the principal invariants of two symmetric tensors $\boldsymbol{A}$ and $\boldsymbol{B}$ coincide if and only if there is an orthogonal tensor $\boldsymbol{Q}$ such that $\boldsymbol{B} = \boldsymbol{Q}\boldsymbol{A}\boldsymbol{Q}^T$. It follows from the definition (8.10) that $\varepsilon(\boldsymbol{A}) = \varepsilon(\boldsymbol{B})$ for any invariant ε whenever $\boldsymbol{A}$ and $\boldsymbol{B}$ have the same principal invariants. This proposition gives rise to the following *representation theorem for invariants*:

A scalar-valued function $\varepsilon(\boldsymbol{A})$ of a symmetric tensor variable is an invariant if and only if it can be expressed as a function of the principal invariants of $\boldsymbol{A}$. We use the same symbol ε for this function, writing

$$\varepsilon(\boldsymbol{A}) = \varepsilon(I_1, I_2, \ldots, I_n). \tag{10.5}$$

The principal invariants $I_k = I_k(\boldsymbol{A})$ are the elementary symmetric functions of the proper numbers a_k of $\boldsymbol{A}$. The fundamental theorem on symmetric functions[1] states that if ε is a polynomial in the a_k, then $\varepsilon(I_1, \ldots, I_n)$ is a polynomial in the I_k. On the other hand, the proper numbers a_k are the elements of the diagonal matrix corresponding to $\boldsymbol{A}$ relative to the basis $\boldsymbol{e}_k$ mentioned before. Hence the representation theorem for invariants may be supplemented by the following proposition:

If $\varepsilon = \varepsilon(\boldsymbol{A})$ is a polynomial invariant, i.e. an invariant that is a polynomial in the components of $\boldsymbol{A}$, then it can be expressed as a polynomial in the principal invariants I_j of $\boldsymbol{A}$.

It would be desirable to have theorems relating the continuity and differentiability properties of $\varepsilon(I_1, \ldots, I_n)$ to those of the corresponding $\varepsilon(\boldsymbol{A})$. These properties can be carried over easily from $\varepsilon(\boldsymbol{A})$ to $\varepsilon(I_1, \ldots, I_n)$ only so long as the proper numbers of $\boldsymbol{A}$ remain distinct. For the general case, no such theorems are known at the present time. That the situation can be involved is shown by the following example in two dimensions ($n = 2$): The invariant $\varepsilon(\boldsymbol{A}) = |a_1 - a_2|^{\frac{3}{2}} = |I^2 - 4II|^{\frac{3}{4}}$ is a continuously differentiable function of the proper numbers a_1 and a_2 of $\boldsymbol{A}$, even when $a_1 = a_2$. But, as a function of the principal invariants I and II, it is not differentiable when $I^2 - 4II = 0$. For a related representation, a theorem of continuity is stated following Eq. (12.7), below.

Instead of the principal invariants we can use also the first n moments $\bar{I}_j(\boldsymbol{A})$ in the representation (10.5). This is the case because the moment $\bar{I}_j$ is the sum of the j'th powers of the proper numbers a_k of $\boldsymbol{A}$, and because the elementary symmetric functions I_j of the a_k can be expressed as polynomials in the first n

[1] Cf. VAN DER WAERDEN [1937, *3*, § 26], WEYL [1939, *2*, Ch. II, § 2].

of the power sums $\bar{I}_j$ by means of NEWTON'S formulae[1]. Thus *any invariant* $\varepsilon(\boldsymbol{A})$ *is expressible as a function of the first n moments* $\bar{I}_k(\boldsymbol{A}) = \operatorname{tr} \boldsymbol{A}^k$ $(k=1, 2, \ldots, n)$ *of* $\boldsymbol{A}$:

$$\varepsilon(\boldsymbol{A}) = \varepsilon(\bar{I}_1, \bar{I}_2, \ldots, \bar{I}_n). \tag{10.6}$$

If $\varepsilon = \varepsilon(\boldsymbol{A})$ *is a polynomial invariant, then* $\varepsilon(\bar{I}_1, \ldots, \bar{I}_n)$ *is a polynomial in the* $\bar{I}_k$.

A linear invariant $\lambda(\boldsymbol{A})$ is necessarily a scalar multiple of the trace:

$$\lambda(\boldsymbol{A}) = c \operatorname{tr} \boldsymbol{A}. \tag{10.7}$$

11. Representation of simultaneous invariants. The following ***basic representation theorem*** for simultaneous invariants of vectors is due to CAUCHY[2]:

A scalar-valued function $\gamma(\boldsymbol{v}_1, \ldots, \boldsymbol{v}_m)$ *of the vectors* $\boldsymbol{v}_1, \ldots, \boldsymbol{v}_m$ *is isotropic, i.e. invariant under the full orthogonal group, if and only if it can be expressed as a function of the inner products* $\boldsymbol{v}_i \cdot \boldsymbol{v}_j$, $i, j = 1, \ldots, m$.

To sketch the proof of CAUCHY'S theorem we consider two sets of vectors $\boldsymbol{v}_1, \ldots, \boldsymbol{v}_m$ and $\boldsymbol{u}_1, \ldots, \boldsymbol{u}_m$ such that

$$\boldsymbol{v}_i \cdot \boldsymbol{v}_j = \boldsymbol{u}_i \cdot \boldsymbol{u}_j, \quad i, j = 1, \ldots, m. \tag{11.1}$$

Let $\mathscr{V}$ be the subspace spanned by $\boldsymbol{v}_1, \ldots, \boldsymbol{v}_m$. After a suitable rearrangement, we may assume that $\mathscr{V}$ is spanned by the linearly independent set $\boldsymbol{v}_1, \ldots, \boldsymbol{v}_p$, where $p \leqq m$, $p \leqq n$. Since

$$\det \|\boldsymbol{v}_i \cdot \boldsymbol{v}_j\|_{i, j = 1, \ldots, p} \neq 0 \tag{11.2}$$

is a necessary and sufficient condition for the linear independence of $\boldsymbol{v}_1, \ldots, \boldsymbol{v}_p$, it follows from (11.1) that the subspace $\mathscr{U}$ spanned by $\boldsymbol{u}_1, \ldots, \boldsymbol{u}_m$ is spanned by the linearly independent set $\boldsymbol{u}_1, \ldots, \boldsymbol{u}_p$. Now, there is a unique linear transformation $\boldsymbol{Q}$ such that

$$\boldsymbol{Q} \boldsymbol{v}_i = \boldsymbol{u}_i, \quad i = 1, \ldots, p, \tag{11.3}$$

and such that $\boldsymbol{Q}$ reduces to the identity on the orthogonal complement of $\mathscr{V}$. Since, by (11.1) and (11.3), we have

$$\boldsymbol{Q} \boldsymbol{v}_i \cdot \boldsymbol{Q} \boldsymbol{v}_j = \boldsymbol{u}_i \cdot \boldsymbol{u}_j = \boldsymbol{v}_i \cdot \boldsymbol{v}_j, \quad i, j = 1, \ldots, p, \tag{11.4}$$

we infer that $\boldsymbol{Q}$ is orthogonal. If $p < k \leqq m$, we have a relation

$$\boldsymbol{v}_k = \sum_{i=1}^{p} \alpha_i \boldsymbol{v}_i. \tag{11.5}$$

Taking the inner product of (11.5) with $\boldsymbol{v}_j$, we obtain

$$\boldsymbol{v}_k \cdot \boldsymbol{v}_j = \sum_{i=1}^{p} \alpha_i (\boldsymbol{v}_i \cdot \boldsymbol{v}_j), \quad j = 1, \ldots, p, \tag{11.6}$$

which is a system of p equations for the p coefficients α_i. These coefficients are uniquely determined from (11.6) because (11.2) holds. Substituting (11.1) into (11.6), we find that

$$\boldsymbol{u}_k = \sum_{i=1}^{p} \alpha_i \boldsymbol{u}_i, \tag{11.7}$$

and hence, by (11.5) and (11.1), that $\boldsymbol{Q} \boldsymbol{v}_k = \boldsymbol{u}_k$. Therefore, if (11.1) holds, there is an orthogonal transformation $\boldsymbol{Q}$ such that $\boldsymbol{Q} \boldsymbol{v}_i = \boldsymbol{u}_i$ for all $i = 1, \ldots, m$. Con-

---

[1] Cf. VAN DER WAERDEN [1937, *3*, § 26, Aufg. 4], WEYL [1939, *2*, Ch. II, § 3].

[2] CAUCHY [1850, *1*], for the case when $n = 3$.

versely, if $\boldsymbol{Q}\boldsymbol{v}_i = \boldsymbol{u}_i$ for $i=1, \ldots, m$, then (11.1) holds. It follows from the definition of isotropy that

$$\gamma(\boldsymbol{u}_1, \ldots, \boldsymbol{u}_m) = \gamma(\boldsymbol{v}_1, \ldots, \boldsymbol{v}_m) \tag{11.8}$$

for any invariant γ whenever (11.1) holds, which shows that the value of γ depends only on the inner products (11.1).

CAUCHY's theorem has the following supplement[1]. *If $\gamma(\boldsymbol{v}_1, \ldots, \boldsymbol{v}_m)$ is a polynomial invariant, then it can be expressed as a polynomial in the inner products $\boldsymbol{v}_i \cdot \boldsymbol{v}_j$, $i, j = 1, \ldots, m$.*

When $\gamma(\boldsymbol{v}_1, \ldots, \boldsymbol{v}_m)$ is not isotropic but only hemitropic, i.e. invariant only under the proper orthogonal group, then the "determinant-products"[2] $[\boldsymbol{v}_{i_1}, \ldots, \boldsymbol{v}_{i_n}]$ must be added[3] to the list of the inner products $\boldsymbol{v}_i \cdot \boldsymbol{v}_j$ in the theorems above.

There is a large literature, mostly recent, devoted to deriving representations for various kinds of tensor functions that are assumed to be isotropic, hemitropic, or isotropic relative to certain subgroups of the orthogonal group. Many of these representations may be related to CAUCHY's theorem. Some of them will be described below.

For simultaneous invariants of several symmetric tensors only partial results, restricted to the case of three dimensions, are known. We give a list of some of these results; for the proofs, which are very involved, and for the numerous known further results of like kind, we refer to the original papers.

(α) Assume that $\lambda = \lambda[\boldsymbol{A}_1, \boldsymbol{A}_2, \ldots, \boldsymbol{A}_r]$ is a *multilinear invariant* of the r symmetric tensors $\boldsymbol{A}_k \in \mathscr{S}$. Then λ can be expressed as a sum of the form

$$\lambda = \sum c\, \varphi_1 \varphi_2 \cdots \varphi_s \tag{11.9}$$

where, in each term, c is a constant and $\varphi_1, \ldots, \varphi_s$ are basic multilinear invariants, taken from the list (11.10)—(11.13) below, such that the product $\varphi_1\varphi_2\cdots\varphi_s$ contains each $\boldsymbol{A}_j$ precisely once.

List of basic multilinear invariants:

$$\operatorname{tr}\boldsymbol{A}_i, \quad \operatorname{tr}(\boldsymbol{A}_i\boldsymbol{A}_j) \quad (i<j), \quad \operatorname{tr}(\boldsymbol{A}_i\boldsymbol{A}_j\boldsymbol{A}_k) \quad (i<j<k); \tag{11.10}$$

$$\operatorname{tr}(\boldsymbol{A}_i\boldsymbol{A}_j\boldsymbol{A}_k\boldsymbol{A}_l), \quad \operatorname{tr}(\boldsymbol{A}_i\boldsymbol{A}_j\boldsymbol{A}_l\boldsymbol{A}_k) \quad (i<j<k<l); \tag{11.11}$$

$$\operatorname{tr}(\boldsymbol{A}_i\boldsymbol{A}_j\boldsymbol{A}_k\boldsymbol{A}_l\boldsymbol{A}_m) \quad (i<j<k<l<m); \tag{11.12}$$

invariants obtained from (11.12) by changing the order (i, j, k, l, m) into (i, j, l, m, k), (i, j, m, k, l), (i, k, l, j, m), (i, k, j, m, l), or (i, l, j, k, m);

$$\operatorname{tr}(\boldsymbol{A}_i\boldsymbol{A}_j\boldsymbol{A}_k\boldsymbol{A}_l\boldsymbol{A}_m\boldsymbol{A}_n) \quad (i<j<k<l<m<n); \tag{11.13}$$

and invariants obtained from (11.13) by changing the order (i, j, k, l, m, n) into (i, n, j, m, k, l), (i, n, j, k, m, l), (i, n, k, m, j, l), (i, n, k, j, m, l), (i, l, m, n, j, k), (i, l, j, m, n, k), (i, l, j, n, m, k), (i, l, n, m, j, k), or (i, l, n, j, m, k). This theorem is an immediate consequence of the theorem stated under (β) below.

For a bilinear invariant $\lambda = \lambda[\boldsymbol{A}, \boldsymbol{B}]$ the representation (11.9) specializes to

$$\lambda[\boldsymbol{A}, \boldsymbol{B}] = c_1 \operatorname{tr}(\boldsymbol{A}\boldsymbol{B}) + c_2 (\operatorname{tr}\boldsymbol{A})(\operatorname{tr}\boldsymbol{B}). \tag{11.14}$$

(β) Assume that $\varepsilon = \varepsilon(\boldsymbol{A}_1, \boldsymbol{A}_2, \ldots, \boldsymbol{A}_r)$ is a *polynomial invariant*, i.e. an invariant that is a polynomial in the components of the r symmetric tensors $\boldsymbol{A}_k$.

[1] Cf. WEYL [1939, *2*, Ch. II, § 9].

[2] If $n=3$, then $[\boldsymbol{v}_{i_1}, \boldsymbol{v}_{i_2}, \boldsymbol{v}_{i_3}] = \boldsymbol{v}_{i_1} \cdot (\boldsymbol{v}_{i_2} \times \boldsymbol{v}_{i_3})$.

[3] For the case when $n=3$ this was shown by CAUCHY [1850, *1*]. For the case of general n cf. WEYL [1939, *2*, Ch. II, § 9].

Then ε can be expressed as a polynomial in the basic invariants listed in (11.10)—(11.13) and in the following

List of non-linear basic invariants:

$$\operatorname{tr}(\boldsymbol{A}_{\mathfrak{i}}^2), \quad \operatorname{tr}(\boldsymbol{A}_{\mathfrak{i}}^3); \tag{11.15}$$

$$\operatorname{tr}(\boldsymbol{A}_{\mathfrak{i}}\boldsymbol{A}_{\mathfrak{j}}^2), \quad \operatorname{tr}(\boldsymbol{A}_{\mathfrak{j}}\boldsymbol{A}_{\mathfrak{i}}^2), \quad \operatorname{tr}(\boldsymbol{A}_{\mathfrak{i}}^2\boldsymbol{A}_{\mathfrak{j}}^2) \quad (\mathfrak{i}<\mathfrak{j}); \tag{11.16}$$

$$\operatorname{tr}(\boldsymbol{A}_{\mathfrak{i}}\boldsymbol{A}_{\mathfrak{j}}\boldsymbol{A}_{\mathfrak{k}}^2), \quad \operatorname{tr}(\boldsymbol{A}_{\mathfrak{i}}\boldsymbol{A}_{\mathfrak{k}}^2\boldsymbol{A}_{\mathfrak{j}}^2) \quad (\mathfrak{i}<\mathfrak{j}<\mathfrak{k}); \tag{11.17}$$

invariants obtained from (11.17) by permuting $\mathfrak{i}$, $\mathfrak{j}$, $\mathfrak{k}$ cyclically;

$$\operatorname{tr}(\boldsymbol{A}_{\mathfrak{i}}\boldsymbol{A}_{\mathfrak{j}}\boldsymbol{A}_{\mathfrak{k}}\boldsymbol{A}_{\mathfrak{l}}^2), \quad \operatorname{tr}(\boldsymbol{A}_{\mathfrak{i}}\boldsymbol{A}_{\mathfrak{k}}\boldsymbol{A}_{\mathfrak{j}}\boldsymbol{A}_{\mathfrak{l}}^2) \quad (\mathfrak{i}<\mathfrak{j}<\mathfrak{k}<\mathfrak{l}); \tag{11.18}$$

invariants obtained from (11.18) by permuting $\mathfrak{i}$, $\mathfrak{j}$, $\mathfrak{k}$, $\mathfrak{l}$ cyclically;

$$\left.\begin{array}{lll} \operatorname{tr}(\boldsymbol{A}_{\mathfrak{i}}\boldsymbol{A}_{\mathfrak{j}}\boldsymbol{A}_{\mathfrak{k}}^2\boldsymbol{A}_{\mathfrak{l}}^2), & \operatorname{tr}(\boldsymbol{A}_{\mathfrak{i}}\boldsymbol{A}_{\mathfrak{k}}\boldsymbol{A}_{\mathfrak{j}}^2\boldsymbol{A}_{\mathfrak{l}}^2), & \operatorname{tr}(\boldsymbol{A}_{\mathfrak{i}}\boldsymbol{A}_{\mathfrak{l}}\boldsymbol{A}_{\mathfrak{j}}^2\boldsymbol{A}_{\mathfrak{k}}^2), \\ \operatorname{tr}(\boldsymbol{A}_{\mathfrak{j}}\boldsymbol{A}_{\mathfrak{k}}\boldsymbol{A}_{\mathfrak{i}}^2\boldsymbol{A}_{\mathfrak{l}}^2), & \operatorname{tr}(\boldsymbol{A}_{\mathfrak{j}}\boldsymbol{A}_{\mathfrak{l}}\boldsymbol{A}_{\mathfrak{i}}^2\boldsymbol{A}_{\mathfrak{k}}^2), & \operatorname{tr}(\boldsymbol{A}_{\mathfrak{k}}\boldsymbol{A}_{\mathfrak{l}}\boldsymbol{A}_{\mathfrak{i}}^2\boldsymbol{A}_{\mathfrak{j}}^2), \end{array}\right. \quad (\mathfrak{i}<\mathfrak{j}<\mathfrak{k}<\mathfrak{l}); \tag{11.19}$$

$$\operatorname{tr}(\boldsymbol{A}_{\mathfrak{i}}\boldsymbol{A}_{\mathfrak{l}}\boldsymbol{A}_{\mathfrak{j}}\boldsymbol{A}_{\mathfrak{k}}\boldsymbol{A}_{\mathfrak{l}}^2) \quad (\mathfrak{i}<\mathfrak{j}<\mathfrak{k}<\mathfrak{l}); \tag{11.20}$$

invariants obtained from (11.20) by permuting $\mathfrak{i}$, $\mathfrak{j}$, $\mathfrak{k}$, $\mathfrak{l}$ cyclically;

$$\left.\begin{array}{ll} \operatorname{tr}(\boldsymbol{A}_{\mathfrak{i}}\boldsymbol{A}_{\mathfrak{j}}\boldsymbol{A}_{\mathfrak{k}}\boldsymbol{A}_{\mathfrak{l}}\boldsymbol{A}_{\mathfrak{m}}^2), & \operatorname{tr}(\boldsymbol{A}_{\mathfrak{i}}\boldsymbol{A}_{\mathfrak{j}}\boldsymbol{A}_{\mathfrak{l}}\boldsymbol{A}_{\mathfrak{k}}\boldsymbol{A}_{\mathfrak{m}}^2), \\ \operatorname{tr}(\boldsymbol{A}_{\mathfrak{i}}\boldsymbol{A}_{\mathfrak{l}}\boldsymbol{A}_{\mathfrak{k}}\boldsymbol{A}_{\mathfrak{j}}\boldsymbol{A}_{\mathfrak{m}}^2), & \operatorname{tr}(\boldsymbol{A}_{\mathfrak{j}}\boldsymbol{A}_{\mathfrak{i}}\boldsymbol{A}_{\mathfrak{k}}\boldsymbol{A}_{\mathfrak{l}}\boldsymbol{A}_{\mathfrak{m}}^2) \end{array}\right. \quad (\mathfrak{i}<\mathfrak{j}<\mathfrak{k}<\mathfrak{l}<\mathfrak{m}); \tag{11.21}$$

invariants obtained from (11.21) by permuting $\mathfrak{i}$, $\mathfrak{j}$, $\mathfrak{k}$, $\mathfrak{l}$, $\mathfrak{m}$ cyclically. Moreover, if $\mathfrak{r}=1, 2, 3, 4$, or 5, the corresponding basic invariants are obtained from the lists (11.10)—(11.13) and (11.15)—(11.21) by deleting all entries in which more than $\mathfrak{r}$ tensors occur.

The theorem just stated is due to SPENCER and RIVLIN[1]. That the lists of basic invariants cannot be reduced was proved later by SMITH and by SPENCER[2].

For two symmetric tensors $\boldsymbol{A}$, $\boldsymbol{B}$, the list of basic invariants reduces to

$$\left.\begin{array}{l} \operatorname{tr}\boldsymbol{A},\ \operatorname{tr}\boldsymbol{A}^2,\ \operatorname{tr}\boldsymbol{A}^3,\ \operatorname{tr}\boldsymbol{B},\ \operatorname{tr}\boldsymbol{B}^2,\ \operatorname{tr}\boldsymbol{B}^3, \\ \operatorname{tr}(\boldsymbol{A}\boldsymbol{B}),\ \operatorname{tr}(\boldsymbol{A}\boldsymbol{B}^2),\ \operatorname{tr}(\boldsymbol{B}\boldsymbol{A}^2),\ \operatorname{tr}(\boldsymbol{A}^2\boldsymbol{B}^2). \end{array}\right\} \tag{11.22}$$

(γ) Let $\varepsilon(\boldsymbol{A},\boldsymbol{B})$ be a simultaneous invariant of two symmetric tensors $\boldsymbol{A}$ and $\boldsymbol{B}$. Assume that the proper numbers of one of them are distinct. Then $\varepsilon(\boldsymbol{A},\boldsymbol{B})$ can be expressed as a function of the ten basic invariants (11.22)[3].

(δ) Let $\varepsilon(\boldsymbol{A}_1, \boldsymbol{A}_2, \ldots, \boldsymbol{A}_{\mathfrak{r}})$ be a simultaneous invariant of the $\mathfrak{r}$ symmetric tensors $\boldsymbol{A}_1, \ldots, \boldsymbol{A}_{\mathfrak{r}}$. Assume that the proper numbers of $\boldsymbol{A}_1$ are distinct and that the off-diagonal elements of the matrix of $\boldsymbol{A}_2$ are non-zero when an orthonormal basis is used relative to which $\boldsymbol{A}_1$ corresponds to a diagonal matrix. Then ε can be expressed as a single-valued function of the $6\mathfrak{r}-2$ invariants which are obtained by adding to the list given by (11.22) for $\boldsymbol{A}=\boldsymbol{A}_1$, $\boldsymbol{B}=\boldsymbol{A}_2$ the $6\mathfrak{r}-12$ invariants[4]

$$\left.\begin{array}{l} \operatorname{tr}\boldsymbol{A}_{\mathfrak{l}},\ \operatorname{tr}(\boldsymbol{A}_1\boldsymbol{A}_{\mathfrak{l}}),\ \operatorname{tr}(\boldsymbol{A}_1^2\boldsymbol{A}_{\mathfrak{l}}), \\ \operatorname{tr}(\boldsymbol{A}_2\boldsymbol{A}_{\mathfrak{l}}),\ \operatorname{tr}(\boldsymbol{A}_1\boldsymbol{A}_2\boldsymbol{A}_{\mathfrak{l}}),\ \operatorname{tr}(\boldsymbol{A}_1^2\boldsymbol{A}_2\boldsymbol{A}_{\mathfrak{l}}), \end{array}\right. \quad \mathfrak{l}=3, 4, \ldots, \mathfrak{r}. \tag{11.23}$$

---

[1] SPENCER and RIVLIN [1960, *53*, §§ 9—10]. Earlier work on the subject was done by RIVLIN [1955, *21*], PIPKIN and RIVLIN [1958, *34*], and SPENCER and RIVLIN [1959, *33* and *34*].

[2] SMITH [1960, *52*] treated the cases when $\mathfrak{r}<6$; SPENCER [1961, *59*], those when $\mathfrak{r}\geq 6$.

[3] RIVLIN and ERICKSEN [1955, *23*, § 34].

[4] RIVLIN and ERICKSEN [1955, *23*, § 35].

Lists of basic simultaneous invariants of an arbitrary number of vectors and symmetric tensors, again when $n=3$, have been constructed by SPENCER and RIVLIN[1].

The following result is valid for arbitrary dimension n: *A scalar-valued function* $\varepsilon(\boldsymbol{A}, \boldsymbol{u})$ *of one symmetric tensor* $\boldsymbol{A}$ *and one vector* $\boldsymbol{u}$ *is an orthogonal invariant if and only if it can be expressed as a function of the* $2n$ *special invariants*[2]

$$I_1(\boldsymbol{A}), \ldots, I_n(\boldsymbol{A}), u^2, \boldsymbol{u}\cdot\boldsymbol{A}\boldsymbol{u}, \ldots, \boldsymbol{u}\cdot\boldsymbol{A}^{n-1}\boldsymbol{u}. \tag{11.24}$$

If the dependence of $\varepsilon(\boldsymbol{A}, \boldsymbol{u})$ *on* $\boldsymbol{A}$ *and* $\boldsymbol{u}$ *is polynomial, then the dependence of the corresponding function of the invariants* (11.24) *is also polynomial*[3].

12. Tensor functions of one variable. According to the definition (8.7), a symmetric tensor function $\mathfrak{f}(\boldsymbol{A})$ of one variable is isotropic if

$$\boldsymbol{Q}\,\mathfrak{f}(\boldsymbol{A})\,\boldsymbol{Q}^T = \mathfrak{f}(\boldsymbol{Q}\boldsymbol{A}\boldsymbol{Q}^T) \tag{12.1}$$

holds for every orthogonal $\boldsymbol{Q}$. The following fundamental representation theorem shows that any such function can be expressed in terms of n invariants[4]:

A tensor function $\boldsymbol{D}=\mathfrak{f}(\boldsymbol{A})$, $\boldsymbol{A}, \boldsymbol{D}\in\mathscr{S}$, *is isotropic if and only if it has a representation of the form*

$$\boldsymbol{D}=\mathfrak{f}(\boldsymbol{A})=\varphi_0\boldsymbol{1}+\varphi_1\boldsymbol{A}+\varphi_2\boldsymbol{A}^2+\cdots+\varphi_{n-1}\boldsymbol{A}^{n-1}, \tag{12.2}$$

where the φ_k *are invariants of* $\boldsymbol{A}$ *and hence can be expressed as functions* $\varphi_k=\varphi_k(I_1, \ldots, I_n)$ *of the principal invariants of* $\boldsymbol{A}$. To outline the proof we consider a proper vector $\boldsymbol{e}$ of $\boldsymbol{A}$ and define an orthogonal transformation $\boldsymbol{Q}$ by

$$\boldsymbol{Q}\boldsymbol{e}=-\boldsymbol{e}, \quad \boldsymbol{Q}\boldsymbol{f}=\boldsymbol{f} \quad \text{if} \quad \boldsymbol{f}\cdot\boldsymbol{e}=0.$$

$\boldsymbol{Q}$ is a reflection on the plane normal to $\boldsymbol{e}$. It is easy to see that $\boldsymbol{Q}\boldsymbol{A}\boldsymbol{Q}^T=\boldsymbol{A}$ and hence, by (12.1), that $\boldsymbol{Q}\boldsymbol{D}\boldsymbol{Q}^T=\boldsymbol{D}$, i.e., $\boldsymbol{Q}\boldsymbol{D}=\boldsymbol{D}\boldsymbol{Q}$. We therefore have

$$\boldsymbol{Q}(\boldsymbol{D}\boldsymbol{e})=\boldsymbol{D}(\boldsymbol{Q}\boldsymbol{e})=-(\boldsymbol{D}\boldsymbol{e}),$$

and we see that $\boldsymbol{Q}$ transforms the vector $\boldsymbol{D}\boldsymbol{e}$ into its opposite. Since the only vectors transformed by the reflection $\boldsymbol{Q}$ into their opposites are the multiples of $\boldsymbol{e}$, it follows that $\boldsymbol{D}\boldsymbol{e}=d\boldsymbol{e}$, i.e. that $\boldsymbol{e}$ is a proper vector of $\boldsymbol{D}$. Thus every proper vector of $\boldsymbol{A}$ is also a proper vector of $\boldsymbol{D}$.

Denote the distinct proper numbers of $\boldsymbol{A}$ by $a_1, a_2, \ldots, a_m$ $(m\leqq n)$, and a set of corresponding proper vectors by $\boldsymbol{e}_1, \ldots, \boldsymbol{e}_m$. They are also proper vectors of $\boldsymbol{D}$ corresponding to proper numbers $d_1, d_2, \ldots, d_m$, which need not be distinct. The system of m equations

$$d_k=\varphi_0+\varphi_1 a_k+\varphi_2 a_k^2+\cdots+\varphi_{m-1}a_k^{m-1}, \qquad k=1, \ldots, m, \tag{12.3}$$

[1] SPENCER and RIVLIN [1962, *63*]. SMITH [1965, *32*] has constructed minimal integrity bases for any number of vectors and second-order tensors for the full orthogonal group; SPENCER [1965, *33*], for the proper orthogonal group.

[2] NOLL [1965, *28*].

[3] PIPKIN and RIVLIN [1958, *35*, § 6]. Later [1959, *21*, §§ 5—8] they gave integrity bases for polynomial functions of one symmetric tensor and any number of vectors, for various subgroups of the orthogonal group. Basic simultaneous invariants relative to certain subgroups of the orthogonal group were obtained by ADKINS [1960, *2* and *3*], [1962, *1*]. For all the crystallographic groups, SMITH, SMITH and RIVLIN [1963, *66*] [1964, *79*] found integrity bases for polynomial functions of one vector and one symmetric tensor, and for any number of vectors. Results for more general functional dependence are obtained by WINEMAN and PIPKIN [1964, *97*].

[4] For the case when $n=3$, this theorem was first proved by RIVLIN and ERICKSEN [1955, *23*, § 39]. An incomplete proof had been given earlier by RICHTER [1948, *10*, § 2]. Our proof is similar to the one given by SERRIN [1959, *27*, § 59] for the case when $n=3$.

has a unique solution for the m unknowns $\varphi_0, \varphi_1, \ldots, \varphi_{m-1}$, because its determinant is $\prod_{j<k}(a_j - a_k) \neq 0$. It is not hard to see that then

$$D = \mathfrak{f}(A) = \varphi_0 1 + \varphi_1 A + \cdots + \varphi_{m-1} A^{m-1}, \qquad m \leq n, \tag{12.4}$$

and, since $QDQ^T = \mathfrak{f}(QAQ^T)$ for every orthogonal Q, that

$$\mathfrak{f}(QAQ^T) = \varphi_0 1 + \varphi_1 (QAQ^T) + \cdots + \varphi_{m-1}(QAQ^T)^{m-1}.$$

Therefore the coefficients φ_k in (12.4) are not changed when A is replaced by an orthogonal conjugate QAQ^T; hence they are invariants of A, and the theorem is proved.

The coefficients $\varphi_k = \varphi_k(A)$ in (12.2) are uniquely determined by $\mathfrak{f}(A)$ only if the proper numbers are distinct, in which case $n = m$ in (12.3) and (12.4). As long as the proper numbers of A remain distinct, the coefficients $\varphi_k = \varphi_k(I_1, \ldots, I_n)$ inherit the continuity and differentiability properties of $\mathfrak{f}(A)$. If $m < n$ in (12.4), the last $n - m$ of the φ_k may be chosen to be zero. However, this choice may cause some φ_k to become discontinuous even when $\mathfrak{f}(A)$ remains continuous. It may happen that there is no way to choose the φ_k such as to remain continuous, even if $\mathfrak{f}(A)$ remains differentiable when the proper numbers of A coalesce, as may easily be seen by investigating the example of the function $\mathfrak{f}(A) = |I^2 - 4II|^{\frac{1}{4}}(I 1 - 2A)$ in two dimensions ($n = 2$)[1]. A theorem showing that such examples are rather pathological is stated overleaf.

If $\mathfrak{f}(A)$ is a tensor power series of the form (7.7), then a representation (12.2) may be obtained explicitly, with coefficients φ_k in the form of power series in the principal invariants. To do this, express A^n and all higher powers A^{n+h} in terms of $1, A, A^2, \ldots, A^{n-1}$ and the principal invariants by means of the HAMILTON-CAYLEY formula (9.10)[2]. This method may be applied even when the coefficients in (7.7) are invariants rather than constants. For details, see Sect. 42 of the Appendix to CFT. It should be kept in mind that tensor power series are only very special cases of isotropic tensor functions and that the representation theorem applies also to more general functions such as $(7.8)_3$.

The representation theorem has the following supplement, which we state without proof[3]:

If $D = \mathfrak{f}(A)$ is a polynomial isotropic tensor function, i.e., if the components of D are polynomials in the components of A, then the coefficients φ_k in (12.2) *may be expressed as polynomials in the principal invariants of A.*

The following result is a special case of this supplement:

Every linear isotropic tensor function $\mathsf{L}[A]$ has a representation

$$\mathsf{L}[A] = c_0 (\operatorname{tr} A) 1 + c_1 A, \tag{12.5}$$

where c_0 and c_1 are constants.

In the case when $n = 3$ the representation theorem (12.2) reads

$$\mathfrak{f}(A) = \varphi_0 1 + \varphi_1 A + \varphi_2 A^2, \qquad \varphi_k = \varphi_k(I_A, II_A, III_A). \tag{12.6}$$

If the tensor variable A is restricted to be invertible (i.e. $\det A \neq 0$), we can use $(9.11)_4$ to obtain the alternative representation

$$\mathfrak{f}(A) = \psi_0 1 + \psi_1 A + \psi_{-1} A^{-1}, \qquad \psi_k = \psi_k(I_A, II_A, III_A). \tag{12.7}$$

[1] An example in three dimensions was given by SERRIN [1959, *28*, § 4].

[2] This reduction method was employed by REINER [1945, *3*, § 4] and PRAGER [1945, *2*] for particular theories of materials, and it was through their papers that the representation (12.6) became known to workers in continuum mechanics.

[3] For the case when $n = 3$, WEITZENBÖCK [1921, *2*] obtained a more general result, in which neither D nor A need be symmetric. While he did not remark on the simplification when A and D are symmetric, it can be derived from formulae given by him. An elegant modern proof when $n = 3$ has been given by SERRIN [1959, *27*, § 60] [1959, *28*, § 4]. SERRIN's proof is easily generalized to the case when n is arbitrary.

The representation theorem (12.6) can be supplemented by the following proposition of SERRIN, stated here without proof[1]:

If the components of $\mathfrak{f}(\boldsymbol{A})$ are three times continuously differentiable functions of the components of $\boldsymbol{A}$, then the coefficients φ_k in (12.6) *are continuous functions of the principal invariants of $\boldsymbol{A}$.*

Assume that $\varepsilon(\boldsymbol{A})=\varepsilon(I_1,\ldots,I_n)$ is an invariant and that it has continuous derivatives with respect to the principal invariants I_k. The gradient $\varepsilon_{\boldsymbol{A}}(\boldsymbol{A})$ is an isotropic tensor function. Applying the formula (9.9) for the gradients of the principal invariants and the chain rule of differentiation, we obtain the following *representation of the gradient of an invariant* $\varepsilon(\boldsymbol{A})$:

$$\frac{\partial\varepsilon}{\partial\boldsymbol{A}}=\varepsilon_{\boldsymbol{A}}(\boldsymbol{A})=\left[\sum_{l=0}^{n-1}\left(\sum_{k=l+1}^{n}\frac{\partial\varepsilon}{\partial I_k}I_{k-l-1}\right)(-1)^l\boldsymbol{A}^l\right]^T \tag{12.8}$$

which has the form (12.2). Using the representation $\varepsilon(\boldsymbol{A})=\varepsilon(\bar{I}_1,\ldots,\bar{I}_n)$ in terms of the moments $\bar{I}_k$, we obtain from (9.22) in a similar way the representation

$$\frac{\partial\varepsilon}{\partial\boldsymbol{A}}=\varepsilon_{\boldsymbol{A}}(\boldsymbol{A})=\left[\sum_{k=1}^{n}k\frac{\partial\varepsilon}{\partial\bar{I}_k}\boldsymbol{A}^{k-1}\right]^T. \tag{12.9}$$

In three dimensions ($n=3$) (12.8) reduces to

$$\frac{\partial\varepsilon}{\partial\boldsymbol{A}}=\left[\left(\frac{\partial\varepsilon}{\partial I}+\frac{\partial\varepsilon}{\partial II}I+\frac{\partial\varepsilon}{\partial III}II\right)\boldsymbol{1}-\left(\frac{\partial\varepsilon}{\partial II}+\frac{\partial\varepsilon}{\partial III}I\right)\boldsymbol{A}+\frac{\partial\varepsilon}{\partial III}\boldsymbol{A}^2\right]^T. \tag{12.10}$$

Using $(9.11)_4$, we get[2]

$$\frac{\partial\varepsilon}{\partial\boldsymbol{A}}=\left[\left(\frac{\partial\varepsilon}{\partial I}+\frac{\partial\varepsilon}{\partial II}I\right)\boldsymbol{1}-\frac{\partial\varepsilon}{\partial II}\boldsymbol{A}+\frac{\partial\varepsilon}{\partial III}III\boldsymbol{A}^{-1}\right]^T, \tag{12.11}$$

$$\frac{\partial\varepsilon}{\partial A_{km}}=\left(\frac{\partial\varepsilon}{\partial I}+\frac{\partial\varepsilon}{\partial II}I\right)g^{mk}-\frac{\partial\varepsilon}{\partial II}A^{mk}+\frac{\partial\varepsilon}{\partial III}III(A^{-1})^{mk}. \tag{12.12}$$

In three dimensions the formula (12.9) has the simple form

$$\frac{\partial\varepsilon}{\partial\boldsymbol{A}}=\left[\frac{\partial\varepsilon}{\partial\bar{I}}\boldsymbol{1}+2\frac{\partial\varepsilon}{\partial\overline{II}}\boldsymbol{A}+3\frac{\partial\varepsilon}{\partial\overline{III}}\boldsymbol{A}^2\right]^T, \tag{12.13}$$

$$\frac{\partial\varepsilon}{\partial A_{km}}=\frac{\partial\varepsilon}{\partial\bar{I}}g^{mk}+2\frac{\partial\varepsilon}{\partial\overline{II}}A^{mk}+3\frac{\partial\varepsilon}{\partial\overline{III}}A^m{}_pA^{pk}. \tag{12.14}$$

13. Tensor and vector functions of several variables. As in the case of simultaneous invariants, so also for isotropic tensor functions of several tensor variables in $\mathscr{S}$ only partial results for three dimensions ($n=3$) are known. We state, without proof, the most important of these results.

(α) Every *multilinear isotropic tensor function* $\boldsymbol{D}=\mathfrak{l}[\boldsymbol{A}_1,\boldsymbol{A}_2,\ldots,\boldsymbol{A}_{\mathfrak{r}}]$, $\boldsymbol{D},\boldsymbol{A}_{\mathfrak{k}}\in\mathscr{S}$, has a representation of the form

$$\boldsymbol{D}=\sum\psi(\boldsymbol{P}+\boldsymbol{P}^T) \tag{13.1}$$

where, in each term, $\boldsymbol{P}$ is a product taken from the list (13.2) below and ψ is a multilinear invariant of those tensors among $\boldsymbol{A}_1,\ldots,\boldsymbol{A}_{\mathfrak{r}}$ that do not occur in the product $\boldsymbol{P}$.

---

[1] SERRIN [1959, *28*, § 6].

[2] This formula plays a central role in the theory of hyperelastic materials (cf. Chapter DII); in principle it is due to FINGER [1894, *3*, Eq. (35)].

List of basic multilinear products:

$$1,\ \boldsymbol{A}_{\mathfrak{i}},\ \boldsymbol{A}_{\mathfrak{i}}\boldsymbol{A}_{\mathfrak{j}},\ \boldsymbol{A}_{\mathfrak{i}}\boldsymbol{A}_{\mathfrak{j}}\boldsymbol{A}_{\mathfrak{k}},\ \boldsymbol{A}_{\mathfrak{i}}\boldsymbol{A}_{\mathfrak{j}}\boldsymbol{A}_{\mathfrak{k}}\boldsymbol{A}_{\mathfrak{l}},\ \boldsymbol{A}_{\mathfrak{i}}\boldsymbol{A}_{\mathfrak{j}}\boldsymbol{A}_{\mathfrak{k}}\boldsymbol{A}_{\mathfrak{l}}\boldsymbol{A}_{\mathfrak{m}}. \tag{13.2}$$

Here the indices $\mathfrak{i}$, $\mathfrak{j}$, $\mathfrak{k}$, $\mathfrak{l}$, $\mathfrak{m}$ are all different and are chosen in all possible ways from *1, 2*, ..., $\mathfrak{r}$. This theorem is a special case of the theorem stated under (β) below.

For a bilinear isotropic tensor function $\boldsymbol{D}=\mathfrak{b}[\boldsymbol{A},\boldsymbol{B}]$ the representation (13.1) can be written explicitly:

$$\boldsymbol{D}=[c_1(\operatorname{tr}\boldsymbol{A})(\operatorname{tr}\boldsymbol{B})+c_2\operatorname{tr}(\boldsymbol{A}\boldsymbol{B})]\,1+c_3(\operatorname{tr}\boldsymbol{B})\boldsymbol{A}+c_4(\operatorname{tr}\boldsymbol{A})\boldsymbol{B}+c_5(\boldsymbol{A}\boldsymbol{B}+\boldsymbol{B}\boldsymbol{A}). \tag{13.3}$$

(β) Let $\boldsymbol{D}=\mathfrak{f}(\boldsymbol{A}_1,\boldsymbol{A}_2,\ldots,\boldsymbol{A}_{\mathfrak{r}}),\ \boldsymbol{D},\boldsymbol{A}_{\mathfrak{k}}\in\mathscr{S}$, be a *polynomial isotropic tensor function*, i.e., assume that the components of $\boldsymbol{D}$ are polynomials in the components of the $\boldsymbol{A}_{\mathfrak{k}}$. Then $\boldsymbol{D}$ has a representation of the form (13.1) in which each $\boldsymbol{P}$ is a product taken from the list (13.2) or the lists (13.4)—(13.6) below and in which ψ is a polynomial invariant.

List of non-linear basic products:

$$\boldsymbol{A}_{\mathfrak{i}}^2,\ \boldsymbol{A}_{\mathfrak{i}}^2\boldsymbol{A}_{\mathfrak{j}},\ \boldsymbol{A}_{\mathfrak{i}}^2\boldsymbol{A}_{\mathfrak{j}}^2, \tag{13.4}$$

$$\boldsymbol{A}_{\mathfrak{i}}^2\boldsymbol{A}_{\mathfrak{j}}\boldsymbol{A}_{\mathfrak{k}},\ \boldsymbol{A}_{\mathfrak{i}}\boldsymbol{A}_{\mathfrak{j}}^2\boldsymbol{A}_{\mathfrak{k}},\ \boldsymbol{A}_{\mathfrak{i}}^2\boldsymbol{A}_{\mathfrak{j}}^2\boldsymbol{A}_{\mathfrak{k}},\ \boldsymbol{A}_{\mathfrak{i}}\boldsymbol{A}_{\mathfrak{j}}\boldsymbol{A}_{\mathfrak{k}}\boldsymbol{A}_{\mathfrak{i}}^2, \tag{13.5}$$

$$\boldsymbol{A}_{\mathfrak{i}}^2\boldsymbol{A}_{\mathfrak{j}}\boldsymbol{A}_{\mathfrak{k}}\boldsymbol{A}_{\mathfrak{l}},\ \boldsymbol{A}_{\mathfrak{i}}\boldsymbol{A}_{\mathfrak{j}}^2\boldsymbol{A}_{\mathfrak{k}}\boldsymbol{A}_{\mathfrak{l}}. \tag{13.6}$$

In (13.4)—(13.6), as in (13.2), the indices $\mathfrak{i}$, $\mathfrak{j}$, $\mathfrak{k}$, $\mathfrak{l}$ are all different and are chosen in all possible ways from *1, 2*, ..., $\mathfrak{r}$. This theorem is due to SPENCER and RIVLIN[1].

For a polynomial isotropic tensor function $\boldsymbol{D}=\mathfrak{f}(\boldsymbol{A},\boldsymbol{B})$ of two variables $\boldsymbol{A},\boldsymbol{B}\in\mathscr{S}$ we have the explicit representation[2]

$$\left.\begin{aligned}\boldsymbol{D}&=\psi_0 1+\psi_1\boldsymbol{A}+\psi_2\boldsymbol{B}+\psi_3\boldsymbol{A}^2+\psi_4\boldsymbol{B}^2+\psi_5(\boldsymbol{A}\boldsymbol{B}+\boldsymbol{B}\boldsymbol{A})+\\&\quad+\psi_6(\boldsymbol{A}^2\boldsymbol{B}+\boldsymbol{B}\boldsymbol{A}^2)+\psi_7(\boldsymbol{A}\boldsymbol{B}^2+\boldsymbol{B}^2\boldsymbol{A})+\psi_8(\boldsymbol{A}^2\boldsymbol{B}^2+\boldsymbol{B}^2\boldsymbol{A}^2)\end{aligned}\right\} \tag{13.7}$$

where $\psi_0,\ldots,\psi_8$ are polynomials in the ten basic invariants (11.22). RIVLIN and ERICKSEN[3] had shown previously that (13.7) is a valid representation for any isotropic tensor function, not necessarily polynomial, of two variables but, while ψ_Γ is of course an isotropic function of $\boldsymbol{A}$ and $\boldsymbol{B}$, no proof that it equals a function of the ten basic invariants (11.22) has yet been found.

The following results are valid for any dimension n: *A vector-valued function* $\boldsymbol{v}=\mathfrak{v}(\boldsymbol{A},\boldsymbol{u})$ *of one symmetric tensor* $\boldsymbol{A}$ *and one vector* $\boldsymbol{u}$ *is isotropic if and only if it has a representation of the form*

$$\boldsymbol{v}=\mathfrak{v}(\boldsymbol{A},\boldsymbol{u})=(\varphi_0 1+\varphi_1\boldsymbol{A}+\cdots+\varphi_{n-1}\boldsymbol{A}^{n-1})\,\boldsymbol{u}, \tag{13.8}$$

where the φ_k *are simultaneous invariants of* $\boldsymbol{A}$ *and* $\boldsymbol{u}$ *and hence can be expressed as functions of the special invariants* (11.24)[4].

If the dependence of $\boldsymbol{v}$ on $\boldsymbol{A}$ and $\boldsymbol{u}$ is polynomial, then the coefficients φ_k in (13.8) may be taken as polynomials in the invariants (11.24)[5].

Various further representation theorems, mainly for cases of polynomial dependence, have been derived by RIVLIN and others[6].

---

[1] SPENCER and RIVLIN [1960, *53*, § 8]. Earlier work on the subject was done by RIVLIN [1955, *21*] and by SPENCER and RIVLIN [1959, *33*].

[2] RIVLIN [1955, *21*, § 10]. The form of the result, based on insufficient considerations, had been given earlier by TRUESDELL [1951, *17*, § 26]. Cf. the discussion by NOVOZHILOV [1963, *59*].

[3] RIVLIN and ERICKSEN [1955, *23*, § 27]. They considered also certain shorter representations valid in some but not all circumstances.

[4] NOLL [1965, *28*]. When $\boldsymbol{v}=\mathfrak{r}(\boldsymbol{u})$, (13.8) yields $\mathfrak{r}(\boldsymbol{u})=\varphi_0(u)\,\boldsymbol{u}$. This result, with an inadequate proof, had been given by TRUESDELL [1950, *16*, § 14].

[5] PIPKIN and RIVLIN [1958, *35*, § 23]. A corresponding representation for hemitropic polynomial functions was obtained by GREEN and ADKINS [1960, *26*, §§ 86—87].

[6] PIPKIN and RIVLIN [1958, *35*] [1959, *21*], RIVLIN [1960, *46*] [1961, *51*], ADKINS [1958, *2*, § 6] [1960, *2* and *3*], LOKHIN [1963, *52* and *53*], LOKHIN and SEDOV [1963, *54* and *62*].

C. The general theory of material behavior.

14. Scope and plan of the chapter. This chapter presents the *general principles* governing the mechanical behavior of materials. These are three: the *principle of material frame-indifference,* which asserts that the response of a material is the same for all observers, and the *principles of determinism* and *local action,* which assert that the present stress at a particle is determined by the history of an arbitrarily small neighborhood of that particle. Subchapter I defines the concept of body and reviews the dynamical quantities associated with its motion; these suffice for a statement of the principle of material indifference. Subchapter II collects the kinematical definitions and formulae needed to describe the deformation and flow of a material. Subchapter III opens with the *general constitutive equation,* which defines mathematically the class of ideal materials obeying the two basic principles. Each such material is characterized by a particular *response functional.* Emphasis is placed on the *simple material,* in which the collection of kinematical facts needed to determine the stress reduces to the history of the deformation gradient alone. The theory so obtained, while special in respect to the three general principles, is still a very broad one: Included are not only the theories of finite elastic strain and of non-linear viscosity but also theories of visco-elasticity and stress relaxation. "Isotropic", "fluid", "solid", "materially uniform", and "homogeneous" are defined in terms of invariance properties of the response functionals. A brief discussion of the structure of materially uniform bodies, i.e. of bodies that, while not necessarily homogeneous, are nevertheless of uniform material, is given. In this way the non-Euclidean geometry of the body manifold, nowadays taken as the basis of the theory of "continuous distributions of dislocations," is shown to follow directly from an assumed relation between response functionals at different particles. The fourth subchapter defines various special kinds of materials, mainly so as to make contact with older and more formal theories. The last subchapter states a fourth principle appropriate to many but not all materials, the *principle of fading memory,* according to which the influence of long past events is weaker than that of recent ones in determining stress. By means of this principle, the general constitutive equation may sometimes be replaced, approximately, by a simpler one of more familiar type: In particular, if the deformation has been very small at all past times, or if the motion is "slow", it is shown that the behavior of the material is described by the classical infinitesimal theory of visco-elasticity or by the classical theory of linearly viscous fluids.

In a word, this chapter presents in explicit, mathematical terms a concept of material response that subsumes as special cases or as approximations all the classical special theories of continuous media and most of the recently proposed non-linear theories. The considerations of this chapter are incomplete in that mechanical interactions not reducible to forces, such as body couples and couple stresses, and thermo-energetic effects, are neglected for the most part. Correspondingly general principles governing such effects are just coming to be known; the direction toward which modern research is pointing may be inferred from the theories sketched in Sects. 96 and 98.

While the contents of this chapter is to be regarded in part as a unification and summary of all the purely mechanical theories of materials developed in earlier years, in the form and generality given here it is due to NOLL and COLEMAN (1957—1961). Ideas similar to some of those presented have been developed also by RIVLIN and his collaborators (1956—1961).

There is no other collected exposition of the material forming the subject of this chapter.

I. Basic principles.

15. Bodies and motions. In this subchapter we give a brief outline of those concepts and principles of mechanics which are of basic importance for the non-linear field theories[1]. For a detailed discussion of the kinematics and dynamics of continuous media the reader is referred to Chapters B, C, and D of CFT.

A *body* $\mathscr{B}$ is a three-dimensional differentiable manifold, the elements of which are called *particles* X. This manifold may be referred to a system of co-ordinates which establishes a one-to-one correspondence between particles and triples (X^1, X^2, X^3) of real numbers:

$$X = X(X^1, X^2, X^3), \qquad X^\alpha = X^\alpha(X), \qquad \alpha = 1, 2, 3. \tag{15.1}$$

Any two systems of co-ordinates are related by a continuously differentiable transformation

$$\overline{X}^\beta = \overline{X}^\beta(X^\alpha), \qquad X^\alpha = X^\alpha(\overline{X}^\beta). \tag{15.2}$$

We refer to the X^α as *material co-ordinates* of the particle X. In general, no particular geometric structure is to be imputed to a body.

A *configuration* $\boldsymbol{\chi}$ of a body $\mathscr{B}$ is a smooth homeomorphism of $\mathscr{B}$ onto a region of three-dimensional Euclidean space:

$$\boldsymbol{x} = \boldsymbol{\chi}(X), \qquad X = \overset{-1}{\boldsymbol{\chi}}(\boldsymbol{x}). \tag{15.3}$$

Here $\overset{-1}{\boldsymbol{\chi}}$ denotes the inverse of the mapping $\boldsymbol{\chi}$. The point $\boldsymbol{x} = \boldsymbol{\chi}(X)$ is called the *place* occupied by the particle X, and $X = \overset{-1}{\boldsymbol{\chi}}(\boldsymbol{x})$ is the particle whose place is $\boldsymbol{x}$.

If a system of general co-ordinates in space is given, and if $\boldsymbol{x}$ has the co-ordinates x^k $(k = 1, 2, 3)$, then (15.3) corresponds to the three co-ordinate equations

$$x^k = \chi^k(X^\alpha), \qquad X^\alpha = \overset{-1}{\chi}{}^\alpha(x^k). \tag{15.4}$$

The co-ordinates $x^k = x^k(X^\alpha)$ are called the *spatial co-ordinates* of the particle X in the configuration $\boldsymbol{\chi}$. The functions χ^k and $\overset{-1}{\chi}{}^\alpha$ are assumed to be continuously differentiable.

The transformation (15.4) may also be viewed in the following two ways: (i) It may be regarded as a transformation of the co-ordinates X^α *of the manifold of the body*[2] to the new co-ordinates x^k, with the effect of endowing the body with a Euclidean metric. If g_{km} are the components of the metric tensor in space, then the components $C_{\mu\nu}$ of the metric tensor in the body are related to the g_{km} by the transformation law

$$C_{\mu\nu} = \frac{\partial x^k}{\partial X^\mu} \frac{\partial x^m}{\partial X^\nu} g_{km}. \tag{15.5}$$

Different configurations will give rise to different metric tensors in the body. The configuration of a body is determined by the metric tensor only up to a rigid displacement. (ii) It may be regarded as a transformation of the co-ordinates x^k *in space* to the new co-ordinates X^α

[1] The development follows NOLL [1958, *31*, Ch. I].

[2] In much of the older work the body manifold is confused with one of its configurations, although from EULER's earliest researches onward it was seen that choice of such a configuration is arbitrary. Apparently LODGE [1951, *6*, § 2] was the first to introduce the body manifold explicitly: "... it is clear that we have to deal with one continuous geometric manifold (the medium) immersed in and moving through another one (space); we shall refer to these as the 'body manifold' and the 'space manifold', and we shall call points of the two manifolds 'particles' and 'places' respectively."

Other works of about the same time seem to make the distinction implicitly. They have given rise to the unfortunate misimpression that some special choice of co-ordinates is necessary or desirable as a result, leading not only to a burdensome formalism but also to much confusion and some error in the literature. TADJBAKHSH and TOUPIN [1964, *81*] have cleared some of the pitfalls for the user of convected co-ordinates.

The components of the metric tensor relative to the new co-ordinates are then given by (15.5). Here the co-ordinates X^α in space depend on the choice of the configuration $\boldsymbol{\chi}$.

Although these two points of view are popular in the literature, we shall adopt neither of them. The reason is that (15.4) is a transformation with a definite physical meaning. To regard assignment of spatial positions to the particles of a body as a co-ordinate transformation can cause confusion, because co-ordinate transformations usually are only relations between different possible mathematical descriptions of the same physical situation.

A body $\mathscr{B}$ is assumed to be also a measure space, i.e., it is endowed with a non-negative scalar measure, m, which is called the *mass distribution* of the body.

The region of space, $\boldsymbol{\chi}(\mathscr{B})$, into which the body is mapped by (15.3) is called the region *occupied* by the body $\mathscr{B}$ in the configuration $\boldsymbol{\chi}$. In cases where confusion would be unwarranted, we refer to this region or a part of it, $\mathscr{P}$, simply as "the body $\mathscr{B}$" or "the part $\mathscr{P}$ of the body", leaving it to the reader to bear in mind that a particular configuration, $\boldsymbol{\chi}$, is intended.

For any configuration $\boldsymbol{\chi}$ of $\mathscr{B}$, one can define a measure over the region $\boldsymbol{\chi}(\mathscr{B})$ in space occupied by $\mathscr{B}$ by assigning the same measure to corresponding subsets of $\mathscr{B}$ and $\boldsymbol{\chi}(\mathscr{B})$. We assume that, for any configuration $\boldsymbol{\chi}$, this induced measure over $\boldsymbol{\chi}(\mathscr{B})$ is absolutely continuous and hence has a density ϱ, which is called the *mass density* of the body in the configuration $\boldsymbol{\chi}$. If the dependence of ϱ on $\boldsymbol{\chi}$ is to be stressed, we write $\varrho=\varrho_{\boldsymbol{\chi}}$.

At this point, therefore, we specialize our formalism explicitly to continuous media. In systems of mass-points, the mass measure m is discrete and thus has no density. A fully general kinematics would allow not only discrete and continuous masses but also more complicated mass distributions such as line masses and surface masses.

The mass of a body is a fundamental physical property and hence is assigned *a priori* as a part of the specification of the body. Not so with the mass density ϱ, which expresses a relation between the physical mass $m(\mathscr{P})$ of every measurable part $\mathscr{P}$ of $\mathscr{B}$ and the volume $v(\mathscr{P})$ of the region $\boldsymbol{\chi}(\mathscr{P})$ in Euclidean space which the part $\mathscr{P}$ happens to occupy in the configuration $\boldsymbol{\chi}$. I.e.,

$$m(\mathscr{P})=\int_{\boldsymbol{\chi}(\mathscr{P})}\varrho_{\boldsymbol{\chi}}\,dv=\int_{\mathscr{P}}\varrho\,dv. \tag{15.6}$$

In this formula the two integrals have exactly the same meaning, the second being only a shorter way of writing the first. The element "dv" is sufficient indication that the integral is defined in terms of Lebesgue measure in Euclidean space. Since we shall never need to define integration over the body manifold itself, we shall give no just cause for confusion by writing $\mathscr{P}$ for $\boldsymbol{\chi}(\mathscr{P})$ in all integrals. From the definition (15.6) it follows that the density is the ultimate ratio of mass to volume; precisely, for a sequence of suitable measurable parts $\mathscr{P}_k$ having only the point $\boldsymbol{x}$ in common and such that $\lim_{k\to\infty} v(\mathscr{P}_k)=0$,

$$\varrho(\boldsymbol{x})=\varrho_{\boldsymbol{\chi}}(\boldsymbol{x})=\lim_{k\to\infty}\frac{m(\mathscr{P}_k)}{v(\mathscr{P}_k)}. \tag{15.7}$$

A *motion* of a body $\mathscr{B}$ is a one-parameter family $\boldsymbol{\chi}_t$ of configurations. The real parameter t is the *time*. We write

$$\left.\begin{aligned}\boldsymbol{x}&=\boldsymbol{\chi}_t(X)=\boldsymbol{\chi}(X,t), & X&=\overset{-1}{\boldsymbol{\chi}}_t(\boldsymbol{x})=\overset{-1}{\boldsymbol{\chi}}(\boldsymbol{x},t),\\ x^k&=\chi^k(X^\alpha,t), & X^\alpha&=\overset{-1}{\chi}{}^\alpha(x^k,t).\end{aligned}\right\} \tag{15.8}$$

The point $\boldsymbol{x}=\boldsymbol{\chi}(X,t)$ is the *place occupied by the particle X at time t*. We shall assume that a sufficient number of derivatives of the χ^k and the $\overset{-1}{\chi}{}^\alpha$ exist and are continuous.

If the point of view (i) of p. 37 is adopted, a motion (15.8) will have the effect of endowing the body with a metric $C_{\alpha\beta}$ which is variable in time. In the point of view (ii) a motion will give rise to a moving co-ordinate system in space, i.e., a co-ordinate system deforming with the body as the body deforms during its motion.

The *velocity*,

$$\dot{\boldsymbol{x}} \equiv \frac{d}{dt}\, \boldsymbol{\chi}(X, t), \tag{15.9}$$

and the *acceleration*,

$$\ddot{\boldsymbol{x}} \equiv \frac{d}{dt}\, \dot{\boldsymbol{x}}(X, t) \equiv \frac{d^2}{dt^2}\, \boldsymbol{\chi}(X, t), \tag{15.10}$$

are, respectively, the first and second time-derivatives of the motion $\boldsymbol{\chi}(X, t)$. These time-dependent vector fields, and in fact any time-dependent scalar, vector, or tensor field Ψ, may be regarded either as a function $\Psi(X, t)$ of the particle and the time, or as a function $\Psi(\boldsymbol{x}, t)$ of the place and the time, provided that a definite motion $\boldsymbol{x} = \boldsymbol{\chi}(X, t)$ be given. If we regard X and t as the independent variables, we adopt the *material description*; if $\boldsymbol{x}$ and t, the *spatial description*. *Material derivatives* and *spatial derivatives (spatial gradients)* are defined accordingly. Material time derivatives are denoted by superposing dots or by superposing $(\mathfrak{n})$, where $\mathfrak{n}$ is the order of the differentiation:

$$\dot{\Psi} \equiv \frac{d}{dt}\, \Psi(X, t), \qquad \ddot{\Psi} \equiv \frac{d^2}{dt^2}\, \Psi(X, t), \qquad \overset{(\mathfrak{n})}{\Psi} \equiv \frac{d^{\mathfrak{n}}}{dt^{\mathfrak{n}}}\, \Psi(X, t). \tag{15.11}$$

Spatial time derivatives are denoted as follows:

$$\frac{\partial \Psi}{\partial t} \equiv \frac{d}{dt}\, \Psi(\boldsymbol{x}, t), \qquad \frac{\partial^{\mathfrak{n}} \Psi}{\partial t^{\mathfrak{n}}} \equiv \frac{d^{\mathfrak{n}}}{dt^{\mathfrak{n}}}\, \Psi(\boldsymbol{x}, t). \tag{15.12}$$

The material derivative $\dot{\Psi}$ is related to the spatial derivative $\partial \Psi / \partial t$ by

$$\left.\begin{aligned} \dot{\Psi} &= \frac{\partial \Psi}{\partial t} + (\operatorname{grad} \Psi)\, \dot{\boldsymbol{x}}, \\ \dot{\Psi}^{k_1 \ldots k_r}{}_{m_1 \ldots m_r} &= \frac{\partial}{\partial t}\, \Psi^{k_1 \ldots k_r}{}_{m_1 \ldots m_r} + \Psi^{k_1 \ldots k_r}{}_{m_1 \ldots m_r, p}\, \dot{x}^p, \end{aligned}\right\} \tag{15.13}$$

where $\operatorname{grad} \Psi$ denotes the spatial gradient of Ψ. I.e., $\operatorname{grad} \Psi = \nabla \Psi(\boldsymbol{x}, t)$. If $\Psi(X, t)$ is the function on the left-hand side of (15.13), it is understood that on the right-hand side Ψ stands for the function $\Psi\big(\overset{-1}{\boldsymbol{\chi}}(\boldsymbol{x}, t), t\big)$. Material gradients have no intrinsic physical meaning, except in relation to a particular reference configuration (cf. Sect. 21).

The mass density ϱ of a body $\mathscr{B}$ in motion, regarded as a function either of X and t or of $\boldsymbol{x}$ and t, will be assumed to have a sufficient number of continuous derivatives. From its definition the mass density may be shown to satisfy the *equation of continuity*:

$$\dot{\varrho} + \varrho \operatorname{div} \dot{\boldsymbol{x}} = 0, \qquad \dot{\varrho} + \varrho\, \dot{x}^k{}_{,k} = 0. \tag{15.14}$$

For *isochoric motions*[1] (15.14) reduces to

$$\operatorname{div} \dot{\boldsymbol{x}} = 0, \qquad \dot{x}^k{}_{,k} = 0, \tag{15.15}$$

and, in the material description, ϱ is time-independent:

$$\varrho(X, t) = \varrho(X), \qquad \dot{\varrho} = 0. \tag{15.16}$$

16. Forces and stresses. The concept of force describes the action of the outside world on a body in motion and the interaction between the different parts of the body. In this treatise, except in Sects. 98 and 127/128, we confine our attention to the non-polar case, i.e., we assume that there are neither body couples nor couple

[1] I.e. volume-preserving motions, cf. CFT, Sect. 77.

stresses[1]. Also for simplicity we assume that mutual body forces are absent and that all forces are continuously distributed. We may then summarize the essential definitions and theorems concerning forces as follows:

Assume that a body $\mathscr{B}$ and a motion of $\mathscr{B}$ are given. A *system of forces* for the body in motion is characterized by the following conditions:

(α) At each time t a vector field $\boldsymbol{b}(\boldsymbol{x}, t)$, defined for $\boldsymbol{x}$ in the region occupied by $\mathscr{B}$ at time t, is given. It is called the density per unit mass of the external body force, or simply the *external body force*, acting on $\mathscr{B}$. The vector $\boldsymbol{f}_b(\mathscr{P})$ defined by the volume integral

$$\boldsymbol{f}_b(\mathscr{P}) = \int_{\mathscr{P}} \boldsymbol{b}\,\varrho\, dv \tag{16.1}$$

over the part $\mathscr{P}$ of $\mathscr{B}$ in the configuration χ_t is called the *resultant external body force* exerted on the part $\mathscr{P}$ at time t.

(β) At any given time t, to each part $\mathscr{P}$ of the body $\mathscr{B}$ corresponds a vector field $\boldsymbol{t}(\boldsymbol{x}; \mathscr{P})$, defined for the points $\boldsymbol{x}$ on the boundary $\partial\mathscr{P}$ of $\mathscr{P}$. It is called the *density of the contact force*, or simply the *stress*, acting on the part $\mathscr{P}$ of $\mathscr{B}$. The *resultant contact force* $\boldsymbol{f}_c(\mathscr{P})$ exerted on $\mathscr{P}$ at time t is defined by the surface integral

$$\boldsymbol{f}_c(\mathscr{P}) = \int_{\partial\mathscr{P}} \boldsymbol{t}(\boldsymbol{x}; \mathscr{P})\, ds \tag{16.2}$$

extended over the boundary $\partial\mathscr{P}$ of $\mathscr{P}$ in the configuration χ_t.

(γ) *The total resultant force* $\boldsymbol{f}(\mathscr{P})$ exerted on the part $\mathscr{P}$ of $\mathscr{B}$ is defined as the sum of the resultant body force and the resultant contact force:

$$\boldsymbol{f}(\mathscr{P}) \equiv \boldsymbol{f}_b(\mathscr{P}) + \boldsymbol{f}_c(\mathscr{P}). \tag{16.3}$$

(δ) *The stress principle*: There is a vector-valued function $\boldsymbol{t}(\boldsymbol{x}, \boldsymbol{n})$, defined for all points $\boldsymbol{x}$ in $\mathscr{B}$ and for all unit vectors $\boldsymbol{n}$, such that the stress acting on any part $\mathscr{P}$ of $\mathscr{B}$ is given by

$$\boldsymbol{t}(\boldsymbol{x}; \mathscr{P}) = \boldsymbol{t}(\boldsymbol{x}, \boldsymbol{n}), \tag{16.4}$$

where $\boldsymbol{n}$ is the exterior unit normal vector at the point $\boldsymbol{x}$ on the boundary of $\mathscr{P}$. The vector $\boldsymbol{t}(\boldsymbol{x}, \boldsymbol{n})$ is called the *stress vector* at $\boldsymbol{x}$ acting across the oriented surface element with normal $\boldsymbol{n}$. If $\boldsymbol{x} \in \partial\mathscr{B}$ and if $\boldsymbol{n}$ is the exterior unit normal to $\partial\mathscr{B}$, then $\boldsymbol{t}(\boldsymbol{x}, \boldsymbol{n})$ reduces to a function of position, $\boldsymbol{t}(\boldsymbol{x})$, which is called the *surface traction* upon the boundary $\partial\mathscr{B}$.

If a body in motion and a corresponding system of forces specified in accord with (α)—(δ) are given, we speak of a *dynamical process* provided the fundamental laws of mechanics, i.e., the *principle of momentum* and the *principle of moment of momentum*[2], are satisfied.

Under suitable continuity conditions these laws imply the existence of a *stress-tensor field* $\boldsymbol{T}$ such that the stress vector $\boldsymbol{t}(\boldsymbol{x}, \boldsymbol{n})$ postulated by the stress principle (δ) may be expressed in terms of $\boldsymbol{T}(\boldsymbol{x})$ as follows[3]:

$$\boldsymbol{t}(\boldsymbol{x}, \boldsymbol{n}) = \boldsymbol{T}(\boldsymbol{x})\,\boldsymbol{n}, \qquad t^k(x^q, n^r) = T^{km}(x^q)\, n_m. \tag{16.5}$$

The fundamental principles are then equivalent to *Cauchy's laws of motion*[4]:

$$\operatorname{div} \boldsymbol{T} + \varrho \boldsymbol{b} = \varrho \ddot{\boldsymbol{x}}, \qquad T^{km}{}_{,m} + \varrho b^k = \varrho \ddot{x}^k, \tag{16.6}$$

$$\boldsymbol{T} = \boldsymbol{T}^T, \qquad T^{km} = T^{mk}. \tag{16.7}$$

[1] Cf. CFT, Sect. 200.
[2] Cf. CFT, Sect. 196.
[3] Cf. CFT, Sect. 203.
[4] Cf. CFT, Sect. 205.

Eq. (16.7), asserting that the stress tensor is symmetric, reflects the absence of body couples and couple stresses.

Suppose we have given a symmetric stress field $\boldsymbol{T}$, a motion $\boldsymbol{\chi}$, and a density ϱ defined over a body $\mathscr{B}$. From (16.6) it is plain that the body force $\boldsymbol{b}$ is then uniquely determined over $\mathscr{B}$. That is, it is always possible to find a body force such that any given pair $\{\boldsymbol{\chi}, \boldsymbol{T}\}$ may be so completed as to constitute a dynamical process for the body. While the body forces thus obtained may be artificial in nature, there is no reason to exclude them. When, henceforth, we speak of a *dynamical process*, or simply a *process*, we shall refer only to $\boldsymbol{\chi}$ and $\boldsymbol{T}$, leaving $\boldsymbol{b}$ to be thereby determined, and we shall use the notation $\{\boldsymbol{\chi}, \boldsymbol{T}\}$ to signify this understanding. In effect, then, when the principle of moment of momentum is satisfied, the principle of linear momentum imposes no restriction whatever upon the motions and stresses possible[1].

17. Changes of frame. Indifference. The fundamental measurable quantities of classical kinematics are distances and time-intervals. The position of an event can be specified only if a frame of reference, or observer, is given. Physically, a frame of reference is a set of objects whose mutual distances change comparatively little in time, like the walls of a laboratory, the fixed stars, or the wooden horses on a merry-go-round. Only if such a frame is given for all times does it make sense to compare the positions of a particle at different times, and only then can we speak about velocities, accelerations, etc., of a particle. The time at which an event takes place can be specified physically only with reference to a particular event in the history of the universe, like the birth of CHRIST. We regard such a reference event as part of a frame of reference. A frame of reference may be described as a possible way of relating physical reality to a three-dimensional Euclidean point space and a real time axis.

An *event* is a pair $\{\boldsymbol{x}, t\}$ consisting of a point $\boldsymbol{x}$ in space and a time t. The totality of all events is called *space-time*. A *change of frame* is a one-to-one mapping of space-time onto itself such that distances, time intervals, and temporal order are preserved. Only the concept of a change of frame, not the concept of frame of reference itself, is regarded as having a mathematical meaning. In particular, frame of reference should *not* be regarded as a synonym for co-ordinate system.

Note that transformations involving a reversal of the sense of time are not considered. While it is commonly asserted that classical mechanics represents only phenomena reversible in time, this is true only for certain special disciplines such as analytical mechanics. As we shall see below, the mechanical behavior of materials is specified in terms of *past* experiences only and thus is fully capable of representing irreversibility.

It can be shown[2] that an event $\{\boldsymbol{x}, t\}$ and its image $\{\boldsymbol{x}^*, t^*\}$ under a change of frame are related by rigid transformations combined with a time-shift. More precisely, if we choose an arbitrary origin in space and then identify points with their position vectors, we can express a change of frame by the formulae

$$\boldsymbol{x}^* = \boldsymbol{c}(t) + \boldsymbol{Q}(t)\,\boldsymbol{x}, \tag{17.1}$$

$$t^* = t - a, \tag{17.2}$$

where $\boldsymbol{c}(t)$ is a point, $\boldsymbol{Q}(t)$ an orthogonal tensor, and a a real number. We assume that $\boldsymbol{c}(t)$ and $\boldsymbol{Q}(t)$ are smooth functions of t. The transformation $\boldsymbol{Q}(t)$ is uniquely determined by the change of frame, but the point $\boldsymbol{c}(t)$ depends also on the choice of the origin.

---

[1] In more general mechanics allowing for body couples and couple stresses, the principle of moment of momentum likewise fails to impose any restriction upon the class of possible stresses and motions.

[2] For a proof of (17.1) see, e.g., NOLL [1964, *64*]. The proof of (17.2) is trivial.

A change of frame induces transformations, for each time t, on scalars, vectors, and tensors. These transformations are uniquely determined by the requirement that the "point-difference" relation (6.1) between points and vectors and the various relations between scalars, vectors, and tensors be preserved under a change of frame. In particular, we have the following transformation laws:

(i) Scalars remain unchanged under changes of frame.

(ii) A vector $\boldsymbol{v}$ is transformed into

$$\boldsymbol{v}^* = \boldsymbol{Q}(t)\,\boldsymbol{v}. \tag{17.3}$$

(iii) Let $\boldsymbol{S}$ be a second-order tensor, regarded as a linear transformation of vectors. Let $\boldsymbol{w} = \boldsymbol{S}\boldsymbol{v}$, and let $\boldsymbol{v}^* = \boldsymbol{Q}(t)\,\boldsymbol{v}$, $\boldsymbol{w}^* = \boldsymbol{Q}(t)\,\boldsymbol{w}$, be the transforms (17.3) of $\boldsymbol{v}$ and $\boldsymbol{w}$ under the change of frame. We then have $\boldsymbol{w}^* = \boldsymbol{Q}(t)\,\boldsymbol{w} = \boldsymbol{Q}(t)\,(\boldsymbol{S}\boldsymbol{v}) = \left(\boldsymbol{Q}(t)\,\boldsymbol{S}\,\boldsymbol{Q}(t)^T\right)\boldsymbol{v}^*$. Hence $\boldsymbol{w}^*$ and $\boldsymbol{v}^*$ are related by the linear transformation

$$\boldsymbol{S}^* = \boldsymbol{Q}(t)\,\boldsymbol{S}\,\boldsymbol{Q}(t)^T, \tag{17.4}$$

which is the transform of $\boldsymbol{S}$ under the change of frame.

Functions and fields whose values are scalars, vectors, or tensors will be called *frame-indifferent* or *objective* if both the dependent and independent variables transform according to the laws (17.1)—(17.4) under a change of frame.

The transformation laws (17.1)—(17.4) have no unambiguous co-ordinate form, and it is best to look at them geometrically, without reference to co-ordinates. It is necessary to distinguish sharply between changes of frame and transformations of co-ordinates. The former are transformations of space-time; the latter do not affect space-time, but only the co-ordinates of points. It is often useful to combine a co-ordinate transformation with a change of frame in such a way that corresponding points $\boldsymbol{x}$ and $\boldsymbol{x}^*$ have the same co-ordinates. Then also corresponding vector and tensor fields, if indifferent, will have the same components in both frames. The equations involving components and co-ordinates in the text below should be understood in this sense.

Under the transformations considered here, the units of length and time are kept fixed. A fully invariant formulation of a physical theory must specify transformation laws under changes of units as well. In this article we rarely present dimensional considerations, because, not offering any peculiar difficulties in mechanics, they may easily be included by the method explained in Part II of the Appendix to CFT. Cf., e.g., the treatment of electromagnetism in Sects. 273 and 274 of CFT.

We consider now the motion of a body $\mathscr{B}$. We assume that, relative to some frame of reference, this motion is given by

$$\boldsymbol{x} = \boldsymbol{\chi}(X, t). \tag{17.5}$$

According to (17.1) and (17.2), the same motion is described in a new frame by

$$\boldsymbol{x}^* = \boldsymbol{\chi}^*(X, t^*) = \boldsymbol{c}(t) + \boldsymbol{Q}(t)\,\boldsymbol{\chi}(X, t), \qquad t^* = t - a. \tag{17.6}$$

Physically, $\boldsymbol{\chi}$ and $\boldsymbol{\chi}^*$ describe the same motion, referred to different frames. In mathematical interpretation, we have to regard $\boldsymbol{\chi}^*$ as the motion obtained from $\boldsymbol{\chi}$ by the superposition of a time-dependent rigid transformation and by a shift in the time scale. We shall say that any two motions related by an equation of the form (17.6) are *equivalent motions*.

We now write (17.6) in the abbreviated form

$$\boldsymbol{x}^* = \boldsymbol{c} + \boldsymbol{Q}\boldsymbol{x}. \tag{17.7}$$

Material time differentiation gives the following transformation formulae for the velocity and the acceleration:

$$\overline{\boldsymbol{x}^*}^{\boldsymbol{\cdot}} - (\dot{\boldsymbol{x}})^* = \dot{\boldsymbol{c}} + \boldsymbol{A}(\boldsymbol{x}^* - \boldsymbol{c}), \tag{17.8}$$

$$\overline{\boldsymbol{x}^*}^{\boldsymbol{\cdot\cdot}} - (\ddot{\boldsymbol{x}})^* = \ddot{\boldsymbol{c}} + 2\boldsymbol{A}\left(\overline{\boldsymbol{x}^*}^{\boldsymbol{\cdot}} - \dot{\boldsymbol{c}}\right) + \left(\dot{\boldsymbol{A}} - \boldsymbol{A}^2\right)(\boldsymbol{x}^* - \boldsymbol{c}) \tag{17.9}$$

where $(\dot{\boldsymbol{x}})^*$ and $(\ddot{\boldsymbol{x}})^*$ are related to $\dot{\boldsymbol{x}}$ and $\ddot{\boldsymbol{x}}$ by (17.3), and where $\boldsymbol{A}$ is defined by

$$\boldsymbol{A} = \dot{\boldsymbol{Q}}\boldsymbol{Q}^T, \quad A_{km} = \dot{Q}_k{}^p Q_{mp}. \tag{17.10}$$

(The overbars cover the quantity to which the dot operation is applied, while parentheses delimit the quantities affected by the asterisks.) $\boldsymbol{A}$, the angular velocity of the original frame with respect to the new one, is a skew tensor: $\boldsymbol{A} = -\boldsymbol{A}^T$, $A_{km} = -A_{mk}$. If we denote by p^k the components of the position vector of $\boldsymbol{x}$ relative to the origin[1], then (17.8) and (17.9) have the forms

$$\overline{x^{*k}}^{\boldsymbol{\cdot}} - \dot{x}^k = \dot{c}^k + A^k{}_m p^m, \tag{17.11}$$

$$\overline{x^{*k}}^{\boldsymbol{\cdot\cdot}} - \ddot{x}^k = \ddot{c}^k + 2A^k{}_m\left(\overline{x^{*m}}^{\boldsymbol{\cdot}} - \dot{c}^m\right) + \left(\dot{A}^k{}_m - A^k{}_p A^p{}_m\right)p^m, \tag{17.12}$$

in general co-ordinates.

18. Equivalent processes. In Sect. 16 we have defined forces in terms of vector functions. If we wish to assume that forces have a frame-indifferent meaning, these vector functions should transform under changes of frame according to the law (17.3). However, when this assumption is made, the fundamental laws of mechanics are not invariant under changes of frame. A dynamical process would not always transform into a dynamical process. The usual way of resolving this difficulty is to introduce the concept of absolute space, usually attributed to NEWTON, and to require the validity of the dynamical laws only when all quantities are referred to absolute space. This procedure was justly criticized by MACH[2]. It is more natural to reinterpret the fundamental laws and to render them invariant under arbitrary changes of frame in the following manner[3]: All forces are frame-indifferent. Inertia is to be regarded as that mutual body force which describes the interaction between the bodies in the solar system and the masses occupying the remainder of the universe. The term $\varrho\ddot{\boldsymbol{x}}$ in (16.6) must then be replaced by $\varrho\boldsymbol{a}$, where $\boldsymbol{a}$ is the *acceleration relative to the fixed stars*. $\ddot{\boldsymbol{x}}$ is not frame-indifferent, but $\boldsymbol{a}$ is, for not only the position of the particle under consideration but also the positions of the fixed stars must be transformed according to (17.1) under a change of frame. Similar remarks apply to all other kinds of external body forces, such as gravity exerted by external masses. The dependence of these body forces on the configuration of the world *outside* of the body $\mathscr{B}$ is usually not made explicit. It is for this reason that these forces appear to be formally not invariant under changes of frame. CAUCHY's first law of motion has the form (16.6) only when the frame is chosen so that $\ddot{\boldsymbol{x}} = \boldsymbol{a}$.

Mutual body forces and internal contact forces depend only on the configuration of the body $\mathscr{B}$, and not on the configuration of the masses of the outside world. Therefore, mutual body forces and internal contact forces are formally

[1] The p^k differ from the co-ordinates x^k of $\boldsymbol{x}$, except when the co-ordinate system is rectilinear.

[2] MACH [1883, *1*, Ch. II, § 6].

[3] Perhaps MACH [1883, *1*, Ch. II, § 6] had this view in mind. A mathematical treatment has been given by NOLL [1963, *58*].

invariant under changes of frame. Since we have assumed (cf. Sect. 16) for the purposes of this treatise that there are no mutual body forces, we have to consider only the requirement that stresses be frame-indifferent. By (17.4) this requirement means that the stress tensor transforms under change of frame according to the law

$$\boldsymbol{T}^*(t^*) = \boldsymbol{Q}(t)\,\boldsymbol{T}(t)\,\boldsymbol{Q}(t)^T, \qquad T^*_{km}(t^*) = T_{km}(t). \tag{18.1}$$

As before, we have assumed that a change of co-ordinates accompanies the change of frame in such a manner that corresponding points have the same co-ordinates.

Two dynamical processes $\{\boldsymbol{\chi}, \boldsymbol{T}\}$ and $\{\boldsymbol{\chi}^*, \boldsymbol{T}^*\}$ that are related to each other by a change of frame of the form (17.6), (18.1) will be called *equivalent processes*. According to the viewpoint used here, they are only different mathematical descriptions of the same physical process.

19. The principle of material frame-indifference. A body was described in Sect. 15 by its possible configurations and by its mass distribution. Experience shows that two physical bodies with the same shape and the same mass distribution may behave quite differently under the influence of the same forces. We describe this difference by saying that the two bodies consist of different materials. For a given body, not every dynamical process is appropriate to a physically natural problem. In the first place, the body force $\boldsymbol{b}$ is usually assigned in advance rather than left to be determined by $\boldsymbol{\chi}$ and $\boldsymbol{T}$ as is done in Sect. 16. Secondly, the possible processes are those that satisfy further conditions defining the material properties of the body[1]. These conditions are called *constitutive assumptions* or *constitutive equations*[2]. A constitutive equation defines an *ideal material*[3]. We say that a certain dynamical process is *admissible* for a given ideal material if the constitutive equation of that material is satisfied. In this Chapter we consider constitutive equations involving only the motion and the stress. Note that, in accord with the observation at the end of Sect. 16, the class of processes consists of the pairs $\{\boldsymbol{\chi}, \boldsymbol{T}\}$ where $\boldsymbol{\chi}$ is an arbitrary motion and $\boldsymbol{T}$ is an arbitrary time-dependent symmetric tensor field.

We assert as a fundamental principle of classical physics that material properties are indifferent, i.e., independent of the frame of reference or observer. This means, mathematically, that constitutive equations are subject to the following

Principle of material frame-indifference:

Constitutive equations must be invariant under changes of frame of reference. If a constitutive equation is satisfied for a process with a motion and a symmetric stress tensor given by

$$\boldsymbol{x} = \boldsymbol{\chi}(X, t), \qquad \boldsymbol{T} = \boldsymbol{T}(X, t), \tag{19.1}$$

then it must be satisfied also for any equivalent process $\{\boldsymbol{\chi}^*, \boldsymbol{T}^*\}$. That is, the constitutive equation must be satisfied also for the motion and stress tensor given by

$$\left.\begin{aligned} \boldsymbol{x}^* &= \boldsymbol{\chi}^*(X, t^*) = \boldsymbol{c}(t) + \boldsymbol{Q}(t)\,\boldsymbol{\chi}(X, t),\\ \boldsymbol{T}^* &= \boldsymbol{T}^*(X, t^*) = \boldsymbol{Q}(t)\,\boldsymbol{T}(X, t)\,\boldsymbol{Q}(t)^T,\\ t^* &= t - a, \end{aligned}\right\} \tag{19.2}$$

where $\boldsymbol{c}(t)$ is an arbitrary point function, $\boldsymbol{Q}(t)$ an arbitrary orthogonal tensor function of the time t, and a an arbitrary number.

---

[1] See also Sects. 7 and 293 of CFT.

[2] The terms "equation of state" and, in special cases, "stress-strain relation" are also frequently used.

[3] Cf. the discussion in Sects. 7 and 293 of CFT.

For brevity we shall refer to this axiom simply as the "principle of material indifference". It is called also the "principle of material objectivity", or simply "objectivity".

From (19.2) and the foregoing considerations it seems that the requirements resulting from the principle of material indifference have little connection with the laws of mechanics. Rather, they express the view of material response that seems to be the only reasonable one when events are regarded as occurring in a space-time consisting of a uniform time and instantaneous Euclidean point spaces. The only consequence of the laws of mechanics actually used is the fact that any symmetric stress tensor $\boldsymbol{T}$ and any motion χ satisfy the dynamical equations for some body force $\boldsymbol{b}$. For any other set of mechanical laws having this property, materials in Euclidean space-time would have to satisfy the principle of material indifference in exactly the same form. However, it can be shown that the laws of mechanics are consequences of another principle of indifference, a principle that postulates the invariance of *work* under changes of frame[1].

19 A. Appendix. History of the principle of material frame-indifference. In 1675 Hooke had published in an anagram the principle of "A new sort of *Philosophical-Scales*, of great use in Experimental Philosophy". When, in 1678, he explained his spring scale [1678, *1*, pp. 5—6], proposing its use as an instrument for absolute rather than merely relative measure of the weights of bodies, he reported that with its aid he had sought to determine the variation of the earth's gravity with altitude, but neither on church towers nor in deep mines was any effect discerned. Thus he regarded it as obvious that while a rigid displacement might alter the weight of a body, it could have no effect on the response of a given spring to a given force. Otherwise, the phenomenon he hoped to detect and could indeed have found by ascending a sufficiently high mountain, namely, that a given suspended body may extend a given spring by different amounts when the experiment is performed in places at very different altitudes, would not afford means of measuring the difference of the forces exerted by the earth upon that same body.

A more striking and familiar example of tacit belief in the principle is afforded by the experiment usually adduced in verification of the formula for the magnitude of centripetal force. A body of known weight, say one pound, when suspended by a given spring is observed to extend it by a given amount, say one inch. The spring and weight, still connected, are then laid upon a horizontal disc, to the center of which the free end of the spring is attached. The disc is then caused to spin at a steady speed such as to extend the spring again by one inch. The spectators are expected to agree that the centripetal force required to hold the weight from flying off is exactly one poundal. That is, *the response of the spring* is unaffected by a rigid motion, even though, in the usual view of classical mechanics, the laws of motion are not. Indeed, if we were to allow the possibility that the material response of the spring may be altered by spinning it rigidly, the experiment would show nothing.

Of course the constitutive equation for the simple spring does satisfy, automatically, the principle of material indifference. So do the constitutive equations of the other better-known classical theories of materials. Thus there is little reference, even oblique, to the principle itself in the older literature.

In a memoir dated October 12, 1829, Poisson [1831, *1*, p. 83], in order to conclude that the stress in the theory of infinitesimal deformation from an unstressed state must be independent of the infinitesimal rotation, appealed to the following principle:

"Si l'on fait tourner le corps autour de l'axe des x, et que chacun de ses points décrive un très-petit angle ..., pour un tel déplacement, le corps demeure dans son état naturel, et les pressions intérieures doivent encore être nulles ...".

In proposing a theory of initially stressed elastic materials, to which we shall refer in more detail in Sects. 68 and 100, Cauchy [1829, *1*, pp. 350—353], acknowledging the idea as being due to Poisson, assumed that "le premier état du corps continuera de subsister, si dans le passage du premier état au second, on a déplacé tous les points, en les faisant tourner simultanément autour de l'un des axes coordonnés". By imposing this requirement, he showed that certain terms could not appear in a constitutive equation of the type he proposed for study. These are the earliest examples of use of the principle to get a reduced form for a constitutive equation.

In the work of Hooke, Poisson, and Cauchy, the *body* is subjected to a rigid deformation. The first reference to the invariance of response of a material for all *observers* occurs implicitly in the work of Zaremba [1903, *11* and *12*] [1937, *4*, Chap. I, § 2]. Without giving any reason, Zaremba chose to demand that the time rates entering a constitutive equation be calculated in a co-rotational frame [CFT, Sect. 148]; the equations so obtained he then transformed to a general frame, remarking [1903, *11*, p. 603] of invariance under arbitrary time-dependent rigid transformation, "C'est là une propriété que doit avoir tout système d'équations exprimant les relations qui existent entre les forces intérieures dans une substance et les circonstances de son mouvement." Later [1903, *12*, pp. 619—620] he suggested, in effect, as an alternative

[1] Noll [1963, *58*].

method the use of convected co-ordinates: "l'emploi des variables introduites en hydrodynamique par Lagrange permet d'éviter, sans introduire aucune complication dans les équations, l'usage d'équations incompatibles avec le principe des mouvements relatifs ..."

Another formulation of the principle is due to Jaumann [1906, *2*, § IV]:

"Wir gehen von dem Axiom aus, daß jeder physikalische Vorgang, sofern er Objekt (Problem) einer Theorie sein kann, im *Endlichen begrenzt* sein muß.

"Es ist dies freilich eine Abstraktion, denn in Wirklichkeit hängen alle Vorgänge im ganzen Weltraume zusammen, aber dieser Zusammenhang darf außer acht gelassen werden. Es ist diese Abstraktion notwendig und die Voraussetzung jeder Theorie, denn es kann keineswegs das Ziel der Theorie sein, alle Naturvorgänge gleichzeitig zu untersuchen.

"Bei der Beschreibung jedes physikalischen Vorganges muß eine bestimmte Grenzschale desselben angegeben werden und der Nachweis erbracht werden, daß alle außerhalb dieser Grenzschale verlaufenden Vorgänge ohne Einfluß auf den betrachteten Vorgang innerhalb der Schale sind. Es ist diese Angabe immer so leicht, daß sie meist nicht ausdrücklich erwähnt zu werden braucht.

"Ferner gehen wir von dem *Nahewirkungsprinzip* aus, welches besagt, daß die Werte der physikalischen Variablen und ihrer räumlichen und zeitlichen Derivationen in einem sehr kleinen Teile des Mediums sich gegenseitig vollkommen bestimmen. Die Fortpflanzung einer Wirkung auf größere Entfernung kann also nicht ohne wesentliche Veränderungen in dem dazwischenliegenden Medium gedacht werden.

"Nun ersehen wir die Bedingung, welche die Grenzschale eines physikalischen Problems immer erfüllen muß. Sie kann nicht als eine mathematische, geschlossene Grenzfläche gedacht werden, denn von einer solchen könnte man niemals behaupten, daß sie alle äußeren Wirkungen abhält. Die Grenzschale muß vielmehr eine bestimmt anzugebende *materielle* Schale von *endlicher* Dicke sein. Für dieselbe muß nachgewiesen werden, daß alle für den betrachteten Vorgang wesentlichen Eigenschaftsveränderungen ihres Materials Null sind. Dann pflanzt die Grenzschale keine Wirkung der äußeren Vorgänge in den Innenraum und keine Wirkung des untersuchten Vorganges im Innenraum nach außen fort.

"Da die Deformation der Teile des Mediums jeden physikalischen Vorgang wesentlich beeinflußt, muß die Grenzschale jedenfalls hinreichend starr sein. Sie bildet also eine geeignete *Grundlage für alle Orts- und Richtungsmessungen.*

"Alle physikalischen Gesetze müssen invariante Form haben, so daß die Lage des Koordinatenursprunges und die Wahl der Koordinatenrichtungen gleichgültig ist. Es reicht also jedenfalls aus, einen bestimmten starren Körper aufzuweisen, auf welchen sich alle Messungen zu beziehen haben, und es ist gleichgültig, wie man das Koordinatensystem in diesem Körper festlegt ...

"Nun halte ich es für ganz unrichtig, deshalb zu verlangen, daß dieser starre Körper für jedes Problem *derselbe Körper* sein müsse. Dann bleibt nichts übrig, als den Fixsternhimmel zur Festlegung des Koordinatensystems zu benützen. Diese Lösung ist aber durchaus nicht universeller, als wenn man das Koordinatensystem in irgend einem anderen starren System festlegt, denn außerhalb unseres Fixsternsystems gibt es andere anders bewegte Fixsternsysteme und man kann doch die Vorgänge in diesen nicht auf unseren Fixsternhimmel beziehen.

"Die einzig denkbare, wirklich universelle Festsetzung ist, daß man das Koordinatensystem immer in der Grenzschale des betrachteten Problems festzulegen hat; dies ist einleuchtend und man tut es ohnehin meist unwillkürlich."

Later Jaumann [1911, *4*, § III] summarized his form of the principle as follows: "Die Orts- und Richtungsangaben des Gleichungssystems müssen meiner Auffassung nach bezogen werden auf ein Koordinatensystem, welches in der indifferenten (also auch undeformierten) Grenzschale des betrachteten Vorganges festgelegt ist."

A landmark in the history of principles of invariance is the Cosserats' *Theory of Euclidean Action.* Considering only materials of essentially elastic type, they laid down the requirement that the energy be invariant under arbitrary proper orthogonal transformations, and they systematically developed the consequences of this requirement for oriented media of one, two, and three dimensions [1907, *1*] [1909, *1*, §§ 3, 8, 33, 51, 61 et passim]. (Their theory is presented, in generalized form, in Sect. 98, below.)

Although several succeeding authors in effect used the principle of indifference, the next statement of it seems to be that of Oldroyd [1950, *11*]: "The form of the completely general equations must be restricted by the requirement that the equations describe properties independent of the frame of reference ... *Moreover, only those tensor quantities need be considered which have a significance for the material element independent of its motion as a whole in space.*" Oldroyd in effect repeated the observation of Zaremba that formulation of all equations in convected co-ordinates *suffices* to satisfy this requirement. Attributing the idea to Cauchy and acknowledging indebtedness also to Ericksen, Truesdell [1953, *25*, § 55bis] wrote: "Now in a dynamical model of actual matter, a rigid rotation should be insufficient to change the state of stress relative to co-ordinates fixed in the body."

While some modern students seem to regard the principle as obvious, the variety of forms and motivations put forward for it suggests that it may be somewhat subtle, as illustrated by the fact that at least two of the persons who ought to have known better, namely, ZAREMBA [1903, *9*] and TRUESDELL [1952, *20*, § 56], had first proposed non-invariant theories that they themselves later corrected by use of it. ZAREMBA [1903, *8* and *10*] had entered the field by criticizing for failure of invariance NATANSON's three-dimensional generalization of MAXWELL's one-dimensional theory of visco-elasticity. NATANSON, a well known physicist in his day, had considered his work important enough to write accounts of it in three languages, published in three countries [1901, *5—8*] [1902, *2—5*], and in a vigorous polemic literature [1903, *5—7*] he never admitted the justice of ZAREMBA's criticism. While the decision of time has been wholly for ZAREMBA, it has come late, and the vast literature on "plasticity" ignores it. ~~We should mention here that the constitutive equations of visco-elastic type that follow from MAXWELL's kinetic theory of gases satisfy the principle of indifference provided not too many "approximations" are introduced. Cf. IKENBERRY and TRUESDELL [1956, *14*, § 16]. Indeed, the intermolecular force, being a function only of mutual distances, is indifferent, so any rigorous consequences of the theory *must* satisfy the principle.~~*

In fact, *two* principles have been stated and studied. According to the first, which may be called the "Hooke-Poisson-Cauchy form", constitutive equations must be invariant under a superimposed rigid rotation of the *body*. According to the second, which may be called the "Zaremba-Jaumann form", an arbitrary change of *observer* is allowed. Since a body may be subjected only to *proper* rotations, while the transformation from one observer's frame to another may be an improper orthogonal transformation, the second form is more restrictive than the first. A constitutive equation satisfying the Zaremba-Jaumann form satisfies also the Hooke-Poisson-Cauchy form, but not conversely. However, as we shall see in Sect. 29, for the purely mechanical theory of simple materials, to which the larger part of this treatise is devoted, the difference is of no consequence.

In this treatise we adopt the Zaremba-Jaumann form, put into precise mathematical terms and given a new motivation by NOLL [1958, *31*, § 11], who called it "the principle of objectivity of material properties". Earlier, NOLL [1955, *19*, § 4] had given a precise formulation of the Hooke-Poisson-Cauchy form, which he called "the principle of isotropy of space", and had clarified the distinction between the invariances corresponding to frame-indifference and material isotropy. Cf. also the cases considered by RIVLIN and ERICKSEN [1955, *23*, § 12], THOMAS [1955, *27* and *29*], COTTER and RIVLIN [1955, *8*]. The Hooke-Poisson-Cauchy form is preferred by RIVLIN and his collaborators.

As has been remarked by DILL [1965, *15*], the claim of ZIEGLER [1963, *89*, § 4] that "the principle of material indifference ... [is] not valid in a continuum moving with respect to a rotating coordinate system" is a mere pronouncement, for which ZIEGLER advanced no basis beyond the phrase, "it is to be expected that". Since *every* continuum moves with respect to rotating as well as stationary co-ordinate systems, we fail to understand what meaning ZIEGLER intended.

20. The main open problem of the theory of material behavior. Apart from the invariance requirement of the preceding section, constitutive equations must satisfy other basic requirements. Unfortunately, we are not able to put these requirements into precise form at the present time. Roughly speaking, they should insure that physically reasonable problems have physically reasonable solutions. For example, a constitutive equation, together with the equation of continuity (15.14) and the equation of motion (16.6), should lead to a definite theory in the sense that the existence, uniqueness and stability of solutions of meaningful boundary-value and initial-value problems can be proved, except in cases where instability or failure of uniqueness can be expected on physical grounds. Another requirement is that a deformation of a thermally isolated body from an unstressed state should never produce mechanical energy. Also, if a body is thermally isolated and held at rest from a certain time on, the stresses should decay to their equilibrium values. This phenomenon is called *stress relaxation*. The requirements mentioned here are probably intimately connected with each other. To find a fully general, complete, and precise statement of these requirements we regard as the *main open problem of the theory of material behavior*. It seems likely that a satisfactory solution of this main problem can be given only after a unified theory of

---

* [The barred passage is incorrect.]

mechanics, thermodynamics, and electromagnetism has been formulated[1], and that the requirements mentioned above will come out as consequences of thermodynamic principles[2].

The work done so far toward the solution of the main problem is almost entirely confined to the special case of hyperelastic materials. A discussion will be given in Sects. 79, 80, 81, and 96 in Chapter D.

II. Kinematics.

21. Deformation. While the body $\mathscr{B}$ is not to be confused with any of its spatial configurations, nevertheless it is available to us only in those configurations. Physical observations can never be made on a body *except* in some configuration. For many purposes it is convenient to reflect this fact by using positions in a certain fixed configuration as a means of specifying the particles of the body. This *reference configuration* $\boldsymbol{\varkappa}$ may be, but need not be, one actually occupied by the body in the course of its motion[3]. The place of a particle in $\boldsymbol{\varkappa}$ will be denoted by

$$\boldsymbol{X} = \boldsymbol{\varkappa}(X). \tag{21.1}$$

If X^α is a system of material co-ordinates, we write

$$\boldsymbol{X} = \boldsymbol{\varkappa}(X^\alpha), \qquad X^\alpha = \overset{-1}{\varkappa}{}^\alpha(\boldsymbol{X}). \tag{21.2}$$

We shall regard $X^\alpha = \overset{-1}{\varkappa}{}^\alpha(\boldsymbol{X})$ as a definition of a co-ordinate system in space, so that a particle X and its place $\boldsymbol{X} = \boldsymbol{\varkappa}(X)$ in the reference configuration have the same co-ordinates X^α. The vectors of the natural basis corresponding to this co-ordinate system will be denoted by $\boldsymbol{d}_\alpha$, where

$$\boldsymbol{d}_\alpha(X) = \partial_\alpha \boldsymbol{\varkappa}(X). \tag{21.3}$$

The unit tensor **1** has the components

$$g_{\alpha\beta} = \boldsymbol{d}_\alpha \cdot \boldsymbol{d}_\beta. \tag{21.4}$$

We distinguish them from the components g_{km} of the unit tensor in the co-ordinate system x^k of the spatial description by using Greek indices.

If $\boldsymbol{\chi}$ is a motion of $\mathscr{B}$, then

$$\boldsymbol{x} = \boldsymbol{\chi}(\overset{-1}{\boldsymbol{\varkappa}}(\boldsymbol{X}), t) \equiv \boldsymbol{\chi}(\boldsymbol{X}, t) \tag{21.5}$$

defines a family of deformations[4] from the reference configuration. We have used the same symbol $\boldsymbol{\chi}$ for the motion and for the corresponding deformation function because ordinarily no confusion should follow, but it must be remembered that

[1] Classical thermodynamics deals mainly with "quasi-static" processes and hence is inadequate for the purposes of continuum mechanics.

[2] While this treatise was in press, an important step toward solution of the main problem was made by COLEMAN, as is explained in Sect. 96 bis.

[3] In the older literature on continuum mechanics the body $\mathscr{B}$ and some reference configuration $\boldsymbol{\varkappa}(\mathscr{B})$ are often confused. Usually it is assumed that $\boldsymbol{\varkappa}(\mathscr{B}) = \boldsymbol{\chi}(\mathscr{B}, t_0)$ for some initial time t_0; with no historical justification, the material co-ordinates are then called "Lagrangean". The fact that the reference configuration need never be assumed has also been recognized; e.g., LAMB [1932, *5*, § 16] wrote, "It is to be remarked that the quantities *a*, *b*, *c*, need not be restricted to mean the initial co-ordinates of the particle; they may be any three quantities which serve to identify a particle, and which vary continuously from one particle to another." Cf. CFT, Sects. 66, 66A.

[4] Here "deformation" is used in a general sense which includes changes both of shape and orientation in space. A rigid rotation is regarded as a special kind of deformation.

the form of $\boldsymbol{\chi}(\boldsymbol{X}, t)$ depends upon the choice of $\boldsymbol{\varkappa}$. The representation of (21.5) in co-ordinates is the same as for $(15.8)_1$, i.e.,

$$x^k = \chi^k(X^\alpha, t). \tag{21.6}$$

Let S be a tensor field we wish to consider with reference to the motion $\boldsymbol{\chi}$. One may wish to employ the components of S relative to the natural basis $\boldsymbol{e}_k$ at $\boldsymbol{x} = \boldsymbol{\chi}(\boldsymbol{X}, t)$ of the co-ordinate system x^k, or those relative to the basis $\boldsymbol{d}_\alpha$ at $\boldsymbol{X}$ of the co-ordinate system X^α. It is possible also to define components of S such that some of the indices correspond to $\boldsymbol{e}_k$ and the others to $\boldsymbol{d}_\alpha$. We distinguish the various types of components by using Greek and Latin indices. Introducing the corresponding *shifters*

$$g_k^\alpha = \boldsymbol{e}_k \cdot \boldsymbol{d}^\alpha, \qquad g_\alpha^k = \boldsymbol{e}^k \cdot \boldsymbol{d}_\alpha, \tag{21.7}$$

we have transformation formulae of the following type[1]:

$$\mathsf{S}^{\cdot\cdot\cdot\cdot\cdot}_{\cdot\cdot\alpha\cdot\cdot}\, g_k^\alpha = \mathsf{S}^{\cdot\cdot\cdot\cdot\cdot}_{\cdot\cdot k\cdot\cdot}, \qquad \mathsf{S}^{\cdot\cdot\cdot\cdot\cdot}_{\cdot\cdot k\cdot\cdot}\, g_\alpha^k = \mathsf{S}^{\cdot\cdot\cdot\cdot\cdot}_{\cdot\cdot\alpha\cdot\cdot}. \tag{21.8}$$

It must be noted that the basis $\boldsymbol{e}_k$ is in general different from the basis $\boldsymbol{d}_\alpha$ even when x^k and X^α define the same co-ordinate system in space. The reason is, of course, that $\boldsymbol{e}_k$ is the basis at $\boldsymbol{x}$ while $\boldsymbol{d}_\alpha$ is the basis at $\boldsymbol{X}$, which is in general different from $\boldsymbol{x}$. The distinction between Greek and Latin indices is superfluous only if x^k and X^α define the same *rectilinear* co-ordinate system in space.

The *deformation gradient* of the motion $\boldsymbol{\chi}$ is a second-order tensor field which we denote by $\boldsymbol{F}$:

$$\boldsymbol{F} \equiv \nabla \boldsymbol{\chi}(\boldsymbol{X}, t), \qquad F^k{}_\alpha \equiv x^k{}_{,\alpha} \equiv \frac{\partial}{\partial X^\alpha} \chi^k(X^\gamma, t). \tag{21.9}$$

The dependence of $\boldsymbol{F}$ on $\boldsymbol{X}$ and t in (21.9) is understood.

In some cases it is convenient to use the configuration $\boldsymbol{x} = \boldsymbol{\chi}_t(X) = \boldsymbol{\chi}(X, t)$ for some particular fixed time t, rather than $\boldsymbol{\varkappa}$, as a reference. The variable time will then be denoted by τ, and the place of X at time τ by

$$\boldsymbol{\xi} = \boldsymbol{\chi}(X, \tau). \tag{21.10}$$

The corresponding deformation function will be denoted by $\boldsymbol{\chi}_{(t)}$, so that

$$\boldsymbol{\xi} = \boldsymbol{\chi}_{(t)}(\boldsymbol{x}, \tau) = \boldsymbol{\chi}(\overset{-1}{\boldsymbol{\chi}}_t(\boldsymbol{x}), \tau). \tag{21.11}$$

We call $\boldsymbol{\chi}_{(t)}$ the *relative deformation function*; its value $\boldsymbol{\xi} = \boldsymbol{\chi}_{(t)}(\boldsymbol{x}, \tau)$ is the place at time τ of that particle which, at time t, occupies the place $\boldsymbol{x}$. The gradient of $\boldsymbol{\chi}_{(t)}$ will be called the *relative deformation gradient*, and it will be denoted by $\boldsymbol{F}_{(t)}$:

$$\boldsymbol{F}_{(t)}(\tau) \equiv \nabla \boldsymbol{\chi}_{(t)}(\boldsymbol{x}, \tau). \tag{21.12}$$

The dependence of $\boldsymbol{F}_{(t)}(\tau)$ on $\boldsymbol{x}$ is understood. The components of $\boldsymbol{F}_{(t)}(\tau)$ are[2]

$$F_{(t)}(\tau)^k{}_m = \xi^k{}_{,m} = \frac{\partial}{\partial x^m} \chi^k{}_{(t)}(x^q, \tau), \tag{21.13}$$

where the first index k corresponds to the natural basis $\boldsymbol{\varepsilon}_k$ at $\boldsymbol{\xi}$ and the second index m to the natural basis $\boldsymbol{e}_m$ at $\boldsymbol{x}$ of the spatial co-ordinate system $\boldsymbol{x}^k$. We have

$$\boldsymbol{F}_{(t)}(t) = \mathbf{1}, \qquad F_{(t)}{}^k{}_m(t) = \delta^k_m. \tag{21.14}$$

---

[1] A detailed discussion of the formalism of the "double tensor" notation used here is given in Part III of the Appendix to CFT.

[2] The $F_{(t)}{}^k{}_m$ are subject to the double tensor transformation law $F_{(t)}{}^k{}_m = \frac{\partial \bar{\xi}^k}{\partial \xi^p} \frac{\partial x^q}{\partial \bar{x}^m} F_{(t)}{}^p{}_q$ under a spatial co-ordinate transformation $\bar{x}^m = \bar{x}^m(x^q)$, $\bar{\xi}^k = \bar{x}^k(\xi^p)$.

Assume that the velocity field $\dot{\boldsymbol{x}}(\boldsymbol{x}, t)$ in the spatial description is assigned. The relative deformation function (21.11) may then be obtained by solving the differential equations

$$\frac{d\boldsymbol{\xi}}{d\tau} = \dot{\boldsymbol{x}}(\boldsymbol{\xi}, \tau), \qquad \frac{d\xi^k}{d\tau} = \dot{x}^k(\xi^q, \tau) \tag{21.15}$$

with the initial conditions

$$\boldsymbol{\xi}\big|_{\tau=t} = \boldsymbol{x}, \qquad \xi^k\big|_{\tau=t} = x^k. \tag{21.16}$$

The deformation function (21.5) cannot be determined from a prescription of the velocity field alone. It is necessary to assign, in addition, the place $\boldsymbol{x}_0 = \boldsymbol{\chi}(\boldsymbol{X}, t_0)$ of the particle for some particular time t_0.

The determinant of the deformation gradient is the Jacobian of the transformation which defines the deformation from the reference configuration to the actual configuration. We use the notation

$$J = J(t) \equiv |\det \boldsymbol{F}(t)|, \qquad J_{(t)}(\tau) = \det \boldsymbol{F}_{(t)}(\tau). \tag{21.17}$$

Note that $J(t)$ and $J_{(t)}(\tau)$ are always positive because $\boldsymbol{F}(t)$ and $\boldsymbol{F}_{(t)}(\tau)$ are invertible and because $J_{(t)}(t) = 1$. We have the following relations for the mass density ϱ_{R} in the reference configuration and the densities $\varrho(t)$ and $\varrho(\tau)$ at times t and τ:

$$J(t)\,\varrho(t) = \varrho_{\mathrm{R}}, \qquad J_{(t)}(\tau)\,\varrho(\tau) = \varrho(t). \tag{21.18}$$

These are integrated forms[1] of the continuity equation (15.14). For isochoric motions ϱ is independent of t, and $J_{(t)}(\tau) = 1$.

When we wish to consider changes of the reference configuration, we cannot continue to use the same symbol for the motion and the deformation function. Let $\boldsymbol{\varkappa}$ and $\hat{\boldsymbol{\varkappa}}$ be two reference configurations, and denote the deformation that carries $\boldsymbol{\varkappa}$ into $\hat{\boldsymbol{\varkappa}}$ by

$$\hat{\boldsymbol{X}} = \hat{\boldsymbol{\varkappa}}(\overset{-1}{\boldsymbol{\varkappa}}(\boldsymbol{X})) \equiv \boldsymbol{\lambda}(\boldsymbol{X}). \tag{21.19}$$

For a given motion, let $\boldsymbol{\chi}$ and $\hat{\boldsymbol{\chi}}$ define the deformations from the reference configurations $\boldsymbol{\varkappa}$ and $\hat{\boldsymbol{\varkappa}}$, respectively. $\boldsymbol{\chi}$ and $\hat{\boldsymbol{\chi}}$ are related by

$$\hat{\boldsymbol{\chi}}(\hat{\boldsymbol{X}}, t) = \hat{\boldsymbol{\chi}}(\boldsymbol{\lambda}(\boldsymbol{X}), t) = \boldsymbol{\chi}(\boldsymbol{X}, t). \tag{21.20}$$

The tensor

$$\boldsymbol{P} = \nabla \boldsymbol{\lambda}(\boldsymbol{X}) \tag{21.21}$$

is the gradient at $\boldsymbol{X}$ of the deformation $\boldsymbol{\lambda}$ relating the two reference configurations. Let

$$\boldsymbol{F} = \nabla \boldsymbol{\chi}(\boldsymbol{X}, t), \qquad \hat{\boldsymbol{F}} = \nabla \hat{\boldsymbol{\chi}}(\hat{\boldsymbol{X}}, t) \tag{21.22}$$

denote the deformation gradients corresponding to use of the two reference configurations $\boldsymbol{\varkappa}$ and $\hat{\boldsymbol{\varkappa}}$. It follows from Eqs. (21.20)—(21.22) and the chain rule for the differentiation of composite functions that

$$\boldsymbol{F} = \hat{\boldsymbol{F}} \boldsymbol{P}. \tag{21.23}$$

In particular, when $\hat{\boldsymbol{\varkappa}}$ becomes the configuration at a particular time t, (21.23) reads

$$\boldsymbol{F}(\tau) = \boldsymbol{F}_{(t)}(\tau)\,\boldsymbol{F}(t). \tag{21.24}$$

If we select as the reference configuration $\boldsymbol{\varkappa}$ the configuration of the material at some fixed time t', (21.24) yields

$$\boldsymbol{F}_{(t')}(\tau) = \boldsymbol{F}_{(t)}(\tau)\,\boldsymbol{F}_{(t')}(t). \tag{21.25}$$

[1] Cf. the discussion in Sect. 156 of CFT.

22. Localizations. Local configurations. Histories. We now give a mathematical description of the local behavior of a motion. Let $\mathcal{N}(X)$ be a small neighborhood of X, and let Z denote a typical particle in $\mathcal{N}(X)$. Let $\boldsymbol{\varkappa}$ be any configuration of $\mathcal{N}(X)$. The vector-valued function $\boldsymbol{\varkappa}_X$ defined by[1]

$$\boldsymbol{Z} = \boldsymbol{\varkappa}_X(Z) = \boldsymbol{\varkappa}(Z) - \boldsymbol{\varkappa}(X) \tag{22.1}$$

is then called the *localization* in $\mathcal{N}(X)$ of the configuration $\boldsymbol{\varkappa}$. The value $\boldsymbol{Z} = \boldsymbol{\varkappa}_X(Z)$ is the position vector of the particle Z in the neighborhood $\mathcal{N}(X)$ when the position of X is taken as origin. We have

$$\boldsymbol{\varkappa}_X(X) = \boldsymbol{0}. \tag{22.2}$$

In particular, if $\boldsymbol{\varkappa}$ is replaced by the configuration at time τ in a motion $\boldsymbol{\chi}$, we define the localization $\boldsymbol{\chi}_X$ at X of the motion $\boldsymbol{\chi}$ by

$$\boldsymbol{\zeta} = \boldsymbol{\chi}_X(Z, \tau) = \boldsymbol{\chi}(Z, \tau) - \boldsymbol{\chi}(X, \tau). \tag{22.3}$$

The localized motion $\boldsymbol{\chi}_X$ describes the motion of a neighborhood of X as it appears to an observer moving with the particle X. The equation

$$\boldsymbol{\chi}_X(X, \tau) = \boldsymbol{0} \tag{22.4}$$

expresses the fact that the particle X is always at rest relative to itself. If a reference configuration $\boldsymbol{\varkappa}$ of $\mathcal{N}(X)$ is given, and if $\boldsymbol{\varkappa}_X$ is its localization at X, then

$$\boldsymbol{\zeta} = \boldsymbol{\chi}_X(\overset{-1}{\boldsymbol{\varkappa}}_X(\boldsymbol{Z}), \tau) = \boldsymbol{\chi}_X(\boldsymbol{Z}, \tau) \tag{22.5}$$

defines the deformation from the localization $\boldsymbol{\varkappa}_X$ of $\boldsymbol{\varkappa}$ to the localization $\boldsymbol{\chi}_X$ of $\boldsymbol{\chi}$. As in (21.5), we have used the same symbol $\boldsymbol{\chi}_X$ for the localized motion and the corresponding deformation function. We observe that

$$\boldsymbol{\chi}_X(\boldsymbol{0}, \tau) = \boldsymbol{0} \tag{22.6}$$

for all times τ.

Since $\boldsymbol{\chi}_X$ and $\boldsymbol{\varkappa}_X$ differ from $\boldsymbol{\chi}$ and $\boldsymbol{\varkappa}$ only by translations, it is clear that the gradient of $\boldsymbol{\chi}$ at $\boldsymbol{X}$ coincides with the gradient of $\boldsymbol{\chi}_X$ at $\boldsymbol{0}$, i.e.

$$\boldsymbol{F}(\boldsymbol{X}, \tau) = \nabla\boldsymbol{\chi}(\boldsymbol{X}, \tau) = \nabla\boldsymbol{\chi}_X(\boldsymbol{0}, \tau). \tag{22.7}$$

If no confusion can arise, we may omit the variable $\boldsymbol{X}$ and write

$$\boldsymbol{F}(\tau) = \nabla\boldsymbol{\chi}_X(\boldsymbol{0}, \tau). \tag{22.8}$$

The deformation from the localization

$$\boldsymbol{z} = \boldsymbol{\chi}_X(Z, t) \tag{22.9}$$

of $\boldsymbol{\chi}$ at a fixed time t to the localization

$$\boldsymbol{\zeta} = \boldsymbol{\chi}_X(Z, \tau) \tag{22.10}$$

of $\boldsymbol{\chi}$ at time τ will be denoted by

$$\boldsymbol{\zeta} = \boldsymbol{\chi}_{X,(t)}(\boldsymbol{z}, \tau). \tag{22.11}$$

Its gradient at $\boldsymbol{z} = \boldsymbol{0}$ is the same as the gradient of the relative deformation function $\boldsymbol{\chi}_{(t)}$ at $\boldsymbol{x}$, i.e.

$$\boldsymbol{F}_{(t)}(\tau) = \nabla\boldsymbol{\chi}_{(t)}(\boldsymbol{x}, \tau) = \nabla\boldsymbol{\chi}_{X,(t)}(\boldsymbol{0}, \tau). \tag{22.12}$$

For each particle X we can define an equivalence relation "$\sim$" among all configurations $\boldsymbol{\varkappa}$ by the condition that $\boldsymbol{\varkappa} \sim \hat{\boldsymbol{\varkappa}}$ if the gradient (21.21) of the

[1] The "$-$" symbol is to be understood in the sense of the point-difference (6.1).

deformation $\boldsymbol{\lambda}$ carrying $\boldsymbol{\varkappa}$ into $\hat{\boldsymbol{\varkappa}}$ is the unit tensor $\mathbf{1}$:

$$\boldsymbol{\varkappa} \sim \hat{\boldsymbol{\varkappa}} \quad \text{if} \quad \nabla \boldsymbol{\lambda}(\boldsymbol{X}) = \mathbf{1}, \tag{22.13}$$

where $\boldsymbol{\lambda}$ is given by (21.19). The resulting equivalence classes are called *local configurations* $\boldsymbol{K}$ at X. We write

$$\boldsymbol{K} = \nabla \boldsymbol{\varkappa}(X) \tag{22.14}$$

for the equivalence class to which $\boldsymbol{\varkappa}$ belongs. $\boldsymbol{K}$ may be viewed also as a linear transformation of the tangent space of $\mathscr{B}$ at X onto the space of ordinary vectors. The gradient (21.21) of the deformation $\boldsymbol{\lambda}$ carrying the configuration $\boldsymbol{\varkappa}$ into the configuration $\hat{\boldsymbol{\varkappa}}$ depends only on the equivalence classes $\boldsymbol{K}$ and $\hat{\boldsymbol{K}}$ to which $\boldsymbol{\varkappa}$ and $\hat{\boldsymbol{\varkappa}}$ belong. Hence, two local configurations $\boldsymbol{K}$ and $\hat{\boldsymbol{K}}$ at X determine a unique tensor $\boldsymbol{P}$, for which we write

$$\boldsymbol{P} = \hat{\boldsymbol{K}} \boldsymbol{K}^{-1} \tag{22.15}$$

and which we call the *local deformation* from the local configuration $\boldsymbol{K}$ to the local configuration $\hat{\boldsymbol{K}}$. If $\boldsymbol{K} = \nabla \boldsymbol{\varkappa}(X)$ and $\hat{\boldsymbol{K}} = \nabla \hat{\boldsymbol{\varkappa}}(X)$, then $\boldsymbol{P}$ coincides with the deformation gradient (21.21).

Let Ψ be any function of a real variable τ whose values are points, vectors, or tensors. For each real t we then define a function $\Psi^{(t)}$ of a non-negative real variable by

$$\Psi^{(t)}(s) = \Psi(t-s) \quad \text{when} \quad s \geqq 0 \tag{22.16}$$

and call it the *history up to time t of the function* Ψ. We have

$$\Psi^{(t)}(0) = \Psi(t). \tag{22.17}$$

It is important to note that $\Psi^{(t)}(s)$ is regarded as undefined for $s<0$, even though $\Psi(\tau)$ may be defined for all real τ. For example, the history $\boldsymbol{\chi}_X{}^{(t)}$ up to time t of a localized motion $\boldsymbol{\chi}_X$ (cf. (22.3)) is defined by

$$\boldsymbol{\chi}_X{}^{(t)}(Z, s) = \boldsymbol{\chi}_X(Z, t-s) \quad \text{when} \quad s \geqq 0. \tag{22.18}$$

$\Psi^{(t)}(s)$ is the value of Ψ at a time s units before the present time t, and the variable s is the *time-lapse* from the past instant τ to the present instant t.

In this section, our notation has been designed so as to make a clear distinction between functions and function values. Such a distinction is necessary for the discussion of functionals whose domains are classes of histories (cf. Subchapter III).

23. Stretch and Rotation. The analysis in the rest of this subchapter in based on the *polar decomposition theorem*[1]:

Any invertible linear transformation $\boldsymbol{F}$ *has two unique multiplicative decompositions*

$$\boldsymbol{F} = \boldsymbol{R}\boldsymbol{U}, \quad \boldsymbol{F} = \boldsymbol{V}\boldsymbol{R}, \tag{23.1}$$

in which $\boldsymbol{R}$ *is orthogonal and* $\boldsymbol{U}$ *and* $\boldsymbol{V}$ *are symmetric and positive-definite. The following relations are valid*:

$$\boldsymbol{U}^2 = \boldsymbol{F}^T \boldsymbol{F}, \qquad \boldsymbol{V}^2 = \boldsymbol{F}\boldsymbol{F}^T, \tag{23.2}$$

$$\boldsymbol{V} = \boldsymbol{R}\boldsymbol{U}\boldsymbol{R}^T, \qquad \boldsymbol{V}^2 = \boldsymbol{R}\boldsymbol{U}^2\boldsymbol{R}^T. \tag{23.3}$$

This theorem, when applied to the deformation gradient $\boldsymbol{F} = \boldsymbol{F}(t) = \boldsymbol{F}(\boldsymbol{X}, t)$ of a motion, gives rise to the following terminology: $\boldsymbol{R}$ is the *rotation tensor*, $\boldsymbol{U}$ the

[1] Cf. Sect. 43 of the Appendix to CFT.

right stretch tensor, and $\boldsymbol{V}$ the *left stretch tensor* of the deformation. The square of the right stretch tensor:

$$\boldsymbol{C}=\boldsymbol{U}^2=\boldsymbol{F}^T\boldsymbol{F}, \tag{23.4}$$

is called the *right Cauchy-Green tensor*, and the square of the left stretch tensor[1]:

$$\boldsymbol{B}=\boldsymbol{V}^2=\boldsymbol{F}\boldsymbol{F}^T=\boldsymbol{R}\boldsymbol{C}\boldsymbol{R}^T, \tag{23.5}$$

is called the *left Cauchy-Green tensor* of the deformation. Components of $\boldsymbol{C}$ and $\boldsymbol{B}$ are

$$C_{\alpha\beta}=F^k{}_\alpha F^m{}_\beta g_{km}=x^k{}_{,\alpha}x^m{}_{,\beta}g_{km}, \tag{23.6}$$

$$B^{km}=F^k{}_\alpha F^m{}_\beta g^{\alpha\beta}=x^k{}_{,\alpha}x^m{}_{,\beta}g^{\alpha\beta}, \tag{23.7}$$

polynomials in the components of $\boldsymbol{F}$. The components of the stretch tensors $\boldsymbol{U}$ and $\boldsymbol{V}$ are complicated irrational functions of the components of $\boldsymbol{F}$, and hence in solutions of special problems it is usually better to use $\boldsymbol{C}$ and $\boldsymbol{B}$ rather than $\boldsymbol{U}$ and $\boldsymbol{V}$ as measures of strain, although these latter are often more suitable for handling general considerations.

The proper numbers of $\boldsymbol{U}$ and $\boldsymbol{V}$ are the principal stretches[2] of the deformation. The proper numbers of $\boldsymbol{C}$ and $\boldsymbol{B}$ are the squares of these principal stretches. Hence, if the principal stretches are denoted by v_a, we have

$$\left.\begin{aligned} I_{\boldsymbol{C}}=I_{\boldsymbol{B}}=v_1^2+v_2^2+v_3^2, \qquad II_{\boldsymbol{C}}=II_{\boldsymbol{B}}=v_1^2v_2^2+v_2^2v_3^2+v_3^2v_1^2,\\ III_{\boldsymbol{C}}=III_{\boldsymbol{B}}=v_1^2v_2^2v_3^2. \end{aligned}\right\} \tag{23.8}$$

If $\boldsymbol{F}$ is replaced by the relative deformation gradient $\boldsymbol{F}_{(t)}$, then the notations $\boldsymbol{R}_{(t)}$, $\boldsymbol{U}_{(t)}$, $\boldsymbol{V}_{(t)}$, $\boldsymbol{C}_{(t)}$, and $\boldsymbol{B}_{(t)}$ are used, respectively, for the corresponding *relative rotation tensor*, *relative stretch tensors*, and *relative Cauchy-Green tensors*. It follows from (21.14) that

$$\boldsymbol{R}_{(t)}(t)=\boldsymbol{U}_{(t)}(t)=\boldsymbol{V}_{(t)}(t)=\boldsymbol{B}_{(t)}(t)=\boldsymbol{C}_{(t)}(t)=\boldsymbol{1}. \tag{23.9}$$

By (23.6) and (21.13) $\boldsymbol{C}_{(t)}$ has the components

$$C_{(t)km}=F_{(t)}{}^p{}_k F_{(t)}{}^q{}_m \gamma_{pq}=\xi^p{}_{,k}\xi^q{}_{,m}\gamma_{pq}, \tag{23.10}$$

where the γ_{pq} are the components of the unit tensor at the point ξ in the spatial co-ordinate system x^k. Both indices of $C_{(t)km}$ correspond to the natural basis $\boldsymbol{e}_k$ at $\boldsymbol{x}$. Hence, the $C_{(t)km}$ are spatial tensor components in the ordinary sense.

The determinants of the stretch tensors and Cauchy-Green tensors are related to the Jacobians (21.17) by

$$\left.\begin{aligned} \det\boldsymbol{U}=\det\boldsymbol{V}=J, \qquad \det\boldsymbol{B}=\det\boldsymbol{C}=J^2,\\ \det\boldsymbol{U}_{(t)}=\det\boldsymbol{V}_{(t)}=J_{(t)}, \qquad \det\boldsymbol{B}_{(t)}=\det\boldsymbol{C}_{(t)}=J_{(t)}^2. \end{aligned}\right\} \tag{23.11}$$

From (23.4) and (21.24) we obtain the following relations between the Cauchy-Green tensors:

$$\boldsymbol{C}(\tau)=\boldsymbol{F}(t)^T\boldsymbol{C}_{(t)}(\tau)\boldsymbol{F}(t), \qquad \boldsymbol{B}(\tau)=\boldsymbol{F}_{(t)}(\tau)\boldsymbol{B}(t)\boldsymbol{F}_{(t)}(\tau)^T. \tag{23.12}$$

24. Stretching and spin. Consider the gradient $\boldsymbol{F}_{(t)}(\tau)$ of the deformation from the configuration at time t to the configuration at time τ. Its derivative with

---

[1] $\boldsymbol{B}$ is the reciprocal of the tensor $\mathbf{c}$ defined by (26.1) of CFT: $\boldsymbol{B}=\mathbf{c}^{-1}$

[2] Cf. CFT, Sect. 25.

respect to τ, evaluated at $\tau = t$, will be denoted by $\boldsymbol{L}(t)$, so that[1]

$$\boldsymbol{L}(t) = \dot{\boldsymbol{F}}_{(t)}(t) = -\dot{\boldsymbol{F}}^{(t)}_{(t)}(0) = \dot{\boldsymbol{F}}(t)\,\boldsymbol{F}(t)^{-1}, \tag{24.1}$$

where $\boldsymbol{F}^{(t)}_{(t)}$ denotes the history of $\boldsymbol{F}_{(t)}$ up to time t, defined according to (22.16). It follows from (21.11) that $\boldsymbol{L}(t)$ is nothing but the *spatial gradient*[2] *of the velocity* $\dot{\boldsymbol{x}}$, i.e.

$$\boldsymbol{L} = \operatorname{grad}\dot{\boldsymbol{x}}, \qquad L_{km} = \dot{x}_{k,m}. \tag{24.2}$$

Similarly the spatial gradient

$$\boldsymbol{L}_2 = \operatorname{grad}\ddot{\boldsymbol{x}}, \qquad L_{2\,km} = \ddot{x}_{k,m}, \tag{24.3}$$

of the acceleration $\ddot{\boldsymbol{x}}$ is related to $\boldsymbol{F}_{(t)}$ by

$$\boldsymbol{L}_2(t) = \ddot{\boldsymbol{F}}_{(t)}(t) = \ddot{\boldsymbol{F}}^{(t)}_{(t)}(0). \tag{24.4}$$

More generally, the $\mathfrak{n}$'*th acceleration gradient*, defined by

$$\boldsymbol{L}_{\mathfrak{n}} = \operatorname{grad}\overset{(\mathfrak{n})}{\boldsymbol{x}}, \qquad L_{\mathfrak{n}\,km} = \overset{(\mathfrak{n})}{x}_{k,m} \tag{24.5}$$

for $\mathfrak{n} = 0, 1, 2, \ldots$, is related to $\boldsymbol{F}_{(t)}$ by

$$\boldsymbol{L}_{\mathfrak{n}}(t) = \overset{(\mathfrak{n})}{\boldsymbol{F}}_{(t)}(t) = (-1)^{\mathfrak{n}}\,\overset{(\mathfrak{n})}{\boldsymbol{F}}{}^{(t)}_{(t)}(0). \tag{24.6}$$

We have, with this notation,

$$\boldsymbol{L}_0 = \boldsymbol{1}, \qquad \boldsymbol{L}_1 = \boldsymbol{L}. \tag{24.7}$$

The polar decomposition

$$\boldsymbol{F}_{(t)}(\tau) = \boldsymbol{R}_{(t)}(\tau)\,\boldsymbol{U}_{(t)}(\tau) \tag{24.8}$$

defines the relative rotation tensor $\boldsymbol{R}_{(t)}$ and the relative right stretch tensor $\boldsymbol{U}_{(t)}$ (cf. Sect. 23). The histories $\boldsymbol{R}^{(t)}_{(t)}$ and $\boldsymbol{U}^{(t)}_{(t)}$ are defined according to (22.16). The tensor

$$\boldsymbol{W}(t) = \dot{\boldsymbol{R}}_{(t)}(t) = -\dot{\boldsymbol{R}}^{(t)}_{(t)}(0) \tag{24.9}$$

is called the *spin tensor*, and

$$\boldsymbol{D}(t) = \dot{\boldsymbol{U}}_{(t)}(t) = -\dot{\boldsymbol{U}}^{(t)}_{(t)}(0) \tag{24.10}$$

is called the *stretching tensor* of the motion. Similarly, the $\mathfrak{n}$'*th spin* $\boldsymbol{W}_{\mathfrak{n}}$ and the $\mathfrak{n}$'*th stretching* $\boldsymbol{D}_{\mathfrak{n}}$ are defined by

$$\boldsymbol{W}_{\mathfrak{n}}(t) = \overset{(\mathfrak{n})}{\boldsymbol{R}}_{(t)}(t) = (-1)^{\mathfrak{n}}\,\overset{(\mathfrak{n})}{\boldsymbol{R}}{}^{(t)}_{(t)}(0) \tag{24.11}$$

and

$$\boldsymbol{D}_{\mathfrak{n}}(t) = \overset{(\mathfrak{n})}{\boldsymbol{U}}_{(t)}(t) = (-1)^{\mathfrak{n}}\,\overset{(\mathfrak{n})}{\boldsymbol{U}}{}^{(t)}_{(t)}(0) \tag{24.12}$$

for $\mathfrak{n} = 0, 1, 2, \ldots$. We have, with this notation,

$$\boldsymbol{W}_0 = \boldsymbol{D}_0 = \boldsymbol{1}, \qquad \boldsymbol{W}_1 = \boldsymbol{W}, \qquad \boldsymbol{D}_1 = \boldsymbol{D}. \tag{24.13}$$

The tensor $\boldsymbol{C}_{(t)}(\tau)$ is the right Cauchy-Green tensor relative to the configuration at time t. The $\mathfrak{n}$'*th Rivlin-Ericksen tensor* $\boldsymbol{A}_{\mathfrak{n}}$ is defined by[3]

$$\boldsymbol{A}_{\mathfrak{n}}(t) = \overset{(\mathfrak{n})}{\boldsymbol{C}}_{(t)}(t) = (-1)^{\mathfrak{n}}\,\overset{(\mathfrak{n})}{\boldsymbol{C}}{}^{(t)}_{(t)}(0). \tag{24.14}$$

---

[1] Variables indicated by subscripts or superscripts are held constant in differentiations denoted by superimposed dots or German minuscules.

[2] The symbol "∇" denotes the gradient operator. When the arguments of the function on which it operates are indicated, its meaning is plain. In those cases, the present being one, in which we adopt the brief though inaccurate notation that confuses functions with their values, we shall reserve "∇" for functions of the material variables $\boldsymbol{X}$, t and use "grad" to inform the reader that the spatial variables $\boldsymbol{x}$, t are being used. Thus, e.g., $\nabla f = \boldsymbol{F}^T \operatorname{grad} f$.

[3] Cf. CFT, Sect. 104.

For *isochoric motions* $J=1$, and hence $\det \boldsymbol{C}_{(t)}(\tau)=1$. Differentiating this formula repeatedly with respect to τ, using (9.3), (9.19), and then putting $\tau=t$, we find relations[1] between the invariants of the $\boldsymbol{A}_{\mathfrak{n}}$. The first three of these are

$$\left.\begin{aligned} &\operatorname{tr}\boldsymbol{A}_1=0, \qquad \operatorname{tr}\boldsymbol{A}_1^2-\operatorname{tr}\boldsymbol{A}_2=0,\\ &2\operatorname{tr}\boldsymbol{A}_1^3-3\operatorname{tr}(\boldsymbol{A}_2\boldsymbol{A}_1)+\operatorname{tr}\boldsymbol{A}_3=0.\end{aligned}\right\} \tag{24.15}$$

More generally, $\operatorname{tr}\boldsymbol{A}_{\mathfrak{n}}$ in an isochoric motion can always be expressed as a linear combination of traces of products formed from $\boldsymbol{A}_1, \boldsymbol{A}_2, \ldots, \boldsymbol{A}_{\mathfrak{n}-1}$.

In the above definitions of $\boldsymbol{W}_{\mathfrak{n}}$, $\boldsymbol{D}_{\mathfrak{n}}$, and $\boldsymbol{A}_{\mathfrak{n}}$, the present configuration has been chosen as the reference configuration. If a fixed reference configuration is used, more complicated formulae ensue. For example,

$$\left.\begin{aligned} \boldsymbol{W}&=\dot{\boldsymbol{R}}\boldsymbol{R}^T+\tfrac{1}{2}\boldsymbol{R}(\dot{\boldsymbol{U}}\boldsymbol{U}^{-1}-\boldsymbol{U}^{-1}\dot{\boldsymbol{U}})\boldsymbol{R}^T,\\ \boldsymbol{D}&=\tfrac{1}{2}\boldsymbol{R}(\dot{\boldsymbol{U}}\boldsymbol{U}^{-1}+\boldsymbol{U}^{-1}\dot{\boldsymbol{U}})\boldsymbol{R}^T.\end{aligned}\right\} \tag{24.16}$$

25. Polynomial expressions for the rates. It is possible to express the components of the tensors $\boldsymbol{W}_{\mathfrak{n}}$, $\boldsymbol{D}_{\mathfrak{n}}$, and $\boldsymbol{A}_{\mathfrak{n}}$, defined in the previous section, as polynomials in the components $L_{\mathfrak{n}}{}^k{}_m=\overset{(\mathfrak{n})}{x}{}^k{}_{,m}$ of the acceleration gradients $\boldsymbol{L}_{\mathfrak{n}}$.

Differentiation of $\boldsymbol{C}_{(t)}(\tau)=\boldsymbol{F}_{(t)}(\tau)^T\boldsymbol{F}_{(t)}(\tau)$ (cf. (23.4)) according to the product rule gives

$$\overset{(\mathfrak{n})}{\boldsymbol{C}}_{(t)}(\tau)=\sum_{\mathfrak{i}=0}^{\mathfrak{n}}\binom{\mathfrak{n}}{\mathfrak{i}}\overset{(\mathfrak{i})}{\boldsymbol{F}}_{(t)}(\tau)^T\overset{(\mathfrak{n}-\mathfrak{i})}{\boldsymbol{F}}_{(t)}(\tau). \tag{25.1}$$

By (24.6) and (24.7) this reduces for $\tau=t$ to

$$\boldsymbol{A}_{\mathfrak{n}}=\boldsymbol{L}_{\mathfrak{n}}+\boldsymbol{L}_{\mathfrak{n}}^T+\sum_{\mathfrak{i}=1}^{\mathfrak{n}-1}\binom{\mathfrak{n}}{\mathfrak{i}}\boldsymbol{L}_{\mathfrak{i}}^T\boldsymbol{L}_{\mathfrak{n}-\mathfrak{i}}, \tag{25.2}$$

which has the component form

$$A_{\mathfrak{n}\,km}=\overset{(\mathfrak{n})}{x}{}_{k,m}+\overset{(\mathfrak{n})}{x}{}_{m,k}+\sum_{\mathfrak{i}=1}^{\mathfrak{n}-1}\binom{\mathfrak{n}}{\mathfrak{i}}\overset{(\mathfrak{i})}{x}{}^q{}_{,k}\,\overset{(\mathfrak{n}-\mathfrak{i})}{x}{}_{q,m}. \tag{25.3}$$

Since $\boldsymbol{C}_{(t)}(\tau)=[\boldsymbol{U}_{(t)}(\tau)]^2$ (cf. (23.4)), one can derive in a similar manner the formula

$$\left.\begin{aligned} \boldsymbol{D}_{\mathfrak{n}}&=\tfrac{1}{2}\Big\{\boldsymbol{A}_{\mathfrak{n}}-\sum_{\mathfrak{i}=1}^{\mathfrak{n}-1}\binom{\mathfrak{n}}{\mathfrak{i}}\boldsymbol{D}_{\mathfrak{i}}\boldsymbol{D}_{\mathfrak{n}-\mathfrak{i}}\Big\},\\ D_{\mathfrak{n}\,km}&=\tfrac{1}{2}\Big\{A_{\mathfrak{n}\,km}-\sum_{\mathfrak{i}=1}^{\mathfrak{n}-1}\binom{\mathfrak{n}}{\mathfrak{i}}D_{\mathfrak{i}\,k}{}^q D_{\mathfrak{n}-\mathfrak{i}\,qm}\Big\}.\end{aligned}\right\} \tag{25.4}$$

This is a recursion formula which can be used to find explicit expressions for the $D_{\mathfrak{n}\,km}$ as polynomials in the $A_{\mathfrak{i}\,km}$, $\mathfrak{i}=1, 2, \ldots, \mathfrak{n}$, and hence, by (25.3), also as polynomials in the $\overset{(\mathfrak{i})}{x}{}_{k,m}$, $\mathfrak{i}=1, 2, \ldots, \mathfrak{n}$.

From $\boldsymbol{F}_{(t)}(\tau)=\boldsymbol{R}_{(t)}(\tau)\,\boldsymbol{U}_{(t)}(\tau)$ (cf. $(23.1)_1$) one can derive the formula

$$\left.\begin{aligned} \boldsymbol{W}_{\mathfrak{n}}&=\boldsymbol{L}_{\mathfrak{n}}-\boldsymbol{D}_{\mathfrak{n}}-\sum_{\mathfrak{i}=1}^{\mathfrak{n}-1}\binom{\mathfrak{n}}{\mathfrak{i}}\boldsymbol{W}_{\mathfrak{i}}\boldsymbol{D}_{\mathfrak{n}-\mathfrak{i}},\\ W_{\mathfrak{n}\,km}&=L_{\mathfrak{n}\,km}-D_{\mathfrak{n}\,km}-\sum_{\mathfrak{i}=1}^{\mathfrak{n}-1}\binom{\mathfrak{n}}{\mathfrak{i}}W_{\mathfrak{i}\,kq}D_{\mathfrak{n}-\mathfrak{i}}{}^q{}_m,\end{aligned}\right\} \tag{25.5}$$

which, again, is a recursion formula showing that the $W_{\mathfrak{n}\,km}$ can be expressed as polynomials in the $\overset{(\mathfrak{i})}{x}{}_{k,m}$, $\mathfrak{i}=1, 2, \ldots, \mathfrak{n}$.

---

[1] Rivlin [1962, 57].

For $\mathfrak{n}=1$ one finds that

$$\left.\begin{aligned} \boldsymbol{D}&=\tfrac{1}{2}\boldsymbol{A}_1=\tfrac{1}{2}(\boldsymbol{L}+\boldsymbol{L}^T), & \boldsymbol{W}&=\tfrac{1}{2}(\boldsymbol{L}-\boldsymbol{L}^T),\\ D_{km}&=\tfrac{1}{2}A_{1\,km}=\tfrac{1}{2}(\dot{x}_{k,m}+\dot{x}_{m,k}), & W_{km}&=\tfrac{1}{2}(\dot{x}_{k,m}-\dot{x}_{m,k}).\end{aligned}\right\}\tag{25.6}$$

We note that the spin $\boldsymbol{W}$ is skew:

$$\boldsymbol{W}=-\boldsymbol{W}^T,\qquad W_{km}=-W_{mk}.\tag{25.7}$$

The higher spins $\boldsymbol{W}_\mathfrak{n}$, $\mathfrak{n}>1$, are not necessarily skew.

For $\mathfrak{n}=2$ one finds that

$$\left.\begin{aligned} \boldsymbol{A}_2&=\boldsymbol{L}_2+\boldsymbol{L}_2^T+2\boldsymbol{L}^T\boldsymbol{L},\\ \boldsymbol{D}_2&=\tfrac{1}{2}(\boldsymbol{A}_2-\tfrac{1}{2}\boldsymbol{A}_1^2)=\tfrac{1}{2}(\boldsymbol{L}_2+\boldsymbol{L}_2^T)-\tfrac{1}{4}(\boldsymbol{L}+\boldsymbol{L}^T)^2+\boldsymbol{L}^T\boldsymbol{L},\\ \boldsymbol{W}_2&=\boldsymbol{L}_2-\boldsymbol{D}_2-2\boldsymbol{W}\boldsymbol{L}=\tfrac{1}{2}(\boldsymbol{L}_2-\boldsymbol{L}_2^T)-\tfrac{1}{4}(\boldsymbol{L}+\boldsymbol{L}^T)^2+(\boldsymbol{L}^T)^2.\end{aligned}\right\}\tag{25.8}$$

The stretching tensors $\boldsymbol{D}_\mathfrak{n}$ and the Rivlin-Ericksen tensors $\boldsymbol{A}_\mathfrak{n}$ are *indifferent* in the sense of Sect. 17, but the spins $\boldsymbol{W}_\mathfrak{n}$ are not.

III. The general constitutive equation.

26. The principle of determinism. The general constitutive equation. We have seen in Sect. 16 that the contact forces acting on the parts of a body $\mathscr{B}$ are determined by the stress tensor field $\boldsymbol{T}$. The physical concept of a contact force implies that the circumstances in the immediate neighborhood of the point of application of such a force determine it. Hence the stress $\boldsymbol{T}(X, t)$ is determined by the state of an arbitrarily small neighborhood of the particle X. Furthermore, the causality of natural processes may be interpreted as implying that the conditions in a body at time t are determined by the past history of the body, and that no aspect of its future behavior need be known in order to determine all of them. In this chapter we shall assume that all mechanical interactions in the body can be described by the stress alone, we shall disregard all non-mechanical influences, and we shall assume that the state of the body is determined by its *kinematical history* alone. We thus lay down as postulates[1] for a theory of purely mechanical phenomena in continuous media the following

(*a*) ***Principle of determinism for the stress:*** *The stress in a body is determined by the history of the motion of that body.*

(*b*) ***Principle of local action:*** *In determining the stress at a given particle X, the motion outside an arbitrary neighborhood of X may be disregarded.*

---

[1] NOLL [1958, *31*, § 12]. The earliest, and in fact the only early statement of the principle of determinism we have found, is that of CAUCHY [1828, *1*, § III]: "Dans un corps solide non élastique, les pressions ou tensions ne dépendent pas seulement du changement de forme que le corps éprouve en passant de l'etat natural à un nouvel état, mais aussi des états intermédiaires et du temps pendant lequel le changement de forme s'effectue." CAUCHY gave equations, however, only for the case he described as being that in which "l'élasticité disparaît entièrement," leading to an extremely special theory. We know of no early statement of the principle of local action as such. Of course, it is a possible counterpart for continua of the old and common idea that intermolecular forces are short-range forces. E.g., STOKES [1845, *1*, § 1] in partial justification of his (in fact extremely special) concept of fluid friction (cf. Sect. 119) wrote, "It is an undoubted result of observation that the molecular forces, whether in solids, liquids, or gases, are forces of enormous intensity, but which are sensible only at insensible distances." Until fairly recently there was not sufficient mathematical equipment to make possible an enunciation both general and precise. In *all* work before NOLL'S, the constitutive equations proposed involved extraneous assumptions of a purely formal nature, having no physical content.

Translated into mathematical language, these principles assert the existence of a *functional* $\mathfrak{F}_t$ with the following properties:

(a) In every kinematically possible process the stress $\boldsymbol{T}(t)$ at time t is related to the motion $\boldsymbol{\chi}$ of the body $\mathscr{B}$ by

$$\boldsymbol{T}(t)=\mathfrak{F}_t(\boldsymbol{\chi}). \tag{26.1}$$

(b) For any two motions $\boldsymbol{\chi}$ and $\overline{\boldsymbol{\chi}}$ that coincide in some neighborhood $\mathscr{N}(X)$ for all times $\tau\leqq t$, the value of $\mathfrak{F}_t$ is the same. Formally,

$$\mathfrak{F}_t(\boldsymbol{\chi})=\mathfrak{F}_t(\overline{\boldsymbol{\chi}}), \tag{26.2}$$

provided there be a neighborhood $\mathscr{N}(X)$ such that

$$\boldsymbol{\chi}(Z,\tau)=\overline{\boldsymbol{\chi}}(Z,\tau) \quad \text{for all} \quad Z\in\mathscr{N}(X) \quad \text{and all} \quad \tau\leqq t. \tag{26.3}$$

Equation (26.1), as restricted by (26.2), is the *general constitutive equation* for purely mechanical theories of continuous media. We note that $\mathfrak{F}_t$ is a functional of a function $\boldsymbol{\chi}$ of two variables, the time t and the variable particle Z in the neighborhood of the fixed particle X. The value of the functional $\mathfrak{F}_t$ is the stress tensor at the place $\boldsymbol{\chi}(X,t)$ occupied by the particle X at the time t. The nature of the functional $\mathfrak{F}_t$ depends upon the choice of X, but we do not indicate this dependence explicitly.

If the constitutive assumptions include restrictive conditions on the local behavior of the allowable motions, the principles just stated must be modified. The effect of such conditions, called *internal constraints*, will be discussed in Sect. 30.

Not any arbitrary functional $\mathfrak{F}_t$ with the property (b) is admissible, because all constitutive equations are subject to the principle of material frame-indifference laid down in Sect. 19. According to that principle,

$$\boldsymbol{T}^*(t^*)=\mathfrak{F}_{t^*}(\boldsymbol{\chi}^*) \tag{26.4}$$

for any process $\{\boldsymbol{\chi}^*, \boldsymbol{T}^*\}$ equivalent to $\{\boldsymbol{\chi}, \boldsymbol{T}\}$, i.e., related to $\{\boldsymbol{\chi}, \boldsymbol{T}\}$ by a change of frame of the form (19.2). The functional equation (26.4) is to hold identically in $\boldsymbol{\chi}$ and t.

In order to discover the restrictions on the form of the functionals $\mathfrak{F}_t$ implied by the principle of indifference, we consider successively three choices of the change of frame (19.2). We shall transform away, in turn, the position, the time, and the orientation of a moving observer.

The first choice is $\boldsymbol{Q}(\tau)\equiv\boldsymbol{1}$, $a=0$, and

$$\boldsymbol{c}(\tau)=-\boldsymbol{\chi}(X,\tau). \tag{26.5}$$

This choice corresponds to a relative translation of the two frames, such that, after the change of frame, the particle X remains at rest at the origin. Here, as in (26.3), we use τ to denote the time variable and reserve t for the particular instant at which the stress is evaluated. We obtain from (19.2)

$$\tau^*=\tau, \qquad \boldsymbol{T}^*(t)=\boldsymbol{T}(t), \qquad \boldsymbol{\chi}^*(Z,\tau)=\boldsymbol{\chi}(Z,\tau)-\boldsymbol{\chi}(X,\tau), \tag{26.6}$$

and we see that $\boldsymbol{\chi}^*$ is nothing but the localization $\boldsymbol{\chi}_X$ at X of the motion $\boldsymbol{\chi}$ as defined by (22.3). Hence, by (26.4) and (26.6), the general constitutive equation must reduce to

$$\boldsymbol{T}(t)=\mathfrak{F}_t(\boldsymbol{\chi}_X). \tag{26.7}$$

Our second choice is $\boldsymbol{Q}(\tau)\equiv\boldsymbol{1}$, $\boldsymbol{c}(\tau)\equiv\boldsymbol{0}$, and $a=t$, shifting the time scale so that the present time t is the reference time after the change of frame. We have then

$$\left.\begin{array}{l}\tau^* = \tau - t, \quad t^* = t - t = 0, \quad \boldsymbol{T}^*(0) = \boldsymbol{T}(t), \\ \boldsymbol{\chi}_X^*(Z, \tau^*) = \boldsymbol{\chi}_X(Z, \tau) = \boldsymbol{\chi}_X(Z, t+\tau^*),\end{array}\right\} \tag{26.8}$$

and hence, by (26.4),

$$\boldsymbol{T}(t) = \boldsymbol{T}^*(0) = \mathfrak{F}_0(\boldsymbol{\chi}_X^*). \tag{26.9}$$

It follows from (22.18) and (26.8) that

$$\boldsymbol{\chi}_X^*(Z, -s) = \boldsymbol{\chi}_X(Z, t-s) = \boldsymbol{\chi}_X^{(t)}(Z, s) \tag{26.10}$$

for $s \geqq 0$. Hence, for non-positive $\tau^* = -s$, $\boldsymbol{\chi}_X^*(Z, \tau^*)$ is determined by the history $\boldsymbol{\chi}_X^{(t)}$ up to time t of the localized motion $\boldsymbol{\chi}_X$. On the other hand, it follows from (26.2) and (26.3) that the value $\mathfrak{F}_0(\boldsymbol{\chi}_X^*)$ does not depend on the values of $\boldsymbol{\chi}_X^*(Z, -s)$ when $s < 0$. Therefore we may drop the index 0 in (26.9), reducing the general constitutive equation to the form

$$\boldsymbol{T}(t) = \mathfrak{F}(\boldsymbol{\chi}_X^{(t)}), \tag{26.11}$$

where $\boldsymbol{\chi}_X^{(t)}$ is the history up to time t of the localization $\boldsymbol{\chi}_X$ at X of the motion $\boldsymbol{\chi}$.

Finally, we consider an equivalent process $\{\boldsymbol{\chi}^*, \boldsymbol{T}^*\}$ that corresponds to the choice $\boldsymbol{c}(\tau) \equiv \boldsymbol{0}$, $a = 0$ in (19.2), leaving $\boldsymbol{Q}(\tau)$ arbitrary. The equation (26.4) then implies that the functional $\mathfrak{F}$ of (26.11) must satisfy the identity

$$\boldsymbol{Q}_0 \mathfrak{F}(\boldsymbol{\chi}_X^{(t)}) \boldsymbol{Q}_0^T = \mathfrak{F}(\boldsymbol{Q}\boldsymbol{\chi}_X^{(t)}), \tag{26.12}$$

where $\boldsymbol{Q}(s)$ is an arbitrary smooth orthogonal tensor function, and where $\boldsymbol{Q}_0$ and $\boldsymbol{Q}\boldsymbol{\chi}_X^{(t)}$ are defined by

$$\boldsymbol{Q}_0 \equiv \boldsymbol{Q}(0), \quad (\boldsymbol{Q}\boldsymbol{\chi}_X^{(t)})(Z, s) \equiv \boldsymbol{Q}(s)\boldsymbol{\chi}_X^{(t)}(Z, s). \tag{26.13}$$

The functional identity (26.12) expresses the indifference of the material response with respect to arbitrary time-dependent rotations of the frame of reference.

It follows from (26.2) and (26.3) that

$$\mathfrak{F}(\boldsymbol{\chi}_X^{(t)}) = \mathfrak{F}(\overline{\boldsymbol{\chi}}_X^{(t)}) \tag{26.14}$$

provided there be a neighborhood $\mathcal{N}(X)$ such that

$$\boldsymbol{\chi}_X^{(t)}(Z, s) = \overline{\boldsymbol{\chi}}_X^{(t)}(Z, s) \quad \text{for all} \quad Z \in \mathcal{N}(X). \tag{26.15}$$

The relation (26.14) is a mathematical statement of the assumption that only the history of the motion of an arbitrarily small neighborhood of the particle X can influence the stress at $\boldsymbol{\chi}(X, t)$.

It is easy to see that, conversely, any constitutive equation of the form (26.11) satisfies the principle of material frame-indifference provided $\mathfrak{F}$ has the properties (26.12) and (26.14). This is so because the general change of frame (17.6) may be obtained by a succession of three changes, one of each of the three special types just considered. Hence (26.11), subject to the requirements (26.12) and (26.14), is the ***most general constitutive equation***[1] for purely mechanical theories of non-polar[2] continuous materials. Any equation of this form restricts the class of all possible dynamical processes of neighborhoods of the particle X and thus characterizes the local material properties of the particle X. We call $\mathfrak{F}$ the *response functional* at the particle X.

27. Material isomorphisms. Homogeneity[3]. We assume now that a reference configuration $\boldsymbol{\varkappa}$ of a neighborhood $\mathcal{N}(X)$ of the particle X is given (cf. Sect. 21). In

[1] NOLL [1958, *31*, § 13].

[2] A non-polar material is one in which body couples and couple stresses are not present (CFT, Sect. 200). A particular theory of polar materials is outlined in Sect. 98, below.

[3] NOLL [1958, *31*, § 14].

accordance with the convention stated in Sect. 22 we replace the particles $Z \in \mathcal{N}(X)$ by $\boldsymbol{Z} = \boldsymbol{\varkappa}_X(Z) = \boldsymbol{\varkappa}(Z) - \boldsymbol{\varkappa}(X)$, i.e., by their position vectors in the reference configuration $\boldsymbol{\varkappa}$ when the position of X in $\boldsymbol{\varkappa}$ is taken as origin. We retain the notation and terminology of the previous section, but we now regard $\boldsymbol{\chi}_X$ as a function of the reference position vector $\boldsymbol{Z}$ and the time τ, rather than as a function of the particle Z and the time τ. It must be noted that the form of the response functional in (26.11) then depends not only on the particle X but also *on the choice of the reference configuration* $\boldsymbol{\varkappa}$. We write $\mathfrak{F} = \mathfrak{F}_{\varkappa}$ if we wish to stress this dependence. The functionals $\mathfrak{F}_{\varkappa}$ and $\mathfrak{F}_{\hat{\varkappa}}$ corresponding to two different reference configurations, $\boldsymbol{\varkappa}$ and $\hat{\boldsymbol{\varkappa}}$, are related by the identity

$$\mathfrak{F}_{\varkappa}(\boldsymbol{\chi}_X^{(t)}) = \mathfrak{F}_{\hat{\varkappa}}(\hat{\boldsymbol{\chi}}_X^{(t)}), \tag{27.1}$$

where

$$\hat{\boldsymbol{\chi}}_X^{(t)}(\hat{\boldsymbol{Z}}, s) = \boldsymbol{\chi}_X^{(t)}(\boldsymbol{Z}, s), \qquad \hat{\boldsymbol{Z}} = \boldsymbol{\lambda}_X(\boldsymbol{Z}), \tag{27.2}$$

$\boldsymbol{\lambda}_X$ being the deformation from $\boldsymbol{\varkappa}_X$ to $\hat{\boldsymbol{\varkappa}}_X$, as defined by (21.19).

A reference configuration $\boldsymbol{\varkappa}$ is said to have constant density if $\varrho_{\varkappa}(\boldsymbol{X})$ is a constant, $\varrho_{\varkappa}$, independent of the particle X.

Two particles X and $\overline{X}$ are said to be *materially isomorphic* if it is possible to find a reference configuration $\boldsymbol{\varkappa}$ of a neighborhood of X and a reference configuration $\bar{\boldsymbol{\varkappa}}$ of a neighborhood of $\overline{X}$ such that

(a) $\boldsymbol{\varkappa}$ and $\bar{\boldsymbol{\varkappa}}$ have the same uniform density, i.e.

$$\varrho_{\varkappa} = \varrho_{\bar{\varkappa}} = \text{const.}, \tag{27.3}$$

(b) the response functional $\mathfrak{F}_{\varkappa}$ of X coincides with the response functional $\mathfrak{F}_{\bar{\varkappa}}$ of $\overline{X}$, i.e.,

$$\mathfrak{F}_{\varkappa}(\boldsymbol{\chi}) = \mathfrak{F}_{\bar{\varkappa}}(\boldsymbol{\chi}) \tag{27.4}$$

holds for all smooth functions $\boldsymbol{\chi}(\boldsymbol{Z}, s)$ defined for all $s \geqq 0$ and all $\boldsymbol{Z}$ in a neighborhood of the origin $\boldsymbol{0}$.

The physical meaning of this definition is that two particles are materially isomorphic if and only if their response to deformation histories, described with respect to suitable reference configurations, is identical. We may express this condition more simply by saying that the mechanical properties of the materials at the two particles are the same. The concept of material isomorphism expresses mathematically the idea that two particles belong to the same material.

A body $\mathscr{B}$ is called *materially uniform* if all of its particles are materially isomorphic to one another. Such a body may be regarded as consisting of a definite ideal material. The body $\mathscr{B}$ is materially uniform, then, if one can find, for each particle $X \in \mathscr{B}$, a reference configuration $\boldsymbol{\varkappa}^X$ of a neighborhood of X such that $\varrho_{\varkappa^X}$ is uniform and independent of X and such that the response functional $\mathfrak{F}$ corresponding to X and $\boldsymbol{\varkappa}^X$ is the same for all X. In general it will be necessary to choose different reference configurations $\boldsymbol{\varkappa}^X$ for different particles X. If it is possible to choose a *single* reference configuration $\boldsymbol{\varkappa}$ of the whole body $\mathscr{B}$ so that $\mathfrak{F}_{\varkappa}$ is the same for all particles, we say that the body is *homogeneous*. In physical terms, a body is homogeneous if and only if there is a reference configuration *of the whole body* such that every particle responds in just the same way as every other to past deformations described with respect to this configuration. For a homogeneous body, the configurations $\boldsymbol{\varkappa}^X$ may be taken to be the localizations $\boldsymbol{\varkappa}_X$ of a global reference configuration, but in general the $\boldsymbol{\varkappa}^X$ are *not* localizations. A body may be materially uniform without being homogeneous.

28. Simple materials, materials of grade $\mathfrak{n}$, dimensional invariance. The condition expressed by (26.14) and (26.15) implies that the value $\mathfrak{F}(\boldsymbol{\chi}_X^{(t)})$ of the functional $\mathfrak{F}$ depends on $\boldsymbol{\chi}_X^{(t)}(\boldsymbol{Z}, s)$ only for $\boldsymbol{Z}$ in an arbitrarily small neighborhood of the origin $\mathbf{0}$. But for small $\boldsymbol{Z}$ one can approximate $\boldsymbol{\chi}_X^{(t)}(\boldsymbol{Z}, s)$ by its directional derivative in the direction of $\boldsymbol{Z}$, evaluated at the origin:

$$\boldsymbol{\chi}_X^{(t)}(\boldsymbol{Z}, s) \approx [\nabla \boldsymbol{\chi}_X^{(t)}(\mathbf{0}, s)]\, \boldsymbol{Z}. \tag{28.1}$$

Hence, by (22.8) and (22.18),

$$\boldsymbol{\chi}_X^{(t)}(\boldsymbol{Z}, s) \approx \boldsymbol{F}^{(t)}(s)\, \boldsymbol{Z} \tag{28.2}$$

where $\boldsymbol{F}^{(t)}$ is the history up to time t of the deformation gradient $\boldsymbol{F}$ at $\boldsymbol{X}$. The approximation (28.2) can be made as precise as is desired by confining $\boldsymbol{Z}$ to a small enough neighborhood of $\mathbf{0}$. Hence it may suffice, in order to describe the behavior of many real materials, to assume that the value $\mathfrak{F}(\boldsymbol{\chi}_X^{(t)})$ depends only on the history $\boldsymbol{F}^{(t)}$ of the deformation gradient.

With this much as physical motivation, we now agree to consider the *special case* when the general constitutive equation (26.11) reduces to

$$\boldsymbol{T}(t) = \mathfrak{G}(\boldsymbol{F}^{(t)}), \tag{28.3}$$

where $\mathfrak{G}$ is a functional which assigns to the tensor-valued function $\boldsymbol{F}^{(t)}(s)$ the symmetric tensor $\boldsymbol{T}(t)$. While, for a fixed particle X, $\mathfrak{F}$ is a functional of a vector-valued function of the two variables $\boldsymbol{Z}$ and s, $\mathfrak{G}$ is a functional of a tensor-valued function of the single variable s. It is often convenient to indicate the domain $0 \leqq s < \infty$ of s by using the more explicit notation $\mathop{\mathfrak{G}}\limits_{s=0}^{\infty}$ instead of $\mathfrak{G}$. In this notation (28.3) becomes

$$\boldsymbol{T}(t) = \mathop{\mathfrak{G}}\limits_{s=0}^{\infty}\big(\boldsymbol{F}^{(t)}(s)\big) = \mathop{\mathfrak{G}}\limits_{s=0}^{\infty}\big(\boldsymbol{F}(t-s)\big). \tag{28.4}$$

The condition (26.12) restricting the form of $\mathfrak{F}$ implies the following condition for the functional $\mathfrak{G}$: For every orthogonal tensor function $\boldsymbol{Q}(s)$ and every history $\boldsymbol{F}^{(t)}(s)$ in a suitable class the relation

$$\boldsymbol{Q}_0 \mathop{\mathfrak{G}}\limits_{s=0}^{\infty}\big(\boldsymbol{F}^{(t)}(s)\big)\, \boldsymbol{Q}_0^T = \mathop{\mathfrak{G}}\limits_{s=0}^{\infty}\big(\boldsymbol{Q}(s)\, \boldsymbol{F}^{(t)}(s)\big), \qquad \boldsymbol{Q}_0 \equiv \boldsymbol{Q}(0), \tag{28.5}$$

must hold. Conversely, if (28.5) holds for a particular functional $\mathfrak{G}$, it is easily seen that this functional satisfies all the conditions so far laid down and thus serves to define a particular material.

Since the special case we are now considering is defined in terms of $\boldsymbol{F}$, implicit in all that we have said is the use of a reference configuration as in Sect. 27. Let us suppose that (26.11) does reduce to the form (28.3) when the motion is referred to $\boldsymbol{\varkappa}$. We may use (27.1) and (27.2) to get the constitutive equation for the same material when a different reference configuration $\hat{\boldsymbol{\varkappa}}$ is employed. It follows that

$$\mathop{\mathfrak{G}_{\varkappa}}\limits_{s=0}^{\infty}\big(\boldsymbol{F}^{(t)}(s)\big) = \mathop{\mathfrak{G}_{\hat{\varkappa}}}\limits_{s=0}^{\infty}\big(\boldsymbol{F}^{(t)}(s)\, \boldsymbol{P}^{-1}\big), \tag{28.6}$$

where

$$\boldsymbol{P} = \nabla \boldsymbol{\lambda}_X(\mathbf{0}) = \nabla \boldsymbol{\lambda}(\boldsymbol{X}) \tag{28.7}$$

is the gradient at $\boldsymbol{X}$ of the deformation $\boldsymbol{\lambda}$ from the configuration $\boldsymbol{\varkappa}$ to the configuration $\hat{\boldsymbol{\varkappa}}$, given by (21.19). From (28.6) we see that if a material is such that its response depends on the history of the motion only through the history of the deformation gradient when the reference configuration $\boldsymbol{\varkappa}$ is used, the material has this same property when we choose to refer its motion to another reference configuration $\hat{\varkappa}$. Moreover, if (28.5) is satisfied by $\mathfrak{G}_{\varkappa}$, it is satisfied also by $\mathfrak{G}_{\hat{\varkappa}}$. Consequently the property expressed by (28.3) is invariant under change of reference configuration, so that no mention of a particular configuration need be made in referring to it. It is clear from (28.6) that $\mathfrak{G}_{\varkappa}=\mathfrak{G}_{\hat{\varkappa}}$ if $\boldsymbol{P}=\mathbf{1}$. Hence $\mathfrak{G}_{\varkappa}$ depends only on the equivalence class (22.14) to which $\varkappa$ belongs. Therefore, it is sufficient to specify a *local* reference configuration $\boldsymbol{K}$, rather than a global reference configuration $\varkappa$, to determine the functional $\mathfrak{G}$. When local, rather than global, reference configurations are used, then (28.6) and (28.7) must be replaced by

$$\mathop{\mathfrak{G}_{\boldsymbol{K}}}_{s=0}^{\infty}\left(\boldsymbol{F}^{(t)}(s)\right)=\mathop{\mathfrak{G}_{\hat{\boldsymbol{K}}}}_{s=0}^{\infty}\left(\boldsymbol{F}^{(t)}(s)\,\boldsymbol{P}^{-1}\right), \tag{28.8}$$

where $\boldsymbol{P}$ is given by (22.15).

Materials defined by a constitutive equation of the form (28.3) are called *simple materials*[1]; the corresponding functionals $\mathfrak{G}$, $\mathfrak{G}_{\varkappa}$, $\mathfrak{G}_{\boldsymbol{K}}$, etc, are the *response functionals* of the simple material for the global or local reference configuration employed or indicated; and (28.5) is the *fundamental functional equation* for those response functionals.

For inhomogeneous bodies, the functional $\mathfrak{G}$ depends also on the particle X. For materially uniform or homogeneous bodies, $\mathfrak{G}$ will be independent of X only if appropriate local reference configurations $\boldsymbol{K}_X$ are employed. These local reference configurations $\boldsymbol{K}_X$ need not fit together to form a global reference configuration (cf. Sect. 34).

The basic physical property of a simple material is immediately seen from (28.3). If, for each fixed X, $\boldsymbol{T}(t)$ is known for every deformation history in which the strain is always homogeneous at X, then $\boldsymbol{T}(t)$ is known for all deformation histories, and conversely. That is, *a material is simple at the particle X if and only if its response to deformations homogeneous in a neighborhood of X determines uniquely its response to every deformation at X.* In the case of a homogeneous body, the qualification "at X" may be omitted. For the general theory of homogeneous strain, see CFT, Sects. 42—46.

In general, the foregoing characterization is not a practical one, since in order to effect a homogeneous strain in a given material suitable body forces, and

[1] NOLL [1957, *15*, §§ 16—17] [1958, *31*, §§ 17—18]. A formally similar but less general theory was proposed earlier by GREEN and RIVLIN [1956, *12*, Eq. (4.3)] [1957, *8*, Eq. (4.3)]. While NOLL used the term "functional" in its usual mathematical sense, GREEN and RIVLIN seem to have regarded a "functional" as a quantity that can be approximated by a sequence of polynomials or a sum of repeated integrals, and they considered only deformation histories that correspond to a state of rest prior to some fixed instant. Thus in order to allow for possible explicit dependence on $\boldsymbol{F}$ and its time derivatives at the present time t, they approached the whole subject anew in later papers [1959, *10*] [1960, *27*]. Such dependence was allowed from the start in NOLL's theory, which does not presume any smoothness in the functional $\mathfrak{G}$. A remark implying the contrary by GREEN and RIVLIN [1960, *27*, § 8] seems to arise from misunderstanding. PIPKIN and WINEMAN [1963, *61*] [1964, *97*] have given arguments to indicate that forms obtained under the hypothesis of polynomial dependence are valid more generally. To us, assuming polynomial dependence seems not only unnecessary, as the reader of this treatise will verify again and again, but unphysical. We see no sign that nature loves a polynomial, and polynomial dependence is not even invariant under change of strain measure. By an appeal to the Stone-Weierstraß theorem, CHACON and RIVLIN [1964, *13*], generalizing and simplifying earlier work by GREEN and RIVLIN [1956, *12*, §§ 4—8] [1957, *8*, §§ 4—8] [1960, *27*] and by GREEN, RIVLIN, and SPENCER [1959, *10*], have found conditions under which a constitutive functional may be uniformly approximated by a sequence of polynomials. Such an approach may remove the latter of the foregoing objections but not the former, for an approximation theorem is valid only subject to certain assumptions of smoothness, while properties of invariance, being algebraic in nature, afford simpler arguments leading to reductions and representations of greater generality.

generally quite artificial ones, must be supplied. Indeed, for a *homogeneous* simple body subject to any homogeneous strain history, the stress $\boldsymbol{T}(t)$ as given by (28.3) is spatially homogeneous, and hence div $\boldsymbol{T}(t)=\boldsymbol{0}$. By (16.6), then, $\boldsymbol{b}=\ddot{\boldsymbol{x}}$. The case of greatest interest in continuum mechanics is that of a uniform gravitational field, $\boldsymbol{b}=\boldsymbol{g}=$const. In order for the homogeneous strain history to be possible subject to this body force, it is necessary and sufficient that $\ddot{\boldsymbol{x}}=\boldsymbol{g}$. That is, *in order for a homogeneous strain history to be possible in a single homogeneous simple body in a uniform gravitational field* $\boldsymbol{g}$, *it must correspond to a deformation with constant acceleration* $\boldsymbol{g}$; *if it does, it may occur in any homogeneous simple body placed in the same gravitational field.*

Now homogeneous strains of this kind are easy to characterize[1]. Denoting the position vectors to $\boldsymbol{x}$ and $\boldsymbol{X}$ by those same symbols, so that a general homogeneous strain may be expressed in the form $\boldsymbol{x}=\boldsymbol{F}\boldsymbol{X}+\boldsymbol{c}$, where $\boldsymbol{F}$ and $\boldsymbol{c}$ depend on t only, we see that $\ddot{\boldsymbol{x}}=\boldsymbol{g}$ if and only if $\ddot{\boldsymbol{F}}=\boldsymbol{0}$ and $\ddot{\boldsymbol{c}}=\boldsymbol{g}$; hence

$$\boldsymbol{F}(t)=\boldsymbol{F}_0(\boldsymbol{1}+t\boldsymbol{F}_1),\qquad \boldsymbol{c}=\tfrac{1}{2}t^2\boldsymbol{g}+t\boldsymbol{e}+\boldsymbol{f},\tag{28.9}$$

where $\boldsymbol{F}_0$ and $\boldsymbol{F}_1$ are constant tensors and where $\boldsymbol{e}$ and $\boldsymbol{f}$ are constant vectors. On the assumption that $\boldsymbol{F}_0$ is invertible, we see that $\boldsymbol{F}(t)$ is invertible only so long as $\det(\boldsymbol{1}+t\,\boldsymbol{F}_1)\neq 0$. This condition restricts $-\frac{1}{t}$ to an interval delimited by such real proper numbers as $\boldsymbol{F}_1$ may have. Substitution of $(28.9)_1$ into (24.1) gives the velocity gradient:

$$\boldsymbol{L}(t)=\boldsymbol{F}_0\boldsymbol{F}_1(\boldsymbol{1}+t\boldsymbol{F}_1)^{-1}\boldsymbol{F}_0^{-1}.\tag{28.10}$$

The initial value $\boldsymbol{L}_0=\boldsymbol{L}(0)$ is $\boldsymbol{F}_0\boldsymbol{F}_1\boldsymbol{F}_0^{-1}$, a tensor having the same proper numbers as $\boldsymbol{F}_1$. We may express the result as follows. Let $-1/t_+$ be the least negative proper number of the initial velocity gradient $\boldsymbol{L}_0$; let $-1/t_-$ be the greatest positive proper number; if $\boldsymbol{L}_0$ has no negative proper number, set $t_+=\infty$, and if no positive proper number, set $t_-=-\infty$. *Then the homogeneous uniformly accelerated deformations having initial velocity gradient* $\boldsymbol{L}_0$ *are non-singular only in the interval* $t_-<t<t_+$. In order for this interval to be $-\infty<t<+\infty$ it is necessary that $\boldsymbol{F}_1$ have as proper numbers 0 and two non-real numbers, or 0 thrice repeated. That is, $III_{\boldsymbol{F}}=0$ and either $I_{\boldsymbol{F}}^2<4II_{\boldsymbol{F}}$ or $I_{\boldsymbol{F}}=II_{\boldsymbol{F}}=0$. If $1/t_0$ is the greatest among the absolute values of the proper numbers of $\boldsymbol{L}_0$, then $\boldsymbol{F}(t)$ is analytic when $|t|<t_0$.

Of course, we are not to conclude that a body of any simple material cast in free fall will *necessarily* be subject to a deformation included in (28.9). We have shown only that (28.9) is compatible with the differential equations of motion and the constitutive equation. When (28.9) is put into (28.4), a certain spatially homogeneous but time-dependent stress field $\boldsymbol{T}(t)$ is determined uniquely, and this stress field in turn determines by (16.5) a surface traction field $\boldsymbol{t}(\boldsymbol{x},t)$ on the boundary $\partial\mathscr{B}$ of the body $\mathscr{B}$. Unless tractions of this amount are supplied upon $\partial\mathscr{B}$, the homogeneous motion corresponding to (28.9) is *not possible*. Thus in any experimental application of the foregoing results it is of the essence *to supply whatever surface tractions may be necessary to maintain homogeneous deformation.*

The foregoing results show that the time-dependent homogeneous strains dynamically possible in simple materials without application of specially adjusted body forces are limited in nature and are non-singular in intervals of time that are, in general, finite. These facts place considerable difficulty in the way of determining $\mathfrak{G}$, even in principle, from a sequence of measurements. In an important if mathematically trivial special case, that of a static homogeneous strain, the difficulties disappear, since the time interval becomes $-\infty<t<\infty$. That is,

[1] We simplify and generalize the analyses of TRUESDELL [1955, *32*] and NOLL [1955, *19*, § 20]; the former's is reproduced in CFT, Sect. 142, where the kinematics of these motions is presented.

any static homogeneous strain is a possible configuration of equilibrium for any[1] *homogeneous simple body, and the general static response of any homogeneous simple body is determinable from its response to homogeneous strain.* Such response is described by the theory of elasticity, the subject of Chapter D.

Eq. (28.3) affords the simplest special case of the general constitutive equation (26.11). In the somewhat more general case when the value (26.11) of the functional $\mathfrak{F}$ depends only on the first $\mathfrak{n}$ gradients of $\boldsymbol{\chi}_X^{(t)}(\boldsymbol{Z}, s)$ with respect to $\boldsymbol{Z}$, evaluated at $\boldsymbol{Z}=\boldsymbol{0}$, we say that the material is of *grade* $\mathfrak{n}$[2]. These gradients, which we denote by ${}_{\mathfrak{j}}\boldsymbol{F}^{(t)}(s) = {}_{\mathfrak{j}}\boldsymbol{F}(t-s)$, $\mathfrak{j}=1, 2, \ldots, \mathfrak{n}$, have the components

$$ {}_{\mathfrak{j}}F^k{}_{\alpha_1\alpha_2\ldots\alpha_{\mathfrak{j}}} = x^k{}_{,\alpha_1\alpha_2\ldots\alpha_{\mathfrak{j}}}, \tag{28.11}$$

evaluated at the particle X and the time $t-s$. The constitutive equation of a material of grade $\mathfrak{n}$ is of the form

$$\boldsymbol{T}(t) = \underset{s=0}{\overset{\infty}{\mathfrak{G}}}\left({}_{1}\boldsymbol{F}(t-s),\ {}_{2}\boldsymbol{F}(t-s),\ \ldots,\ {}_{\mathfrak{n}}\boldsymbol{F}(t-s)\right). \tag{28.12}$$

Simple materials are those of grade *1*. The condition (26.12) restricting the form of $\mathfrak{F}$ implies the following condition for the response functional $\mathfrak{G}$: For every orthogonal tensor function $\boldsymbol{Q}(s)$ and every set ${}_{1}\boldsymbol{F}^{(t)}(s), \ldots, {}_{\mathfrak{n}}\boldsymbol{F}^{(t)}(s)$ of histories in a suitable class the relation

$$\boldsymbol{Q}_0 \underset{s=0}{\overset{\infty}{\mathfrak{G}}}\left({}_{1}\boldsymbol{F}^{(t)}(s), \ldots, {}_{\mathfrak{n}}\boldsymbol{F}^{(t)}(s)\right) \boldsymbol{Q}_0^T = \underset{s=0}{\overset{\infty}{\mathfrak{G}}}\left({}_{1}\overline{\boldsymbol{F}}^{(t)}(s), \ldots, {}_{\mathfrak{n}}\overline{\boldsymbol{F}}^{(t)}(s)\right) \tag{28.13}$$

---

[1] We presume here that there are no kinematical or dynamical constraints.

[2] While the early and classical studies of continuum mechanics concern simple materials almost exclusively, there are a few exceptions. Cauchy [1851, *1*] once suggested that the stress components may be "fonctions linéaires des déplacements ... et de leurs dérivées des divers ordres". After discussing a theory of this type resulting from a molecular model, he stated that in a continuum theory of small motions "one will find" an expression for the stress of the following form [1851, *1*, Eqs. (6), (7)]:

$$\boldsymbol{T} = k\tilde{\boldsymbol{E}} + K I_{\tilde{\boldsymbol{E}}}\boldsymbol{1} + J \operatorname{grad}\operatorname{curl}\boldsymbol{u} + \text{const},$$

where $\tilde{\boldsymbol{E}}$ is the infinitesimal strain tensor [defined by Eq. (41.12), below], $\boldsymbol{u}$ is the infinitesimal displacement vector, while the operators k, K, and J are entire functions of the operator Δ. Apparently guided by Cauchy's work, Lévy [1869, *2*] proposed a theory of fluids in which the stress is supposed to be a function of spatial derivatives of the velocity field of all orders, but only in certain invariant combinations. St. Venant [1869, *1*] thereupon took up Cauchy's idea for solids and expressed it in a form similar to Lévy's for liquids. Both these latter theories were made to rest upon Navier's molecular notions. Apart from the empty formal classifications of the rheologists in the 1920's and 1930's, according to which all sorts of derivatives may occur here, there, and everywhere, usually in one dimension, there have been few further proposals of theories of non-simple materials except those we summarize in some detail in Sects. 98, 124—125.

An interesting attempt to formulate a theory of anelastic response was made by Eckart [1948, *3*], who proposed to let the stress be determined as a function of the deformation gradient $\boldsymbol{F}$ calculated with respect to a varying reference configuration. A differential equation for determining that configuration was then laid down as an additional constitutive equation. The attempt of Truesdell [1952, *20*, § 82] to place this theory on a sound footing with respect to the principles of mechanics must be regarded as a failure. Cf. also Manfredi [1957, *12*]. A rational basis for specifying the change of preferred reference configuration in time remains to be found.

Restrictions imposed by the principle of material indifference on the constitutive equations for materials of grade $\mathfrak{n}$ were found by Noll [1957, *15*, § 15].

Investigations by Gurtin [1965, *20* and *21*], in progress as this treatise goes to press, indicate that theories of materials of grade $\mathfrak{n}>1$ are in conflict with thermodynamic principles unless they are modified so as to take into account couple-stresses and other more general mechanical interactions. Cf. the indications to this effect furnished by Toupin's second theory of polar-elastic materials, presented in Sect. 98γ.

holds, where $\boldsymbol{Q}_0 = \boldsymbol{Q}(0)$ and

$$ {}_{\mathfrak{j}}\bar{F}^{(t)}(s)^k{}_{\alpha_1 \ldots \alpha_{\mathfrak{j}}} = Q(s)^k{}_m \, {}_{\mathfrak{j}}F^{(t)}(s)^m{}_{\alpha_1 \ldots \alpha_{\mathfrak{j}}} . \tag{28.14} $$

A response functional $\mathfrak{F}$ depends not only on the material and on the reference configuration, but also on the choice of dimensional units used to measure masses, lengths, times, forces, stresses, etc. We wish to investigate how $\mathfrak{F}$ changes when a new system of units is employed. Suppose the original units are chosen from one particular system, for example the cm-g-sec system. Let us now consider a new system of units. The units of stress, time, and length of the new system will have certain measures E_0, s_0, and l_0, respectively, in terms of the units of the old system. Let $\boldsymbol{T}$ be a stress tensor, s a time lapse, and $\boldsymbol{p}$ a position vector, all measured in units of the old system. In the units of the new system, they will become

$$ \boldsymbol{T}^* = \frac{1}{E_0} \boldsymbol{T}, \qquad s^* = \frac{1}{s_0} s, \qquad \boldsymbol{p}^* = \frac{1}{l_0} \boldsymbol{p}. \tag{28.15} $$

Let $\boldsymbol{\chi}_X^{(t)}$ be the history of the localization of a given motion. Its arguments $\boldsymbol{Z}$ and s are position vectors and time lapses, and its values $\boldsymbol{\chi}_X^{(t)}(\boldsymbol{Z}, s)$ are position vectors. In the units of the new system, the history $\boldsymbol{\chi}_X^{(t)}$ therefore becomes $\boldsymbol{\chi}_X^{*(t^*)}$, defined by

$$ \boldsymbol{\chi}_X^{*(t^*)}(\boldsymbol{Z}^*, s^*) = \frac{1}{l_0} \boldsymbol{\chi}_X^{(t)}(\boldsymbol{Z}, s), \quad \boldsymbol{Z} = l_0 \boldsymbol{Z}^*, \quad s = s_0 s^*, \quad t = s_0 t^*. \tag{28.16} $$

Since the value of the response functional $\mathfrak{F}$ is the stress $\boldsymbol{T}$, that value in the new system is given by $(28.15)_1$. Therefore, the functional $\mathfrak{F}^*$ in the new system is related to $\mathfrak{F}$ by

$$ \mathfrak{F}^*(\boldsymbol{\chi}_X^{*(t^*)}) = \frac{1}{E_0} \mathfrak{F}(\boldsymbol{\chi}_X^{(t)}), \tag{28.17} $$

where $\boldsymbol{\chi}_X^{*(t^*)}$ is defined by (28.16).

Often it is possible to define dimensional units in terms of the responses of the material itself in certain special deformation processes (cf. the examples given in Sects. 43, 54, 57, and 108). If $\mathfrak{F}^*$ is the functional corresponding to such *internal* units, it is said to be *dimensionless*[1], because then $\mathfrak{F}^*$ does not depend on the *a priori* choice of *external* dimensional units. The measure of an internal unit in terms of an external system of units is called a *dimensional modulus* or *dimensional material constant*. E_0, s_0, and l_0, in this interpretation, are called a *natural elasticity*, *natural time lapse*, and *natural length* for the material. For example, in elasticity, the stress that must be applied to double the length of a bar may be taken as the internal unit of stress. The measure E_0 of this stress in some conventional system of external units is usually called the modulus of elasticity. It is clear from (28.16) and (28.17) that *a general material can have at most three independent dimensional moduli, having the dimensions of stress, time, and length.* The existence of the natural elasticity, natural time lapse, and the natural length suggests the possibility of representing, in an extremely general way, the aspects of material response commonly called *spring*, *relaxation*, and *absorption*, and all aspects obtainable from them by combination and interaction, e.g., *viscosity*, *phase shift*, and *dispersion*. Special materials may have fewer dimensional

---

[1] Dimensional reductions of the constitutive equations of various non-linear materials, in increasing degrees of generality, were given by TRUESDELL and SCHWARTZ [1947, *12*, § 65], TRUESDELL [1952, *22*, pp. 89—91] and other papers, culminating in [1964, *85*, § 4]. See Sects. 119A—121, 125, below. In the work of TRUESDELL, dimension-bearing constants are supposed given *a priori* as part of the definition of the material and are admitted as additional arguments in the constitutive equations.

moduli, because it may happen that there exists no response of the material which can serve to define an internal dimensional unit for one or the other physical dimension.

It follows from differentiation of (28.16) with respect to $\boldsymbol{Z}$ that the gradients ${}_{\mathfrak{i}}\boldsymbol{F}^{(t)}$ of $\boldsymbol{\chi}_X^{(t)}$ are related to the gradients ${}_{\mathfrak{i}}\boldsymbol{F}^{*(t^*)}$ of $\boldsymbol{\chi}_X^{*(t^*)}$ by

$$ {}_{\mathfrak{i}}\boldsymbol{F}^{*(t^*)}(s^*) = l_0^{\mathfrak{i}-1}\,{}_{\mathfrak{i}}\boldsymbol{F}^{(t)}(s), \qquad s = s_0 s^*, \qquad t = s_0 t^*. \tag{28.18} $$

Therefore, in the special case of a material of grade $\mathfrak{n}$, (28.17) reduces to the following relation between the two defining response functionals $\mathfrak{G}$ and $\mathfrak{G}^*$ for the two systems of units:

$$ \mathop{\mathfrak{G}^*}_{s^*=0}^{\infty}\left({}_1\boldsymbol{F}^{*(t^*)}(s^*), \ldots, {}_{\mathfrak{n}}\boldsymbol{F}^{*(t^*)}(s^*)\right) = \frac{1}{E_0} \mathop{\mathfrak{G}}_{s=0}^{\infty}\left({}_1\boldsymbol{F}^{(t)}(s), \ldots, {}_{\mathfrak{n}}\boldsymbol{F}^{(t)}(s)\right). \tag{28.19} $$

Substitution of (28.18) into (28.19) gives the following *dimensionally reduced form for the constitutive equation of a material of grade* $\mathfrak{n}$:

$$ \left.\begin{aligned} \boldsymbol{T}(t) &= \mathop{\mathfrak{G}}_{s=0}^{\infty}\left({}_1\boldsymbol{F}^{(t)}(s), \ldots, {}_{\mathfrak{n}}\boldsymbol{F}^{(t)}(s)\right) \\ &= E_0 \mathop{\mathfrak{G}^*}_{s^*=0}^{\infty}\left({}_1\boldsymbol{F}^{(t)}(s_0 s^*),\, l_0 \cdot {}_2\boldsymbol{F}^{(t)}(s_0 s^*), \ldots, l_0^{\mathfrak{n}-1} \cdot {}_{\mathfrak{n}}\boldsymbol{F}^{(t)}(s_0 s^*)\right). \end{aligned}\right\} \tag{28.20} $$

In the special case of a simple material (28.20) reduces to

$$ \boldsymbol{T}(t) = \mathop{\mathfrak{G}}_{s=0}^{\infty}\left(\boldsymbol{F}^{(t)}(s)\right) = E_0 \mathop{\mathfrak{G}^*}_{s^*=0}^{\infty}\left(\boldsymbol{F}^{(t)}(s_0 s^*)\right) = E_0 \mathop{\mathfrak{G}^*}_{s^*=0}^{\infty}\left(\boldsymbol{F}(t - s_0 s^*)\right). \tag{28.21} $$

Since l_0 does not occur in (28.21), it follows that *a simple material can have at most two independent dimensional moduli, one having the dimension of stress, the other of time.* Thus, in contrast to materials of higher grade, the simple material is not general enough to represent a physical material in which there is an internal unit of length. Special simple materials have even fewer dimensional moduli. For example, the constitutive equation of an elastic material (Sect. 43) involves at most one dimensional modulus E_0, which has the dimension of a stress; that of an incompressible linearly viscous fluid (Sect. 41), only a viscosity μ, which has the dimensions of a stress multiplied by a time; that of an incompressible elastic fluid, none at all[1].

A theory of scaling may be constructed along classical lines, but it is useless. Only for the simplest of theories are there any non-trivial scaling laws. The more complicated the constitutive equation, the greater the number of dimensionless numbers that must be controlled in order to assure dynamical similarity. Just as the dimensionless material constant called "Poisson's modulus" is a scaling parameter in the infinitesimal theory of elasticity, so that two materials with different Poisson moduli generally fail to exhibit dynamical similarity even in that extremely simple theory, in more general theories a great number of such dimensionless ratios, or, ultimately, the dimensionless response functional $\mathfrak{G}^*$ itself, must be the same for the two materials in order for scaling to be possible. To see the point at issue, the reader may study the example worked out by Serrin in Sect. 66 of *Mathematical Principles of Classical Fluid Mechanics*, this Encyclopedia, Volume VIII/1, where it is proved that even for the classical linearly viscous compressible fluid, dynamical similarity is *impossible* unless the viscosities depend upon pressure and density according to a compound power law.

The strongest experimental evidence in favor of the classical theories comes, not from the so-called "fundamental experiments" or the imaginary "operational definitions", but from the millions of successful if rough or even crude uses of the scaling laws based upon those theories, scaling laws which in fact *come close to characterizing the classical theories.*

---

[1] These materials require for their full definition an additional modulus which enters the equations of motion (16.6) but does not affect the constitutive equation, namely, the density ϱ.

Except in passages where the contrary is stated explicitly, we shall confine attention to simple materials, or to equations that are appropriate at least in some cases to simple materials.

29. Reduced constitutive equations. The property (28.5) permits further reductions of the constitutive equation (28.4). $\boldsymbol{F}^{(t)}$ has the polar decomposition

$$\boldsymbol{F}^{(t)} = \boldsymbol{R}^{(t)}\,\boldsymbol{U}^{(t)}, \tag{29.1}$$

where $\boldsymbol{R}^{(t)}$ and $\boldsymbol{U}^{(t)}$ are the histories up to time t of the rotation tensor and the right stretch tensor, respectively. With the special choice[1] $\boldsymbol{Q}(s) = \left(\boldsymbol{R}^{(t)}(s)\right)^T$ equation (28.5) reduces to

$$\boldsymbol{R}(t)^T \mathop{\mathfrak{G}}_{s=0}^{\infty}\left(\boldsymbol{F}^{(t)}(s)\right)\boldsymbol{R}(t) = \mathop{\mathfrak{G}}_{s=0}^{\infty}\left(\boldsymbol{U}^{(t)}(s)\right). \tag{29.2}$$

Thus (28.4) can be written in the form

$$\boldsymbol{T}(t) = \boldsymbol{R}(t) \mathop{\mathfrak{G}}_{s=0}^{\infty}\left(\boldsymbol{U}(t-s)\right)\boldsymbol{R}(t)^T. \tag{29.3}$$

Conversely, the invariance requirement (28.5) is automatically satisfied for a constitutive equation of the form (29.3) with an arbitrary functional $\mathfrak{G}$ having as its domain of definition a class of positive-definite symmetric-tensor valued functions $\boldsymbol{U}^{(t)}(s)$ of a non-negative real variable. Hence (29.3) is a ***general form for the constitutive equation of a simple material***[2]. This result is of major importance for two reasons. First, it gives *the general solution of the functional equation* (28.5), so that we may deal with a concise yet full and invariant mathematical statement of the constitutive equation of a simple material, namely, (29.3). Second, it shows that *while the present rotation generally affects the stress in a simple material, past rotations are without influence*; moreover, the dependence of $\boldsymbol{T}(t)$ on $\boldsymbol{R}(t)$ is given explicitly and is *exactly the same for all simple materials.* The reduced form (29.3) and the equivalent forms we shall presently develop will be used again and again and may be regarded as statements of the ***fundamental theorem on simple materials.*** In view of the fundamental theorem and of the main theorem of Sect. 28, we may summarize the definition of a simple material in physical terms: *A material is simple if and only if its response to any deformation history is known as soon as its response to all homogeneous pure-stretch histories is specified.*

In order to determine $\mathfrak{G}$ in general it is sufficient, by (29.2), to know its value for all positive-definite symmetric arguments. The homogeneous pure-stretch histories afford one means of exhausting this class. By $(24.16)_1$, we may instead set $\boldsymbol{W} = \boldsymbol{0}$ and with an assigned $\boldsymbol{U}(t)$ determine $\boldsymbol{R}(t)$ uniquely to within its initial value. That is, the class of homogeneous stretch-histories $\boldsymbol{U}^{(t)}(s)$ is exhausted also by the class of irrotational motions. Therefore, the last theorem may be rephrased

---

[1] Here we note the effect that would follow from adopting the Hooke-Poisson-Cauchy form rather than the Zaremba-Jaumann form of the principle of material frame-indifference (Sect. 19A). Eq. (28.5) would then be required to hold identically in $\boldsymbol{F}$ and in the *proper* orthogonal tensor $\boldsymbol{Q}$. Setting $\varepsilon \equiv \left(\text{sign}\det \boldsymbol{F}^{(t)}(s)\right)1$, which by continuity is either $+1$ for all s and t or -1 for all s and t, we should then make the special choice $\boldsymbol{Q}(s) = \varepsilon\left(\boldsymbol{R}^{(t)}(s)\right)^T$. As a result, in (29.3) the argument $\boldsymbol{U}(t-s)$ is replaced by $\varepsilon\boldsymbol{U}(t-s)$. For a given deformation process, however, ε remains constant, either always $+1$ or always -1, and so in all applications the results would be just the same as according to the apparently more restrictive Zaremba-Jaumann form adopted in this treatise.

In the secondary literature material frame-indifference is still sometimes confused with properties of material symmetry. Cf. footnote 1, p. 78, and the end of Sect. 19A.

[2] NOLL [1957, *15*, § 17] [1958, *31*, § 18], broadly generalizing a classic result in the theory of finite elastic strain.

as follows[1]: *A material is simple if and only if its response to any deformation history is known as soon as its response to all homogeneous irrotational histories is specified.*

It is often useful to introduce another functional, $\mathfrak{J}$, by the definition

$$\mathop{\mathfrak{J}}_{s=0}^{\infty}\left([\boldsymbol{U}^{(t)}(s)]^2\right) \equiv \boldsymbol{U}_0^{(t)} \mathop{\mathfrak{G}}_{s=0}^{\infty}\left(\boldsymbol{U}^{(t)}(s)\right) \boldsymbol{U}_0^{(t)}, \qquad \boldsymbol{U}_0^{(t)} = \boldsymbol{U}^{(t)}(0). \tag{29.4}$$

Note that

$$[\boldsymbol{U}^{(t)}(s)]^2 = [\boldsymbol{U}(t-s)]^2 = \boldsymbol{C}(t-s) = \boldsymbol{C}^{(t)}(s) \tag{29.5}$$

is the history, up to time t, of the right Cauchy-Green tensor, defined by (23.4), and note the relation $\boldsymbol{F}(t) = \boldsymbol{R}(t)\,\boldsymbol{U}(t) = \boldsymbol{R}(t)\,\boldsymbol{U}_0^{(t)}$. It follows from (29.4) that the general constitutive equation (29.3) may be written in the alternative form[2]

$$\overline{\boldsymbol{T}}(t) = \mathop{\mathfrak{J}}_{s=0}^{\infty}\left(\boldsymbol{C}^{(t)}(s)\right) = \mathop{\mathfrak{J}}_{s=0}^{\infty}\left(\boldsymbol{C}(t-s)\right), \tag{29.6}$$

where $\overline{\boldsymbol{T}}$, the *convected stress tensor*, is defined by

$$\overline{\boldsymbol{T}} \equiv \boldsymbol{F}^T \boldsymbol{T} \boldsymbol{F}, \qquad \overline{T}_{\alpha\beta} = x^k{}_{,\alpha} x^m{}_{,\beta} T_{km}. \tag{29.7}$$

Equation (29.6) has the co-ordinate form

$$\overline{T}_{\alpha\beta}(t) = \mathop{\mathfrak{J}_{\alpha\beta}}_{s=0}^{\infty}\left(C_{\mu\nu}(t-s)\right). \tag{29.8}$$

The functional form of the component functionals $\mathfrak{J}_{\alpha\beta}$ depends, of course, on the natural basis $\boldsymbol{d}_\alpha$ at X of the material co-ordinates X^α in the reference state. Since in general $\boldsymbol{d}_\alpha$ will change with X, the forms of the functionals $\mathfrak{J}_{\alpha\beta}$ will change from particle to particle, even if the body is homogeneous. It is therefore sometimes useful to stress the dependence on the $\boldsymbol{d}_\alpha$ and to rewrite (29.8) in the more explicit form

$$\overline{T}_{\alpha\beta}(t) = \mathop{\mathfrak{J}_{\alpha\beta}}_{s=0}^{\infty}\left(C_{\mu\nu}(t-s);\, \boldsymbol{d}_\sigma\right). \tag{29.9}$$

If the point of view (ii) mentioned in Sect. 15 is used, then the $C_{\alpha\beta}$ are the time-dependent components of the metric tensor in a co-ordinate system which deforms with the body. The components $\overline{T}_{\alpha\beta}$ of $\overline{\boldsymbol{T}}$ coincide with the components of the stress tensor in the deforming co-ordinate system[3]. However, confusion may result from this viewpoint, since it seems to allow $\overline{\boldsymbol{T}}$ and $\boldsymbol{T}$ to be used interchangeably, contrary to the fact, obvious from the general formulae (29.7), that *neither determines the other*. CAUCHY's tensor $\boldsymbol{T}$ specifies *the actual contact forces* in the material, irrespective of reference configurations and deformation histories. Corresponding to each choice of reference configuration, a different $\overline{\boldsymbol{T}}$ results from the same stress field $\boldsymbol{T}$, and the connection between them depends not only on the strain but also on the rotation from the reference configuration to the present configuration.

If we set

$$\mathop{\mathfrak{L}}_{s=0}^{\infty}\left(\boldsymbol{C}^{(t)}(s)\right) \equiv \boldsymbol{C}^{-1}(t) \mathop{\mathfrak{J}}_{s=0}^{\infty}\left(\boldsymbol{C}^{(t)}(s)\right) \boldsymbol{C}^{-1}(t), \tag{29.10}$$

then (29.7) can be written in the form

$$\boldsymbol{T} = \boldsymbol{F} \mathop{\mathfrak{L}}_{s=0}^{\infty}\left(\boldsymbol{C}^{(t)}(s)\right) \boldsymbol{F}^T, \tag{29.11}$$

---

[1] COLEMAN and TRUESDELL [1965, *14*].

[2] GREEN and RIVLIN [1956, *12*, § 2] [1957, *8*, § 2]. The *formal* equivalence of (29.3) with GREEN and RIVLIN's constitutive equation was noticed by NOLL [1958, *31*, § 22] and has been reiterated by LIANIS [1963, *49*]. GREEN and RIVLIN's hypotheses were far more restrictive than NOLL's, as has been remarked in footnote 1, p. 61. Cf. the further discussion by RIVLIN [1964, *72*].

[3] This is the point of view stressed by GREEN and RIVLIN [1957, *8*].

or, in components,

$$T^{km} = x^k{}_{,\alpha}\, x^m{}_{,\beta}\, \underset{s=0}{\overset{\infty}{\mathfrak{L}}}{}^{\alpha\beta}\big(C_{\mu\nu}(t-s);\, \boldsymbol{d}_\sigma\big). \tag{29.12}$$

While these forms are not so illuminating as (29.3), they are more useful when it comes to calculation of cases.

Another form of the general constitutive equation (29.3) is[1]

$$\boldsymbol{T}(t) = \boldsymbol{R}_{(t')}(t)\, \boldsymbol{R}(t')\, \underset{s=0}{\overset{\infty}{\mathfrak{K}}}\big(\boldsymbol{U}^*_{(t')}(t-s);\, \boldsymbol{U}(t')\big)\, \boldsymbol{R}(t')^T \boldsymbol{R}_{(t')}(t)^T, \tag{29.13}$$

where $\boldsymbol{R}_{(t)}$ is the relative rotation tensor, where $\boldsymbol{U}^*_{(t)}$ is defined in terms of the relative right stretch tensor $\boldsymbol{U}_{(t)}$ by

$$\boldsymbol{U}^*_{(t)}(\tau) = \boldsymbol{R}(t)^T\, \boldsymbol{U}_{(t)}(\tau)\, \boldsymbol{R}(t), \tag{29.14}$$

and where $\mathfrak{K}$ is a functional of the function $\boldsymbol{U}^*_{(t')}(t-s)$, $s \geqq 0$, and a function of the tensor parameter $\boldsymbol{U}(t')$. In order to prove (29.13), we substitute the polar decompositions $\boldsymbol{F}(t) = \boldsymbol{R}(t)\, \boldsymbol{U}(t)$ and $\boldsymbol{F}_{(t)}(\tau) = \boldsymbol{R}_{(t)}(\tau)\; \boldsymbol{U}_{(t)}(\tau)$ into (21.24), obtaining

$$\boldsymbol{F}(\tau) = \boldsymbol{R}_{(t)}(\tau)\, \boldsymbol{R}(t)\, \boldsymbol{U}^*_{(t)}(\tau)\, \boldsymbol{U}(t), \tag{29.15}$$

where $\boldsymbol{U}^*_{(t)}$ is defined by (29.14). Replacing t by t' and putting $\tau = t-s$ in (29.15), we find that

$$\boldsymbol{Q}(s)\, \boldsymbol{F}^{(t)}(s) = \boldsymbol{U}^*_{(t')}(t-s)\, \boldsymbol{U}(t'), \tag{29.16}$$

where

$$\boldsymbol{Q}(s) = [\boldsymbol{R}_{(t')}(t-s)\, \boldsymbol{R}(t')]^T. \tag{29.17}$$

Substitution of (29.16) into (28.5) shows that

$$\boldsymbol{T}(t) = \underset{s=0}{\overset{\infty}{\mathfrak{G}}}\big(\boldsymbol{F}^{(t)}(s)\big) = \boldsymbol{Q}_0^T\, \underset{s=0}{\overset{\infty}{\mathfrak{G}}}\big(\boldsymbol{U}^*_{(t')}(t-s)\, \boldsymbol{U}(t')\big)\, \boldsymbol{Q}_0. \tag{29.18}$$

This equation is of the form (29.13).

With the special choice $t' = t$ the general constitutive equation (29.13) becomes

$$\boldsymbol{R}(t)^T\, \boldsymbol{T}(t)\, \boldsymbol{R}(t) = \underset{s=0}{\overset{\infty}{\mathfrak{K}}}\big(\boldsymbol{U}^*_{(t)}(t-s);\, \boldsymbol{U}(t)\big). \tag{29.19}$$

If we introduce a functional $\mathfrak{J}$ by a definition parallel to (29.4), we see that (29.13) and (29.19) have the alternative forms

$$\boldsymbol{F}_{(t')}(t)^T\, \boldsymbol{T}(t)\, \boldsymbol{F}_{(t')}(t) = \boldsymbol{R}(t')\, \underset{s=0}{\overset{\infty}{\mathfrak{J}}}\big(\boldsymbol{C}^*_{(t')}(t-s);\, \boldsymbol{C}(t')\big)\, \boldsymbol{R}(t')^T \tag{29.20}$$

and

$$\boldsymbol{R}(t)^T\, \boldsymbol{T}(t)\, \boldsymbol{R}(t) = \underset{s=0}{\overset{\infty}{\mathfrak{J}}}\big(\boldsymbol{C}^*_{(t)}(t-s);\, \boldsymbol{C}(t)\big), \tag{29.21}$$

where $\boldsymbol{C}^*_{(t)}$ is related to the relative right Cauchy-Green tensor $\boldsymbol{C}_{(t)}$ (cf. Sect. 23) by

$$\boldsymbol{C}^*_{(t)}(\tau) = \boldsymbol{R}(t)^T\, \boldsymbol{C}_{(t)}(\tau)\, \boldsymbol{R}(t). \tag{29.22}$$

The equations (29.13)—(29.22) have no simple representation in terms of co-ordinates, except, of course, when rectilinear co-ordinates are used.

The same methods that have been used to reduce the constitutive equations for simple materials can also be applied to those for materials of grade $\mathfrak{n} > 1$.

---

[1] NOLL [1958, *31*, Theorem 4].

A result generalizing (29.3) is[1]

$$\boldsymbol{T}(t) = \boldsymbol{R}(t) \underset{s=0}{\overset{\infty}{\mathfrak{G}}} \left({}_1\boldsymbol{U}(t-s), \ldots, {}_n\boldsymbol{U}(t-s)\right) \boldsymbol{R}(t)^T, \tag{29.23}$$

where the tensors ${}_i\boldsymbol{U}$ are related to the gradient tensors (28.11) by

$${}_iU^k{}_{\alpha_1 \ldots \alpha_i} = R_p{}^k \, {}_iF^p{}_{\alpha_1 \ldots \alpha_i}. \tag{29.24}$$

A reduced constitutive equation generalizing (29.6) is

$$\overline{\boldsymbol{T}}(t) = \underset{s=0}{\overset{\infty}{\mathfrak{J}}} \left({}_1\boldsymbol{C}(t-s), \ldots, {}_n\boldsymbol{C}(t-s)\right), \tag{29.25}$$

where $\overline{\boldsymbol{T}}$ is the convected stress tensor (29.7) and where the tensors ${}_i\boldsymbol{C}$ are related to the gradient tensors (28.11) by

$${}_iC_{\beta\alpha_1 \ldots \alpha_i} = F_{k\beta} \, {}_iF^k{}_{\alpha_1 \ldots \alpha_i}. \tag{29.26}$$

Reduced constitutive equations generalizing (29.11), (29.13), and (29.21) are also easily obtained.

It is often useful to put the general constitutive equation (29.21) of a simple material into a slightly different form by writing the right-hand side as a sum of an "equilibrium term" $\mathfrak{f}(\boldsymbol{C}(t))$ and a term that vanishes when the material has always been at rest. Also, we may use the function

$$\boldsymbol{G}^*(s) = \boldsymbol{C}^*_{(t)}(t-s) - \boldsymbol{1} = \boldsymbol{R}(t)^T \boldsymbol{C}_{(t)}(t-s) \boldsymbol{R}(t) - \boldsymbol{1} \tag{29.27}$$

to describe the deformation history; the rest history then corresponds to $\boldsymbol{G}^*(s) \equiv \boldsymbol{0}$. Suppressing the dependence on the present time t, we see that (29.21) may be written in the form

$$\boldsymbol{R}^T \boldsymbol{T} \boldsymbol{R} = \mathfrak{f}(\boldsymbol{C}) + \underset{s=0}{\overset{\infty}{\mathfrak{F}}} \left(\boldsymbol{G}^*(s); \boldsymbol{C}\right), \tag{29.28}$$

where the functional $\mathfrak{F}$ vanishes when $\boldsymbol{G}^*(s) \equiv \boldsymbol{0}$:

$$\underset{s=0}{\overset{\infty}{\mathfrak{F}}} (\boldsymbol{0}; \boldsymbol{C}) = \boldsymbol{0}. \tag{29.29}$$

A form similar to (29.28) is

$$\overline{\boldsymbol{T}} = \mathfrak{s}(\boldsymbol{C}) + \underset{s=0}{\overset{\infty}{\mathfrak{S}}} \left(\overline{\boldsymbol{G}}(s); \boldsymbol{C}\right), \tag{29.30}$$

where $\overline{\boldsymbol{T}}$ is the convected stress (29.7), where, instead of (29.27),

$$\overline{\boldsymbol{G}}(s) = \boldsymbol{F}^T(t) \boldsymbol{C}_{(t)}(t-s) \boldsymbol{F}(t) - \boldsymbol{C}(t), \tag{29.31}$$

and where the response functional $\mathfrak{S}$ has the value $\boldsymbol{0}$ when $\overline{\boldsymbol{G}}(s) \equiv \boldsymbol{0}$.

There are many more reduced forms for the constitutive equations of simple materials and materials of grade $\mathfrak{n}$. We have given here only a selection[2].

30. Internal constraints, incompressibility, inextensibility. α) *General theory.* It was mentioned in Sect. 26 that the principle of determinism must be modified for bodies subject to internal constraints. A *simple internal constraint* is defined by a scalar-valued function $\gamma(\boldsymbol{F})$ of a tensor variable $\boldsymbol{F}$. *We say that a particle X*

---

[1] Noll [1957, *15*, § 15].

[2] Reduced forms for constitutive equations of various types not presented here have been given by Pipkin and Rivlin [1958, *35*, § 2] [1959, *21*, § 2], Rivlin [1960, *46*] [1961, *51*]. Cf. also Sect. 96.

in a body is subject to the constraint defined by γ if the possible motions of the body are restricted to those for which

$$\gamma(\boldsymbol{F}(\tau))=0, \quad -\infty<\tau<\infty, \tag{30.1}$$

where $\boldsymbol{F}(\tau)$ is the deformation gradient at X and τ. The constraint function γ depends on the choice of the local reference configuration. The constraint functions γ and $\hat{\gamma}$ corresponding to two local reference configurations $\boldsymbol{K}$ and $\hat{\boldsymbol{K}}$ are related by the following identity in $\boldsymbol{F}$:

$$\gamma(\boldsymbol{F})=\gamma(\hat{\boldsymbol{F}}\boldsymbol{P})=\hat{\gamma}(\hat{\boldsymbol{F}}), \tag{30.2}$$

where $\boldsymbol{P}$ is the local deformation from $\boldsymbol{K}$ to $\hat{\boldsymbol{K}}$.

A constraint (30.1) is a constitutive equation and hence is subject to the principle of material frame-indifference. A consideration analogous to one given in Sect. 26 shows that this principle is satisfied if and only if (30.1) is equivalent to an equation of the form

$$\lambda(\boldsymbol{C}(\tau))=0, \quad -\infty<\tau<\infty, \tag{30.3}$$

where $\boldsymbol{C}(\tau)$ is the right Cauchy-Green tensor (23.4). Differentiation of (30.3) with respect to τ gives

$$\operatorname{tr}(\lambda_{\boldsymbol{C}}(\boldsymbol{C})\dot{\boldsymbol{C}})=\frac{\partial\lambda}{\partial C_{\alpha\beta}}\dot{C}^{\alpha\beta}=0. \tag{30.4}$$

Differentiating $(23.12)_1$ with respect to τ and then putting $\tau=t$ enables us to express $\dot{\boldsymbol{C}}$ in terms of the stretching tensor $\boldsymbol{D}$:

$$\dot{\boldsymbol{C}}(t)=2\boldsymbol{F}(t)^T\boldsymbol{D}(t)\boldsymbol{F}(t). \tag{30.5}$$

In deriving this equation (24.14) and $(25.6)_1$ have been used (cf. CFT, Eq. (95.7)). Substitution of (30.5) into (30.4) gives

$$\operatorname{tr}[\boldsymbol{F}\lambda_{\boldsymbol{C}}(\boldsymbol{C})\boldsymbol{F}^T\boldsymbol{D}]=\frac{\partial\lambda}{\partial C_{\alpha\beta}}x^k{}_{,\alpha}x^m{}_{,\beta}D_{km}=0. \tag{30.6}$$

Of course, it is impossible to *derive* from our earlier analysis constitutive equations for materials subject to constraints. Constraints are maintained by forces, and it is conceivable that infinitely many different systems of forces may suffice to maintain any given constraint. Just as in the theory of discrete mechanical systems[1], it is necessary to make some kind of additional *assumption*, beyond those sufficient for unconstrained systems, in order to get a definite constitutive equation for a constrained material. We choose here a generalized form of an assumption grown familiar through many special cases in the earlier literature. Let the general principles of motions for unconstrained systems be expressed by a work principle. The same work principle has meaning for constrained systems, and it may be postulated as a generalized axiom[2].

Such an axiom is mathematically equivalent to a statement that *the forces maintaining the constraints do no work*. While more complicated systems of forces might be found, workless forces seem to be the simplest imaginable that would suffice to maintain the constraints. Once this fact has been seen, there is no need to mention variations or a formal work principle, and we can simply lay down as our constitutive assumption the following

Principle of determinism for simple materials subject to internal constraints: *The stress* $\boldsymbol{T}$ *at time* t *is determined by the history* $\boldsymbol{F}^{(t)}(s)$ *of the deformation gradient only to within a stress* $\boldsymbol{N}$ *that does no work in any motion satisfying the constraints.*

---

[1] While this matter is often obscure in elementary treatments, the critical reader will easily find the points where tacit assumptions are slipped in, whatever the line of attack followed.

[2] Such is the approach in Sect. 233 of CFT, generalizing results of ERICKSEN and RIVLIN [1954, *7*, §§ 3—4] for hyperelasticity. Cf. also ADKINS [1958, *2*, § 9].

As shown in Sect. 217 of CFT, the rate at which the stresses do work, per unit volume, is given by the stress power,

$$P \equiv \operatorname{tr}(\boldsymbol{T}\boldsymbol{D}) = T^{km} D_{km}. \tag{30.7}$$

Hence the principle of determinism requires that the indeterminate stress $\boldsymbol{N}$ be such that

$$\operatorname{tr}(\boldsymbol{N}\boldsymbol{D}) = 0 \tag{30.8}$$

for all symmetric tensors $\boldsymbol{D}$ satisfying (30.6). Since $\operatorname{tr}(\boldsymbol{A}\boldsymbol{B})$ defines an inner product in the 6-dimensional space $\mathscr{S}$ of symmetric tensors, the condition on $\boldsymbol{N}$ may be expressed in geometric language as follows: $\boldsymbol{N}$ must be orthogonal to all $\boldsymbol{D}$ that are orthogonal to $\boldsymbol{F}\lambda_{\boldsymbol{C}}(\boldsymbol{C})\boldsymbol{F}^T$. Clearly, this can be the case only if $\boldsymbol{N}$ is a scalar multiple of $\boldsymbol{F}\lambda_{\boldsymbol{C}}(\boldsymbol{C})\boldsymbol{F}^T$:

$$\boldsymbol{N} = q\boldsymbol{F}\lambda_{\boldsymbol{C}}(\boldsymbol{C})\boldsymbol{F}^T. \tag{30.9}$$

If there are several constraints, defined by the constraint functions $\lambda^{\mathfrak{i}}$, $\mathfrak{i} = 1, 2, \ldots, \mathfrak{p}$, we have instead of (30.9)

$$\boldsymbol{N} = \sum_{\mathfrak{i}=1}^{\mathfrak{p}} q_{\mathfrak{i}} \boldsymbol{F}\lambda^{\mathfrak{i}}_{\boldsymbol{C}}(\boldsymbol{C})\boldsymbol{F}^T, \tag{30.10}$$

the quantities $q_{\mathfrak{i}}$ being scalar coefficients. Therefore, a mathematical statement of the principle of determinism for simple materials subject to constraints is the following: The history $\boldsymbol{F}^{(t)}(s)$ of the deformation gradient determines the *extra stress*[1]

$$\left.\begin{aligned} \boldsymbol{T}_{\mathrm{E}} &= \boldsymbol{T} + \sum_{\mathfrak{i}=1}^{\mathfrak{p}} q_{\mathfrak{i}} \boldsymbol{F}\lambda^{\mathfrak{i}}_{\boldsymbol{C}}(\boldsymbol{C})\boldsymbol{F}^T, \\ T^{km}_{\mathrm{E}} &= T^{km} + x^k{}_{,\alpha} x^m{}_{,\beta} \sum_{\mathfrak{i}=1}^{\mathfrak{p}} q_{\mathfrak{i}} \frac{\partial \lambda^{\mathfrak{i}}}{\partial C_{\alpha\beta}}, \end{aligned}\right\} \tag{30.11}$$

by a constitutive equation of the form

$$\boldsymbol{T}_{\mathrm{E}}(t) = \mathop{\mathfrak{G}}_{s=0}^{\infty}\big(\boldsymbol{F}(t-s)\big) = \boldsymbol{R}(t) \mathop{\mathfrak{G}}_{s=0}^{\infty}\big(\boldsymbol{U}(t-s)\big)\boldsymbol{R}(t)^T. \tag{30.12}$$

The response functional $\mathfrak{G}$ need be defined only when $\boldsymbol{F}(\tau)$ satisfies the constraints:

$$\lambda^{\mathfrak{i}}\big(\boldsymbol{F}(\tau)\big) = 0 \quad \tau \leqq t,\ \mathfrak{i} = 1, 2, \ldots, \mathfrak{p}. \tag{30.13}$$

The scalars $q_{\mathfrak{i}}$ in (30.11) are indeterminate, and the functional $\mathfrak{G}$ is determined only up to a functional whose values are of the form (30.10). Of course, $\mathfrak{G}$ must satisfy (28.5).

β) *Special cases.* We now consider certain special constraints:

a) Incompressibility. For isochoric motions, $|\det \boldsymbol{F}(\tau)| = J(\tau) = 1$, and hence

$$\det \boldsymbol{C}(\tau) = 1, \tag{30.14}$$

provided that the reference configuration be chosen suitably. A material is *incompressible* if it is susceptible only of isochoric motions. A corresponding constraint function is

$$\lambda(\boldsymbol{C}) = \det \boldsymbol{C} - 1. \tag{30.15}$$

[1] Note that the extra stress is always symmetric if $\boldsymbol{T}$ is symmetric. In the discussion of more general constraints by POINCARÉ [1889, *2*, § 152] [1892, *1*, § 33], the requirement (30.2), reflecting the principle of material indifference, is not imposed. As a result, the stress tensors he obtained as corresponding to constraints need not be symmetric.

From (9.3) and the fact that $\boldsymbol{C}^{-1}=\boldsymbol{F}^{-1}(\boldsymbol{F}^T)^{-1}$, it follows that $\boldsymbol{F}\lambda_{\boldsymbol{C}}(\boldsymbol{C})\boldsymbol{F}^T=(\det\boldsymbol{C})\boldsymbol{1}=\boldsymbol{1}$ when λ is given by (30.15) and when (30.14) holds. Hence, when there are no constraints beyond that of incompressibility, the extra stress reduces to

$$\boldsymbol{T}_{\mathrm{E}}=\boldsymbol{T}+p\boldsymbol{1}. \tag{30.16}$$

Therefore, *for incompressible materials, the stress is determined by the deformation history only to within a hydrostatic pressure* p. The response functional is determined only up to an arbitrary functional whose values are scalar multiples of $\boldsymbol{1}$, and it need be defined only for arguments $\boldsymbol{F}$ that are unimodular: $\det\boldsymbol{F}=\pm 1$. One can remove this indeterminacy by a normalization such as

$$\operatorname{tr}\overset{\infty}{\underset{s=0}{\mathfrak{G}}}(\boldsymbol{F}(t-s))=\operatorname{tr}\boldsymbol{T}_{\mathrm{E}}=0. \tag{30.17}$$

When (30.17) holds, the extra stress becomes the stress deviator, and the indeterminate pressure p of (30.16) coincides with the mean pressure (cf. CFT, Sect. 204):

$$p=\bar{p}=-\tfrac{1}{3}\operatorname{tr}\boldsymbol{T}. \tag{30.18}$$

b) Inextensibility[1]. The directional derivative of the deformation function $\boldsymbol{\chi}(\boldsymbol{X},\tau)$ in the direction of the unit vector $\boldsymbol{e}$ at $\boldsymbol{X}$ is

$$\boldsymbol{g}(\tau)=\nabla\boldsymbol{\chi}(\boldsymbol{X},\tau)\,\boldsymbol{e}=\boldsymbol{F}(\tau)\,\boldsymbol{e}. \tag{30.19}$$

Inextensibility of the material at $\boldsymbol{X}$ in a direction defined by $\boldsymbol{e}$ in the reference configuration corresponds to the restriction to motions in which the length of the vector $\boldsymbol{g}(\tau)$ is independent of the time τ:

$$[\boldsymbol{g}(\tau)]^2=(\boldsymbol{F}(\tau)\,\boldsymbol{e})^2=\boldsymbol{e}\cdot\boldsymbol{F}(\tau)^T\boldsymbol{F}(\tau)\,\boldsymbol{e}=\boldsymbol{e}\cdot\boldsymbol{C}(\tau)\,\boldsymbol{e}=\text{const}. \tag{30.20}$$

Thus for a suitable reference configuration, *inextensibility in the* **e**-*direction* is defined by the constraint function

$$\lambda(\boldsymbol{C})=\boldsymbol{e}\cdot\boldsymbol{C}\boldsymbol{e}-1. \tag{30.21}$$

For this case, (30.9) reduces to

$$\boldsymbol{N}=q\boldsymbol{F}\boldsymbol{e}\otimes\boldsymbol{F}\boldsymbol{e}, \tag{30.22}$$

and the extra stress (30.11) becomes

$$\left.\begin{aligned}\boldsymbol{T}_{\mathrm{E}}&=\boldsymbol{T}+q\boldsymbol{F}\boldsymbol{e}\otimes\boldsymbol{F}\boldsymbol{e},\\ T_{\mathrm{E}}^{km}&=T^{km}+q\,x^k{}_{,\alpha}\,x^m{}_{,\beta}\,e^\alpha e^\beta.\end{aligned}\right\} \tag{30.23}$$

Thus, *for materials that are inextensible in the direction of* **e** *in the reference configuration, the stress is determined by the deformation history only to within a uniaxial tension in the direction of* $\boldsymbol{F}\boldsymbol{e}$. The response functional $\mathfrak{G}$ is determined only up to an arbitrary functional whose values are scalar multiples of $\boldsymbol{F}\boldsymbol{e}\otimes\boldsymbol{F}\boldsymbol{e}$. As in the case of incompressibility, one can remove this indeterminacy by a suitable normalization. If inextensibility in two or three directions is assumed, two or three terms of the form (30.22) must be added in determining the extra stress, according to the general formula (30.11).

c) Rigidity. For rigid motions the deformation gradient $\boldsymbol{F}(\tau)$ must always be a proper orthogonal tensor, when the reference configuration is suitably selected,

---

[1] The constraint of inextensibility, in the context of a sheet of hyperelastic material reinforced by a network of inextensible cords, seems first to have been considered by ADKINS and RIVLIN [1955, *4*]. Cf. also GREEN and ADKINS [1960, *26*, Ch. VII]. References to special solutions for constrained materials are given in Sect. 43.

and hence

$$C(\tau) = F(\tau)^T F(\tau) = 1. \tag{30.24}$$

Since the space of symmetric tensors $\mathscr{S}$ is 6-dimensional, the restriction (30.24) corresponds to 6 independent scalar constraints. Rigidity is defined by these constraints. If we take as the 6 constraint functions the components $C_{\alpha\beta} - g_{\alpha\beta}$ when $\alpha \leqq \beta$, in (30.11) we obtain 6 indeterminate scalars $q^{\alpha\beta}$, $\alpha \leqq \beta$, one for each component of $\boldsymbol{T}$. Thus for a rigid material the stress is completely indeterminate. The response functional may be taken to be identically zero, and there is no relation at all between the stress and the deformation history, which must always reduce to a rotation history. The concept of stress is in fact of little if any use in the theory of rigid bodies. The dynamics of rigid bodies is treated in the article by SYNGE in Vol. III/1 of this Encyclopedia.

All the reduced forms obtained in Sect. 29 are valid also for constrained materials, provided $\boldsymbol{T}$ be replaced by $\boldsymbol{T}_{\mathrm{E}}$.

γ) *Motions possible in all homogeneous incompressible simple bodies*[1]. In Sect. 28 we have determined all homogeneous motions that can occur in a general and unconstrained simple material when a uniform field of force is applied, and we have found them very limited in kind. In a material subject to internal constraints, the *kinematically* possible motions are restricted, by definition, but the arbitrary functions occurring in the extra stress generally render a great many more motions *dynamically* possible. We illustrate this difference in the case of an incompressible material. We consider only homogeneous bodies, describing their motions in terms of a homogeneous reference configuration, and we assume the density is uniform in the reference configuration and hence constant in space and time forever. We assume the body force $\boldsymbol{b}$ to be conservative, with single-valued potential v:

$$\boldsymbol{b} = -\operatorname{grad} v. \tag{30.25}$$

By (30.16), CAUCHY's first law of motion (16.6) assumes the form

$$\operatorname{div} \boldsymbol{T}_{\mathrm{E}} - \varrho \operatorname{grad} \varphi = \varrho \ddot{\boldsymbol{x}}, \tag{30.26}$$

where

$$\varphi = \frac{p}{\varrho} + v \tag{30.27}$$

and where $\boldsymbol{T}_{\mathrm{E}}$ is obtained from (30.12). If a motion and a body are given, both $\boldsymbol{T}_{\mathrm{E}}$ and $\varrho\ddot{\boldsymbol{x}}$ are determined, while p remains arbitrary. If for a given motion of a given body subject to a given body force, a pressure p satisfying (30.26) exists, *then such a p exists for that body in the same motion, subject to any other conservative body force:* We have only to adjust p so that φ, as given by (30.27), has the same value in each case. In particular, the motion is possible when $\boldsymbol{b} = \boldsymbol{0}$, and conversely. Thus *for a given motion to be possible in a given homogeneous incompressible simple body subject to an arbitrary conservative body force, it is necessary and sufficient that that motion be possible in the same body subject to suitable surface tractions alone.* Notice that the stress systems giving rise to the same motion of the same body subject to different conservative body forces *differ from each other only by a hydrostatic pressure field.*

If a single motion is to be possible in *all* incompressible simple bodies, it must be possible, in particular, in an incompressible perfect fluid. By KELVIN's theorem [CFT, Sect. 297], it must be circulation-preserving:

$$\ddot{\boldsymbol{x}} = -\operatorname{grad} \zeta, \tag{30.28}$$

[1] From this point to the end of the section, all results for which no other reference is given are due to COLEMAN and TRUESDELL [1965, *14*].

where $\zeta(\boldsymbol{x}, t)$ is a single-valued scalar field. For such motions the main theorems of classical hydrodynamics, named after BERNOULLI, CAUCHY, HELMHOLTZ, KELVIN, etc., are valid. A full account of these theorems will be found in CFT, Sects. 105—138; we leave it to the reader to bear in mind the relevance of all this classical apparatus to the subject at hand. From (30.26) and (30.12) we see that *a circulation-preserving motion is possible in a homogeneous incompressible simple body if and only if, for that motion,*

$$\operatorname{div} \underset{s=0}{\overset{\infty}{\mathfrak{G}}}\left(\boldsymbol{F}^{(t)}(s)\right) = -\varrho \operatorname{grad} \lambda, \tag{30.29}$$

where $\lambda(\boldsymbol{x}, t)$ is a single-valued scalar field. If (30.29) is satisfied, the stress system is determined to within an arbitrary time-dependent pressure $h(t)$:

$$\boldsymbol{T} = -\varrho(\zeta - \lambda - v + h)\boldsymbol{1} + \underset{s=0}{\overset{\infty}{\mathfrak{G}}}\left(\boldsymbol{F}^{(t)}(s)\right). \tag{30.30}$$

From (30.29) we have the necessary condition

$$\operatorname{curl} \operatorname{div} \underset{s=0}{\overset{\infty}{\mathfrak{G}}}\left(\boldsymbol{F}^{(t)}(s)\right) = \boldsymbol{0}. \tag{30.31}$$

Thus, in particular, a circulation-preserving motion which renders $\underset{s=0}{\overset{\infty}{\mathfrak{G}}}\left(\boldsymbol{F}^{(t)}(s)\right)$ a quadratic function of the spatial position vector is always a possible motion for the material defined by $\mathfrak{G}$.

When the motion is homogeneous, so that $\boldsymbol{F}^{(t)}(s)$ is constant in space at each time, then $\mathfrak{G}\left(\boldsymbol{F}^{(t)}(s)\right)$ is also a function of time alone for every homogeneous body, and (30.29) is satisfied with $\lambda = 0$. As in Sect. 28, we write such a motion in the form $\boldsymbol{x} = \boldsymbol{F}\boldsymbol{X} + \boldsymbol{c}$, where now $|\det \boldsymbol{F}| = 1$. Then

$$\boldsymbol{L} \equiv \operatorname{grad} \dot{\boldsymbol{x}} = \dot{\boldsymbol{F}}\boldsymbol{F}^{-1}, \quad \boldsymbol{L}_2 \equiv \operatorname{grad} \ddot{\boldsymbol{x}} = \ddot{\boldsymbol{F}}\boldsymbol{F}^{-1}, \tag{30.32}$$

and the motion is circulation-preserving if and only if[1]

$$\boldsymbol{L}_2 = \boldsymbol{L}_2^T. \tag{30.33}$$

When this condition is satisfied,

$$-\zeta = \boldsymbol{x} \cdot \left(\tfrac{1}{2}\boldsymbol{L}_2\boldsymbol{x} + \ddot{\boldsymbol{c}} - \boldsymbol{L}_2\boldsymbol{c}\right). \tag{30.34}$$

We have proved the following ***theorem on homogeneous motions***: *A homogeneous isochoric motion is possible in every homogeneous incompressible simple body subject to surface tractions alone (and hence also subject to any conservative field of body force) if and only if (30.33) holds. The stress in such a motion is given by*

$$\boldsymbol{T} = \varrho\left[\boldsymbol{x} \cdot \left(\tfrac{1}{2}\boldsymbol{L}_2\boldsymbol{x} + \ddot{\boldsymbol{c}} - \boldsymbol{L}_2\boldsymbol{c}\right) + v - h\right]\boldsymbol{1} + \underset{s=0}{\overset{\infty}{\mathfrak{G}}}\left(\boldsymbol{F}^{(t)}(s)\right), \tag{30.35}$$

where $h(t)$ is an arbitrary function.

Notice that while the extra stress is spatially homogeneous in a homogeneous motion, the pressure, in general, is not, even if the body force vanishes.

Since (30.33) is satisfied when $\ddot{\boldsymbol{x}} = \boldsymbol{0}$ or when $\boldsymbol{W} = \boldsymbol{0}$, we have also the following ***corollary***: *Every homogeneous isochoric motion that is either accelerationless or irrotational is possible, subject to surface tractions alone, in every homogeneous incompressible simple body.*

---

[1] TRUESDELL [1962, *65*, Eq. (17)]. The result is easy to derive directly; also, it follows as a special case from the D'Alembert-Euler vorticity equation [CFT, Eq. (130.1)].

By use of any of the reduced forms in Sect. 29, applied to $\boldsymbol{T}_E$, we see that the form of $\mathfrak{G}$ is determined for all admissible histories if it is known for every admissible homogeneous pure-stretch history. As was shown in Sect. 29, the class of irrotational motions includes motions with arbitrary $\boldsymbol{U}(t)$. Hence follows the ***determination theorem**: By supplying suitable time-dependent tractions on the bounding surface of any homogeneous incompressible simple body, subject to any conservative field of body force, it is possible to produce a variety of motions sufficient to determine the form of the response functional for the material by measurement of the corresponding stresses. Irrotational motions suffice.*

To find all $\boldsymbol{F}$ consistent with (30.33), we may note that CAUCHY's criterion for circulation-preserving motion [CFT, Eq. (134.1)] becomes $\boldsymbol{F}^T\boldsymbol{W}\boldsymbol{F}=\boldsymbol{W}_0\equiv\boldsymbol{W}(0)$. By $(24.16)_1$ CAUCHY's criterion is equivalent to[1]

$$\dot{\boldsymbol{R}}=\boldsymbol{R}\boldsymbol{Y} \quad \text{where} \quad \boldsymbol{Y}=\tfrac{1}{2}(\boldsymbol{U}^{-1}\dot{\boldsymbol{U}}-\dot{\boldsymbol{U}}\boldsymbol{U}^{-1})+\boldsymbol{U}^{-1}\boldsymbol{W}_0\boldsymbol{U}^{-1}. \tag{30.36}$$

Therefore, *if a unimodular right stretch tensor $\boldsymbol{U}(t)$ is prescribed arbitrarily as a function of time, and if the initial spin $\boldsymbol{W}_0$ and the initial rotation $\boldsymbol{R}_0$ are given, there exists a unique rotation $\boldsymbol{R}(t)$ such that the homogeneous motion with $\boldsymbol{F}=\boldsymbol{R}\boldsymbol{U}$ is a possible motion, subject to surface tractions alone, in every homogeneous incompressible simple body.*

We now consider some examples of isochoric homogeneous motions, all for the special case when $\ddot{\boldsymbol{c}}=\boldsymbol{0}$. It is then possible by choice of origin to annul $\boldsymbol{c}$ also. The motion is irrotational if and only if $\boldsymbol{W}_0=\boldsymbol{0}$ (as follows also by application of the Lagrange-Cauchy velocity-potential theorem [CFT, Sect. 134]). By $(24.16)_1$, in an irrotational motion the finite rotation $\boldsymbol{R}(t)$ generally cannot be constant in time. In order that $\dot{\boldsymbol{R}}=\boldsymbol{0}$ in an irrotational motion, it is necessary and sufficient that the right stretch tensor satisfy the condition

$$\boldsymbol{U}\dot{\boldsymbol{U}}=\dot{\boldsymbol{U}}\boldsymbol{U}. \tag{30.37}$$

In a suitably selected Cartesian co-ordinate system this differential equation is satisfied by the following *homogeneous irrotational motions with constant rotation tensor*:*

$$\left.\begin{aligned} &\dot{x}_k=\alpha_k(t)\,x_k, \qquad \alpha_1+\alpha_2+\alpha_3=0,\\ &[\boldsymbol{F}]=[\boldsymbol{U}]=\operatorname{diag}\left(e^{\int\alpha_1\,dt}, e^{\int\alpha_2\,dt}, e^{\int\alpha_3\,dt}\right),\\ &[\boldsymbol{L}]=[\boldsymbol{D}]=\operatorname{diag}(\alpha_1,\alpha_2,\alpha_3),\\ &[\boldsymbol{L}_2]=\operatorname{diag}(\dot{\alpha}_1+\alpha_1^2, \dot{\alpha}_2+\alpha_2^2, \dot{\alpha}_3+\alpha_3^2). \end{aligned}\right\} \tag{30.38}$$

Substitution into (30.35) yields the following stress system:

$$\boldsymbol{T}=\varrho\left[\tfrac{1}{2}\sum_{k=1}^{3}(\dot{\alpha}_k+\alpha_k^2)\,x_k^2+v-h\right]\boldsymbol{1}+\mathop{\mathfrak{F}}_{s=0}^{\infty}\left(\alpha_1(t-s),\alpha_2(t-s),\boldsymbol{e}_\sigma\right), \tag{30.39}$$

where $\mathfrak{F}$ is a tensor-valued functional and where $\boldsymbol{e}_\sigma$ is the orthonormal basis of vectors in the directions of the co-ordinate axes.

Another important special case of homogeneous motion is that in which the velocity field is steady[2]. By (30.32), we see that $\boldsymbol{L}_2=\dot{\boldsymbol{L}}+\boldsymbol{L}^2$ in general, so that $\boldsymbol{L}_2=\boldsymbol{L}^2$ in steady homogeneous motion. Substitution in (30.33) yields the following criterion for *steady circulation-preserving homogeneous motion:*

$$\boldsymbol{L}^2=(\boldsymbol{L}^2)^T. \tag{30.40}$$

---

[1] This result may be found also by substituting $(23.1)_1$ in (30.33) and integrating once; cf. TRUESDELL [1962, *65*, p. 427].

[2] An extensive kinematical analysis of these flows has been made by GIESEKUS [1962, *21*].

* [There are other solutions.]

By (25.6), $\boldsymbol{L}=\boldsymbol{D}+\boldsymbol{W}$; hence (30.40) is equivalent to

$$\boldsymbol{D}\boldsymbol{W}+\boldsymbol{W}\boldsymbol{D}=\boldsymbol{0}. \tag{30.41}$$

We now consider three special cases in which (30.41) is satisfied.

1. $\boldsymbol{D}=\boldsymbol{0}$. The motion is then a steady rigid rotation.

2. $\boldsymbol{W}=\boldsymbol{0}$. The motion is then irrotational. If we assume in addition that $\dot{\boldsymbol{R}}=\boldsymbol{0}$, then Eqs. (30.38) hold with $\dot{\alpha}_k=0$. In such a motion, which is called a *steady extension*, (30.39) reduces to

$$\boldsymbol{T}=\varrho\left[\tfrac{1}{2}\sum_{k=1}^{3}\alpha_k^2 x_k^2+v-h\right]\boldsymbol{1}+\boldsymbol{f}(\alpha_1,\alpha_2,\boldsymbol{e}_\sigma), \tag{30.42}$$

where $\boldsymbol{f}$ is a tensor-valued function.

3. $\boldsymbol{L}^2=\boldsymbol{0}$. In this case it is possible to choose Cartesian co-ordinates such that

$$[\boldsymbol{L}]=\varkappa\begin{Vmatrix}0&0&0\\1&0&0\\0&0&0\end{Vmatrix}, \tag{30.43}$$

where $\varkappa$ is a constant. Since we have assumed that $\boldsymbol{c}=\boldsymbol{0}$, this motion is accelerationless, in fact a *rectilinear shearing*, which with properly selected origin of co-ordinates may be written in the form

$$x=X,\qquad y=Y+\varkappa t X,\qquad z=Z. \tag{30.44}$$

Since

$$[\boldsymbol{F}^{(t)}(s)]=\begin{Vmatrix}1&0&0\\(t-s)\varkappa&1&0\\0&0&0\end{Vmatrix}, \tag{30.45}$$

substitution into (30.35) yields

$$\boldsymbol{T}=\varrho(v-h)\boldsymbol{1}+\boldsymbol{f}(\varkappa,t,\boldsymbol{e}_\sigma), \tag{30.46}$$

where $\boldsymbol{f}$ is a tensor-valued function.

Some important types of homogeneous motion are *not* circulation-preserving and hence *not possible* in *any* homogeneous incompressible simple material, unless, of course, suitable non-conservative body force be applied. For example, consider a time-dependent lineal flow. Such a flow is described by Eqs. (30.44) with $\varkappa t$ replaced by $f(t)$. Since $\ddot{y}=f''(t)X$, this motion is circulation-preserving if and only if $f''(t)=0$, i.e., if and only if it reduces to (30.44).

Important special cases of the results of this section will be developed in Sects. 61 and 110ff.

31. The isotropy group[1]. It may happen that a particle X is materially isomorphic to itself in a non-trivial manner. According to the definition of Sect. 27, this means that there are two different configurations $\boldsymbol{\varkappa}$ and $\hat{\boldsymbol{\varkappa}}$, with the same constant density, of a neighborhood of X such that the corresponding response functionals $\mathfrak{F}_{\varkappa}$ and $\mathfrak{F}_{\hat{\varkappa}}$ coincide. The physical meaning of such a coincidence is that the material at the particle X in the configuration $\boldsymbol{\varkappa}$ is indistinguishable in its response from the same material after it has been deformed into the configuration $\hat{\boldsymbol{\varkappa}}$. For simple materials, it follows from (28.8) that the response functionals $\mathfrak{G}_K$

[1] Noll [1957, *15*, § 14] [1958, *31*, § 19].

and $\mathfrak{G}_{\hat{K}}$ coincide if and only if for all histories $\boldsymbol{F}^{(t)}$ in the domain of $\mathfrak{G}$

$$\underset{s=0}{\overset{\infty}{\mathfrak{G}_{K}}}\left(\boldsymbol{F}^{(t)}(s)\,\boldsymbol{P}^{-1}\right)=\underset{s=0}{\overset{\infty}{\mathfrak{G}_{\hat{K}}}}\left(\boldsymbol{F}^{(t)}(s)\,\boldsymbol{P}^{-1}\right)=\underset{s=0}{\overset{\infty}{\mathfrak{G}_{K}}}\left(\boldsymbol{F}^{(t)}(s)\right), \tag{31.1}$$

where $\boldsymbol{P}=\hat{\boldsymbol{K}}\boldsymbol{K}^{-1}$ is a unimodular tensor (i.e., det $\boldsymbol{P}=\pm 1$) since $\boldsymbol{K}$ and $\hat{\boldsymbol{K}}$ have the same density. It is not hard to see that, conversely, any unimodular tensor $\boldsymbol{H}$ such that for all $\boldsymbol{F}^{(t)}$

$$\underset{s=0}{\overset{\infty}{\mathfrak{G}}}\left(\boldsymbol{F}^{(t)}(s)\right)=\underset{s=0}{\overset{\infty}{\mathfrak{G}}}\left(\boldsymbol{F}^{(t)}(s)\,\boldsymbol{H}\right) \tag{31.2}$$

gives rise to a material isomorphism of X onto itself.

It is clear that the set of all unimodular tensors $\boldsymbol{H}$ for which (31.2) holds forms a group, which is called the *isotropy group* of the response functional $\mathfrak{G}$ and is denoted by g. It is a subgroup of the group u of all unimodular tensors: $g \subset u$.

From its definition, the isotropy group generally depends upon the choice of the local reference configuration $\boldsymbol{K}$. Let $\boldsymbol{K}$ and $\hat{\boldsymbol{K}}$ be two local reference configurations, and denote the corresponding response functionals and isotropy groups by $\mathfrak{G}$, g, and $\hat{\mathfrak{G}}$, $\hat{g}$, respectively. In the present notation (28.8) assumes the form

$$\underset{s=0}{\overset{\infty}{\mathfrak{G}}}\left(\boldsymbol{F}^{(t)}(s)\right)=\underset{s=0}{\overset{\infty}{\hat{\mathfrak{G}}}}\left(\boldsymbol{F}^{(t)}(s)\,\boldsymbol{P}^{-1}\right), \tag{31.3}$$

and this relation holds for all histories $\boldsymbol{F}^{(t)}(s)$. The tensor $\boldsymbol{H}$ belongs to the isotropy group g if and only if (31.2) holds for all $\boldsymbol{F}^{(t)}(s)$. This is the case, by (31.3), if and only if the relation

$$\underset{s=0}{\overset{\infty}{\hat{\mathfrak{G}}}}\left(\boldsymbol{F}^{(t)}(s)\,\boldsymbol{P}^{-1}\right)=\underset{s=0}{\overset{\infty}{\hat{\mathfrak{G}}}}\left(\boldsymbol{F}^{(t)}(s)\,\boldsymbol{H}\boldsymbol{P}^{-1}\right) \tag{31.4}$$

is an identity in the history $\boldsymbol{F}^{(t)}(s)$, or, equivalently, if and only if

$$\underset{s=0}{\overset{\infty}{\hat{\mathfrak{G}}}}\left(\boldsymbol{F}^{(t)}(s)\right)=\underset{s=0}{\overset{\infty}{\hat{\mathfrak{G}}}}\left(\boldsymbol{F}^{(t)}(s)\,\boldsymbol{P}\boldsymbol{H}\boldsymbol{P}^{-1}\right) \tag{31.5}$$

is an identity in $\boldsymbol{F}^{(t)}(s)$. Therefore, $\boldsymbol{H}$ is a member of the isotropy group g if and only if $\boldsymbol{P}\boldsymbol{H}\boldsymbol{P}^{-1}$ is a member of the isotropy group $\hat{g}$. We have thus proved the following theorem[1]:

Let g and $\hat{g}$ be the isotropy groups of a particle relative to the local reference configurations $\mathbf{K}$ and $\hat{\mathbf{K}}$, and let $\mathbf{P}=\hat{\mathbf{K}}\mathbf{K}^{-1}$. Then

$$\hat{g}=\mathbf{P}\,g\,\mathbf{P}^{-1}. \tag{31.6}$$

Thus the isotropy groups of a given particle with respect to various reference configurations are conjugate to one another.

In the foregoing theorem, the two local reference configurations need not have the same density. In particular, one may be obtained from the other by a dilatation: $\mathbf{P}=\alpha\mathbf{1}$, $\alpha\neq 0$. Then (31.6) yields $\hat{g}=g$. That is, *the isotropy group is unchanged by a dilatation.*

In many cases we are primarily or exclusively interested in such members of the isotropy group as are also orthogonal. These are characterized by the

[1] Noll [1958, *31*, § 19].

following theorem: *An orthogonal tensor* $\boldsymbol{Q}$ *belongs to the isotropy group* g *if and only if*

$$\boldsymbol{Q}\,\underset{s=0}{\overset{\infty}{\mathfrak{G}}}\left(\boldsymbol{F}^{(t)}(s)\right)\boldsymbol{Q}^T=\underset{s=0}{\overset{\infty}{\mathfrak{G}}}\left(\boldsymbol{Q}\boldsymbol{F}^{(t)}(s)\,\boldsymbol{Q}^T\right) \tag{31.7}$$

holds identically in $\boldsymbol{F}^{(t)}$. To prove this theorem, let $\boldsymbol{Q}$ be a constant orthogonal tensor, replace $\boldsymbol{F}^{(t)}(s)$ by $\boldsymbol{Q}\boldsymbol{F}^{(t)}(s)$ and $\boldsymbol{H}$ by $\boldsymbol{Q}^T$ in (31.2), and use (28.5).

The relation (31.7) is equivalent to the following relations for the functional $\mathfrak{F}$ and the function $\mathfrak{f}$ of (29.28):

$$\left.\begin{aligned}\boldsymbol{Q}\,\underset{s=0}{\overset{\infty}{\mathfrak{F}}}\left(\boldsymbol{G}^*(s),\boldsymbol{C}\right)\boldsymbol{Q}^T&=\underset{s=0}{\overset{\infty}{\mathfrak{F}}}\left(\boldsymbol{Q}\boldsymbol{G}^*(s)\,\boldsymbol{Q}^T,\boldsymbol{Q}\boldsymbol{C}\boldsymbol{Q}^T\right),\\ \boldsymbol{Q}\,\mathfrak{f}(\boldsymbol{C})\,\boldsymbol{Q}^T&=\mathfrak{f}(\boldsymbol{Q}\boldsymbol{C}\boldsymbol{Q}^T).\end{aligned}\right\} \tag{31.8}$$

The identity (31.7) continues to be valid for all $\boldsymbol{Q}\in g$ if the functional $\mathfrak{G}$ is replaced by the functional $\mathfrak{I}$ of (29.6) or the functional $\mathfrak{L}$ of (29.11). The identity $(31.8)_1$ remains valid if $\mathfrak{F}$ is replaced by the functional $\mathfrak{K}$ of (29.13), (29.19) or by the functional $\mathfrak{J}$ of (29.20), (29.21). Also, $\mathfrak{F}$ and $\mathfrak{f}$ in (31.8) may be replaced by $\mathfrak{S}$ and $\mathfrak{s}$ of (29.30).

It follows from (31.7) that g always contains the identity $\mathbf{1}$ and the inversion $-\mathbf{1}$. Thus if g contains a certain proper orthogonal transformation $\boldsymbol{Q}$, it contains also the improper transformation $-\boldsymbol{Q}$. In particular, g cannot contain the group of proper orthogonal transformations unless it contains also the group of all orthogonal transformations, proper or not.

A material is called *isotropic* if there is at least one local reference configuration such that the isotropy group of its corresponding response functional contains the full orthogonal group. A local reference configuration with this property is called an *undistorted state* of the material. For an isotropic material in an undistorted state, a physical test cannot detect whether or not the material has been rotated arbitrarily before the test is made. This property may be described by saying that the material has no preferred directions when in an undistorted state[1].

From the definitions of "simple fluid" and "simple solid" to be given in Sects. 32 and 33 it will follow that *every isotropic simple material is either a simple fluid or a simple solid, and all simple fluids are isotropic.* In the remainder of the present section we shall derive only such results as are stated more economically without specifying whether a fluid or solid is intended.

For an isotropic material, (31.7) and (31.8) are valid for all orthogonal transformations $\boldsymbol{Q}$, provided that the reference configuration is an undistorted state. From (23.3) we see that $\boldsymbol{U}=\boldsymbol{R}^T\boldsymbol{V}\boldsymbol{R}$ and $\boldsymbol{U}^2=\boldsymbol{R}^T\boldsymbol{V}^2\boldsymbol{R}$; using these facts in (31.7) and (31.8), respectively, shows that for isotropic materials the general constitutive equations (29.19) and (29.28) may be written in the following equivalent but simpler forms:

$$\boldsymbol{T}(t)=\underset{s=0}{\overset{\infty}{\mathfrak{K}}}\left(\boldsymbol{U}_{(t)}(t-s);\,\boldsymbol{V}(t)\right), \tag{31.9}$$

$$\boldsymbol{T}(t)=\mathfrak{f}(\boldsymbol{B})+\underset{s=0}{\overset{\infty}{\mathfrak{F}}}\left(\boldsymbol{G}(s);\,\boldsymbol{B}\right),\quad \boldsymbol{G}(s)\equiv\boldsymbol{C}_{(t)}(t-s)-\mathbf{1}, \tag{31.10}$$

[1] We could define a material as being *hemitropic* if, for some reference configuration, g contains the *proper* orthogonal group. Because of the theorem stated in the preceding paragraph of the text, it follows that every *hemitropic simple material is isotropic,* and conversely, so the distinction is immaterial here . In view of the confusion between frame-indifference and material symmetry still current in some secondary literature, however, we state with emphasis that *no such result has ever been claimed for other theories,* in particular, for those in which tensors of odd order appear. Cf. the theory of heat conduction in Sect. 96.

where $\boldsymbol{V}(t)$ is the left stretch tensor and $\boldsymbol{B}(t)$ the left Cauchy-Green tensor (cf. Sect. 23). The *reduced forms*[1] (31.9) and (31.10), broadly generalizing results long known in the theory of elasticity, are of major importance because, as expected from the idea of isotropy, *the rotation is altogether eliminated*. They state, in rough terms, that the stress is determined by the strain history and the present strain alone, provided that as measures of present strain we select $\boldsymbol{V}$ or $\boldsymbol{B}$ (or any single-valued, uniquely invertible isotropic function of $\boldsymbol{V}$)[2].

It must be emphasized that while the various reduced forms in Sect. 29 hold for any local reference configuration, the reduced forms (31.9) and (31.10) are valid, in general, *only* for isotropic materials and *only* when an undistorted state is used as the local reference configuration. If another local reference configuration is used, as of course is entirely legitimate, the constitutive equation of an isotropic material does not appear to have any particularly simple form.

A simplification for the component form (29.9) results from (31.7), with $\mathfrak{G}$ replaced by $\mathfrak{F}$ and $\boldsymbol{F}^{(t)}$ by $\boldsymbol{C}^{(t)}$, i.e. from

$$\boldsymbol{Q} \mathop{\mathfrak{F}}_{s=0}^{\infty} (\boldsymbol{C}^{(t)}(s))\, \boldsymbol{Q}^T = \mathop{\mathfrak{F}}_{s=0}^{\infty} (\boldsymbol{Q}\, \boldsymbol{C}^{(t)}(s)\, \boldsymbol{Q}^T). \tag{31.11}$$

Using an argument similar to the one leading to the condition (8.12), one can show that $\boldsymbol{Q}$ belongs to the isotropy group g if and only if the component functionals $\mathfrak{F}_{\alpha\beta}$ of (29.9) satisfy the identity

$$\mathop{\mathfrak{F}_{\alpha\beta}}_{s=0}^{\infty} (C_{\lambda\mu}(t-s)\,;\, \boldsymbol{d}_\sigma) = \mathop{\mathfrak{F}_{\alpha\beta}}_{s=0}^{\infty} (C_{\lambda\mu}(t-s)\,;\, \boldsymbol{Q}\boldsymbol{d}_\sigma). \tag{31.12}$$

The condition for the isotropy of tensor functions stated at the end of Sect. 8 is analogous to the following condition for material isotropy: A simple material is isotropic if and only if its constitutive equation (29.9), when the local reference configuration is an undistorted state, reduces to

$$\overline{T}_{\alpha\beta}(t) = \mathop{\mathfrak{F}_{\alpha\beta}}_{s=0}^{\infty} (C_{\mu\nu}(t-s)\,;\, g_{\sigma\tau}). \tag{31.13}$$

Equation (31.10) has the component form

$$T_{km}(t) = \mathfrak{f}_{km}(B_{ru}(t), g_{vw}(t)) + \mathop{\mathfrak{F}_{km}}_{s=0}^{\infty} (C_{(t)pq}(t-s) - g_{pq};\, B_{ru}(t);\, g_{vw}(t)). \tag{31.14}$$

The components $C_{(t)\,km}$ of the relative right Cauchy-Green tensor $\boldsymbol{C}_{(t)}$ are related to the deformation gradients by (23.10). The components $g_{km}(t)$ of the metric tensor at $\boldsymbol{x}$ may vary with t because the place $\boldsymbol{x} = \boldsymbol{\chi}(X, t)$ occupied by the particle X may vary with t.

32. Simple fluids[3]. The physical concept of a "fluid" is vague, but it may be interpreted as including the idea that a fluid should not alter its material response after an arbitrary deformation that leaves the density unchanged. Guided by this idea, we call a simple material a *simple fluid* if, for some one reference configuration, *the isotropy group is the full group of unimodular transformations*, i.e., if $g = u$. It is a trivial consequence of this definition that *all simple fluids are*

---

[1] Noll [1957, *15*, § 17] [1958, *31*, § 20 and § 22]. An important step toward this reduction had been made by Green and Rivlin [1956, *12*, § 3] [1957, *8*, § 3].

[2] In the older terminology, to achieve this simplification we must use a spatial rather than a material measure of present strain.

[3] This concept of a fluid, generalizing and including earlier attempts by several authors, was put forward by Noll [1958, *31*, § 21]. References to older theories will be given in Sect. 119.

isotropic. But this is not all. From Noll's theorem in Sect. 31 we conclude that if $\boldsymbol{K}$ is the undistorted state for which $g=u$, the isotropy group $\hat{g}$ corresponding to the local reference configuration $\hat{\boldsymbol{K}}$ is given by $\hat{g}=\boldsymbol{P}u\boldsymbol{P}^{-1}=u$. That is, *every configuration of a simple fluid is an undistorted state.* (31.3) is valid for all $\boldsymbol{H}$ in u, irrespective of the choice of the reference configuration. It then follows from (31.2) and (31.3) that

$$\underset{s=0}{\overset{\infty}{\mathfrak{G}_{\boldsymbol{K}}}}\left(\boldsymbol{F}^{(t)}(s)\right)=\underset{s=0}{\overset{\infty}{\mathfrak{G}_{\hat{\boldsymbol{K}}}}}\left(\boldsymbol{F}^{(t)}(s)\right) \tag{32.1}$$

holds whenever

$$\boldsymbol{P}=\hat{\boldsymbol{K}}\boldsymbol{K}^{-1} \tag{32.2}$$

is unimodular, i.e., whenever $|\det \boldsymbol{P}|=1$. Now the density $\varrho_{\varkappa}(X)$ at X depends only on the equivalence class $\boldsymbol{K}=\nabla\varkappa$ to which the configuration $\varkappa$ belongs. Hence we can associate with each local configuration $\boldsymbol{K}$ at X a density $\varrho_{\boldsymbol{K}}$. The densities $\varrho_{\boldsymbol{K}}$ and $\varrho_{\hat{\boldsymbol{K}}}$ corresponding to two local configurations $\boldsymbol{K}$ and $\hat{\boldsymbol{K}}$ are related by an equation similar to (21.18), namely,

$$|\det \boldsymbol{P}|\,\varrho_{\hat{\boldsymbol{K}}}=\varrho_{\boldsymbol{K}}. \tag{32.3}$$

It follows that (32.1) holds whenever

$$\varrho_{\hat{\boldsymbol{K}}}=\varrho_{\boldsymbol{K}}, \tag{32.4}$$

i.e., whenever the mass density at X is the same in the local configurations $\boldsymbol{K}$ and $\hat{\boldsymbol{K}}$. We can conclude that the form of the functional $\mathfrak{G}_{\boldsymbol{K}}$ depends only on the density $\varrho_{\boldsymbol{K}}$ and not on any other properties of $\boldsymbol{K}$. Therefore, in the case of a fluid, the general constitutive equation (29.3) reduces to

$$\boldsymbol{T}(t)=\boldsymbol{R}(t)\underset{s=0}{\overset{\infty}{\mathfrak{H}}}\left(\boldsymbol{U}^{(t)}(s);\,\varrho_{\boldsymbol{K}}\right)\boldsymbol{R}(t)^{T}, \tag{32.5}$$

where $\mathfrak{H}$ is a functional of the history $\boldsymbol{U}^{(t)}(s)$, $s\geqq 0$, and a function of the scalar parameter $\varrho_{\boldsymbol{K}}$. The functional $\mathfrak{H}$ does not depend on the choice of the local reference configuration $\boldsymbol{K}$. Since the isotropy group u contains the orthogonal group, it follows from (31.4) that $\mathfrak{H}$ satisfies the relation

$$\boldsymbol{Q}\underset{s=0}{\overset{\infty}{\mathfrak{H}}}\left(\boldsymbol{U}^{(t)}(s);\,\varrho\right)\boldsymbol{Q}^{T}=\underset{s=0}{\overset{\infty}{\mathfrak{H}}}\left(\boldsymbol{Q}\,\boldsymbol{U}^{(t)}(s)\,\boldsymbol{Q}^{T};\,\varrho\right) \tag{32.6}$$

for *all* orthogonal transformations $\boldsymbol{Q}$.

If the configuration at time t is taken as the reference configuration, then $\boldsymbol{R}(t)=\mathbf{1}$, and (32.5) becomes

$$\boldsymbol{T}(t)=\underset{s=0}{\overset{\infty}{\mathfrak{H}}}\left(\boldsymbol{U}_{(t)}(t-s);\,\varrho(t)\right), \tag{32.7}$$

which could also have been obtained by specialization from (31.9).

For a simple fluid, the dependence on $\boldsymbol{B}$ in (31.10) reduces to a dependence on the density ϱ. The equilibrium term $\mathfrak{f}(\boldsymbol{B})$ then reduces to a hydrostatic pressure, and hence the general constitutive equation of a simple fluid may be written in the form

$$\boldsymbol{T}=-p(\varrho)\,\mathbf{1}+\underset{s=0}{\overset{\infty}{\mathfrak{Q}}}\left(\boldsymbol{G}(s);\,\varrho\right), \tag{32.8}$$

where $p(\varrho)$ is a scalar function of ϱ, and where $\boldsymbol{G}(s)$ is given by $(31.10)_2$. The response functional $\mathfrak{Q}$ of (32.8) has the value $\mathbf{0}$ when $\boldsymbol{G}(s)\equiv\mathbf{0}$ and satisfies the isotropy relation

$$\boldsymbol{Q}\,\underset{s=0}{\overset{\infty}{\mathfrak{Q}}}\left(\boldsymbol{G}(s);\,\varrho\right)\boldsymbol{Q}^T=\underset{s=0}{\overset{\infty}{\mathfrak{Q}}}\left(\boldsymbol{Q}\,\boldsymbol{G}(s)\,\boldsymbol{Q}^T;\,\varrho\right) \tag{32.9}$$

identically in $\boldsymbol{G}(s)$, ϱ, and the orthogonal tensor $\boldsymbol{Q}$.

The response functionals $\mathfrak{H}$ and $\mathfrak{Q}$ of (32.7) and (32.8) do not depend on any reference configuration. This fact may be described in physical terms by saying that simple fluids have no preferred configurations; i.e., they have no permanent memory for any particular state. Eqs. (32.8) and (32.9), expressing NOLL's ***fundamental theorem on simple fluids***, delimit the peculiar kind of memory a simple fluid may have. A simple fluid may remember everything that ever happened to it, yet it cannot recall any one configuration as being physically different from any other except in regard to its mass density. It reconciles these two almost contradictory qualities by remembering the past only insofar as it is reflected by the tensor $\boldsymbol{G}(s)$, which measures the deformation of the configuration at time $t-s$ with respect to that at the ever-changing present time t.

Sometimes a "fluid" is defined as a substance capable of flow in any direction. Formally, a "fluid" in this sense cannot support a shear stress when in equilibrium. The simple fluid has this property, since $\boldsymbol{T}=-p\boldsymbol{1}$ when $\boldsymbol{G}(s)\equiv\boldsymbol{0}$, but, as is clear from (29.28), the converse is false, and indeed we shall see in Sect. 126 that certain anisotropic solids can flow.

For incompressible simple fluids, the stress $\boldsymbol{T}$, in the various forms of the constitutive equation given in this section, must be replaced by the extra-stress $\boldsymbol{T}_E=\boldsymbol{T}+p\boldsymbol{1}$. Moreover, the density ϱ at a particle cannot depend on time and hence can be omitted in the constitutive equations for incompressible simple fluids. For example, the constitutive equation (32.8) must be replaced by

$$\boldsymbol{T}=-p\boldsymbol{1}+\underset{s=0}{\overset{\infty}{\mathfrak{Q}}}\left(\boldsymbol{G}(s)\right). \tag{32.10}$$

The functional $\mathfrak{Q}$ need here be defined only when $\boldsymbol{C}_{(t)}(t-s)=\boldsymbol{G}(s)+\boldsymbol{1}$ is unimodular for all $s\geqq 0$, i.e., if

$$\det\left(\boldsymbol{G}(s)+\boldsymbol{1}\right)=\det\boldsymbol{C}_{(t)}(t-s)=1\,. \tag{32.11}$$

As we explained in Sect. 26, the functional $\mathfrak{Q}$ of (32.10) is determined only up to an arbitrary scalar-valued functional of $\boldsymbol{C}_{(t)}(t-s)$. We often shall use the normalization (30.17), i.e.,

$$\operatorname{tr}\underset{s=0}{\overset{\infty}{\mathfrak{Q}}}\left(\boldsymbol{G}(s)\right)=\operatorname{tr}\boldsymbol{T}_E(t)=0, \tag{32.12}$$

so that $\boldsymbol{T}_E$ reduces to the stress deviator.

A detailed discussion of the properties of simple fluids will be given in Chapter E.

33. Simple solids. Real substances commonly thought of as "solid" have preferred configurations such that any pure deformation from one of these brings the material into a configuration from which its response is different. For example, if an unloaded bar of iron or rubber is stretched and then held at rest, the stretching force, or at least some part of it, must continue to be applied, even for very long periods of time, while to maintain the body in its initial configuration, no force at all was needed. This behavior is entirely different from that of a physical substance regarded as a "fluid", in which the response to a given deformation history is just the same, starting from any two configurations with the same density. For solids, more generally, while any pure strain from a preferred configuration affects the subsequent behavior of the material, rotations may or may not have an influence. This characterization leads us to call a simple material a *simple solid*[1] if there is a local reference configuration $\boldsymbol{K}$ such that the corresponding isotropy group g is a subgroup of the orthogonal group o, i.e. if $g \subset o$. The particular local reference configurations having this property are called the *undistorted states* of the solid. Note that the term "undistorted state" has been defined distinctly for isotropic simple materials in Sect. 31 and here for simple solids. Later in this section, the two definitions will be shown to reduce to one and the same in any case in which both are applicable.

A simple solid is called *aeolotropic*, or anisotropic, if its isotropy group, taken relative to an undistorted reference state, is a *proper* subgroup g of the orthogonal group. The type of aeolotropy is characterized by the type of the group g. A simple solid has the aeolotropy characterized by the isotropy group g if and only if (31.7) holds for all $\boldsymbol{Q}$ in g and all deformation histories $\boldsymbol{F}^{(t)}$. As remarked already in Sect. 31, every possible isotropy group g contains the two-element group $\{\boldsymbol{1}, -\boldsymbol{1}\}$ as a subgroup. In fact, it is easily seen that g is the direct product of this two-element group and a group g_0 which consists only of proper orthogonal transformations, i.e. rotations. Consequently the type of aeolotropy is characterized by the type of the group g_0. Although there are an infinite number of types of rotation groups g_0, twelve of them seem to exhaust the kinds of symmetries occurring in theories proposed up to now as being appropriate to describe the behavior of real aeolotropic solids[2]. We shall define these twelve types.

We use the notation $\boldsymbol{R}_{\boldsymbol{n}}^{\varphi}$ for the right-handed rotation through the angle φ, $0<\varphi<2\pi$, about an axis in the direction of the unit vector $\boldsymbol{n}$. To specify the isotropy group g of a solid, it is sufficient to list a set of generators for the corresponding rotation group g_0, i.e., a set of elements of g_0 which, when they and their inverses are multiplied among themselves in various combinations, yield all the elements of g_0 as products.

The first eleven types of aeolotropy correspond to the thirty-two crystal classes[3]. We assume that the isotropy group of a crystalline solid is generated by the crystallographic point group of the crystal and the inversion, $-\boldsymbol{1}$. There are thirty-two such point groups but only eleven corresponding types of isotropy groups. In the following table, $\boldsymbol{i}, \boldsymbol{j}, \boldsymbol{k}$ denote a right-handed orthonormal basis, and $\boldsymbol{p} \equiv \sqrt{\tfrac{1}{3}}(\boldsymbol{i}+\boldsymbol{j}+\boldsymbol{k})$.

The last type of aeolotropy, called *transverse isotropy*, is characterized by the condition that g_0 consist of $\boldsymbol{1}$ and all rotations $\boldsymbol{R}_{\boldsymbol{k}}^{\varphi}$, $0<\varphi<2\pi$, about the axis

[1] NOLL [1958, *31*, § 20].

[2] As far as the theory is concerned, *any* subgroup of the orthogonal group may be an isotropy group. The impulse to concentrate attention upon these particular subgroups, like the emphasis on isotropic materials, comes from experience with real materials rather than from any distinguishing feature within the theory, and hence in this treatise we do not attempt to explain the origins and properties of the cristallographic groups. Neither do we claim that these groups are the only ones leading to theories of solids with important physical applications.

[3] A definitive mathematical treatment of the thirty-two crystal classes was given by SCHOENFLIESS [1891, *1*]. The subject was developed from a more physical standpoint by VOIGT [1910, *2*, §§ 1—58, esp. §§ 47—55]. A mineralogical description, using the now current terminology, was given by DANA and HURLBUT [1959, *5*, Ch. 2, §§ B—F]. Tensor bases for the corresponding *anisotropic tensors*, i.e., second-order tensors $\boldsymbol{S}$ such that $\boldsymbol{Q}\boldsymbol{S}\boldsymbol{Q}^T = \boldsymbol{S}$ and higher-order tensors satisfying analogous identities for all $\boldsymbol{Q}$ that belong to an assigned subgroup of the orthogonal group, were constructed by SMITH and RIVLIN [1957, *22*], PIPKIN and RIVLIN [1958, *35*, § 5] [1959, *21*, § 4], and SIROTIN [1960, *50* and *51*]. Cf. also the works cited in Sect. 13.

determined by a unit vector $\boldsymbol{k}$. Transverse isotropy is appropriate to real materials having a laminated or a bundled structure.

A material is called *orthotropic* if its isotropy group contains reflections on three mutually perpendicular planes. Such a triple of reflections is $-\boldsymbol{R}_{\boldsymbol{i}}^{\pi}$, $-\boldsymbol{R}_{\boldsymbol{j}}^{\pi}$, $-\boldsymbol{R}_{\boldsymbol{k}}^{\pi}$. Since $\boldsymbol{R}_{\boldsymbol{i}}^{\pi}\boldsymbol{R}_{\boldsymbol{j}}^{\pi}=\boldsymbol{R}_{\boldsymbol{k}}^{\pi}$ and $(\boldsymbol{R}_{\boldsymbol{i}}^{\frac{1}{2}\pi})^2=\boldsymbol{R}_{\boldsymbol{i}}^{\pi}$, it follows easily that the crystals listed under 3, 5, 6, and 7 in the above table are particular orthotropic materials.

	Crystal class	Generators of g_0	Order of g
1.	*Triclinic system* all classes	$\mathbf{1}$	2
2.	*Monoclinic system* all classes	$\boldsymbol{R}_{\boldsymbol{k}}^{\pi}$	4
3.	*Rhombic system* all classes	$\boldsymbol{R}_{\boldsymbol{i}}^{\pi}, \boldsymbol{R}_{\boldsymbol{j}}^{\pi}$	8
4.	*Tetragonal system* tetragonal-disphenoidal tetragonal-pyramidal tetragonal-dipyramidal	$\boldsymbol{R}_{\boldsymbol{k}}^{\frac{1}{2}\pi}$	8
5.	tetragonal-scalenohedral ditetragonal-pyramidal tetragonal-trapezohedral ditetragonal-dipyramidal	$\boldsymbol{R}_{\boldsymbol{k}}^{\frac{1}{2}\pi}, \boldsymbol{R}_{\boldsymbol{i}}^{\pi}$	16
6.	*Cubic system* tetartoidal diploidal	$\boldsymbol{R}_{\boldsymbol{i}}^{\pi}, \boldsymbol{R}_{\boldsymbol{j}}^{\pi}, \boldsymbol{R}_{\boldsymbol{p}}^{\frac{2}{3}\pi}$	24
7.	hextetrahedral gyroidal hexoctahedral	$\boldsymbol{R}_{\boldsymbol{i}}^{\frac{1}{2}\pi}, \boldsymbol{R}_{\boldsymbol{j}}^{\frac{1}{2}\pi}, \boldsymbol{R}_{\boldsymbol{k}}^{\frac{1}{2}\pi}$	48
8.	*Hexagonal system* trigonal-pyramidal rhombohedral	$\boldsymbol{R}_{\boldsymbol{k}}^{\frac{2}{3}\pi}$	6
9.	ditrigonal-pyramidal trigonal-trapezohedral hexagonal-scalenohedral	$\boldsymbol{R}_{\boldsymbol{i}}^{\pi}, \boldsymbol{R}_{\boldsymbol{k}}^{\frac{2}{3}\pi}$	12
10.	trigonal-dipyramidal hexagonal-pyramidal hexagonal-dipyramidal	$\boldsymbol{R}_{\boldsymbol{k}}^{\frac{1}{3}\pi}$	12
11.	ditrigonal-dipyramidal dihexagonal-pyramidal hexagonal-trapezohedral dihexagonal-dipyramidal	$\boldsymbol{R}_{\boldsymbol{i}}^{\pi}, \boldsymbol{R}_{\boldsymbol{k}}^{\frac{1}{3}\pi}$	24

As was pointed out in Sect. 31, both the response functional and the isotropy group depend not only on the material but also on the reference configuration[1]. Let g and $\hat{g}$ be the isotropy groups of a simple solid which correspond to two *undistorted* states $\boldsymbol{K}$ and $\hat{\boldsymbol{K}}$. By Eq. (31.6), the elements of $\hat{g}$ must be of the form

$$\hat{\boldsymbol{Q}}=\boldsymbol{P}\boldsymbol{Q}\boldsymbol{P}^{-1}, \tag{33.1}$$

where $\boldsymbol{Q}$ belongs to g and $\boldsymbol{P}$ is the deformation from $\boldsymbol{K}$ to $\hat{\boldsymbol{K}}$. Our assumptions ensure that both $\boldsymbol{Q}$ and $\hat{\boldsymbol{Q}}$ are orthogonal. Let

$$\boldsymbol{P}=\boldsymbol{R}\boldsymbol{U} \tag{33.2}$$

[1] The results from this point on to the end of the section are due to COLEMAN and NOLL [1964, *19*].

be the right polar decomposition of $\boldsymbol{P}$, so that $\boldsymbol{R}$ is the rotation tensor and $\boldsymbol{U}$ the right stretch tensor of the deformation from $\boldsymbol{K}$ to $\hat{\boldsymbol{K}}$. By (33.2), (33.1) is equivalent to

$$\hat{\boldsymbol{Q}}\boldsymbol{R}\boldsymbol{U}=\boldsymbol{R}\boldsymbol{U}\boldsymbol{Q}=\boldsymbol{R}\boldsymbol{Q}(\boldsymbol{Q}^T\boldsymbol{U}\boldsymbol{Q}). \tag{33.3}$$

The assertion of uniqueness in the polar decomposition theorem yields

$$\hat{\boldsymbol{Q}}\boldsymbol{R}=\boldsymbol{R}\boldsymbol{Q}, \quad \boldsymbol{U}=\boldsymbol{Q}^T\boldsymbol{U}\boldsymbol{Q}. \tag{33.4}$$

Eq. $(33.4)_1$ shows that $\hat{\boldsymbol{Q}}=\boldsymbol{R}\boldsymbol{Q}\boldsymbol{R}^T$, i.e. that the group $\hat{g}$ is the conjugate $\hat{g}=\boldsymbol{R}\,g\boldsymbol{R}^{-1}$ of g *within the orthogonal group* o. Two subgroups of o are considered to be of the same type if they are conjugate within o, and this is the precise meaning for the term "type of rotation group" used above. Only such a type of isotropy group, not the group itself, represents an intrinsic property of a simple solid. In the table given above, a change from a group g_0 to a conjugate within o merely corresponds to a change of the orthonormal basis $\boldsymbol{i}, \boldsymbol{j}, \boldsymbol{k}$ to another.

In order to exploit the result $(33.4)_2$ we need the following theorem from geometrical linear algebra[1]: The symmetric tensor $\boldsymbol{S}$ commutes with the orthogonal tensor $\boldsymbol{Q}$ (i.e., $\boldsymbol{Q}\boldsymbol{S}=\boldsymbol{S}\boldsymbol{Q}$) if and only if $\boldsymbol{Q}$ leaves each of the characteristic spaces of $\boldsymbol{S}$ invariant. Characteristic spaces are defined as being the maximal subspaces of the three-dimensional ordinary vector space which consist only of proper vectors of $\boldsymbol{S}$. If the proper numbers of $\boldsymbol{S}$ are distinct, there are three such spaces, all one-dimensional and mutually perpendicular. If $\boldsymbol{S}$ has only two distinct proper numbers, then there is one two-dimensional and one one-dimensional characteristic space, and $\boldsymbol{S}$ has the form

$$\boldsymbol{S}=\alpha\boldsymbol{1}+\beta\boldsymbol{k}\otimes\boldsymbol{k}. \tag{33.5}$$

If all proper numbers of $\boldsymbol{S}$ are equal, then the space of all vectors is the only characteristic space, and $\boldsymbol{S}$ has the form $\boldsymbol{S}=\alpha\boldsymbol{1}$. The theorem above and $(33.4)_2$ imply that *a local deformation of a simple solid from an undistorted state brings it into another undistorted state if and only if the characteristic spaces of the right stretch tensor of the local deformation are invariant under all transformations of the isotropy group* relative to the original undistorted state. As should be expected, the rotation plays no part in the result: Rotation always carries one undistorted state into another. However, the isotropy groups corresponding to two different undistorted states of a solid are generally different from one another; in particular, rotating a solid, even from an undistorted state, generally brings it into a configuration having a different isotropy group.

For isotropic solids, simpler statements may be made. In the two preceding paragraphs, the term "undistorted state" has been used in the sense of its definition for solids, given at the beginning of this section. Namely, $\boldsymbol{K}$ is an undistorted state if and only if $g_{\boldsymbol{K}} \subset o$. The solid is isotropic in the sense defined in Sect. 31 if and only if there exists a $\hat{\boldsymbol{K}}$ such that $g_{\hat{\boldsymbol{K}}} \supset o$. Now the orthogonal group is maximal in the unimodular group[2]. Equivalently, *every isotropic simple material is either a fluid or a solid.* In particular, for an isotropic simple solid $g_{\hat{\boldsymbol{K}}}=o$. Consequently, $\hat{\boldsymbol{K}}$ is an undistorted state in *both* senses of that term. Since the only orthogonal conjugate of o is o itself, $g_{\boldsymbol{K}}=o$, and thus $\boldsymbol{K}$, too, is an undistorted state in both senses of the term. Therefore, the two definitions of

---

[1] This theorem is a corollary to Theorem 2, p. 77 and Theorem 3, p. 157 of HALMOS [1958, *22*].

[2] That is, $o \subset g \subset u$ if and only if $g=o$ or $g=u$. Elementary proofs of this theorem have been given by BRAUER [1965, *4*] and NOLL [1965, *26*].

"undistorted state" coalesce in case both are applicable, and *for an isotropic simple solid, the isotropy group corresponding to any undistorted state is the full orthogonal group.*

Consider now an isotropic solid. No subspace of the ordinary vector space is invariant under all transformations in $\mathscr{o}$ other than the full vector space itself. In this case, the right stretch tensor $\boldsymbol{U}$ must reduce to a scalar multiple of the identity $\mathbf{1}$. COLEMAN and NOLL's theorem above, therefore, has the following corollary: *Any undistorted state of an isotropic simple solid may be obtained from any other by a conformal deformation* $\boldsymbol{P}=\alpha\boldsymbol{R}$, *and every conformal deformation carries an undistorted state into another.*

For the various types of aeolotropy, the conditions listed in the table below are easily seen to be necessary and sufficient to insure that $\boldsymbol{U}$ be the right stretch tensor of a deformation carrying a given undistorted state into another. The vectors $\boldsymbol{i}, \boldsymbol{j}, \boldsymbol{k}$ are the same as those employed in the table listing the generators of the isotropy groups. The numbers in the left-hand column of the table below refer to the enumeration of isotropy groups in the earlier table.

Type of aeolotropy	Restrictions on $\boldsymbol{U}$
Triclinic system (1)	no restriction
Monoclinic system (2)	$\boldsymbol{k}$ is a proper vector of $\boldsymbol{U}$
Rhombic system (3)	$\boldsymbol{i}, \boldsymbol{j}, \boldsymbol{k}$ are proper vectors of $\boldsymbol{U}$
Tetragonal system (4, 5) Hexagonal system (8, 9, 10, 11) Transverse isotropy	$\boldsymbol{U}=\alpha\mathbf{1}+\beta\boldsymbol{k}\otimes\boldsymbol{k}$
Cubic system (6, 7)	$\boldsymbol{U}=\alpha\mathbf{1}$

Note that triclinic simple solids share one property with simple fluids: *Any configuration is an undistorted state, and the isotropy group is the same for every reference configuration.* While $\mathscr{g}=\mathscr{u}$ for simple fluids, $\mathscr{g}=\{\mathbf{1}, -\mathbf{1}\}$ for triclinic simple solids. These are the largest and smallest, respectively, of all possible isotropy groups. That is, simple fluids and triclinic simple solids *have no preferred configurations*, but for the opposite reasons, the fluids because they have the maximum possible material symmetry in every configuration, and the triclinic solids because they have no non-trivial material symmetries in any configuration. Moreover, according to a known theorem of group theory, the proper unimodular group $\mathscr{u}_+$ is a simple group. That is, if $\mathscr{g}\subset\mathscr{u}_+$ and if $\boldsymbol{H}\mathscr{g}\boldsymbol{H}^{-1}=\mathscr{g}$ for every $\boldsymbol{H}\in\mathscr{u}_+$, then either $\mathscr{g}=\{\mathbf{1}\}$ or $\mathscr{g}=\mathscr{u}_+$. By NOLL's theorem (31.6), we may interpret this property of $\mathscr{u}_+$ in terms of the theory of simple materials as follows: *The only simple materials having no preferred configurations are fluids and triclinic solids.*

No particular simplification of the various forms of the constitutive equation given in Sect. 29 has been shown to result for simple solids[1]. The general constitutive equation for *isotropic* simple solids may be written in any of the forms (29.3), (29.6), (29.8), (29.10), (29.11), (29.28), (29.30), (31.9), (31.10), (31.13), or (31.14). The response functionals involved must satisfy relations of the type (31.7) or (31.8).

[1] Reduced forms appropriate to invariance under some or all of the orthogonal subgroups listed here have been obtained for tensor-valued or vector-valued functions or functionals of various kinds of variables by SMITH and RIVLIN [1957, *23*], PIPKIN and RIVLIN [1958, *35*, §§ 6—10] [1959, *21*, §§ 5—9], RIVLIN [1960, *46*, §§ 4—5] [1961, *51*, §§ 4—5], and SMITH, SMITH, and RIVLIN [1963, *66*] [1964, *76*].

The definition of a simple solid may be rephrased by saying that the isotropy group, relative to an undistorted state as reference, should have no element which is not orthogonal. That is, non-trivial strain always results in a different material response.

It is clear that simple solids and simple fluids are mutually exclusive[1]. However, they do not exhaust the possibilities for simple materials. There exist simple materials that are neither solids nor fluids in the sense in which these terms are used here. For such materials, the isotropy group is a proper subgroup of the unimodular group which is not conjugate to any subgroup of the orthogonal group. As shown above, such materials cannot be isotropic.

33 bis. Simple subfluids (added in proof). While of course unimodular sub-groups not comparable to the orthogonal group have long been known, the first specific study of simple materials that are neither fluids not solids has been made by C.-C. WANG[2]. His work aims at explaining the phenomena observed in liquid crystals. (A different class of theories for them was proposed earlier by ERICKSEN and is summarized in Sects. 127—129.) These materials, typically, are able to flow, at least in certain preferred directions, although they may support non-vanishing shear stress on certain planes when in equilibrium. Thus they have some properties in common with fluids and others with solids.

Specifically, WANG defines a *simple subfluid* as a *simple material whose isotropy group contains a dilatation group*. In his terms, *a dilatation group* is the group of all unimodular tensors having three fixed linearly independent vectors as their proper vectors. The directions of these fixed vectors are called the *axes* of the dilatation group. While this definition seems to depend upon choice of the reference configuration, in fact it is easily shown not to do so. Clearly, *a subfluid is never a solid*, and WANG showed that *a subfluid is a fluid if and only if it is isotropic*. More generally, WANG proved[3] that *there are exactly fourteen distinct types of subfluids*.

If $\boldsymbol{H}\in g$, let $H_i^{\,j}$ be its components relative to a basis of unit vectors along the axes of a dilatation group contained, by hypothesis, in g. Then, according to WANG's results, for given g the matrices $\|H_i^{\,j}\|$ of these components must be of some one of the following fourteen types, where $a, b, c, \ldots, i$ are arbitrary real numbers, subject only to the requirement that $|\det\|H_i^{\,j}\||=1$.

From the table it is easily verified that for a subfluid, two local configurations with the same density are isomorphic if and only if the corresponding isotropy groups are the same. Therefore the constitutive equation of a subfluid reduces to

$$\boldsymbol{T}(t)=\underset{s=0}{\overset{\infty}{\mathfrak{G}}}\left(\boldsymbol{C}_{(t)}(t-s),\,\varrho(t),\,\boldsymbol{e}_1(t),\,\boldsymbol{e}_2(t),\,\boldsymbol{e}_3(t)\right) \tag{33b.1}$$

where the $\boldsymbol{e}_i(t)$ are unit vectors along the axes of a dilatation group contained in the isotropy group of the present configuration. From the principle of material

[1] Note that simple solids and fluids are not defined by particular kinds of functional dependence but rather by properties of invariance. For example, consider a functional relation $\boldsymbol{T}=\mathfrak{f}(\boldsymbol{F})$. For some functions $\mathfrak{f}$, this constitutive equation defines a kind of simple solid; however, in the special case when $\mathfrak{f}(\boldsymbol{F})=\hat{\mathfrak{f}}(\det\boldsymbol{F})$, it defines a kind of simple fluid. Thus our terminology departs from that of the older works on hydrodynamics, in which the "elastic fluid" (piezotropic fluid) was regarded as a special case of the elastic solid. [Cf. CFT, Sect 218.] In our terminology, elastic materials include both elastic solids and elastic fluids (cf. Sect. 50).

[2] WANG [1965, *42*].

[3] The proof rests upon a theorem of group theory, apparently new, found by WANG: Any two dilatation groups with independent axes generate the unimodular group.

indifference, the functional $\mathfrak{G}$ is isotropic. By CAUCHY's theorem in Sect. 11, Eq. (33b.1) has the following representation:

$$\boldsymbol{T}(t)=\underset{s=0}{\overset{\infty}{\mathfrak{G}^{ij}}}\left(C_{(t)}^{kl}(t-s),\ \varrho(t),\ g_{12}(t),\ g_{23}(t),\ g_{31}(t)\right)\boldsymbol{e}_i(t)\otimes\boldsymbol{e}_j(t). \tag{33b.2}$$

Since the axes generally fail to be unique, for each type of subfluid WANG has selected a convenient particular set of axes such as to minimize the numbers of arguments in the functionals $\mathfrak{G}^{ij}$.

Reference No.	$\|H_i^j\|\cdot(\lvert\det\|H_i^j\|\rvert=1)$	Reference No.	$\|H_i^j\|\cdot(\lvert\det\|H_i^j\|\rvert=1)$
1 (Simple fluid)	$\begin{Vmatrix} a & b & c \\ d & e & f \\ g & h & i \end{Vmatrix}$	8	$\begin{Vmatrix} a & 0 & b \\ 0 & c & d \\ 0 & 0 & e \end{Vmatrix}$ or $\begin{Vmatrix} 0 & a & b \\ c & 0 & d \\ 0 & 0 & e \end{Vmatrix}$
2	$\begin{Vmatrix} a & b & 0 \\ c & d & 0 \\ e & f & g \end{Vmatrix}$	9	$\begin{Vmatrix} a & 0 & b \\ 0 & c & d \\ 0 & 0 & e \end{Vmatrix}$
3	$\begin{Vmatrix} a & b & c \\ d & e & f \\ 0 & 0 & g \end{Vmatrix}$	10	$\begin{Vmatrix} a & 0 & b \\ 0 & c & 0 \\ 0 & 0 & d \end{Vmatrix}$
4	$\begin{Vmatrix} a & b & c \\ 0 & d & e \\ 0 & 0 & f \end{Vmatrix}$	11	Generated by the union of Types 12—14.
5	$\begin{Vmatrix} a & b & c \\ 0 & d & 0 \\ 0 & 0 & e \end{Vmatrix}$ or $\begin{Vmatrix} a & b & c \\ 0 & 0 & d \\ 0 & e & 0 \end{Vmatrix}$	12	$\begin{Vmatrix} a & 0 & 0 \\ 0 & b & 0 \\ 0 & 0 & c \end{Vmatrix}$ or $\begin{Vmatrix} 0 & a & 0 \\ 0 & 0 & b \\ c & 0 & 0 \end{Vmatrix}$ or $\begin{Vmatrix} 0 & 0 & a \\ b & 0 & 0 \\ 0 & c & 0 \end{Vmatrix}$
6	$\begin{Vmatrix} a & b & c \\ 0 & d & 0 \\ 0 & 0 & e \end{Vmatrix}$	13	$\begin{Vmatrix} a & 0 & 0 \\ 0 & b & 0 \\ 0 & 0 & c \end{Vmatrix}$ or $\begin{Vmatrix} 0 & a & 0 \\ b & 0 & 0 \\ 0 & 0 & c \end{Vmatrix}$
7	$\begin{Vmatrix} a & b & 0 \\ c & d & 0 \\ 0 & 0 & e \end{Vmatrix}$	14	$\begin{Vmatrix} a & 0 & 0 \\ 0 & b & 0 \\ 0 & 0 & c \end{Vmatrix}$

From (33b.2), we see that the static stress in a subfluid is of the form

$$\boldsymbol{T}(t)=\mathfrak{g}^{ij}\left(\varrho(t),\ g_{12}(t),\ g_{23}(t),\ g_{31}(t)\right)\boldsymbol{e}_i(t)\otimes\boldsymbol{e}_j(t). \tag{33b.3}$$

Since the constitutive equation is indifferent to reflections in the directions of the $\boldsymbol{e}_i$, we easily verify that

$$\mathfrak{g}^{ij}(\varrho(t),0,0,0)=0 \quad \text{if} \quad i\neq j. \tag{33b.4}$$

WANG defines the *undistorted states* of a subfluid by the conditions

$$g_{ij}=0 \quad \text{if} \quad i\neq j, \tag{33b.5}$$

i.e., as those configurations for which the $\boldsymbol{e}_i$ form an orthonormal basis. Every subfluid has undistorted states, and WANG determines for each type the class of deformations that carry one undistorted state into another. As for simple solids, this class always includes all rotations. From (33b.4), the static stress (33b.3) in an undistorted state has the representation

$$\boldsymbol{T}=\alpha\boldsymbol{e}_1\otimes\boldsymbol{e}_1+\beta\boldsymbol{e}_2\otimes\boldsymbol{e}_2+\gamma\boldsymbol{e}_3\otimes\boldsymbol{e}_3, \tag{33b.6}$$

where α, β, γ are functions of the density. Additional restrictions on α, β, γ for each type of subfluid can be found from the table above. The results are as follows:

Reference No.	Representation of the static stress in an undistorted state
1, 11, 12	$\boldsymbol{T} = \alpha(\mathbf{e}_1 \otimes \mathbf{e}_1 + \mathbf{e}_2 \otimes \mathbf{e}_2 + \mathbf{e}_3 \otimes \mathbf{e}_3) = \alpha \mathbf{1}$
2, 3, 7, 8, 13	$\boldsymbol{T} = \alpha(\mathbf{e}_1 \otimes \mathbf{e}_1 + \mathbf{e}_2 \otimes \mathbf{e}_2) + \gamma \mathbf{e}_3 \otimes \mathbf{e}_3 = \alpha \mathbf{1} + \gamma^* \mathbf{e}_3 \otimes \mathbf{e}_3$
4, 6, 9, 10, 14	$\boldsymbol{T} = \alpha \mathbf{e}_1 \otimes \mathbf{e}_1 + \beta \mathbf{e}_2 \otimes \mathbf{e}_2 + \gamma \mathbf{e}_3 \otimes \mathbf{e}_3$
5	$\boldsymbol{T} = \alpha \mathbf{e}_1 \otimes \mathbf{e}_1 + \beta(\mathbf{e}_2 \otimes \mathbf{e}_2 + \mathbf{e}_3 \otimes \mathbf{e}_3) = \alpha^* \mathbf{e}_1 \otimes \mathbf{e}_1 + \beta \mathbf{1}$

Thus all simple subfluids share with simple fluids the property of *supporting no shear stress* when in equilibrium in an undistorted state, but generally not all configurations are undistorted. That is, a subfluid can flow in any direction from rest in an undistorted state, but in other configurations it can generally withstand certain shear stresses. Types 1, 11, and 12, and only these, share with the simple fluid the property of *being unable to support a difference of normal pressures* when in equilibrium in an undistorted state, but again, not all configurations are undistorted. It is easy to verify that subfluids of Type 1—4 share with the simple fluid the property that *every configuration is undistorted*, but in general for these types unequal normal pressures are required to maintain equilibrium, even in an undistorted state. Thus the definition of "undistorted state" by WANG for simple subfluids is consistent with those introduced in Sects. 31 and 33 for isotropic materials and for simple solids. Subfluids are materials in some ways similar to solids, in other ways to fluids.

In work done before WANG's but made available to us only later, COLEMAN[1] has proposed the name *simple liquid crystal* for simple materials that are neither fluids nor solids. He has singled out two special classes of isotropy groups for further study. First, he shows that if g is the group of all unimodular tensors that preserve or reverse all vectors in a *one-dimensional or two-dimensional subspace*, the material is a simple liquid crystal. Groups of this type are not included in WANG's theory. Second, if g is the group of all unimodular tensors *mapping a one-dimensional or two-dimensional subspace onto itself*, COLEMAN proves the material to be a simple liquid crystal. Groups of this type are included in WANG's theory also (Nos. 2 and 3 in the table above). COLEMAN obtains several representations of the stress tensor and characterizations, and he compares his results with some of those obtained on the basis of a different theory by ERICKSEN (Sects. 127 to 129).

34. Material connections, inhomogeneity, curvilinear aeolotropy, continuous dislocations[2]. Let $\mathscr{B}$ be a materially uniform simple body, i.e. a materially uniform

[1] COLEMAN [1965, *6*].

[2] NOLL [1965, *27*]. NOLL's analysis generalizes and gives a rigorous foundation for the continuum theory of dislocations proposed and developed in various forms by KONDO [1949, *7*] [1950, *6—9*] [1951, *5*] [1952, *13*] [1954, *12*] [1955, *16*] [1958, *27*] [1962, *43*], NYE [1953, *21*], BILBY, BULLOUGH and collaborators [1955, *5* and *6*] [1956, *2*] [1958, *8—10*], SEEGER [1956, *26*], GÜNTHER [1958, *21*], KRÖNER and SEEGER [1959, *17*], KRÖNER [1960, *35*] [1961, *34*], and AMARI [1962, *2*]. Expositions have been given by KRÖNER [1958, *28*], BILBY [1960, *13*] and SEEGER [1961, *54*]. These authors either lay down *a priori* some geometric structure on the basis of considerations of lattice defects in crystals or else, and often also, they employ some very special constitutive equation, usually a linear approximation to some standard equation of elasticity or plasticity. NOLL's theory, on the other hand, is free from *ad hoc* assumptions. As will be seen from the outline given above in the text, he shows that once a con-

body consisting of a simple material (cf. Sect. 27 and 28). Then for each particle $X \in \mathscr{B}$ there is a local reference configuration $\boldsymbol{K}(X)$ such that the response functional $\mathfrak{G} = \mathfrak{G}_{\boldsymbol{K}(X)}$ corresponding to $\boldsymbol{K}(X)$ does not depend on X (cf. Sect. 28). Such a distribution $\boldsymbol{K}(X)$ of local reference configurations will be called a *uniform reference* for the body.

Consider a particular particle X. Recalling that $\boldsymbol{K}(X)$ is an equivalence class of configurations (cf. Sect. 22), let $\boldsymbol{\varkappa}$ be a configuration of a neighborhood of X belonging to the class $\boldsymbol{K}(X)$. The vectors[1]

$$\boldsymbol{d}_\alpha(X) = \partial_\alpha \boldsymbol{\varkappa}(X), \tag{34.1}$$

i.e. the derivatives of $\boldsymbol{\varkappa}$ at X taken with respect to the material co-ordinates X^α, clearly depend only on the class $\boldsymbol{K}(X)$ to which $\boldsymbol{\varkappa}$ belongs. Moreover, they form a basis for the space of ordinary vectors. We assume that $\boldsymbol{d}_\alpha(X) = \boldsymbol{d}_\alpha(X^\lambda)$ depends smoothly on the material co-ordinates X^λ. Thus, given a uniform reference $\boldsymbol{K}(X)$ and a system of material co-ordinates X^α, we can define three vector-valued functions $\boldsymbol{d}_\alpha = \boldsymbol{d}_\alpha(X)$ on $\mathscr{B}$ whose values, for each particle X, form a vector basis. Let $\boldsymbol{u} = \boldsymbol{u}(X)$ be any vector-valued function defined on $\mathscr{B}$. Let $u^\alpha = u^\alpha(X)$ be the components of $\boldsymbol{u}(X)$ with respect to the basis $\boldsymbol{d}_\alpha(X)$, so that

$$\boldsymbol{u}(X) = u^\alpha(X)\, \boldsymbol{d}_\alpha(X). \tag{34.2}$$

From the definition (34.1) of the $\boldsymbol{d}_\alpha$ it follows easily that the u^α obey the transformation law for components of a contravariant vector field on the body-manifold $\mathscr{B}$. Conversely, if the u^α are the components of a contravariant vector field on $\mathscr{B}$, then the vector function $\boldsymbol{u}(X)$ on $\mathscr{B}$ defined by (34.2) can be shown to be independent of the choice of the co-ordinate system X^α. Thus, when a uniform reference $\boldsymbol{K}(X)$ is given, (34.2) establishes a one-to-one correspondence between contravariant vector fields on $\mathscr{B}$ and functions on $\mathscr{B}$ whose values are ordinary spatial vectors.

A uniform reference $\boldsymbol{K}(X)$ of $\mathscr{B}$ determines two geometric structures on $\mathscr{B}$, as follows:

(i) The functions $g_{\alpha\beta} = g_{\alpha\beta}(X)$ given by

$$g_{\alpha\beta} = \boldsymbol{d}_\alpha \cdot \boldsymbol{d}_\beta \tag{34.3}$$

are the components of a positive-definite symmetric covariant tensor field on $\mathscr{B}$ and hence define a *Riemannian metric* $\boldsymbol{g}$ on $\mathscr{B}$.

(ii) The functions $\Gamma_{\mu\lambda}{}^\nu = \Gamma_{\mu\lambda}{}^\nu(X)$, determined by

$$\partial_\mu \boldsymbol{d}_\lambda = \Gamma_{\mu\lambda}{}^\nu \boldsymbol{d}_\nu, \tag{34.4}$$

define a *linear connection* $\boldsymbol{\Gamma}$ on $\mathscr{B}$. Indeed, it is easily verified that the $g_{\alpha\beta}$ and the $\Gamma_{\mu\lambda}{}^\nu$ obey the transformation laws of a covariant tensor field and of a linear connection, respectively.

stitutive equation such as to define a materially uniform simple body is laid down, the geometric structure in the body is determined. Thus the geometric structure is the *outcome*, not the first assumption, of the physical theory.

In one respect, however, NOLL's theory is less general than some of those above referenced, since couple-stresses are excluded from the start. That couple-stresses are really needed for an adequate description of the behavior of real solids has not yet been shown, in our opinion. In any case, a treatment of the polar-elastic media of TOUPIN (Sect. 98) along the lines of NOLL's theory of simple materials can surely be constructed, though as yet it has not been, and such a treatment must yield the geometry of the body manifold from the forms of the constitutive equations of the particles. The geometrical structure resulting must contain as a special case the one developed in our text above.

[1] In contrast to (21.3), where $\boldsymbol{\varkappa}$ is a global configuration independent of X, the configuration $\boldsymbol{\varkappa}$ to be used in (34.1) must belong to the class $\boldsymbol{K}(X)$ and hence in general varies with X.

Differentiation of (34.2) with respect to X^μ and use of (34.4) yields

$$\partial_\mu \boldsymbol{u} = (\partial_\mu u^\alpha)\boldsymbol{d}_\alpha + u^\alpha \partial_\mu \boldsymbol{d}_\alpha = (\nabla_\mu u^\alpha)\boldsymbol{d}_\alpha, \tag{34.5}$$

where the quantities $\nabla_\mu u^\alpha$, given by

$$\nabla_\mu u^\alpha = \partial_\mu u^\alpha + u^\sigma \Gamma_{\mu\sigma}{}^\alpha, \tag{34.6}$$

are the covariant derivatives of the u^α with respect to the connection $\boldsymbol{\Gamma}$. A corollary of (34.5) is the following result: *A vector-valued function* $\boldsymbol{u}$ *on* $\mathscr{B}$ *is constant if and only if the contravariant vector field on* $\mathscr{B}$ *corresponding to* $\boldsymbol{u}$ *is covariantly constant with respect to the material connection* $\boldsymbol{\Gamma}$.

It can be shown that $\boldsymbol{\Gamma}$ defines a distant parallelism, i.e. that the parallel displacement corresponding to $\boldsymbol{\Gamma}$ is independent of path. It follows from this fact that the curvature tensor $\boldsymbol{R}$ based on the material connection $\boldsymbol{\Gamma}$ is zero:

$$\tfrac{1}{2} R_{\nu\mu\lambda}{}^\varkappa = \partial_{[\nu} \Gamma_{\mu]\lambda}{}^\varkappa + \Gamma_{[\nu|\varrho|}{}^\varkappa \Gamma_{\mu]\lambda}{}^\varrho = 0. \tag{34.7}$$

If $\boldsymbol{\Gamma}$ is a symmetric connection, i.e. if $\Gamma_{[\nu\mu]}{}^\lambda = 0$, it follows from (34.7) that $\boldsymbol{\Gamma}$ is in fact a *flat* connection, i.e. that the co-ordinates can be so chosen that $\Gamma_{\nu\mu}{}^\lambda \equiv 0$. In general, several co-ordinate patches will be necessary to cover the whole body if $\Gamma_{\nu\mu}{}^\lambda$ is to vanish. If the body is simply connected, however, a single system suffices. For such a choice of co-ordinates, by (34.4), we then have $\boldsymbol{d}_\alpha = \text{const}$. Defining a global configuration $\boldsymbol{\varkappa}$ by $\boldsymbol{\varkappa}(X) = \boldsymbol{c} + X^\alpha \boldsymbol{d}_\alpha$, where $\boldsymbol{c}$ is an arbitrary point, we have $\partial_\alpha \boldsymbol{\varkappa}(X) = \boldsymbol{d}_\alpha$. Therefore, by (34.1), $\boldsymbol{\varkappa}$ belongs to the equivalence class $\boldsymbol{K}(X)$ for *every* $X \in \mathscr{B}$; i.e. in the notation of (22.14), the condition $\boldsymbol{K}(X) = \nabla\boldsymbol{\varkappa}(X)$ holds for every $X \in \mathscr{B}$. We conclude (cf. Sect. 28), that $\mathfrak{G}_{\boldsymbol{\varkappa}} = \mathfrak{G}_{\boldsymbol{K}(X)}$ and hence that the body is homogeneous. Conversely, if the body is homogeneous and if $\boldsymbol{\varkappa}$ is one of its homogeneous configurations, then $\boldsymbol{K}(X) = \nabla\boldsymbol{\varkappa}(X)$ is a uniform reference that gives rise to a symmetric material connection. To summarize: *A simply connected materially uniform body is homogeneous if and only if it has a symmetric material connection.* This result motivates the following definition: The torsion tensor $\boldsymbol{S}$ of a material connection $\boldsymbol{\Gamma}$, with components

$$S_{\mu\lambda}{}^\gamma = \Gamma_{[\mu\lambda]}{}^\gamma, \tag{34.8}$$

is called the *inhomogeneity* of the connection $\boldsymbol{\Gamma}$ or of the uniform reference from which $\boldsymbol{\Gamma}$ is defined. The inhomogeneity is zero if not only the body, but also the uniform reference used, is homogeneous.

In order to relate the connection $\boldsymbol{\Gamma}$ defined by (34.4) to the metric (34.3) we differentiate (34.3) and obtain

$$\partial_\gamma g_{\alpha\beta} = (\partial_\gamma \boldsymbol{d}_\alpha) \cdot \boldsymbol{d}_\beta + \boldsymbol{d}_\alpha \cdot (\partial_\gamma \boldsymbol{d}_\beta). \tag{34.9}$$

Substitution of (34.4) into (34.9) gives

$$\partial_\gamma g_{\alpha\beta} = \Gamma_{\gamma\beta\alpha} + \Gamma_{\gamma\alpha\beta}, \tag{34.10}$$

where $\Gamma_{\gamma\alpha\beta} = g_{\beta\sigma}\Gamma_{\gamma\alpha}{}^\sigma$. (34.10) states that the covariant derivatives of $g_{\alpha\beta}$ are zero: $\nabla_\gamma g_{\alpha\beta} = 0$. Raising and lowering indices by means of the $g_{\alpha\beta}$ in the usual manner, we find that the equations (34.8) and (34.10) can be solved for $\Gamma_{\alpha\beta\gamma}$, with the result that

$$\Gamma_{\alpha\beta\gamma} = \{\alpha\gamma\beta\} + S_{\alpha\beta\gamma} + 2S_{\gamma(\alpha\beta)}, \tag{34.11}$$

where $\{\alpha\gamma\beta\} = \tfrac{1}{2}(\partial_\alpha g_{\gamma\beta} + \partial_\beta g_{\gamma\alpha} - \partial_\gamma g_{\alpha\beta})$ are the Christoffel symbols based on the Riemannian metric $g_{\alpha\beta}$.

The curvature tensor $\overset{*}{\boldsymbol{R}}$ of the Riemannian metric $\boldsymbol{g}$, in contrast to the curvature tensor $\boldsymbol{R}$ of the material connection $\boldsymbol{\Gamma}$, is in general not zero. Consider the special case when $\overset{*}{\boldsymbol{R}}$ is zero. Then the metric $\boldsymbol{g}$ is Euclidean, and we can choose a co-ordinate system X^α such that

$$g_{\alpha\beta}(X) = \boldsymbol{d}_\alpha(X)\cdot\boldsymbol{d}_\beta(X) = \delta_{\alpha\beta}. \tag{34.12}$$

If the body is simply connected, this co-ordinate system will cover all of it. Hence, for such a choice of co-ordinates, the $\boldsymbol{d}_\alpha(X)$ form an *orthonormal basis* for each $X \in \mathscr{B}$. Assume now that the body is isotropic and that the local reference configurations $\boldsymbol{K}(X)$ are undistorted (cf. Sect. 31). Choose a fixed particle X_0, and define a global reference configuration γ by

$$\gamma(X) = \boldsymbol{c} + X^\alpha \boldsymbol{d}_\alpha(X_0). \tag{34.13}$$

Denoting the vectors (34.1) corresponding to the configuration γ by $\hat{\boldsymbol{d}}_\alpha(X) = \partial_\alpha \gamma(X)$, we find that $\hat{\boldsymbol{d}}_\alpha(X) = \boldsymbol{d}_\alpha(X_0)$. Hence, by (34.12) and (34.3), the metric $\hat{\boldsymbol{g}}$ corresponding to the global configuration γ is the same as the metric $\boldsymbol{g}$ corresponding to the uniform reference $\boldsymbol{K} = \boldsymbol{K}(X)$. But, by (31.12), the components of the response functional $\mathfrak{J}$ of an isotropic material depend only on the metric $\boldsymbol{g}$. Therefore, the functional $\mathfrak{J}$ of (29.6) is not altered when the uniform reference $\boldsymbol{K}(X)$ is replaced by the global reference configuration γ. We have obtained the following result: *If $\boldsymbol{K} = \boldsymbol{K}(X)$ is an undistorted uniform reference in a simply connected isotropic body and if the curvature $\overset{*}{\boldsymbol{R}}$ of the Riemannian metric corresponding to $\boldsymbol{K}$ vanishes, then the body is homogeneous.*

An aeolotropic body can be inhomogeneous even when $\overset{*}{\boldsymbol{R}}$ vanishes. A special type of such inhomogeneous aeolotropy is *curvilinear aeolotropy*[1], defined by the following condition: It is possible to choose a *global* reference configuration $\boldsymbol{\varkappa}$ and a system of orthogonal curvilinear co-ordinates $\hat{X}^\alpha$ in space such that the $\boldsymbol{d}_\alpha(X)$, for each $\boldsymbol{X} = \boldsymbol{\varkappa}(X)$ are the unit direction vectors of the co-ordinate lines through $\boldsymbol{X} = \boldsymbol{\varkappa}(X)$. In this case the unit vectors $\boldsymbol{d}_\alpha$ are proportional to the vectors of the natural basis of the system $\hat{X}^\alpha$, and components of vectors and tensors relative to $\boldsymbol{d}_\alpha$ are identical with their *physical* components in the orthogonal curvilinear system $\hat{X}^\alpha$. Thus, in the case of curvilinear aeolotropy, the constitutive equation (29.9) may be written

$$\overline{T}^{\langle\alpha\beta\rangle}(t) = \mathop{\mathfrak{J}^{\langle\alpha\beta\rangle}}_{s=0}^{\infty}\left(C^{\langle\mu\nu\rangle}(t-s)\right), \tag{34.14}$$

where the forms of the component functionals $\mathfrak{J}^{\langle\alpha\beta\rangle}$ do not depend on the particle X, and where $\overline{T}^{\langle\alpha\beta\rangle}$ and $C^{\langle\mu\nu\rangle}$ are physical components with respect to the special orthogonal co-ordinate system $\hat{X}^\alpha$.

We consider now the general case, when the Riemannian metric $\boldsymbol{g}$ is not necessarily Euclidean, i.e. when $\overset{*}{\boldsymbol{R}}$ does not necessarily vanish. The following identity can be obtained from (34.7) and (34.10) by straightforward calculation:

$$\tfrac{1}{2}\overset{*}{R}_{\nu\mu\lambda\varkappa} + V_{[\nu} T_{\mu]\lambda\varkappa} + T_{[\nu|\varkappa|\varrho} T_{\mu]\lambda}{}^{\varrho} + S_{\nu\mu}{}^{\varrho} T_{\varrho\lambda\varkappa} = 0 \tag{34.15}$$

where

$$T_{\mu\lambda\varkappa} = S_{\mu\lambda\varkappa} + 2 S_{\varkappa(\mu\lambda)}. \tag{34.16}$$

It is easily seen that each of the terms in (34.15) is alternating in both ν, μ and $\lambda, \varkappa$. Therefore, (34.15) is equivalent to its contraction,

$$\tfrac{1}{2}\overset{*}{R}_{\sigma\mu\lambda}{}^{\sigma} + V_{[\sigma} T_{\mu]\lambda}{}^{\sigma} + T_{[\sigma}{}^{\sigma}{}_{|\varrho|} T_{\mu]\lambda}{}^{\varrho} + S_{\sigma\mu}{}^{\varrho} T_{\varrho\lambda}{}^{\sigma} = 0. \tag{34.17}$$

---

[1] In the context of hyperelasticity, curvilinear aeolotropy of finitely deformable bodies was first discussed by ADKINS [1955, *1*, § 3]. Cf. also GREEN and ADKINS [1960, *26*, § 1.16].

After some calculation one finds that (34.17) is equivalent to the following two identities:

$$\nabla_\sigma S_{\mu\lambda}{}^{\sigma} + 2\nabla_{[\mu} S_{\lambda]\varrho}{}^{\varrho} + 2 S_{\mu\lambda}{}^{\sigma} S_{\sigma\varrho}{}^{\varrho} = 0, \tag{34.18}$$

$$\left.\begin{aligned}\overset{*}{R}_{\sigma\mu\lambda}{}^{\sigma} = &- 2\nabla_\sigma S^{\sigma}{}_{(\mu\lambda)} + 2\nabla_{(\mu} S_{\lambda)\sigma}{}^{\sigma} + \\ &+ 4 S_{\varrho\sigma}{}^{\sigma} S^{\varrho}{}_{(\mu\lambda)} + (2 S_{\mu}{}^{(\sigma\varrho)} + S^{\sigma\varrho}{}_{\mu})(2 S_{\lambda(\sigma\varrho)} + S_{\sigma\varrho\lambda}).\end{aligned}\right\} \tag{34.19}$$

Eq. (34.18) is a differential identity that every possible inhomogeneity tensor $\boldsymbol{S}$ must satisfy. Eq. (34.19) shows that the curvature $\overset{*}{\boldsymbol{R}}$ of the Riemannian metric $\boldsymbol{g}$ can be expressed in terms of the inhomogeneity tensor $\boldsymbol{S}$ and its covariant derivatives with respect to the material connection $\boldsymbol{\Gamma}$.

Let $\boldsymbol{e}$ be an alternating third-order tensor field on $\mathscr{B}$ with the property

$$e_{\nu\mu\lambda} e^{\nu\mu\lambda} = 3! = 6. \tag{34.20}$$

This property determines $\boldsymbol{e}$ to within sign. The tensor field $\boldsymbol{A}$ on $\mathscr{B}$ with components

$$A^{\alpha\beta} = - e^{\alpha\lambda\mu} S_{\lambda\mu}{}^{\beta}, \qquad S_{\lambda\mu}{}^{\varkappa} = -\tfrac{1}{2} e_{\sigma\mu\lambda} A^{\sigma\varkappa} \tag{34.21}$$

is called the *dislocation density* for the uniform reference $\boldsymbol{K}(X)$ from which $\boldsymbol{g}$ and $\boldsymbol{S}$ are obtained. The term "dislocation density" stems from the interpretation that can be given to $\boldsymbol{A}$ in the theory of lattice defects in crystals. Substituting $(34.21)_2$ into (34.18) yields the following relation satisfied by the dislocation density:

$$\nabla_\sigma A^{\sigma\alpha} = A^{\alpha\sigma} a_\sigma, \qquad a_\sigma = - e_{\sigma\mu\lambda} A^{\mu\lambda} = -2 S_{\sigma\lambda}{}^{\lambda}, \tag{34.22}$$

which states that the divergence of $\boldsymbol{A}^T$ with respect to the material connection is equal to the result of operating with $\boldsymbol{A}$ on the "axial" vector $\boldsymbol{a}$ of $\boldsymbol{A}$. (34.22) is a differential identity for the dislocation density $\boldsymbol{A}$. It is *not* a differential equation for the determination of $\boldsymbol{A}$, because the covariant differential operator ∇ and $\boldsymbol{A}$ are not independent.

IV. Special classes of materials.

The older definitions of various kinds of materials rest upon special functional dependences rather than properties of invariance. Originally these concepts of material response were fashioned by "superposition" of elastic and viscous effects, but it was soon seen that there are infinitely many possible combinations[1], and since there is no physical reason to prefer one to another, the whole idea now

[1] The simplest "superpositions" lead to the visco-elastic theories of MEYER [1874, *2* and *3*] [1875, *2*] and MAXWELL [1867, *1*, pp. 30—31], respectively. The former, studied especially by VOIGT [1889, *4*] [1892, *2* and *3*], is generalized by our *material of the differential type*, defined in Sect. 35. The latter, along with various generalizations by BUTCHER [1876, *1*, pp. 103—111], PEARSON [1889, *1*], REYNOLDS [1901, *9*], NATANSON [1901, *5*—*8*], and ZAREMBA [1903, *8*—*12*], is included as a special case in our material *of the rate type*, defined in Sect. 36. Superposition of infinitely many "elastic elements" leads to the linear accumulative theory of BOLTZMANN [1874, *1*], generalized by our *material of the integral type*, defined in Sect. 37.

We have cited only the earliest representatives of the vast literature on these theories. Classifications according to the numbers and types of derivatives occurring, etc., have been frequent: e.g. v. MISES [1930, *2*], WEISSENBERG [1931, *7*], HOHENEMSER and PRAGER [1932, *3* and *4*], REINER [1943, *2*, Lect. VI]; there is also an enormous number of papers applying theories of this kind to plane infinitesimal ultrasonic waves. In all this work the principle of material frame-indifference is disregarded, so that only motion "small" in some sense is envisaged.

In the following sections we shall cite explicitly those few proposals that have led to properly invariant theories. We mention here, however, that some of these rest on an appeal to "simplicity" and hence fail to be invariant under change of strain measure or stress rate.

seems misleading. In this subchapter we give some rather general functional definitions, but we regard them as subservient to the invariant definitions given in the previous subchapter. While we shall apply the results already attained so as to reduce the special functional dependence to forms as explicit as possible, it is not always clear from these functional forms whether a material satisfying them is a fluid, a solid, or neither, or even whether, for materials of the rate type, any solutions really correspond to materials in the sense specified in Sects. 26 and 28.

The reader interested in modern continuum mechanics should turn at once to Sect. 38, using the intermediate text only as a lexicon.

35. Materials of the differential type. According to Sect. 29 the stress

$$\boldsymbol{T}(t) = \underset{s=0}{\overset{\infty}{\mathfrak{G}}}\left(\boldsymbol{F}^{(t)}(s)\right) \tag{35.1}$$

in a simple material is determined by the values of $\boldsymbol{F}^{(t)}(s)$ for all $s \geqq 0$, i.e., by the values of the deformation gradient $\boldsymbol{F}(\tau) = \boldsymbol{F}^{(t)}(t-\tau)$ for all times $\tau \leqq t$. It may happen that only a very short part of the history of the deformation gradient $\boldsymbol{F}$ has an influence on the stress i.e., that the value of $\mathfrak{G}$ depends only on the values of $\boldsymbol{F}^{(t)}(s)$ for s very near to zero. If the deformation gradient $\boldsymbol{F}$ has sufficiently many continuous time derivatives, then $\boldsymbol{F}^{(t)}(s)$ may be approximated, for s near zero, by its Taylor expansion up to some order $\mathfrak{r}$. This expansion is determined by the value of $\boldsymbol{F}^{(t)}$ and its derivatives up to the order $\mathfrak{r}$ at $s=0$, i.e., by

$$\boldsymbol{F}^{(t)}(0) = \boldsymbol{F}(t), \quad \dot{\boldsymbol{F}}^{(t)}(0) = \dot{\boldsymbol{F}}(t), \ \ldots, \quad \overset{(\mathfrak{r})}{\boldsymbol{F}^{(t)}}(0) = \overset{(\mathfrak{r})}{\boldsymbol{F}}(t). \tag{35.2}$$

A material in which the stress depends on only a finite number of these time derivatives is called a *material of the differential type*.

In order to find the constitutive equation for such materials, in reduced form so that the principle of material indifference is satisfied, we start from the general constitutive equation (29.19). By (24.12) the derivatives at $\tau = t$ of the relative right stretch tensor $\boldsymbol{U}_{(t)}(\tau)$, are the stretchings,

$$\boldsymbol{D}_{\mathfrak{k}}(t) = \frac{d^{\mathfrak{k}}}{d t^{\mathfrak{k}}} \boldsymbol{U}_{(t)}(\tau)\bigg|_{\tau=t}. \tag{35.3}$$

In a material of the differential type, the stress and hence the value of the constitutive functional $\underset{s=0}{\overset{\infty}{\mathfrak{K}}}\left(\boldsymbol{U}^*_{(t)}(t-s);\, \boldsymbol{U}(t)\right)$ of (29.19) can depend on $\boldsymbol{U}^*_{(t)}(t-s)$ only through its first $\mathfrak{r}$ derivatives with respect to s, evaluated at $s=0$, because $\boldsymbol{U}^*_{(t)}(t-s)$ depends on the data (35.2) only through these derivatives. Thus the functional $\mathfrak{K}$ must reduce to a function $\mathfrak{k}$ of these derivatives, which differ only by the factor $(-1)^{\mathfrak{k}}$ from the tensors

$$\boldsymbol{D}^*_{\mathfrak{k}}(t) = \boldsymbol{R}(t)^T \boldsymbol{D}_{\mathfrak{k}}(t)\, \boldsymbol{R}(t), \tag{35.4}$$

as is easily seen from (35.3) and (29.14). Recall that $\boldsymbol{R}$ is the rotation tensor defined in Sect. 23. Hence the *general constitutive equation of the differential type*[1] is

$$\boldsymbol{R}^T \boldsymbol{T} \boldsymbol{R} = \mathfrak{k}\left(\boldsymbol{D}^*_1, \boldsymbol{D}^*_2, \ldots, \boldsymbol{D}^*_{\mathfrak{r}};\, \boldsymbol{U}\right), \tag{35.5}$$

where $\mathfrak{k}$ is a function of $\mathfrak{r}+1$ symmetric tensors, the $\boldsymbol{D}^*_{\mathfrak{k}}$ and the right stretch tensor $\boldsymbol{U}$. $\mathfrak{r}$ is called the *complexity* of the material.

[1] Noll [1958, *31*, § 23].

The same reasoning, applied to (29.21), leads to the following *alternative form of the general constitutive equation for materials of the differential type*[1]:

$$\boldsymbol{R}^T \boldsymbol{T} \boldsymbol{R} = \mathfrak{j}(\boldsymbol{A}_1^*, \ldots, \boldsymbol{A}_{\mathfrak{r}}^*; \boldsymbol{C}). \tag{35.6}$$

Here $\boldsymbol{C}$ is the right Cauchy-Green tensor (23.4), and the $\boldsymbol{A}_{\mathfrak{k}}^*$ are defined in terms of the Rivlin-Ericksen tensors (24.14) by

$$\boldsymbol{A}_{\mathfrak{k}}^* = \boldsymbol{R}^T \boldsymbol{A}_{\mathfrak{k}} \boldsymbol{R}. \tag{35.7}$$

In the case when the material is isotropic the functions $\mathfrak{k}$ and $\mathfrak{j}$ of (35.5) and (35.6) must be isotropic tensor functions (cf. Sect. 8), as is evident from (31.8) and the fact that the function $\mathfrak{k}$ is obtained from the functional $\mathfrak{K}$ by specialization. From the definition (8.7) of an isotropic tensor function, it follows that $\boldsymbol{R}$ can be drawn into the arguments in (35.5) and (35.6) and hence that the *general constitutive equation for an isotropic material of the differential type* may be written also in either of the following two forms[1]:

$$\boldsymbol{T} = \mathfrak{k}(\boldsymbol{D}, \boldsymbol{D}_2, \ldots, \boldsymbol{D}_{\mathfrak{r}}; \boldsymbol{V}), \tag{35.8}$$

$$\boldsymbol{T} = \mathfrak{j}(\boldsymbol{A}_1, \boldsymbol{A}_2, \ldots, \boldsymbol{A}_{\mathfrak{r}}; \boldsymbol{B}). \tag{35.9}$$

Here $\boldsymbol{V} = \boldsymbol{R}\boldsymbol{U}\boldsymbol{R}^T$ is the left stretch tensor and $\boldsymbol{B} = \boldsymbol{R}\boldsymbol{C}\boldsymbol{R}^T$ the left Cauchy-Green tensor. Both $\mathfrak{k}$ and $\mathfrak{j}$ are isotropic tensor functions.

It follows from the theorem on components of isotropic functions stated in Sect. 8 that (35.9) has the co-ordinate form[2]

$$T_{km} = \mathfrak{j}_{km}(A_{1pq}, A_{2pq}, \ldots, A_{\mathfrak{r}pq}; B_{ru}; g_{vw}), \tag{35.10}$$

where the forms of the component functions $\mathfrak{j}_{km}$ are independent of the choice of co-ordinate system.

A material which is a simple fluid as defined in Sect. 32 and is also a material of the differential type as just defined will be called a *fluid of the differential type*. A *solid of the differential type* is defined analogously.

Starting from the constitutive equations (32.7) or (32.8) for simple fluids and applying the same reasoning as we used in the general case, we obtain the following two forms for the *constitutive equation of a fluid of the differential type*[2]:

$$\boldsymbol{T} = \mathfrak{h}(\boldsymbol{D}, \boldsymbol{D}_2, \ldots, \boldsymbol{D}_{\mathfrak{r}}; \varrho), \tag{35.11}$$

$$\boldsymbol{T} = -p(\varrho)\boldsymbol{1} + \mathfrak{q}(\boldsymbol{A}_1, \boldsymbol{A}_2, \ldots, \boldsymbol{A}_{\mathfrak{r}}; \varrho), \tag{35.12}$$

Here $\mathfrak{h}$ and $\mathfrak{q}$ are isotropic tensor functions of $\mathfrak{r}$ symmetric tensor variables and one scalar variable, the density ϱ. $\mathfrak{q}$ vanishes when all $\boldsymbol{A}_{\mathfrak{k}}$ are zero. Co-ordinate forms of (35.11) and (35.12) are

$$T_{km} = \mathfrak{h}_{km}(D_{pq}, D_{2pq}, \ldots, D_{\mathfrak{r}pq}; \varrho; g_{ru}), \tag{35.13}$$

$$T_{km} = -p(\varrho)\, g_{km} + \mathfrak{q}_{km}(A_{1pq}, A_{2pq}, \ldots, A_{\mathfrak{r}pq}; \varrho; g_{ru}), \tag{35.14}$$

where the forms of the functions $\mathfrak{h}_{km}$ and $\mathfrak{q}_{km}$ are independent of the choice of co-ordinate system.

Looking back at (35.8) and (35.9), we see that if those equations do not reduce to the forms (35.11) and (35.12), respectively, then the isotropy group is not the

[1] NOLL [1958, *31*, § 23].

[2] This form was first derived by RIVLIN and ERICKSEN [1955, *23*, § 15], who used a different method. Special but properly invariant theories of the differential type were proposed by KILCHEVSKI [1938, *3*, Ch. I, §§ 11—14] and YAMAMOTO [1959, *40*, §§ 13—15].

unimodular group. Hence (35.8) and (35.9), when they do not degenerate into (35.11) and (35.12), are constitutive equations of an isotropic simple solid.

We have given here only some of the reduced forms for materials of the differential type; as in the case of general simple materials, one could list many more.

36. Materials of the rate type. The general constitutive equation

$$\boldsymbol{T}(t)=\underset{s=0}{\overset{\infty}{\mathfrak{G}}}\left(\boldsymbol{F}(t-s)\right) \tag{36.1}$$

for simple materials relates the deformation gradient $\boldsymbol{F}(t)$ as a function of time to the stress $\boldsymbol{T}(t)$ as a function of time. For a particular constitutive equation (36.1) it may happen that for *every* process compatible with (36.1) the functions $\boldsymbol{T}=\boldsymbol{T}(t)$ and $\boldsymbol{F}=\boldsymbol{F}(t)$ satisfy a differential equation of the form

$$\overset{(\mathfrak{p})}{\boldsymbol{T}}=\mathfrak{g}(\boldsymbol{T}, \dot{\boldsymbol{T}}, \ldots, \overset{(\mathfrak{p}-1)}{\boldsymbol{T}}; \boldsymbol{F}, \dot{\boldsymbol{F}}, \ldots, \overset{(\mathfrak{r})}{\boldsymbol{F}}). \tag{36.2}$$

Materials for which this is the case will be called *materials of the rate type*. We assume that $\mathfrak{g}$ is a tensor-valued function with such smoothness properties as to ensure that, for each prescribed sufficiently smooth function $\boldsymbol{F}(t)$ and prescribed initial data

$$\boldsymbol{T}(t_0), \dot{\boldsymbol{T}}(t_0), \ldots, \overset{(\mathfrak{p}-1)}{\boldsymbol{T}}(t_0), \tag{36.3}$$

the differential equation (36.2) has a unique solution $\boldsymbol{T}(t)$. The stress $\boldsymbol{T}(t)$ depends not only on the values of $\boldsymbol{F}(\tau)$ for $t_0\leqq\tau\leqq t$, but also on the initial data (36.3). In general it is not possible to reconstruct from the relation (36.2) the corresponding constitutive equation (36.1). In fact, it is conceivable that a single relation of the form (36.2) be satisfied for several different simple materials, and that some solutions not correspond to any material at all. Thus, we must regard a particular differential equation (36.2) as defining a *class*, possibly empty, of materials of the rate type rather than a single such material. Nevertheless, we shall refer to (36.2) as a *constitutive equation of the rate type*.

In order to reduce the constitutive equation of the rate type (36.2) so as to satisfy the principle of material indifference, we start from the reduced form (29.13) of the general constitutive equation for simple materials. This equation (29.13) may be rewritten as

$$\boldsymbol{T}^*_{(t')}(t)=\underset{s=0}{\overset{\infty}{\mathfrak{K}}}\left(\boldsymbol{U}^*_{(t')}(t-s);\ \boldsymbol{U}(t')\right) \tag{36.4}$$

when the abbreviation

$$\boldsymbol{T}^*_{(t')}(t)=\boldsymbol{R}(t')^T\boldsymbol{R}_{(t')}(t)^T\boldsymbol{T}(t)\boldsymbol{R}_{(t')}(t)\boldsymbol{R}(t') \tag{36.5}$$

is used. It is apparent from (36.4) that the function $\boldsymbol{T}^*_{(t')}(t)$ is completely determined for all times t if the function

$$\boldsymbol{U}^*_{(t')}(\tau)=\boldsymbol{R}(t')^T\,\boldsymbol{U}_{(t')}(\tau)\,\boldsymbol{R}(t') \tag{36.6}$$

is prescribed for all τ. If the material is of the rate type, it is readily seen that the functions $\boldsymbol{T}^*_{(t')}(t)$ and $\boldsymbol{U}^*_{(t')}(t)$ must be related by a differential equation of the form

$$\left.\begin{aligned}\overset{(\mathfrak{p})}{\boldsymbol{T}^*}_{(t')}(t)=\mathfrak{k}\big(&\boldsymbol{T}^*_{(t')}(t), \dot{\boldsymbol{T}}^*_{(t')}(t), \ldots, \overset{(\mathfrak{p}-1)}{\boldsymbol{T}^*}_{(t')}(t);\\ &\boldsymbol{U}^*_{(t')}(t), \dot{\boldsymbol{U}}^*_{(t)}(t), \ldots, \overset{(\mathfrak{r})}{\boldsymbol{U}^*}_{(t')}(t); \boldsymbol{U}(t')\big).\end{aligned}\right\} \tag{36.7}$$

Since t' is arbitrary, we may choose $t'=t$ in (36.7), without loss of generality. By (36.6) and (35.3) we have

$$\overset{(\mathfrak{k})}{U}{}^{*}_{(t)}(t)=\boldsymbol{R}(t)^{T}\boldsymbol{D}_{\mathfrak{k}}(t)\boldsymbol{R}(t), \tag{36.8}$$

where $\boldsymbol{D}_{\mathfrak{k}}$ is the $\mathfrak{k}$'th stretching tensor and $\boldsymbol{R}$ is the rotation tensor (cf. Sects. 23, 24). The tensor

$$\overset{\circ}{\boldsymbol{T}}_{\mathfrak{k}}(t)=\frac{d^{\mathfrak{k}}}{dt^{\mathfrak{k}}}\left[\boldsymbol{R}_{(t')}(t)^{T}\boldsymbol{T}(t)\boldsymbol{R}_{(t')}(t)\right]\Big|_{t'=t} \tag{36.9}$$

will be called the *$\mathfrak{k}$'th co-rotational stress rate.* It follows from (36.5) and (36.9) that (36.7), for $t'=t$, takes the form[1]

$$\overset{\circ}{\boldsymbol{T}}{}^{*}_{\mathfrak{p}}=\mathfrak{k}(\boldsymbol{T}^{*},\overset{\circ}{\boldsymbol{T}}{}^{*},\ldots,\overset{\circ}{\boldsymbol{T}}{}^{*}_{\mathfrak{p}-1};\ \boldsymbol{D}^{*},\ldots,\boldsymbol{D}^{*}_{\mathfrak{r}};\ \boldsymbol{U}), \tag{36.10}$$

where

$$\overset{\circ}{\boldsymbol{T}}{}^{*}_{\mathfrak{k}}=\boldsymbol{R}^{T}\overset{\circ}{\boldsymbol{T}}_{\mathfrak{k}}\boldsymbol{R}. \tag{36.11}$$

Eq. (36.10) is a form of the *general constitutive equation of the rate type.*

Comparison with (35.5) shows that materials of the differential type are included as specially simple kinds of materials of the rate type, namely, those for which $\mathfrak{p}=0$ in (36.10).

After carrying out the differentiation on the right-hand side of (36.9) according to the product rule, we obtain

$$\overset{\circ}{\boldsymbol{T}}_{\mathfrak{k}}=\sum_{\substack{\mathfrak{a}+\mathfrak{b}+\mathfrak{c}=\mathfrak{k}\\ \mathfrak{a},\mathfrak{b},\mathfrak{c}=0,1,\ldots,\mathfrak{k}}}\frac{\mathfrak{k}!}{\mathfrak{a}!\,\mathfrak{b}!\,\mathfrak{c}!}\boldsymbol{W}_{\mathfrak{a}}^{T}\overset{(\mathfrak{b})}{\boldsymbol{T}}\boldsymbol{W}_{\mathfrak{c}}, \tag{36.12}$$

where $\boldsymbol{W}_{\mathfrak{k}}$ is the $\mathfrak{k}$'th spin (cf. (24.11)). For $\mathfrak{k}=1$ we get the *co-rotational stress rate*[2]

$$\overset{\circ}{\boldsymbol{T}}_{1}=\overset{\circ}{\boldsymbol{T}}=\dot{\boldsymbol{T}}-\boldsymbol{W}\boldsymbol{T}+\boldsymbol{T}\boldsymbol{W}, \tag{36.13}$$

$$\overset{\circ}{T}_{km}=\dot{T}_{km}-W_{kp}T^{p}{}_{m}+T_{k}{}^{p}W_{pm}. \tag{36.14}$$

The time derivatives $\overset{(\mathfrak{k})}{\boldsymbol{T}}$ of the stress are not indifferent in the sense of Sect. 17, but the co-rotational stress rates $\overset{\circ}{\boldsymbol{T}}_{\mathfrak{k}}$ are.

From (29.20), using the method that led to (36.10), one can derive the following alternative form of the *general constitutive equation of the rate type*[3]:

$$\overset{\vartriangle}{\boldsymbol{T}}{}^{*}_{\mathfrak{p}}=\mathfrak{j}(\boldsymbol{T}^{*},\overset{\vartriangle}{\boldsymbol{T}}{}^{*},\ldots,\overset{\vartriangle}{\boldsymbol{T}}{}^{*}_{\mathfrak{p}-1};\ \boldsymbol{A}^{*}_{1},\ldots,\boldsymbol{A}^{*}_{\mathfrak{r}};\ \boldsymbol{C}). \tag{36.15}$$

Here the tensors $\boldsymbol{A}^{*}_{\mathfrak{k}}$ are defined by (35.7), $\boldsymbol{C}$ is the right Cauchy-Green tensor (23.4), and

$$\overset{\vartriangle}{\boldsymbol{T}}{}^{*}_{\mathfrak{k}}=\boldsymbol{R}^{T}\overset{\vartriangle}{\boldsymbol{T}}_{\mathfrak{k}}\boldsymbol{R}, \tag{36.16}$$

where $\overset{\vartriangle}{\boldsymbol{T}}_{\mathfrak{k}}$ is the *$\mathfrak{k}$'th convected stress rate,* defined by

$$\overset{\vartriangle}{\boldsymbol{T}}_{\mathfrak{k}}(t)=\frac{d^{\mathfrak{k}}}{dt^{\mathfrak{k}}}\left[\boldsymbol{F}_{(t')}(t)^{T}\boldsymbol{T}(t)\boldsymbol{F}_{(t')}(t)\right]\Big|_{t'=t}. \tag{36.17}$$

---

[1] NOLL [1958, *31*, § 24].

[2] It was first used by ZAREMBA [1903, *11* and *12*], and it is the co-rotational time flux of the stress tensor in the sense of CFT, Sect. 148.

[3] NOLL [1958, *31*, § 24].

It is related to the time derivatives $\overset{(\mathfrak{k})}{\boldsymbol{T}}$ of the stress $\boldsymbol{T}$ and to the acceleration gradients $\boldsymbol{L}_{\mathfrak{k}}$ by

$$\overset{\triangle}{\boldsymbol{T}}_{\mathfrak{k}} = \sum_{\substack{\mathfrak{a}+\mathfrak{b}+\mathfrak{c}=\mathfrak{k}\\ \mathfrak{a},\mathfrak{b},\mathfrak{c}=0,1,\dots,\mathfrak{k}}} \frac{\mathfrak{k}!}{\mathfrak{a}!\,\mathfrak{b}!\,\mathfrak{c}!}\, \boldsymbol{L}_{\mathfrak{a}}^{T} \overset{(\mathfrak{b})}{\boldsymbol{T}} \boldsymbol{L}_{\mathfrak{c}}, \tag{36.18}$$

$$\overset{\triangle}{T}_{\mathfrak{k}ij} = \sum_{\substack{\mathfrak{a}+\mathfrak{b}+\mathfrak{c}=\mathfrak{k}\\ \mathfrak{a},\mathfrak{b},\mathfrak{c}=0,1,\dots,\mathfrak{k}}} \frac{\mathfrak{k}!}{\mathfrak{a}!\,\mathfrak{b}!\,\mathfrak{c}!}\, \overset{(\mathfrak{a})}{x}{}^{u}{}_{,i}\, \overset{(\mathfrak{b})}{T}_{uv}\, \overset{(\mathfrak{c})}{x}{}^{v}{}_{,j}, \tag{36.19}$$

as can be seen by carrying out the differentiation on the right-hand side of (36.17) and observing (24.6). For $\mathfrak{k}=1$ we get the *convected stress rate*

$$\overset{\triangle}{\boldsymbol{T}}_{1} = \overset{\triangle}{\boldsymbol{T}} = \dot{\boldsymbol{T}} + \boldsymbol{L}^{T}\boldsymbol{T} + \boldsymbol{T}\boldsymbol{L}, \tag{36.20}$$

$$\overset{\triangle}{T}_{ij} = \dot{T}_{ij} + \dot{x}^{u}{}_{,i}\, T_{uj} + T_{iu}\, \dot{x}^{u}{}_{,j}. \tag{36.21}$$

The convected stress rates $\overset{\triangle}{\boldsymbol{T}}_{\mathfrak{k}}$, like the co-rotational stress rates $\overset{\circ}{\boldsymbol{T}}_{\mathfrak{k}}$, are indifferent.

The fluxes $\overset{\circ}{\boldsymbol{T}}$ and $\overset{\triangle}{\boldsymbol{T}}$ are but two of the infinitely many possible invariant time fluxes that may be used. Clearly the properties of a material are *independent of the choice of flux*, which, like the choice of a measure of strain, is *absolutely immaterial*. Before the invariance to be required of constitutive equations was fully understood, there was some discussion of this point among the major theorists, but it has ceased. Thus we leave intentionally uncited the blossoming literature on invariant time fluxes subjected to various arbitrary requirements.

Let us assume that the constitutive equation (36.1) is satisfied for some functions $\boldsymbol{T}(t)$ and $\boldsymbol{F}(t)$. If the material is isotropic, the relation (31.4) holds for all orthogonal $\boldsymbol{Q}$. Hence the constitutive equation must remain satisfied if $\boldsymbol{T}(t)$ is replaced by $\boldsymbol{Q}\boldsymbol{T}(t)\boldsymbol{Q}^{T}$ and $\boldsymbol{F}(t)$ by $\boldsymbol{Q}\boldsymbol{F}(t)\boldsymbol{Q}^{T}$, for an arbitrary constant orthogonal tensor $\boldsymbol{Q}$. If the material is of the rate type, it follows that the relation (36.2) must remain valid if $\overset{(\mathfrak{k})}{\boldsymbol{T}}$ is replaced by $\boldsymbol{Q}\overset{(\mathfrak{k})}{\boldsymbol{T}}\boldsymbol{Q}^{T}$ and $\overset{(\mathfrak{k})}{\boldsymbol{F}}$ by $\boldsymbol{Q}\overset{(\mathfrak{k})}{\boldsymbol{F}}\boldsymbol{Q}^{T}$. Hence $\mathfrak{g}$ must be an isotropic tensor function of its variables [cf. (8.7)]. The same argument applies to the reduced constitutive equations of the rate type (36.10) and (36.15). Thus, finally, the *constitutive equation of the rate type*[1] *satisfied by an isotropic material* may be written in either of the following two forms[2]:

$$\overset{\circ}{\boldsymbol{T}}_{\mathfrak{p}} = \mathfrak{f}(\boldsymbol{T}, \overset{\circ}{\boldsymbol{T}}, \dots, \overset{\circ}{\boldsymbol{T}}_{\mathfrak{p}-1};\ \boldsymbol{D}, \boldsymbol{D}_{2}, \dots, \boldsymbol{D}_{\mathfrak{r}};\ \boldsymbol{V}), \tag{36.22}$$

$$\overset{\triangle}{\boldsymbol{T}}_{\mathfrak{p}} = \mathfrak{j}(\boldsymbol{T}, \overset{\triangle}{\boldsymbol{T}}, \dots, \overset{\triangle}{\boldsymbol{T}}_{\mathfrak{p}-1};\ \boldsymbol{A}_{1}, \boldsymbol{A}_{2}, \dots, \boldsymbol{A}_{\mathfrak{r}};\ \boldsymbol{B}), \tag{36.23}$$

where $\overset{\circ}{\boldsymbol{T}}_{\mathfrak{k}}$ is the $\mathfrak{k}$'th co-rotational stress rate (36.12), $\overset{\triangle}{\boldsymbol{T}}_{\mathfrak{k}}$ is the $\mathfrak{k}$'th convected stress rate (36.18), $\boldsymbol{D}_{\mathfrak{k}}$ is the stretching (25.4), $\boldsymbol{A}_{\mathfrak{k}}$ is the Rivlin-Ericksen tensor (25.2), $\boldsymbol{V}=\boldsymbol{R}\boldsymbol{U}\boldsymbol{R}^{T}$ is the left stretch tensor, and $\boldsymbol{B}=\boldsymbol{R}\boldsymbol{C}\boldsymbol{R}^{T}$ the left Cauchy-Green tensor (cf. Sect. 23). A co-ordinate form of (36.23) is

$$\overset{\triangle}{T}_{\mathfrak{p}km} = \mathfrak{j}_{km}(T_{pq}, \overset{\triangle}{T}_{pq}, \dots, \overset{\triangle}{T}_{(\mathfrak{p}-1)pq};\ A_{1ru}, \dots, A_{\mathfrak{r}ru};\ B_{vw};\ g_{ij}), \tag{36.24}$$

---

[1] The term is somewhat misleading, as is shown by the remarks in the next-to-last paragraph of this section.

[2] The form (36.23) was first derived by COTTER and RIVLIN [1955, *8*], using a different method. Various special but properly invariant theories of the rate type for isotropic materials have been proposed by ZAREMBA [1903, *11—12*] [1937, *4*], FROMM [1933, *2*], OLDROYD [1950, *11*, Eqs. (59), (63)—(65)] [1951, *9*], TORRE [1954, *25* and *26*], NOLL [1955, *19*], YAMAMOTO [1959, *40*, §§ 5—7], GIESEKUS [1961, *21* A]. Cf. also the literature on hypo-elasticity cited in Sect. 99 and the literature on fluids of the rate type cited in Sect. 119.

where the forms of the component functions $\mathfrak{j}_{km}$ do not depend on the choice of co-ordinate system.

A simple fluid of the rate type will be called a *fluid of the rate type*. A *solid of the rate type* is defined analogously.

The constitutive equation for a *fluid of the rate type* may be reduced to either of the following two forms:

$$\overset{\circ}{\boldsymbol{T}}_{\mathfrak{p}} = \mathfrak{h}(\boldsymbol{T}, \overset{\circ}{\boldsymbol{T}}, \ldots, \overset{\circ}{\boldsymbol{T}}_{\mathfrak{p}-1}; \boldsymbol{D}, \boldsymbol{D}_2, \ldots, \boldsymbol{D}_{\mathfrak{r}}; \varrho), \tag{36.25}$$

$$\overset{\circ}{T}_{\mathfrak{p}\,km} = \mathfrak{h}_{km}(T_{pq}, \overset{\circ}{T}_{pq}, \ldots, \overset{\circ}{T}_{(\mathfrak{p}-1)pq}; D_{ru}, \ldots, D_{\mathfrak{r}ru}; \varrho; g_{ij}), \tag{36.26}$$

or

$$\overset{\vartriangle}{\boldsymbol{T}}_{\mathfrak{p}} = \mathfrak{q}(\boldsymbol{T}, \overset{\vartriangle}{\boldsymbol{T}}, \ldots, \overset{\vartriangle}{\boldsymbol{T}}_{\mathfrak{p}-1}; \boldsymbol{A}_1, \ldots, \boldsymbol{A}_{\mathfrak{r}}; \varrho), \tag{36.27}$$

$$\overset{\vartriangle}{T}_{\mathfrak{p}\,km} = \mathfrak{q}_{km}(T_{pq}, \overset{\vartriangle}{T}_{pq}, \ldots, \overset{\vartriangle}{T}_{(\mathfrak{p}-1)pq}; A_{1ru}, \ldots, A_{\mathfrak{r}ru}; \varrho; g_{ij}). \tag{36.28}$$

The tensor functions $\mathfrak{h}$ and $\mathfrak{q}$ are isotropic, and the forms of the component functions $\mathfrak{h}_{km}$ and $\mathfrak{q}_{km}$ do not depend on the choice of the co-ordinate system. (36.25) and (36.28) can be derived from (36.22) and (36.23), respectively, using the method that led to (32.7) in the case of general simple fluids.

To every fluid of the rate type correspond constitutive equations of the forms (36.25)—(36.28). The converse, however, is not true. It may happen that a differential equation of the form (36.25) or (36.27) is satisfied for every process in some simple material which is not a fluid. For example, if we differentiate the constitutive equation (36.22) for an isotropic material of the rate type and then eliminate $\boldsymbol{V}$, we obtain a differential equation of the form (36.25). Hence it is possible that equations of the form (36.25)—(36.28) are satisfied by the constitutive equations of certain isotropic simple solids. We should remember that a constitutive equation of the rate type defines a class of simple materials, rather than only a single material[1]. As far as is known today, a particular class of this kind may contain both fluids and solids, or even both isotropic and anisotropic materials.

37. Materials of the integral type. Consider the general constitutive equation of a simple material in the form (29.28), i.e.

$$\boldsymbol{R}^T \boldsymbol{T} \boldsymbol{R} = \mathfrak{f}(\boldsymbol{C}) + \mathop{\mathfrak{F}}_{s=0}^{\infty}(\boldsymbol{G}^*(s); \boldsymbol{C}), \quad \boldsymbol{G}^*(s) = \boldsymbol{C}^*_{(t)}(t-s) - \boldsymbol{1}. \tag{37.1}$$

It may happen that the response functional $\mathfrak{F}$ can be expressed in terms of an integral polynomial over $\boldsymbol{G}^*(s)$. More precisely, $\mathfrak{F}$ may be a functional of the type

$$\mathop{\mathfrak{F}}_{s=0}^{\infty}(\boldsymbol{G}^*(s);\ \boldsymbol{C}) = \sum_{\mathfrak{k}=1}^{\mathfrak{m}} \int_0^\infty \cdots \int_0^\infty \mathfrak{g}_{\mathfrak{k}}(s_1, \ldots, s_{\mathfrak{k}}; \boldsymbol{C})\,[\boldsymbol{G}^*(s_1), \ldots, \boldsymbol{G}^*(s_{\mathfrak{k}})]\, ds_1 \cdots ds_{\mathfrak{k}}, \tag{37.2}$$

where $\mathfrak{g}_{\mathfrak{k}}(s_1, \ldots, s_{\mathfrak{k}}; \boldsymbol{C})\,[\ ]$, for each choice of $s_1, \ldots, s_{\mathfrak{k}}$, and $\boldsymbol{C}$, is a multilinear tensor function of $\mathfrak{k}$ tensor variables (cf. Sect. 7). The values $\mathfrak{g}_{\mathfrak{k}}(s_1, \ldots, s_{\mathfrak{k}}; \boldsymbol{C})$ may be regarded as tensors of order $2\mathfrak{k}$. Also, the tensor functions $\mathfrak{g}_{\mathfrak{k}}$ may be taken to be symmetric in the sense that $\mathfrak{g}_{\mathfrak{k}}(s_1, \ldots, s_{\mathfrak{k}}; \boldsymbol{C})\,[\boldsymbol{G}^*_1, \ldots, \boldsymbol{G}^*_{\mathfrak{k}}]$ does not change its value under a permutation of the indices $1, \ldots, \mathfrak{k}$. Then the $\mathfrak{g}_{\mathfrak{k}}$ are uniquely[2] determined by the response functional $\mathfrak{F}$.

[1] A discussion of the difficulties arising in the definition of materials of the rate type has been given by BERNSTEIN [1961, *3*]. For the special case of hypo-elastic materials, this subject will be considered again in Chapter D IV.

[2] While the proof of uniqueness is straightforward, it depends on concepts from functional analysis and thus is too technical to be included here.

We say that a response functional is of the *integral type* if it is of the form (37.2). A simple material is called a *material of the integral type* if the response functional $\mathfrak{F}$ in its defining constitutive equation (37.1) is of the integral type. The largest number $\mathfrak{m}$ of iterated integrations occuring in (37.2) is called the *order* of the material[1].

For (37.2) to have meaning it is necessary that the integrals on the right-hand side converge. In general, this will be the case only when the history $\boldsymbol{G}^*(s)$ is restricted to a suitable class. Arbitrary continuous histories are permitted only when the $\mathfrak{g}_{\mathfrak{k}}(s_1, \ldots, s_{\mathfrak{k}}; \boldsymbol{C})$ vanish for large values of $s_1, \ldots, s_{\mathfrak{k}}$. In this case, only a finite part of the history of the motion has an effect on the present stress.

For *isotropic* materials the general constitutive equation (37.1) reduces to (31.10), i.e.

$$\boldsymbol{T} = \mathfrak{f}(\boldsymbol{B}) + \mathop{\mathfrak{F}}_{s=0}^{\infty}(\boldsymbol{G}(s); \boldsymbol{B}), \quad \boldsymbol{G}(s) = \boldsymbol{C}_{(t)}(t-s) - \mathbf{1}. \tag{37.3}$$

The tensor function $\mathfrak{f}(\boldsymbol{B})$ and the response functional $\mathfrak{F}$ in (37.3) satisfy the isotropy relations (31.8). If the isotropic material defined by (37.3) is of the integral type, then $\mathfrak{F}$ is of the form (37.2), with $\boldsymbol{C}$ replaced by $\boldsymbol{B}$. In this case, each of the tensor functions $\mathfrak{g}_{\mathfrak{k}}$ that occur as an integrand in (37.2) is an isotropic function of its $\mathfrak{k}+1$ tensor variables, i.e., the relations

$$\left.\begin{aligned} &\boldsymbol{Q}\,\mathfrak{g}_{\mathfrak{k}}(s_1, \ldots, s_{\mathfrak{k}}; \boldsymbol{B})\,[\boldsymbol{G}_1, \ldots, \boldsymbol{G}_{\mathfrak{k}}]\,\boldsymbol{Q}^T \\ &\quad = \mathfrak{g}_{\mathfrak{k}}(s_1, \ldots, s_{\mathfrak{k}}; \boldsymbol{Q}\boldsymbol{B}\boldsymbol{Q}^T)\,[\boldsymbol{Q}\boldsymbol{G}_1\boldsymbol{Q}^T, \ldots, \boldsymbol{Q}\boldsymbol{G}_{\mathfrak{k}}\boldsymbol{Q}^T] \end{aligned}\right\} \tag{37.4}$$

hold identically in the symmetric tensor variables $\boldsymbol{B}, \boldsymbol{G}_1, \ldots, \boldsymbol{G}_{\mathfrak{k}}$ and the orthogonal tensor variable $\boldsymbol{Q}$. The identity (37.4) follows from $(31.8)_1$ and from the fact that the integral representation (37.2) of $\mathfrak{F}$ is unique.

According to (32.8) and (32.10) the general constitutive equation of a *simple fluid, compressible* or *incompressible*, may be written in the form

$$\boldsymbol{T} = -p\mathbf{1} + \mathop{\mathfrak{Q}}_{s=0}^{\infty}(\boldsymbol{G}(s)), \quad \boldsymbol{G}(s) = \boldsymbol{C}_{(t)}(t-s) - \mathbf{1}, \tag{37.5}$$

where, for a compressible fluid, the dependence of p and $\mathfrak{Q}$ on the density is understood, and where the functional $\mathfrak{Q}$ satisfies the isotropy relation (32.9). The fluid is of the integral type if the response functional $\mathfrak{Q}$ is of the form

$$\mathop{\mathfrak{Q}}_{s=0}^{\infty}(\boldsymbol{G}(s)) = \sum_{\mathfrak{k}=1}^{\mathfrak{m}} \int_0^{\infty} \cdots \int_0^{\infty} \mathfrak{g}_{\mathfrak{k}}(s_1, \ldots, s_{\mathfrak{k}})\,[\boldsymbol{G}(s_1), \ldots, \boldsymbol{G}(s_{\mathfrak{k}})]\,ds_1 \cdots ds_{\mathfrak{k}}. \tag{37.6}$$

Here the tensor functions $\mathfrak{g}_{\mathfrak{k}}(s_1, \ldots, s_{\mathfrak{k}})\,[\boldsymbol{G}_1, \ldots, \boldsymbol{G}_{\mathfrak{k}}]$ are multilinear and isotropic in the $\mathfrak{k}$ tensor variables $\boldsymbol{G}_1, \ldots, \boldsymbol{G}_{\mathfrak{k}}$. Also, they may be taken to be invariant under permutations of $1, \ldots, \mathfrak{k}$. By use of this fact and the results of Sect. 11 (α) and Sect. 13 (α), it is not hard to see that each integrand in (37.6) may be replaced by a sum of terms of the form

$$\left.\begin{aligned} &\psi(s_1, \ldots, s_{\mathfrak{k}})\,\mathrm{tr}\,[\boldsymbol{G}(s_1) \cdots \boldsymbol{G}(s_{\mathfrak{l}_1})]\,\mathrm{tr}\,[\boldsymbol{G}(s_{\mathfrak{l}_1+1}) \cdots \boldsymbol{G}(s_{\mathfrak{l}_2})] \\ &\qquad \cdots \mathrm{tr}\,[\boldsymbol{G}(s_{\mathfrak{l}_{r-1}+1}) \cdots \boldsymbol{G}(s_{\mathfrak{l}_r})]\,\boldsymbol{G}(s_{\mathfrak{l}_r+1}) \cdots \boldsymbol{G}(s_{\mathfrak{k}}) \end{aligned}\right\} \tag{37.7}$$

where $1 \leqq \mathfrak{l}_1 < \mathfrak{l}_2 < \cdots < \mathfrak{l}_r \leqq \mathfrak{k}$ and where $\mathfrak{l}_1 \leqq 6$, $\mathfrak{l}_p - \mathfrak{l}_{p-1} \leqq 6$, $\mathfrak{k} - \mathfrak{l}_r \leqq 5$. If $\mathfrak{k} = \mathfrak{l}_r$, then the product at the end of (37.7) must be replaced by the unit tensor, $\mathbf{1}$.

[1] Materials of order $\mathfrak{m}$ that are not necessarily of the integral type will be defined in Sect. 40. Apparently the first frame-indifferent theories of particular fluids of the integral type are those of Oldroyd [1950, *11*, §§ 3 (c), 4], Green and Rivlin [1956, *12*, §§ 4—9] [1957, *8*, §§ 4—9], and Pao [1957, *16*].

For *fluids of the first order*, which correspond to $\mathfrak{m}=1$ in (37.6), the constitutive equation (37.5) may be written in the explicit form[1]

$$\boldsymbol{T}=-p\boldsymbol{1}+\int_0^\infty \psi(s)\operatorname{tr}\boldsymbol{G}(s)\,ds\,\boldsymbol{1}+\int_0^\infty \zeta(s)\,\boldsymbol{G}(s)\,ds. \tag{37.8}$$

The constitutive equation for *second-order fluids* of the integral type is obtained from (37.8) by adding to the right-hand side the expression

$$\left.\begin{aligned}&\int_0^\infty\int_0^\infty \{\gamma_1(s_1,s_2)\operatorname{tr}\boldsymbol{G}(s_1)\operatorname{tr}\boldsymbol{G}(s_2)+\gamma_2(s_1,s_2)\operatorname{tr}[\boldsymbol{G}(s_1)\,\boldsymbol{G}(s_2)]\}\,ds_1\,ds_2\,\boldsymbol{1}+\\ &\quad+\int_0^\infty\int_0^\infty\{\alpha(s_1,s_2)\,[\operatorname{tr}\boldsymbol{G}(s_1)]\,\boldsymbol{G}(s_2)+\beta(s_1,s_2)\,\boldsymbol{G}(s_1)\,\boldsymbol{G}(s_2)\}\,ds_1\,ds_2.\end{aligned}\right\} \tag{37.9}$$

For a compressible fluid the functions ψ, ζ, γ_1, γ_2, α, and β depend on the density ϱ.

For an *incompressible simple fluid* all terms in (37.7) that are scalar multiples of the unit tensor may be absorbed in the pressure term $-p\boldsymbol{1}$. For example, the constitutive equation of an *incompressible second-order fluid of the integral type* is

$$\left.\begin{aligned}\boldsymbol{T}=&-p\boldsymbol{1}+\int_0^\infty\zeta(s)\,\boldsymbol{G}(s)\,ds+\\ &+\int_0^\infty\int_0^\infty\{\alpha(s_1,s_2)\,[\operatorname{tr}\boldsymbol{G}(s_1)]\,\boldsymbol{G}(s_2)+\beta(s_1,s_2)\,\boldsymbol{G}(s_1)\,\boldsymbol{G}(s_2)\}\,ds_1\,ds_2.\end{aligned}\right\} \tag{37.10}$$

Explicit representations may also be obtained for general isotropic materials of the integral type, provided that the tensor functions $\mathfrak{g}_{\mathfrak{k}}(s_1,\ldots,s_{\mathfrak{k}};\boldsymbol{B})\,[\boldsymbol{G}_1,\ldots,\boldsymbol{G}_{\mathfrak{k}}]$, which appear as integrands in (37.2), are polynomial functions of $\boldsymbol{B}$. These representations are easily inferred from the results of Sect. 11 (β) and Sect. 13 (β), but they are too cumbersome to be stated here in general[2].

For the constitutive equation of an *isotropic material of the first order* one obtains

$$\left.\begin{aligned}\boldsymbol{T}=\mathfrak{f}(\boldsymbol{B})+\int_0^\infty\{&\mathfrak{k}_1(s,\boldsymbol{B})\,\boldsymbol{G}(s)+\boldsymbol{G}(s)\,\mathfrak{k}_1(s,\boldsymbol{B})+\operatorname{tr}[\boldsymbol{G}(s)\,\mathfrak{k}_2(s,\boldsymbol{B})]\,\boldsymbol{1}+\\ &+\operatorname{tr}[\boldsymbol{G}(s)\,\mathfrak{k}_3(s,\boldsymbol{B})]\,\boldsymbol{B}+\operatorname{tr}[\boldsymbol{G}(s)\,\mathfrak{k}_4(s,\boldsymbol{B})]\,\boldsymbol{B}^2\}\,ds,\end{aligned}\right\} \tag{37.11}$$

where $\mathfrak{f}$, $\mathfrak{k}_1$, $\mathfrak{k}_2$, $\mathfrak{k}_3$ and $\mathfrak{k}_4$ are isotropic tensor functions of only one tensor variable $\boldsymbol{B}$ and hence have explicit representations of the form (12.6) or (12.7). It is likely that (37.11) continues to be valid even when the dependence on $\boldsymbol{B}$ is not polynomial.

In the considerations of this section we have used the relative right Cauchy-Green tensor $\boldsymbol{C}_{(t)}$ to describe the deformation history. Instead of $\boldsymbol{C}_{(t)}$ we could also have used the relative right stretch tensor $\boldsymbol{U}_{(t)}=\sqrt{\boldsymbol{C}_{(t)}}$ or any other equivalent measure of deformation, such as $\boldsymbol{C}_{(t)}^{-1}$ or $\log\boldsymbol{C}_{(t)}$. The concept of a material of the integral type depends on the choice of the deformation measure. For example, a material which is of the integral type with respect to $\boldsymbol{C}_{(t)}$ is *not* of the integral type with respect to $\boldsymbol{U}_{(t)}$.

There are various ways of motivating the consideration of materials of the integral type. If the domain of the response functional $\mathfrak{F}$ is a suitable function space endowed with a suitable topology, and if $\mathfrak{F}$ is continuous with respect to

---

[1] A special theory of this kind has been discussed by BERNSTEIN, KEARSLEY, and ZAPAS [1963, *5*] [1964, *7*].

[2] A description of such representations was given by SPENCER and RIVLIN [1960, *53*, § 11], using the approach developed by GREEN, RIVLIN and SPENCER [1956, *12*] [1957, *8*] [1959, *10*]. These authors assume a special dependence of $\mathfrak{g}_{\mathfrak{k}}(s_1,\ldots,s_{\mathfrak{k}};\boldsymbol{B})$ on the scalar variables $s_1,\ldots,s_{\mathfrak{k}}$, which results in a slight simplification for the representations.

that topology, one can often use the Stone-Weierstraß theorem[1] to show that $\mathfrak{F}$ may be uniformly approximated by integral polynomials of the form exhibited on the right-hand side of (37.2). This is the view of GREEN and RIVLIN[2].

It is possible to consider materials whose response functional is of a form similar to (37.2), except that the integrands depend explicitly not only on the present value $\boldsymbol{C}$ of the right Cauchy-Green tensor, but also on the values at $s=0$ of the first $\mathfrak{r}$ derivatives of $\boldsymbol{G}(s)$. Such materials may be called materials of a *mixed integral-differential type*[3]. It follows from (29.27), (29.22), (35.7), and (24.14) that the derivatives of $\boldsymbol{G}(s)$ at $s=0$ differ only by a sign factor from the tensors $\boldsymbol{A}_{\mathfrak{k}}^*$ defined by (35.7). Hence, for materials of a mixed integral-differential type, the response functional has the form (37.2) with $\mathfrak{g}_{\mathfrak{k}}(s_1, \ldots, s_{\mathfrak{k}}; \boldsymbol{C})$ replaced by

$$\mathfrak{g}_{\mathfrak{k}}(s_1, \ldots, s_{\mathfrak{k}}; \boldsymbol{C}, \boldsymbol{A}_1^*, \ldots, \boldsymbol{A}_{\mathfrak{r}}^*). \tag{37.12}$$

For materials that are also isotropic, we may take $\boldsymbol{G}(s)$ to be defined by $(37.3)_2$, replace $\boldsymbol{C}$ by $\boldsymbol{B}$ and $\mathfrak{g}_{\mathfrak{k}}(s_1, \ldots, s_{\mathfrak{k}}; \boldsymbol{C})$ by

$$\mathfrak{g}_{\mathfrak{k}}(s_1, \ldots, s_{\mathfrak{k}}; \boldsymbol{B}, \boldsymbol{A}_1, \ldots, \boldsymbol{A}_{\mathfrak{r}}), \tag{37.13}$$

where $\boldsymbol{A}_{\mathfrak{i}}$ is the $\mathfrak{i}$'th Rivlin-Ericksen tensor (25.2). In this case, $\mathfrak{g}_{\mathfrak{k}}(s_1, \ldots, s_{\mathfrak{k}}; \boldsymbol{B}, \boldsymbol{A}_1, \ldots, \boldsymbol{A}_{\mathfrak{r}})[\boldsymbol{G}_1, \ldots, \boldsymbol{G}_{\mathfrak{k}}]$ is an isotropic function of its $\mathfrak{r}+\mathfrak{k}+1$ tensor variables, multilinear in $\boldsymbol{G}_1, \ldots, \boldsymbol{G}_{\mathfrak{k}}$. For fluids, the dependence on $\boldsymbol{B}$ reduces to a dependence on the density ϱ.

Materials of a *mixed integral-rate* type, corresponding to the case when $\mathfrak{j}$ in (36.15) is replaced by an integral polynomial involving the histories of the arguments, have been considered[4].

V. Fading memory.

38. The principle of fading memory[5]. Simple materials were defined in Sect. 29 as follows:

The present stress is determined by the history of the first spatial gradient of the deformation function.

A theory based on this assumption alone can scarcely have predictive value, for the *entire* history of a body can never be known. The interpretation of the results of an experiment in terms of the theory of simple materials can be justified only if additional assumptions are made, at least tacitly. One such assumption would be that the history of the body previous to the start of an experiment has no appreciable influence on its result. Under suitable circumstances, this assumption will be satisfied as a consequence of the *principle of fading memory*:

Deformations that occurred in the distant past should have less influence in determining the present stress than those that occurred in the recent past.

We shall assume that this principle is appropriate to most simple materials. A notable exception is furnished by materials that are hypo-elastic but not elastic (see Subchapter D IV). When we seek to render this intuitive principle of fading memory precise, we find several different possibilities. We select one particular mathematical interpretation and investigate it in detail.

[1] Cf., e.g., STONE [1948, *17*].

[2] For references and for a further comment on the insufficiency of such an approach, see the first footnote to Sect. 28.

[3] Materials of this type were first discussed by GREEN and RIVLIN [1960, *27*, §§ 6—9].

[4] OLDROYD [1950, *11*, § 4], GREEN and RIVLIN [1960, *27*, § 5].

[5] The theory presented in this subchapter is due to COLEMAN and NOLL [1960, *17*] [1961, *5* and *7*] [1964, *20*].

The general constitutive equation of a simple material may be written in the form (29.28), i.e.

$$\boldsymbol{R}^T \boldsymbol{T} \boldsymbol{R} = \mathfrak{f}(\boldsymbol{C}) + \underset{s=0}{\overset{\infty}{\mathfrak{F}}} (\boldsymbol{G}^*(s); \boldsymbol{C}). \tag{38.1}$$

Here $\boldsymbol{T}$ is the stress tensor, $\boldsymbol{R}$ is the rotation tensor, $\boldsymbol{C}$ is the right Cauchy-Green tensor and $\boldsymbol{G}^*(s)$ is defined, in terms of the relative right Cauchy-Green tensor $\boldsymbol{C}_{(t)}(s)$, by (29.27), namely,

$$\boldsymbol{G}^*(s) = \boldsymbol{R}(t)^T \boldsymbol{C}_{(t)}(t-s) \boldsymbol{R}(t) - \boldsymbol{1}, \quad \boldsymbol{G}^*(0) = \boldsymbol{0}. \tag{38.2}$$

The value of the functional $\mathfrak{F}$ of (38.1) is zero when $\boldsymbol{G}^*(s) \equiv \boldsymbol{0}$:

$$\underset{s=0}{\overset{\infty}{\mathfrak{F}}} (\boldsymbol{0}; \boldsymbol{C}) = \boldsymbol{0}. \tag{38.3}$$

Recall that $\boldsymbol{R}\mathfrak{f}(\boldsymbol{C})\boldsymbol{R}^T$ is the *equilibrium value of the stress*, i.e., the value of $\boldsymbol{T}$ corresponding to the rest history, $\boldsymbol{G}^*(s) \equiv \boldsymbol{0}$.

We interpret the principle of fading memory as a requirement of smoothness for the response functional $\mathfrak{F}$. In order to do so, we first introduce the concept of an obliviator, intended to characterize the rate at which the memory fades. A function h is called an *obliviator*[1] *of order* $\mathfrak{r}$ (positive but not necessarily integral) fi it satisfies the following conditions:

a) $h(s)$ is defined for $0 \leqq s < \infty$ and has positive real values: $h(s) > 0$.

b) $h(s)$ is normalized by the condition $h(0) = 1$. (38.4)

c) $h(s)$ decays to zero in such a way that $\lim_{s\to\infty} s^{\mathfrak{r}} h(s) = 0$, monotonically for large s. (38.5)

For example,

$$h(s) = \frac{1}{(s+1)^p} \tag{38.6}$$

is an obliviator of order $\mathfrak{r}$ for $\mathfrak{r} < p$. An exponential,

$$h(s) = e^{-\beta s}, \quad \beta > 0, \tag{38.7}$$

is an obliviator of arbitrary order.

One can consider various normed linear function spaces of histories[2] $\boldsymbol{G}(s)$ such that, in computing the norm of $\boldsymbol{G}(s)$, the values of $\boldsymbol{G}(s)$ are weighted by the obliviator $h(s)$. We confine our attention to a particular norm of this type[3]: The *recollection* of a history $\boldsymbol{G}(s)$ is defined by

$$\|\boldsymbol{G}(s)\|_h = \left(\int_0^\infty [h(s)\,|\boldsymbol{G}(s)|]^2\, ds \right)^{\frac{1}{2}}, \tag{38.8}$$

where $|\boldsymbol{G}(s)| = \sqrt{\operatorname{tr}[\boldsymbol{G}(s)^2]}$ is the magnitude of the symmetric tensor $\boldsymbol{G}(s)$ (cf. (6.15)). Here and subsequently it is understood that the admissible histories $\boldsymbol{G}(s)$ are measurable functions of s and that histories differing only on a set of measure zero on the interval $0 \leqq s < \infty$ are regarded as the same.

---

[1] COLEMAN and NOLL's term is "influence function".

[2] Throughout the rest of this section $\boldsymbol{G}(s)$ an arbitrary function of s ($0 \leqq s < \infty$) whose values are symmetric tensors. For the interpretation in the theory of simple materials, $\boldsymbol{G}(s)$ is to be taken as the particular history $\boldsymbol{G}^*(s)$ defined by (38.2),.

[3] COLEMAN and NOLL [1960, *17*, § 2] study also the rorresponding L^p norms when $p \neq 2$.

The collection of all histories with finite recollection (38.8) forms a Hilbert space $\mathscr{H}$. The *inner product* of two histories $\boldsymbol{G}(s)$ and $\boldsymbol{H}(s)$ in $\mathscr{H}$ is given by

$$\langle \boldsymbol{G}(s), \boldsymbol{H}(s)\rangle_h = \int_0^\infty \operatorname{tr}\big(\boldsymbol{G}(s)\boldsymbol{H}(s)\big)\, h(s)^2\, ds. \tag{38.9}$$

The *a priori* domain of definition of $\mathfrak{F}$ contains only histories $\boldsymbol{G}(s)$ that are continuous and that satisfy the conditions $\boldsymbol{G}(0)=\boldsymbol{0}$ and $\det(\boldsymbol{G}(s)+\boldsymbol{1}) = \det \boldsymbol{C}_{(t)}(t-s) > 0$. We assume that this domain of definition of $\mathfrak{F}$ can be extended so as to render valid the following ***weak principle of fading memory:***

(*C*) *There exists an obliviator $h(s)$ of order $\mathfrak{r} > \frac{1}{2}$ such that the response functional $\underset{s=0}{\overset{\infty}{\mathfrak{F}}}\big(\boldsymbol{G}(s)\big)$ is defined and continuous for histories $\boldsymbol{G}(s)$ in a neighborhood of the zero history in the function space $\mathscr{H}$.*

The continuity of $\mathfrak{F}$ is to be understood, of course, in terms of the topology defined by the norm (38.8) in the function space $\mathscr{H}$. It is this continuity that expresses the assumption of fading memory: Two histories differ very little in norm if their values are close to each other for small s (recent past), though they may be far apart for large s (distant past), because the norm is weighted with a decaying obliviator. The continuity of $\mathfrak{F}$ then insures that the corresponding stresses differ but little. In the statement of (*C*) we have suppressed the dependence on $\boldsymbol{C}$, which plays the role of a tensor parameter. We assume that (*C*) is satisfied for each value of $\boldsymbol{C}$ in an appropriate domain. We note that one may have to select different obliviators $h(s)$ in order to satisfy the requirement (*C*) for different response functionals, i.e. for different materials. For a given material, however, there is no unique way of choosing $h(s)$, and $h(s)$ is not a material function, although the existence of an $h(s)$ of the required type is a material property.

For incompressible simple materials the kinematically possible histories must satisfy the condition $\det\big(\boldsymbol{G}(s)+\boldsymbol{1}\big)\equiv 1$. This condition severely restricts the *a priori* domain of definition of $\mathfrak{F}$. However, by putting

$$\underset{s=0}{\overset{\infty}{\mathfrak{F}}}\big(\boldsymbol{G}(s)\big) = \underset{s=0}{\overset{\infty}{\mathfrak{F}}}\big([\det(\boldsymbol{G}(s)+\boldsymbol{1})]^{-1}[\boldsymbol{G}(s)+\boldsymbol{1}]-\boldsymbol{1}\big) \tag{38.10}$$

when $\det\big(\boldsymbol{G}(s)+\boldsymbol{1}\big)\not\equiv 1$, we can extend the domain of definition of $\mathfrak{F}$. We assume that this extension of $\mathfrak{F}$ can be extended further so as to satisfy the requirement (*C*) stated above.

For many purposes it is necessary to replace the requirement (*C*) that $\mathfrak{F}$ be continuous by the stronger requirement ($C^{\mathfrak{n}}$) that $\mathfrak{F}$ be $\mathfrak{n}$ times differentiable in the sense of FRÉCHET[1]. Before defining the Fréchet differential of a functional $\mathfrak{F}$, we recall some preliminary concepts.

A functional

$$\boldsymbol{P} = \underset{s=0}{\overset{\infty}{\mathfrak{P}}}\big(\boldsymbol{G}_1(s), \ldots, \boldsymbol{G}_{\mathfrak{j}}(s), \ldots, \boldsymbol{G}_{\mathfrak{k}}(s)\big) \tag{38.11}$$

with $\mathfrak{k}$ argument functions $\boldsymbol{G}_{\mathfrak{j}}(s)$ in $\mathscr{H}$ and symmetric tensor values $\boldsymbol{P}$ is called a *bounded $\mathfrak{k}$-linear functional* if it is a linear functional in each of its arguments $\boldsymbol{G}_{\mathfrak{j}}(s)$, and if there is a constant M, independent of the functions $\boldsymbol{G}_{\mathfrak{j}}(s)$, such that

$$\Big|\underset{s=0}{\overset{\infty}{\mathfrak{P}}}\big(\boldsymbol{G}_1(s), \ldots, \boldsymbol{G}_{\mathfrak{k}}(s)\big)\Big| \leq M \|\boldsymbol{G}_1(s)\|_h \cdots \|\boldsymbol{G}_{\mathfrak{k}}(s)\|_h. \tag{38.12}$$

---

[1] For an exposition of properties of the Fréchet differential, cf. e.g. VAINBERG [1964, 89, § 3ff.].

The functional $\mathfrak{P}$ is said to be *symmetric* if any permutation of the argument functions $\boldsymbol{G}_{\mathfrak{i}}(s)$ leaves its value unchanged.

A functional obtained by substituting the same function for the argument functions of a symmetric bounded $\mathfrak{k}$-linear functional,

$$\mathop{\mathfrak{P}}_{s=0}^{\infty}(\boldsymbol{G}(s))=\mathop{\mathfrak{P}}_{s=0}^{\infty}(\boldsymbol{G}(s),\ldots,\boldsymbol{G}(s)), \tag{38.13}$$

is called a *bounded homogeneous polynomial functional of degree* $\mathfrak{k}$. We use the same symbol $\mathfrak{P}$ for the polynomial functional and the corresponding symmetric multilinear functional because one is uniquely determined by the other.

A functional $\mathfrak{F}$ defined on a neighborhood of the function $\boldsymbol{H}(s)$ is said to be $\mathfrak{n}$ *times Fréchet-differentiable* at $\boldsymbol{H}(s)$ if there exist bounded homogeneous polynomial functionals

$$\delta^{\mathfrak{k}}\mathop{\mathfrak{F}}_{s=0}^{\infty}(\boldsymbol{H}(s);\boldsymbol{G}(s)) \quad \text{of degree} \quad \mathfrak{k}=0,1,\ldots,\mathfrak{n} \quad \text{in} \quad \boldsymbol{G}(s)$$

such that

$$\mathop{\mathfrak{F}}_{s=0}^{\infty}(\boldsymbol{H}(s)+\boldsymbol{G}(s))=\sum_{\mathfrak{k}=0}^{\mathfrak{n}}\frac{1}{\mathfrak{k}!}\,\delta^{\mathfrak{k}}\mathop{\mathfrak{F}}_{s=0}^{\infty}(\boldsymbol{H}(s);\boldsymbol{G}(s))+o(\|\boldsymbol{G}(s)\|_h^{\mathfrak{n}}). \tag{38.14}$$

The polynomial functional $\delta^{\mathfrak{k}}\mathop{\mathfrak{F}}\limits_{s=0}^{\infty}(\boldsymbol{H}(s);\boldsymbol{G}(s))$ or its corresponding symmetric $\mathfrak{k}$-linear functional is called the $\mathfrak{k}$*'th Fréchet-differential* or the $\mathfrak{k}$*'th variation* of $\mathfrak{F}$ at $\boldsymbol{H}(s)$ in $\mathscr{H}$.

In the special case when $\boldsymbol{H}(s)$ is the zero function, $\boldsymbol{H}(s)\equiv\boldsymbol{0}$, we write

$$\delta^{\mathfrak{k}}\mathop{\mathfrak{F}}_{s=0}^{\infty}(\boldsymbol{0};\boldsymbol{G}(s))=\delta^{\mathfrak{k}}\mathop{\mathfrak{F}}_{s=0}^{\infty}(\boldsymbol{G}(s)), \tag{38.15}$$

so that (38.14) reduces to

$$\mathop{\mathfrak{F}}_{s=0}^{\infty}(\boldsymbol{G}(s))=\sum_{\mathfrak{k}=1}^{\mathfrak{n}}\frac{1}{\mathfrak{k}!}\,\delta^{\mathfrak{k}}\mathop{\mathfrak{F}}_{s=0}^{\infty}(\boldsymbol{G}(s))+o(\|\boldsymbol{G}(s)\|_h^{\mathfrak{n}}). \tag{38.16}$$

We are now able to state the following ***stronger principles of fading memory***:

($C^{\mathfrak{n}}$) *There exists an obliviator of order greater than* $\mathfrak{n}+\frac{1}{2}$ *such that the response functional* $\mathop{\mathfrak{F}}\limits_{s=0}^{\infty}(\boldsymbol{G}(s))$ *is defined and* $\mathfrak{n}$ *times Fréchet-differentiable in a neighborhood of the zero history of the function space* $\mathscr{H}$.

The remarks made after (C) apply here too.

Of course, if ($C^{\mathfrak{n}}$) holds, so does ($C^{\mathfrak{k}}$) for every $\mathfrak{k}$ less than $\mathfrak{n}$. We note that the statements (C) or ($C^{\mathfrak{n}}$) do not require the admissible histories $\boldsymbol{G}(s)$ to be continuous. In the discussion of stress relaxation given in the following section we shall consider deformations that are discontinuous in time.

The theory of materials of the differential type, defined by the constitutive equations (35.5) or (35.6), is meaningful only for motions that are sufficiently differentiable. These materials do not obey the smoothness requirements (C) or ($C^{\mathfrak{n}}$). However, since the present stretchings $\boldsymbol{D}, \boldsymbol{D}_2, \ldots, \boldsymbol{D}_{\mathfrak{n}}$ do not depend on the motion in the distant past, materials of the differential type do have, in a different sense, a fading memory. In Sect. 40 we shall see that the theory of materials of the differential type results as an asymptotic approximation to the theory of general simple materials obeying the strong principle of fading memory ($C^{\mathfrak{n}}$).

[Added in proof. Two theories of fading memory have been constructed in recent studies by C.-C. WANG. His first theory[1] generalizes that of COLEMAN and NOLL, outlined above. WANG calls a Lebesgue-Stieltjes measure μ defined on the closed half-line (0, ∞) *obliviating* if it is generated by a non-vanishing lower semi-continuous function $\sigma(s)$ such that

$$\left.\begin{aligned} &\sigma(s)=0 \quad \text{if} \quad s\leqq 0,\\ &\lim_{s\to+\infty}\sigma(s) \quad \text{exists.}\end{aligned}\right\} \tag{38.17}$$

In terms of an obliviating measure, the recollection of a history $\boldsymbol{G}(s)$ is then defined as follows:

$$\|\boldsymbol{G}(s)\|_{\mu}=\Big(\int_{[0,\infty)}|\boldsymbol{G}(s)|^2\,d\mu\Big)^{\frac{1}{2}}. \tag{38.18}$$

If $h(s)$ is an obliviator as defined at the beginning of this section, set

$$\mu(\mathscr{I})\equiv\int_{\mathscr{I}}[h(s)]^2\,ds \tag{38.19}$$

for all Borel sets $\mathscr{I}$; then μ is an obliviating measure, and (38.18) reduces to (38.8). Thus COLEMAN and NOLL's theory is included as a special case in WANG's first theory. The greater generality of the obliviating measure includes a broader class of materials as being endowed with fading memory and as enjoying, therefore, the consequences of having that property.

WANG's second theory[2] of fading memory is based upon a different idea. Instead of using the recollection as a measure of the departure of a deformation history from the rest history, WANG introduces a topology of compact convergence. In his terms, the *order* p of a simple material is the largest integer such that $\mathfrak{F}$ depends explicitly on $d^p\boldsymbol{G}(s)/ds^p$ at some $s\in[0,\infty)$. He then assumes that the domain $\mathscr{D}$ of $\mathfrak{F}$ is the set of all p-times continuously differentiable functions whose values are positive-definite symmetric tensors. A sequence $\boldsymbol{G}_n(s)\in\mathscr{D}$ is said to *converge* to a function $\boldsymbol{G}(s)$ if $d^q\boldsymbol{G}_n(s)/ds^q\to d^q\boldsymbol{G}(s)/ds^q$ uniformly on every finite interval in [0, ∞), for every $q\leqq p$. Weak and strong principles of fading memory can then be stated as before, as requirements of continuity and differentiability to a specified order, but the meaning and the consequences of these assumptions are now different because the topology is a different one.

WANG's second theory implies for any material that obeys the corresponding weak principle of fading memory the existence of two material parameters: a *time of sentience* η_0 and a *grade of sentience* δ_0. The time of sentience of a material is a measure of the weight of the effects on the value of $\mathfrak{F}$ of variations of $\boldsymbol{G}(s)$ at different s, and the grade of sentience characterizes the effects on the value of $\mathfrak{F}$ of variations of $\boldsymbol{G}(s)$ for times from the present time back through an interval equal to the time of sentience. If the material has been kept in a configuration sufficiently near to a state of rest from the time $t-\eta_0$ up to the present time t, then it is impossible to find any deformation history for the times prior to $t-\eta_0$ that renders the present stress arbitrarily different from the static stress. The time interval $[t-\eta_0, t]$ may thus be regarded as the "major memory" of the material, while $(-\infty, t-\eta_0]$ is the "minor memory". In these terms, if a simple material obeys the principle of fading memory, then it has a finite major memory. The grade of sentience is the largest possible deviation of a deformation history (including also a suitable number of its derivatives, depending upon the nature of $\mathfrak{F}$) from the rest history such that the collection of all possible present stresses remains bounded.

[1] WANG [1965, *40*].
[2] WANG [1965, *41*].

While linearly viscous fluids and general materials of the differential type, defined by Eqs. (35.5), (35.9), etc., do not satisfy the weak principle of fading memory as formulated by COLEMAN and NOLL, they do satisfy that principle in WANG's second theory. However, WANG's second theory is *not* a generalization of COLEMAN and NOLL's. In the first place, the postulated domains of the response functionals are different for the two theories. But more importantly, even if these domains are restricted in such a way that the concepts of continuity of both theories are meaningful, a response functional continuous in the sense of WANG's second theory need not be continuous in the sense of COLEMAN and NOLL's theory and vice versa. In other words, there are materials having fading memory in the sense of COLEMAN and NOLL but not in the sense of WANG's second theory, and there are also materials having fading memory in the sense of WANG's second theory but not in the sense of COLEMAN and NOLL.]

39. Stress relaxation. Consider the deformation of a simple material which, up to the time 0, is described by a prescribed deformation history $\boldsymbol{G}(s)$. Given any positive real number t, we define a new deformation history $\boldsymbol{G}_{(t)}(s)$ as follows[1]:

$$\boldsymbol{G}_{(t)}(s) = \begin{cases} \boldsymbol{0} & \text{if} \quad 0 \leqq s \leqq t. \\ \boldsymbol{G}(s-t) & \text{if} \quad s > t. \end{cases} \tag{39.1}$$

The history $\boldsymbol{G}_{(t)}(s)$ is called the *statical continuation* of $\boldsymbol{G}(s)$. Physically, $\boldsymbol{G}_{(t)}(s)$ describes a deformation which, up to the time 0, coincides with the one described by $\boldsymbol{G}(s)$, while for all times between 0 and t the material is held at rest.

We assume now that the original history $\boldsymbol{G}(s)$ is bounded in magnitude, $|\boldsymbol{G}(s)| \leqq M < \infty$. For the recollection (38.8) of $\boldsymbol{G}_{(t)}(s)$ we then obtain

$$\|\boldsymbol{G}_{(t)}(s)\|_h^2 = \int_t^\infty |\boldsymbol{G}(s-t)|^2 h(s)^2\, ds \leqq M^2 \int_t^\infty h(s)^2\, ds. \tag{39.2}$$

If the order $\mathfrak{r}$ of the obliviator $h(s)$ is greater than $\frac{1}{2}$, it follows from (38.5) that $\lim_{t\to\infty} \int_t^\infty h(s)^2\, ds = 0$ and hence, by (39.2), that[2]

$$\lim_{t\to\infty} \|\boldsymbol{G}_{(t)}(s)\|_h = 0. \tag{39.3}$$

From (39.3) and (38.1) we infer the following ***theorem on stress relaxation***[3]: *Assume that the weak principle of fading memory (C) holds for a particle. Let $\boldsymbol{G}(s)$ be any bounded deformation history at that particle, let $\boldsymbol{G}_{(t)}(s)$ be the statical continuation of $\boldsymbol{G}(s)$, and let $\boldsymbol{T}(t)$ be the stress corresponding to $\boldsymbol{G}_{(t)}(s)$. Then*

$$\lim_{t\to\infty} \boldsymbol{T}(t) = \boldsymbol{R}\,\mathfrak{f}(\boldsymbol{C})\,\boldsymbol{R}^T, \tag{39.4}$$

where $\boldsymbol{R}\,\mathfrak{f}(\boldsymbol{C})\,\boldsymbol{R}^T$ is the equilibrium value of the stress.

This theorem states that in a simple material with fading memory, if it is held in a fixed configuration from some time on, the stress will decay to its equilibrium value, which depends only on the present values of the Cauchy-Green tensor $\boldsymbol{C}$ and the rotation tensor $\boldsymbol{R}$ and not on the previous history, provided this previous history was bounded.

Elastic materials will be defined in Sect. 43 and will be studied in great detail, presentation of their theory furnishing in fact the bulk of the contents of this treatise. Early in Sect. 43 it will be shown that for an elastic material,

$$\boldsymbol{T}(t) = \boldsymbol{R}\,\mathfrak{f}(\boldsymbol{C})\,\boldsymbol{R}^T, \tag{39.5}$$

---

[1] COLEMAN and NOLL (1962) [1964, *20*, Eq. (3.8)].

[2] COLEMAN and NOLL (1962) [1964, *20*, Eq. (3.9)].

[3] An only slightly less general theorem was obtained by COLEMAN and NOLL (1962) [1964, *20*, § 3].

for all t, no matter what the history $\boldsymbol{G}(s)$ may be. I.e., for an elastic material $\underset{s=0}{\overset{\infty}{\mathfrak{F}}}(\boldsymbol{G}(s); \boldsymbol{C}) \equiv \boldsymbol{0}$. A certain general position for the theory of elastic materials is given by the following reformulation of the ***theorem on stress relaxation***: *In equilibrium, every simple material has the response of an elastic material, and moreover, if the weak form of the principle of fading memory holds, then also the response to the statical continuation of any bounded deformation history is, ultimately, that of an elastic material.*

We note that the assumption that the previous history is bounded may be dropped when the obliviator is an exponential of the form (38.7). In this case, it suffices to assume that $\boldsymbol{G}(s)$ belongs to the function space $\mathscr{H}$. [Added in proof. In WANG's first theory, the theorem on stress relaxation, freed of the restriction to bounded deformation histories, is proved to hold for any obliviating measure that satisfies the following condition:

$$\mu(\mathscr{I}) \geqq \mu(\mathscr{I}_\alpha) \tag{39.5 A}$$

for every Borel set $\mathscr{I}$ on $[M, \infty)$, where M is any positive constant and where $\mathscr{I}_\alpha$ is the Borel set on $[M, \infty)$ obtained by translating $\mathscr{I}$ by a non-negative number α, namely, $\mathscr{I}_\alpha \equiv \{x | x = y + \alpha,\ y \in \mathscr{I}\}$. Thus, in a very general way, the obliviating measure μ assigns less weight to the part of the history $\boldsymbol{G}(s)$ occuring in the distant past than to the part in the recent past. (From (38.19), the obliviating measure corresponding to an obliviator clearly satisfies (39.5 A). Consequently, the theorem on stress relaxation in the preceding text is included as a special case in WANG's results.)

In WANG's second theory, the statical continuation $\boldsymbol{G}_{(t)}(s)$ obviously converges to the rest history with respect to the topology mentioned at the end of Sect. 38. Thus the theorem on stress-relaxation holds trivially for materials obeying the weak principle of fading memory in the sense of WANG's second theory.]

The mode of decay of the stress to its equilibrium value depends, in general, on the previous history. We investigate this mode of decay in the case when the previous history consists of a sudden jump deformation, or strain impulse. Assume, then, that the material is at rest at all times $\tau < 0$, that it is subjected to a sudden deformation at time $\tau = 0$, and that it is held at rest again for all times $\tau > 0$. We take the initial rest configuration as reference and denote the gradient of the jump deformation at $\tau = 0$ by $\boldsymbol{F}$. It is easily seen that the relative deformation gradient $\boldsymbol{F}_{(t)}(\tau)$ at a time $t > 0$ is given by

$$\boldsymbol{F}_{(t)}(\tau) = \begin{cases} \boldsymbol{F}^{-1} & \text{if} \quad \tau < 0, \\ \boldsymbol{1} & \text{if} \quad \tau > 0. \end{cases} \tag{39.6}$$

The corresponding relative right Cauchy-Green tensor is

$$\boldsymbol{C}_{(t)}(\tau) = \boldsymbol{F}_{(t)}(\tau)^T \boldsymbol{F}_{(t)}(\tau) = \begin{cases} (\boldsymbol{F}^{-1})^T \boldsymbol{F}^{-1} = \boldsymbol{B}^{-1} & \text{if} \quad \tau < 0, \\ \boldsymbol{1} & \text{if} \quad \tau > 0, \end{cases} \tag{39.7}$$

where

$$\boldsymbol{B} = \boldsymbol{F}\boldsymbol{F}^T \tag{39.8}$$

is the left Cauchy-Green tensor of the jump deformation. Using the relation (23.5) between $\boldsymbol{B}$ and $\boldsymbol{C}$, we see that the deformation history (38.2) corresponding to (39.7) is given by

$$\boldsymbol{G}(s) = (\boldsymbol{C}^{-1} - \boldsymbol{1})\,\sigma(s - t), \quad t > 0, \tag{39.9}$$

where σ is the step function defined by

$$\sigma(\tau)=\begin{cases}0 & \text{if} \quad \tau<0, \\ 1 & \text{if} \quad \tau>0.\end{cases} \tag{39.10}$$

Substitution of the history (39.9) into the functional $\mathfrak{F}$ of (38.1) shows that the value of this functional depends only on $\boldsymbol{C}$ and the time t ($t>0$). Hence the entire right-hand side of (38.1) reduces to a function of $\boldsymbol{C}$ and t only:

$$\boldsymbol{R}^T \boldsymbol{T} \boldsymbol{R}=\bar{\mathfrak{f}}(\boldsymbol{C}; t), \qquad t>0. \tag{39.11}$$

This equation has the same form as the constitutive equation (39.5) for an elastic material, except that the right-hand side of (39.11) depends explicitly on the time t. Therefore, *the stress arising from a strain impulse is the same as if the material were an elastic one, except that the form of the response function* $\bar{\mathfrak{f}}$ *depends upon the time since the impulse occurred*[1]. It follows that the special solutions of the theory of elastostatics to be discussed in Subchapter DI apply also to any simple material subjected to a sudden deformation. The only difference is that the material functions occurring in those solutions must be regarded as depending explicitly on the time.

In the special case of a sudden deformation, the theorem on stress relaxation states that

$$\lim_{t\to\infty} \bar{\mathfrak{f}}(\boldsymbol{C}, t)=\mathfrak{f}(\boldsymbol{C}). \tag{39.12}$$

This limit relation implies similar relations for the various material functions occurring in the special solutions of Subchapter DI.

40. Asymptotic approximations. In the remainder of this subchapter we deal with simple materials which satisfy the strong principle of fading memory ($C^{\mathfrak{n}}$) (cf. Sect. 38), and we add the following additional smoothness requirements. (i) The Fréchet-differentiability of the response functional $\overset{\infty}{\underset{s=0}{\mathfrak{F}}}(\boldsymbol{G}(s), \boldsymbol{C})$ postulated in ($C^{\mathfrak{n}}$) is uniform in the tensor parameter $\boldsymbol{C}$. (ii) The tensor function $\mathfrak{f}(\boldsymbol{C})$ of (38.1) is $\mathfrak{n}$ times continuously differentiable. By the requirement (i) we mean that the variations $\delta^{\mathfrak{k}} \overset{\infty}{\underset{s=0}{\mathfrak{F}}}(\boldsymbol{G}(s); \boldsymbol{C})$ depend continuously on $\boldsymbol{G}(s)$ and $\boldsymbol{C}$ jointly and that the error term in (38.16), which of course depends on the parameter $\boldsymbol{C}$ also, is of the order $o(\|\boldsymbol{G}(s)\|_h^{\mathfrak{n}})$ *uniformly in* $\boldsymbol{C}$. The two requirements (i) and (ii) are of a technical nature. Together with ($C^{\mathfrak{n}}$) they suffice for justifying the approximations to be discussed here. Substitution of the approximation formula (38.16) into the constitutive equation (38.1) yields

$$\boldsymbol{R}^T \boldsymbol{T} \boldsymbol{R}=\mathfrak{f}(\boldsymbol{C})+\sum_{\mathfrak{k}=1}^{\mathfrak{n}} \frac{1}{\mathfrak{k}!} \delta^{\mathfrak{k}} \overset{\infty}{\underset{s=0}{\mathfrak{F}}}(\boldsymbol{G}^*(s); \boldsymbol{C})+o(\|\boldsymbol{G}^*(s)\|_h^{\mathfrak{n}}). \tag{40.1}$$

---

[1] This idea has developed by degrees. NOLL [1955, *19*, § 19] proved a theorem of this kind for certain materials of the rate type. RIVLIN [1956, *23*] then proposed the result as being physically natural; later RIVLIN [1960, *47*, § VII] [1959, *23*, § 6] derived the above theorem under certain special assumptions about the material. In these papers RIVLIN also explored solutions for particular cases; his results may be obtained by specialization or approximation from those we shall present below in Sects. 54—57. Relevant experiments on strain impulse and stress relaxation have been reported by BERGEN [1960, *6*], BERGEN, MESSERSMITH, and RIVLIN [1960, *7*], and BERNSTEIN, KEARSLEY and ZAPAS [1963, *5*].

This result shows that the general constitutive equation of a simple material with fading memory can be approximated by a constitutive equation of the form

$$\boldsymbol{R}^T\boldsymbol{T}\boldsymbol{R}=\mathfrak{f}(\boldsymbol{C})+\sum_{\mathfrak{k}=1}^{\mathfrak{n}}\underset{s=0}{\overset{\infty}{\mathfrak{P}_{\mathfrak{k}}}}(\boldsymbol{G}^*(s);\boldsymbol{C}), \tag{40.2}$$

where $\mathfrak{P}_{\mathfrak{k}}$ is a bounded homogeneous polynomial functional of $\boldsymbol{G}^*(s)$ of degree $\mathfrak{k}$, and a continuous function of the tensor parameter $\boldsymbol{C}$. The error made in replacing the general equation by its approximation (40.2) approaches zero faster than the $\mathfrak{n}$'th power of the recollection (38.8) of the deformation history (38.2). A material obeying a constitutive equation of the form (40.2) will be called a *simple material of order* $\mathfrak{n}$.

When $\mathfrak{n}=1$, the sum in (40.2) reduces to a single term involving the bounded linear functional $\mathfrak{P}_1$. A theorem of the theory of Hilbert spaces states that any bounded linear functional may be represented by an inner product (38.9). Allowing for the fact that the values of $\mathfrak{P}_1$ are tensors rather than scalars, we find that (40.2) for $\mathfrak{n}=1$ reduces to

$$\boldsymbol{R}^T\boldsymbol{T}\boldsymbol{R}=\mathfrak{f}(\boldsymbol{C})+\int_0^\infty \mathsf{K}(\boldsymbol{C};s)\,[\boldsymbol{G}^*(s)]\,ds, \tag{40.3}$$

where $\mathsf{K}(\boldsymbol{C};s)\,[\;]$, for each choice of $s\geqq 0$ and $\boldsymbol{C}$, is a linear function of one tensor variable. $\mathsf{K}(\boldsymbol{C};s)$ may be identified with a fourth-order tensor (cf. (7.11)) whose magnitude, as defined by (6.14), satisfies

$$\int_0^\infty |\mathsf{K}(\boldsymbol{C};s)|^2 h(s)^{-2}\,ds<\infty. \tag{40.4}$$

Equation (40.3) is a constitutive equation of the integral type of order *1*. The considerations of Sect. 37, in particular those pertaining to the special cases of isotropic materials and simple fluids of the first order, apply. We call the theory of simple materials of order *1*, based on the constitutive equation (40.3), *finite linear visco-elasticity*.

Every simple material of order *1* is of the integral type, but simple materials of order $\mathfrak{n}>1$ are in general not of the integral type, because not all bounded polynomial functionals in Hilbert space have integral representations[1].

Roughly speaking, the constitutive equation (40.2) approximates the general constitutive equation of a simple material with fading memory when the recollection $\|\boldsymbol{G}^*(s)\|_h$ of the history (38.2) is small. The definition (38.8) shows that $\|\boldsymbol{G}^*(s)\|_h$ will be small when $\boldsymbol{G}^*(s)$ is small for small values of s even though it may be large for large values of s. In physical terms, the recollection of the history $\boldsymbol{G}^*(s)$ is small when the deformation, taking the present configuration as reference, was small in the recent past even though it may have been large in the distant past. This condition will be satisfied if, in a sense to be made precise, the deformation is "slow". We assume a particular history $\boldsymbol{G}(s)$ in the space $\mathscr{H}$ given, and we consider the histories

$$\boldsymbol{G}^\alpha(s)=\boldsymbol{G}(\alpha s),\quad 0<\alpha<1, \tag{40.5}$$

obtained from $\boldsymbol{G}(s)$ by a *retardation* with retardation factor α. Physically, the deformations corresponding to $\boldsymbol{G}^\alpha(s)$ are "the same" as those corresponding to $\boldsymbol{G}(s)$ but take place at a slower rate. Under the assumption that the obliviator h is of order $\mathfrak{r}>\mathfrak{n}+\frac{1}{2}$ and that $\boldsymbol{G}(s)$ has $\mathfrak{n}$ derivatives at $s=0$, it follows from

[1] A different process of approximation, introduced by Chacon and Rivlin [1964, *13*] (cf. our remarks in footnote 1, p. 61), was shown by Lianis and De Hoff [1964, *56*] to yield (40.3) at the first stage and similar integral representations at all stages.

the ***retardation theorem*** of COLEMAN and NOLL[1] that

$$G^\alpha(s) = \sum_{j=0}^{n} \frac{s^j}{j!} \overset{(j)}{G^\alpha} + o(\alpha^n), \tag{40.6}$$

where

$$\overset{(j)}{G^\alpha} = \frac{d^j}{ds^j} G^\alpha(s)\Big|_{s=0} = \alpha^j \frac{d^j}{ds^j} G(s)\Big|_{s=0}. \tag{40.7}$$

The order symbol $o(\alpha^n)$ in (40.6) means that the recollection of $\alpha^{-n} o(\alpha^n)$ tends to zero as $\alpha \to 0$. Let $\boldsymbol{G}^*(s)$ be related by (38.2) to a relative Cauchy-Green tensor $\boldsymbol{C}_{(t)}(\tau)$ which is continuous and has n derivatives at the present time $\tau = t$. Then, since $\boldsymbol{C}_{(t)}(t) = \mathbf{1}$, we have

$$\boldsymbol{G}^*(0) = \mathbf{0}, \qquad \frac{d^j}{ds^j} \boldsymbol{G}^*(s)\Big|_{s=0} = (-1)^j \boldsymbol{R}^T(t) \overset{(j)}{\boldsymbol{C}}_{(t)}(t) \boldsymbol{R}(t) = (-1)^j \boldsymbol{A}_j^*(t), \tag{40.8}$$

where $\boldsymbol{A}_j^*$ is related to the j'th Rivlin-Ericksen tensor $\boldsymbol{A}_j$ by $\boldsymbol{A}_j^* = \boldsymbol{R}^T \boldsymbol{A}_j \boldsymbol{R}$ (cf. (35.7)). If we put

$$\boldsymbol{A}_j^\alpha = \alpha^j \boldsymbol{A}_j, \qquad \boldsymbol{A}_j^{*\alpha} = \alpha^j \boldsymbol{A}_j^*, \tag{40.9}$$

we may regard $\boldsymbol{A}_j^\alpha$ as the j'th Rivlin-Ericksen tensor corresponding to the retarded history $\boldsymbol{G}^{*\alpha}(s)$. Substituting (40.9), (40.8) and (40.7) into (40.6), we obtain

$$\boldsymbol{G}^{*\alpha}(s) = \sum_{j=1}^{n} \frac{s^j}{j!} (-1)^j \boldsymbol{A}_j^{*\alpha} + o(\alpha^n). \tag{40.10}$$

The j'th term in the sum in (40.10) is of order $O(\alpha^j)$. Since there is no term with $j = 0$, we have, in particular,

$$\boldsymbol{G}^{*\alpha}(s) = O(\alpha). \tag{40.11}$$

Hence, if we substitute $\boldsymbol{G}^{*\alpha}(s)$ for $\boldsymbol{G}^*(s)$ in (40.1), the error term is of order $o(\alpha^n)$. Thus, up to terms of order $o(\alpha^n)$, the general equation (40.1) may be replaced, for $\boldsymbol{G}^{*\alpha}(s)$, by the constitutive equation (40.2) for simple materials of order n, and we obtain

$$\boldsymbol{R}^T \boldsymbol{T}_\alpha \boldsymbol{R} = \mathfrak{f}(\boldsymbol{C}) + \sum_{\mathfrak{k}=1}^{n} \underset{s=0}{\overset{\infty}{\mathfrak{P}_{\mathfrak{k}}}} (\boldsymbol{G}^{*\alpha}(s); \boldsymbol{C}) + o(\alpha^n), \tag{40.12}$$

where $\boldsymbol{T}_\alpha$ denotes the stress corresponding to the retarded history $\boldsymbol{G}^{*\alpha}(s)$. Using (40.10) and the fact that $\mathfrak{P}_{\mathfrak{k}}$ is a bounded homogeneous polynomial functional of degree $\mathfrak{k}$, one can prove[2] that

$$\underset{s=0}{\overset{\infty}{\mathfrak{P}_{\mathfrak{k}}}} (\boldsymbol{G}^{*\alpha}(s); \boldsymbol{C}) = \sum_{\substack{(j_1, \ldots, j_{\mathfrak{k}}) \\ \mathfrak{k}\ \text{fixed}}} \mathfrak{l}_{j_1, \ldots, j_{\mathfrak{k}}}(\boldsymbol{C}) [\boldsymbol{A}_{j_1}^{*\alpha}, \ldots, \boldsymbol{A}_{j_{\mathfrak{k}}}^{*\alpha}] + o(\alpha^n), \tag{40.13}$$

where each $\mathfrak{l}_{j_1, \ldots, j_{\mathfrak{k}}}(\boldsymbol{C})[\;]$, for each choice of $\boldsymbol{C}$, is a multilinear tensor function of $\mathfrak{k}$ tensor variables. The summation is to be extended over all sets of $\mathfrak{k}$ indices $(j_1, \ldots, j_{\mathfrak{k}})$ satisfying the inequalities

$$1 < j_1 \leqq \cdots \leqq j_{\mathfrak{k}} \leqq n, \quad j_1 + \cdots + j_{\mathfrak{k}} \leqq n. \tag{40.14}$$

The results (40.12) and (40.13), which may be called COLEMAN and NOLL's ***explicit retardation formulae***, show that, for slow motion, the general constitutive equation of a simple material with fading memory can be approximated by a

[1] COLEMAN and NOLL [1960, *17*, Theorem 1]. The proof of this theorem is too difficult to be included here.

[2] The proof, straightforward but tedious, is given by COLEMAN and NOLL [1960, *17*, Theorem 2].

constitutive equation of the form

$$\boldsymbol{R}^T\boldsymbol{T}\boldsymbol{R}=\mathfrak{f}(\boldsymbol{C})+\sum_{(\mathfrak{j}_1,\ldots,\mathfrak{j}_\mathfrak{k})}\mathfrak{l}_{\mathfrak{j}_1,\ldots,\mathfrak{j}_\mathfrak{k}}(\boldsymbol{C})[\boldsymbol{A}^*_{\mathfrak{j}_1},\ldots,\boldsymbol{A}^*_{\mathfrak{j}_\mathfrak{k}}], \tag{40.15}$$

where the summation is to be extended over all sets of indices $(\mathfrak{j}_1,\ldots,\mathfrak{j}_\mathfrak{k})$, $\mathfrak{k}=1,\ldots\mathfrak{n}$, that satisfy the inequalities (40.14). If (40.15) is applied to a motion which is related to some "model motion" by a retardation, then the error made in using (40.15) instead of the general equation approaches zero faster than the $\mathfrak{n}$'th power of the retardation factor α.

Comparison of (40.15) with (35.6) shows that (40.15) is a special case of the constitutive equation for materials of the differential type of complexity $\mathfrak{n}$; in particular, $\boldsymbol{A}_\mathfrak{n}$ occurs linearly. A material obeying (40.15) will be called a *material of the differential type of grade* $\mathfrak{n}$. The derivation of (40.15) shows that, *for slow motions, the theory of general simple materials with fading memory is approximated by the theory of materials of the differential type of grade* $\mathfrak{n}$.

Note that the right-hand side of (40.15) is a polynomial tensor function in $\boldsymbol{A}^*_1,\ldots,\boldsymbol{A}^*_\mathfrak{k}$. The $\mathfrak{l}_{\mathfrak{j}_1,\ldots,\mathfrak{j}_\mathfrak{k}}(\boldsymbol{C})$ may be identified with tensors of order $2\mathfrak{k}+2$ (cf. Sect. 7), whose components may be regarded as the coefficients of the polynomial tensor function. Since the physical dimensions of $\boldsymbol{A}^*_\mathfrak{j}$ and $\boldsymbol{R}^T\boldsymbol{T}\boldsymbol{R}$ are given by

$$\dim \boldsymbol{A}^*_\mathfrak{j}=\dim \boldsymbol{A}_\mathfrak{j}=T^{-\mathfrak{j}},\qquad \dim \boldsymbol{R}^T\boldsymbol{T}\boldsymbol{R}=\dim\boldsymbol{T}=ML^{-1}T^{-2}, \tag{40.16}$$

we see that the constitutive equation of a material of the differential type of grade $\mathfrak{n}$ may be described as follows: The rotated stress $\boldsymbol{R}^T\boldsymbol{T}\boldsymbol{R}$ is given by a polynomial tensor function of the rotated Rivlin-Ericksen tensors $\boldsymbol{A}^*_\mathfrak{j}=\boldsymbol{R}^T\boldsymbol{A}_\mathfrak{j}\boldsymbol{R}$, such that the coefficients, which depend on the right Cauchy-Green tensor $\boldsymbol{C}$, have physical dimensions $ML^{-1}T^\mathfrak{q}$, $-2\leqq\mathfrak{q}\leqq\mathfrak{n}-2$.

In the case of a isotropic material (cf. Sect. 35), the constitutive equation (40.15) reduces to

$$\boldsymbol{T}=\mathfrak{f}(\boldsymbol{B})+\sum_{(\mathfrak{j}_1,\ldots,\mathfrak{j}_\mathfrak{k})}\mathfrak{l}_{\mathfrak{j}_1,\ldots,\mathfrak{j}_\mathfrak{k}}(\boldsymbol{B})[\boldsymbol{A}_{\mathfrak{j}_1},\ldots,\boldsymbol{A}_{\mathfrak{j}_\mathfrak{k}}], \tag{40.17}$$

where $\boldsymbol{B}$ is the left Cauchy-Green tensor and the $\boldsymbol{A}_\mathfrak{j}$ are the Rivlin-Ericksen tensors; moreover, all terms on the right side of (40.17) are isotropic functions of their tensor-arguments.

In the case of simple fluids, the dependence on $\boldsymbol{B}$ in (40.17) reduces to a dependence on the density ϱ only, and the equilibrium term $\mathfrak{f}(\boldsymbol{B})$ becomes $-p(\varrho)\mathbf{1}$. Hence the *constitutive equation of a fluid of the differential type of grade* $\mathfrak{n}$ is of the form

$$\boldsymbol{T}=-p(\varrho)\mathbf{1}+\sum_{(\mathfrak{j}_1,\ldots,\mathfrak{j}_\mathfrak{k})}\mathfrak{l}_{\mathfrak{j}_1,\ldots,\mathfrak{j}_\mathfrak{k}}(\varrho)[\boldsymbol{A}_{\mathfrak{j}_1},\ldots,\boldsymbol{A}_{\mathfrak{j}_\mathfrak{k}}], \tag{40.18}$$

where the summation extends over all sets of indices satisfying (40.14).

The case when $\mathfrak{n}=0$ in (40.12), corresponding to the weak principle of fading memory, deserves special mention:

$$\boldsymbol{R}^T\boldsymbol{T}_\alpha\boldsymbol{R}=\mathfrak{f}(\boldsymbol{C})+o(1). \tag{40.19}$$

That is, *in sufficiently retarded motion the response of any simple material is that of an elastic material*[1].

[1] Coleman and Noll [1964, *19*, p. 88]. (Added in proof. This theorem, too, has been generalized by Wang [1965, *40*, § 5] in his first study. There he uses the concept of obliviating measure to define on the space of deformation histories a topology more general than that given by his own generalized recollection (38.18). Formulating a weak principle of fading memory in terms of this more general topology, Wang finds that (40.19) remains valid. Thus approximately elastic response in sufficiently slow motion seems to be an extremely general property of simple materials.)

[Added in proof. In WANG's second theory[1], described at the end of Sect. 38, other expansions similar to (40.12) are constructed. WANG defines the *λ-relaxation* of a simple material by the constitutive functional

$$ {}_{\lambda}\overset{\infty}{\underset{s=0}{\mathfrak{F}}}\big(\boldsymbol{G}(s)\big)=\overset{\infty}{\underset{s=0}{\mathfrak{F}}}\big(\lambda\,\boldsymbol{G}(s)\big). \tag{40.20} $$

The strong principle of fading memory is then stated as the Taylor-series expansion

$$ {}_{\lambda}\overset{\infty}{\underset{s=0}{\mathfrak{F}}}\big(\boldsymbol{G}(s)\big)=\sum_{k=1}^{n}\frac{\lambda^k}{k!}\,\delta^k\overset{\infty}{\underset{s=0}{\mathfrak{F}}}\big(\boldsymbol{G}(s)\big)+\overset{\infty}{\underset{s=0}{\boldsymbol{o}}}\big(\lambda^n,\boldsymbol{G}(s)\big) \tag{40.21} $$

where the functional $\boldsymbol{o}$ is of order $o(\lambda^n)$ for each fixed $\boldsymbol{G}(s)$. Similarly, WANG defines the *α-retardation* of a simple material by the constitutive functional

$$ \overset{\infty}{\underset{s=0}{\mathfrak{F}_{\alpha}}}\big(\boldsymbol{G}(s)\big)\equiv\overset{\infty}{\underset{s=0}{\mathfrak{F}}}\big(\boldsymbol{G}^{\alpha}(s)\big)=\overset{\infty}{\underset{s=0}{\mathfrak{F}}}\big(\boldsymbol{G}(\alpha s)\big). \tag{40.22} $$

From the strong principle of fading memory (40.21), he proves that for each fixed $\boldsymbol{G}(s)$, $\overset{\infty}{\underset{s=0}{\mathfrak{F}_{\alpha}}}\big(\boldsymbol{G}(s)\big)$ has a Taylor series expansion in α about $\alpha=0$. The coefficient of α^k $(k\leqq n)$, generally, is a continuous homogeneous polynomial functional of $\boldsymbol{G}(s)$ of degree k which is local[2] at $s=0$. In particular, if the deformation history $\boldsymbol{G}(s)$ is smooth near $s=0$, then the homogeneous polynomial functional reduces to a continuous polynomial function of the Rivlin-Ericksen tensors of $\boldsymbol{G}(s)$, and the expansion becomes[3]

$$ \boldsymbol{R}^{T}\boldsymbol{T}_{\alpha}\boldsymbol{R}=\mathfrak{f}(\boldsymbol{C})+\sum_{\mathfrak{k}=1}^{\mathfrak{n}}\alpha^{\mathfrak{k}}\sum_{(\mathfrak{j}_1,\ldots\mathfrak{j}_{\mathfrak{k}})}\mathfrak{t}_{\mathfrak{j}_1\ldots\mathfrak{j}_{\mathfrak{k}}}(\boldsymbol{C})\,[\boldsymbol{A}^{*}_{\mathfrak{j}_1},\ldots,\boldsymbol{A}^{*}_{\mathfrak{j}_k}]+\boldsymbol{o}(\alpha^{n}), \tag{40.23} $$

where

$$ \boldsymbol{T}_{\alpha}=\mathfrak{f}(\boldsymbol{C})+\overset{\infty}{\underset{s=0}{\mathfrak{F}_{\alpha}}}\big(\boldsymbol{G}(s)\big). \tag{40.24} $$

While COLEMAN and NOLL's retardation process considers response of a given material to *a family of deformation histories* of like kind, described more and more slowly, WANG's λ-relaxations and α-retardations are *families of materials*. As $\lambda\to 0$, the λ-relaxations respond to a given deformation in just the same way as the original material would respond to deformation described in the same way in time, but of proportionately lesser magnitude at each instant. The α-retardations, on the other hand, respond in just the same way as the original material responds to a progressively retarded deformation history. If the times and grades of sentience (Sect. 38) of the respective materials are denoted by η_0, δ_0, ${}^{(\lambda)}\eta_0$, ${}^{(\lambda)}\delta_0$, $\eta_0^{(\alpha)}$, $\delta_0^{(\alpha)}$, then

$$ \left.\begin{aligned} {}^{(\lambda)}\eta_0&=\eta_0, & {}^{(\lambda)}\delta_0&=\frac{1}{\lambda}\,\delta_0,\\ \eta_0^{(\alpha)}&=\alpha\,\eta_0, & \delta_0^{(\alpha)}&=\delta_0. \end{aligned}\right\} \tag{40.25} $$

Thus, as $\alpha\to 0$, the α-retardations are materials with a fixed grade of sentience but proportionately decreased time of sentience, while the λ-relaxations have fixed time of sentience but proportionately increased grade of sentience. The α-retardations have smaller and smaller major memories but are neither softer nor harder

---

[1] WANG [1965, *41*].

[2] I.e., the value of the functional is the same if the argument functions coincide in a neighborhood of $s=0$.

[3] This expansion gives a rigorous basis for the heuristic theory of TRUESDELL, presented for fluids in Sect. 121, below.

than the given material, while the λ-relaxations are materials successively less and less sentient in that they show less and less variation in stress when subjected to various deformations, but their major memories remain constant. Of course, COLEMAN and NOLL's process could have been interpreted as one of α-retardation, and WANG's two processes, conversely, could be interpreted as approximating the response of a given material to families of deformation histories.

For all materials of the differential type, including elastic materials and linearly viscous fluids, WANG's asymptotic approximations apply and reduce to the Taylor series expansion of the response functions of these materials about the rest history along a certain fixed curve (parametrized in α) in the Cartesian space of $(\boldsymbol{C}^*, \boldsymbol{A}_1^*, \ldots, \boldsymbol{A}_n^*)$.]

41. Position of the classical theories of viscosity and visco-elasticity. Various classical theories of materials may be shown to furnish approximations to the theory of simple materials, or to that of isotropic simple materials, or to that of simple fluids, when the motion is *slow* or the deformation is *small*. We consider these two possibilities in turn, restricting attention to materials free from internal constraints, since the reader may easily modify the results in accord with the principle laid down in Sect. 30.

α) *Slow motions.* In the special case when $n=0$, the constitutive equations (40.15), (40.17) and (40.18) reduce to

$$\boldsymbol{R}^T\boldsymbol{T}\boldsymbol{R}=\mathfrak{f}(\boldsymbol{C}), \quad \boldsymbol{T}=\mathfrak{f}(\boldsymbol{B}), \quad \text{and} \quad \boldsymbol{T}=-p(\varrho)\,\boldsymbol{1}, \tag{41.1}$$

respectively. These are the classical constitutive equations of finite elasticity. Eq. $(41.1)_1$, which has been stated already as Eq. (39.5), will be derived as Eq. (43.6); its special case $(41.1)_2$ for isotropic materials will be derived as Eq. (47.4); while $(41.1)_3$ is the constitutive equation for elastic fluids, on which a large portion of classical hydrodynamics is based. The approximation theorem of the previous section shows that the classical equations (41.1) can be expected to apply to general simple materials with fading memory when only very slow motions are considered.

In the case when $n=1$, the equations (40.15), (40.17), and (40.18) become

$$\boldsymbol{R}^T\boldsymbol{T}\boldsymbol{R}=\mathfrak{f}(\boldsymbol{C})+\mathsf{V}(\boldsymbol{C})\,[\boldsymbol{R}^T\boldsymbol{A}_1\boldsymbol{R}], \tag{41.2}$$

$$\boldsymbol{T}=\mathfrak{f}(\boldsymbol{B})+\mathsf{V}(\boldsymbol{B})\,[\boldsymbol{A}_1], \tag{41.3}$$

and

$$\boldsymbol{T}=-p(\varrho)\,\boldsymbol{1}+\mathsf{V}(\varrho)\,[\boldsymbol{A}_1], \tag{41.4}$$

respectively. These equations may be regarded as supplying the first-order correction, still for slow motions, to the classical equations (41.1).

The representation (12.5) and Eq. $(25.6)_1$ show that (41.4) may be given the explicit form

$$\boldsymbol{T}=(-p(\varrho)+\lambda\,\mathrm{tr}\boldsymbol{D})\,\boldsymbol{1}+2\mu\boldsymbol{D} \tag{41.5}$$

where $\boldsymbol{D}$ is the stretching tensor, i.e., the symmetric part of the velocity gradient. Now (41.5) is the classical constitutive equation of a linearly viscous fluid (CFT, Sect. 298). Hence the classical theory of linear viscosity applies to general simple fluids with fading memory in the limit of slow flow. The degree of "slowness" required to make (41.5) applicable depends on the particular material and on the particular flow being retarded. The classical theory based on (41.5) is generally regarded as giving a remarkably good description of the mechanical behavior of some real fluids, such as water and air, for all motions ordinarily encountered.

With other real fluids, such as molten plastics, it is easy to produce in the laboratory phenomena which cannot be described by Eqs. (41.5).

The constitutive equations (41.2) and (41.3) are analogues, applying to non-fluids, of the classical equation (41.5) for fluids. We call materials obeying (41.2) *linearly viscous materials*. The following equation, easily seen to be equivalent to (41.2), is an alternative form of the constitutive equation for linearly viscous materials:

$$\overline{\boldsymbol{T}}=\mathfrak{g}(\boldsymbol{C})+\mathsf{P}(\boldsymbol{C})[\boldsymbol{F}^T\boldsymbol{D}\boldsymbol{F}], \tag{41.6}$$

where $\overline{\boldsymbol{T}}=\boldsymbol{F}^T\boldsymbol{T}\boldsymbol{F}$ is the convected stress tensor. A component form of (41.6) is

$$T_{km}=[\mathfrak{g}_{\alpha\beta}(C_{\mu\nu})+P_{\alpha\beta}{}^{\gamma\delta}x^p{}_{,\gamma}x^q{}_{,\delta}D_{pq}]X^\alpha{}_{,k}X^\beta{}_{,m}. \tag{41.7}$$

The representation theorems of Sects. 12 and 13 may be used to render (41.3) explicit. When the dependence of $\mathsf{V}(\boldsymbol{B})$ on $\boldsymbol{B}$ is polynomial, we obtain the following explicit equation of an *isotropic linearly viscous material*[1]:

$$\left.\begin{aligned}\boldsymbol{T}=&\aleph_0\boldsymbol{1}+\aleph_1\boldsymbol{B}+\aleph_2\boldsymbol{B}^2+[\mathfrak{d}_1\operatorname{tr}\boldsymbol{D}+\mathfrak{d}_2\operatorname{tr}(\boldsymbol{D}\boldsymbol{B})+\mathfrak{d}_3\operatorname{tr}(\boldsymbol{D}\boldsymbol{B}^2)]\boldsymbol{1}+\\&+[\mathfrak{d}_4\operatorname{tr}\boldsymbol{D}+\mathfrak{d}_5\operatorname{tr}(\boldsymbol{D}\boldsymbol{B})+\mathfrak{d}_6\operatorname{tr}(\boldsymbol{D}\boldsymbol{B}^2)]\boldsymbol{B}+\\&+[\mathfrak{d}_7\operatorname{tr}\boldsymbol{D}+\mathfrak{d}_8\operatorname{tr}(\boldsymbol{D}\boldsymbol{B})+\mathfrak{d}_9\operatorname{tr}(\boldsymbol{D}\boldsymbol{B}^2)]\boldsymbol{B}^2+\\&+\mathfrak{d}_{10}\boldsymbol{D}+\mathfrak{d}_{11}(\boldsymbol{D}\boldsymbol{B}+\boldsymbol{B}\boldsymbol{D})+\mathfrak{d}_{12}(\boldsymbol{D}\boldsymbol{B}^2+\boldsymbol{B}^2\boldsymbol{D}),\end{aligned}\right\} \tag{41.8}$$

where $\aleph_0, \aleph_1, \aleph_2$, and $\mathfrak{d}_1, \ldots, \mathfrak{d}_{12}$ are functions of the principal invariants $I_{\boldsymbol{B}}$, $II_{\boldsymbol{B}}$, $III_{\boldsymbol{B}}$. It is likely that (41.8) applies even when the dependence on $\boldsymbol{B}$ is not polynomial. The terms involving $\mathfrak{d}_1, \ldots, \mathfrak{d}_{12}$ in (41.8) may be interpreted as representing linear internal friction.

β) *Small deformations.* We now investigate the behavior of simple materials with fading memory for "small" deformations. To make the meaning of "small" precise, assume that $\boldsymbol{F}=\boldsymbol{F}(\tau)$ is the deformation gradient at some particle, taken relative to a fixed reference configuration, and put

$$\boldsymbol{H}\equiv\boldsymbol{F}-\boldsymbol{1}. \tag{41.9}$$

The tensor $\boldsymbol{H}$ is the gradient of the vector $\boldsymbol{u}=\boldsymbol{u}(\tau)$ of the displacement from the reference configuration. We put

$$\varepsilon=\sup_{t\geqq\tau}|\boldsymbol{H}(\tau)|=\sup_{t\geqq\tau}\sqrt{u^i{}_{,j}(\tau)\,u^j{}_{,i}(\tau)} \tag{41.10}$$

and regard ε as the measure of "smallness" of the deformation history. An easy analysis[2] shows that the history (38.2) is given by

$$\boldsymbol{G}^*(s)=2[\tilde{\boldsymbol{E}}(t-s)-\tilde{\boldsymbol{E}}(t)]+O(\varepsilon^2)=O(\varepsilon), \tag{41.11}$$

where

$$\tilde{\boldsymbol{E}}=\tfrac{1}{2}(\boldsymbol{H}+\boldsymbol{H}^T), \tag{41.12}$$

---

[1] The first theory of this type, restricted to infinitesimal deformations and, presumably, sufficiently slow motions, so that (41.8) may be approximated by neglecting all squares and products of $\boldsymbol{D}$ and $\boldsymbol{B}$, was proposed by O. E. MEYER [1874, *2* and *3*] [1875, *2*]; it has a large literature, in which it is usually associated with the names of KELVIN and VOIGT. A possible but special theory for finite deformations results by supposing that the coefficients of all terms containing the products of $\boldsymbol{B}$ and $\boldsymbol{D}$ in (41.8) vanish, so that a linearly viscous response is superimposed upon the classical theory of finite elastic strain. This special theory was proposed by DUHEM [1903, *1* and *2*] [1904, *1*]. Cf. also MANACORDA [1957, *11*].

[2] Cf. COLEMAN and NOLL [1961, *7*, § 2].

the symmetric part of the displacement gradient $\boldsymbol{H}$. We call $\tilde{\boldsymbol{E}}$ the *infinitesimal strain tensor*[1] because it is used as the measure of strain in the classical theory of infinitesimal elasticity. The order term $O(\varepsilon^2)$ in (41.11) has the property that $\varepsilon^{-2}O(\varepsilon^2)$ is bounded uniformly in s. Since the obliviator h is square-integrable in $0 \leqq s < \infty$, it follows from the definition (38.8) that the recollection of $\varepsilon^{-2}O(\varepsilon^2)$ is bounded. Hence, the order symbol in (41.11) may also be interpreted in terms of recollections. It is this interpretation that will be relevant. It follows from $(41.11)_2$ that the error term in (40.1), when $\mathfrak{n}=1$, is of order $o(\varepsilon)$. Hence, the constitutive equation (40.3) of finite linear visco-elasticity will give the correct stress up to terms of order $o(\varepsilon)$, and, since the error term $O(\varepsilon^2)$ in $(41.11)_1$ gives only another contribution of order $o(\varepsilon)$, we obtain

$$\boldsymbol{R}^T\boldsymbol{T}\boldsymbol{R} = \mathfrak{f}(\boldsymbol{C}) + 2\int_0^\infty \mathsf{K}(\boldsymbol{C}; s)\,[\tilde{\boldsymbol{E}}(t-s) - \tilde{\boldsymbol{E}}(t)]\,ds + o(\varepsilon). \tag{41.13}$$

On the other hand, the right Cauchy-Green tensor $\boldsymbol{C}=\boldsymbol{C}(t)$ is related to $\tilde{\boldsymbol{E}}(t)$ by

$$\boldsymbol{C}(t) = \mathbf{1} + 2\tilde{\boldsymbol{E}}(t) + O(\varepsilon^2) = \mathbf{1} + O(\varepsilon), \tag{41.14}$$

and the rotation tensor $\boldsymbol{R}(t)$ is related to the *infinitesimal rotation tensor*

$$\tilde{\boldsymbol{R}}(t) = \tfrac{1}{2}(\boldsymbol{H}(t) - \boldsymbol{H}^T(t)) \tag{41.15}$$

by the formula

$$\boldsymbol{R}(t) = \mathbf{1} + \tilde{\boldsymbol{R}}(t) + O(\varepsilon^2) = \mathbf{1} + O(\varepsilon) \tag{41.16}$$

(cf. CFT, Sects. 54 and 57). The assumed smoothness of $\mathfrak{f}(\boldsymbol{C})$ and (41.14) imply that

$$\mathfrak{f}(\boldsymbol{C}(t)) = \boldsymbol{T}_0 + \mathsf{L}[\tilde{\boldsymbol{E}}(t)] + o(\varepsilon), \tag{41.17}$$

where $\mathsf{L}[\ ]$ is a linear tensor function (cf. Sect. 7), and where

$$\boldsymbol{T}_0 = \mathfrak{f}(\mathbf{1}) \tag{41.18}$$

is the *residual stress*, i.e., the stress the material would sustain if it had been held in the reference configuration at all past times.

Substitution of (41.14), (41.16) and (41.17) into (41.13) yields

$$\left.\begin{aligned} \boldsymbol{T}(t) - \boldsymbol{T}_0 = \tilde{\boldsymbol{R}}(t)\,\boldsymbol{T}_0 - \boldsymbol{T}_0\tilde{\boldsymbol{R}}(t) + \mathsf{L}[\tilde{\boldsymbol{E}}(t)] + {}\\ + 2\int_0^\infty \mathsf{K}(\mathbf{1}; s)\,[\tilde{\boldsymbol{E}}(t-s) - \tilde{\boldsymbol{E}}(t)]\,ds + o(\varepsilon). \end{aligned}\right\} \tag{41.19}$$

Here, to show that replacing $\boldsymbol{C}$ by $\mathbf{1}$ in $\mathsf{K}(\boldsymbol{C}; s)$ results in an error of order $o(\varepsilon)$ only, one must use the assumptions of smoothness stated in the beginning of Sect. 40. If we define the *stress-relaxation function* $\mathsf{M}(s)$ by

$$\mathsf{M}(s) = -2\int_s^\infty \mathsf{K}(\mathbf{1}; \sigma)\,d\sigma, \qquad \dot{\mathsf{M}}(s) = \frac{d}{ds}\mathsf{M}(s) = 2\mathsf{K}(\mathbf{1}; s), \tag{41.20}$$

we may rewrite (41.19) in the form[2]

$$\left.\begin{aligned} \boldsymbol{T}(t) - \boldsymbol{T}_0 = \tilde{\boldsymbol{R}}(t)\,\boldsymbol{T}_0 - \boldsymbol{T}_0\tilde{\boldsymbol{R}}(t) + (\mathsf{L} + \mathsf{M}(0))\,[\tilde{\boldsymbol{E}}(t)] + {}\\ + \int_0^\infty \dot{\mathsf{M}}(s)\,[\tilde{\boldsymbol{E}}(t-s)]\,ds + o(\varepsilon). \end{aligned}\right\} \tag{41.21}$$

---

[1] It is called the "elongation tensor" in CFT, Sect. 31, where it is interpreted also in finite strains.

[2] A survey of various ideas and results of the types presented above and in Sect. 40, including explicit forms for incompressible materials, has been published by PIPKIN [1964, *68 A*].

When ε, as given by (41.10), is small enough, the error term $o(\varepsilon)$ can be neglected in comparison with the other terms on the right-hand side of (41.21), which are of order $O(\varepsilon)$. A component form of (41.21) is

$$\left.\begin{aligned} T^{km}(t)-T_0^{km} &= \tilde{R}^k{}_p(t)\,T_0^{pm}-T_0^{kp}\tilde{R}_p{}^m(t)+\left(\mathsf{L}^{kmpq}+\mathsf{M}^{kmpq}(0)\right)\tilde{E}_{pq}(t)+\\ &+\int_0^\infty \dot{\mathsf{M}}^{kmpq}(s)\,\tilde{E}_{pq}(t-s)\,ds+o(\varepsilon).\end{aligned}\right\}\tag{41.22}$$

The equations (41.21) and (41.22) describe the theory of infinitesimal visco-elastic deformations superimposed upon a state of arbitrary strain[1].

When the residual stress (41.18) is zero, the reference configuration is called a *natural state.* In this case, if the error term $o(\varepsilon)$ is omitted, (41.21) reduces to

$$\boldsymbol{T}(t)=\left(\mathsf{L}+\mathsf{M}(0)\right)[\tilde{\boldsymbol{E}}(t)]+\int_0^\infty \dot{\mathsf{M}}(s)\,[\tilde{\boldsymbol{E}}(t-s)]\,ds.\tag{41.23}$$

This equation is the constitutive equation of the classical theory of *infinitesimal visco-elasticity*, first proposed by BOLTZMANN[2]. BOLTZMANN considered only the case when the material is isotropic. In this case, the linear tensor functions $\mathsf{L}[\;]$ and $\mathsf{M}(s)[\;]$ are isotropic and hence have representations of the form (12.5). Thus the constitutive equation of *isotropic infinitesimal visco-elasticity* is

$$\left.\begin{aligned} \boldsymbol{T}(t) &= \left\{\left(\lambda+\bar{\lambda}(0)\right)\operatorname{tr}\tilde{\boldsymbol{E}}(t)+\int_0^\infty \dot{\bar{\lambda}}(s)\operatorname{tr}\tilde{\boldsymbol{E}}(t-s)\,ds\right\}\mathbf{1}+\\ &+2\left(\mu+\bar{\mu}(0)\right)\tilde{\boldsymbol{E}}(t)+2\int_0^\infty \dot{\bar{\mu}}(s)\,\tilde{\boldsymbol{E}}(t-s)\,ds,\end{aligned}\right\}\tag{41.24}$$

where λ and μ are material constants[3] and $\bar{\lambda}(s)$ and $\bar{\mu}(s)$ are scalar material functions. The constants λ and μ are Lamé coefficients of the material in equilibrium. The functions $\lambda+\bar{\lambda}(t)$ and $\mu+\bar{\mu}(t)$ may be regarded as time-dependent Lamé coefficients for the stress-relaxation response to a sudden deformation at time $t=0$ (cf. Sect. 39).

The theory of infinitesimal visco-elasticity is treated in Vol. VI of this Encyclopedia.

The general equation (41.21) or (41.22), with $\boldsymbol{T}_0\neq\mathbf{0}$, applies to small deformations superimposed on a large deformation from an unstressed state, when the configuration resulting from the large deformation is taken as reference. It should be noted that the linear tensor functions L and $\mathsf{M}(s)$ in (41.21) depend not only on the material but also on the reference configuration, i.e., on the original large deformation. In general, when $\boldsymbol{T}_0\neq\mathbf{0}$, L and $\mathsf{M}(s)$ are not isotropic even when the material is isotropic.

When $\mathsf{M}(s)\equiv\mathbf{0}$, (41.21) and (41.22) become

$$\left.\begin{aligned} \boldsymbol{T}(t)-\boldsymbol{T}_0 &= \tilde{\boldsymbol{R}}(t)\,\boldsymbol{T}_0-\boldsymbol{T}_0\tilde{\boldsymbol{R}}(t)+\mathsf{L}[\tilde{\boldsymbol{E}}(t)]+o(\varepsilon),\\ T^{km}(t)-T_0^{km} &= \tilde{R}^k{}_p(t)\,T_0^{pm}-T_0^{kp}\tilde{R}_p{}^m(t)+\mathsf{L}^{kmpq}\tilde{E}_{pq}(t)+o(\varepsilon),\end{aligned}\right\}\tag{41.25}$$

---

[1] The derivation of this theory as a consequence of the general theory of simple materials with fading memory is due to COLEMAN and NOLL [1961, *7*]. A similar equation was derived later by PIPKIN and RIVLIN [1961, *45*] on the basis of a formal perturbation argument starting from the theory of materials of a mixed integral-differential type (cf. Sect. 37). A corresponding representation formula for isotropic materials was obtained by LIANIS [1963, *48*], who considered also some particular deformations [1963, *50*] and some special cases suggested by experimental data [1963, *51*] and by thermodynamic conjectures [1964, *55*]. A similar study, accompanied by experimental tests, has been made by KO and BLATZ [1964, *51*]. A special theory of this kind has been set up by ZAHORSKI [1963, *86*].

[2] BOLTZMANN [1874, *1*]. The theory is often associated with the name of VOLTERRA, who developed it extensively in connection with integral equations [1909, *3* and *4*].

[3] These coefficients are not to be confused with the λ and μ occurring in (41.5).

which describe small *elastic* deformations superimposed on a fixed large deformation, which will be treated in detail in Sects. 68, 69, and 70. When $\boldsymbol{T}_0 = \boldsymbol{0}$ and when the error term $o(\varepsilon)$ is omitted, (41.25) reduces to the constitutive equation of the *classical theory of infinitesimal elasticity* (cf. CFT, Sect. 301):

$$\boldsymbol{T} = \mathsf{L}[\tilde{\boldsymbol{E}}], \qquad T^{km} = \mathsf{L}^{kmpq}\tilde{E}_{pq}. \tag{41.26}$$

The tensor L is the *linear elasticity* of the material described, approximately, by the classical infinitesimal theory. For an isotropic material, (41.26) becomes the familiar law of CAUCHY (CFT, Eq. (301.2)):

$$\boldsymbol{T} = (\lambda \operatorname{tr} \tilde{\boldsymbol{E}})\,\mathbf{1} + 2\mu\tilde{\boldsymbol{E}}, \qquad T^{km} = \lambda \tilde{E}^p_p g^{km} + 2\mu \tilde{E}^{km}. \tag{41.27}$$

We remark that the infinitesimal theories based on (41.23), (41.24), (41.26), and (41.27) are physically meaningless for finite deformations, because they do not have the invariance properties required by the principle of material indifference (cf. Sect. 19). Thus, these infinitesimal theories cannot possibly apply to *any* material in general finite deformation. The other approximating theories discussed here, however, do satisfy the principle of material indifference. It is possible, therefore, that these theories describe the behavior of *some* materials for arbitrary finite deformations, even though only in limiting cases can they be expected to apply to *all* simple materials with fading memory.

To obtain approximations better than those afforded by the classical linear or infinitesimal theories, we may resort to the simple materials of higher order, or to the materials of the differential type of higher grade, defined in Sect. 40. Examples will be considered in Sects. 119—122.

D. Elasticity.

42. Scope and plan of the chapter. This chapter is a treatise on theories of *elastic response*. Most of it concerns *elastic materials*, defined as those in which the stress is determined by the strain and rotation from a single, fixed reference configuration. Our objectives are two: First, to present everything known concerning the fully general theory, and, second, to derive and explain all known exact solutions of special problems in it and all approximate methods that promise any breadth of application. We emphasize static problems, not only because they are mathematically easier but also because, since they pertain in common to all simple materials, of howsoever various dynamic response, they enjoy greater physical breadth.

This program is largely achieved in the first subchapter (p. 119), where energetic assumptions are expressly avoided. The second subchapter (p. 294) concerns *hyperelastic materials*, namely, those possessing a stored-energy function. This short exposition develops those relatively few results for which the existence of such a function seems to be necessary, chief among them being work theorems.

The third subchapter (p. 355) discusses briefly theories of thermo-elastic and electro-elastic response, as well as theories of polar elastic media.

The fourth subchapter (p. 401) explains the newer concept of *hypo-elasticity*, defined by constitutive equations that specify the stressing of a material as a function of the stretching and the stress. While this theory rests upon relations among rates, it describes response elastic in one sense, namely, that the results are independent of the time scale of the deformation process occurring, since the stressing depends linearly on the stretching. Relations between elasticity and hypo-elasticity are developed; while the two theories overlap to some extent, neither is included within the other.

After important work on non-linear one-dimensional elasticity by several savants of the eighteenth century, notably JAMES BERNOULLI (1691—1705) and EULER (1727—1778), the main ideas of the general theories of elasticity and hyper-elasticity, respectively, were suggested by CAUCHY (1823—1828) and GREEN (1839—1841). Subsequent developments were restricted mainly to GREEN'S theory, which was given definitive mathematical formulation by KIRCHHOFF (1852—1859), KELVIN (1863), and other great physicists of the last century. Important special developments were added by FINGER (1894), HADAMARD (1901—1903), DUHEM (1903—1906), SIGNORINI (1930), RIVLIN (1947—1951), and others. To the last named are due the two major ideas ruling the notable resurgence of the subject in the last decade: For incompressible materials more explicit and handy results can often be found, and certain special problems can be solved exactly without supposing any particular or approximate form for the response functions.

While a special case of hypo-elasticity had been proposed by JAUMANN (1911), the general theory, the motivation for it, and the first solutions of particular problems were given by TRUESDELL (1952—1955).

For parts of the subject treated here, there are numerous other expositions[1], to some of which we gladly acknowledge our debt. In comparison with them, however, we trust this new presentation will be found worthwhile, on the one hand because the mathematics, while more general in scope, is simpler and more direct and hence easier to follow, and, on the other, because the treatment stays closer to the physics than the earlier approaches, more formal, elaborate, and special, could allow. Despite the length of this treatise, the continuing rush of new work on elasticity has made it impossible for us, in reasonable time, to include all the facets we wished to. Intentionally we have left not only unmentioned but even uncited the flood of papers on unsystematic approximations or special theories.

---

[1] The classic exposition, unfortunately omitting the work of FINGER, is the treatise by E. and F. COSSERAT [1896, *1*]. The exposition most widely read in recent years is that of MURNAGHAN [1937, *2*], which is brief, easy, and so incomplete as scarcely to be representative. Our treatment is closest to those of TRUESDELL [1952, *20*, Ch. IV] [1953, *25*] and NOLL [1955, *19*, §§ 15—16], which rest partly upon the two earlier works just cited but otherwise are drawn directly from the sources. Other expositions, some much narrower in scope but some covering topics not treated by us, have been given by HAMEL [1912, *1*, § 60], HELLINGER [1914, *1*, Ch. III], BRILLOUIN [1925, *2*] [1938, *2*, Ch. X], ARIANO [1925, *1*] [1928, *1*] [1929, *1*] [1930, *1*] [1933, *1*], SIGNORINI [1936, *3*] [1943, *3*] [1945, *4*] [1949, *21*] [1955, *25*], KAPPUS [1939, *1*], MURNAGHAN [1941, *1*] [1951, *8*], KUTILIN [1947, *6*], RICHTER [1948, *10*] [1949, *14*] [1952, *16*], MILNE-THOMSON [1949, *8*], GREEN & ZERNA [1950, *2*] [1954, *11*, Chs. II—IV], RIVLIN [1951, *13*] [1956, *24*] [1960, *48*], NIORDSEN [1953, *19*], MIŞICU [1953, *17*], NOVOZHILOV [1953, *20*] [1961, *42*, Chaps. I—III], SHIMAZU [1954, *21*, Chs. 2—4], SHENG [1955, *24*], DOYLE and ERICKSEN [1956, *5*], KOPPE [1956, *15*], ANGLES D'AURIAC [1958, *4*], JOHN [1958, *25*, Chs. II—IV], GREEN and ADKINS [1960, *26*], PRAGER [1961, *47*, Chs. IX—X], GRIOLI [1962, *24*], ERINGEN [1962, *18*, Chs. 6, 8], ZAHORSKI [1962, *77* and *78*], and GUO ZHONG-HENG [1963, *29*].

In the period between the two great wars, knowledge of the classical theory of finite elastic strain sank so far that "engineering" papers sprouted here and there with "new" theories, all either pointlessly special or wrong. Only in Italy, due to the teaching and writing of SIGNORINI, was the true theory still widely known, though little advanced. The experts of the older generation, such as HADAMARD, HILBERT, and HAMEL, seem to have lost interest. While to this day references to "MURNAGHAN'S theory" are published in the American "applied" literature, SIGNORINI in a conversation with TRUESDELL in 1949 expressed surprise that a major journal had accepted MURNAGHAN'S paper [1937, *2*], adding „en tout cas il est simplement un mémoire qu'on n'a pas besoin de citer." SIGNORINI, however, had meanwhile stopped reading, so that he took no notice of the great work of RIVLIN, just then coming into print, and despite a starting knowledge of finite elasticity unequalled anywhere else, the younger Italian specialists let at least a decade pass before beginning to follow the new British and American work, which, as the reader will see, furnishes some three quarters of the contents of the complete exposition of the subject we give in the present treatise.

I. Elastic materials.

a) General considerations.

43. Definition of an elastic material. A material is called *elastic* if it is simple (cf. Sect. 29) and if the stress at time t depends only on the local configuration at time t, and not on the entire past history of the motion. We may say that elastic materials are simple materials with a perfect memory of a very special and limited kind. A change in stress arises solely in response to a change in configuration, and the material is outright oblivious to the manner in which the change of configuration has occurred in space and time. The change in stress is completely determined by comparison of the present configuration with any other. The common statement that an elastic material has a perfect memory of a single state and no memory of any other is misleading if interpreted as imputing any special physical properties to the material in that state. While the response of an elastic material may exhibit certain symmetries with respect to one configuration which it does not show with respect to some others, in principle any smooth configuration may be used as a reference, and it would be more accurate to say that an elastic material has an equally perfect awareness of *every* configuration, including even those it never occupies.

For an elastic material, the functional $\mathfrak{G}$ of the general constitutive equation (28.4) of a simple material reduces to a function $\mathfrak{g}$, and the *constitutive equation of an elastic material*[1] is of the form

$$\boldsymbol{T}=\mathfrak{g}(\boldsymbol{F}), \tag{43.1}$$

where $\boldsymbol{F}$ is the deformation gradient at the present time, taken relative to a fixed, but arbitrary, local reference configuration. We call $\mathfrak{g}$ the *response function* of the elastic material. A formula such as (43.1) is often called a "stress-strain relation", but for reasons to be made clear in Sect. 80 we shall refer to it simply as a *stress relation.*

Most of our considerations in this section and in Sects. 45—53 are *local,* i.e., they refer to a single particle. It is not necessary to assume that there exists for the whole body a reference configuration $\varkappa$ such that the elastic response function $\mathfrak{g}_\varkappa$ is the same for all particles. In the senses defined in Sect. 27, a body composed of elastic material may be materially uniform yet not homogeneous. The theory of strain for such bodies has been outlined in Sect. 34; the general differential equations of the corresponding theory will be presented in Sect. 44.

Direct specialization of the results of Sects. 28 and 29 yields the following theorems:

---

[1] The concept of elasticity here embodied is due to CAUCHY [1823, *1*] [1828, *1*, § 11] but was put in mathematical form by him only in the case of linear response to infinitesimal strain. Most of the researches on finite strain concern the more special theory to be presented in Subchapter II, p. 294. The first attempts at mathematical treatment of CAUCHY's idea, apparently, are those of REINER [1948, *9*, §§ 1—2], RICHTER [1948, *10*, § 2], and GLEYZAL [1949, *4*, § 2]; a formula equivalent to (47.8) for the isotropic case, but for infinitesimal strain only, had been proposed by PRAGER [1945, *2*, § 2] and by HANDELMAN, LIN, and PRAGER [1947, *5*, § 2]. The physical circumstances and interpretations envisaged by these authors are various.

In most studies the starting point has been one of the reduced forms such as (43.10) or (47.8). Proofs of the reduction of (43.1) to (43.7), under unnecessary and complicating restrictions, are due to CELLERIER [1893, *3*, § 4], E. and F. COSSERAT [1896, *1*, § 26], and MURNAGHAN [1937, *2*, § 3]. RICHTER [1952, *16*, § 2] was the first to observe that the reduction follows at once from a simple and natural requirement of invariance, which is in fact a special case of the principle of material frame-indifference. Cf. NOLL [1955, *19*, § 15a].

a) The response function $\mathfrak{g}$ of an elastic material satisfies the relation

$$\boldsymbol{Q}\,\mathfrak{g}(\boldsymbol{F})\,\boldsymbol{Q}^T = \mathfrak{g}(\boldsymbol{Q}\boldsymbol{F}), \tag{43.2}$$

identically in the tensor variable $\boldsymbol{F}$ and the orthogonal tensor variable $\boldsymbol{Q}$.

b) The form of the response function depends upon the choice of the local reference configuration. If $\mathfrak{g}_{\boldsymbol{K}}$ and $\mathfrak{g}_{\hat{\boldsymbol{K}}}$ are the response functions corresponding to the reference configurations $\boldsymbol{K}$ and $\hat{\boldsymbol{K}}$, then[1]

$$\mathfrak{g}_{\boldsymbol{K}}(\boldsymbol{F}) = \mathfrak{g}_{\hat{\boldsymbol{K}}}(\boldsymbol{F}\boldsymbol{P}^{-1}), \tag{43.3}$$

where

$$\boldsymbol{P} = \hat{\boldsymbol{K}}\boldsymbol{K}^{-1} \tag{43.4}$$

is the local deformation from the local configuration $\boldsymbol{K}$ to the local configuration $\hat{\boldsymbol{K}}$. As a corollary to this result it follows that if a relation having the form (43.1) holds for a particular fixed local reference configuration, then a relation of the same form holds for every fixed local reference configuration. That is, if $\boldsymbol{F}^* = \boldsymbol{F}\boldsymbol{P}$, then $\boldsymbol{T} = \mathfrak{g}^*(\boldsymbol{F}^*)$, where $\mathfrak{g}^*(\boldsymbol{F}^*) = \mathfrak{g}(\boldsymbol{F}^*\boldsymbol{P}^{-1})$. Consequently, *the definition of an elastic material is invariant under change of the local reference configuration.*

c) The general constitutive equation of an elastic material may be written in numerous different forms, reduced so as to satisfy the principle of material indifference (Sect. 19). Some of these forms[2] follow at once from more general results already given:

$$\boldsymbol{T} = \boldsymbol{R}\,\mathfrak{g}(\boldsymbol{U})\,\boldsymbol{R}^T, \tag{43.5}$$

$$\boldsymbol{T} = \boldsymbol{R}\,\mathfrak{f}(\boldsymbol{C})\,\boldsymbol{R}^T, \tag{43.6}$$

$$\overline{\boldsymbol{T}} = \mathfrak{s}(\boldsymbol{C}). \tag{43.7}$$

Here $\boldsymbol{U}$ is the right stretch tensor; $\boldsymbol{R}$, the rotation tensor; $\boldsymbol{C}$, the right Cauchy-Green tensor (cf. Sect. 23); and $\overline{\boldsymbol{T}}$, the convected stress tensor, defined by (29.7). Equation (43.5) follows from (29.3) or (29.19); (43.6), from (29.28); and (43.7), from (29.6). When specializing (29.19) and (29.28), one must take into account that, for elastic materials, $\boldsymbol{T}$ can depend only on the values of $\boldsymbol{U}_{(t)}(t-s)$ or $\boldsymbol{C}_{(t)}(t-s)$ for $s=0$, while by (23.9), these values are the unit tensor, $\mathbf{1}$. It is clear that elastic materials are special materials of the differential type (cf. Sect. 35). The equations (43.5) and (43.6) can be obtained also by specialization of (35.5) and (35.6), respectively.

Other forms of the general constitutive equation of elasticity employ, instead of the Cauchy stress, $\boldsymbol{T}$, the first or second Piola-Kirchhoff tensors, $\boldsymbol{T}_{\mathrm{R}}$ and $\tilde{\boldsymbol{T}}$. While these two tensors are well known, for convenience we list their major properties in an appendix at the end of this section[3]. From (43.1) and $(43\mathrm{A}.11)_1$ below we see that the constitutive equation of an elastic material may be put into the form

$$\boldsymbol{T}_{\mathrm{R}} = \mathfrak{h}(\boldsymbol{F}), \tag{43.8}$$

---

[1] Special results of this kind were discussed by SIGNORINI [1943, *3*, Ch. III, ¶ 6] and MANACORDA [1957, *11*].

[2] RICHTER [1952, *16*, §§ 3—4]. Many other reduced forms also are correct, but not all. From a correct starting point [our Eqs. (43.1) and (43.2)], by misconstruing a classical but irrelevant theorem on polynomials ERINGEN [1962, *18*, § 46] was able to derive no less than two generally incorrect forms: $\boldsymbol{T} = \mathfrak{f}(\boldsymbol{C}, \boldsymbol{d}_\sigma)$ [his Eqs. (46.8) or (46.14), which are right only for pure strains] and $\boldsymbol{T} = \mathfrak{g}(\boldsymbol{B}, \boldsymbol{d}_\sigma)$ [his Eqs. (46.21), right only for isotropic materials, cf. footnote 1, p. 140].

[3] A fuller treatment, with references, is given in Sect. 210 of CFT.

where, by (43.2), $\mathfrak{h}$ must satisfy the functional equation

$$\mathfrak{h}(\boldsymbol{Q}\boldsymbol{F})=\boldsymbol{Q}\,\mathfrak{h}(\boldsymbol{F}), \tag{43.9}$$

identically in the tensor variable $\boldsymbol{F}$ and the orthogonal tensor variable $\boldsymbol{Q}$. Putting $\boldsymbol{Q}=\boldsymbol{R}^T$ in (43.9) shows that (43.8) may be expressed in the first of the following two forms, reduced so as to satisfy automatically the principle of material indifference:

$$\boldsymbol{T}_{\mathrm{R}}=\boldsymbol{R}\,\mathfrak{h}(\boldsymbol{U}), \qquad \tilde{\boldsymbol{T}}=\mathfrak{t}(\boldsymbol{C}), \tag{43.10}$$

while the second is derived from (29.11) and $(43\text{A}.11)_5$, below. Equations (43.8) and (43.10) give forms of the stress relation very useful in the general theory of elasticity.

The tensor functions occurring in the various forms of the constitutive equation are related by the following identities:

$$\left.\begin{aligned}
\mathfrak{g}(\boldsymbol{F})&=\frac{1}{|\det \boldsymbol{F}|}\,\mathfrak{h}(\boldsymbol{F})\boldsymbol{F}^T, & \mathfrak{h}(\boldsymbol{F})&=|\det\boldsymbol{F}|\,\mathfrak{g}(\boldsymbol{F})(\boldsymbol{F}^{-1})^T,\\
\mathfrak{f}(\boldsymbol{U}^2)&=\mathfrak{g}(\boldsymbol{U}), & \boldsymbol{U}\,\mathfrak{g}(\boldsymbol{U})\,\boldsymbol{U}&=\mathfrak{s}(\boldsymbol{U}^2),\\
\sqrt{III_{\boldsymbol{U}}}\,\mathfrak{s}(\boldsymbol{U})&=\boldsymbol{U}\,\mathfrak{t}(\boldsymbol{U})\,\boldsymbol{U}, & \mathfrak{h}(\boldsymbol{F})&=\boldsymbol{F}\,\mathfrak{t}(\boldsymbol{F}^T\boldsymbol{F}),
\end{aligned}\right\} \tag{43.11}$$

identically in the tensor $\boldsymbol{F}$ and the positive-definite symmetric tensor $\boldsymbol{U}$, respectively.

Among the corresponding co-ordinate forms of the stress relation, the following are most useful:

$$\left.\begin{aligned}
T_{km}&=\mathfrak{s}_{\alpha\beta}(C_{\mu\nu})\,X^{\alpha}{}_{,k}X^{\beta}{}_{,m},\\
T^{km}&=J^{-1}\mathfrak{t}^{\alpha\beta}(C_{\mu\nu})\,x^{k}{}_{,\alpha}x^{m}{}_{,\beta},\\
T^{k}_{m}&=J^{-1}x^{k}{}_{,\alpha}\mathfrak{h}_{m}{}^{\alpha}(x^{q}{}_{,\gamma}),\\
\tilde{T}^{\alpha\beta}&=\mathfrak{t}^{\alpha\beta}(C_{\mu\nu}),\\
T_{\mathrm{R}k}{}^{\alpha}&=\mathfrak{h}_{k}{}^{\alpha}(x^{m}{}_{,\beta})=g_{km}x^{m}{}_{,\beta}\mathfrak{t}^{\alpha\beta}(C_{\mu\nu}).
\end{aligned}\right\} \tag{43.12}$$

Here, J is given by (43A.4) below. Of course, the various component functions depend also on the natural basis, at the particle under consideration, of the material co-ordinates X^α in the reference configuration. The component functions $\mathfrak{h}_k{}^\alpha$ depend, in addition, on the natural basis of the spatial co-ordinates x^k. These dependences will be made explicit only when needed (cf. Sect. 29).

For *incompressible elastic materials*, the stress tensor $\boldsymbol{T}$ in (43.1) must be replaced by the extra stress, $\boldsymbol{T}_{\mathrm{E}}=\boldsymbol{T}+p\boldsymbol{1}$, where p is an indeterminate pressure (cf. Sect. 30). The response function $\mathfrak{g}(\boldsymbol{F})$ need be defined only if $|\det\boldsymbol{F}|=1$, and it is determined only up to a scalar function of $\boldsymbol{F}$. Thus, for example, the counterpart of (43.8) is

$$\boldsymbol{T}_{\mathrm{R}}+p(\boldsymbol{F}^{-1})^T=\mathfrak{h}(\boldsymbol{F}), \tag{43.13}$$

where (43.9) still holds, while in place of (43.1), (43.5), (43.7), and (43.10) we now have the following reduced forms of the constitutive equation[1]:

$$\left.\begin{aligned}
\boldsymbol{T}&=-p\boldsymbol{1}+\mathfrak{g}(\boldsymbol{F})=-p\boldsymbol{1}+\boldsymbol{R}\,\mathfrak{g}(\boldsymbol{U})\boldsymbol{R}^T=-p\boldsymbol{1}+(\boldsymbol{F}^{-1})^T\mathfrak{s}(\boldsymbol{C})\boldsymbol{F}^{-1},\\
\overline{\boldsymbol{T}}&=-p\boldsymbol{C}+\mathfrak{s}(\boldsymbol{C}),\\
\tilde{\boldsymbol{T}}&=-p\boldsymbol{C}^{-1}+\mathfrak{t}(\boldsymbol{C}),
\end{aligned}\right\} \tag{43.14}$$

where $\mathfrak{g}(\boldsymbol{U})$, $\mathfrak{s}(\boldsymbol{C})$, and $\mathfrak{t}(\boldsymbol{C})$ need be defined only when $\det\boldsymbol{U}=1$, $\det\boldsymbol{C}=1$.

[1] Noll [1955, *19*, § 15].

It is clear from contrast of (43.14) with (43.1) that for a given isochoric deformation, the dynamic response of an incompressible elastic material is not at all the same as that of a compressible one. The appearance of the arbitrary hydrostatic pressure in an incompressible material is somewhat mysterious, since it has absolutely no counterpart for compressible materials in isochoric deformations or in deformations with very small changes of volume. So as to render possible an isochoric deformation in a compressible material, a certain definite pressure must be *supplied*; in an incompressible material, no pressure of any amount can produce *anything but* isochoric deformations. If the two theories are to be reconciled, it must be by some limiting process[1], giving the theory for incompressible materials a status as an asymptotic approximation. However, the value of the theory for incompressible materials does not grow from such a reconciliation, any more than the value of the theory of rigid bodies need be reduced to a (still unproved) theorem of approximation giving it asymptotic status in respect to elasticity theory or some other theory of deformable media. Any theory simplifies the phenomena it is designed to represent. The mathematical simplicity of the theory for incompressible materials justifies its study, leading to simple results fit for direct comparison with experimental data, a comparison (Sect. 55) that seems to substantiate present confidence in it for rubber-like materials.

There is a considerable literature[2] on elastic materials reinforced by a network of inextensible cords. The constitutive equation may be written down at once from the results given in Sect. 30; specifically, from (30.11) and (30.21). The literature concerns mainly special solutions. Since these are of interest primarily to the extent that they are useful in the design of rubber tires, we do not present anything further regarding reinforced elastic materials in this treatise.

In addition to the forms of the stress relation given here, others are possible. The literature of the subject, particularly in the older studies by engineers, contains many arguments purporting to show that one or another form, sometimes not generally correct and usually more special than any given here, is better than the rest. As REINER and RICHTER[3] have stated, these futile wrangles arise from arbitrary preferences for one or another way of representing stress and strain and thus reflect failure to grasp the basic physical idea of elasticity, which is independent of the mathematical means used to express it. Indeed, as RIVLIN[4] has remarked, "strain" need never be defined at all. Our use of particular measures of strain such as $\boldsymbol{U}$ and $\boldsymbol{C}$ is only so as to be able to write down certain explicit formulae which are often useful. As is plain from our treatment, any correct and sufficiently general mathematical statement of the theory of elasticity is equivalent to the rest[5]. For special problems or for special elastic materials one form may have some advantage over others.

By substituting (43.1) into (31.2) we see that the *isotropy group* of an elastic material is the set of unimodular tensors $\boldsymbol{H}$ such that

$$\mathfrak{g}(\boldsymbol{F}) = \mathfrak{g}(\boldsymbol{F}\boldsymbol{H}) \tag{43.15}$$

---

[1] Such processes have been discussed by OLDROYD [1950, *12*, § 6] and SPENCER [1964, *80*].

[2] RIVLIN [1955, *22*] [1959, *24*], ADKINS and RIVLIN [1955, *4*], ADKINS [1956, *1*] [1957, *2*] [1958, *3*] [1961, *2*, § *18*], GENENSKY and RIVLIN [1959, *8*], GREEN and ADKINS [1960, *26*, Ch. VII], JONES and ADKINS [1962, *38*]. While these authors assumed the material to be hyperelastic, little if any simplification results from the existence of a stored-energy function.

[3] REINER [1948, *9*, § 3] [1951, *11*], RICHTER [1952, *16*, § 7].

[4] RIVLIN [1950, *13*, § 1].

[5] Cf. also the discussion of the equivalence of strain measures in CFT, Sect. 32.

for all invertible tensors $\boldsymbol{F}$. By (31.7) and $(31.8)_2$, an orthogonal tensor $\boldsymbol{Q}$ belongs to the isotropy group of an elastic material if and only if

$$\left.\begin{aligned} \boldsymbol{Q}\,\mathfrak{g}(\boldsymbol{F})\,\boldsymbol{Q}^T &= \mathfrak{g}(\boldsymbol{Q}\boldsymbol{F}\boldsymbol{Q}^T), \\ \boldsymbol{Q}\,\mathfrak{f}(\boldsymbol{C})\,\boldsymbol{Q}^T &= \mathfrak{f}(\boldsymbol{Q}\boldsymbol{C}\boldsymbol{Q}^T), \end{aligned}\right\} \tag{43.16}$$

for all invertible tensors $\boldsymbol{F}$ and all positive-definite symmetric tensors $\boldsymbol{C}$, respectively. In these equations $\mathfrak{g}$ is the response function occurring in (43.1) and (43.5), while $\mathfrak{f}$ is the response function occurring in (43.6). Similar results hold for the response functions of constrained elastic materials.

While $\boldsymbol{F}$ is dimensionless, $\dim \boldsymbol{T} = M L^{-1} T^{-2}$, and hence the response function $\mathfrak{g}$ in (43.1) must have these same physical dimensions. If E is any modulus having these dimensions, then $\mathfrak{g} = E\,\mathfrak{g}_0$, where $\mathfrak{g}_0$ is dimensionless. I.e.,

$$\boldsymbol{T}/E = \mathfrak{g}_0(\boldsymbol{F}), \tag{43.17}$$

as follows also by specialization of (28.21). The value of the modulus E, in any particular system of units, is fixed by a convention; e.g., it may be taken as a particular physical component of $\boldsymbol{T}$ occurring in a particular deformation, such as one of the homogeneous strains discussed in Sect. 54. When E has been fixed, the dimensionless response function $\mathfrak{g}_0$ is then determined. On the understanding that, as in most of the present treatment, possible dependence of $\mathfrak{g}$ upon temperature or entropy is not rendered explicit, we may rephrase the above conclusion thus: *Every elastic material has a single dimensional modulus, which bears the dimensions of stress.* I.e., all other elastic moduli are either themselves dimensionless or else dimensionless multiples of any given one, just as in the infinitesimal theory.

There may well be no real materials that obey precisely an elastic constitutive equation (43.1), for, as we shall see in Sect. 80, dissipation of energy, although it seems to be universally observed in nature, is impossible in at least the commonest theories of elastic materials. There are a few real materials which, in many physical situations, dissipate but negligible energy and may be represented by an elastic ideal material for a wide range of experimental conditions.

However, the theory of elasticity applies to all simple materials in the case of statics. We mean by statics that branch of mechanics dealing with bodies which are at rest at the present time and which may be regarded as having been at rest at all times in the past. In statics, therefore, the history $\boldsymbol{F}^{(t)}(s) = \boldsymbol{F}(t-s) = \boldsymbol{F}$ of the deformation gradient is independent of s, so that (28.4) reduces to (43.1). This is a result of capital importance for the theory: *In static deformation, all simple materials obey the theory of elasticity*[1]. Solutions of the equations of static elasticity theory thus have a greater relevance than might at first be thought. Moreover, since there are no real bodies which are eternally at rest, to apply statics to a real situation one must impute to real materials only a limited memory. It should make no difference how a body has been deformed into the present configuration after a sufficiently long time has elapsed. This expectation is confirmed, subject to certain assumptions, by the theorem on stress relaxation in Sect. 39. Thus further relevance for static elasticity is afforded by that theorem and still more by the theorems on slow motions in Sect. 41. It is to be noted especially that those theorems refer to elasticity in general, not necessarily to hyperelasticity.

---

[1] In passing it may be remarked that this fact affords a particularly strong example of the falsity of one of the common statements of "D'ALEMBERT's principle". If dynamical equations could be obtained from statical ones by adding "inertial forces" to the body force, all simple materials would be elastic! Except in the most primitive special theories of mechanics, from statics little or nothing regarding dynamics may be inferred.

43 A. Appendix. The Piola-Kirchhoff stress tensors. So as to state problems in which the boundary of a body is deformed in a prescribed way or is subject to assigned forces, it is convenient to use the material description. A material formulation may be attained as follows. Let $\partial\mathscr{P}$ be the boundary of a part $\mathscr{P}$ of a body in some configuration χ defined by its deformation $\boldsymbol{x}=\chi(\boldsymbol{X})$ from a fixed global reference configuration. By (16.2) and (16.5), the contact force acting on $\mathscr{P}$ is given by

$$\boldsymbol{f}_c=\int_{\partial\mathscr{P}}\boldsymbol{T}\boldsymbol{n}\,ds, \tag{43 A.1}$$

where $\boldsymbol{n}$ is the outward unit vector normal to $\partial\mathscr{P}$ in the configuration χ. Let ds_R be the surface element and $\boldsymbol{n}_R$ the outward unit normal to $\partial\mathscr{P}$ in the reference configuration. By the laws of transformation of surface integrals[1],

$$\boldsymbol{f}_c=\int_{\partial\mathscr{P}}\boldsymbol{T}\boldsymbol{n}\,ds=\int_{\partial\mathscr{P}}\boldsymbol{T}_R\boldsymbol{n}_R\,ds_R, \tag{43 A.2}$$

where $\boldsymbol{T}_R$ is defined by the relation

$$\boldsymbol{T}=J^{-1}\boldsymbol{T}_R\boldsymbol{F}^T,\qquad T^{km}=J^{-1}T_R^{k\alpha}x^m{}_{,\alpha}, \tag{43 A.3}$$

in which

$$J\equiv|\det\boldsymbol{F}|=\sqrt{\det\boldsymbol{C}}=\sqrt{III_{\boldsymbol{C}}}=\frac{\varrho_R}{\varrho}. \tag{43 A.4}$$

The subscript R recalls use of the reference configuration. Thus ϱ_R is the density in the reference configuration, and the tensor $\boldsymbol{T}_R$, called the *first Piola-Kirchhoff stress tensor*[2], measures the actual contact force when it is taken per unit area in the reference configuration.

From (43 A.2) we see that the surface traction $\boldsymbol{t}_R$ acting on the boundary is given by[3]

$$\boldsymbol{t}_R=\boldsymbol{T}_R\boldsymbol{n}_R, \tag{43 A.5}$$

while CAUCHY's first law of motion (16.6) assumes the form[4]

$$\operatorname{Div}\boldsymbol{T}_R+\varrho_R\boldsymbol{b}=\varrho_R\ddot{\boldsymbol{x}}, \tag{43 A.6}$$

where "Div" is the divergence operator with respect to the place $\boldsymbol{X}$ in the reference configuration. A co-ordinate form of (43 A.6) is

$$T_{Rk}{}^{\alpha}{}_{;\alpha}+\varrho_R b_k=\varrho_R\ddot{x}_k, \tag{43 A.7}$$

where ";" indicates the total covariant derivative, explained in Sect. 20 of the Appendix to CFT.

The first Piola-Kirchhoff tensor $\boldsymbol{T}_R$ is not, in general, a symmetric tensor. The symmetry of the Cauchy stress tensor follows from the requirement that the moments be balanced in the actual configuration. The first Piola-Kirchhoff stress tensor is symmetric if and only if the moments remain balanced when the neighborhood of the particle is deformed into the reference configuration with the contact forces held constant in magnitude and direction. More generally, CAUCHY's second law (16.7) expressed in terms of the first Piola-Kirchhoff tensor has the form

$$\boldsymbol{T}_R\boldsymbol{F}^T=\boldsymbol{F}\boldsymbol{T}_R{}^T,\qquad T_R^{k\alpha}x^m{}_{,\alpha}=T_R^{m\alpha}x^k{}_{,\alpha}. \tag{43 A.8}$$

The second Piola-Kirchhoff tensor, $\tilde{\boldsymbol{T}}$, is introduced so as to provide a representation of stress that, while transformed back to the reference configuration, is still a symmetric tensor. By definition

$$\tilde{\boldsymbol{T}}\equiv\boldsymbol{F}^{-1}\boldsymbol{T}_R,\qquad \tilde{T}^{\alpha\beta}=X^\alpha{}_{,k}T_R^{k\beta}. \tag{43 A.9}$$

---

[1] Cf. CFT, Eq. (20.8).

[2] For no good reason, its components are sometimes called "pseudo-stresses".

[3] The relation between $\boldsymbol{t}_R$ and the stress vector $\boldsymbol{t}$ occurring in (16.5) is

$$\boldsymbol{t}_R\,ds_R=\boldsymbol{t}\,ds.$$

Cf. CFT, Sect. 210. From Eq. (29.2) of CFT we see that

$$ds=J\sqrt{\boldsymbol{n}_R\cdot(\boldsymbol{C}^{-1}\boldsymbol{n}_R)}\,ds_R.$$

Thus the relation between $\boldsymbol{t}_R$ and the actual stress vector $\boldsymbol{t}$ acting upon the boundary of the loaded material depends in a complicated way on the deformation itself.

[4] CFT, Sect. 210.

From (43 A.8) we then verify at once that $\tilde{\boldsymbol{T}}$ is symmetric if and only if CAUCHY's second law holds. That is,

$$\tilde{\boldsymbol{T}}^T = \tilde{\boldsymbol{T}}, \qquad \tilde{T}^{\alpha\beta} = \tilde{T}^{\beta\alpha}, \tag{43 A.10}$$

if and only if $\boldsymbol{T}^T = \boldsymbol{T}$. CAUCHY's first law, however, assumes a form more complicated than (43 A.6); that form is not needed here.

From (43 A.3), (43 A.9), and (29.7) we see that

$$\left.\begin{aligned}
\boldsymbol{T}_{\mathrm{R}} &= J\,\boldsymbol{T}(\boldsymbol{F}^{-1})^T, & T_{\mathrm{R}}^{k\alpha} &= J\,T^{km} X^{\alpha}{}_{,m},\\
&= \boldsymbol{F}\tilde{\boldsymbol{T}}, & &= x^k{}_{,\beta}\,\tilde{T}^{\beta\alpha},\\
\tilde{\boldsymbol{T}} &= J\,\boldsymbol{F}^{-1}\boldsymbol{T}(\boldsymbol{F}^{-1})^T, & \tilde{T}^{\alpha\beta} &= J\,X^{\alpha}{}_{,k} X^{\beta}{}_{,m}\,T^{km},\\
&= \boldsymbol{F}^{-1}\boldsymbol{T}_{\mathrm{R}}, & &= X^{\alpha}{}_{,k}\,T_{\mathrm{R}}^{k\beta},\\
\boldsymbol{T} &= J^{-1}\boldsymbol{F}\tilde{\boldsymbol{T}}\boldsymbol{F}^T, & T^{km} &= J^{-1} x^k{}_{,\alpha} x^m{}_{,\beta}\,\tilde{T}^{\alpha\beta},\\
J\,\overline{\boldsymbol{T}} &= \boldsymbol{C}\tilde{\boldsymbol{T}}\boldsymbol{C}, & J\,\overline{T}^{\alpha\beta} &= C^{\alpha}{}_{\gamma}\,\tilde{T}^{\gamma\delta}\,C^{\beta}_{\delta}.
\end{aligned}\right\} \tag{43 A.11}$$

44. Formulation of boundary-value problems. Let $\mathscr{B}$ be an *elastic body*, i.e., assume that the material at *each* particle X of $\mathscr{B}$ is elastic. We employ a fixed reference configuration of $\mathscr{B}$ and describe all other configurations $\boldsymbol{\chi}$ by their deformations $\boldsymbol{x} = \boldsymbol{\chi}(\boldsymbol{X})$ from the reference configuration. For each $\boldsymbol{X}$ we then have a constitutive equation of the form (43.8), i.e.,

$$\boldsymbol{T}_{\mathrm{R}} = \mathfrak{h}(\boldsymbol{F}, \boldsymbol{X}), \tag{44.1}$$

where $\boldsymbol{F} = \nabla\boldsymbol{\chi}(\boldsymbol{X})$. If not only the body but also the reference configuration is homogeneous, then $\mathfrak{h}$ in (44.1) reduces to a function of $\boldsymbol{F}$ only, independent of $\boldsymbol{X}$.

Substituting (44.1) into (43 A.6) yields the following *differential equation of finite elasticity*:

$$\operatorname{Div}\mathfrak{h}(\boldsymbol{F}, \boldsymbol{X}) + \varrho_{\mathrm{R}}\boldsymbol{b} = \varrho_{\mathrm{R}}\ddot{\boldsymbol{x}}. \tag{44.2}$$

When the external body force $\boldsymbol{b}$ acting on each particle $\boldsymbol{X}$ is given as a function of place $\boldsymbol{x}$ and time t, $\boldsymbol{b} = \boldsymbol{b}(\boldsymbol{x}, \boldsymbol{X}, t)$, (44.2) becomes a quasi-linear differential equation of second order for determining the deformation $\boldsymbol{x} = \boldsymbol{\chi}(\boldsymbol{X}, t)$ when suitable initial conditions and boundary conditions are satisfied.

We set (cf. (9.13), (9.14))

$$\mathsf{A} = \mathsf{A}(\boldsymbol{F}, \boldsymbol{X}) \equiv \mathfrak{h}_{\boldsymbol{F}}(\boldsymbol{F}, \boldsymbol{X}), \qquad \mathsf{A}_k{}^{\alpha}{}_m{}^{\beta} = \frac{\partial \mathfrak{h}_k{}^{\alpha}}{\partial x^m{}_{,\beta}} \tag{44.3}$$

and

$$\boldsymbol{q} = \boldsymbol{q}(\boldsymbol{F}, \boldsymbol{X}) \equiv [\operatorname{Div}\mathfrak{h}(\boldsymbol{F}, \boldsymbol{X})]_{(\boldsymbol{F})}, \qquad q_k = \frac{\partial \mathfrak{h}_k{}^{\alpha}}{\partial X^{\alpha}}, \tag{44.4}$$

where the index $(\boldsymbol{F})$ in $(44.4)_1$ indicates that $\boldsymbol{F}$ must be held constant when applying the operator Div.[1] We then have the following co-ordinate form for (44.2):

$$\mathsf{A}_k{}^{(\alpha}{}_m{}^{\beta)}\, x^m{}_{,\alpha;\beta} + q_k + \varrho_{\mathrm{R}} b_k = \varrho_{\mathrm{R}}\ddot{x}_k, \tag{44.5}$$

where $\mathsf{A}_k{}^{(\alpha}{}_m{}^{\beta)} \equiv \frac{1}{2}(\mathsf{A}_k{}^{\alpha}{}_m{}^{\beta} + \mathsf{A}_k{}^{\beta}{}_m{}^{\alpha})$. Here the quantities $x^m{}_{,\alpha;\beta}$ are total covariant derivatives, given explicitly by

$$x^m{}_{,\alpha;\beta} = \frac{\partial^2 x^m}{\partial X^{\alpha}\partial X^{\beta}} + \left\{ {m \atop s\,r} \right\} x^s{}_{,\alpha} x^r{}_{,\beta} - \left\{ {\sigma \atop \beta\,\alpha} \right\} x^m{}_{,\sigma}, \tag{44.6}$$

where the braces $\{\ \}$ denote Christoffel symbols based upon g_{km} and $g_{\alpha\beta}$, respectively. The vector $\boldsymbol{q}$ is the density of *resultant force due to inhomogeneity*, per unit volume in the reference configuration.

[1] When the derivatives of the component functions $\mathfrak{h}_k{}^{\alpha}$ with respect to $x^m{}_{,\beta}$ and X^{α} are taken, the basis vectors $\boldsymbol{e}_r$ and $\boldsymbol{d}_{\sigma}$ must be held fixed. The natural basis vectors at $\boldsymbol{x}$ and $\boldsymbol{X}$ should be substituted for $\boldsymbol{e}_r$ and $\boldsymbol{d}_{\sigma}$ only after the differentiations have been performed.

If the body and the reference configuration are homogeneous, then $\boldsymbol{q}=\mathbf{0}$, and A does not depend explicitly on $\boldsymbol{X}$. If the body is *materially uniform but not homogeneous*, there are no homogeneous configurations. In this case, it is useful to employ, instead of a global reference configuration, a uniform reference in the sense of Sect. 34. When this is done, then again $\boldsymbol{q}=\mathbf{0}$, and A does not depend explicitly on $\boldsymbol{X}$. However, $\boldsymbol{F}=\boldsymbol{F}(\boldsymbol{X}, t)$ is no longer the gradient of a deformation; rather, it is a field of local deformations (cf. Sect. 22). It can be shown[1] that under these circumstances (44.5) and (44.6) may be replaced by

$$\left.\begin{aligned} \mathsf{A}_{k\ m}^{\ \alpha\ \beta} F^m{}_{\alpha;\beta} - \mathfrak{h}_k{}^\alpha S_{\alpha\sigma}{}^\sigma + \varrho_{\mathrm{R}} b_k &= \varrho_{\mathrm{R}} \ddot{x}_k, \\ F^m{}_{\alpha;\beta} = \frac{\partial^2 x^m}{\partial X^\alpha \partial X^\beta} + \left\{ {m \atop s\,r} \right\} x^s{}_{,\alpha} x^r{}_{,\beta} &- \Gamma_{\beta\alpha}{}^\sigma x^m{}_{,\sigma} \end{aligned}\right\} \tag{44.7}$$

where $\boldsymbol{\Gamma}$ is the material connection and $\boldsymbol{S}$ the inhomogeneity tensor associated with the uniform reference employed.

A boundary-value problem is the problem of finding solutions of the differential equations (44.2) that satisfy certain *boundary conditions*. For finite elasticity it is not so easy to decide what boundary conditions are reasonable as it is in theories that are governed by linear partial differential equations. For example, if we attempt to prescribe the stress vector $\boldsymbol{t}$ acting on the deformed boundary, $\boldsymbol{t}=\boldsymbol{f}(\boldsymbol{x}, t)$, we cannot generally do so in an unambiguous manner since the places $\boldsymbol{x}$ to be occupied by the boundary particles $\boldsymbol{X}$ are unknown. In order to obtain a definite boundary condition, we have to prescribe a function $\boldsymbol{t}=\boldsymbol{f}_{\mathscr{C}}(\boldsymbol{x}, t)$, $\boldsymbol{x}\in\mathscr{C}$, for *every conceivable* configuration $\mathscr{C}$ of the boundary $\partial\mathscr{B}$ of the body.

Such a boundary condition may be described as an advance strategy that tells us how to load the body no matter how it is deformed. The simplest such strategy is to prescribe surface tractions $\boldsymbol{t}_{\mathrm{R}}(\boldsymbol{X})$, $\boldsymbol{X}\in\partial\mathscr{B}$, measured per unit area in the reference configuration, and to keep these tractions fixed in magnitude and direction no matter what the configurations of $\partial\mathscr{B}$ in the deformed states may be. The resulting *boundary condition of traction* is

$$\boldsymbol{t}_{\mathrm{R}}(\boldsymbol{X}) = \boldsymbol{T}_{\mathrm{R}} \boldsymbol{n}_{\mathrm{R}} = \mathfrak{h}(\boldsymbol{F}) \boldsymbol{n}_{\mathrm{R}}. \tag{44.8}$$

Another boundary condition results when the traction vector $\boldsymbol{t}$ is required to be normal to the surface occupied by $\partial\mathscr{B}$ in every configuration and to have magnitude p, a prescribed constant. This condition,

$$-p\boldsymbol{n} = \boldsymbol{T}\boldsymbol{n} = \mathfrak{g}(\boldsymbol{F})\boldsymbol{n}, \tag{44.9}$$

is called a *boundary condition of pressure*. Instead of assuming p to be a constant, one could also assume p to be a function of the volume of $\mathscr{B}$ in the deformed states.

A *boundary condition of place* corresponds to constraining the boundary $\partial\mathscr{B}$ of $\mathscr{B}$, or a part $\mathscr{S}$ of it, to assume a given shape. This shape may be described by a function $\boldsymbol{x}=\boldsymbol{\chi}_0(\boldsymbol{X}, t)$, $\boldsymbol{X}\in\partial\mathscr{B}$, and the condition may be written

$$\boldsymbol{\chi}(\boldsymbol{X}, t) = \boldsymbol{\chi}_0(\boldsymbol{X}, t), \quad \boldsymbol{X}\in\mathscr{S}. \tag{44.10}$$

---

[1] NOLL [1965, *27*]. Eq. (44.7) and the constitutive equation $\boldsymbol{T}_{\mathrm{R}}=\mathfrak{h}(\boldsymbol{F})$, using a uniform reference in the sense of Sect. 34, yield a continuum model for an elastic material with "dislocations". Some writers on dislocation theory, however, contend that the concept of simple material is not inclusive enough for their purposes. According to them, a theory allowing for couple stresses and oriented material elements is needed. Cf. GÜNTHER [1958, *21*], KRÖNER [1962, *44*] [1963, *43* and *44*] [1965, *23*]. In Sect. 98 we present a theory of polar elastic materials. While only homogeneous bodies have been considered so far in the applications of this theory, the concept of an inhomogeneous but materially uniform body could be developed for it also.

Mixed boundary-value problems result when different types of boundary conditions are imposed for different portions of the boundary $\partial\mathcal{B}$.

The presently known theorems of existence and uniqueness of the solutions to partial differential equations are not sufficiently general to cover the problems defined here. Some special cases will be presented below in Sects. 46 and 68. In general, it is clear that the results will depend strongly on the nature of the coefficients $\mathsf{A}_k{}^\alpha{}_m{}^\beta$ in (44.5).

For problems of *elastic equilibrium* we must put $\ddot{\boldsymbol{x}}=\mathbf{0}$ in (44.2) and (44.5)$_1$. We then obtain the *differential equations of finite elastostatics*:

$$\left.\begin{aligned} \operatorname{Div} \mathfrak{h}(\boldsymbol{F}, \boldsymbol{X})+\varrho_{\mathrm{R}} \boldsymbol{b}&=\mathbf{0}, \\ \mathsf{A}_k{}^{(\alpha}{}_m{}^{\beta)} x^m{}_{,\alpha;\beta}+q_k+\varrho_{\mathrm{R}} b_k&=0. \end{aligned}\right\} \tag{44.11}$$

Any solution of a boundary-value problem in statics must be such that the external loads are equilibrated in the deformed state. That is, the total force and total torque acting upon the body in the configuration of equilibrium must vanish[1]:

$$\int_{\partial\mathcal{B}} \boldsymbol{t}\, ds+\int_{\mathcal{B}} \varrho\, \boldsymbol{b}\, dv=\mathbf{0}, \tag{44.12}$$

$$\int_{\partial\mathcal{B}} \boldsymbol{p}\times \boldsymbol{t}\, ds+\int_{\mathcal{B}} \boldsymbol{p}\times \varrho\, \boldsymbol{b}\, dv=\mathbf{0}, \tag{44.13}$$

where $\boldsymbol{p}$ is the position vector of the place $\boldsymbol{x}$ from a fixed origin. Transforming variables so as to integrate over the reference configuration, we have the equivalent conditions

$$\int_{\partial\mathcal{B}} \boldsymbol{t}_{\mathrm{R}}\, ds_{\mathrm{R}}+\int_{\mathcal{B}} \varrho_{\mathrm{R}} \boldsymbol{b}\, dv_{\mathrm{R}}=\mathbf{0}, \tag{44.14}$$

$$\int_{\partial\mathcal{B}} \boldsymbol{p}\times \boldsymbol{t}_{\mathrm{R}}\, ds_{\mathrm{R}}+\int_{\mathcal{B}} \boldsymbol{p}\times \varrho_{\mathrm{R}} \boldsymbol{b}\, dv_{\mathrm{R}}=\mathbf{0}, \tag{44.15}$$

where $\boldsymbol{p}$ is still the position vector of $\boldsymbol{x}$ but now is expressed as a function of $\boldsymbol{X}$.

Consider the case when the body force $\boldsymbol{b}$ is assigned as a function of $\boldsymbol{X}\in\mathcal{B}$ and when a boundary condition of traction (44.8) is imposed. The condition (44.14) then becomes a necessary condition that the boundary data $\boldsymbol{t}_{\mathrm{R}}=\boldsymbol{t}_{\mathrm{R}}(\boldsymbol{X})$, $\boldsymbol{X}\in\partial\mathcal{B}$, must satisfy if the problem is to have a solution. The condition (44.15), however, since it involves the position vector of the place $\boldsymbol{x}$ occupied by $\boldsymbol{X}$ *after the body has been deformed*, cannot be checked directly when only the differential equation and the boundary data are known. Rather, as was first observed by SIGNORINI[2], (44.15) is to be regarded as a *condition of compatibility* to be satisfied by any solution of the traction boundary-value problem that is defined by (44.11) and (44.8).

The conditions (44.14) and (44.15) may be written more compactly by means of Stieltjes integrals in the form

$$\int_{\mathcal{B}} d\boldsymbol{f}=\mathbf{0}, \quad \int_{\mathcal{B}} \boldsymbol{p}\times d\boldsymbol{f}=\mathbf{0}, \tag{44.16}$$

where $d\boldsymbol{f}$ is the element of the external force, equal to $\varrho_{\mathrm{R}} \boldsymbol{b}\, dv_{\mathrm{R}}$ in the interior of $\mathcal{B}$ and equal to $\boldsymbol{t}_{\mathrm{R}}\, ds_{\mathrm{R}}$ on the boundary $\partial\mathcal{B}$. Introducing the *astatic load*[3] tensor

$$\boldsymbol{A}=\int_{\mathcal{B}} \boldsymbol{p}\otimes d\boldsymbol{f}=\int_{\partial\mathcal{B}} \boldsymbol{p}\otimes \boldsymbol{t}_{\mathrm{R}}\, ds_{\mathrm{R}}+\int_{\mathcal{B}} \boldsymbol{p}\otimes \varrho_{\mathrm{R}} \boldsymbol{b}\, dv_{\mathrm{R}}, \tag{44.17}$$

---

[1] Cf. CFT, Sect. 196.

[2] SIGNORINI [1930, *3*, § 6], and numerous later papers. GRIOLI [1962, *24*, Ch. IV, § 3] derived equivalent results as formal conditions of compatibility for the systems (63.17) of Sect. 63, below.

[3] Cf. CFT, Sect. 220.

we see that the compatibility condition (44.15) is satisfied if and only if $\boldsymbol{A}$ is a symmetric tensor.

The following theorem of DA SILVA[1] is of interest in connection with the compatibility condition (44.15): *The total torque of any system of loads acting on a body can be made to vanish by subjecting the body to a suitable rigid rotation about the origin.* To prove this theorem we observe first that a uniform rotation $\boldsymbol{Q}$ about the origin changes the position vectors $\boldsymbol{p}$ into $\boldsymbol{Q}\boldsymbol{p}$. Hence, the astatic load (44.17), after the rotation, is given by

$$\int_{\mathscr{B}} (\boldsymbol{Q}\boldsymbol{p})\otimes d\boldsymbol{f} = \boldsymbol{Q}\left[\int_{\mathscr{B}} \boldsymbol{p}\otimes d\boldsymbol{f}\right] = \boldsymbol{Q}\boldsymbol{A}. \tag{44.18}$$

The torque vanishes after the rotation if and only if $\boldsymbol{Q}\boldsymbol{A}$ is symmetric, i.e., if and only if

$$\boldsymbol{Q}\boldsymbol{A} - \boldsymbol{A}^T\boldsymbol{Q}^T = \boldsymbol{0}. \tag{44.19}$$

The theorem is proved if it is shown that (44.19) has a proper orthogonal solution $\boldsymbol{Q}$ when $\boldsymbol{A}$ is given. But such a solution can easily be found by considering a polar decomposition $\boldsymbol{A} = \boldsymbol{P}\boldsymbol{S}$ of $\boldsymbol{A}$, where $\boldsymbol{P}$ is orthogonal and $\boldsymbol{S}$ symmetric and positive semi-definite. We need only take $\boldsymbol{Q}$ to be $+\boldsymbol{P}^T$ or $-\boldsymbol{P}^T$ depending on whether $\boldsymbol{P}$ is proper or improper. Actually it is easy to see that (44.19) has at least four solutions.

Applying DA SILVA's theorem to the reference configuration, we see that a change of reference configuration by a rigid rotation can always be found such that, after the change, the equation

$$\int_{\partial\mathscr{B}} \boldsymbol{p}_R \times \boldsymbol{t}_R \, ds_R + \int_{\mathscr{B}} \boldsymbol{p}_R \times \varrho_R \boldsymbol{b} \, dv_R = \boldsymbol{0} \tag{44.20}$$

is satisfied, where $\boldsymbol{p}_R$ is the position vector in the reference configuration. A rigid rotation preserves any significant property the reference configuration might have, such as the property of being homogeneous, undistorted, or natural. For an isotropic material even the response functions remain unchanged after a rigid rotation of an undistorted homogeneous reference configuration. Therefore, it is usually no loss of generality to assume, for a traction boundary-value problem, that the boundary data satisfy not only the condition (44.14), but also the condition (44.20). Alternatively, we may leave the reference configuration fixed and rotate the loads so as to satisfy (44.20).

From the physical standpoint it is clear that uniqueness of solution to the traction boundary-value problem of equilibrium is not expected or desired except in special circumstances. If the ends of a bar are subjected to equal but opposite tractions t, two solutions, trivially, are possible: In addition to the one that comes to mind at once, a second is obtained by rotating the bar through a straight angle and applying tractions of amount $-t$ to the ends after rotation. That is, tension and compression correspond to two different solutions of the same traction boundary-value problem. An elastic hemisphere may be turned inside out; corresponding to vanishing assigned forces on the boundary and in the interior of the undeformed hemisphere, then, there are two solutions, one in which there is no deformation, and one corresponding to eversion[2]. Likewise, one end of a long hollow cylinder may be drawn through the other, or a long cylindrical annular segment may be snapped into reversed curvature[3]. Exact solutions of

[1] DA SILVA [1851, *2*, §§ 153—158] constructed an exhaustive classification of systems of forces with resultant force zero. Cf. also SIGNORINI [1932, *7*], STOPPELLI [1954, *22*].

[2] For a special theory of elasticity, the problem of eversion of an entire sphere was solved by ARMANNI [1915, *1*].

[3] ALMANSI [1916, *1*].

this kind, for a general incompressible material, are presented in Sect. 57. Such problems belong in the finite theory of elasticity, and no condition sufficient to secure a theorem of uniqueness strong enough to exclude them can be right.

If we were to follow the lead of analysis, we might be tempted to impose the condition that the differential equation (44.11) be *strongly elliptic* for the case of equilibrium:

$$A_k{}^{(\alpha}{}_m{}^{\beta)} \lambda^k \lambda^m \mu_\alpha \mu_\beta > 0 \tag{44.21}$$

for all non-vanishing vectors $\boldsymbol{\lambda}$ and $\boldsymbol{\mu}$. We shall refer to (44.21) as the *S-E condition*; we do not lay it down as a postulate, but we propose it for study, and at several points in the following pages we shall determine the consequences that would follow if it were assumed.

If the equations (44.11) were linear partial differential equations with constant coefficients, the S-E condition would suffice to ensure uniqueness of the place boundary-value problem. However, Eqs. (44.11) are not linear, and exploration so far tends to suggest rather the reverse, namely, that even if (44.21) holds for arbitrary $\boldsymbol{F}$, it does not guarantee a unique solution $\boldsymbol{x} = \boldsymbol{\chi}(\boldsymbol{X})$ assuming prescribed values upon the boundary of an arbitrary body.

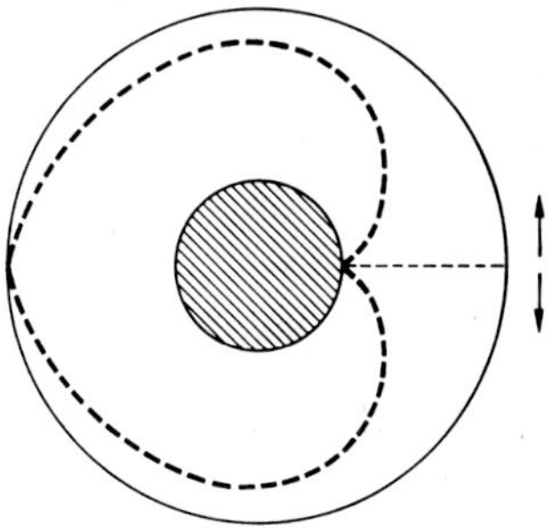

Fig. 1. Example to show that uniqueness of solution to the boundary-value problem of place cannot generally hold.

Indeed, an example due to JOHN[1] shows that general uniqueness of solution to the boundary-value problem of place is not to be expected, no matter what restrictions be imposed upon the form of the stress relation. Consider the material confined between concentric cylinders. Let the inner cylinder be held fixed, and let the outer be rotated through a straight angle, so that each point is carried into its antipode. If such a deformation is possible when the rotation is effected in one sense, it is equally possible, at least in a homogeneous and isotropic body, for the rotation in the opposite sense. Formally, let the deformation be $x = \bar{x}(X, Y)$ $y = \bar{y}(X, Y)$, $z = Z$, with the boundary conditions $\bar{x}(X, Y) = X$ and $\bar{y}(X, Y) = Y$ when $X^2 + Y^2 = a^2$, while $\bar{x}(X, Y) = -X$ and $\bar{y}(X, Y) = -Y$ when $X^2 + Y^2 = b^2$. Then the deformation $x = \bar{x}(X, Y)$, $y = -\bar{y}(X, -Y)$ satisfies these same boundary conditions of place but is not the same as the first. In Fig. 1, the dashed curves show the places occupied, in these two deformations, by the particles lying initially on the dashed straight segment. If $q_k = 0$, one of these deformations satisfies (44.11) if and only if the other does. A similar example may be constructed for a spherical shell. Notice that if we regard the given finite rotation of the outer boundary cylinder as built up by any number of smaller rotations, there is no reason to expect that the solutions corresponding to these successive problems fail to be unique, or that further small displacement of the boundary points fail to yield unique further deformations from either of the two solutions. A similar failure of uniqueness without loss of stability occurs in the theory of minimal surfaces.

For the place boundary-value problem of equilibrium the intuitive picture is not so clear as for the traction boundary-value problem. Consider, for example, an initially cylindrical isotropic[2] body which has been severely buckled by compressing plates which assign the positions of the particles on its two ends; keeping these two plates fixed, let tractions be applied to its curved surface so as to bring that surface slowly back to the form of a cylinder, moreover, the very cylinder that bounds the unstable homogeneous deformation known to be possible subject

[1] JOHN [1964, *47*].

[2] Isotropic elastic materials are defined in Sect. 47.

to the same end conditions. Will the interior particles be carried to the positions they would have in the homogeneous strain? This seems unlikely: It would mean that the surface tractions to be applied on the mantle so as to impel the buckled cylinder into a straight form would have to increase from zero to maximum values and then fall off again to zero, that being their value in the homogeneous strain. Rather, it seems to us that there should be, corresponding to these assigned surface displacements, in addition to the homogeneous solution a second and stable one, with non-uniform tractions applied to the mantle, just as in the solution which is stable when the mantle is free there is a non-homogeneous deformation. Such a solution would represent a kind of *interior buckling*.

There is also a problem of *free shape:* Let the deformed shape of the body be given, subject to the action of given loads, and find the shape the body will assume when unloaded. This is an inverse problem[1] rather than a boundary-value problem, and it has been studied little. Clearly it has solutions in special cases[2].

For certain problems it is convenient to introduce the vector $\boldsymbol{u}(\boldsymbol{X})$ giving the *displacement* of the particle X from its place $\boldsymbol{X}$ in the reference configuration to its place $\boldsymbol{x}=\boldsymbol{\chi}(\boldsymbol{X})$ in the deformed state:

$$\boldsymbol{u}(\boldsymbol{X})=\boldsymbol{\chi}(\boldsymbol{X})-\boldsymbol{X}. \tag{44.22}$$

Thus

$$\ddot{\boldsymbol{u}}=\ddot{\boldsymbol{x}} \quad \text{and} \quad \boldsymbol{F}=\boldsymbol{1}+\boldsymbol{H}, \quad \boldsymbol{H}=\nabla\boldsymbol{u}(\boldsymbol{X}). \tag{44.23}$$

$\boldsymbol{H}$, already defined by Eq. (41.9), is called the *displacement gradient* (cf. CFT, Sect. 19).

By (44.23) the constitutive equation (43.8) may be written in the form

$$\boldsymbol{T}_{\mathrm{R}}=\mathfrak{h}(\boldsymbol{F})=\mathfrak{k}(\boldsymbol{H}). \tag{44.24}$$

The differential equation (44.2) and the boundary condition of traction (44.8) assume the forms

$$\operatorname{Div}\mathfrak{k}(\boldsymbol{H})+\varrho_{\mathrm{R}}\boldsymbol{b}=\varrho_{\mathrm{R}}\ddot{\boldsymbol{u}}, \tag{44.25}$$

$$\boldsymbol{t}_{\mathrm{R}}(\boldsymbol{X})=\mathfrak{k}(\boldsymbol{H})\boldsymbol{n}_{\mathrm{R}}, \tag{44.26}$$

respectively.

While the formulation in terms of the displacement and its gradient played a considerable part in older studies, more recent and concrete exact work rests on the essentially simpler forms expressed in terms of the deformation $\boldsymbol{x}=\boldsymbol{\chi}(\boldsymbol{X},t)$, except for problems which are approached by perturbation procedures.

For incompressible materials it is easy to write down counterparts of all equations given in this section. For example, by substituting (43.13) into (43A.6)

---

[1] For the elastica, it was formulated by JAMES BERNOULLI in 1696, solved in one special case by EULER, in another by TRUESDELL [1954, *27*]; later TRUESDELL [1963, *73*] called attention to his having suggested the general problem in his lecture of 1953, just cited, and WESOŁOWSKI [1963, *83* and *85*] analyzed the problem in some detail.

[2] Given a stress $\boldsymbol{T}=\boldsymbol{T}(\boldsymbol{x})$ such that CAUCHY's laws (16.6) and (16.7) are satisfied with $\ddot{\boldsymbol{x}}=\boldsymbol{0}$, consider the stress relation (43.1):

$$\boldsymbol{T}(\boldsymbol{x})=\mathfrak{g}(\boldsymbol{F})=\mathfrak{l}(\boldsymbol{F}^{-1}), \tag{A}$$

say. In general, $\mathfrak{l}$ will have many inverse relations, so that

$$\boldsymbol{F}^{-1}=\boldsymbol{f}(\boldsymbol{x}), \qquad \frac{\partial X^\alpha}{\partial x^j}=f^\alpha{}_j(\boldsymbol{x}). \tag{B}$$

If one of these inverses satisfies the integrability conditions

$$\frac{\partial f^\alpha{}_j}{\partial x^k}=\frac{\partial f^\alpha{}_k}{\partial x^j}, \tag{C}$$

the system (B) can be solved to yield solutions $\boldsymbol{X}=\overline{\boldsymbol{X}}(\boldsymbol{x})$. These give the desired free shapes.

and using Eq. (18.2) of CFT, we find that

$$-p_{,k}+\mathsf{A}_{k\ m}^{(\alpha\ \beta)}x^m_{,\alpha;\beta}+q_k+\varrho_R b_k=\varrho_R\ddot{x}_k, \tag{44.27}$$

where $\mathsf{A}_{k\ m}^{\alpha\ \beta}$ and q_k are defined by (44.3) and (44.4), $\mathfrak{h}$ being understood now as the function occurring in (43.13). Eq. (44.27) replaces $(44.5)_1$, and other formulae for incompressible materials are similarly obtained.

45. The elasticities of an elastic material. As has already been remarked, the constitutive equation (43.8), on which all the foregoing analysis is based, is more general than that of elasticity. The fourth-order tensor A, which plays a great role throughout the theory, is defined by (44.3) directly in terms of the function $\mathfrak{h}$ and hence need not enjoy any properties of symmetry. When, as will usually be the case in this chapter, we wish to consider elasticity specifically, then we must exploit the reduction resulting from the principle of material indifference. While we have noted already the functional equation (43.9), more generally useful are certain necessary conditions following from the fact that $\mathfrak{h}$ is given by $(43.11)_6$ in terms of the symmetric-tensor-valued function $\mathfrak{t}$ of the symmetric argument $\boldsymbol{C}$. First, from (44.3) and $(43.11)_6$ we easily show that[1]

$$\left.\begin{aligned}\mathsf{A}_{k\ m}^{\alpha\ \beta}&=g_{km}\mathfrak{t}^{\alpha\beta}+g_{kq}x^q_{,\lambda}\frac{\partial\mathfrak{t}^{\alpha\lambda}}{\partial C_{\mu\nu}}\frac{\partial C_{\mu\nu}}{\partial x^m_{,\beta}},\\&=g_{km}\mathfrak{t}^{\alpha\beta}+2g_{kp}g_{mq}x^p_{,\gamma}x^q_{,\delta}\frac{\partial\mathfrak{t}^{\alpha\gamma}}{\partial C_{\beta\delta}}.\end{aligned}\right\} \tag{45.1}$$

Second, we introduce another fourth-order tensor $\mathsf{C}=\mathsf{C}(\boldsymbol{F},\boldsymbol{X})$ of coefficients[2]:

$$\mathsf{C}^{kmpq}\equiv 2J^{-1}x^k_{,\alpha}x^m_{,\beta}x^p_{,\gamma}x^q_{,\delta}\frac{\partial\mathfrak{t}^{\alpha\beta}}{\partial C_{\gamma\delta}}; \tag{45.2}$$

in view of the above-mentioned symmetry of $\mathfrak{t}$ and $\boldsymbol{C}$, these coefficients satisfy the following conditions of symmetry:

$$\mathsf{C}^{kmpq}=\mathsf{C}^{mkpq}=\mathsf{C}^{kmqp}, \tag{45.3}$$

reflecting the principle of material indifference. If we set[3]

$$\mathsf{B}^{kmpq}\equiv J^{-1}x^m_{,\alpha}x^q_{,\beta}g^{kr}g^{ps}\mathsf{A}_{r\ s}^{\alpha\ \beta}, \tag{45.4}$$

so that

$$\mathsf{A}_{k\ m}^{\alpha\ \beta}=J\,g_{kr}g_{ms}X^\alpha_{,p}X^\beta_{,q}\mathsf{B}^{rpsq}, \tag{45.5}$$

from (45.1), (45.2), (43.10), and $(43\text{A}.11)_{10}$ we see that

$$\mathsf{B}^{kmpq}=g^{kp}T^{mq}+\mathsf{C}^{kmpq}. \tag{45.6}$$

By (45.3), then

$$\left.\begin{aligned}\mathsf{B}^{kmpq}-\mathsf{B}^{mkpq}&=g^{kp}T^{mq}-g^{mp}T^{kq},\\\mathsf{B}^{kmpq}-\mathsf{B}^{kmqp}&=g^{kp}T^{mq}-g^{kq}T^{mp},\\\mathsf{B}^{kmpq}-\mathsf{B}^{mkqp}&=g^{kp}T^{mq}-g^{mq}T^{kp}.\end{aligned}\right\} \tag{45.7}$$

These identities, relating the coefficients B^{kmpq} to the stresses T^{rs}, express a restriction imposed upon the coefficients B^{kmpq} by the principle of material indifference. From them we see that a necessary and sufficient condition that $\boldsymbol{T}=\boldsymbol{0}$ is

$$\mathsf{B}^{kmpq}=\mathsf{B}^{mkpq}=\mathsf{B}^{kmqp}. \tag{45.8}$$

[1] We use the dual of Eq. $(29.4)_1$ of CFT.

[2] Murnaghan [1949, *9*, p. 333], Truesdell [1952, *20*, § 55].

[3] Truesdell [1961, *63*, § 6], slightly generalizing a definition of Toupin and Bernstein [1961, *62*, § 2].

If, on the other hand, we set[1]

$$D^{kmpq} \equiv \tfrac{1}{2}(B^{kpmq} + B^{kqmp}) = D^{kmqp}, \tag{45.9}$$

so that, by $(45.7)_2$,

$$B^{kmpq} = g^{km} T^{pq} - g^{kq} T^{mp} + D^{kpmq} + D^{kqpm} - D^{kmpq}, \tag{45.10}$$

then

$$\left.\begin{aligned} D^{kmpq} - D^{pqkm} &= \tfrac{1}{2}(g^{km} T^{pq} - g^{pq} T^{km}) + \tfrac{1}{2}(B^{kqmp} - B^{pmqk}), \\ D^{kmpq} - D^{mkpq} &= \tfrac{1}{2}(B^{kpmq} - B^{mqkp}) + \tfrac{1}{2}(B^{kqmp} - B^{mpkq}). \end{aligned}\right\} \tag{45.11}$$

In general states of stress, the tensors B and D share some but not all of the symmetries of the elasticity L occuring in the constitutive equation (41.26) of the infinitesimal theory[2].

The fourth-order tensors A, B, C, and D, being determined by the gradient of the stress with respect to the deformation gradient, measure in various ways the sensitivity of the stress to a change of strain. Therefore they may be called *elasticities* of the elastic material defined by $\mathfrak{h}$. Alternatively, when no confusion results, we shall call the individual components $A_k{}^\alpha{}_m{}^\beta$, B^{kmpq}, C^{kmpq}, and D^{kmpq} "elasticities".

By (45.4), we can express the identities (45.7) and the conditions (45.8) in terms of the original coefficients $A_k{}^\alpha{}_m{}^\beta$ if desired, but the results are not illuminating.

Since $J = |\det \boldsymbol{F}|$, by differentiating $(43.12)_3$ with respect to $x^p{}_{,\beta}$ we have

$$\frac{\partial T^k{}_m}{\partial x^p{}_{,\beta}} = J^{-1}[A_m{}^\alpha{}_p{}^\beta x^k{}_{,\alpha} + \mathfrak{h}_m{}^\beta \delta^k_p - \mathfrak{h}_m{}^\alpha X^\beta{}_{,p} x^k{}_{,\alpha}], \tag{45.12}$$

where (44.3) has been used. By (45.4) it follows that

$$B^{km}{}_p{}^q = T^{km}\delta^q_p - T^{kq}\delta^m_p + \frac{\partial T^{km}}{\partial x^p{}_{,\beta}} x^q{}_{,\beta}. \tag{45.13}$$

In terms of the tensors B and D, the S-E condition (44.21) takes on a very simple form, namely[3],

$$B^{kmpq}\lambda_k\lambda_p\nu_m\nu_q = D^{kmpq}\lambda_k\lambda_m\nu_p\nu_q > 0 \tag{45.14}$$

for arbitrary vectors $\boldsymbol{\lambda}$ and $\boldsymbol{\nu}$. The equivalence of (45.14) to (44.21) follows from the fact that since $\boldsymbol{F}$ is invertible, the vector $\boldsymbol{\nu} = \boldsymbol{F}^{-1}\boldsymbol{\mu}$ is arbitrary if the vector $\boldsymbol{\mu}$ is arbitrary.

The tensor C is of main importance in calculating the stress rate. Putting $(43.10)_2$ into $(43\text{A}.11)_9$ yields the following reduced form for the stress relation of elasticity:

$$\boldsymbol{T} = J^{-1}\boldsymbol{F}\mathfrak{t}(\boldsymbol{C})\boldsymbol{F}^T. \tag{45.15}$$

Material differentiation yields

$$\dot{\boldsymbol{T}} = -(\dot{J}/J)\boldsymbol{T} + J^{-1}(\dot{\boldsymbol{F}}\mathfrak{t}(\boldsymbol{C})\boldsymbol{F}^T + \boldsymbol{F}\mathfrak{t}(\boldsymbol{C})\dot{\boldsymbol{F}}^T) + J^{-1}\boldsymbol{F}\mathfrak{t}_{\boldsymbol{C}}(\boldsymbol{C})[\dot{\boldsymbol{C}}]\boldsymbol{F}^T, \tag{45.16}$$

in the notation (9.15). From (21.18), (15.14), and $(25.6)_1$ it follows that $\dot{J}/J = \operatorname{tr}\boldsymbol{D}$; by use of this result, (24.1), (30.5), and (45.2) we find from (45.16) a simple formula

[1] As noted by THURSTON [1965, *34*], analogous coefficients have been used by HUANG [1950, *5*, § 4] and other writers on lattice dynamics, but interpretation of their definitions seems uncertain.

[2] While HUANG [1950, *5*, § 6] claimed that the symmetry relations of BRILLOUIN [1938, *2*, Eq. (X.65) on p. 232] are incorrect in stressed states, it seems to us that these conditions (apart from those ensuring the existence of a stored-energy function) are in fact equivalent to (45.3) and that HUANG has misunderstood BRILLOUIN'S definitions.

[3] TOUPIN and BERNSTEIN [1961, *62*, Eq. (3.4)] for hyperelastic materials, and in general by TRUESDELL [1961, *63*, Eq. (10.3)].

for the stress rate[1]:

$$\left.\begin{aligned}\dot{\boldsymbol{T}} &= -\boldsymbol{T}\operatorname{tr}\boldsymbol{D} + \boldsymbol{L}\boldsymbol{T} + \boldsymbol{T}\boldsymbol{L}^T + \mathsf{C}[\boldsymbol{D}],\\ \dot{T}^{km} &= -T^{km}\dot{x}^r{}_{,r} + \dot{x}^k{}_{,q}T^{qm} + T^{kq}\dot{x}^m{}_{,q} + \mathsf{C}^{kmpq}D_{pq}.\end{aligned}\right\} \tag{45.17}$$

It is easy to express this result in terms of the invariant stress rates $\overset{\circ}{\boldsymbol{T}}$ and $\overset{\triangle}{\boldsymbol{T}}$, introduced in Sect. 36.

As we have said in Sect. 26, for the most part in this treatise we disregard all non-mechanical influence, but here we shall take note of a more general and physically flexible concept of elasticity in which possible dependence of the stress upon the specific entropy η or the absolute temperature θ is taken into account. The stress relation (43.8) is replaced by the more general relations

$$\boldsymbol{T}_{\mathrm{R}} = \hat{\mathfrak{h}}(\boldsymbol{F}, \boldsymbol{X}, \eta) = \tilde{\mathfrak{h}}(\boldsymbol{F}, \boldsymbol{X}, \theta), \tag{45.18}$$

and η and θ are assumed to be related by an invertible *equation of state*

$$\eta = \tilde{\eta}(\boldsymbol{F}, \boldsymbol{X}, \theta), \quad \theta = \hat{\theta}(\boldsymbol{F}, \boldsymbol{X}, \eta). \tag{45.19}$$

If we set

$$\left.\begin{aligned}\hat{\mathsf{A}}_k{}^\alpha{}_m{}^\beta(\boldsymbol{F}, \boldsymbol{X}, \eta) &\equiv \frac{\partial \hat{\mathfrak{h}}_k{}^\alpha}{\partial x^m{}_{,\beta}}, & \tilde{\mathsf{A}}_k{}^\alpha{}_m{}^\beta(\boldsymbol{F}, \boldsymbol{X}, \theta) &\equiv \frac{\partial \tilde{\mathfrak{h}}_k{}^\alpha}{\partial x^m{}_{,\beta}},\\ \hat{q}_k(\boldsymbol{F}, \boldsymbol{X}, \eta) &\equiv \frac{\partial \hat{\mathfrak{h}}_k{}^\alpha}{\partial X^\alpha}, & \tilde{q}_k(\boldsymbol{F}, \boldsymbol{X}, \theta) &\equiv \frac{\partial \tilde{\mathfrak{h}}_k{}^\alpha}{\partial X^\alpha},\end{aligned}\right\} \tag{45.20}$$

the equations of motion (44.5) are generalized as follows:

$$\left.\begin{aligned}\hat{\mathsf{A}}_k{}^{(\alpha}{}_m{}^{\beta)} x^m{}_{,\alpha;\beta} + \hat{q}_k + \frac{\partial \hat{\mathfrak{h}}_k{}^\alpha}{\partial \eta}\eta_{,m}\, x^m{}_{,\alpha} + \varrho_{\mathrm{R}} b_k &= \varrho_{\mathrm{R}}\ddot{x}_k,\\ \tilde{\mathsf{A}}_k{}^{(\alpha}{}_m{}^{\beta)} x^m{}_{,\alpha;\beta} + \tilde{q}_k + \frac{\partial \tilde{\mathfrak{h}}_k{}^\alpha}{\partial \theta}\theta_{,m}\, x^m{}_{,\alpha} + \varrho_{\mathrm{R}} b_k &= \varrho_{\mathrm{R}}\ddot{x}_k.\end{aligned}\right\} \tag{45.21}$$

Using the chain rule, we obtain from (45.18) — (45.20) the relation

$$\hat{\mathsf{A}}_k{}^\alpha{}_m{}^\beta = \tilde{\mathsf{A}}_k{}^\alpha{}_m{}^\beta + \frac{\partial \tilde{\mathfrak{h}}_k{}^\alpha}{\partial \theta}\,\frac{\partial \hat{\theta}}{\partial x^m{}_{,\beta}}. \tag{45.22}$$

From (45.18) and $(45.19)_2$ we have $\hat{\mathfrak{h}}(\boldsymbol{F}, \eta) = \tilde{\mathfrak{h}}(\boldsymbol{F}, \hat{\theta}(\boldsymbol{F}, \eta))$; hence

$$\frac{\partial \hat{\mathfrak{h}}_k{}^\alpha}{\partial \eta} = \frac{\partial \tilde{\mathfrak{h}}_k{}^\alpha}{\partial \theta}\,\frac{\partial \hat{\theta}}{\partial \eta} = \frac{\theta}{\varkappa_{(\boldsymbol{F})}}\,\frac{\partial \tilde{\mathfrak{h}}_k{}^\alpha}{\partial \theta}, \tag{45.23}$$

where $\varkappa_{(\boldsymbol{F})}$ is the specific heat at constant deformation [CFT, Sect. 249]. Hence (45.22) may be written also in the form

$$\tilde{\mathsf{A}}_k{}^\alpha{}_m{}^\beta = \hat{\mathsf{A}}_k{}^\alpha{}_m{}^\beta - \frac{\varkappa_{(\boldsymbol{F})}}{\theta}\,\frac{\partial \hat{\mathfrak{h}}_k{}^\alpha}{\partial \eta}\,\frac{\partial \hat{\theta}}{\partial x^m{}_{,\beta}}. \tag{45.24}$$

In (45.24) we see a *general relation* giving the *isothermal elasticity* $\tilde{\mathsf{A}}$ in terms of the *isentropic elasticity* $\hat{\mathsf{A}}$, for the elastic material defined by (45.18). Thus if we know $\varkappa_{(\boldsymbol{F})}$ and the form of the equation of state $(45.19)_2$, the isothermal elasticity is determined entirely from the description in terms of entropy, and the reader may easily derive a converse relation.

From (45.21) we see that in general it is necessary to call in additional physical principles if definite problems are to be solved. In the first place, the balance of energy is to be used [CFT, Sect. 241], but it does not suffice, for it introduces as additional variables the specific internal energy ε and the flux of energy $\boldsymbol{h}$, which must be related to the motion by additional constitutive assumptions. Examples of the use of these principles are given below in Sects. 80, 82, 96 and 96 bis.

46. STOPPELLI's theorems of the existence, uniqueness, and analyticity of the solution of a class of traction boundary-value problems. A theorem of existence and uniqueness for certain traction boundary-value problems of equilibrium has

[1] MURNAGHAN [1949, *9*, p. 333], TRUESDELL [1952, *20*, § 55].

been proved by STOPPELLI[1]. He considered a family of problems depending upon a parameter ε; the loads are assumed to be of the form

$$\boldsymbol{t}_{\mathrm{R}} = \varepsilon \boldsymbol{t}_{\mathrm{R}1}, \qquad \boldsymbol{b} = \varepsilon \boldsymbol{b}_1, \tag{46.1}$$

where $\boldsymbol{t}_{\mathrm{R}1}$ and $\boldsymbol{b}_1$ are prescribed functions of $\boldsymbol{X}$ for $\boldsymbol{X}\in\partial\mathscr{B}$ and $\boldsymbol{X}\in\mathscr{B}$, respectively. As we remarked in Sect. 44, the condition (44.14) is necessary for the existence of a solution, and the condition (44.20) may be assumed without loss of generality. When (46.1) holds, these two conditions are equivalent to

$$\left.\begin{aligned} \int_{\partial\mathscr{B}} \boldsymbol{t}_{\mathrm{R}1}\, d s_{\mathrm{R}} + \int_{\mathscr{B}} \varrho_{\mathrm{R}} \boldsymbol{b}_1\, d v_{\mathrm{R}} &= \boldsymbol{0}, \\ \int_{\partial\mathscr{B}} \boldsymbol{p}_{\mathrm{R}} \times \boldsymbol{t}_{\mathrm{R}1}\, d s_{\mathrm{R}} + \int_{\mathscr{B}} \boldsymbol{p}_{\mathrm{R}} \times \varrho_{\mathrm{R}} \boldsymbol{b}_1\, d v_{\mathrm{R}} &= \boldsymbol{0}. \end{aligned}\right\} \tag{46.2}$$

The idea is to reduce the problem of existence and uniqueness in the general theory, provided $|\varepsilon|$ be sufficiently small, to that for the theory obtained by linearization.

STOPPELLI assumed that the reference configuration is a homogeneous natural state, i.e., that the function $\mathfrak{h}$ in (44.1) is a function of $\boldsymbol{F}$ only, independent of $\boldsymbol{X}$, and that the body is free of stresses when in the reference configuration, so that

$$\boldsymbol{0} = \mathfrak{h}(\boldsymbol{1}). \tag{46.3}$$

Under these assumptions, linearization of (44.8) and (44.11) yields the traction boundary-value problem of equilibrium in the classical theory of infinitesimal elasticity corresponding to the loads $\boldsymbol{t}_{\mathrm{R}1}$ and $\boldsymbol{b}_1$. The conditions (46.2) are necessary in order that this problem have a solution. *We shall assume that the body* $\mathscr{B}$ *and the elasticities* (44.3), *evaluated at* $\boldsymbol{F}=\boldsymbol{1}$, *are such that the conditions* (46.2) *are sufficient for the existence and uniqueness* (to within a rigid displacement) *of the solution of the problem in the infinitesimal theory*[2].

In the classical infinitesimal theory, the solution of this problem is unique only to within an arbitrary infinitesimal rigid displacement, but in the finite theory we expect the rotation to be determined in the course of the solution, leaving the possibility of an arbitrary rigid translation but no more. Thus we expect to determine from the problem defined by (44.8) and (44.11), i.e., from

$$\left.\begin{aligned} \operatorname{Div} \mathfrak{h}(\boldsymbol{F}) &= -\varrho_{\mathrm{R}} \boldsymbol{b}, \\ \mathfrak{h}(\boldsymbol{F})\, \boldsymbol{n}_{\mathrm{R}} &= \boldsymbol{t}_{\mathrm{R}}, \end{aligned}\right\} \tag{46.4}$$

a unique deformation $\boldsymbol{x}=\boldsymbol{\chi}(\boldsymbol{X})$ such that a single assigned particle suffers no displacement. Taking the place of this particle as the origin, $\boldsymbol{0}$, and identifying spatial points with their position vectors, we have

$$\boldsymbol{\chi}(\boldsymbol{0}) = \boldsymbol{0}. \tag{46.5}$$

If we superimpose a rigid rotation $\boldsymbol{Q}$ about $\boldsymbol{0}$ upon a deformation $\boldsymbol{\chi}^*$, we obtain a new deformation $\boldsymbol{\chi}=\boldsymbol{Q}\boldsymbol{\chi}^*$ with gradient $\boldsymbol{F}=\boldsymbol{Q}\boldsymbol{F}^*$. It follows from (43.9) that $\mathfrak{h}(\boldsymbol{F})=\boldsymbol{Q}\mathfrak{h}(\boldsymbol{F}^*)$ and $\operatorname{Div}\mathfrak{h}(\boldsymbol{F})=\boldsymbol{Q}\operatorname{Div}\mathfrak{h}(\boldsymbol{F}^*)$. Hence, if $\boldsymbol{\chi}^*$ is a solution of

[1] The analysis of STOPPELLI [1954, *23*] refers to hyperelastic materials but applies equally well to the more general theory considered here. An exposition of parts of STOPPELLI's work has been given by GRIOLI [1962, *24*, Ch. V, §§ 1—5].

[2] Sufficient conditions are given in the classical literature. Even for hyperelastic materials, the problem is still under study. Unfortunately, there is no adequate exposition of the analytical theory of solutions in the classical infinitesimal theory as it is known today.

(46.4), then χ is a solution of

$$\left.\begin{aligned}\operatorname{Div}\mathfrak{h}(\boldsymbol{F}) &= -\varrho_R \boldsymbol{Q}\boldsymbol{b},\\ \mathfrak{h}(\boldsymbol{F})\,\boldsymbol{n}_R &= \boldsymbol{Q}\boldsymbol{t}_R.\end{aligned}\right\} \tag{46.6}$$

Conversely, if χ is a solution of (46.6), then $\chi^* = \boldsymbol{Q}^T\chi$ is a solution of (46.4). STOPPELLI preferred to consider the system (46.6) and to determine the rigid rotation $\boldsymbol{Q}$ so as to make (46.6) have a unique solution χ such that (46.5) holds and such that $\boldsymbol{F}(\boldsymbol{0})$ is a pure stretch, i.e., a positive-definite and symmetric tensor. Then (46.4) has a unique solution $\chi^* = \boldsymbol{Q}^T\chi$ which also satisfies (46.5); for this solution $\boldsymbol{Q}^T$ is the local rotation tensor at $\boldsymbol{X} = \boldsymbol{0}$.

So as to solve (46.6), let us consider a deformation χ and introduce the additional load which would be required to equilibrate it:

$$\left.\begin{aligned}\operatorname{Div}\mathfrak{h}(\boldsymbol{F}) + \varrho_R \boldsymbol{Q}\boldsymbol{b} &= \boldsymbol{h},\\ -\mathfrak{h}(\boldsymbol{F})\,\boldsymbol{n}_R + \boldsymbol{Q}\boldsymbol{t}_R &= \boldsymbol{g}.\end{aligned}\right\} \tag{46.7}$$

Regarding this as a functional transformation, we seek to determine both $\boldsymbol{Q}$ and χ in such a way that $\boldsymbol{g} = \boldsymbol{0}$ and $\boldsymbol{h} = \boldsymbol{0}$; if we can do so, our boundary-value problem will be solved. First we shall determine $\boldsymbol{Q}$ in such a way that the loads $\boldsymbol{g}$ and $\boldsymbol{h}$, for given χ, are equilibrated. By use of GREEN's transformation and (46.2) we see at once that

$$\int_{\partial\mathscr{B}} \boldsymbol{g}\, ds_R + \int_{\mathscr{B}} \boldsymbol{h}\, dv_R = \boldsymbol{0} \tag{46.8}$$

for any constant $\boldsymbol{Q}$, but in order that

$$\int_{\partial\mathscr{B}} \boldsymbol{p}_R \times \boldsymbol{g}\, ds_R + \int_{\mathscr{B}} \boldsymbol{p}_R \times \boldsymbol{h}\, dv_R = \boldsymbol{0}, \tag{46.9}$$

a straightforward calculation using (46.1) shows it to be necessary and sufficient that

$$\varepsilon(\boldsymbol{A}_1\boldsymbol{Q}^T - \boldsymbol{Q}\boldsymbol{A}_1) = \boldsymbol{M} \equiv \int_{\mathscr{B}} [\mathfrak{h}(\boldsymbol{F})^T - \mathfrak{h}(\boldsymbol{F})]\, dv_R, \tag{46.10}$$

where $\boldsymbol{A}_1$ is the astatic load [cf. (44.17)] of the forces $\boldsymbol{t}_{R1}$, $\boldsymbol{b}_1$ in the reference configuration:

$$\boldsymbol{A}_1 \equiv \int_{\partial\mathscr{B}} \boldsymbol{p}_R \otimes \boldsymbol{t}_{R1}\, ds_R + \int_{\mathscr{B}} \boldsymbol{p}_R \otimes \varrho_R\, \boldsymbol{b}_1\, dv_R. \tag{46.11}$$

The condition $(46.2)_2$ ensures that $\boldsymbol{A}_1$ is a symmetric tensor. Of course, the tensor $\boldsymbol{M}$ in $(46.10)_2$ is skew.

For given χ, (46.10) is an equation to be solved for $\boldsymbol{Q}$. This equation has the form

$$\boldsymbol{A}_1\boldsymbol{Q}^T - \boldsymbol{Q}\boldsymbol{A}_1 \equiv \boldsymbol{\Phi}(\boldsymbol{Q}) = \frac{1}{\varepsilon}\boldsymbol{M}, \tag{46.12}$$

in which $\boldsymbol{\Phi}$ is a transformation which maps the orthogonal group $\mathscr{o}$ into the three-dimensional space $\mathscr{W}$ of all skew tensors. Clearly, $\boldsymbol{\Phi}(\boldsymbol{1}) = \boldsymbol{0}$. By the inverse function theorem, $\boldsymbol{\Phi}$ is invertible near $\boldsymbol{Q} = \boldsymbol{1}$ if the differential $\delta\boldsymbol{\Phi}$ of $\boldsymbol{\Phi}$ at $\boldsymbol{Q} = \boldsymbol{1}$ is an invertible linear transformation. Now, $\delta\boldsymbol{\Phi}(\boldsymbol{W})$ is defined for $\boldsymbol{W} \in \mathscr{W}$ by[1]

$$\delta\boldsymbol{\Phi}(\boldsymbol{W}) = \boldsymbol{A}_1\boldsymbol{W}^T - \boldsymbol{W}\boldsymbol{A}_1 = -(\boldsymbol{A}_1\boldsymbol{W} + \boldsymbol{W}\boldsymbol{A}_1), \tag{46.13}$$

and $\delta\boldsymbol{\Phi}$ is invertible if and only if the equation

$$\boldsymbol{A}_1\boldsymbol{W} + \boldsymbol{W}\boldsymbol{A}_1 = \boldsymbol{0} \tag{46.14}$$

---

[1] We use the fact that the tangent-space at $\boldsymbol{Q} = \boldsymbol{1}$ of the manifold of the orthogonal group may be identified with the space $\mathscr{W}$ of skew tensors.

has only the trivial solution $\boldsymbol{W}=\mathbf{0}$ in the space $\mathscr{W}$. It is easily seen that (46.14) is equivalent to

$$(\boldsymbol{A}_1-\mathbf{1}\operatorname{tr}\boldsymbol{A}_1)\,\boldsymbol{w}=\mathbf{0}, \tag{46.15}$$

where $\boldsymbol{w}$ is the axial vector[1] equivalent to the skew tensor $\boldsymbol{W}$. Thus $\delta\boldsymbol{\Phi}$ is invertible if and only if

$$\det(\boldsymbol{A}_1-\mathbf{1}\operatorname{tr}\boldsymbol{A}_1)\neq 0. \tag{46.16}$$

It follows that $\boldsymbol{\Phi}$ is locally invertible near $\boldsymbol{Q}=\mathbf{1}$ if (46.16) holds. If, on the other hand,

$$\det(\boldsymbol{A}_1-\mathbf{1}\operatorname{tr}\boldsymbol{A}_1)=0, \tag{46.17}$$

then, as is well known, there exists an axis such that any rotation of $\mathscr{B}$ about it while the vectors $\boldsymbol{t}_{R1}$ and $\boldsymbol{b}_1$ are held constant in space does not destroy the equilibrium of the load system $\boldsymbol{t}_{R1}$, $\boldsymbol{b}_1$. The load acting on $\mathscr{B}$ in the reference configuration is then said to possess an *axis of equilibrium*. Accordingly, so as to continue the procedure initiated here, *we shall assume henceforth that the load* $\boldsymbol{t}_{R1}$, $\boldsymbol{b}_1$ *does not possess an axis of equilibrium*[2]. Some remarks on the excluded case are given at the end of this section.

In view of the local invertibility of the transformation $\boldsymbol{\Phi}$ of (46.12) we have the following precise result[3]: There are positive numbers α and β, independent of ε, such that if

$$|\boldsymbol{M}|\leqq\alpha|\varepsilon|, \tag{46.18}$$

the equation $(46.10)_1$ has a unique solution $\boldsymbol{Q}$ with the property

$$|\boldsymbol{Q}-\mathbf{1}|\leqq\beta. \tag{46.19}$$

We now specify conditions of regularity sufficient to ensure the necessary degree of smoothness of the functional transformation defined by (46.7).

1. The region occupied by $\mathscr{B}$ in the reference configuration is compact. We identify $\mathscr{B}$ with the set of position vectors of the particles in the reference configuration.

2. The boundary surface $\partial\mathscr{B}$ in the reference configuration has Hölder-continuous[4] principal curvatures with exponent λ, and it is possible to cover $\partial\mathscr{B}$ by a finite number of regions $\mathscr{R}^{(p)}$, each of which can be represented parametrically on a disc $\mathscr{C}^{(p)}$, and every point of $\partial\mathscr{B}$ is interior to at least one $\mathscr{R}^{(p)}$.

3. $\boldsymbol{b}_1$ is Hölder-continuous in $\mathscr{B}$ with exponent λ; $\boldsymbol{t}_{R1}$ and its derivatives with respect to the parameters of some parametric representation of $\partial\mathscr{B}$ (as in No. 2) are Hölder-continuous with exponent λ.

4. The response function $\mathfrak{h}(\boldsymbol{F})$ and its derivatives up to the third order are continuous in the region of the 9-dimensional space $\mathscr{L}$ of tensors $\boldsymbol{F}$ that is defined by the inequality $|\boldsymbol{F}-\mathbf{1}|\leqq R$.

Let Σ'' be the class of all vector-valued functions $\boldsymbol{u}$ on $\mathscr{B}$ which have Hölder-continuous second derivatives, with exponent λ, on $\mathscr{B}$, and which satisfy $\boldsymbol{u}(\mathbf{0})=\mathbf{0}$.

[1] Cf. Eq. (3.3) of the Appendix to CFT.

[2] The special difficulty of the case when the load has an axis of equilibrium was first noticed by SIGNORINI in connection with the formal expansion procedure we shall describe in Sects. 63 and 64.

[3] STOPPELLI [1954, *23*].

[4] In a space with metric ϱ, the function F is Hölder-continuous with exponent λ in a domain $\mathscr{D}$ if

$$\frac{|F(X')-F(X)|}{[\varrho(X',X)]^\lambda}$$

is bounded as X and X' vary over $\mathscr{D}$. The least upper bound of the quotient is the Hölder coefficient of F in the domain $\mathscr{D}$.

Using a single Cartesian co-ordinate system, we set

$$\|\boldsymbol{u}\| \equiv \sum_{k=1}^{3} \operatorname*{Max}_{\boldsymbol{X}\in\mathscr{B}} |u^k(\boldsymbol{X})| + \sum_{k,\alpha=1}^{3} \operatorname*{Max}_{\boldsymbol{X}\in\mathscr{B}} |u^k{}_{,\alpha}(\boldsymbol{X})| + \sum_{k,\alpha,\beta=1}^{3} [\operatorname*{Max}_{\boldsymbol{X}\in\mathscr{B}} |u^k{}_{,\alpha\beta}(\boldsymbol{X})| + B^k{}_{\alpha\beta}], \tag{46.20}$$

where $B^k{}_{\alpha\beta}$ is the Hölder coefficient of $u^k{}_{,\alpha\beta}$ in $\mathscr{B}$. With (46.20) as a norm the class Σ'' becomes a Banach space, i.e. a complete normed linear space.

Let Σ' be the class of all pairs $(\boldsymbol{g}, \boldsymbol{h})$ of vector-valued functions with the following properties:

(a) $\boldsymbol{h}$ is defined on $\mathscr{B}$ and is Hölder-continuous with exponent λ in $\mathscr{B}$.

(b) $\boldsymbol{g}$ is defined on the boundary surface $\partial\mathscr{B}$; $\boldsymbol{g}$ and its derivatives $g^k{}_{,\Gamma}$ with respect to the parameters π_Γ, $\Gamma=1, 2$, of some parametric representation of $\partial\mathscr{B}$ (as in No. 2, above) are Hölder-continuous with exponent λ.

(c) The equations (46.8) and (46.9) are satisfied.

We observe that the pair $(\boldsymbol{t}_{R1}, \varrho_R \boldsymbol{b}_1)$ is a member of Σ', by Hypothesis 3. We set

$$\|(\boldsymbol{g}, \boldsymbol{h})\| = \sum_{k=1}^{3} [\operatorname*{Max}_{\boldsymbol{X}\in\mathscr{B}} |h^k(\boldsymbol{X})| + \operatorname*{Max}_{\boldsymbol{X}\in\partial\mathscr{B}} |g^k(\boldsymbol{X})| + C^k + D^k], \tag{46.21}$$

where C^k is the Hölder coefficient of h^k, and where D^k is the greatest of the Hölder coefficients of the derivatives $g^k{}_{,\Gamma}$ in the various discs $\mathscr{C}^{(p)}$ (No. 2, above). It can be verified that Σ', with (46.21) as a norm, is also a Banach space.

Finally, let Σ be the subspace of Σ'' obtained by adding the requirement that

$$\boldsymbol{H}(0) = \boldsymbol{H}(0)^T, \qquad \boldsymbol{H} = \nabla \boldsymbol{u}. \tag{46.22}$$

It can be shown that Σ is a closed subspace of Σ'' and hence is also a Banach space.

A member $\boldsymbol{u}$ of Σ can be regarded as a possible displacement (44.22) of $\mathscr{B}$, giving rise to a deformation $\boldsymbol{\chi}$, a deformation gradient $\boldsymbol{F} = \nabla\boldsymbol{u} + \boldsymbol{1}$, and hence, by means of $(46.10)_2$, to a skew tensor $\boldsymbol{M}$. From Hypotheses 4 and 1 STOPPELLI infers that there is a number K, independent of $\boldsymbol{u}$, such that

$$|\boldsymbol{M}| \leqq K \|\boldsymbol{u}\|^2 \tag{46.23}$$

provided that $\|\boldsymbol{u}\| \leqq R$. Hence (46.18) is satisfied if

$$\|\boldsymbol{u}\| \leqq \sqrt{\frac{\alpha}{K}}\, |\varepsilon| \tag{46.24}$$

and

$$|\varepsilon| \leqq \frac{K}{\alpha} R^2. \tag{46.25}$$

It follows that if (46.24) and (46.25) hold, then there is a unique orthogonal tensor $\boldsymbol{Q}$ such that (46.19) and $(46.10)_1$ are satisfied. With this choice of $\boldsymbol{Q}$, the load $(\boldsymbol{g}, \boldsymbol{h})$ determined by (46.7) is equilibrated; in view of the hypotheses, it defines a point in the Banach space Σ'. In this manner, (46.7) can be seen to define a mapping $\mathfrak{T}$ of a neighborhood Λ of $\boldsymbol{0} \in \Sigma$ into the space Σ':

$$(\boldsymbol{g}, \boldsymbol{h}) = \mathfrak{T}[\boldsymbol{u}], \qquad \boldsymbol{u} \in \Lambda. \tag{46.26}$$

The neighborhood Λ is defined by (46.24). In view of (46.3) and $(46.10)_2$ we have $\boldsymbol{M} = \boldsymbol{0}$ if $\boldsymbol{u} = \boldsymbol{0}$. The corresponding solution $\boldsymbol{Q}$ of $(46.12)_1$ is $\boldsymbol{Q} = \boldsymbol{1}$. Hence (46.7) and (46.3) yield

$$(\boldsymbol{t}_R, \varrho_R \boldsymbol{b}) = \mathfrak{T}(\boldsymbol{0}). \tag{46.27}$$

Through a long chain of estimates, STOPPELLI proved that the transformation $\mathfrak{T}$ has a continuous Fréchet differential[1] $\delta\mathfrak{T}(\boldsymbol{u})$ at all $\boldsymbol{u}\in\Lambda$. For each $\boldsymbol{u}\in\Lambda$, $\delta\mathfrak{T}(\boldsymbol{u})$ is a continuous linear transformation of the Banach space Σ into the Banach space Σ'.

According to a known inverse mapping theorem of functional analysis, the mapping $\mathfrak{T}$ is locally invertible in a neighborhood of $\mathbf{0}\in\Sigma$ if the Fréchet differential $\delta\mathfrak{T}(\mathbf{0})=\mathfrak{L}$ is an invertible linear transformation of Σ onto Σ'. We write

$$(\delta\boldsymbol{g},\, \delta\boldsymbol{h})=\mathfrak{L}(\delta\boldsymbol{u}) \tag{46.28}$$

for the result of operating with $\mathfrak{L}$ on $\delta\boldsymbol{u}\in\Sigma$. The pair $(\delta\boldsymbol{g},\, \delta\boldsymbol{h})$ is a member of Σ'. STOPPELLI showed that (46.28) has the explicit form

$$\left.\begin{aligned} \mathsf{A}^{(0)\alpha\ \beta}_{k\ \ m}\,\delta u^m{}_{,\alpha;\beta}&=\delta h_k,\\ -\mathsf{A}^{(0)\alpha\ \beta}_{k\ \ m}\,\delta u^m{}_{,\beta}\, n_{\mathrm{R}\alpha}&=\delta g_k,\end{aligned}\right\} \tag{46.29}$$

where $\mathsf{A}^{(0)}$ is the elasticity tensor (44.3) evaluated for the natural reference state, which corresponds to $\boldsymbol{F}=\mathbf{1}$.

The equations (46.29) coincide with the equations defining the traction boundary-value problem of classical infinitesimal elasticity, with the load $\delta\boldsymbol{g}$, $\delta\boldsymbol{h}$. If $(\delta\boldsymbol{g},\, \delta\boldsymbol{h})\in\Sigma'$, by part (c) of the definition of Σ', the load is equilibrated. By our initial hypothesis, these equations then possess a solution $\delta\boldsymbol{u}$ which is unique to within a rigid displacement. The solution $\delta\boldsymbol{u}$ is rendered unique by imposing the requirements $\delta\boldsymbol{u}(\mathbf{0})=\mathbf{0}$ and (46.22), with $\boldsymbol{u}$ replaced by $\delta\boldsymbol{u}$. This unique solution can be shown to be a member of the Banach space Σ. It follows that the linear transformation $\mathfrak{L}$ of (46.28) is indeed invertible and hence that the mapping $\mathfrak{T}$ of (46.26) is invertible in a neighborhood Λ_1 of $\mathbf{0}\in\Sigma$ contained in Λ.

STOPPELLI then showed that the zero element $(\mathbf{0},\mathbf{0})$ of Σ' belongs to the image of Λ_1 under the mapping $\mathfrak{T}$, provided $|\varepsilon|$ be sufficiently small. Accordingly, for small $|\varepsilon|$ the equation

$$(\mathbf{0},\, \mathbf{0})=\mathfrak{T}(\boldsymbol{u}) \tag{46.30}$$

has a unique solution $\boldsymbol{u}$ in Λ_1. This solution $\boldsymbol{u}$, since it belongs to Σ, satisfies $\boldsymbol{u}(\mathbf{0})=\mathbf{0}$ and (46.22). Now, $\boldsymbol{H}=\nabla\boldsymbol{u}$ can be made as small as desired by making $|\varepsilon|$ small enough. Hence, for $|\varepsilon|$ sufficiently small, $\boldsymbol{F}=\boldsymbol{H}+\mathbf{1}$ is invertible at all particles $\boldsymbol{X}\in\mathscr{B}$ and, by (46.22), the value of $\boldsymbol{F}$ at $\boldsymbol{X}=\mathbf{0}$ is symmetric and positive-definite. Therefore, the deformation $\boldsymbol{\chi}$ defined by $\boldsymbol{x}=\boldsymbol{\chi}(\boldsymbol{X})=\boldsymbol{X}+\boldsymbol{u}(\boldsymbol{X})$ satisfies all the requirements for the solution of (46.6) mentioned in the beginning, and $\boldsymbol{\chi}^*=\boldsymbol{Q}^T\boldsymbol{\chi}$ is a solution of the original problem defined by (46.4).

In summary, we have the following ***theorem of existence and uniqueness of Stoppelli***: *Let Hypotheses 1—4 be satisfied, and suppose the load* $\boldsymbol{t}_{\mathrm{R}1}$, $\boldsymbol{b}_1$ *has no axis of equilibrium; then it is possible to find positive numbers* γ, $\varkappa$ *such that for all* ε *for which* $|\varepsilon|\leqq\gamma$, *the system* (46.4) *has a unique solution* $\boldsymbol{x}=\boldsymbol{\chi}(\boldsymbol{X})=\boldsymbol{X}+\boldsymbol{u}(\boldsymbol{X})$ *satisfying the conditions*

$$\|\boldsymbol{u}\|\leqq\varkappa,\qquad \boldsymbol{\chi}(\mathbf{0})=\mathbf{0}. \tag{46.31}$$

It is easily seen that $\boldsymbol{u}$ tends to zero in norm as $\varepsilon\to 0$.

STOPPELLI next[2] considered the case when Hypothesis 4 above is replaced by the stronger

Hypothesis 4a: The response function $\mathfrak{h}(\boldsymbol{F})$ is *analytic* in the region of $\mathscr{L}$ defined by $|\boldsymbol{F}-\mathbf{1}|\leqq R$.

---

[1] The concept of Fréchet differential, in another context, is explained in Sect. 38.

[2] STOPPELLI [1955, *26*].

Under this hypothesis, STOPPELLI showed that *the solution* χ *whose existence and uniqueness was established above depends analytically on* ε *when* $|\varepsilon|<\gamma$. Again the proof employs known theorems of functional analysis. Methods for constructing the coefficients of a power series expansion of $\boldsymbol{u}$ with respect to ε will be presented in Sects. 63 and 64.

In a sequence of further papers STOPPELLI[1] has investigated the nature of solutions in the case when the loads have an axis of equilibrium. The results are very complicated. In some cases, a unique solution exists and is an analytic function of ε. In others, there are two or three distinct solutions, all or some of which are analytic functions of ε, or of $\sqrt{\varepsilon}$ when $\varepsilon>0$, or of $\sqrt{-\varepsilon}$ when $\varepsilon<0$. It is interesting that in some cases a solution may exist for given forces but fail to exist when those forces are reversed. For details the reader should consult STOPPELLI's papers.

The results of STOPPELLI serve, in the main, to give the classical infinitesimal theory a position with respect to the general theory of elasticity. In the same spirit, JOHN[2] has shown that for an isotropic elastic body, the boundary-value problem of place does not have more than one solution, provided the assigned displacements be not too large.

47. Isotropic elastic materials, I. General properties. In the discussion of simple solids in Sect. 33, the types of aeolotropy corresponding to the various crystal classes and to some other requirements of symmetry were defined, and in Sect. 34 the concept of curvilinear aeolotropy was presented. For elastic materials, to which these definitions apply *a fortiori*, no important simplification results from them except, of course, when it comes to the solution of specific problems. In the present treatise we must be content to touch but lightly upon all kinds of special elastic symmetries except full isotropy, which now we proceed to consider in detail.

If a material is elastic in the sense of the definition of Sect. 43 and isotropic in the sense of the definition of Sect. 31, it is called an *isotropic elastic material.*

The following theorem is a direct consequence of $(43.16)_1$ and the definitions of an isotropic material and of an undistorted state (Sect. 31):

An elastic material is isotropic if and only if its response function $\mathfrak{g}(\boldsymbol{F})$, *taken relative to an undistorted state as reference configuration, is an isotropic tensor function of* $\boldsymbol{F}$, i.e., if and only if $\mathfrak{g}$ satisfies the relation

$$\boldsymbol{Q}\,\mathfrak{g}(\boldsymbol{F})\,\boldsymbol{Q}^T=\mathfrak{g}(\boldsymbol{Q}\boldsymbol{F}\boldsymbol{Q}^T), \tag{47.1}$$

identically in the tensor $\boldsymbol{F}$ and the orthogonal tensor $\boldsymbol{Q}$ (cf. Sect. 8).

Other necessary and sufficient conditions for response functions $\mathfrak{g}$ and $\mathfrak{h}$ to define an isotropic elastic material are the identities

$$\mathfrak{g}(\boldsymbol{F})=\mathfrak{g}(\boldsymbol{F}\boldsymbol{Q}), \quad \mathfrak{h}(\boldsymbol{F}\boldsymbol{Q})=\mathfrak{h}(\boldsymbol{F})\,\boldsymbol{Q}. \tag{47.2}$$

The former identity follows from (43.15), since if and only if the material is isotropic may $\boldsymbol{H}$ in (43.15) be taken as any orthogonal tensor $\boldsymbol{Q}$. The latter follows from the former by $(43.11)_1$.

Reduced forms of the constitutive equation for isotropic elastic materials may be obtained from (43.5) and (43.6) with use of (47.1):

$$\boldsymbol{T}=\mathfrak{g}(\boldsymbol{V}), \tag{47.3}$$

$$\boldsymbol{T}=\mathfrak{f}(\boldsymbol{B}). \tag{47.4}$$

---

[1] STOPPELLI [1957, *25* and *26*] [1958, *45*].

[2] JOHN [1958, *25*, Ch. IV].

Here $\boldsymbol{V}=\boldsymbol{R}\boldsymbol{U}\boldsymbol{R}^T$ is the left stretch tensor and $\boldsymbol{B}=\boldsymbol{V}^2=\boldsymbol{R}\boldsymbol{C}\boldsymbol{R}^T=\boldsymbol{F}\boldsymbol{F}^T$ the left Cauchy-Green tensor (cf. Sect. 23). For isotropic materials, the response functions $\mathfrak{g}$, $\mathfrak{f}$, and $\mathfrak{h}$ [cf. (43.7)] are isotropic tensor functions (cf. Sect. 8). A co-ordinate form of (47.4) is

$$T_{km}=\mathfrak{f}_{km}(B_{pq};\, g_{ru}), \tag{47.5}$$

where g_{ru} are the components of the metric tensor, and where the forms of the component functions $\mathfrak{f}_{km}$ do not depend on the choice of the co-ordinate system.

The equations (47.3), (47.4), (47.5) also follow directly by specialization from (31.9). (31.10), (31.14), or from (35.8), (35.9), (35.10).

The representation theorems (12.6) and (12.7) for isotropic tensor functions, applied to (47.3) and (47.4), yield the following explicit forms of the general constitutive equation for an isotropic elastic material[1]:

$$\boldsymbol{T}=\text{ש}_0\boldsymbol{1}+\text{ש}_1\boldsymbol{V}+\text{ש}_2\boldsymbol{V}^2, \tag{47.6}$$

$$\boldsymbol{T}=\gimel_0\boldsymbol{1}+\gimel_1\boldsymbol{V}+\gimel_{-1}\boldsymbol{V}^{-1}, \tag{47.7}$$

$$\left.\begin{aligned}\boldsymbol{T}&=\aleph_0\boldsymbol{1}+\aleph_1\boldsymbol{B}+\aleph_2\boldsymbol{B}^2,\\ T^k_m&=\aleph_0\,\delta^k_m+\aleph_1\,B^k_m+\aleph_2\,B^k_p\,B^p_m,\end{aligned}\right\} \tag{47.8}$$

$$\left.\begin{aligned}\boldsymbol{T}&=\beth_0\boldsymbol{1}+\beth_1\boldsymbol{B}+\beth_{-1}\boldsymbol{B}^{-1},\\ T^k_m&=\beth_0\,\delta^k_m+\beth_1\,B^k_m+\beth_{-1}(B^{-1})^k_m,\end{aligned}\right\} \tag{47.9}$$

where the *response coefficients*

$$\left.\begin{aligned}&\text{ש}_\Gamma=\text{ש}_\Gamma(\boldsymbol{V}), &&\aleph_\Gamma=\aleph_\Gamma(\boldsymbol{B}), &&\Gamma=0,1,2,\\ &\gimel_\Gamma=\gimel_\Gamma(\boldsymbol{V}), &&\beth_\Gamma=\beth_\Gamma(\boldsymbol{B}), &&\Gamma=0,1,-1,\end{aligned}\right\} \tag{47.10}$$

are scalar invariant functions of the left stretch tensor $\boldsymbol{V}$ and of the left Cauchy-Green tensor $\boldsymbol{B}$, respectively. These invariants may be expressed alternatively as functions of the principal invariants of $\boldsymbol{V}$ or of $\boldsymbol{B}$:

$$\left.\begin{aligned}&\text{ש}_\Gamma=\text{ש}_\Gamma(I_{\boldsymbol{V}}, II_{\boldsymbol{V}}, III_{\boldsymbol{V}}), &&\aleph_\Gamma=\aleph_\Gamma(I_{\boldsymbol{B}}, II_{\boldsymbol{B}}, III_{\boldsymbol{B}}), &&\Gamma=0,1,2;\\ &\gimel_\Gamma=\gimel_\Gamma(I_{\boldsymbol{V}}, II_{\boldsymbol{V}}, III_{\boldsymbol{V}}), &&\beth_\Gamma=\beth_\Gamma(I_{\boldsymbol{B}}, II_{\boldsymbol{B}}, III_{\boldsymbol{B}}), &&\Gamma=0,1,-1;\end{aligned}\right\} \tag{47.11}$$

or also as symmetric functions of the proper numbers of $\boldsymbol{V}$ or of $\boldsymbol{B}$. Denoting the principal stretches, which are the proper numbers of $\boldsymbol{V}$, by v_a, we may thus write, alternatively,

$$\text{ש}_\Gamma=\bar{\text{ש}}_\Gamma(v_1, v_2, v_3), \qquad \aleph_\Gamma=\bar{\aleph}_\Gamma(v_1, v_2, v_3), \qquad \text{etc.}, \tag{47.12}$$

where $\bar{\text{ש}}_\Gamma$, $\bar{\aleph}_\Gamma$, etc., are symmetric functions of their arguments. The directions defined by the proper vectors of $\boldsymbol{V}$, which coincide with the proper vectors of $\boldsymbol{B}$, are called the *principal axes of strain in the deformed state* (cf. CFT, Sects. 27, 28).

[1] REINER [1948, *9*, § 2], RICHTER [1948, *10*, § 2] [1952, *16*, § 6]; an equivalent result was derived from a more complicated definition of isotropy by MANACORDA [1953, *16*, § 5]. That formulae of this kind with *special* values for the response coefficients, e.g. (86.9) and (86.10), hold for isotropic hyperelastic materials had been known since 1894, and such formulae appear often in the Italian literature, e.g. SIGNORINI [1930, *3*, § 8].

From our presentation it is immediately clear that (47.3) and (47.4) are correct *only* for isotropic materials. This fact was first observed, for hyperelastic materials, by MURNAGHAN [1928, *3*] [1931, *6*]. A simple direct proof is as follows: Since both $\boldsymbol{T}$ and $\boldsymbol{B}$ are indifferent tensors, in order for (47.4) to satisfy the principle of material indifference it is necessary that

$$\boldsymbol{Q}\,\mathfrak{f}(\boldsymbol{B})\,\boldsymbol{Q}^T=\mathfrak{f}(\boldsymbol{Q}\boldsymbol{B}\boldsymbol{Q}^T),$$

showing that (47.4) is not indifferent *unless* $\mathfrak{f}$ is isotropic. This fact notwithstanding, PRAGER [1961, *47*, Ch. X, § 1] took (47.4) as the definition of an elastic material in general.

Since the response coefficients $\aleph_\Gamma$ and $\beth_\Gamma$ are used equally frequently in modern work, it is helpful to have relations connecting them:

$$\beth_0 = \aleph_0 - II_{\boldsymbol{B}}\aleph_2, \qquad \beth_1 = \aleph_1 + I_{\boldsymbol{B}}\aleph_2, \qquad \beth_{-1} = III_{\boldsymbol{B}}\aleph_2, \tag{47.13}$$

$$\aleph_0 = \beth_0 + \frac{II_{\boldsymbol{B}}}{III_{\boldsymbol{B}}}\beth_{-1}, \qquad \aleph_1 = \beth_1 - \frac{I_{\boldsymbol{B}}}{III_{\boldsymbol{B}}}\beth_{-1}, \qquad \aleph_2 = \frac{1}{III_{\boldsymbol{B}}}\beth_{-1}, \tag{47.14}$$

as follows at once by application of the Hamilton-Cayley theorem (9.10) so as to convert (47.8) to the form (47.9), or vice versa.

When the material is isotropic, $\tilde{\boldsymbol{T}}$ is given by an isotropic function of $\boldsymbol{C}$, and hence the function $\mathfrak{t}$ in (43.10) has a representation of the type (12.7), namely,

$$\left.\begin{aligned}\tilde{\boldsymbol{T}} &= \mathfrak{t}(\boldsymbol{C}) = \daleth_0 \boldsymbol{1} + \daleth_1 \boldsymbol{C} + \daleth_{-1}\boldsymbol{C}^{-1},\\ \tilde{T}^{\alpha\beta} &= \mathfrak{t}^{\alpha\beta}(\boldsymbol{C}) = \daleth_0 g^{\alpha\beta} + \daleth_1 C^{\alpha\beta} + \daleth_{-1}(C^{-1})^{\alpha\beta}.\end{aligned}\right\} \tag{47.15}$$

So as to specialize general results to isotropic materials, it is often useful to have the explicit form of that representation in terms of the response coefficients occuring in (47.9). To derive it, we need the easy identity[1]

$$III_{\boldsymbol{C}}\boldsymbol{F}^{-1}\boldsymbol{B}^{-1}(\boldsymbol{F}^{-1})^T = \boldsymbol{C} - I_{\boldsymbol{C}}\boldsymbol{1} + II_{\boldsymbol{C}}\boldsymbol{C}^{-1}. \tag{47.16}$$

From $(43\,\mathrm{A}.11)_5$, (47.9), (23.4), and (23.5) we then find that[2]

$$\left.\begin{aligned}\daleth_0 &= \sqrt{III_{\boldsymbol{C}}}\,\beth_1 - \frac{I_{\boldsymbol{C}}}{\sqrt{III_{\boldsymbol{C}}}}\beth_{-1},\\ \daleth_1 &= \frac{\beth_{-1}}{\sqrt{III_{\boldsymbol{C}}}},\\ \daleth_{-1} &= \sqrt{III_{\boldsymbol{C}}}\,\beth_0 + \frac{II_{\boldsymbol{C}}}{\sqrt{III_{\boldsymbol{C}}}}\beth_{-1}.\end{aligned}\right\} \tag{47.17}$$

The arguments $I_{\boldsymbol{B}}, II_{\boldsymbol{B}}, III_{\boldsymbol{B}}$ of the functions $\beth_\Gamma$ are to be replaced by $I_{\boldsymbol{C}}, II_{\boldsymbol{C}}, III_{\boldsymbol{C}}$, to which they are numerically equal, as follows from the remark about the proper numbers of $\boldsymbol{C}$ and $\boldsymbol{B}$ in Sect. 23.

An explicit form for the elasticity C of an isotropic elastic material may be found as follows. First, by use of (23.4) and the Hamilton-Cayley equation we derive the identity

$$\boldsymbol{F}\boldsymbol{C}\boldsymbol{F}^T = \boldsymbol{B}^2 = III\,\boldsymbol{B}^{-1} - II\,\boldsymbol{1} + I\,\boldsymbol{B}, \tag{47.18}$$

where $I \equiv I_{\boldsymbol{C}} = I_{\boldsymbol{B}}$, etc. Second, for the derivative of a function $f(I, II, III)$, by use of (9.11), (23.5), and (47.18) we find the formula

$$\left.\begin{aligned}\boldsymbol{F} f_{\boldsymbol{C}} \boldsymbol{F}^T &= \boldsymbol{F}\left[\frac{\partial f}{\partial I}\boldsymbol{1} + \frac{\partial f}{\partial II}(I\boldsymbol{1} - \boldsymbol{C}) + \frac{\partial f}{\partial III}III\,\boldsymbol{C}^{-1}\right]\boldsymbol{F}^T,\\ &= \frac{\partial f}{\partial I}\boldsymbol{B} + \frac{\partial f}{\partial II}(II\boldsymbol{1} - III\,\boldsymbol{B}^{-1}) + III\frac{\partial f}{\partial III}\boldsymbol{1}.\end{aligned}\right\} \tag{47.19}$$

We now differentiate (47.15) so as to obtain $\partial\mathfrak{t}^{\alpha\lambda}/\partial C_{\mu\nu}$, substitute the result into (45.2), and simplify it by using (47.18) and (47.19). The expression so obtained is linear in the coefficients $\partial\daleth_\Gamma/\partial I$, $\partial\daleth_\Gamma/\partial II$, and $\partial\daleth_\Gamma/\partial III$. Straightforward calculation starting from (47.17) converts this expression into the desired formula

---

[1] It is the dual of Eq. (29.7) of CFT. To derive it, note that from (23.4) and (23.5) we have $\boldsymbol{C}^{-2} = \boldsymbol{F}^{-1}\boldsymbol{B}^{-1}(\boldsymbol{F}^{-1})^T$, and then use the Hamilton-Cayley equation.

[2] Formulae of this kind, for hyperelastic materials, seem first to have been given by SIGNORINI [1943, *3*, Chap. III, ¶ 11].

for C in terms of the response coefficients $\beth_\Gamma$:

$$\left.\begin{aligned}
\mathsf{C}^{kmpq} = {} & \frac{1}{III}\beth_{-1}(B^{kp}B^{mq}+B^{mp}B^{kq}) - \left(\beth_0+\frac{II}{III}\beth_{-1}\right)(g^{kp}g^{mq}+g^{mp}g^{kq}) + \\
& + 2g^{km}\left[\frac{\partial\beth_0}{\partial I}B^{pq}+\left(\frac{1}{2}\beth_0+\frac{II}{III}\beth_{-1}+II\frac{\partial\beth_0}{\partial II}+III\frac{\partial\beth_0}{\partial III}\right)g^{pq}-\right.\\
& \left.-\left(\beth_{-1}+III\frac{\partial\beth_0}{\partial II}\right)(B^{-1})^{pq}\right]+\\
& + 2B^{km}\left[\left(-\frac{1}{III}\beth_{-1}+\frac{\partial\beth_1}{\partial I}\right)B^{pq}+\right.\\
& \left.+\left(\frac{1}{2}\beth_1+II\frac{\partial\beth_1}{\partial II}+III\frac{\partial\beth_1}{\partial III}\right)g^{pq}-III\frac{\partial\beth_1}{\partial II}(B^{-1})^{pq}\right]+\\
& + 2(B^{-1})^{km}\left[\frac{\partial\beth_{-1}}{\partial I}B^{pq}+\left(-\frac{1}{2}\beth_{-1}+II\frac{\partial\beth_{-1}}{\partial II}+III\frac{\partial\beth_{-1}}{\partial III}\right)g^{pq}-\right.\\
& \left.-III\frac{\partial\beth_{-1}}{\partial II}(B^{-1})^{pq}\right].
\end{aligned}\right\}\quad(47.20)$$

It is commonly and loosely said that a severely strained isotropic elastic material loses its isotropy. Simple sense is easily given to this misleadingly dramatic statement. Of course, no physical change occurs: The properties of an elastic material are defined, once and for all, by its response function $\mathfrak{g}$, the form of which, by hypothesis, remains unchanged by whatever deformation the material may suffer. Moreover, those properties are independent of the choice of local reference configuration and strain measure. However, the property of isotropy is defined in terms of *particular local reference configurations*, namely, those that are undistorted states. If the local reference configuration is *distorted*, the corresponding response function will not satisfy (47.1) and (47.2). For example, if we subject an elastic solid to a local deformation $\boldsymbol{P}$ carrying it from an undistorted state $\boldsymbol{K}$ into a local configuration $\hat{\boldsymbol{K}}$, then $\hat{\boldsymbol{K}}$ will be distorted unless $\boldsymbol{P}$ is conformal (cf. Sect. 33). If we then, as it is always perfectly legitimate to do, choose to measure strain taking $\hat{\boldsymbol{K}}$ as the local reference configuration, isotropic stress relations of the forms (47.6)—(47.9) will fail to hold. This purely formal difference is all that is meant by the "loss" of isotropy. In a word, the stress relation of an isotropic elastic material is particularly simple for strains measured from undistorted states, but not so for strains measured from distorted states[1]. For an isotropic elastic material, this remark will be illustrated by explicit formulae in Sect. 69.

Henceforth, without further comment, when dealing with isotropic materials *we shall always measure strains from an undistorted state as local reference configuration, unless the contrary is stated explicitly.*

In the reference configuration, the tensors $\boldsymbol{V}$, $\boldsymbol{B}$, and $\boldsymbol{B}^{-1}$ reduce to $\mathbf{1}$. Since here the local reference configuration has been assumed to be an undistorted state, it follows from any one of Eqs. (47.6)—(47.9) that *the stress in any undistorted state of an isotropic elastic material is hydrostatic.*

48. Isotropic elastic materials, II. The principal stresses and principal forces. The proper numbers t_a of the stress tensor $\boldsymbol{T}$ are called the *principal stresses.* The directions defined by the proper vectors of $\boldsymbol{T}$ are called the *principal axes of stress* [cf. CFT, Sect. 208].

[1] Similarly, in analytic geometry a given figure "loses" the simplicity of its equation if inappropriate co-ordinate systems are used.

As has been remarked by BONDAR [1963, *9*], we may replace the effect of change of reference configuration by that of a body force of a suitable (and very artificial) kind.

The following theorem may be read off from Eqs. (47.6)—(47.9): *In an isotropic elastic material any principal axis of strain in the deformed state is also a principal axis of stress*[1]. *The principal stresses t_a are related to the principal stretches v_b by*

$$\left.\begin{aligned} t_a &= \bar{\mathfrak{w}}_0 + \bar{\mathfrak{w}}_1 v_a + \bar{\mathfrak{w}}_2 v_a^2, \\ &= \bar{\gimel}_0 + \bar{\gimel}_1 v_a + \bar{\gimel}_{-1} v_a^{-1}, \\ &= \bar{\aleph}_0 + \bar{\aleph}_1 v_a^2 + \bar{\aleph}_2 v_a^4, \\ &= \bar{\beth}_0 + \bar{\beth}_1 v_a^2 + \bar{\beth}_{-1} v_a^{-2}, \qquad a = 1, 2, 3 \end{aligned}\right\} \tag{48.1}$$

where (47.12) has been used.

The forms (47.8), (47.9), and $(48.1)_4$ for the stress relation are the ones most useful in solutions of special problems for isotropic materials.

Components of tensors taken with respect to an orthonormal basis of proper vectors of stress and strain are called *principal components*.

The stress relation of an isotropic elastic material, referred to principal components, has the form

$$t_a = \bar{t}_a(v_1, v_2, v_3), \tag{48.2}$$

where, as is easily seen from (48.1), the three functions $\bar{t}_a$ must all be expressible in terms of a single function $\bar{t}$, as follows[2]:

$$\left.\begin{aligned} \bar{t}_1(v_1, v_2, v_3) &= \bar{t}\,(v_1, v_2, v_3), \\ \bar{t}_2(v_1, v_2, v_3) &= \bar{t}\,(v_2, v_3, v_1), \\ \bar{t}_3(v_1, v_2, v_3) &= \bar{t}\,(v_3, v_1, v_2), \end{aligned}\right\} \tag{48.3}$$

where

$$\bar{t}\,(\xi, \eta, \zeta) = \bar{t}\,(\xi, \zeta, \eta). \tag{48.4}$$

Conversely, *if the three principal stresses t_a are known as functions $\bar{t}_a(v_1, v_2, v_3)$ of the principal stretches v_b, and if they satisfy the symmetry conditions* (48.3) *and* (48.4), *then a unique stress relation for an isotropic elastic material is determined.* Indeed, (48.2) is the component form, in principal components, of a relation $\boldsymbol{T} = \mathfrak{g}(\boldsymbol{V})$. The condition (48.4) ensures that $\mathfrak{g}$ is an isotropic function. If the principal stretches v_a are distinct, the response coefficients can be determined explicitly in terms of the $\bar{t}_a$ by eliminating t_a between (48.1) and (48.2). E.g.,

$$\bar{\mathfrak{w}}_0 = \frac{\begin{vmatrix} \bar{t}_1(v_1, v_2, v_3) & v_1 & v_1^2 \\ \bar{t}_2(v_1, v_2, v_3) & v_2 & v_2^2 \\ \bar{t}_3(v_1, v_2, v_3) & v_3 & v_3^2 \end{vmatrix}}{\begin{vmatrix} 1 & v_1 & v_1^2 \\ 1 & v_2 & v_2^2 \\ 1 & v_3 & v_3^2 \end{vmatrix}}. \tag{48.5}$$

---

[1] The coincidence of the principal axes of stress and strain plays a great part in older treatments of isotropic elasticity from FRESNEL's time down to the present and is made much of in expositions for engineers. However, *it does not characterize the isotropic case*, since stress relations of the form (47.6)—(47.9) always yield coaxial stress and strain, regardless of the form of the response coefficients $\bar{\mathfrak{w}}_0, \ldots, \bar{\beth}_{-1}$. Conversely, stress relations of the form (47.6)—(47.9) characterize an isotropic elastic material referred to an undistorted state *only* if $\bar{\mathfrak{w}}_0, \ldots, \bar{\beth}_{-1}$, are invariant functions of $\boldsymbol{B}$.

These and related matters have been discussed by FINGER [1894, *2*] [1894, *3*, pp. 1085—1086], ARIANO [1924, *1*, § 8] [1930, *1*, pp. 742—743] [1933, *1*], SETH [1935, *2*, § 1], MURNAGHAN [1937, *2*, Introd.].

[2] NOLL [1955, *19*, § 19]. See also BARTA [1957, *3*].

If the v_a are not all distinct, the response coefficients are not uniquely determined by the material, but the $\bar{t}_a$ are (cf. Sect. 12).

In describing the response of isotropic elastic materials the first Piola-Kirchhoff tensor $\boldsymbol{T}_\mathrm{R}$ offers often a special advantage. From (23.1) and $(43\,\mathrm{A}.11)_1$ we see that

$$\boldsymbol{T}_\mathrm{R} = (J\boldsymbol{T}\boldsymbol{V}^{-1})\boldsymbol{R}. \tag{48.6}$$

For an isotropic elastic material, the proper vectors of $\boldsymbol{V}$ are also proper vectors of $\boldsymbol{T}$. Hence the tensor $J\boldsymbol{T}\boldsymbol{V}^{-1}$ is symmetric and has the same proper vectors as $\boldsymbol{V}$ and $\boldsymbol{T}$. We denote the proper numbers of $J\boldsymbol{T}\boldsymbol{V}^{-1}$ by T_a. Since $J = \det \boldsymbol{V} = v_1 v_2 v_3$, we have

$$T_a = v_1 v_2 v_3 \frac{t_a}{v_a}, \qquad a = 1, 2, 3. \tag{48.7}$$

To interpret the quantities T_a, consider a block of homogeneous isotropic elastic material subject to a homogeneous strain [cf. Sect. 55 and also CFT, Sect. 42] with principal stretches v_a. Let the faces of the block be normal to the principal stress-strain axes. Then the faces of this block are subject to normal tractions of amounts t_a. If the block before deformation was a unit cube, the area of the face normal to the a'th principal axis in the deformed block is $(v_1 v_2 v_3)/v_a$. By (48.7) the tensile force acting upon this face is T_a. Alternatively, we may derive this same result from the considerations at the beginning of Sect. 43 A. In an arbitrary deformation, the quantities T_a may thus be called the *principal forces*[1] acting at a given place and time in the isotropic elastic material. As we shall see later, the behavior of the principal forces is of primary importance in the theory of elastic response. From (48.7) and (48.2) we see that

$$T_a = \bar{T}_a(v_1, v_2, v_3) = v_1 v_2 v_3 \frac{\bar{t}_a(v_1, v_2, v_3)}{v_a}; \tag{48.8}$$

thus the principal forces are uniquely determined by the principal stretches. If we set

$$J_{ab} \equiv \frac{\partial \bar{T}_a}{\partial v_b}, \qquad j_{ab} \equiv \frac{\partial \bar{t}_a}{\partial v_b}, \tag{48.9}$$

then in general

$$J_{ab} \neq J_{ba}, \qquad j_{ab} \neq j_{ba}. \tag{48.10}$$

However, from (48.3) we easily show that[1]

$$J_{ab} = J_{ba} \quad \text{and} \quad j_{ab} = j_{ba} \quad \text{if} \quad v_a = v_b. \tag{48.11}$$

In particular, the matrices $\|J_{ab}\|$ and $\|j_{ab}\|$ are symmetric whenever all three principal stretches are equal.

It is sometimes useful to have an explicit form of the equations of motion for homogeneous isotropic elastic bodies. It is possible, of course, to calculate the elasticity A occurring in (44.5) by first obtaining the special form that the function $\mathfrak{h}$ in (43.8) assumes in the special case of an isotropic material. This long calculation can be avoided[2] by proceeding directly from (47.4) and (47.8), substitution of which into (16.6) yields

$$F^{km}{}_{pq} B^{pq}{}_{,m} + \varrho b^k = \varrho \ddot{x}^k, \tag{48.12}$$

---

[1] TRUESDELL and TOUPIN [1963, *76*, § 2].

[2] The approach is due to ERICKSEN [1953, *6*, § 2] for incompressible hyperelastic materials. Eq. (48.13) was given by TRUESDELL [1961, *63*, § 7].

where

$$\left.\begin{aligned}
\mathsf{F}^{km}{}_{pq} &\equiv \frac{\partial \mathfrak{f}^{km}(B^{rs})}{\partial B^{pq}} \\
&= \frac{1}{2}\aleph_1(\delta^k_p\delta^m_q+\delta^m_p\delta^k_q)+\frac{1}{2}\aleph_2(\delta^k_p B^m_q+\delta^k_q B^m_p+\delta^m_p B^k_q+\delta^m_q B^k_p)+\\
&\quad+g^{km}\left[\frac{\partial\aleph_0}{\partial I}g_{pq}+\frac{\partial\aleph_0}{\partial II}(I g_{pq}-B_{pq})+III\frac{\partial\aleph_0}{\partial III}(B^{-1})_{pq}\right]+\\
&\quad+B^{km}\left[\frac{\partial\aleph_1}{\partial I}g_{pq}+\frac{\partial\aleph_1}{\partial II}(I g_{pq}-B_{pq})+III\frac{\partial\aleph_1}{\partial III}(B^{-1})_{pq}\right]+\\
&\quad+B^k_r B^{rm}\left[\frac{\partial\aleph_2}{\partial I}g_{pq}+\frac{\partial\aleph_2}{\partial II}(I g_{pq}-B_{pq})+III\frac{\partial\aleph_2}{\partial III}(B^{-1})_{pq}\right],
\end{aligned}\right\} \tag{48.13}$$

the formulae (9.12) having been used in the calculation. Eq. $(29.4)_4$ of CFT assumes the following form in the present notation:

$$B^{pq}{}_{,m}=X^\alpha{}_{,m}g^{\beta\gamma}\left[x^p{}_{,\gamma}x^q{}_{,\alpha;\beta}+x^q{}_{,\gamma}x^p{}_{,\alpha;\beta}\right]. \tag{48.14}$$

The desired explicit form of the equations of motion results from substituting (48.14) and (48.13) into (48.12). Forms alternative to (48.13) may be obtained by replacing $\boldsymbol{B}^{-1}$ by its expression in terms of $\boldsymbol{B}$ and $\boldsymbol{B}^2$, by using the identities (47.14), etc.

We have seen that the principal components of the stress relation reduce to the especially simple forms (48.1). Correspondingly simple forms for the components of the tensor F are easily calculated[1] if we note from (23.8) that for any function $f(I, II, III)$ we have

$$\frac{\partial f(I,II,III)}{\partial v_1^2}=\frac{\partial f}{\partial I}+(v_2^2+v_3^2)\frac{\partial f}{\partial II}+v_2^2v_3^2\frac{\partial f}{\partial III}. \tag{48.15}$$

From (48.13) we easily show that the principal components of F are

$$\mathsf{F}\langle 1111\rangle=\frac{\partial t_1}{\partial v_1^2}, \qquad \mathsf{F}\langle 1122\rangle=\frac{\partial t_1}{\partial v_2^2},\ \dots, \qquad \mathsf{F}\langle 1212\rangle=\frac{t_1-t_2}{2(v_1^2-v_2^2)},\ \dots, \tag{48.16}$$

while all other principal components vanish. The physical interpretation of these terms is immediate. For the direction in which the stretch is v_1, the quantity $2\mathsf{F}\langle 1111\rangle$ is a *tangent modulus of longitudinal extension*, while the quantity $2\mathsf{F}\langle 1122\rangle$ is a *tangent modulus of transverse extension* corresponding to the direction in which the stretch is v_2. In view of CFT, Eqs. (28.12) and Appendix (46.9), if τ_3 is the shearing stress in the plane normal to the direction in which the stretch is v_3, and if γ_3 is the corresponding principal shear, then from $(48.16)_3$ we see that

$$\mathsf{F}\langle 1212\rangle=\frac{\tau_3}{(v_1^2+v_2^2)\sin\gamma_3}. \tag{48.17}$$

Thus $2\mathsf{F}\langle 1212\rangle$ is a kind of *secant modulus of shear* for the plane normal to the direction in which the stretch is v_3.

In general, the 12 moduli (48.16) are distinct, although, if regarded as functions $\bar{\mathsf{F}}\langle kmpq\rangle$ of v_1, v_2, v_3, they are subject to various identities. From (48.16) it follows that

$$\left.\begin{aligned}
v_1\frac{\partial\bar{\mathsf{F}}\langle 1111\rangle}{\partial v_2}&=v_2\frac{\partial\bar{\mathsf{F}}\langle 1122\rangle}{\partial v_1}, \qquad v_2\frac{\partial\bar{\mathsf{F}}\langle 1122\rangle}{\partial v_3}=v_3\frac{\partial\bar{\mathsf{F}}\langle 1133\rangle}{\partial v_2},\\
v_3\frac{\partial\bar{\mathsf{F}}\langle 1133\rangle}{\partial v_1}&=v_1\frac{\partial\bar{\mathsf{F}}\langle 1111\rangle}{\partial v_3}.
\end{aligned}\right\} \tag{48.18}$$

[1] Truesdell [1961, *63*, § 12].

If three functions $\bar{F}^{\langle 1111\rangle}$, $\bar{F}^{\langle 1122\rangle}$, and $\bar{F}^{\langle 1133\rangle}$ satisfying these three conditions are given, the function $\bar{t}$ in $(48.3)_1$ is determined to within a constant, but it does not necessarily satisfy $(48.3)_{2,3}$ or (48.4). These identities require that

$$\left.\begin{aligned}
\bar{F}^{\langle 2222\rangle}(v_1, v_2, v_3) &= \bar{F}^{\langle 1111\rangle}(v_2, v_1, v_3),\\
\bar{F}^{\langle 3333\rangle}(v_1, v_2, v_3) &= \bar{F}^{\langle 1111\rangle}(v_3, v_2, v_1),\\
\bar{F}^{\langle 2211\rangle}(v_1, v_2, v_3) &= \bar{F}^{\langle 1122\rangle}(v_2, v_1, v_3),\\
\bar{F}^{\langle 3311\rangle}(v_1, v_2, v_3) &= \bar{F}^{\langle 1133\rangle}(v_3, v_2, v_1),\\
\bar{F}^{\langle 3322\rangle}(v_1, v_2, v_3) &= \bar{F}^{\langle 1122\rangle}(v_3, v_2, v_1),\\
\bar{F}^{\langle 2233\rangle}(v_1, v_2, v_3) &= \bar{F}^{\langle 1122\rangle}(v_2, v_3, v_1).
\end{aligned}\right\} \tag{48.19}$$

If these conditions, conversely, are used to define $\bar{F}^{\langle 22kk\rangle}$ and $\bar{F}^{\langle 33kk\rangle}$, then $(48.3)_{2,3}$ are satisfied. The further conditions

$$\left.\begin{aligned}
\bar{F}^{\langle 1111\rangle}(v_1, v_2, v_3) &= \bar{F}^{\langle 1111\rangle}(v_1, v_3, v_2)\\
\bar{F}^{\langle 1122\rangle}(v_1, v_2, v_3) &= \bar{F}^{\langle 1133\rangle}(v_1, v_3, v_2)
\end{aligned}\right\} \tag{48.20}$$

are necessary and sufficient that (48.4) be satisfied. Accordingly, for a set of 9 functions $\bar{F}^{\langle kkmm\rangle}(v_1, v_2, v_3)$ to define a unique stress relation for an isotropic material, to within an arbitrary constant stress, it is necessary and sufficient that the 8 algebraic conditions (48.19), (48.20) and the 3 differential conditions (48.18) be satisfied. The corresponding further components $F^{\langle 1212\rangle}$, $F^{\langle 2323\rangle}$, and $F^{\langle 3131\rangle}$ are then determined from those already given. The process of determining them requires integration of the system (48.16) but may be shortened by use of three identities of the form

$$\frac{1}{v_1}\frac{\partial}{\partial v_1}\left[(v_1^2 - v_2^2)\,\bar{F}^{\langle 1212\rangle}\right] = \bar{F}^{\langle 1111\rangle} - \bar{F}^{\langle 2211\rangle}, \ldots, \tag{48.21}$$

which follow from (48.16) by inspection and by use of the trivial algebraic identity[1]

$$(v_1^2 - v_2^2)\,F^{\langle 1212\rangle} + (v_2^2 - v_3^2)\,F^{\langle 2323\rangle} + (v_3^2 - v_1^2)\,F^{\langle 3131\rangle} = 0. \tag{48.22}$$

In an isotropic material, by using $(48.13)_1$ and (23.7) we see that

$$\left.\begin{aligned}
\frac{\partial T^{km}}{\partial x^p{}_{,\beta}}\, x^q{}_{,\beta} &= F^{km}{}_{rs}\,\frac{\partial B^{rs}}{\partial x^p{}_{,\beta}}\, x^q{}_{,\beta},\\
&= F^{km}{}_{rs}\,(\delta^r_p\, x^s{}_{,\gamma} + \delta^s_p\, x^r{}_{,\gamma})\, g^{\beta\gamma}\, x^q{}_{,\beta} = 2F^{km}{}_{ps}\, B^{qs}.
\end{aligned}\right\} \tag{48.23}$$

By (45.13) we then obtain the following expression for the elasticity $B^{km}{}_p{}^q$ of an isotropic material:

$$B^{km}{}_p{}^q = T^{km}\,\delta^q_p - T^{kq}\,\delta^m_p + 2F^{km}{}_{ps}\, B^{qs}. \tag{48.24}$$

---

[1] As an application of the above ideas, consider the infinitesimal theory, in which all components $F^{\langle kmpq\rangle}$ are constants. Thus the conditions (48.18) are satisfied, while from (48.19) it follows that at most two components of the type $F^{\langle kkmm\rangle}$ can differ from one another:

$$F^{\langle 1111\rangle} = F^{\langle 2222\rangle} = F^{\langle 3333\rangle} \equiv \tfrac{1}{2}(\lambda + 2\mu), \qquad F^{\langle 1122\rangle} = F^{\langle 2211\rangle} = \cdots \equiv \tfrac{1}{2}\lambda,$$

say, for conformity with the classical notations. From (48.21) we have

$$F^{\langle 1212\rangle} = \tfrac{1}{2}\left(F^{\langle 1111\rangle} - F^{\langle 2211\rangle}\right) = \tfrac{1}{2}\mu = F^{\langle 2323\rangle} = F^{\langle 3131\rangle}.$$

Of course (48.22) is satisfied. These same results are easily calculated directly from (48.16) and the stress relation of the infinitesimal theory.

It follows by (45.5) and (23.7) that

$$\mathsf{A}_{k\ m}^{\ \alpha\ \beta} = 2J g_{kp}(X^{[\alpha}{}_{,q} X^{\beta]}{}_{,m} T^{pq} + g^{\beta\gamma} X^{\alpha}{}_{,q} x^{r}{}_{,\gamma} \mathsf{F}^{pq}{}_{mr}). \tag{48.25}$$

Either from this result and (44.21), or from (48.24) and (45.14), we see that the *S-E condition for isotropic materials* takes the form[1]

$$\mathsf{F}^{km}{}_{ps} B^{sq} v_q v_m \lambda_k \lambda^p > 0, \tag{48.26}$$

where $\boldsymbol{v}$ and $\boldsymbol{\lambda}$ are arbitrary vectors.

Let $\boldsymbol{v}$ be a unit proper vector of $\boldsymbol{B}$, so that $\boldsymbol{B}\boldsymbol{v} = v_1^2 \boldsymbol{v}$, say. Then (48.26) becomes

$$\mathsf{F}^{km}{}_{ps} \lambda_k \lambda^p v_m v^s > 0 \tag{48.27}$$

for all non-vanishing vectors $\boldsymbol{\lambda}$. In principal components, by the remark following (48.16), this condition assumes the form[1]

$$\mathsf{F}\langle 1111\rangle \lambda_1^2 + \mathsf{F}\langle 1212\rangle \lambda_2^2 + \mathsf{F}\langle 1313\rangle \lambda_3^2 > 0; \tag{48.28}$$

equivalently,

$$\mathsf{F}\langle 1111\rangle > 0, \quad \mathsf{F}\langle 1212\rangle > 0, \quad \mathsf{F}\langle 1313\rangle > 0. \tag{48.29}$$

That is, the secant moduli of shear are all positive. Interpretations for these simple inequalities will be given in Sects. 52 and 74. We note here that they do not express the full force of the S-E condition: While (48.29) is equivalent to (48.27), that inequality is only the corollary of the S-E condition that results when $\boldsymbol{v}$ is taken as parallel to a principal axis of stress and strain. It is easy to write out the form assumed by the general condition (48.26) in principal components, but the result has no evident interpretation.

By substituting (48.24) into $(45.7)_2$ we find the identity[1]

$$2(\mathsf{F}^{km}{}_{ps} B^s_q - \mathsf{F}^{km}{}_{qs} B^s_p) = \delta^k_p T^m_q - \delta^k_q T^m_p + \delta^m_p T^k_q - \delta^m_q T^k_q, \tag{48.30}$$

which, in the abstract notation of Sect. 7, is equivalent to the identity[2]

$$\mathsf{F}[\boldsymbol{Z}\boldsymbol{B} - \boldsymbol{B}\boldsymbol{Z}] = \boldsymbol{Z}\boldsymbol{T} - \boldsymbol{T}\boldsymbol{Z}, \tag{48.31}$$

valid for all skew tensors $\boldsymbol{Z}$. (48.31) can also be obtained directly from the isotropy relation $\boldsymbol{Q}\mathfrak{f}(\boldsymbol{B})\boldsymbol{Q}^T = \mathfrak{f}(\boldsymbol{Q}\boldsymbol{B}\boldsymbol{Q}^T)$ by differentiation with respect to $\boldsymbol{Q}$.

49. Incompressible isotropic elastic materials. For incompressible materials only isochoric deformations are possible, so that

$$III_{\boldsymbol{V}} = \det \boldsymbol{V} = 1, \quad III_{\boldsymbol{B}} = \det \boldsymbol{B} = 1. \tag{49.1}$$

Hence for isotropic elastic materials that are incompressible the dependence on $III_{\boldsymbol{V}}$ or $III_{\boldsymbol{B}}$ in formulae such as (47.11) may be omitted. Corresponding to the general constitutive equations (43.14), in the reduced forms (47.6) through (47.9) the stress $\boldsymbol{T}$ must be replaced by the extra stress, $\boldsymbol{T} + p\boldsymbol{1}$, while the right-hand sides of those relations, which give explicit forms for the response function, are determined only up to a scalar function of $\boldsymbol{V}$ or $\boldsymbol{B}$. We remove this indeterminacy by omitting the scalar terms $\mathfrak{v}_0\boldsymbol{1}$, $\gimel_0\boldsymbol{1}$, $\aleph_0\boldsymbol{1}$, and $\beth_0\boldsymbol{1}$, respectively. It follows that *the constitutive equation of an incompressible isotropic elastic material* may be written

---

[1] Truesdell and Toupin [1963, *76*, § 7].
[2] Noll [1955, *19*, § 15].

in any of the following forms:

$$\boldsymbol{T} = -p\mathbf{1} + \text{ש}_1 \boldsymbol{V} + \text{ש}_2 \boldsymbol{V}^2, \tag{49.2}$$

$$\boldsymbol{T} = -p\mathbf{1} + \gimel_1 \boldsymbol{V} + \gimel_{-1} \boldsymbol{V}^{-1}, \tag{49.3}$$

$$\left.\begin{aligned} \boldsymbol{T} &= -p\mathbf{1} + \aleph_1 \boldsymbol{B} + \aleph_2 \boldsymbol{B}^2, \\ T^k_m &= -p\,\delta^k_m + \aleph_1 B^k_m + \aleph_2 B^k_p B^p_m, \end{aligned}\right\} \tag{49.4}$$

$$\left.\begin{aligned} \boldsymbol{T} &= -p\mathbf{1} + \beth_1 \boldsymbol{B} + \beth_{-1} \boldsymbol{B}^{-1}, \\ T^k_m &= -p\,\delta^k_m + \beth_1 B^k_m + \beth_{-1} (B^{-1})^k_m. \end{aligned}\right\} \tag{49.5}$$

The response coefficients ש_1, ש_2, $\gimel_1$, $\gimel_{-1}$, $\aleph_1$, $\aleph_2$, $\beth_1$, $\beth_{-1}$ are functions of the two invariants $I_{\boldsymbol{V}}$ and $II_{\boldsymbol{V}}$ or of the two invariants $I_{\boldsymbol{B}}$ and $II_{\boldsymbol{B}}$. The scalar p is an indeterminate pressure. We note that the quantities p occurring in the different equations (49.2) through (49.5) are not the same, because these equations correspond to different normalizations of the response function which determines the extra stress, $\boldsymbol{T} + p\mathbf{1}$.

The coefficients $\aleph_\Gamma$ in (49.4) are related to the coefficients $\beth_\Gamma$ in (49.5) as follows:

$$\beth_1 = \aleph_1 + I_{\boldsymbol{B}} \aleph_2, \quad \beth_{-1} = \aleph_2, \quad \aleph_1 = \beth_1 - I_{\boldsymbol{B}} \beth_{-1}. \tag{49.6}$$

From $(43\text{A}.11)_5$ and (49.5) it is easy to show that the tensor $\tilde{\boldsymbol{T}}$ for an isotropic incompressible material is given by an expression of the form

$$\tilde{\boldsymbol{T}} = \daleth_0 \mathbf{1} + \daleth_1 \boldsymbol{C} - p\boldsymbol{C}^{-1}, \quad \daleth_0 = \beth_1 - I_{\boldsymbol{C}} \beth_{-1}, \quad \daleth_1 = \beth_{-1} \tag{49.7}$$

where $\beth_\Gamma = \beth_\Gamma(I_{\boldsymbol{C}}, II_{\boldsymbol{C}}) = \beth_\Gamma(I_{\boldsymbol{B}}, II_{\boldsymbol{B}})$, $\Gamma = \pm 1$.

For an explicit form of the equations of motion[1], we may proceed as at the end of the previous section, obtaining

$$-p^{,k} + \mathsf{F}^{km}{}_{pq} B^{pq}{}_{,m} + \varrho b^k = \varrho \ddot{x}^k, \tag{49.8}$$

where

$$\left.\begin{aligned} \mathsf{F}^{km}{}_{pq} = {} & \frac{1}{2} \aleph_1 (\delta^k_p \delta^m_q + \delta^m_p \delta^k_q) + \frac{1}{2} \aleph_2 (\delta^k_p B^m_q + \delta^k_q B^m_p + \delta^m_p B^k_q + \delta^m_q B^k_p) + \\ & + B^{km} \left[\frac{\partial \aleph_1}{\partial I} g_{pq} + \frac{\partial \aleph_1}{\partial II} (I g_{pq} - B_{pq}) \right] + \\ & + B^k_r B^{rm} \left[\frac{\partial \aleph_2}{\partial I} g_{pq} + \frac{\partial \aleph_2}{\partial II} (I g_{pq} - B_{pq}) \right]. \end{aligned}\right\} \tag{49.9}$$

In these formulae, I and II are written for $I_{\boldsymbol{B}}$ and $II_{\boldsymbol{B}}$. Substituting (49.9) and (48.14) in (49.8) yields a second-order differential equation to be solved for $\boldsymbol{x} = \boldsymbol{\chi}(\boldsymbol{X}, t)$, but it is not fully explicit, since p is to be eliminated by imposing the condition $|\det \boldsymbol{F}| = 1$, or, in co-ordinate form, $\sqrt{\det \|g_{km}\|}\, |\det \|x^p{}_{,\alpha}\|| = \sqrt{\det \|g_{\beta\gamma}\|}$. An equation for $\boldsymbol{\chi}$ alone is obtained by taking the curl of (49.8). The procedure is more easily understood by looking at the derivation of particular solutions in Sects. 55—57.

50. Elastic fluids and solids. Natural states. When a simple fluid, as defined in Sect. 32, is also an elastic material, it is called an *elastic fluid*. The constitutive equation of an elastic fluid may be obtained easily by specializing the general constitutive equation (32.8) of a simple fluid to the case when the stress $\boldsymbol{T}$ does not depend on the history $\boldsymbol{G}(s)$. Since the value of the functional $\mathfrak{Q}$ of (32.8) vanishes for $\boldsymbol{G}(s) \equiv \boldsymbol{0}$, it must then always vanish; hence the *general constitutive*

[1] ERICKSEN [1953, *6*, § 2], for hyperelastic materials.

equation of an elastic fluid must have the form

$$T = -p(\varrho)\,\mathbf{1}, \tag{50.1}$$

where $p(\varrho)$ is a scalar function of the density ϱ, called the *pressure*. The response functions $\mathfrak{g}$, $\mathfrak{f}$, $\mathfrak{h}$, $\mathfrak{s}$, $\mathfrak{t}$ (cf. Sects. 43 and 47) are related to the pressure function $p(\varrho)$ by

$$\left.\begin{aligned}
&\mathfrak{g}(F) = -p\left(\frac{\varrho_R}{|\det F|}\right)\mathbf{1}, \qquad \mathfrak{f}(B) = -p\left(\frac{\varrho_R}{\sqrt{III_B}}\right)\mathbf{1},\\
&\mathfrak{h}(F) = -|\det F|\,p\left(\frac{\varrho_R}{|\det F|}\right)(F^{-1})^T,\\
&\mathfrak{s}(C) = -p\left(\frac{\varrho_R}{\sqrt{III_C}}\right)C,\\
&\mathfrak{t}(C) = -\sqrt{III_C}\,p\left(\frac{\varrho_R}{\sqrt{III_C}}\right)C^{-1},
\end{aligned}\right\} \tag{50.2}$$

where ϱ_R is the density in the reference configuration.

We recall (Sect. 32) that every simple fluid, and hence every elastic fluid, is isotropic and that every one of its local configurations is an undistorted state. Therefore, in specializing results from the general theory of isotropic elastic materials to the case of fluids, the choice of local reference configuration is immaterial.

For an incompressible elastic fluid[1], we may take the extra stress $T + p\mathbf{1}$ to be zero, so that the constitutive equation degenerates into

$$T = -p\,\mathbf{1}, \tag{50.3}$$

where p, as for any incompressible material, is a pressure which remains indeterminate until a dynamical problem with suitable specific boundary conditions has been set. The response functions of (43.13) and (43.14) may be taken to be identically zero in the case of an incompressible elastic fluid.

A large part of what is usually called hydrodynamics is the theory of elastic fluids, which are called perfect fluids when treated in the framework of the thermodynamical theory to be discussed in Sect. 80. A detailed exposition of the subject is given in Serrin's article, "The mathematical principles of fluid dynamics", Vol. VIII/1 of this Encyclopedia; see also CFT, Sect. 297. We have mentioned elastic fluids here only so as to show how they fit into the general framework.

An elastic material that is also a solid according to the definition of Sect. 33 is called an *elastic solid*. In terms of the response function $\mathfrak{g}(F)$ corresponding to an undistorted state, an elastic solid is characterized by the property that a non-orthogonal H cannot satisfy (43.15): For each non-orthogonal unimodular tensor H there is at least one tensor F such that

$$\mathfrak{g}(F) \neq \mathfrak{g}(FH). \tag{50.4}$$

In physical terms, the response of the material in an undistorted local reference configuration will be different, for at least one deformation gradient F, from its response in any configuration related to the reference configuration by a distortion H.

[1] The classical term is "incompressible perfect fluid". The at first surprising name used in the text results by condensing the term, "incompressible simple fluid which is also an elastic material", in which the formal definitions of Sects. 32 and 43 are combined.

The definitions and results of Sect. 33 apply to elastic solids in the same way as they do to general simple solids. For elastic solids, we can derive certain additional results[1] concerning the stress corresponding to an undistorted state. When an undistorted state is taken as the local reference configuration, all members of the isotropy group are orthogonal, so that $(43.16)_1$ characterizes them all. That is, a tensor belongs to the isotropy group g if and only if it is an orthogonal tensor $\boldsymbol{Q}$ such that

$$\boldsymbol{Q}\,\mathfrak{g}(\boldsymbol{F})\,\boldsymbol{Q}^T=\mathfrak{g}(\boldsymbol{Q}\boldsymbol{F}\boldsymbol{Q}^T) \tag{50.5}$$

holds for all $\boldsymbol{F}$. Using (50.5) for $\boldsymbol{F}=\mathbf{1}$, we find that

$$\boldsymbol{Q}\,\boldsymbol{T}^{(0)}=\boldsymbol{T}^{(0)}\boldsymbol{Q} \tag{50.6}$$

holds for all members $\boldsymbol{Q}$ of the isotropy group, where $\boldsymbol{T}^{(0)}=\mathfrak{g}(\mathbf{1})$ is the stress in the undistorted reference state. Using the algebraic theorem mentioned in Sect. 33, we find that *the stress $\boldsymbol{T}^{(0)}$ on an undistorted state of an elastic solid must be such that the characteristic spaces of $\boldsymbol{T}^{(0)}$ are invariant under all transformations of the isotropy group.* For isotropic elastic solids, it follows that the stress on an undistorted state must be hydrostatic, which is a special case of the result mentioned at the end of Sect. 47. For elastic solids having one of the eleven types of aeolotropy described in Sect. 33, the restrictions on the possible stresses $\boldsymbol{T}^{(0)}$ on undistorted states can be read off from the second table of Sect. 33 by replacing $\boldsymbol{U}$ there by $\boldsymbol{T}^{(0)}$. For example, the stress $\boldsymbol{T}^{(0)}$ on an undistorted state of an elastic solid that belongs to the tetragonal system or the hexagonal system, or is transversely isotropic, must be of the form

$$\boldsymbol{T}^{(0)}=-p\mathbf{1}+q\boldsymbol{k}\otimes\boldsymbol{k}, \tag{50.7}$$

i.e., it must consist of a uniaxial tension q in the direction of $\boldsymbol{k}$ superimposed upon an ambient pressure p. For elastic solids belonging to the cubic system, as for isotropic elastic materials, the stress on an undistorted state must be hydrostatic.

When the response function $\mathfrak{g}$ of (43.5) is a polynomial tensor function[2], it is possible to derive certain representations for $\mathfrak{g}$ from the condition that (50.5) holds for all $\boldsymbol{Q}$ in the isotropy group. The same is true for the response functions $\mathfrak{f}$ and $\mathfrak{S}$ of (43.6) and (43.7). For some types of aeolotropy, such representations have been derived in the literature. These results are too complicated to be included here. While the resulting reduced forms of the stress relation for the various crystal classes are doubtless valid for more general functional dependence, it cannot be said that proper proofs have yet been given[3].

The following theorem gives a criterion which can often be used to decide whether a given response function defines a solid. *If the response function $\mathfrak{g}(\boldsymbol{U})$ of* (43.5), *restricted to positive-definite and symmetric argument tensors $\boldsymbol{U}$, has a unique inverse $\overset{-1}{\mathfrak{g}}$, and if $\mathfrak{g}(\mathbf{1})=-p_0\mathbf{1}$, then $\mathfrak{g}$ defines an elastic solid with respect to an undistorted local reference configuration.* To prove this theorem it is sufficient to show that (50.4) is in fact satisfied for $\boldsymbol{F}=\mathbf{1}$, i.e., that $\mathfrak{g}(\boldsymbol{H})\neq-p_0\mathbf{1}$ for all non-orthogonal unimodular tensors $\boldsymbol{H}$. Non-orthogonality of $\boldsymbol{H}$ implies that the positive-definite and symmetric tensor $\overline{\boldsymbol{U}}$ in the polar decomposition $\boldsymbol{H}=\overline{\boldsymbol{R}}\,\overline{\boldsymbol{U}}$ does

[1] COLEMAN and NOLL [1964, *19*].

[2] Cf. our remarks on polynomial dependence in footnote 1, p. 61.

[3] For hyperelastic materials, where scalar invariants specify the stress relation, references will be given in Sect. 85. For elastic materials in general, subject to the assumption of polynomial dependence, a method was given by SMITH and RIVLIN [1957, *23*]; they worked out the details for the monoclinic and rhombic systems. Materials of the differential type, with special reference to orthotropic and transversely isotropic materials, were considered by ADKINS [1958, *2*, §§ 4—8] [1960, *2* and *3*].

not reduce to $\mathbf{1}$. It follows from (43.2) that $\mathfrak{g}(\boldsymbol{H})=\overline{\boldsymbol{R}}\mathfrak{g}(\overline{\boldsymbol{U}})\overline{\boldsymbol{R}}^T$. But $\mathfrak{g}(\overline{\boldsymbol{U}})\neq-p_0\mathbf{1}$ because $\mathfrak{g}(\boldsymbol{U})$ is one-to-one and $\overline{\boldsymbol{U}}\neq\mathbf{1}$; hence $\mathfrak{g}(\boldsymbol{H})\neq-p_0\mathbf{1}$. The criterion remains valid when the response function $\mathfrak{g}(\boldsymbol{U})$ of (43.5) is replaced by any of the response functions $\mathfrak{f}(\boldsymbol{C})$, $\mathfrak{s}(\boldsymbol{C})$, or $\mathfrak{t}(\boldsymbol{C})$ of (43.6), (43.7), and $(43.10)_2$.

In summary, we have the following ***theorem on the undistorted states of isotropic elastic solids***: *All undistorted states are obtained from a given one by subjecting it to all possible conformal deformations*[1]. *The reduced forms* (47.6)—(47.9) *of the stress relation are valid when and only when the reference configuration is undistorted. The stress on an undistorted state is always hydrostatic.*

Intuitively, one would expect the converse of the last statement to hold, also. However, there are exceptional stress relations which give a hydrostatic stress for certain distorted states. For example, if $\beth_1=\frac{1}{2}\beth_{-1}$ in (47.9), then the stress is easily seen to be hydrostatic whenever two of the proper numbers of $\boldsymbol{B}$ are equal and the third is twice the reciprocal of the first. A left Cauchy-Green tensor $\boldsymbol{B}$ with this property cannot correspond to a conformal deformation.

In Sect. 52 we shall see that a *non-positive* hydrostatic pressure on a distorted state is impossible if the stress relation satisfies the restrictive GCN condition, to be explained in Sect. 52.

As is clear from the general properties of fluids and solids presented in Sects. 32 and 33, elastic solids and elastic fluids constitute mutually exclusive classes of elastic materials, but there are also elastic materials, necessarily anisotropic, that are neither fluids nor solids. For example, *elastic subfluids* have been characterized by WANG[2].

A local configuration of an elastic material is called a *natural state* if it is undistorted and if it corresponds to zero stress. Experience shows that natural states do not exist at all in elastic fluids because cavitation tends to occur unless the pressure is positive. A theorem based on the GCN condition of Sect. 52 asserts that the pressure is positive for every non-zero density and hence implies that a fluid has no natural state[3]. While elastic solids, by virtue of the definition of a simple solid in Sect. 33, necessarily have infinitely many undistorted states, they need not have a natural state. Since, however, many real solids seem to be more or less free of internal tensions when subject to no external loads, in applications of the theory it is usual to assume the existence of a natural state.

When the local reference configuration is a natural state, the various response functions introduced in Sect. 43 are subject to the conditions

$$\mathfrak{g}(\mathbf{1})=\mathfrak{h}(\mathbf{1})=\mathfrak{f}(\mathbf{1})=\mathfrak{s}(\mathbf{1})=\mathfrak{t}(\mathbf{1})=\mathbf{0}, \tag{50.8}$$

while the response coefficients for isotropic materials introduced in Sect. 47 must satisfy the conditions

$$q_0(3,3,1)+q_1(3,3,1)+q_{\mathfrak{a}}(3,3,1)=0, \tag{50.9}$$

where q may be taken as $\aleph$, $\beth$, $\gimel$, or $\dot{\text{ש}}$ and where $\mathfrak{a}=2$ or -1, as appropriate.

Contact with the classical theory of infinitesimal elasticity may be made in various ways. That theory[4] is defined by the constitutive equation (41.26), i.e.

$$\boldsymbol{T}=\mathsf{L}[\tilde{\boldsymbol{E}}], \qquad T^{km}=\mathsf{L}^{kmpq}\tilde{E}_{pq}, \tag{50.10}$$

---

[1] For hyperelastic materials this seems to be a result aimed at by SIGNORINI [1943, *3*, Cap. III, ¶¶ 13—16] and CARICATO [1958, *12*].

[2] WANG [1965, *42*].

[3] From the second table in Sect. 33bis it seems that subfluids may have natural states.

[4] Here and until further notice by "the classical infinitesimal theory" we mean that of CAUCHY, in which there may be 36 independent elasticities, rather than the hyperelastic theory of GREEN, in which there are at most 21.

where $\tilde{\boldsymbol{E}}$ is the classical infinitesimal strain tensor (41.12) and where L is the linear elasticity tensor. Applying the approximation procedure described in Sect. 41 to the stress relation in the form $(43.10)_2$ and using (50.8), we find that[1]

$$\left.\begin{aligned} \mathsf{L} &= 2\left.\frac{\partial \mathfrak{t}(\boldsymbol{C})}{\partial \boldsymbol{C}}\right|_{\boldsymbol{C}=\mathbf{1}}, \\ \mathsf{L}^{\alpha\beta\gamma\delta} &= 2\left.\frac{\partial \mathfrak{t}^{\alpha\beta}}{\partial C_{\gamma\delta}}\right|_{\boldsymbol{C}=\mathbf{1}}, \qquad \mathsf{L}^{kmpq} = g^k_\alpha g^m_\beta g^p_\gamma g^q_\delta \mathsf{L}^{\alpha\beta\gamma\delta}. \end{aligned}\right\} \tag{50.11}$$

Alternative formulae for L may be read off from $(45.1)_2$, (45.2), and (45.6):

$$\mathsf{L}^{kmpq} = g^{mr} g^{qs} g^k_\alpha g^p_\beta \mathsf{A}^{(0)}{}_r{}^\alpha{}_s{}^\beta = \mathsf{C}^{kmpq}_{(0)} = \mathsf{B}^{kmpq}_{(0)}, \tag{50.12}$$

where the index (0) indicates that the elasticities must be evaluated for $\boldsymbol{F}=\mathbf{1}$, i.e. in the natural reference state. Thus the elasticities A, B, and C are different generalizations of the linear elasticity L, reducing to it in the natural reference state. Since L represents a linear transformation of the 6-dimensional space $\mathscr{S}$ of symmetric tensors, its components have the symmetry properties (7.12). By the definition (45.2) the components of C enjoy the same symmetry properties. The components of B, however, have these properties only in the natural state, and (45.7) takes their place in a deformed state. While the formulae (50.11) and (50.12) for the linear elasticities are meaningful for any elastic material having a response function $\mathfrak{t}$ which is differentiable at the reference state (where $\boldsymbol{C}=\mathbf{1}$, by definition, and where the stress is assumed to vanish), it is introduced here only for use in comparison of results from the non-linear theory with their counterparts from the linearized. The classical moduli $\mathsf{L}^{\alpha\beta\gamma\delta}$, as such, are of little use in the general theory.

The assumption that the stress relation is differentiable at and near the natural reference state renders the classical linear formulae a first approximation, for small strains, to those of the finite theory. In the last century it was common to "prove" the approximate validity of the infinitesimal theory in this way. Of course no proof at all is involved. A certain amount of smoothness of the response function may seem reasonable, but there is no way to assure it except by *assuming* it. We formulated a far more general assumption of this kind when we developed the classical linear theories of viscosity and visco-elasticity in Sects. 40—41, where the more sophisticated mathematics required makes the point a little less obvious[2] than it is here. The general theories of mechanical response rest on *physical concepts of determinism and invariance alone*; to obtain (short of assuming outright) the linearity nineteenth-century researches have accustomed us to expecting, some *purely mathematical assumption of smoothness* must be added. We emphasize this distinction here, partly because it is so easy to understand in the context of elasticity, but more because few of the specific results we shall develop in the later sections of this chapter require the assumption of differentiability, except, of course, when it is desired to see how they reduce to their known predecessors in the infinitesimal theory. To a great extent, our special solutions express the consequences of elasticity alone, without purely analytic extra assumptions.

[1] We employ component transformations of the form (21.8), involving the shifters $(21.7)_2$.

[2] Nevertheless PEARSON [1886, *1*, §§ 928—929] managed to make it seem profound and deeply empirical.

For isotropic materials, counterparts of (50.11) are the formulae

$$\left.\begin{aligned}\lambda &= 2\left(\frac{\partial}{\partial I_{\boldsymbol{B}}}+2\frac{\partial}{\partial II_{\boldsymbol{B}}}+\frac{\partial}{\partial III_{\boldsymbol{B}}}\right)(\beth_0+\beth_1+\beth_{-1})\Big|_{\boldsymbol{B}=\mathbf{1}},\\ &= 2\left(\frac{\partial}{\partial I_{\boldsymbol{B}}}+2\frac{\partial}{\partial II_{\boldsymbol{B}}}+\frac{\partial}{\partial III_{\boldsymbol{B}}}\right)(\aleph_0+\aleph_1+\aleph_2)\Big|_{\boldsymbol{B}=\mathbf{1}},\\ \mu &= (\beth_1-\beth_{-1})\,|_{\boldsymbol{B}=\mathbf{1}}=(\aleph_1+2\aleph_2)\,|_{\boldsymbol{B}=\mathbf{1}},\\ &=\beth_1(3,3,1)-\beth_{-1}(3,3,1),\\ &=\aleph_1(3,3,1)+2\aleph_2(3,3,1),\end{aligned}\right\}\qquad(50.13)$$

where λ and μ are the coefficients in the stress relation (41.27) for isotropic infinitesimal elasticity.

For an incompressible isotropic elastic material, the stress is determined only up to a hydrostatic pressure. Thus, any undistorted state may correspond to zero stress and hence may be regarded as a natural state. The requirement that the local reference configuration be a natural state imposes no restrictions upon the response coefficients occurring in (49.2)—(49.5). Contact with the infinitesimal theory of isotropic incompressible materials may be made by setting

$$\mu=\beth_1(3,3)-\beth_{-1}(3,3)=\aleph_1(3,3)+2\aleph_2(3,3);\qquad(50.14)$$

cf. $(50.13)_{5,6}$.

51. Restrictions upon the response functions, I. Isotropic compressible materials. The definition of an elastic material as embodied in (43.1) and its various consequences is too general in that nothing regarding the direction of response is said. Within the framework given so far, it is conceivable that an elastic material may grow shorter rather than longer when pulled, or may swell when subjected to pressure.

There are two simple requirements universally accepted as just and complete in order that the classical infinitesimal theory of isotropic elastic materials be a physically appropriate one for small strains from a stable natural state:

$$\mu>0,\qquad 3\lambda+2\mu>0,\qquad(51.1)$$

where λ and μ are the moduli occurring in (41.27). Using "C" as a mnemonic for "classical," we shall call (51.1) the *C-inequalities*. They arise in connection with several different kinds of argument in the infinitesimal theory.

A. Plausibility in statics.

A 1. In special strains. The condition $(51.1)_1$ is necessary and sufficient that in simple shear, the shearing stress be directed in the same sense as the shear effected. The condition $(51.1)_2$ is necessary and sufficient that tension be required to produce a uniform dilatation, while pressure be required to produce a uniform condensation.

A 2. In general strains. In any strain whatever, the condition $(51.1)_1$ is necessary and sufficient that the shearing stress on any plane shall have the same direction as the shear of that plane. The condition $(51.1)_2$ is necessary and sufficient that the mean tension have the same sign as the increment in volume.

B. Work theorem. The C-inequalities are necessary and sufficient that the work done in any non-rigid local deformation be positive.

C. Stability. The C-inequalities are sufficient but not necessary in order that for every homogeneous body a displacement satisfying the conditions of equilibrium shall correspond to a lesser total stored energy than does any other displacement having the same boundary values. Necessary and sufficient conditions are given by the weaker inequalities[1]

$$\mu > 0, \qquad \lambda + 2\mu > 0. \tag{51.2}$$

It is important to distinguish between this condition and the preceding. In essence, the work theorem imposes the *local* requirement that the energy stored in *every* finite portion of the deformed body be positive, while the stability theorem is *global*, referring only to the total energy stored by a (necessarily non-equilibrated if non-trivial) deformation within a fixed boundary.

D. Uniqueness theorems. The C-inequalities are sufficient in order that the mixed boundary-value problem of equilibrium, for smooth enough boundaries and boundary data, have a solution that is unique, at least to within a rigid displacement. Moreover, the C-inequalities or their opposites are also necessary in order that, for a general region, *both* the place boundary-value problem and the traction boundary-value problem have unique solutions, although they are not necessary for uniqueness of solution to either of these problems by itself[2]. They are sufficient but not necessary for the uniqueness of solution to the equations of motion corresponding to assigned initial displacements and velocities upon the bounding surface[3].

E. Wave speeds. The C-inequalities are sufficient but not necessary in order that the squared speeds of all possible weak waves[4] be positive. Necessary and sufficient conditions, again[5], are the less restrictive inequalities (51.2), provided $\varrho > 0$.

For the finite theory, whether a full set of restrictions such as to ensure physically reasonable behavior in any possible deformation may be established once and for all is a matter still under discussion[6]. Certainly, arguments of all these types can be given; those of types *B* through *E* will be presented in Sects. 68, 71, 83, 87, 89, but here we shall consider those of type *A*, the simplest of all, in

---

[1] KELVIN [1888, *3*].

[2] In terms of the Poisson modulus,

$$\sigma \equiv \frac{\lambda}{2(\lambda+\mu)}, \tag{A}$$

the inequalities

$$-1 < \sigma < \tfrac{1}{2}, \qquad \mu > 0, \tag{B}$$

are equivalent to (51.1). The inequality $(\mathrm{B})_1$ alone implies the validity of either the C-inequalities or their opposites, *viz* $\mu < 0$, $3\lambda + 2\mu < 0$. The solution of the place boundary-value problem is unique, within arbitrary smooth regions, *if and only if*

$$\sigma < \tfrac{1}{2}, \quad \text{or} \quad \sigma > 1; \tag{C}$$

of the traction boundary-value problem, *if and only if*

$$-1 < \sigma < 1. \tag{D}$$

The range of common validity of (D) and (C) is given by $(\mathrm{B})_1$. Thus follows the result stated in the text above.

[3] It suffices that the two squared wave-speeds be non-negative: $\mu/\varrho \geqq 0$ and $(\lambda + 2\mu)/\varrho \geqq 0$. If $\varrho > 0$ and if "$\geqq$" is replaced by "$>$", this condition reduces to (51.2); if $\varrho \neq 0$, it is equivalent to (C) in the preceding footnote.

[4] I.e., singular surfaces of second or higher order, or plane infinitesimal oscillations.

[5] A connection between wave propagation and stability as defined under *C* was seen by KELVIN [1888, *3*]: "Surely, then, if there is a real finite propagational velocity for each of the two kinds of wave-motion, the equilibrium *must* be stable!" The matter is not really so simple as this. In the finite theory, there are various plausible but different definitions of stability (Sects. 68bis, 89), and there are many kinds of waves (Sects. 71—78).

[6] This is the "Hauptproblem" of TRUESDELL [1956, *32*]. The reader will see that much has been learned about the matter since 1956.

which physically natural requirements are laid down, scrutinized, and connected with each other[1].

In the theory of finite elastic strain it is clearly of no use for the present end to consider special deformations such as simple tension, simple shear, and uniform dilatation, since special relations hold among the stretches occurring in those deformations, so no restriction upon general functions of those stretches can possibly result by laying down conditions for these cases[2]. Thus the approach numbered *A 1*, effective in the classical infinitesimal theory only because the response of the material is specified in terms of constants, *viz* λ and μ, instead of functions of the stretches, would be fruitless here.

Not so for *A 2*, as we shall see now. At the beginning we consider only isotropic materials because for materials of less complete symmetry the intuitive picture is no longer clear.

α) The P-C inequality. The condition that the volume of a compressible isotropic material should be decreased by pressure but increased by tension is expressed by requiring the hydrostatic tension t to be a strictly increasing function $t=\bar{t}(v)$ of the stretch, $v=v_1=v_2=v_3$:

$$(t^*-t)(v^*-v)>0, \qquad v^*\neq v, \tag{51.3}$$

where $t^*=\bar{t}(v^*)$. In particular, if the reference state is a natural state, so that $\bar{t}(1)=0$, (51.3) implies that $t(v-1)>0$ if $v\neq 1$. For a general deformation, there are many inequalities that are invariant and that reduce to this latter one in the special case of hydrostatic stress. An example is a set of three inequalities $t_a(v_a-1)>0$, $a=1, 2, 3$, stating that each principal stress is a pressure or a tension according as the corresponding principal stretch is a contraction or an elongation, but these inequalities are too strong. It should be possible in general, as it is in the infinitesimal theory of elasticity, that a severe pull in one direction may result in transverse contraction even when some transverse tension is applied. It is reasonable, however, to demand an inequality of this type in mean, and we are led to consider either or both of the following[3]:

$$\sum_{a=1}^{3} t_a(v_a-1)>0, \qquad \sum_{a=1}^{3} T_a(v_a-1)>0, \tag{51.4}$$

if not all $v_a=1$. Here the t_a are the principal stresses, while the T_a are the principal forces, defined by (48.7). We shall see later that $(51.4)_2$ is implied by certain more general and plausible conditions that do not always lead to $(51.4)_1$. In terms of principal stresses $(51.4)_2$ can be written

$$\sum_{a=1}^{3} t_a \frac{v_a-1}{v_a}>0 \quad \text{if not all} \quad v_a=1. \tag{51.5}$$

Thus our intuition does not suggest, in this instance, which of the following two strain measures to select as a basis for inequalities:

$$\left.\begin{aligned} v_a-1 &= \frac{\text{change in length}}{\text{initial length}}, \\ \frac{v_a-1}{v_a} &= \frac{\text{change in length}}{\text{final length}}. \end{aligned}\right\} \tag{51.6}$$

---

[1] We follow TRUESDELL and TOUPIN [1963, *76*], to whom are due all results in this section and the next not specifically attributed to others.

[2] To substantiate this statement the reader may consider the special solutions given in Sects. 54 and 55.

[3] For hyperelastic materials the inequality $(51.4)_2$ was first mentioned by NOLL in 1960 (unpublished), who derived it as a consequence of the C-N condition, which we shall discuss in Sect. 87.

Using the letters P-C to recall "pressure-compression", we shall refer to $(51.4)_2$ as the *P-C inequality*. An invariant form of it is

$$\operatorname{tr}\left[\boldsymbol{T}(\mathbf{1}-\boldsymbol{V}^{-1})\right]>0 \quad \text{if} \quad \boldsymbol{V}\neq\mathbf{1}. \tag{51.7}$$

Another form is

$$\operatorname{tr}\boldsymbol{T}=\sum_{a=1}^{3} t_a > \operatorname{tr}(\boldsymbol{T}\boldsymbol{V}^{-1}) = J^{-1}\sum_{a=1}^{3} T_a \quad \text{if not all} \quad v_b=1. \tag{51.8}$$

In particular, in a state of pressure in mean (i.e., $\sum t_a \leqq 0$), the mean of the principal forces is negative ($\sum T_a<0$ unless all $v_a=1$); conversely, if the mean of the principal forces is positive, there is a state of tension in mean.

It is easy to express the P-C inequality as a restriction upon the three response coefficients $\beth_\Gamma$. From (47.9) we easily show that $(51.4)_2$ is equivalent to the condition

$$\left(3-\frac{II_{\boldsymbol{V}}}{III_{\boldsymbol{V}}}\right)\beth_0+(I_{\boldsymbol{B}}-I_{\boldsymbol{V}})\beth_1>\left[\frac{II_{\boldsymbol{V}}II_{\boldsymbol{B}}}{III_{\boldsymbol{V}}^3}-\frac{II_{\boldsymbol{B}}+I_{\boldsymbol{V}}II_{\boldsymbol{V}}-3III_{\boldsymbol{V}}}{III_{\boldsymbol{B}}}\right]\beth_{-1} \tag{51.9}$$

when $\boldsymbol{B}\neq\mathbf{1}$. A condition expressed entirely in terms of invariants of $\boldsymbol{B}$ would be preferable, but unfortunately $I_{\boldsymbol{V}}$, $II_{\boldsymbol{V}}$, and $III_{\boldsymbol{V}}$ are irrational functions of those invariants, and in practice probably the most useful form will be $(51.4)_2$ itself.

Note that only for materials having a natural state and only when the stretches are measured from that natural state is the P-C inequality to be expected. E.g., for an elastic fluid, while (51.3) ought to be satisfied, the P-C inequality $(51.4)_2$ certainly is not. Neither is the inequality $(51.4)_1$, in general.

β) *The T-E inequalities.* It is natural to expect that when a cube of isotropic material is lengthened along one principal direction while its faces parallel to that direction are kept fixed, the tensile force must be increased, but to shorten it, the tensile force must be reduced. This condition may be expressed in either of the following equivalent forms:

$$(T_a^*-T_a)(v_a^*-v_a)>0, \qquad (t_a^*-t_a)(v_a^*-v_a)>0, \tag{51.10}$$

provided $v_b^*=v_b$ if $b\neq a$. Here we are using the notations $T_a^*=\overline{T}_a(v_1^*, v_2^*, v_3^*)$, etc. [cf. (48.8)]. From these conditions it follows that

$$\frac{\partial\overline{T}_a}{\partial v_a}>0, \qquad \frac{\partial\overline{t}_a}{\partial v_a}>0, \tag{51.11}$$

except perhaps upon a nowhere dense set of values v_a. Conversely, if we require (51.11) to hold everywhere, then (51.10) follows. Using the letters T-E to recall "tension-extension", we shall refer to (51.10) as the *T-E inequalities*, while the slightly stronger conditions (51.11) will be called the *T-E⁺ inequalities*. From $(48.16)_1$ and (48.13) we see at once two equivalent alternative forms of the T-E⁺ condition (51.11):

$$F\langle 1111\rangle>0, \qquad F\langle 2222\rangle>0, \qquad F\langle 3333\rangle>0, \tag{51.12}$$

and

$$\aleph_1+2v_a^2\aleph_2+\sum_{\Gamma=0}^{2} v_a^{2\Gamma}\left[\frac{\partial\aleph_\Gamma}{\partial I}+(I-v_a^2)\frac{\partial\aleph_\Gamma}{\partial II}+\frac{III}{v_a^2}\frac{\partial\aleph_\Gamma}{\partial III}\right]>0. \tag{51.13}$$

γ) *The IFS condition.* It is reasonable to expect that if given pairs of opposing normal forces of magnitudes T_a are applied to the faces of a unit cube of isotropic elastic material, one and only one pure homogeneous deformation will result. Thus we require that the equations $T_a=\overline{T}_a(v_1, v_2, v_3)$ be uniquely invertible. Using "IFS" to recall "invertibility of force-stretch", we may name this requirement the *IFS condition*.

In terms of the quantities J_{ab} defined by $(48.9)_1$, the IFS condition implies that

$$\det \|J_{ab}\| \neq 0, \tag{51.14}$$

except perhaps upon a nowhere dense set. Conversely, if (51.14) holds everywhere, the IFS condition follows. The slightly stronger requirement (51.14) will be called the *IFS⁺ condition*.

For an isotropic elastic solid with an invertible response function $\mathfrak{t}$ it is necessary, again with the exception of points constituting a nowhere dense set, that

$$\det \|j_{ab}\| \neq 0, \tag{51.15}$$

where j_{ab} is defined by $(48.9)_2$. Neither of the conditions (51.14) and (51.15) implies the other.

The difference between invertibility of the force-stretch functions $\overline{T}_a$ and invertibility of the stress-stretch functions $\overline{t}_a$ is illustrated by the case of an elastic fluid, for which, in view of $(50.2)_2$, we have $t_a = -p\left(\varrho_R/(v_1 v_2 v_3)\right)$. Obviously no such relation can be inverted to give v_1 as a function of the t_a, yet the corresponding formulae for the T_a may be invertible. E.g., in an "ideal" gas $p = K/(v_1 v_2 v_3)$, leading to the unique inverse relations $v_a = -K/T_a$. Thus the IFS condition (51.14) may be imposed on elastic fluids, but (51.15) may not.

δ) The E-T inequalities. Granted that the IFS condition is satisfied, i.e., that we may solve for the principal stretches as functions $v_a = \overline{v}_a(T_1, T_2, T_3)$ of the principal forces T_a, it is reasonable to expect that if one pair of opposing normal forces on a homogeneously strained block with faces perpendicular to the principal axes are increased in magnitude, while the remaining pairs are kept fixed, the block will lengthen in the corresponding direction. This condition may be expressed in the form

$$(T_a^* - T_a)(v_a^* - v_a) > 0 \quad \text{if} \quad T_b^* = T_b \quad \text{for} \quad b \neq a, \tag{51.16}$$

where we are using the notation $v_a^* = \overline{v}_a(T_1^*, T_2^*, T_3^*)$. This is a condition on the three functions $\overline{v}_a$, while the formally similar condition $(51.10)_1$ restricts the three functions $\overline{T}_a$. Thus, neither condition can be expected to imply the other in general. From (51.16) we infer that

$$\frac{\partial \overline{v}_a}{\partial T_a} > 0 \tag{51.17}$$

except perhaps on a nowhere dense set of values T_a. Conversely, if we require (51.17) to hold everywhere, (51.16) follows. Using the letters E-T to recall "extension-tension", we refer to (51.16) as the *E-T inequalities*, while the slightly stronger conditions (51.17) will be called the *E-T⁺ inequalities*.

ε) The O-F inequalities. Consider again a block of isotropic material supposed to be in equilibrium subject to pairs of equal and oppositely directed normal forces acting upon its faces. It is reasonable to expect that the greater stretch will occur in the direction of the greater force. Thus we are led to require that

$$(T_a - T_b)(v_a - v_b) > 0 \quad \text{if} \quad v_a \neq v_b. \tag{51.18}$$

For hyperelastic materials this inequality was first mentioned by COLEMAN and NOLL[1], who derived it as a consequence of the C-N condition, which we shall discuss below in Sect. 87. Using the letters O-F to recall "ordered forces," we shall refer to (51.18) as the *O-F inequalities*. An alternative form is

$$\frac{t_a}{v_a} > \frac{t_b}{v_b} \quad \text{if and only if} \quad v_a > v_b. \tag{51.19}$$

---

[1] COLEMAN and NOLL [1959, *3*, § 12].

To express this requirement in terms of the response coefficients $\beth_\Gamma$, note from $(48.1)_4$ that

$$\frac{t_a}{v_a}-\frac{t_b}{v_b}=(v_a-v_b)\left[-\frac{\bar{\beth}_0}{v_a v_b}+\bar{\beth}_1-\frac{v_a^2+v_a v_b+v_b^2}{v_a^3 v_b^3}\bar{\beth}_{-1}\right]. \tag{51.20}$$

Hence (51.19) is equivalent to the former of the conditions

$$\left.\begin{aligned} &-\frac{\bar{\beth}_0}{v_a v_b}+\bar{\beth}_1-\frac{v_a^2+v_a v_b+v_b^2}{v_a^3 v_b^3}\bar{\beth}_{-1}>0 \quad \text{if} \quad v_a\neq v_b,\\ &-\frac{\bar{\beth}_0}{v_a^2}+\bar{\beth}_1-\frac{3}{v_a^4}\bar{\beth}_{-1}\geqq 0 \qquad\qquad \text{if} \quad v_a=v_b, \end{aligned}\right\} \tag{51.21}$$

while the latter follows from the former by the assumption that the response coefficients $\beth_\Gamma$ are continuous functions.

Conditions expressed in terms of the principal invariants of $\boldsymbol{B}$ would be preferable if they were simple, but they are not. The reader will easily derive simple conditions expressed in terms of the response coefficients $\dot{\mathfrak{w}}_\Gamma$ occurring in (47.6); while these have the virtue of restricting $\dot{\mathfrak{w}}_0$ and $\dot{\mathfrak{w}}_2$ only, the difficulty of calculating $\boldsymbol{V}$ renders them little likely to be useful in practice.

ζ) The B-E inequalities. Inequalities similar to (51.18), but referring to principal stresses rather than to principal forces, were first suggested by BAKER and ERICKSEN[1]. Specifically, they proposed that the greater principal stress occur always in the direction of the greater principal stretch:

$$(t_a-t_b)(v_a-v_b)>0 \quad \text{if} \quad v_a\neq v_b. \tag{51.22}$$

We shall refer to these as the *B-E inequalities.*

From (48.16) and (48.13) we see that equivalent forms of the B-E inequalities are

$$F\langle 1212\rangle>0, \qquad F\langle 2323\rangle>0, \qquad F\langle 3131\rangle>0, \tag{51.23}$$

or, alternatively,

$$\aleph_1+(I_{\boldsymbol{B}}-v_a^2)\,\aleph_2>0. \tag{51.24}$$

Since, by $(48.1)_4$,

$$t_a-t_b=(v_a^2-v_b^2)\left(\bar{\beth}_1-\frac{1}{v_a^2 v_b^2}\bar{\beth}_{-1}\right), \tag{51.25}$$

the B-E inequalities are equivalent to[1]

$$\left.\begin{aligned} &\bar{\beth}_1-\frac{1}{v_a^2 v_b^2}\bar{\beth}_{-1}>0 \quad \text{if} \quad v_a\neq v_b,\\ &\bar{\beth}_1-\frac{1}{v_a^4}\bar{\beth}_{-1}\geqq 0 \quad\;\; \text{if} \quad v_a=v_b. \end{aligned}\right\} \tag{51.26}$$

η) The E-inequalities. From (51.21) and (51.26) we see that a simple set of inequalities implying both the B-E inequalities and the O-F inequalities is

$$\beth_0\leqq 0, \qquad \beth_1>0, \qquad \beth_{-1}\leqq 0. \tag{51.27}$$

To the limited extent that experimental data are available (Sects. 55, 57), they seem to support these inequalities, which shall therefore be called the *E-inequalities*, where "E" is a mnemonic for "empirical". No theoretical motivation has been found for them, however, beyond their logical relation to the B-E and O-F inequalities, which may be abbreviated as follows:

$$\text{E} \Rightarrow \text{B-E \& O-F}. \tag{51.28}$$

---

[1] BAKER and ERICKSEN [1954, *3*].

We now establish some less obvious connections between the various inequalities stated.

If the reference state is a natural state, then the linear approximations to the functions $\overline{T}_a$ near $v_1 = v_2 = v_3 = 1$ are given by

$$T_a = \lambda(v_1 + v_2 + v_3 - 3) + 2\mu(v_a - 1), \tag{51.29}$$

expressing in principal components the stress relation (41.27) of the infinitesimal theory. Indeed, for infinitesimal strains the proper numbers of $\widetilde{\boldsymbol{E}}$ are $v_a - 1$, and the distinction between principal forces and principal stresses disappears. From (51.29) we infer easily that, in the infinitesimal theory, the various inequalities reduce to the following forms[1]:

$$\left.\begin{array}{rl} \text{P-C:} & \mu > 0,\ 3\lambda + 2\mu > 0. \\ \text{T-E, T-E}^+: & \lambda + 2\mu > 0. \\ \text{IFS, IFS}^+: & \mu(3\lambda + 2\mu) \neq 0. \\ \text{E-T, E-T}^+: & \dfrac{\mu(3\lambda + 2\mu)}{\lambda + \mu} > 0. \\ \text{O-F, B-E:} & \mu > 0. \\ \text{E:} & \mu > 0. \end{array}\right\} \tag{51.30}$$

From these forms we may read off several non-implications:

$$\left.\begin{array}{l} \text{O-F} \not\Rightarrow \text{P-C or T-E or E-T.} \\ \text{B-E} \not\Rightarrow \text{P-C or T-E or E-T.} \\ \text{(T-E or E-T) \& (O-F or B-E)} \not\Rightarrow \text{P-C.} \end{array}\right\} \tag{51.31}$$

Either from the table (51.30) or directly from $(51.4)_2$ we see that in the infinitesimal theory the P-C inequality is equivalent to the requirement that the work done in effecting any non-rigid deformation be positive. Hence the P-C inequality implies all the requirements customarily imposed in the infinitesimal theory, as may be verified also from the table (51.30): P-C$\Rightarrow$C. In a finite deformation, however, the P-C inequality does not have any evident energetic meaning, nor does it imply any of the others considered here. Neither do the E-inequalities (51.27) imply the P-C inequalities.

The inequalities listed fall into three types: those that relate a single stress or force to the corresponding stretch, those that relate pairs of stresses to pairs of stretches, and those that relate all three of each. We shall discuss now the three types separately, beginning with the second, consisting only in the O-F and B-E inequalities.

While the O-F and B-E inequalities are equivalent in the infinitesimal theory, for finite strains

$$\text{O-F} \not\Rightarrow \text{B-E} \quad \text{and} \quad \text{B-E} \not\Rightarrow \text{O-F}. \tag{51.32}$$

For example, if $T < 0$, the triples $(T, 2T, 3T)$ and $(v, \frac{1}{2}v, \frac{1}{4}v)$ satisfy the O-F inequalities, but the principal stresses are $(8T/v^2, 8T/v^2, 6T/v^2)$, which do not satisfy the B-E inequalities. There is, however, an intimate connection between the two inequalities. To see it, we observe that, by (48.7), the O-F inequality

[1] As might be expected, the E-T inequality leads to the requirement that the modulus of extension be positive. On this point, TRUESDELL and TOUPIN's paper [1963, *76*, § 4[1]] is incorrect.

(51.18) is equivalent to

$$\left(\frac{t_a}{v_a} - \frac{t_b}{v_b}\right)(v_a - v_b) > 0 \quad \text{if} \quad v_a \neq v_b. \tag{51.33}$$

A simple calculation shows that (51.33), in turn, is equivalent to

$$(t_a - t_b)(v_a - v_b) > \frac{t_a}{v_a}(v_a - v_b)^2 \quad \text{if} \quad v_a \neq v_b. \tag{51.34}$$

Since interchange of a and b does not alter the left-hand sides of (51.33) and (51.34), we infer that

$$(t_a - t_b)(v_a - v_b) > (v_a - v_b)^2 \operatorname{Max}\left(\frac{t_a}{v_a}, \frac{t_b}{v_b}\right) \quad \text{if} \quad v_a \neq v_b. \tag{51.35}$$

Now, it is clear that the right-hand side of (51.35) is non-negative for all choices of a and b if at least two principal stresses are non-negative. Comparing (51.35) with (51.22), we infer that *if two of the three principal stresses t_a are non-negative, then* O-F$\Rightarrow$B-E. States of pure tension are included as a special case[1].

By (48.7), the B-E inequality (51.22) is equivalent to

$$(v_a T_a - v_b T_b)\left(\frac{1}{v_a} - \frac{1}{v_b}\right) < 0 \quad \text{if} \quad v_a \neq v_b. \tag{51.36}$$

Replacing t_a by T_a, v_a by v_a^{-1}, and $>$ by $<$ in the argument that led from (51.33) to (51.35), we find that

$$(T_a - T_b)\left(\frac{1}{v_a} - \frac{1}{v_b}\right) < \left(\frac{1}{v_b} - \frac{1}{v_a}\right)^2 \operatorname{Min}(v_a T_a, v_b T_b) \quad \text{if} \quad v_a \neq v_b, \tag{51.37}$$

or, equivalently,

$$(T_a - T_b)(v_a - v_b) > (v_a - v_b)^2 \operatorname{Max}\left(-\frac{T_a}{v_b}, -\frac{T_b}{v_a}\right) \quad \text{if} \quad v_a \neq v_b. \tag{51.38}$$

The right-hand side of (51.38) is non-negative for all choices of a and b if two of the principal forces are non-positive. Of course, a principal stress has the same sign as the corresponding principal force. In view of (51.18), we conclude that *if two of the three principal stresses are non-positive, then* B-E$\Rightarrow$O-F. States of pure pressure are included as a special case. In any state of stress, one or the other hypothesis holds, and hence one or the other of the implications B-E$\Rightarrow$O-F and O-F$\Rightarrow$B-E follows, but generally not both. In a state of stress such that $t_1 > 0$, $t_2 = 0$, and $t_3 < 0$, either inequality implies the other: B-E$\Leftrightarrow$O-F.

Further analysis may reveal other connections between the B-E and O-F inequalities, but it is clear that they express, in general, distinct requirements.

A summary of several of the preceding statical inequalities may be obtained by demanding that $(51.10)_1$ hold in mean even when the requirement that $v_b^* = v_b$ is abandoned. That is[2],

$$\sum_{a=1}^{3} (T_a^* - T_a)(v_a^* - v_a) > 0 \quad \text{if} \quad v_a^* \neq v_a \quad \text{for some } a. \tag{51.39}$$

For reasons that will shortly appear, this condition is called the *GCN_0 condition.* It is a statement that *the transformation from principal stretches to principal forces*

---

[1] That O-F$\Rightarrow$B-E for states of pure tension was remarked by COLEMAN and NOLL [1959, *3*, § 12]. TRUESDELL and TOUPIN [1963, *76*, § 4] proved that O-F$\Rightarrow$B-E when $t_1 \geqq 0$, $t_2 \geqq 0$, $t_2 + t_3 \geqq 0$, and that B-E$\Rightarrow$O-F when $t_3 \leqq 0$, $t_2 \leqq 0$, $t_1/v_1 + t_2/v_2 \leqq 0$, the t_a being ordered so that $v_1 > v_2 > v_3$. The generalization, here published for the first time, is due to NOLL.

[2] For hyperelastic materials, this inequality was first given by NOLL in 1960 (unpublished), who derived it from the C-N condition, to be given Sect. 87. NOLL also supplied the derivation of (51.40) from (51.39).

is a monotone[1] *transformation*. Obviously it includes $(51.10)_1$ as a special case. Moreover, if $v_a^* \neq v_a$ for any a, (51.39) cannot be satisfied if $T_a^* = T_a$ for all a. Thus different values of the v_a always lead to different values of the T_a, or, in other words, the force-stretch relations have unique inverses. This being so, we may set $T_b^* = T_b$ for $b \neq a$ and infer (51.16). If we employ a natural state as the reference configuration, by setting $v_a^* = 1$ we obtain $T_a^* = 0$ and so reduce (51.39) to $(51.4)_2$. But this is not all. If we choose the v_a^* as a permutation of the v_a, because of (48.3), (48.4), and (48.7) the T_a^* are the corresponding permutation of the T_a, so that (51.39) yields

$$\sum_{a=1}^{3} (T_{\pi(a)} - T_a)(v_{\pi(a)} - v_a) > 0, \tag{51.40}$$

provided only that π is not the identity permutation of (1, 2, 3) and that $v_a \neq v_b$ if $a \neq b$. Permuting any two of the v_c but leaving the other unchanged reduces (51.40) to (51.18). In summary of the results in the sentences just preceding,

$$\text{GCN}_0 \Rightarrow \left\{\begin{array}{l} \text{P-C} \;\; \& \\ \text{IFS} \;\; \& \\ \text{T-E} \;\; \& \\ \text{E-T} \;\; \& \\ \text{O-F}, \end{array}\right\}, \tag{51.41}$$

so that the GCN_0 inequality implies a full set of simple statical conditions laid down earlier as plausible. [It is necessary to employ a natural state as the local reference configuration in order to infer the P-C inequality, but the other implications in (51.41) hold also for materials without a natural state.]

If (51.39) holds, the corresponding Hessian form can never be negative. That is, for arbitrary numbers w^c,

$$\sum_{a,b=1}^{3} J_{ab}\, w^a w^b \geqq 0, \tag{51.42}$$

where J_{ab} is defined by $(48.9)_1$. From (51.42) it follows that the principal minors of the symmetric matrix $\|J_{(ab)}\|$ must be non-negative; in particular,

$$\left(\frac{\partial T_a}{\partial v_b} + \frac{\partial T_b}{\partial v_a}\right)^2 \leqq 4\, \frac{\partial T_a}{\partial v_a}\, \frac{\partial T_b}{\partial v_b}. \tag{51.43}$$

Since $\alpha^2 + \beta^2 \geqq 2\alpha\beta$, we see that

$$\text{GCN}_0 \Rightarrow \frac{\partial T_a}{\partial v_b}\, \frac{\partial T_b}{\partial v_a} \leqq \frac{\partial T_a}{\partial v_a}\, \frac{\partial T_b}{\partial v_b}, \qquad a \neq b. \tag{51.44}$$

This is a reciprocal inequality, stating that the product of transverse tangent moduli can never exceed the product of the corresponding tangent moduli of extension. As a corollary, the lesser transverse modulus can never exceed the greater modulus of extension.

According to a theorem of Bernstein and Toupin[2], equality can hold in (51.42) only on a nowhere dense set. Conversely, if the Hessian form is positive-

---

[1] We call a transformation $\boldsymbol{f}$ of a Euclidean space (of any dimension) into itself *monotone* if $[\boldsymbol{f}(\boldsymbol{x}) - \boldsymbol{f}(\boldsymbol{y})] \cdot (\boldsymbol{x} - \boldsymbol{y}) > 0$ holds for every pair of distinct points $\boldsymbol{x}, \boldsymbol{y}$. When the dimension of the space is one, this concept of monotonicity reduces to the usual monotonicity for functions of one variable. Transformations of this kind were called "convex" by Truesdell and Toupin [1963, *76*, § 6].

[2] Bernstein and Toupin [1962, *4*].

definite, i.e. if

$$\sum_{a,b=1}^{3} J_{ab}\, w^a\, w^b > 0 \tag{51.45}$$

except when $w^1 = w^2 = w^3 = 0$, the transformation from principal stretches to principal forces is monotone. That is, if we call (51.45) the *GCN$_0^+$ condition*[1], then

$$\mathrm{GCN}_0^+ \Rightarrow \mathrm{GCN}_0, \tag{51.46}$$

but of course

$$\mathrm{GCN}_0 \not\Rightarrow \mathrm{GCN}_0^+ \tag{51.47}$$

in general. Combining (51.46) and (51.41) yields an implication connecting GCN$_0^+$ with the separate statical inequalities discussed above, but in fact a somewhat stronger implication holds[2]:

$$\mathrm{GCN}_0^+ \Rightarrow \left\{\begin{matrix} \text{P-C} & \& \\ \text{IFS}^+ & \& \\ \text{T-E}^+ & \& \\ \text{E-T}^+ & \& \\ \text{O-F.} & \end{matrix}\right\}. \tag{51.48}$$

In the results of this section the use of principal forces T_a rather than principal stresses t_a has led to simpler as well as more plausible restrictions. For example, we have seen that the B-E inequalities cannot hold for elastic fluids, yet the O-F inequalities can.

52. Restrictions upon the response functions, II. Compressible materials in general. With this much information about isotropic materials gained, we may now approach materials of arbitrary symmetry. Since it generalizes a condition for hyperelastic materials first proposed by COLEMAN and NOLL, which we shall present in Sect. 87, the following condition of restricted monotonicity will be called the *GCN condition*[3]:

$$\operatorname{tr}\{[\mathfrak{h}(\boldsymbol{F}^*) - \mathfrak{h}(\boldsymbol{F})][\boldsymbol{F}^{*T} - \boldsymbol{F}^T]\} > 0, \tag{52.1}$$

provided $\boldsymbol{F}^* = \boldsymbol{S}\boldsymbol{F}$, where $\boldsymbol{S}$ is positive-definite, symmetric, and not equal to $\mathbf{1}$. In view of (43.8), the GCN condition asserts that *the transformation from defor-*

---

[1] Inequalities leading to the condition

$$\sum_{a,b=1}^{3} j_{ab}\, w^a\, w^b > 0$$

were proposed by BARTA [1957, *3*] but were shown to be unsatisfactory by TRUESDELL and TOUPIN [1963, *76*, § 4].

[2] The first and last implications follow as indicated in the text (and we repeat the remark that a natural state is chosen as local reference configuration so as to infer P-C but is not needed for the rest). The three intermediate implications were obtained by TRUESDELL and TOUPIN [1963, *76*, § 6]. Their argument rests on the fact that if a 3×3 matrix (whether or not it be symmetric) is positive-definite, its principal minors of orders 1, 2, 3 are positive. They also showed that

$$\mathrm{GCN}_0^+ \Rightarrow \frac{\partial T_a}{\partial v_b}\,\frac{\partial T_b}{\partial v_a} < \frac{\partial T_a}{\partial v_a}\,\frac{\partial T_b}{\partial v_b} \qquad a \neq b.$$

[3] TRUESDELL and TOUPIN [1963, *76*, § 5]. COLEMAN and NOLL [1964, *19*, § 4] called it the "Weakened Thermostatic Inequality". BRAGG [1964, *11*] has shown that the GCN condition holds at $\boldsymbol{F}$ if and only if the virtual work on every one-parameter family of pure stretches from $\boldsymbol{F}$ is a monotone function of the parameter.

mation gradient to first Piola-Kirchhoff stress tensor shall be monotone[1] *with respect to pairs of deformations differing from one another by a pure stretch.*

First, we motivate necessity for comparing only states differing by a pure stretch. If we were to demand that (52.1) should hold for all $\boldsymbol{F}$ and $\boldsymbol{F}^*$, we could choose $\boldsymbol{F}=\mathbf{1}$, $\boldsymbol{F}^*=\boldsymbol{Q}$, where $\boldsymbol{Q}$ is orthogonal. For a material having a natural state, we have $\mathfrak{h}(\mathbf{1})=\mathbf{0}$. Moreover, by (43.9) we conclude that $\mathfrak{h}(\boldsymbol{Q})=\boldsymbol{Q}\mathbf{0}=\mathbf{0}$. Hence, for this choice of $\boldsymbol{F}^*$ and $\boldsymbol{F}$ the tensor in braces in (52.1) is the zero tensor, so (52.1) itself cannot hold[2]. This fact, while showing that (52.1) cannot be asserted for arbitrary pairs $\boldsymbol{F}, \boldsymbol{F}^*$, does not prove that the deformations compared must satisfy the relation $\boldsymbol{F}^*=\boldsymbol{S}\boldsymbol{F}$. This specific restriction we can justify only indirectly by verifying that it leads to acceptable consequences.

Second, consider an infinitesimal deformation from a natural state taken as reference configuration. Set $\boldsymbol{F}=\mathbf{1}$, $\boldsymbol{F}^*=\mathbf{1}+\tilde{\boldsymbol{E}}+\tilde{\boldsymbol{R}}$, where $\tilde{\boldsymbol{E}}$ and $\tilde{\boldsymbol{R}}$ are tensors of infinitesimal strain and rotation, respectively. Since $\mathfrak{h}(\boldsymbol{F})=\mathbf{0}$ and $\mathfrak{h}(\boldsymbol{F}^*)\approx\boldsymbol{T}$, (52.1) reduces to the requirement that $\operatorname{tr}(\boldsymbol{T}\tilde{\boldsymbol{E}})>0$ if $\tilde{\boldsymbol{E}}\neq\mathbf{0}$. Thus *the GCN condition includes and generalizes all the requirements customarily imposed in the classical infinitesimal theory*[3]. In particular, for an isotropic material, GCN$\Rightarrow$C.

Third, consider an arbitrary deformation of an isotropic elastic material. Assume that

$$\boldsymbol{F}=\boldsymbol{V}\boldsymbol{R}, \quad \boldsymbol{F}^*=\boldsymbol{V}^*\boldsymbol{R}, \quad \boldsymbol{V}\boldsymbol{V}^*=\boldsymbol{V}^*\boldsymbol{V}, \tag{52.2}$$

i.e., that $\boldsymbol{F}$ and $\boldsymbol{F}^*$ have the same rotation tensor $\boldsymbol{R}$ and that the left stretch tensors $\boldsymbol{V}$ and $\boldsymbol{V}^*$ commute. Then $\boldsymbol{V}$ and $\boldsymbol{V}^*$ have an orthonormal basis of proper vectors in common, and $\boldsymbol{F}^*$ has the form $\boldsymbol{F}^*=\boldsymbol{S}\boldsymbol{F}$, where $\boldsymbol{S}=\boldsymbol{V}^*\boldsymbol{V}^{-1}$ is symmetric and positive-definite. By use of (48.6) we find that

$$(\boldsymbol{T}_{\mathrm{R}}^*-\boldsymbol{T}_{\mathrm{R}})(\boldsymbol{F}^{*T}-\boldsymbol{F}^T)=(J^*\boldsymbol{T}^*\boldsymbol{V}^{*-1}-J\boldsymbol{T}\boldsymbol{V}^{-1})(\boldsymbol{V}^*-\boldsymbol{V}). \tag{52.3}$$

---

[1] The term "monotone," in the sense of footnote 1 on p. 161, is applied here to the 9-dimensional space $\mathscr{L}$ of second-order tensors with $\operatorname{tr}(\boldsymbol{A}\boldsymbol{B}^T)$ as the inner product (cf. Sect. 7).

[2] TRUESDELL and TOUPIN [1963, *76*, § 8]. Indeed, if $\boldsymbol{F}^*=\boldsymbol{Q}\boldsymbol{F}$, by $(43\mathrm{A}.11)_1$ we see that

$$\operatorname{tr}\{[\mathfrak{h}(\boldsymbol{F}^*)-\mathfrak{h}(\boldsymbol{F})][\boldsymbol{F}^{*T}-\boldsymbol{F}^T]\}=J\operatorname{tr}[(\boldsymbol{Q}-\mathbf{1})\boldsymbol{T}(\boldsymbol{Q}^T-\mathbf{1})]. \tag{*}$$

If we now choose for $\boldsymbol{Q}$ the tensor having the matrix diag $(-1, 1, 1)$ relative to an orthonormal basis of proper vectors of $\boldsymbol{T}$, (52.1) becomes

$$4\,J\,t_1>0.$$

Thus, as remarked in effect by COLEMAN and NOLL [1959, *3*, § 8], (52.1) cannot hold for *all* pairs $\boldsymbol{F}^*, \boldsymbol{F}$ if a single principal stress is a pressure. In the example just given, $\boldsymbol{Q}$ is improper. If only proper rotations are allowed, we may take tensors $\boldsymbol{Q}$ having the matrices diag $(-1, 1, -1)$, diag $(1, -1, -1)$ and diag $(-1, -1, 1)$ in turn. These choices yield

$$t_1+t_2>0, \quad t_2+t_3>0, \quad t_3+t_1>0. \tag{**}$$

That is, if $\boldsymbol{F}^*=\boldsymbol{Q}\boldsymbol{F}$ where $\boldsymbol{Q}$ is proper, (52.1) cannot hold except at a point where two principal stresses are tensions and are numerically greater than the third. Moreover, as HILL pointed out to us in a letter, this last result holds even when $\boldsymbol{F}^*$ is allowed to differ from $\boldsymbol{F}$ only by an infinitesimal deformation, since then (*) becomes $-J\operatorname{tr}\tilde{\boldsymbol{R}}\boldsymbol{T}\tilde{\boldsymbol{R}}$, which is positive for all non-vanishing infinitesimal rotations $\tilde{\boldsymbol{R}}$ if and only if (**) holds.

[3] As remarked in Sect. 50, by "the infinitesimal theory" we mean the theory of CAUCHY, in which there need be no stored-energy function. In hyperelasticity, $\operatorname{tr}(\boldsymbol{T}\tilde{\boldsymbol{E}})$ is the work done per unit volume, but in the more general theory of elasticity it may or may not have that significance. By (50.10), condition $\operatorname{tr}(\boldsymbol{T}\tilde{\boldsymbol{E}})>0$ demands that $L^{kmpq}\tilde{E}_{km}\tilde{E}_{pq}>0$ for every symmetric, non-vanishing $\tilde{\boldsymbol{E}}$. This requirement restricts the symmetric part of the tensor of linear elasticities, namely, $\frac{1}{2}(L^{kmpq}+L^{pqkm})$, leaving the skew-symmetric part, namely, $\frac{1}{2}(L^{kmpq}-L^{pqkm})$, entirely unrestricted. For there to be a stored-energy function, in the infinitesimal theory, it is necessary and sufficient that $L^{kmpq}=L^{pqkm}$. In hyperelasticity, the GCN condition implies that the stored energy be positive.

Since the principal axes of stress and strain in the deformed state coincide, the symmetric tensors $\boldsymbol{V}, \boldsymbol{V}^*$, $J\boldsymbol{T}\boldsymbol{V}^{-1}$, and $J^*\boldsymbol{T}^*\boldsymbol{V}^{*-1}$ have an orthonormal basis of proper vectors in common. In conformity with (48.7) we denote the corresponding proper numbers by v_a, v_a^*, T_a, and T_a^*, respectively. We then infer from (52.3) that $(\boldsymbol{T}_R^* - \boldsymbol{T}_R)(\boldsymbol{F}^{*T} - \boldsymbol{F}^T)$ is a symmetric tensor with proper numbers $(T_a^* - T_a) \cdot (v_a^* - v_a)$. It follows that (52.1) reduces to the form (51.39) when (52.2) holds. Thus we have shown that

$$\mathrm{GCN} \Rightarrow \mathrm{GCN}_0 \tag{52.4}$$

for isotropic materials. In fact, the argument given above shows that the GCN_0 condition is the form assumed by the GCN condition when applied to a pair of coaxial strains of an isotropic material. It is clear that the implication in (52.4) cannot be reversed, since the GCN_0 condition gives no relation whatever between states of strain that are not coaxial.

Fourth, it follows from (52.1) that

$$\mathfrak{h}(\boldsymbol{S}\boldsymbol{F}) \neq \mathfrak{h}(\boldsymbol{F}) \tag{52.5}$$

for any $\boldsymbol{F}$ and any positive-definite, symmetric $\boldsymbol{S}$ other than $\boldsymbol{1}$. That is, if a pure stretch $\boldsymbol{S} \neq \boldsymbol{1}$ is superimposed upon an arbitrary state of strain, the first Piola-Kirchhoff tensor $\boldsymbol{T}_R$ cannot remain unchanged. It does not follow from (52.1) that the Cauchy stress $\boldsymbol{T}$ is necessarily changed by a pure stretch. For example, in an elastic fluid, an isochoric pure stretch leaves $\boldsymbol{T} = -p\boldsymbol{1}$ unchanged, but (52.1) may nevertheless be satisfied. For an isotropic material, it follows from (52.4) and (51.41) that GCN ⇒ IFS, but no other kind of easily interpreted invertibility has been shown to follow[1] from the GCN condition.

The GCN condition suffices to carry through the argument toward a general theorem of uniqueness, formally similar but conceptually far weaker than the usual one in the classical infinitesimal theory. As remarked by HILL[2], a formula of KIRCHHOFF's type follows at once from CAUCHY's law (43A.6) and from (43A.5) and (21.9) in the case of equilibrium:

$$\left.\begin{aligned}&\int_{\mathscr{B}} \operatorname{tr}\{(\boldsymbol{T}_R^* - \boldsymbol{T}_R)(\boldsymbol{F}^{*T} - \boldsymbol{F}^T)\}\, dv_R \\ &\quad = \int_{\mathscr{B}} \varrho_R (\boldsymbol{b}^* - \boldsymbol{b}) \cdot (\boldsymbol{x}^* - \boldsymbol{x})\, dv_R + \int_{\partial\mathscr{B}} (\boldsymbol{x}^* - \boldsymbol{x}) \cdot (\boldsymbol{t}_R^* - \boldsymbol{t}_R)\, ds_R.\end{aligned}\right\} \tag{52.6}$$

Suppose now that there are two solution-pairs, $(\boldsymbol{F}, \boldsymbol{T}_R)$ and $(\boldsymbol{F}^*, \boldsymbol{T}_R^*)$, to the same mixed boundary-value problem of place and traction for a homogeneous elastic body. Then $\boldsymbol{b}^* = \boldsymbol{b}$, and on $\partial\mathscr{B}$ we have either $\boldsymbol{x}^* = \boldsymbol{x}$ or $\boldsymbol{t}_R^* = \boldsymbol{t}_R$. From (52.6) it follows that

$$\int_{\mathscr{B}} \operatorname{tr}\{(\boldsymbol{T}_R^* - \boldsymbol{T}_R)(\boldsymbol{F}^{*T} - \boldsymbol{F}^T)\}\, dv_R = 0. \tag{52.7}$$

This condition may be satisfied in several ways: (1) the two solutions are identical, (2) the stress relation is such as to allow more than one deformation to correspond to identical stress fields, so that $\boldsymbol{F}^* \neq \boldsymbol{F}$ but $\boldsymbol{T}_R^* = \boldsymbol{T}_R$, (3) the two solutions are distinct in such a way as to give the integrand both positive and negative values and zero mean value.

The GCN condition (52.1) contradicts (52.7) if $\boldsymbol{F}^* = \boldsymbol{S}\boldsymbol{F}$, where $\boldsymbol{S}$ is positive-definite and symmetric. In order for such a relation to hold everywhere in a homogeneous body, it is necessary and sufficient, since $\boldsymbol{F}^*$ and $\boldsymbol{F}$ are both deformation gradients, that $\boldsymbol{S}$ also be a deformation gradient. Since $\boldsymbol{S}$ is symmetric,

---

[1] TRUESDELL and TOUPIN [1963, *76*, § 5] erred in asserting that GCN implies invertibility of (43.8) in a pure strain.

[2] HILL [1957, *9*, § 2]. HILL's uniqueness theorem will be given in Sect. 68bis; the theorem proved here is due to TRUESDELL and TOUPIN [1963, *76*, § 8].

the deformation $\boldsymbol{x}^* = \boldsymbol{\varphi}(\boldsymbol{x})$ which gives rise to it must be derivable from a potential: $\boldsymbol{x}^* = \boldsymbol{x} + \nabla\Phi$, where $\Phi = \Phi(\boldsymbol{x})$ is a scalar field. Such deformations are called *potential deformations* [CFT, Sects. 36, 38]. We have thus established the following ***uniqueness theorem***: *In a homogeneous elastic body such that the GCN condition holds, a given mixed boundary-value problem of place and traction cannot have two distinct solutions that differ from each other at each point by a pure potential deformation.* This type of uniqueness is consistent with the general view of buckling, since the common and familiar buckled states certainly differ from the corresponding ground states by large and non-uniform local rotations. As a trivial corollary of the above theorem, or directly from the fact that GCN$\Rightarrow$IFS, we may infer that, for a homogeneous elastic body as specified in the theorem, *a given mixed boundary-value problem cannot be satisfied by more than one homogeneous strain.*

To cast further light upon the GCN condition, we relate it to a differential inequality involving the fourth-order tensor B of elasticities (45.4). Let $\boldsymbol{D}$ be an arbitrary non-zero symmetric tensor, and define $\bar{\boldsymbol{F}}(\tau)$ by

$$\bar{\boldsymbol{F}}(\tau) = (\mathbf{1} + \tau\boldsymbol{D})\boldsymbol{F}, \quad \bar{\boldsymbol{F}}(0) = \boldsymbol{F}. \tag{52.8}$$

For small but non-vanishing values of $|\tau|$, the tensor $\mathbf{1} + \tau\boldsymbol{D}$ is symmetric, positive-definite, and not $\mathbf{1}$. Thus we may substitute $\bar{\boldsymbol{F}}(\tau)$ for $\boldsymbol{F}^*$ in (52.1), obtaining

$$\tau[f(\tau) - f(0)] > 0, \tag{52.9}$$

where

$$f(\tau) = \operatorname{tr}\{(\boldsymbol{D}\boldsymbol{F})^T \mathfrak{h}(\bar{\boldsymbol{F}}(\tau))\}. \tag{52.10}$$

The derivative $f'(\tau)$ of (52.10) may be expressed in terms of the gradient $\mathsf{A}(\bar{\boldsymbol{F}}(\tau)) = \mathfrak{h}_{\boldsymbol{F}}(\bar{\boldsymbol{F}}(\tau))$ by

$$f'(\tau) = \operatorname{tr}\{(\boldsymbol{D}\boldsymbol{F})^T \mathsf{A}(\bar{\boldsymbol{F}}(\tau))[\boldsymbol{D}\boldsymbol{F}]\}, \tag{52.11}$$

where the notation of Sect. 7 has been used. Now, if we divide (52.9) by τ^2 and take the limit as $\tau \to 0$, we obtain $f'(0) \geqq 0$. It follows from (52.11), $(52.8)_2$, and (45.4) that

$$\left.\begin{aligned} J^{-1} f'(0) &= J^{-1} \operatorname{tr}\{(\boldsymbol{D}\boldsymbol{F})^T \mathsf{A}(\boldsymbol{F})[\boldsymbol{D}\boldsymbol{F}]\} = J^{-1} \mathsf{A}_k{}^\alpha{}_p{}^\beta x^m{}_{,\alpha} x^q{}_{,\beta} D^p_q D^k_m \\ &= \mathsf{B}^{kmpq} D_{km} D_{pq} = \mathsf{B}^{(km)(pq)} D_{km} D_{pq} = \beta[\boldsymbol{D}, \boldsymbol{D}]. \end{aligned}\right\} \tag{52.12}$$

Thus the GCN condition implies the differential inequality[1]

$$\beta[\boldsymbol{D}, \boldsymbol{D}] = \mathsf{B}^{(km)(pq)} D_{km} D_{pq} \geqq 0 \tag{52.13}$$

for all symmetric tensors $\boldsymbol{D}$.

In the infinitesimal theory the classical requirement that the stress work be non-negative in every strain may be expressed alternatively in dynamic terms: The working of the stress rate cannot be negative, $\operatorname{tr}(\dot{\boldsymbol{T}}\boldsymbol{D}) \geqq 0$, where $\boldsymbol{D}$ is now the stretching tensor. Since, as shown at the beginning of this section, the GCN condition reduces in the infinitesimal theory to the usual requirement, in the motion of a finitely deformed body it may be expected to imply a corresponding dynamic statement:

$$\operatorname{tr}(\overset{*}{\boldsymbol{T}}\boldsymbol{D}) \geqq 0, \tag{52.14}$$

---

[1] For hyperelastic materials (52.13) was asserted to follow from the C-N condition by Toupin and Bernstein [1961, *62*, § 3]; for general elastic materials, by Truesdell and Toupin [1963, *76*, § 6]. The proof given here is adapted from that of Coleman and Noll [1964, *19*, § 6].

where $\overset{*}{\boldsymbol{T}}$ is some invariant stress rate. COLEMAN and NOLL[1] have shown that (52.14) follows from GCN if

$$\overset{*}{\boldsymbol{T}} = \dot{\boldsymbol{T}} + \boldsymbol{T}\boldsymbol{W} - \boldsymbol{W}\boldsymbol{T} + \boldsymbol{T}\,\mathrm{tr}\,\boldsymbol{D} - \tfrac{1}{2}(\boldsymbol{T}\boldsymbol{D} + \boldsymbol{D}\boldsymbol{T}). \tag{52.15}$$

For proof, we need only put (45.6) into (45.17) so as to express $\mathsf{B}[\boldsymbol{D}]$ in terms of $\dot{\boldsymbol{T}}$, then put the result into (52.13).

Let us call the strict inequality corresponding to (52.13), viz

$$\beta[\boldsymbol{D}, \boldsymbol{D}] = \mathsf{B}^{(km)(pq)} D_{km} D_{pq} > 0 \tag{52.16}$$

for all non-zero symmetric $\boldsymbol{D}$, the *GCN⁺ condition*. The notation is motivated by the implication[2]

$$\mathrm{GCN}^+ \Rightarrow \mathrm{GCN}. \tag{52.17}$$

In order to prove (52.17), assume that $\boldsymbol{F}$ and a symmetric positive-definite $\boldsymbol{S} \neq \boldsymbol{1}$ are given. We put $\boldsymbol{D} = \boldsymbol{S} - \boldsymbol{1}$ and define $\bar{\boldsymbol{F}}(\tau)$ and $f(\tau)$ as in (52.8) and (52.10), respectively. It is easily seen, for $0 < \tau \leqq 1$, that the tensor $\boldsymbol{1} + \tau\boldsymbol{D}$ is invertible and that $\boldsymbol{D}(\boldsymbol{1} + \tau\boldsymbol{D})^{-1}$ is symmetric. Writing the GCN⁺ inequality (52.16) with $\boldsymbol{F}$ replaced by $\bar{\boldsymbol{F}}(\tau)$ and $\boldsymbol{D}$ by $\boldsymbol{D}(\boldsymbol{1} + \tau\boldsymbol{D})^{-1}$, we obtain

$$\beta|_{\boldsymbol{F} = \bar{\boldsymbol{F}}(\tau)}[\boldsymbol{D}(\boldsymbol{1} + \tau\boldsymbol{D})^{-1}, \boldsymbol{D}(\boldsymbol{1} + \tau\boldsymbol{D})^{-1}] > 0 \tag{52.18}$$

when $0 < \tau \leqq 1$. On the other hand, it is easily seen from $(52.12)_5$ and (52.11) that the left-hand side of (52.18) differs only by the factor J^{-1} from $f'(\tau)$. Therefore, the GCN⁺ condition implies that $f'(\tau) > 0$ when $0 < \tau \leqq 1$. Integration of this inequality gives $f(1) - f(0) > 0$, which by (52.10) is just the GCN condition (52.1) with $\boldsymbol{F}^* = \boldsymbol{S}\boldsymbol{F}$. Q.e.d.

For isotropic materials we can put the GCN⁺ condition into an immediately plausible static form[3]. The condition (52.16) asserts that the quadratic form $\beta[\boldsymbol{D}, \boldsymbol{D}]$ on the 6-dimensional space $\mathscr{S}$ of symmetric tensors $\boldsymbol{D}$ is positive-definite. For an isotropic material, the principal components $\mathsf{B}^{(km)(pq)}$ of the bilinear form β on $\mathscr{S}$ can be calculated explicitly by aid of (48.16), (48.18), and (48.24). These principal components are the elements of the following 6 by 6 matrix:

$$[\beta] = \left\|\begin{matrix} \frac{v_1}{v_2 v_3}\frac{\partial T_1}{\partial v_1} & \frac{1}{v_3}\frac{\partial T_1}{\partial v_2} & \frac{1}{v_2}\frac{\partial T_1}{\partial v_3} & 0 & 0 & 0 \\ \frac{1}{v_3}\frac{\partial T_2}{\partial v_1} & \frac{v_2}{v_1 v_3}\frac{\partial T_2}{\partial v_2} & \frac{1}{v_1}\frac{\partial T_2}{\partial v_3} & 0 & 0 & 0 \\ \frac{1}{v_2}\frac{\partial T_3}{\partial v_1} & \frac{1}{v_1}\frac{\partial T_3}{\partial v_2} & \frac{v_3}{v_1 v_2}\frac{\partial T_3}{\partial v_3} & 0 & 0 & 0 \\ 0 & 0 & 0 & A_1 & 0 & 0 \\ 0 & 0 & 0 & 0 & A_2 & 0 \\ 0 & 0 & 0 & 0 & 0 & A_3 \end{matrix}\right\| \tag{52.19}$$

where

$$A_1 = \frac{(t_2 - t_3)(v_2^2 + v_3^2)}{2(v_2^2 - v_3^2)} - \frac{1}{4}(t_2 + t_3), \tag{52.20}$$

[1] COLEMAN and NOLL [1964, *19*, § 6] regard this inequality as expressing the "loi du déplacement isothermique de l'équilibre" asserted by DUHEM [1905, *1*, p. 193].

[2] For hyperelastic materials, the implication (52.17) was asserted without proof by TOUPIN and BERNSTEIN [1961, *62*, § 3], who determined some consequences of (52.16). For general elastic materials a proof of (52.17) was published by TRUESDELL and TOUPIN [1963, *76*, § 6]. The proof in the text above, here published for the first time, is due to NOLL.

[3] TRUESDELL and TOUPIN [1963, *76*, § 6].

etc. The elements of the 3 by 3 matrix in the upper left-hand corner of (52.19) are proportional to $v_a v_b J_{ab}$; that matrix, accordingly, is positive-definite if and only if the GCN_0^+ condition (51.45) holds. Thus

$$\text{GCN}^+ \Leftrightarrow \text{GCN}_0^+ \ \& \ (A_\Gamma > 0). \tag{52.21}$$

In the previous section we have analysed the GCN_0^+ condition. It is natural to ask if the inequalities $A_\Gamma > 0$ follow from it. L. BRAGG has shown us analysis proving that such is not the case. Thus a full understanding of the GCN^+ condition would require a statical interpretation of the inequalities $A_\Gamma > 0$. While we are unable to construct such an interpretation, we can go some way toward it[1]. The inequalities $A_\Gamma > 0$ are clearly equivalent to

$$(v_a - v_b)\left[(t_a - t_b)(v_a^2 + v_b^2) - \tfrac{1}{2}(t_a + t_b)(v_a^2 - v_b^2)\right] > 0 \quad \text{if} \quad a \neq b. \tag{52.22}$$

Rearranging the expression on the left-hand side and observing (48.7), we find that (52.22) is equivalent to either of the following two inequalities:

$$\left.\begin{aligned} (v_a - v_b)(t_a - t_b) &> t_b \frac{2(v_a - v_b)^2 (v_a + v_b)}{v_a^2 + 3v_b^2} \quad &\text{if} \quad a \neq b,\\ (v_a - v_b)(T_a - T_b) &> -T_b \frac{(v_a - v_b)^4}{v_a (v_a^2 + 3v_b^2)} \quad &\text{if} \quad a \neq b. \end{aligned}\right\} \tag{52.23}$$

Since (52.22) and the left-hand sides of (52.23) remain unaltered if a and b are interchanged, it follows that we can also interchange a and b on the right-hand sides of (52.23) without destroying the inequalities. Therefore, if one of t_a and t_b is non-positive and if $(v_a - v_b)(t_a - t_b) > 0$, then $(52.23)_1$ and hence (52.22) holds. Considering all possibilities for a and b, $a \neq b$, we find that $A_\Gamma > 0$, $\Gamma = 1, 2, 3$, if the B-E inequalities hold and if two of the principal stresses are non-positive. The same reasoning applied to $(52.23)_2$ shows that $A_\Gamma > 0$, $\Gamma = 1, 2, 3$, if the O-F inequalities hold and if two of the principal forces are non-negative. Now, a principal force has the same sign as the corresponding principal stress. Since there are only three of each, there must always be two that are either non-positive or non-negative. Hence if *both* the B-E and the O-F inequalities hold, we can infer that $A_\Gamma > 0$, $\Gamma = 1, 2, 3$, without further assumptions; i.e.,

$$\text{O-F} \ \& \ \text{B-E} \Rightarrow A_\Gamma > 0. \tag{52.23a}$$

The implication cannot be reversed. From (51.48), (52.21), and (52.23a) it follows that

$$\text{GCN}_0^+ \ \& \ \text{B-E} \Rightarrow \text{GCN}^+, \tag{52.24}$$

where again the implication cannot be reversed.

Returning to the complete result (52.21), we remark that the formal conditions GCN_0^+ and $A_\Gamma > 0$ deserve notice, since they afford an immediate and simple test for finding whether or not any given stress relation for isotropic materials satisfies the GCN^+ condition. The two parts of the test refer to different aspects of the stress relation. The GCN_0^+ condition, asserting that *the transformation from principal stretches to principal forces is strongly monotone*, refers to a comparison of the forms of the stress relation at pairs of points, while the independent conditions $A_\Gamma > 0$ restrict those forms at a single point. To make this latter fact particularly clear, observe from (51.28) and (52.23a) that

$$\text{E} \Rightarrow A_\Gamma > 0, \tag{52.25}$$

[1] Most of the analysis from this point through Eq. (52.32) was given by TRUESDELL and TOUPIN [1963, *76*, § 6]; the proof of implication (52.23a) in the text, here published for the first time, is due to NOLL.

and hence by (52.21) that

$$\mathrm{GCN}_0^+ \ \& \ \mathrm{E} \Rightarrow \mathrm{GCN}^+. \tag{52.26}$$

It has already been mentioned that what empirical evidence there is seems to favor the E-inequalities (51.27), which are no more than statements of sign for the coefficients in a particular representation of the stress. In a case where these simple and immediate inequalities are known to hold, one has only to test the Jacobian matrix $\|J_{ab}\|$ in order to see if the GCN$^+$ condition be satisfied.

From (52.21) and (51.48) we see that[1]

$$\mathrm{GCN}^+ \Rightarrow \begin{Bmatrix} \text{P-C} & \& \\ \text{IFS}^+ & \& \\ \text{T-E}^+ & \& \\ \text{E-T}^+ & \& \\ \text{O-F} & \end{Bmatrix} \tag{52.27}$$

(As usual, a natural state is used as the reference configuration so as to include P-C among the consequences, but the rest of the implications do not presume the existence of a natural state.) The implication cannot be reversed.

We now note some static consequences of the S-E condition (48.26) for isotropic materials. From (48.29) and (48.16), after inspecting (51.11) and (51.23), we see immediately that[2]

$$\text{S-E} \Rightarrow \text{T-E}^+ \ \& \ \text{B-E}. \tag{52.28}$$

Now in the classical infinitesimal theory of isotropic materials the S-E condition reduces to (51.2), weaker than the C-inequalities (51.1). Thus S-E$\not\Rightarrow$C. Hence

$$\text{S-E} \not\Rightarrow \text{P-C or IFS}, \tag{52.29}$$

since the C-inequalities are necessary in order that either condition on the right shall hold [cf. (51.30)]. Further statical consequences of the S-E condition are easy to derive but hard to understand[3]. The main significance of the S-E condition will appear in connection with wave propagation in Sect. 90.

At the beginning of this section we showed that the GCN condition includes and generalizes all the requirements customarily imposed upon the infinitesimal theory. We have seen above that the S-E condition does not have this property. Thus we have proved the former of the following two denials:

$$\text{S-E} \not\Rightarrow \text{GCN}, \quad \text{GCN}^+ \not\Rightarrow \text{S-E}, \tag{52.30}$$

while the latter follows[4] by considering the case of an elastic fluid, since the S-E condition is never satisfied by a fluid[5], while the GCN$^+$ condition may be.

The similarities and differences between the S-E and GCN$^+$ conditions are reflected when they are written in terms of a common differential form. We may

---

[1] The implication GCN$^+ \Rightarrow$T-E$^+$ is equivalent to a theorem on wave speeds asserted by TOUPIN [1961, *61*, Th. XII], proved by TRUESDELL [1961, *63*, § 10], and given below in Sect. 74.

[2] TRUESDELL [1956, *32*, § 8] attributed an equivalent result to ERICKSEN and TOUPIN, for hyperelastic materials. The second part of the implication was rediscovered by HAYES and RIVLIN [1961, *24*, § 6].

[3] Some examples were given by ZORSKI [1964, *98*, § 2]. An interpretation in terms of linear velocity fields was found by HILL [1962, *34*, § 4 (i)].

[4] Alternatively, one may use a counter-example given by TOUPIN and BERNSTEIN [1961, *62*, § 3].

[5] By (52.28), S-E$\Rightarrow$B-E, but the B-E inequalities can never be satisfied by an elastic fluid, since $t_1 = t_2 = t_3$ always.

summarize both (52.16) and (45.14) as follows:

$$B^{kmpq} H_{km} H_{pq} > 0, \tag{52.31}$$

where it is only the domain of non-vanishing tensors $\boldsymbol{H}$ that differs in the two cases, namely,

$$\left.\begin{array}{rl} \text{GCN}^+: & \text{all symmetric } \boldsymbol{H} \\ \text{S-E}: & \text{all } \boldsymbol{H} \text{ of rank } 1. \end{array}\right\} \tag{52.32}$$

The effects of the GCN, GCN$^+$, and S-E conditions, and various of their corollaries derived in this section, will be considered again and again in the following pages. While there remains uncertainty as to what is the true general condition to be imposed, some restriction of one or both of these types seems necessary if the theory of elasticity is to yield results physically acceptable when applied to materials ordinarily regarded as "solids" or "fluids".

While the GCN condition seems to be the most plausible so far proposed, a disturbing consequence of it has been noticed by BRAGG and COLEMAN[1]. Let $\bar{p}$ be the mean pressure, $\bar{p} \equiv -\frac{1}{3} \operatorname{tr} \boldsymbol{T}$, and consider its value in a uniform dilatation of the configuration in which the deformation gradient is $\boldsymbol{F}$. By (43A.3) and (43.8),

$$-3\bar{p}(\nu \boldsymbol{F}) J = \frac{1}{\nu^3} \operatorname{tr} [\mathfrak{h}(\nu \boldsymbol{F}) \nu \boldsymbol{F}^T] = \frac{1}{\nu^2} \operatorname{tr} [\mathfrak{h}(\nu \boldsymbol{F}) \boldsymbol{F}^T], \tag{52.33}$$

where $J \equiv |\det \boldsymbol{F}|$. If $\nu^* \neq \nu$, we may apply (52.1) with $\boldsymbol{F}$ and $\boldsymbol{F}^*$ replaced by $\nu \boldsymbol{F}$ and $\nu^* \boldsymbol{F}$, respectively. By (52.33), it follows that

$$-(\nu^* - \nu)[\nu^{*2} \bar{p}(\nu^* \boldsymbol{F}) - \nu^2 \bar{p}(\nu \boldsymbol{F})] > 0. \tag{52.34}$$

This inequality asserts that $-\nu^2 \bar{p}(\nu \boldsymbol{F})$ is a strictly increasing function of ν, for every fixed $\boldsymbol{F}$. The quantity

$$k(\boldsymbol{F}) \equiv -\left.\frac{\partial}{\partial \alpha} \bar{p}(\alpha^{\frac{1}{3}} \boldsymbol{F})\right|_{\alpha=1} \tag{52.35}$$

reduces in the case of an elastic fluid to the quantity called the compressibility, $\varrho\, d\bar{p}/d\varrho$. From what was just established, it follows that

$$\frac{\partial}{\partial \alpha} [\alpha^{\frac{2}{3}} \bar{p}(\alpha^{\frac{1}{3}} \boldsymbol{F})] \leqq 0 \tag{52.36}$$

if $\alpha > 0$. Writing out this condition when $\alpha = 1$ and using the definition (52.35) yields

$$k(\boldsymbol{F}) \geqq \frac{2}{3} \bar{p}(\boldsymbol{F}), \tag{52.37}$$

which is BRAGG and COLEMAN's result. In the special case of an elastic fluid, it reduces to the form[2]

$$\varrho \frac{dp}{d\varrho} \geqq \frac{2}{3} p. \tag{52.38}$$

We conclude this section by discussing some limitations that the GCN condition places on the possible states of hydrostatic pressure in an elastic material[3]. Consider two local configurations which both correspond to hydrostatic pressures:

$$\boldsymbol{T}_0 = -p_0 \boldsymbol{1}, \quad \boldsymbol{T} = -p \boldsymbol{1}. \tag{52.39}$$

Take the first of these configurations as the reference configuration, and denote the deformation gradient of the second by $\boldsymbol{F} = \boldsymbol{R}\boldsymbol{U}$. The first Piola-Kirchhoff

---

[1] BRAGG and COLEMAN [1963, *11*].
[2] COLEMAN [1962, *9*, § 4]. Cf. also COLEMAN and NOLL [1959, *3*, § 11].
[3] COLEMAN and NOLL [1964, *19*, § 5].

tensors corresponding to (52.39) are, by $(43\text{A}.11)_1$, (52.39), and (43.8)

$$\boldsymbol{T}_{\mathrm{R}\,0}=\boldsymbol{T}_0=-p_0\mathbf{1}=\mathfrak{h}(\mathbf{1}), \tag{52.40}$$

$$\boldsymbol{T}_{\mathrm{R}}=J\,\boldsymbol{T}(\boldsymbol{F}^{-1})^T=-p\,J\,(\boldsymbol{F}^{-1})^T=\mathfrak{h}(\boldsymbol{F}). \tag{52.41}$$

By $(43.10)_1$ we infer from (52.41) that

$$\mathfrak{h}(\boldsymbol{U})=-p\,J\,\boldsymbol{U}^{-1}=-p\,(\det \boldsymbol{U})\,\boldsymbol{U}^{-1}. \tag{52.42}$$

If we replace $\boldsymbol{F}^*$ and $\boldsymbol{F}$ in (52.1) by $\boldsymbol{U}$ and $\mathbf{1}$, respectively, we obtain from (52.40) and (52.42)

$$p_0(\operatorname{tr}\boldsymbol{U}-3)+p\det\boldsymbol{U}(\operatorname{tr}\boldsymbol{U}^{-1}-3)>0 \tag{52.43}$$

if $\boldsymbol{U}\neq\mathbf{1}$.

Consider first the case when the two states for which (52.39) holds have the same density, so that $\det\boldsymbol{F}=\det\boldsymbol{U}=\det\boldsymbol{U}^{-1}=1$. The arithmetic mean of a set of positive numbers is strictly greater than the geometric mean unless the numbers are all equal to one another. Applying this result to the proper numbers of $\boldsymbol{U}$ and $\boldsymbol{U}^{-1}$, whose geometric mean is equal to 1, we obtain

$$\operatorname{tr}\boldsymbol{U}-3>0, \qquad \operatorname{tr}\boldsymbol{U}^{-1}-3>0, \tag{52.44}$$

whenever $\boldsymbol{U}\neq\mathbf{1}$. Now if $p_0\leqq 0$ and $p\leqq 0$, the inequalities (52.43) and (52.44) are inconsistent. Therefore, $p_0\leqq 0$ and $p\leqq 0$ are possible only if $\boldsymbol{U}=\mathbf{1}$, in which case $\boldsymbol{F}=\boldsymbol{R}$, and hence is orthogonal. Substituting this result into (52.41) and using the fact that $\mathfrak{h}(\boldsymbol{R})=\boldsymbol{R}\,\mathfrak{h}(\mathbf{1})$, we find that we must also have $p=p_0$. Thus we have proved COLEMAN and NOLL's ***fundamental theorem on states of hydrostatic tension***: *If the GCN condition holds, then two states of strain with the same density can correspond to hydrostatic tensions only if these states differ only by a rotation and if the tensions are the same in each.* As a corollary, it follows that

$$-p\,\mathbf{1}=\mathfrak{g}(\mathbf{1})=\mathfrak{g}(\boldsymbol{H}), \tag{52.45}$$

where $\mathfrak{g}$ is the response function of (43.1), is possible for $p\leqq 0$ only if $\boldsymbol{H}$ is orthogonal. Thus, if $\boldsymbol{H}$ is unimodular but not orthogonal, then (50.4) holds with $\boldsymbol{F}=\mathbf{1}$, and the material must be solid. Therefore, *if the GCN condition holds, and if the material has a natural state or a state of hydrostatic tension, then the material must be an elastic solid, and every natural state or state of hydrostatic tension must be undistorted.* This theorem shows that if the GCN condition holds, NOLL's concept of a solid as a material having "undistorted states", for which the isotropy group is contained in the orthogonal group, subsumes the older concept of a solid as a material having a stress-free configuration. As a corollary to this theorem, the GCN condition requires that *in an elastic fluid the stress must always be a positive hydrostatic pressure*[1].

Other aspects of the GCN condition are discussed in Sect. 87.

It is possible to draw a conclusion from the inequality (52.43) also in the case when p and p_0 are equal *positive* pressures; i.e. when $p=p_0>0$. For this case, (52.43) is equivalent to

$$g\equiv(\operatorname{tr}\boldsymbol{U}-3)+\det\boldsymbol{U}(\operatorname{tr}\boldsymbol{U}^{-1}-3)>0. \tag{52.46}$$

Since the proper numbers of $\boldsymbol{U}$ are the principal stretches, v_a, we have

$$g\equiv(v_1+v_2+v_3-3)+(v_2v_3+v_1v_3+v_1v_2-3v_1v_2v_3). \tag{52.47}$$

A simple analysis[2] shows that $g\leqq 0$ if the principal stretches v_a obey the inequalities

$$v_a\geqq 1 \quad \text{for} \quad a=1,2,3 \tag{52.48}$$

or

$$v_a\leqq 1 \quad \text{for} \quad a=1,2,3. \tag{52.49}$$

---

[1] COLEMAN and NOLL [1959, *3*, § 11].
[2] COLEMAN and NOLL [1964, *19*, § 5].

Thus the inequalities (52.48) or (52.49) are inconsistent with (52.46). Hence, *if the GCN condition holds, then extensions in three perpendicular directions or compressions in three perpendicular directions cannot be effected under constant, positive hydrostatic pressure.* This result applies in particular to uniform dilatations and compressions and is a complement to the implication GCN ⇒ P-C for isotropic elastic materials having a natural state.

53. Restrictions upon the response functions, III. Incompressible materials. For incompressible materials the requirement of isochoric deformation and its consequence, the presence of the arbitrary hydrostatic pressure in the stress relations (43.14), render inappropriate nearly all the considerations of the last two sections. The P-C inequality (51.4) is wrong because change of volume is impossible, the T-E inequality (51.10) cannot be applied because it is impossible to hold v_2 and v_3 fixed while varying v_1; and more general conditions such as (51.42), since they are not invariant under change of the arbitrary hydrostatic pressure, are meaningless.

Of all conditions stated in Sect. 51, the only one free from the foregoing objections is given by the B-E inequalities (51.22). The equivalent conditions (51.26) may be derived in the same way, but in view of the fact that $v_1 v_2 v_3 = 1$ they assume the following simpler forms[1]:

$$\left.\begin{aligned} \beth_1 &> v_b^2 \beth_{-1} \quad \text{if} \quad v_a \neq v_c, \\ \beth_1 &\geqq v_b^2 \beth_{-1} \quad \text{if} \quad v_a = v_c, \end{aligned}\right\} \tag{53.1}$$

where a, b, c are any permutation of 1, 2, 3. By combination of these inequalities we infer that

$$\left.\begin{aligned} &\beth_1 \geqq \tfrac{1}{3} I_{\boldsymbol{B}} \beth_{-1} \geqq \beth_{-1}, \\ &I_{\boldsymbol{B}} \beth_1 \geqq II_{\boldsymbol{B}} \beth_{-1} \geqq 3 \beth_{-1} \\ &II_{\boldsymbol{B}} \beth_1 \geqq 3 \beth_{-1}. \end{aligned}\right\} \tag{53.2}$$

By inspection of (53.1) we see that the conditions

$$\beth_1 > 0, \quad \beth_{-1} \leqq 0 \tag{53.3}$$

are sufficient that (53.1) hold. Strong experimental evidence for rubber (see Sects. 55 and 57) supports (53.3), which we shall call the *E-inequalities*[2]. The result derived just above may be stated symbolically as

$$\text{E} \Rightarrow \text{B-E}. \tag{53.4}$$

For incompressible materials no fundamental studies of inequalities, comparable to those presented for compressible materials in the last two sections, seem to have been undertaken.

b) Exact solutions of special problems of equilibrium[3].

54. Homogeneous strain of compressible elastic bodies. In a body subject to homogeneous strain, all measures of strain and rotation are constant over the body. While the geometry of homogeneous strains is analysed in detail in CFT,

---

[1] The former inequality, for hyperelastic materials, was obtained by TRUESDELL [1952, *20*, § 41] on the basis of an energy argument. His inference that $(53.3)_2$ follows from (53.1) is false. Earlier RIVLIN and SAUNDERS [1951, *14*, § 16] had obtained a weaker inequality. Cf. also the discussion of TRUESDELL [1956, *32*].

[2] For hyperelastic materials, they were offered as a conjecture by TRUESDELL [1952, *20*, § 41].

[3] The following exposition is complete for special problems we consider to have broad interest. Subsidiary work is cited in the footnotes. See also footnote 2, p. 122.

Sects. 42 to 46, here one need only remark, observing (43.1) or any of its numerous alternative forms such as (43.6), that in a homogeneous body div $\boldsymbol{T}=\boldsymbol{0}$, so that CAUCHY's first law (16.6) is satisfied when $\boldsymbol{b}=\boldsymbol{0}$, $\ddot{\boldsymbol{x}}=\boldsymbol{0}$. That is, *any homogeneous elastic body, whether or not it be isotropic, may be maintained in equilibrium in any desired state of homogeneous strain by application of suitable tractions alone.*

For a given deformation to be possible, as is any homogeneous strain, in *every* kind of homogeneous elastic body, is an altogether singular property. In Sect. 91 we shall see that even for hyperelastic materials, the only such deformations are homogeneous. It follows *a fortiori* that *other than homogeneous deformations, there are no deformations possible in all kinds of homogeneous elastic bodies free of body force.* This negative result still holds when only isotropic materials are considered. In other words, when two bodies of the same shape but of different elastic materials are held in equilibrium subject to like displacements upon their boundaries, some of the corresponding interior deformations certainly will be different unless the boundary values happen to be consistent with a homogeneous strain. Stated in this way, the result is not surprising at all.

In the classical infinitesimal theory of elasticity, for example, it is plain from the fact that the Poisson ratio occurs as a coefficient in NAVIER's statical equation that the solution of a boundary-value problem of place generally will depend upon that ratio. There exist, however, certain exceptional deformations, all the St. Venant torsion solutions being examples, that are possible in every infinitesimally elastic material[1], whatever its elastic moduli, even though, as is obvious from (43.6), the stresses required to effect one of these deformations vary from one material to another. In the finite theory it is only natural to expect this special class of deformations to be much more restricted, and for compressible materials it turns out to be the class of homogeneous strains, as already stated. However, we shall see in Sect. 57 that for incompressible materials, there are several important non-homogeneous deformations possible for all isotropic materials and some aeolotropic ones.

Little is known about solutions for inhomogeneous elastic bodies. For them, even homogeneous deformation need not always be possible subject to surface tractions alone.

Despite its rare and degenerate nature, the simple exact solution for homogeneous strain of compressible materials is deeply illuminating upon the nature of finite elastic strain. Not going into the bewildering complications of aeolotropy[2], we consider here only homogeneous isotropic bodies[3]. We employ the stress relation (47.9), and for most of the results it is presumed without further remark that a natural state exists and is taken as the reference configuration.

In this and in the following section we employ Cartesian co-ordinates x, y, z. The matrix of a tensor $\boldsymbol{A}$ with respect to these co-ordinates is denoted by $[\boldsymbol{A}]$. The components of the stress tensor $\boldsymbol{T}$ are denoted by $T\langle xx\rangle$, $T\langle xy\rangle$, etc.

---

[1] When $\boldsymbol{b}=\boldsymbol{0}$, NAVIER's statical equation is

$$(1-2\sigma)\,\Delta\boldsymbol{u}+\operatorname{grad}\operatorname{div}\boldsymbol{u}=\boldsymbol{0},$$

σ being POISSON's ratio and Δ the Laplacean operator. According to the infinitesimal theory, then, the displacement fields that can be produced in *any* homogeneous isotropic body by applying suitable surface tractions are those for which

$$\Delta\boldsymbol{u}=\boldsymbol{0} \quad \text{and} \quad \operatorname{grad}\operatorname{div}\boldsymbol{u}=\boldsymbol{0}.$$

[2] Some discussion of homogeneous strain in aeolotropic materials is given by GREEN and ADKINS [1960, *26*, § 27].

[3] We generalize slightly the results of RIVLIN [1948, *12*, § 6].

For a *simple extension* with stretch v along the z-direction, by Eq. (44.1) of CFT we have for the left Cauchy-Green tensor $\boldsymbol{B}$

$$[\boldsymbol{B}]=\operatorname{diag}(\alpha^2 v^2, \alpha^2 v^2, v^2), \tag{54.1}$$

where α is a constant to be determined, so that by (47.9) follows

$$\left.\begin{aligned} T\langle x x\rangle = T\langle y y\rangle &= \tilde{\beth}_0 + \alpha^2 v^2 \tilde{\beth}_1 + \alpha^{-2} v^{-2} \tilde{\beth}_{-1}, \\ T\langle z z\rangle &= \tilde{\beth}_0 + v^2 \tilde{\beth}_1 + v^{-2} \tilde{\beth}_{-1}, \\ \tilde{\beth}_\Gamma(v^2, \alpha^2) &\equiv \beth_\Gamma\left((1+2\alpha^2) v^2, (2+\alpha^2)\alpha^2 v^4, \alpha^4 v^6\right), \quad \Gamma = 0, \pm 1. \end{aligned}\right\} \tag{54.2}$$

If we consider a block whose lateral faces $x=$const. and $y=$const. are to be free of traction, we must satisfy the equation

$$\tilde{\beth}_0 + \alpha^2 v^2 \tilde{\beth}_1 + \alpha^{-2} v^{-2} \tilde{\beth}_{-1} = 0. \tag{54.3}$$

Use of (50.13) shows that the derivatives of the left-hand side of (54.3) with respect to α^2 and v^2, evaluated at $\alpha^2 = v^2 = 1$, are given by $\lambda+\mu$ and $\frac{3}{2}\lambda+\mu$, respectively. By the implicit function theorem it follows that if the response coefficients $\beth_\Gamma$ are continuously differentiable near $\boldsymbol{B}=\boldsymbol{1}$ and if $\lambda+\mu\neq 0$, a solution α of (54.3) exists if v^2 is sufficiently near to 1, and that this solution is consistent with the classical result of the infinitesimal theory, viz

$$\frac{1-\alpha v}{v-1} = \sigma + O(v-1), \quad \text{where} \quad \sigma \equiv \frac{\lambda}{2(\lambda+\mu)}, \tag{54.4}$$

the classical Poisson modulus. If the E-T inequalities (51.16) are satisfied, (54.3) has a unique solution T_1 corresponding to an arbitrary stretch v, since then $v = \bar{v}_1(T_1, 0, 0)$ is a monotonically increasing function of T_1. The contraction ratio α is then given by

$$\alpha = v^{-1}\bar{v}_2(T_1, 0, 0) = v^{-1}\bar{v}_3(T_1, 0, 0).$$

The solution α may be substituted into $(54.2)_2$ so as to yield a definite stress-stretch relation $T\langle zz\rangle = \varphi(v)$ for simple extension. The existence of a unique stress-stretch relation of this form does not imply, however, the uniqueness of solution to the corresponding boundary-value problem. Even if only one homogeneous simple extension is compatible with the assigned uniform end loading $T\langle zz\rangle$, it is possible, as in problems of buckling, that several non-homogeneous strains may also correspond to that same load. The function φ in the relation $T\langle zz\rangle = \varphi(v)$ for simple extension affords one possible generalization of the concept of modulus of extension, E, of the infinitesimal theory, since $E = \varphi'(1)$.

We now consider uniform dilatation (CFT, Sect. 43). We see at once from (47.9) that *only hydrostatic stress can produce uniform dilatation, and if the dilatational stretch is v, the corresponding pressure is given by*

$$p = \tilde{\varkappa}(v), \tag{54.5}$$

where

$$\tilde{\varkappa}(v) \equiv -\beth_0(3v^2, 3v^4, v^6) - v^2 \beth_1(3v^2, 3v^4, v^6) - v^{-2}\beth_{-1}(3v^2, 3v^4, v^6). \tag{54.6}$$

Hence it is clear that for any given function $\tilde{\varkappa}(v)$, there are infinitely many different isotropic elastic materials such that (54.5) holds in dilatation. That is, *any law of static compression is consistent with the theory of isotropically elastic materials*[1]. The physically natural requirement that pressure is needed in order to effect compression from the natural state, but tension to effect expansion, has already been expressed in the generalized forms (51.4), (51.9), etc.

The most striking characteristics of finite elastic deformation are revealed in simple shear[2] (CFT, Sect. 45). A shear of amount K is given by $x = X + KY$, $y = Y$, $z = Z$; by (47.9) we have then

$$\left.\begin{aligned}[\boldsymbol{T}] = (\hat{\beth}_0 + \hat{\beth}_1 + \hat{\beth}_{-1})\,[\mathbf{1}] + (\hat{\beth}_1 - \hat{\beth}_{-1})\,K \begin{Vmatrix} 0 & 1 & 0 \\ \cdot & 0 & 0 \\ \cdot & \cdot & 0 \end{Vmatrix} + \\ + \hat{\beth}_1 K^2 \begin{Vmatrix} 1 & 0 & 0 \\ \cdot & 0 & 0 \\ \cdot & \cdot & 0 \end{Vmatrix} \qquad + \hat{\beth}_{-1} K^2 \begin{Vmatrix} 0 & 0 & 0 \\ \cdot & 1 & 0 \\ \cdot & \cdot & 0 \end{Vmatrix},\end{aligned}\right\} \tag{54.7}$$

where

$$\hat{\beth}_\Gamma(K^2) \equiv \beth_\Gamma(3 + K^2, 3 + K^2, 1), \qquad \Gamma = 0, \pm 1. \tag{54.8}$$

This result will now be interpreted.

First, to consider the relation between shear stress and the amount of shear, put

$$\left.\begin{aligned}\hat{\mu}(K^2) &\equiv \hat{\beth}_1(K^2) - \hat{\beth}_{-1}(K^2), \\ &= \beth_1(3 + K^2, 3 + K^2, 1) - \beth_{-1}(3 + K^2, 3 + K^2, 1).\end{aligned}\right\} \tag{54.9}$$

By $(50.13)_4$, $\hat{\mu}(0) = \mu$, the shear modulus of the infinitesimal theory, and in finite simple shear

$$T\langle xy\rangle / K = \hat{\mu}(K^2), \tag{54.10}$$

so that $\hat{\mu}(K^2)$ is a generalized shear modulus. That $\hat{\mu}$, as defined by (54.9), is an even function of K manifests itself in (54.10) as the physically obvious fact that the *shear stress is an odd function of the amount of shear.* In case $\hat{\mu}$ is differentiable at $K^2 = 0$, *any departure from the classical linear proportionality of shear stress to shear strain is necessarily an effect of third or higher odd order in the amount of shear.*

The experimental fact that shear stress remains proportional to shear strain for strains of magnitude far greater than the limit of linearity for simple extension has long been known, but until fairly recently it was taken as justifying some particular, special theory of elasticity[3], rather than recognized as an inevitable consequence of the laws of mechanics when applied to elasticity.

It is well known [CFT, Sect. 45] that the principal stretches in a simple shear of amount K are given by the formulae

$$\left.\begin{aligned} v_1^2 &= 1 + \tfrac{1}{2} K^2 + K\sqrt{1 + \tfrac{1}{4} K^2}, \\ v_2^2 &= 1 + \tfrac{1}{2} K^2 - K\sqrt{1 + \tfrac{1}{4} K^2} = \frac{1}{v_1^2}, \\ v_3^2 &= 1. \end{aligned}\right\} \tag{54.11}$$

---

[1] In particular, there is no foundation for the widespread belief that according to the theory of elasticity, pressure and tension have equal but opposite effects. Such a symmetry requires that $\tilde{\varkappa}(1/v) = -\tilde{\varkappa}(v)$; if $\tilde{\varkappa}$ is differentiable at $v = 1$, this relation always holds when $v - 1$ is small, and for certain special elastic materials it holds for all v, but there is no reason for expecting it to be valid generally.

[2] RIVLIN [1948, *12*, § 13], for hyperelastic materials, cf. also the discussion of REINER [1948, *9*, § 5]. A superposition principle for the second-order effects in a succession of shears was derived from (54.7) by TRUESDELL [1952, *20*, § 42G].

[3] Cf., e.g., MOONEY [1940, *4*, p. 582]. That the theory of MOONEY, described below in Sect. 95, covers fairly well the large deformation of various rubbers must be regarded as a fact of *experiment*, independent of the formal procedures originally and subsequently introduced so as to infer it mathematically.

By substituting these expressions into $(48.1)_4$ and then using (54.9) we conclude that[1]

$$t_1 - t_2 = 2K\sqrt{1 + \tfrac{1}{4} K^2 \hat{\mu}(K^2)}, \tag{54.12}$$

or

$$\hat{\mu}(K^2) = \frac{t_1 - t_2}{v_1^2 - v_2^2}. \tag{54.13}$$

From (51.22) we see that *if the B-E inequalities hold, the shear modulus $\hat{\mu}$ is a positive function.* Conversely, however, a positive shear modulus does not suffice for the truth of the B-E inequalities, since the magnitude of t_3, the stress corresponding to the stretch $v_3 = 1$, is unaffected by the sign of $\hat{\mu}$. Rather, *the shear modulus is positive if and only if, in simple shear, the greater of the principal stresses in the plane of shear occurs in the direction of the greatest principal stretch*[2]. The physically natural condition that the shear stress point in the same direction as the shear effected, expressed by the inequality

$$\hat{\mu}(K^2) > 0, \tag{54.14}$$

can be shown *not* to follow, for all values of K, from the GCN or GCN$^+$ conditions, although it *does* follow from the S-E condition, and also, by (51.28), from the E-inequalities.

From (54.9) it is clear that infinitely many different isotropically elastic materials correspond to any given relation $T\langle xy\rangle = f(K)$, where f is an odd function, for simple shear.

In the infinitesimal theory, shear stress suffices to produce shear, but a glance at (54.7) shows that *this simple property can never hold exactly, for any isotropic elastic material.* In fact, *normal stresses, negligible only in small shears, are required* if a simple shear is to be effected[3]. These normal stresses are given by the first, third and fourth terms in (54.7). In order for these to vanish exactly, it is necessary that $\hat{\beth}_1 = \hat{\beth}_{-1} = \hat{\beth}_0 = 0$, whence follows $\hat{\mu} = 0$, so that the material is highly degenerate, having a natural state and yet being susceptible of simple shear subject to no stress at all. In particular, such a material violates nearly all the inequalities laid down as plausible in Sects. 51–52: P-C, IFS, E-T, O-F, B-E, etc. This proves the exceptional quality of the infinitesimal theory in its results

---

[1] TRUESDELL [1965, *36*].

[2] TRUESDELL [1965, *36*] has investigated conditions under which there exists a value K_* of K such that $\hat{\mu}(K_*^2) = 0$. At such a value of K, by (47.9),

$$\frac{t_1 - t_3}{v_1^2 - 1} = \frac{1}{v_1^2}(v_1^2 - 1)\,\hat{\beth}_1(K_*^2),$$

$$\frac{t_2 - t_3}{v_2^2 - 1} = -(v_1^2 - 1)\,\hat{\beth}_1(K_*^2).$$

Unless $\hat{\beth}_1(K_*^2) = \hat{\beth}_{-1}(K_*^2) = 0$, the expressions on the right-hand side are of opposite sign. Consequently, if $\hat{\mu}(K_*^2) = 0$, then either the stress is hydrostatic when $K = K_*$ or else the B-E inequalities have been violated at some lesser amount of shear.

For discussion of the possible surfaces of discontinuity associated with this phenomenon, see Sect. 74.

Since $\hat{\mu}(0) > 0$, it follows that if there exists an amount of shear K_* such that $\hat{\mu}(K_*^2) = 0$, at some lesser amount of shear the shear stress $T\langle xy\rangle$ as a function of K shall have experienced a maximum or point of inflection. Examples adduced by TRUESDELL [1965, *36*] show that no general rule connecting discontinuity surfaces with a maximum value of the shear stress can be expected.

[3] RIVLIN [1948, *15*, § 4] mentioned that from failure to supply the right surface tractions, many so-called shear mountings do not produce a state of simple shear in the material.

We may remark from (54.13) that $\hat{\mu}(K^2)$ is the ratio of an extremal shear stress to an extremal value of a shear component of $\boldsymbol{C}$ (CFT, Sect. App. 46). Of course, the appearance of a particular measure of strain reflects our choice of K as the particular measure of finite shear to be used in defining $\hat{\mu}$.

regarding shear. The infinitesimal theory does not show us even how a particular, special sort of elastic material *may* behave. The phenomenon of isotropic elastic simple shear, for large shear, is different not only in the form of the stress-strain relations *but also in kind* from its counterpart in the infinitesimal theory.

We now analyse the normal stresses which must be supplied. By (54.7), these are

$$\left.\begin{aligned} T\langle zz\rangle/K^2 &= \tau(K^2) \equiv (\hat{\beth}_0+\hat{\beth}_1+\hat{\beth}_{-1})/K^2,\\ (T\langle xx\rangle - T\langle zz\rangle)/K^2 &= \hat{\beth}_1,\\ (T\langle yy\rangle - T\langle zz\rangle)/K^2 &= \hat{\beth}_{-1}.\end{aligned}\right\} \tag{54.15}$$

From (50.9) we see that if $\hat{\beth}_0+\hat{\beth}_1+\hat{\beth}_{-1}$ is differentiable at $K=0$, the function $\tau(K^2)$ is finite at $K=0$. That the moduli τ, $\hat{\beth}_1$, $\hat{\beth}_{-1}$ are all even functions of K reflects the fact, physically obvious, that if the direction of shear is reversed, the normal forces required to effect the shear are unchanged. From it follows the centrally important conclusion that while departures from the linear relation between shear stress and shear strain are generally of third order in the amount of shear, *the second order effects in shear are normal stresses*, not present at all in the infinitesimal theory. *The effects of non-linearity manifest themselves by a different quality of response at shears not large enough to indicate noticeable change in the classical quantity.*

Since, in general, all three normal stresses are different from zero, normal forces must act upon all faces of a block in order to maintain it in a state of simple shear. Unless these normal forces be supplied, it is natural to presume that a cube of the material if subjected to shear stress alone upon its faces will tend to contract or expand according as the mean tension given by Eq. (54.15), namely, $[\tau+\frac{1}{3}(\hat{\beth}_1+\hat{\beth}_{-1})]K^2$, is positive or negative. Since this conclusion follows from considering finite strain according to any possible stress relation for an isotropic elastic material, it is not surprising that KELVIN and TAIT[1] remarked upon it long before the general equations of Sect. 47 had been derived: "But it is possible that a distorting stress may produce, in a truly isotropic solid, condensation or dilatation in proportion to the square of its value: and it is probable that such effects may be sensible in India rubber, or cork, or other bodies susceptible of great deformations or compressions, with persistent elasticity." Accordingly, effects associated with the presence of a hydrostatic stress which vanishes in the corresponding deformation according to the infinitesimal theory are called *Kelvin effects*. There seems to be no indication of any general rule of sign concerning them.

Equally present in any isotropically elastic material is the *inequality of the normal stresses* indicated by $(54.15)_{3,4}$. For if $T\langle xx\rangle = T\langle yy\rangle$, it follows that $\hat{\beth}_1-\hat{\beth}_{-1}=0$, whence $\hat{\mu}=0$ by (54.9), so that again the material is of a highly degenerate kind. The necessity of these unequal normal pressures in order to effect a simple shear suggests that unless they be supplied, an initially cubical specimen when subjected to shear force will tend to dilate unequally. On the basis of a special theory of elasticity, this phenomenon was noticed by POYNTING, who then by a sequence of classical experiments verified its occurrence in nature[2].

[1] KELVIN and TAIT [1867, *2*, footnote to § 679].

[2] POYNTING [1905, *3*, p. 338] [1909, *2*, §§ I—II] [1913, *2*, pp. 397—412, 415—418]. The experiments of POYNTING refer to torsion rather than simple shear, but POYNTING was not in possession of any solution for non-linear torsion, let alone the general one given in Sect. 57. POYNTING's interpretation of results on torsion in terms of the theory of simple shear has a heuristic basis in the geometry of the two problems, as explained in CFT, Sect. 48, but such neglect of the dynamical conditions, unfortunately a common practice in "physical" reasoning, can easily lead to error. A rigorous relation between torsion and shear is derived in Sect. 57, Case 3a, below.

Accordingly, the existence of the unequal normal pressures is called the *Poynting effect.*

In the general theory, the meaning of the three *moduli of normal stress* may be read off from (54.15). The modulus τ determines the tension which must be applied to the planes of shear (cf. CFT, Sect. 45, Fig. 7); $\hat{\beth}_{-1}$, the amount by which the tension on the shearing planes must exceed it; $\hat{\beth}_{1}$, the amount by which the tension upon the planes normal to the direction of shear must exceed it. Two special cases, distinguished by qualitative differences, are determined by any one of the following equivalent conditions:

$$\text{Case 1:}\quad T\langle xx\rangle = T\langle zz\rangle,\quad \hat{\beth}_{1}=0,\ \hat{\mu}=-\hat{\beth}_{-1},\quad T\langle zz\rangle - T\langle yy\rangle = K\,T\langle xy\rangle$$

$$\text{Case 2:}\quad T\langle yy\rangle = T\langle zz\rangle,\quad \hat{\beth}_{-1}=0,\ \hat{\mu}=\hat{\beth}_{1},\quad T\langle xx\rangle - T\langle zz\rangle = K\,T\langle xy\rangle .$$

In general, the three moduli of normal stresses are independent of each other, but the shear modulus is determined from them by the relation (54.9). In fact, from (54.7) we see that[1]

$$T\langle xx\rangle - T\langle yy\rangle = K\,T\langle xy\rangle . \qquad (54.16)$$

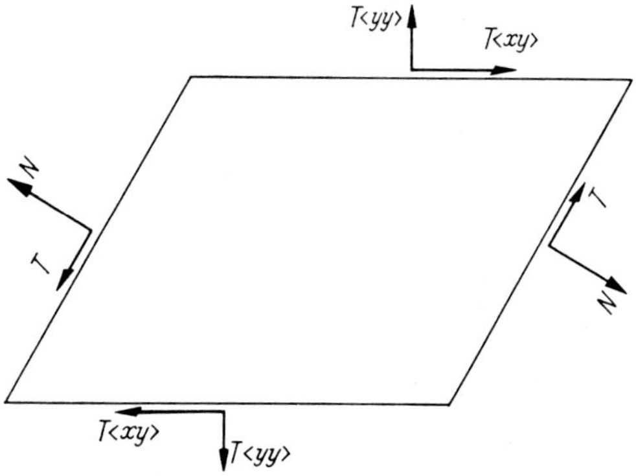

Fig. 2. Tractions on the faces of a sheared block.

This relation is characteristic of isotropic elasticity in that it need not hold for anisotropic materials or for materials that are not elastic, but it is a *universal relation* in that it pertains to all kinds of isotropic elastic materials in simple shear, irrespective of their response functions $\beth_{\Gamma}$, which need not be determined in order to test it. We have seen above that the normal stresses cannot all be equal, but from (54.16) we infer the stronger result that *in an isotropically elastic body in simple shear, it is impossible*[2] *that all planes parallel to the axis of shear sustain equal tension,* since $T\langle xx\rangle = T\langle yy\rangle$ implies that $T\langle xy\rangle = 0$. Thus any *isotropically elastic body, no matter what its response functions, exhibits the Poynting effect*[3]. In the special materials such that $\hat{\beth}_{-1}=0$, we have $T\langle yy\rangle = T\langle zz\rangle$; it is thus not impossible that all planes parallel to the direction of shear suffer equal normal tensions.

The stresses acting upon the planes $x=$const are less accessible to measurement in that the particles comprising them change as K increases, so that if they are to be made available by cutting away material upon one side, a different specimen has to be used for each K. So as to follow the tractions acting upon the faces of a single specimen (Fig. 2), we calculate the tangential and normal tensions T and N acting upon the faces $X=$const., which in the deformed state occupy the slanted planes $x-Ky=$const. Since the unit normal $\boldsymbol{n}$ and unit tangent $\boldsymbol{\tau}$ have the components

$$n_x = \pm(1+K^2)^{-\frac{1}{2}},\quad n_y = \mp K(1+K^2)^{-\frac{1}{2}},\quad \tau_x = -n_y,\quad \tau_y = n_x,$$

[1] Rivlin [1948, *15*, Eq. (12)], for incompressible hyperelastic materials. Eq. (54.16) is a necessary condition that $\boldsymbol{T}$ and $\boldsymbol{B}$ be co-axial. As such, it was remarked in a different context by St. Venant [1843, *1*].

[2] Again we exclude elastic materials having a natural state but a vanishing shear modulus, or, if we prefer, materials violating one or more of the conditions P-C, IFS, E-T, O-F, B-E, etc.

[3] Reiner [1948, *9*, § 5].

from the statical relation (16.5) we have

$$\left.\begin{aligned}(1+K^2)\,T &= (1+K^2)\,\boldsymbol{\tau}\cdot(\boldsymbol{T}\boldsymbol{n}),\\ &= K\,(T_{\langle xx\rangle}-T_{\langle yy\rangle})+(1-K^2)\,T_{\langle xy\rangle},\\ (1+K^2)\,N &= (1+K^2)\,\boldsymbol{n}\cdot(\boldsymbol{T}\boldsymbol{n}),\\ &= T_{\langle xx\rangle}-2K\,T_{\langle xy\rangle}+K^2\,T_{\langle yy\rangle},\end{aligned}\right\} \tag{54.17}$$

Thus the universal relation (54.16) may be put into the alternative forms

$$T_{\langle xy\rangle}=(1+K^2)\,T,\quad K\,T_{\langle xy\rangle}=(1+K^2)\,(T_{\langle yy\rangle}-N),\qquad N=T_{\langle yy\rangle}-K\,T. \tag{54.18}$$

These universal relations connect the tractions that have to be applied upon the faces of an initially rectangular block if it is to be maintained in a simple shear of amount K. In particular, $|T|<|T_{\langle xy\rangle}|$, and $N<T_{\langle yy\rangle}$. Hence if $T_{\langle yy\rangle}$ is negative, so is N; *if the normal traction on the shearing planes is a pressure, so is that on the inclined faces,* but the converse need not hold. From $(54.18)_2$ it follows that there may be special elastic materials such that $N=0$ for all shears. There are no elastic solids, however, such that $N=T_{\langle yy\rangle}$ since, as is plain from (54.18), this would again imply that $\hat{\mu}=0$. Thus the omnipresence of the Poynting effect as concluded above still holds when referred to the actual faces of the sheared block.

From the formulae (54.10) and (54.15) we see that there are, in general, four non-vanishing stresses in simple shear, and that these stresses are given by four functions of the amount of shear: $\hat{\mu}$, $\hat{\beth}_1$, $\hat{\beth}_{-1}$, and τ. One relation, namely, (54.9), connects these functions. This relation has already been given a physical interpretation in the equivalent form (54.16). Looking upon it formally, we infer that *full knowledge of the behavior of the shear stress in simple shear gives no information whatever about the normal stresses.* Thus it is no wonder that the numerous attempts, particularly in the engineering literature, to predict behavior in large shears on the basis of no more information than is contained in the classical infinitesimal theory are inconsistent with each other, being based on one or another hidden and unjustified assumption about the functions $\beth_\Gamma$. The "intuition" gained from the infinitesimal theory is specious: In any elastic material, *it is not the shear stress but the normal stresses that characterize the phenomenon of simple shear,* since, while the shear stress cannot determine the normal stresses, (54.16) shows that *the normal stresses determine the shear stress,* and moreover, they determine it in exactly the same way for *every* isotropic elastic material, whether the shear be large or small[1]. Tnis is one of the most striking results in the exact theory.

In the examples worked out, the arguments of the response functions were specialized, so that experimental measurements of the relations between stresses and deformations could not suffice to determine fully the forms of those functions. In principle, however, the functions may be inferred, once and for all, from experimental measurements upon an appropriate sequence of homogeneous strains, since such strains suffice to yield every possible combination of principal stretches, so that fully general equations of the form (48.2) may be obtained.

---

[1] In small shear, the quantities on each side of (54.16) are ultimately proportional to K^2 and hence are of the order neglected in the infinitesimal theory. However, neither can justly be said to be negligible in respect to the other . The true result for small shear is expressed by the formula

$$\mu=\hat{\mu}(0)=\lim_{K\to 0}\frac{T_{\langle xy\rangle}}{K}=\lim_{K\to 0}\frac{T_{\langle xx\rangle}-T_{\langle yy\rangle}}{K^2}.$$

According to the infinitesimal theory, the limit at the right-hand end is always zero. In this respect the infinitesimal theory does *not* give results correct in the limit as $K\to 0$.

For an isotropic material, there are various natural ways of specifying the elastic modulus E_0 mentioned in Sect. 43. For example, we may take E_0 as the shear modulus μ, defined by $(50.13)_3$; to interpret it, we observe that since $\hat{\mu}(0)=\mu$, by (54.10) we have

$$\mu = \lim_{K\to 0} \frac{T\langle xy\rangle}{K} \tag{54.19}$$

in simple shear. A modulus of this kind is often called a "tangent modulus".

We have seen that the form of the stress relation for an isotropic material may be determined, in principle, by measurements carried out on homogeneous strains. To the best of our knowledge, no sufficient experiments of this kind have ever been done. Instead, the experiments on compressible materials[1] concern either special homogeneous strains such as simple extension, or non-homogeneous deformations such as torsion. As we have remarked, there is no non-homogeneous deformation that can be maintained in any compressible, isotropic, elastic material by surface tractions alone. Thus in order to obtain a solution appropriate to torsion, for example, it is *first* necessary to determine the form of the response coefficients. In other words, check of theory against experiments on torsion of compressible materials is *impossible* unless preceded by experiments on homogeneous strain sufficient to determine the form of the stress relation. In the experimental literature we have found no trace of understanding of this simple fact. Worse, some experimenters seem content to assume a particular deformation, calculate the corresponding stresses from the general constitutive equation, and proceed to "check" these, oblivious of the fact that the assumed deformation cannot yield stresses that satisfy the conditions of equilibrium unless specially adjusted body force be supplied.

55. Incompressible elastic materials, I. Homogeneous strain. Incompressible materials are simpler than compressible ones in two ways: Not only are all possible deformations isochoric, but also the stress is determinate from the deformation only to within an arbitrary hydrostatic pressure, as may be seen from Eqs. (43.14). Thus the response of an incompressible material in a given isochoric deformation generally is not the same as would be that of a compressible material in precisely the same deformation. For example, simple shear is a deformation which may be undergone by compressible and incompressible bodies alike, but the stress system in the latter may not be read off from the solution for the former, since the arbitrary pressure makes possible a greater variety in the solution. Moreover, various non-homogeneous isochoric deformations which may not be maintained as states of equilibrium for arbitrary isotropic compressible bodies free from the action of body force are amenable to exact solution for all incompressible isotropic bodies and for certain aeolotropic ones, subject to suitable tractions applied upon the boundary. The radically simpler nature of the incompressible material in finite deformation was seen by RIVLIN and exploited in a sequence of remarkable papers in 1948—1954; his work opened new land in non-linear continuum mechanics. The solutions obtained enjoy a particular aptness since rubber, the first example of an elastic material that comes to mind, even under very great strain is highly elastic but only slightly compressible. Moreover, rubber had long been regarded as a rebel to the theory of elasticity, and, so long as "elasticity" was confused with "infinitesimal elasticity", various elaborate physical and chemical conjectures were put forward so as to explain the "anomalous" behavior of the most truly elastic of real materials. As soon as the properly

[1] Among the more recent and general programs may be mentioned those of ANGLES D'AURIAC [1958, *4*, 3me Partie] and of FOUX and REINER [1960, *23*].

general theory of mechanical elasticity was developed and interpreted, these anomalies disappeared.

First we consider homogeneous strain. From (43.14) and (16.6) it is clear that any homogeneous isochoric strain is a configuration of equilibrium subject to surface tractions only, for any homogeneous incompressible elastic body, and that the pressure p is an arbitrary constant, $p=p_0$.

In a simple extension with stretch v in the direction of z, the transverse stretch αv is now determined by the isochoric requirement, $\alpha=v^{-\frac{3}{2}}$ (cf. CFT, Sect. 44). For an isotropic incompressible material, from (49.5) we have

$$\left.\begin{aligned} T_{\langle xx\rangle}=T_{\langle yy\rangle}&=-p_0+v^{-1}\tilde{\beth}_1+v\tilde{\beth}_{-1}\\ T_{\langle zz\rangle}&=-p_0+v^2\tilde{\beth}_1+v^{-2}\tilde{\beth}_{-1}\end{aligned}\right\} \tag{55.1}$$

where

$$\tilde{\beth}_\Gamma(v)\equiv\beth_\Gamma(2v^{-1}+v^2,\ 2v+v^{-2}),\qquad \Gamma=\pm 1. \tag{55.2}$$

By choice of the constant p_0, for any v, we may satisfy the condition $T_{\langle xx\rangle}=T_{\langle yy\rangle}=0$. That is, *for any incompressible isotropic elastic material, there exists a unique simple tension $T_{\langle zz\rangle}$ such as to effect any given simple extension*, the corresponding stress-stretch relation being

$$T_{\langle zz\rangle}=\left(v^2-\frac{1}{v}\right)\tilde{\beth}_1+\left(\frac{1}{v^2}-v\right)\tilde{\beth}_{-1}; \tag{55.3}$$

equivalently, in terms of the principal force T in the direction of extension we have[1]

$$T=\alpha^2 v^2\,T_{\langle zz\rangle}=v^{-1}\,T_{\langle zz\rangle}=\left(v-\frac{1}{v^2}\right)\left(\tilde{\beth}_1-\frac{1}{v}\,\tilde{\beth}_{-1}\right). \tag{55.4}$$

Since the principal stresses satisfy $t_1=t_2=0$, $t_3=T_{\langle zz\rangle}$, and since the principal stretches are given by $v_1=v_2=\alpha v=v^{-\frac{1}{2}}$, $v_3=v$, from (55.3) we find that

$$\frac{t_3-t_1}{v_3-v_1}=\frac{t_3-t_2}{v_3-v_2}=\frac{t_3}{v-\frac{1}{\sqrt{v}}}. \tag{55.5}$$

If $v\neq 1$, the sign of the right-hand side of (55.5) is the same as that of $T_{\langle zz\rangle}(v-1)$ and hence as that of $T(v-1)$ as determined by (55.4). Comparison of these facts with (51.22) shows that *in simple extension of an isotropic incompressible material from its natural state, the B-E inequalities are equivalent to the requirement that tension produces lengthening, while compression produces shortening.* Schematically,

$$\text{B-E}\Leftrightarrow\text{T-E}, \tag{55.6}$$

as far as concerns simple extension alone[2].

More generally, for any homogeneous strain of an incompressible isotropic material the principal stresses and principal stretches are related by the simple equations[3]

$$t_a=-p_0+v_a^2\beth_1+v_a^{-2}\beth_{-1}, \tag{55.7}$$

where $v_1v_2v_3=1$. By choice of p_0, any one of the stresses t_a may be made to vanish, so that homogeneous plane stress is always possible. If $t_1=0$, say, then for

[1] RIVLIN [1956, *23*, § 4], the special case for hyperelastic materials having been obtained earlier [1948, *12*, § 6].

[2] Note that "T-E" does not have the same significance here as in Sect. 51. Since T_1 is not in general a function of the principal stretches alone for an incompressible material, an inequality such as $(51.10)_1$ cannot even be stated.

[3] A detailed discussion of (55.7), for hyperelastic materials, was given by RIVLIN and SAUNDERS [1951, *14*, § 2].

$a, b = 2, 3$, and $a \neq b$ we have

$$\left.\begin{aligned} t_a = \left(v_a^2 - \frac{1}{v_2^2 v_3^2}\right) \times \\ \times (\beth_I - v_b^2 \beth_{-I}). \end{aligned}\right\} \quad (55.8)$$

Here, moreover, the arguments $I_{\boldsymbol{B}}$ and $II_{\boldsymbol{B}}$ of the response functions $\beth_\Gamma$ may assume any values consistent with the requirement of isochoric deformation, $III_{\boldsymbol{B}} = 1$ (cf. the discussion in Sect. 40 of CFT). Hence *a sequence of experiments on plane biaxial stress of a sheet of isotropic incompressible elastic material suffices to determine the response functions uniquely.* For "a pure gum mix of natural rubber", in which effects of hysteresis are slight even for stretches as great as 4, such experiments were carried out by RIVLIN and SAUNDERS[1] for the range $5 \leqq I_{\boldsymbol{B}} \leqq 11$ and $5 \leqq II_{\boldsymbol{B}} \leqq 30$. All recorded data was obtained in tests in which one principal invariant was held constant while the other was increased. As will be seen from Fig. 3 and the accompanying tables of their data, they found that $\frac{1}{2}\beth_I$ is very nearly constant, staying close to 1.7 kgwt/cm², while $-\frac{1}{2}\beth_{-I}$ varies between 0.06 and 0.28 kgwt/cm², and they inferred that a satisfactory account of the results of an extensive program of experiments on several kinds of deformation[2] is given by taking

$$\left.\begin{aligned} \beth_I &= \text{const.}, \\ \beth_{-I} &= -f(II_{\boldsymbol{B}} - 3), \end{aligned}\right\} \quad (55.9)$$

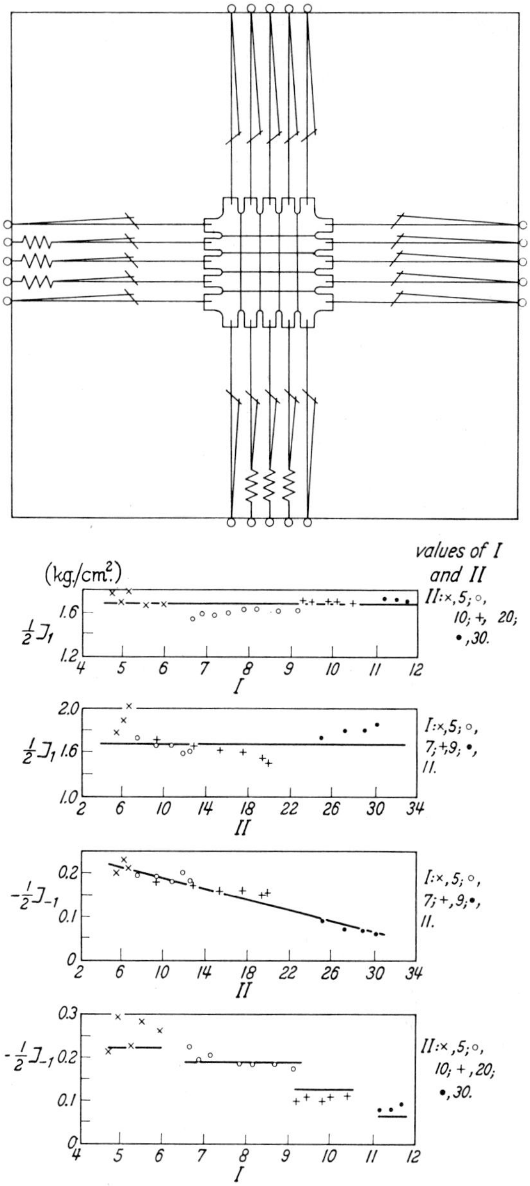

Fig. 3. Schema of RIVLIN and SAUNDERS' experiment on biaxial tension of a sheet, and their values of the response coefficients of a certain rubber.

---

[1] RIVLIN and SAUNDERS [1951, *14*, §§ 3—5] [1952, *17*].

[2] Pure shear [§ 6], pure shear superimposed upon simple extension [§§ 7—8], simple extension [§§ 9—10], compression [§§ 11—13]. References to experiments on non-homogeneous strains will be given in Sects. 57 and 59. Forms for $\beth_I$ and $\beth_{-I}$ somewhat different from (55.9) have been proposed by GENT and THOMAS [1958, *18*] to account for the same and further data. A survey of experimental results for rubber and their relation to the theory is given by GREEN and ADKINS [1960, *26*, Ch. X]. According to KLINGBEIL and SHIELD [1964, *50*], experiments on membranes involving values of II up to 1000 support RIVLIN and SAUNDERS' formulae (55.9) and controvert those proposed by GENT and THOMAS.

where $f(x)$ is a positive function. The inequalities $\beth_1 > 0$, $\beth_{-1} < 0$, even stronger than the E-inequalities (53.3), are satisfied in all cases. Some more recent experimental results do not confirm $(55.9)_1$ but indicate rather a non-constant value for $\beth_1$. In this treatise we refrain from commitment toward forms of the response coefficients appropriate for particular materials.

Table 1.

I	v_1	v_2	II	$\frac{1}{2}\beth_1$ (kgwt/cm²)	$-\frac{1}{2}\beth_{-1}$ (kgwt/cm²)
5	1.90	1.07	5.28	1.77	0.20
	1.80	1.25	6.01	1.89	0.23
	1.70	1.39	6.45	2.01	0.21
	1.60	1.51	6.67	—	—
7	2.40	1.04	7.33	1.72	0.19
	2.30	1.25	9.10	1.66	0.19
	2.20	1.43	10.59	1.66	0.18
	2.10	1.58	11.63	1.58	0.20
	2.00	1.71	12.27	1.60	0.18
	1.90	1.82	12.54	—	—
9	2.80	1.02	9.24	1.70	0.18
	2.70	1.28	12.69	1.64	0.17
	2.60	1.47	15.22	1.61	0.16
	2.50	1.64	17.34	1.59	0.16
	2.40	1.80	19.14	1.53	0.15
	2.30	1.91	19.76	1.49	0.15
	2.20	2.03	20.40	—	—
11	2.80	1.76	24.74	1.73	0.09
	2.70	1.91	27.01	1.79	0.07
	2.60	2.05	28.79	1.80	0.07
	2.50	2.18	30.07	1.85	0.06
	2.40	2.28	30.31	—	—

Table 2.

II	v_1	v_2	I	$\frac{1}{2}\beth_1$ (kgwt/cm²)	$-\frac{1}{2}\beth_{-1}$ (kgwt/cm²)
5	2.40	0.69	6.60	—	—
	2.30	0,76	6.20	—	—
	2.20	0.83	5.92	1.66	0.26
	2.10	0.89	5.49	1.65	0.28
	2.00	0.96	5.19	1.78	0.22
	1.90	1.02	4.92	1.68	0.29
	1.80	1.09	4.69	1.76	0.21
	1.70	1.17	4.51	—	—
	1.60	1.24	4.35	—	—
	1.50	1.33	4.27	—	—
10	2.80	1.07	9.10	1.61	0.17
	2.70	1.11	8.63	1.61	0.18
	2.60	1.16	8.13	1.63	0.18
	2.50	1.21	7.82	1.61	0.18
	2.40	1.26	7.46	1.59	0.19
	2.30	1.32	7.14	1.56	0.20
	2.20	1.38	6.85	1.57	0.19
	2.10	1.45	6.62	1.52	0.22
	2.00	1.53	6.45	—	—
	1.90	1.61	6.31	—	—
	1.80	1.70	6.24	—	—
20	2.80	1.58	10.39	1.67	0.11
	2.70	1.63	10.00	1.68	0.11
	2.60	1.70	9.80	1.69	0.10
	2.50	1.76	9.40	1.68	0.11
	2.40	1.84	9.20	1.69	0.10
	2.30	1.92	9.03	—	—
	2.20	2.01	8.95	—	—
30	2.80	1.94	11.64	1.69	0.09
	2.70	2.02	11.40	1.71	0.08
	2.60	2.09	11.16	1.72	0.08
	2.50	2.18	11.04	—	—
	2.40	2.27	10.95	—	—

While all possible homogeneous strains are included in the results just given, it is instructive to look directly at simple shear. The tractions required to produce this particular isochoric strain are different from those which effect the same strain in a compressible material. From (49.5), for the simple shear $x = X + KY$, $y = Y$, $z = Z$ we have[1]

$$\left.\begin{aligned}[\boldsymbol{T}] = (-p_0 + \hat{\beth}_1 + \hat{\beth}_{-1})[\mathbf{1}] + (\hat{\beth}_1 - \hat{\beth}_{-1}) K \begin{Vmatrix} 0 & 1 & 0 \\ \cdot & 0 & 0 \\ \cdot & \cdot & 0 \end{Vmatrix} + \\ + \hat{\beth}_1 K^2 \begin{Vmatrix} 1 & 0 & 0 \\ \cdot & 0 & 0 \\ \cdot & \cdot & 0 \end{Vmatrix} + \hat{\beth}_{-1} K^2 \begin{Vmatrix} 0 & 0 & 0 \\ \cdot & 1 & 0 \\ \cdot & \cdot & 0 \end{Vmatrix},\end{aligned}\right\} \tag{55.10}$$

where

$$\hat{\beth}_\Gamma(K^2) \equiv \beth_\Gamma(3 + K^2, 3 + K^2), \quad \Gamma = \pm 1. \tag{55.11}$$

The relation between shear stress and shear is still given by (54.10), with $\hat{\mu}$ given by $(54.9)_1$. Eqs. (54.12) and (54.13) hold here, too, and again it follows that *the*

[1] RIVLIN [1948, *12*, § 12], for hyperelastic materials.

shear modulus $\hat{\mu}$ is positive if the B-E inequalities hold[1]:

$$\text{B-E} \Rightarrow \hat{\mu}(K^2) > 0. \tag{55.12}$$

As for compressible materials, the sign of $\hat{\mu}$ is not affected by the magnitude of the normal stress on the plane of shear, so that if $\hat{\mu}(K^2) > 0$, it does not necessarily follow that the B-E inequalities hold in simple shear.

Despite the superficial resemblance to (54.7), the presence of the arbitrary pressure p_0 shows that in place of the Kelvin effect for compressible materials, for incompressible materials there is a greater variety of Poynting effects, since the normal traction on any pair of opposite faces of a sheared block may be assigned at will. In particular, *it is possible that any pair of faces be free of traction.* Let it be supposed, for example, that the planes of shear are free: $T\langle zz\rangle = 0$. In this case, $p_0 = \hat{\beth}_1 + \hat{\beth}_{-1}$, and

$$T\langle yy\rangle / K^2 = \hat{\beth}_{-1}, \qquad T\langle xx\rangle / K^2 = \hat{\beth}_1, \tag{55.13}$$

so that $\hat{\beth}_{-1}$ and $\hat{\beth}_1$ may be interpreted directly in terms of the normal stresses in simple shear. If the E-inequalities (53.3) are satisfied, as seems to be the case for rubber, *the stress on the shearing planes is always a pressure, while the stress on the planes normal to the direction of shear is always a tension.* If only the B-E inequalities (53.1) are imposed, this rule of sign does not necessarily hold[2]. This example does not characterize the difference between the E and B-E inequalities, since only the restricted functions $\hat{\beth}_\Gamma$ which are defined by (55.11), not the general response coefficients $\beth_\Gamma$, make their appearance in simple shear. Nevertheless, simple shear serves as an example giving the difference a physical interpretation.

The universal relation (54.16) remains valid, and the relations (54.17) and (54.18) give the tractions on the inclined faces $x - Ky = \text{const.}$ in the present case also. Again it follows that if $T\langle yy\rangle < 0$, i.e. if $\beth_{-1} < 0$, then $N < 0$ also; i.e., *the normal force on the inclined faces is also a pressure.* Unless these pressures be supplied, an incompressible specimen in shear will tend to *lengthen and broaden in the plane of shear* and, consequently, to contract normally to the plane of shear. It is in this form that the Poynting effect is most commonly observed. The general considerations of orders of magnitude of the normal stresses and of the non-linear correction to the shear stress as discussed in Sect. 54 are applicable here, too.

56. Incompressible elastic materials, II. Preliminaries for the non-homogeneous solutions for arbitrary materials. By substituting (49.5) into (16.6) we obtain the following differential equations of equilibrium for a homogeneous isotropic body of incompressible elastic material:

$$-\operatorname{grad}(p + \varrho v) + \operatorname{div}(\beth_1 \boldsymbol{B} + \beth_{-1} \boldsymbol{B}^{-1}) = \boldsymbol{0}, \tag{56.1}$$

where we have assumed that the external body force $\boldsymbol{b}$ is conservative (CFT, Sect. 218), so that $\boldsymbol{b} = -\operatorname{grad} v$. Thus the presence of conservative body force does not make the solution of problems for incompressible materials more difficult[3], provided only that v be single-valued, for all that is needed is to make a corresponding adjustment in the pressure p. Of course, the surface tractions

[1] That $\hat{\mu}(K^2) > 0$ was asserted, in effect, as a postulate by RIVLIN [1949, *16*, § 16].

[2] It is easy to verify that the B-E inequalities are equivalent, for simple shear, to the following:

$$\hat{\beth}_1 > \max(\hat{\beth}_{-1}, v_1^2 \hat{\beth}_{-1}),$$

where v_1^2 is given by $(54.11)_1$.

[3] This was remarked, in effect, in CFT, Sect. 207.

that must be applied in order to effect the deformation will depend upon v, but the compatibility of any given deformation is independent of whether or not a conservative body force is present. From (43.14) we see that this conclusion holds also for anisotropic materials.

Given a field $\boldsymbol{B}=\boldsymbol{B}(\boldsymbol{x})$ of left Cauchy-Green tensors, we see from (56.1) that the deformation from which it is calculated is a possible configuration of equilibrium only if

$$\left.\begin{aligned} &\operatorname{curl}\operatorname{div}(\beth_1\boldsymbol{B}+\beth_{-1}\boldsymbol{B}^{-1})=\boldsymbol{0},\\ &(\beth_1 B^q_k+\beth_{-1}(B^{-1})^q_k)_{,qm}=(\beth_1 B^q_m+\beth_{-1}(B^{-1})^q_m)_{,qk}.\end{aligned}\right\} \tag{56.2}$$

Given a particular tensor field $\boldsymbol{B}$, whether or not (56.2) is satisfied will depend, in general, upon the functions $\beth_\Gamma$. As already mentioned in Sect. 54, however, there are certain forms for $\boldsymbol{B}$ such that (56.2) is satisfied no matter what functions $\beth_1$ and $\beth_{-1}$ are. The deformations leading to these forms of $\boldsymbol{B}$ *can be effected in every homogeneous, incompressible, isotropic, elastic body by application of suitable surface tractions*, although of course, as is plain from (49.5), the tractions required and the resulting interior stresses in a particular body depend very strongly upon the material properties.

There are, in fact, four simple families of such deformations; moreover, results to be presented in Sect. 91 make it seem probable, though some details of proof are still lacking, that no further such solutions exist. These families are so special that it would be cumbrous to approach them via (56.2). We shall instead be content to exhibit them[1] and to verify that they satisfy all the required conditions. To this end we first consider their purely statical aspects, which will do service also in connection with more general deformations and anisotropic materials mentioned later. The statical results, when regarded in the light of the elastic constitutive equations, go far to motivate the assumptions leading to the special solutions and to provide some rationale for what might otherwise seem to be, and perhaps originally was, guesswork.

Statical preliminaries. All solutions, it turns out, are such that appropriate physical components of stress are constant on each member of a family of parallel planes, coaxial cylinders, or concentric spheres. In all cases, we consider only equilibrium subject to no body force, i.e., $\ddot{\boldsymbol{x}}=\boldsymbol{0}$, $\boldsymbol{b}=\boldsymbol{0}$.

Case A. Planar problems. In a system of Cartesian co-ordinates x, y, z, suppose that $\boldsymbol{T}+p\boldsymbol{1}$ is a function of x only. Then CAUCHY's first law (16.6) assumes the form

$$\partial_x T\langle xx\rangle=0,\quad \partial_x T\langle xy\rangle-\partial_y p=0,\quad \partial_x T\langle xz\rangle-\partial_z p=0. \tag{56.3}$$

If we assume that

$$T\langle xy\rangle=0,\quad T\langle xz\rangle=0, \tag{56.4}$$

it follows from (56.3) that

$$p=p(x),\quad T\langle xx\rangle=\text{const.}=-p_0. \tag{56.5}$$

To obtain this result it suffices to assume the weaker conditions $T\langle xy\rangle=\text{const.}$, $T\langle xz\rangle=\text{const.}$ in place of (56.4). From (56.5) we see then that the planes $x=\text{const.}$ are subject to a stress of uniform magnitude and direction. If (56.4) holds, these planes can be rendered altogether free of traction by setting $p_0=0$.

Case B. Cylindrical problems. In a system of cylindrical co-ordinates r, θ, z, we assume that the physical components of the extra-stress, $\boldsymbol{T}+p\boldsymbol{1}$, are functions

[1] References for discovery of these results are given below. The use of independently selected co-ordinate systems to obtain such solutions is due to TRUESDELL [1952, *20*, § 42].

of r only. We denote the physical components of $\boldsymbol{T}$ by $T\langle rr\rangle$, $T\langle r\theta\rangle$, etc. By (6A.4), CAUCHY's first law (with $\ddot{\boldsymbol{x}}=\boldsymbol{b}=\boldsymbol{0}$) reduces to the form

$$\left.\begin{aligned} \partial_r T\langle rr\rangle+\frac{1}{r}(T\langle rr\rangle-T\langle\theta\theta\rangle)=0,\\ \frac{1}{r^2}\partial_r(r^2 T\langle r\theta\rangle)-\frac{1}{r}\partial_\theta p=0,\\ \frac{1}{r}\partial_r(r\,T\langle rz\rangle)-\partial_z p=0.\end{aligned}\right\}\tag{56.6}$$

If we assume that

$$T\langle r\theta\rangle=0,\quad T\langle rz\rangle=0,\tag{56.7}$$

it follows from (56.6) that p and hence $(T\langle rr\rangle+p)-p$ depend only on r and that

$$T\langle rr\rangle=-\int\frac{T\langle rr\rangle-T\langle\theta\theta\rangle}{r}\,dr,\qquad T\langle\theta\theta\rangle=\frac{d}{dr}(r\,T\langle rr\rangle).\tag{56.8}$$

By superimposing a uniform pressure, the cylinder $r=r_1$ may be rendered free of traction, so that $T\langle rr\rangle=0$ on $r=r_1+0$. The resultant normal force R per unit height on the part of the axial plane $\theta=$ const. cut off by the cylinders $r=r_1$ and $r=r_2$ is given by

$$R=\int_{r_1}^{r_2}T\langle\theta\theta\rangle\,dr=r\,T\langle rr\rangle\big|_{r=r_2},\tag{56.9}$$

where we exclude the case when $r_2=0$. From (56.9) we infer that *the faces* $\theta=$const. *are free of resultant normal traction if and only if the cylinder $r=r_2$ is free of normal traction.* Independently of this condition, the resultant moment M per unit height, with respect to a point on the axis $r=0$, exerted by the normal stress acting upon these faces, is given by

$$\left.\begin{aligned} M&=\int_{r_1}^{r_2}r\,T\langle\theta\theta\rangle\,dr=\int_{r_1}^{r_2}r\frac{d}{dr}(r\,T\langle rr\rangle)\,dr,\\ &=r^2\,T\langle rr\rangle\Big|_{r_1}^{r_2}-\int_{r_1}^{r_2}r\,T\langle rr\rangle\,dr,\\ &=\frac{1}{2}r^2\,T\langle rr\rangle\Big|_{r_1}^{r_2}+\frac{1}{2}\int_{r_1}^{r_2}r^2\frac{d\,T\langle rr\rangle}{dr}\,dr.\end{aligned}\right\}\tag{56.10}$$

If $T\langle rr\rangle$ vanishes on $r=r_1$ and $r=r_2$, the first term in each of the latter expressions vanishes also.

The resultant normal traction N acting upon the plane annular wedge $r_1\leqq r\leqq r_2$, $z=$const., $|\theta|\leqq\theta_0$ may be calculated as follows. First, from $(56.8)_2$ we see that $2\int r\,T\langle rr\rangle\,dr=r^2\,T\langle rr\rangle-\int r(T\langle\theta\theta\rangle-T\langle rr\rangle)\,dr$. Hence

$$\left.\begin{aligned} N&=2\theta_0\int_{r_1}^{r_2}T\langle zz\rangle\,r\,dr,\\ &=2\theta_0\int_{r_1}^{r_2}(T\langle zz\rangle-T\langle rr\rangle)\,r\,dr+2\theta_0\int_{r_1}^{r_2}r\,T\langle rr\rangle\,dr,\\ &=2\theta_0\int_{r_1}^{r_2}(T\langle zz\rangle-T\langle rr\rangle)\,r\,dr+\theta_0 r^2\,T\langle rr\rangle\Big|_{r_1}^{r_2}-\theta_0\int_{r_1}^{r_2}r(T\langle\theta\theta\rangle-T\langle rr\rangle)\,dr,\\ &=\theta_0 r^2\,T\langle rr\rangle\Big|_{r_1}^{r_2}+\theta_0\int_{r_1}^{r_2}(2\,T\langle zz\rangle-T\langle rr\rangle-T\langle\theta\theta\rangle)\,r\,dr.\end{aligned}\right\}\tag{56.11}$$

Thus N is independent of z.

The solution (56.8) and its consequences (56.9) through (56.11) remain valid when (56.7)is replaced by the weaker conditions $r^2 T\langle r\theta\rangle = \text{const.}$, $r T\langle rz\rangle = \text{const.}$, but some of the interpretations must be modified a little.

Case C. Spherical problems. In a system of spherical co-ordinates r, θ, φ, assume that the physical components of $\boldsymbol{T}+p\boldsymbol{1}$ are functions of r only, and that

$$T\langle r\theta\rangle = T\langle r\varphi\rangle = T\langle \theta\varphi\rangle = 0, \quad \text{and} \quad T\langle\theta\theta\rangle = T\langle\varphi\varphi\rangle. \tag{56.12}$$

By (6A.8), CAUCHY's first law (16.6) then reduces to the form

$$r\,\partial_r T\langle rr\rangle = 2(T\langle\theta\theta\rangle - T\langle rr\rangle), \quad \partial_\theta p = \partial_\varphi p = 0, \tag{56.13}$$

where we have assumed that $\boldsymbol{b}=\boldsymbol{0}$ and $\ddot{\boldsymbol{x}}=\boldsymbol{0}$. It follows that

$$\left.\begin{aligned} p &= p(r), \\ T\langle rr\rangle &= 2\int \frac{T\langle\theta\theta\rangle - T\langle rr\rangle}{r}\,dr. \end{aligned}\right\} \tag{56.14}$$

By superimposition of a uniform pressure, one sphere $r=\text{const.}$ may be rendered free of traction.

For the purposes of the next section it is essential to observe that *all the statical requirements* have been eliminated in the three cases just presented. A stress system satisfying the conditions defining any one of them is thus compatible, and there is no need to have recourse to more complicated general conditions such as (56.2).

57. Incompressible elastic materials, III. The non-homogeneous solutions for arbitrary isotropic materials. We are now ready to apply the statical results of the previous section to particular deformations which render them compatible with the response of an isotropic homogeneous elastic body. Except when the contrary is stated explicitly, we consider only the case of equilibrium subject to no body force. Thus $\boldsymbol{b}=\boldsymbol{0}$, $\ddot{\boldsymbol{x}}=\boldsymbol{0}$.

Family 1 (Bending, stretching, and shearing of a rectangular block)[1]. Let the co-ordinates in the natural state be Cartesian (X, Y, Z); those in the loaded configuration, cylindrical (r, θ, z). Consider the deformation

$$r=\sqrt{2AX}, \quad \theta = BY, \quad z=\frac{Z}{AB} - BCY, \tag{57.1}$$

where A, B, C are constants, $AB \neq 0$. If $C=0$, this deformation carries the block bounded by the planes $X=X_1$, $X=X_2$, $Y=\pm Y_0$, $Z=\pm Z_0$ into the annular wedge bounded by the cylinders $r=r_1=\sqrt{2AX_1}$ and $r=r_2=\sqrt{2AX_2}$, and the planes $\theta=\pm\theta_0=\pm BY_0$, $z=\pm z_0=\pm Z_0/(AB)$. Cf. CFT, Sect. 50. If we think of B as given, then an arbitrary axial stretch $1/(AB)$ is allowed, and the radial stretch is so adjusted as to render the deformation isochoric. In the general case, the deformation may be effected in two steps, the first of which is the bending and axial stretch just described, while the second is a homogeneous strain, in fact a simple shear, which carries the body into the solid bounded by the cylindrical surfaces $r=r_1$ and $r=r_2$, the planes $\theta=\pm\theta_0$, and the helicoidal surfaces $z+C\theta=\pm z_0$.

---

[1] In the major special case when $C=0$, corresponding to pure bending, this family of solutions was discovered by RIVLIN [1949, *15*] [1949, *16*, §§ 14—16]. The generalization is due to ERICKSEN [1954, *5*, § 4]. Both these authors considered only hyperelastic materials, for which somewhat simpler forms for some of the results are possible; cf. Sect. 92.

Substituting (57.1) into (23.7) yields

$$\|B^{km}\| = \left\| \begin{matrix} \frac{A^2}{r^2} & 0 & 0 \\ \cdot & B^2 & -B^2 C \\ \cdot & \cdot & B^2 C^2 + \frac{1}{A^2 B^2} \end{matrix} \right\|, \quad \|(B^{-1})_{km}\| = \left\| \begin{matrix} \frac{r^2}{A^2} & 0 & 0 \\ \cdot & \frac{1}{B^2} + A^2 B^2 C^2 & A^2 B^2 C \\ \cdot & & A^2 B^2 \end{matrix} \right\|, \tag{57.2}$$

$$I \equiv I_{\boldsymbol{B}} = \frac{A^2}{r^2} + B^2 r^2 + B^2 C^2 + \frac{1}{A^2 B^2},$$

$$II \equiv II_{\boldsymbol{B}} = I_{\boldsymbol{B}^{-1}} = \frac{r^2}{A^2} + \frac{1}{r^2}\left(\frac{1}{B^2} + A^2 B^2 C^2\right) + A^2 B^2.$$

From (49.5) we see at once that the physical components of $\boldsymbol{T} + p\mathbf{1}$ are functions of r only, and that (56.7) is satisfied. The stress system thus falls under Case B of the previous section, so that Eqs. (56.8) through (56.11) are applicable. Substituting into $(56.8)_2$ the value of $T\langle rr\rangle - T\langle\theta\theta\rangle$ as given by (57.2) and (49.5), we see that

$$\left.\begin{aligned} T\langle rr\rangle &= -\int \left\{ \left[\frac{A^2}{r^3} - B^2 r\right] \beth_1 - \left[\frac{1}{r^3}\left(\frac{1}{B^2} + A^2 B^2 C^2\right) - \frac{r}{A^2}\right] \beth_{-1} \right\} dr, \\ &= \int \left(\frac{1}{2} \beth_1 \frac{dI}{dr} - \frac{1}{2} \beth_{-1} \frac{dII}{dr} \right) dr. \end{aligned}\right\} \tag{57.3}$$

The rest of the stress components are obtained by use of this result combined with (49.5) and (57.2):

$$\left.\begin{aligned} T\langle\theta\theta\rangle &= T\langle rr\rangle - \left[\frac{A^2}{r^2} - B^2 r^2\right] \beth_1 + \left[\frac{1}{r^2}\left(\frac{1}{B^2} + A^2 B^2 C^2\right) - \frac{r^2}{A^2}\right] \beth_{-1}, \\ T\langle zz\rangle &= T\langle rr\rangle + \left(B^2 C^2 + \frac{1}{A^2 B^2} - \frac{A^2}{r^2}\right) \beth_1 - \left(\frac{r^2}{A^2} - A^2 B^2\right) \beth_{-1}, \\ T\langle\theta z\rangle &= -B^2 C r \beth_1 + \frac{A^2 B^2 C}{r} \beth_{-1}, \end{aligned}\right\} \tag{57.4}$$

where $\beth_1(I, II)$ and $\beth_{-1}(I, II)$ are functions of r in view of $(57.2)_{3,4,5,6}$.

By choice of the constant of integration in (57.3), the cylinder $r = r_1$ may be rendered free of traction. In order for the surface $r = r_2$ also to be free of traction, so that the block may be bent and sheared by forces applied to its plane and helicoidal faces only, by (57.3) it is necessary that

$$\int_{r_1}^{r_2} \left(\frac{1}{2} \beth_1 \frac{dI}{dr} - \frac{1}{2} \beth_{-1} \frac{dII}{dr} \right) dr = 0. \tag{57.5}$$

For given r_1 and r_2, this is a relation among the constants A, B, C. Whether or not this equation has any real roots depends on the nature of $\beth_1$ and $\beth_{-1}$.

Independently of whether or not (57.5) can be satisfied, it is obvious from (57.4) that the helicoidal faces cannot be free of traction. From (16.5) we readily calculate the normal and tangential tractions N and T that have to be applied on these faces to maintain the deformation:

$$\left.\begin{aligned} N &= \frac{1}{1 + \frac{C^2}{r^2}} \left[T\langle zz\rangle + 2\frac{C}{r} T\langle\theta z\rangle + \frac{C^2}{r^2} T\langle\theta\theta\rangle \right], \\ T &= \frac{1}{1 + \frac{C^2}{r^2}} \left[\left(1 - \frac{C^2}{r_2}\right) T\langle\theta z\rangle + \frac{C}{r}\left(T\langle\theta\theta\rangle - T\langle zz\rangle\right) \right] \end{aligned}\right\} \tag{57.6}$$

where the values of $T\langle km\rangle$ are to be taken from (57.4). In the case of pure bending, $C = 0$, and then $T = 0$ and $N = T\langle zz\rangle$, but in general both normal and tangential tractions must be applied. The presence of these tractions is the *Poynting effect for bending*. Like the Poynting effect in simple shear (Sect. 54), it is an inevitable accompaniment of large strain. In general, the only practical way to effect a deformation of the kind described would be within cheeks having the form $z+C\theta=\text{const.}$, since the required surface tractions depend upon r, A, B, C and the response coefficients in a complicated way. Even when $C=0$, the normal traction is a function of r. The resultant normal force may be calculated from (56.11) but is difficult to estimate in general terms.

The tractions that must be applied upon the plane ends $\theta=\pm\theta_0$ may be read off from (57.4), and the resultant normal force on them is given by (56.9). When $C=0$, the condition $R=0$ was shown in Sect. 56 to be equivalent to (57.5); that is, in the case of pure bending ($C=0$), *in order that the bending be produced by terminal couples it is necessary and sufficient that the inner and outer curved faces be free.* Independently of this condition, the resultant moment on the plane ends is obtained by substituting (57.4) into (56.10):

$$M=\frac{1}{2}\,r^2\,T\langle r r\rangle\Big|_{r_1}^{r_2}-\frac{1}{2}\int_{r_1}^{r_2}\left\{\left[\frac{A^2}{r}-B^2 r^3\right]\beth_1+\left[\frac{r^3}{A^2}-\frac{1}{r}\left(\frac{1}{B^2}+A^2B^2C^2\right)\right]\beth_{-1}\right\}dr. \tag{57.7}$$

All these results are referred to the deformed body but can easily be expressed in terms of the undeformed body by use of (57.1).

Family 2 (Straightening, stretching, and shearing a sector of a hollow cylinder)[1]. Using a cylindrical co-ordinate system (R,Θ,Z), consider a body bounded by the cylinders $R=R_1$ and $R=R_2$, the planes $\Theta=\pm\Theta_0$, and the planes $Z=\pm Z_0$. We use Cartesian co-ordinates (x, y, z) for deformed states. Let the body first be straightened by the deformation $\bar{x}=\frac{1}{2}A R^2$, $\bar{y}=\Theta/A$, $\bar{z}=Z$; it then becomes the block bounded by the planes $\bar{x}=\frac{1}{2}A R_1^2$, $\bar{x}=\frac{1}{2}A R_2^2$, $\bar{y}=\pm\Theta_0/A$, $\bar{z}=\pm Z_0$. Now stretch the block along the x-axis with equal transverse contractions: $x'=B^2\bar{x}$, $y'=\bar{y}/B$, $z'=\bar{z}/B$. Finally, effect a shear in the z-y planes: $x=x'$, $y=y'$, $z=z'+Cy'$. The resulting deformation is

$$x=\frac{1}{2}A\,B^2R^2,\qquad y=\frac{\Theta}{A\,B},\qquad z=\frac{Z}{B}+\frac{C\Theta}{A\,B}, \tag{57.8}$$

where $A\,B\neq 0$.

Substituting (57.8) into (23.7) yields

$$\left.\begin{aligned}
\|B^{km}\|&=\left\|\begin{matrix}2A\,B^2x & 0 & 0\\ \cdot & \dfrac{1}{2A\,x} & \dfrac{C}{2A\,x}\\ \cdot & \cdot & \dfrac{1}{B^2}+\dfrac{C^2}{2A\,x}\end{matrix}\right\|,\\
\|(B^{-1})_{km}\|&=\left\|\begin{matrix}\dfrac{1}{2A\,B^2x} & 0 & 0\\ \cdot & 2A\,x+B^2C^2 & -B^2C\\ \cdot & \cdot & B^2\end{matrix}\right\|,\\
I&=\frac{1}{B^2}+2A\,B^2x+\frac{1}{2A\,x}(1+C^2),\\
II&=B^2(1+C^2)+2A\,x+\frac{1}{2A\,B^2x}.
\end{aligned}\right\} \tag{57.9}$$

---

[1] ERICKSEN [1954, 5, § 4], for hyperelastic materials.

Since $\boldsymbol{B}^{-1}$ is positive-definite, we must have $A x > 0$ and $B \neq 0$. From (57.9) and (49.5) it is plain that $\boldsymbol{T} + p\mathbf{1}$ depends upon x only and that the conditions (56.4) are satisfied. Thus this deformation falls under Case A of Sect. 56, so that (56.5) holds. Therefore, by (49.5) and (57.9) we see that

$$p = p_0 + 2A B^2 x \beth_1 + \frac{1}{2A B^2 x} \beth_{-1}. \tag{57.10}$$

Hence the full set of stresses is given by the explicit formulae

$$\left.\begin{aligned}
T\langle xx\rangle &= -p_0,\\
T\langle yy\rangle &= -p_0 + \left(\frac{1}{2Ax} - 2AB^2 x\right)\beth_1 - \left(\frac{1}{2AB^2x} - B^2C^2 - 2Ax\right)\beth_{-1},\\
T\langle zz\rangle &= -p_0 + \left(\frac{1}{B^2} + \frac{C^2}{2Ax} - 2AB^2x\right)\beth_1 - \left(\frac{1}{2AB^2x} - B^2\right)\beth_{-1},\\
T\langle yz\rangle &= \frac{C}{2Ax}\beth_1 - B^2 C \beth_{-1}.
\end{aligned}\right\} \tag{57.11}$$

The faces $x = \text{const.}$ may be rendered free of traction by taking $p_0 = 0$. On the slanted faces, $z - Cy = \text{const.}$, must be supplied, in general, both normal and tangential tractions, this fact being a counterpart of the Poynting effect. When $C = 0$, so that the faces in the deformed state are parallel to the co-ordinate planes, normal tractions suffice, but these are necessarily functions of x, even for materials in which the response coefficients are constant. The resultant normal force on the planes $y = \text{const.}$ vanishes if and only if

$$\int_{x_1}^{x_2} \left[\left(\frac{1}{2Ax} - 2AB^2x\right)\beth_1 + \left(B^2C^2 + 2Ax - \frac{1}{2AB^2x}\right)\beth_{-1}\right] dx = 0, \tag{57.12}$$

where $x_a = \frac{1}{2} A B^2 R_a^2$, $a = 1, 2$. This condition imposes a relation between A, B, and C for a given material; whether it can be satisfied depends on the response coefficients $\beth_1, \beth_{-1}$; a solution is known to exist in some simple cases.

Family 3 (Inflation, bending, torsion, extension, and shearing of an annular wedge)[1]. The deformation which we now consider carries the particle with cylindrical co-ordinates R, Θ, Z into the place with cylindrical co-ordinates r, θ, z as follows:

$$r = \sqrt{A R^2 + B}, \quad \theta = C\Theta + DZ, \quad z = E\Theta + FZ, \quad A(CF - DE) = 1, \tag{57.13}$$

where B, C, D, E, F are arbitrary constants, and A and B have values such that $AR^2 + B > 0$ when R is in some interval $R_1 \leqq R \leqq R_2$. This deformation may be regarded as resulting from a succession of four simpler ones. First the cylinder defined by $R_1 \leqq R \leqq R_2$ is inflated uniformly: $r' = \sqrt{AR^2 + B}$, $\theta' = \Theta$, $z' = Z/A$. Second, the inflated cylinder is subjected to a uniform longitudinal stretch of amount $1/C$, so that $\bar{r} = r'$, $\bar{\theta} = C\theta'$, $\bar{z} = z'/C$. Third, the inflated and stretched

[1] For pure torsion, the solution was found by RIVLIN [1947, *9*, p. 837] [1948, *12*, § 14] [1949, *18*]. Later [1949, *16*, §§ 3—8] he gave the solution for the case when $C = 1$ and $E = 0$; this includes inflation, torsion, and extension. GREEN and SHIELD [1950, *3*, § 6] calculated the stresses (57.27) induced by steady rotation of a solid cylinder about its axis. ADKINS, GREEN, and SHIELD [1953, *1*, § 7] found the solution when $D = E = 0$; this corresponds to bending without torsion or shear. The solution for the unrestricted deformation (57.13) was first given by ERICKSEN and RIVLIN [1954, *7*, §§ 8—11]. All these authors considered only hyperelastic materials, for which somewhat simpler forms of the results are possible; cf. Sect. 92. To ERICKSEN and RIVLIN is due the dislocation solution discussed under our Case 3c. That a stored-energy function is unnecessary to get these results was observed by RIVLIN [1956, *23*, § 4] in the special case when $B = 0$, $A = D = E = 1$.

cylinder is twisted with a torsion of amount ACD; thus $r' = \bar{r}$, $\theta' = \bar{\theta} + ACD\bar{z}$, $z' = \bar{z}$. Fourth, a kind of shear of the azimuthal planes is effected: $r = r'$, $\theta = \theta'$, $z = z' + K\theta'$. As the result of this last deformation, the planes $Z = \text{const.}$ are deformed into the helicoidal surfaces $z - E\theta/C = \text{const.}$ The constants E and F of the resulting deformation are given by $E = CK$, $F = DK + 1/(AC)$. Note that if $A > 0$, r is an increasing function of R, while if $A < 0$, r is a decreasing function of R; thus the case when $A < 0$ represents a hollow cylinder, or a part of one, which is turned inside out. If $D = E = 0$, the deformation is a plane strain superposed upon a uniform extension perpendicular to the plane.

Either directly from (23.7) or by use of results in Sect. 49 of CFT we find that

$$\left.\begin{aligned}
\|B^{km}\| &= \left\|\begin{matrix} \frac{A^2R^2}{r^2} & 0 & 0 \\ \cdot & \frac{C^2}{R^2} + D^2 & \frac{CE}{R^2} + DF \\ \cdot & \cdot & \frac{E^2}{R^2} + F^2 \end{matrix}\right\|, \\
\|(B^{-1})_{km}\| &= \left\|\begin{matrix} \frac{r^2}{A^2R^2} & 0 & 0 \\ \cdot & A^2(E^2 + F^2R^2) & -A^2(CE + DFR^2) \\ \cdot & \cdot & A^2(C^2 + D^2R^2) \end{matrix}\right\|, \\
I &= \frac{A^2R^2}{r^2} + r^2\left(\frac{C^2}{R^2} + D^2\right) + \frac{E^2}{R^2} + F^2, \\
II &= \frac{r^2}{A^2R^2} + \frac{A^2}{r^2}(E^2 + F^2R^2) + A^2(C^2 + D^2R^2).
\end{aligned}\right\} \tag{57.14}$$

Putting (57.14) into (49.5) shows that the physical components of the extra-stress are functions of r only and that (56.7) is satisfied. The stress thus falls under Case B of Sect. 56, so that Eqs. (56.8) through (56.11) apply. Substituting into $(56.8)_3$ the value of $T\langle rr\rangle - T\langle\theta\theta\rangle$ as given by (57.14) and (49.5), we see that

$$\left.\begin{aligned}
T\langle rr\rangle &= -\int\left\{\left[\frac{A^2R^2}{r^3} - r\left(\frac{C^2}{R^2} + D^2\right)\right]\beth_1 - \left[\frac{A^2}{r^3}(E^2 + F^2R^2) - \frac{r}{A^2R^2}\right]\beth_{-1}\right\}dr, \\
T\langle\theta\theta\rangle &= T\langle rr\rangle - \left[\frac{A^2R^2}{r^2} - r^2\left(\frac{C^2}{R^2} + D^2\right)\right]\beth_1 + \left[\frac{A^2}{r^2}(E^2 + F^2R^2) - \frac{r^2}{A^2R^2}\right]\beth_{-1}, \\
T\langle zz\rangle &= T\langle rr\rangle - \left[\frac{A^2R^2}{r^2} - \frac{E^2}{R^2} - F^2\right]\beth_1 + \left[A^2(C^2 + D^2R^2) - \frac{r^2}{A^2R^2}\right]\beth_{-1}, \\
T\langle\theta z\rangle &= (CE + DFR^2)\left(\frac{r}{R^2}\beth_1 - \frac{A^2}{r}\beth_{-1}\right),
\end{aligned}\right\} \tag{57.15}$$

where $R^2 = (r^2 - B)/A$. By choice of the constant of integration in $(57.15)_1$, one cylinder $r = r_1$ may be rendered free of traction. In some cases a suitable relation between the constants $A, B, \ldots F$ yields a solution in which a second cylinder $r = r_2$ is free of traction.

Formulae for resultant forces and torques on the boundaries are easy to calculate from (57.15) but are elaborate. At this level of generality we calculate only the actions upon the annulus $r_1 \leqq r \leqq r_2$ in the plane $z = \text{const.}$, taking care to recall, however, that only when $E = 0$ is such a plane a free boundary. The resultant twisting moment T is then found as follows from $(57.15)_4$:

$$T = \int_{r_1}^{r_2} r\,T\langle\theta z\rangle\,2\pi r\,dr = 2\pi A\int_{R_1}^{R_2}\left(DF + \frac{CE}{R^2}\right)\left[(AR^2 + B)\beth_1 - A^2R^2\beth_{-1}\right]R\,dR, \tag{57.16}$$

where $R_a=\sqrt{(r_a^2-B)/A}$, $a=1, 2$. The resultant normal force N is gotten by substituting (57.15) into (56.11):

$$\left.\begin{aligned} N=&\,\pi(A R^2+B)\, T\langle rr\rangle \Big|_{R_1}^{R_2} - \\ &-\pi A\int_{R_1}^{R_2}\Big\{\Big[\frac{A^2R^2}{A R^2+B}-2\Big(\frac{E^2}{R^2}+F^2\Big)+(A R^2+B)\Big(\frac{C^2}{R^2}+D^2\Big)\Big]\beth_1 - \\ &-\Big[2A^2(C^2+D^2R^2)-\frac{A R^2+B}{A^2R^2}-\frac{A^2(E^2+F^2R^2)}{A R^2+B}\Big]\beth_{-1}\Big\} R\,dR. \end{aligned}\right\} \tag{57.17}$$

Important conclusions are drawn from these results in various special cases we now consider.

Case 3a (Pure torsion). Put $A=1$, $B=0$, $C=1$, $E=0$, $F=1$. The constant D is then the twist. We have $I=II=3+D^2r^2$, and the formulae (57.15) reduce to

$$\left.\begin{aligned} T\langle rr\rangle &= D^2\int r\,\hat{\beth}_1\,dr, & T\langle\theta\theta\rangle &= T\langle rr\rangle + D^2r^2\hat{\beth}_1, \\ T\langle zz\rangle &= T\langle rr\rangle + D^2r^2\hat{\beth}_{-1}, & T\langle\theta z\rangle &= D\,r(\hat{\beth}_1-\hat{\beth}_{-1}), \end{aligned}\right\} \tag{57.18}$$

where $\hat{\beth}_\Gamma=\hat{\beth}_\Gamma(r^2D^2)$ is defined by (55.11). If the surface $r=r_2$ is free of traction, it is generally impossible for any other surface $r=$const. to be free; for example, if the E-inequalities (53.3) hold, then $T\langle rr\rangle<0$ if $r<r_2$. This result means that pure torsion of a hollow cylinder can be effected only by aid of a suitable internal pressure if the outside is free. The analogy between torsion and shear (CFT, Sect. 48) makes it natural that the shear stress in torsion should be determined by the shear modulus alone: $T\langle\theta z\rangle=Dr\hat{\mu}(D^2r^2)$, where $\hat{\mu}$ is defined by $(54.9)_1$. The presence of the normal stress $T\langle zz\rangle$, proportional to D^2 for small values of D, is evidence of the *Poynting effect for torsion*, for if this normal stress is not supplied, the corresponding part of the cylinder being twisted will tend to shorten or lengthen according as $T\langle zz\rangle>0$ or $T\langle zz\rangle<0$. Note that $T\langle zz\rangle$ is a function of r even when the response coefficients are constants. A separation of effects much like that in simple shear is seen in (57.18). The normal stresses are even functions of D, while the shear stress is an odd function. The coefficient $\hat{\beth}_1(D^2r^2)$ determines $T\langle rr\rangle$ and $T\langle\theta\theta\rangle$, but, in distinction to the case of shear, both $\hat{\beth}_1$ and $\hat{\beth}_{-1}$ are required to determine $T\langle zz\rangle$. Since the functions $\hat{\beth}_\Gamma$ occurring in (57.18) are the same as those occurring in (55.10), we infer the following capital result: *Measurement of the shear stress and any one normal stress in the simple shear of an isotropic incompressible elastic material when the planes of shear are free suffices to determine the behavior of a cylinder of the same material in simple torsion.* This fact gives a rational position to the common loose claim that torsion is "locally equivalent" to shear, for isotropic incompressible elastic materials.

The formulae (57.16) and (57.17) for the resultant torque and force are specialized as follows:

$$\left.\begin{aligned} T&=2\pi D\int_{r_1}^{r_2} r^3\hat{\mu}(D^2r^2)\,dr=\frac{\pi}{D}\int_{r_1}^{r_2} r^2\Big(\hat{\beth}_1\frac{dI}{dr}-\hat{\beth}_{-1}\frac{dII}{dr}\Big)\,dr, \\ N&=-\pi\Big[-r^2\,T\langle rr\rangle\Big|_{r_1}^{r_2}+D^2\int_{r_1}^{r_2} r^3(\hat{\beth}_1-2\hat{\beth}_{-1})\,dr\Big]. \end{aligned}\right\} \tag{57.19}$$

It is worth noting that *the torsional couple T is determined by the generalized shear modulus, $\hat{\mu}$.* The normal force, however, does not seem to be related in any simple way to any single normal traction in simple shear. When the outer bounding cylinder $r=r_2$ is free of traction, by putting $(57.18)_1$ into $(57.19)_3$ we show that

$$N = -\pi D^2 \int_{r_1}^{r_2} \left[r(r^2 - r_1^2)\,\hat{\beth}_1 - 2r^3\,\hat{\beth}_{-1}\right] dr. \tag{57.20}$$

If the E-inequalities (53.3) are assumed, it follows from (57.20) that $N<0$; that is, the resultant force on the ends required to effect the pure torsion is *always a pressure*, not a tension. This fact suggests explanation of the Poynting effect in its classic form: *When an incompressible cylinder, free on its outer surface, is twisted, it experiences an elongation ultimately proportional to the square of the twist.* What has been shown here is a little less, namely, that pressure ultimately proportional to the square of the twist must be supplied in order to prevent elongation; we now proceed to set up a more general problem, from whose solution we can prove the italicized statement above.

Case 3b (Extension, torsion, and inflation). Put $C=1$, $E=0$; hence $AF=1$. We may regard the deformation as being a longitudinal stretch of amount F, followed by a torsion in which the twist is D/F. Instead of the extension, we may think of an inflation, since either must be accompanied by the other if there is to be no change of volume.

Two results of interest are included here. First, there is the possibility of explaining the fact, known since the classical experiments of COULOMB, that subjecting a cylinder to large tensile force changes its torsional modulus, even for twists small enough that the torque-twist relation remains linear. We consider a solid rod of radius R_2 in the undeformed state and hence take $B=0$, so that also $A>0$. Putting $r_1=R_1=0$ now reduces Eqs. (57.16) and (57.17) to

$$\left.\begin{aligned} \frac{T}{D} &= \frac{2\pi}{F}\int_0^{R_2}\left(\beth_1 - \frac{1}{F}\beth_{-1}\right) R^3\, dR,\\ N &= 2\pi\left(F - \frac{1}{F^2}\right)\int_0^{R_2}\left(\beth_1 - \frac{1}{F}\beth_{-1}\right) R\, dR - \frac{\pi D^2}{F^2}\int_0^{R_2}\left(\beth_1 - \frac{2}{F}\beth_{-1}\right) R^3\, dR. \end{aligned}\right\} \tag{57.21}$$

Now let D become small, and set $\tilde{\beth}_\Gamma \equiv \tilde{\beth}_\Gamma(F) = \beth_\Gamma(F^2+2/F,\ 1/F^2+2F)$. Since $\tilde{\beth}_\Gamma = \text{const.}$, we have from (57.21)

$$\left.\begin{aligned} \lim_{D\to 0}\frac{T}{D/F} &= \frac{1}{2}\pi R_2^4\left(\tilde{\beth}_1 - \frac{1}{F}\tilde{\beth}_{-1}\right) \equiv \tau(F),\\ \lim_{D\to 0} N &\equiv N_0 = \pi R_2^2\left(F - \frac{1}{F^2}\right)\left(\tilde{\beth}_1 - \frac{1}{F}\tilde{\beth}_{-1}\right). \end{aligned}\right\} \tag{57.22}$$

Since the twist superimposed upon the stretch F is D/F, the result $(57.22)_1$ gives the modulus of torsion $\tau(F)$ for this twist. The E-inequalities (53.3) are sufficient that this modulus be positive. In the present notation, the classical modulus is $\tau(1) = \frac{1}{2}\pi R_2^4 \mu = \frac{1}{2}\pi R_2^4 \hat{\mu}(0) = \frac{1}{2}\pi R_2^4 (\tilde{\beth}_1(1) - \tilde{\beth}_{-1}(1))$. The inequalities given in Sect. 53 do not seem to offer any information regarding the behavior of $\tilde{\beth}_1 - \tilde{\beth}_{-1}/F$ as a function of F. The example when $\beth_1$ and $-\beth_{-1}$ are positive constants, which is consistent with those inequalities, shows that it is possible for extension to have the effect of softening the incompressible material as far as torsion is concerned, while contraction stiffens it. Perhaps this example is typical.

Taking the ratio of $(57.22)_2$ to $(57.22)_4$ yields a remarkable formula of RIVLIN:

$$\frac{R_2^2 N_0}{\tau(F)} = 2\left(F - \frac{1}{F^2}\right), \tag{57.23}$$

whereby the torsional modulus is related directly to the tensile force N_0 and the stretch F. This result is a *universal relation*, valid for all kinds of isotropic incompressible elastic materials, independently of the form of the response coefficients $\beth_\Gamma$. The quantity on the right-hand side of (57.23) is an increasing function of F; *the greater the stretch, the greater must be the magnitude of the stretching force in proportion to the torsional modulus.* Note that this conclusion, not dependent on the various inequalities tentatively considered in Sect. 53, is universal.

The exact problem of POYNTING may be set up formally as follows: for given D, set $N=0$. Then $(57.21)_2$ may be regarded as an equation for the elongation F which renders the plane ends of the cylinder free of resultant normal tension. If upon rigid planes bonded to these ends we apply a torsional couple sufficient to produce a twist D with respect to the undeformed cylinder, then F is the elongation that will result. Under what conditions a solution of the resulting equation exists in general is difficult to see, but the inequalities (53.3) are obviously sufficient that $F>1$ for any solution F, proving that, subject to the assumption that the E-inequalities hold, *the cylinder will always elongate when twisted.* Since $F\to 1$ as $D\to 0$, from $(57.21)_2$ we easily show that

$$\left.\begin{aligned}\lim_{D\to 0}\frac{F-1}{(D/F)^2}&=\frac{1}{3}\lim_{D\to 0}\frac{F^3-1}{D^2}=\frac{1}{12}R_2^2\,\frac{\tilde\beth_1(1)-2\tilde\beth_{-1}(1)}{\tilde\beth_1(1)-\tilde\beth_{-1}(1)}\\&=\frac{1}{12}R_2^2\left(1-\frac{\tilde\beth_{-1}(1)}{\mu}\right),\end{aligned}\right\}\qquad(57.24)$$

where μ is the classical shear modulus. The limit so obtained might be called the *Poynting modulus*, since it gives the ratio of the extension to the square of the twist. If $-\tilde\beth_{-1}(1)/\mu\geqq 0$, and this is the case if the E-inequalities (53.3) hold, the Poynting modulus is positive and is never less than $\frac{1}{12}R_2^2$. The formula (57.24) may be used to give a direct interpretation for the material function $\beth_{-1}$ in small strain, since $\tilde\beth_{-1}(1)=\beth_{-1}(3,3)$. In the infinitesimal theory of incompressible materials, no such modulus as $\beth_{-1}$ appears, since the material properties are exhausted by the shear modulus, μ. Thus, as might be expected, *the Poynting effect is governed by a modulus which cannot be explained in terms of the concepts and formalism of the infinitesimal theory*[1].

Now set $D=0$, so that only expansion without torsion takes place. The condition that the cylinders $R=R_1$ and $R=R_2$ be deformed into cylinders free of traction follows at once from $(57.15)_1$:

$$0=\int_{R_1}^{R_2}\left[\frac{R^2}{(R^2+B/A)^2}-\frac{1}{R^2}\right]\left(\beth_1-\frac{1}{A^2}\beth_{-1}\right)R\,dR,\qquad(57.25)$$

while by (57.17) the condition that also the plane ends be free of resultant axial force is

$$\left.\begin{aligned}0=\int_{R_1}^{R_2}\Big[&\left(\frac{A^2R^2}{AR^2+B}-\frac{2}{A^2}+\frac{AR^2+B}{R^2}\right)\beth_1+\\&+\left(\frac{AR^2+B}{A^2R^2}-2A^2+\frac{R^2}{AR^2+B}\right)\beth_{-1}\Big]R\,dR.\end{aligned}\right\}\qquad(57.26)$$

These equations are to be solved for A and B. If, in addition to the trivial solution $A=1, B=0$, there is a solution in which $A<0$, it will be appropriate to the eversion of a tube subject to no applied forces. Since $(57.13)_4$ reduces to $AF=1$,

[1] Historical references and further discussion are given in Sect. 70, where the corresponding problem for compressible materials is solved for sufficiently small twists.

in eversion necessarily $F<0$. For special elastic materials (Sect. 95), it is known that a unique solution to (57.25) and (57.26) exists, but in general, little can be said about the matter. If we assume that the E-inequalities (53.3) hold, some inequalities restricting the inner and outer radii after eversion may be derived by considering values of B/A and A for which (57.25) and (57.26) have no solution.

While dynamical problems will be treated more generally in Sect. 61, here we consider one that reduces immediately to a case of the static problem just solved: a solid cylindrical shaft rotating steadily about its axis at angular speed ω. At the beginning of the previous section it was remarked that any conservative body force may be included by simply adjusting the pressure; since centrifugal force may be regarded as a body force with potential $v=-\frac{1}{2}\omega^2 r^2$, the stresses arising from rotation are known at once and may be added to any solution, but we consider only the case of a spinning cylinder since it is one in which the results correspond to a natural problem and are easy to assess. We put $B=D=E=0$, $C=1$, $AF=1$ in (57.13), and by adding a pressure of amount $\frac{1}{2}\varrho\,\omega^2 r^2$ to (57.15) obtain the stresses

$$\left.\begin{aligned} T\langle rr\rangle &= T\langle\theta\theta\rangle = -\frac{1}{2}\varrho\omega^2 r^2 + \text{const.} = \frac{1}{2}\frac{\varrho\omega^2}{F}(R_0^2-R^2),\\ T\langle zz\rangle &= \frac{1}{2}\frac{\varrho\omega^2}{F}(R_0^2-R^2) + \left(F^2-\frac{1}{F}\right)\left(\tilde{\beth}_1-\frac{1}{F}\tilde{\beth}_{-1}\right), \end{aligned}\right\} \tag{57.27}$$

where $R=R_0$ is the cylindrical boundary before the rotation is started, and where the functions $\tilde{\beth}_\Gamma$ are those occurring in (57.22). While $T\langle zz\rangle$ is a function of R for all materials, the resultant tension N on the plane ends may be made to vanish if

$$\frac{1}{4}\frac{\varrho\omega^2 R_0^2}{F^2} + \left(F-\frac{1}{F^2}\right)\left(\tilde{\beth}_1-\frac{1}{F}\tilde{\beth}_{-1}\right)=0. \tag{57.28}$$

This is an equation for F, the longitudinal extension incident upon the rotation. In terms of the modulus of torsion $(57.22)_2$, we may write (57.28) in the universal form

$$\frac{\pi\varrho\omega^2 R_0^6}{8F^2\tau(F)} = \frac{1}{F^2} - F. \tag{57.29}$$

Thus the physically natural condition that the torsional modulus be positive, no matter how great is F, is seen to be equivalent to the equally natural condition that spinning a shaft shall always make it shorten and thicken. Both these conditions are satisfied if the E-inequalities (53.3) are assumed.

Case 3c (Pure bending). Put $D=E=0$. The resulting deformation represents the bending of one annular wedge into another. If $A<0$, the curvature is reversed in the bending. By $(57.15)_1$ the condition that the curved faces $r=r_1$ and $r=r_2$ of the deformed block be free of traction is

$$\int_{r_1}^{r_2}\left[\beth_1\left(\frac{A^2R^2}{r^3}-\frac{C^2 r}{R^2}\right)+\beth_{-1}\left(\frac{r}{A^2R^2}-\frac{R^2}{C^2r^3}\right)\right]dr=0. \tag{57.30}$$

If the constants A, B, C can be selected so as to satisfy this relation, by putting $(57.15)_1$ into (56.10) we see that the resultant couple acting upon the ends $\theta=\pm\theta_0$, per unit height in the deformed body, is given by

$$M=\frac{1}{2}\int_{r_1}^{r_2} r^2\frac{d}{dr}T\langle rr\rangle\,dr=\frac{1}{2}\int_{r_1}^{r_2}\left[\left(\frac{C^2r^3}{R^2}-\frac{A^2R^2}{r}\right)\beth_1-\left(\frac{r^3}{A^2R^2}-\frac{R^2}{C^2r}\right)\beth_{-1}\right]dr. \tag{57.31}$$

All these results are referred to the deformed body but of course can easily be expressed in terms of the undeformed body.

A simple "dislocation" is also included in the present case. From a solid right circular cylinder of radius R_0, remove the material filling the wedge between the planes $\Theta=2\pi$ and $\Theta=2\pi-G$, where $0<G<2\pi$, and let these faces be brought together and joined in such a way that the body again becomes a right circular cylinder. The body is materially uniform, but after the faces are joined it does not remain homogeneous in the sense of Sect. 34, since a singularity is produced along the plane $r=0$. In addition to putting $D=E=0$ as before, we now put $B=0$, $C=2\pi/G$, so that $A=G/(2\pi F)$. Since by $(57.14)_{3,4}$

$$I=\frac{G}{2\pi F}\left(1+\frac{4\pi^2}{G^2}\right)+F^2,\qquad II=\frac{G}{2\pi F}\left(F^2+\frac{2\pi}{FG}\right)+\frac{2\pi F}{G},\tag{57.32}$$

the response coefficients $\beth_\Gamma(I, II)$ are independent[1] of r. Denoting these constants by $\hat{\beth}_\Gamma(F, G)$, by specializing $(57.15)_1$ and (57.17) to the present case we obtain

$$\left.\begin{aligned}T\langle rr\rangle&=-\frac{G}{2\pi F}\left(\frac{4\pi^2}{G^2}-1\right)\left(\hat{\beth}_1-F^2\hat{\beth}_{-1}\right)\log\frac{R_0}{R},\\ N&=\frac{R_0^2 G}{4F}\left\{\left[2F^2-\frac{1}{F}\left(\frac{G}{2\pi}+\frac{2\pi}{G}\right)\right]\hat{\beth}_1+\left[-F\left(\frac{G}{2\pi}+\frac{2\pi}{G}\right)+\frac{2}{F^2}\right]\hat{\beth}_{-1}\right\},\end{aligned}\right\}\tag{57.33}$$

where the constant of integration has been chosen so that $T\langle rr\rangle=0$ when $R=R_0$. G is regarded as given, and the requirement $N=0$ is set up as an equation to determine the stretch F resulting from the dislocation. It is not clear under what conditions this equation always has a root F. If we adopt the E-inequalities (53.3), then a necessary condition for the existence of a root is that the coefficients of $\hat{\beth}_1$ and $\hat{\beth}_{-1}$ in $(57.33)_2$ be of the same sign. The condition is easily shown to be equivalent to

$$\left[\frac{1}{2}\left(\frac{G}{2\pi}+\frac{2\pi}{G}\right)\right]^{-\frac{1}{3}}<F<\left[\frac{1}{2}\left(\frac{G}{2\pi}+\frac{2\pi}{G}\right)\right]^{\frac{1}{3}},\tag{57.34}$$

but this inequality does not discriminate between elongations and contractions.

Family 4 (Inflation or eversion of a sector of a spherical shell)[2]. Let the point with spherical co-ordinates R, Θ, Φ be deformed into one with spherical co-ordinates r, θ, φ, where

$$r=(\pm R^3+A)^{\frac{1}{3}},\qquad \theta=\pm\Theta,\qquad \varphi=\Phi.\tag{57.35}$$

This deformation carries the region between two concentric spheres into a region between two other concentric spheres; any angular boundaries are preserved. The $\pm$ signs are associated; if both are taken as $+$, the curvature of the shell is unchanged in sign, while if both are taken as $-$, the shell is everted. Either by use of the formulae in Sect. 52 of CFT or directly from (23.7) we find that

$$\left.\begin{aligned}\|B^k_m\|&=\operatorname{diag}\left(\frac{R^4}{r^4},\frac{r^2}{R^2},\frac{r^2}{R^2}\right),&\qquad \|(B^{-1})^k_m\|&=\operatorname{diag}\left(\frac{r^4}{R^4},\frac{R^2}{r^2},\frac{R^2}{r^2}\right),\\ I&=\frac{R^4}{r^4}+\frac{2r^2}{R^2},&\qquad II&=\frac{r^4}{R^4}+\frac{2R^2}{r^2}.\end{aligned}\right\}\tag{57.36}$$

Putting this result into (49.5) shows that the conditions (56.12) are satisfied and that $T\langle rr\rangle+p$ and $T\langle\theta\theta\rangle+p$ are functions of r only. Therefore, the stress system falls under Case C of Sect. 56, so that the results (56.14) may be applied. From

[1] This fact was overlooked by ERICKSEN and RIVLIN [1954, *7*, § 11], who, accordingly, did not render the solution as explicit as that given above.

[2] The solution for inflation, for hyperelastic materials, was obtained by GREEN and SHIELD [1950, *3*, § 7]; that for eversion, by ERICKSEN [1955, *10*].

(49.5) and $(56.14)_2$ we thus conclude that

$$\left.\begin{aligned} T\langle rr\rangle &= 2\int\left[\left(\frac{r^2}{R^2}-\frac{R^4}{r^4}\right)\beth_1-\left(\frac{r^4}{R^4}-\frac{R^2}{r^2}\right)\beth_{-1}\right]\frac{dr}{r}, \\ T\langle\theta\theta\rangle &= T\langle\varphi\varphi\rangle = T\langle rr\rangle+\left(\frac{r^2}{R^2}-\frac{R^4}{r^4}\right)\beth_1-\left(\frac{r^4}{R^4}-\frac{R^2}{r^2}\right)\beth_{-1}. \end{aligned}\right\} \tag{57.37}$$

Consider first the case when the + signs are used in (57.35), so that r is an increasing function of R. If the inner and outer radii of the undeformed shell are R_1 and R_2, we have then, if we set $\varpi(r) = -T\langle rr\rangle =$ the radial pressure,

$$\varpi(R_1)-\varpi(R_2) = 2\int_{R_1}^{R_2}\left[\left(\frac{r^2}{R^2}-\frac{R^4}{r^4}\right)\beth_1-\left(\frac{r^4}{R^4}-\frac{R^2}{r^2}\right)\beth_{-1}\right]\frac{R^2\,dR}{r^3}, \tag{57.38}$$

where $r^3 = R^3 + A$. If $\varpi(R_1)-\varpi(R_2)$ is given, Eq. (57.38) is to be solved for A. If a unique solution exists, then the inner and outer radii of the deformed shell are determined at once from $(57.35)_1$. If, on the other hand, we assign the inner radius $r_1 = \sqrt[3]{R_1^3+A}$ of the deformed shell, then A is known, and (57.38) gives directly the unique difference of pressures which suffices to maintain this radius.

Fig. 4. RIVLIN's torsion apparatus (1947).

Finally, consider the case when the − signs are used in (57.35), so that the sphere is everted. The most interesting deformation of this kind is that in which both surfaces of the deformed shell are free of traction; to effect such a deformation in a complete spherical shell, it is necessary to suppose that first a cut is made somewhere, then the shell is turned inside out and afterwards joined together again. The condition that $T\langle rr\rangle = 0$ on both deformed surfaces is

$$0 = \int_{R_1}^{R_2}\left[\left(\frac{r^2}{R^2}-\frac{R^4}{r^4}\right)\beth_1+\left(\frac{R^2}{r^2}-\frac{r^4}{R^4}\right)\beth_{-1}\right]\frac{R^2\,dR}{r^3}, \tag{57.39}$$

where $r^3 = A - R^3$. Whether or not this equation has a solution A such that $A > R_2^3$ depends on the response coefficients; in important special cases it does (see Sect. 95).

If the deformations of this family are to be produced in portions of a spherical shell rather than a complete shell, it is easy to calculate the forces that must be supplied upon the edges.

Fairly extensive experiments confirming some of the results in this section have been carried out on rubber specimens. While the classic experiment of RIVLIN[1] on torsion of a short, solid, circular cylinder (Fig. 4) was interpreted in terms of the special theory in which the response coefficients are constants, later and more accu-

[1] RIVLIN [1947, 9].

rate work[1] on combined torsion, extension, and inflation of circular-cylindrical tubes gives data in good agreement with the results of the general theory and with the values of the response coefficients obtained by measurements on homogeneous strain, mentioned in Sect. 55. In particular, the formula (57.23) is confirmed very closely. Experiments on the eversion of a tube[2] are harder to interpret because the theoretical formulae are not fully explicit, but here, too, the agreement seems to be satisfactory. A summary of the experimental results and a discussion of their relation to the theory has been given by GREEN and ADKINS[3].

58. Incompressible elastic materials, IV. Non-homogeneous solutions for bodies with certain kinds of aeolotropy and inhomogeneity[4]. While in the next section we shall look at the problem of finding solutions for aeolotropic and inhomogeneous bodies from a somewhat broader viewpoint, here we remark that the four families of deformations just shown to be possible in all homogeneous incompressible isotropic bodies are possible also in some incompressible aeolotropic bodies possessing certain kinds of inhomogeneity. The guiding idea is that the aeolotropy and inhomogeneity shall be restricted in such a way that:

α) The physical components $T\langle km\rangle + p\,\delta_{km}$ of the extra stress, $\boldsymbol{T}+p\boldsymbol{1}$, do not depend on more of the co-ordinates than they do in the case of an isotropic material.

β) A physical component of $\boldsymbol{T}+p\boldsymbol{1}$ vanishes if it does so for an isotropic material in the particular deformation considered.

Under these conditions, the analysis of Sect. 56 is again applicable; not only are the deformations thus proved compatible, but also the statical formulae there given remain in force. The generalization is not a trivial one, however, since the stresses themselves are more complicated functions of position and of material properties, and hence the surface tractions required to maintain the deformations are also more elaborate.

The appropriate stress relation is $(43.14)_4$, a component form of which is

$$\left.\begin{aligned}\sqrt{g_{kk}g_{mm}}\,(T\langle km\rangle + p\,\delta_{km})\,x^k{}_{,\alpha}\,x^m{}_{,\beta} = (T_{km}+p\,g_{km})\,x^k{}_{,\alpha}\,x^m{}_{,\beta} = \overline{T}_{\alpha\beta}+p\,C_{\alpha\beta},\\ = \mathfrak{g}_{\alpha\beta}(C_{\mu\nu},\,\boldsymbol{d}_\varrho,\,X^\sigma)\,.\end{aligned}\right\}\tag{58.1}$$

By calculation from (57.1), (57.8), (57.13), and (57.35) we see that for each of the families of deformations considered, the derivatives $x^k{}_{,\alpha}$, and hence also the components $C_{\mu\nu}$, are functions of just the same single co-ordinate as are the components B^k_m. For Families 1, 2, and 3, the metric components g_{kk}, in the co-ordinates used, depend only upon that same variable. Therefore, Condition α is satisfied for these three families when the dependence of $\mathfrak{g}_{\alpha\beta}$ upon X^σ and the natural basis $\boldsymbol{d}_\varrho$ of the material co-ordinates reduces to a dependence upon that same variable only. Family 4 requires special consideration because the metric components g_{kk}, which are used in transformations from covariant to physical components, depend upon θ as well as upon r. We now consider the four families in detail.

[1] RIVLIN and SAUNDERS [1951, *14*, §§ 15–19], GENT and RIVLIN [1952, *8*].

[2] GENT and RIVLIN [1952, *7*].

[3] GREEN and ADKINS [1960, *26*, Ch. X]. The more recent work of ZAHORSKI [1962, *79*] seems to be compromised by a predisposition toward a special form of stress relation.

[4] The presentation by GREEN and ADKINS [1960, *26*, §§ 2.8–2.10] gives the misleading impression that their results are valid for all kinds of aeolotropic materials. This is not true. Up to the present, the classes of deformations considered here have been shown to be possible, subject to surface tractions alone, *only* for bodies whose aeolotropies are such as to ensure that conditions (58.3), (58.5), and (58.7), respectively, are satisfied. For more general incompressible elastic bodies, solution of the problems of torsion of a cylinder, bending of a block, etc., requires introduction of additional deformations determined by the nature of the material, as is shown in Sect. 59.

Family 1. The deformation is given by (57.1). In the co-ordinates used to specify (57.1), the $X^\alpha{}_{,k}$ and $C_{\mu\nu}$ depend only on X in the material description, or on r in the spatial description. The vectors $\boldsymbol{d}_\alpha$ of the natural basis are constant. Assume that the elastic response of the material is the same on each of the planes $X=$ const.; these are the planes bent into the concentric circular cylinders $r=$ const. In the direction normal to them, the elastic response may vary arbitrarily. More precisely, let the inhomogeneity be such that in the Cartesian co-ordinates X, Y, Z already used

$$\mathfrak{g}_{\alpha\beta}=\mathfrak{g}_{\alpha\beta}(C_{\mu\nu}, X). \tag{58.2}$$

Condition α is now satisfied. To satisfy Condition β also, we need to make certain that (56.7) is satisfied. From (57.1) it is easy to show that this condition is equivalent to

$$\mathfrak{g}_{XY}(C_{\mu\nu}, X)=0, \quad \mathfrak{g}_{XZ}(C_{\mu\nu}, X)=0, \tag{58.3}$$

where for the components $C_{\mu\nu}$ the specific values calculated from (57.1) are to be substituted.

Various materials are known to satisfy these conditions. For the special case when $C=0$ in (57.1), the simplest example[1] is a material transversely isotropic with respect to the axes in the Z-direction.

Family 2. The deformation is given by (57.8). In the co-ordinates there used, the $X^\alpha{}_{,k}$ and the $C_{\mu\nu}$ depend only on R in the material description, or on r in the spatial description. The vectors $\boldsymbol{d}_\alpha$ of the natural basis depend on both R and Θ. Assume that the elastic response of the material is uniform on each of the cylinders $R=$ const., which are bent into the planes $x=$ const. More precisely, in the cylindrical co-ordinates R, Θ, Z already used, we assume that the physical components of $\overline{\boldsymbol{T}}+p\boldsymbol{C}$ depend only on the physical components of $\boldsymbol{C}$ and on R. This is the case, in particular, if the body is materially uniform and possesses the kind of curvilinear aeolotropy corresponding to the cylindrical co-ordinate system R, Θ, Z [cf. (34.14)]. In terms of covariant components, our assumption may be written

$$\overline{T}_{\alpha\beta}+p\,C_{\alpha\beta}=\mathfrak{g}_{\alpha\beta}(C_{\mu\nu}, R). \tag{58.4}$$

Condition α is now satisfied. To satisfy Condition β also, we need to make certain that (56.4) is satisfied. From (57.11) it is easy to show that this condition is equivalent to

$$\mathfrak{g}_{R\Theta}(C_{\mu\nu}, R)=0, \quad \mathfrak{g}_{RZ}(C_{\mu\nu}, R)=0, \tag{58.5}$$

where for $C_{\mu\nu}$ the specific values calculated from (57.8) are to be substituted.

Aeolotropies such as to render these equations valid do not seem yet to have been studied in detail.

Family 3. The deformation is given by (57.13). In the co-ordinates there used, the $X^\alpha{}_{,k}$ and the $C_{\mu\nu}$ depend only on R in the material description, or on r in the spatial description. As for Family 2, the $\boldsymbol{d}_\varrho$ depend on both R and Θ. Assume that the elastic response of the material is uniform on each of the cylinders $R=$ const., which are inflated into the cylinders $r=$ const. Thus, in the cylindrical co-ordinates R, Θ, Z already used, a result of the form (58.4) is assumed to hold.

[1] For hyperelastic materials, this solution is due to ERICKSEN and RIVLIN [1954, *7*, § 12], who gave the stress relation and stress system explicitly. Some results valid in a more general class of deformations, irrespective of material symmetries, were obtained by ADKINS [1955, *1*, § 8]. Some crystal classes satisfying (58.3) were determined by GREEN and ADKINS [1960, *26*, §§ 2.6 and 2.8]. However, their treatment of the general aeolotropic material [ibid., § 2.8] is partly insufficient, since their Eqs. (2.8.5) do not satisfy the conditions of equilibrium for all kinds of crystals.

Condition α is now satisfied. To satisfy Condition β also, we need to make sure that (56.7) is satisfied. From (57.13) it is easy to show that this condition is equivalent[1] to equations of the form (58.5), except, of course, that the arguments of the functions $\mathfrak{s}_{R\Theta}$ and $\mathfrak{s}_{RZ}$ are to be calculated from (57.13) in the present case.

Various materials are known to satisfy these conditions. The simplest example[2] is a material transversely isotropic with respect to the axis $r=0$.

Family 4. The deformation is given by (57.35). In the co-ordinates there used, the $X^\alpha{}_{,k}$ and the $C_{\mu\nu}$ depend only on R in the material description, or on r in the spatial description. The basis vectors $\boldsymbol{d}_\varrho$ depend on all three co-ordinates R, Θ, Φ. Assume that the elastic response of the material is uniform on each of the spheres $R=$ const., which are inflated into the spheres $r=$ const. More precisely, in the spherical co-ordinates R, Θ, Φ already used, we assume that the physical components of $\boldsymbol{T}+p\boldsymbol{C}$ depend only on the physical components of $\boldsymbol{C}$ and on R. This is the case, in particular, if the body possesses curvilinear aeolotropy appropriate to the spherical co-ordinates. We cannot conclude, as we did in the considerations for Families 1, 2, and 3, that Condition α is satisfied, because the transformation from covariant to physical components, both for the material and spatial co-ordinates, involves the metric components $g_{\Phi\Phi}$ and $g_{\varphi\varphi}$, which depend not only on R or r, but also on Θ or θ [cf. (6A.5)]. The symmetry of the material must be such that, for the particular values of $x^m{}_{,\alpha}$ calculated from (57.35), the dependence of $T\langle km\rangle+p\,\delta_{km}$ on θ drops out [as it has already been shown to do for isotropic materials, cf. (57.37)]. Condition α will then be satisfied. To satisfy Condition β, we need to make sure that (56.12) is satisfied. From (57.35) it is easy to show that this condition is equivalent to

$$\mathfrak{s}_{R\Theta}=\mathfrak{s}_{\Theta\Phi}=\mathfrak{s}_{\Phi R}=0, \qquad \sin^2\Theta\,\mathfrak{s}_{\Theta\Theta}=\mathfrak{s}_{\Phi\Phi}. \tag{58.6}$$

The last of these equations may be written in the form $\mathfrak{s}^\Theta_\Theta=\mathfrak{s}^\Phi_\Phi$. The problem of determining whether there be aeolotropies consistent with these conditions does not appear to have been studied.

59. Semi-inverse methods: Reduction of certain static deformations that depend upon material properties. As already stated, the class of static deformations that are possible in all homogeneous, isotropic, elastic bodies subject to surface tractions only is likely to be exhausted by the cases studied in Sects. 54—55 and 57. We expect that any other deformation which is possible for one isotropic material will fail to be possible for some other[3], unless suitable external body force be supplied. Some progress toward determining such deformations may be made if we replace the purely inverse method of Sects. 54—57 by a semi-inverse one, considering a family of deformations involving one or more arbitrary functions which may be determined so as to render the deformation possible for a particular material. In this section we shall sketch the beginning of the analysis for a few cases where the problem can be reduced to one of solving ordinary differential equations. At the same time, we abandon the restriction to isotropic materials, so

---

[1] The proof employs the fact that $FC-DE\neq 0$, as follows from (57.13).

[2] Ericksen and Rivlin [1954, *7*, § 5]. Green and Wilkes [1954, *10*, §§ 5, 7] extended the results to orthotropic materials. Adkins [1955, *1*, §§ 3—7] studied a more general problem including this one as a special case and determined some results valid irrespective of the symmetries of the material. Inflation, torsion, and extension of a cylinder of curvilinearly orthotropic material has been studied by Urbanowski [1959, *38*, § 3]. Various cases were discussed by Green and Adkins [1960, *26*, §§ 2.1 to 2.9], who determined also some crystal classes such that (58.5) is satisfied. Cf. their note of correction [1961, *1*].

[3] This fact makes an experimental program for compressible materials harder to plan and carry out. The work of Blatz and Ko [1962, *7*] presumes a simple class of response functions for compressible materials.

that the stress relation for compressible and incompressible materials, respectively, may be taken as $(43.12)_1$ or (58.1). In the special cases to be considered, we shall add assumptions of symmetry just sufficient to get a simple result, but never so strong as to require full isotropy. First we consider generalizations of Families 2 and 1 of Sect. 57.

Family 2. Suppose that the point with cylindrical co-ordinates (R, Θ, Z) in the natural state is deformed into that with cylindrical co-ordinates (r, θ, z), where

$$\left.\begin{aligned} &r=f(R), \quad \theta=g(R)+A\Theta+BZ, \quad z=h(R)+C\Theta+DZ, \\ &R=\overset{-1}{f}(r), \quad \Theta=\frac{Bh(\overset{-1}{f}(r))-Dg(\overset{-1}{f}(r))+D\theta-Bz}{AD-BC}, \\ &Z=\frac{-Ah(\overset{-1}{f}(r))+Cg(\overset{-1}{f}(r))-C\theta+Az}{AD-BC}, \quad (AD-BC)f'>0. \end{aligned}\right\} \tag{59.1}$$

This deformation[1], generalizing (57.13), includes torsion, extension, inflation, eversion, shearing, etc., of a part of a cylindrical tube. By (23.6) we find from (59.1) that

$$\|C_{\alpha\beta}\| = \begin{Vmatrix} f'^2+r^2g'^2+h'^2 & Ar^2g'+Ch' & Br^2g'+Dh' \\ \cdot & A^2r^2+C^2 & ABr^2+CD \\ \cdot & \cdot & B^2r^2+D^2 \end{Vmatrix}. \tag{59.2}$$

Hence $\sqrt{III_{\mathbf{C}}}=(AD-BC)ff'/R$, so that the isochoric subclass of (59.1) is characterized by

$$r=f(R)=\sqrt{\frac{R^2+E}{AD-BC}}, \tag{59.3}$$

while A, B, C, D, $g(R)$ and $h(R)$ remain arbitrary. Both in the isochoric and in the general case, then, the components $C_{\alpha\beta}$ and the derivatives $X^\alpha{}_{,k}$ are functions of r only.

We now assume the symmetries of the material to be such that the elastic response is uniform on each cylinder $r=$const. Various kinds of inhomogeneity and aeolotropy are included. In consequence, the stress components T_{km} as given by $(43.12)_1$ are functions of r only, and the equations of equilibrium for compressible materials reduce to the forms

$$\frac{d}{dr}T\langle rr\rangle+\frac{T\langle rr\rangle-T\langle\theta\theta\rangle}{r}=0, \quad \frac{d}{dr}(r^2T\langle r\theta\rangle)=0, \quad \frac{d}{dr}(rT\langle rz\rangle)=0, \tag{59.4}$$

while for incompressible materials they reduce to the forms (56.6). Since $(59.4)_1$ and $(56.6)_1$ are identical, all the statical formulae (56.8) through (56.10), and the conclusions drawn from them, are valid for these problems, whether the material be compressible or incompressible. In the case of a compressible material, substitution of the appropriate specialization of $(43.12)_1$ into (59.4) yields a system of differential equations, generally non-linear, for the three unknown functions $f(R)$,

[1] The case when $f(R)=\sqrt{\alpha R^2+\beta}$, $B=C=0$, $A=1$, which represents extension, inflation, and a double shear of a cylindrical tube, was considered by RIVLIN [1949, *16*, §§ 9—10] for isotropic incompressible materials. The case when $g(R)=h(R)=C=0$, $A=1$, which represents torsion, inflation, and extension, was set up for isotropic compressible materials by GREEN [1955, *13*, § 3]. For orthotropic materials, the case when also $B=0$ was studied by GREEN and WILKES [1954, *10*, § 7]. The case when $g(R)=h(R)=0$, and also the general deformation (59.1), were studied by ADKINS [1955, *1*, §§ 4—5] [1955, *2*, §§ 3—7]. The case when $g(R)=h(R)=0$ was studied for transversely isotropic materials by GREEN and ADKINS [1960, *26*, §§ 2.11—2.13, 2.15, 2.17]. All these authors considered only hyperelastic materials.

$g(R)$, and $h(R)$[1]. The form of this system depends, in general, upon those of the response functions $\mathfrak{s}_{\alpha\beta}$. For incompressible materials, $(56.8)_2$ may be used to eliminate p from $(56.6)_{2,3}$, so that a system of two equations for the unknown functions $g(R)$ and $h(R)$ results.

Only one special case within this class of deformations will be considered here, that of a homogeneous isotropic incompressible cylinder which is extended, inflated, and sheared:

$$r=\sqrt{A R^2+B}, \quad \theta=\Theta+g(R), \quad z=Z/A+h(R). \tag{59.5}$$

By the results in Sect. 49 of CFT we have formulae for B_{km} and $(B^{-1})_{km}$, I and II which need not be written here, for we only have to remark that these quantities are functions of r alone, and that by substitution into (49.5) we obtain, among other results, the formulae

$$T\langle r\theta\rangle=A R\left(\beth_1-\frac{1}{A^2}\beth_{-1}\right) g', \quad T\langle rz\rangle=A\left(\frac{R}{r}\beth_1-\frac{r}{R}\beth_{-1}\right) h'. \tag{59.6}$$

Since by $(56.6)_{2,3}$ we know that $r^2 T\langle r\theta\rangle=$ const., $r T\langle rz\rangle=$ const., from (59.6) we obtain the differential system

$$g'(R)=\frac{K}{R(A R^2+B)\left(\beth_1-\frac{1}{A^2}\beth_{-1}\right)}, \quad h'(R)=\frac{L R}{R^2\beth_1-(A R^2+B)\beth_{-1}}, \tag{59.7}$$

where K and L are constants, and where the response functions $\beth_\Gamma$ are given, explicit functions of I and II and hence of g', h', A, and B. For certain special elastic materials, these equations are known to have a solution.

Family 1. We consider the deformation which carries the particle with Cartesian co-ordinates (X, Y, Z) in the natural state into the place with cylindrical co-ordinates (r, θ, z) as follows:

$$\left.\begin{aligned} r=f(X), \quad \theta=g(X)+A Y+B Z, \quad z=h(X)+C Y+D Z,\\ (A D-B C) f'>0. \end{aligned}\right\} \tag{59.8}$$

This deformation[2] represents bending, extension, and variable shears of a rectangular block, generalizing (57.1). The isochoric subfamily is obtained by taking $f(X)=\sqrt{2X/(AD-BC)}$. Since

$$\|C_{\alpha\beta}\|=\left\|\begin{matrix} f'^2+r^2 g'^2+h'^2 & A r^2 g'+C h' & B r^2 g'+D h' \\ \cdot & A^2 r^2+C^2 & A B r^2+C D \\ \cdot & & B^2 r^2+D^2 \end{matrix}\right\|, \tag{59.9}$$

the components $C_{\alpha\beta}$ are functions of r only, and so are the derivatives $X^\alpha{}_{,k}$. Therefore if the elastic response of the material is uniform on each cylinder $r=$ const., the stress components as given by $(43.12)_1$ are functions of r only; the

---

[1] The claim of Reiner [1955, *20*], "the general stress-strain equation of an isotropic elastic material ... is solved exactly in closed form for the case of torsion of a circular cylinder," is not borne out by his analysis, which is limited to the constitutive equation of a very special elastic (and not hyperelastic) material. Indeed from a theorem of Ericksen, to be given below in Sect. 91, no single, fixed torsional deformation is possible in all compressible isotropic elastic materials.

[2] For hyperelastic materials, a deformation of this kind was studied by Adkins [1955, *1*, § 6] [1955, *2*, § 8] and by Adkins and Green [1961, *1*]. A special case for isotropic compressible materials had been considered earlier by Green [1955, *13*, § 3]; one for orthotropic materials, by Green and Wilkes [1954, *10*, § 5]. The case when $g(X)=h(X)=0$, for compressible or incompressible hyperelastic materials with various kinds of aeolotropy, was considered by Green and Adkins [1960, *26*, §§ 2.10, 2.14, 2.16, 2.18].

equations of equilibrium are again (59.4) or (56.6), the relations (56.8) through (56.11) are again applicable, etc., and the problem of determining f, g, and h for compressible materials, or p, g, and h for incompressible materials, is again reduced to the solution of a system of three explicitly given and generally non-linear ordinary differential equations.

Both of the deformations considered so far are members of a class for which the same method applies. Let there be given an orthogonal curvilinear co-ordinate system such that the components g_{kk} of the metric tensor are functions of x^1 only, and consider a deformation such that the derivatives $x^k{}_{,\alpha}$ are functions of x^1 only. For materials whose elastic response is uniform on each surface $x^1=\text{const.}$, the equations of equilibrium reduce to a system of ordinary differential equations. If a sufficient number of arbitrary functions are included in the deformation considered, this system may be compatible.

An example is furnished by the deformation

$$x=f(R), \quad y=g(R)+A\Theta+BZ, \quad z=h(R)+C\Theta+DZ, \tag{59.10}$$

in a body which is homogeneous or whose response function varies with R only. Deformations of this kind, representing the straightening out, extension, and variable shear of an annular wedge, include (57.8) as a special case.

Solutions for a slightly more general class of problems may be reduced to solving ordinary differential equations if a more restricted class of elastic bodies is considered. More general derivatives $x^k{}_{,\alpha}$ and metric components g_{kk} are allowed, but the symmetries of the material must be such that the combinations of stress components $T\langle km\rangle$ and metric components g_{kk} which occur in the equations of equilibrium [Eq. (205 A.1) of CFT] depend on one co-ordinate only. It is hard to formulate this method in general terms, but the idea is made clear by a simple example[1]. Consider the uniform inflation of a spherical shell, given in spherical co-ordinates by

$$r=f(R), \quad \theta=\Theta, \quad \varphi=\Phi, \tag{59.11}$$

where f is a function to be determined. Since the co-ordinate directions determine principal axes of strain at each point, in an isotropic material the shear stresses $T\langle km\rangle$, $k\neq m$ will vanish. Moreover, a simple calculation with (23.7) shows that the physical components and invariants of $\boldsymbol{B}$ and $\boldsymbol{B}^{-1}$ are functions of r only. It follows by (47.9) that the normal stresses $T\langle mm\rangle$ in an isotropic compressible material are functions of r only, and that $T\langle\theta\theta\rangle=T\langle\varphi\varphi\rangle$. Hence the three equations of equilibrium reduce to the single Eq. $(56.13)_1$; substitution of the appropriate constitutive equations (47.9) into it yields

$$\frac{d}{dr}\left(\beth_0+f'^2\beth_1+\frac{1}{f'^2}\beth_{-1}\right)=\frac{2}{r}\left[\left(\frac{r^2}{R^2}-f'^2\right)\beth_1-\left(\frac{1}{f'^2}-\frac{R^2}{r^2}\right)\beth_{-1}\right]. \tag{59.12}$$

Since $R=\overset{-1}{f}(r)$, and since $\beth_0$, $\beth_1$, and $\beth_{-1}$ are explicitly given functions of R, r, and f', Eq. (59.12) is an ordinary differential equation for determining the function $\overset{-1}{f}$ that is compatible with the material properties.

A slightly more complicated kind of deformation for isotropic incompressible materials has been considered by ADKINS[2]:

$$x=AX, \quad y=AY, \quad z=\frac{Z}{A^2}+f(AX, AY), \tag{59.13}$$

---

[1] GREEN [1955, *13*, § 4], for hyperelastic materials; further discussion was given by GREEN and ADKINS [1960, *26*, § 2.19]. Generalization to certain aeolotropic materials was effected by ADKINS [1955, *1*, § 7].

[2] ADKINS [1954, *1*] (for hyperelastic materials). A generalization to anisotropic materials was considered by ADKINS [1955, *2*, § 9].

where both systems of co-ordinates are Cartesian, where A is an assigned constant, and where $f(x, y)$ is a function to be determined. From (23.7) we obtain the formulae

$$\|B_{km}\| = \left\| \begin{matrix} A^2 & 0 & A^2 f_x \\ \cdot & A^2 & A^2 f_y \\ \cdot & \cdot & A^2(f_x^2+f_y^2)+\dfrac{1}{A^4} \end{matrix} \right\|, \quad \|(B^{-1})_{km}\| = \left\| \begin{matrix} A^4 f_x^2 + \dfrac{1}{A^2} & A^4 f_x f_y & -A^4 f_x \\ \cdot & A^4 f_y^2 + \dfrac{1}{A^2} & -A^4 f_y \\ \cdot & \cdot & A^4 \end{matrix} \right\|, \tag{59.14}$$

$$I = 2A^2 + \frac{1}{A^4} + A^2(f_x^2+f_y^2), \qquad II = \frac{2}{A^2} + A^4 + A^4(f_x^2+f_y^2).$$

If we set $h \equiv A^2 \beth_1 - A^4 \beth_{-1}$, $p' \equiv p - \beth_1/A^4 - A^4 \beth_{-1}$, the stress system given by (47.9) may be written in the form

$$\left.\begin{aligned} T\langle xx\rangle &= -p' + \left(1 - \frac{1}{A^6}\right) h + A^4 f_x^2 \beth_{-1}, \\ T\langle yy\rangle &= -p' + \left(1 - \frac{1}{A^6}\right) h + A^4 f_y^2 \beth_{-1}, \\ T\langle zz\rangle &= -p' + A^2 (f_x^2 + f_y^2) \beth_1, \\ T\langle xy\rangle &= A^4 f_x f_y \beth_{-1}, \qquad T\langle yz\rangle = f_y h, \qquad T\langle zx\rangle = f_x h. \end{aligned}\right\} \tag{59.15}$$

The conditions of integrability for (56.1), i.e., the appropriate specializations of (56.2), imply

$$\frac{\partial p}{\partial z} = \text{const.}, \qquad \left(\frac{\partial^2}{\partial x^2} - \frac{\partial^2}{\partial y^2}\right)(f_x f_y \beth_{-1}) = \frac{\partial^2}{\partial x\, \partial y}\left[(f_x^2 - f_y^2)\beth_{-1}\right]. \tag{59.16}$$

The results can be simplified by writing $J \equiv f_x^2 + f_y^2$ and noting that $\beth_1$ and $\beth_{-1}$ are functions of J, so that their derivatives are simply related. If we write $h = H(J)$ and put (59.15) and $(59.16)_1$ into the z-component of (56.1), we obtain

$$(\Delta f) H + \bar{J} H' = \text{const.}, \qquad \Delta f \equiv f_{xx} + f_{yy}, \tag{59.17}$$

where $\bar{J} \equiv f_x \dfrac{\partial J}{\partial x} + f_y \dfrac{\partial J}{\partial y}$. If we write $\beth_{-1} = H_{-1}(J)$, then it is possible to show that $(59.16)_2$ assumes the form

$$\frac{\partial(\Delta f, f)}{\partial(x, y)} H_{-1} + \frac{\partial(\bar{J}, f)}{\partial(x, y)} H'_{-1} + \frac{\partial(J, f)}{\partial(x, y)}\left[(\Delta f) H'_{-1} + \bar{J} H''_{-1}\right] = 0. \tag{59.18}$$

For a given material, the functions H and H_{-1} are given; thus (59.17) and (59.18) constitute a system for determination of f. It is possible, however, to find functions f such that all three Jacobians occurring in (59.18) are zero. Such functions include

$$f = \varphi(k_1 x + k_2 y) \quad \text{and} \quad f = \psi(x^2 + y^2). \tag{59.19}$$

With these choices of f, (59.17) becomes an ordinary differential equation for determining φ or ψ. A still simpler choice of f is

$$f = f(x). \tag{59.20}$$

Then (59.16) is satisfied, while (59.17) may be integrated once to yield a relatively simple ordinary differential equation[1] for f:

$$f' H = A + B x. \tag{59.21}$$

The authors who have studied these classes of deformations have derived various relations connecting resultant forces and moments, and in some cases have rendered the solutions more explicit, but all this further work depends upon additional specializing assumptions. These are of two kinds. Either special forms are assumed for $\beth_1$ and $\beth_{-1}$, or certain integrals are assumed to be functions of certain parameters. Some of the results from the former type of specialization will be mentioned in Sect. 95.

[1] Adkins, Green, and Shield [1953, *1*, § 8].

60. Plane problems[1]. We first consider a plane strain combined with a uniform stretch λ normal to its plane. Leaving both the material and the spatial co-ordinate systems for the plane arbitrary (X^α and x^k, respectively), for such a deformation we may write

$$x^k = x^k(X^1, X^2), \qquad k=1,2, \qquad x^3=\lambda X^3, \tag{60.1}$$

where the x^3 and X^3 co-ordinates are distances normal to the plane, so that for the metric components we have $g_{3k}=0$ $(k\neq 3)$, $g_{33}=1$, $g_{3\alpha}=0$ $(\alpha\neq 3)$, $g_{33}=1$. In this section, all indices have the range 1, 2, while all 3-components are written expressly. Calculating the Cauchy-Green tensor $\boldsymbol{C}$ from (23.6) shows that $C_{33}=\lambda^2$, $C_{3\alpha}=0$ $(\alpha\neq 3)$, and that $C_{\alpha\beta}$ is a function of x^1, x^2 only. The explicit stress relation $(43.12)_1$ now reduces to

$$\left.\begin{aligned} T_{km} &= \mathfrak{S}_{\alpha\beta}(C_{\gamma\delta}, \lambda^2, X^\alpha, X^3)\, X^\alpha{}_{,k}\, X^\beta{}_{,m}, \\ T_{3m} &= T_{m3} = \frac{1}{\lambda}\,\mathfrak{S}_{3\alpha}(C_{\gamma\delta}, \lambda^2, X^\alpha, X^3)\, X^\alpha{}_{,m} \\ T_{33} &= \frac{1}{\lambda^2}\,\mathfrak{S}_{33}(C_{\gamma\delta}, \lambda^2, X^\alpha, X^3). \end{aligned}\right\} \tag{60.2}$$

CAUCHY's first law (16.6), for the case of equilibrium, may be written in the form

$$T^{km}{}_{,m} + \frac{\partial T^{k3}}{\partial x^3} = 0, \tag{60.3}$$

where the comma denotes covariant differentiation based on the plane Euclidean metric tensor g_{km}. It has been assumed here that $\boldsymbol{b}=\boldsymbol{0}$, but conservative body force may be included by the usual device of redefining $\boldsymbol{T}$ (CFT, Sect. 207). Consider only materials whose elastic response is the same in each plane $X^3=$const. Then the dependence of $\mathfrak{S}_{\alpha\beta}$ on X^3 drops out in (60.2), and (60.3) reduces to the fully plane condition

$$T^{km}{}_{,m} = 0. \tag{60.4}$$

The general solution of this equation is given in terms of AIRY's stress function A as follows:

$$T^{km} = -e^{kp}\,e^{mq}\,A_{,pq}, \tag{60.5}$$

where e^{kp} are the components of the plane determinant tensor[2]. (For a derivation of this result and a development of the properties of A, see Sect. 224 of CFT.) Hence, by (60.2),

$$\left.\begin{aligned} A_{,rs} &= -e_{rk}\,e_{sm}\,T^{km}, \\ &= -e_{rk}\,e_{sm}\,g^{kp}\,g^{mq}\,X^\alpha{}_{,p}\,X^\beta{}_{,q}\,\mathfrak{S}_{\alpha\beta}(C_{\gamma\delta}, \lambda^2, X^\alpha). \end{aligned}\right\}. \tag{60.6}$$

If the Eqs. (60.1) defining the deformation are given, the result (60.6) may be regarded as a system of three partial differential equations to be satisfied by the single function A. If a solution can be found, the stresses required to effect this deformation are then given by (60.2). In general, of course, the system will not be compatible if so regarded. Instead, we should look upon it as a system of three equations to be satisfied by the three functions $A(x^1, x^2)$, $X^1(x^1, x^2)$, $X^2(x^1, x^2)$, for a prescribed constant λ and prescribed material properties[3].

---

[1] A fuller exposition of this subject, for hyperelastic materials, has been given by GREEN and ADKINS [1960, *26*, Ch. III].

[2] The components e^{kp} are $e^{12} = -e^{21} = 1/\sqrt{g}$, $e^{11} = e^{22} = 0$.

[3] Of course a similar formulation would be possible for general deformations in three-dimensional space, by means of the Beltrami (Gwyther-Finzi) stress functions [CFT, Eq. (227.10)], but it would seem even less promising for use.

The specialization to isotropic materials is perhaps worth writing down[1]. By (23.7) we have $B_{3k}=0$, $B_{33}=\lambda^2$, while B_{km} is a function of x^1, x^2 only. Hence

$$\left.\begin{aligned} I &= b_1+b_2+\lambda^2 = I_0+\lambda^2,\\ II &= b_1 b_2+\lambda^2(b_1+b_2) = II_0+\lambda^2 I_0,\\ III &= b_1 b_2 \lambda^2 = II_0\lambda^2, \end{aligned}\right\} \tag{60.7}$$

where we write b_1 and b_2 for the proper numbers of the plane tensor $\boldsymbol{B}_0$ with components B^k_m, and I_0 and II_0 for its principal invariants. Regarding λ as given, we may replace the response coefficients $\beth_\Gamma$ in (47.9) by equivalent functions of I_0 and II_0 only. From (47.9) we see that not only

$$T^3_k = T^k_3 = 0, \quad T^3_3 = \beth_0 + \lambda^2\beth_1 + \frac{1}{\lambda^2}\beth_{-1}, \tag{60.8}$$

but also either the term in $\boldsymbol{B}_0^{-1}$ or the term in $\boldsymbol{B}_0$ may be eliminated by use of the Hamilton-Cayley equation, $\boldsymbol{B}_0 - I_0\boldsymbol{1} + II_0\boldsymbol{B}_0^{-1} = \boldsymbol{0}$, so that

$$T^k_m = \text{ל}_0\,\delta^k_m + \text{ל}_1\,B^k_m = \text{ה}_0\,\delta^k_m + \text{ה}_1\,(B^{-1})^k_m, \tag{60.9}$$

where ל_0, ל_1, ה_0, ה_1, the effective response co-efficients for the plane strain, are related to the general response coefficients $\beth_\Gamma$ as follows:

$$\left.\begin{aligned} \text{ל}_0(I_0, II_0, \lambda) &= \beth_0(I_0+\lambda^2, II_0+\lambda^2 I_0, II_0\lambda^2) +\\ &\quad + \frac{I_0}{II_0}\beth_{-1}(I_0+\lambda^2, II_0+\lambda^2 I_0, II_0\lambda^2),\\ \text{ל}_1(I_0, II_0, \lambda) &= \beth_1(I_0+\lambda^2, II_0+\lambda^2 I_0, II_0\lambda^2) -\\ &\quad - \frac{1}{II_0}\beth_{-1}(I_0+\lambda^2, II_0+\lambda^2 I_0, II_0\lambda^2),\\ \text{ה}_0 = \text{ל}_0 + I_0\text{ל}_1 &= \beth_0(I_0+\lambda^2, II_0+\lambda^2 I_0, \lambda^2 II_0) +\\ &\quad + I_0\beth_1(I_0+\lambda^2, II_0+\lambda^2 I_0, \lambda^2 II_0),\\ \text{ה}_1 = -II_0\text{ל}_1 &= -II_0\beth_1(I_0+\lambda^2, II_0+\lambda^2 I_0, \lambda^2 II_0) +\\ &\quad + \beth_{-1}(I_0+\lambda^2, II_0+\lambda^2 I_0, \lambda^2 II_0). \end{aligned}\right\} \tag{60.10}$$

For incompressible materials, it is necessary that $III=1$, so that $II_0=1/\lambda^2$. Instead of (60.8), (60.9), and (60.10), from (49.5) we have

$$\left.\begin{aligned} &T^3_3 = -p + \lambda^2\beth_1 + \frac{1}{\lambda^2}\beth_{-1}, \quad T^k_3 = T^3_k = 0,\\ &T^k_m = -p\,\delta^k_m + \text{ל}_1 B^k_m = -p'\,\delta^k_m + \text{ה}_1(B^{-1})^k_m,\\ &\text{ל}_1(I_0, \lambda) = \beth_1\left(I_0+\lambda^2, \frac{1}{\lambda^2}+\lambda^2 I_0\right) - \lambda^2\beth_{-1}\left(I_0+\lambda^2, \frac{1}{\lambda^2}+\lambda^2 I_0\right),\\ &\text{ה}_1 = -\text{ל}_1/\lambda^2 = -\beth_1\left(I_0+\lambda^2, \frac{1}{\lambda^2}+\lambda^2 I_0\right)/\lambda^2 +\\ &\quad + \beth_{-1}\left(I_0+\lambda^2, \frac{1}{\lambda^2}+\lambda^2 I_0\right). \end{aligned}\right\} \tag{60.11}$$

By $(60.6)_1$ it follows that for compressible materials

$$\left.\begin{aligned} A_{,rs} &= -e_{rk}e_{sm}(\text{ל}_0 g^{km} + \text{ל}_1 B^{km}),\\ &= -\text{ל}_0 g_{rs} - \text{ל}_1 II_0 (B^{-1})_{rs}, \end{aligned}\right\} \tag{60.12}$$

---

[1] ADKINS, GREEN, and SHIELD [1953, *1*, §§ 3–4], for hyperelastic materials. GREEN and WILKES [1954, *10*] generalized the formulation so as to apply to orthotropic hyperelastic materials.

while for incompressible materials

$$A_{,rs} = p g_{rs} - \frac{\text{ל}_1}{\lambda^2}(B^{-1})_{rs}. \tag{60.13}$$

By forming the product $g^{rs} A_{,rs}$ we may eliminate ל_0 from (60.12), obtaining the result

$$A_{,rs} = \tfrac{1}{2}(A_{;q}^{;q} + \text{ל}_1 I_0) g_{rs} - \text{ל}_1 II_0 (B^{-1})_{rs}, \tag{60.14}$$

which may be regarded as the definitive system of three partial differential equations for the three unknowns A, X^1, and X^2 as functions of x^1 and x^2. For incompressible materials we may eliminate p from (60.13) in just the same way; the result is of the same form, except, of course, that ל_1 is given by $(60.11)_6$ rather than by $(60.10)_2$, and that $II_0 = 1/\lambda^2$.

An alternative formulation[1] may be given in terms of complex variables related to real Cartesian co-ordinates as follows:

$$z = x + iy, \quad \bar{z} = x - iy, \quad Z = X + iY, \quad \bar{Z} = X - iY. \tag{60.15}$$

We consider only incompressible materials; hence we must have

$$\frac{\partial(z, \bar{z})}{\partial(Z, \bar{Z})} = \frac{1}{\lambda}, \quad \frac{\partial(Z, \bar{Z})}{\partial(z, \bar{z})} = \lambda. \tag{60.16}$$

If we introduce "complex stresses" by the definitions[2]

$$T^{zz} = \overline{T^{\bar{z}\bar{z}}} = T\langle xx\rangle - T\langle yy\rangle + 2i\,T\langle xy\rangle, \quad T^{z\bar{z}} = T\langle xx\rangle + T\langle yy\rangle, \tag{60.17}$$

then (60.5) assumes the form

$$T^{zz} = \overline{T^{\bar{z}\bar{z}}} = 4\frac{\partial^2 A}{\partial \bar{z}^2}, \quad T^{z\bar{z}} = -4\frac{\partial^2 A}{\partial z\, \partial \bar{z}}. \tag{60.18}$$

Since $g_{zz} = g_{\bar{z}\bar{z}} = 0$, $g_{z\bar{z}} = \frac{1}{2}$, while $\frac{1}{4}(B_0^{-1})^{zz} = (B_0^{-1})_{\bar{z}\bar{z}} = \frac{\partial Z}{\partial \bar{z}}\frac{\partial \bar{Z}}{\partial \bar{z}}$, from $(60.9)_2$ we obtain the following complex constitutive equation for isotropic incompressible materials:

$$T^{zz} = 4\frac{\partial Z}{\partial \bar{z}}\frac{\partial \bar{Z}}{\partial \bar{z}}\,\text{ה}_1. \tag{60.19}$$

Comparison with $(60.18)_1$ shows that

$$\frac{\partial^2 A}{\partial \bar{z}^2} = \frac{\partial Z}{\partial \bar{z}}\frac{\partial \bar{Z}}{\partial \bar{z}}\,\text{ה}_1. \tag{60.20}$$

A condition that this equation shall be compatible with its complex conjugate is

$$\frac{\partial^2}{\partial \bar{z}^2}\left(\frac{\partial Z}{\partial z}\frac{\partial \bar{Z}}{\partial z}\,\text{ה}_1\right) = \frac{\partial^2}{\partial z^2}\left(\frac{\partial Z}{\partial \bar{z}}\frac{\partial \bar{Z}}{\partial \bar{z}}\,\text{ה}_1\right). \tag{60.21}$$

For a given isotropic incompressible material, ה_1 will be a given function of λ and I_0; in the present co-ordinates

$$I_0 = \frac{4}{\lambda^2}(B_0^{-1})_{z\bar{z}} = \frac{2}{\lambda^2}\left(\frac{\partial Z}{\partial z}\frac{\partial \bar{Z}}{\partial \bar{z}} + \frac{\partial \bar{Z}}{\partial z}\frac{\partial Z}{\partial \bar{z}}\right) = 2\left(\frac{2}{\lambda^2}\frac{\partial \bar{Z}}{\partial z}\frac{\partial Z}{\partial \bar{z}} + \frac{1}{\lambda}\right), \tag{60.22}$$

where we have used (60.16). Hence ה_1 is a given function of λ and of $\frac{\partial \bar{Z}}{\partial z}\frac{\partial Z}{\partial \bar{z}}$. Therefore (60.21) is a partial differential equation for determining all plane strains $Z = Z(z, \bar{z})$ which may be superimposed upon a given uniform normal stretch λ, for the material in question, subject to conservative body force.

[1] ADKINS, GREEN, and SHIELD [1953, *1*, §§ 9 and 11], ADKINS [1955, *3*].

[2] For a brief explanation of formulae such as these, see CFT, Appendix, Sects. 5—6.

Several of the problems solved in Sect. 57 are examples of plane strain superimposed on uniform normal extension and hence may be approached by way of the formulation just given. For other cases, it may be made the basis of methods of successive approximation[1].

By use of $(60.16)_1$ and of the usual formulae for inversion of partial differentiations, we have

$$\frac{\partial \overline{Z}}{\partial z} = -\lambda \frac{\partial \overline{z}}{\partial Z}, \qquad \frac{\partial Z}{\partial \overline{z}} = -\lambda \frac{\partial z}{\partial \overline{Z}}. \tag{60.23}$$

By (60.22), then,

$$I_0 = 2\left(2 \frac{\partial z}{\partial \overline{Z}} \frac{\partial \overline{z}}{\partial Z} + \frac{1}{\lambda}\right). \tag{60.24}$$

Suppose now that a solution

$$Z = f(z, \overline{z}, \lambda) \tag{60.25}$$

of (60.21) has been found, and consider the deformation defined by

$$z = \frac{1}{\lambda} f(Z, \overline{Z}, \lambda), \tag{60.26}$$

where f is the same function as in (60.25). By (60.24) and (60.22), I_0 as calculated from (60.26) is the same function of Z, $\overline{Z}$, and λ as is I_0 calculated from the original deformation (60.25). Consequently the material function $\beth_1$ for the new deformation is the same function of Z, $\overline{Z}$, and λ as it was for the original one. From the homogeneity of (60.21), it follows that the deformation (60.26) is compatible for the same material as that for which (60.25) is compatible. This result constitutes the *reciprocal theorem* of ADKINS[2]. It enables us, when a single plane strain superimposed on a uniform normal extension is known for a given isotropic incompressible material, to find a second such strain superposable upon the same normal extension for the same material. A second application of the reciprocal theorem leads back to the original solution.

We cannot expect to find in this way any new deformation possible in an arbitrary isotropic elastic material, since it is conjectured that all such deformations are known. As illustration, however, we consider the appropriate special case of (57.1), viz

$$X = \frac{1}{2} \lambda B (x^2 + y^2), \qquad Y = \frac{1}{B} \arctan \frac{y}{x}, \qquad x^3 = \lambda X^3. \tag{60.27}$$

The reciprocal deformation given by (60.26) is

$$x = \frac{1}{2} B (X^2 + Y^2), \qquad y = \frac{1}{\lambda B} \arctan \frac{Y}{X}, \qquad x^3 = \lambda X^3. \tag{60.28}$$

This is a special case of (57.8), for which the general solution has already been obtained in Sect. 57. The plane case of (57.13) is, of course, self-reciprocal in the sense of ADKINS' theorem, the usefulness of which appears for special rather than general materials.

It is clear from the example of homogeneous strain, studied in Sects. 54 and 55, that a strictly plane stress, in which $T^{33} = 0$, $T^{3i} = 0$, and T^{ij} is independent of x^3, is to be expected only in trivial cases. A kind of generalized plane stress may be approached by the classical method of averaging along the direction normal to the plane. The portion of the theory that can be carried through in exact terms, namely, the general theory of stress resultants and stress functions for a flat plate, may be read off from Sects. 212, 213, and 229 of CFT. In particular, the mean stresses are given in terms of AIRY's function by formulae of the same form as (60.5). Further progress is made by assuming[3] that a deformation of the form (60.1)

[1] ADKINS, GREEN, and SHIELD [1953, *1*, §§ 14—15] introduced series expansions and illustrated their use in the problem of an infinite plate containing a circular hole and subject to uniform tension at infinity, and in the problem of an infinite plate containing a rigid circular inclusion. More general results of this kind were obtained by ADKINS and GREEN [1957, *1*]. An exposition is given by GREEN and ADKINS [1960, *26*, Ch. VI]. A solution for a plate with an elliptic hole was obtained by LIANIS [1963, *47*].

[2] ADKINS [1958, *1*] showed that the result holds also for transversely isotropic materials.

[3] ADKINS, GREEN, and NICHOLAS [1954, *2*, § 5]. These authors gave also a formulation in terms of complex variables and a unified second-order theory including both plane strain and generalized plane stress, which was developed further by ADKINS and GREEN [1957, *1*]. Cf. also the theory of flat plates formulated by GUO ZHONG-HENG [1962, *25*, *26*, *28*, and *29*]. Earlier RIVLIN and THOMAS [1951, *15*] had found an approximate solution for a deformation of this kind, namely, that produced in a thin sheet containing a circular hole on which uniform radial traction is applied, and had found good agreement between their results and experimental values for rubber.

is effected in the plate initially bounded by the surfaces $X^3 = \pm h_0$, where h_0 is a constant, so that the deformed plate is bounded by the surfaces $x^3 = \pm h_0 \lambda$, but λ is no longer restricted to be constant. If $\lambda(X^1, X^2)$ is assumed to vary slowly enough that its derivatives may be neglected, the averaging of the stresses may be carried out, leading to results of the forms (60.12) and (60.13), except that the coefficients $\beth_0$ and $\beth_1$ are related to the response coefficients of the material in a somewhat different way. Thus the formal theory for this kind of approximately plane stress is essentially the same as that for plane strain superimposed upon uniform normal extension.

As in the infinitesimal theory, it is possible to develop theories of rods[1] and membranes[2] by truncating appropriate expansions of the general constitutive equations or their averages over the cross-section or thickness.

c) Exact solutions of special problems of motion.

61. Quasi-equilibrated motions of incompressible bodies[3]. A very simple idea, illustrated in Sect. 57 in the problem of finding the inflation due to whirling of a shaft, serves to reduce a great many cases of motion of incompressible homogeneous bodies to corresponding problems of equilibrium. For a given incompressible elastic body and a given body force $\boldsymbol{b} = \boldsymbol{b}(\boldsymbol{x}, t)$, a motion $\boldsymbol{x} = \boldsymbol{\chi}(\boldsymbol{X}, t)$ is dynamically possible, in view of $(43.14)_1$ and (16.6), if there exists a pressure $p = p(\boldsymbol{x}, t)$ such that

$$-\operatorname{grad} p + \operatorname{div} \mathfrak{g}(\boldsymbol{F}) + \varrho \boldsymbol{b} = \varrho \ddot{\boldsymbol{x}}, \tag{61.1}$$

where $\boldsymbol{F} = \boldsymbol{F}(\boldsymbol{x}, t)$ is the deformation gradient. Now consider t not as the time but as a time-independent parameter. Then $\boldsymbol{x} = \boldsymbol{\chi}(\boldsymbol{X}, t)$, for each value of t, defines a static deformation. We say that $\boldsymbol{\chi}$ is *quasi-equilibrated* for the given body if each of the deformations $\boldsymbol{x} = \boldsymbol{\chi}(\boldsymbol{X}, t)$ satisfies the conditions of equilibrium for the given incompressible elastic body subject to the body force $\boldsymbol{b} = \boldsymbol{b}(\boldsymbol{x}, t)$. These conditions of equilibrium are satisfied for each value of t if there exists a pressure $p_0 = p_0(\boldsymbol{x}, t)$ such that

$$-\operatorname{grad} p_0 + \operatorname{div} \mathfrak{g}(\boldsymbol{F}) + \varrho \boldsymbol{b} = \boldsymbol{0}. \tag{61.2}$$

Now, if (61.2) is valid, then (61.1) is also valid if and only if

$$\operatorname{grad}(p_0 - p) = \varrho \ddot{\boldsymbol{x}}, \tag{61.3}$$

or[4]

$$\ddot{\boldsymbol{x}} = -\operatorname{grad} \zeta, \quad \varrho \zeta = p - p_0. \tag{61.4}$$

Eq. (61.4) states that the motion is circulation-preserving, with single-valued acceleration-potential ζ. For such motions the main theorems of classical hydrodynamics, named after BERNOULLI, CAUCHY, HELMHOLTZ, KELVIN, etc., are valid (cf. CFT, Sects. 105–138). We have proved the following ***theorem on quasi-equilibrated motions***: *Given a quasi-equilibrated motion of a particular incompressible elastic body subject to a particular body force, this motion is dynamically possible, subject to the same body force, if and only if the motion is circulation-preserving with a single-valued acceleration-potential ζ. If this condition is satisfied, the stress tensor is given by the formula*

$$\boldsymbol{T} = -\varrho \zeta \boldsymbol{1} + \boldsymbol{T}_0 \tag{61.5}$$

where $\boldsymbol{T}_0$ is the most general stress corresponding to equilibrium in the configuration at time t.

---

[1] KÖTTER [1910, *1*, § 2], HAY [1942, *3*].

[2] RIVLIN and THOMAS [1951, *15*], ADKINS and RIVLIN [1952, *1*], CORNELIUSSEN and SHIELD [1961, *8*], GUO ZHONG-HENG [1962, *32*].

[3] TRUESDELL [1962, *65*].

[4] Since p and p_0 are determined only to within an arbitrary function of time, the same indeterminacy follows for ζ.

Thus a quasi-equilibrated motion is dynamically possible for an incompressible elastic material if and only if it is a possible motion of a very special case of such a material, namely, an incompressible elastic fluid subject to no body force at all. Such a motion, moreover, is one that can be brought instantly to rest by applying a suitable pressure impulse upon the boundary of the body experiencing it. Notice also from (61.5) that *the shear stresses in the motion are the same as those that correspond to equilibrium in the configuration occupied by the body at time t.*

The theorem is not one of general dynamics, for it applies neither, on the one hand, to compressible materials nor, on the other, to incompressible ones for which the stress field is not determined by the present configuration alone. Among simple materials, it applies only to incompressible elastic ones. It is not to be confused with the more general but less specific results in Sect. 30, although it may be derived by using the criterion (30.29). It applies also, obviously, to non-simple incompressible materials in which the extra stress at $\boldsymbol{x}=\boldsymbol{\chi}(\boldsymbol{X}, t)$ is determined by the values $\boldsymbol{z}=\boldsymbol{\chi}(\boldsymbol{Z}, t)$ for $\boldsymbol{Z}$ in a neighborhood of $\boldsymbol{X}$ (cf. Sect. 26). In physical terms, it applies only to materials not showing memory effects.

While the foregoing theorem may find its main future application to particular quasi-equilibrated motions of particular materials, a major general corollary is found once and for all. In order for a quasi-equilibrated motion $\boldsymbol{\chi}$ to be dynamically possible, subject to a given field of body force $\boldsymbol{b}$, in *all* elastic materials, it is necessary that at each instant the mapping $\boldsymbol{x}=\boldsymbol{\chi}(\boldsymbol{X}, t)$ gives a possible configuration of equilibrium, subject to the body force $\boldsymbol{b}(\boldsymbol{x}, t)$, in all materials. ~~For isotropic materials subject to no body force, results of ERICKSEN to be presented in Sect. 91 make it seem unlikely that there are any such deformations beyond homogeneous strains and the four families studied in Sect. 57.~~* The corresponding quasi-equilibrated motions of an isotropic material subject to surface tractions alone are obtained by letting the arbitrary constants in those deformations be arbitrary functions of time. It is then a straightforward matter to see if the acceleration satisfies (61.4). First, it is necessary that

$$\operatorname{curl} \ddot{\boldsymbol{x}}=\mathbf{0}; \tag{61.6}$$

next it is necessary and sufficient that the potential ζ, existence of which is a consequence of (61.6), be single-valued.

First there are those homogeneous motions that are possible in all homogeneous incompressible simple materials. These motions must satisfy (30.33); the acceleration-potential is given by (30.34).

Next consider the family (57.1). Note that arbitrary constants which can be annulled by choice of origin for static deformations cannot be annulled in general, since the dynamical equations are not invariant under arbitrary time-dependent shifts of origin. Thus some quasi-equilibrated motions corresponding to (57.1) are of the form

$$r^2=A(2X+D), \qquad \theta=B(Y+E), \qquad z=\frac{Z}{AB}-BCY+F, \tag{61.7}$$

where $A, B, \ldots, F$ are functions of time, and $AB\neq 0$. The contravariant components of velocity, namely, $\dot{r}, \dot{\theta}, \dot{z}$, are calculated straight off, and from them the contravariant components of acceleration follow[1] by (6A.3). Thus

$$\begin{aligned}\dot{r}&=\frac{1}{2r}(2\dot{A}X+\dot{A}D+\dot{D}A)=\frac{1}{2}\left(\frac{\dot{A}}{A}r+\frac{\dot{D}A}{r}\right),\\ \dot{\theta}&=\dot{B}Y+\dot{B}E+\dot{E}B=\frac{\dot{B}}{B}\theta+B\dot{E},\end{aligned}$$

[1] Alternatively,

$$\ddot{x}^k=(\dot{x}^k)^{\cdot}+\left\{{k \atop pq}\right\}\dot{x}^p\dot{x}^q,$$

and the same result follows by (6A.2).

* [The conjecture is false.]

$$\left.\begin{aligned}
\dot{z} &= \left(\frac{1}{AB}\right)^{\cdot} Z - (BC)^{\cdot}\, Y + \dot{F} = A\,B\left(\frac{1}{AB}\right)^{\cdot} z + \\
&\quad + \left[A\,B\left(\frac{1}{AB}\right)^{\cdot} - \frac{(BC)^{\cdot}}{BC}\right](C\theta - BCE) + \dot{F} - FA\,B\left(\frac{1}{AB}\right)^{\cdot},\\
\ddot{r} &= \frac{1}{2}\left[\frac{\ddot{A}}{A} - \frac{1}{2}\left(\frac{\dot{A}}{A}\right)^{2}\right] r + \frac{1}{2}\,\frac{(A\dot{D})^{\cdot}}{r} - \frac{1}{4}\,\frac{(A\dot{D})^{2}}{r^{3}},\\
\ddot{x}_r &= \ddot{x}^r = \ddot{r} - r\left(\frac{\dot{B}}{B}\,\theta + B\dot{E}\right)^{2},\\
\ddot{x}_\theta &= r^2 \ddot{x}^\theta = \left(\frac{\ddot{B}}{B} + \frac{\dot{A}\dot{B}}{AB}\right) r^2\theta + \frac{A\dot{B}\dot{D}}{B}\,\theta + \\
&\quad + \left(B\ddot{E} + 2\dot{B}\dot{E} + \frac{\dot{A}}{A}\,B\dot{E}\right) r^2 + A\,B\dot{D}\dot{E},\\
\ddot{x}_z &= \ddot{z} = A\,B\left(\frac{1}{AB}\right)^{\cdot\cdot} z + \left[A\,B\left(\frac{1}{AB}\right)^{\cdot\cdot} - \frac{(BC)^{\cdot\cdot}}{BC}\right](C\theta - BCE) + \\
&\quad + \ddot{F} - FA\,B\left(\frac{1}{AB}\right)^{\cdot\cdot}.
\end{aligned}\right\} \tag{61.8}$$

The condition (61.6) assumes the form $\partial_z \ddot{x}_r = \partial_r \ddot{x}_z$, $\partial_\theta \ddot{x}_z = \partial_z \ddot{x}_\theta$, $\partial_\theta \ddot{x}_r = \partial_r \ddot{x}_\theta$. The first of these equations is satisfied identically; the second reduces to

$$A\,B\left(\frac{1}{AB}\right)^{\cdot\cdot} = \frac{(BC)^{\cdot\cdot}}{BC}, \tag{61.9}$$

the general solution of which is given by

$$BC = \frac{1}{AB}\left(K^* + K^{**}\int A^2 B^2\, dt\right), \tag{61.10}$$

where K^* and K^{**} are constants. The condition $\partial_\theta \ddot{x}_r = \partial_r \ddot{x}_\theta$ yields

$$\theta\left(\frac{\ddot{B}}{B} + \frac{\dot{A}\dot{B}}{AB}\right) + B\ddot{E} + 2\dot{B}\dot{E} + \frac{\dot{A}}{A}\,B\dot{E} = -\left(\frac{\dot{B}}{B}\,\theta + B\dot{E}\right)\frac{\dot{B}}{B}, \tag{61.11}$$

equivalent to

$$\frac{\ddot{B}}{B} + \frac{\dot{B}^2}{B^2} + \frac{\dot{A}\dot{B}}{AB} = 0, \qquad B\ddot{E} + 3\dot{B}\dot{E} + \frac{\dot{A}}{A}\,B\dot{E} = 0. \tag{61.12}$$

Hence $A\,B\dot{B} = \frac{1}{2}K$, where $K = \text{const.}$, so that

$$B^2 = K\int \frac{dt}{A} + B_0^2, \tag{61.13}$$

while (61.10) yields

$$C = \left\{\begin{array}{ll}
\dfrac{\dot{B}}{B}\left(K' + K''\displaystyle\int \frac{dt}{\dot{B}^2}\right) & \text{if} \quad K \neq 0,\\[2ex]
\dfrac{1}{A}\left(K' + K''\displaystyle\int A^2\, dt\right) & \text{if} \quad K = 0,
\end{array}\right\} \tag{61.14}$$

and $(61.12)_2$ can be integrated also. From the specializations so obtained, simpler forms for $\ddot{x}_r$, $\ddot{x}_\theta$, and $\ddot{x}_z$ result by substitution into $(61.8)_{6-11}$. The form of ζ can then be seen by inspection.

Similar calculations, lengthier in the case of the third family discussed in Sect. 57, may be carried out for the remaining three families. We omit the details and present the final result. ~~If, as seems likely from the analysis of Ericksen to be presented in Sect. 91, there are no further static deformations that can be maintained in any homogeneous, incompressible, isotropic elastic material loaded by surface tractions alone, then the following list gives *all dynamically possible quasi-equilibrated motions that can be produced in*~~ *every* ~~*isotropic incompressible material by the application of surface tractions*~~ *only*.*

Family 0. Homogeneous motions:

$$\left.\begin{aligned} \boldsymbol{x} &= \boldsymbol{F}\boldsymbol{X}+\boldsymbol{c}, \quad |\det \boldsymbol{F}|=1, \\ \ddot{\boldsymbol{F}}^T &= \boldsymbol{F}^T\ddot{\boldsymbol{F}}\boldsymbol{F}^{-1}, \\ -\zeta &= \tfrac{1}{2}(\ddot{\boldsymbol{F}}\boldsymbol{F}^{-1}\boldsymbol{x})\cdot\boldsymbol{x}+(\ddot{\boldsymbol{c}}-\ddot{\boldsymbol{F}}\boldsymbol{F}^{-1}\boldsymbol{c})\cdot\boldsymbol{x}. \end{aligned}\right\} \tag{61.15}$$

Family 1. Bending, extension, and shear of a block:

$$r^2=A\,(2X+D), \qquad \theta=B\,(Y+E), \qquad z=\frac{Z}{AB}-BCY+F, \qquad AB\neq 0, \tag{61.16}$$

$$\left.\begin{aligned} A &= A\,(t) = \text{arbitrary function}, \\ B^2 &= K\int\frac{dt}{A}+B_0^2, \qquad K,\, B_0=\text{const.}, \\ C &= \left\{\begin{aligned} &\frac{\dot{B}}{B}\left(K'+K''\int\frac{dt}{\dot{B}^2}\right) && \text{if}\quad K\neq 0, \\ &\frac{1}{A}\left(K'+K''\int A^2\,dt\right) && \text{if}\quad K=0, \end{aligned}\right\} K',\, K''=\text{const.}, \\ D &= D\,(t) = \text{arbitrary function}, \\ E &= K'''\int\frac{\dot{B}}{B^2}\,dt+E_0, \qquad K''',\, E_0=\text{const.}, \\ F &= F\,(t) = \text{arbitrary function}, \\ -\zeta &= \frac{1}{2}\left[\frac{1}{2}\frac{\ddot{A}}{A}-\frac{1}{4}\frac{\dot{A}^2}{A^2}-K'''\frac{\dot{B}^2}{B^2}\right]r^2+\frac{1}{2}(A\dot{D})^{\boldsymbol{\cdot}}\log r+ \\ &\quad +\frac{A^2\dot{D}^2}{6r^3}-\frac{1}{2}\frac{\dot{B}^2}{B^2}r^2\theta^2-2K'''\frac{\dot{B}^2}{B^2}r^2\theta+\frac{1}{4}\frac{K\dot{D}}{B^2}\theta^2+ \\ &\quad +\frac{1}{2}KK'''\frac{\dot{D}}{D^2}\theta+\frac{1}{2}\frac{\dddot{B}}{\dot{B}}z^2+\left(\ddot{F}-F\frac{\dddot{B}}{\dot{B}}\right)z, \end{aligned}\right\} \tag{61.17}$$

where in the case when $K=0$, the coefficient $\dddot{B}/\dot{B}$ is to be replaced by $A\,(1/A)^{\boldsymbol{\cdot\cdot}}$.

Family 2. Straightening, extension, and shear of a sector of a cylindrical tube:

$$x=\frac{1}{2}AB^2R^2+D, \qquad y=\frac{\Theta}{AB}+E, \qquad z=\frac{Z}{B}+\frac{C\Theta}{AB}+F, \qquad AB\neq 0, \tag{61.18}$$

* [The conjecture is false.]

$$\left.\begin{aligned}
&A = A(t) = \text{arbitrary function},\\
&B = B(t) = \text{arbitrary function},\\
&C = A\,(K + K'\textstyle\int B^2\,dt),\ K = \text{const.},\ K' = \text{const.},\\
&D = D(t) = \text{arbitrary function},\\
&E = E(t) = \text{arbitrary function},\\
&F = F(t) = \text{arbitrary function},\\
&-\zeta = \frac{1}{2}\,\frac{(A B^2)^{\cdot\cdot}}{A B^2}\,(x - D)^2 + \frac{1}{2}\,A B\left(\frac{1}{A B}\right)^{\cdot\cdot}(y - E)^2 +\\
&\qquad + \frac{1}{2}\,B\left(\frac{1}{B}\right)^{\cdot\cdot}(z - F)^2 + \ddot{D}\,x + \ddot{E}\,y + \ddot{F}\,z.
\end{aligned}\right\} \tag{61.19}$$

Family 3. Inflation, bending, torsion, extension, and shear of a sector of a cylindrical tube:

$$\left.\begin{aligned}
r^2 = A R^2 + B, \qquad \theta = C\Theta + D Z + G, \qquad z = E\Theta + F Z + H,\\
A\,(C F - D E) = 1.
\end{aligned}\right\} \tag{61.20}$$

General case: $CF \neq 0$. Put

$$\left.\begin{aligned}
&\alpha \equiv 1 - K\left(K' + K''\int\frac{dt}{F^2}\right), \qquad \beta \equiv K^* - K'''\,\frac{G}{C},\\
&\gamma \equiv \dot{B} - \frac{\dot{A}}{A}\,B, \qquad K,\ K^*,\ K',\ K'',\ K''' = \text{const.}
\end{aligned}\right\} \tag{61.21}$$

Then

$$\left.\begin{aligned}
&A = \frac{1}{C F \alpha}, \qquad \alpha \neq 0,\\
&B = B(t) = \text{arbitrary function},\\
&C = K'''\textstyle\int F\alpha\,dt + C_0, \qquad C_0 = \text{const.},\\
&D = K C,\\
&E = F\left(K' + K''\displaystyle\int\frac{dt}{F^2}\right),\\
&F = F(t) = \text{arbitrary function},\\
&G = K^*\textstyle\int F\alpha\,dt + G_0, \qquad G_0 = \text{const.},\\
&H = H(t) = \text{arbitrary function},\\
&-\zeta = \frac{1}{2}\left[\frac{1}{2}\left(\frac{\dot{A}}{A}\right)^{\cdot} + \frac{1}{4}\,\frac{\dot{A}^2}{A^2} - F^2\alpha^2\beta^2\right] r^2 + \frac{1}{2}\,\dot{\gamma}\log r +\\
&\qquad + \frac{1}{8}\,\frac{\gamma^2}{r^2} - \frac{1}{2}\,\frac{\dot{C}^2}{C^2}\,r^2\theta^2 - \frac{\dot{C}}{C}\,F\alpha\beta r^2\theta + \frac{1}{2}\,\gamma\,\frac{\dot{C}}{C}\,\theta^2 +\\
&\qquad + F\alpha\beta\gamma\theta + \frac{1}{2}\,\frac{\ddot{F}}{F}\,z^2 + \left(\frac{\ddot{H}}{H} - \frac{\ddot{F}}{F}\right) H z.
\end{aligned}\right\} \tag{61.22}$$

Special case: $CF=0$, hence $ADE=-1\neq 0$.

$$\left.\begin{aligned}
&A=-\frac{1}{DE},\\
&B=B(t)=\text{arbitrary function},\\
&C=KD,\ K=\text{const., where } K=0 \text{ if } F\neq 0,\\
&D=K^*\left[\int E\,dt+K^{**}\right],\quad K^*, K^{**}=\text{const.},\quad K^*\neq 0,\\
&E=\begin{cases} K'F\left(1+K''\int\frac{dt}{F^2}\right) & \text{if } F\neq 0,\quad K', K''=\text{const.},\quad K'\neq 0,\\ E(t)=\text{arbitrary function if } F=0, \text{ but always } E(t)\neq 0,\end{cases}\\
&F=F(t)=\text{arbitrary function, and } F=0 \text{ if } K\neq 0,\\
&G=K'''D+G_0,\quad K''', G_0=\text{const.},\\
&H=H(t)=\text{arbitrary function},\\
&-\zeta=\frac{1}{2}\left[\frac{1}{2}\left(\frac{\dot A}{A}\right)^{\bullet}+\frac{1}{4}\frac{\dot A^2}{A^2}-G_0^2\frac{\dot D^2}{D^2}\right]r^2+\frac{1}{2}\dot\gamma\log r+\\
&\quad+\frac{1}{8}\frac{\gamma^2}{r^2}-\frac{1}{2}\frac{\dot D^2}{D^2}r^2\theta^2+G_0\frac{\dot D^2}{D^2}r^2\theta+\frac{1}{2}\gamma\frac{\dot D}{D}\theta^2-\gamma G_0\frac{\dot D}{D}\theta+\\
&\quad+\frac{1}{2}\frac{\ddot E}{E}z^2+\left(\ddot H-\frac{\ddot E}{E}H\right)z,
\end{aligned}\right\}\tag{61.23}$$

where $\gamma\equiv\dot B-\dot A\,B/A$.

Family 4. Inflation or eversion of a sphere:

$$r^3=\pm R^3+A,\qquad \theta=\pm\Theta+B,\qquad \varphi=\Phi+C,\tag{61.24}$$

$$\left.\begin{aligned}
&A=A(t)=\text{arbitrary function},\\
&B=\text{const.},\\
&C=\text{const.},\\
&-\zeta=\frac{1}{3r}\left(-\ddot A+\frac{1}{6}\frac{\dot A^2}{r^3}\right).
\end{aligned}\right\}\tag{61.25}$$

In each case, the stress system is given by (61.5); the value of ζ has been calculated explicitly for each family, and the equilibrium stress $\boldsymbol{T}_0$ may be read off by appropriate specialization of the static results in Sects. 55 and 57.

For example, in (61.22) let the arbitrary constants be chosen as follows: $K=K'=K''=K'''=0$, $C_0=1$, and take $B(t)\equiv H(t)\equiv 0$, $F(t)\equiv\text{const}$. Then it follows from (61.22) that $A=1/F=\text{const.}$, $C=1$, $D=0$, $\alpha=1$, $\beta=K^*$, $\gamma=0$, $G=K^*Ft+G_0$. The motion is thus a longitudinal extension of amount F effected in a cylinder spinning at angular speed $\omega=K^*F$. From the last of Eqs. (61.22) we find that

$$-\zeta=-\tfrac{1}{2}\omega^2r^2.\tag{61.26}$$

Substitution of this result into (61.5), using the appropriate specialization of (57.15), leads to the solution (57.27) we have found earlier for this same problem.

More interesting applications are reserved for the next section. Here we remark only certain general features of the four new families of motions found.

Family 0. Cf. Eqs. (30.32)—(30.42) for discussion and examples.

Family 1. While these motions represent a sequence of deformations of a block into a segment of a hollow circular cylinder, all the successive configurations of what was a block in the reference configuration are bent. At no finite time is the natural state assumed. Thus no problems of free oscillation are included in this family. All solutions require the application of suitable surface tractions on at least a part of the bounding surface at all times.

Family 2. Remarks analogous to those for Family 1 may be made.

Family 3. This is the most interesting and important family of all. A major though very special case included within it will be discussed in the next section. Some problems one might hope to treat by its aid are excluded, however. Consider the case of torsional-longitudinal oscillation, for which we should have to have F and D oscillatory functions, and $C\neq 0$. Suppose, for example, that $F=F_0+J(t)$, where $F_0>0$ and $|J(t)|<F_0-\varepsilon$, where $\varepsilon>0$; the cylinder is then set into longitudinal motion about a given length, which may be its length in the natural state ($F_0=1$). If the cross-sections remain plane, $E=0$, and hence $\alpha=1$. Then

$$\left.\begin{aligned} C&=K'''\int_0^t[F_0+J(\tau)]\,d\tau+C_0\\ &=K'''\left[F_0 t+\int_0^t J(\tau)\,d\tau\right]+C_0.\end{aligned}\right\}\qquad(61.27)$$

Since $\left|\int_0^t J(\tau)\,d\tau\right|<(F_0-\varepsilon)t$, it follows that $C\to\infty$ as $t\to\infty$. But $D=KC$. Thus, if there is any torsional motion at all, D is not bounded and hence not oscillatory. That is, a *longitudinal-torsional oscillation is not possible.* This conclusion is easily reached also on the basis of purely mechanical reasoning. If a torsional oscillation begins, the rings at greater distance from the fixed plane are turning with greater angular speed and hence require greater centripetal force in order to maintain their radius constant. The radial tensions arising from the elasticity of the material in such a deformation are independent of z, as shown in Sect. 57. Thus, in order to maintain the radius of the cylinder uniform, pressures varying with distance from the fixed plane $\left(\text{given explicitly by the term } \frac{1}{2}\frac{\ddot F}{F}z^2 \text{ in the last of Eqs. (61.22)}\right)$ must be applied. The surface of the cylindrical rod, then, cannot be left free if purely torsional oscillations are to be produced.

A similar result, with similar explanation, holds for radial-longitudinal oscillations, for which A and F are oscillatory functions, while $B=D=E=0$ and $C=1$. The corresponding strain is homogeneous and may be analysed alternatively as a special case of Family 0.

Family 4. The motions described are analogous to those for purely radial motion of a cylinder, which is a very special case of Family 3 and is the subject of the next section. Choice of the minus sign leads to motions about the everted state, as does taking A to be negative in Family 3. A double stability is to be expected: For sufficiently small initial deformation from either the natural or the everted state, and for sufficiently small initial velocity, bounded motion should result, but no matter how large the initial velocity and deformation, it will be impossible to carry one kind of motion smoothly into the other.

Some motions of undulatory type, which generally are not quasi-equilibrated, are discussed at the end of Sect. 74.

62. Radial oscillations of isotropic cylinders and spheres. To consider radial oscillations of a cylinder[1], in (61.20) put $A=C=F=1$, $D=G=E=H=0$, so that

[1] The analysis here slightly simplifies and slightly generalizes that of KNOWLES [1960, *34*] [1962, *42*], given for hyperelastic materials. TADJBAKHSH and TOUPIN [1964, *81*] have shown that KNOWLES' derivation of (62.3), which rests on use of convected co-ordinates, contains two errors which cancel one another.

the motion is of the form

$$\left.\begin{aligned} \theta&=\Theta, \quad z=Z, \\ r^2&=R^2-R_1^2+r_1^2, \\ r_2^2&=R_2^2-R_1^2+r_1^2, \end{aligned}\right\} \tag{62.1}$$

where $r_1=r_1(t)$ is the inner radius at time t of the cylindrical tube whose inner and outer radii in the reference state are R_1 and R_2. The function $r_1(t)$, which determines the entire motion, is itself to be determined from suitable initial conditions. The choices leading to (62.1) correspond in (61.21) and (61.22) to $B=r_1^2-R_1^2$ (hence $\frac{1}{2}\gamma=\frac{1}{2}\dot{B}=r_1\dot{r}_1$), $K'''=0$, $C_0=1$ (hence $C=1$), $K=0$ (hence $D=0$, $\alpha=1$), $K'=K''=0$ (hence $E=0$), $F=1$, $K^*=G_0=0$ (hence $G=0$, $B=0$), $H=0$. These specializations reduce the last of Eqs. (61.22) to the form

$$-\zeta=(r_1\dot{r}_1)^{\cdot}\log r+\frac{r_1^2\dot{r}_1^2}{2r^2}\,. \tag{62.2}$$

By making the corresponding specializations in the static stress system (57.15), from (61.5) we conclude that

$$\left.\begin{aligned} T\langle rr\rangle=\varrho\left[(r_1\ddot{r}_1+\dot{r}_1^2)\log r+\frac{r_1^2\dot{r}_1^2}{2r^2}\right]+\psi(t)-\\ -\int\frac{1}{r}\left(\frac{R^2}{r^2}-\frac{r^2}{R^2}\right)\left(\hat{\beth}_1-\hat{\beth}_{-1}\right)dr, \end{aligned}\right\} \tag{62.3}$$

where both the arguments of the functions $\beth_\Gamma$ are $1+\frac{r^2}{R^2}+\frac{R^2}{r^2}$, as follows from $(57.14)_{3,4}$. Let the pressures on the inner and outer faces of the cylinder be $P_1(t)$ and $P_2(t)$. By evaluating (62.3) at $r=r_1$ and $r=r_2$, and then subtracting the results, we obtain a differential equation for the unknown inner radius $r_1(t)$. If we set

$$\left.\begin{aligned} x&\equiv\frac{r_1(t)}{R_1}, \qquad u\equiv\frac{r^2}{R^2}, \\ \gamma&\equiv\frac{R_2^2}{R_1^2}-1>0, \\ f(x,\gamma)&\equiv\frac{1}{\varrho R_1^2}\int\limits_{\frac{\gamma+x^2}{\gamma+1}}^{x^2}(1+u)\,\hat{\mu}\left(u+\frac{1}{u}-2\right)\frac{du}{u^2}, \end{aligned}\right\} \tag{62.4}$$

where $\hat{\mu}\left(u+\frac{1}{u}-2\right)$ is the generalized shear modulus (54.9) with $K^2=u+\frac{1}{u}-2$, we may write that differential equation in the form

$$\left.\begin{aligned} x\log\left(1+\frac{\gamma}{x^2}\right)\ddot{x}+\left[\log\left(1+\frac{\gamma}{x^2}\right)-\frac{\gamma}{\gamma+x^2}\right]\dot{x}^2+\\ +f(x,\gamma)=\frac{P_1(t)-P_2(t)}{\frac{1}{2}\varrho R_1^2}\,. \end{aligned}\right\} \tag{62.5}$$

This is the basic equation obtained by Knowles. Notice that the effect of material properties upon the motion is expressed entirely through the form of the function $f(x,\gamma)$, which is determined by the generalized shear modulus, $\hat{\mu}$, alone. If we set

$$\left.\begin{aligned} F(x,\gamma)&\equiv\int\limits_1^x\xi f(\xi,\gamma)\,d\xi, \\ &=\frac{1}{\varrho R_1^2}\int\limits_1^x\xi\,d\xi\int\limits_{\frac{\gamma+\xi^2}{\gamma+1}}^{\xi^2}(1+u)\,\hat{\mu}\left(u+\frac{1}{u}-2\right)\frac{du}{u^2}, \end{aligned}\right\} \tag{62.6}$$

we see that the integrand is positive if and only if the generalized shear modulus $\hat{\mu}$ is positive. Assuming that $\hat{\mu}>0$, we infer that in the range when $0<x<\infty$, $F(x,\gamma)>0$ except when $x=1$, and that $F(x,\gamma)$ is monotone decreasing when $0<x<1$, zero when $x=1$, and monotone increasing when $x>1$.

Consider first the problem of free oscillations. Then $P_1(t)=P_2(t)=0$, and (62.5) may be integrated at once to yield the rate of change $\dot{x}$ of the inner radius (in units of R_1):

$$\frac{1}{2}\dot{x}^2 x^2 \log\left(1+\frac{\gamma}{x^2}\right)+F(x,\gamma)=C, \tag{62.7}$$

where the constant of integration may be evaluated by setting $x(0)=x_0$, $\dot{x}(0)=\dot{x}_0$, say. Thus $F(x,\gamma)<C$ except at values of x for which $\dot{x}=0$, and

$$\dot{x}=\pm\sqrt{\frac{2C-2F(x,\gamma)}{x^2\log\left(1+\frac{\gamma}{x^2}\right)}}. \tag{62.8}$$

From the theory of vibrations it is known that the motion is periodic if and only if the energy curves (62.7) are closed curves in the phase plane. The curves (62.7) are closed if and only if the equation

$$F(x,\gamma)=C, \tag{62.9}$$

for a given C, has exactly two roots in the interval $0<x<\infty$. If those roots are $x=a$ and $x=b>a$, the period T is then

$$T=2\int_a^b\frac{dx}{\dot{x}}=2\int_a^b\sqrt{\frac{x^2\log\left(1+\frac{\gamma}{x^2}\right)}{2C-2F(x,\gamma)}}\,dx. \tag{62.10}$$

Now if $\hat{\mu}\geqq A$, where $A>0$, then from (62.10) it follows that $F(x,\gamma)$ is unbounded both as $x\to 0+$ and as $x\to\infty$. In this case, from the properties of F derived above we infer that for any non-zero value of C, (62.9) has exactly one root in each of the intervals $0<x<1$ and $1<x<\infty$, and each root is simple. These roots, a and b, give the minimum and maximum values of the ratio x and are to be substituted into (62.10); since a and b are simple roots of (62.9), the integral (62.10) converges and yields a finite period corresponding to any given initial conditions. Without assuming that $\hat{\mu}$ is uniformly bounded below, we can infer the existence of periodic solutions only for initial velocities and amplitudes such as to render C sufficiently small.

To obtain results appropriate for a thin shell, we calculate the forms of various expressions as $\gamma\to 0+$. First, from (62.6) we easily show that

$$\left.\begin{aligned}\lim_{\gamma\to 0+}\frac{F(x,\gamma)}{\gamma}&=\frac{1}{2\varrho R^2}\int_0^{x^2+\frac{1}{x^2}-2}\hat{\mu}(\xi)\,d\xi,\\&\equiv\frac{1}{2}W(x^2),\end{aligned}\right\} \tag{62.11}$$

say, where $R=R_1=R_2$. Writing (62.7) in the explicit form

$$F(x,\gamma)=\frac{1}{2}\dot{x}_0^2 x_0^2\log\left(1+\frac{\gamma}{x_0^2}\right)+F(x_0,\gamma), \tag{62.12}$$

by dividing by γ and then calculating the limits as $\gamma\to 0+$, by use of (62.11) we see that for the thin shell the amplitude equation becomes

$$W(x^2)=\dot{x}_0^2+W(x_0^2). \tag{62.13}$$

Since, as it is plain from (62.11), $W(x^2) = W\left(\frac{1}{x^2}\right)$, the two roots a_0 and b_0 of this equation satisfy the condition $a_0 b_0 = 1$, or $(r)_{\min}(r)_{\max} = R^2$. That is, the radius in the natural state is the geometric mean of the greatest and least radii in the oscillatory motion. The expression (62.10) for the period now becomes

$$T = 2\int_{a_0}^{1/a_0} [\dot{x}_0^2 + W(x_0^2) - W(x^2)]^{-\frac{1}{2}} dx. \tag{62.14}$$

Another problem that can be solved explicitly is furnished by letting a pressure difference be applied impulsively to the walls of the tube:

$$\frac{P_1(t) - P_2(t)}{\frac{1}{2}\varrho R_1^2} = \begin{cases} 0 & t \leqq 0, \\ p & t > 0, \end{cases} \tag{62.15}$$

where p is a constant. Again (62.5) can be integrated easily. If we assume that $\dot{x}(0) = 0$, $x(0) = 1$, so that the pressure impulse is applied to the tube when it is at rest in the natural state, then

$$\frac{1}{2}\dot{x}^2 x^2 \log\left(1 + \frac{\gamma}{x^2}\right) + F(x, \gamma) = \frac{1}{2} p(x^2 - 1), \quad t > 0. \tag{62.16}$$

From (62.5) we see that $\ddot{x}(0) > 0$ if $p > 0$, while $\ddot{x}(0) < 0$ if $p < 0$, as is to be expected. For a periodic motion it is necessary that the equation

$$F(x, \gamma) = \tfrac{1}{2} p(x^2 - 1) \tag{62.17}$$

shall have a positive root $x = a$ other than $x = 1$. If $p > 0$, then $a > 1$, while if $p < 0$, then $a < 1$, as follows from the fact that x initially increases or decreases according as $p > 0$ or $p < 0$. Since, by (62.6), $F'(x, \gamma) = x f(x, \gamma)$, the equation $f(x, \gamma) = p$ must have a root, $x = b$, between $x = 1$ and $x = a$. From (62.5) we see that this root yields an equilibrium solution subject to the assigned pressure difference. That is, *the tube oscillates about its equilibrium amplitude, $x = b$, from a minimum inner radius such that $x = a$, where $a > 1$ or $a < 1$ according as net outward or inward pressure is applied; b lies between a and 1; and the period of oscillation is given by*

$$T = 2\left|\int_1^a \sqrt{\frac{\log\left(1 + \frac{\gamma}{x^2}\right)}{p(x^2 - 1) - 2F(x, \gamma)}}\, x\, dx\right|. \tag{62.18}$$

It is easy to formulate conditions on the shear modulus $\hat{\mu}$ sufficient that (62.17) shall have a root for arbitrary values of p, yielding periodic motions, as shown above. However, a special case, which we shall present in Sect. 95, indicates that such a condition is unnatural. Rather, we should expect periodic motion only for sufficiently small values of the applied pressure difference p. For combined radial and axial vibrations we replace the assumption $A = F = 1$ by $AF = 1$, where the axial stretch $F(t)$, like $B(t)$, is a function to be determined. The resulting system of two ordinary differential equations for, say, the inner radius and the length has been studied in special cases and approximately by Wesołowski[1].

To consider radial oscillations of a sphere[2], given by $(61.25)_{1,2,3}$ (with the upper sign) when $B = C = 0$, we write $(61.25)_8$ in terms of the varying radius $r(R, t)$ of a typical sphere:

$$-\zeta = -(r\ddot{r} + \tfrac{3}{2}\dot{r}^2). \tag{62.19}$$

---

[1] Wesołowski [1964, *94*].

[2] The analysis here, generalizing slightly that of Guo Zhong-Heng and Solecki [1963, *33* and *34*], is essentially that given by Knowles and Jakub [1965, *22*] and Wang [1965, *43*]. That such a solution could easily be obtained had been remarked by Knowles [1960, *42*].

By (61.5) and $(57.37)_1$ we obtain the following expression for the radial tension:

$$T\langle rr\rangle = -\varrho\left(r\ddot{r}+\frac{3}{2}\dot{r}^2\right)+2\int\left(\frac{r^2}{R^2}-\frac{R^4}{r^4}\right)\left(\beth_1-\frac{r^2}{R^2}\beth_{-1}\right)\frac{dr}{r} \tag{62.20}$$

the arguments of $\beth_1$ and $\beth_{-1}$ being $(57.36)_{3,4}$. If the pressures on the inner and outer spherical boundaries are $P_1(t)$ and $P_2(t)$, respectively, and if we set[1]

$$\left.\begin{aligned} &x\equiv\frac{r_1(t)}{R_1}, \qquad u\equiv\frac{r^3}{R^3}, \qquad \gamma\equiv\frac{R_2^3}{R_1^3}-1>0, \\ &g(x,\gamma)\equiv\frac{4}{3\varrho R_1^2}\int\limits_{\frac{\gamma+x^3}{\gamma+1}}^{x^3} u^{-\frac{7}{3}}(1+u)\left(\beth_1-u^{\frac{2}{3}}\beth_{-1}\right)du, \end{aligned}\right\} \tag{62.21}$$

where the arguments of $\beth_1$ and $\beth_{-1}$ are $2u^{\frac{2}{3}}+u^{-\frac{4}{3}}$ and $2u^{-\frac{2}{3}}+u^{\frac{4}{3}}$, then from (62.20) we find that

$$\frac{2x^2}{\varrho R_1^2}\left(P_1(t)-P_2(t)\right)=x^2 g(x,\gamma)+\frac{d}{dx}\left\{\left[1-\left(1+\frac{\gamma}{x^3}\right)^{-\frac{1}{3}}\right]x^3\dot{x}^2\right\}. \tag{62.22}$$

This is an ordinary differential equation similar to (62.5), and it may be studied by similar means. Before going into detail we remark from comparison of $(62.4)_5$ with $(62.21)_5$ that while the laws of finite oscillations of incompressible cylinders are governed by the generalized shear modulus $\hat{\mu}$, those of spheres are not. Therefore, observation of one of these two kinds of vibrations cannot determine information sufficient to predict all results for the other. Similar but not identical analysis applies in each case. For example, by $(57.36)_1$ the B-E inequalities (53.1) suffice for the integrand in the integral defining $g(x,\gamma)$ to be positive. Thus if we set

$$G(x,\gamma)\equiv\int\limits_1^x \xi^2 g(\xi,\gamma)\,d\xi=\frac{4}{3\varrho R_1^2}\int\limits_1^x \xi^2\,d\xi\int\limits_{\frac{\gamma+\xi^3}{\gamma+1}}^{\xi^3} u^{-\frac{7}{3}}(1+u)\left(\beth_1-u^{\frac{2}{3}}\beth_{-1}\right)du, \tag{62.23}$$

the B-E inequalities imply for $G(x,\gamma)$ the same growth properties as they do for $F(x,\gamma)$, as set forth above.

In free oscillations, $P_1(t)-P_2(t)\equiv 0$, and (62.22) yields at once

$$\left[1-\left(1+\frac{\gamma}{x^3}\right)^{-\frac{1}{3}}\right]x^3\dot{x}^2+G(x,\gamma)=C, \tag{62.24}$$

where C may be evaluated by setting $x(0)=x_0$, $\dot{x}(0)=\dot{x}_0$. It is easy to formulate conditions on the nature of the function $\beth_1-u^{\frac{2}{3}}\beth_{-1}$ near $u=0$ and near $u=\infty$ sufficient that $\lim\limits_{x\to 0} G(x,\gamma)=\lim\limits_{x\to\infty} G(x,\gamma)=\infty$. When the B-E inequalities hold, these conditions are sufficient that for any initial conditions other than $x_0=\dot{x}_0=0$, the equation $G(x,\gamma)=C$ shall have exactly two roots, $x=a<1$ and $x=b>1$. These roots give the least and greatest excursions of the inner surface of the shell. The speed is obtained from the equation

$$\dot{x}=\pm\sqrt{\frac{C-G(x,\gamma)}{x^3\left[1-\left(1+\frac{\gamma}{x^3}\right)^{-\frac{1}{3}}\right]}}, \tag{62.25}$$

[1] Notice that $\frac{1}{2}\varrho R_1^2 g(x,\gamma)=T_0\big|_{r_1}^{r_2}$, where T_0 is the radial tension $(57.37)_1$ corresponding to equilibrium of the shell in its instantaneous configuration.

while the period T is given by

$$T = 2\int_a^b \frac{dx}{\dot{x}} = 2\int_a^b \sqrt{\frac{x^3\left[1-\left(1+\frac{\gamma}{x^3}\right)^{-\frac{1}{3}}\right]}{C-G(x,\gamma)}}\,dx. \tag{62.26}$$

GUO ZHONG-HENG and SOLECKI, after obtaining the foregoing results, discussed the motion arising from a pressure impulse. The equations are similar to (62.15) — (62.18).

KNOWLES and JAKUB[1] have considered in detail the radial vibration of a spherical cavity in an infinite medium, which corresponds to the case when $R_2=\infty$, $\gamma=\infty$ in the foregoing analysis.

WANG[2] has found the simplifications resulting when the shell is thin. We may obtain his result by linearizing the right-hand side of (62.22) with respect to γ:

$$\varrho\ddot{x} + \frac{2}{R^2}\,\frac{x^6-1}{x^5}\,(\beth_I - x^2\beth_{-I}) = \frac{(P_1(t)-P_2(t))\,x^2}{\frac{1}{3}\gamma R^2}, \tag{62.27}$$

the arguments of $\beth_I$ and $\beth_{-I}$ being $2x^{\frac{2}{3}}+x^{-\frac{4}{3}}$ and $2x^{-\frac{2}{3}}+x^{\frac{4}{3}}$. This equation governs the radial oscillations of a spherical shell whose thickness is the small quantity $\frac{1}{3}\gamma R$ and whose radius at time t is $R\,x(t)$, where R is a constant radius. For the thin shell, the conditions of free vibration are easier to determine than for a shell of arbitrary thickness, since (62.24) may be seen to reduce to the more explicit form (62.13), except that now W is given by

$$W(x^2) = \frac{4}{\varrho R^2}\int_1^x \frac{\xi^2-1}{\xi^5}\,(\beth_I - \xi^2\beth_{-I})\,d\xi \tag{62.28}$$

in place of $(62.11)_2$. If the B-E inequalities hold, the sign of the integrand is that of ξ^2-1, so that it is easy to formulate simple conditions on the behavior of the function $\beth_I - \xi^2\beth_{-I}$ at $\xi=0$ and $\xi=\infty$ sufficient that (62.13) have exactly two roots x, no matter what the value of the initial condition $\dot{x}_0$.

WANG has studied also the response of a thin shell to pressure impulse, and the oscillations of a shell filled with a polytropic gas.

d) Systematic methods of approximation.

63. SIGNORINI's expansion. All presently known systematic methods of approximate solution in the theory of finite elastic deformation refer to the boundary-value problem of traction set in the material description, as defined in Sect. 44, although in applications sometimes semi-inverse devices are used to lighten the calculation. The differential equations and boundary conditions are (44.25) and (44.26), respectively, from which the displacement vector $\boldsymbol{u}$ is to be determined, either as a function of $\boldsymbol{X}$ in $\mathscr{B}$ in the case of equilibrium, or more generally as a function of $\boldsymbol{X}$ and t when suitable initial values are prescribed. The surface tractions $\boldsymbol{t}_R$ are given functions of $\boldsymbol{X}$ on $\partial\mathscr{B}$ and of t, and the body forces $\boldsymbol{b}$ are given functions of $\boldsymbol{X}$ in $\mathscr{B}$ and of t. We assume here that the reference configuration is a natural state and that the response function $\mathfrak{k}(\boldsymbol{H})$ is an *analytic* function of $\boldsymbol{H}$. We then have an expansion

$$\mathfrak{k}(\boldsymbol{H}) = \sum_{s=1}^{\infty} \mathfrak{k}_s(\boldsymbol{H}), \tag{63.1}$$

where $\mathfrak{k}_s(\boldsymbol{H})$ is a homogeneous polynomial of degree s in $\boldsymbol{H}$. It is not necessary to assume that the body is homogeneous, i.e., both $\mathfrak{k}$ and ϱ_R may depend on $\boldsymbol{X}$.

The expansion to be discussed does not apply to a single boundary-value problem but rather to a one-parameter family of such problems. It is assumed that the prescribed surface tractions $\boldsymbol{t}_R$ and body forces $\boldsymbol{b}$ depend analytically on

[1] KNOWLES and JAKUB [1965, *22*].

[2] WANG [1965, *43*].

the parameter ε and that they vanish when $\varepsilon=0$. We then have series expansions

$$\boldsymbol{b}=\sum_{n=1}^{\infty}\varepsilon^n\,\boldsymbol{b}_n,\quad \boldsymbol{t}_{\mathrm{R}}=\sum_{n=1}^{\infty}\varepsilon^n\,\boldsymbol{t}_{\mathrm{R}\,n}. \tag{63.2}$$

We *assume* that there exists a solution $\boldsymbol{u}$ to the boundary-value problem defined by the loads (63.2) which depends analytically on ε and vanishes when $\varepsilon=0$. In other words, we assume the existence of a solution of the form

$$\boldsymbol{u}=\sum_{n=1}^{\infty}\varepsilon^n\,\boldsymbol{u}_n. \tag{63.3}$$

In the specification of a particular problem the parameter ε may arise naturally as a quantity which may be increased from zero, such as a twist or some other strain.

We seek methods to determine the successive terms $\boldsymbol{u}_n$ in the expansion (63.3).

The first method of this kind is that of SIGNORINI[1], which has been considered extensively by the Italian school. In this scheme, $\boldsymbol{t}_{\mathrm{R}\,n}$ and $\boldsymbol{b}_n$ are set equal to $\mathbf{0}$ if $n\geqq 2$, so that, by assumption, the load is of the form (46.1). As has been indicated in Sect. 46, for the static problem the existence of a unique convergent series solution of the form (63.3) has been proved, subject to stated restrictions, in the case when the load does not possess an axis of equilibrium, and also in some cases when it does. GREEN and SPRATT[2] introduced the more general expansion (63.2). We shall now obtain results equivalent to theirs, without, however, adopting their choice of variables and notations.

By (63.3) the displacement gradient $\boldsymbol{H}=\nabla\boldsymbol{u}$ has a series expansion:

$$\boldsymbol{H}=\sum_{r=1}^{\infty}\varepsilon^r\boldsymbol{H}_r,\quad \boldsymbol{H}_r=\nabla\boldsymbol{u}_r. \tag{63.4}$$

Substitution of (63.4) into (63.1) shows that the Piola-Kirchhoff stress $\boldsymbol{T}_{\mathrm{R}}$ has a series expansion

$$\boldsymbol{T}_{\mathrm{R}}=\sum_{s=1}^{\infty}\mathfrak{k}_s\Big(\sum_{r=1}^{\infty}\varepsilon^r\boldsymbol{H}_r\Big)=\sum_{n=1}^{\infty}\varepsilon^n\,\boldsymbol{T}_{\mathrm{R}\,n}. \tag{63.5}$$

Since $\mathfrak{k}_s$ is a homogeneous polynomial of degree s, by collecting the terms proportional to ε^n in the first sum in (63.5) we see that the coefficient $\boldsymbol{T}_{\mathrm{R}\,n}$ must be of the form

$$\boldsymbol{T}_{\mathrm{R}\,n}=\mathfrak{k}_1(\boldsymbol{H}_n)+\mathfrak{h}_n(\boldsymbol{H}_1,\boldsymbol{H}_2,\ldots,\boldsymbol{H}_{n-1}), \tag{63.6}$$

where $\mathfrak{h}_n$ is a polynomial tensor function of $n-1$ tensor variables that is uniquely determined by the response function $\mathfrak{k}$ of the material. In particular, $\mathfrak{h}_1=\mathbf{0}$.

The coefficient polynomials $\mathfrak{k}_s$ in (63.1), and hence the polynomials $\mathfrak{h}_n$ in (63.6), cannot be arbitrary, since the response function $\mathfrak{k}$ must satisfy the principle of material indifference. In order to find the restrictions imposed by this principle we consider the response function $\mathfrak{t}$ of the reduced constitutive equation (43.10), assuming that $\mathfrak{t}$ has the expansion:

$$\tilde{\boldsymbol{T}}=\mathfrak{t}(\boldsymbol{C})=\mathfrak{l}_1(\boldsymbol{E})+\mathfrak{l}_2(\boldsymbol{E})+\cdots, \tag{63.7}$$

---

[1] SIGNORINI [1930, *3*, § 6] [1936, *3*, pp. 19—21] [1945, *4*, pp. 156—158] [1949, *21*, Cap. I, §§ 2—3] [1950, *15*] [1955, *25*, Cap. I, § 2]. In these papers, however, the process is rather described than presented explicitly. A detailed exposition in Cartesian tensor notation was given by GRIOLI [1962, *24*, Chapter IV].

[2] GREEN and SPRATT [1954, *9*]. A scheme of perturbation was given also by MIŞICU [1953, *17*, § 20]. References to works on second-order effects will be given in Sect. 66. In addition, there is an extensive literature on special non-linear theories, some of which yield results similar to those resulting at the second or third stage of a systematic procedure.

where $\boldsymbol{E}$ is the Green-St. Venant strain tensor[1]:

$$\boldsymbol{E} \equiv \tfrac{1}{2}(\boldsymbol{C}-\mathbf{1}) = \tfrac{1}{2}(\boldsymbol{H}+\boldsymbol{H}^T+\boldsymbol{H}^T\boldsymbol{H}), \tag{63.8}$$

and where $\mathfrak{l}_s$ is a homogeneous polynomial of degree s. In view of (50.11), we have

$$\mathfrak{l}_1(\boldsymbol{E}) = \mathsf{L}[\boldsymbol{E}], \tag{63.9}$$

where L is the fourth-order tensor of linear elasticities. A component form of (63.7) is

$$\mathfrak{t}_{\alpha\beta}(\boldsymbol{C}) = \mathsf{L}_{\alpha\beta}{}^{\gamma\delta} E_{\gamma\delta} + \mathsf{L}_{\alpha\beta}{}^{\gamma\delta\varepsilon\varrho} E_{\gamma\delta} E_{\varepsilon\varrho} + \cdots. \tag{63.10}$$

The coefficients $\mathsf{L}_{\alpha\beta}{}^{\gamma\delta}$, $\mathsf{L}_{\alpha\beta}{}^{\gamma\delta\varepsilon\varrho}$, ..., are the *elasticities*[2] of order 1, 2, ... Those of order one are the linear elasticities. The elasticities of all orders are symmetric in each successive pair of indices, and any two pairs after the first may be exchanged with each other:

$$\mathsf{L}_{\alpha\beta}{}^{\gamma\delta\varepsilon\varrho} = \mathsf{L}_{\beta\alpha}{}^{\gamma\delta\varepsilon\varrho} = \mathsf{L}_{\alpha\beta}{}^{\delta\gamma\varepsilon\varrho} = \mathsf{L}_{\alpha\beta}{}^{\gamma\delta\varrho\varepsilon} = \mathsf{L}_{\alpha\beta}{}^{\varepsilon\varrho\gamma\delta}. \tag{63.11}$$

With the help of $(43.11)_6$ we can express the terms $\mathfrak{t}_s(\boldsymbol{H})$ of the expansion (63.1) of $\mathfrak{t}(\boldsymbol{H}) = \mathfrak{h}(\boldsymbol{F})$ by means of the terms $\mathfrak{l}_s(\boldsymbol{E})$ of the expansion (63.7). By using also $(63.8)_2$ we find for the first two terms:

$$\mathfrak{t}_1(\boldsymbol{H}) = \mathsf{L}[\tilde{\boldsymbol{E}}], \tag{63.12}$$

$$\mathfrak{t}_2(\boldsymbol{H}) = \mathsf{L}[\tfrac{1}{2}\boldsymbol{H}^T\boldsymbol{H}] + \boldsymbol{H}\,\mathsf{L}[\tilde{\boldsymbol{E}}] + \mathfrak{l}_2(\tilde{\boldsymbol{E}}), \tag{63.13}$$

where $\tilde{\boldsymbol{E}} = \tfrac{1}{2}(\boldsymbol{H}+\boldsymbol{H}^T)$, the infinitesimal strain tensor. Substitution of (63.12) into (63.6) shows that

$$\boldsymbol{T}_{\mathrm{R}\,n} = \mathsf{L}[\tilde{\boldsymbol{E}}_n] + \mathfrak{h}_n(\boldsymbol{H}_1, \ldots, \boldsymbol{H}_{n-1}), \tag{63.14}$$

where $\tilde{\boldsymbol{E}}_n \equiv \tfrac{1}{2}(\boldsymbol{H}_n + \boldsymbol{H}_n^T)$. From (63.13), (63.5), and (63.14) we derive

$$\mathfrak{h}_2(\boldsymbol{H}_1) = \mathfrak{t}_2(\boldsymbol{H}_1). \tag{63.15}$$

The formulae (63.12), (63.13), and (63.15) illustrate how the $\mathfrak{t}_s$ and $\mathfrak{h}_n$ can be expressed in terms of the $\mathfrak{l}_s$. When $s, n > 2$, the corresponding formulae become increasingly complex. The $\mathfrak{t}_s$ and $\mathfrak{h}_n$ are restricted by the condition that they must be expressible by such formulae in terms of arbitrary homogeneous polynomial tensor functions $\mathfrak{l}_s$, i.e., in terms of elasticities of higher order, which are subject only to the symmetry conditions mentioned above.

We have chosen to define the "elasticities" of various orders in terms of the expansion (63.7), or, equivalently, (63.10). By using, instead of $\boldsymbol{E}$, any equivalent measure of strain (CFT, Sect. 32) that vanishes in the natural state, we should obtain an equally legitimate expansion, with, of course, different coefficients, which merit equally the name "elasticities". The previous paragraph has been included so as to illustrate the definiteness of the method. Starting with any other formal series expression for the stress relation using any other strain measure, we should find the same form of the functions $\mathfrak{h}_n$ in (63.6). For the method we now present it is (63.6) that is essential, no matter what means be used to calculate $\mathfrak{h}_n$. Indeed, whether or not the form of $\mathfrak{t}_1$ and $\mathfrak{h}_n$ be such as to satisfy the principle of material indifference is of little or no moment in what follows.

Substitution of the series (63.2), (63.3), and (63.5) into (44.25) and (44.26) and then equating like powers of ε yields the following successive systems of

[1] Cf. CFT, Sect. 31.

[2] A simplified notation for the elasticities of a hyperelastic material was proposed by BRUGGER [1964, *12*].

differential equations and associated boundary conditions:

$$\left.\begin{aligned} \operatorname{Div} \boldsymbol{T}_{\mathrm{R}\,n} + \varrho_{\mathrm{R}} \boldsymbol{b}_n &= \varrho_{\mathrm{R}} \ddot{\boldsymbol{u}}_n, \\ \boldsymbol{T}_{\mathrm{R}\,n} \boldsymbol{n}_{\mathrm{R}} &= \boldsymbol{t}_{\mathrm{R}\,n}, \end{aligned}\right\} n=1, 2, \ldots . \tag{63.16}$$

When (63.14) is used, these systems take the form

$$\left.\begin{aligned} \operatorname{Div} \mathsf{L}\,[\tilde{\boldsymbol{E}}_n] + \varrho_{\mathrm{R}} \boldsymbol{b}_n^* &= \varrho_{\mathrm{R}} \ddot{\boldsymbol{u}}_n, \\ \mathsf{L}\,[\tilde{\boldsymbol{E}}_n] \boldsymbol{n}_{\mathrm{R}} &= \boldsymbol{t}_{\mathrm{R}\,n}^*, \end{aligned}\right\} \tag{63.17}$$

where

$$\left.\begin{aligned} \varrho_{\mathrm{R}} \boldsymbol{b}_n^* &= \varrho_{\mathrm{R}} \boldsymbol{b}_n + \operatorname{Div} \mathfrak{h}_n(\boldsymbol{H}_1, \ldots, \boldsymbol{H}_{n-1}), \\ \boldsymbol{t}_{\mathrm{R}\,n}^* &= \boldsymbol{t}_{\mathrm{R}\,n} - \mathfrak{h}_n(\boldsymbol{H}_1, \ldots, \boldsymbol{H}_{n-1}) \boldsymbol{n}_{\mathrm{R}}. \end{aligned}\right\} \tag{63.18}$$

When $n=1$, since $\boldsymbol{b}_1^* \equiv \boldsymbol{b}_1$, $\boldsymbol{t}_{\mathrm{R}1}^* \equiv \boldsymbol{t}_{\mathrm{R}1}$, the equations (63.17) are precisely those which define the boundary-value problem of traction in the classical infinitesimal theory of elasticity, where $\boldsymbol{b}_1$ and $\boldsymbol{t}_{\mathrm{R}1}$ are the assigned external body force and surface traction. More generally, let it be supposed that fields $\boldsymbol{u}_1, \boldsymbol{u}_2, \ldots, \boldsymbol{u}_{m-1}$ satisfying (63.17) for $n=1, 2, \ldots, m-1$ have been determined. Then for $n=m$ (63.17) has the form of the equations defining the traction boundary-value problem of the classical infinitesimal theory *for the same material and the same boundary,* though subject to assigned external body force and surface traction which depend in an explicitly known way upon the previously determined fields $\boldsymbol{u}_1, \boldsymbol{u}_2, \ldots, \boldsymbol{u}_{m-1}$. Thus *the calculation of m terms in the expansion* (63.3) *is reduced, formally, to the solution of m boundary-value problems of traction in the classical infinitesimal theory of elasticity for the same body.*

The process applies to anisotropic and isotropic materials alike. For the latter, the details of calculation of the $\mathfrak{h}_n$ are more explicit if one replaces (63.7) by an expansion arising from a reduced form such as (47.9). In place of the invariants of $\boldsymbol{B}$, which coincide with those of $\boldsymbol{C}$, it is preferable to use others which vanish when $\boldsymbol{u} \equiv \boldsymbol{0}$. For example, we may use the invariants of $\boldsymbol{E}$ [cf. (63.8)], which are related to those of $\boldsymbol{B}$, used in the presentation of exact solutions in Sect. 66, as follows[1]:

$$\left.\begin{aligned} 2I_{\boldsymbol{E}} &= I_{\boldsymbol{B}} - 3, \quad 4II_{\boldsymbol{E}} = II_{\boldsymbol{B}} - 2I_{\boldsymbol{B}} + 3, \\ & \qquad 8III_{\boldsymbol{E}} = III_{\boldsymbol{B}} - II_{\boldsymbol{B}} + I_{\boldsymbol{B}} - 1, \\ I_{\boldsymbol{B}} &= 2I_{\boldsymbol{E}} + 3, \quad II_{\boldsymbol{B}} = 4II_{\boldsymbol{E}} + 4I_{\boldsymbol{E}} + 3, \\ & \qquad III_{\boldsymbol{B}} = 8III_{\boldsymbol{E}} + 4II_{\boldsymbol{E}} + 2I_{\boldsymbol{E}} + 1. \end{aligned}\right\} \tag{63.19}$$

The expansion corresponding to (63.7) is so complicated that there is no point in attempting to exhibit it in general; of course, the calculation of the first few terms offers no difficulty. For a somewhat different method, the details of the second stage are presented in Sect. 66.

It will be recalled that in the classical infinitesimal theory the three stress tensors $\boldsymbol{T}$, $\boldsymbol{T}_{\mathrm{R}}$, and $\tilde{\boldsymbol{T}}$ are indistinguishable. As soon as terms quadratic in the deformation gradients are taken into account, such confusion is no longer legitimate. For example, from $(43.11)_1$, (63.1), (63.12), and (63.13) we obtain the expansion

$$\left.\begin{aligned} \boldsymbol{T} &= \mathfrak{g}(\boldsymbol{F}) = [\det(\boldsymbol{1}+\boldsymbol{H})]^{-1} \mathfrak{t}(\boldsymbol{H}) (\boldsymbol{1}+\boldsymbol{H}^T) \\ &= (1 - \operatorname{tr} \boldsymbol{H} + \cdots) (\mathfrak{t}_1(\boldsymbol{H}) + \mathfrak{t}_2(\boldsymbol{H}) + \cdots\cdots) (\boldsymbol{1}+\boldsymbol{H}^T) \\ &= \mathsf{L}[\tilde{\boldsymbol{E}}] + \boldsymbol{H}\mathsf{L}[\tilde{\boldsymbol{E}}] + \mathsf{L}[\tilde{\boldsymbol{E}}]\boldsymbol{H}^T + \mathsf{L}[\tfrac{1}{2}\boldsymbol{H}^T\boldsymbol{H}] - \operatorname{tr}\boldsymbol{H}\,\mathsf{L}[\tilde{\boldsymbol{E}}] + \mathfrak{l}_2(\tilde{\boldsymbol{E}}) + \cdots, \end{aligned}\right\} \tag{63.20}$$

[1] These formulae are immediate consequences of the duals of Eqs. (31.5) of CFT.

where all terms of degree 1 and 2 in $\boldsymbol{H}$ are shown explicitly. The terms in this expansion, for $\boldsymbol{T}$, are obviously different from those in (63.7), the expansion for $\tilde{\boldsymbol{T}}$; likewise, they differ from those in the expansion (63.5) for $\boldsymbol{T}_{\mathrm{R}}$ by the quadratic term $\mathsf{L}[\tilde{\boldsymbol{E}}]\boldsymbol{H}^T - \operatorname{tr}\boldsymbol{H}\mathsf{L}[\tilde{\boldsymbol{E}}]$ and, of course, also by terms of higher order. A component form of (63.20) is[1]

$$T^{km} = L^{km\alpha\beta} u_{\alpha,\beta} + Q^{km\alpha\beta\gamma\delta} u_{\alpha,\beta} u_{\gamma,\delta} + \cdots, \tag{63.21}$$

where the components of Q are expressed as follows in terms of the linear elasticities and second-order elasticities, defined by (63.10):

$$Q^{km\alpha\beta\gamma\delta} = L^{km\alpha\beta\gamma\delta} + g^{k\alpha} L^{m\beta\gamma\delta} + g^{m\alpha} L^{k\beta\gamma\delta} + \tfrac{1}{2} g^{\alpha\gamma} L^{km\beta\delta} - g^{\alpha\beta} L^{km\gamma\delta}. \tag{63.22}$$

The coefficients of terms of higher degree depend in increasingly complicated ways upon the elasticities of various orders.

The expansion (63.21) has been used by TOUPIN and RIVLIN[2] as a basis for determining changes in the dimensions of bodies in whose interior there are "dislocations", i.e. bodies that can be brought back to their homogeneous natural reference configurations only after a finite number of cuts are effected. Specifically, they consider a body free of body force and surface loads. In such a body, whatever be its constitution, the average stress components $\overline{T\langle km\rangle}$ in any Cartesian co-ordinate system vanish [CFT, Sect. 220]. In the infinitesimal theory, since the strain components are linear functions of the stress components, the average strains vanish also, and, in particular, there is no overall change of volume. This result is not characteristic of the general theory, in which it is possible for there to be non-vanishing stresses and strains in the interior of a body subject to no body force or boundary tractions, with a corresponding change of volume. Using a bar to denote a mean value, from (63.21) we see that

$$L^{km\alpha\beta}\,\overline{u_{\alpha,\beta}} = -\,Q^{km\alpha\beta\gamma\delta}\,\overline{u_{\alpha,\beta} u_{\gamma,\delta}} + \cdots. \tag{63.23}$$

Since L is the tensor of linear elasticities, under the usual assumptions of the infinitesimal theory it has an inverse M such that $M_{\lambda\mu km} L^{km\alpha\beta} = \frac{1}{2}(\delta^\alpha_\lambda \delta^\beta_\mu + \delta^\beta_\lambda \delta^\alpha_\mu)$. From (63.23), then,

$$\overline{u_{(\lambda,\mu)}} = -\,M_{\lambda\mu km} Q^{km\alpha\beta\gamma\delta}\,\overline{u_{\alpha,\beta} u_{\gamma,\delta}} + \cdots. \tag{63.24}$$

Thus the mean linearized strains are not zero in general, and they may be determined as means of certain quadratic expressions. An example is given in Sect. 66.

64. SIGNORINI'S theorems of compatibility and uniqueness of the equilibrium solution. In the case of equilibrium ($\ddot{\boldsymbol{u}} \equiv \boldsymbol{0}$) the terms involving the unknown function $\boldsymbol{u}_n$ in (63.17) are of exactly the same form as those involving the displacement in the classical infinitesimal theory of elasticity for the same material. The functions $\boldsymbol{u}_1, \boldsymbol{u}_2, \ldots, \boldsymbol{u}_{n-1}$ affect only the apparent forces giving rise to the displacement $\boldsymbol{u}_n$. If those apparent forces are compatible with equilibrium, then *every theorem concerning solution of the traction boundary-value problem in the classical infinitesimal theory for bodies of the material under consideration may be carried over at once to the displacement $\boldsymbol{u}_n$ in the formal series expansion* (63.3). In particular, *for a body such that the solution of the traction boundary-value problem is unique in the infinitesimal theory*[3], *to within a uniform infinitesimal rotation, the solution $\boldsymbol{u}_n$ is also unique, for all n, to within a uniform infinitesimal rotation.*

Here must be raised two objections. First, in the finite theory, there is no reason why an arbitrary rotation should remain in the solution. Second, in order for a solution to exist in the infinitesimal theory, it is necessary that the applied

---

[1] $L^{km\alpha\beta} = g^k_\gamma g^m_\delta L^{\gamma\delta\alpha\beta}$, etc.

[2] TOUPIN and RIVLIN [1960, *56*].

[3] As is well known, it is sufficient but not necessary for such uniqueness that the stress be derivable from a definite strain energy, positive or negative. For isotropic materials a necessary and sufficient condition, in a general region, is that the Poisson modulus σ satisfy the condition $|\sigma| < 1$. References to the literature of which this conclusion is a summary are given by TRUESDELL and TOUPIN [1963, *76*, § 1].

loads be equilibrated. Since the effective loads giving rise to the displacement $\boldsymbol{u}_n$ satisfying (63.17) depend upon $\boldsymbol{u}_1, \boldsymbol{u}_2, \ldots, \boldsymbol{u}_{n-1}$, that condition here restricts the displacements previously found. Both these objections will now be met.

As we have seen in Sect. 44, in order that a boundary-value problem of traction have a solution, it is necessary that the applied loads (63.2) satisfy the condition (44.14). Also, by DA SILVA's theorem, we may assume without loss of generality that (44.20) holds. The assumed solution (63.3) must satisfy the condition of compatibility (44.15). Subtracting (44.20) from (44.15), we see that this condition, in terms of the displacement $\boldsymbol{u}=\boldsymbol{x}-\boldsymbol{X}=\boldsymbol{p}-\boldsymbol{p}_{\mathrm{R}}$, may be written in the form[1]

$$\int_{\partial\mathscr{B}} \boldsymbol{u}\times \boldsymbol{t}_{\mathrm{R}}\, d s_{\mathrm{R}} + \int_{\mathscr{B}} \boldsymbol{u}\times \varrho_{\mathrm{R}} \boldsymbol{b}\, d v_{\mathrm{R}} = \boldsymbol{0}. \tag{64.1}$$

If we substitute (63.2) and (63.3) into (64.1) and expand, we obtain

$$\sum_{m=1}^{n-1} \left[\int_{\partial\mathscr{B}} \boldsymbol{u}_{n-m}\times \boldsymbol{t}_{\mathrm{R}m}\, d s_{\mathrm{R}} + \int_{\mathscr{B}} \boldsymbol{u}_{n-m}\times \varrho_{\mathrm{R}} \boldsymbol{b}_m\, d v_{\mathrm{R}}\right] = \boldsymbol{0}, \qquad n=2, 3, \ldots. \tag{64.2}$$

Thus, in order that the expansion (63.3) *shall give a solution, it is necessary not only that the* $\boldsymbol{u}_n$ *be solutions of the system* (63.17) *but also that they satisfy the conditions of compatibility* (64.2). As remarked at the beginning of this section, the system (63.17), since it has the form of the equations governing the traction boundary-value problem of infinitesimal elasticity, determines $\boldsymbol{u}_n$ at most to within a uniform infinitesimal rotation. We shall see that, in general, this rotation is uniquely determined by the compatibility conditions (64.2). Thus the $\boldsymbol{u}_n$, and hence $\boldsymbol{u}$, are unique, and the indeterminacy of the rotation, a notorious feature of the classical infinitesimal theory which obviously cannot carry over to the finite theory, is removed, in general, in the method of series solution[2].

In order to determine the $\boldsymbol{u}_n$ inductively, suppose that solutions $\boldsymbol{u}_1, \boldsymbol{u}_2, \ldots, \boldsymbol{u}_{p-1}$ of (63.17) for $n=1, 2, \ldots, p-1$ have already been found such that the compatibility conditions (64.2) are satisfied for $n=2, 3, \ldots, p$. Let $\boldsymbol{u}_p$ be any solution of (63.17) for $n=p$, and put

$$\boldsymbol{r}_p \equiv \sum_{m=1}^{p} \left[\int_{\partial\mathscr{B}} \boldsymbol{u}_{p+1-m}\times \boldsymbol{t}_{\mathrm{R}m}\, d s_{\mathrm{R}} + \int_{\mathscr{B}} \boldsymbol{u}_{p+1-m}\times \varrho_{\mathrm{R}} \boldsymbol{b}_m\, d v_{\mathrm{R}}\right]. \tag{64.3}$$

If $\boldsymbol{\omega}_p$ is any constant vector, then $\boldsymbol{u}_p + \boldsymbol{\omega}_p\times\boldsymbol{p}_{\mathrm{R}}$ is also a solution of (63.17) for $n=p$. This solution will satisfy the compatibility condition (64.2) if

$$\int_{\partial\mathscr{B}} (\boldsymbol{\omega}_p\times\boldsymbol{p}_{\mathrm{R}})\times \boldsymbol{t}_{\mathrm{R}1}\, d s_{\mathrm{R}} + \int_{\mathscr{B}} (\boldsymbol{\omega}_p\times\boldsymbol{p}_{\mathrm{R}})\times \varrho_{\mathrm{R}} \boldsymbol{b}_1\, d v_{\mathrm{R}} = -\boldsymbol{r}_p. \tag{64.4}$$

By use of the vectorial identity $(\boldsymbol{a}\times\boldsymbol{b})\times\boldsymbol{c}=(\boldsymbol{a}\cdot\boldsymbol{c})\,\boldsymbol{b}-(\boldsymbol{b}\cdot\boldsymbol{c})\,\boldsymbol{a}$ we may express (64.4) in the form

$$(\boldsymbol{A}_1 - \boldsymbol{1}\operatorname{tr}\boldsymbol{A}_1)\,\boldsymbol{\omega}_p = -\boldsymbol{r}_p, \tag{64.5}$$

where $\boldsymbol{A}_1$ is the astatic load (46.11) corresponding to $\boldsymbol{t}_{\mathrm{R}1}$ and $\boldsymbol{b}_1$. In order that (64.5) possess a unique solution, it is necessary and sufficient that (46.16) holds; that is, the load $\boldsymbol{t}_{\mathrm{R}1}$, $\boldsymbol{b}_1$ must not possess an axis of equilibrium. Under this condition, therefore, the $\boldsymbol{u}_n$ are unique.

---

[1] SIGNORINI [1930, *3*, § 6], and numerous later papers.

[2] Thus is answered, in principle, an objection stated by ERICKSEN [1955, *10*, § 1], though that objection was directed toward the work of RIVLIN and TOPAKOGLU (Sect. 65), for which no conditions of compatibility were given. Cf. the way in which the rotation enters the analytical existence theory in Sect. 46.

In order that the system (63.17) have a solution $\boldsymbol{u}_n$ at all, it is necessary that the effective loads $\boldsymbol{t}^*_{\mathrm{R}n}$, $\boldsymbol{b}^*_{\mathrm{R}n}$ be equilibrated, i.e., that (44.14) and (44.20) hold if $\boldsymbol{t}_{\mathrm{R}}$ and $\boldsymbol{b}$ there are replaced by $\boldsymbol{t}^*_{\mathrm{R}n}$ and $\boldsymbol{b}^*_n$. Now, if we substitute the expansion (63.2) into (44.14) and (44.20), we see that $\boldsymbol{t}_{\mathrm{R}n}$ and $\boldsymbol{b}_n$ are equilibrated by themselves. Hence the conditions for the existence of a solution $\boldsymbol{u}_n$ of (63.17) are equivalent to

$$\left.\begin{aligned}&\int_{\partial\mathscr{B}}(\boldsymbol{t}^*_{\mathrm{R}n}-\boldsymbol{t}_{\mathrm{R}n})\,d s_{\mathrm{R}}+\int_{\mathscr{B}}\varrho_{\mathrm{R}}(\boldsymbol{b}^*_n-\boldsymbol{b}_n)\,d v_{\mathrm{R}}=0,\\&\int_{\partial\mathscr{B}}\boldsymbol{p}_{\mathrm{R}}\times(\boldsymbol{t}^*_{\mathrm{R}n}-\boldsymbol{t}_{\mathrm{R}n})\,d s_{\mathrm{R}}+\int_{\mathscr{B}}\boldsymbol{p}_{\mathrm{R}}\times\varrho_{\mathrm{R}}(\boldsymbol{b}^*_n-\boldsymbol{b}_n)\,d v_{\mathrm{R}}=\boldsymbol{0}.\end{aligned}\right\}\tag{64.6}$$

Use of (63.18) and the divergence theorem shows easily that $(64.6)_1$ is automatically satisfied. In order to analyse $(64.6)_2$ we shall make repeated use of the identity

$$\int_{\mathscr{B}}(\nabla\boldsymbol{v})\boldsymbol{S}^T\,d v_{\mathrm{R}}=\int_{\partial\mathscr{B}}\boldsymbol{v}\otimes(\boldsymbol{S}\boldsymbol{n}_{\mathrm{R}})\,d s_{\mathrm{R}}-\int_{\mathscr{B}}(\boldsymbol{v}\otimes\operatorname{div}\boldsymbol{S})\,d v_{\mathrm{R}}.\tag{64.7}$$

Applying (64.7) with the choice $\boldsymbol{v}=\boldsymbol{p}_{\mathrm{R}}$ and $\boldsymbol{S}=\mathfrak{h}_n(\boldsymbol{H}_1,\ldots,\boldsymbol{H}_{n-1})$, we see from (63.18) that $(64.6)_2$ is equivalent to the requirement that

$$\int_{\mathscr{B}}\mathfrak{h}_n(\boldsymbol{H}_1,\ldots,\boldsymbol{H}_{n-1})\,d v_{\mathrm{R}}=\text{a symmetric tensor}.\tag{64.8}$$

On the other hand, since the Cauchy stress $\boldsymbol{T}$ is a symmetric tensor, it follows from (43 A.3) that

$$J\boldsymbol{T}=\boldsymbol{T}_{\mathrm{R}}(\boldsymbol{1}+\boldsymbol{H}^T)=\text{a symmetric tensor}.\tag{64.9}$$

Substituting the expansions (63.4) and (63.5) into (64.9) and collecting the terms multiplied by like powers of ε, we find that (64.9) is equivalent to

$$\boldsymbol{T}_{\mathrm{R}n}+\sum_{m=1}^{n-1}\boldsymbol{T}_{\mathrm{R}m}\boldsymbol{H}^T_{n-m}=\text{a symmetric tensor}.\tag{64.10}$$

Integrating (64.10) over $\mathscr{B}$ and observing that, by (63.12), the first term, $\mathfrak{t}_1(\boldsymbol{H}_n)$, in (63.14) is symmetric, we conclude that (64.8) is satisfied if and only if

$$\sum_{m=1}^{n-1}\int_{\mathscr{B}}\boldsymbol{H}_{n-m}\boldsymbol{T}^T_{\mathrm{R}m}\,d v_{\mathrm{R}}=\text{a symmetric tensor}.\tag{64.11}$$

Applying the identity (64.7) with the choice $\boldsymbol{v}=\boldsymbol{u}_{n-m}$, $\boldsymbol{S}=\boldsymbol{T}_{\mathrm{R}m}$ and observing (63.16), we see that (64.11) is actually equivalent to the compatibility condition (64.2) for n. Thus these compatibility conditions are just the right restrictions to be imposed in order that (63.17) can have any solutions at all[0].

The result established so far in this section may then be stated as the following ***theorem of compatibility and uniqueness***, due to SIGNORINI[1]: *Let it be supposed that the loads $\boldsymbol{t}_{\mathrm{R}1}$, $\boldsymbol{b}_1$ do not possess an axis of equilibrium. Then*

1. The system (63.17), which must be satisfied if (63.3) is to be a solution, is compatible; i.e., if solutions exist for the traction boundary-value problem in the infinitesimal theory, then there exist solutions $\boldsymbol{u}_1, \boldsymbol{u}_2, \ldots$ of the system (63.17).

2. If the classical uniqueness theorem for the traction boundary-value problem in the infinitesimal theory holds, then the solution $\boldsymbol{u}_1, \boldsymbol{u}_2, \ldots$ of the system (63.17) is unique to within a uniform translation[2].

[0] GRIOLI [1962, *24*, Ch. IV, § 3] derived equivalent equations as formal conditione of compatibility for (63.17).

[1] The original assertion of uniqueness by SIGNORINI [1930, *3*, § 6] is too broad; it was corrected in [1949, *21*, Cap. I, ¶ 6], where a proof was given. Modifications appropriate to incompressible materials were made in [1955, *25*, Cap. I, ¶¶ 6—8].

[2] In the case when the hypotheses for STOPPELLI's theorem in Sect. 46 are satisfied, SIGNORINI's theorem is of course a corollary of it. However, SIGNORINI's theorem, being weaker in statement, holds under more general conditions. It is possible that the coefficients in (63.3) may be uniquely determined in circumstances when the series is not convergent.

It has not been shown, however, that the system (63.17) is incompatible in the case when there is an axis of equilibrium; what has been shown is that in this case (64.5) does not determine *unique* infinitesimal rotations $\boldsymbol{\omega}_p$; (64.5) may now have no solution, or more than one. For example, it may turn out that (64.2) for $n=2$, is satisfied by a particular solution $\boldsymbol{u}_1$:

$$\int_{\partial\mathscr{B}} \boldsymbol{u}_1 \times \boldsymbol{t}_{R1}\, d s_R + \int_{\mathscr{B}} \boldsymbol{u}_1 \times \varrho_R \boldsymbol{b}_1 d v_R = \boldsymbol{0}. \tag{64.12}$$

If there is an axis of equilibrium, then there are also infinitely many possible rotations $\boldsymbol{\omega}_1$. These possibilities have been subjected to much study by the Italian school. First, as remarked by SIGNORINI[1], uniqueness continues to hold, subject to the same proviso regarding the infinitesimal theory, if we prescribe rotations $\boldsymbol{\omega}_n^0$ about each axis of equilibrium; the remaining part, if any, of the rotation $\boldsymbol{\omega}_n$ is then determined by the conditions (64.5). Second, TOLOTTI[2] has given an exhaustive analysis of the various possible cases when there is an axis of equilibrium and the successive systems (63.17) are compatible for every n, for n up to a certain value, or for no n.

The numerous papers of the Italian school on series solution of the traction boundary-value problem[3] seem to confine themselves to determination and illustration of incompatible cases. In the still more numerous researches of the British school, some of which will be summarized in the next section, we have found no mention at all of the compatibility problem and no attempt to determine the rotation in any of the many special solutions to which most of that literature is devoted. A just view lies somewhere between these extremes.

First, the Italian school regards the incompatibility of (63.17) for any n, as *an objection against the classical infinitesimal theory* for the solution of the corresponding traction boundary-value problem for small loads. What is shown, precisely, is that in the incompatible cases the solution given by the infinitesimal theory cannot be the first term in a series of the type (63.3) satisfying formally the particular problem of the finite theory which has been defined by (63.2). While SIGNORINI[4] asserted, "The classical theory tacitly but systematically presupposes expansions of the form (63.3), with the intention of retaining only the first term," and while TOLOTTI[5] considered that in the incompatible cases "the application of the usual theory is never justified," TRUESDELL[6] replied, "It does not seem to me that there is any reason to expect even a formal power series ... to exist: rather, the infinitesimal theory might be expected to have an elaborate asymptotic character with respect to the general non-linear theory." Indeed, if we presuppose both finite elasticity and series of the forms (63.2) and (63.3), incompatibility of the infinitesimal theory follows inevitably for certain problems. Neither presumption, however, is a necessary one. First, the theory of finite elastic strain is not the only possible basis for the classical infinitesimal theory; other plausible theories, such as hypo-elasticity, serve equally well as a starting point from which to derive the classical infinitesimal theory as a first approximation (cf. Sect. 100). Second, if we agree to start from the finite theory, to presume the loads given by series of the forms (63.3) is unnecessarily restrictive. Because of the occurrence of the arbitrary parameter ε, the *one* boundary-value problem of the infinitesimal theory is embedded in a special way in a *family* of boundary-value problems of the finite theory, and there is no reason to suppose this the only way the infinitesimal theory may emerge. To establish the position of the infinitesimal theory, it would be quite enough to show that the solution of the linear problem is the asymptotic form of *some* family of solutions of the finite theory as the loads tend to zero, and to lay down a power series as the only way of defining such a family is merely arbitrary. Moreover, the researches of the Italian school are limited to the case when (46.1) holds, rather than the more general expansions (63.2). Even if we adopt the viewpoint of the Italian school, the major part of the compatibility problem vanishes if we agree to accept an expansion of the form (63.2) instead of (46.1) as defining the problem, for the linearized problem gives no information about $\boldsymbol{t}_{R2}, \boldsymbol{t}_{R3}, \ldots, \boldsymbol{b}_2, \boldsymbol{b}_3, \ldots$. In any case

[1] SIGNORINI [1949, *21*, Cap. I, ¶ 6].

[2] TOLOTTI [1943, *4*].

[3] SIGNORINI [1930, *3*, § 6] [1936, *3*, pp. 19—21] [1942, *6*, pp. 64—65] [1945, *4*, pp. 156—158] [1949, *21*, Cap. I, § 3] [1950, *15*] [1955, *25*, Cap. I, ¶ 11]; TOLOTTI [1943, *4*]; CAPRIZ [1959, *1*]; GRIOLI [1962, *24*, Chap. IV].

[4] SIGNORINI [1949, *21*, Cap. I, ¶ 6].

[5] TOLOTTI [1943, *4*, p. 1141].

[6] TRUESDELL [1952, *20*, § 39, footnote 1].

when the infinitesimal theory is compatible with the second approximation, i.e., in any problem where (64.12) is satisfied, we can satisfy all higher conditions of compatibility by proper choice of $\boldsymbol{t}_{R2}, \boldsymbol{t}_{R3}, \ldots, \boldsymbol{b}_2, \boldsymbol{b}_3, \ldots$. For example, setting $n=3$ in (64.2), we have

$$\int_{\partial\mathscr{B}} (\boldsymbol{u}_1 \times \boldsymbol{t}_{R2} + \boldsymbol{u}_2 \times \boldsymbol{t}_{R1})\, ds_R + \int_{\mathscr{B}} (\boldsymbol{u}_1 \times \varrho_R \boldsymbol{b}_2 + \boldsymbol{u}_2 \times \varrho_R \boldsymbol{b}_2)\, dv_R = \mathbf{0}; \tag{64.13}$$

with $\boldsymbol{u}_1, \boldsymbol{u}_2, \boldsymbol{t}_{R1}$, and $\boldsymbol{b}_1$ given, there are infinitely many choices of $\boldsymbol{t}_{R2}$ and $\boldsymbol{b}_2$ such that (64.13) is satisfied. Similarly, all conditions of compatibility of higher order may be satisfied by infinitely many choices of $\boldsymbol{t}_{R3}, \boldsymbol{t}_{R4}, \ldots \boldsymbol{b}_3, \boldsymbol{b}_4, \ldots$. If all that is desired is to exhibit a problem of the finite theory whose solution, as $\varepsilon \to 0$, becomes that of the corresponding problem of the infinitesimal theory, this expedient suffices. What has been shown, then, is that *the only essential condition of compatibility, if we are to embed the solution of an infinitesimal traction boundary-value problem as the first term in a formal series expansion of the solution of a corresponding problem in the finite theory, is the first one, namely,* (64.12).

However, the objections raised by the Italian school cannot be dismissed altogether. It is natural to expect that the solution of a problem of the infinitesimal theory should be the common limit of the solutions of a class of problems of the finite theory, as the loads tend to zero. From what has been shown above, it is clear that this class is much smaller in the incompatible cases than in the compatible ones. Moreover, breakdown of a formal series method, however arbitrary, usually indicates some analytic distinction. Such is in fact the case here, as has been seen from the results of STOPPELLI, summarized in Sect. 46.

Finally, it is important to see clearly that no perturbation method can be general, for perturbation is relevant only to those problems of finite deformation that reduce smoothly to problems of infinitesimal deformation as the loads vanish. Some of the most typical and striking problems of finite strain fail to be of this kind. For example, in the solutions for eversion of a tube or of a sphere, presented in Sect. 57, the assigned loads $\boldsymbol{t}_R$ and $\boldsymbol{b}$ are identically zero; for these loads, two distinct solutions are possible, one being that in which the stresses vanish identically and the other being that which corresponds to eversion. Only the former and trivial solution can follow from any method of perturbation about the solution given by the classical infinitesimal theory[1]. The same objection holds in reference to the analytical theory of STOPPELLI summarized in Sect. 46, where the objective is rather to extend the results of the classical type than to find the true nature of solutions according to the finite theory.

65. RIVLIN and TOPAKOGLU'S interpretation. We now present an interpretation of the expansion of Sect. 63 due to RIVLIN and TOPAKOGLU[2], which provides a simple and direct physical motivation: Given a solution of the equations of displacement for the $(n-1)$'st step, we substitute it into the equations for the n'th step, thus finding the surface tractions and body forces that would be required in order to maintain this displacement if the n'th-order theory were exact. Next we find the displacement which these surface tractions and body forces, if reversed, would produce according to the infinitesimal theory. To within terms of the order neglected at the n'th stage, a solution of the equations at that stage is given by adding this new displacement to the one previously known. This scheme grows out of the simple idea that the step from any stage to the next is a small one and hence nearly "infinitesimal"; the discrepancy of a solution at the $(n-1)$'st stage from being one at the n'th stage is equivalent to the effect of a set of forces, which

[1] As shown by the calculation at the end of Sect. 66, even in the second-order theory obtained by considering simultaneously the first two stages of the perturbation process, without splitting the problem into two, this objection is avoided.

[2] Like the works of SIGNORINI, the memoir of RIVLIN and TOPAKOGLU [1954, *19*] refers only to hyperelastic materials. RIVLIN and TOPAKOGLU were not aware that their expansion is but an adjustment of SIGNORINI'S. After this had been remarked by TRUESDELL [1959, *37*, p. 77], GRIOLI [1962, *24*, Ch. IV, § 6] showed the identity of the expansion for the first two terms.

The generalization to materials not necessarily having a stored-energy function is due to SHENG [1955, *24*, § 10], who considered explicitly the mixed boundary-value problem.

are cancelled by supplying the displacement that corresponds, according to the infinitesimal theory, to their negatives.

To make this idea precise, let us denote the partial sums of the expansions (63.1), (63.2), (63.3), (63.4), and (63.5) by $\mathfrak{t}^{(n)}$, $\boldsymbol{b}^{(n)}$, $\boldsymbol{t}_{\mathrm{R}}^{(n)}$, $\boldsymbol{u}^{(n)}$, $\boldsymbol{H}^{(n)}$, and $\boldsymbol{T}_{\mathrm{R}}^{(n)}$, respectively. For example

$$\mathfrak{t}^{(n)}(\boldsymbol{H}) \equiv \sum_{s=1}^{n} \mathfrak{t}_s(\boldsymbol{H}), \qquad \boldsymbol{H}^{(n)} \equiv \sum_{r=1}^{n} \varepsilon^r \boldsymbol{H}_r. \tag{65.1}$$

It is clear from (63.5) that

$$\mathfrak{t}^{(n)}(\boldsymbol{H}^{(n)}) = \boldsymbol{T}_{\mathrm{R}}^{(n)} + O(\varepsilon^{n+1}) \tag{65.2}$$

where $O(\varepsilon^{n+1})$ indicates terms of order ε^{n+1} and higher. It follows from (63.16) that

$$\left.\begin{aligned} \operatorname{Div} \boldsymbol{T}_{\mathrm{R}}^{(n)} + \varrho_{\mathrm{R}} \boldsymbol{b}^{(n)} &= \varrho_{\mathrm{R}} \ddot{\boldsymbol{u}}^{(n)}, \\ \boldsymbol{T}_{\mathrm{R}}^{(n)} \boldsymbol{n}_{\mathrm{R}} &= \boldsymbol{t}_{\mathrm{R}}^{(n)}. \end{aligned}\right\} \tag{65.3}$$

Hence, by (65.2),

$$\left.\begin{aligned} \operatorname{Div} \mathfrak{t}^{(n)}(\boldsymbol{H}^{(n)}) + \varrho_{\mathrm{R}} \boldsymbol{b}^{(n)} &= \varrho_{\mathrm{R}} \ddot{\boldsymbol{u}}^{(n)} + O(\varepsilon^{n+1}), \\ \mathfrak{t}^{(n)}(\boldsymbol{H}^{(n)}) \boldsymbol{n}_{\mathrm{R}} &= \boldsymbol{t}_{\mathrm{R}}^{(n)} + O(\varepsilon^{n+1}). \end{aligned}\right\} \tag{65.4}$$

We shall say we have solved *the equations of n'th-order elasticity* corresponding to the response function (63.1) if we have found a polynomial solution

$$\boldsymbol{u}^{(n)} = \sum_{r=1}^{n} \varepsilon^r \boldsymbol{u}_r \tag{65.5}$$

of the system (65.4). The partial sum $\boldsymbol{u}^{(n)}$ of SIGNORINI's expansion (63.3) gives such a solution, and every solution is of this form. It follows from (63.6) and (65.2) that

$$\mathfrak{t}^{(n)}(\boldsymbol{H}^{(n)}) = \sum_{r=1}^{n} \varepsilon^r [\mathfrak{t}_1(\boldsymbol{H}_r) + \mathfrak{h}_r(\boldsymbol{H}_1, \ldots, \boldsymbol{H}_{r-1})] + O(\varepsilon^{n+1}). \tag{65.6}$$

Now, $\mathfrak{t}^{(n)}(\boldsymbol{H}^{(n-1)})$ is obtained from $\mathfrak{t}^{(n)}(\boldsymbol{H}^{(n)})$ by replacing $\boldsymbol{H}_n$ by $\boldsymbol{0}$. Hence, by (65.6) and (65.2),

$$\mathfrak{t}^{(n)}(\boldsymbol{H}^{(n-1)}) = \boldsymbol{T}_{\mathrm{R}}^{(n-1)} + \varepsilon^n \mathfrak{h}_n(\boldsymbol{H}_1, \ldots, \boldsymbol{H}_{n-1}) + O(\varepsilon^{n+1}). \tag{65.7}$$

Writing (65.3) with n replaced by $n-1$ and substituting (65.7) into the result, we find

$$\left.\begin{aligned} \operatorname{Div} \mathfrak{t}^{(n)}(\boldsymbol{H}^{(n-1)}) + \varrho_{\mathrm{R}}(\boldsymbol{b}^{(n)} - \varepsilon^n \boldsymbol{b}_n^*) &= \varrho_{\mathrm{R}} \ddot{\boldsymbol{u}}^{(n-1)} + O(\varepsilon^{n+1}), \\ \mathfrak{t}^{(n)}(\boldsymbol{H}^{(n-1)}) \boldsymbol{n}_{\mathrm{R}} &= \boldsymbol{t}_{\mathrm{R}}^{(n)} - \varepsilon^n \boldsymbol{t}_{\mathrm{R}\,n}^* + O(\varepsilon^{n+1}) \end{aligned}\right\} \tag{65.8}$$

where $\boldsymbol{b}_n^*$ and $\boldsymbol{t}_{\mathrm{R}\,n}^*$ are given by (63.18). The result (65.8) holds whenever $\boldsymbol{u}^{(n-1)}$ is a solution of the equations of the $(n-1)$'st-order theory.

Suppose that we have a solution $\boldsymbol{u}^{(n-1)}$ of the equations of $(n-1)$'st-order elasticity in the sense defined above. Equation (65.8) shows that $-\varepsilon^n \boldsymbol{t}_{\mathrm{R}\,n}^*$ is the surface traction and $-\varepsilon^n \boldsymbol{b}_n^*$ the body force that would be required, in addition to the assigned loads $\boldsymbol{t}_{\mathrm{R}}^{(n)}$ and $\boldsymbol{b}_n$, in order to maintain the displacement $\boldsymbol{u}^{(n-1)}$ in equilibrium according to the n'th-order theory of elasticity. Equations (63.17) show that $\varepsilon^n \boldsymbol{u}_n$ is a solution in the infinitesimal theory corresponding to the loads $\varepsilon^n \boldsymbol{t}^*_{\mathrm{R}\,n}$, $\varepsilon^n \boldsymbol{b}_n^*$. Hence $\boldsymbol{u}^{(n)} = \boldsymbol{u}^{(n-1)} + \varepsilon^n \boldsymbol{u}_n$ is a solution in the n'th-order theory.

We may summarize the procedure as follows:

(i) Assume that a solution $\boldsymbol{u}^{(n-1)}$ of the equations $(n-1)$'st-order elasticity has been found. Substituting $\boldsymbol{u}^{(n-1)}$ into the equations of n'th-order elasticity, find the loads $-\varepsilon^n \boldsymbol{t}_{\mathrm{R}\,n}^*, -\varepsilon^n \boldsymbol{b}_n^*$ required to maintain the displacement in equilibrium.

(ii) Find the solution $\varepsilon^n \boldsymbol{u}_n$ of the infinitesimal theory corresponding to the reversed loads $\varepsilon^n \boldsymbol{t}_{\mathrm{R}\,n}^*$, $\varepsilon^n \boldsymbol{b}_n^*$.

(iii) Then a solution of the equations of n'th-order elasticity is given by the sum $\boldsymbol{u}^{(n)} = \boldsymbol{u}^{(n-1)} + \varepsilon^n \boldsymbol{u}_n$.

According to the considerations of the previous section, the procedure cannot be carried out unless the solution in (ii), at each step, is chosen such that the compatibility condition (64.2), with n replaced by $n+1$, is satisfied.

We have assumed so far that the solution $\boldsymbol{u}^{(n)}$ of (65.4) is a polynomial of degree n in ε. However, the procedure described above can easily be seen to apply also when by a solution $\boldsymbol{u}^{(n)}$ of the equations of n-th order elasticity we agree to mean *any* analytic function of ε that satisfies (65.4). Of course, the coefficients of order higher than n in the expansion of $\boldsymbol{u}^{(n)}$ are not determined by (65.4), but a judicious choice of these coefficients often lightens the solution in (ii) at the later steps.

Furthermore, since each stage requires solution of a problem in the infinitesimal theory, all the devices used in that theory, such as BETTI's theorem on the average strains, are at our disposal to ease the work.

The expansion is not directly applicable to incompressible bodies, since the forces required at step (ii) are generally infinite, because a deformation which is isochoric to within terms of order $n+1$ is generally not isochoric to a higher order and thus not possible subject to finite forces. To obtain results for bodies of incompressible material, a method of iteration may be set up directly for them[1], or, as will be seen in Sect. 67, an incompressible material may be regarded as the limit of a compressible one.

66. Second-order effects in compressible isotropic materials[2]. In order to apply the method given in the last section, we now derive the explicit equations of second-order elasticity for isotropic compressible materials[3].

[1] GREEN and SPRATT [1954, *9*, § 3], SIGNORINI [1955, *25*, Cap. I, ¶¶ 7—11].

[2] Explicit equations of second-order elasticity for various kinds of material symmetries more general than full isotropy have been developed and discussed by several authors, usually for hyperelastic materials: BORN and MISRA [1940, *2*], BIRCH [1947, *2*], BHAGAVANTAM and SURYANARAYANA [1947, *1*] [1949, *1*], JAHN [1949, *6*], MURNAGHAN [1951, *7*, Ch. 5, § 4], FUMI [1951, *1*] [1952, *4*—*6*], HEARMON [1953, *11*], SHIMAZU [1954, *21*, Ch. 5], BHAGAVANTAM [1960, *9*].

Particular notice is deserved by the detailed study of second-order elasticity given by SHENG [1955, *24*], obscured, unfortunately, by misprints, some of them serious, in nearly every line. He determined the maximum number of independent components of the elasticities of first and second order and their explicit forms for elastic and hyperelastic materials enjoying various symmetries [§§ 2—6, 9]; for example, the most general elastic material may have 126 independent second-order elasticities, while a hyperelastic material can have but 56. He developed and interpreted the iterative scheme we have presented in Sect. 65 [§ 10]. Finally, he obtained and discussed in great detail three particular second-order solutions: (1) torsion, extension, and inflation of a circular-cylindrical tube of "axomootropic" material [§ 13]; (2) pure bending of a rectangular plate of orthotropic hyperelastic material [§ 14]; (3) torsion and extension of an isotropic cylinder of arbitrary cross-section [§ 15]. Results apparently equivalent to some of his are included in our text below, but because of misprints and differences of notation it has seemed more economical for us to develop them independently.

To the extent that the stress relations in the works just cited are correct, they could be derived by expansion of the exact forms of the stress relation for the classes of materials in question. Literature on the exact theory is cited in Sects. 33—34.

[3] FINGER [1894, *1*, Part 1, Eqs. (47), (50), and (54)], for hyperelastic materials. While formulation of the second-order theory now seems a straightforward matter, it was not always so. The earlier work, some of which will be mentioned in Sect. 94, now seems to us merely special, resting upon arbitrary preferences for one or another particular strain measure and hence lacking any basis in physical principle. To the faulty attempts in this century may be added the less discreditable one of VOIGT [1893, *5*], which FINGER corrected [1894, *1*, pp. 197—200]; VOIGT was the first to try to formulate a theory containing all terms of second-order in the displacement gradient, but he did not succeed.

Among the many authors who have worked out equations for second-order elasticity for isotropic materials may be mentioned KÖTTER [1910, *1*, § 1], BRILLOUIN [1925, *2*, §§ 5—7].

To the second order in the displacement gradient $\boldsymbol{H}$ we have the expansions

$$\left.\begin{aligned} &\boldsymbol{B} = 1 + 2\tilde{\boldsymbol{E}} + \boldsymbol{H}\boldsymbol{H}^T, \\ &\boldsymbol{B}^{-1} = 1 - 2\tilde{\boldsymbol{E}} - \boldsymbol{H}\boldsymbol{H}^T + 4\tilde{\boldsymbol{E}}^2 + \cdots, \\ &I_{\boldsymbol{E}} = I_{\tilde{\boldsymbol{E}}} + \tfrac{1}{2} I_{\boldsymbol{H}\boldsymbol{H}^T}, \quad II_{\boldsymbol{E}} = II_{\tilde{\boldsymbol{E}}} + \cdots, \quad III_{\boldsymbol{E}} = 0 + \cdots, \end{aligned}\right\} \tag{66.1}$$

[1938, *2*, Ch. X, § 12], (the work of KAPLAN [1931, *3*] and CERUTI [1932, *2*] is partly incorrect), MURNAGHAN [1937, *2*, pp. 250—252], SAKADI [1949, *20*, § II], RIVLIN [1953, *24*, § 2], HUGHES and KELLY [1953, *13*], SHIMAZU [1954, *21*, Chap. 3], SHENG [1955, *24*, § 6 (v)], TOUPIN and BERNSTEIN [1961, *62*, § 5], GUO ZHONG-HENG [1962, *30*]. An exposition has been given by GREEN and ADKINS [1960, *26*, Ch. V]. JOHN [1958, *25*, Chs. III, V, VI] obtained equations of the third-order theory and derived from them a theory of the bending of thin plates. The theory in our text corresponds to that first given by SAKADI [1949, *20*, § II] and SHENG [1955, *24*, § 6 (vii)], the only ones among the authors just cited who did not restrict attention to hyperelastic materials.

To express the results of the text in terms of the notations used by several of these authors, use the following table:

Present notation	BRILLOUIN	MURNAGHAN	RIVLIN	TOUPIN and BERNSTEIN
$\mu\alpha_1$	λ	λ	$4(a_1 + 2a_2)$	λ
$\mu\alpha_2 = \mu$	μ	μ	$-2a_1$	μ
$\mu\alpha_3$	$-\lambda + 8A + 24B$	$-\lambda + 4(6l + 2m)$	$4(6a_4 + 2a_3 - a_1 - 2a_2)$	$-\lambda + \nu_2 + \frac{1}{2}\nu_1$
$\mu\alpha_4$	$-16A$	$8(m + n)$	$8(a_3 + a_5)$	$-2\nu_2$
$\mu\alpha_5$	$2(\lambda - \mu + 8A)$	$2(\lambda - \mu) - 8(m + n)$	$4(3a_1 + 4a_2 - 2a_3 - 2a_5)$	$2(\lambda - \mu + \nu_2)$
$\mu\alpha_6$	$4(\mu + 6C)$	$4\mu - 8n$	$-8(a_1 - a_5)$	$4(\mu + \nu_3)$

Our results, being more general than those with which they are compared in the table, cannot be deduced from them for general values of the 6 coefficients $\mu\alpha_1, \mu, \mu\alpha_3, \ldots, \mu\alpha_6$. While always

$$a_1 = -\tfrac{1}{2}\mu\alpha_2 = -\tfrac{1}{2}\mu, \qquad a_2 = \tfrac{1}{8}\mu(\alpha_1 + 2\alpha_2) = \tfrac{1}{8}(\lambda + 2\mu),$$

the relations connecting the second-order coefficients cannot be solved for a_3, a_4, a_5, or l, m, n, or A, B, C, unless the condition

$$\mu(\alpha_4 + \alpha_5) = 2(\lambda - \mu)$$

is satisfied. This condition, reducing the number of second-order elasticities from 4 to 3 in such a way as to express the existence of a stored-energy function in the second-order theory, specializes elasticity to hyperelasticity in second approximation, as will be shown in Sect. 93. If this condition is satisfied, then MURNAGHAN's and RIVLIN's coefficients are given in terms of ours by the formulae

$$\begin{aligned} l &= \tfrac{1}{24}\mu(4 + \alpha_1 + \alpha_3 - \alpha_4 + \alpha_6), & a_3 &= \tfrac{1}{8}\mu(4 + \alpha_4 - \alpha_6), \\ m &= \tfrac{1}{8}\mu(-4 + \alpha_4 - \alpha_6), & a_4 &= \tfrac{1}{24}\mu(-4 + \alpha_1 + \alpha_3 - \alpha_4 + \alpha_6), \\ n &= \tfrac{1}{8}\mu(4 + \alpha_6), & a_5 &= \tfrac{1}{8}\mu(-4 + \alpha_6), \end{aligned}$$

and corresponding expressions for the coefficients used by the other authors cited are easily derived.

A good deal of experimental work can be interpreted as yielding measured values of second-order elasticities. A survey and comparison of this work, regarded as relevant to hyperelastic materials, mainly isotropic, is given by SEEGER and BUCK [1960, *49*]. E.g., from one interpretation they infer the following values of MURNAGHAN's constants l, m, n for copper and iron:

	m kp/mm²	n kp/mm²	l kp/mm²
Cu	$(-6.2 \pm 0.1) \cdot 10^4$	$(-15.9 \pm 0.2) \cdot 10^4$	$(-1.6 \pm 0.7) \cdot 10^4$
Fe	$(-7.7 \pm 0.1) \cdot 10^4$	$(-15.2 \pm 0.1) \cdot 10^4$	$(-1.7 \pm 0.4) \cdot 10^4$

but they remark that other interpretations of the data lead to somewhat different values. Cf. also SEEGER [1964, *74*]. The literature of the subject multiplies.

where $\boldsymbol{B}$ is the left Cauchy-Green tensor; $\tilde{\boldsymbol{E}} \equiv \frac{1}{2}(\boldsymbol{H}+\boldsymbol{H}^T)$, the infinitesimal strain tensor; and $I_{\boldsymbol{E}}$, $II_{\boldsymbol{E}}$, and $III_{\boldsymbol{E}}$, the invariants of $\boldsymbol{E}$, which are determined from those of $\boldsymbol{B}$ by (63.19).

We assume that the response coefficients $\beth_\Gamma$ in the stress-relation (47.9) have expansions

$$\beth_\Gamma = \mu(\alpha_{\Gamma 0} + \alpha_{\Gamma 1} I_{\boldsymbol{E}} + \alpha_{\Gamma 2} I_{\boldsymbol{E}}^2 + \alpha_{\Gamma 3} II_{\boldsymbol{E}} + \cdots), \qquad \Gamma = 0, \pm 1. \tag{66.2}$$

Substitution of (66.1) and (66.2) into (47.9) shows that, to within terms of the third order in $\boldsymbol{H}$, the stress relation for isotropic compressible materials becomes

$$\left.\begin{aligned} \boldsymbol{T}/\mu = \alpha_1 I_{\tilde{\boldsymbol{E}}} \boldsymbol{1} + 2\alpha_2 \tilde{\boldsymbol{E}} + (\tfrac{1}{2}\alpha_1 I_{\boldsymbol{H}\boldsymbol{H}^T} + \alpha_3 I_{\tilde{\boldsymbol{E}}}^2 + \alpha_4 II_{\tilde{\boldsymbol{E}}})\boldsymbol{1} + \\ + \alpha_5 I_{\tilde{\boldsymbol{E}}} \tilde{\boldsymbol{E}} + \alpha_2 \boldsymbol{H}\boldsymbol{H}^T + \alpha_6 \tilde{\boldsymbol{E}}^2 + \cdots, \end{aligned}\right\} \tag{66.3}$$

where we have assumed that the reference configuration is a natural state, so that $\alpha_{00} = -\alpha_{10} - \alpha_{-10}$, and where

$$\left.\begin{aligned} &\alpha_1 = \alpha_{01} + \alpha_{11} + \alpha_{-11}, \quad &&\alpha_2 = \alpha_{10} - \alpha_{-10}, \quad &&\alpha_3 = \alpha_{02} + \alpha_{12} + \alpha_{-12}, \\ &\alpha_4 = \alpha_{03} + \alpha_{13} + \alpha_{-13}, \quad &&\alpha_5 = 2(\alpha_{11} - \alpha_{-11}), \quad &&\alpha_6 = 4\alpha_{-10}. \end{aligned}\right\} \tag{66.4}$$

If we interpret μ as the classical shear modulus, as we shall do, then

$$\alpha_2 = 1, \tag{66.5}$$

and the other classical elastic coefficients are given by

$$\lambda = \alpha_1 \mu, \qquad E = \frac{3\alpha_1 + 2}{\alpha_1 + 1}\mu, \qquad \sigma = \frac{\alpha_1}{2(\alpha_1 + 1)}, \qquad \alpha_1 = \frac{2\sigma}{1 - 2\sigma}; \tag{66.6}$$

hence the condition $\infty > \alpha_1 > -\frac{2}{3}$ is equivalent to the classical requirement that $\infty > 3\lambda + 2\mu > 0$. Of the coefficients of the 8 terms in (66.3), only 6 are independent; these 6, however, do not suffice to determine uniquely the 11 coefficients in the expansions of the response coefficients (66.2) through the relations (66.4). The four coefficients α_3, α_4, α_5, and α_6 are the *second-order elasticities* of an isotropic elastic material.

We have seen after (63.20) that the expansion of $\boldsymbol{T}$ is obtained from the expansion of $\boldsymbol{T}_{\mathrm{R}}$ by adding the quadratic term $\mathsf{L}[\tilde{\boldsymbol{E}}]\boldsymbol{H}^T - (\mathrm{tr}\,\boldsymbol{H})\mathsf{L}[\tilde{\boldsymbol{E}}]$ and terms of higher order. Since, for isotropic materials,

$$\mathsf{L}[\tilde{\boldsymbol{E}}] = 2\mu\left(\tilde{\boldsymbol{E}} + \frac{\sigma}{1 - 2\sigma} I_{\tilde{\boldsymbol{E}}} \boldsymbol{1}\right), \tag{66.7}$$

we derive from (66.3), (66.5), and (66.6) the expansion

$$\left.\begin{aligned} \boldsymbol{T}_{\mathrm{R}}/\mu = 2\left(\frac{\sigma}{1-2\sigma} I_{\tilde{\boldsymbol{E}}} \boldsymbol{1} + \tilde{\boldsymbol{E}}\right) + \left[\frac{\sigma}{1-2\sigma}(I_{\boldsymbol{H}\boldsymbol{H}^T} + 2 I_{\tilde{\boldsymbol{E}}}^2) + \alpha_3 I_{\tilde{\boldsymbol{E}}}^2 + \alpha_4 II_{\tilde{\boldsymbol{E}}}\right]\boldsymbol{1} + \\ + (\alpha_5 + 2) I_{\tilde{\boldsymbol{E}}} \tilde{\boldsymbol{E}} - \frac{2\sigma}{1-2\sigma} I_{\tilde{\boldsymbol{E}}} \boldsymbol{H}^T - (\boldsymbol{H}^T)^2 + \alpha_6 \tilde{\boldsymbol{E}}^2 + \cdots. \end{aligned}\right\} \tag{66.8}$$

In the preceding section it was shown that a stress boundary-value problem in the second-order theory is equivalent to two such boundary-value problems in the infinitesimal theory. A direct attack on the second linear problem, however, often leads to complications which may be avoided if we have some idea in advance of what the solution should be. We now present two examples in which use of a semi-inverse method from the start simplifies the analysis.

Example 1. Torsion and extension of a cylindrical tube[1]. We consider a deformation of a cylinder obtained by superimposing a twist of amount ε upon an extension of amount δ accompanied by a lateral contraction of amount $\sigma\delta$. The Cartesian components of the displacement vector $\boldsymbol{u}^{(1)}$ representing this deformation are given by

$$\left.\begin{aligned} u^{(1)} &= (1-\sigma\delta)\{X\cos[(1+\delta)\varepsilon Z] - Y\sin[(1+\delta)\varepsilon Z]\} - X \\ &= -[\sigma\delta X + \varepsilon Z Y] - [\tfrac{1}{2}\varepsilon^2 X Z^2 + (1-\sigma)\delta\varepsilon Y Z] + \cdots, \\ v^{(1)} &= (1-\sigma\delta)\{X\sin[(1+\delta)\varepsilon Z] + Y\cos[(1+\delta)\varepsilon Z]\} - Y \\ &= -[\sigma\delta Y - \varepsilon Z X] - [\tfrac{1}{2}\varepsilon^2 Y Z^2 - (1-\sigma)\delta\varepsilon X Z] + \cdots, \\ w^{(1)} &= \delta Z, \end{aligned}\right\} \tag{66.9}$$

where in the second forms the terms of order greater than two in ε and δ have not been written. If we assume that δ and ε are related by an expansion

$$\delta = \delta_1\varepsilon + \delta_2\varepsilon^2 + \cdots, \tag{66.10}$$

then (66.9) represents a one-parameter family of displacements.

We assume that the contraction ratio σ in (66.9) is equal to the Poisson modulus $(66.6)_3$. Comparison of the leading terms in (66.9) with the known solutions in the infinitesimal theory then shows that, to within terms of order two in ε, the stress corresponding to (66.9) is equilibrated and the tractions on the cylinders $R=\sqrt{X^2+Y^2}=\text{const.}$ vanish. Thus (66.9) is a solution, in the first-order theory of elasticity, of the problem of extending and twisting a cylindrical tube whose surfaces are free of traction. Of course, this would remain true if all terms of second-order in ε were omitted in (66.9). These second-order terms are retained only to ease the calculation of the second-order solution $\boldsymbol{u}^{(2)}$ later.

From this point onward in this section, the sign "=" is to be interpreted as equality to within an error of order three or greater in ε.

The matrix of the gradient $\boldsymbol{H}^{(1)}$ of the displacement (66.9) is given by

$$\left.\begin{aligned} [\boldsymbol{H}^{(1)}] = &-\left\|\begin{matrix} \sigma\delta & \varepsilon Z & \varepsilon Y \\ -\varepsilon Z & \sigma\delta & -\varepsilon X \\ 0 & 0 & -\delta \end{matrix}\right\| - \\ &-\left\|\begin{matrix} \tfrac{1}{2}\varepsilon^2 Z^2 & (1-\sigma)\delta\varepsilon Z & \varepsilon^2 X Z + (1-\sigma)\delta\varepsilon Y \\ -(1-\sigma)\delta\varepsilon Z & \tfrac{1}{2}\varepsilon^2 Z^2 & \varepsilon^2 Y Z - (1-\sigma)\delta\varepsilon X \\ 0 & 0 & 0 \end{matrix}\right\|. \end{aligned}\right\} \tag{66.11}$$

Substitution of (66.11) into (66.8) gives the following expansion for the Piola-Kirchhoff stress $\boldsymbol{T}_{\mathrm{R}}^{(1)}$ corresponding to the displacement (66.9):

$$\left.\begin{aligned} \boldsymbol{T}_{\mathrm{R}}^{(1)}/\mu &= 2(\sigma\delta\boldsymbol{1} + \tilde{\boldsymbol{E}}_1) + \\ &+ \xi\boldsymbol{1} + \delta(1-2\sigma)(\alpha_5+2)\tilde{\boldsymbol{E}}_1 + \boldsymbol{K} + \alpha_6\tilde{\boldsymbol{E}}_1^2, \end{aligned}\right\} \tag{66.12}$$

[1] RIVLIN [1953, *24*, § 5].

where

$$\left.\begin{aligned}
&R^2 = X^2 + Y^2,\\
&\xi = \delta^2\left[\frac{\sigma(1+2\sigma^2)}{1-2\sigma} + 2\sigma(1-2\sigma) + \alpha_3(1-2\sigma)^2 - \alpha_4\sigma(2-\sigma)\right] +\\
&\quad + \left(\frac{\sigma}{1-2\sigma} - \frac{\alpha_4}{4}\right)\varepsilon^2 R^2,\\
&[\tilde{\boldsymbol{E}}_1] = \begin{Vmatrix} -\sigma\delta & 0 & -\frac{1}{2}\varepsilon Y\\ 0 & -\sigma\delta & \frac{1}{2}\varepsilon X\\ -\frac{1}{2}\varepsilon Y & \frac{1}{2}\varepsilon X & \delta \end{Vmatrix},\\
&[\boldsymbol{K}] = \begin{Vmatrix} \sigma^2\delta^2 & 0 & -\varepsilon^2 XZ - (1-\sigma)\delta\varepsilon Y\\ 0 & \sigma^2\delta^2 & -\varepsilon^2 YZ + (1-\sigma)\delta\varepsilon X\\ 2\sigma\delta\varepsilon Y & -2\sigma\delta\varepsilon X & -(2\sigma+1)\delta^2 \end{Vmatrix}.
\end{aligned}\right\} \tag{66.13}$$

It follows from (66.12) and (66.13) that

$$\left.\begin{aligned}
\operatorname{Div}\boldsymbol{T}_{\mathrm{R}}^{(1)} &= \mu\{\nabla\xi + \operatorname{Div}\boldsymbol{K} + \alpha_6\operatorname{Div}(\tilde{\boldsymbol{E}}_1^2)\},\\
&= -\varepsilon^2\mu\left(\frac{1-4\sigma}{1-2\sigma} + \frac{\alpha_4}{2} + \frac{\alpha_6}{4}\right) R\,\boldsymbol{e}_{\mathrm{R}}
\end{aligned}\right\} \tag{66.14}$$

where $\boldsymbol{e}_{\mathrm{R}}$ is the unit vector in the radial direction, having the components X/R and Y/R.

The tractions corresponding to $\boldsymbol{T}_{\mathrm{R}}^{(1)}$ on the cylindrical surfaces $R=\text{const.}$ are easily obtained from (66.12) and (66.13):

$$\left.\begin{aligned}
\boldsymbol{T}_{\mathrm{R}}^{(1)}\boldsymbol{e}_R &= \mu\left[\left(\frac{\sigma}{1-2\sigma} - \frac{1}{4}\alpha_4\right)\varepsilon^2 R^2 + \gamma\delta^2\right]\boldsymbol{e}_R,\\
\gamma &= \frac{\sigma(1+\sigma)}{1-2\sigma} + \alpha_3(1-2\sigma)^2 - \alpha_4\sigma(2-\sigma) - \alpha_5\sigma(1-2\sigma) + \alpha_6\sigma^2.
\end{aligned}\right\} \tag{66.15}$$

We now consider a tube bounded by the cylinders $R=R_I$ and $R=R_{II}>R_I$. We wish to determine a solution $\boldsymbol{u}^{(2)}$ of the equations of second-order elasticity that corresponds to twisting and extending such a tube. We require that $\boldsymbol{u}^{(2)}$ be a solution subject to no body forces and no tractions on the bounding surfaces $R=R_I$ and $R=R_{II}$. It follows from (66.14) and (66.15) that the loads required to maintain the displacement $\boldsymbol{u}^{(1)}$ in equilibrium according to the second-order theory consist of a radial body force $-\boldsymbol{b}_{(1)}^*$, where

$$\varrho_{\mathrm{R}}\boldsymbol{b}_{(1)}^* = -\mu\varepsilon^2\left(\frac{1-4\sigma}{1-2\sigma} + \frac{\alpha_4}{2} + \frac{\alpha_6}{4}\right) R\,\boldsymbol{e}_R, \tag{66.16}$$

and radial surface tractions of amounts $-t_I^*$ and $-t_{II}^*$ at $R=R_I$ and $R=R_{II}$, respectively, where

$$\left.\begin{aligned}
t_I^* &= -\mu\left[\left(\frac{\sigma}{1-2\sigma} - \frac{\alpha_4}{4}\right)\varepsilon^2 R_I^2 + \gamma\delta^2\right],\\
t_{II}^* &= -\mu\left[\left(\frac{\sigma}{1-2\sigma} - \frac{\alpha_4}{4}\right)\varepsilon^2 R_{II}^2 + \gamma\delta^2\right].
\end{aligned}\right\} \tag{66.17}$$

According to the results of the previous section, the displacement $\boldsymbol{u}_{(2)}$ to be added to $\boldsymbol{u}^{(1)}$ is obtained by solving the equations of the infinitesimal theory corresponding to the negatives of the loads required to maintain $\boldsymbol{u}^{(1)}$. That is, $\boldsymbol{u}_{(2)}$ must be a solution in the infinitesimal theory, corresponding to the radial loads $\boldsymbol{b}_{(1)}^*, t_I^*, t_{II}^*$ given by (66.16) and (66.17). We expect a radial displacement

$$\boldsymbol{u}_{(2)} = U(R)\,\boldsymbol{e}_R \tag{66.18}$$

to be such a solution. After routine calculation, the differential equations of the infinitesimal theory for a displacement of the form (66.18) and body forces (66.16) lead to the condition

$$\left.\begin{aligned} &U'' + \frac{U'}{R} - \frac{U}{R^2} = 8\varkappa\varepsilon^2 R,\\ &\varkappa = \frac{1}{16(1-\sigma)}\left[1 - 4\sigma + \frac{1}{2}\left(\alpha_4 + \frac{1}{2}\alpha_6\right)(1-2\sigma)\right].\end{aligned}\right\} \tag{66.19}$$

The general solution of (66.19) is

$$U = \varkappa\varepsilon^2 R^3 + aR + \frac{b}{R}, \tag{66.20}$$

where a and b are constants of integration. The condition that the radial tensions at $R = R_I$ and $R = R_{II}$ must be given by (66.17) serves to determine appropriate values for a and b. After more routine calculation we finally obtain

$$\left.\begin{aligned} &U = U(R) = \varepsilon^2\left\{\varkappa R^3 - \beta\left[(R_I{}^2 + R_{II}^2) R + \frac{R_I^2 R_{II}^2}{(1-2\sigma)R}\right]\right\} - \delta^2 \frac{1-2\sigma}{2}\gamma R,\\ &\beta = \frac{1-2\sigma}{16(1-\sigma)}\left[3 - \frac{1}{2}(1-2\sigma)\alpha_4 + \frac{3-2\sigma}{4}\alpha_6\right].\end{aligned}\right\} \tag{66.21}$$

The complete solution $\boldsymbol{u}^{(2)} = \boldsymbol{u}^{(1)} + \boldsymbol{u}_{(2)}$ of the equations of second-order elasticity is obtained from (66.9), (66.18), and (66.21). $\boldsymbol{u}^{(2)}$ has the components

$$\left.\begin{aligned} &u^{(2)} = -[\sigma\delta X + \varepsilon Z Y] - \left[\frac{1}{2}\varepsilon^2 X Z^2 + (1-\sigma)\delta\varepsilon Y Z - \frac{X}{R}U(R)\right],\\ &v^{(2)} = -[\sigma\delta Y - \varepsilon Z X] - \left[\frac{1}{2}\varepsilon^2 Y Z^2 - (1-\sigma)\delta\varepsilon X Z - \frac{Y}{R}U(R)\right],\\ &w^{(2)} = \delta Z,\end{aligned}\right\} \tag{66.22}$$

where $U(R)$ is given by (66.21).

It requires only a routine calculation, similar to the one leading from (66.9) to (66.11) and (66.12), to obtain the Piola-Kirchhoff stress $\boldsymbol{T}_R^{(2)}$ corresponding to the displacement (66.22). The result of this calculation may be used to determine the surface tractions that must be supplied at the plane ends of the tube in order to maintain the deformation according to the second-order theory. These tractions consist in an azimuthal shear stress of amount

$$t_\Theta = \mu\varepsilon R\{1 + \delta[(2-3\sigma) + \tfrac{1}{2}\alpha_5(1-2\sigma) + \tfrac{1}{2}\alpha_6(1-\sigma)]\} \tag{66.23}$$

and a normal traction of amount

$$t_Z = \mu\left\{2(1+\sigma)\delta + \varepsilon^2\left[\beta' R^2 - \frac{4\sigma\beta}{1-2\sigma}(R_I^2 + R_{II}^2)\right] + \delta^2\gamma'\right\}, \tag{66.24}$$

where

$$\left.\begin{aligned} &\beta' = \frac{1}{8(1-\sigma)}[12\sigma + 2(2\sigma-1)\alpha_4 + (2-\sigma)\alpha_6],\\ &\gamma' = (\sigma+1)(1-4\sigma) + \alpha_3(1-2\sigma)^3 + \alpha_4\sigma(2-\sigma)(2\sigma-1) +\\ &\quad + \alpha_5(1+2\sigma^2)(1-2\sigma) + \alpha_6(1-2\sigma^3).\end{aligned}\right\} \tag{66.25}$$

These tractions are taken per unit area in the undeformed natural state.

The total torque M necessary to produce the twist is obtained from (66.23):

$$M = \frac{\pi}{2}\mu\varepsilon(R_{II}^4 - R_I^4)\left\{1 + \delta\left[(2-3\sigma) + \frac{1}{2}\alpha_5(1-2\sigma) + \frac{1}{2}\alpha_6(1-\sigma)\right]\right\}. \tag{66.26}$$

The quantity in braces is the ratio of of the modulus of torsion in the rod subject to extension δ to that in the unstretched rod. Eq. (66.26) solves the classical problem of finding *the effect of a small stretch upon the torsional rigidity*, for a circular cylinder of compressible isotropic material.

The total longitudinal tension N needed to maintain the extension can be calculated from (66.24):

$$\left.\begin{aligned} N &= \pi\mu(R_{II}^2 - R_I^2)\{[2\delta(1+\sigma) + \delta^2\gamma'] + \varepsilon^2\beta''(R_I^2 + R_{II}^2)\},\\ \beta'' &= \tfrac{1}{8}[-(1-2\sigma)\alpha_4 + (1-\sigma)\alpha_6]. \end{aligned}\right\} \tag{66.27}$$

If no longitudinal tension is supplied, i.e., if $N=0$, then it follows from (66.27) and (66.10) that $\delta_1=0$ and[1]

$$\delta = \delta_2\varepsilon^2, \qquad \delta_2 = -\frac{\beta''}{2(1+\sigma)}(R_I^2 + R_{II}^2), \tag{66.28}$$

when terms of an order higher than two are omitted. It is an extension of this kind, proportional to the square of the twist, that was found in the experiments of POYNTING on torsion (below, p. 237). If (66.28) holds, then (66.26), to within terms of order three, reduces to

$$M = \frac{\pi}{2}\mu\varepsilon(R_{II}^4 - R_I^4), \tag{66.29}$$

which is identical with the formula for the torque given by the infinitesimal theory. To find deviation from this classical formula, a theory taking into account terms of third order is needed, as may be expected from COULOMB's experiments on torsion (below, p. 236).

The deformation (66.22) can be carried out in two steps: first an extension produced by a tensile force N; second a twist produced by a torque M, the tensile force N being held constant during the twist. The extension δ^* of the first step can be computed from the result of putting $\varepsilon=0$ in (66.27). The twist ε of the second step is obtained by substituting δ^* into (66.26). If we then put

$$\delta = \delta_1\varepsilon + \delta_2\varepsilon^2, \qquad \delta_1 = \delta^*/\varepsilon, \tag{66.30}$$

we see that the torque (66.26) and the tension (66.27) corresponding to (66.30) differ from the ones originally given only by terms of order higher than two. Therefore, (66.30) gives the extension of the total deformation, the quadratic term being produced by the twist alone. Since this quadratic term $\delta_2\varepsilon^2$ gives a contribution of higher order to (66.21), it follows that the radial displacement resulting from the twist alone is independent of the tensile force N; it is given by

$$U^*(R) = \varepsilon^2\left\{\varkappa R^3 - \beta\left[(R_I^2 + R_{II}^2)R + \frac{R_I^2 R_{II}^2}{(1-2\sigma)R}\right]\right\}. \tag{66.31}$$

Since the extension (66.28) produced by the twist is also independent of N, it follows that the change of volume resulting from the twist is independent of the tensile force N applied before the twist was effected. This change of volume, per unit original volume, is given by[2]

$$\left.\begin{aligned} \frac{\Delta V}{V} &= \frac{2\pi[R_{II}U^*(R_{II}) - R_I U^*(R_I)] + \pi(R_{II}^2 - R_I^2)\delta_2\varepsilon^2}{\pi(R_{II}^2 - R_I^2)},\\ &= \pi\varepsilon^2\beta'''(R_{II}^2 + R_I^2),\\ \beta''' &= -\frac{1}{4} + \frac{1}{16(1+\sigma)}[(3 - 4\sigma - 4\sigma^2)\alpha_4 + (2\sigma^2 + 2\sigma - 2)\alpha_6]. \end{aligned}\right\} \tag{66.32}$$

---

[1] Results equivalent to (66.26) and (66.28), for the case when $R_I=0$, were obtained by SAKADI [1949, *20*, § IV] and MURNAGHAN [1951, *8*, Ch. 7, § 3]; the latter considered only hyperelastic materials.

[2] This formula differs from the one given by RIVLIN [1953, *24*, Eq. (5.38)], who assumed that the initial extension, rather than the tensile force N, is held constant during the superimposed twist.

It is a change of bulk of this kind, proportional to the square of the twist and independent of the tensile force (provided it be small enough), that was found in the experiments of POYNTING (below, p. 237). This change is an example of the KELVIN effect (Sect. 54).

The general problem of combined torsion and extension of a solid hyperelastic cylinder has been set up by GREEN and WILKES[1]. The deformation they considered is a special case of (59.1). They expanded all quantities in powers of the twist and calculated the results explicitly up to terms of third order. Thus they found the third-order correction to the classical linear relation (66.29) between torque and twist; this correction is determined by the third-order elasticities. Corresponding results could easily be gotten for general elastic materials by pushing the above-explained scheme one stage further.

While, as we have seen in Sect. 57, a cylinder of incompressible material will always elongate when twisted (granted certain basic inequalities), from (66.28) we can infer no general rule of sign for δ_2. Note that the moduli α_4 and α_6 determine the extension. Likewise, α_4 and α_6 determine the change of volume through Eq. (66.32), while α_5 and α_6 determine the charge in torsional rigidity through Eq. (66.27). *It is impossible to predict the magnitude of the second-order effects in terms of the infinitesimal elasticities, λ and μ.* Some distinguished savants have deluded themselves into a contrary opinion.

The theory of elasticity grew mainly from four special problems: the deformation of flexible cords and sheets, the bending of a beam, the torsion of a rod, the bending and twisting of plates. Non-linear theories were developed for the second and third, but the special forms and special successes of the early work on non-linear bending were not at all typical (cf. the history by TRUESDELL [1960, *57*]). *Non-linear* elasticity, in turn, grew mainly from considerations of generality, based upon obvious faults of the three-dimensional linear theory. One special problem, however, did have an influence: combined torsion and tension of a circular cylinder. This is the problem completely solved, as far as terms of the second order in the twist, by the results of RIVLIN, given in the text preceding. The earlier history of the problem is as follows.

The first theory of torsion was proposed, hesitantly, by EULER in 1763—1764 [1769, *1*]; he regarded torsional elasticity as resulting from the longitudinal extension of linearly elastic fibres, but the effect of all these he represented as being that of a single straight spring, so that the torque was found to be proportional to the sine of the angle of torsion. The first experiments were reported by COULOMB [1780, *1*, Ch. III, § 44]. The "force of torsion" was inferred to be proportional to the angle of torsion in hairs and in silk threads stretched taught by a weight. The amount of that weight "does not at all influence the force of torsion. One must remark nevertheless that if the weight of the body is very much increased, and if the hairs or silk threads are ready to break, this same law does not hold exactly. The force of torsion seems then much diminished, the oscillations are no longer isochrone, the times of the large ones then being much greater than those of the small ones. It happens in this case that the thread, from too great tension, loses its elasticity ..." I.e., for moderate tensions the torsional moment is independent of the amount of tension, but large tension first diminishes the torsional rigidity and then invalidates the linear response. His later experiments on brass and steel wires [1787, *1*, § XI] do not emphasize departures from the linear relation, but again COULOMB seems to have inferred that great tension decreases the torsional rigidity. (More details regarding this early work are given by TRUESDELL [1960, *57*, §§ 50, 61].)

YOUNG [1807, *1*, Vol. 1, Lecture XIII (see especially pp. 140—141)] wrote, "We might consider a wire as composed of a great number of minute threads, extending through its length, and closely connected together; if we twisted such a wire, the external threads would be extended, and, in order to preserve the equilibrium, the internal ones would be contracted; and it may be shown that the whole wire would be shortened one fourth as much as the external fibres would be extended if the length remained undiminished; and that the force would vary as the cube of the angle through which the wire is twisted. But the force of torsion, as it is determined by experiment, varies simply as the angle of torsion; it cannot, therefore, be explained by the action of longitudinal fibres only; but it appears to depend principally, if not intirely, on the rigidity, or lateral adhesion, which resists the detrusion of the particles." Whatever may have been YOUNG's method of calculation, he ends by *rejecting the model* of a rod as a bundle of slippery wires, so he cannot be accused of having himself claimed that a twisted wire necessarily shortens, although the passage just quoted seems to be the source whence the erroneous belief to that effect has diffused through much of the profession of mechanical engineers.

[1] GREEN and WILKES [1953, *8*].

MAXWELL [1853, *1*, Case VII] set and solved the problem of finding "the conditions of torsion of a cylinder composed of a great number of parallel wires bound together without adhering to each other". According to his result, such a cylinder will shorten. Indeed, this is evident, since in order for the rod to remain of the same length, each filament must lengthen, while the effect of the longitudinal elasticity is to diminish that lengthening to some extent. MAXWELL did not consider this model of slippery wires as adequate by itself. Rather, he thought that the force of torsion in a wire having "fibrous texture" would be the sum of that in the bundle and that obtained by the linear theory.

According to PEARSON [1893, *4*, § 735], KUPFFER in 1851 published as being due to F. NEUMANN a formula for the effect of a small extension on the torsional rigidity. In the great memoir on torsion, ST. VENANT [1856, *1*, Ch. XII, § 119], referring to experimental results just published by WERTHEIM, included an analysis of the second-order shortening of a twisted bundle of fibres and of the corresponding third-order effect on the torsional rigidity. While he saw that another elasticity beyond the classical λ and μ was needed to describe the phenomenon, he took it to be merely the change in effective modulus of extension arising from the transverse contraction of the fibres. He seemed to believe that all effects of the adherence of the fibers to one another were taken into account by the linear theory of torsion. PEARSON [1893, *4*, §§ 51, 735] claimed to correct the analysis and results of NEUMANN and ST. VENANT; he, too, found that a twisted wire always shortens. According to PEARSON, [1893, *4*, §§ 735, 803—806, 811—818], KUPFFER measured M as a function of δ and ε for wires of brass and steel, and WERTHEIM in 1855 reported experimental determination of the second-order change of the volume of the cavity in a hollow cylinder. Since such a change might result from a change of shape without a change of total volume, we cannot attribute discovery of the Kelvin effect to WERTHEIM. In explanation of the failure of his own formulae to agree with the values measured, PEARSON dismissed the data: The wires of KUPFFER he regarded as being "really aeolotropic", and he suspected "difficulty in WERTHEIM's apparatus".

The experiments of POYNTING [1905, *3*] [1909, *2*] [1913, *2*] showed that wires of steel, copper, brass, and rubber lengthen when twisted.

In bland disregard of the experimental facts, an engineering literature arose so as to explain in suitably practical and intuitive terms the YOUNG-MAXWELL-ST. VENANT theory of shortening. Some of this literature was cited by TRUESDELL [1953, *25*, p. 600]. An equivalent argument was published in a widely diffused textbook (TIMOSHENKO [1941, *4*, § 55]) even within the memory of living men.

That the assertion is false appears at once from the result of the theory of elasticity, expressed in (66.28). Its falsity may be seen directly, however. Indeed, if one of two parallel bars whose tips are connected by two parallel inextensible wires is turned about an axis through their centers, the bars will be drawn closer to each other. In the same way, a thin strip if twisted will shorten. But in a circular rod, the longitudinal fibres are not inextensible, slippery wires; they adhere to one another by tangential forces, and also they may swell or shrink in thickness, changing the normal forces they exert on each other. The laws of this adherence and the amount of this swelling are not determined by the modulus of extension of the fibres. Lengthening, shortening, or neither occurs according as $\beta''<0$, $\beta''>0$, or $\beta''=0$; thus the sign of the effect is an *independent material property*, not included among the concepts forming the basis of the infinitesimal theory. The dimensionless coefficient β'' is determined by $(66.27)_2$ in terms of the second-order elasticities α_6 and α_4 in addition to the Poisson modulus. Only for incompressible materials is there any reason for a general rule of sign, as shown in Sect. 57, cases 3a and 3b; it indicates *lengthening*.

The results we have just summarized show that even the giants of mechanics have stumbled when they placed faith in resolutions of a continuous medium into material elements of a special kind, or when they leaned upon the infinitesimal theory as a conceptual guide rather than just a peculiarly simple approximation to elastic response.

POYNTING interpreted his results by means of a fragile analogy to shear, using a special non-linear theory of three-dimensional elasticity. From the 1930's onward, solutions of the torsion-extension problem in special theories of elasticity have appeared; some of them were cited by TRUESDELL [1952, *20*, §§ 49—53] [1953, *25*, p. 602]. Most contain only the linear moduli and thus are as wrong as the Young-Maxwell-St. Venant theory, and for essentially the same reason. Some contain adjustable constants.

The experimental results on compressible materials are not clear. While SWIFT [1947, *11*] found lengthening in all seven metals he tested, the recent experiments of FOUX and REINER [1960, *23*] and of FOUX [1964, *29*] lead to no definite conclusion.

Example 2. Simultaneous extension and torsion of a prism of arbitrary cross-section[1]. For a prism of arbitrary cross-section $\mathscr{A}$, a deformation corresponding

[1] RIVLIN [1953, *24*, § 6].

to torsion superimposed upon an extension is obtained from (66.9) by modifying $(66.9)_5$ to read

$$w^{(1)} = \delta Z + \varepsilon \varphi(X, Y). \tag{66.33}$$

Again we assume that δ and ε are related by an expansion (66.10). If φ in (66.33) is chosen to be the classical warping function of the cross-section, then $(66.9)_{1,3}$ and (66.33) give a solution, in the first-order theory of elasticity, of the problem of extending and twisting the prism while keeping its mantle free of traction. The warping function φ satisfies Laplace's equation

$$\varphi_{,XX} + \varphi_{,YY} = 0 \tag{66.34}$$

and the boundary condition

$$(\varphi_{,X} - Y) n_X + (\varphi_{,Y} + X) n_Y = 0, \tag{66.35}$$

on the boundary $\partial\mathscr{A}$, on which $\boldsymbol{n}$ is the outward unit normal. Eq. (66.34) results from the differential equations of infinitesimal elasticity, and (66.35) from the condition that the mantle of the prism be free of traction. The Z-axis is chosen so as to pass through the centroids of the cross-sections.

The procedure for formulation of the second-order problem is just the same as in the preceding example, so we do not give details or restate assumptions. Again we have (66.12), where now

$$\left.\begin{aligned}
\xi &= \delta^2\left[\frac{\sigma(1+2\sigma)^2}{1-2\sigma} + 2\sigma(1-2\sigma) + \alpha_3(1-2\sigma)^2 - \alpha_4\sigma(2-\sigma)\right] + \\
&\quad + \varepsilon^2\left[\frac{\sigma}{1-2\sigma}(X^2+Y^2+\varphi_{,X}^2+\varphi_{,Y}^2) - \frac{\alpha_4}{4}\left((X+\varphi_{,Y})^2 + (Y-\varphi_{,X})^2\right)\right], \\
[\boldsymbol{E}_1] &= \left\|\begin{matrix} -\sigma\delta & 0 & -\frac{1}{2}\varepsilon(Y-\varphi_{,X}) \\ 0 & -\sigma\delta & \frac{1}{2}\varepsilon(X+\varphi_{,Y}) \\ -\frac{1}{2}\varepsilon(Y-\varphi_{,X}) & \frac{1}{2}\varepsilon(X+\varphi_{,Y}) & \delta \end{matrix}\right\|, \\
[\boldsymbol{K}] &= \left\|\begin{matrix} \sigma^2\delta^2+\varepsilon^2 Y\varphi_{,X} & -\varepsilon^2 X\varphi_{,X} & -\varepsilon^2 Z(X+\varphi_{,Y}) - \delta\varepsilon[(1-\sigma)Y+(1+\sigma)\varphi_{,X}] \\ \varepsilon^2 Y\varphi_{,Y} & \sigma^2\delta^2-\varepsilon^2 X\varphi_{,Y} & -\varepsilon^2 Z(Y-\varphi_{,X}) + \delta\varepsilon[(1-\sigma)X-(1+\sigma)\varphi_{,Y}] \\ 2\sigma\delta\varepsilon Y & -2\sigma\delta\varepsilon X & -(2\sigma+1)\delta^2 - \varepsilon^2(X\varphi_{,Y} - Y\varphi_{,X}) \end{matrix}\right\|.
\end{aligned}\right\} \tag{66.36}$$

A straightforward but lengthy calculation based on (66.12) and (66.36) will yield the body forces

$$\boldsymbol{b}^*_{(1)} = \varrho_R^{-1}\,\mathrm{Div}\,\boldsymbol{T}_R^{(1)} \tag{66.37}$$

and the surface tractions

$$\boldsymbol{t}^*_{(1)} = -\boldsymbol{T}_R^{(1)}\boldsymbol{n}_R, \tag{66.38}$$

the negatives of which must be supplied in order to maintain the displacement $\boldsymbol{u}^{(1)}$ in equilibrium according to the second-order theory. The additional displacement $\boldsymbol{u}_{(2)}$ necessary to obtain the second-order solution $\boldsymbol{u}^{(2)} = \boldsymbol{u}^{(1)} + \boldsymbol{u}_{(2)}$ for extension and twist of the prism is the solution of the equations of the infinitesimal theory corresponding to the body forces (66.37) and surface tractions (66.38).

We cannot expect to solve this problem explicitly for a cross-section of arbitrary form. However, some features of the deformation are determined directly by the load. For example, the mean strains may be calculated by BETTI's formulae[1] in the infinitesimal theory. It is a straightforward matter

[1] For any material in equilibrium, the mean values of the stresses are determined from the assigned loads by Eq. (220.2) of CFT. In the infinitesimal theory of elasticity, the strains are linear functions of the components of stress, and this fact yields BETTI's formulae at once. For their explicit form, see e.g. LOVE [1927, *1*, § 121].

to show in this way that the mean extension $\bar{\delta}$ resulting from the twist is given by the expression

$$\bar{\delta} = \frac{\varepsilon^2}{2A_0}\left[-\frac{2\beta''}{1+\sigma}S_0 - (I_0 - S_0)\right] \tag{66.39}$$

where β'' is given by $(66.27)_2$, where A_0 is the area of the cross-section $\mathscr{A}$, where I_0 is its polar moment of inertia, and where μS_0 is its classical torsional rigidity:

$$S_0 \equiv \iint\limits_{\mathscr{A}} [(X+\varphi_{,Y})^2 + (Y-\varphi_{,X})^2]\,dX\,dY = I_0 - \iint\limits_{\mathscr{A}} [\varphi_{,X}^2 + \varphi_{,Y}^2]\,dX\,dY. \tag{66.40}$$

This result generalizes (66.28).

BLACKBURN and GREEN[1] have formulated problems of second-order torsion and bending in terms of two complex functions satisfying a relatively simple boundary-value problem. They show also how (66.39) can be obtained by their procedure.

The problem of combined torsion and tension of a circular cylinder, or sometimes of a prism of more general form, has been a favorite for proponents of special or approximate theories of elasticity. Much of the early literature concerning the problem was summarized by TRUESDELL[2].

Other examples. SAKADI[3] obtained second-order solutions for the bending and torsion of an elliptic cylinder, and for radial oscillation of a sphere, and for torsional vibration of a circular cylinder. MURNAGHAN[4] obtained second-order solutions for purely radial deformation of a circular-cylindrical tube or a spherical shell of hyperelastic material. More general results of this kind for spheres and cylinders can be obtained by specializing the general theory of the same problems, starting from (59.5) and (59.11), respectively.

[*Added in proof.* WANG[5] has used the second-order theory to calculate the change in the infinitesimal shear and bulk moduli for a small dilatation from the natural state. If $\boldsymbol{F} = v\boldsymbol{1}$, then $\boldsymbol{H} = (v-1)\boldsymbol{1}$, and (66.3) yields

$$-p_0 = \tfrac{1}{3}\operatorname{tr}\boldsymbol{T} = \mu(v-1)\left[2+3\alpha_1+(v-1)\left(1+\tfrac{3}{2}\alpha_1+9\alpha_3+3\alpha_4+3\alpha_5+\alpha_6\right)\right]. \tag{66.41}$$

First, if an infinitesimal dilatation with stretch $1+\delta$ is superimposed, then $\boldsymbol{F} = (1+\delta)v\boldsymbol{1}$, so that by (66.41)

$$\left.\begin{aligned} -p = \tfrac{1}{3}\operatorname{tr}\boldsymbol{T} &= \mu(v-1+v\delta)\,[2+3\alpha_1+ \\ &+ (v-1+v\delta)\left(1+\tfrac{3}{2}\alpha_1+9\alpha_3+3\alpha_4+3\alpha_5+\alpha_6\right)].\end{aligned}\right\} \tag{66.42}$$

The apparent bulk modulus $\varkappa_{\text{app}}(v)$ is given by

$$\left.\begin{aligned} \varkappa_{\text{app}}(v) &\equiv \lim_{\delta\to 0} -\frac{(p-p_0)}{3\delta}, \\ &= \mu v\left[\tfrac{2}{3}+\alpha_1+(v-1)\left(\tfrac{2}{3}+\alpha_1+6\alpha_3+2\alpha_4+2\alpha_5+\tfrac{2}{3}\alpha_6\right)\right].\end{aligned}\right\} \tag{66.43}$$

Hence, of course, $\varkappa_{\text{app}}(1) = \mu\left(\tfrac{2}{3}+\alpha_1\right) = \tfrac{2}{3}\mu+\lambda \equiv \varkappa$, and also

$$\varkappa'_{\text{app}}(1) = \mu\left[\tfrac{4}{3}+2\alpha_1+6\alpha_3+2\alpha_4+2\alpha_5+\tfrac{2}{3}\alpha_6\right], \tag{66.44}$$

where the prime denotes differentiation. Second, superimpose an infinitesimal shear, so that

$$[\boldsymbol{F}] = \begin{Vmatrix} v & v\delta & 0 \\ 0 & v & 0 \\ 0 & 0 & v \end{Vmatrix}. \tag{66.45}$$

[1] BLACKBURN and GREEN [1957, *4*].
[2] TRUESDELL [1952, *20*, §§ 49—53].
[3] SAKADI [1949, *20*, §§ III—VI].
[4] MURNAGHAN [1951, *8*, Ch. 7, §§ 1—2].
[5] WANG [1965, *44*].

By (66.3),

$$T\langle x y\rangle=\mu v \delta[1+(v-1)(1+\tfrac{3}{2}\alpha_5+\alpha_6). \tag{66.46}$$

The shear modulus $\mu_{\rm app}(v)$ is given by

$$\mu_{\rm app}(v)\equiv \lim_{\delta\to 0}\frac{T\langle x y\rangle}{\delta}=\mu v\left[1+(v-1)\left(1+\frac{3}{2}\alpha_5+\alpha_6\right)\right]. \tag{66.47}$$

Hence

$$\mu'_{\rm app}(1)=\mu[2+\tfrac{3}{2}\alpha_5+\alpha_6]. \tag{66.48}$$

If we prefer to use the ratio of volumes rather than the corresponding stretch as argument, we can write

$$\mu_{\rm app}(v)=\mu^*_{\rm app}(v^3),\qquad \varkappa_{\rm app}(v)=\varkappa^*_{\rm app}(v^3), \tag{66.49}$$

and obtain

$$\left.\begin{aligned}\varkappa^{*\prime}_{\rm app}(1)&=\tfrac{1}{3}\varkappa'_{\rm app}(1)=\tfrac{2}{3}\mu(\varkappa/\mu+3\alpha_3+\alpha_4+\alpha_5+\tfrac{1}{3}\alpha_6),\\ \mu^{*\prime}_{\rm app}(1)&=\tfrac{1}{3}\mu'_{\rm app}(1)=\mu(\tfrac{2}{3}+\tfrac{1}{2}\alpha_5+\tfrac{1}{3}\alpha_6).\end{aligned}\right\} \tag{66.50}$$

These formulae give the second-order changes in the shear modulus and bulk modulus for dilatations from the natural state. As they show, some but not all of the second-order moduli can be interpreted by means of variations of the first-order moduli in a dilatation.

TOUPIN and RIVLIN's theory of mean strains in a body subject to no load has been mentioned at the end of Sect. 63. WANG[1] has found a simple means of calculating the change of volume. Since

$$V=\int_{\mathscr{B}}|\det \boldsymbol{F}|\,dv_{\rm R}=\int_{\mathscr{B}}|\det(\boldsymbol{1}+\boldsymbol{H})|\,dv_{\rm R},$$

by Eq. (8.2)

$$\left.\begin{aligned}V-V_{\rm R}&=\int_{\mathscr{B}}|I_{\boldsymbol{H}}+II_{\boldsymbol{H}}+III_{\boldsymbol{H}}|\,dv_{\rm R},\\ &=\int_{\mathscr{B}}I_{\tilde{\boldsymbol{E}}}\,dv_{\rm R}+\tfrac{1}{2}\int_{\mathscr{B}}(I^2_{\tilde{\boldsymbol{E}}}-I_{\boldsymbol{H}^2})\,dv_{\rm R},\end{aligned}\right\} \tag{66.51}$$

as far as terms of second-order in the displacement gradient. The bar in Sect. 63 denotes an average over the deformed configuration. Thus in the second-order theory, given by (66.3), the trace of the general equation $\overline{T\langle km\rangle}=0$ assumes the form

$$\left.\begin{aligned}0&=\int_{\mathscr{B}}I_{\boldsymbol{T}}\,dv=\int_{\mathscr{B}}I_{\boldsymbol{T}}(1+I_{\tilde{\boldsymbol{E}}})\,dv_{\rm R},\\ &=\mu\int_{\mathscr{B}}[(3\alpha_1+2)I_{\tilde{\boldsymbol{E}}}+(\tfrac{3}{2}\alpha_1+1)I_{\boldsymbol{H}\boldsymbol{H}^T}+(3\alpha_3+\alpha_5)I^2_{\tilde{\boldsymbol{E}}}+\\ &\quad+3\alpha_4 II_{\tilde{\boldsymbol{E}}}+\alpha_6 I_{\tilde{\boldsymbol{E}}^2}](1+I_{\tilde{\boldsymbol{E}}})\,dv_{\rm R}.\end{aligned}\right\} \tag{66.52}$$

This equation, of course, is a special case of (63.24). If $\varkappa\neq 0$, we may solve (66.52) for $\int I_{\tilde{\boldsymbol{E}}}\,dv_{\rm R}$ and substitute the result into (66.51). Use of the identities $2I_{\tilde{\boldsymbol{E}}^2}=I_{\boldsymbol{H}\boldsymbol{H}^T}+I_{\boldsymbol{H}^2}$ and $4II_{\tilde{\boldsymbol{E}}}=2I^2_{\tilde{\boldsymbol{E}}}-I_{\boldsymbol{H}\boldsymbol{H}^T}-I_{\boldsymbol{H}^2}$ allows us to write the result in the form

$$\left.\begin{aligned}V-V_{\rm R}=-\frac{1}{\frac{2}{3}+\alpha_1}&\left\{\left(\frac{10}{9}+\frac{5}{3}\alpha_1+2\alpha_3+\frac{2}{3}\alpha_4+\frac{2}{3}\alpha_5+\frac{2}{9}\alpha_6\right)\int_{\mathscr{B}}\frac{1}{2}I^2_{\tilde{\boldsymbol{E}}}\,dv_{\rm R}+\right.\\ &\left.+\left(\frac{2}{3}+\alpha_1-\frac{1}{2}\alpha_4+\frac{1}{3}\alpha_6\right)\int_{\mathscr{B}}\left[\frac{1}{2}(I_{\boldsymbol{H}\boldsymbol{H}^T}+I_{\boldsymbol{H}^2})-\frac{1}{2}I^2_{\tilde{\boldsymbol{E}}}\right]dv_{\rm R}\right\}.\end{aligned}\right\} \tag{66.53}$$

[1] WANG [1965, *44*].

This formula may be interpreted partly in terms of the changes of the shear and bulk moduli, as given by (66.50):

$$\left.\begin{aligned} V-V_{\mathrm{R}}=&-\left(1+\frac{\varkappa^{*\prime}_{\mathrm{app}}(1)}{\varkappa}\right)\int_{\mathscr{B}}\frac{1}{2}\,I_{\tilde{E}}^{2}\,dv_{\mathrm{R}}-\\ &-\frac{\lambda+\mu^{*\prime}_{\mathrm{app}}(1)-\frac{1}{2}\mu(\alpha_4+\alpha_5)}{\varkappa}\int_{\mathscr{B}}\left[\frac{1}{2}\,(I_{\boldsymbol{H}\boldsymbol{H}^T}+I_{\boldsymbol{H}^2})-\frac{1}{3}\,I_{\tilde{E}}^{2}\right]dv_{\mathrm{R}}.\end{aligned}\right\}\tag{66.54}$$

An application of this form of the result will be presented in Sect. 93.

In the special case when the first-order change of volume vanishes, $I_{\tilde{E}}=0$, and (66.51) yields for the second-order change

$$V-V_{\mathrm{R}}=-\tfrac{1}{2}\int_{\mathscr{B}}I_{\boldsymbol{H}^2}\,dv_{\mathrm{R}}.\tag{66.55}$$

By (66.53), we obtain a relation between the mean values of $I_{\boldsymbol{H}^2}$ and $I_{\boldsymbol{H}\boldsymbol{H}^T}$ for such a deformation, still, of course, on the assumption that the body is free of load:

$$(\tfrac{1}{2}\alpha_4-\tfrac{1}{3}\alpha_6)\int_{\mathscr{B}}I_{\boldsymbol{H}^2}\,dv_{\mathrm{R}}=(\tfrac{2}{3}+\alpha_1-\tfrac{1}{2}\alpha_4+\tfrac{1}{3}\alpha_6)\int_{\mathscr{B}}I_{\boldsymbol{H}\boldsymbol{H}^T}\,dv_{\mathrm{R}}.\tag{66.56)]}$$

67. Second-order effects in incompressible isotropic materials. In an isochoric deformation the invariants $I_{\boldsymbol{B}}-3$ and $II_{\boldsymbol{B}}-3$ are both of the second order in the extensions [CFT, Eq. (40.6)]. Hence the most general second-order stress relation for isotropic incompressible materials is of the form obtained by approximating the response coefficients $\beth_1$ and $\beth_{-1}$ in (49.5) by constants. Writing

$$(\tfrac{1}{2}+\beta)\mu\equiv\beth_1(3,3),\qquad(\tfrac{1}{2}-\beta)\mu\equiv-\beth_{-1}(3,3),\tag{67.1}$$

to within terms of third order in the displacement gradient $\boldsymbol{H}$ we find[1] with the help of (66.1) that

$$\left.\begin{aligned}\boldsymbol{T}/\mu&=-\varpi\boldsymbol{1}+(\tfrac{1}{2}+\beta)(2\tilde{\boldsymbol{E}}+\boldsymbol{H}\boldsymbol{H}^T)+(\tfrac{1}{2}-\beta)(2\tilde{\boldsymbol{E}}+\boldsymbol{H}\boldsymbol{H}^T-4\tilde{\boldsymbol{E}}^2)+\cdots\\&=-\varpi\boldsymbol{1}+2\tilde{\boldsymbol{E}}+\boldsymbol{H}\boldsymbol{H}^T-4(\tfrac{1}{2}-\beta)\tilde{\boldsymbol{E}}^2+\cdots,\end{aligned}\right\}\tag{67.2}$$

where ϖ is an undetermined dimensionless scalar and β is a dimensionless material modulus. In the infinitesimal theory, $I_{\tilde{E}}$ is the change of volume, so in the passage to the case of an incompressible material $I_{\tilde{E}}\to 0$. At the same time the classical bulk modulus must become infinitely large in order that an incompressible material may support non-zero hydrostatic pressure. That is, $3\lambda+2\mu=(3\alpha_1+2)\mu\to\infty$ in such a way that $(3\lambda+2\mu)I_{\tilde{E}}$ tends to a finite, undetermined limit. Hence $\alpha_1\to\infty$, or, equivalently, $\sigma\to\frac{1}{2}$. In a limit process of this kind, (66.3) reduces to the form (67.2) if we put

$$\alpha_6=4(\beta-\tfrac{1}{2}).\tag{67.3}$$

This is not all, however, since the second-order condition for isochoric deformation, namely,

$$I_{\boldsymbol{H}\boldsymbol{H}^T}+4II_{\tilde{E}}=0,\tag{67.4}$$

must be considered[2]. [Cf. Eq. (66.51).] From (66.3) we see that if $I_{\boldsymbol{H}\boldsymbol{H}^T}\to-4II_{\tilde{E}}$ while $\alpha_1\to\infty$, in order for a finite indeterminate pressure to result it is necessary that $\alpha_4\to 2\alpha_1$. From $(66.6)_4$, since $\sigma\to\frac{1}{2}$, we see that

$$(1-2\sigma)\alpha_4\to 2.\tag{67.5}$$

---

[1] The result in this context but in slightly lesser generality was obtained by RIVLIN and SAUNDERS [1951, *14*, § 21].

[2] This discussion differs somewhat from that of RIVLIN [1953, *24*, § 7], who obtained equivalent results by demanding that the coefficient written as α_3' in his notation shall remain finite as $\sigma\to\frac{1}{2}$.

Putting (67.3) and (67.5) into $(66.27)_2$ and (66.28) now yields the following formula[1] for the extension of an incompressible cylindrical tube resulting from a twist of magnitude ε:

$$\delta = \frac{1}{24}(3-2\beta)(R_I^2 + R_{II}^2)\,\varepsilon^2. \tag{67.6}$$

Similarly, from (66.39) we obtain the corresponding result for an incompressible prism of arbitrary simply connected cross-section[2]:

$$\bar{\delta} = \frac{\varepsilon^2}{6A_0}\left[\left(\frac{3}{2}-\beta\right)S_0 - 3(I_0 - S_0)\right]. \tag{67.7}$$

If the E-inequalities (53.3) are adopted, it follows that $-\frac{1}{2}<\beta\leqq+\frac{1}{2}$. From (67.6), then, we see that an *incompressible isotropic cylinder always elongates when twisted*, as was shown on the basis of the exact solution in Sect. 57. For non-circular cross-sections, however, no such general conclusion holds. Since $S_0 < I_0$ unless $\mathscr{A}$ is a circle, from (67.7) we see that lengthening rather than shortening will result if the material and the cross-section satisfy the relation

$$\frac{3}{2} - \beta < \frac{3(I_0 - S_0)}{S_0}. \tag{67.8}$$

For example, if $I_0 > \frac{5}{3}S_0$, (67.8) holds for all materials satisfying the E-inequalities (53.3).

There is uncertainty as well as indirectness in the limit process just discussed. We shall now present a direct perturbation method which presumes from the outset that the material is incompressible.

This method, due to GREEN and SPRATT[3], starts from an assumed expansion of the form (63.3) but differs from SIGNORINI's method in that it is formulated in terms of the second Piola-Kirchhoff tensor $\tilde{\boldsymbol{T}}$, defined by (43A.9), rather than the first, $\boldsymbol{T}_R$. To within terms of order two in $\boldsymbol{H}$ we have

$$\boldsymbol{F}^{-1} = (\mathbf{1}+\boldsymbol{H})^{-1} = \mathbf{1} - \boldsymbol{H} + \boldsymbol{H}^2 + \cdots. \tag{67.9}$$

Substituting (67.9) and (67.2) into $(43\text{A}.11)_5$ and noting that $J=1$ for isochoric deformations, we obtain the expansion

$$\left.\begin{aligned}\tilde{\boldsymbol{T}}/\mu = &-\varpi(\mathbf{1} - 2\tilde{\boldsymbol{E}} + \boldsymbol{H}^2 + \boldsymbol{H}^{T2} + \boldsymbol{H}\boldsymbol{H}^T) + 2\tilde{\boldsymbol{E}} + \boldsymbol{H}\boldsymbol{H}^T - \\ &- 2(\boldsymbol{H}\tilde{\boldsymbol{E}} + \tilde{\boldsymbol{E}}\boldsymbol{H}^T) - 4(\tfrac{1}{2}-\beta)\tilde{\boldsymbol{E}}^2 + \cdots.\end{aligned}\right\} \tag{67.10}$$

As in Sect. 63, we assume that we have a boundary-value problem depending on a parameter ε and that a solution $\boldsymbol{u}$ having an expansion of the form (63.3) exists. In the case of an incompressible material, the determination of the pressure ϖ is a part of the solution of the boundary-value problem. We assume that the pressure ϖ, as well as the displacement, has a convergent expansion

$$\varpi = \varepsilon\,\varpi_1 + \varepsilon^2\varpi_2 + \cdots. \tag{67.11}$$

Substitution of (63.4) and (67.11) into (67.10) yields

$$\left.\begin{aligned}\tilde{\boldsymbol{T}} &= \varepsilon\,\tilde{\boldsymbol{T}}_1 + \varepsilon^2\,\tilde{\boldsymbol{T}}_2 + O(\varepsilon^3),\\ \tilde{T}^{\alpha\beta} &= \varepsilon\,\tilde{T}^{\alpha\beta}_{(1)} + \varepsilon^2\,\tilde{T}^{\alpha\beta}_{(2)} + O(\varepsilon^3),\end{aligned}\right\} \tag{67.12}$$

---

[1] RIVLIN [1953, *24*, § 7].

[2] RIVLIN [1953, *24*, § 7]. For an elliptic cylinder, in the case when $\beta = \frac{1}{2}$ the result was obtained by GREEN and SHIELD [1951, *2*, § 11].

[3] GREEN and SPRATT [1954, *9*, § 3].

where

$$\tilde{\boldsymbol{T}}_1/\mu = -\varpi_1 \mathbf{1} + 2\tilde{\boldsymbol{E}}_1,$$
$$\tilde{\boldsymbol{T}}_2/\mu = -\varpi_2 \mathbf{1} + 2\varpi_1 \tilde{\boldsymbol{E}}_1 + \boldsymbol{H}_1 \boldsymbol{H}_1^T - 2(\boldsymbol{H}_1 \tilde{\boldsymbol{E}}_1 + \tilde{\boldsymbol{E}}_1 \boldsymbol{H}_1^T) - 4(\tfrac{1}{2} - \beta)\tilde{\boldsymbol{E}}_1^2 + 2\tilde{\boldsymbol{E}}_2, \quad (67.13)$$
$$\tilde{\boldsymbol{E}}_n \equiv \tfrac{1}{2}(\boldsymbol{H}_n + \boldsymbol{H}_n^T), \qquad n = 1, 2.$$

The condition for isochoric deformation, $J = \det(\mathbf{1} + \boldsymbol{H}) = 1$, is easily shown to yield the equations

$$\left.\begin{aligned} \operatorname{tr} \boldsymbol{H}_1 &= u^\alpha_{(1),\alpha} = 0, \\ 2\operatorname{tr} \boldsymbol{H}_2 - \operatorname{tr} \boldsymbol{H}_1^2 &= 2u^\alpha_{(2),\alpha} - u^\alpha_{(1),\beta} u^\beta_{(1),\alpha} = 0 \end{aligned}\right\} \quad (67.14)$$

[cf. (67.4)]. From (43 A.7) and $(43\,\text{A}.11)_4$ it follows that the equations of motion, in terms of the second Piola-Kirchhoff tensor $\tilde{\boldsymbol{T}}$, assume the form[1]

$$\tilde{T}^{\alpha\beta}{}_{,\beta} + X^\alpha{}_{,k}\, g^k_\delta\, u^\delta{}_{,\beta\gamma}\, \tilde{T}^{\beta\gamma} + \varrho X^\alpha{}_{,k}(b^k - \ddot{u}^k) = 0. \quad (67.15)$$

By (67.9) we have $X^\alpha{}_{,k} = g^\alpha_k + O(\varepsilon)$, so that

$$X^\alpha{}_{,k}\, g^k_\delta\, u^\delta{}_{,\beta\gamma} = \varepsilon\, u^\alpha_{(1),\beta\gamma} + O(\varepsilon^2). \quad (67.16)$$

Substitution of (67.16) and the expansion $(67.12)_2$ into (67.15) gives the following differential equations to be satisfied by the successive terms

$$\left.\begin{aligned} &\tilde{T}^{\alpha\beta}_{(1),\beta} + \varrho(b^{(1)\alpha} - \ddot{u}^{(1)\alpha}) = 0, \\ &\tilde{T}^{\alpha\beta}_{(2),\beta} + u^\alpha_{(1),\beta\gamma}\, \tilde{T}^{\beta\gamma}_{(1)} + \varrho(b^{(2)\alpha} - \ddot{u}^{(2)\alpha}) = 0, \ldots, \end{aligned}\right\} \quad (67.17)$$

where it has been assumed that[2]

$$X^\alpha{}_{,k}(b^k - \ddot{u}^k) = \varepsilon(b^{(1)\alpha} - \ddot{u}^{(1)\alpha}) + \varepsilon^2(b^{(2)\alpha} - \ddot{u}^{(2)\alpha}) + O(\varepsilon^3). \quad (67.18)$$

The dots after (67.17) do not indicate any rule for forming higher-order equations; as in the method of SIGNORINI, the terms grow more and more complicated at the successive stages.

In the place of $(63.2)_2$, suppose that the assigned surface traction $\boldsymbol{t}_{\mathrm{R}}$ has an expansion of the form

$$X^\alpha{}_{,k}\, t^k_{\mathrm{R}} = \varepsilon\, t^{(1)\alpha}_{\mathrm{R}} + \varepsilon^2 t^{(2)\alpha}_{\mathrm{R}} + O(\varepsilon^3). \quad (67.19)$$

Then the conditions at the bounding surface are easily shown to be

$$\tilde{T}^{\alpha\beta}_{(1)}\, n_{\mathrm{R}\beta} = t^{(1)\alpha}_{\mathrm{R}}, \qquad \tilde{T}^{\alpha\beta}_{(2)}\, n_{\mathrm{R}\beta} = t^{(2)\alpha}_{\mathrm{R}}. \quad (67.20)$$

Eqs. (67.13), when substituted into (67.17) and combined with (67.14) and (67.20), yield a system of partial differential equations and boundary conditions for the second-order theory of incompressible materials, formulated entirely in material co-ordinates. The first displacement field $\boldsymbol{u}_1$ is determined from the classical infinitesimal theory, governed by a single modulus, the shear modulus μ.

---

[1] Cf. CFT, Eq. (210.10).

[2] The coefficients $\boldsymbol{b}^n$, $\boldsymbol{u}^n$, and $\overset{n}{\boldsymbol{t}}_{\mathrm{R}}$ are not the same as $\boldsymbol{b}_n$, $\boldsymbol{u}_n$, and $\boldsymbol{t}_{\mathrm{R}n}$ in (63.2) and (63.3). We have, in fact, from (67.9),

$$t^{(1)\alpha}_{\mathrm{R}} = g^\alpha_k\, t^k_{\mathrm{R}(1)},$$
$$t^{(2)\alpha}_{\mathrm{R}} = g^\alpha_k\, t^k_{\mathrm{R}(2)} - g^\beta_k\, u^\alpha_{(1),\beta}\, t^k_{\mathrm{R}(1)}, \ldots$$

along with similar relations connecting $\boldsymbol{b}^n - \ddot{\boldsymbol{u}}^n$ to $\boldsymbol{b}_n - \ddot{\boldsymbol{u}}_n$.

The second displacement field, $\boldsymbol{u}_2$, is thus determined by the first and by equations involving a single additional material constant, the dimensionless quantity β. The second-order equations, of course, are also of the same form as those of the first-order theory, but the effective additional body force required to maintain the deformation is a complicated combination of derivatives of the first-order solution having no evident physical interpretation.

Example. Torsion of a solid of revolution[1]. To represent torsion of a solid of revolution subject to no body force, select a cylindrical co-ordinate system R, Θ, Z, and consider a first-order purely azimuthal displacement $\varepsilon \boldsymbol{u}_1$ of amount $\varepsilon R v(R, Z)$. Then

$$u^R_{(1)} = 0, \qquad u^\Theta_{(1)} = v(R, Z) = v, \qquad u^Z_{(1)} = 0. \tag{67.21}$$

A straightforward calculation based on (67.13) yields

$$\tilde{T}^{RR}_{(1)} = R^2 \tilde{T}^{\Theta\Theta}_{(1)} = \tilde{T}^{ZZ}_{(1)} = -\varpi_1 \mu, \qquad \tilde{T}^{RZ}_{(1)} = 0, \qquad \tilde{T}^{R\Theta}_{(1)} = \mu \frac{\partial v}{\partial R}, \qquad \tilde{T}^{\Theta Z}_{(1)} = \mu \frac{\partial v}{\partial Z}. \tag{67.22}$$

Aside from the hydrostatic part, this is a purely torsional state of stress; it is equilibrated if and only if $\varpi_1 = \text{const.}$ and[2]

$$\tilde{T}^{R\Theta}_{(1)} = -\frac{1}{R^3} \frac{\partial W}{\partial Z}, \qquad \tilde{T}^{\Theta Z}_{(1)} = \frac{1}{R^3} \frac{\partial W}{\partial R}, \tag{67.23}$$

where $W = W(R, Z)$. Comparison of $(67.22)_{5,6}$ with (67.23) yields a system of equations for v and W, which are compatible if and only if

$$\frac{\partial^2 W}{\partial R^2} - \frac{3}{R} \frac{\partial W}{\partial R} + \frac{\partial^2 W}{\partial Z^2} = 0, \qquad \frac{\partial}{\partial R}\left(R^3 \frac{\partial v}{\partial R}\right) + \frac{\partial}{\partial Z}\left(R^3 \frac{\partial v}{\partial Z}\right) = 0. \tag{67.24}$$

The bounding surface is taken as having the equation $R = f(Z)$. It is easy to show that this surface is free of traction, according to the infinitesimal theory, if and only if $\varpi_1 = 0$ and

$$W|_{R=f(Z)} = \text{const.} \tag{67.25}$$

The tractions acting upon the plane end $Z = Z_0$ are equipollent to a torsional couple of magnitude M, where

$$M = \int_0^a 2\pi R (\varepsilon R \tilde{T}^{\Theta Z}_{(1)}) R\, dR = 2\pi \varepsilon [W(a, Z_0) - W(0, Z_0)], \tag{67.26}$$

a being the radius of the section, $a = f(Z_0)$. The parameter ε, which has appeared here in virtue of (67.12), is a measure of the amount of torsion. This completes the first-order solution, since v may be determined by integrating $(67.22)_{5,6}$.

For the second-order solution, consider a deformation of the form

$$u^R_{(2)} = G - \tfrac{1}{2} R v^2, \qquad u^\Theta_{(2)} = 0, \qquad u^Z_{(2)} = K, \tag{67.27}$$

where G and K are functions of R and Z. A straightforward calculation using $(67.13)_2$ then yields the following formulae for the second-order stresses:

$$\left.\begin{aligned}
\tilde{T}^{RR}_{(2)}/\mu &= -\varpi_2 + 2\frac{\partial G}{\partial R} - \left(\frac{1}{2} - \beta\right) R^2 \left(\frac{\partial v}{\partial R}\right)^2, \\
R^2 \tilde{T}^{\Theta\Theta}_{(2)}/\mu &= -\varpi_2 + 2\frac{G}{R} - \left(\frac{3}{2} + \beta\right) R^2 \left[\left(\frac{\partial v}{\partial R}\right)^2 + \left(\frac{\partial v}{\partial Z}\right)^2\right], \\
\tilde{T}^{ZZ}_{(2)}/\mu &= -\varpi_2 + 2\frac{\partial K}{\partial Z} - \left(\frac{1}{2} - \beta\right) R^2 \left(\frac{\partial v}{\partial Z}\right)^2, \\
\tilde{T}^{RZ}_{(2)}/\mu &= \frac{\partial G}{\partial Z} + \frac{\partial K}{\partial R} - \left(\frac{1}{2} - \beta\right) R^2 \frac{\partial v}{\partial R} \frac{\partial v}{\partial Z}, \\
\tilde{T}^{R\Theta}_{(2)} &= \tilde{T}^{\Theta Z}_{(2)} = 0.
\end{aligned}\right\} \tag{67.28}$$

---

[1] GREEN and SPRATT [1954, 9, §§ 5—6].

[2] Eq. (228.5) of CFT.

The Θ-component of $(67.17)_2$ is satisfied identically; the Z-component assumes the form

$$\frac{\partial}{\partial R}(R\,\tilde{T}^{RZ}_{(2)}) + \frac{\partial}{\partial Z}(R\,\tilde{T}^{ZZ}_{(2)}) = 0. \tag{67.29}$$

Hence there exists a function $Y = Y(R, Z)$ such that

$$\tilde{T}^{RZ}_{(2)} = -\frac{\mu}{R}\frac{\partial Y}{\partial Z}, \qquad \tilde{T}^{ZZ}_{(2)} = \frac{\mu}{R}\frac{\partial Y}{\partial R}. \tag{67.30}$$

The second-order criterion of isochoric deformation $(67.14)_2$ assumes the form

$$\frac{\partial}{\partial R}(R\,G) + \frac{\partial}{\partial Z}(R\,K) = 0. \tag{67.31}$$

Regarding this equation as a condition of integrability for a plane vector field, we may write the general solution in the form

$$G = -\frac{\partial^3 H}{\partial R\,\partial Z^2}, \qquad K = \nabla_1^2\frac{\partial H}{\partial Z}, \tag{67.32}$$

where $\nabla_1^2 = \frac{\partial^2}{\partial R^2} + \frac{1}{R}\frac{\partial}{\partial R}$ and H is a scalar potential. Putting (67.32) into $(67.28)_{3,4}$ and comparing the result with (67.30), we obtain

$$\left.\begin{aligned}
\frac{\partial Y}{\partial Z} &= -R\frac{\partial^2}{\partial R\,\partial Z}\left(\nabla_1^2 - \frac{\partial^2}{\partial Z^2}\right)H + \left(\frac{1}{2} - \beta\right)R^3\frac{\partial v}{\partial R}\frac{\partial v}{\partial Z},\\
\frac{\partial Y}{\partial R} &= -\varpi_2 R + 2R\nabla_1^2\frac{\partial^2 H}{\partial Z^2} - \left(\frac{1}{2} - \beta\right)R^3\left(\frac{\partial v}{\partial Z}\right)^2.
\end{aligned}\right\} \tag{67.33}$$

If we set

$$\varpi_2 = \nabla_1^2\nabla^2 H - \frac{1}{2}\left(\frac{1}{2} - \beta\right)R^2\left[\left(\frac{\partial v}{\partial R}\right)^2 + \left(\frac{\partial v}{\partial Z}\right)^2\right], \tag{67.34}$$

where $\nabla^2 = \nabla_1^2 + \partial^2/\partial Z^2$, and if we let λ be any solution of the system

$$\frac{\partial \lambda}{\partial Z} = R^3\frac{\partial v}{\partial R}\frac{\partial v}{\partial Z}, \qquad \frac{\partial \lambda}{\partial R} = \frac{1}{2}R^3\left[\left(\frac{\partial v}{\partial R}\right)^2 - \left(\frac{\partial v}{\partial Z}\right)^2\right], \tag{67.35}$$

the system (67.33) can be integrated to yield

$$Y = -R\frac{\partial}{\partial R}\left(\nabla_1^2 - \frac{\partial^2}{\partial Z^2}\right)H + \left(\frac{1}{2} - \beta\right)\lambda. \tag{67.36}$$

In view of $(67.24)_2$, the system (67.35) is compatible.

Putting $(67.32)_1$, (67.34), and (67.36) into $(67.28)_{1,2}$ and $(67.30)_1$ yields

$$\left.\begin{aligned}
\tilde{T}^{RR}_{(2)}\big/\mu &= -\nabla_1^2\nabla^2 H - 2\frac{\partial^4 H}{\partial Z^2\,\partial R^2} - \frac{1}{2}\left(\frac{1}{2} - \beta\right)R^2\left[\left(\frac{\partial v}{\partial R}\right)^2 - \left(\frac{\partial v}{\partial Z}\right)^2\right],\\
R^2\,\tilde{T}^{\Theta\Theta}_{(2)}\big/\mu &= -\nabla_1^2\nabla^2 H - \frac{2}{R}\frac{\partial^3 H}{\partial Z^2\,\partial R} - \frac{1}{2}\left(\frac{5}{2} - \beta\right)R^2\left[\left(\frac{\partial v}{\partial R}\right)^2 + \left(\frac{\partial v}{\partial Z}\right)^2\right],\\
\tilde{T}^{RZ}_{(2)}\big/\mu &= \frac{\partial}{\partial R}\left(\nabla_1^2 - \frac{\partial^2}{\partial Z^2}\right)\frac{\partial H}{\partial Z} - \left(\frac{1}{2} - \beta\right)R^2\frac{\partial v}{\partial R}\frac{\partial v}{\partial Z}.
\end{aligned}\right\} \tag{67.37}$$

It remains to consider the R-component of $(67.17)_2$; by a straightforward calculation we easily show that

$$\frac{\partial \tilde{T}^{RR}_{(2)}}{\partial R} + \frac{\partial \tilde{T}^{RZ}_{(2)}}{\partial Z} + \frac{\tilde{T}^{RR}_{(2)} - R^2\,\tilde{T}^{\Theta\Theta}_{(2)}}{R} - 2\mu R\left[\left(\frac{\partial v}{\partial R}\right)^2 + \left(\frac{\partial v}{\partial Z}\right)^2\right] = 0. \tag{67.38}$$

Substituting (67.37) into this equation yields, after use of $(67.24)_2$, some reduction, and one integration, the following partial differential equation for determining H:

$$\nabla^4 H = -2\beta\int R\left[\left(\frac{\partial v}{\partial R}\right)^2 + \left(\frac{\partial v}{\partial Z}\right)^2\right]dR, \tag{67.39}$$

where the arbitrary function of Z implied by the quadrature contributes nothing to the displacements or stresses. From (67.20) it is possible to show that for the bounding surface $R=f(Z)$ to be free of second-order tractions it is necessary that

$$\left.\begin{aligned} &Y=D=\text{const.},\\ &\frac{dS}{dZ}\frac{d}{dS}\left[\frac{\partial}{\partial Z}\left(\nabla_1^2-\frac{\partial^2}{\partial Z^2}\right)H\right]+\nabla^4 H-\frac{2}{R}\frac{\partial^3 H}{\partial R\,\partial Z^2}\\ &\qquad=\frac{1}{2}\left(\frac{1}{2}-\beta\right)R^2\left[\left(\frac{\partial v}{\partial R}\right)^2+\left(\frac{\partial v}{\partial Z}\right)^2\right]\end{aligned}\right\} \tag{67.40}$$

when $R=f(Z)$, where S is arc-length along the curve $R=f(Z)$ in a plane $\Theta=$ const. To the first order the planes $Z=$ const. are free of normal traction but are subject to a couple given by (67.26). The second-order solution of course contributes nothing to the couple, but in order to maintain it we must apply a normal tension of amount

$$\left.\begin{aligned} N&=\varepsilon^2\int_0^a \tilde{T}_{(2)}^{ZZ}\,2\pi R\,dR,\\ &=2\pi\mu\,\varepsilon^2\int_0^a\frac{\partial Y}{\partial R}\,dR=2\pi\mu\,\varepsilon^2\,[D-Y(0,Z)].\end{aligned}\right\} \tag{67.41}$$

GREEN and SHIELD[1] have given a formulation of the theory of second-order effects in the torsion and extension of isotropic incompressible prisms in terms of functions of a complex variable and have specialized it to several forms of cross-section. LIANIS[2] has investigated the second-order effects in the biaxial stretching of an infinite plate with an elliptical hole. FOSDICK and SHIELD[3] have analyzed the small bending of a circular bar subject to finite extension or compression.

Note added in proof. In work which was disclosed to us too late for detailed presentation in this treatise, CARLSON and SHIELD[4] have applied ADKINS' reciprocal theorem (Sect. 60) to show that the second-order solution of a plane problem may be read off from the first-order one, and they have found similar results for the third-order and fourth-order effects. As illustrations they give the second-order solution for an elliptic inclusion in a slab subject to uniform stress at infinity, the third-order solution for a circular inclusion, and the fourth-order solution for a concentrated couple in a slab subject to no stress or displacement at infinity.

68. Infinitesimal strain superimposed upon a given strain. I. General equations[5]. Consider now an elastic body $\mathscr{B}$, not necessarily homogeneous, in an *arbitrary* reference configuration $\boldsymbol{\varkappa}$. This reference configuration need not be a natural state and hence may be any state of strain for $\mathscr{B}$. We consider a small displace-

---

[1] GREEN and SHIELD [1951, *2*, §§ 5—11]. GREEN [1954, *8*] showed that this theory yields RIVLIN's result (67.7). The method was generalized to transversely isotropic materials by BLACKBURN [1958, *11*].

[2] LIANIS [1963, *47*].

[3] FOSDICK and SHIELD [1963, *24*].

[4] CARLSON and SHIELD [1965, *5*].

[5] The theory is a special case of one due to CAUCHY [1829, *1*, Eqs. (36) (37)], who derived (68.15) from different hypotheses, as explained below, p. 250. Cf. also ST. VENANT [1868, *2*]. CAUCHY's results were not understood and were reported obscurely or even incorrectly by nineteenth-century expositors. There have been many subsequent treatments in various notations and subject to various restrictive assumptions, e.g. BRILLOUIN [1925, *2*, § 10]. The recent paper most often cited is by GREEN, RIVLIN and SHIELD [1952, *11*]; cf. GREEN and ZERNA [1954, *11*, Ch. IV], ZORSKI [1964, *98*, § 1]. The presentation in the text, influenced by that given by TOUPIN and BERNSTEIN [1961, *62*, § 2] for hyperelastic materials, seeks clarity through directness and generality.

ment $\boldsymbol{u}=\boldsymbol{u}(\boldsymbol{x}, t)$ superimposed upon such a state of strain. Places in the reference configuration are denoted by $\boldsymbol{x}$, and their co-ordinates by x^k. The displacement gradient

$$\boldsymbol{H} \equiv \nabla \boldsymbol{u} \tag{68.1}$$

is assumed to be infinitesimal, so that all terms of an order higher than one in $\boldsymbol{H}$ can be neglected. All subsequent formulae are valid only up to an error which goes to zero like $|\boldsymbol{H}|^2$. The components of $\boldsymbol{H}$ are the covariant derivatives $u^k{}_{,m}$. The constitutive equation of the material may be written in the form (44.24), where the response function $\mathfrak{k}$ may depend on $\boldsymbol{x}$ because the reference configuration is not necessarily homogeneous. We denote the stress in the reference configuration by

$$\boldsymbol{T}_0 = \boldsymbol{T}_0(\boldsymbol{x}) = \mathfrak{k}(\boldsymbol{0}, \boldsymbol{x}). \tag{68.2}$$

The linear approximation to (44.24) then becomes

$$\boldsymbol{T}_{\mathrm{R}} = \mathfrak{k}(\boldsymbol{H}) = \boldsymbol{T}_0 + \mathsf{A}_0[\boldsymbol{H}], \tag{68.3}$$

where

$$\mathsf{A}_0 = \left.\frac{\partial \mathfrak{k}}{\partial \boldsymbol{H}}\right|_{\boldsymbol{H}=\boldsymbol{0}}, \qquad \mathsf{A}_0^{k\,m\,p\,q} = \left.\frac{\partial \mathfrak{k}^{k\,m}}{\partial H_{p\,q}}\right|_{H_{rs}=0} \tag{68.4}$$

[cf. (44.3)]. Substitution of (68.3) into (44.25) gives *the equations of motion for the superimposed infinitesimal displacement,*

$$\operatorname{div} \mathsf{A}_0[\boldsymbol{H}] = \varrho_0(\ddot{\boldsymbol{u}} - \boldsymbol{b}^*), \tag{68.5}$$

where

$$\boldsymbol{b}^* = \boldsymbol{b} + \frac{1}{\varrho_0} \operatorname{div} \boldsymbol{T}_0. \tag{68.6}$$

In (68.5) and (68.6) "div" denotes the divergence operator with respect to the place $\boldsymbol{x}$ in the reference configuration, and ϱ_0 denotes the density in the reference configuration.

The traction $\boldsymbol{t}_0$ on the boundary $\partial\mathscr{B}$ produced by the stress $\boldsymbol{T}_0$ in the reference configuration is

$$\boldsymbol{t}_0 = \boldsymbol{T}_0\, \boldsymbol{n}, \tag{68.7}$$

where $\boldsymbol{n}$ is the outer unit normal to $\partial\mathscr{B}$ in the reference configuration. The displacement $\boldsymbol{u}$ produces additional tractions $\boldsymbol{t}_{\mathrm{R}}^*$ given by

$$\boldsymbol{t}_{\mathrm{R}}^* = \boldsymbol{T}_{\mathrm{R}}\boldsymbol{n} - \boldsymbol{t}_0 = (\boldsymbol{T}_{\mathrm{R}} - \boldsymbol{T}_0)\boldsymbol{n} = \mathsf{A}_0[\boldsymbol{H}]\boldsymbol{n}. \tag{68.8}$$

In co-ordinate form, the equations just derived may be written

$$\left.\begin{aligned} (\mathsf{A}_0^{k\,p\,m\,q} u_{m,q})_{,p} &= \varrho_0(\ddot{u}^k - b^{*k}), \\ \mathsf{A}_0^{k\,p\,m\,q} u_{m,q} n_p &= t_{\mathrm{R}}^{*k}, \\ b^{*k} &= b^k + \frac{1}{\varrho_0} T_0^{k\,m}{}_{,m}. \end{aligned}\right\} \tag{68.9}$$

In the special case when the strained reference configuration is homogeneous the response function $\mathfrak{k}$, and hence also the tensor A_0, is independent of $\boldsymbol{x}$, and $\boldsymbol{b}^* = \boldsymbol{b}$. In this case $(68.9)_1$ reduces to

$$\mathsf{A}_0^{k\,(p\,|m|\,q)} u_{m,qp} = \varrho_0(\ddot{u}^k - b^k). \tag{68.10}$$

If A_0 is constant, then its components $\mathsf{A}_0^{k\,p\,m\,q}$ are constant in Cartesian, but not necessarily in curvilinear co-ordinates.

Since $\boldsymbol{F}=\mathbf{1}+\boldsymbol{H}$ and $J^{-1}=1-\operatorname{tr}\boldsymbol{H}$ when quadratic terms are neglected, it follows from substitution of (68.3) into (43A.3) that

$$\left.\begin{aligned}\boldsymbol{T}&=(1-\operatorname{tr}\boldsymbol{H})\,\boldsymbol{T}_0+\boldsymbol{T}_0\boldsymbol{H}^T+\mathsf{A}_0[\boldsymbol{H}]\\&=\boldsymbol{T}_{\mathrm{R}}-(\operatorname{tr}\boldsymbol{H})\,\boldsymbol{T}_0+\boldsymbol{T}_0\boldsymbol{H}^T,\end{aligned}\right\}\tag{68.11}$$

expressing the Cauchy-stress $\boldsymbol{T}$ in terms of the first Piola-Kirchhoff stress $\boldsymbol{T}_{\mathrm{R}}$.

If the S-E condition (44.21) is satisfied throughout the body in the reference configuration, i.e. if

$$\mathsf{A}_0^{kpmq}\lambda_k\lambda_m\mu_p\mu_q>0 \tag{68.12}$$

for all non-zero vectors $\boldsymbol{\lambda}$ and $\boldsymbol{\mu}$, then the system $(68.9)_1$ is strongly elliptic when $\ddot{\boldsymbol{u}}=\mathbf{0}$. The coefficients A_0^{kpmq} are known functions of $\boldsymbol{x}$. Thus the existing body of theorems on strongly elliptic linear partial differential systems may be applied. Unfortunately, these seem to be rather incomplete. While the S-E condition should ensure existence and uniqueness of solution to the boundary-value problem of place for the superimposed infinitesimal deformation, we have not been able to find this proved anywhere[1].

The condition of *uniformly strong ellipticity* in a bounded open set $\mathscr{R}$ requires that $\mathsf{A}_0^{kpmq}\lambda_k\lambda_m\mu_p\mu_q\geqq\varrho\lambda^2\mu^2$, for some positive constant ϱ. If A_0 is a continuous function of $\boldsymbol{x}$ in the closure of $\mathscr{R}$, assumed compact, this condition reduces to the S-E condition. Within a certain class of functions, a theorem of BROWDER[2] implies that for the boundary-value problem of place, when uniformly strong ellipticity is assumed, uniqueness implies existence.

The work of MORREY[3] for uniformly strongly elliptic systems proves both existence and uniqueness for a one-parameter family of equations including (68.9) in the case of equilibrium but allows uniqueness to fail for exceptional values of the parameter. Thus it is not certain that his result applies to elasticity at all.

The work on strongly elliptic systems fails also to show that solutions are smooth at the boundary. A weak solution to the boundary-value problem of place that yields boundary displacements not having interior derivatives is unlikely to be of interest, since it does not correspond to any possible tractions on the boundary.

The whole matter of existence and uniqueness in the theory of infinitesimal strain superimposed upon a given strain is elaborate, as shown by the results in the next section and by further work of ERICKSEN[4].

The condition (68.12) restricts only the symmetric part of the coefficient tensor, namely, $\frac{1}{4}(\mathsf{A}_0^{kpmq}+\mathsf{A}_0^{mpkq}+\mathsf{A}_0^{kqmp}+\mathsf{A}_0^{mqkp})$. Some physical implications of the condition of strong ellipticity will be discussed in Sect. 71.

---

[1] The broader claims in the literature, e.g., those of TRUESDELL and TOUPIN [1963, *76*, § 8] and ZORSKI [1964, *98*], are not borne out by the analytic sources. ZORSKI's claim that the S-E condition suffices for uniqueness of solution to the traction boundary-value problem is contradicted by a known theorem of the infinitesimal theory for isotropic materials.

Counter-examples given by HAYES [1963, *37*] [1964, *42*] show that the S-E condition is not *necessary* for uniqueness of the boundary-value problems of superimposed infinitesimal deformation. HAYES classified cases when uniqueness does and does not hold, correlating them with the signs of the squared wave speeds.

HILL [1957, *9*, § 3] remarked that the condition

$$\mathsf{A}_0^{kpmq}F_{km}F_{pq}>0$$

for arbitrary non-zero $\boldsymbol{F}$ is sufficient for unqualified uniqueness of solution to the problem of small deformation superimposed upon an arbitrary deformation. Like the unrestricted monotonicity of $\mathfrak{h}$ (Sect. 52), this inequality contradicts the principle of material indifference unless all principal stresses of $\boldsymbol{T}_0$ are positive. HILL, however, proposed the inequality only as a sufficient condition for stability and only for $\boldsymbol{F}$ differing sufficiently little from $\mathbf{1}$. In this case, the condition is compatible with frame-indifference if and only if $t_1+t_2>0$, $t_2+t_3>0$, $t_3+t_1>0$. See also Sect. 68bis.

[2] BROWDER [1954, *4*].

[3] MORREY [1954, *14*].

[4] ERICKSEN [1965, *17*].

The equations so far derived are valid for theories more general[1] than elasticity, since in the formula (44.24) taken as the starting point the principle of material indifference has not been imposed. To find properly invariant expressions for the stress, we may begin from any of the reduced forms (43.5) through (43.7) or (43.10).

At the reference configuration we have $\boldsymbol{F}=\mathbf{1}$, $J=1$, and (45.2) reduces to

$$\mathsf{C}_0^{kmpq} = 2\left.\frac{\partial \mathfrak{t}^{km}}{\partial C_{pq}}\right|_{C_{rs}=g_{rs}}, \tag{68.13}$$

i.e., $\frac{1}{2}\mathsf{C}_0$ is the gradient of the response function $\mathfrak{t}$ in the reference configuration $\boldsymbol{\varkappa}$. The components C_0^{kmpq} satisfy the symmetry conditions (45.3). From (45.4) and (45.6) we obtain

$$\mathsf{B}_0^{kmpq} = \mathsf{A}_0^{kmpq} = g^{kp}\, T_0^{mq} + \mathsf{C}_0^{kmpq}. \tag{68.14}$$

Substitution of (68.14) into (68.11) gives the reduced stress-relation:

$$\left.\begin{aligned} \boldsymbol{T} &= \boldsymbol{T}_0 + \boldsymbol{T}_0\boldsymbol{H}^T + \boldsymbol{H}\boldsymbol{T}_0 - (\operatorname{tr}\tilde{\boldsymbol{E}})\,\boldsymbol{T}_0 + \mathsf{C}_0[\tilde{\boldsymbol{E}}],\\ T^{km} &= T_0^{km} + T_0^{kp}\,u^m{}_{,p} + u^k{}_{,p}\,T_0^{pm} - u^p{}_{,p}\,T_0^{km} + \mathsf{C}_0^{kmpq}\tilde{E}_{pq}, \end{aligned}\right\} \tag{68.15}$$

where $\tilde{\boldsymbol{E}} \equiv \frac{1}{2}(\boldsymbol{H}+\boldsymbol{H}^T)$ is the infinitesimal strain tensor. Introducing the infinitesimal rotation tensor $\tilde{\boldsymbol{R}} \equiv \frac{1}{2}(\boldsymbol{H}-\boldsymbol{H}^T)$, we see that (68.15) may also be written in the form

$$\boldsymbol{T} = \boldsymbol{T}_0 + \tilde{\boldsymbol{R}}\boldsymbol{T}_0 - \boldsymbol{T}_0\tilde{\boldsymbol{R}} + \mathsf{L}_0[\tilde{\boldsymbol{E}}], \tag{68.16}$$

where

$$\mathsf{L}_0[\tilde{\boldsymbol{E}}] = \tilde{\boldsymbol{E}}\boldsymbol{T}_0 + \boldsymbol{T}_0\tilde{\boldsymbol{E}} - (\operatorname{tr}\tilde{\boldsymbol{E}})\,\boldsymbol{T}_0 + \mathsf{C}_0[\tilde{\boldsymbol{E}}], \tag{68.17}$$

a result which has already been obtained by another method as Eq. (41.25).

On the basis of these formulae we can now establish a uniqueness theorem somewhat different from that given above. It is the specialization to infinitesimal strain of the theorem given in Sect. 52, but now instead of requiring the GCN condition to hold for arbitrary strains, we demand that the stronger GCN$^+$ condition (52.16) hold throughout the body in the particular strained reference configuration $\boldsymbol{\varkappa}$. Supposing that the displacement gradient $\boldsymbol{H}$ is symmetric, so that $\tilde{\boldsymbol{E}}=\boldsymbol{H}$, we see from (68.3) and $(68.14)_1$ that

$$\operatorname{tr}\{(\boldsymbol{T}_{\mathrm{R}} - \boldsymbol{T}_0)\,(\boldsymbol{F}^T - \mathbf{1})\} = \operatorname{tr}\{\mathsf{A}_0[\boldsymbol{H}]\,\boldsymbol{H}^T\} = \mathsf{B}_0^{kmpq}\tilde{E}_{km}\tilde{E}_{pq}. \tag{68.18}$$

By the GCN$^+$ condition (52.16), this quantity is positive unless $\tilde{\boldsymbol{E}}=\mathbf{0}$. Since this result contradicts (52.7) unless $\boldsymbol{T}_{\mathrm{R}}=\boldsymbol{T}_0$, we infer the following ***uniqueness theorem***[2]: *Let an elastic material be subject to a particular strain such that the GCN$^+$ condition is satisfied everywhere in a certain region; then the mixed boundary-value problem for that region has at most one solution such that the difference between the resulting strain and the given strain is an infinitesimal pure strain.*

Note that this theorem does not yield a condition of stability. It remains possible that there may exist, corresponding to the same boundary conditions, other infinitesimal additional strains that are not pure. Thus, even for infinitesimal deformations (provided they be from a state of equilibrium subject

---

[1] E.g., for MacCullagh's theory of the rotationally elastic aether (CFT, Sect. 302).

[2] Truesdell and Toupin [1963, *76*, § 8].

to non-vanishing stress[1]), the GCN$^+$ condition is not so restrictive as to disallow infinitesimal rotation as a possible mechanism for instability.

CAUCHY approached the problem from a more general standpoint, not using the theory of elasticity directly. He showed, in effect, that an expression of the form (68.15) is the most general bilinear function of $\boldsymbol{H}$ and $\boldsymbol{T}_0$ which satisfies the principle of material frame-indifference. The stress $\boldsymbol{T}_0$ is then not necessarily elastic, and the coefficients C_0^{kmpq} need satisfy no other conditions beyond (45.3). A result equivalent to CAUCHY's is easily derived by use of the corotational or the convected time flux (CFT, Sects. 148, 150, 304).*

When the initial stress $\boldsymbol{T}_0$ vanishes, the various forms (68.11), (68.15), and (68.16) all reduce to the stress-strain relation of the classical infinitesimal theory of elasticity, and $C_0^{kmpq} = L^{kmpq}$, the classical tensor of linear elasticities (50.11). For a general value of $\boldsymbol{T}_0$, they show how the stress in the strained reference state affects the response of the material to the superimposed infinitesimal strain. Thus the present theory may be regarded as a generalization of the classical one in that an arbitrarily stressed and strained configuration, rather than a natural state, is taken as the initial state. Inspection of the result shows that the response of a severely strained elastic material to a small further deformation is, while of course linear and again elastic in the sense of (43.1), *not generally elastic in the sense of the classical infinitesimal theory*, for the additional stress depends not only on the additional infinitesimal strain $\tilde{\boldsymbol{E}}$ but also on the additional infinitesimal rotation $\tilde{\boldsymbol{R}}$. The principle of material frame-indifference, which allowed (43.1) to be reduced to (43.10) in general, does not allow the effect of the rotation to be neglected when the second deformation, however small, is considered by itself. Since $\boldsymbol{H} = \tilde{\boldsymbol{R}} + \tilde{\boldsymbol{E}}$, from (68.16) we see that a necessary and sufficient condition for the stress increment to depend *only* on the superimposed infinitesimal strain $\tilde{\boldsymbol{E}}$ is that $\tilde{\boldsymbol{R}}\boldsymbol{T}_0 = \boldsymbol{T}_0\tilde{\boldsymbol{R}}$. Let $\boldsymbol{n}_1$ and $\boldsymbol{n}_2$ be orthogonal proper vectors of $\boldsymbol{T}_0$. Multiplying the preceding equation on the left by $\boldsymbol{n}_1$ and on the right by $\boldsymbol{n}_2$ yields the condition $t_1\boldsymbol{n}_1\cdot(\tilde{\boldsymbol{R}}\boldsymbol{n}_2) = t_2\boldsymbol{n}_1\cdot(\tilde{\boldsymbol{R}}\boldsymbol{n}_2)$, whence we derive one or both of the conditions $t_1 = t_2$ and $\tilde{R}\langle 12\rangle = 0$. Hence we may replace the previous italicized statement by the following, more explicit, ***theorem on superimposed elastic strain***[2]: *The response of a strained elastic material to an arbitrary further infinitesimal deformation is linearly elastic if and only if the given initial strain is maintained by a state of hydrostatic stress. If two but not three principal initial stresses are equal, the response of the material is linearly elastic for those further infinitesimal deformations, and only for those, in which the axis of infinitesimal rotation, if any, is perpendicular to the plane of the two corresponding principal directions of stress. If all three principal initial stresses are distinct, the response of the material is linearly elastic in further pure infinitesimal deformations, and only in them.*

But this is not all. Even if the body is homogeneous, the strained state which we employed as the reference configuration need not be homogeneous, and the tensors A_0, C_0, and L_0 in (68.3), (68.15), and (68.16) may vary with the place $\boldsymbol{x}$. This dependence on $\boldsymbol{x}$ can be made explicit by introducing another *homogeneous* reference configuration $\varkappa_0$, for example a global natural state of $\mathscr{B}$.

[1] This qualification is explained as follows. To derive (68.18) we have assumed that $\boldsymbol{H}$ is symmetric. This restriction, however, is sufficient, not necessary. While (68.18) does not hold *in general* if $\boldsymbol{H}$ fails to be symmetric, it does so in the special case when $\boldsymbol{T}_0 = \boldsymbol{0}$, as the reader will easily verify. Thus the GCN$^+$ condition suffices for the usual uniqueness theorem of the infinitesimal theory. This same conclusion follows also from a number of arguments given in Sect. 52; e.g., GCN$^+ \Rightarrow$ C $\Rightarrow$ S-E in the infinitesimal theory, and S-E $\Rightarrow$ the existence and uniqueness theorem given earlier in this section.

[2] While this theorem is rather obvious, the nearest to a statement of it we have seen in the literature is that of ZORSKI [1964, *98*, § 1].

* CAUCHY in developing his molecular theory [1828, *2*, eq. (30) (31)] first obtained (68.16). The reader will note that for (68.16) to hold the stress need not be equilibrated.

Let the strained reference configuration $\varkappa$ be related to $\varkappa_0$ by the deformation of $\boldsymbol{x}=\chi_0(\boldsymbol{X})$. The response function $\mathfrak{t}$ of (43.10) depends, of course, on the choice of the reference configuration.* Denoting the function corresponding to $\varkappa$ by $\mathfrak{t}$ and the one corresponding to $\varkappa_0$ by $\mathfrak{t}_0$, we easily see (cf. Eq. (43.3)) that the two are related by

$$\mathfrak{t}(\boldsymbol{C}) = J_0^{-1}\boldsymbol{F}_0\,\mathfrak{t}_0(\boldsymbol{F}_0^T\boldsymbol{C}\,\boldsymbol{F}_0)\,\boldsymbol{F}_0^T, \qquad J_0=|\det \boldsymbol{F}_0|, \tag{68.19}$$

where $\boldsymbol{F}_0=\nabla\chi_0(\boldsymbol{X})$. Taking the derivative of (68.19) at $\boldsymbol{C}=\boldsymbol{1}$ in the direction of the symmetric tensor $\tilde{\boldsymbol{E}}$ and using (68.13), we obtain (cf. Sect. 9)

$$\mathsf{C}_0[\tilde{\boldsymbol{E}}] = 2\,J_0^{-1}\boldsymbol{F}_0\left.\frac{\partial\,\mathfrak{t}_0}{\partial\boldsymbol{C}}\right|_{\boldsymbol{C}_0}[\boldsymbol{F}_0^T\tilde{\boldsymbol{E}}\boldsymbol{F}_0]\,\boldsymbol{F}_0^T, \tag{68.20}$$

where $\boldsymbol{C}_0=\boldsymbol{F}_0^T\boldsymbol{F}_0$. This equation shows that C_0 is identical with the elasticity tensor C defined by (45.2) when it is evaluated for the response function $\mathfrak{t}_0$ and the deformation gradient $\boldsymbol{F}_0$. Comparing $(68.14)_2$ with (45.6), we see that an analogous statement is true for the tensor B_0; i.e. we have

$$\mathsf{C}_0=\mathsf{C}(\boldsymbol{F}_0), \qquad \mathsf{B}_0=\mathsf{B}(\boldsymbol{F}_0), \tag{68.21}$$

when the tensors C and B on the right-hand sides are evaluated with respect to $\varkappa_0$. The tensor $\mathsf{A}_0=\mathsf{B}_0$, however, differs in general from the value of A at $\boldsymbol{F}_0$ when $\varkappa_0$ is taken as the reference configuration (cf. (45.4)). Since $\varkappa_0$ is a homogeneous configuration, the response function $\mathfrak{t}_0$ is the same for all particles. The elasticity tensors C_0 and B_0, however, can depend on $\boldsymbol{X}$ and hence $\boldsymbol{x}$ through $\boldsymbol{F}_0$. Thus the preceding theorem may be supplemented as follows: *The response of a severely deformed homogeneous elastic material to a further pure infinitesimal deformation, while linearly elastic, is not generally that of a homogeneous material according to the infinitesimal theory.* This rather obvious statement is a rational substitute for the common loose claim that an elastic material "loses" its homogeneity when severely strained. In the special case when the first deformation is homogeneous, so that $\boldsymbol{F}_0$ is constant, no dependence on $\boldsymbol{x}$ is introduced by it. In this case, as we have seen, the further deformation of a homogeneous material is governed by (68.10), an equation with constant coefficients, and the coefficients occurring in other equations such as (68.15) are constants when Cartesian co-ordinates are used. Thus it is only when the first deformation is homogeneous that a simple boundary-value problem can be expected to result from linearizing the general equations in the neighborhood of a given large strain. Such problems will be studied in Sect. 70.

However, those who insist upon talking about "apparent elasticities" of a strained material may do so by aid of the device of writing the stress relation (50.10) of the infinitesimal theory in the form

$$T^{km}-T_0^{km}=\mathsf{L}^{kmpq}u_{p,q}, \tag{68.22}$$

where $\boldsymbol{T}_0=\boldsymbol{0}$. One may reconcile (68.22) formally with (68.15) by saying that the *apparent elasticities* $\mathsf{L}_{\text{app}}^{kmpq}$ are given by the expression

$$\mathsf{L}_{\text{app}}^{kmpq}=-T_0^{km}g^{pq}+T_0^{kq}g^{mp}+T_0^{mq}g^{kp}+\mathsf{C}_0^{kmpq}. \tag{68.23}$$

As noted in the statement of the theorem on superimposed strain, the one case when the concept of "apparent elasticity" is a natural one is that in which the initial stress is a hydrostatic pressure, $\boldsymbol{T}_0=-p_0\boldsymbol{1}$. When, furthermore, the body is homogeneous and isotropic and the configuration corresponding to $\boldsymbol{F}_0$ is undistorted, $\mathsf{C}_0[\tilde{\boldsymbol{E}}]=\lambda_0(\operatorname{tr}\tilde{\boldsymbol{E}})\boldsymbol{1}+2\mu_0\tilde{\boldsymbol{E}}$, and (68.15) yields

$$\left.\begin{aligned}\boldsymbol{T}&=(1-\operatorname{tr}\tilde{\boldsymbol{E}})(-p_0\boldsymbol{1})-2p_0\tilde{\boldsymbol{E}}+\lambda_0(\operatorname{tr}\tilde{\boldsymbol{E}})\boldsymbol{1}+2\mu_0\tilde{\boldsymbol{E}},\\&=-p_0\boldsymbol{1}+(\lambda_0+p_0)(\operatorname{tr}\tilde{\boldsymbol{E}})\boldsymbol{1}+2(\mu_0-p_0)\tilde{\boldsymbol{E}},\end{aligned}\right\} \tag{68.24}$$

* [The subscript zero on this page is used not only as on the preceding pages but also to denote quantities related to $\varkappa_0$. The kernel letters to which there zeroes are attached differ.]

or, in the notation suggested by (68.23),

$$\lambda_{\rm app}=\lambda_0+p_0, \qquad \mu_{\rm app}=\mu_0-p_0. \tag{68.25}$$

This classical result[1] is sometimes applied in seismology. Formulae for $\mu_{\rm app}$ and $\varkappa_{\rm app}$ in the second-order theory have been given as Eqs. (66.43) and (66.47); of course $\lambda_{\rm app}=\varkappa_{\rm app}-\frac{2}{3}\mu_{\rm app}$. Explicit formulae for $\mu_{\rm app}$ and $\lambda_{\rm app}$ in general are given below as Eqs. (69.9).

The second-order theory of elasticity may be derived by taking the initially stressed state as corresponding to an infinitesimal strain from the natural state. Then, approximately,

$$\mathsf{C}_0^{kmpq}=\mathsf{L}^{kmpq}+\mathsf{M}^{kmpqrs}u_{r,s}, \tag{68.26}$$

where

$$\mathsf{M}^{kmpqrs}=\left.\frac{\partial \mathsf{C}^{kmpq}}{\partial u_{r,s}}\right|_{u_{v,w}=0}. \tag{68.27}$$

The tensor M^{kmpqrs} may thus be regarded, formally, as expressing the change of the linear elasticities L^{kmpq} when quadratic terms in the displacement gradient are taken into account[2]. To relate these results with those previously given, we may use (45.2), carry out the differentiations, and simplify the results by use of Eqs. (17.8) and (29.4) of CFT, so obtaining the formula

$$\mathsf{M}^{kmpqrs}=\mathsf{L}^{kmpq}g^{rs}+\mathsf{L}^{smpq}g^{kr}+\mathsf{L}^{kspq}g^{mr}+\mathsf{L}^{kmsq}g^{pr}+\mathsf{L}^{kmps}g^{qr}+\mathsf{L}^{kmpqrs}, \tag{68.28}$$

where L^{kmpqrs} is the tensor of second-order elasticities defined by (63.10). Substituting (68.28), (68.26), and (68.23) into (68.22), where $\boldsymbol{T}_0$ is given by the classical infinitesimal theory, yields a result of the same form as (63.21).

68bis. Infinitesimal strain superimposed upon a given strain. II. Infinitesimal stability[3]. (Mainly added in proof.) There is a vast literature on elastic stability, and to little purpose. Most of it rests upon improper, or at best unduly special, formulations of the principles of elasticity. Whole volumes have been devoted to presenting ostensible solutions to particular problems by means of criteria that are never even clearly stated, and the morass of equations spouted forth on the subject can be regarded as little else than rhetoric. Stability theory is, necessarily, an application of some theory of finite deformation, such as elasticity, but most of the specialists in stability analysis show no evidence of having troubled to learn the theory they claim to be applying. Here or in Sect. 89 we present everything solid we have been able to find concerning this unhappy subject.

A static deformation of a body subject to boundary conditions of place and traction is said to be *infinitesimally stable*[4] if the work done in every further infinitesimal deformation compatible with the boundary conditions is not less than that required to effect the same infinitesimal deformation subject to "dead loading", i.e., at the same state of stress as in the ground state of strain. That is,

$$\int_{\mathscr{B}} \operatorname{tr}[(\boldsymbol{T}_{\rm R}-\boldsymbol{T}_0)\boldsymbol{H}^T]\,dv \geqq 0, \tag{68b.1}$$

---

[1] It was attributed by L. BRILLOUIN [1938, *2*, Ch. X, XIII] [1940, *3*, § 6] to POINCARÉ [1892, *1*]; a different proof was given by TRUESDELL [1952, *20*, § 43].

[2] We generalize the analysis of TOUPIN and RIVLIN [1960, *56*, § 5].

[3] An exposition of some aspects of the subject has been given by GREEN and ADKINS [1960, *26*, Ch. 9].

[4] The criterion (68b.1), (68b.2) was derived by HADAMARD [1903, *4*, ¶ 269] from a more general and rigorous definition of finite stability, which we shall give in Sect. 89. Despite an abundant literature, including whole treatises, on elastic stability for engineers, the first writers to rediscover correctly this aspect of HADAMARD's work seem to be PEARSON [1956, *21*, Eq. (10)] and HILL [1957, *9*, Eq. (15)], to whom Eq. (68b.1) is sometimes attributed, e.g., by GREEN and ADKINS [1960, *26*, §§ 9.2—9.3]. Cf. also CHELAM [1960, *16*]. HADAMARD considered only the boundary conditions of place; all the authors just cited considered only hyperelastic materials, but the existence of a stored-energy function does not seem to simplify the results in any way.

where the integration is extended over the body in the (possibly strained) ground state, and where gradients $\boldsymbol{H}$ of all superimposed infinitesimal displacements $\boldsymbol{u}$ compatible with the boundary conditions of place and traction are to be considered. If we drop the subscript 0 and write A for the elasticity (68.4) of the material in the strained reference configuration, by substituting (68.3) into (68b.1) we obtain as the condition for infinitesimal stability[1]

$$\int_{\mathscr{B}} \operatorname{tr}\{\mathsf{A}[\boldsymbol{H}]\boldsymbol{H}^T\}\, dv = \int_{\mathscr{B}} \mathsf{A}_k{}^\alpha{}_m{}^\beta H^k{}_\alpha H^m{}_\beta\, dv \geqq 0, \tag{68b.2}$$

where, again, *the field $\boldsymbol{H}$ is the gradient of an infinitesimal displacement compatible with the boundary conditions.* The following **basic stability theorem** was found by HADAMARD[2]: *In order that a configuration of an elastic body be infinitesimally stable for boundary conditions of place and traction, it is necessary that the inequality*

$$\mathsf{A}_k{}^{(\alpha}{}_m{}^{\beta)} \lambda^k \lambda^m \mu_\alpha \mu_\beta \geqq 0 \tag{68b.3}$$

hold for all vectors $\boldsymbol{\lambda}$ and $\boldsymbol{\mu}$ at each particle. It is understood that A is evaluated at the position of the particle in $\boldsymbol{\chi}$. We shall refer to (68b.3) as *Hadamard's inequality.*

To prove HADAMARD's theorem, assume that $\boldsymbol{\chi}$ is an infinitesimally stable configuration. Then the inequality (68b.2) holds, in particular, for all $\boldsymbol{H}$ that vanish on the boundary $\partial\mathscr{B}$. Let $\boldsymbol{\lambda}$ be any constant vector, and consider infinitesimal displacements of the form $\boldsymbol{u} = \varphi\boldsymbol{\lambda}$, where φ is a scalar field that vanishes on $\partial\mathscr{B}$. With this choice of, (68b.2) gives

$$\left.\begin{aligned} \int_{\mathscr{B}} \mathsf{A}_k{}^\alpha{}_m{}^\beta \lambda^k \lambda^m \varphi_{,\alpha}\varphi_{,\beta}\, dm &= \int_{\mathscr{B}} S^{\alpha\beta}\varphi_{,\alpha}\varphi_{,\beta}\, dm \\ &= \int_{\mathscr{B}} \nabla\varphi\cdot\boldsymbol{S}\nabla\varphi\, dm \geqq 0, \end{aligned}\right\} \tag{68b.4}$$

where the symmetric tensor $\boldsymbol{S}$ is defined by

$$S^{\alpha\beta} = \mathsf{A}_k{}^{(\alpha}{}_m{}^{\beta)} \lambda^k \lambda^m. \tag{68b.5}$$

Let $\boldsymbol{X}_0$ be an arbitrary point in the interior of the body $\mathscr{B}$, and let $\boldsymbol{S}_0$ denote the value of $\boldsymbol{S}$ at $\boldsymbol{X}_0$. Since $\mathsf{A}_k{}^\alpha{}_m{}^\beta$, and hence $\boldsymbol{S}$, is assumed to depend continuously on $\boldsymbol{X}$, if $\varepsilon > 0$ is given we have

$$\int_{\mathscr{B}_0} \nabla\varphi\cdot\boldsymbol{S}\nabla\varphi\, dm \leqq \int_{\mathscr{B}_0} [\nabla\varphi\cdot\boldsymbol{S}_0\nabla\varphi + \varepsilon(\nabla\varphi)^2]\, dm, \tag{68b.6}$$

provided $\mathscr{B}_0$ be a sufficiently small neighborhood of $\boldsymbol{X}_0$. We now employ a Cartesian co-ordinate system X^α whose origin is $\boldsymbol{X}_0$ and whose axes point in the directions of three orthogonal proper vectors of $\boldsymbol{S}_0$. We then have

$$\int_{\mathscr{B}_0} [\nabla\varphi\cdot\boldsymbol{S}_0\nabla\varphi + \varepsilon(\nabla\varphi)^2]\, dm = \sum_{\alpha=1}^{3} (s_\alpha + \varepsilon) \int_{\mathscr{B}_0} \left(\frac{\partial\varphi}{\partial X^\alpha}\right)^2 dm, \tag{68b.7}$$

where s_1, s_2, s_3 are the proper numbers of $\boldsymbol{S}_0$.

---

[1] Since $\boldsymbol{H}$ is the gradient of an infinitesimal displacement, we may, if we wish, confuse co-ordinates in the resulting deformation with those in the strained reference configuration:

$$\mathsf{A}_k{}^\alpha{}_m{}^\beta H^k{}_\alpha H^m{}_\beta = \mathsf{A}_k{}^p{}_m{}^q u^k{}_{,p} u^m{}_{,q}.$$

[2] HADAMARD [1903, *4*, ¶ 270]. HADAMARD's sketchy proof was criticized by DUHEM [1905, *1*, Ch. III, § IV]. DUHEM did not succeed in giving a rigorous proof, but he reduced it to a "point litigieux" which he characterized as follows: "Cette supposition que l'intuition fait apparaître come vraisemblable, mais qu'il est malaisé de justifier par un raisonnement rigoureux, permet seule de poursuivre la démonstration de M. HADAMARD." A rigorous proof was given by CATTANEO [1946, *1*, pp. 732—734]. The proof in the text, not previously published, is due to NOLL.

We now consider a rectangular parallelepiped $\mathscr{B}_0$ whose center is at $\boldsymbol{X}_0$ and whose edges are parallel to the co-ordinate axes and have lengths $2d^\alpha$, $\alpha=1, 2, 3$. We denote the volume of $\mathscr{B}_0$ by $v_0 = 8d^1 d^2 d^3$. The region $\mathscr{B}_0$ is the union of three subregions $\mathscr{P}_\alpha$, $\alpha=1, 2, 3$, defined by the condition that $\boldsymbol{X}\in\mathscr{P}_\alpha$ if and only if $\left|\frac{X^\beta}{d^\beta}\right| \leqq \left|\frac{X^\alpha}{d^\alpha}\right| \leqq 1$ for all $\beta=1, 2, 3$. Each $\mathscr{P}_\alpha$ consists of two opposite pyramids which touch at $\boldsymbol{X}_0$. The $\mathscr{P}_\alpha$ all have the same volume, $\frac{1}{3}v_0$. We define a function φ, non-zero only in $\mathscr{B}_0$, by

$$\varphi = \left\{\begin{array}{ll} 1-\left|\dfrac{X^\alpha}{d^\alpha}\right| & \text{if} \quad \boldsymbol{X}\in\mathscr{P}_\alpha \\ 0 & \text{if} \quad \boldsymbol{X}\notin\mathscr{B}_0 \end{array}\right\}. \tag{68b.8}$$

This function is continuous; it has a piecewise continuous gradient, and

$$\left(\frac{\partial\varphi}{\partial X^\alpha}\right)^2 = \left\{\begin{array}{ll} (d^\alpha)^{-2} & \text{if} \quad \boldsymbol{X}\in\mathscr{P}_\alpha \\ 0 & \text{if} \quad \boldsymbol{X}\notin\mathscr{P}_\alpha \end{array}\right\}. \tag{68b.9}$$

The inequality (68b.6) applies to the function φ given by (68b.8) and to the rectangular parallelepiped $\mathscr{B}_0$ if the d^α are sufficiently small. Since φ vanishes outside $\mathscr{B}_0$, we may replace $\mathscr{B}_0$ by $\mathscr{B}$ on the left-hand side of (68b.6). Using (68b.7) and (68b.9), we thus obtain

$$\int_{\mathscr{B}} \nabla\varphi\cdot\boldsymbol{S}\nabla\varphi\, dm \leqq \tfrac{1}{3}\varrho_R v_0 \sum_{\alpha=1}^{3} (s_\alpha+\varepsilon)(d^\alpha)^{-2}. \tag{68b.10}$$

Combining (68b.4) and (68b.10), we find that $\sum_{\alpha=1}^{3}(s_\alpha+\varepsilon)(d^\alpha)^{-2}$ must be non-negative for all $\varepsilon>0$ and all sufficiently small d^α. It is clear that this can be the case only when $s_\alpha \geqq 0$, $\alpha=1, 2, 3$. Therefore, $\boldsymbol{S}_0$ must be positive semi-definite. Since $\boldsymbol{X}_0$ was chosen arbitrarily, it follows that the tensor $\boldsymbol{S}$ must be positive semi-definite everywhere in $\mathscr{B}$. It is clear from (68b.5) that this condition is equivalent to (68b.3), which proves HADAMARD's stability theorem. The proof has made use of infinitesimal displacements whose gradients are only piecewise continuous. A slight modification of (68b.8) would show that the proof can be carried through even if only infinitesimal displacements with continuous gradients are admitted.

While stability is a condition on the deformed body as a whole, HADAMARD's inequality is a local necessary condition, offering a convenient and easy check for stability. If it is not satisfied by the elasticity A in the ground state, that state cannot be infinitesimally stable for any of the boundary conditions. At the end of this section will be found statements of what may be inferred if HADAMARD's inequality holds throughout a region.

Comparison with (44.21) shows that HADAMARD's inequality (68b.3) is only slightly weaker than the S-E condition. Many of the consequences of the S-E condition (cf. Sects. 45, 48, 52, and 71) follow also from (68b.3) provided that in the conclusions strict in equalities "$>$" are everywhere replaced by non-strict inequalities "$\geqq$". On the other hand, (68b.3) does not seem to be strong enough to ensure some of the most important results that follow from the S-E condition, such as the existence of solution to the place boundary-value problem under certain further assumptions.

A uniqueness theorem may be proved directly from a strengthened requirement of stability. In effect, we have only to notice that the classical argument of KIRCHHOFF for the infinitesimal theory rests on an integral inequality:

$$\int_{\mathscr{B}} \operatorname{tr}[(\boldsymbol{T}_R-\boldsymbol{T}_0)\boldsymbol{H}^T]\,dv = \int_{\mathscr{B}} \operatorname{tr}\{\mathsf{A}[\boldsymbol{H}]\boldsymbol{H}^T\}\,dv = \int_{\mathscr{B}} \mathsf{A}_{k\ m}^{\ \alpha\ \beta} H^k{}_\alpha H^m{}_\beta\, dv > 0, \tag{68b.11}$$

unless, perhaps, $\boldsymbol{H}$ is an infinitesimal rotation. We shall call a configuration in which the inequality (68 b.11) holds *infinitesimally superstable*[1] for the boundary conditions being considered. Suppose that we have two solutions to a boundary-value problem of place and traction. The difference of the two displacement fields then yields a solution pair $(\boldsymbol{T}_R, \boldsymbol{H})$ satisfying null boundary data. By adjusting the argument leading to (52.7), or by suitably changing notations in (52.7) itself, we infer that

$$\int_{\mathscr{B}} \operatorname{tr}[(\boldsymbol{T}_R - \boldsymbol{T}_0)\boldsymbol{H}^T]\, dv = 0. \tag{68b.12}$$

This contradicts (68 b.11) unless $\boldsymbol{H}$ is a pure rotation. Schematically,

$$\text{superstability} \Rightarrow \text{uniqueness}, \tag{68b.13}$$

where the infinitesimal displacements allowed in the condition of stability respect the boundary conditions for which uniqueness is to hold. Specifically, we have the following ***uniqueness theorem***[2] of KIRCHHOFF's type: *If a configuration of an elastic body is infinitesimally superstable for mixed boundary conditions of place and traction, the mixed boundary-value problem of place and traction for superimposed infinitesimal strain has at most one solution, to within an arbitrary infinitesimal rotation.* In many cases, as in the infinitesimal theory, the boundary conditions themselves remove the ambiguity just mentioned.

The converse of this theorem is false: uniqueness does not imply stability. Indeed, to secure uniqueness by the same argument it suffices to require that "$>$" be replaced by "$<$" in (68 b.11). Examples given by ERICKSEN and TOUPIN[3] show that *stability*, rather than superstability, *does not suffice* as premise for the uniqueness theorem.

The foregoing theorem constitutes the only known basis for the "method of adjacent equilibria" or "bifurcation", in which the critical load is defined as that for which uniqueness ceases to hold[4]. By (68 b.13), the configuration at the bifurcation strain cannot be infinitesimally superstable. It may, however, be infinitesimally stable. Moreover, failure of stability does not necessarily imply bifurcation. Thus the method of adjacent equilibria is insecure.

In a similar way, the classic argument for uniqueness of the problem of elastic vibrations may be extended to an arbitrary stable ground state. In view of the

---

[1] Stability of this kind is to be distinguished from "ordinary" stability, for which (68 b.11) must hold for *all* $\boldsymbol{H}$ other than $\boldsymbol{0}$, while if the integral may vanish for some non-vanishing $\boldsymbol{H}$ the stability is said to be "neutral." The concept of ordinary stability does not seem to be useful in elasticity.

[2] For the boundary-value problem of place in a hyperelastic body, the theorem was obtained by ERICKSEN and TOUPIN [1956, *8*, Th. 1]. The more general result is due, in effect, to HILL [1957, *9*, § 3]. BEATTY [1965, *1*, § 2.1 A] has shown that the result still holds if stress waves are allowed.

Notice that the variations allowed in the criterion of superstability respect the boundary conditions for which uniqueness is asserted to follow. An example given by ERICKSEN and TOUPIN [1956, *8*, Th. 2] shows that if (68 b.11) holds for $\boldsymbol{H}$ appropriate to the place boundary-value problem, uniqueness of solution to the traction boundary-value problem generally does *not* follow.

[3] ERICKSEN and TOUPIN [1956, *8*, Th. 3].

[4] WILKES [1955, *33*] considered a hollow circular cylinder, of arbitrary isotropic incompressible material, subject to end thrust. GREEN and SPENCER [1959, *11*] considered a solid circular cylinder subject to extension and torsion but got detailed results only for a special case of the Mooney-Rivlin material (Sect. 95). The results of these authors have been generalized somewhat by WESOŁOWSKI [1963, *82* and *84*] [1964, *95*]. FOSDICK and SHIELD [1963, *24*] analysed the condition of bifurcation for small bending of a solid circular cylinder, subject to finite extension or compression. WESOŁOWSKI [1962, *73* and *74*] set up and studied the problem for homogeneous strain. PEARSON [1956, *21*, § 8] attempted to formulate a general method.

linearity of the differential equations (68.5), it suffices to consider a solution $\boldsymbol{u}$ corresponding to null boundary and initial data. Instead of (68b.11) we find for such a solution the condition

$$\int_{\mathscr{B}} \varrho_0 \dot{u}^2 dv + \int_{\mathscr{B}} \operatorname{tr}[(\boldsymbol{T}_R - \boldsymbol{T}_0)\boldsymbol{H}^T]\, dv = 0, \tag{68b.14}$$

where ϱ_0 is the density in the ground state. Because of the assumption of stability (68b.1), not only the first but also the second integral is non-negative. To satisfy (68b.14), each integral must vanish separately. Hence $\boldsymbol{u}$ = const. in time, and since $\boldsymbol{u}=\boldsymbol{0}$ initially throughout the body, it follows that $\boldsymbol{u}=\boldsymbol{0}$ always. Thus follows the ***uniqueness theorem for free vibrations***[1]: *Corresponding to prescribed initial velocity and infinitesimal displacement from an infinitesimally stable configuration of a body subject to prescribed mixed boundary conditions, there is at most one solution.* Note that only stability, not superstability, is assumed here.*

Consider now an infinitesimal free sinusoidal oscillation about an arbitrarily stressed state. By assumption, $\boldsymbol{b}^* = \boldsymbol{0}$. Therefore, there is an approximate solution of (68.5) of the form $\boldsymbol{u}(\boldsymbol{x}, t) = \boldsymbol{U}(\boldsymbol{x}) \sin \omega t$. Taking the inner product of (68.5) by $\boldsymbol{U}(\boldsymbol{x})$ yields

$$\boldsymbol{U} \cdot \operatorname{div}(\boldsymbol{T}_R - \boldsymbol{T}_0) = -\varrho_0 \omega^2 U^2 \sin \omega t. \tag{68b.15}$$

Integrating over $\mathscr{B}$ and then using the divergence theorem, we obtain

$$-\sin \omega t \int_{\partial\mathscr{B}} \boldsymbol{U} \cdot (\boldsymbol{t}_R - \boldsymbol{t}_0)\, ds + \int_{\mathscr{B}} \operatorname{tr}[(\boldsymbol{T}_R - \boldsymbol{T}_0)\boldsymbol{H}^T]\, dv = \omega^2 \sin^2 \omega t \int_{\mathscr{B}} \varrho_0 U^2\, dv. \tag{68b.16}$$

Under various boundary conditions, the first integral vanishes. When it does, and when the condition of stability (68b.1) or superstability (68b.11) holds, it follows that the expression on the right-hand side of (68b.16) is non-negative, or, in the latter case, positive. Hence we infer the following ***theorem***[2]: *In infinitesimal free sinusoidal motion about a stable configuration of an elastic body, subject to boundary conditions such that*

$$\int_{\partial\mathscr{B}} \boldsymbol{U} \cdot (\boldsymbol{t}_R - \boldsymbol{t}_0)\, ds = 0, \tag{68b.17}$$

every possible frequency of oscillation is real. If the configuration is superstable, no frequency can vanish.

A connection between stability and the speeds of propagation of plane progressive waves will be found in Sect. 71.

Since the boundary conditions do not enter the S-E condition (44.21) or the weaker condition of HADAMARD (68b.3) at all, neither would seem sufficient for stability. Nevertheless, suppose that A is *constant*, as it is for a homogeneous body subject to homogeneous strain from any homogeneous configuration, and suppose it satisfies the S-E condition. Let $\mathscr{B}$ be a bounded region of space, and let its boundary $\partial\mathscr{B}$ consist in a finite number of non-intersecting surfaces, regular in the sense of KELLOGG. According to a theorem of VAN HOVE[3], for the gradient $\boldsymbol{H}$ of any field $\boldsymbol{u}$ that is twice continuously differentiable in $\mathscr{B} \cup \partial\mathscr{B}$ and that vanishes on $\partial\mathscr{B}$, HADAMARD's local inequality (68b.3) is *equivalent* to the condition (68b.2) for infinitesimal stability. Moreover, the S-E condition is *sufficient* that (68b.11) holds, unless, of course, $\boldsymbol{u}$ vanishes identically in $\mathscr{B} \cup \partial\mathscr{B}$. That is, *a homogeneous elastic body in a homogeneously strained configuration is infinitesimally stable for the boundary condition of place if and only if Hadamard's inequality* (68b.3) *holds; the S-E condition is sufficient that the configuration be superstable*[4]. Schematically, for the boundary condition of place superimposed

[1] BEATTY [1964, *4*] [1965, *1*, § 2.3A], generalizing a classic theorem of the infinitesimal theory.

[2] BEATTY [1964, *4*] [1965, *1*, § 2.2], generalizing a classic theorem of the infinitesimal theory.

[3] VAN HOVE [1947, *13*].

[4] HILL [1962, *34*, p. 10, footnote] mentioned that this interpretation of VAN HOVE's theorem is due to TOUPIN. In the infinitesimal theory of isotropic materials, the result reduces to KELVIN's stability theorem.

* [The conclusions obtained in the text following (68b.14) are valid for hyperelastic bodies but not for all elastic bodies.]

upon homogeneous strain of an elastic body *from a homogeneous configuration*,

$$\text{HADAMARD's inequality} \Leftrightarrow \text{stability}, \qquad \text{S-E} \Rightarrow \text{superstability}. \tag{68b.18}$$

The result (68b.18) is particularly interesting since it shows that for the boundary condition of place, *the shape of the body is irrelevant in the question of stability*. On the other hand, for stability for the traction boundary-value problem the shape is quite clearly of paramount importance.

By (68b.13) and BROWDER's theorem, summarized above, it follows that *for a homogeneously strained homogeneous elastic body in which the S-E condition is satisfied, the boundary-value problem of place for superimposed infinitesimal deformation has a solution, and that solution is unique*[1]. The conclusion is subject, of course, to the assumptions of smoothness laid down for the three theorems which are combined to yield it. This rather weak theorem of existence and uniqueness seems to be the most general so far known for the boundary-value problem of place.

For several years attempts have been made to extend this theorem, at least in regard to uniqueness, to a more general ground state, or even to finite strain, subject to appropriate restriction (cf. Sect. 44), but they have failed.

Let us now examine more closely the condition (68b.2) itself. As we have seen in the discussion following (68.18), the GCN$^+$ condition (52.16) suffices that the integrand in (68b.2) be everywhere positive if $\boldsymbol{H}$ is symmetric, i.e., if the superimposed deformation is a pure strain. This fact is not related in any obvious way to the criterion of stability, which requires only that the *mean value* of $\operatorname{tr}[(\boldsymbol{T}_R - \boldsymbol{T}_0)\boldsymbol{H}^T]$ be positive, but requires it for *all* $\boldsymbol{H}$, whether symmetric or not. The GCN$^+$ condition restricts, in effect, only the work corresponding to the pure strain. In the classical infinitesimal theory, this is all the work there is, but non-zero dead loads can do work even in a pure rotation.

The role of the given stress and the superimposed infinitesimal rotation is made clear by using (68.14) to express (68b.2) in the form

$$\int_{\mathscr{B}} [T_0^{mq} H^k{}_q H_{km} + C_0^{kmpq} \tilde{E}_{km} \tilde{E}_{pq}]\, dv \geqq 0, \tag{68b.19}$$

where C_0 is the value of C in the ground state. Alternatively, by use of (45.6) and the relation $\boldsymbol{H} = \tilde{\boldsymbol{E}} + \tilde{\boldsymbol{R}}$, we obtain

$$\int_{\mathscr{B}} T_0^{mq} (\tilde{E}^k{}_q \tilde{R}_{km} + \tilde{R}^k{}_q \tilde{E}_{km} + \tilde{R}^k{}_q \tilde{R}_{km})\, dv + \int_{\mathscr{B}} \mathsf{B}_0^{(km)(pq)} \tilde{E}_{km} \tilde{E}_{pq}\, dv \geqq 0, \tag{68b.20}$$

where B_0 is the value of B in the ground state. If superstability rather than stability is required, the sign "$\geqq$" in (68b.19) and (68b.20) is to be replaced by "$>$", with the restriction to non-vanishing $\tilde{\boldsymbol{E}}$. If $\boldsymbol{T}_0 = \boldsymbol{0}$, the condition of superstability reduces to the standard requirement of the infinitesimal theory: the work done in any infinitesimal pure strain is positive. Thus, as is well known, every solution according to the infinitesimal theory is a stable one. If $\boldsymbol{T}_0 \neq \boldsymbol{0}$, this universal stability is lost, as it should be. From (68b.19) and (68b.20) it is clear that the condition of stability is two-fold: it restricts *both* the stress in the ground state and the form of the stress relation for small deformations from that state. In particular, *no condition on the form of the stress relation can ensure infinitesimal stability for all states of stress*. For example, the GCN$^+$ condition (52.16) suffices to make the second integral in (68b.19) and (68b.20) positive if $\tilde{\boldsymbol{E}} \neq \boldsymbol{0}$ but has no effect on the value of the first integral.

---

[1] For the case of the classical infinitesimal theory ($\boldsymbol{T}_0 = \boldsymbol{0}$), this observation seems to be due to MIŞICU [1953, *17*, § 16]; some unsubstantiated broader claims have been cited in footnote 1, p. 248.

Indeed, consider the case of pure infinitesimal rotation[1]: $\tilde{\boldsymbol{E}}=\boldsymbol{0}$. If $T_0^{mq}\tilde{R}^k{}_q\tilde{R}_{km}<0$, the condition of stability is violated, *no matter what is the form of the elastic response function.* In particular, if the sum of two principal stresses in the ground state is negative, any rotation about the third principal axis of stress makes $T_0^{mq}\tilde{R}^k{}_q\tilde{R}_{km}<0$. For example, a state of uniaxial pressure is *always unstable,* no matter what the pressure is; in classical terms, the critical load is zero. Fig. 5 makes this instability clear. In the particular problem of buckling of a strut, such instability has long been recognized[2] but cast aside as being of no interest.

So as to exclude undesirable instabilities such as the one just noticed, BEATTY[3] has proposed to restrict the class of allowed deformations to those preserving the equilibrium of moments. Only fields $\boldsymbol{H}$ that satisfy this *moment condition* are to be allowed in (68b.1). By assumption, in the ground state we have

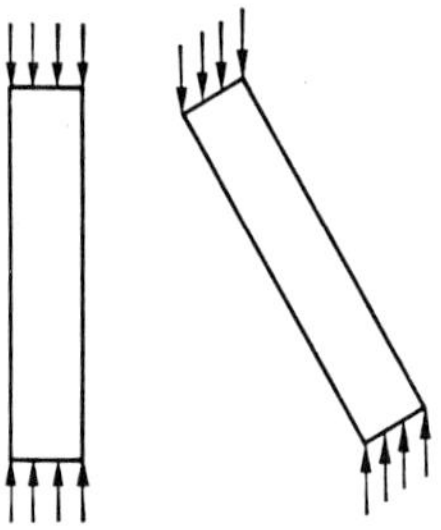

Fig. 5. Rotational instability of a strut.

$$\int_{\partial\mathscr{B}} \boldsymbol{p}\times\boldsymbol{t}_0\,ds+\int_{\mathscr{B}}\boldsymbol{p}\times\varrho_0\boldsymbol{b}\,dv=\boldsymbol{0}, \tag{68b.21}$$

where $\boldsymbol{t}_0$ is the traction vector. If $\boldsymbol{t}$ is the traction vector after the superimposed displacement $\boldsymbol{u}$ has occurred, the moment condition requires that

$$\int_{\partial\mathscr{B}}(\boldsymbol{p}+\boldsymbol{u})\times\boldsymbol{t}\,ds+\int_{\mathscr{B}}(\boldsymbol{p}+\boldsymbol{u})\times\varrho_0\boldsymbol{b}\,dv=\boldsymbol{0}. \tag{68b.22}$$

Subtraction of (68b.21) from (68b.22) yields

$$\int_{\partial\mathscr{B}}[(\boldsymbol{p}+\boldsymbol{u})\times(\boldsymbol{t}-\boldsymbol{t}_0)+\boldsymbol{u}\times\boldsymbol{t}_0]\,ds+\int_{\mathscr{B}}\boldsymbol{u}\times\varrho_0\boldsymbol{b}\,dv=\boldsymbol{0}. \tag{68b.23}$$

On the part of the boundary $\partial\mathscr{B}$ where the traction vector is prescribed, the condition of dead loading requires that $\boldsymbol{t}=\boldsymbol{t}_0$. Calling the remaining part $\mathscr{S}_2$, by use of the divergence theorem and (16.6) we find the following form for the moment condition[4]:

$$\int_{\mathscr{S}_2}(\boldsymbol{p}+\boldsymbol{u})\times(\boldsymbol{t}-\boldsymbol{t}_0)\,ds+\int_{\mathscr{B}}(\boldsymbol{H}\boldsymbol{T}_0)_\times\,dv=\boldsymbol{0}, \tag{68b.24}$$

where the subscript $\times$ denotes the axial vector formed from the tensor $\boldsymbol{H}\boldsymbol{T}_0$. The case of main interest is the boundary-value problem of traction. For this case, $\mathscr{S}_2$ vanishes, and the moment condition becomes

$$\int_{\mathscr{B}}\boldsymbol{H}\boldsymbol{T}_0\,dv=\text{a symmetric tensor}. \tag{68b.25}$$

For infinitesimal displacements, this condition has already been analysed in Sect. 64 [cf. Eqs. (64.2) and (64.11)], with a different end in view. The equation to be solved now is (64.5) with $\boldsymbol{r}_1=\boldsymbol{0}$. From the results given in Sect. 64 we infer the following ***theorem on the moment condition***[5]: *If the loads acting on the ground state possess no axis of equilibrium, the moment condition excludes all infinitesimal rotations as competing deformations for the criterion of stability in the traction boundary-value problem; if there is an axis of equilibrium, only rotations*

[1] Cf. also the remarks on monotone transformations in footnote 1, p. 248. In fact, $T_0^{mq}\tilde{R}^k{}_q\tilde{R}_{km}>0$ for all non-vanishing infinitesimal rotations $\tilde{\boldsymbol{R}}$ if and only if $t_1+t_2>0$, $t_2+t_3>0$, $t_3+t_1>0$.

[2] In effect, it was hinted obscurely by D'ALEMBERT (1780) as an objection against EULER's theory of buckling.

[3] BEATTY [1964, *4*] [1965, *1*, § 1.4.1].

[4] BEATTY [1964, *4*] [1965, *1*, § 1.4.1].

[5] BEATTY [1964, *4*] [1965, *1*, § 1.4.2] obtained this result by considering finite rotations $\boldsymbol{H}=\boldsymbol{Q}$ satisfying (68b.25). A rotation through any straight angle is a solution; otherwise, the only solutions are arbitrary rotations about an axis of equilibrium, if one exists.

about that axis are allowed. In many of the most interesting stress boundary-value problems there is an axis of equilibrium. For eversion of a sphere or cylinder, every line is an axis of equilibrium. For a state of uniaxial tension, the direction of the non-vanishing principal stress is an axis of equilibrium; the moment condition excludes rotations about any other axis, and in particular the rotations illustrated in Fig. 5. Thus the undesired trivial instability of the Euler strut is removed by the moment condition. On the basis of this new condition, HOLDEN[1] has obtained the first estimate of the critical load of a strut in compression according to a correct three-dimensional theory.

In the foregoing paragraphs we have attempted to present in detail all that is known regarding infinitesimal stability in the general theory of elasticity, insofar as the concept of stored-energy function is not necessary. A few further results are given later in connection with finite static stability of hyperelastic materials (Sect. 89). The reader will see that these static criteria fail of the mark. The concept of stability most immediate to understanding and nearest to application is dynamic: A configuration is *kinetically stable* if the motion resulting from imparting any sufficiently small initial velocities and displacements to it, while holding the loads sufficiently near to assigned values, results in a displacement sufficiently small for a time. Usually also certain derivatives and other subsidiary quantities are required to be small. According to a classic theorem of DIRICHLET in analytical dynamics, static stability of a conservative system having a finite number of degrees of freedom implies kinetic stability. No such result holds, in general, for systems of infinitely many degrees of freedom. As shown above, the proper modes of infinitesimal free oscillation about a configuration infinitesimally stable in the static sense remain small if they are small initially, but a general disturbance is an infinite sum of such modes, and counter-examples show that not all such sums can be kept small at all places and times by reducing their initial values. Thus *the criterion of real proper frequencies is merely necessary, not sufficient, for infinitesimal kinetic stability*[2].

GUO ZHONG-HENG and URBANOWSKI[3] have attempted to formulate a general kinetic method.

It would be naive to assume, however, that the kinetic concept of stability is necessarily superior to static ones. To discuss kinetic stability, obviously, the theorist must commit himself in regard to the behavior of the material in motion. In particular, the attempts so far presume the material to be elastic when in motion, i.e., to be devoid of internal friction, and such an idealization may be quite inappropriate for some real materials. In statics, all simple materials behave like elastic materials (Sects. 39, 40), so a static criterion has, potentially, wider applicability. It is conceivable that a configuration which is unstable for a particular

[1] HOLDEN [1964, *44*]. The earlier attempt by PEARSON [1956, *21*, § 4] must be discarded as faulty, since PEARSON did not impose any condition on the competing deformations. As we have seen above, the correct value of the critical load according to PEARSON's definition of stability is zero.

[2] HELLINGER [1914, *1*, § 7d] referred to unpublished work of HAAR showing that for systems of more than one dimension, the assertion of DIRICHLET's theorem becomes false, although there is a counterpart for certain one-dimensional bodies. Correcting an erroneous statement by GREEN and ADKINS [1960, *26*, § 9.1], SHIELD and GREEN [1963, *65*] exhibited a family of solutions in the infinitesimal theory for isotropic bodies (where, of course, all conditions for static stability are satisfied, and all squared wave speeds are positive) in which the motion does not grow small everywhere when the initial displacements, velocities, and their derivatives, are reduced arbitrarily. Cf. also ZIEGLER's illustrations of different kinds of stability in the theory of the elastica [1956, *37*, §§ 1—12].

[3] GUO ZHONG-HENG and URBANOWSKI [1963, *35*]. In earlier papers GUO ZHONG-HENG [1962, *27* and *31*] had considered the problem of infinitesimal motion of an isotropic incompressible cylinder subject to inflation and extension. GUO and URBANOWSKI found that the Mooney-Rivlin material (Sect. 95) does not show typical behavior. As far as stability is concerned, the forms of the response coefficients are decisive.

elastic material according to a static criterion could be kinetically stable for materials having some kinds of internal friction but unstable for others.

69. Infinitesimal strain superimposed upon a given strain, III. General theory for isotropic materials. We assume now that the body is homogeneous and isotropic. Then an undistorted homogeneous configuration $\varkappa_0$, usually a natural state, exists, and, with respect to $\varkappa_0$, the response of the body is given by any of the stress relations (47.6) to (47.9). The elasticities C_0 or L_0 occurring in (68.15) or (68.16) can be expressed in terms of the response coefficients occurring in any of (47.6) to (47.9). Throughout the analysis, $\boldsymbol{B} \equiv \boldsymbol{B}_0 = \boldsymbol{F}_0 \boldsymbol{F}_0^T$, the left Cauchy-Green tensor of the deformation from $\varkappa_0$ to the strained reference configuration $\varkappa$ upon which the infinitesimal deformation is to be superimposed; I, II, III stand for the principal invariants of $\boldsymbol{B}$. We shall not use $\boldsymbol{B}$ in any other sense in this section.

Combining (68.14) with (48.24), we find that

$$C_0^{kmpq} = T_0^{km} g^{pq} - T_0^{kq} g^{mp} - T_0^{mq} g^{kp} + 2 \mathsf{F}_0^{kmps} B_s^q, \tag{69.1}$$

where F_0 is the gradient tensor (48.13), evaluated at $\boldsymbol{B}$. Substitution of (69.1) into (68.17) gives

$$\mathsf{L}_0[\tilde{\boldsymbol{E}}] = \mathsf{F}_0[\boldsymbol{B}\tilde{\boldsymbol{E}} + \tilde{\boldsymbol{E}}\boldsymbol{B}], \tag{69.2}$$

so that the stress relation (68.16) becomes

$$\left.\begin{aligned} \boldsymbol{T} &= \boldsymbol{T}_0 + \tilde{\boldsymbol{R}}\boldsymbol{T}_0 - \boldsymbol{T}_0\tilde{\boldsymbol{R}} + \mathsf{F}_0[\boldsymbol{B}\tilde{\boldsymbol{E}} + \tilde{\boldsymbol{E}}\boldsymbol{B}], \\ T^{km} &= T_0^{km} + T_0^{kp}\tilde{R}^m{}_p + T_0^{mp}\tilde{R}^k{}_p + 2\mathsf{F}_0^{km}{}_{pq} B^{pr} \tilde{E}^q_r. \end{aligned}\right\} \tag{69.3}$$

The formula (48.13) gives the tensor F_0 explicitly in terms of the response coefficients $\aleph_\Gamma$, $\Gamma = 0, 1, 2,$ of (47.8). Often more useful for special problems is a formula in terms of the response coefficients $\beth_\Gamma$, $\Gamma = 0, +1, -1,$ of (47.9). Differentiation of (47.9) in the direction of a symmetric tensor $\boldsymbol{S}$ can easily be carried out with the help of the formulae of Sect. 9. We find

$$\left.\begin{aligned} &\frac{\partial \mathfrak{f}(\boldsymbol{B})}{\partial \boldsymbol{B}}[\boldsymbol{S}] = \mathsf{F}_0[\boldsymbol{S}] = \beth_1 \boldsymbol{S} - \beth_{-1}\boldsymbol{B}^{-1}\boldsymbol{S}\boldsymbol{B}^{-1} + \\ &+ \sum_{\Gamma=-1}^{+1} \left\{ \frac{\partial \beth_\Gamma}{\partial I} \operatorname{tr} \boldsymbol{S} + \frac{\partial \beth_\Gamma}{\partial II} [I \operatorname{tr} \boldsymbol{S} - \operatorname{tr}(\boldsymbol{B}\boldsymbol{S})] + \frac{\partial \beth_\Gamma}{\partial III} III \operatorname{tr}(\boldsymbol{B}^{-1}\boldsymbol{S}) \right\} \boldsymbol{B}^\Gamma. \end{aligned}\right\} \tag{69.4}$$

Putting $\boldsymbol{S} = \boldsymbol{B}\tilde{\boldsymbol{E}} + \tilde{\boldsymbol{E}}\boldsymbol{B}$ in (69.4), using the Hamilton-Cayley equation $\boldsymbol{B}^2 = I\boldsymbol{B} - II\boldsymbol{1} + III\boldsymbol{B}^{-1}$, and substituting the result into (69.3) yields an explicit formula for the stress[1]:

$$\left.\begin{aligned} &\boldsymbol{T} = \boldsymbol{T}_0 + \tilde{\boldsymbol{R}}\boldsymbol{T}_0 - \boldsymbol{T}_0\tilde{\boldsymbol{R}} + \beth_1(\boldsymbol{B}\tilde{\boldsymbol{E}} + \tilde{\boldsymbol{E}}\boldsymbol{B}) - \beth_{-1}(\boldsymbol{B}^{-1}\tilde{\boldsymbol{E}} + \tilde{\boldsymbol{E}}\boldsymbol{B}^{-1}) + \\ &+ 2\sum_{\Gamma=-1}^{+1} \left\{ \left(II \frac{\partial \beth_\Gamma}{\partial II} + III \frac{\partial \beth_\Gamma}{\partial III} \right) \operatorname{tr} \tilde{\boldsymbol{E}} + \frac{\partial \beth_\Gamma}{\partial I} \operatorname{tr}(\boldsymbol{B}\tilde{\boldsymbol{E}}) - III \frac{\partial \beth_\Gamma}{\partial II} \operatorname{tr}(\boldsymbol{B}^{-1}\tilde{\boldsymbol{E}}) \right\} \boldsymbol{B}^\Gamma. \end{aligned}\right\} \tag{69.5}$$

[1] Using different methods, GREEN, RIVLIN and SHIELD [1952, *11*, § 3], URBANOWSKI [1961, *66*, § 3], GUO ZHONG-HENG and URBANOWSKI [1963, *36*] and ZORSKI [1964, *98*, § 1] obtained corresponding results in different notations, for hyperelastic materials. (The analysis of BERG [1958, *5*] rests on a conjectured special form for the stress relation.) Corresponding results for a curvilinearly orthotropic material were obtained by URBANOWSKI [1959, *38*, § 3].

Since $\tilde{\boldsymbol{R}}=\boldsymbol{H}-\tilde{\boldsymbol{E}}=\tilde{\boldsymbol{E}}-\boldsymbol{H}^T$ and $\boldsymbol{T}_0=\beth_0\boldsymbol{1}+\beth_1\boldsymbol{B}+\beth_{-1}\boldsymbol{B}^{-1}$, the following formula is equivalent to (69.5):

$$\left.\begin{aligned}
\boldsymbol{T}&=\boldsymbol{T}_0+\boldsymbol{H}\boldsymbol{T}_0+\boldsymbol{T}_0\boldsymbol{H}^T-2[\beth_0\tilde{\boldsymbol{E}}+\beth_{-1}(\tilde{\boldsymbol{E}}\boldsymbol{B}^{-1}+\boldsymbol{B}^{-1}\tilde{\boldsymbol{E}})]+\\
&+2\sum_{\Gamma=-1}^{+1}\left\{\left(II\frac{\partial\beth_\Gamma}{\partial II}+III\frac{\partial\beth_\Gamma}{\partial III}\right)\operatorname{tr}\tilde{\boldsymbol{E}}+\frac{\partial\beth_\Gamma}{\partial I}\operatorname{tr}(\boldsymbol{B}\tilde{\boldsymbol{E}})-III\frac{\partial\beth_\Gamma}{\partial II}\operatorname{tr}(\boldsymbol{B}^{-1}\tilde{\boldsymbol{E}})\right\}\boldsymbol{B}^\Gamma,\\
T^{km}&=T_0^{km}+T_0^{kp}u^m{}_{,p}+T_0^{mp}u^k{}_{,p}-2\beth_{-1}[(B^{-1})^{kp}\tilde{E}_p^m+(B^{-1})^{mp}\tilde{E}_p^k]-2\beth_0\tilde{E}^{km}+\\
&+2B^{pq}\tilde{E}_{pq}\left[\frac{\partial\beth_0}{\partial I}g^{km}+\frac{\partial\beth_1}{\partial I}B^{km}+\frac{\partial\beth_{-1}}{\partial I}(B^{-1})^{km}\right]+\\
&+2\tilde{E}_q^q\left[\left(II\frac{\partial\beth_0}{\partial II}+III\frac{\partial\beth_0}{\partial III}\right)g^{km}+\left(II\frac{\partial\beth_1}{\partial II}+III\frac{\partial\beth_1}{\partial III}\right)B^{km}+\right.\\
&\left.+\left(II\frac{\partial\beth_{-1}}{\partial II}+III\frac{\partial\beth_{-1}}{\partial III}\right)(B^{-1})^{km}\right]-\\
&-2III(B^{-1})^{pq}\tilde{E}_{pq}\left[\frac{\partial\beth_0}{\partial II}g^{km}+\frac{\partial\beth_1}{\partial II}B^{km}+\frac{\partial\beth_{-1}}{\partial II}(B^{-1})^{km}\right].
\end{aligned}\right\}\quad(69.6)$$

Alternatively, one may derive this result from the general equation (68.15) by substituting in it the special form (47.20) for the elasticity C of an isotropic material.

The counterpart of (69.6) for *incompressible materials* is easily obtained on the basis of (49.5). Observing that $\operatorname{tr}\tilde{\boldsymbol{E}}=0$ for isochoric infinitesimal deformations and that (47.9) formally reduces to (49.5) if $\boldsymbol{T}$ is replaced by $\boldsymbol{T}_{\mathrm{E}}=\boldsymbol{T}+p\boldsymbol{1}$, III by 1, and $\beth_0$ and $\frac{\partial\beth_\Gamma}{\partial III}$ by 0, we can read off the result from (69.6):

$$\left.\begin{aligned}
\boldsymbol{T}&=-p\boldsymbol{1}+\boldsymbol{T}_{\mathrm{E}0}+\boldsymbol{H}\boldsymbol{T}_{\mathrm{E}0}+\boldsymbol{T}_{\mathrm{E}0}\boldsymbol{H}^T-2\beth_{-1}(\tilde{\boldsymbol{E}}\boldsymbol{B}^{-1}+\boldsymbol{B}^{-1}\tilde{\boldsymbol{E}})+\\
&+2\left\{\left[\frac{\partial\beth_1}{\partial I}\boldsymbol{B}+\frac{\partial\beth_{-1}}{\partial I}\boldsymbol{B}^{-1}\right]\operatorname{tr}(\boldsymbol{B}\boldsymbol{E})-\left[\frac{\partial\beth_1}{\partial II}\boldsymbol{B}+\frac{\partial\beth_{-1}}{\partial II}\boldsymbol{B}^{-1}\right]\operatorname{tr}(\boldsymbol{B}^{-1}\boldsymbol{E})\right\},\\
T^{km}&=-p\,g^{km}+T_{\mathrm{E}0}^{km}+T_{\mathrm{E}0}^{kp}u^m{}_{,p}+T_{\mathrm{E}0}^{mp}u^k{}_{,p}-2\beth_{-1}[(B^{-1})^{kp}\tilde{E}_p^m+\\
&+(B^{-1})^{mp}\tilde{E}_p^k]+2B^{pq}\tilde{E}_{pq}\left[\frac{\partial\beth_1}{\partial I}B^{km}+\frac{\partial\beth_{-1}}{\partial I}(B^{-1})^{km}\right]-\\
&-2(B^{-1})^{pq}\tilde{E}_{pq}\left[\frac{\partial\beth_1}{\partial II}B^{km}+\frac{\partial\beth_{-1}}{\partial II}(B^{-1})^{km}\right],
\end{aligned}\right\}\quad(69.7)$$

where $\boldsymbol{T}_{\mathrm{E}0}=\beth_1\boldsymbol{B}+\beth_{-1}\boldsymbol{B}^{-1}$, and where, as in (69.6), $\boldsymbol{B}=\boldsymbol{B}_0$, the left Cauchy-Green tensor of the given deformation, upon which the infinitesimal deformation is superimposed.

When $\boldsymbol{B}_0=\boldsymbol{1}$, the various forms just derived for the stress relation reduce to their counterparts in the classical infinitesimal theory, use being made of (50.13) and (50.14)

From (69.6) and (69.7) we see that *the response of a severely deformed isotropic elastic material to further infinitesimal deformation is not that of an isotropic elastic material according to the classical infinitesimal theory.* This simple comment provides a concrete and specific replacement for the common vague claim that an "initially" isotropic elastic material "loses" its isotropy if severely deformed[1].

[1] Cf. the corresponding remarks on homogeneity in the previous section. According to the definitions used in the present work, homogeneity and isotropy are properties determined *once and for all* from the constitutive equation of a material and cannot be lost or created unless the constitutive equation itself is changed. This is to be contrasted with the usage in the majority of works, even recent ones, on continuum mechanics, where "homogeneous" and "isotropic" are used in an undefined, intuitive, physical sense which may change from case to case. In the literature, the point seems first to have been raised in a footnote of Rivlin and Ericksen [1955, *23*, § 11].

For the simplest application of the theory, consider a compressible isotropic material subject to hydrostatic pressure p_0, so that the initial state is obtained by a uniform dilatation from the natural state: $\boldsymbol{B}_0 = v_0^2 \mathbf{1}$, $\boldsymbol{B}_0^{-1} = v_0^{-2} \mathbf{1}$, while p_0 is related to v_0 through (54.5), or, if we prefer, through an equivalent relation of the form $p_0 = \tilde{p}(\varrho_0)$, since $\varrho_0 v_0^3 = \varrho_R$. From (69.6) we easily show that

$$\boldsymbol{T} + p_0 \mathbf{1} = \lambda_{\text{app}} (\operatorname{tr} \tilde{\boldsymbol{E}}) \mathbf{1} + 2\mu_{\text{app}} \tilde{\boldsymbol{E}}, \tag{69.8}$$

where

$$\left.\begin{aligned} \mu_{\text{app}} &= v_0^2 \beth_1 (3 v_0^2, 3 v_0^4, v_0^6) - \frac{1}{v_0^2} \beth_{-1} (3 v_0^2, 3 v_0^4, v_0^6), \\ \lambda_{\text{app}} &= \left. \frac{d\tilde{p}}{d \log \varrho} \right|_{\varrho = \varrho_0} - \frac{2}{3} \mu_{\text{app}}. \end{aligned}\right\} \tag{69.9}$$

These formulae render (68.25) more explicit. Indeed, Eqs. (69.9) allow us to derive explicit formulae for the coefficient λ_0 and μ_0 appearing in (68.25).

In the previous section we have seen that in order for the response of an elastic material to arbitrary further infinitesimal strain to be linearly elastic it is *necessary* that the underlying strain be maintained by hydrostatic pressure. From (69.8) we see that in such states, and hence in such states only, the further response of an isotropic elastic material is that given by the infinitesimal theory of isotropic elastic materials. Consequently, only for states of hydrostatic initial stress can one justly speak of apparent moduli for further strains, in the sense of the *general* classical infinitesimal stress relation. Of course, one may extend the sense of modulus to cover any linear coefficient, without its being a strict counterpart of a modulus in the infinitesimal theory. For example, one can consider a simple shear: $u_x = K y$, $u_y = 0$, $u_z = 0$. Then from (69.3) we see that

$$T\langle x y\rangle - T_0\langle x y\rangle = K\left[\tfrac{1}{2}(T_0\langle y y\rangle - T_0\langle x x\rangle) + \sum_{q=x,y,z} (\mathsf{F}\langle x y x q\rangle B\langle y q\rangle + \mathsf{F}\langle x y y q\rangle B\langle x q\rangle)\right]. \tag{69.10}$$

If we regard as the apparent shear modulus the ratio of the increment of shear stress to the corresponding amount of shear *in a simple shear*, then that modulus is given by the quantity in brackets in (69.10). In the infinitesimal theory, however, the shear modulus is not merely the above special ratio: It is the ratio of shear stress to the corresponding amount of shear in *any* infinitesimal deformation. For a state of initial stress other than a hydrostatic one, no such modulus exists. Results of this kind may be seen more explicitly and generally from the formulae of the next section.

There is literature on apparent moduli of various kinds[1].

SPENCER[2] has observed that the general theory developed here may be given a different interpretation if slightly generalized. Suppose that a deformation leading to a particular deformation gradient $\boldsymbol{F}_0$ is possible for a material having a particular response function $\mathfrak{h}$. As mentioned in Sect. 54, this same deformation is not a possible one, in general, subject to the same body force, for other elastic materials. We may consider, however, a material having the response function $\mathfrak{h} + \mathfrak{h}'$, where the magnitude of $\mathfrak{h}'$, for the deformation considered, is small in comparison with the magnitude of $\mathfrak{h}$, and we may expect that a small change in the deformation will render it compatible for the slightly perturbed material. The formal apparatus is just the same as above, except that the initial stress $\boldsymbol{T}_0$ must be calculated from $\mathfrak{h} + \mathfrak{h}'$ rather than from $\mathfrak{h}$. For example, for isotropic materials the response coefficients are $\beth_0 + \beth_0'$, $\beth_1 + \beth_1'$, $\beth_{-1} + \beth_{-1}'$; the formula (69.6) remains valid, except that $\boldsymbol{T}_0$ must be replaced by

$$\boldsymbol{T}_0 + \beth_0' \mathbf{1} + \beth_1' \boldsymbol{B}_0 + \beth_{-1}' \boldsymbol{B}_0^{-1}, \tag{69.11}$$

where $\boldsymbol{T}_0$ is calculated from $\boldsymbol{B}_0$ and where $\beth_\Gamma' = \beth_\Gamma'(\boldsymbol{B}_0)$. As an example, SPENCER has worked out the details for the deformation mentioned in footnote 1, p. 200 in the case when the response coefficients $\beth_\Gamma$ are those for the Mooney-Rivlin material, defined in Sect. 95, below.

Following a suggestion of OLDROYD[3], SPENCER[4] has constructed, for isotropic hyperelastic materials, a perturbation procedure in which the change of volume and the superimposed

[1] SHIMAZU [1954, *21*, Chap. 3], BHAGAVANTAM and CHELAM [1960, *10* and *11*], CHELAM [1960, *15*].

[2] SPENCER [1959, *35*].

[3] OLDROYD [1950, *12*, § 6].

[4] SPENCER [1964, *80*]. He applied the results to a study of the simultaneous extension, inflation, and torsion of a cylindrical tube.

infinitesimal displacement are both taken as of the same order of smallness. In this way a solution for incompressible materials may be made the starting point and a small change of volume calculated incidentally in finding the deformation corresponding to the perturbed response coefficients. Various schemes of this sort will result from various choices of expansion parameter. It would seem simpler to us to proceed as follows: (1) Given a particular solution for an incompressible elastic material, calculate the additional body and surface forces required to maintain it in a compressible material having elastic response of the desired type; (2) from the general theory given above, find the additional small deformation resulting when the negatives of these additional forces are applied. If the elastic response is rather insensitive to changes of volume, the resulting combined deformation will be nearly isochoric and nearly the one arising from application of the given forces to the slightly compressible material.

70. Infinitesimal strain superimposed upon a given strain. IV. Solutions of special problems. As remarked in the last section, the theory given there can be expected to lead to relatively simple differential equations of equilibrium or motion only in case the given large deformation is homogeneous, so that the left Cauchy-Green tensor $\boldsymbol{B}_0$ is independent of $\boldsymbol{x}$. It is then natural to choose Cartesian co-ordinates x, y, z with axes in the direction of the proper vectors of $\boldsymbol{B}_0$, thus making the matrices $[\boldsymbol{B}_0]$ and $[\boldsymbol{B}_0^{-1}]$ of $\boldsymbol{B}_0$ and $\boldsymbol{B}_0^{-1}$ diagonal:

$$[\boldsymbol{B}_0]=\operatorname{diag}(v_x^2,\, v_y^2,\, v_z^2), \qquad [\boldsymbol{B}_0^{-1}]=\operatorname{diag}(v_x^{-2},\, v_y^{-2},\, v_z^{-2}), \tag{70.1}$$

where v_x, v_y, v_z are the principal stretches of the initial large deformation. By straightforward calculation, from (69.6) we may show that the normal and shear components of the stress tensor have the forms

$$T\langle ll\rangle = T_0\langle ll\rangle + \sum_{m=1}^{3} a_{lm} u_{m,m}, \qquad T\langle lm\rangle = b_{lm} u_{l,m} + b_{ml} u_{m,l}, \qquad (l \neq m), \tag{70.2}$$

where the constant coefficients a_{lm} and b_{lm} are expressed in terms of the response functions as follows. Set

$$\left.\begin{aligned} c_k &\equiv \frac{\partial \beth_0}{\partial I} + v_k^2 \frac{\partial \beth_1}{\partial I} + v_k^{-2} \frac{\partial \beth_{-1}}{\partial I}, \\ d_k &= II \frac{\partial \beth_0}{\partial II} + III \frac{\partial \beth_0}{\partial III} + v_k^2 \left(II \frac{\partial \beth_1}{\partial II} + III \frac{\partial \beth_1}{\partial III}\right) + \\ &\quad + v_k^{-2}\left(II \frac{\partial \beth_{-1}}{\partial II} + III \frac{\partial \beth_{-1}}{\partial III}\right), \\ e_k &= -\, III \left(\frac{\partial \beth_0}{\partial II} + v_k^2 \frac{\partial \beth_1}{\partial II} + v_k^{-2} \frac{\partial \beth_{-1}}{\partial II}\right), \end{aligned}\right\} \tag{70.3}$$

where of course $I = v_x^2 + v_y^2 + v_z^2$, $II = v_x^2 v_y^2 + v_y^2 v_z^2 + v_z^2 v_x^2$, $III = v_x^2 v_y^2 v_z^2$. Then

$$\tfrac{1}{2} a_{lm} = (v_l^2 \beth_1 - v_l^{-2} \beth_{-1})\,\delta_{lm} + v_m^2 c_l + d_l + v_m^{-2} e_l, \qquad b_{lm} = v_m^2 \beth_1 - v_l^{-2} \beth_{-1}. \tag{70.4}$$

While the coefficients in (70.2) depend upon the material and the given homogeneous deformation in a somewhat complicated way, the form (70.2) itself shows us that some properties of the classical infinitesimal theory of elasticity are carried over to the present case. First, the normal stresses depend upon the normal strains only, but through the medium of 9 rather than 2 coefficients. Second, the shear stresses are independent of all the normal strains and of the shears in other planes; each depends, however, upon the local rotation in its own plane as well as upon the shear, and the coefficients giving one shear stress are different from those for another. Moreover, the 6 coefficients determining the shear stresses are independent, in general, of the 9 that determine the normal stresses. The resulting theory, resting upon 15 constant coefficients instead of the classical 2, might be thought of as representing a kind of quasi-elastic anisotropic response. Writing u_x, u_y, u_z for the components of the displacement vector $\boldsymbol{u}$, from (70.2) and (16.6) we obtain the following differential equations of motion:

$$\left.\begin{aligned}\left(a_{xx}\frac{\partial^2}{\partial x^2}+b_{xy}\frac{\partial^2}{\partial y^2}+b_{xz}\frac{\partial^2}{\partial z^2}\right)u_x+(a_{xy}+b_{yx})\frac{\partial^2 u_y}{\partial x\,\partial y}+\\ +(a_{xz}+b_{zx})\frac{\partial^2 u_z}{\partial x\,\partial z}+\varrho_0 b_x=\varrho_0\ddot{u}_x,\end{aligned}\right\}\tag{70.5}$$

and two others which may be gotten from it by cyclic permutation of the letters x, y, z.

Since these equations are so similar in form to those occurring in the classical infinitesimal theories of anisotropic elastic media, it is to be expected that the method of displacement potentials should be applicable. As yet, however, little has been done in this direction.

GREEN, RIVLIN, and SHIELD[1] have specialized the above formulae to the case when $v_x=v_y$ and have explored the properties of certain special given solutions in terms of a single displacement-potential. In particular, they have applied the results to problems of small indentation of the surface of a semi-infinite incompressible material by a rigid punch. They also discussed the small bending and small plane stress of a stretched plate.

For incompressible materials, WOO and SHIELD[2] have considered in some generality the partial analogy to infinitesimally deformed anisotropic materials. Since $v_x^2=1/v_z$ when $v_x=v_y$, the equations are more tractable in that case; WOO and SHIELD established the general solution in terms of potential functions and expressed the total traction on a surface in terms of line integrals around its boundary curve. They obtained explicit solutions for two problems: (i) a concentrated force points along a principal axis of extension in an infinite medium, (ii) in a semi-infinite medium with a plane boundary normal to the z axis, a concentrated force either normal or parallel to the plane acts at one point of the plane.

By far the most interesting case of the foregoing results is that which represents the small torsion of a prism which has been subjected to a large extension[3] from a natural state $\boldsymbol{\varkappa}_0$. The extension is described in Cartesian co-ordinates by the equations

$$x=\alpha v X,\qquad y=\alpha v Y,\qquad z=vZ,\tag{70.6}$$

in which v and αv are the longitudinal and transverse stretches. The superimposed infinitesimal torsional displacement $\boldsymbol{u}$ is assumed to have the components

$$u_x=-\varepsilon y z,\qquad u_y=\varepsilon x z,\qquad u_z=\varepsilon\Phi(x,y)\tag{70.7}$$

[cf. (66.9) and (66.33)]. From (70.2) and (70.3) we have at once

$$\left.\begin{aligned}&T\langle xx\rangle=T\langle yy\rangle=T_0\langle xx\rangle=T_0\langle yy\rangle=\tilde{\beth}_0+\alpha^2v^2\tilde{\beth}_1+\alpha^{-2}v^{-2}\tilde{\beth}_{-1},\\ &T\langle zz\rangle=T_0\langle zz\rangle=\tilde{\beth}_0+v^2\tilde{\beth}_1+v^{-2}\tilde{\beth}_{-1},\\ &T\langle xy\rangle=0,\\ &T\langle yz\rangle/\varepsilon=(v^2\tilde{\beth}_1-\alpha^{-2}v^{-2}\tilde{\beth}_{-1})(\alpha^2\Phi_{,y}+x),\\ &T\langle zx\rangle/\varepsilon=(v^2\tilde{\beth}_1-\alpha^{-2}v^{-2}\tilde{\beth}_{-1})(\alpha^2\Phi_{,x}-y),\end{aligned}\right\}\tag{70.8}$$

where $\tilde{\beth}_\Gamma\equiv\beth_\Gamma(v^2(1+2\alpha^2),\ \alpha^2v^4(2+\alpha^2),\ \alpha^4v^6))$, as in $(54.2)_3$. The condition that the mantle of the prism be free of normal tension again yields (54.3); on the assumption that this condition can be satisfied, from $(70.8)_6$ we obtain the same

[1] GREEN, RIVLIN and SHIELD [1952, *11*, §§ 5—9], for hyperelastic materials.

[2] WOO and SHIELD [1962, *76*], for hyperelastic materials.

[3] GREEN and SHIELD [1951, *2*, § 3]. Cf. also the treatment by GREEN and ZERNA [1954, *11*, § 4.3]. That a stored-energy function, the existence of which had been presumed by these authors, is unnecessary for the results was observed by RIVLIN [1956, *23*, § 4]. A generalization to the case of a rhombic crystal has been given by GREEN [1962, *23*, § 8].

formula for the tensile stress as that which holds when there is no torsion, namely,

$$T = T\langle zz\rangle = (1-\alpha^2)\,(v^2\,\tilde{\beth}_1 - \alpha^{-2}v^{-2}\,\tilde{\beth}_{-1}). \tag{70.9}$$

The conditions of equilibrium (70.5) then reduce to the single equation

$$\Phi_{,xx} + \Phi_{,yy} = 0. \tag{70.10}$$

The theory given here is somewhat more general but not essentially more difficult than that given in Example 2 of Sect. 66, where the longitudinal stretch was assumed to be of the second order of smallness, but also terms of the second order in ε were retained.

For incompressible materials, all the analysis goes through in just the same way, except of course that $\alpha = v^{-\frac{3}{2}}$ and $\tilde{\beth}_\Gamma \equiv \beth_\Gamma(v^2 + 2/v,\, 2v + 1/v^2)$, as in (55.2).

Let the boundary $\partial\mathscr{A}$ of the cross-section $\mathscr{A}$ of the prism be given by

$$F(x, y) = F(\alpha v X, \alpha v Y) = 0. \tag{70.11}$$

The components n_x, n_y, n_z of the outer unit normal $\boldsymbol{n}$ to $\partial\mathscr{B}$, after the displacement (70.7) has been carried out, are given by

$$\left.\begin{aligned} &G n_x = F_{,x} - \varepsilon z F_{,y}, \quad G n_y = F_{,y} + \varepsilon z F_{,x}, \quad G n_z = \varepsilon(y F_{,x} - x F_{,y}),\\ &G^2 = F_{,x}^2 + F_{,y}^2, \end{aligned}\right\} \tag{70.12}$$

to the first order in ε. Application of $(70.8)_{8,9}$ to (16.5) thus yields, again to the first order in ε, the condition

$$T\langle xz\rangle F_{,x} + T\langle yz\rangle F_{,y} + T\langle zz\rangle (y F_{,x} - x F_{,y}) = 0, \tag{70.13}$$

or, equivalently,

$$(\Phi_{,x} - y)\,F_{,x} + (\Phi_{,y} + x)\,F_{,y} = 0 \quad \text{when} \quad F(x, y) = 0, \tag{70.14}$$

as the condition that the mantle of the prism be free of traction. Since $F_{,x}$ and $F_{,y}$ are the components of a vector normal to the boundary $\partial\mathscr{B}$ in the strained reference state $\boldsymbol{\varkappa}$, comparison of (70.10), (70.14) with (66.34), (66.35) shows that Φ must be the classical warping function for the cross-section $\mathscr{A}$ in the strained reference state $\boldsymbol{\varkappa}$. Since the deformation (70.6) effects only a uniform dilatation of the cross-section and hence leaves directions in the cross-section unchanged, it is easily seen that Φ can be obtained from the warping function $\varphi = \varphi(X, Y)$ for the natural state $\boldsymbol{\varkappa}_0$ by

$$\Phi(x, y) = \alpha^2 v^2 \varphi\left(\frac{x}{\alpha v}, \frac{y}{\alpha v}\right). \tag{70.15}$$

Thus the entire theory of small twist for a severely extended prism may be read off from the classical infinitesimal theory for the same prism unextended.

The system of forces and moments acting upon the initially plane ends $Z = \text{const.}$ is easily obtained. As in the classical infinitesimal theory, if the origin is chosen at the centroid, these reduce to a tensile force N and a twisting couple M. The magnitudes of N and M are calculated as follows from $(70.8)_{8,9}$ and (70.9):

$$\left.\begin{aligned} N &= \iint T\langle zz\rangle\, dx\, dy = T A = (1-\alpha^2)\,(v^2\,\tilde{\beth}_1 - \alpha^{-2}v^{-2}\,\tilde{\beth}_{-1})\,\alpha^2 v^2 A_0,\\ M &= \iint (x\,T\langle yz\rangle - y\,T\langle xz\rangle)\, dx\, dy = \varepsilon\alpha^2 (v^2\,\tilde{\beth}_1 - \alpha^{-2}v^{-2}\,\tilde{\beth}_{-1})\, S + \varepsilon T I, \end{aligned}\right\} \tag{70.16}$$

where A and I are the area and the polar moment of inertia of the cross-section $\mathscr{A}$ in the strained reference state, where A_0 is the area of the cross-section $\mathscr{A}_0$ in the natural state, and where

$$S = \iint_{\mathscr{A}} (x^2 + y^2 + x\Phi_{,y} - y\Phi_{,x})\, dx\, dy. \tag{70.17}$$

From this formula and (70.15) we see that

$$\left.\begin{aligned} S &= \alpha^4 v^4 \iint_{\mathscr{A}_0} (X^2 + Y^2 + X\varphi_{,Y} - Y\varphi_{,X})\, dX\, dY \\ &= \alpha^4 v^4 S_0 \end{aligned}\right\} \tag{70.18}$$

where S_0 is the classical torsional rigidity for the undeformed prism. Since $I = \alpha^4 v^4 I_0$, where I_0 is the polar moment of inertia of the undeformed cross-section $\mathscr{A}_0$, substitution of (70.18) into (70.16) yields GREEN and SHIELD's solution of the classical problem[1] of finding the ***torsional rigidity*** $\tau(v)$ ***of a prism subject to large extension***:

$$\tau(v) \equiv M/\varepsilon = \alpha^4 v^4 (v^2 \tilde{\beth}_1 - \alpha^{-2} v^{-2} \tilde{\beth}_{-1}) [I_0 - \alpha^2 (I_0 - S_0)]. \tag{70.19}$$

For a solid or hollow circular cylinder, $I_0 = S_0$. For other prisms, $S_0 < I_0$; thus (70.19) shows that the twisting couple vanishes, at least to within terms of third order in ε, when

$$\frac{1}{\alpha^2} = \frac{I_0 - S_0}{I_0} < 1. \tag{70.20}$$

The number α as a function of v is to be determined by solving the equation $T = T\langle zz\rangle = 0$. Physical experience suggests that a compressed cylinder thickens; that is, $\alpha > 1$ when $v < 1$. Thus (70.20) indicates that sufficient compression may have the effect of weakening a prism so much that a very slight couple will suffice to twist it.

Dividing $(70.16)_3$ by (70.19) yields GREEN and SHIELD's elegant universal relation

$$\frac{N}{\tau(v)} = \frac{1 - \alpha^2}{\alpha^2 v^2} \cdot \frac{A_0}{I_0 - \alpha^2 (I_0 - S_0)}, \tag{70.21}$$

which connects the force required to produce the stretch v with the torsional modulus for a superimposed small twist. Since it is independent of the response functions, it may be used as a general test of the theory of isotropic elasticity; that is, a material for which this relation between directly measurable quantities does not hold cannot be isotropically elastic.

For incompressible materials, since $\alpha = v^{-\frac{1}{2}}$, the relations $(70.16)_3$ and (70.19) through (70.21) assume the forms

$$\left.\begin{aligned} N &= \left(v - \frac{1}{v^2}\right)\left(\tilde{\beth}_1 - \frac{1}{v}\tilde{\beth}_{-1}\right) A_0, \\ \tau(v) &\equiv M/\varepsilon = \left(\tilde{\beth}_1 - \frac{1}{v}\tilde{\beth}_{-1}\right)\left[I_0 - \frac{I_0 - S_0}{v^3}\right], \\ v^3 &= \frac{I_0 - S_0}{I_0}, \\ \frac{N}{\tau(v)} &= \left(v - \frac{1}{v^2}\right) \frac{A_0}{I_0 - (I_0 - S_0)/v^3}. \end{aligned}\right\} \tag{70.22}$$

When the prism is a solid circular cylinder, the first, second, and fourth of these formulae reduce to (57.22) and (57.23). Some of the relations derived here have been tested experimentally and found good[2].

[1] We do not cite the abundant literature in which this problem has been attacked by use of special theories of elasticity or for special cross-sections. Some of the oldest attempts, which are best interpreted by assuming the extension to be small, have been discussed in Sect. 66.

[2] GENT and RIVLIN [1952, *9*], BERGEN, MESSERSMITH and RIVLIN [1960, *7*, § 7].

The results have been generalized to the case when a hydrostatic pressure is superimposed[1]; in particular, the torsional couple is unaffected[2].

A solution for infinitesimal torsional vibrations of a circular rod subject to finite longitudinal extension has been found by GREEN[3]. As was remarked in Sect. 69, the effect of centrifugal forces makes purely torsional oscillations impossible, in the strict theory, unless certain radial pressures are applied.

e) Wave propagation.

71. General theory of acceleration waves[4]. A singular surface of second order with respect to the deformation $\boldsymbol{x}=\boldsymbol{\chi}(\boldsymbol{X}, t)$ is defined as a surface across which the functions χ^k and their first derivatives are continuous, but at least one of the second derivatives $x^k{}_{,\alpha\beta}$ suffers a jump discontinuity[5]. We recall the geometrical and kinematical conditions of compatibility for such singularities[6]:

$$\left.\begin{aligned} [x^k{}_{,\alpha;\beta}] &= a^k x^m{}_{,\alpha} x^p{}_{,\beta} n_m n_p, \\ [\dot{x}^k{}_{,m}] &= -U a^k n_m, \\ [\ddot{x}^k] &= U^2 a^k, \end{aligned}\right\} \tag{71.1}$$

---

[1] GREEN and SHIELD [1951, *2*, § 4].

[2] GENT and RIVLIN [1952, *9*, App.].

[3] GREEN [1961, *20*]. Cf. also ZORSKI [1964, *98*, § 2 (b)].

[4] Regarding *shock waves* and *vortex sheets* in elastic materials, we have been unable to find definitive literature. For the various jump conditions that are valid for such discontinuities in arbitrary media, see CFT, Sects. 189, 205, 241, 258. Studies of shock waves in elastic solids seem to contain little more than repetition of these necessary but insufficient conditions; most work concerns the limiting forms as the shock strength is made small. Some authors also presume that shock waves are either longitudinal or transverse, although there is no reason to expect waves of these kinds to exist in general, as is shown by the complications developed in our text for acceleration waves, which may surely be expected to enjoy simpler properties than do shock waves. In order to get a definite theory of shock propagation in elastic fluids, strong thermodynamic inequalities must be presumed; indeed, few treatments go beyond the case of the perfect gas. The authors writing on shocks in elastic solids do not show themselves aware of the corresponding, and of course more complicated, requirements on the form of the internal-energy function of the solid. In fact, the constitutive equation of the elastic material seems to be used little if at all, although, as our development shows, it is the form of the constitutive equation that determines the nature and speeds of weak waves, and hence, presumably, of strong ones also.

Cf. JOUGUET [1920, *1*—*4*] [1921, *1*, 1st note], DEWEY [1959, *6*], BLAND [1964, *8*]. VERMA [1964, *90*] has considered special kinds of shock waves according to the special theory of ST. VENANT and KIRCHHOFF (cf. Sect. 94).

It may be mentioned also that the "physical" literature on shock waves in solids (e.g. RICE, MCQUEEN and WALSH [1958, *40*]) assumes, with no basis in experiment or theory, that *hydrodynamics* suffices to describe them. As the following pages show (see especially Sect. 75), even for weak compression waves the results following from elasticity are very much different from their counterparts in gas dynamics.

[Note added in proof: Some recent and important work on shock waves in dissipative materials is mentioned in Sect. 98ter.]

[5] The reader is presumed familiar with the general theory of singular surfaces, as explained, for example, in CFT, Chapter C, where precise conditions of regularity are laid down.

[6] Set $a^k \equiv s_0^k$ to obtain $(71.1)_{2,3}$ from Eqs. (190.5) of CFT. From $(183.3)_3$ and (183.4) of CFT we see that

$$\frac{U}{U_N} = \frac{\sqrt{F_{,\alpha} F^{,\alpha}}}{\sqrt{f_{,k} f^{,k}}}.$$

By this result and (190.4) of CFT,

$$s^k = a^k \frac{F_{,\alpha} F^{,\alpha}}{f_{,p} f^{,p}}.$$

Use of this result and of Eqs. (182.8) and $(190.1)_3$ of CFT yields our $(71.1)_1$, if we recall that $[x^k{}_{,\alpha\beta}] = [x^k{}_{,\alpha;\beta}]$.

where the bracket denotes the jump occasioned; where the vector $\boldsymbol{a}$, characterizing the strength of the discontinuity, is called the *amplitude* of the singularity and is assumed different from zero; where $\boldsymbol{n}$ is a unit normal to the singular surface; and where U is the *local speed of propagation*. If $U \neq 0$, the surface propagates and is therefore called a *wave*; since, by $(71.1)_3$, such a wave must carry a jump of the acceleration, it is called an *acceleration wave*. It is customary to identify such waves physically with sound waves. If $\boldsymbol{a}$ is parallel to $\boldsymbol{n}$, the wave is said to be *longitudinal*; if normal, *transverse*. A typical wave is neither.

We now consider an elastic body and assume that we can choose a reference configuration such that the response functions depend continuously on the place $\boldsymbol{X}$ in it. If the body is homogeneous, for example, we may choose the reference configuration to be homogeneous; the response functions will then be independent of $\boldsymbol{X}$. Since the deformation gradient $\boldsymbol{F}$ is continuous across an acceleration wave, it follows then that the stress $\boldsymbol{T}$ is also continuous. That is,

$$[\boldsymbol{T}] = \boldsymbol{0}, \quad \text{and } a\ fortiori \quad [\boldsymbol{T}]\,\boldsymbol{n} = \boldsymbol{0}. \tag{71.2}$$

Therefore, the dynamical condition (205.5) of CFT is automatically satisfied.

The differential equations $(44.5)_1$ must hold on each side of the surface. Subtracting the limit values of these two equations as the surface is approached from the one side or the other, and assuming that the external body force and the derivatives of the response function $\mathfrak{h}(\boldsymbol{F}, \boldsymbol{X})$ with respect to $\boldsymbol{X}$ are continuous across the wave, we obtain from $(71.1)_{1,3}$ a relation of the form

$$(\boldsymbol{Q}(\boldsymbol{n}) - \varrho U^2 \boldsymbol{1})\,\boldsymbol{a} = \boldsymbol{0}, \tag{71.3}$$

where

$$Q_{km}(\boldsymbol{n}) = \frac{\varrho}{\varrho_R} \mathsf{A}_k{}^\alpha{}_m{}^\beta x^p{}_{,\alpha} x^q{}_{,\beta} n_p n_q. \tag{71.4}$$

In terms of the elasticities B and D, defined by (45.4) and (45.9), we may write (71.4) in the forms[1]:

$$Q^{km}(\boldsymbol{n}) = \mathsf{B}^{kpmq} n_p n_q = \mathsf{D}^{kmpq} n_p n_q. \tag{71.5}$$

We shall refer to (71.3) as the *propagation condition*. From $(71.1)_3$ and (71.3) follows a result which may be called the ***Fresnel-Hadamard theorem***[2]: *the amplitude* $\boldsymbol{a}$ *of an acceleration wave travelling in the direction of* $\boldsymbol{n}$ *must be a proper vector of* $\boldsymbol{Q}(\boldsymbol{n})$; *the speed* U *of propagation of the wave must be such that* ϱU^2 *is the corresponding proper number.*

The tensor $\boldsymbol{Q}(\boldsymbol{n})$, which for each value of $\boldsymbol{n}$ is determined by the response function of the material and by the deformation gradient $\boldsymbol{F}$, is the *acoustical tensor*[3] for the direction of $\boldsymbol{n}$ in the actual state of strain; the directions of its proper vectors are the *acoustical axes* for waves travelling in the direction of $\boldsymbol{n}$ through the strained material. The speeds U are determined from the three pairs of roots of the real bi-cubic

$$\left.\begin{aligned} 0 &= \det(\boldsymbol{Q} - \varrho U^2 \boldsymbol{1}), \\ &= -(\varrho U^2)^3 + I_{\boldsymbol{Q}}(\varrho U^2)^2 - II_{\boldsymbol{Q}}\varrho U^2 + III_{\boldsymbol{Q}}. \end{aligned}\right\} \tag{71.6}$$

---

[1] TRUESDELL [1961, *63*, § 6], generalizing slightly results of TOUPIN and BERNSTEIN [1961, *62*, § 3]. The second form is due to THURSTON [1965, *34*].

[2] HADAMARD [1901, *1*, § 7] [1903, *4*, ¶ 267], for hyperelastic materials, generalizing results of FRESNEL, CAUCHY, POISSON, CHRISTOFFEL, and HUGONIOT. The result in the text was given by TRUESDELL [1961, *63*, § 2]. The method used earlier by FINZI [1942, *2*], although applied to arbitrary states of strain, is not sufficiently general, being limited in principle to infinitesimal deformations.

[3] In more special theories the quadric of $\boldsymbol{Q}(\boldsymbol{n})$ is called the *Fresnel quadric* or the *quadric of polarization*, but at the present level of generality $\boldsymbol{Q}(\boldsymbol{n})$, since it is not necessarily symmetric, is not determined by its quadric.

However, if it is known that $\boldsymbol{a}$ is a real proper vector of $\boldsymbol{Q}(\boldsymbol{n})$, the speed of a wave travelling in the direction of $\boldsymbol{n}$ and having amplitude $\boldsymbol{a}$ is gotten more easily from the explicit formula

$$\varrho U^2 = (\boldsymbol{a} \cdot \boldsymbol{Q}(\boldsymbol{n})\,\boldsymbol{a})/a^2 = (Q_{km} a^k a^m)/a^2, \qquad a \equiv \sqrt{\boldsymbol{a}^2}, \tag{71.7}$$

immediate from (71.3). The quantity ϱU^2 is uniquely determined but need not be positive even though $\boldsymbol{a}$ is real; thus U is not always real. The simplest general statement we can infer from (71.7) is: *In a given material, for a wave of given amplitude and direction there is at most one possible squared speed of propagation, determined by the existing deformation from the reference state.* In particular, the speed is unaffected by the shape of the wave, the path over which it has travelled, the external body force, and all other factors — unaffected, that is, except through whatever influence those factors may have had in bringing about the existing deformation.

All these results remain unchanged if $\boldsymbol{n}$ is replaced by $-\boldsymbol{n}$ or if $\boldsymbol{a}$ is replaced by $k\boldsymbol{a}$, where k is any non-zero real number, or if U is replaced by $-U$. That is, *if the propagation condition is satisfied for a wave of amplitude* $\boldsymbol{a}$ *travelling in the direction of* $\boldsymbol{n}$ *at speed* U, *it is satisfied also by any parallel amplitude in the same or the opposite direction with the same or the opposite velocity of propagation.* For simplicity of stating results it is customary to confound all the possible waves having direction $\pm\boldsymbol{n}$ and amplitude $k\boldsymbol{a}$; setting $\boldsymbol{m} \equiv \boldsymbol{a}/a$, one speaks of "the" wave travelling in the direction of $\boldsymbol{n}$ with amplitude $\boldsymbol{m}$ and speed $\pm U$ or $|U|$.

While according to the equations given, neither the acoustical axes nor the speeds are necessarily real, only waves of real amplitude and real speed have physical significance, so that we take care to phrase all our propositions on waves with only the real case intended. Thus, for example, most of the foregoing propositions are put as necessary conditions only, not as sufficient. In the general case, the possible nature of the amplitude and speeds is various. Since (71.6) is a real cubic in U^2, $\boldsymbol{Q}(\boldsymbol{n})$ has at least one real proper number, and the corresponding unit proper vector is real. However, by (71.7), U^2 may be negative. Thus in the generality maintained here, we cannot even assert that at least one wave can propagate in each direction. The best we can do is apply some theorems on the proper numbers of second-order tensors (e.g. CFT, Appendix, Sect. 37) so as to infer the following results, which do not, however, exhaust all the possibilities in this complicated situation.

Case 1: $\boldsymbol{Q}(\boldsymbol{n})$ is symmetric and positive-definite with distinct proper numbers. Then there exists a uniquely determined orthogonal triad of acoustical axes. Corresponding to each of these axes there is a distinct speed.

Case 2: $\boldsymbol{Q}(\boldsymbol{n})$ is symmetric and positive-definite, but two proper numbers coincide. Then one acoustical axis is uniquely determined, and any direction orthogonal to it is also an acoustical axis. All waves corresponding to the latter axes have a common speed, different from that corresponding to the uniquely determined axis aforementioned.

Case 3: $\boldsymbol{Q}(\boldsymbol{n})$ has three coincident positive numbers. Then there is but one speed of propagation, while the amplitude is arbitrary.

Case 4: $\boldsymbol{Q}(\boldsymbol{n})$ is symmetric and has distinct proper numbers, one of which is negative, the other two, positive. Then (since the speeds of propagation corresponding to the negative proper numbers are imaginary) only two amplitudes are possible, and these are orthogonal.

Case 5: $\boldsymbol{Q}(\boldsymbol{n})$ has only one positive proper number. Then only one amplitude is possible.

Case 6: $\boldsymbol{Q}(\boldsymbol{n})$ is singular. Then $U=0$ is one solution of (71.6). Corresponding to this solution, by (71.1), there may exist a non-propagating singular surface. Across such a surface, which permanently divides the material into two parts, it follows from (71.1) that the acceleration and the velocity gradient must be continuous, the only possible discontinuities being those of $x^k{}_{,\alpha\beta}$.

Case 7: $\boldsymbol{Q}(\boldsymbol{n})$ is not symmetric, but at least three distinct waves may propagate. Then the three possible amplitudes cannot be mutually orthogonal.

We repeat that these cases are not exhaustive. In the generality maintained here, the full diversity of positive proper numbers and real proper vectors of a real tensor is possible, and each case has a different physical significance in terms of waves. Moreover, the results are all local, so that in general the nature of $\boldsymbol{Q}(\boldsymbol{n})$ will vary from point to point in a given strained material. Even if one kind of wave is possible at one point, it may fail to be so at neighboring points and thus may not propagate in the large.

We expect that a superimposed rotation, though it may well turn the acoustic axes, cannot affect the speeds of propagation, which must be determinate from the strain alone for given $\boldsymbol{n}$. That this is so may be verified as follows. Setting

$$\left.\begin{aligned} \boldsymbol{S}(\boldsymbol{\nu}) &\equiv \frac{\varrho_{\mathrm{R}}}{\varrho}\boldsymbol{F}^{-1}\boldsymbol{Q}(\boldsymbol{n})\boldsymbol{F}, & \boldsymbol{\nu} &\equiv \boldsymbol{F}^T\boldsymbol{n}, \\ S^{\alpha}{}_{\beta}(\boldsymbol{\nu}) &= \frac{\varrho_{\mathrm{R}}}{\varrho} X^{\alpha}{}_{,k}\, Q^k{}_m(\boldsymbol{n})\, x^m{}_{,\beta}, & \nu_\alpha &= x^k{}_{,\alpha} n_k, \end{aligned}\right\} \tag{71.8}$$

by use of (71.4) and $(45.1)_2$ we see that[1]

$$\left.\begin{aligned} S^{\alpha}{}_{\beta}(\boldsymbol{\nu}) &= X^{\alpha}{}_{,k}\, x^p{}_{,\beta}\, g_{pm}\, \mathsf{A}^{k\gamma m\delta}\nu_\gamma\nu_\delta \\ &= \left(\delta^{\alpha}_{\beta}\mathfrak{t}^{\gamma\delta} + 2X^{\alpha}{}_{,k}\, x^p{}_{,\beta}\, x^k{}_{,\mu}\, x^m{}_{,\lambda}\, g_{pm}\frac{\partial \mathfrak{t}^{\gamma\mu}}{\partial C_{\delta\lambda}}\right)\nu_\gamma\nu_\delta \\ &= \left(\delta^{\alpha}_{\beta}\mathfrak{t}^{\gamma\delta} + 2C_{\beta\lambda}\frac{\partial \mathfrak{t}^{\alpha\gamma}}{\partial C_{\lambda\delta}}\right)\nu_\gamma\nu_\delta; \end{aligned}\right\} \tag{71.9}$$

or

$$\left.\begin{aligned} \boldsymbol{S}(\boldsymbol{\nu}) &= \mathsf{D}[\boldsymbol{\nu}\otimes\boldsymbol{\nu}], \\ S^{\alpha\beta}(\boldsymbol{\nu}) &= \mathsf{D}^{\alpha\beta\gamma\delta}\nu_\gamma\nu_\delta, \end{aligned}\right\} \tag{71.10}$$

where

$$\mathsf{D}^{\alpha\beta\gamma\delta} = g^{\alpha\beta}\mathfrak{t}^{\gamma\delta} + 2C^{\beta}_{\lambda}\frac{\partial \mathfrak{t}^{\alpha\gamma}}{\partial C_{\lambda\delta}}. \tag{71.11}$$

Thus, for a given material and a given reference configuration, D is a function of the right Cauchy-Green tensor $\boldsymbol{C}$ alone and hence remains unchanged if a local rotation is superimposed. From (71.8) we see that the propagation condition (71.3) may be written in the form

$$(\boldsymbol{S}(\boldsymbol{\nu}) - \varrho_{\mathrm{R}} U^2 \boldsymbol{1})\,\boldsymbol{a}_{\mathrm{R}} = \boldsymbol{0}, \tag{71.12}$$

where[2]

$$\boldsymbol{a}_{\mathrm{R}} = \boldsymbol{F}^{-1}\boldsymbol{a}. \tag{71.13}$$

Superimposing a rotation $\bar{\boldsymbol{Q}}$ corresponds to changing $\boldsymbol{F}$ into $\bar{\boldsymbol{Q}}\boldsymbol{F}$ and $\boldsymbol{n}$ into $\bar{\boldsymbol{Q}}\boldsymbol{n}$. It follows from $(71.8)_2$ that the vector $\boldsymbol{\nu}$ and hence, by (71.10), also $\boldsymbol{S}(\boldsymbol{\nu})$, remains unchanged. Therefore *the speeds of propagation may be determined from the response*

---

[1] TRUESDELL [1961, *63*, § 6]. The result on the wave speeds was stated and proved in a different way by TOLOTTI [1943, *5*, § 1, ¶ 1] and by TOUPIN and BERNSTEIN [1961, *62*, § 4].

[2] The corresponding formula of TRUESDELL [1961, *63*, § 6] is wrong.

function $\mathfrak{t}$, *the right Cauchy-Green tensor* $\boldsymbol{C}$, *and the vector* $\boldsymbol{v}$ *through the equations*[1]

$$\det\left(\boldsymbol{S}(\boldsymbol{v})-\varrho_{\mathrm{R}}U^2\boldsymbol{1}\right)=0. \tag{71.14}$$

From (71.13) we see that the acoustical axes are obtained by transforming the principal directions of $\boldsymbol{S}$ like material elements under the deformation.

As we have noted, in the fully general theory it is possible that real speeds of propagation may fail to exist, so that there are no possible waves. Necessary and sufficient conditions for the existence of waves of various kinds are not known. Two special sufficient conditions are as follows. First, either from (71.4) and (44.21) or from (71.5) and (45.11) we see that the S-E condition takes the form

$$\boldsymbol{\lambda}\cdot\boldsymbol{Q}(\boldsymbol{n})\,\boldsymbol{\lambda}=Q_{km}(\boldsymbol{n})\,\lambda^k\lambda^m>0 \tag{71.15}$$

for arbitrary non-vanishing vectors $\boldsymbol{\lambda}$ and $\boldsymbol{n}$. That is, *the S-E condition is equivalent to the assertion that the symmetric part of the acoustical tensor is positive-definite for every direction of propagation.* Comparison with (71.7) shows that $U^2>0$ if $\boldsymbol{a}$ is real. Thus *if the S-E condition holds, then, for any given direction of propagation the squared speeds corresponding to each real acoustical axis are positive.* Since (71.15) restricts only the symmetric part of $\boldsymbol{Q}(\boldsymbol{n})$, it does not guarantee the existence of real proper vectors of $\boldsymbol{Q}(\boldsymbol{n})$, that is, of real acoustical axes. However, as was already remarked, $\boldsymbol{Q}(\boldsymbol{n})$ always has at least one real proper vector, and therefore in particular it follows that *the S-E condition ensures that in any given direction the propagation condition is satisfied by at least one real amplitude with positive speed.* By HADAMARD's stability theorem in Sect. 68bis, in an infinitesimally stable configuration (71.15) holds with "$>$" replaced by "$\geqq$". Thus follows ***Hadamard's fundamental theorem on waves***[2]: *In an infinitesimally stable configuration, for any given direction of propagation the speeds corresponding to each real acoustical axis are real.* Non-propagating singular surfaces are not excluded by the condition of stability, while by the S-E condition they are.

If $\boldsymbol{n}$ is a proper vector of $\boldsymbol{Q}(\boldsymbol{n})$, the corresponding wave is longitudinal. Putting $\boldsymbol{a}=\boldsymbol{n}$ in (71.7) and then using (71.5) yields a formula for the speed U_{nn} of such a wave:

$$\varrho\,U_{\mathrm{nn}}^2=B^{kmpq}n_k n_p n_m n_q. \tag{71.16}$$

By (52.16), the GCN⁺ condition is sufficient that $U_{\mathrm{nn}}^2>0$. Thus *the GCN⁺ condition implies that if one acoustical axis is normal to the wave, longitudinal waves in that direction have a positive squared speed of propagation.* In general, however, $\boldsymbol{n}$ will not be a right proper vector of $\boldsymbol{Q}(\boldsymbol{n})$, and the statement remains vacuous.

Since it rests upon the constitutive equation (43.1), the foregoing theory takes no account of the effects of thermal conditions varying throughout the body in which the waves are propagated. DUHEM[3] has shown that when more general constitutive equations such as $(45.18)_1$ or $(45.18)_2$, allowing for variation of entropy or temperature, are considered, something may be said nevertheless. Clearly the results hold if $\eta=$ const. or if $\theta=$ const., but since the functions $\hat{\mathfrak{h}}$ and $\tilde{\mathfrak{h}}$ are generally different, so are the corresponding acoustical tensors $\hat{\boldsymbol{Q}}(\boldsymbol{n})$ and $\tilde{\boldsymbol{Q}}(\boldsymbol{n})$. Moreover, if we calculate the jump of the equations of motion (45.21) at the singular surface, we see that if $[\operatorname{grad}\eta]=\boldsymbol{0}$, we still get the same propagation condition (71.3) as when $\eta=$ const.,

[1] In general, $\boldsymbol{v}$ is not a unit vector. THURSTON and BRUGGER [1964, *83*, §§ 3–4] call the unit vector in the direction of $\boldsymbol{v}$ "the natural direction" and the speed νU, where ν is the magnitude of $\boldsymbol{v}$, "the natural velocity" of the wave.

[2] The result was obtained by HADAMARD [1901, *1*, § 7] [1903, *4*, ¶ 271] in the somewhat stronger form appropriate to hyperelastic materials (Sect. 90). As remarked by CATTANEO [1946, *1*, § 2] and TRUESDELL [1961, *63*, § 10], the criticism of HADAMARD's work by SIGNORINI [1945, *4*, pp. 154—155] and TOLOTTI [1943, *5*] misses the point by failing to mention the question of stability.

[3] While the results were obtained by DUHEM [1903, *2*] [1906, *1*, Ch. I, § 2], the proofs in the text are those of TRUESDELL [1961, *63*, § 13].

while if $[\operatorname{grad}\theta]=\mathbf{0}$, we still get the same propagation condition as when $\theta=$ const. Following DUHEM, we now formulate conditions such as to ensure physical circumstances leading to these jump conditions.

First, consider the case of isentropic motion[1]: $\dot{\eta}=0$. Assume, moreover, that the singular surface is of at least first order with respect to η, so that $[\eta]=0$. By HADAMARD's kinematical conditions [CFT, Eq. (180.5)] it follows that

$$[\operatorname{grad}\eta]=B\,\mathbf{n},\qquad [\dot{\eta}]=-U B. \tag{71.17}$$

But since $\dot{\eta}=0$, it is necessary that $B=0$ in the case of a wave, and thus $[\operatorname{grad}\eta]=\mathbf{0}$. The conclusion may be phrased as ***Duhem's first theorem***: *the laws of propagation of isentropic acceleration waves are the same as those of homentropic waves*, generalizing a familiar proposition in gas dynamics [cf. CFT, Sect. 297].

Second, consider a material that conducts heat according to FOURIER's law:

$$\mathbf{h}=\mathbf{K}\operatorname{grad}\theta, \tag{71.18}$$

where $\mathbf{K}$ is a positive-definite tensor. Assume that the wave is at least of first order with respect to θ, so that $[\theta]=0$. By MAXWELL's theorem [CFT, Eq. (175.8)],

$$[\operatorname{grad}\theta]=B\,\mathbf{n}, \tag{71.19}$$

but also it is necessary for conservation of energy at the wave front that FOURIER's condition be satisfied, viz [CFT, Eq. (241.8)]:

$$[\mathbf{h}]\cdot\mathbf{n}=0. \tag{71.20}$$

From (71.18), (71.19), and (71.20) it follows that

$$B\,\mathbf{n}\cdot\mathbf{K}\mathbf{n}=0, \tag{71.21}$$

where $\mathbf{K}$ has been assumed continuous. Since $\mathbf{K}$ is positive-definite, $B=0$. Thus we have established ***Duhem's second theorem***[2]: *In a material that conducts heat according to Fourier's law with continuous and positive-definite conductivity, the laws of propagation of waves of second order are the same as those of homothermal waves.*

The two foregoing theorems are essentially thermal in character. Their relevance for elastic materials arises only from the fact, mentioned at the outset, that for the constitutive equations (45.18) the resulting conditions $[\operatorname{grad}\eta]=\mathbf{0}$ and $[\operatorname{grad}\theta]=\mathbf{0}$ are sufficient to obtain the same laws of propagation as hold under the more stringent assumptions that $\eta=$ const. or $\theta=$ const., respectively.

72. Waves of higher order. For a singular surface of third order, the geometrical and kinematical conditions may be written as follows (CFT, Sect. 191):

$$\left.\begin{aligned}[][x^k{}_{,\alpha\beta\gamma}] &= a^k x^m{}_{,\alpha} x^p{}_{,\beta} x^q{}_{,\gamma} n_m n_p n_q,\\ [\dot{x}^k{}_{,mp}] &= -U a^k n_m n_p,\\ [\ddot{x}^k{}_{,m}] &= U^2 a^k n_m,\qquad [\dddot{x}^k]=-U^3 a^k.\end{aligned}\right\} \tag{72.1}$$

Differentiation of (44.5) with respect to x^p yields

$$\left.\begin{aligned}&\left\{\mathsf{A}_k{}^{(\alpha}{}_m{}^{\beta)} x^m{}_{,\alpha;\beta\gamma}+\frac{\partial \mathsf{A}_k{}^{(\alpha}{}_m{}^{\beta)}}{\partial x^q{}_{,\delta}} x^q{}_{,\delta;\gamma} x^m{}_{,\alpha;\beta}+\frac{\partial q_k}{\partial x^q{}_{,\delta}} x^q{}_{,\delta;\gamma}+\right.\\ &\left.+\frac{\partial \mathsf{A}_k{}^{(\alpha}{}_m{}^{\beta)}}{\partial X^\gamma} x^m{}_{,\alpha;\beta}+\frac{\partial q_k}{\partial X^\gamma}\right\} X^\gamma{}_{,p}+(\varrho_R b_k)_{,p}=(\varrho_R \ddot{x}_k)_{,p}.\end{aligned}\right\} \tag{72.2}$$

We assume that the response function $\mathfrak{h}(\mathbf{F},\mathbf{X})$ has continuous second derivatives with respect to $\mathbf{F}$ and to $\mathbf{X}$, that the density ϱ_R in the reference configuration has continuous first derivatives with respect to $\mathbf{X}$, and that the external body force $\varrho_R\mathbf{b}$ has a continuous gradient. Then the only term on the left-hand side of (72.2) that is not continuous across the surface is the first one. Calculating the jump of (72.2) at the surface and substituting $(72.1)_{1,3}$ into the result yields

$$\mathsf{A}_k{}^{(\alpha}{}_m{}^{\beta)} X^\gamma{}_{,p} a^m x^q{}_{,\alpha} x^r{}_{,\beta} x^s{}_{,\gamma} n_q n_r n_s=\varrho_R U^2 a_k n_p, \tag{72.3}$$

[1] For perfect materials this condition results when heat conduction and heat addition are altogether negligible, as follows from (79.9) and (80.8).

[2] The argument derives from FOURIER in a some what different and entwined context; *cf.* CFT, Sect. 296.

which reduces easily to the form (71.3). The same process may be applied to waves of fourth or higher order, with the same result. Thus waves of all orders greater than 1 satisfy a propagation condition of the same form. More precisely, we have the following ***equivalence theorem***[1] for the propagation condition: *Let it be assumed that at the point in question the response function* $\mathfrak{h}(\boldsymbol{F}, \boldsymbol{X})$ *has continuous derivatives of orders* $0, 1, \ldots, q$, *with respect to* $\boldsymbol{F}$ *and* $\boldsymbol{X}$ *and that* ϱ_R *and* $\boldsymbol{b}$ *have continuous derivatives of orders* $0, 1, \ldots, q-1$ *with respect to* $\boldsymbol{X}$. *Then* $\boldsymbol{a}$ *and* U *are a possible amplitude and a corresponding speed of propagation for a singular surface of order* k *if and only if they are likewise possible for a singular surface of order* m, *where* k *and* m *are any pair of integers between* 2 *and* $q+1$.

This theorem makes it superfluous to give special consideration, beyond the various known general kinematical properties, to waves of order greater than 2. When, henceforth, we speak of acceleration waves, we shall expect the reader to recall that all results are immediately applicable to waves of higher order, provided only the response function be sufficiently smooth.

The two theorems of DUHEM given at the end of the preceding section may be extended to waves of higher order, provided appropriate additional continuity of the thermal variables be assumed[2]. Consider, for example, a third-order wave across which the internal energy rate $\dot{\varepsilon}$, the gradient of the conductivity $K^{pq}{}_{,r}$, and the temperature gradient $\theta_{,k}$ are assumed continuous. By the differential form of the energy equation [CFT, Eq. (241.4)] it follows that

$$[\operatorname{div} \boldsymbol{h}] = 0. \tag{72.4}$$

By MAXWELL's theorem [CFT, Eq. (175.8)] there exists a vector $\boldsymbol{k}$ such that

$$[\theta_{,pq}] = k_p n_q = k_q n_p. \tag{72.5}$$

Hence $k_q = n^p k_p n_q$. Setting $B \equiv n^p A_p$ yields

$$[\theta_{,pq}] = B\, n_p n_q. \tag{72.6}$$

From (72.4),

$$K^{pq} [\theta_{,pq}] = 0, \tag{72.7}$$

whence by (72.6) we conclude that in fact $[\theta_{,pq}] = 0$. Thus the propagation condition for the conducting elastic material is again the same as for homothermal waves.

73. General theory of plane infinitesimal progressive waves. A displacement of the form

$$\boldsymbol{u} = \boldsymbol{a} \sin\left[\frac{2\pi}{l}(\boldsymbol{n}\cdot\boldsymbol{p} - Ut)\right] \tag{73.1}$$

where $\boldsymbol{n}$ is a constant vector and $\boldsymbol{p}$ is the position vector, is said to represent a *sinusoidal plane progressive wave* of amplitude $\boldsymbol{a}$ and wave length l, propagating in the direction $\boldsymbol{n}$ at speed U. In general, such displacements are not possible except when the equations of motion are linear partial differential equations with constant coefficients, as is the case in some of the best known classical special theories of materials. From (44.25) we see at once that plane progressive waves cannot generally subsist in an elastic body. There is, however, an important special case in which (44.25) may be replaced, approximately, by a linear system with constant coefficients, namely, when $\boldsymbol{u}$ is an *infinitesimal displacement from a homogeneous configuration*.

[1] TRUESDELL [1961, *63*, § 3]; for isotropic incompressible hyperelastic materials the result is due to ERICKSEN [1953, *6*, § 6]; for compressible materials, it generalizes a familiar proposition in gas dynamics.
[2] TRUESDELL [1961, *63*, § 13].

The equations of motion for such a displacement are (68.10). On the assumption that the external body force is zero[1], they reduce to

$$B_0^{kpmq} u_{m,pq} = \varrho_0 \ddot{u}^k, \tag{73.2}$$

where $(68.14)_1$ has been used. As already remarked in Sect. 68, the tensor B_0, and hence also its components in a Cartesian system, are constant. Substituting (73.1) into (73.2) thus yields the algebraic conditions[2]

$$(B_0^{kpmq} n_p n_q - \varrho_0 U^2 g^{km}) a_m = 0, \tag{73.3}$$

Note that, by (68.21), B_0 may be computed from any reference state, not necessarily the one upon which the infinitesimal displacement (73.1) is superimposed. We often assume that the homogeneous initial strain is obtained from a homogeneous natural state by a deformation with uniform gradient $\boldsymbol{F}_0$. It follows from (71.5) that (73.3) has the same form as does (71.3). *Thus, the laws of propagation of infinitesimal plane waves in a homogeneously deformed body are the same as those of acceleration waves*[3]. More, precisely, at a point in an elastic material where $\boldsymbol{F} = \boldsymbol{F}_0$, *the propagation condition* (71.3) *for an acceleration wave (or a wave of higher order) in the direction* $\boldsymbol{n}$ *is satisfied by the amplitude* $\boldsymbol{a}$ *if and only if a plane infinitesimal sinusoidal wave of amplitude* $\boldsymbol{a}$ *may propagate in the direction* $\boldsymbol{n}$ *through a specimen of the material subjected to a homogeneous strain with gradient* $\boldsymbol{F}_0$; *the absolute speeds of propagation of all these waves are equal.* All the formulae and results of the two previous sections on acceleration waves or waves of higher order may thus be interpreted also in terms of infinitesimal oscillations, and conversely.

While it seems to be believed widely that singular surfaces of order 2 or greater, such as acceleration waves, and infinitesimal oscillations are physically equivalent, such an idea is justified only in the simpler, classical theories. For general elastic materials, the identical behavior established above results only from the special hypotheses made. The laws of propagation of acceleration waves hold for waves of any form, in bodies subject to any kind of strain. The results for infinitesimal oscillations are more special, being restricted to plane disturbances from a homogeneous configuration. From (68.9) it is clear that if the strain is not homogeneous, plane waves cannot generally persist. Another difference between acceleration waves and infinitesimal oscillations lies in the effect of body forces. Only a *continuous* body force has no effect on an acceleration wave; only a *time-independent* body force has no effect on an oscillation. It is known from experience with special theories that infinitesimal waves of certain curved forms are sometimes possible, but no general theory has yet been worked out for them.

On the other hand, the relevance of the theory of acceleration waves is uncertain. That theory refers, strictly, only to necessary conditions for solutions. If there is a solution in which a singular surface with unit normal $\boldsymbol{n}$ carries a jump of amplitude $\boldsymbol{a}$, then (71.3) must be satisfied at each point, but a solution of (71.3) yielding an $\boldsymbol{a}$ and an $\boldsymbol{n}$ varying smoothly over a set of points does not generally

[1] Since the equations are linear, body force may be taken into account by addition of a particular integral. If the body force is independent of time, so is the corresponding integral; therefore, time-independent body forces have no effect on the propagation of infinitesimal waves.

[2] Results of this type, in slowly increasing generality, have been obtained ever since the beginning of the theory of elasticity. Our presentation simplifies that of TRUESDELL [1961, *63*, § 4], who generalized slightly the approach of TOUPIN and BERNSTEIN [1961, *62*, § 3].

[3] Results of the type given here, but of course more special, first appear in the work of HUGONIOT and DUHEM. Cf. CFT, Sect. 194A. The general statement, foreshadowed by some extent by CATTANEO [1946, *1*, § 1], was given by TRUESDELL [1961, *63*, § 4].

yield a singular surface. First, it is necessary that at each time the unit normals $\boldsymbol{n}$ correspond to surface elements that sweep out a surface. Second, as is clear from known examples in the classical infinitesimal theory, the dynamical equations impose conditions upon the motion of the surface and upon the variation of $\boldsymbol{a}$ in time. The only deformation possible in all uniformly aeolotropic elastic bodies is a homogeneous strain (Sects. 54, 91). Since $x^k{}_{,\alpha;\beta}=0$ in any such state, it follows that $[x^k{}_{,\alpha;\beta}]=0$, so that no non-trivial singular surface of second or higher order can travel through such a state. Thus, at present, no solution corresponding to any propagating singular surface is known for a general elastic material.

These facts raise the disquieting thought that the whole theory of singular surfaces may turn out to be an empty formalism as far as the general theory of elasticity is concerned. We doubt, however, that this danger be real. It is clear that solutions consisting of two different homogeneous strains connected by a suitable plane shock wave exist. As the initial strength of this shock wave is taken smaller and smaller, we expect that the relation between its amplitude and its speed will tend to satisfy the propagation condition for waves of second order. Analysis towards this end has been given by JOUGUET and BLAND[1], and doubtless a general theorem of equivalence corresponding to more modern views on the thermodynamics of deformation could be established. In this way should result a universal status for plane waves of second order, and possibly also for waves of certain curved forms, as limiting cases of weak shocks. [Note added in proof: See the memoirs cited in Sect. 96ter.]

In this chapter, no more is attempted than a fairly complete *analysis of the propagation condition* (71.3). So as to avoid lengthy wording, the criteria refer only to the propagation condition, with no implication that singular surfaces corresponding to the results actually exist.

The interpretation in the context of infinitesimal waves offers one advantage in that the nature of the roots of (71.6) is reflected obviously in a criterion of stability. First, a negative root leads to a purely imaginary value of U, say $U=\pm i\alpha$ where $\alpha>0$; if the amplitude $\boldsymbol{a}$ is real, a suitable linear combination of two solutions of the type (73.1) leads to a particular real solution of the form

$$\boldsymbol{u}=\boldsymbol{c}e^{\alpha t}\sin\frac{2\pi}{l}(\boldsymbol{n}\cdot\boldsymbol{p}), \tag{73.4}$$

the magnitude of which increases without limit as $t\to\infty$. Thus any disturbance of this type if compatible with the boundary conditions, as it is, for example, in an infinite medium, will tend to build up[2]. Similarly, since $\boldsymbol{Q}(\boldsymbol{n})$ is real, complex roots of (71.6) and corresponding complex amplitudes must occur in conjugate pairs, suitable combination of which again leads to solutions increasing exponentially with time. Thus *unless all the proper numbers of $\boldsymbol{Q}(\boldsymbol{n})$ are non-negative*[3], *in an infinite medium there are certain infinitesimal disturbances which amplify with time*, according to the infinitesimal theory. Conversely, as we have seen in Sect. 68bis, in an infinitesimally stable configuration every mode of plane oscillation has a real frequency.

This does not mean that, necessarily, we should impose as a condition on the constitutive equation the requirement that $\boldsymbol{Q}(\boldsymbol{n})$ be such as to have only positive proper numbers in all states of strain. Rather, it indicates that for a given material any homogeneous strain giving $\boldsymbol{Q}(\boldsymbol{n})$ a negative or complex proper number is not likely to be encountered in practice, since in such a strain there are certain kinds of disturbances which, however small in magnitude initially and in whatever way excited, begin[4] to grow at once and hence to destroy the given state of strain. Perhaps in this way physical instabilities such as buckling ultimately may come

---

[1] JOUGUET [1920, *1*—*4*] [1921, *1*]; BLAND [1964, *8*].
[2] HAYES and RIVLIN [1961, *24*, § 5].
[3] If 0 is a proper number of $\boldsymbol{Q}(\boldsymbol{n})$, the corresponding solutions are independent of time.
[4] The infinitesimal wave theory is only an approximation having a range of validity as yet unknown.

to be explained. It may be the occurrence of certain waves, presumably transverse, that disturbs a sufficiently severe homogeneous compression of a bar so much as to carry the body over into a state of small oscillation about a different configuration subject to the same resultant terminal loads.

In any case, under circumstances making possible solutions of the form (73.1), the constants $\boldsymbol{a}$ and l cancel out of the propagation equation (73.3). In particular, the independence of $\boldsymbol{a}$ and U from $\boldsymbol{p}$ and l, respectively, shows that *according to the finite theory of elasticity, plane sinusoidal free oscillations are propagated without absorption or dispersion.* This is only to be expected on dimensional grounds, since the elastic response function does not depend on any characteristic length (cf. Sect. 43), nor is any such length provided by the specification of the particular problem. The speed U does depend very markedly, however, on the direction of propagation, the direction of the amplitude $\boldsymbol{a}$, and the magnitudes of the stretches v_1, v_2, v_3. The theory of singular surfaces, presented in Sect. 71, is not limited to deformations in which the strain is homogeneous. Accordingly, the speed of a singular surface need not be the same at all points or times, so the above statement that in elasticity theory there is no dispersion needs to be hedged about with qualifications. Likewise the amplitude $\boldsymbol{a}$ of a singular surface will generally be subjected to conditions beyond those given in Sect. 71 if it is to divide two regions in which the equations of the theory have continuously differentiable solutions. Thus in the interpretation of singular surfaces also the statement that there is no absorption or amplification needs to be qualified. Indeed, as already mentioned, it has never been shown that for general elastic materials any solutions with singular surfaces exist.

Infinitesimal surface waves in an isotropic elastic body subject to finite homogeneous strain have been studied by HAYES and RIVLIN[1]. Their results are similar to those found by RAYLEIGH and LOVE for an unstressed material.

One of the best ways to determine the elasticities of a material by measurement is afforded by its response to infinitesimal waves. For a given direction of propagation $\boldsymbol{n}$, let it be supposed possible to measure the speed of plane waves and to determine their directions of vibration. If three linearly independent amplitudes may subsist, there is a simple formula[2] expressing the matrix of $\boldsymbol{Q}(\boldsymbol{n})$ in terms of those amplitudes and the corresponding speeds. Let us adopt as a working hypothesis the supposition that, for initial states only slightly different from the natural state, the acoustical tensor $\boldsymbol{Q}(\boldsymbol{n})$ is known. $\boldsymbol{S}(\boldsymbol{v})$, defined by (71.8), will then be known also.

Under this hypothesis, the problem of determining the elasticities from measured data on wave propagation is equivalent to that of calculating the elasticities in terms of $\boldsymbol{S}(\boldsymbol{v})$ and its derivatives. Although many authors have considered this problem, TOUPIN and BERNSTEIN[3] were the first to find a general method, which we now present. Substituting the expansion (63.10) into (71.11) yields

$$D_{\alpha\beta\gamma\delta} = g_{\alpha\beta}(L_{\gamma\delta\lambda\mu}E^{\lambda\mu} + \cdots) + C^{\mu}_{\beta}(L_{\gamma\alpha\delta\mu} + 2L_{\gamma\alpha\delta\mu\varepsilon\varrho}E^{\varepsilon\varrho} + \cdots). \tag{73.5}$$

If we write

$$\left.\begin{aligned} S^{(0)}_{\alpha\beta}(\boldsymbol{v}) &= S_{\alpha\beta}(\boldsymbol{v})\big|_{\boldsymbol{C}=\boldsymbol{1}},\\ S^{(0)}_{\alpha\beta\varepsilon\varrho}(\boldsymbol{v}) &= \frac{\partial}{\partial E^{\varepsilon\varrho}}\left(S_{\alpha\beta}(\boldsymbol{v})\right)\Big|_{\boldsymbol{C}=\boldsymbol{1}},\end{aligned}\right\} \tag{73.6}$$

[1] HAYES and RIVLIN [1961, *25*]. References to thermo-elastic oscillations are given in Sect. 96.

[2] E.g. Eq. (App. 37.12)$_3$ of CFT.

[3] TOUPIN and BERNSTEIN [1961, *62*, § 4].

where $\boldsymbol{E} \equiv \frac{1}{2}(\boldsymbol{C}-\mathbf{1})$, then from (73.5) and (71.10) we see that[1]

$$\left.\begin{aligned} S^{(0)}_{\alpha\beta}(\boldsymbol{v}) &= L_{\gamma\alpha\delta\beta} v^\gamma v^\delta, \\ S^{(0)}_{\alpha\beta\varepsilon\varrho}(\boldsymbol{v}) &= [2L_{\gamma\alpha\delta\beta\varepsilon\varrho} + g_{\alpha\beta} L_{\gamma\delta\varepsilon\varrho} + g_{\varrho\beta} L_{\gamma\alpha\delta\varepsilon} + g_{\varepsilon\beta} L_{\gamma\alpha\delta\varrho}] v^\gamma v^\delta, \ldots \end{aligned}\right\} \tag{73.7}$$

Similar relations hold for the elasticities of higher order.

In order that the relations of the form (73.7) may be solved for $L_{\alpha\beta\gamma\delta}$, $L_{\alpha\beta\gamma\delta\varepsilon\varrho}, \ldots,$ the quantities $S^{(0)}_{\alpha\beta}(\boldsymbol{v})$, $S^{(0)}_{\alpha\beta\varepsilon\varrho}(\boldsymbol{v}), \ldots$ must satisfy certain conditions of compatibility. For the linear coefficients $L_{\alpha\beta\gamma\delta}$, these may be found as follows. Let $\boldsymbol{v}_\Omega$, $\Omega = 1, 2, \ldots, 6$, be any set of six vectors such that the six symmetric tensors

$$P^{\gamma\delta}_\Omega = v^\gamma_\Omega v^\delta_\Omega \tag{73.8}$$

are linearly independent. Then an inverse set of six symmetric tensors $R^\Omega_{\lambda\mu}$ remains:

$$\sum_{\Omega=1}^{6} R^\Omega_{\lambda\mu} P^{\gamma\delta}_\Omega = \tfrac{1}{2}(\delta^\gamma_\lambda \delta^\delta_\mu + \delta^\delta_\lambda \delta^\gamma_\mu). \tag{73.9}$$

For six directions of propagation such that the corresponding vectors $\boldsymbol{v}_\Omega$ satisfy the above condition of independence, consider the corresponding equations $(73.7)_1$:

$$L_{\gamma\alpha\delta\beta} P^{\gamma\delta}_\Omega = S^{(0)}_{\alpha\beta}(\boldsymbol{v}_\Omega). \tag{73.10}$$

In terms of the acoustical tensors $S^{(0)}_{\alpha\beta}(\boldsymbol{v}_\Omega)$ for the six directions $\boldsymbol{v}_\Omega$ we define a tensor $K_{\alpha\lambda\beta\mu}$ as follows:

$$K_{\alpha\lambda\beta\mu} = \sum_{\Omega=1}^{6} S^{(0)}_{\alpha\beta}(\boldsymbol{v}_\Omega) R^\Omega_{\lambda\mu}. \tag{73.11}$$

Multiplying (73.10) by $R^\Omega_{\lambda\mu}$, summing on Ω, and using (73.9) yields

$$\tfrac{1}{2}(L_{\lambda\alpha\mu\beta} + L_{\mu\alpha\lambda\beta}) = K_{\alpha\lambda\beta\mu}. \tag{73.12}$$

Interchanging first λ and α, then α and μ, yields

$$\tfrac{1}{2}(L_{\alpha\lambda\mu\beta} + L_{\mu\lambda\alpha\beta}) = K_{\lambda\alpha\beta\mu}, \quad \tfrac{1}{2}(L_{\lambda\mu\alpha\beta} + L_{\alpha\mu\lambda\beta}) = K_{\mu\lambda\beta\alpha}. \tag{73.13}$$

Since $L_{\pi\sigma\mu\beta} = L_{\sigma\pi\mu\beta}$, by adding (73.12) to $(73.13)_1$ and subtracting $(73.13)_2$ from the sum we see that

$$L_{\lambda\alpha\mu\beta} = K_{\alpha\lambda\beta\mu} + K_{\lambda\alpha\beta\mu} - K_{\mu\lambda\beta\alpha}. \tag{73.14}$$

Thus follows a major theorem, due essentially to TOUPIN and BERNSTEIN[2]: *If the acoustical tensors for six direction of propagation $\boldsymbol{v}_\Omega$ leading to a linearly independent set of tensors $\boldsymbol{P}_\Omega = \boldsymbol{v}_\Omega \otimes \boldsymbol{v}_\Omega$ are known in the natural state, the linear elasticities $L_{\lambda\alpha\mu\beta}$, at most 36 of which are independent, are uniquely determined.*

The acoustical tensors may be regarded as available directly from experiment and thus not restricted to any particular material. To see if measured wave speeds and amplitudes are consistent with the theory of elasticity, it is necessary that unique values of the coefficients $L_{\lambda\alpha\mu\beta}$ result from the process indicated. Therefore, the right-hand side of (73.14) must be independent of the choice of the six vectors $\boldsymbol{v}_\Omega$. Thus by (73.11) follows TOUPIN and BERNSTEIN's ***first condition of compatibility***: *For every set of six vectors $\boldsymbol{v}_\Omega$ such that the six symmetric tensors $\boldsymbol{P}_\Omega = \boldsymbol{v}_\Omega \otimes \boldsymbol{v}_\Omega$ are linearly independent, the tensor*

$$\sum_{\Omega=1}^{6} [S^{(0)}_{\alpha\beta}(\boldsymbol{v}_\Omega) R^\Omega_{\lambda\mu} + S^{(0)}_{\lambda\beta}(\boldsymbol{v}_\Omega) R^\Omega_{\alpha\mu} - S^{(0)}_{\mu\beta}(\boldsymbol{v}_\Omega) R^\Omega_{\lambda\alpha}] \tag{73.15}$$

has the same value.

Since $R^\Omega_{\lambda\mu} = R^\Omega_{\mu\lambda}$, from (73.11) we see that $K_{\alpha\lambda\beta\mu} = K_{\alpha\mu\beta\lambda}$; hence by (73.13) it follows that the one condition of symmetry $L_{\lambda\alpha\mu\beta} = L_{\alpha\lambda\mu\beta}$ is always satisfied automatically. Not so for the other condition, $L_{\lambda\alpha\mu\beta} = L_{\lambda\alpha\beta\mu}$, which leads to the ***second condition of compatibility***[3]:

$$K_{\alpha\lambda\beta\mu} - K_{\alpha\lambda\mu\beta} + K_{\lambda\alpha\beta\mu} - K_{\lambda\alpha\mu\beta} + K_{\beta\alpha\mu\lambda} - K_{\mu\lambda\beta\alpha} = 0. \tag{73.16}$$

---

[1] Eq. $(73.7)_2$ corrects the corresponding formula of TOUPIN and BERNSTEIN [1961, *62*, § 4].

[2] The argument of TOUPIN and BERNSTEIN, carried through under the assumption that there are at most 21 independent linear elasticities, is here seen to hold more generally.

[3] This is slightly more general than the corresponding result of TOUPIN and BERNSTEIN [1961, *62*, § 4], who considered only hyperelastic materials.

If these two conditions are not satisfied, no symmetry or lack of symmetry ascribed to the material can make the acoustical tensor $\boldsymbol{Q}$ consistent with elasticity theory. If the two conditions are satisfied, then linear elasticities $L_{\alpha\beta\gamma\delta}$, at most 36 of which are distinct, are uniquely determined, and the classical infinitesimal theory may therefore account for the data, as far as it goes.

Similar analysis may be applied so as to determine the higher elasticities $L_{\alpha\beta\gamma\delta\mu\nu}$, $L_{\alpha\beta\gamma\delta\mu\nu\pi\varrho}$, ..., but it has been worked out only for isotropic materials. The results will be obtained by a simpler method in Sect. 77.

74. Waves in isotropic compressible materials. I. General properties. In an isotropic material the tensor B may be expressed in the form (48.24) when the reference configuration is undistorted; substitution into (71.5) gives the following simple formula for the acoustical tensor[1]:

$$Q^k{}_m(\boldsymbol{n}) = 2\mathsf{F}^{kp}{}_{mq} B^{qr} n_r n_p, \tag{74.1}$$

where the tensor F is given in terms of the response coefficients $\aleph_\Gamma$ by means of (48.13). The equation to be satisfied is still (71.3). Since F, for a given material, is a function of $\boldsymbol{B}$ alone, it follows that for isotropic materials not only the possible wave speeds but also the corresponding jumps or amplitudes depend only on $\boldsymbol{B}$ and hence remain unaltered if the (undistorted) reference state is changed by a rotation, as is only to be expected.

In the classical infinitesimal theory of isotropic elastic materials, every wave must be either longitudinal or transverse[2]. In order for such a decomposition to follow, it is necessary and sufficient that $\boldsymbol{n}$ itself be a proper vector of $\boldsymbol{Q}(\boldsymbol{n})$, for every $\boldsymbol{n}$, and that the other two proper vectors be orthogonal to $\boldsymbol{n}$. From (71.4) it is clear that $\boldsymbol{Q}(\boldsymbol{n})$ does not generally have these properties[3]. In the infinitesimal theory, the material is effectively unstrained as regards the passage of waves; only from properties of the material is there reason for the laws of propagation to differ from one direction to another, and in an isotropic material there are no distinguished directions[4]. In a severely strained body, such indifference holds no longer. If the material is anisotropic, there are three different sets of directions having especially simple properties and physical significance: the principal axes of stress, the principal axes of strain, and such directions as specify intrinsic symmetries of the material. These three sets of directions, along with the local rotation, must influence in some way the acoustical axes for the direction $\boldsymbol{n}$, and the relation cannot be expected to be simple; its complexity has been shown by the results in Sect. 71. In an isotropic material, the third of the sets consists of all

---

[1] TRUESDELL [1961, *63*, § 7].

[2] E.g., Sect. 301 of CFT.

[3] HADAMARD [1901, *1*, § 8] [1903, *4*, ¶¶ 274—277], who was not in possession of reduced forms for the stress relation of an isotropic material, reached this same conclusion in a somewhat different way. For a hyperelastic material he found the most general form of strain energy such that $\boldsymbol{Q}(\boldsymbol{n})$, which is then symmetric (see below, Sect. 90), shall have $\boldsymbol{n}$ as a proper vector; the energy function turns out to be the sum of a special kind of quadratic polynomial in the components of $\boldsymbol{E}$, plus an arbitrary linear function of $\boldsymbol{E}$, plus an arbitrary function of $III_{\boldsymbol{C}}$. Since the stress relation of an isotropic material may involve terms of arbitrarily high degree in $I_{\boldsymbol{C}}$ and $II_{\boldsymbol{C}}$, it is generally not of the requisite form. A more special result, following from still more restrictive hypotheses, was obtained by TOLOTTI [1943, *5*, Eq. (3)]. Further results of this kind have been presented by BRESSAN [1963, *12*, Parte II].

[4] In fact, from (74.1), (48.13), and (50.13) we easily show that

$$\boldsymbol{Q}(\boldsymbol{n})|_{\boldsymbol{B}=\boldsymbol{1}} = (\lambda+\mu)\,\boldsymbol{n}\otimes\boldsymbol{n} + \mu\,\boldsymbol{1}, \qquad Q_{km}(\boldsymbol{n})\,|_{\boldsymbol{B}=\boldsymbol{1}} = (\lambda+\mu)\,n_k n_m + \mu g_{km}.$$

$\boldsymbol{Q}(\boldsymbol{n})$ has thus the form of the most general tensor-valued isotropic function of $\boldsymbol{n}$; *a fortiori*, it is a symmetric tensor. This observation might be used to give physical motivation for a direct treatment of the laws of propagation of waves in slightly deformed isotropic materials, without use of the constitutive equations and laws of motion.

directions and the former two coincide; the only distinguished directions are the principal axes of stress or strain, induced by the fact of the body's being strained; the reception given to the passing wave will be of much the same kind as that offered by an unstrained but suitably aeolotropic body. The wave itself has one distinguished direction, that in which it propagates. If this direction is itself that of a principal axis, the only other distinguished directions are those of the other principal axes. Calling a wave a *principal wave*[1] if $\boldsymbol{n}$ is parallel to a principal axis of stress, we may expect that *in an isotropic material, the accoustic axes for principal waves coincide with the principal axes; in particular, every principal wave is either longitudinal or transverse.* That such is the case is proved formally as follows.

Let $\boldsymbol{n}_1$ be a proper vector of $\boldsymbol{B}$ corresponding to the stretch v_1, so that $\boldsymbol{B}\boldsymbol{n}_1 = v_1^2 \boldsymbol{n}_1$, $\boldsymbol{B}^{-1}\boldsymbol{n}_1 = \boldsymbol{n}_1/v_1^2$. Then from (48.13) and (74.1) we easily show that[2]

$$\left.\begin{aligned}\frac{1}{v_1^2}\boldsymbol{Q}(\boldsymbol{n}_1) = (\aleph_1 + v_1^2\aleph_2)\,\mathbf{1} + \aleph_2\boldsymbol{B} + 2\boldsymbol{n}_1\otimes\boldsymbol{n}_1\Big\{\frac{1}{2}\aleph_1 + v_1^2\aleph_2 + {}\\ + \sum_{\Gamma=0}^{2} v_1^{2\Gamma}\left[\frac{\partial\aleph_\Gamma}{\partial I} + (v_2^2+v_3^2)\frac{\partial\aleph_\Gamma}{\partial II} + v_2^2v_3^2\frac{\partial\aleph_\Gamma}{\partial III}\right]\Big\}.\end{aligned}\right\} \tag{74.2}$$

The tensor $\boldsymbol{Q}(\boldsymbol{n}_1)$ is symmetric, and $\boldsymbol{n}_1$ is a proper vector. If $\boldsymbol{n}_2$ is a proper vector of $\boldsymbol{B}$ along a different principal axis, it is necessarily normal to $\boldsymbol{n}_1$; from (74.2) it is immediate that $\boldsymbol{n}_2$ is also a proper vector of $\boldsymbol{Q}(\boldsymbol{n}_1)$. We have thus established the theorem stated at the end of the previous paragraph.

Thus far we have considered only the directions, not the speeds. By substituting (74.2) into (71.7) we see that the speed U_{11} of a longitudinal wave in the direction of the principal axis for which the stretch is v_1 is given by

$$\frac{\varrho U_{11}^2}{2v_1^2} = \aleph_1 + 2v_1^2\aleph_2 + \sum_{\Gamma=0}^{2} v_1^{2\Gamma}\left[\frac{\partial\aleph_\Gamma}{\partial I} + (I - v_1^2)\frac{\partial\aleph_\Gamma}{\partial II} + \frac{III}{v_1^2}\frac{\partial\aleph_\Gamma}{\partial III}\right]. \tag{74.3}$$

Similarly, the speed U_{12} of a transverse wave whose amplitude is parallel to $\boldsymbol{n}_2$ is given by

$$\left.\begin{aligned}\frac{\varrho U_{12}^2}{v_1^2} &= \aleph_1 + (v_1^2+v_2^2)\,\aleph_2,\\ &= \beth_1 - \frac{1}{v_1^2v_2^2}\beth_{-1},\end{aligned}\right\} \tag{74.4}$$

where we have used $(47.14)_{2,3}$.

Alternatively, we may calculate the explicit forms of the principal components of $\boldsymbol{Q}(\boldsymbol{n})$ (cf. Sect. 48). To this end we need only substitute (48.16) into (74.1), obtaining

$$\left.\begin{aligned}Q\langle 11\rangle(\boldsymbol{n}) &= \frac{\partial t_1}{\partial\log v_1}\cos^2\theta_1 + \frac{t_1-t_2}{v_1^2-v_2^2}v_2^2\cos^2\theta_2 + \frac{t_1-t_3}{v_1^2-v_3^2}v_3^2\cos^2\theta_3, \ldots.\\ Q\langle 12\rangle(\boldsymbol{n}) &= \left[\frac{\partial t_1}{\partial\log v_2} + \frac{t_1-t_2}{v_1^2-v_2^2}v_1^2\right]\cos\theta_1\cos\theta_2, \ldots\end{aligned}\right\} \tag{74.5}$$

where $\boldsymbol{n}$ has the principal components $\cos\theta_1$, $\cos\theta_2$, $\cos\theta_3$. If $\boldsymbol{n} = \boldsymbol{n}_1$, then $\cos\theta_2 = \cos\theta_3 = 0$ and $Q\langle 12\rangle(\boldsymbol{n}_1) = Q\langle 21\rangle(\boldsymbol{n}_1) = 0$, but for non-principal directions of

[1] Manacorda [1959, *19*, § 5], for special elastic materials.

[2] This result and its various consequences presented below, including the explicit formulae for the wave speeds, were obtained by Truesdell [1961, *63*, § 7]. In the context of infinitesimal oscillation of a hyperelastic material, an equivalent formula in a special co-ordinate system may be obtained by combining three equations given by Hayes and Rivlin [1961, *24*, Eqs. (4.8), (4.7), and (3.10)], and earlier Tolotti [1943, *5*, Eqs. (18′) (19′)] had virtually calculated $\boldsymbol{Q}(\boldsymbol{n})$ for isotropic hyperelastic materials, but these authors did not classify the waves or obtain explicit formulae for the speeds of propagation.

propagation $\boldsymbol{Q}(\boldsymbol{n})$ generally fails to be symmetric. From $(74.5)_1$ we see that

$$Q\langle 11\rangle(\boldsymbol{n}_1) = \frac{\partial t_1}{\partial \log v_1}, \qquad Q\langle 22\rangle(\boldsymbol{n}_1) = v_1^2 \frac{t_1 - t_2}{v_1^2 - v_2^2}. \tag{74.6}$$

From (71.7) it follows that

$$\left.\begin{aligned} \varrho\, U_{11}^2 &= \frac{\partial t_1}{\partial \log v_1}, \\ \frac{\varrho\, U_{12}^2}{v_1^2} &= \frac{t_1 - t_2}{v_1^2 - v_2^2}. \end{aligned}\right\} \tag{74.7}$$

These elegant formulae, which are due to ERICKSEN[1], may be obtained also by interpretation of (74.3) and (74.4); since they make no direct use of the response coefficients $\aleph_\Gamma$, they give insight into the general nature of wave propagation in all kinds of elastic materials.

From (74.4) we see that

$$\frac{\varrho}{v_1^2}\left(U_{12}^2 - U_{13}^2\right) = \left(v_2^2 - v_3^2\right)\aleph_2 = \frac{v_2^2 - v_3^2}{III}\,\beth_{-1}. \tag{74.8}$$

From this formula follows the ***fundamental theorem on transverse principal waves***: *In an isotropic elastic material such that $\beth_{-1} = 0$ for the strain in question, both kinds of transverse waves that may travel down a given principal axis of strain have the same absolute speed of propagation; if, however, $\beth_{-1} \neq 0$, then these two kinds of transverse waves travel at the same absolute speed if and only if the corresponding principal stretches are equal; and in this latter case, any transverse amplitude is possible, and all these transverse waves have the same absolute speeds of propagation.* From (74.8), moreover, we see that $\beth_{-1} < 0$ *if and only if $U_{12}^2 - U_{13}^2$ and $v_2^2 - v_3^2$ are of opposite sign; that is, if and only if transverse waves with amplitudes parallel to the axis of lesser transverse stretch travel at greater absolute speeds than the others.* Thus we have obtained a physical interpretation for the sign of $\beth_{-1}$ and, in particular, for the third of the E-inequalities (51.27), in slightly strengthened form[2].

From (74.3) it is clear that $U_{11}^2 = U_{22}^2$ if $v_1 = v_2$. A converse condition is immediate from $(74.7)_1$, leading to the ***fundamental theorem on longitudinal principal waves***: *In an isotropic elastic material, the squared speeds of propagation of longitudinal waves in the directions of two equal principal stretches are equal; conversely, if $U_{11}^2 = U_{22}^2$, it is necessary that*

$$\frac{\partial t_1}{\partial \log v_1} = \frac{\partial t_2}{\partial \log v_2}. \tag{74.9}$$

---

[1] According to TRUESDELL [1961, *63*, § 7]. A result close to $(74.7)_1$ had been attributed earlier to ERICKSEN and TOUPIN by TRUESDELL [1956, *32*, Eq. (11)]. GREEN [1963, *26*, § 2] showed that Eqs. (74.7) hold for certain aeolotropic materials. BELL [1964, *5*] has reported confirmation of these formulae in experiments on high-velocity impact of metal bars.

[2] The wave speeds in a material subject to simple shear have been calculated by TRUESDELL [1965, *36*]. From (54.13) and $(74.7)_2$ we see that

$$\frac{\varrho\, U_{12}^2}{v_1^2} = \frac{\varrho\, U_{21}^2}{v_2^2} = \hat{\mu}(K^2),$$

while more complicated expressions hold for the other principal shear waves. Vanishing of $\hat{\mu}(K^2)$ is thus equivalent to the vanishing of both U_{12}^2 and U_{21}^2, that is, to the possibility of weak transverse discontinuities permanently dividing the material on opposite sides of either principal plane of stress and strain normal to the plane of shear. It is but reinterpretation of the theorem in footnote 2, p. 175, to state that if $\hat{\mu}(K_*^2) = 0$, then either all transverse wave speeds vanish when $K = K_*$, or else at some lesser value of K either U_{23}^2 or U_{31}^2 vanishes. Therefore, in gradually increasing the amount of shear, the first principal slip planes that become possible are parallel to the plane of shear rather than perpendicular to it, except in the case when any principal plane may be a slip plane if $K = K_*$.

These results show that the state of affairs in the infinitesimal theory of isotropic materials, where all transverse waves travel at one absolute speed and all longitudinal waves at another, is not typical but rather degenerate for isotropic elastic materials.

Either direct application of (48.27) to (74.3) and (74.4), or alternatively, specialization of the general theorem stated after (71.15), or again, inspection of (48.16), (48.29) and (74.7), shows the S-E condition to imply that all six speeds $U_{11}, U_{12}, \ldots, U_{33}$ are real and not zero. That is, *the S-E condition is sufficient that the squared speeds of all principal waves in an isotropic material be positive.* In the present connection, the GCN⁺ and B-E inequalities lead to weaker results. Since $\boldsymbol{n}_1$ is a proper vector of $\boldsymbol{Q}(\boldsymbol{n}_1)$, we may apply to the present case the theorem stated just after (71.16) and conclude *that if the GCN⁺ condition holds, the squared speeds of propagation of all longitudinal principal waves are positive*[1].

ERICKSEN's formulae (74.6) and (74.7) make possible the statement of simple and immediate criteria of reality, which may be summarized in the ***fundamental theorem on the speeds of principal waves***: *In an isotropic elastic material the squared speeds of propagation of the principal longitudinal waves are positive if and only if each principal tension is an increasing function of the corresponding principal stretch while the other principal stretches are held constant; the squared speeds of the principal transverse waves are positive if and only if the greater principal tension occurs always in the direction of the greater principal stretch.* By rephrasing the statements in the theorem, or by comparing (74.3) and (74.4) with (51.13) and (51.26), respectively, or directly from (74.7), we see that *the T-E⁺ inequalities are necessary and sufficient that the squared speeds of propagation of all principal longitudinal waves be positive, while the B-E inequalities are necessary and sufficient that the squared speeds of propagation of all principal transverse waves be positive*[2].

Not much has been found out about the behavior of non-principal waves. First we notice that only two of the three types of non-vanishing principal elasticities of an isotropic material may be interpreted in terms of the speeds of principal waves. Indeed, from (74.7) and (48.16) we see that

$$F\langle abab\rangle = \frac{\varrho U_{ab}^2}{2v_a^2}. \tag{74.10}$$

The third kind of elasticity, $F\langle aabb\rangle$ when $a \neq b$, can be connected to undulatory phenomena only if non-principal waves are considered. Indeed, by use of (74.7) we easily put the explicit formula (74.5) for the acoustical tensor for the direction $\boldsymbol{n}$ into the following not inelegant form:

$$\left.\begin{aligned} Q\langle aa\rangle &= \varrho \sum_{c=1}^{3} U_{ca}^2 \cos^2\theta_c, \\ Q\langle ab\rangle &= (2v_b^2 F\langle aabb\rangle + \varrho U_{ab}^2)\cos\theta_a \cos\theta_b, \quad a \neq b, \end{aligned}\right\} \tag{74.11}$$

where $\boldsymbol{n}$ has the principal components $\cos\theta_a$. These formulae, which generalize (74.6), suggest that it is possible for the squared speeds of all principal waves to be positive yet in some non-principal directions only certain waves may propagate, since for any given values of U_{ab}^2, the off-diagonal terms $Q\langle ab\rangle$ may be given any value by choice of the corresponding elasticities $F\langle aabb\rangle$. However, from $(74.11)_1$

[1] The result is due to TOUPIN [1961, *61*, Th. XII], with an insufficiently general proof. The argument in the text was given by TRUESDELL [1961, *63*, § 10].

[2] A weaker result was obtained by HAYES and RIVLIN [1961, *24*, § 5]. Specifically, they showed that the S-E condition implies the B-E inequalities; in the present treatment, this fact was proved directly from the definitions in Sect. 52, above.

we see that $\mathrm{tr}[\boldsymbol{Q}(\boldsymbol{n})] > 0$ for every $\boldsymbol{n}$ if and only if[1]

$$\sum_{b=1}^{3} U_{ab}^2 > 0, \qquad a = 1, 2, 3. \tag{74.12}$$

Now the condition $\mathrm{tr}[\boldsymbol{Q}(\boldsymbol{n})] > 0$ is sufficient that at least one of the proper numbers of $\boldsymbol{Q}(\boldsymbol{n})$ be positive. Thus follows ***Bressan's theorem***: *The condition* (74.12) *is sufficient that at least one squared speed be positive for each* $\boldsymbol{n}$. *In particular, if all squared speeds of principal waves are positive, so is at least one squared speed for every direction of propagation.* According to the former statement in the theorem, the existence of one very great positive squared speed of propagation for each principal direction, even though the other two squared speeds may be negative (but of not too large absolute value), suffices for the existence of at least one real wave in every direction.

The growth of plane acceleration waves propagating into an isotropic body in a state of homogeneous strain has been examined by W. A. GREEN[2]. While, as we see directly from the propagation condition, the speed and direction of propagation are constant, the amplitude generally varies as the wave progresses. If a is the magnitude of the amplitude $\boldsymbol{a}$ pointing in the direction of a unit vector $\boldsymbol{l}$, on the assumption that there is a real wave for which a is the same at all points on the plane wave front GREEN calculated the displacement derivative [CFT, Sect. 179] of a and integrated the resulting differential equation to get a as a function of the distance, $s = U\,t$, traversed by the wave. Assuming that there be real amplitudes pointing in the directions $\bar{\boldsymbol{l}}$ and $\tilde{\boldsymbol{l}}$, and that $\boldsymbol{l}, \bar{\boldsymbol{l}}, \tilde{\boldsymbol{l}}$ form a linearly independent set, he found that

$$\frac{a(s)}{a(0)} = \frac{1}{1 + s\,\dfrac{K\,a(0)}{\varrho\,U^2}}, \tag{74.13}$$

where

$$\left.\begin{aligned} K &= \frac{\boldsymbol{A}\cdot\bar{\boldsymbol{l}}\times\tilde{\boldsymbol{l}}}{\boldsymbol{l}\cdot\bar{\boldsymbol{l}}\times\tilde{\boldsymbol{l}}}, \\ A^p &= B^{wu} n_w n_u F^{pq}{}_{rs} n_q l^r l^s + 2\frac{\partial F^{pq}{}_{rs}}{\partial B_{uv}} B_{mv} n^m B^{ws} n_w n_q l_u . \end{aligned}\right\} \tag{74.14}$$

When the vectors $\boldsymbol{l}, \bar{\boldsymbol{l}}, \tilde{\boldsymbol{l}}$ form an orthonormal set, as is the case for principal waves, $(74.14)_1$ reduces of the simpler form $K = \boldsymbol{A}\cdot\boldsymbol{l}$. For a given wave in a given state of strain, K is a constant. If $K > 0$, the amplitude of an acceleration wave of rarefaction ($a < 0$) increases steadily until it becomes infinite at the time $t = -\varrho|U|/[K\,a(0)]$. The amplitude of a wave of condensation, however, decays steadily to 0 as $t \to \infty$. If $K < 0$, the behaviors of condensation and rarefaction waves are interchanged. If $K = 0$, the wave is propagated without change of amplitude.

In the case of principal waves, K is easy to interpret. The values K_{ab} of K corresponding to the nine kinds of principal waves were shown by W. A. GREEN to be

$$\left.\begin{aligned} K_{11} &= \frac{1}{2}\,\frac{v_1}{\varrho}\,\frac{\partial}{\partial v_1}\left(\varrho^2 U_{11}^2\right), \\ &= v_1^2\,\frac{\partial t_1}{\partial v_1^2} + 2 v_1^4\,\frac{\partial^2 t_1}{\partial v_1^2\,\partial v_1^2}, \ldots, \\ K_{ab} &= 0 \quad \text{if} \quad a \neq b. \end{aligned}\right\} \tag{74.15}$$

Thus the transverse principal waves are propagated unchanged, while the longitudinal principal waves either decay or amplify.

[1] BRESSAN [1963, *12*, § 6], for hyperelastic materials.
[2] W. A. GREEN [1964, *40*], corrected [1965, *19*].

W. A. GREEN found that waves of order 3 or more are always propagated without change of amplitude.

In further work[1], W. A. GREEN shows that a plane acceleration wave cannot generally exist alone but must be accompanied by third-order waves propagating at the same speed (rather than at their own characteristic speeds) and having amplitudes with non-zero components in the directions of $\bar{\boldsymbol{l}}$ and $\tilde{\boldsymbol{l}}$. A plane wave of order greater than two, likewise, is accompanied by waves of order one higher, with amplitudes parallel to that of the primary wave.

A. E. GREEN[2] has discussed wave propagation in elastic materials from the standpoint of plane waves of finite amplitude. First consider a deformation representing a plane longitudinal displacement superimposed upon a homogeneous strain having stretches V_1, v_2, v_3:

$$\left.\begin{aligned} x &= \xi + u(\xi, t), & \xi &= V_1 X, \\ y &= v_2 Y, & z &= v_3 Z. \end{aligned}\right\} \tag{74.16}$$

The co-ordinate axes are principal axes, and the principal stretches are v_1, v_2, v_3, where $v_1 = V_1(1+u_\xi)$. Since $I = V_1^2(1+u_\xi)^2 + v_2^2 + v_3^2$, $II = V_1^2(v_2^2+v_3^2)(1+u_\xi)^2 + v_2^2 v_3^2$, $III = V_1^2 v_2^2 v_3^2 (1+u_\xi)^2$, from (47.9) we have

$$\left.\begin{aligned} T\langle xx\rangle &= \beth_0 + V_1^2(1+u_\xi)^2 \beth_1 + V_1^{-2}(1+u_\xi)^{-2}\beth_{-1}, \\ &= t_1(v_1) = -F(\varrho), \end{aligned}\right\} \tag{74.17}$$

say, since $\varrho(1+u_\xi) = \varrho_0$, the density in the homogeneously deformed configuration. The only component of CAUCHY's first law (16.6) not trivially satisfied is $T\langle xx\rangle_{,x} = \varrho u_{tt}$. Hence the equation of motion assumes the following various forms:

$$u_{tt} = \frac{V_1^2}{\varrho v_1}\frac{\partial t_1}{\partial v_1} u_{\xi\xi} = \frac{V_1}{\varrho_0}\frac{\partial t_1}{\partial v_1} u_{\xi\xi} = \frac{\varrho^2}{\varrho_0^2} F'(\varrho)\, u_{\xi\xi} = g(u_\xi)\, u_{\xi\xi}, \tag{74.18}$$

where g is a given function. The second of these shows that when u_ξ is infinitesimal, so that $v_1 \approx V_1$, the disturbance is propagated according to the linear wave equation with speed given by $(74.7)_1$, as already shown by our more general analysis in Sects. 73—74. The third form is identical with that governing propagation of plane waves in an elastic (piezotropic) fluid whose pressure function is $F(\varrho)$. Therefore the extensive known theory[3] of such waves may be carried over bodily to isotropic elastic materials. In the process it must be recalled both that the form of $F(\varrho)$ depends on that of the constitutive equation (47.5) and on the values of the constants V_1, v_2, v_3, and also that the transverse tensions $T\langle yy\rangle$ and $T\langle zz\rangle$, while determined explicitly and uniquely by u_ξ, are not equal to $T\langle xx\rangle$ or to each other. In fact,

$$\left.\begin{aligned} T\langle yy\rangle &= t_2 = \beth_0 + v_2^2 \beth_1 + v_2^{-2}\beth_{-1}, \\ T\langle zz\rangle &= t_3 = \beth_0 + v_3^2 \beth_1 + v_3^{-2}\beth_{-1}. \end{aligned}\right\} \tag{74.19}$$

---

[1] W. A. GREEN [1965, *19*, § 3].

[2] A. E. GREEN [1963, *26*, §§ 2 and 4]. A special case had been investigated earlier by JAIN [1959, *13*]. After the theory for certain anisotropic materials had been discussed by GREEN, BLAND [1964, *9*, § 3] showed that $(74.18)_4$ holds in a material of arbitrary symmetry, provided longitudinal waves be possible at all, according to a condition of compatibility. BLAND [1964, *9*, §§ 3—5] has discussed the theory of simple waves and characteristics in this context and has derived the conditions for "similarity solutions". For the case when $V_1 = V_2 = v_3 = 1$, some specific problems with particular reference to the formation of shocks have been analyzed by CHU [1964, *14*].

[3] E.g. LAMB [1932, *5*, §§ 281—284], COURANT and FRIEDRICHS [1948, *2*, §§ 40—48]. More recent work on one-dimensional fluid motions, emphasizing the formation and interaction of shock waves, rests on other formulations. The complexities encountered there will be found in aggravated form in elasticity.

Thus it would be wrong to regard finite longitudinal waves in general isotropic elastic materials as being just like those in fluids.

A. E. GREEN has given a similar analysis for transverse waves of finite amplitude in an incompressible isotropic material[1]. Let the underlying homogeneous strain have principal stretches V_1, V_2, v_3, where $V_1 V_2 v_3 = 1$, and consider deformations of the form

$$x = V_1 X, \qquad y = V_2 Y + u(x, t), \qquad z = v_3 Z. \tag{74.20}$$

The co-ordinate axes are no longer principal axes, and V_1, V_2 are not principal stretches; neither is the motion a quasi-equilibrated one[2], so that the analysis of Sect. 61 does not apply here. Since, by (49.5), the matrix $[\boldsymbol{T}]$ of $\boldsymbol{T}$ is

$$[\boldsymbol{T}] = -p[1] + \beth_1 \begin{Vmatrix} V_1^2 & V_1^2 u_x & 0 \\ \cdot & V_2^2 + V_1^2 u_x^2 & 0 \\ \cdot & \cdot & v_3^2 \end{Vmatrix} + \beth_{-1} \begin{Vmatrix} v_3^2 (V_2^2 + V_1^2 u_x^2) & -V_1^2 v_3^2 u_x & 0 \\ \cdot & V_1^2 v_3^2 & 0 \\ \cdot & \cdot & V_1^2 V_2^2 \end{Vmatrix}; \tag{74.21}$$

thus $T\langle xz\rangle = T\langle yz\rangle = 0$; and since $I = V_1^2 + V_2^2 + v_3^2 + V_1^2 u_x^2$, $II = V_1^2 V_2^2 + V_2^2 v_3^2 + v_3^2 V_1^2 + V_1^2 v_3^2 u_x^2$, it follows that $T\langle xz\rangle = T\langle yz\rangle = 0$, that all stresses are functions of u_x alone, that by choice of p we can make $T\langle xx\rangle$ vanish, and that the motion is then governed by the following single partial differential equation:

$$\left.\begin{aligned} u_{tt} &= \frac{V_1^2}{\varrho} \frac{\partial}{\partial x} \left[(\beth_1 - v_3^2 \beth_{-1}) u_x \right], \\ &= \frac{V_1^2}{\varrho} \frac{\partial}{\partial x} \left[\frac{t_1 - t_2}{V_1^2 - V_2^2} u_x \right], \\ &= g(u_x) u_{xx}, \quad \text{where} \quad g(u_x) = \frac{1}{\varrho} \frac{d T\langle xy\rangle}{d u_x}. \end{aligned}\right\} \tag{74.22}$$

This equation, like (74.18), is of the same form as that governing the propagation of plane waves of finite amplitude in a compressible elastic fluid. When linearized with respect to u_x, it becomes the ordinary linear wave equation with speed of propagation given by $(74.7)_2$, as expected.

BLAND[3] has derived a partial differential equation for spherical waves.

75. Waves in isotropic compressible materials. II. The case of hydrostatic pressure[4]. From (74.4) we see that in a material for which $\beth_{-1} \neq 0$, if the six kinds of transverse principal waves travel at the same speed, then $v_1 = v_2 = v_3 = v$, say, and the stress is hydrostatic. In this case, conversely, follow still stronger results, of just the same kind as those to which we have grown accustomed in the classical infinitesimal theory.

First, since every vector is a proper vector of $\boldsymbol{B}$, from (74.2) we conclude that the acoustical tensor $\boldsymbol{Q}(\boldsymbol{n})$ is an isotropic function of $\boldsymbol{n}$. Thus the acoustical properties of the deformed body are the same in every direction. The quadric of $\boldsymbol{Q}(\boldsymbol{n})$ is a quadric of revolution with its axis of symmetry along $\boldsymbol{n}$ and with form independent of $\boldsymbol{n}$. Thus for a wave travelling in any direction a longitudinal

---

[1] The reader will easily see that in compressible materials such waves are not possible unless suitable time-dependent transverse tractions, or, of course, body forces, are applied. Thus the theory, while equally easy to work out, is of little interest.

[2] It is easy to show that for general forms of the response coefficients $\beth_\Gamma$ this motion is quasi-equilibrated if and only if u is linear in x; in this special case, the deformation remains homogeneous, and the motion given here is included in (61.19).

[3] While BLAND [1964, *9*, § 6] considered only hyperelastic materials, his results are easily generalized.

[4] TRUESDELL [1961, *63*, § 8], clearing and unifying results of BIRCH [1938, *1*] and BIOT [1940, *1*].

amplitude or any transverse amplitude is possible, and all waves are either transverse or longitudinal. Also, the squared speed $U_\perp^2$ of all transverse waves is the same, since (74.4) reduces to the form

$$U_\perp^2 = \frac{v^5}{\varrho_R}(\aleph_1 + 2v^2\aleph_2) = \frac{v^5}{\varrho_R}\left(\beth_1 - \frac{1}{v^4}\beth_{-1}\right), \tag{75.1}$$

where (47.13) has been used in order to get the second form. Since $I = 3v^2$, $II = 3v^4$, and $III = v^6$, it follows that

$$\frac{\partial}{\partial I} + 2v^2\frac{\partial}{\partial II} + v^4\frac{\partial}{\partial III} = \frac{1}{3}\frac{d}{dv^2}; \tag{75.2}$$

taking the stress-relation in the form (54.5), from (74.3) we conclude that all longitudinal waves travel at a speed $U_\parallel$ such that

$$U_\parallel^2 = \frac{4}{3}\frac{v^5}{\varrho_R}(\aleph_1 + 2v^2\aleph_2) + \frac{dp}{d\varrho}. \tag{75.3}$$

Elimination of $\aleph_1 + 2v^2\aleph_2$ between (75.1) and (75.3) yields a ***universal identity*** connecting the two wave speeds with the bulk modulus:

$$U_\parallel^2 = \frac{4}{3}U_\perp^2 + \frac{dp}{d\varrho}. \tag{75.4}$$

The linearized case of this result is familiar, since in the infinitesimal theory $\varrho U_\parallel^2 = \lambda + 2\mu$, $\varrho U_\perp^2 = \mu$, and $\varrho\, dp/d\varrho = \lambda + \frac{2}{3}\mu$. Another case is equally familiar, that of an elastic fluid, for (50.1) is equivalent to $\aleph_1 = \aleph_2 = 0$, $\aleph_0 = -p(\varrho)$, whence it follows from $(75.1)_1$ that $U_\perp = 0$: In an elastic fluid, a transversal singular surface of second or higher order is necessarily a material surface; in other words, all waves are necessarily longitudinal [cf. CFT, Sect. 297]. From (75.4) then follows the classical formula for the speed of sound in fluids: $U_\parallel^2 = dp/d\varrho$. Eq. (75.4) in generality is implied by the results on effective moduli in Sect. 69; in particular, by Eq. $(69.9)_2$.

Various general inequalities, including the S-E condition and the GCN$^+$ condition, when specialized to the case of hydrostatic stress yield the conclusion that $dp/d\varrho > 0$ always, as is to be expected intuitively. From (75.4) it follows that

$$U_\parallel^2 > \tfrac{4}{3}U_\perp^2. \tag{75.5}$$

In particular, *longitudinal waves always travel faster than transverse waves.* Whether or not transverse waves may propagate at all is a more delicate matter. The B-E inequalities $(51.26)_2$ are necessary and sufficient that $U_\perp^2 \geqq 0$; the possibility that $U_\perp = 0$, as is the case in an elastic fluid, is not excluded. If the B-E inequalities are satisfied, then from (75.4) we see that

$$U_\parallel^2 \geqq \frac{dp}{d\varrho}; \tag{75.6}$$

that is, *in an elastic material subject to hydrostatic pressure, if transverse waves are possible, so also are longitudinal waves, which necessarily propagate faster than sound waves in an elastic fluid with the same law of compression, and also faster than the transverse waves.* If, on the other hand, $U_\perp^2 < 0$, then transverse waves are impossible, while the speed of the longitudinal waves, if it is real, is less than the hydrodynamical speed of sound. It still remains possible, of course, that no waves at all may propagate.

76. Waves in isotropic compressible materials. III. Determination of the stress relation from wave speeds[1]. When the form of the stress relation is known,

[1] TRUESDELL [1961, *63*, § 9].

formulae such as (74.3) and (74.4) give unique values for the speeds of the 9 possible kinds of waves travelling along the principal axes of stress and strain. Conversely, if these 9 speeds are known as functions of the three principal stretches v_a, as for example from the results of experiments, then the three response coefficients $\aleph_0$, $\aleph_1$, and $\aleph_2$, or $\beth_0$, $\beth_1$, and $\beth_{-1}$, are uniquely determined.

Indeed, they are overdetermined. From (74.3) and (74.4) we may derive a sequence of *universal relations* connecting the wave speeds, relations which must hold in any isotropic elastic material regardless of the forms of $\aleph_1$, $\aleph_2$, and $\aleph_3$. Only if these relations are satisfied can there exist functions $\aleph_1$, $\aleph_2$, and $\aleph_3$ such that (74.3) and (74.4) hold. Thus these universal relations serve as *conditions of compatibility* for measured values of the wave speeds: unless these relations are satisfied by the data, those data are not consistent with the theory of isotropic elasticity[1].

Compatibility of the transverse waves. From (74.4) we derive at once two sets of three universal relations:

$$\left.\begin{aligned}
&\frac{U_{12}^2}{v_1^2}=\frac{U_{21}^2}{v_2^2}, \qquad \frac{U_{23}^2}{v_2^2}=\frac{U_{32}^2}{v_3^2}, \qquad \frac{U_{31}^2}{v_3^2}=\frac{U_{13}^2}{v_1^2},\\
&\frac{1}{v_1^2-v_3^2}\left(\frac{U_{12}^2}{v_1^2}-\frac{U_{23}^2}{v_2^2}\right)=\frac{1}{v_2^2-v_1^2}\left(\frac{U_{23}^2}{v_2^2}-\frac{U_{31}^2}{v_3^2}\right)=\frac{1}{v_3^2-v_2^2}\left(\frac{U_{31}^2}{v_3^2}-\frac{U_{12}^2}{v_1^2}\right),\\
&=\varphi(v_1, v_2, v_3),
\end{aligned}\right\} \tag{76.1}$$

where φ is a symmetric function of its arguments. Of these 6 conditions, the first 3 are independent, but from any 1 of the last 3 the other 2 may be derived by use of the first set, so in all there are 4 independent conditions. If we please, we may take as the independent set $(76.1)_{1,2,3}$ and

$$U_{23}^2=\frac{v_2^2}{v_1^2(v_2^2-v_3^2)}\left[(v_2^2-v_1^2)\,U_{12}^2-(v_3^2-v_1^2)\,U_{13}^2\right]. \tag{76.2}$$

If these conditions are satisfied, the functions $\aleph_1$ and $\aleph_2$ are uniquely determined by relations of the type (74.4). For example,

$$\left.\begin{aligned}
\aleph_2&=\varrho\,\varphi=\varrho\,\frac{U_{12}^2-U_{13}^2}{v_1^2(v_2^2-v_3^2)},\\
\aleph_1&=\frac{1}{3}\,\varrho\left[\frac{U_{12}^2}{v_1^2}+\frac{U_{23}^2}{v_2^2}+\frac{U_{31}^2}{v_3^2}-2I\varphi\right],\\
&=\varrho\,\frac{(v_2^2-I)\,U_{12}^2-(v_3^2-I)\,U_{13}^2}{v_1^2(v_2^2-v_3^2)}.
\end{aligned}\right\} \tag{76.3}$$

In each case the second form follows from the first by use of the identities (76.1) and (76.2). From the form of the results we see that *in an isotropic elastic material, the response coefficients $\aleph_1$ and $\aleph_2$ are uniquely determined by the speeds of the two kinds of transverse waves that may travel down any one principal axis of strain.* The experimenter willing to take compatibility for granted may content himself with measuring U_{12}^2 and U_{13}^2 as functions of v_1, v_2, v_3.

---

[1] TOUPIN and BERNSTEIN [1961, *62*, § 4] were the first to consider problems of this kind. They derived necessary and sufficient conditions for the infinitesimal theory of hyperelastic materials. The results presented above for the finite theory are sufficient only as regards waves travelling along a principal axis of stress and strain. If waves travelling in other directions are to be considered, additional conditions of compatibility must be satisfied. Earlier HUGHES and KELLY [1953, *13*, p. 1146], after obtaining more than sufficient formulae for determining the second-order elasticities of isotropic hyperelastic materials, had remarked that "any additional measurements give checks on the theory."

By (47.13) we may calculate $\beth_I$ and $\beth_{-I}$:

$$\beth_I = \varrho \frac{v_2^2 U_{12}^2 - v_3^2 U_{13}^2}{v_1^2 (v_2^2 - v_3^2)}, \qquad \beth_{-I} = \varrho v_1^2 v_2^2 v_3^2 \frac{U_{12}^2 - U_{13}^2}{v_1^2 (v_2^2 - v_3^2)}. \tag{76.4}$$

With the help of $(74.7)_2$, the relations (76.1) may be expressed alternatively in terms of the stresses:

$$\left.\begin{aligned} \varrho (U_{12}^2 - U_{21}^2) = t_1 - t_2 = 2\tau_3, \qquad \varrho (U_{23}^2 - U_{32}^2) = t_2 - t_3 = 2\tau_1, \\ \varrho (U_{31}^2 - U_{13}^2) = t_3 - t_1 = 2\tau_2, \end{aligned}\right\} \tag{76.5}$$

where τ_1, τ_2, τ_3 are the principal shear stresses (CFT, Sect. App. 46). Hence follows a *universal identity* connecting the six transverse wave speeds, no matter what the state of stress and strain:

$$U_{12}^2 + U_{23}^2 + U_{31}^2 = U_{21}^2 + U_{32}^2 + U_{13}^2. \tag{76.6}$$

Compatibility of the longitudinal waves. Let us assume that the conditions of compatibility for the transverse waves are satisfied, so that $\aleph_I$ and $\aleph_2$ are uniquely determined. We now consider the compatibility of the speeds of the 3 longitudinal waves, from which the single function $\aleph_0$ is to be determined.

If we set

$$A_1 \equiv \frac{\varrho U_{11}^2}{2 v_1^2} - \aleph_I - 2 v_1^2 \aleph_2 - \sum_{\Gamma=1}^{2} v_1^{2\Gamma} \left[\frac{\partial \aleph_\Gamma}{\partial I} + (v_2^2 + v_3^2) \frac{\partial \aleph_\Gamma}{\partial II} + v_2^2 v_3^2 \frac{\partial \aleph_\Gamma}{\partial III} \right], \tag{76.7}$$

etc., then because of (76.3) we can regard the 3 quantities A_a as explicitly given functions of U_{12}^2, U_{13}^2, U_{11}^2, U_{22}^2, U_{33}^2, v_1, v_2, v_3. In view of (74.3) we then have 3 relations of the type

$$A_1 = \frac{\partial \aleph_0}{\partial I} + (v_2^2 + v_3^2) \frac{\partial \aleph_0}{\partial II} + v_2^2 v_3^2 \frac{\partial \aleph_0}{\partial III}, \tag{76.8}$$

from which $\aleph_0$ is to be determined. If the three principal stretches v_a are distinct, we may solve (76.8) as a linear system, obtaining the following results:

$$\left.\begin{aligned} \frac{\partial \aleph_0}{\partial I} &= -\frac{A_1 v_1^4 (v_2^2 - v_3^2) + A_2 v_2^4 (v_3^2 - v_1^2) + A_3 v_3^4 (v_1^2 - v_2^2)}{(v_1^2 - v_2^2)(v_2^2 - v_3^2)(v_3^2 - v_1^2)} \equiv B_1, \\ \frac{\partial \aleph_0}{\partial II} &= \frac{A_1 v_1^2 (v_2^2 - v_3^2) + A_2 v_2^2 (v_3^2 - v_1^2) + A_3 v_3^2 (v_1^2 - v_2^2)}{(v_1^2 - v_2^2)(v_2^2 - v_3^2)(v_3^2 - v_1^2)} \equiv B_2, \\ \frac{\partial \aleph_0}{\partial III} &= -\frac{A_1 (v_2^2 - v_3^2) + A_2 (v_3^2 - v_1^2) + A_3 (v_1^2 - v_2^2)}{(v_1^2 - v_2^2)(v_2^2 - v_3^2)(v_3^2 - v_1^2)} \equiv B_3. \end{aligned}\right\} \tag{76.9}$$

The quantities B_a are known functions of v_1, v_2, v_3; they must be symmetric functions and hence equal to functions of I, II, III. In order that $\aleph_0$ may be determined from the system (76.9) ,the functions B_a must satisfy the following conditions of integrability:

$$\frac{\partial B_1}{\partial II} = \frac{\partial B_2}{\partial I}, \qquad \frac{\partial B_1}{\partial III} = \frac{\partial B_3}{\partial I}, \qquad \frac{\partial B_2}{\partial III} = \frac{\partial B_3}{\partial II}. \tag{76.10}$$

If these conditions are satisfied, then

$$\aleph_0 = \int (B_1 dI + B_2 dII + B_3 dIII). \tag{76.11}$$

A summary of these results is given by the ***compatibility theorem:*** *Let the 9 speeds of propagation for principal waves be known functions of the principal stretches. These data are consistent with the theory of isotropic elastic materials if and only if:*

(a) *The quantities* φ, B_1, B_2, B_3, *as defined by* $(76.1)_{4,5,6}$ *and* (76.9), *are symmetric functions of the principal stretches.*

(b) *The 4 algebraic conditions* $(76.1)_{1,2,3}$ *and* (76.2) *and the 3 differential conditions* (76.10) *are satisfied.*

If conditions (a) *and* (b) *hold, by means of* (76.3) *and* (76.11) *the data determine the form of the stress relation, uniquely to within an arbitrary uniform pressure.*

Thus we have shown that in the theory of finite elastic strain of isotropic materials, just as in the infinitesimal theory, a suitable set of wave speeds determines the form of the stress relation. In the case of the infinitesimal theory, the experimental difficulties connected with reconciling static and undular moduli are well known, and in the finite theory we may expect them to be aggravated.

77. Waves in isotropic compressible materials. IV. Second-order effects[1]. We now find the first terms in the expansions of the exact formulae (74.3) and (74.4) in powers of the extensions δ_a, where $\delta_a = v_a - 1$. Exactly the same results would follow from using the second-order theory presented in Sect. 66; accordingly, we shall use the notations laid down there, although we shall not directly employ the stress relation (66.3).

We presume the response coefficients $\beth_\Gamma$ to have power-series expansions of the form (66.2). These we substitute into the formulae for the wave speeds, retaining only the terms linear in the δ_a. Using the notation $\doteq$ to denote equality to within an error which is $O(\delta_1^2+\delta_2^2+\delta_3^2)$ as $(\delta_1^2+\delta_2^2+\delta_3^2)\to 0$, from $(74.4)_2$ we have

$$\left.\begin{aligned} \varrho\frac{U_{12}^2}{\mu} &\doteq \frac{1}{\mu}(1+2\delta_1)\beth_1 - \frac{1}{\mu}(1-2\delta_2)\beth_{-1},\\ &\doteq (1+2\delta_1)(\alpha_{10}+\alpha_{11}I_{\tilde{E}}) - (1-2\delta_2)(\alpha_{-10}+\alpha_{-11}I_{\tilde{E}}),\\ &\doteq (1+2\delta_1)(\alpha_{10}-\alpha_{-10}) + (\alpha_{11}-\alpha_{-11})I_{\tilde{E}} + 2\alpha_{-10}(\delta_1+\delta_2),\\ &\doteq 1+2\delta_1+\tfrac{1}{2}\alpha_5(\delta_1+\delta_2+\delta_3)+\tfrac{1}{2}\alpha_6(\delta_1+\delta_2), \end{aligned}\right\} \tag{77.1}$$

where (66.4) and (66.5) have been used. Since $\varrho_R \doteq \varrho(1+\delta_1+\delta_2+\delta_3)$, we may express (77.1) in the form

$$\varrho_R\frac{U_{12}^2}{\mu} \doteq 1+2\delta_1+\left(1+\frac{1}{2}\alpha_5\right)(\delta_1+\delta_2+\delta_3)+\frac{1}{2}\alpha_6(\delta_1+\delta_2), \tag{77.2}$$

giving the *second-order speeds of propagation of transverse principal waves.*

It is not so easy to get the corresponding results for longitudinal waves. First we write (74.3) in the form

$$\varrho\frac{U_{11}^2}{2v_1^2} = A + B, \tag{77.3}$$

[1] TRUESDELL [1961, *63*, § 11] [1964, *88*]. Results of this kind, more general in considering materials of cubic symmetry but less general in restricting attention to a one-parameter family of deformations in a hyperelastic material, are given by THURSTON and BRUGGER [1964, *83*, §§ 5—7]. For hyperelastic materials, an elaborate discussion of second-order effects in wave propagation in isotropic media was given by L. BRILLOUIN [1925, *2*, §§ 6—7, 10—11] [1938, *2*, Ch. XI, §§ VII—XVI]. He obtained the wave speeds in a body subject to hydrostatic pressure [1925, *2*, Eqs. (76) (77)] [1938, *2*, Eqs. (xi.125), (xi.126)] (cf. also JOUGUET [1920, *2*]), or to uniaxial tension [1925, *2*, § 11], and he proposed certain inequalities for the second-order elasticities based on physically natural requirements for the wave speeds and radiation pressures [1925, *2*, § 12] [1938, *2*, Ch. XI, § XVII]. Wave speeds in more general circumstances were calculated by HUGHES and KELLY [1953, *13*, Eqs. (12)]. SHIMAZU [1954, *21*, Chap. 4] calculated the second-order wave speeds in hyperelastic materials when the initial deformation is biaxial or is a simple shear; the former case was studied later by TOUPIN and BERNSTEIN [1961, *62*, Eqs. (5.12)], who gave various equivalent forms and consequences of the results as well as discussion of related experimental methods.

where

$$\left.\begin{aligned} A &\equiv \aleph_1 + 2v_1^2\aleph_2, \\ B &\equiv \sum_{\Gamma=0}^{2} v_1^{2\Gamma}\left[\frac{\partial\aleph_\Gamma}{\partial I} + (v_2^2+v_3^2)\frac{\partial\aleph_\Gamma}{\partial II} + v_2^2 v_3^2 \frac{\partial\aleph_\Gamma}{\partial III}\right]. \end{aligned}\right\} \tag{77.4}$$

Then from (47.14), (66.6), (66.4), and (66.5) we see that

$$\left.\begin{aligned} \frac{A}{\mu} &= \frac{1}{\mu}\left[\beth_1 + \frac{v_1^2-v_2^2-v_3^2}{v_1^2v_2^2v_3^2}\beth_{-1}\right], \\ &\doteq \alpha_{10} + \alpha_{11} I_{\tilde{E}} + \frac{-1-2I_{\tilde{E}}+4\delta_1}{1+2I_{\tilde{E}}}(\alpha_{-10}+\alpha_{-11}I_{\tilde{E}}), \\ &\doteq 1 + \tfrac{1}{2}\alpha_5(\delta_1+\delta_2+\delta_3) + \alpha_6\delta_1. \end{aligned}\right\} \tag{77.5}$$

Also

$$\left.\begin{aligned} B &\doteq \frac{\partial\aleph_0}{\partial I} + 2(1+I_{\tilde{E}}-\delta_1)\frac{\partial\aleph_0}{\partial II} + (1+2I_{\tilde{E}}-2\delta_1)\frac{\partial\aleph_0}{\partial III} + \\ &\quad + (1+2\delta_1)\left[\frac{\partial\aleph_1}{\partial I} + 2(1+I_{\tilde{E}}-\delta_1)\frac{\partial\aleph_1}{\partial II} + (1+2I_{\tilde{E}}-2\delta_1)\frac{\partial\aleph_1}{\partial III}\right] + \\ &\quad + (1+4\delta_1)\left[\frac{\partial\aleph_2}{\partial I} + 2(1+I_{\tilde{E}}-\delta_1)\frac{\partial\aleph_2}{\partial II} + (1+2I_{\tilde{E}}-2\delta_1)\frac{\partial\aleph_2}{\partial III}\right], \\ &\doteq \left(\frac{\partial}{\partial I} + 2\frac{\partial}{\partial II} + \frac{\partial}{\partial III}\right)(\aleph_0+\aleph_1+\aleph_2) + \\ &\quad + 2(I_{\tilde{E}}-\delta_1)\left(\frac{\partial}{\partial II} + \frac{\partial}{\partial III}\right)(\aleph_0+\aleph_1+\aleph_2) + \\ &\quad + 2\delta_1\left(\frac{\partial}{\partial I} + 2\frac{\partial}{\partial II} + \frac{\partial}{\partial III}\right)(\aleph_1+2\aleph_2). \end{aligned}\right\} \tag{77.6}$$

But from (63.19),

$$\frac{\partial}{\partial I} + 2\frac{\partial}{\partial II} + \frac{\partial}{\partial III} = \frac{1}{2}\frac{\partial}{\partial I_E}, \qquad \frac{\partial}{\partial II} + \frac{\partial}{\partial III} = \frac{1}{4}\frac{\partial}{\partial II_E}. \tag{77.7}$$

Hence (77.6) becomes

$$\left.\begin{aligned} B &\doteq \frac{1}{2}\frac{\partial}{\partial I_E}(\aleph_0+\aleph_1+\aleph_2) + \\ &\quad + \frac{1}{2}(I_{\tilde{E}}-\delta_1)\frac{\partial}{\partial II_E}(\aleph_0+\aleph_1+\aleph_2) + \delta_1\frac{\partial}{\partial I_E}(\aleph_1+2\aleph_2). \end{aligned}\right\} \tag{77.8}$$

Now by (47.14) and (63.19),

$$\aleph_0+\aleph_1+\aleph_2 = \beth_0+\beth_1+\beth_{-1} - \frac{8III_E}{1+2I_E+4II_E+8III_E}\beth_{-1}. \tag{77.9}$$

From this formula, by the aid of (66.2) and (66.4),

$$\left.\begin{aligned} \frac{\partial}{\partial I_E}(\aleph_0+\aleph_1+\aleph_2) &\doteq \frac{\partial}{\partial I_E}(\beth_0+\beth_1+\beth_{-1}), \\ &\doteq \mu(\alpha_{01}+\alpha_{11}+\alpha_{-11}) + 2\mu(\alpha_{02}+\alpha_{12}+\alpha_{-12})I_{\tilde{E}} \\ &\doteq \mu(\alpha_1+2\alpha_3 I_{\tilde{E}}). \end{aligned}\right\} \tag{77.10}$$

Likewise

$$\left.\begin{aligned} (I_{\tilde{E}}-\delta_1)\frac{\partial}{\partial II_E}(\aleph_0+\aleph_1+\aleph_2) &\doteq (I_{\tilde{E}}-\delta_1)\mu(\alpha_{03}+\alpha_{13}+\alpha_{-13}), \\ &= (I_{\tilde{E}}-\delta_1)\mu\alpha_4, \end{aligned}\right\} \tag{77.11}$$

and

$$\left.\begin{aligned} \delta_1\frac{\partial}{\partial I_{\tilde{E}}}(\aleph_1+2\aleph_2) &\doteq \delta_1\frac{\partial}{\partial I_{\tilde{E}}}(\beth_1-\beth_{-1}) \\ &\doteq \mu(\alpha_{11}-\alpha_{-11})\delta_1 = \tfrac{1}{2}\mu\alpha_5\delta_1. \end{aligned}\right\} \tag{77.12}$$

Putting (77.12), (77.11), and (77.10) into (77.8) yields

$$\left.\begin{aligned}\frac{B}{\mu} \doteq \frac{1}{2}\alpha_1 + \left(\alpha_3 + \frac{1}{2}\alpha_4\right)(\delta_1+\delta_2+\delta_3) + \\ +\left(-\tfrac{1}{2}\alpha_4 + \tfrac{1}{2}\alpha_5\right)\delta_1.\end{aligned}\right\} \tag{77.13}$$

Finally, substituting (77.13) and (77.5) into (77.3) and observing $(66.6)_1$, we obtain the definite result for the *second-order speeds of propagation of longitudinal principal waves:*

$$\left.\begin{aligned}\varrho_R \frac{U_{11}^2}{\mu} \doteq \frac{\lambda}{\mu} + 2 + \left(\frac{\lambda}{\mu} + 2 + 2\alpha_3 + \alpha_4 + \alpha_5\right)(\delta_1+\delta_2+\delta_3) + \\ +\left(\frac{2\lambda}{\mu} + 4 - \alpha_4 + \alpha_5 + 2\alpha_6\right)\delta_1.\end{aligned}\right\} \tag{77.14}$$

No linear function is positive-definite unless it reduces to a positive constant. Hence there are *no* conditions on the second-order elasticities α_3, α_4, α_5, and α_6 sufficient to render the second-order wave speeds (77.2) and (77.14) positive in all deformations, beyond the trivial one that $\alpha_3 = \alpha_4 = \alpha_5 = \alpha_6 = 0$, reducing the second-order theory to the first-order one. Of course, the second-order theory is designed only to describe, approximately, deformations that are sufficiently small, but for sufficiently small extensions δ_a all squared wave speeds are positive in the second-order theory, provided only their first-order counterparts be positive, so that by this argument *no restrictions* on $\alpha_3, \alpha_4, \alpha_5, \alpha_6$ result. This negative outcome is typical of the failure of perturbation processes to give clear and definite information on matters of principle. In the exact theory, first-order conditions are of course far from sufficient that the squared wave-speeds, given by (74.3) and (74.4), be positive.

With the explicit formulae (77.2) and (77.14) in hand, we are in a position to examine the compatibility of measured values of the wave speeds with the second-order theory for isotropic elastic materials. To express these results simply, we write $U_\perp^2$ and $U_\parallel^2$ for the speeds according to the infinitesimal theory:

$$\varrho_R U_\perp^2 = \mu, \qquad \varrho_R U_\parallel^2 = \lambda + 2\mu, \tag{77.15}$$

corresponding to the terms free of the δ_a in (77.2) and (77.14).

Compatibility of the transverse waves. The general compatibility relations (76.1) may be reduced to the following four independent conditions for the second-order theory:

$$\left.\begin{aligned}\frac{U_{12}^2 - U_{21}^2}{\delta_1-\delta_2} = \frac{U_{23}^2 - U_{32}^2}{\delta_2-\delta_3} = \frac{U_{31}^2 - U_{13}^2}{\delta_3-\delta_1} = 2U_\perp^2, \\ U_{23}^2 = 2(\delta_2-\delta_1)U_\perp^2 + \frac{1}{2}(U_{12}^2 + U_{13}^2) + \frac{1}{2}\frac{U_{12}^2 - U_{13}^2}{\delta_2-\delta_3}(\delta_2+\delta_3-2\delta_1).\end{aligned}\right\} \tag{77.16}$$

The first three of Eqs. (77.16) may be obtained directly from (77.1) and (77.15) or, alternatively, from (76.5) and the relations $t_a - t_b = 2\mu(\delta_a - \delta_b)$. The last of Eqs. (77.16) follows from (76.2) and (77.15). The conditions (77.16) show that at most 2 of the 6 squared speeds of transversal waves are independent. If the conditions of compatibility (77.16) are satisfied, then *unique values for the second-order elasticities α_5 and α_6 are determined by the speeds of the transverse waves.* For example, from (77.2) we derive

$$\left.\begin{aligned}\frac{1}{2}\alpha_6 = \frac{U_{12}^2 - U_{13}^2}{(\delta_2-\delta_3)U_\perp^2}, \\ \frac{1}{2}\alpha_5 = \frac{1}{\delta_1+\delta_2+\delta_3}\left[-2\delta_1 - \frac{1}{4}\alpha_6(2\delta_1+\delta_2+\delta_3) + \frac{U_{12}^2 + U_{13}^2}{2U_\perp^2} - 1\right] - 1.\end{aligned}\right\} \tag{77.17}$$

The second of these is not the simplest expression that might be used, but together the two formulae give the constants α_5 and α_6 in terms of the simplest data, namely the sum and the difference of the squared speeds of the two kinds of transverse waves that can travel down the same principal axis.

Compatibility of the longitudinal waves. The conditions (76.10) are complicated and awkward to apply to the second-order approximation. It is easier to work directly with the system of three equations of the form (77.14). We see at once that in order for this system to be compatible, it is necessary and sufficient that

$$(U_{11}^2 - U_{\|}^2)(\delta_2 - \delta_3) + (U_{22}^2 - U_{\|}^2)(\delta_3 - \delta_1) + (U_{33}^2 - U_{\|}^2)(\delta_1 - \delta_2) = 0. \quad (77.18)$$

If the three longitudinal wave speeds satisfy this condition, then the values of any two of them suffice to determine the values of α_3 and α_4. For example,

$$\left.\begin{aligned} \alpha_3 &= \frac{(U_{11}^2 - U_{\|}^2)\,\delta_2 - (U_{22}^2 - U_{\|}^2)\,\delta_1}{2(\delta_1 + \delta_2 + \delta_3)(\delta_2 - \delta_1)\,U_\perp^2} + \frac{U_{22}^2 - U_{11}^2}{2(\delta_2 - \delta_1)\,U_\perp^2} - \frac{3}{2}\left(2 + \frac{\lambda}{\mu}\right) - \alpha_5 - \alpha_6, \\ \alpha_4 &= -\frac{U_{22}^2 - U_{11}^2}{(\delta_2 - \delta_1)\,U_\perp^2} + 2\left(2 + \frac{\lambda}{\mu}\right) + \alpha_5 + 2\alpha_6. \end{aligned}\right\} \quad (77.19)$$

In summary, *if the 9 speeds of propagation of principal waves are linear functions of the principal extensions, in order to be compatible with the second-order theory of elastic materials they must satisfy the 5 algebraic identities (77.16) and (77.18); if they do satisfy these relations, then any 2 transverse speeds and any 2 longitudinal speeds determine unique values for the 2 first-order and 4 second-order elasticities.*

The formula $(74.15)_1$ when applied to the second-order theory, or, equivalently to the case of a wave propagating in a material that is undeformed before the wave passes, yields[1]

$$K_{11} = \frac{1}{2}\,\mu\left(\frac{\lambda}{\mu} + 2 + 2\alpha_3 + 2\alpha_5 + 2\alpha_6\right). \quad (77.20)$$

This formula, interpreted as in Sect. 74, shows that the discontinuities which grow are those giving rise to speeds of propagation behind the wave which exceed the speed of the wave itself. The discontinuity then waxes by accumulation of disturbances from the region of strain behind the wave; in this region the strain differs from the homogeneous strain at the wave, and this region catches up with the wave as the motion progresses, in the same way as a shock is formed according to gas dynamics. From (77.20) and the result in Sect. 74 it follows that if

$$\frac{\lambda}{\mu} + 2 + 2\alpha_3 + 2\alpha_5 + 2\alpha_6 > 0, \quad (77.21)$$

only rarefaction shocks can result from the growth of an acceleration wave propagating into a body at rest in its natural state. If the inequality in (77.21) is reversed, only condensation shocks can form in this way. The results here are consistent with the criteria of BLAND[2] for propagation of infinitely weak shock waves.

78. Waves in incompressible materials[3]. The theory of waves in incompressible materials is much the same as the foregoing, except that the constraint of incompressibility renders longitudinal waves impossible: *In an incompressible material, all singular surfaces are necessarily transversal* (CFT, Sect. 191):

$$\boldsymbol{a} \cdot \boldsymbol{n} = 0. \quad (78.1)$$

[1] W. A. GREEN [1964, *40*].
[2] BLAND [1964, *8*].
[3] Some of the analysis here, under assumptions partly more general and partly less so, has been given by HILL [1962, *34*, § 6].

The geometrical and kinematical conditions (71.1) remain in force. Substituting into $(71.2)_2$ any of the forms $(43.14)_{1,2,3}$ of the constitutive equations for incompressible elastic materials yields

$$[p]=0, \tag{78.2}$$

provided the response functions depend continuously on the place in the reference configuration. Thus, across any singular surface of order greater than 1, the pressure is continuous. By MAXWELL's theorem [CFT, Eq. (175.8)] it follows that

$$[\operatorname{grad} p]=A\,\boldsymbol{n}. \tag{78.3}$$

Substitution of (71.1) and (78.3) into the equation resulting from calculating the jump of (44.27) across a singular surface of order 2, we have

$$-A\,\boldsymbol{n}+\left(\boldsymbol{Q}(\boldsymbol{n})-\varrho\,U^2\,\mathbf{1}\right)\boldsymbol{a}=\mathbf{0}, \tag{78.4}$$

where $\boldsymbol{Q}(\boldsymbol{n})$ is given by an equation of the form (71.4) or (71.5). Taking the scalar product of (78.4) by $\boldsymbol{n}$ and then using (78.1) yields

$$A=\boldsymbol{n}\cdot\left(\boldsymbol{Q}(\boldsymbol{n})\,\boldsymbol{a}\right). \tag{78.5}$$

Substituting this result into (78.4), we obtain the following definitive equation for the amplitudes and speeds[1]:

$$\left(\boldsymbol{Q}^*(\boldsymbol{n})-\varrho\,U^2\,\mathbf{1}\right)\boldsymbol{a}=\mathbf{0}, \tag{78.6}$$

where

$$\boldsymbol{Q}^*(\boldsymbol{n})=\boldsymbol{Q}(\boldsymbol{n})-\boldsymbol{n}\otimes\boldsymbol{Q}(\boldsymbol{n})^T\boldsymbol{n}, \qquad Q^*_{km}=Q_{km}-n_k\,n^p\,Q_{pm}. \tag{78.7}$$

The speeds may be calculated explicitly as follows. First, the characteristic equation of (78.6) is

$$-(\varrho\,U^2)^3+(\varrho\,U^2)^2\,I_{\boldsymbol{Q}^*}-\varrho\,U^2\,II_{\boldsymbol{Q}^*}+III_{\boldsymbol{Q}^*}=0. \tag{78.8}$$

But, from (78.7), $\boldsymbol{Q}^{*T}\boldsymbol{n}=\mathbf{0}$; hence $\boldsymbol{Q}^*$ is singular; hence $III_{\boldsymbol{Q}^*}=0$, so that (78.8) assumes the form

$$(\varrho\,U^2)^2-\varrho\,U^2\,I_{\boldsymbol{Q}^*}+II_{\boldsymbol{Q}^*}=0, \tag{78.9}$$

where a root $\varrho\,U^2=0$, corresponding to a possible non-propagating discontinuity of the second derivatives $x^k{}_{,\alpha\beta}$ alone, has been cancelled. Hence the two pairs of generally non-zero speeds of propagation are given by the formula

$$2\varrho\,U^2=I_{\boldsymbol{Q}^*}\pm\sqrt{I^2_{\boldsymbol{Q}^*}-4\,II_{\boldsymbol{Q}^*}}, \tag{78.10}$$

first found by ERICKSEN for a more special theory[2]. The corresponding amplitudes may be read off from (78.6); for example, if the x^3 co-ordinate is taken normal to the wave front, then $\boldsymbol{a}$ has the components $(a_1, a_2, 0)$ where $(Q^*_{11}-\varrho\,U^2)\,a_1+Q^*_{12}\,a_2=0$. Of course, if it is known that $\boldsymbol{a}$ is a possible amplitude, the corresponding speed of propagation is given more easily by the formula

$$\varrho\,U^2=\frac{\boldsymbol{a}\cdot(\boldsymbol{Q}^*(\boldsymbol{n})\,\boldsymbol{a})}{a^2}, \qquad a=\sqrt{\boldsymbol{a}^2}, \tag{78.11}$$

which follows at once from (78.6).

The formal correspondence between singular surfaces of any two orders greater than 1, and that between any such surface and a plane progressive infinitesimal

[1] MANACORDA [1959, *19*, § 3].
[2] ERICKSEN [1953, *6*, § 4] considered isotropic hyperelastic materials.

wave, established in Sects. 72 and 73 for compressible materials, are easily extended to incompressible materials as well.

For isotropic materials[1], by substituting (49.9) into (74.1) we find that $\boldsymbol{Q}(\boldsymbol{n})$ is given by

$$\left.\begin{aligned} \boldsymbol{Q}(\boldsymbol{n}) = {} & \aleph_I[(\boldsymbol{n}\cdot\boldsymbol{B}\boldsymbol{n})\mathbf{1}+(\boldsymbol{B}\boldsymbol{n})\otimes\boldsymbol{n}]+ \\ & +\aleph_2[(\boldsymbol{n}\cdot\boldsymbol{B}^2\boldsymbol{n})\mathbf{1}+(\boldsymbol{B}\boldsymbol{n})\otimes(\boldsymbol{B}\boldsymbol{n})+(\boldsymbol{B}^2\boldsymbol{n})\otimes\boldsymbol{n}+(\boldsymbol{n}\cdot\boldsymbol{B}\boldsymbol{n})\boldsymbol{B}]+ \\ & +2\sum_{\Gamma=1}^{2}\boldsymbol{B}^\Gamma\boldsymbol{n}\otimes\left[\left(\frac{\partial\aleph_\Gamma}{\partial I}+I\frac{\partial\aleph_\Gamma}{\partial II}\right)\boldsymbol{B}\boldsymbol{n}-\frac{\partial\aleph_\Gamma}{\partial II}\boldsymbol{B}^2\boldsymbol{n}\right]. \end{aligned}\right\} \quad (78.12)$$

Accordingly, when $\boldsymbol{n}$ is a unit vector in a principal direction of strain, from (78.7) and (78.12) we easily show that

$$v_1^{-2}\boldsymbol{Q}^*(\boldsymbol{n}_1) = (\aleph_I + v_1^2\aleph_2)\mathbf{1} - (\aleph_I + 2v_1^2\aleph_2)\boldsymbol{n}_1\otimes\boldsymbol{n}_1 + \aleph_2\boldsymbol{B}. \quad (78.13)$$

Therefore, just as for a compressible material, the acoustical axes for principal waves are themselves the principal axes. However, longitudinal waves are impossible because the material is incompressible. By taking $\boldsymbol{a}=\boldsymbol{n}_2$, where $\boldsymbol{n}_2$ is a proper vector of $\boldsymbol{B}$ that is orthogonal to $\boldsymbol{n}_1$, from (78.11) and (78.13) we see at once that the speed U_{12} of a transverse wave with amplitude $\boldsymbol{n}_2$ is again given by (74.4), or, alternatively, since $v_1v_2v_3=1$,

$$\frac{\varrho U_{12}^2}{v_1^2} = \beth_I - v_3^2\beth_{-I}. \quad (78.14)$$

Hence *the B-E inequalities* $(53.1)_2$ *are necessary and sufficient conditions that the speeds of all principal waves be real.* After the great stability theorem of HADAMARD (Sect. 71), this now celebrated result of ERICKSEN was the first simple and definite connection found between the existence of waves and physically reasonable static response of an elastic material in finite deformation.

For isotropic incompressible materials the speeds of the (necessarily transverse) principal waves depend upon two variables rather than three, since $v_1v_2v_3=1$, but in general the 6 speeds U_{12}^2, U_{13}^2, ..., U_{32}^2 are distinct. The conditions of compatibility (76.1) and (76.2) remain valid, except, of course, that only those stretches for which $v_1v_2v_3=1$ are possible. The formulae (76.3) and (76.4) for the response coefficients remain valid under the same restriction. Therefore, *the squared speeds of propagation for the two kinds of transverse waves corresponding to any one principal axis of stress and strain determine uniquely the stress relation of an isotropic incompressible elastic material.*

For second-order effects we may proceed just as in Sect. 77, obtaining the following result:

$$\frac{\varrho U_{12}^2}{\mu} \doteq 1 + (\delta_1 - \delta_2) + 2\beta(\delta_1 + \delta_2), \quad (78.15)$$

where we have used (67.1). The 4 conditions of compatibility (77.16) remain in force, but since $\delta_1+\delta_2+\delta_3=0$, $(77.16)_4$ may be expressed more simply as follows:

$$U_{23}^2 = 2(\delta_2-\delta_1)U_\perp^2 + \frac{1}{2}(U_{12}^2+U_{13}^2) - \frac{3}{2}\frac{\delta_1}{\delta_1+2\delta_2}(U_{12}^2-U_{13}^2). \quad (78.16)$$

There is, however, 1 additional condition, namely,

$$\frac{U_{12}^2+U_{13}^2-2U_\perp^2}{\delta_1} - \frac{U_{12}^2-U_{13}^2}{\delta_2-\delta_3} = 4U_\perp^2. \quad (78.17)$$

[1] Eqs. (78.12)—(78.14), and the conclusions drawn from them, were obtained by ERICKSEN [1953, *6*, § 4] for hyperelastic materials.

When all these conditions are satisfied, there remains but one independent squared speed of propagation. Therefore, *the stress relation of the second-order theory of isotropic incompressible elastic materials is uniquely determined if any one principal-wave speed is known as a linear function of the extensions.* Indeed, it is not necessary that two extensions be allowed to vary independently, for in the special case when $\delta_1 = \delta_2 = \delta$ and hence $\delta_3 = -2\delta$, from (78.15) we see that

$$\beta = \frac{1}{4\delta}\left(\frac{U_{12}^2}{U_\perp^2} - 1\right). \tag{78.18}$$

II. Hyperelastic materials.

a) General considerations.

79. Thermodynamic preliminaries. For a detailed treatment of the concepts of the thermodynamics of continuous media we refer the reader to Chapter E of CFT. Here we summarize those definitions and theorems that serve as a basis for the theory of hyperelastic materials.

We associate with each dynamical process a *specific internal energy* $\varepsilon = \varepsilon(X, t)$. If $\mathscr{P}$ is a part of the body under consideration, the integral

$$E(\mathscr{P}) = \int_{\mathscr{P}} \varepsilon\, dm \tag{79.1}$$

defines the *internal energy* of the part $\mathscr{P}$. We introduce a vector field $\boldsymbol{h} = \boldsymbol{h}(X, t)$, called the *heat flux*[1], and a scalar field $q = q(X, t)$, called the specific *heat absorption*. The integral

$$Q = \int_{\partial\mathscr{P}} \boldsymbol{h}\cdot\boldsymbol{n}\, ds + \int_{\mathscr{P}} q\, dm \tag{79.2}$$

where $\partial\mathscr{P}$ is the boundary surface of $\mathscr{P}$ and $\boldsymbol{n}$ the exterior unit normal, defines the *heat* entering the part $\mathscr{P}$ per unit time. We may interpret[2] the first term on the right-hand side of (79.2) as giving the heat entering through the surface of $\mathscr{P}$ by conduction, the second term as giving the heat transmitted into $\mathscr{P}$ by radiation.

The equation of energy balance[3] reads

$$\left.\begin{aligned} \varrho\dot{\varepsilon} &= \operatorname{tr}(\boldsymbol{T}\boldsymbol{D}) + \operatorname{div}\boldsymbol{h} + \varrho q, \\ \varrho\dot{\varepsilon} &= T^{km} D_{km} + h^k{}_{,k} + \varrho q, \end{aligned}\right\} \tag{79.3}$$

where ϱ is the density, $\boldsymbol{T}$ the stress tensor, and $\boldsymbol{D}$ the stretching tensor.

We also associate with each dynamical process a *specific entropy* $\eta = \eta(X, t)$. The entropy H of a part $\mathscr{P}$ of the body is defined by the integral

$$H = \int_{\mathscr{P}} \eta\, dm. \tag{79.4}$$

Finally, we introduce the *temperature field* $\theta = \theta(X, t)$, which is assumed to have positive values:

$$\theta > 0. \tag{79.5}$$

If a dynamical process $\{\boldsymbol{\chi}, \boldsymbol{T}\}$, an entropy field η, and a temperature field θ are given, and if an internal energy field ε, a heat flux field $\boldsymbol{h}$, and a heat absorption

[1] It is more usual to call $-\boldsymbol{h}$ the heat flux, since $\boldsymbol{h}\cdot\boldsymbol{n} < 0$ at points where heat is entering the body.

[2] In more general theories, such as those of polar, oriented, or heterogeneous media, other interpretations are possible. Here we wish to stay close to classical, uncomplicated physical contexts.

[3] This is Eq. (241.4) of CFT in the special case when ${}_m p^{q r} = 0$. It is assumed that momentum and moment of momentum are balanced, i.e., that CAUCHY's laws (16.6) and (16.7) hold.

field q are prescribed such that the equation of energy balance (79.3) is valid, we speak of a *thermodynamic process.*

Suppose that $\boldsymbol{\chi}$, $\boldsymbol{T}$, ε, $\boldsymbol{h}$, and η are given. The specific heat absorption q is then uniquely determined by the equation of energy balance (79.3), just as the body force $\boldsymbol{b}$ is determined by CAUCHY's law (16.6). Thus, any given quintuple $\{\boldsymbol{\chi}, \boldsymbol{T}, \varepsilon, \boldsymbol{h}, \eta\}$ may be completed by a heat absorption q so as to constitute a thermodynamic process. While the heat absorptions obtained in this way may be artificial, there is no reason to exclude them from the general theory.

The *production of entropy* per unit time in the part $\mathscr{P}$ is defined by

$$\Gamma \equiv \dot{H} - \int_{\partial\mathscr{P}} \frac{\boldsymbol{h}}{\theta} \cdot \boldsymbol{n}\, ds - \int_{\mathscr{P}} \frac{q}{\theta}\, dm. \tag{79.6}$$

Here the vector $\boldsymbol{h}/\theta$ is to be interpreted as the flux of entropy leaving $\mathscr{P}$ by conduction through the surface, and q/θ the entropy leaving $\mathscr{P}$ by radiation. Using the divergence theorem and (79.4), one can easily show that Γ is the integral,

$$\Gamma = \int_{\mathscr{P}} \gamma\, dm, \tag{79.7}$$

of the *specific production of entropy*, γ:

$$\gamma = \gamma_{\text{loc}} + \gamma_{\text{con}}, \tag{79.8}$$

where

$$\gamma_{\text{loc}} = \dot{\eta} - \frac{1}{\varrho\theta} \operatorname{div} \boldsymbol{h} - \frac{q}{\theta} \tag{79.9}$$

and

$$\gamma_{\text{con}} = \frac{1}{\varrho\theta^2} \boldsymbol{h} \cdot \operatorname{grad} \theta. \tag{79.10}$$

We call γ_{loc} the *local entropy production* and γ_{con} the *entropy production by conduction of heat.*

In thermodynamic processes often *both types of entropy production are non-negative,* as we here assume[1]:

$$\gamma_{\text{loc}} \geqq 0, \qquad \gamma_{\text{con}} \geqq 0. \tag{79.11}$$

The first of these inequalities corresponds to the physical observation that a substance at uniform temperature free from sources of heat may consume mechanical energy but cannot give it out. The second inequality corresponds to the fact that heat does not flow spontaneously from the colder to the hotter parts of a body. The postulate (79.11) is a special case of the ***Clausius-Duhem inequality*** or the ***principle of dissipation***.

It follows from (79.3), (79.5), and (79.9) that the inequality $(79.11)_1$ is equivalent to

$$\theta\gamma_{\text{loc}} = \frac{1}{\varrho} \operatorname{tr}(\boldsymbol{T}\boldsymbol{D}) + \theta\dot{\eta} - \dot{\varepsilon} \geqq 0. \tag{79.12}$$

We extend the concepts of change of frame of reference (cf. Sect. 17) and of equivalence of processes (cf. Sect. 18) to thermodynamic processes by requiring all thermal variables to be frame-indifferent. Thus, the scalars ε, q, η, and θ are invariant under change of frame, while the heat flux $\boldsymbol{h}$ obeys the vector law of transformation (17.3), i.e.

$$\boldsymbol{h}^*(X, t^*) = \boldsymbol{Q}(t)\, \boldsymbol{h}(X, t). \tag{79.13}$$

[1] In Sect. 96, below, inequalities (79.11) are shown to follow from an apparently less specific assumption, for a certain class of materials.

It is not hard to see that the energy balance condition (79.3) is invariant under changes of frame. This invariance insures that thermodynamic processes are transformed again into thermodynamic processes.

We now formulate an extension of the ***principle of material frame-indifference*** of Sect. 19: *Constitutive equations involving thermodynamic and mechanical variables must be invariant under change of frame.*

80. Perfect materials. The best understood theory of the combined thermal and mechanical behavior of a continuous medium is the theory of perfect materials. The local material properties of a perfect material at a particle X are defined by the following thermo-mechanical constitutive assumption:

(i) *There are response functions $\hat{\varepsilon}$, $\hat{\theta}$, and $\hat{\mathfrak{g}}$ such that, for every admissible process of a neighborhood of X, the specific entropy $\eta(t)$ and the deformation gradient $\boldsymbol{F}(t)$ determine the values of the specific energy*

$$\varepsilon(t)=\hat{\varepsilon}\left(\boldsymbol{F}(t), \eta(t)\right), \tag{80.1}$$

the temperature

$$\theta(t)=\hat{\theta}\left(\boldsymbol{F}(t), \eta(t)\right), \tag{80.2}$$

and the stress tensor

$$\boldsymbol{T}(t)=\hat{\mathfrak{g}}\left(\boldsymbol{F}(t), \eta(t)\right). \tag{80.3}$$

Of course, the functions $\hat{\varepsilon}$, $\hat{\theta}$, and $\hat{\mathfrak{g}}$ will depend on the choice of the reference configuration and, in inhomogeneous bodies, also on the particle. Furthermore, the extended principle of material indifference mentioned in the previous section will impose restrictions on these functions. These matters will be treated in Sect. 84.

By (25.6), the stretching tensor $\boldsymbol{D}$ is the symmetric part of the velocity gradient $\boldsymbol{L}=\operatorname{grad}\dot{\boldsymbol{x}}$. Since the stress tensor $\boldsymbol{T}$ is symmetric and since the trace of a product of a symmetric and a skew tensor is zero, we have

$$\operatorname{tr}(\boldsymbol{T}\boldsymbol{D})=\operatorname{tr}(\boldsymbol{T}\boldsymbol{L}). \tag{80.4}$$

Substitution into (79.12) gives

$$\theta\gamma_{\text{loc}}=\operatorname{tr}\left(\frac{1}{\varrho}\boldsymbol{T}\boldsymbol{L}\right)+\theta\dot{\eta}-\dot{\varepsilon}\geqq 0. \tag{80.5}$$

Differentiating (80.1) yields, in the notation of Sect. 9,

$$\left.\begin{aligned}\dot{\varepsilon}&=\operatorname{tr}\left[\hat{\varepsilon}_{\boldsymbol{F}}(\boldsymbol{F},\eta)\dot{\boldsymbol{F}}^{T}\right]+\hat{\varepsilon}_{\eta}(\boldsymbol{F},\eta)\dot{\eta}\,,\\ &=\frac{\partial\hat{\varepsilon}}{\partial x^{k}_{,\alpha}}\dot{x}^{k}_{,\alpha}+\frac{\partial\hat{\varepsilon}}{\partial\eta}\dot{\eta}.\end{aligned}\right\} \tag{80.6}$$

Combining (80.5), $(24.1)_3$, and (80.6), we find that the inequality $\gamma_{\text{loc}}\geqq 0$ is equivalent to

$$\theta\gamma_{\text{loc}}=\operatorname{tr}\left[\left(\frac{1}{\varrho}\hat{\mathfrak{g}}(\boldsymbol{F},\eta)-\boldsymbol{F}\hat{\varepsilon}_{\boldsymbol{F}}(\boldsymbol{F},\eta)^{T}\right)\boldsymbol{L}\right]+\left(\hat{\theta}(\boldsymbol{F},\eta)-\hat{\varepsilon}_{\eta}(\boldsymbol{F},\eta)\right)\dot{\eta}\geqq 0. \tag{80.7}$$

(ii) We now assume the local entropy production γ_{loc} to be non-negative for *all* thermodynamic processes that are compatible with the constitutive equations (80.1)—(80.3). If $\boldsymbol{F}(t)$ and $\eta(t)$ are prescribed, then (80.1)—(80.3) may be used to determine $\varepsilon(t)$, $\theta(t)$ and $\boldsymbol{T}(t)$. It is clear, therefore, that there are thermodynamic processes compatible with (80.1)—(80.3) for arbitrary $\boldsymbol{F}(t)$ and $\eta(t)$. If the values of $\boldsymbol{F}$ and η at a particular time are given, we can always choose $\boldsymbol{F}$ and η as functions of time such that $\boldsymbol{L}$ and $\dot{\eta}$ have arbitrarily assigned values at the particular time t. It follows easily from this observation that the inequality (80.7) must actually be an equality, i.e., that

$$\gamma_{\text{loc}} = 0, \tag{80.8}$$

and that

$$\hat{\theta}(\boldsymbol{F}, t) = \hat{\varepsilon}_\eta(\boldsymbol{F}, \eta), \qquad \hat{\mathfrak{g}}(\boldsymbol{F}, \eta) = \varrho \boldsymbol{F} \hat{\varepsilon}_{\boldsymbol{F}}(\boldsymbol{F}, \eta)^T. \tag{80.9}$$

Thus, the response functions $\hat{\mathfrak{g}}$ and $\hat{\theta}$ are determined by $\hat{\varepsilon}$. We call $\hat{\varepsilon}$ the *energy function* and (80.1) the *caloric equation of state* of the material. The energy function $\hat{\varepsilon}$ alone characterizes the local thermo-mechanical properties of a perfect material. We summarize: *In a perfect material the deformation gradient* $\boldsymbol{F}$ *and the entropy* η *determine the stress* $\boldsymbol{T}$ *and the temperature* θ *by the stress relation*

$$\boldsymbol{T} = \varrho \boldsymbol{F} \hat{\varepsilon}_{\boldsymbol{F}}(\boldsymbol{F}, \eta)^T, \qquad T^k{}_m = \varrho\, x^k_{,\alpha} \frac{\partial \hat{\varepsilon}}{\partial x^m_{,\alpha}}, \tag{80.10}$$

and the temperature relation

$$\theta = \hat{\varepsilon}_\eta(\boldsymbol{F}, \eta) = \frac{\partial \hat{\varepsilon}}{\partial \eta}. \tag{80.11}$$

For every admissible process in a perfect material, the local entropy production is zero.

For perfect materials that are subject to simple internal constraints, the constitutive Eq. (80.3) must be modified in a manner conforming to the principle of determinism stated in Sect. 30. I.e., for such materials, the stress $\boldsymbol{T}(t)$ is determined by $\boldsymbol{F}(t)$ and $\eta(t)$ only up to a stress $\boldsymbol{N}$ that does no work in motions satisfying the constraints. Thus (80.3) must be replaced by

$$\boldsymbol{T}_{\text{E}}(t) = \boldsymbol{T}(t) + \boldsymbol{N}(t) = \hat{\mathfrak{g}}\left(\boldsymbol{F}(t), \eta(t)\right), \tag{80.12}$$

where $\boldsymbol{T}_{\text{E}}$ is an extra-stress and $\boldsymbol{N}$ is given by (30.10). The response functions $\hat{\varepsilon}(\boldsymbol{F}, \eta)$, $\hat{\theta}(\boldsymbol{F}, \eta)$, $\hat{\mathfrak{g}}(\boldsymbol{F}, \eta)$ are defined only for those $\boldsymbol{F}$ that satisfy the constraints. It is then meaningless to talk about the gradient $\hat{\varepsilon}_{\boldsymbol{F}}(\boldsymbol{F}, \eta)$ of $\hat{\varepsilon}$ with respect to $\boldsymbol{F}$. Let us assume, however, that we have extended the definition of $\hat{\varepsilon}$ so as to render $\hat{\varepsilon}(\boldsymbol{F}, \eta)$ meaningful even when $\boldsymbol{F}$ does not satisfy the constraints. The same argument as the one leading to (80.7) now yields

$$\theta \gamma_{\text{loc}} = \operatorname{tr}\left[\left(\frac{1}{\varrho} \hat{\mathfrak{g}}(\boldsymbol{F}, \eta) - \frac{1}{\varrho} \boldsymbol{N} - \boldsymbol{F} \hat{\varepsilon}_{\boldsymbol{F}}(\boldsymbol{F}, \eta)^T\right) \boldsymbol{L}\right] + \left(\hat{\theta}(\boldsymbol{F}, \eta) - \hat{\varepsilon}_\eta(\boldsymbol{F}, \eta)\right)\dot{\eta} \geqq 0. \tag{80.13}$$

This inequality must be valid for all functions $\eta = \eta(t)$ and all $\boldsymbol{F} = \boldsymbol{F}(t)$ that satisfy the constraints. It follows, first, that (80.13) must always reduce to an equality and that $(80.9)_1$ holds. Instead of $(80.9)_2$, however, by an argument similar to the one leading from (30.8) to (30.9), we obtain the conclusion that

$$\hat{\mathfrak{g}}(\boldsymbol{F}, \eta) - \varrho \boldsymbol{F} \hat{\varepsilon}_{\boldsymbol{F}}(\boldsymbol{F}, \eta)^T = \boldsymbol{N}' \tag{80.14}$$

must have the form (30.10). Replacing $\boldsymbol{N} - \boldsymbol{N}'$ by $\boldsymbol{N}$ and redefining $\boldsymbol{T}_{\text{E}}$ accordingly, we infer from (80.12) and (80.14) the following *stress relation for constrained perfect materials*:

$$\boldsymbol{T}_{\text{E}} = \varrho \boldsymbol{F} \hat{\varepsilon}_{\boldsymbol{F}}(\boldsymbol{F}, \eta)^T, \tag{80.15}$$

where $\boldsymbol{T}_{\text{E}}$ is an extra-stress of the form (30.11). The temperature relation remains given by (80.11). In particular, by (30.16), for *incompressible perfect materials* (80.15) becomes

$$\boldsymbol{T} = -p\boldsymbol{1} + \varrho \boldsymbol{F} \hat{\varepsilon}_{\boldsymbol{F}}(\boldsymbol{F}, \eta)^T. \tag{80.16}$$

The values of the gradient $\hat{\varepsilon}_{\boldsymbol{F}}(\boldsymbol{F}, \eta)$, and hence the values of $\boldsymbol{T}_{\mathrm{E}}$ given by (80.15), depend on the way in which the function $\hat{\varepsilon}(\boldsymbol{F}, \eta)$ has been extended to values of $\boldsymbol{F}$ which do not satisfy the constraints. In practice the extension of the definition of $\hat{\varepsilon}$ will be performed so as to render the functional form of $\hat{\varepsilon}$ as simple as possible (cf. the example given at the end of Sect. 86). A change of the mode of extension of $\hat{\varepsilon}$ corresponds merely to a change of the indeterminate scalars $q_{\mathfrak{i}}$ in the definition (30.11) of $\boldsymbol{T}_{\mathrm{E}}$.

In summary, we have shown that *for perfect materials, consistency with the principle of dissipation requires that the temperature and stress relations be determined uniquely from the energy function by Eqs.* (80.9).

81. Thermal equilibrium of simple materials. The problem of devising a physically correct mathematical theory of the thermo-mechanical behavior of simple materials is unsolved at the present time[1]. The construction of such a theory would be a major step towards the solution of the main open problem of the theory of material behavior, posed in Sect. 20. The thermo-mechanical constitutive assumptions defining simple materials should be generalizations of the constitutive assumptions (i), (ii) for perfect materials of the previous section. In analogy to the mechanical constitutive assumption (28.4) for simple materials, the constitutive assumption for perfect materials of the previous section lends itself readily to the following generalization: In a simple material, the specific energy $\varepsilon(t)$, the temperature $\theta(t)$, and the stress $\boldsymbol{T}(t)$ at a particle X and at the present time t are all given by functionals of the history of the deformation gradient and of the history of the specific entropy at X. In particular, the caloric equation of state (80.1) is generalized by a relation of the form

$$\varepsilon(t) = \mathop{\hat{\varepsilon}}_{\tau=-\infty}^{t}\big(\boldsymbol{F}(\tau), \eta(\tau)\big). \tag{81.1}$$

The functional $\hat{\varepsilon}$ will be called the *energy functional* of the material.

It is not clear, however, how the stress relation (80.10) and the temperature relation (80.11) for perfect materials should be generalized to the case of simple materials. In simple materials the equality (80.8) cannot be expected to hold, i.e., positive local entropy production by dissipation of energy should be the rule. What is needed is a new thermodynamical principle for simple materials which would allow the determination of the stress and temperature functionals from a knowledge of the energy functional (81.1).

Lacking such a general principle, we confine our attention to the case of thermostatics, i.e. to the case when all mechanical and thermal variables are independent of time. The functional $\hat{\varepsilon}$ of (81.1) then reduces to a function, and we obtain the *caloric equation of state*

$$\varepsilon = \hat{\varepsilon}(\boldsymbol{F}, \eta) = \hat{\varepsilon}(x^r{}_{,\lambda}, \eta). \tag{81.2}$$

It has the same form as (80.1) but a different meaning: The caloric equation (81.2) applies to a far wider class of materials under far more limited circumstances.

In order to derive the general stress relation for simple materials in statics, some postulate is necessary. Such a postulate was proposed by COLEMAN and NOLL[2]. We now outline their procedure.

Consider a one-parameter family $\boldsymbol{\chi}_\alpha$, $\alpha_1 \leqq \alpha \leqq \alpha_2$, of configurations of a body $\mathscr{B}$. Let

$$\boldsymbol{x} = \boldsymbol{\chi}(\boldsymbol{X}, \alpha) = \boldsymbol{\chi}_\alpha(\boldsymbol{X}) \tag{81.3}$$

---

[1] Note added in proof: See Sect. 96bis for what seems to be the solution of this problem.

[2] COLEMAN and NOLL [1959, *3*, §§ 5—9].

describe the deformation carrying $\mathscr{B}$ from a reference configuration into the configuration $\boldsymbol{\chi}_\alpha$. Assume, furthermore, that the body in the configuration $\boldsymbol{\chi}_\alpha$ is subject to surface tractions $\boldsymbol{t}_R = \boldsymbol{t}_R(\alpha)$, measured per unit area in the reference configuration. The work done by these surface tractions as $\mathscr{B}$ deforms through the configurations $\boldsymbol{\chi}_\alpha$ is given by

$$W_{12} = \int_{\alpha_1}^{\alpha_2} \left[\int_{\partial\mathscr{B}} \dot{\boldsymbol{\chi}} \cdot \boldsymbol{t}_R \, d s_R \right] d\alpha, \tag{81.4}$$

where the superimposed dot denotes differentiation with respect to α and where $d s_R$ is the element of surface area of the boundary $\partial\mathscr{B}$ of $\mathscr{B}$ in the reference configuration.

We now make the following three assumptions:

(i) The deformation (81.3), for each α, is *homogeneous*, so that the deformation gradient

$$\boldsymbol{F} = \boldsymbol{F}(\alpha) = \nabla \boldsymbol{\chi}(\boldsymbol{X}, \alpha) \tag{81.5}$$

is independent of the particle X.

(ii) The surface tractions $\boldsymbol{t}_R = \boldsymbol{t}_R(\alpha)$ are homogeneous in the sense that they arise from a homogeneous stress field given by the value $\boldsymbol{T}_R = \boldsymbol{T}_R(\alpha)$ for the first Piola-Kirchhoff stress tensor. By (43A.5) we then have

$$\boldsymbol{t}_R(\alpha) = \boldsymbol{T}_R(\alpha)\, \boldsymbol{n}_R, \tag{81.6}$$

where $\boldsymbol{n}_R$ is the outer unit normal to $\partial\mathscr{B}$ in the reference configuration.

(iii) The body $\mathscr{B}$ has unit mass, so that its volume in the reference configuration is ϱ_R^{-1}.

Using (81.6), the divergence theorem, and (81.5), we find

$$\left.\begin{aligned} \int_{\partial\mathscr{B}} \dot{\boldsymbol{\chi}} \cdot \boldsymbol{t}_R \, d s_R &= \int_{\partial\mathscr{B}} \boldsymbol{n}_R \cdot \boldsymbol{T}_R^T \dot{\boldsymbol{\chi}} \, d s_R \\ &= \int_{\mathscr{B}} \operatorname{div}(\boldsymbol{T}_R^T \dot{\boldsymbol{\chi}}) \, d v_R = \varrho_R^{-1} \operatorname{tr}(\boldsymbol{T}_R^T \dot{\boldsymbol{F}}). \end{aligned}\right\} \tag{81.7}$$

Substitution of (81.7) into (81.4) gives the following expression for the *work done on a body of unit mass by homogeneous surface tractions during a homogeneous deformation process:*

$$W_{12} = \int_{\alpha_1}^{\alpha_2} \frac{1}{\varrho_R} \operatorname{tr}(\boldsymbol{T}_R^T \dot{\boldsymbol{F}}) \, d\alpha. \tag{81.8}$$

Suppose, now, that the surface tractions (81.6) are held fixed during the deformation process, i.e., that $\boldsymbol{T}_R$ is independent of α. Then (81.8) may be integrated to yield

$$W_{12} = \frac{1}{\varrho_R} \operatorname{tr}\left(\boldsymbol{T}_R^T \boldsymbol{F}(\alpha_2)\right) - \frac{1}{\varrho_R} \operatorname{tr}\left(\boldsymbol{T}_R^T \boldsymbol{F}(\alpha_1)\right). \tag{81.9}$$

Hence, when the surface tractions are held fixed, they have the potential

$$\hat{\pi}(\boldsymbol{F}) = -\frac{1}{\varrho_R} \operatorname{tr}(\boldsymbol{T}_R^T \boldsymbol{F}). \tag{81.10}$$

Assuming that fixed surface tractions characterized by $\boldsymbol{T}_R$ and a fixed uniform temperature θ are prescribed, we introduce the *thermomechanical potential*,

$$\hat{\lambda}(\boldsymbol{F}, \eta) \equiv \hat{\varepsilon}(\boldsymbol{F}, \eta) + \hat{\pi}(\boldsymbol{F}) - \theta\eta, \tag{81.11}$$

which is the sum of the internal energy, the potential (81.10) of the surface tractions, and the thermal potential $-\eta\theta$ associated with the prescribed temperature θ. The potential (81.11) applies to a homogeneous body of unit mass in homogeneous deformations with uniform entropy η.

A pair $(\boldsymbol{F}, \eta)$, consisting of a value $\boldsymbol{F}$ for the deformation gradient tensor and a value η for the specific entropy, will be called a *local state* of the material. We say that two local states $(\boldsymbol{F}, \eta)$ and $(\boldsymbol{F}^*, \eta^*)$ *differ by a pure stretch* if there is a positive-definite symmetric tensor $\boldsymbol{G}$ such that

$$\boldsymbol{F}^* = \boldsymbol{G}\boldsymbol{F}. \tag{81.12}$$

In accord with Sect. 23, this means that the rotation tensor of the deformation from the configuration corresponding to $\boldsymbol{F}$ to the configuration corresponding to $\boldsymbol{F}^*$ reduces to the identity, $\mathbf{1}$, and that $\boldsymbol{G}$ is the left and right stretch tensor of this deformation.

We are now able to give a precise definition of *local thermal equilibrium: Assuming that a Piola-Kirchhoff tensor* $\boldsymbol{T}_{\mathrm{R}}$ *and a temperature* θ *be prescribed, we say that* $(\boldsymbol{F}, \eta)$ *is a state of thermal equilibrium if it minimizes the thermomechanical potential* (81.11) *within the class of all local states differing from* $(\boldsymbol{F}, \eta)$ *by a pure stretch, and if the stress tensor* $\boldsymbol{T} = \frac{\varrho}{\varrho_{\mathrm{R}}} \boldsymbol{T}_{\mathrm{R}} \boldsymbol{F}^T$ *corresponding to* $\boldsymbol{T}_{\mathrm{R}}$ *and* $\boldsymbol{F}$ *is symmetric.* Recall that the last condition corresponds to the requirement that the moments of the surface tractions determined by $\boldsymbol{T}_{\mathrm{R}}$ be balanced in the configuration determined by $\boldsymbol{F}$.

With the definition above we can state the postulate of COLEMAN and NOLL for the ***thermostatics of simple materials:*** *For every local state* $(\boldsymbol{F}, \eta)$ *there exists a Piola-Kirchhoff tensor* $\boldsymbol{T}_{\mathrm{R}}$ *and a temperature* θ *that make* $(\boldsymbol{F}, \eta)$ *a state of local thermal equilibrium.*

The Piola-Kirchhoff tensor $\boldsymbol{T}_{\mathrm{R}}$ and the temperature θ must have the property that the function

$$\hat{\lambda}(\boldsymbol{G}\boldsymbol{F}, \eta^*) = \hat{\varepsilon}(\boldsymbol{G}\boldsymbol{F}, \eta^*) - \frac{1}{\varrho_{\mathrm{R}}} \operatorname{tr}(\boldsymbol{G}\boldsymbol{F}\boldsymbol{T}_{\mathrm{R}}^T) - \eta^*\theta \tag{81.13}$$

achieves its minimum for $\eta^* = \eta$ and $\boldsymbol{G} = \mathbf{1}$, as is clear from (81.10), (81.11), (81.12), and the definition of thermal equilibrium. It follows from the theory of maxima and minima that the gradient with respect to $\boldsymbol{G}$ and the derivative with respect to η^* of $\hat{\lambda}(\boldsymbol{G}\boldsymbol{F}, \eta^*)$ must vanish for $\boldsymbol{G} = \mathbf{1}$ and $\eta^* = \eta$. Using the rules of Sect. 9 to calculate the gradient with respect to $\boldsymbol{G}$, we obtain the equation

$$\operatorname{tr}\left\{\left[\boldsymbol{F}\hat{\varepsilon}_{\boldsymbol{F}}(\boldsymbol{F}, \eta)^T - \frac{1}{\varrho_{\mathrm{R}}} \boldsymbol{F}\boldsymbol{T}_{\mathrm{R}}^T\right] \boldsymbol{S}\right\} = 0, \tag{81.14}$$

which must be valid for all symmetric tensors $\boldsymbol{S}$. Introducing the symmetric stress tensor $\boldsymbol{T} = \frac{\varrho}{\varrho_{\mathrm{R}}} \boldsymbol{T}_{\mathrm{R}} \boldsymbol{F}^T = \frac{\varrho}{\varrho_{\mathrm{R}}} \boldsymbol{F}\boldsymbol{T}_{\mathrm{R}}^T$, we see that (81.14) is equivalent to

$$\operatorname{tr}\{[\varrho\boldsymbol{F}\hat{\varepsilon}_{\boldsymbol{F}}(\boldsymbol{F}, \eta)^T - \boldsymbol{T}]\boldsymbol{S}\} = 0. \tag{81.15}$$

We shall see in Sect. 84 that the principle of material indifference implies the symmetry of $\boldsymbol{F}\hat{\varepsilon}_{\boldsymbol{F}}(\boldsymbol{F}, \eta)^T$. Thus, the tensor $\varrho\boldsymbol{F}\hat{\varepsilon}_{\boldsymbol{F}}(\boldsymbol{F}, \eta)^T - \boldsymbol{T}$ must be symmetric. Recalling that $\operatorname{tr}(\boldsymbol{A}\boldsymbol{B})$ has the properties of an inner product in the 6-dimensional space $\mathscr{S}$ of all symmetric tensors, we see that (81.15) amounts to the statement that $\varrho\boldsymbol{F}\hat{\varepsilon}_{\boldsymbol{F}}(\boldsymbol{F}, \eta)^T - \boldsymbol{T}$ be orthogonal to every tensor in $\mathscr{S}$. But since only the zero tensor is orthogonal to every tensor in $\mathscr{S}$, we obtain the *stress relation*

$$\boldsymbol{T} = \varrho\boldsymbol{F}\hat{\varepsilon}_{\boldsymbol{F}}(\boldsymbol{F}, \eta)^T, \qquad T^k{}_m = \varrho\, x^k{}_{,\alpha} \frac{\partial\hat{\varepsilon}}{\partial x^m{}_{,\alpha}}. \tag{81.16}$$

The condition that the derivative of (81.13) with respect to η^* be zero for $\eta=\eta^*$ leads to the *temperature relation*

$$\theta=\hat{\varepsilon}_\eta(\boldsymbol{F},\eta)=\frac{\partial\hat{\varepsilon}}{\partial\eta}\,. \tag{81.17}$$

Although the stress relation (81.16) and the temperature relation (81.17) have the same forms as those given by Eqs. (80.10), they have been shown valid under different conditions. They hold for any simple material in thermal equilibrium as defined above.

Motivation for the restriction (81.12) on the class of states $(\boldsymbol{F}^*, \eta^*)$ within which the thermo-mechanical potential is minimized stems from the observation that a body is not mechanically stable with respect to certain rigid rotations under compressive forces held constant in magnitude *and direction* (cf. Sect. 68b, Fig. 5). Hence, if the restriction (81.12) were not imposed, thermal equilibrium under compressive stresses would not be possible, which would contradict the intuitive physical notion of thermal equilibrium. However, it is conceivable that a restriction different from (81.12) might lead to an equally consistent theory of thermostatics of simple materials.

The foregoing considerations can easily be modified so as to apply to the statics of simple materials subject to constraints. The resulting stress relation has the same form, (80.15), as for the dynamics of perfect materials subject to constraints.

82. Definition of a hyperelastic material. The main results of the two previous sections are the *stress relation*

$$\boldsymbol{T}=\varrho\boldsymbol{F}\hat{\varepsilon}_{\boldsymbol{F}}(\boldsymbol{F},\eta)^T,\qquad T^k{}_m=\varrho\, x^k{}_{,\alpha}\frac{\partial\hat{\varepsilon}}{\partial x^m{}_{,\alpha}} \tag{82.1}$$

and the *temperature relation*

$$\theta=\hat{\varepsilon}_\eta(\boldsymbol{F},\eta)=\frac{\partial\hat{\varepsilon}}{\partial\eta}\,. \tag{82.2}$$

These relations are valid (i) for any motion of a perfect material and (ii) for thermal equilibrium of any simple material. We assume that (82.2) can be solved for η for each fixed $\boldsymbol{F}$ to yield the *entropy function*

$$\eta=\tilde{\eta}(\boldsymbol{F},\theta)\,. \tag{82.3}$$

We now consider the *specific free energy* [cf. CFT, (251.1)]

$$\psi=\varepsilon-\eta\theta\,. \tag{82.4}$$

With the help of the caloric equation $\varepsilon=\hat{\varepsilon}(\boldsymbol{F},\eta)$ and of (82.3), ψ may be expressed as a function of the deformation gradient $\boldsymbol{F}$ and the temperature θ:

$$\psi=\tilde{\psi}(\boldsymbol{F},\theta)=\hat{\varepsilon}(\boldsymbol{F},\tilde{\eta}(\boldsymbol{F},\theta))-\theta\tilde{\eta}(\boldsymbol{F},\theta)\,. \tag{82.5}$$

Using (82.2) and the chain rule, we calculate the gradient of $\tilde{\psi}$ with respect to $\boldsymbol{F}$ and the derivative of $\tilde{\psi}$ with respect to θ:

$$\tilde{\psi}_{\boldsymbol{F}}(\boldsymbol{F},\theta)=\hat{\varepsilon}_{\boldsymbol{F}}(\boldsymbol{F},\eta),\qquad \eta=-\tilde{\psi}_\theta(\boldsymbol{F},\theta)\,, \tag{82.6}$$

where η is related to $\boldsymbol{F}$ and θ by (82.3). This result and (82.1) show that when we wish to employ as independent variables the deformation gradient $\boldsymbol{F}$ and the temperature θ rather than $\boldsymbol{F}$ and η, we have the following alternative form of the *stress relation*[1]:

$$\boldsymbol{T} = \varrho \boldsymbol{F} \tilde{\psi}_{\boldsymbol{F}}(\boldsymbol{F}, \theta)^T, \quad T^k{}_m = \varrho x^k{}_{,\alpha} \frac{\partial \tilde{\psi}}{\partial x^m{}_{,\alpha}}, \tag{82.7}$$

and the *entropy relation* becomes

$$\eta = -\tilde{\psi}_\theta(\boldsymbol{F}, \theta) = -\frac{\partial \tilde{\psi}}{\partial \theta}. \tag{82.8}$$

When the thermal variables η and θ, respectively, are ignored, we see that the stress relations (82.1) and (82.7) correspond to a purely mechanical constitutive equation of the form[2]

$$\boldsymbol{T} = \varrho \boldsymbol{F} \sigma_{\boldsymbol{F}}(\boldsymbol{F})^T, \quad T^k{}_m = \varrho x^k{}_{,\alpha} \frac{\partial \sigma}{\partial x^m{}_{,\alpha}}. \tag{82.9}$$

To within consistency with the principle of material indifference, which will be established in Sect. 84, (82.9) is a special case of the constitutive equation (43.1) for elastic materials. Accordingly, *we define a hyperelastic material as an elastic material whose response function* $\mathfrak{g}$ *has the special form*

$$\mathfrak{g}(\boldsymbol{F}) = \varrho \boldsymbol{F} \sigma_{\boldsymbol{F}}(\boldsymbol{F})^T = \varrho \sigma_{\boldsymbol{F}}(\boldsymbol{F}) \boldsymbol{F}^T. \tag{82.10}$$

In other words, a hyperelastic material is characterized by the constitutive equation (82.9), the scalar function $\sigma(\boldsymbol{F})$ being called the *strain-energy function* or the *stored-energy function*. Alternatively, as we see at once from (43.8) and $(43.11)_2$, we have[3]

$$\boldsymbol{T}_{\mathrm{R}} = \mathfrak{h}(\boldsymbol{F}) = \varrho_{\mathrm{R}} \sigma_{\boldsymbol{F}}(\boldsymbol{F}), \quad T_{\mathrm{R}k}{}^{\alpha} = \mathfrak{h}_k{}^{\alpha}(x^m{}_{,\beta}) = \varrho_{\mathrm{R}} \frac{\partial \sigma}{\partial x^k{}_{,\alpha}}. \tag{82.11}$$

Therefore the elasticities $\mathsf{A}_k{}^{\alpha}{}_m{}^{\beta}$, defined by (44.3), have the form

$$\mathsf{A}_k{}^{\alpha}{}_m{}^{\beta} = \varrho_{\mathrm{R}} \frac{\partial^2 \sigma}{\partial x^k{}_{,\alpha} \partial x^m{}_{,\beta}}, \tag{82.12}$$

whence we infer the following *necessary and sufficient condition that an elastic material be hyperelastic:*

$$\mathsf{A}_k{}^{\alpha}{}_m{}^{\beta} = \mathsf{A}_m{}^{\beta}{}_k{}^{\alpha}. \tag{82.13}$$

Here, too there is a gap in the argument, for we defer until Sec. 84 proof ot consistency with the principle of material frame-indifference.

Since the terminology used here is not yet standard, it must be explained with more emphasis. We have shown that if (82.1) holds, so does (82.7), and conversely; these are alternative formulations of the same property of certain materials. It does not follow, however, that these materials are always hyperelastic, or even, for that matter, that they are elastic. According to our definition in Sect. 43, an elastic material is one in which $\boldsymbol{T}$ is a function of $\boldsymbol{F}$ *only*. A perfect material is characterized by a caloric equation of state (80.1) and has been shown to have a stress relation which may be written, for all kinds of deformations, alternatively in the forms (82.1) and (82.7), but neither of these, in general, is subsumed under (82.9). Rather, (82.1) reduces to (82.9) when $\eta = \text{const.}$, provided we put $\sigma = \hat{\varepsilon}$; it reduces to (82.9) also when $\theta = \text{const.}$, but then we must put $\sigma = \tilde{\psi}$. Thus the same perfect material is a hyperelastic material in a homothermal

[1] Note that (82.7) follows from (82.1), and conversely, by functional transformations. Thus it is not necessary at this point to have further recourse to the energy equation or to introduce an isothermal process as in the traditional method, cited in Sect. 82 A, below.

[2] C. NEUMANN [1860, *1*, Eq. (21)], BOUSSINESQ [1872, *1*, Note 3, p. 594].

[3] KIRCHHOFF [1852, *1*, p. 772], GIBBS [1875, *1*, p. 190].

deformation and also a hyperelastic material, usually a different one[1], in homentropic deformation. A similar remark may be made in regard to simple materials in thermal equilibrium.

Thus we may summarize the thermodynamic analysis of this section and the preceding as showing that the theory of hyperelastic materials is appropriate to the following cases[2]:

(i) Perfect materials in any motion at uniform temperature
(ii) Perfect materials in any motion at uniform specific entropy
(iii) Simple materials in thermal equilibrium at uniform temperature
(iv) Simple materials in thermal equilibrium at uniform specific entropy

The strain-energy functions are $\tilde{\psi}$ for cases (i) and (iii), $\hat{\varepsilon}$ for cases (ii) and (iv), with the qualification stated after (82.9). It remains possible that the theory of hyperelastic materials may be shown, eventually, to apply in still further circumstances. [Note added in proof: See Sect. 96bis.]

As an example of the distinctions made here, we consider a given perfect material, or a given simple material subject to static deformation, and we find a relation[3] between the different elasticities $A_k{}^\alpha{}_m{}^\beta$ corresponding to the use of the internal-energy function $\hat{\varepsilon}$ or the free-energy function $\tilde{\psi}$ as the strain-energy function. Comparing (82.11) with (45.18), and (82.2) with $(45.19)_2$, we see that now

$$\frac{\partial\hat{\theta}}{\partial x^m{}_{,\beta}} = \frac{\partial^2\hat{\varepsilon}}{\partial x^m{}_{,\beta}\,\partial\eta} = \frac{1}{\varrho_{\mathrm{R}}}\frac{\partial\hat{\mathfrak{h}}_m{}^\beta}{\partial\eta}, \tag{82.14}$$

so that (45.22) becomes

$$\left.\begin{aligned}\hat{A}_k{}^\alpha{}_m{}^\beta &= \tilde{A}_k{}^\alpha{}_m{}^\beta + \frac{1}{\varrho_{\mathrm{R}}}\frac{\partial\tilde{\mathfrak{h}}_k{}^\alpha}{\partial\theta}\frac{\partial\hat{\mathfrak{h}}_m{}^\beta}{\partial\eta},\\ &= \tilde{A}_k{}^\alpha{}_m{}^\beta + \frac{\varkappa(\boldsymbol{F})}{\varrho_{\mathrm{R}}\theta}\frac{\partial\hat{\mathfrak{h}}_k{}^\alpha}{\partial\eta}\frac{\partial\hat{\mathfrak{h}}_m{}^\beta}{\partial\eta},\\ &= \tilde{A}_k{}^\alpha{}_m{}^\beta + \frac{\theta}{\varrho_{\mathrm{R}}\varkappa(\boldsymbol{F})}\frac{\partial\tilde{\mathfrak{h}}_k{}^\alpha}{\partial\theta}\frac{\partial\tilde{\mathfrak{h}}_m{}^\beta}{\partial\theta},\end{aligned}\right\} \tag{82.15}$$

where (45.23) has been used. When thermodynamic variables are taken into account, (82.11) becomes

$$\boldsymbol{T}_{\mathrm{R}} = \hat{\mathfrak{h}}(\boldsymbol{F},\eta) = \varrho_{\mathrm{R}}\hat{\varepsilon}_{\boldsymbol{F}}(\boldsymbol{F},\eta) = \tilde{\mathfrak{h}}(\boldsymbol{F},\theta) = \varrho_{\mathrm{R}}\tilde{\psi}_{\boldsymbol{F}}(\boldsymbol{F},\theta). \tag{82.16}$$

Hence the two relations $(82.15)_{2,3}$ may be written in the alternative forms

$$\left.\begin{aligned}\hat{A}_k{}^\alpha{}_m{}^\beta - \tilde{A}_k{}^\alpha{}_m{}^\beta &= \frac{\varrho_{\mathrm{R}}\varkappa(\boldsymbol{F})}{\theta}\frac{\partial^2\hat{\varepsilon}}{\partial x^k{}_{,\alpha}\,\partial\eta}\frac{\partial^2\hat{\varepsilon}}{\partial x^m{}_{,\beta}\,\partial\eta},\\ &= \frac{\varrho_{\mathrm{R}}\theta}{\varkappa(\boldsymbol{F})}\frac{\partial^2\tilde{\psi}}{\partial x^k{}_{,\alpha}\,\partial\theta}\frac{\partial^2\tilde{\psi}}{\partial x^m{}_{,\beta}\,\partial\theta},\\ &= \frac{\varrho_{\mathrm{R}}\varkappa(\boldsymbol{F})}{\theta}\frac{\partial\hat{\theta}}{\partial x^k{}_{,\alpha}}\frac{\partial\hat{\theta}}{\partial x^m{}_{,\beta}},\\ &= \frac{\varrho_{\mathrm{R}}\theta}{\varkappa(\boldsymbol{F})}\frac{\partial\tilde{\eta}}{\partial x^k{}_{,\alpha}}\frac{\partial\tilde{\eta}}{\partial x^m{}_{,\beta}},\end{aligned}\right\} \tag{82.17}$$

---

[1] Thus the terminology is different from that of TRUESDELL [1952, *20*, § 36], who preferred to define *elastic response* rather than a hyperelastic material. While the terms are different, the distinction made is the same, and not without reason. Most of the theorems about hyperelastic materials depend upon differentiating $\boldsymbol{T}$ with respect to $\boldsymbol{x}$, as is apparent from the fact that it is div $\boldsymbol{T}$ rather than $\boldsymbol{T}$ itself that enters the equations of motion (16.6). If η is allowed to vary in (82.1), or θ in (82.7), scarcely any of these theorems retain a form simple enough to be definite and enlightening. An exception is given in Sect. 90.

[2] Two more cases may be added by considering motion or equilibrium at uniform internal energy; the strain-energy function is then the specific entropy, taken as a function of $\boldsymbol{F}$ and ε.

[3] TRUESDELL [1961, *63*, § 13].

where we use $\tilde{\eta}(\boldsymbol{F}, \theta)$ to stand for the function $-\tilde{\psi}_\theta(\boldsymbol{F}, \theta)$ occurring in (82.8). These not inelegant expressions show how one set of elasticities may be calculated from the other, provided the specific heat $\varkappa_{(\boldsymbol{F})}$ and the thermal coefficients $\hat{\theta}_{\boldsymbol{F}}$ or $\tilde{\eta}_{\boldsymbol{F}}$ be known.

In accord with (80.16) we define an *incompressible hyperelastic material* by the constitutive equation

$$\boldsymbol{T} + p\boldsymbol{1} = \boldsymbol{T}_{\mathrm{E}} = \varrho \boldsymbol{F} \sigma_{\boldsymbol{F}}(\boldsymbol{F})^{T}, \qquad T^k_m + p\,\delta^k_m = T_{\mathrm{E}}{}^k_m = \varrho\, x^k{}_{,\alpha} \frac{\partial \sigma}{\partial x^m{}_{,\alpha}}. \tag{82.18}$$

As for all incompressible materials, the stress $\boldsymbol{T}$ is determined by the motion only to within a hydrostatic pressure p. The strain-energy function $\sigma(\boldsymbol{F})$ is originally defined only for unimodular $\boldsymbol{F}$ (det $\boldsymbol{F} = \pm 1$). As we already pointed out in Sect. 80, in order to make (82.17) meaningful, the definition of σ must be extended to non-unimodular tensors.

Such an extension is not unique, and $\boldsymbol{F}\sigma_{\boldsymbol{F}}(\boldsymbol{F})^T$ is determined by the material only up to arbitrary scalar multiple of the unit tensor **1**.

Corresponding to the ideas of elastic response proposed by CAUCHY and GREEN (though not fully formulated by them), the literature[1] often refers to elastic materials as "Cauchy-elastic" and to hyperelastic materials as "Green-elastic".

82A. Appendix. History of the theory of hyperelastic materials in finite strain. In the literature the term "perfectly elastic" is used to describe, loosely, the concepts studied in this section and the preceding. The theory of perfect materials originates in the more special theory of perfect (elastic) fluids and thus has a long early history. The first theory of finite elastic deformation of solids was given by GREEN [1839, *1*, pp. 248—255] [1841, *1*, pp. 298—300]. In effect, he supposed that $\sigma(\boldsymbol{F}) = \hat{\sigma}(\boldsymbol{E})$ where $\boldsymbol{E} \equiv \frac{1}{2}(\boldsymbol{C} - \boldsymbol{1})$. After a preliminary attempt [1850, *2*, § 1], KIRCHHOFF [1852, *1*, pp. 770—772] was the first to obtain a stress relation that is fully correct; while he considered only a special form of the energy, his method is general. KELVIN [1855, *1*] [1856, *1*, Chs. XIII, XIV] attempted to connect GREEN's method with the principles of thermodynamics as they were then understood; finally he succeeded [1863, *2*, §§ 61—67] [1867, *2*, § 673 and App. C., §§ (c)—(d)]. Meanwhile C. NEUMANN [1860, *1*, § 4], starting from a molecular hypothesis but not restricting himself to it, had introduced a general stored-energy function and had derived from it the general constitutive equation of hyperelasticity. Some of the many later presentations, essentially repeating the ideas of one or another of the earlier ones, were cited by TRUESDELL [1952, *20*, §§ 33, 35]. The connection between the adiabatic and isothermal theories, given just above, seems to be due to VOIGT [1889, *3*, pp. 943—949], of course in a much less general setting. While his derivation has come to be standard in textbooks and expositions of elasticity, it was criticized for vagueness and for the use of hidden thermodynamic assumptions by TRUESDELL [1952, *20*, § 33].

"Perfect materials" correspond to what is called "the fully recoverable case" in Sect. 256 A of CFT, where an approach suggested by TOUPIN and ERICKSEN (according to TRUESDELL [1952, *20*, § 33]) is developed. The presentation in Sect. 80 above, which is due to NOLL and is here published for the first time, starts from similar ideas but is more explicit and concise. A slightly different treatment, also starting from local inequalities, was found independently by A. E. GREEN and communicated to TRUESDELL in a letter dated May 20, 1962.

A treatment based on global inequalities was given by GIBBS [1875, *1*, pp. 184—192].

The stress relation for incompressible hyperelastic materials was first obtained by POINCARÉ [1889, *2*, § 152] [1892, *1*, § 33].

83. Work theorems. We consider a homogeneous elastic body $\mathscr{B}$ of unit mass whose material properties are defined by a constitutive equation of the form (43.8), i.e.

$$\boldsymbol{T}_{\mathrm{R}} = \mathfrak{h}(\boldsymbol{F}). \tag{83.1}$$

We choose a homogeneous reference configuration, so that the response function $\mathfrak{h}$ is the same for all particles X in $\mathscr{B}$. In this section we deal only with homogeneous deformation processes of $\mathscr{B}$. Every such process is characterized by a one-parameter family of deformation gradients $\boldsymbol{F}(\alpha)$, $\alpha_1 \leqq \alpha \leqq \alpha_2$ (cf. Sect. 81). Such a family

[1] Following TRUESDELL [1952, *20*, § 35].

describes a path $\mathcal{P}$ in the 9-dimensional space $\mathcal{L}$ of all tensors. The path $\mathcal{P}$ joins the two "points"

$$F_1 = F(\alpha_1) \quad F_2 = F(\alpha_2) \tag{83.2}$$

in $\mathcal{L}$. If we put $F = F(\alpha)$ in (83.1) and substitute the result into (81.8), we find the following expression for the work done by the actual surface tractions on $\mathcal{B}$:

$$W_{12} = \frac{1}{\varrho_R} \int_{\alpha_1}^{\alpha_2} \operatorname{tr}[\mathfrak{h}(F)^T \dot{F}]\, d\alpha. \tag{83.3}$$

Recalling that $\operatorname{tr}(A^T B)$ defines an inner product in the tensor space $\mathcal{L}$ (cf. Sect. 7), we see that W_{12} is the line integral of $\varrho_R^{-1}\mathfrak{h}(F)$ in $\mathcal{L}$ along the path $\mathcal{P}$ defined by the deformation process $F(\alpha)$. It is evident from (82.11) that the function $\varrho_R^{-1}\mathfrak{h}(F)$ is the gradient of a scalar function $\sigma(F)$,

$$\varrho_R^{-1}\mathfrak{h}(F) = \sigma_F(F), \tag{83.4}$$

if and only if the elastic material defined by $\mathfrak{h}$ is actually hyperelastic. In this case, the work (83.3) done by the surface tractions is given by

$$W_{12} = \sigma(F_2) - \sigma(F_1). \tag{83.5}$$

We assume always that the domain of definition of $\mathfrak{h}$ consists of simply connected regions in the tensor space $\mathcal{L}$. We say that a deformation process $F(\alpha)$ is closed if the corresponding path $\mathcal{P}$ in $\mathcal{L}$ is closed, i.e. if $F_1 = F_2$. The following ***first work theorem***[1] is a corollary to well known facts about line-integrals, applied to paths in the tensor space $\mathcal{L}$:

The following three conditions on the response function of an elastic material are equivalent:

(1) *The work done by the actual surface tractions in every closed homogeneous deformation process is non-negative.*

(2) *The work done by the actual surface tractions in a homogeneous deformation process depends only on the initial and final configurations.*

(3) *The material is hyperelastic.*

The following ***second work theorem*** is due to CAPRIOLI[2]: *If there is a configuration* $\varkappa$ *such that the actual work done in every homogeneous deformation process from* $\varkappa$ *is non-negative, then the material is hyperelastic, and its strain energy func-*

---

[1] While the content of this theorem will be considered obvious by most students of elasticity today, we cannot find any previous statement and proof of it. First, how work is to be defined in finite deformation may not be clear; in fact, most of the statements made in the literature about work and stored energy in elasticity are meaningless if not wrong from failure to demand that the deformation process be homogeneous. In a general deformation $F(X, t)$, the work done by the surface tractions acting on a portion of a hyperelastic material is not independent of path and is not given by the volume integral of (83.5), although a belief to the contrary is frequently met. Second, without the presumption that the material is elastic, or that some other form of constitutive equation holds, a statement such as (1) or (2), while common enough in the thermodynamic literature, has no definite meaning. The classic treatment of KELVIN [1855, *1*] [1863, *2*, §§ 61—62], heavily influenced by the Lagrangean tradition, really introduces the stored-energy function *a priori*; while E. and F. COSSERAT [1896, *1*, §§ 22—25] saw the need for restricting attention to families of homogeneous strains, they likewise assumed existence of a stored-energy function. More recent presentations vary the formalism but not the essential content. That is, the authors are content to verify that work or virtual work done is independent of the path connecting two deformations, but they do not state this property as a criterion distinguishing the existence of a stored-energy function.

[2] CAPRIOLI [1955, *7*]. CAPRIOLI's theorem generalizes an analogous result for infinitesimal elasticity found by UDESCHINI [1943, *7*].

tion σ satisfies the inequality

$$\sigma(\boldsymbol{F}) \geqq \sigma(\mathbf{1}) \tag{83.6}$$

where $\boldsymbol{F}$ is the gradient, with respect to $\boldsymbol{\varkappa}$ as reference configuration, of any deformation. In addition, the configuration $\boldsymbol{\varkappa}$ must be a natural state, i.e., the stress corresponding to $\boldsymbol{\varkappa}$ is zero.

To prove this theorem, consider any closed deformation process $\boldsymbol{F}(\alpha)$, $\alpha_1 \leqq \alpha \leqq \alpha_2$, $\boldsymbol{F}(\alpha_1) = \boldsymbol{F}(\alpha_2) = \boldsymbol{F}_0$. This process corresponds to a closed path $\mathscr{C}$ in the tensor space $\mathscr{L}$. Join the "point" $\mathbf{1}$ in $\mathscr{L}$ to the "point" $\boldsymbol{F}_0$ on $\mathscr{C}$ by a path $\mathscr{P}$ (see Fig. 6). Consider the closed path in $\mathscr{L}$ obtained by first traversing $\mathscr{P}$ from $\mathbf{1}$ to $\boldsymbol{F}_0$, then going around $\mathscr{C}$ from $\boldsymbol{F}_0$ back to $\boldsymbol{F}_0$, and finally traversing $\mathscr{P}$ again from $\boldsymbol{F}_0$

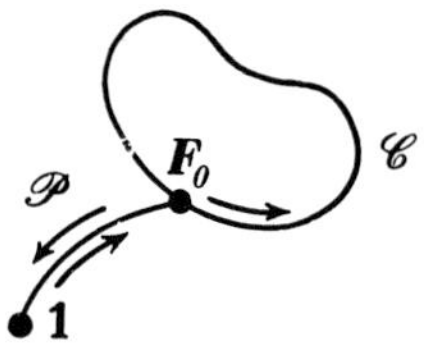

Fig. 6. Construction for proof of CAPRIOLI's theorem.

to $\mathbf{1}$. By hypothesis, the line integral (83.3) along this path must be non-negative. The path $\mathscr{P}$, being traversed twice in opposite senses, makes no contribution to the line integral, and it follows that the integral along the original closed path $\mathscr{C}$ must be non-negative. The first work theorem shows that the material must be hyperelastic. The inequality (83.6) is a consequence of the hypothesis and (83.5). The inequality (83.6) states that σ is a minimum for $\boldsymbol{F} = \mathbf{1}$. Hence the gradient $\sigma_{\boldsymbol{F}}(\mathbf{1})$ must vanish, which, by (82.9), ensures that the stress $\boldsymbol{T}$ in the configuration $\boldsymbol{\varkappa}$ is zero.

The work done by the actual surface tractions in a deformation process of an elastic body is given by (83.3). If the surface tractions do not correspond to the actual response (83.1) of the material but are held at their initial values during the entire deformation process, they will do the *virtual work*

$$W_{12}^* = \frac{1}{\varrho_R} \operatorname{tr}[\mathfrak{h}(\boldsymbol{F}_1)^T (\boldsymbol{F}_2 - \boldsymbol{F}_1)], \tag{83.7}$$

as is seen by substitution of the initial Piola-Kirchhoff tensor $\boldsymbol{T}_R = \mathfrak{h}(\boldsymbol{F}_1)$ into (83.3) and observing (83.2). We recall that the initial and final configurations of the deformation process, which correspond to the deformation gradients $\boldsymbol{F}_1$ and $\boldsymbol{F}_2$, respectively, differ by only a pure stretch if

$$\boldsymbol{G} = \boldsymbol{F}_2 \boldsymbol{F}_1^{-1} \tag{83.8}$$

is positive definite and symmetric [cf. (81.12)].

We can now state the ***third work theorem,*** due to COLEMAN[1]: *The following two conditions on the response function of an elastic material are equivalent:*

(1) *For any homogeneous deformation process whose initial and final configurations are distinct and differ by only a pure stretch, the work done by the actual contact forces is greater than the virtual work done if the contact forces had been held at their initial values (dead loading):*

$$W_{12} > W_{12}^*. \tag{83.9}$$

[1] COLEMAN [1962, *9*, Theorem 3].

(2) *The material is hyperelastic, and its strain energy obeys the inequality*

$$\sigma(\boldsymbol{F}_2)-\sigma(\boldsymbol{F}_1)-\operatorname{tr}[\sigma_{\boldsymbol{F}}(\boldsymbol{F}_1)^T(\boldsymbol{F}_2-\boldsymbol{F}_1)]>0 \tag{83.10}$$

whenever $\boldsymbol{F}_1\neq\boldsymbol{F}_2$ *and* (83.8) *holds with positive-definite and symmetric* $\boldsymbol{G}$.

To prove this theorem assume first that condition (1) is satisfied. A limiting argument will show that $W_{12}\geqq W_{12}^*$ if the initial and final configuration actually coincide. In this case, we also have $\boldsymbol{F}_1=\boldsymbol{F}_2$ and hence, by (83.7), $W_{12}^*=0$. Therefore, the work done in every closed deformation process is non-negative. By the first work theorem, it follows that the material must be hyperelastic. The inequality (83.10) then follows from (83.9), (83.5), (83.7), and (83.4). That condition (2) implies condition (1), becomes evident on looking at (83.5) and (83.7).

While COLEMAN'S work theorem suggests an interpretation in terms of infinitesimal stability (Sect. 68bis), it is not clear what that interpretation should be.

ERICKSEN[1] has noticed that a work theorem is a corollary of the following theorem of CARATHÉODORY: If in every neighborhood of every point P in a Euclidean space there is a point Q such that for no curve joining P and Q the line integral of a given continuous vector field is zero, then the vector field is complex-lamellar [CFT, Appendix, Sect. 33]. Application of this result to the line integral (83.3) in the 9-dimensional tensor space $\mathscr{L}$ yields ***Ericksen's work theorem***: *In order that arbitrarily close to every configuration of an elastic body there be another configuration that cannot be attained by a workless deformation process, it is necessary and sufficient that the constitutive equation* (83.1) *of the material reduce to the more special form*

$$\boldsymbol{T}_{\rm R}=\varphi(\boldsymbol{F})\,\psi_{\boldsymbol{F}}(\boldsymbol{F}),\qquad T_{{\rm R}k}{}^{\alpha}=\varphi\frac{\partial\psi}{\partial x^k{}_{,\alpha}}, \tag{83.11}$$

where φ and ψ are scalar functions of $\boldsymbol{F}$.

The constitutive equation (83.11) is of a generality intermediate between elasticity and hyperelasticity, since only if it is possible to choose φ in such a way that $\varphi=h(\psi)$ does (83.11) reduce to the form (82.11). Thus the condition that work must be done along *some* deformation process leading to each configuration is not sufficient to characterize hyperelasticity.

A work principle has been proposed by RIVLIN and THOMAS[2] in formulating a definite, explicit theory of rupture of sheets of incompressible hyperelastic material. A specimen suffering a slow straight tear is held in such a way that no external work is done upon it; in particular, parts of the boundary where external forces are applied are held fixed. The stored energy is then a certain function of the length, c, of the tear (since nothing else is changing). Call this function Σ_c. Then the principle put forward by RIVLIN and THOMAS is

$$\frac{1}{h}\,\frac{d\Sigma_c}{dc}=E, \tag{83.12}$$

where h is the thickness of the sheet in its natural state and where E, an energy per unit area, is a material function. For two different test situations, in one of which a large part of the sheet is in simple extension, while in the other it is mainly in simple shear, they calculated Σ_c approximately from the general solutions given in Sect. 55. They found good agreement with the results of measurement[3].

This proposal seems to be consonant with that of GRIFFITH[4], which is generally discussed within the framework of the infinitesimal theory.

84. The strain-energy function. As foreshadowed in Sect. 79, the extended principle of material frame-indifference imposes restrictions on the strain-energy

[1] ERICKSEN [1956, *6*].
[2] RIVLIN and THOMAS [1952, *18*].
[3] Further experimental confirmation has been found by GREENSMITH [1963, *27*].
[4] Cf., e.g., GREEN and ZERNA [1954, *11*, § 8.15].

function σ, defined in Sect. 82. When dependence on the time is ignored, it follows from $(19.2)_2$ that a configuration $\boldsymbol{x}=\boldsymbol{\chi}(\boldsymbol{X})$ transforms under a change of frame according to the law

$$\boldsymbol{\chi}^*(\boldsymbol{X})=\boldsymbol{c}+\boldsymbol{Q}\boldsymbol{\chi}(\boldsymbol{X}), \tag{84.1}$$

provided that the reference configuration be unaffected by the change of frame. Taking the gradient of (84.1) with respect to $\boldsymbol{X}$, we obtain the law of transformation for the deformation gradient:

$$\boldsymbol{F}^*=\boldsymbol{Q}\boldsymbol{F}. \tag{84.2}$$

The strain energy σ is assumed to be frame-indifferent. Therefore, by (84.2) it must obey the identity

$$\sigma(\boldsymbol{F}^*)=\sigma(\boldsymbol{Q}\boldsymbol{F})=\sigma(\boldsymbol{F}) \tag{84.3}$$

for all tensors $\boldsymbol{F}$ in the domain of definition of σ and all orthogonal tensors $\boldsymbol{Q}$. Substituting in (84.3) for $\boldsymbol{Q}$ the inverse $\boldsymbol{R}^{-1}=\boldsymbol{R}^T$ of the rotation tensor and using the polar decomposition $(23.1)_1$ of the deformation gradient $\boldsymbol{F}$, we find that

$$\sigma(\boldsymbol{F})=\sigma(\boldsymbol{U}); \tag{84.4}$$

i.e., the strain energy depends upon $\boldsymbol{F}$ only through the right stretch tensor $\boldsymbol{U}$. Instead of $\boldsymbol{U}$, we may also employ the right Cauchy-Green tensor $\boldsymbol{C}=\boldsymbol{U}^2=\boldsymbol{F}^T\boldsymbol{F}$ and write[1]

$$\sigma(\boldsymbol{F})=\bar{\sigma}(\boldsymbol{F}^T\boldsymbol{F})=\bar{\sigma}(\boldsymbol{C})=\bar{\sigma}(C_{\alpha\beta}). \tag{84.5}$$

Taking the gradient of (84.3) with respect to $\boldsymbol{F}$, we obtain the identity

$$\sigma_{\boldsymbol{F}}(\boldsymbol{F})^T|_{\boldsymbol{Q}\boldsymbol{F}}\boldsymbol{Q}=\sigma_{\boldsymbol{F}}(\boldsymbol{F})^T, \tag{84.6}$$

and hence, by (82.10),

$$\mathfrak{g}(\boldsymbol{Q}\boldsymbol{F})=\varrho\boldsymbol{Q}\boldsymbol{F}\,\sigma_{\boldsymbol{F}}(\boldsymbol{Q}\boldsymbol{F})^T=\boldsymbol{Q}\left[\varrho\boldsymbol{F}\sigma_{\boldsymbol{F}}(\boldsymbol{F})^T\right]\boldsymbol{Q}^T=\boldsymbol{Q}\mathfrak{g}(\boldsymbol{F})\,\boldsymbol{Q}^T. \tag{84.7}$$

This relation is but the specialization of the identity (43.2) to the case of the response function of a hyperelastic material. However, it has been derived here from the requirement that the strain energy be frame-indifferent. Equation (84.7) shows that the requirement of indifference for the stress imposes no further restrictions on the strain-energy function.

Let $\boldsymbol{W}$ be an arbitrary skew tensor. We can find functions $\boldsymbol{Q}(\alpha)$ whose values are orthogonal tensors and which have the properties

$$\boldsymbol{Q}(0)=\boldsymbol{1},\qquad \dot{\boldsymbol{Q}}(0)=\frac{d}{d\alpha}\boldsymbol{Q}(\alpha)\,|_{\alpha=0}=\boldsymbol{W}.$$

If we substitute $\boldsymbol{Q}=\boldsymbol{Q}(\alpha)$ into (84.3), differentiate with respect to α and then put $\alpha=0$ we obtain

$$\operatorname{tr}\left[\sigma_{\boldsymbol{F}}(\boldsymbol{F})^T\boldsymbol{W}\boldsymbol{F}\right]=\operatorname{tr}\left[\boldsymbol{F}\sigma_{\boldsymbol{F}}(\boldsymbol{F})^T\boldsymbol{W}\right]=0.$$

This equation can hold for every skew $\boldsymbol{W}$ only if $\boldsymbol{F}\sigma_{\boldsymbol{F}}(\boldsymbol{F})^T$ is symmetric. Hence σ must satisfy the identity

$$\boldsymbol{F}\sigma_{\boldsymbol{F}}(\boldsymbol{F})^T=\sigma_{\boldsymbol{F}}(\boldsymbol{F})\boldsymbol{F}^T,\qquad x^k{}_{,\alpha}\frac{\partial\sigma}{\partial x^m{}_{,\alpha}}=x^m{}_{,\alpha}\frac{\partial\sigma}{\partial x^k{}_{,\alpha}}. \tag{84.8}$$

By (82.10), the identity (84.8) expresses the symmetry of the stress tensor $\boldsymbol{T}$. Hence, for hyperelastic materials, the indifference of the strain energy implies not

[1] This result was assumed by GREEN as his starting point in the works cited on p. 304. Proofs by CELLERIER, the COSSERATS, and MURNAGHAN have been cited in footnote 1, p. 119.

only the indifference of the stress, but also the symmetry of the stress. It can be proved that the indifference of the strain energy, the indifference of the stress, and the symmetry of the stress are actually equivalent[1].

Taking the gradient of (84.5) with respect to $\boldsymbol{F}$ and then using (9.2), the chain rule, and the product rule, we obtain the equation

$$\left.\begin{aligned}\operatorname{tr}[\sigma_{\boldsymbol{F}}(\boldsymbol{F})^T\boldsymbol{A}] &= \operatorname{tr}[\bar{\sigma}_{\boldsymbol{C}}(\boldsymbol{C})(\boldsymbol{A}^T\boldsymbol{F}+\boldsymbol{F}^T\boldsymbol{A})],\\ &= 2\operatorname{tr}[\bar{\sigma}_{\boldsymbol{C}}(\boldsymbol{C})\boldsymbol{F}^T\boldsymbol{A}],\end{aligned}\right\} \tag{84.9}$$

which must be valid for all tensors $\boldsymbol{A}$. It follows that the gradients of $\sigma(\boldsymbol{F})$ and $\bar{\sigma}(\boldsymbol{C})$ are related by

$$\sigma_{\boldsymbol{F}}(\boldsymbol{F})^T = 2\bar{\sigma}_{\boldsymbol{C}}(\boldsymbol{C})\boldsymbol{F}^T, \qquad \frac{\partial\sigma}{\partial x^k{}_{,\alpha}} = 2\frac{\partial\bar{\sigma}}{\partial C_{\alpha\beta}}x^m{}_{,\beta}g_{mk}. \tag{84.10}$$

Substituting this result into (82.9), we find the following *reduced form of the constitutive equation for hyperelastic materials*[2]:

$$\boldsymbol{T} = 2\varrho\boldsymbol{F}\bar{\sigma}_{\boldsymbol{C}}(\boldsymbol{C})\boldsymbol{F}^T, \qquad T^{km} = 2\varrho x^k{}_{,\alpha}x^m{}_{,\beta}\frac{\partial\bar{\sigma}}{\partial C_{\alpha\beta}}. \tag{84.11}$$

This form corresponds to the special case

$$\bar{\boldsymbol{T}} = \mathfrak{G}(\boldsymbol{C}) = 2\varrho\boldsymbol{C}\bar{\sigma}_{\boldsymbol{C}}(\boldsymbol{C})\,\boldsymbol{C}, \qquad \bar{T}_{\alpha\beta} = 2\varrho C_{\alpha\lambda}C_{\beta\mu}\frac{\partial\bar{\sigma}}{\partial C_{\lambda\mu}}, \tag{84.12}$$

of (43.7). Alternatively[3]

$$\boldsymbol{T}_{\mathrm{R}} = 2\varrho_{\mathrm{R}}\boldsymbol{F}\bar{\sigma}_{\boldsymbol{C}}(\boldsymbol{C}), \qquad T_{\mathrm{R}}^{k\alpha} = 2\varrho_{\mathrm{R}}x^k{}_{,\beta}\frac{\partial\bar{\sigma}}{\partial C_{\alpha\beta}}, \tag{84.13}$$

or[4], by $(43.10)_2$,

$$\left.\begin{aligned}\tilde{\boldsymbol{T}} &= \mathfrak{t}(\boldsymbol{C}) = 2\varrho_{\mathrm{R}}\bar{\sigma}_{\boldsymbol{C}}(\boldsymbol{C}),\\ \tilde{T}^{\alpha\beta} &= \mathfrak{t}^{\alpha\beta}(C_{\gamma\delta}) = 2\varrho_{\mathrm{R}}\frac{\partial\bar{\sigma}}{\partial C_{\alpha\beta}}.\end{aligned}\right\} \tag{84.14}$$

If we prefer to use $\boldsymbol{E}=\frac{1}{2}(\boldsymbol{C}-\mathbf{1})$ instead of $\boldsymbol{C}$ to measure the deformation, we may set $\bar{\sigma}(\boldsymbol{C})=\hat{\sigma}(\boldsymbol{E})$, $2\bar{\sigma}_{\boldsymbol{C}}=\hat{\sigma}_{\boldsymbol{E}}$ in (84.11) through (84.14). The formulae resulting by specialization from (43.5) and (43.6) are not very useful, and we omit them[5].

From (82.19), (45.5), and (45.6) we may read off the following alternative *necessary and sufficient conditions* for an elastic material to be hyperelastic[6]:

$$\mathsf{B}^{kmpq} = \mathsf{B}^{pqkm}, \tag{84.15}$$

$$\mathsf{C}^{kmpq} = \mathsf{C}^{pqkm}. \tag{84.16}$$

Comparison with (45.3) and (45.8) shows that *while for elastic materials there may be as many as* 36 *linearly independent elasticities* B^{kmpq} *and* C^{kmpq}, *for hyper-*

---

[1] NOLL [1955, *19*, Theorem I of § 16]. *Cf.* the step in this direction taken by GIBBS [1875, *1*] on p. 190 of the reprint in his *Works*.

[2] BOUSSINESQ [1870, *1*] [1872, *1*, Note 3, p. 591].

[3] KELVIN [1863, *2*, § 62].

[4] E. & F. COSSERAT [1896, *1*, § 24, Eq. (59)], correcting erroneous forms due to GREEN [1841, *1*] and ST. VENANT [1863, *1*, ¶ 2]. Cf. also OLDROYD [1950, *12*, § 2].

[5] A list of various alternative forms, with references, was given by TRUESDELL [1952, *20*, §§ 39—40].

[6] TRUESDELL [1961, *63*, § 12]. By $(45.11)_2$, the condition

$$\mathsf{D}^{kmpq} = \mathsf{D}^{mkpq}$$

is necessary but not sufficient for an elastic material to be hyperelastic. By $(45.11)_1$ and (84.15), in a hyperelastic material the condition

$$\mathsf{D}^{kmpq} = \mathsf{D}^{pqkm}$$

is necessary and sufficient that the stress be hydrostatic.

elastic materials there are at most 21. Since in the infinitesimal theory these coefficients reduce to the linear elasticities $L^{\alpha\beta\gamma\delta}$ occurring in (50.12), we see that the foregoing statement is an extension to the general theory of elasticity of a well known distinction between the infinitesimal theories of CAUCHY and GREEN. In GREEN'S theory, by hypothesis, there is a strain-energy function, while in CAUCHY'S, there need not be [CFT, Sect. 301]; the maximal numbers of independent elasticities in the two cases are 21 and 36.

It is clear that the strain-energy function depends not only on the material but also on the choice of the local reference configuration. An argument similar to the one given at the end of Sect. 28 shows that the strain-energy functions $\sigma_{\boldsymbol{K}}$ and $\sigma_{\hat{\boldsymbol{K}}}$ corresponding to two local reference configurations $\boldsymbol{K}$ and $\hat{\boldsymbol{K}}$ are related by

$$\sigma_{\boldsymbol{K}}(\boldsymbol{F}) = \sigma_{\hat{\boldsymbol{K}}}(\boldsymbol{F}\boldsymbol{G}^{-1}), \tag{84.17}$$

where $\boldsymbol{G}$ is the gradient of the deformation from $\boldsymbol{K}$ to $\hat{\boldsymbol{K}}$, given by (22.15).

The dependence of the internal energy function $\hat{\varepsilon}(\boldsymbol{F}, \eta)$ and the free-energy function $\tilde{\psi}(\boldsymbol{F}, \theta)$ (cf. Sect. 82) on the choice of the local reference configuration has been investigated by TOLOTTI[1]. He assumed that a natural state for any one temperature may be obtained from that for any other by a suitable dilatation. He inferred formulae relating the functions $\tilde{\psi}$ and $\hat{\varepsilon}$ corresponding to a fixed reference state to the analogous functions corresponding to a natural state which varies with θ or η. These formulae involve repeated integrals of the specific heat at constant pressure.

85. The isotropy group of the strain-energy function. Isotropic hyperelastic materials, hyperelastic fluids and solids. The isotropy group g of an elastic material, at a given particle and for a given reference configuration, is obtained by finding all unimodular tensors $\boldsymbol{H}$ such that (43.15) holds, identically in $\boldsymbol{F}$. For a hyperelastic material, by (82.10), this equation takes the form

$$\left.\begin{aligned} \sigma_{\boldsymbol{F}}(\boldsymbol{F})\boldsymbol{F}^T &= \sigma_{\boldsymbol{F}}(\boldsymbol{F}\boldsymbol{H})(\boldsymbol{F}\boldsymbol{H})^T, \\ &= \sigma_{\boldsymbol{F}}(\boldsymbol{F}\boldsymbol{H})\boldsymbol{H}^T\boldsymbol{F}^T. \end{aligned}\right\} \tag{85.1}$$

Since only non-singular $\boldsymbol{F}$ are considered,

$$\sigma_{\boldsymbol{F}}(\boldsymbol{F}) = \sigma_{\boldsymbol{F}}(\boldsymbol{F}\boldsymbol{H})\boldsymbol{H}^T. \tag{85.2}$$

For a fixed $\boldsymbol{H}$ this equation may be integrated to yield

$$\sigma(\boldsymbol{F}) = \sigma(\boldsymbol{F}\boldsymbol{H}) + \psi(\boldsymbol{H}). \tag{85.3}$$

Setting $\boldsymbol{F}=\boldsymbol{1}$ yields $\psi(\boldsymbol{H}) = \sigma(\boldsymbol{1}) - \sigma(\boldsymbol{H})$. Hence (85.3) becomes[2]

$$\sigma(\boldsymbol{F}) = \sigma(\boldsymbol{F}\boldsymbol{H}) + \sigma(\boldsymbol{1}) - \sigma(\boldsymbol{H}). \tag{85.4}$$

The members of g, then, are the unimodular tensors $\boldsymbol{H}$ such that (85.4) is satisfied identically in $\boldsymbol{F}$.

The considerations of Sect. 31, when applied to the strain-energy function σ of a hyperelastic material, suggest defining the *isotropy group* g_σ of σ as the group of all unimodular transformations $\boldsymbol{H}$ for which

$$\sigma(\boldsymbol{F}) = \sigma(\boldsymbol{F}\boldsymbol{H}), \tag{85.5}$$

identically in the tensor $\boldsymbol{F}$. While the elements of g give rise to deformations carrying the material from the reference configuration into others indiscernible from it by any experiment on the *stress*, the elements of g_σ correspond to deformations indiscernible by experiments on the *energy*. In general, these two groups

[1] TOLOTTI [1943, *6*].
[2] TRUESDELL [1964, *86*].

are not the same. If (85.5) is satisfied for a particular $\boldsymbol{H}$, putting $\boldsymbol{F}=\mathbf{1}$ yields $\sigma(\mathbf{1})=\sigma(\boldsymbol{H})$, and (85.4) for general $\boldsymbol{F}$ reduces to (85.5). Thus $\boldsymbol{H}$ satisfies (85.4). That is, $g_\sigma \subset g$. We now consider cases when $g_\sigma = g$ and when $g_\sigma \neq g$.

First, it is easy to see from (84.17) that the isotropy groups g_σ of σ for two difference reference configurations are related by the rule derived in Sect. 31 for g. Thus $g=g_\sigma$ for one reference configuration if and only if $g=g_\sigma$ for all.

Second, note that if $\boldsymbol{H}=\boldsymbol{Q}$, it follows from (84.3) that $\sigma(\boldsymbol{H})=\sigma(\mathbf{1})$, and again (85.4) reduces to (85.5). That is, *the orthogonal transformations in g and g_σ are the same.* If the material is a solid, by definition $g \subset o$ when the reference configuration is undistorted. It follows then that $g_\sigma = g$. *For a hyperelastic solid, the isotropy group of the strain-energy function is the isotropy group of the material.*

Moreover, if we suppose $\boldsymbol{H}\in g$ and set $\boldsymbol{F}=\boldsymbol{H}^{-1}$ in (85.4), we see that if $\sigma(\boldsymbol{H}) > \sigma(\mathbf{1})$, then $\sigma(\boldsymbol{H}^{-1}) < \sigma(\mathbf{1})$. Therefore, if σ has a strict minimum at $\boldsymbol{F}=\mathbf{1}$ in the sense that $\sigma(\boldsymbol{F})>1$ for all non-orthogonal $\boldsymbol{F}$, the isotropy group can consist only of orthogonal tensors. Thus follows the ***fundamental theorem on hyperelastic solids***[1]: *If the strain-energy function of a hyperelastic material has a strict minimum in the sense just described, the material is a simple solid, and any minimizing configuration is a natural state.*

Now if $\boldsymbol{H}=\boldsymbol{Q}$, by use of (84.4) we may put (85.5) into the form

$$\sigma(\boldsymbol{U})=\sigma(\boldsymbol{Q}^T\boldsymbol{U}\boldsymbol{Q}), \tag{85.6}$$

or, by (84.5),

$$\bar{\sigma}(\boldsymbol{C})=\bar{\sigma}(\boldsymbol{Q}^T\boldsymbol{C}\boldsymbol{Q}). \tag{85.7}$$

Therefore, the isotropy group of a hyperelastic solid, taken relative to an undistorted reference state, is the group of all orthogonal transformations that leave the strain energy function invariant, i.e., the group g consists of all orthogonal $\boldsymbol{Q}$ such that (85.6) or (85.7) holds.

The invariance under the isotropy group places limitations on the possible forms of the strain-energy function. For isotropic materials, these limitations will be discussed in the next section. Here we consider aeolotropic hyperelastic solids whose isotropy groups are of one of the types listed in Sect. 33. It turns out[2] that the strain energy is invariant under the isotropy group g if and only if it can be expressed as a function of a certain number of scalars $I_1, \ldots, I_{\mathfrak{k}}$ that are invariant under g:

$$\sigma=\sigma(\boldsymbol{U})=\bar{\sigma}(\boldsymbol{C})=\sigma(I_1, \ldots, I_{\mathfrak{k}}). \tag{85.8}$$

The table overleaf lists the invariant scalars $I_1, \ldots, I_{\mathfrak{k}}$ for each of the isotropy groups in the table for the crystal classes in Sect. 33. The notation C_{ij} is used for the components of $\boldsymbol{C}$ relative to the orthogonal basis $\boldsymbol{i}, \boldsymbol{j}, \boldsymbol{k}$ used to describe the isotropy groups. The components U_{ij} of $\boldsymbol{U}$ may be used instead of the C_{ij} to define $I_1, \ldots, I_{\mathfrak{k}}$.

Smith and Rivlin showed also that when $\bar{\sigma}(\boldsymbol{C})$ is a *polynomial* in the components C_{ij} of $\boldsymbol{C}$, it can be expressed as a *polynomial* in the invariant scalars $I_1, \ldots, I_{\mathfrak{k}}$. Moreover, Smith[3] has shown that those invariant scalars satisfy certain algebraic relations (syzygies). He has used these relations to obtain *unique* representations of polynomial strain-energy functions in terms of $I_1, \ldots, I_{\mathfrak{k}}$.

---

[1] Truesdell [1964, *86*, Coroll. 5].

[2] The proof, straightforward but tedious, was given by Smith and Rivlin [1958, *44*]. Corresponding explicit forms of the stress relation were given by Green and Adkins [1960, *26*, Ch. I].

[3] Smith [1962, *62*].

Crystal Class	Type	List of $I_1, \ldots, I_{\mathfrak{t}}$
Triclinic System	1	$C_{11}, C_{22}, C_{33}, C_{12}, C_{13}, C_{23}$
Monoclinic System	2	$C_{11}, C_{22}, C_{33}, C_{12}, C_{13}^2, C_{23}^2, C_{13}\,C_{23}$
Rhombic System	3	$C_{11}, C_{22}, C_{33}, C_{23}^2, C_{13}^2, C_{12}^2, C_{12}\,C_{23}\,C_{13}$
Tetragonal System	4	$C_{11}+C_{22}, C_{33}, C_{13}^2+C_{23}^2, C_{12}^2, C_{11}\,C_{22}, C_{12}(C_{11}-C_{22}),$ $C_{13}\,C_{23}\,(C_{11}-C_{22}), C_{12}\,C_{23}\,C_{13}, C_{12}(C_{13}^2-C_{23}^2),$ $C_{11}\,C_{23}^2+C_{22}\,C_{13}^2, C_{13}\,C_{12}\,(C_{13}^2-C_{23}^2), C_{13}^2\,C_{23}^2$
	5	$C_{11}+C_{22}, C_{33}, C_{13}^2+C_{23}^2, C_{11}\,C_{22}, C_{12}\,C_{23}\,C_{13},$ $C_{11}\,C_{23}^2+C_{22}\,C_{13}^2, C_{13}^2\,C_{23}^2$
Cubic System	6	$C_{11}+C_{22}+C_{33}, C_{22}\,C_{33}+C_{33}\,C_{11}+C_{11}\,C_{22}, C_{11}\,C_{22}\,C_{33},$ $C_{23}^2+C_{31}^2+C_{12}^2, C_{31}^2\,C_{12}^2+C_{12}^2\,C_{23}^2+C_{23}^2\,C_{31}^2, C_{23}C_{31}C_{12},$ $C_{22}\,C_{12}^2+C_{33}\,C_{23}^2+C_{11}\,C_{31}^2, C_{31}^2\,C_{33}+C_{12}^2\,C_{11}+C_{23}^2\,C_{22},$ $C_{33}\,C_{22}^2+C_{11}\,C_{33}^2+C_{22}\,C_{11}^2, C_{12}^2\,C_{31}^4+C_{23}^2\,C_{12}^4+C_{31}^2\,C_{23}^4,$ $C_{11}\,C_{31}^2\,C_{12}^2+C_{22}\,C_{12}^2\,C_{23}^2+C_{33}\,C_{23}^2\,C_{31}^2, C_{23}^2\,C_{22}\,C_{33}+$ $C_{31}^2\,C_{33}\,C_{11}+C_{12}^2\,C_{11}\,C_{22}, C_{11}\,C_{22}\,C_{31}^2+C_{22}\,C_{33}\,C_{12}^2+$ $C_{33}\,C_{11}\,C_{23}^2, C_{23}^2\,C_{31}^2\,C_{22}+C_{31}^2\,C_{12}^2\,C_{33}+C_{12}^2\,C_{23}^2C_{11}$
	7	$C_{11}+C_{22}+C_{33}, C_{22}\,C_{33}+C_{33}\,C_{11}+C_{11}\,C_{22}, C_{11}\,C_{22}\,C_{33},$ $C_{23}^2+C_{31}^2+C_{12}^2, C_{31}^2C_{12}^2+C_{12}^2C_{23}^2+C_{23}^2C_{31}^2, C_{23}C_{31}C_{12},$ $C_{22}\,C_{12}^2+C_{33}\,C_{31}^2+C_{33}\,C_{23}^2+C_{11}\,C_{12}^2+C_{11}C_{31}^2+C_{22}\,C_{23}^2,$ $C_{11}\,C_{31}^2\,C_{12}^2+C_{22}\,C_{12}^2\,C_{23}^2+C_{33}\,C_{23}^2\,C_{31}^2,$ $C_{23}^2\,C_{22}\,C_{33}+C_{31}^2\,C_{33}\,C_{11}+C_{12}^2\,C_{11}\,C_{22}$
	8	$C_{33}, C_{11}+C_{22}, C_{11}\,C_{22}-C_{12}^2, C_{11}[(C_{11}+3C_{22})^2-12C_{12}^2],$ $C_{31}^2+C_{23}^2, C_{31}(C_{31}^2-3C_{23}^2), (C_{11}-C_{22})\,C_{31}-2C_{12}\,C_{23},$ $(C_{22}-C_{11})\,C_{23}-2C_{12}\,C_{31}, 3C_{12}(C_{11}-C_{22})^2-4C_{12}^3,$ $C_{23}(C_{23}^2-3C_{31}^2), C_{22}C_{31}^2+C_{11}\,C_{23}^2-2C_{23}\,C_{31}\,C_{12},$ $C_{31}[(C_{11}+C_{22})^2+4(C_{12}^2-C_{22}^2)]-8C_{11}\,C_{12}\,C_{23},$ $C_{23}[(C_{11}+C_{22})^2+4(C_{12}^2-C_{22}^2)]+8C_{11}\,C_{12}\,C_{31},$ $(C_{11}-C_{22})\,C_{23}\,C_{31}+C_{12}(C_{23}^2-C_{31}^2)$
	9	$C_{33}, C_{11}+C_{22}, C_{11}\,C_{22}-C_{12}^2, C_{11}[(C_{11}+3C_{22})^2-12C_{12}^2],$ $C_{31}^2+C_{23}^2, C_{23}(C_{23}^2-3C_{31}^2), (C_{11}-C_{22})\,C_{23}+2C_{12}\,C_{31},$ $C_{11}\,C_{31}^2+C_{22}\,C_{23}^2+2C_{23}\,C_{31}\,C_{12},$ $C_{23}[(C_{11}+C_{22})^2-4(C_{22}^2-C_{12}^2)]+8C_{11}\,C_{12}\,C_{31}$
Hexagonal System	10	$C_{33}, C_{11}+C_{22}, C_{11}\,C_{22}-C_{12}^2, C_{11}[(C_{11}+3C_{22})^2-12C_{12}^2],$ $C_{31}^2+C_{23}^2, C_{31}^2(C_{31}^2-3C_{23}^2)^2, C_{11}\,C_{23}^2+C_{22}\,C_{31}^2-2C_{23}C_{31}C_{12},$ $C_{12}(C_{31}^2-C_{23}^2)+(C_{22}-C_{11})\,C_{31}\,C_{23},$ $3C_{12}(C_{11}-C_{22})^2-4C_{12}^3,$ $C_{31}\,C_{23}[3(C_{31}^2-C_{23}^2)^2-4C_{31}^2\,C_{23}^2],$ $C_{11}(C_{31}^4+3C_{23}^4)+2C_{22}C_{31}^2(C_{31}^2+3C_{23}^2)-8C_{12}\,C_{23}\,C_{31}^3,$ $C_{31}^2[(C_{11}+C_{22})^2-4(C_{22}^2-C_{12}^2)]$ $-2C_{11}[(C_{11}+3C_{22})(C_{31}^2+C_{23}^2)-4C_{23}\,C_{31}\,C_{12}],$ $C_{23}C_{31}[(C_{11}+C_{22})^2-4(C_{22}^2-C_{12}^2)]+4C_{11}C_{12}(C_{23}^2-C_{31}^2),$ $C_{12}[(C_{31}^2+C_{23}^2)^2+4C_{23}^2(C_{31}^2-C_{23}^2)]-4C_{31}^3C_{23}(C_{11}-C_{22})$
	11	$C_{33}, C_{11}+C_{22}, C_{11}\,C_{22}-C_{12}^2, C_{11}[(C_{11}+3C_{22})^2-12C_{12}^2],$ $C_{31}^2+C_{23}^2, C_{31}^2(C_{31}^2-3C_{23}^2)^2,$ $C_{11}\,C_{23}^2+C_{22}\,C_{31}^2-2C_{23}\,C_{31}\,C_{12},$ $C_{11}(C_{31}^4+3C_{23}^4)+2C_{22}\,C_{31}^2(C_{31}^2+3C_{23}^2)-8C_{12}C_{23}C_{31}^3,$ $C_{31}^2[(C_{11}+C_{22})^2-4(C_{22}^2-C_{12}^2)]$ $-2C_{11}[(C_{11}+3C_{22})(C_{31}^2+C_{23}^2)-4C_{23}\,C_{31}\,C_{12}]$

Next consider an elastic fluid. Setting $v \equiv 1/\varrho$, we can define a function $\hat{\sigma}(v)$ in terms of the pressure function $p(\varrho)$ occurring in the constitutive equation (50.1):

$$p = p(\varrho) = -\hat{\sigma}_v(v). \tag{85.9}$$

Since $v = v_R|\det \boldsymbol{F}|$, where $v_R \equiv 1/\varrho_R$, we see that $\hat{\sigma}(v) = \hat{\sigma}(v_R|\det \boldsymbol{F}|)$, and by using the formula (9.3) for the gradient of a determinant and then applying the chain rule we find that in this case

$$\left.\begin{aligned} \sigma_{\boldsymbol{F}}(\boldsymbol{F})^T &= \hat{\sigma}_v(v) v_R |\det \boldsymbol{F}| \boldsymbol{F}^{-1} \\ &= \frac{1}{\varrho}\hat{\sigma}_v(v)\boldsymbol{F}^{-1}, \end{aligned}\right\} \tag{85.10}$$

Hence (85.9) is of the form (82.1). That is, *an elastic fluid is hyperelastic*, with stored-energy function

$$\sigma(\boldsymbol{F}) = \varrho \int \frac{p(\xi)\, d\xi}{\xi^2}, \quad \varrho = \varrho_R/|\det \boldsymbol{F}|. \tag{85.11}$$

Either directly from (85.11) or by glancing at the corresponding special case of (85.4) we see that $\mathcal{g} = \mathcal{g}_\sigma = \mathcal{u}$, the unimodular group.

When thermodynamic variables are taken into account, according to the results of Sect. 83, we may identify the strain-energy function $\hat{\sigma}(v)$ with the internal energy function $\hat{\varepsilon}(v, \eta)$ or with the free energy function $\tilde{\psi}(v, \theta)$. Then, (85.9) corresponds to the following *thermal equations of state*:

$$p = -\hat{\varepsilon}_v(v, \eta) = -\tilde{\psi}_v(v, \theta). \tag{85.12}$$

These thermal equations are valid for both statics and dynamics of perfect fluids as well as for the statics of all simple fluids.

Returning to the general consideration of the groups $\mathcal{g}$ and $\mathcal{g}_\sigma$ at the beginning of this section, we note a simple corollary of the fact that all orthogonal transformations in $\mathcal{g}$ belong also to $\mathcal{g}_\sigma$. Namely[1], $\mathcal{g} \supset \mathcal{o}$ if and only if $\mathcal{g}_\sigma \supset \mathcal{o}$. That is, *a hyperelastic material is isotropic if and only if its strain-energy function $\sigma(\boldsymbol{U}) = \bar{\sigma}(\boldsymbol{C})$, where $\boldsymbol{U}$ and $\boldsymbol{C}$ are taken relative to an undistorted state, is an isotropic function, i.e., an orthogonal invariant*[2].

For an isotropic hyperelastic material, then, (85.6) and (85.7) are identities in $\boldsymbol{Q}$. If we substitute for $\boldsymbol{Q}$ in (85.6) and (85.7) the transpose $\boldsymbol{R}^T$ of the rotation tensor $\boldsymbol{R}$, we find, by (23.3) and (23.5), that

$$\sigma(\boldsymbol{F}) = \sigma(\boldsymbol{U}) = \sigma(\boldsymbol{V}) = \bar{\sigma}(\boldsymbol{C}) = \bar{\sigma}(\boldsymbol{B}). \tag{85.13}$$

Thus, for an isotropic material, the strain energy may be expressed as an isotropic function of the right stretch tensor $\boldsymbol{U}$, or of the left stretch tensor $\boldsymbol{V}$, or of the right Cauchy-Green tensor $\boldsymbol{C} = \boldsymbol{F}^T\boldsymbol{F}$, or of the left Cauchy-Green tensor $\boldsymbol{B} = \boldsymbol{F}\boldsymbol{F}^T$ (cf. Sect. 23).

Applying to $\sigma(\boldsymbol{F}) = \bar{\sigma}(\boldsymbol{B}) = \bar{\sigma}(\boldsymbol{F}\boldsymbol{F}^T)$ an argument analogous to the one which led from (84.5) to (84.10), we obtain

$$\sigma_{\boldsymbol{F}}(\boldsymbol{F})^T = 2\boldsymbol{F}^T\bar{\sigma}_{\boldsymbol{B}}(\boldsymbol{B}), \qquad \frac{\partial \sigma}{\partial x^k{}_{,\alpha}} = 2 x^m{}_{,\beta}\, g^{\alpha\beta} \frac{\partial \bar{\sigma}}{\partial B^{mk}}. \tag{85.14}$$

Substituting this result into (82.9) and using the fact, clear from (12.10), that $\bar{\sigma}_{\boldsymbol{B}}(\boldsymbol{B})$ commutes with $\boldsymbol{B}$, we find that the constitutive equation of an isotropic hyperelastic material may be written in the form

$$\left.\begin{aligned} \boldsymbol{T} &= 2\varrho \boldsymbol{B}\bar{\sigma}_{\boldsymbol{B}}(\boldsymbol{B}) = 2\varrho\bar{\sigma}_{\boldsymbol{B}}(\boldsymbol{B})\boldsymbol{B}, \\ T^k{}_m &= 2\varrho B^{kr}\frac{\partial \bar{\sigma}}{\partial B^{mr}}, \end{aligned}\right\} \tag{85.15}$$

---

[1] This result is due to NOLL [1955, *19*, Theorem II of § 16], in whose proof it was assumed that $\sigma_{\boldsymbol{F}}$ is continuous.

[2] In older works this formal statement, or an equivalent one, was taken as the definition of material isotropy. Cf. KELVIN [1863, *2*, § 69], FINGER [1894, *3*, p. 1088], E. and F. COSSERAT [1896, *1*, § 29], MURNAGHAN [1937, *2*, § 3], TRUESDELL [1952, *20*, § 38]. FINGER [1894, *3*, pp. 1084—1088] had attempted first to define an elastic material as isotropic by the condition, in fact insufficient (cf. footnote 1, p. 143), that the principal axes of stress and strain coincide.

which corresponds to the response function $\mathfrak{f}(\boldsymbol{B}) = 2\varrho \boldsymbol{B}\bar{\sigma}_{\boldsymbol{B}}(\boldsymbol{B})$ in (47.4). If we put $\tilde{\sigma}(\boldsymbol{B}^{-1}) = \bar{\sigma}(\boldsymbol{B})$, we may use (9.19) to show that (85.15) is equivalent to[1]

$$\left.\begin{aligned} \boldsymbol{T} &= -2\varrho \boldsymbol{B}^{-1}\tilde{\sigma}_{\boldsymbol{B}^{-1}}(\boldsymbol{B}^{-1}) = -2\varrho\tilde{\sigma}_{\boldsymbol{B}^{-1}}(\boldsymbol{B}^{-1})\,\boldsymbol{B}^{-1}, \\ T^k{}_m &= -2\varrho (B^{-1})^k_p \frac{\partial\tilde{\sigma}}{\partial (B^{-1})^m_p}. \end{aligned}\right\} \tag{85.16}$$

Using the identity $\bar{\sigma}(\boldsymbol{B}) = \bar{\sigma}(\boldsymbol{V}^2) = \sigma(\boldsymbol{V})$, one can easily show that for hyperelastic materials the stress relation (47.3) specializes to

$$\boldsymbol{T} = \varrho \boldsymbol{V}\sigma_{\boldsymbol{V}}(\boldsymbol{V}) = \varrho\sigma_{\boldsymbol{V}}(\boldsymbol{V})\boldsymbol{V}, \tag{85.17}$$

where the second form follows from the first because $\bar{\sigma}_{\boldsymbol{B}}(\boldsymbol{B})$, since it commutes with $\boldsymbol{B}$, commutes also with its square root, which is $\boldsymbol{V}$. Thus the response function occurring in (47.3) is $\mathfrak{g}(\boldsymbol{V}) = \varrho\boldsymbol{V}\sigma_{\boldsymbol{V}}(\boldsymbol{V})$.

At the beginning of this section it was shown that $g_\sigma \neq g$ in general. Indeed[2], by (85.4), in order that $g_\sigma \neq g$ it is necessary and sufficient that g contain an element $\boldsymbol{H}$, necessarily non-orthogonal, such that $\sigma(\boldsymbol{H}) \neq \sigma(\boldsymbol{1})$. In other words, there must exist a deformation that leaves the form of the stress relation unchanged but cannot be effected without loss or gain of work. This criterion is obvious from the very ideas behind the definitions of the two isotropy groups. g gives rise to all deformations, from the given reference configuration, that cannot be detected by experiments measuring contact forces; g_σ, to all that cannot be detected by experiments on work done.

85 bis. Hyperelastic subfluids (added in proof). The class of isotropy groups g that define subfluids has been delimited in Sect. 33 bis. If a subfluid is also a hyperelastic material, it is called a *hyperelastic subfluid*. In the previous section, after seeing that g and g_σ need not be the same in general, we have proved that for hyperelastic fluids and solids they are. The first example of a hyperelastic material with two distinct isotropy groups has been constructed by WANG[3], who has shown that for a *hyperelastic subfluid, in general*, $g \neq g_\sigma$. WANG considers the functional equation (85.4) for a hyperelastic subfluid of Type 14. By definition, the isotropy group g is a dilatation group. He shows that the general solution[4] of (84.4) and (85.4) for which $\boldsymbol{H}$ belongs to the dilatation group g is of the form

$$\bar{\sigma}(\boldsymbol{C}) = \hat{\sigma}(C_{12}\sqrt{C_{33}}, C_{23}\sqrt{C_{11}}, C_{13}\sqrt{C_{22}}, C_{11}C_{22}C_{33}) + K_1 \log C_{11} + K_2 \log C_{22}, \tag{85b.1}$$

where the reference configuration is undistorted, and the components C_{ij} are taken relative to the axes of the subfluid. The function $\hat{\sigma}$ satisfies only the following restrictions:

$$\hat{\sigma}(\alpha, \beta, \gamma, \delta) = \hat{\sigma}(-\alpha, -\beta, \gamma, \delta) = \hat{\sigma}(-\alpha, \beta, -\gamma, \delta) = \hat{\sigma}(\alpha, -\beta, -\gamma, \delta). \tag{85b.2}$$

From (85 b.1), we see that

$$\bar{\sigma}(\boldsymbol{1}) = \hat{\sigma}(0, 0, 0, 1). \tag{85b.3}$$

---

[1] MURNAGHAN [1937, *2*, § 3]. Several other forms for isotropic materials, with references, were listed by TRUESDELL [1952, *20*, § 41].

[2] TRUESDELL [1964, *86*].

[3] WANG [1965, *42*].

TRUESDELL [1964, *86*] and BRAGG [1964, *11*, § 3] remarked that $\sigma(\boldsymbol{H}) = \sigma(\boldsymbol{1})$ on any locally compact subgroup of g, providing σ be integrable and the group integral over the subgroup finite. The isotropy groups of WANG's subfluids are locally compact but of infinite measure with respect to the group integration.

[4] Hence the stored-energy function of *every* hyperelastic subfluid can be identified as an appropriate special case of (85 b.1), and, conversely, every function included in (85 b.1) is the stored-energy function of some hyperelastic subfluid.

By definition, if $\boldsymbol{H}\in g$, then $\|H_i^{\,j}\|=\operatorname{diag}\left(a,\, b,\, \frac{\pm 1}{ab}\right)$. Hence from (85b.1)

$$\bar{\sigma}(\boldsymbol{H}^T\boldsymbol{H})=\hat{\sigma}(0,0,0,1)+K_1\log a^2+K_2\log b^2. \tag{85b.4}$$

Since $\boldsymbol{H}\in g_\sigma$ if and only if $\bar{\sigma}(\boldsymbol{H}^T\boldsymbol{H})=\bar{\sigma}(\mathbf{1})$, from (85b.3) and (85b.4), we see that, if $K_1 \neq K_2$ and $K_1K_2\neq 0$, then $\boldsymbol{H}\in g_\sigma$ if and only if

$$a^2=b^2=1. \tag{85b.5}$$

Thus g_σ consists in the reflections in the axes of the subfluid[1].

Representations of the stored-energy function of other types of hyperelastic subfluids are special cases of (85b.1). The explicit forms, as found by WANG, are listed in the following table:

Reference No.	Representation of the stored-energy function
1 (Elastic fluid)	$\hat{\sigma}(\det \boldsymbol{C})$
2	$\hat{\sigma}(\det \boldsymbol{C})+K_1\log C_{33}$
3	$\hat{\sigma}(\det \boldsymbol{C})+K_1\log(C_{11}C_{22}-C_{12}^2)$
4	$\hat{\sigma}(\det \boldsymbol{C})+K_1\log C_{11}+K_2\log(C_{11}C_{22}-C_{12}^2)$
5	$\hat{\sigma}\left(\frac{(C_{23}C_{11}-C_{13}C_{12})^2}{(C_{11}C_{22}-C_{12}^2)(C_{11}C_{33}-C_{13}^2)},\ \det \boldsymbol{C}\right)+K_1\log C_{11}$
6	$\hat{\sigma}\left(\frac{(C_{23}C_{11}-C_{13}C_{12})^2}{(C_{11}C_{22}-C_{12}^2)(C_{11}C_{33}-C_{13}^2)},\ \det \boldsymbol{C}\right)+K_1\log(C_{11}C_{22}-C_{12}^2)+$ $+K_2\log(C_{11}C_{33}-C_{13}^2)$
7	$\hat{\sigma}(C_{33}(C_{11}C_{22}-C_{12}^2),\ \det \boldsymbol{C})+K_1\log C_{33}$
8	$\hat{\sigma}\left(\frac{C_{12}^2}{C_{11}C_{12}},\ \det \boldsymbol{C}\right)+K_1\log C_{11}C_{22}$
9	$\hat{\sigma}\left(\frac{C_{12}^2}{C_{11}C_{12}},\ \det \boldsymbol{C}\right)+K_1\log C_{11}+K_2\log C_{22}$
10	$\hat{\sigma}\left(\frac{C_{12}^2}{C_{11}C_{22}},\ C_{22}(C_{11}C_{33}-C_{13}^2),\ \det \boldsymbol{C}\right)+K_1\log C_{11}+K_2\log C_{22}$
11* 12*	$\hat{\sigma}\left(\frac{C_{12}}{\sqrt{C_{11}C_{22}}},\ \frac{C_{23}}{\sqrt{C_{22}C_{33}}},\ \frac{C_{13}}{\sqrt{C_{11}C_{33}}},\ \det \boldsymbol{C}\right)$
13**	$\hat{\sigma}\left(\frac{C_{12}}{\sqrt{C_{11}C_{22}}},\ \frac{C_{23}}{\sqrt{C_{22}C_{33}}},\ \frac{C_{13}}{\sqrt{C_{11}C_{33}}},\ \det \boldsymbol{C}\right)+K_1\log C_{33}$
14**	$\hat{\sigma}\left(\frac{C_{12}}{\sqrt{C_{11}C_{22}}},\ \frac{C_{23}}{\sqrt{C_{22}C_{33}}},\ \frac{C_{13}}{\sqrt{C_{11}C_{33}}},\ \det \boldsymbol{C}\right)+K_1\log C_{11}+K_2\log C_{22}$

* For Type 11 $\hat{\sigma}$ satisfies the identities

$$\hat{\sigma}(\alpha,\beta,\gamma,\delta)=\hat{\sigma}(\beta,\gamma,\alpha,\delta)=\hat{\sigma}(\gamma,\alpha,\beta,\delta),$$

while for Type 12 $\hat{\sigma}$ is a symmetric function of α, β, γ.

** For Type 13 $\hat{\sigma}$ satisfies the identity

$$\hat{\sigma}(\alpha,\beta,\gamma,\delta)=\hat{\sigma}(\alpha,\gamma,\beta,\delta).$$

Furthermore, for Types 11—14 $\hat{\sigma}$ satisfies (85b.2).

[1] As noted in Sect. 33, g_σ always contains the orthogonal part of g. For the particular hyperelastic subfluid considered, g_σ coincides with the orthogonal part of g.

The isotropy groups $\mathscr{g}_\sigma$ of the stored-energy functions of the various types of hyperelastic subfluids can be obtained easily from the previous table. The components $H_i^{\,j}$ of $\boldsymbol{H} \in \mathscr{g}_\sigma$ relative to the axes of the subfluid are listed in the following table:

Reference No.	$\|H_i^{\,j}\|$ $(\lvert\det\|H_i^{\,j}\|\rvert = 1)$	Reference No.	$\|H_i^{\,j}\|$ $(\lvert\det\|H_i^{\,j}\|\rvert = 1)$
1 (Elastic fluid)	$\mathscr{g}_\sigma = \mathscr{g} = \mathscr{U}$	8	$\begin{Vmatrix} a & 0 & b \\ 0 & c & d \\ 0 & 0 & \pm 1 \end{Vmatrix}$ or $\begin{Vmatrix} 0 & a & b \\ c & 0 & d \\ 0 & 0 & \pm 1 \end{Vmatrix}$
2	$\begin{Vmatrix} a & b & 0 \\ c & d & 0 \\ e & f & \pm 1 \end{Vmatrix}$	9	$\begin{Vmatrix} \pm 1 & 0 & a \\ 0 & \pm 1 & b \\ 0 & 0 & \pm 1 \end{Vmatrix}$
3	$\begin{Vmatrix} a & b & c \\ d & e & f \\ 0 & 0 & \pm 1 \end{Vmatrix}$	10	$\begin{Vmatrix} \pm 1 & 0 & a \\ 0 & \pm 1 & 0 \\ 0 & 0 & \pm 1 \end{Vmatrix}$
4	$\begin{Vmatrix} \pm 1 & a & b \\ 0 & \pm 1 & c \\ 0 & 0 & \pm 1 \end{Vmatrix}$ (where the ± signs are not associated in any way)	11, 12	$\mathscr{g}_\sigma = \mathscr{g}$
5	$\begin{Vmatrix} \pm 1 & a & b \\ 0 & c & 0 \\ 0 & 0 & d \end{Vmatrix}$ or $\begin{Vmatrix} \pm 1 & a & b \\ 0 & 0 & c \\ 0 & d & 0 \end{Vmatrix}$	13	$\begin{Vmatrix} a & 0 & 0 \\ 0 & b & 0 \\ 0 & 0 & \pm 1 \end{Vmatrix}$ or $\begin{Vmatrix} 0 & a & 0 \\ b & 0 & 0 \\ 0 & 0 & \pm 1 \end{Vmatrix}$
6	$\begin{Vmatrix} \pm 1 & a & b \\ 0 & \pm 1 & 0 \\ 0 & 0 & \pm 1 \end{Vmatrix}$	14	$\begin{Vmatrix} \pm 1 & 0 & 0 \\ 0 & \pm 1 & 0 \\ 0 & 0 & \pm 1 \end{Vmatrix}$
7	$\begin{Vmatrix} a & b & 0 \\ c & d & 0 \\ 0 & 0 & \pm 1 \end{Vmatrix}$		

Comparing the previous table with that given in Sect. 33 bis, we see that only Types 11 and 12 share with the elastic fluid the property of having the same isotropy group for the stress as for the stored-energy function. The elastic fluid, however, is the only type of elastic subfluid that is always hyperelastic. The explicit forms of the response functions of hyperelastic subfluids can be derived easily from representations of the stored-energy function. For Types 1—4 the results are, respectively[1],

$$\left.\begin{aligned} \boldsymbol{T} &= -p(\varrho)\,\mathbf{1}, \\ \boldsymbol{T} &= -p(\varrho)\,\mathbf{1} + 2K_1\varrho\,\boldsymbol{e}_3 \otimes \boldsymbol{e}_3, \\ \boldsymbol{T} &= -p(\varrho)\,\mathbf{1} - 2K_1\varrho\,\boldsymbol{e}_3 \otimes \boldsymbol{e}_3, \\ \boldsymbol{T} &= -p(\varrho)\,\mathbf{1} + 2K_1\varrho\,\boldsymbol{e}_1 \otimes \boldsymbol{e}_1 - 2K_2\varrho\,\boldsymbol{e}_3 \otimes \boldsymbol{e}_3. \end{aligned}\right\} \tag{85 b.6}$$

Since the types are so defined as to be mutually exclusive, neither K_1 nor K_2 can vanish. Therefore, from (85 b.6) we see that *the stress is never hydrostatic for hyperelastic subfluids of Types 2—4. In particular, these subfluids do not have a natural state.* That the elastic fluids also fail to have a natural state follows, however, only if additional thermodynamical restrictions, such as the C-N condition, are imposed, as we shall see in the following sections.

[1] The response functions of Types 2 and 3 take same form, but $\boldsymbol{e}_3$ has different meanings. For Type 2, $\boldsymbol{e}_3$ is a material direction, while for Type 3, $\boldsymbol{e}_3$ is the unit normal of a material plane.

86. Explicit forms of the stress relation for isotropic hyperelastic materials. The strain energy $\sigma(\boldsymbol{V})$, for isotropic materials, is an invariant and hence may be expressed as a symmetric function of the principal stretches v_a:

$$\sigma(\boldsymbol{V}) = \sigma(v_1, v_2, v_3). \tag{86.1}$$

We now consider an orthonormal basis of proper vectors of the left stretch tensor $\boldsymbol{V}$. These vectors define the principal axes of strain in the deformed state, which coincide with the principal axes of stress. Referring the tensors occurring in (85.17) to these axes, we obtain the following relation between the principal stretches v_a and the corresponding principal stresses t_a:

$$t_a = \varrho v_a \frac{\partial \sigma}{\partial v_a} = \varrho v_a \sigma_{v_a}(v_1, v_2, v_3), \qquad a = 1, 2, 3 \text{ (unsummed)}. \tag{86.2}$$

Noting that the density ϱ_{R} in the undistorted reference configuration is related to ϱ by the identity

$$\varrho_{\mathrm{R}} = (\det \boldsymbol{V})\, \varrho = v_1 v_2 v_3\, \varrho, \tag{86.3}$$

we may rewrite (86.2) in the form[1]

$$T_a = t_a v_b v_c = \varrho_{\mathrm{R}} \frac{\partial \sigma}{\partial v_a} = \frac{\partial}{\partial v_a}(\varrho_{\mathrm{R}} \sigma) \qquad (a, b, c \neq), \tag{86.4}$$

which shows that the *principal forces in an isotropic hyperelastic body are the derivatives of the strain energy, per unit volume in the reference configuration, with respect to the corresponding stretches.*

If the reference configuration is a natural state, by (86.2) the function $\sigma(v_1, v_2, v_3)$ satisfies the relations

$$\sigma_{v_a}(1, 1, 1) = 0. \tag{86.5}$$

According to the representation theorems of Sect. 10, the strain energy may be expressed as a function of the principal invariants $I_{\boldsymbol{V}}, II_{\boldsymbol{V}}, III_{\boldsymbol{V}}$ of the stretch tensors $\boldsymbol{V}$ or $\boldsymbol{U}$,

$$\sigma(\boldsymbol{V}) = \sigma(\boldsymbol{U}) = \sigma(I_{\boldsymbol{V}}, II_{\boldsymbol{V}}, III_{\boldsymbol{V}}), \tag{86.6}$$

or as a function of the principal invariants[2] $I_{\boldsymbol{B}}, II_{\boldsymbol{B}}, III_{\boldsymbol{B}}$ of the Cauchy-Green tensors $\boldsymbol{B}$ or $\boldsymbol{C}$:

$$\bar{\sigma}(\boldsymbol{B}) = \bar{\sigma}(\boldsymbol{C}) = \bar{\sigma}(I_{\boldsymbol{B}}, II_{\boldsymbol{B}}, III_{\boldsymbol{B}}). \tag{86.7}$$

Since, as easily verified, $I_{\boldsymbol{B}} = I_{\boldsymbol{V}}^2 - 2 II_{\boldsymbol{V}}$, $II_{\boldsymbol{B}} = II_{\boldsymbol{V}}^2 - 2 I_{\boldsymbol{V}} III_{\boldsymbol{V}}$, $III_{\boldsymbol{B}} = III_{\boldsymbol{V}}^2$, the functions σ and $\bar{\sigma}$ are related by the formula

$$\sigma(I_{\boldsymbol{V}}, II_{\boldsymbol{V}}, III_{\boldsymbol{V}}) = \bar{\sigma}(I_{\boldsymbol{V}}^2 - 2 II_{\boldsymbol{V}}, II_{\boldsymbol{V}}^2 - 2 I_{\boldsymbol{V}} III_{\boldsymbol{V}}, III_{\boldsymbol{V}}^2). \tag{86.8}$$

The strain energy can also be expressed as a function of the first three moments $\bar{I}_{\boldsymbol{V}}, \overline{II}_{\boldsymbol{V}}, \overline{III}_{\boldsymbol{V}}$ or $\bar{I}_{\boldsymbol{B}}, \overline{II}_{\boldsymbol{B}}, \overline{III}_{\boldsymbol{B}}$ of the stretch tensors or Cauchy-Green tensors, respectively, or as a function of various other triples of invariants.

In order to calculate the response coefficients $\aleph_\Gamma$, $\beth_\Gamma$ of (47.8) and (47.9) from the strain-energy function we only have to differentiate $\bar{\sigma}(\boldsymbol{B}) = \bar{\sigma}(I_{\boldsymbol{B}}, II_{\boldsymbol{B}}, III_{\boldsymbol{B}})$ according to the chain rule, use the formulae (9.11), and substitute the result into

[1] Kötter [1910, *1*, § 1], Almansi [1911, *1*, § 7]. Cf. the more elaborate form given by Finger [1894, *1*, Eq. (35)] [1894, *3*, Eq. (39)]. In effect, Almansi [1911, *2*, § 3] showed how to derive (86.9) from (86.4). Signorini [1930, *3*, Eq. (26)] [1933, *4*, Eq. (10)] obtained the corresponding relation between principal components of $\tilde{\boldsymbol{T}}$ and $\boldsymbol{E}$. Similarly explicit forms of the stress relation for anisotropic materials were obtained by Signorini [1958, *43*].

[2] Finger [1894, *3*, pp. 1088—1089, 1092].

(85.15). We obtain[1]

$$\aleph_0 = 2\varrho III_{\boldsymbol{B}} \frac{\partial \bar{\sigma}}{\partial III_{\boldsymbol{B}}}, \quad \aleph_1 = 2\varrho\left(\frac{\partial \bar{\sigma}}{\partial I_{\boldsymbol{B}}} + I_{\boldsymbol{B}} \frac{\partial \bar{\sigma}}{\partial II_{\boldsymbol{B}}}\right), \quad \aleph_2 = -2\varrho \frac{\partial \bar{\sigma}}{\partial II_{\boldsymbol{B}}}, \tag{86.9}$$

$$\left.\begin{aligned} \beth_0 &= 2\varrho\left(II_{\boldsymbol{B}} \frac{\partial \bar{\sigma}}{\partial II_{\boldsymbol{B}}} + III_{\boldsymbol{B}} \frac{\partial \bar{\sigma}}{\partial III_{\boldsymbol{B}}}\right), \quad \beth_1 = 2\varrho \frac{\partial \bar{\sigma}}{\partial I_{\boldsymbol{B}}}, \\ \beth_{-1} &= -2\varrho III_{\boldsymbol{B}} \frac{\partial \bar{\sigma}}{\partial II_{\boldsymbol{B}}}. \end{aligned}\right\} \tag{86.10}$$

The formulae for the response coefficients שׁ_Γ, $\gimel_\Gamma$ of (47.6) and (47.7) have the same form as (86.9) and (86.10), respectively, except that the factor 2 must be omitted and that $\bar{\sigma}(I_{\boldsymbol{B}}, II_{\boldsymbol{B}}, III_{\boldsymbol{B}})$ must be replaced by $\sigma(I_{\boldsymbol{V}}, II_{\boldsymbol{V}}, III_{\boldsymbol{V}})$.

If the isotropic material is a solid and if the reference configuration is a natural state, it follows from (50.9) and (86.9) that $\bar{\sigma}(I_{\boldsymbol{B}}, II_{\boldsymbol{B}}, III_{\boldsymbol{B}})$ satisfies the condition

$$\left[\frac{\partial \bar{\sigma}}{\partial I_{\boldsymbol{B}}} + 2\frac{\partial \bar{\sigma}}{\partial II_{\boldsymbol{B}}} + \frac{\partial \bar{\sigma}}{\partial III_{\boldsymbol{B}}}\right]_{(3,3,1)} = 0. \tag{86.11}$$

For an isotropic elastic material with given response coefficients $\aleph_\Gamma$ or $\beth_\Gamma$ the equations (86.9) or (86.10) may be regarded as a system of partial differential equations for the strain-energy function $\bar{\sigma}(I_{\boldsymbol{B}}, II_{\boldsymbol{B}}, III_{\boldsymbol{B}})$. The integrability conditions, which are of course special cases of the conditions (82.13), (84.15), and (84.16), may be written in the forms[2]

$$\left.\begin{aligned} &\frac{\partial}{\partial II_{\boldsymbol{B}}}\left(\frac{\aleph_0}{\varrho}\right) + III_{\boldsymbol{B}} \frac{\partial}{\partial III_{\boldsymbol{B}}}\left(\frac{\aleph_2}{\varrho}\right) = 0, \\ &III_{\boldsymbol{B}} \frac{\partial}{\partial III_{\boldsymbol{B}}}\left(\frac{\aleph_1}{\varrho}\right) - I_{\boldsymbol{B}} \frac{\partial}{\partial II_{\boldsymbol{B}}}\left(\frac{\aleph_0}{\varrho}\right) - \frac{\partial}{\partial I_{\boldsymbol{B}}}\left(\frac{\aleph_0}{\varrho}\right) = 0, \\ &\frac{\partial}{\partial II_{\boldsymbol{B}}}\left(\frac{\aleph_1}{\varrho}\right) + \frac{\partial}{\partial I_{\boldsymbol{B}}}\left(\frac{\aleph_2}{\varrho}\right) + I_{\boldsymbol{B}} \frac{\partial}{\partial II_{\boldsymbol{B}}}\left(\frac{\aleph_2}{\varrho}\right) = 0, \end{aligned}\right\} \tag{86.12}$$

or

$$\left.\begin{aligned} &\frac{\partial}{\partial I_{\boldsymbol{B}}}\left(\frac{\beth_0}{\varrho}\right) - II_{\boldsymbol{B}} \frac{\partial}{\partial II_{\boldsymbol{B}}}\left(\frac{\beth_1}{\varrho}\right) - III_{\boldsymbol{B}} \frac{\partial}{\partial III_{\boldsymbol{B}}}\left(\frac{\beth_1}{\varrho}\right) = 0, \\ &III_{\boldsymbol{B}}\left[\frac{\partial}{\partial II_{\boldsymbol{B}}}\left(\frac{\beth_0}{\varrho}\right) + \frac{\partial}{\partial III_{\boldsymbol{B}}}\left(\frac{\beth_{-1}}{\varrho}\right)\right] + II_{\boldsymbol{B}} \frac{\partial}{\partial II_{\boldsymbol{B}}}\left(\frac{\beth_{-1}}{\varrho}\right) = 0, \\ &\frac{\partial}{\partial I_{\boldsymbol{B}}}\left(\frac{\beth_{-1}}{\varrho}\right) + III_{\boldsymbol{B}} \frac{\partial}{\partial II_{\boldsymbol{B}}}\left(\frac{\beth_1}{\varrho}\right) = 0. \end{aligned}\right\} \tag{86.13}$$

Either set of equations (86.12) or (86.13) constitutes necessary and sufficient conditions for an isotropic elastic material to be hyperelastic.

In an experimental situation it is often easier to start from measured values of the principal stresses t_a as functions of the principal stretches. In order for such functions to be compatible with elastic response, they must satisfy 6 independent algebraic identities following from (48.3) and (48.4). For such a material to be hyperelastic, from (86.4) we at once infer the following necessary and sufficient conditions[3]:

$$\frac{\partial}{\partial v_b}(v_b t_a) = \frac{\partial}{\partial v_a}(v_a t_b), \quad a \neq b. \tag{86.14}$$

---

[1] The first to derive a stress relation with these coefficients was FINGER [1894, *3*, Eq. (35)]. Various alternative forms, obtained by various later authors, were listed by TRUESDELL [1952, *20*, § 41].

[2] Apparently the first to write down conditions of this kind in terms of principal invariants was GOLDENBLAT [1950, *4*]. Other forms were given by TRUESDELL [1952, *20*, § 41].

[3] Noticed in a special case by HOUSTOUN [1911, *3*], given in the above form by TRUESDELL [1952, *20*, § 41].

For an incompressible isotropic hyperelastic material the strain-energy function $\sigma(\boldsymbol{V})=\bar{\sigma}(\boldsymbol{B})$ is defined *a priori* only for unimodular tensors $\boldsymbol{V}$ or $\boldsymbol{B}$ (cf. the remarks at the end of Sect. 80), i.e. when

$$\det \boldsymbol{V}=III_{\boldsymbol{V}}=1, \qquad \det \boldsymbol{B}=III_{\boldsymbol{B}}=1. \tag{86.15}$$

Hence, in this case, (86.6) and (86.7) reduce to

$$\sigma(\boldsymbol{V})=\sigma(I_{\boldsymbol{V}}, II_{\boldsymbol{V}}), \qquad \bar{\sigma}(\boldsymbol{B})=\bar{\sigma}(I_{\boldsymbol{B}}, II_{\boldsymbol{B}}). \tag{86.16}$$

These formulae, however, are meaningful even when $\boldsymbol{V}$ and $\boldsymbol{B}$ are not unimodular and hence may be used to extend the domain of definition of the strain-energy function in a simple way. We may then use the formulae (86.9) and (86.10), with $III_{\boldsymbol{B}}=1$, for the response coefficients in the stress relations (49.4) and (49.5). Explicitly, for the latter,

$$\beth_1=2\varrho\frac{\partial\bar{\sigma}}{\partial I_{\boldsymbol{B}}}, \qquad \beth_{-1}=-2\varrho\frac{\partial\bar{\sigma}}{\partial II_{\boldsymbol{B}}}, \tag{86.17}$$

so that[1]

$$\left.\begin{aligned} \boldsymbol{T}&=-p\boldsymbol{1}+2\varrho\left(\frac{\partial\bar{\sigma}}{\partial I_{\boldsymbol{B}}}+I_{\boldsymbol{B}}\frac{\partial\bar{\sigma}}{\partial II_{\boldsymbol{B}}}\right)\boldsymbol{B}-2\varrho\frac{\partial\bar{\sigma}}{\partial II_{\boldsymbol{B}}}\boldsymbol{B}^2,\\ &=-p\boldsymbol{1}+2\varrho\frac{\partial\bar{\sigma}}{\partial I_{\boldsymbol{B}}}\boldsymbol{B}-2\varrho\frac{\partial\bar{\sigma}}{\partial II_{\boldsymbol{B}}}\boldsymbol{B}^{-1}. \end{aligned}\right\} \tag{86.18}$$

Note that in these two formulae, as in the more general (49.4) and (49.5), the two indeterminate pressures p differ. These formulae, though now seldom needed, played a great part in the recent renascence of finite elasticity.

b) General Theorems.

87. Consequences of restrictions upon the strain-energy function. We return to the problem of thermal equilibrium of simple materials discussed in Sect. 81. The postulate stated at the end of that section led to the stress relation (81.16) and the temperature relation (81.17). These relations are necessary but not sufficient conditions for local thermal equilibrium. The requirement that the thermomechanical potential (81.13) achieve its minimum for $\eta^*=\eta$ and $\boldsymbol{G}=\boldsymbol{1}$ is the content of the inequality

$$\hat{\lambda}(\boldsymbol{G}\boldsymbol{F}, \eta^*) > \hat{\lambda}(\boldsymbol{F}, \eta), \tag{87.1}$$

valid when $\boldsymbol{G}-\boldsymbol{1}$ and $\eta^*-\eta$ do not both vanish. Substituting into this inequality the expression (81.13) for $\hat{\lambda}$ and (81.16) and (81.17), we obtain the ***inequality of Coleman and Noll***[2]:

$$\hat{\varepsilon}(\boldsymbol{G}\boldsymbol{F}, \eta^*)-\hat{\varepsilon}(\boldsymbol{F}, \eta)-\operatorname{tr}[(\boldsymbol{G}-\boldsymbol{1})\boldsymbol{F}\hat{\varepsilon}_{\boldsymbol{F}}(\boldsymbol{F}, \eta)^T]-(\eta^*-\eta)\hat{\varepsilon}_\eta(\boldsymbol{F}, \eta)>0, \tag{87.2}$$

which must be valid for all positive-definite symmetric tensors $\boldsymbol{G}$ and all η^* such that $\boldsymbol{G}-\boldsymbol{1}$ and $\eta^*-\eta$ do not both vanish.

We now consider perfect materials, treated in Sect. 80. From (80.5), (24.1), and $(43\text{A}.3)_1$ we see that the local entropy production γ_{loc} in a thermodynamical

[1] Eq. (86.18) was first given by RIVLIN [1948, *12*, § 4]. Essentially the same result except for lack of the all-important arbitrary pressure had been obtained by ARIANO [1930, *1*. pp. 753—754].

[2] COLEMAN and NOLL [1959, *3*, § 8].

process may be expressed in the form

$$\left.\begin{aligned} -\theta\gamma_{\mathrm{loc}} &= \dot{\varepsilon} - \frac{1}{\varrho}\,\mathrm{tr}\,(\boldsymbol{T}\boldsymbol{L}) - \theta\dot{\eta},\\ &= \dot{\varepsilon} - \frac{1}{\varrho}\,\mathrm{tr}\,(\boldsymbol{F}^{-1}\boldsymbol{T}\dot{\boldsymbol{F}}) - \theta\dot{\eta},\\ &= \dot{\varepsilon} - \frac{1}{\varrho_{\mathrm{R}}}\,\mathrm{tr}\,(\dot{\boldsymbol{F}}\boldsymbol{T}_{\mathrm{R}}^{T}) - \theta\dot{\eta}. \end{aligned}\right\} \tag{87.3}$$

Consider a neighborhood $\mathcal{N}$ of a particle X. The surface tractions, per unit area in the reference configuration, acting on the boundary $\partial\mathcal{N}$ of $\mathcal{N}$ are given by $\boldsymbol{T}_{\mathrm{R}}\boldsymbol{n}_{\mathrm{R}}$, where $\boldsymbol{T}_{\mathrm{R}}$ is to be evaluated at a typical point of $\partial\mathcal{N}$ and where $\boldsymbol{n}_{\mathrm{R}}$ is the outer unit normal at that point, in the reference configuration. If the neighborhood $\mathcal{N}$ of X is small and if $\boldsymbol{T}_{\mathrm{R}}$ is continuous, its value at a point of $\partial\mathcal{N}$ is approximately the same as its value at X. Thus the Piola-Kirchhoff tensor $\boldsymbol{T}_{\mathrm{R}}$ at a particle characterizes the contact forces acting on a neighborhood of the particle in the limit as this neighborhood vanishes. In a process in which both θ and $\boldsymbol{T}_{\mathrm{R}}$ are held constant in time (87.3) reduces to

$$-\theta\gamma_{\mathrm{loc}} = \frac{d}{dt}\left[\varepsilon - \frac{1}{\varrho_{\mathrm{R}}}\,\mathrm{tr}\,(\boldsymbol{F}\boldsymbol{T}_{\mathrm{R}}^{T}) - \theta\eta\right]. \tag{87.4}$$

Thus the entropy $\int_t^{t^*}\gamma_{\mathrm{loc}}(\tau)\,d\tau$ produced locally in such a process in the time interval from t to t^* is given by

$$-\theta\int_t^{t^*}\gamma_{\mathrm{loc}}(\tau)\,d\tau = \varepsilon^* - \varepsilon - \frac{1}{\varrho_{\mathrm{R}}}\,\mathrm{tr}\,[(\boldsymbol{F}^* - \boldsymbol{F})\,\boldsymbol{T}_{\mathrm{R}}^{T}] - \theta\,(\eta^* - \eta), \tag{87.5}$$

where the asterisk indicates that the corresponding quantity is evaluated at time t^*.

The physical notion that a perfect material will not stretch unless either the forces or the temperature is changed, and the notion that a process with positive entropy production will actually take place, lead to the following ***postulate***: *in a hypothetical process such that*

(i) *the Piola-Kirchhoff tensor* $\boldsymbol{T}_{\mathrm{R}}$ *and the temperature* θ *are held constant*;

(ii) *the deformation effected in the process is a pure stretch*;

(iii) *the caloric equation of state* (81.2) *is always valid*;

(iv) *the stress relation* (81.16) *and the temperature relation* (81.17) *are satisfied initially*;

the entropy produced is negative.

It follows easily from (87.5) that this postulate is equivalent to the requirement that the Coleman-Noll inequality (87.2) be valid for the energy function $\hat{\varepsilon}$ of a perfect material. We note that all the classical inequalities of the thermodynamics of perfect fluids are consequences[1] of inequality (87.2).

Using (82.4) and (82.6), we may express the Coleman-Noll inequality (87.2) also in terms of the free-energy function $\tilde{\psi}$:

$$\tilde{\psi}(\boldsymbol{G}\boldsymbol{F},\theta^*) - \tilde{\psi}(\boldsymbol{F},\theta) - \mathrm{tr}\,[(\boldsymbol{G}-\boldsymbol{1})\,\boldsymbol{F}\tilde{\psi}_{\boldsymbol{F}}(\boldsymbol{F},\theta)^{T}] - (\theta^* - \theta)\,\tilde{\psi}_{\theta}(\boldsymbol{G}\boldsymbol{F},\theta^*) > 0 \tag{87.6}$$

for all positive-definite symmetric $\boldsymbol{G}$ and all θ^* such that $\boldsymbol{G}-\boldsymbol{1}$ and $\theta^*-\theta$ do not both vanish.

[1] The details of this verification were given by COLEMAN and NOLL [1959, *3*, § 11].

In the special case when $\eta^* = \eta$ or $\theta^* = \theta$, the inequalities (87.2) and (87.6) lead to the *C-N condition* for the strain-energy function of a hyperelastic material:

$$\sigma(\boldsymbol{G}\boldsymbol{F}) - \sigma(\boldsymbol{F}) - \operatorname{tr}[(\boldsymbol{G}-\boldsymbol{1})\boldsymbol{F}\sigma_{\boldsymbol{F}}(\boldsymbol{F})^T)] > 0 \tag{87.7}$$

for all tensors $\boldsymbol{F}$ *and all positive-definite symmetric tensors* $\boldsymbol{G} \neq \boldsymbol{1}$.

The inequality (87.7) differs from (83.10) only in notation. Thus, by COLEMAN's theorem in Sect. 83, the C-N condition may be interpreted as the requirement that the work actually done in any pure stretch, starting from an arbitrary configuration, shall exceed the work that would be done if the same stretch could be effected under dead loading. (Since it refers only to pure stretches, the C-N condition does not ensure dead-loading stability in the sense defined in Sect. 68bis.)

As the initials "C-N" suggest, the condition (87.7) was proposed by COLEMAN and NOLL[1]. To see its connection with the GCN condition (52.1), let it be written in the form

$$\sigma(\bar{\boldsymbol{F}}) - \sigma(\boldsymbol{F}) - \frac{1}{\varrho_R}\operatorname{tr}[(\bar{\boldsymbol{F}}^T - \boldsymbol{F}^T)\,\mathfrak{h}(\boldsymbol{F})] > 0, \qquad \bar{\boldsymbol{F}} = \boldsymbol{G}\boldsymbol{F}, \tag{87.8}$$

as follows from the stress relation (82.11). Interchanging $\bar{\boldsymbol{F}}$ and $\boldsymbol{F}$ in (87.8) and adding the result to (87.8), we derive (52.1). Conversely, suppose (52.1) holds for any pair of "points" $\tilde{\boldsymbol{F}}$ and $\boldsymbol{F}$ in a convex domain $\mathcal{D}$ of the space $\mathcal{L}$ of invertible tensors, provided, of course, that $\tilde{\boldsymbol{F}} = \tilde{\boldsymbol{G}}\boldsymbol{F}$, where $\tilde{\boldsymbol{G}}$ is positive-definite and unequal to $\boldsymbol{1}$. The "points" $\bar{\boldsymbol{F}}$ and $\boldsymbol{F}$ may then be connected by a straight line in $\mathcal{D}$:

$$\left.\begin{aligned} \tilde{\boldsymbol{F}} &= \boldsymbol{F} + \lambda(\bar{\boldsymbol{F}} - \boldsymbol{F}), \qquad 0 \leqq \lambda \leqq 1, \\ &= [(1-\lambda)\boldsymbol{1} + \lambda\boldsymbol{G}]\boldsymbol{F}, \\ &= \tilde{\boldsymbol{G}}\boldsymbol{F}, \qquad \tilde{\boldsymbol{G}} = (1-\lambda)\boldsymbol{1} + \lambda\boldsymbol{G}. \end{aligned}\right\} \tag{87.9}$$

Since $\boldsymbol{G}$ is positive-definite and symmetric, so also is $\tilde{\boldsymbol{G}}$. Thus (52.1) may be applied whenever $0 \leqq \lambda \leqq 1$. Therefore integration along the line (87.9), followed by application of (82.11), shows that

$$\left.\begin{aligned} &\sigma(\bar{\boldsymbol{F}}) - \sigma(\boldsymbol{F}) - \operatorname{tr}[(\bar{\boldsymbol{F}} - \boldsymbol{F})^T \sigma_{\boldsymbol{F}}(\boldsymbol{F})] \\ &\quad = \int_0^1 \frac{d}{d\lambda}\,\sigma(\tilde{\boldsymbol{F}})\,d\lambda - \operatorname{tr}[(\bar{\boldsymbol{F}} - \boldsymbol{F})^T \sigma_{\boldsymbol{F}}(\boldsymbol{F})] \\ &\quad = \int_0^1 \{\operatorname{tr}[(\bar{\boldsymbol{F}} - \boldsymbol{F})^T \sigma_{\boldsymbol{F}}(\tilde{\boldsymbol{F}})] - \operatorname{tr}[(\bar{\boldsymbol{F}} - \boldsymbol{F})^T \sigma_{\boldsymbol{F}}(\boldsymbol{F})]\}\,d\lambda \\ &\quad = \frac{1}{\varrho_R}\int_0^1 \operatorname{tr}\{(\tilde{\boldsymbol{F}} - \boldsymbol{F})^T\,[\mathfrak{h}(\tilde{\boldsymbol{F}}) - \mathfrak{h}(\boldsymbol{F})]\}\frac{d\lambda}{\lambda} > 0. \end{aligned}\right\} \tag{87.10}$$

It has been proved, then, that if the GCN condition (52.1) holds in a convex domain, and if the material is hyperelastic, then the C-N condition (87.7) holds[2]. That is, for hyperelastic materials,

$$\text{GCN} \Leftrightarrow \text{C-N}. \tag{87.11}$$

If $\boldsymbol{G}$ were an arbitrary tensor in (87.7), the C-N condition would assert that $\sigma(\boldsymbol{F})$ is a convex function. Such a requirement would imply that (52.1) holds for all pairs $\boldsymbol{F}$, $\bar{\boldsymbol{F}}$, an

[1] COLEMAN and NOLL [1959, *3*, § 8].
[2] TRUESDELL and TOUPIN [1963, *76*, § 5].

assertion we have shown in Sect. 52 to stand in contradiction to the principle of material indifference[1].

Consider in $\mathscr{D}$ the class $\mathscr{C}$ of smooth deformation paths $\boldsymbol{F}(\alpha)$ whose end-points are related by a pure stretch. COLEMAN's work theorem (Sect. 83) states that the C-N condition is equivalent to the assertion that the work inequality (83.9) holds for all curves in $\mathscr{C}$. Dropping the restriction to hyperelasticity, BRAGG and COLEMAN[2] have shown that the GCN condition (52.1) is equivalent to the assertion that (83.9) holds for all *straight lines* in $\mathscr{C}$.

In Sects. 51 and 52 numerous consequences of the GCN condition were derived. By (87.11), for hyperelastic materials these same results follow from the C-N condition. Also it follows from a result stated toward the end of Sect. 52, or by an easy direct argument, that the C-N condition reduces in the infinitesimal theory to the statement that the stored-energy function is a positive-definite quadratic form. Indeed, a more general inequality can be obtained. Assume that the reference configuration is a natural state, and put $\boldsymbol{F}=\mathbf{1}$ and $\overline{\boldsymbol{F}}=\boldsymbol{U}$ in (87.8); since $\boldsymbol{T}_{\mathrm{R}}=\mathfrak{h}(\mathbf{1})=\mathbf{0}$ in a natural state, it follows that

$$\sigma(\boldsymbol{U})=\bar{\sigma}(\boldsymbol{C})>\sigma(\mathbf{1}) \tag{87.12}$$

if $\boldsymbol{C}\neq\mathbf{1}$. That is[3],

$$\text{C-N}\Rightarrow\sigma \text{ is a strict minimum in a natural state.} \tag{87.13}$$

Consequently any two natural states can differ at most by a rigid rotation. In other words,

$$\left.\begin{array}{l}\text{C-N}\Rightarrow\text{there is at most one natural state,}\\ \qquad\quad\text{defined to within a rigid displacement.}\end{array}\right\} \tag{87.14}$$

As was shown in Sect. 52, an elastic material that satisfies the GCN condition and has a natural state is a solid. By (87.13) and the last result stated in Sect. 85 it follows that[4] *in a hyperelastic material (necessarily solid) that satisfies the C-N condition and has a natural state, the isotropy group of the material is the isotropy group of the strain-energy function.*

For isotropic hyperelastic materials, from (87.11), (52.4), and (51.40) we see that

$$\text{C-N}\Rightarrow\text{GCN}_0\Rightarrow\left\{\begin{array}{l}\text{P-C \&}\\ \text{IFS \&}\\ \text{T-E \&}\\ \text{E-T \&}\\ \text{O-F.}\end{array}\right\} \tag{87.15}$$

(In order to include the P-C inequality, the reference state is assumed to be natural, though such an assumption is not necessary for the remaining four implications.) This is not all, however. The GCN_0 condition, asserting that the transformation from the v_a to the T_b is monotone, is now equivalent[5], if it holds in a convex domain of the space of stretches, to the convexity of the stored-energy function:

$$\text{GCN}_0\Leftrightarrow\sigma(v_1, v_2, v_3) \text{ is strictly convex.} \tag{87.16}$$

Hence[6], by $(87.15)_1$,

$$\text{C-N}\Rightarrow\sigma(v_1, v_2, v_3) \text{ is strictly convex.} \tag{87.17}$$

---

[1] TRUESDELL and TOUPIN [1963, *76*, § 8]. Earlier HILL [1957, *9*, § 2] had remarked that if $\sigma(\boldsymbol{F})$ were convex, unqualified uniqueness of solution to the mixed boundary-value problem would result.

[2] BRAGG and COLEMAN [1963, *10*].

[3] The requirement that σ shall be minimized in a natural state is common in the Italian literature, where it is taken more or less directly as a postulate. Cf. e.g., SIGNORINI [1949, *21*, Cap. I, ¶ 3].

[4] TRUESDELL [1964, *86*].

[5] The argument is parallel to that used to derive (87.10), but simpler.

[6] This result was obtained in a different way by COLEMAN and NOLL [1959, *3*, § 12].

COLEMAN and NOLL raised the question whether the converse holds. That is, does every stored energy that is a strictly convex function of the principal stretches satisfy the C-N condition? By means of a counter-example, BRAGG and COLEMAN[1] have shown that the answer is no.

Returning to general hyperelastic materials, we see that the C-N condition (87.7) requires a type of convexity of the stored-energy function $\sigma(\boldsymbol{F})$ with regard to a restricted class of arguments. We may consider replacing this condition by a stronger requirement that the appropriate Hessian form be positive-definite. This condition, which may be called the *C-N⁺ condition*, is expressed formally by (52.16), where, of course, the tensor B^{kmpq} is now restricted by the relations (84.15). From (52.17) it follows that

$$\text{C-N}^+ \Rightarrow \text{C-N}; \tag{87.18}$$

while the reverse implication does not hold.

A statical interpretation of the C-N⁺ condition for isotropic materials is furnished by (52.21). Although the meaning of the inequalities $A_\Gamma > 0$ does not appear to be any simpler for hyperelastic materials than for more general elastic ones, the condition (51.45), which we now call the CN_0^+ condition, may be given an elegant statical expression. The IFS⁺ condition (51.14) follows at once since $\det \|J_{ab}\| > 0$. Since $J_{ab} = J_{ba}$, the CN_0^+ condition (51.45) holds if and only if the principal minors of $\|J_{ab}\|$, i.e.

$$\frac{\partial \overline{T}_a}{\partial v_a}, \quad \frac{\partial \overline{v}_a}{\partial T_a} \det \|J_{bc}\|, \quad \det \|J_{bc}\|, \tag{87.19}$$

are positive. Thus we read off from (51.11) and (51.17) the following equivalence[2]:

$$\left.\begin{array}{l} \text{C} \quad \& \\ \text{IFS}^+ \ \& \\ \text{T-E}^+ \ \& \\ \text{E-T}^+ \end{array}\right\} \Leftrightarrow \text{CN}_0^+ . \tag{87.20}$$

None of the four conditions listed on the left-hand side of the equivalence is redundant; if we leave off C, there remains the possibility that $\det \|J_{ab}\| < 0$,

[1] BRAGG and COLEMAN [1963, *10*] considered the material for which

$$\varrho_R \sigma(v_1, v_2, v_3) = [\tfrac{1}{2}(v_1^2 + v_2^2 + v_3^2) - (v_1 + v_2 + v_3)] K,$$

so that, by (86.4),

$$t_1 = \frac{K}{v_2 v_3}(v_1 - 1), \qquad \text{etc.}$$

Since the Hessian matrix of this σ is proportional to the unit matrix, σ is strongly convex, for all v_a. Using the results derived in Sect. 52, we can replace BRAGG and COLEMAN's argument by a shorter one. From (52.20) it is a straightforward matter to show that $A_3 < 0$ if and only if

$$(v_1 - v_2)^2 > (v_1 + v_2)(v_1^2 + v_2^2).$$

If $0 < v_2 < 1$, there is an interval of values of v_1 satisfying this inequality and such that $v_2 < v_1 < 1$. It was shown in Sect. 52 that in order for the C-N condition to hold, it is necessary that $A_\Gamma \geqq 0$. Therefore, the above stored-energy function, while strongly convex, does not satisfy the C-N condition in sufficiently severe and unequal states of compression.

This same material serves to illustrate the fact that O-F$\not\Rightarrow$B-E, already developed in general in Sect. 52. Indeed, since σ is strongly convex, O-F holds, but B-E takes the form $v_a + v_b > 1$, $a \neq b$. Thus the B-E inequalities are never satisfied by this material in severe enough compression, e.g., if all three stretches are less than $\frac{1}{2}$. Cf. the results at the end of Sect. 51.

[2] From this point on in the present section, the analysis is due to TRUESDELL and TOUPIN [1963, *76*, §§ 4 and 6].

while if we leave off IFS⁺, it becomes possible that $\det \|J_{ab}\| = 0$, etc. Thus we have found a full and minimal set of simple, plausible, purely static conditions equivalent to the CN_0^+ condition, for hyperelastic materials.

Combining (87.20) with (52.21) yields

$$\left.\begin{array}{ll} \mathrm{C} & \& \\ \mathrm{IFS}^+ & \& \\ \mathrm{T\text{-}E}^+ & \& \\ \mathrm{E\text{-}T}^+ & \& \\ A_\Gamma > 0 & \end{array}\right\} \Leftrightarrow \mathrm{C\text{-}N}^+. \tag{87.21}$$

As shown by the counter-example of COLEMAN and BRAGG given in footnote 1, p. 323, this result cannot be weakened.

Since the CN_0^+ condition is equivalent to the strong convexity of $\sigma(v_1, v_2, v_3)$, from (87.16) and (51.40) we see that

$$\mathrm{CN}_0^+ \Rightarrow \mathrm{GCN}_0 \Rightarrow \left\{\begin{array}{l} \mathrm{P\text{-}C}\ \& \\ \mathrm{O\text{-}F}. \end{array}\right\}. \tag{87.22}$$

Hence, by (87.20)

$$\left.\begin{array}{ll} \mathrm{C} & \& \\ \mathrm{IFS}^+ & \& \\ \mathrm{T\text{-}E}^+ & \& \\ \mathrm{E\text{-}T}^+ & \end{array}\right\} \Rightarrow \left\{\begin{array}{l} \mathrm{P\text{-}C}\ \& \\ \mathrm{O\text{-}F}. \end{array}\right\}. \tag{87.23}$$

This implication does not seem to hold for isotropic elastic materials that are not hyperelastic. Finally, from (87.22) and (52.24) we see that

$$\mathrm{CN}_0^+ \ \&\ \mathrm{B\text{-}E} \Rightarrow \mathrm{C\text{-}N}^+. \tag{87.24}$$

Of course, all inequalities for hyperelastic materials may be expressed in terms of the stored-energy function if desired[1].

88. BETTI'S theorem. Variational problems. The differential equations for the deformation $\boldsymbol{x} = \boldsymbol{\chi}(\boldsymbol{X}, t)$ of a hyperelastic body $\mathscr{B}$ from a homogeneous reference state may be obtained by substituting (82.11) into (44.2):

$$\operatorname{Div} \sigma_{\boldsymbol{F}}(\boldsymbol{F}) + \boldsymbol{b} = \ddot{\boldsymbol{x}}, \quad \boldsymbol{F} = \nabla \boldsymbol{\chi}(\boldsymbol{X}), \quad \left(\frac{\partial \sigma}{\partial x^k{}_{,\alpha}}\right)_{;\alpha} + b_k = \ddot{x}_k. \tag{88.1}$$

These equations are the same as (44.5) when $q_k = 0$ and $\mathsf{A}_k{}^\alpha{}_m{}^\beta$ is given by (82.12).

The condition (82.13), necessary and sufficient that the elastic material be hyperelastic, asserts that the equations (88.1) form a self-adjoint system. As is

[1] The concise presentation in the text above does not reflect the tentative way in which such inequalities were first approached. As a criterion that the work done in increasing an arbitrary simple shear be positive RIVLIN [1949, *16*, § 16] found and proposed that $\hat{\mu}(K^2) > 0$; as shown in Sect. 54, we now know that this inequality is equivalent, for simple shear, to the B-E inequalities. TRUESDELL [1952, *20*, § 41] asserted that positive work must be done on an incompressible, isotropic, hyperelastic material in order to increase one principal extension while holding another fixed. He derived the necessary and sufficient condition

$$\frac{\partial \bar{\sigma}}{\partial I} + v_a^2 \frac{\partial \bar{\sigma}}{\partial II} > 0.$$

By (86.17), we recognize this inequality as equivalent, essentially, to (53.1) and hence to the B-E inequalities, as was shown by BAKER and ERICKSEN [1954, *3*]. After the general problem had been presented by TRUESDELL [1956, *32*] in a lecture delivered in 1955, the entire body of researches reported in Sects. 51, 52, and 87 grew from this beginning.

well known, a system is self-adjoint if and only if a certain reciprocity relation holds. We now find and interpret this relation in the context of the theory of infinitesimal strain superimposed upon a given configuration, not necessarily a natural state. To this end, we multiply $(68.9)_1$ by $\bar{u}_k$, where $\bar{\boldsymbol{u}}$ is a displacement vector corresponding to body-force increment $\bar{\boldsymbol{b}}^*$ and surface-traction increment $\bar{\boldsymbol{t}}^*$, thus obtaining

$$\left.\begin{aligned} \varrho_0(\ddot{u}^k - b^{*k})\bar{u}_k &= (\mathsf{A}_0^{kpmq} u_{m,q})_{,p}\bar{u}_k, \\ &= (\mathsf{A}_0^{kpmq} u_{m,q}\bar{u}_k)_{,p} - \mathsf{A}_0^{kpmq} u_{m,q}\bar{u}_{k,p}. \end{aligned}\right\} \tag{88.2}$$

Integration over the body $\mathscr{B}$ in the reference configuration (denoted by subscript 0), followed by use of the divergence theorem and $(68.9)_2$, yields the following *Betti identity:*

$$\int_{\mathscr{B}} \varrho_0 (b^{*k} - \ddot{u}^k)\bar{u}_k\, dv_0 + \int_{\partial\mathscr{B}} t^{*k}\bar{u}_k\, ds_0 = \int_{\mathscr{B}} \mathsf{A}_0^{kpmq} u_{m,q}\bar{u}_{k,p}\, dv_0. \tag{88.3}$$

BETTI's reciprocal theorem of the infinitesimal theory asserts that in the sum on the left-hand side, barred and unbarred quantities may be interchanged. In the usual language of the infinitesimal theory, "the whole work done by the forces of the first set (including kinetic reactions), acting over the displacements produced by the second set, is equal to the whole work done by the forces of the second set, acting over the displacements produced by the first"[1]. In order that BETTI's theorem shall hold for all regions $\mathscr{B}$, it is necessary and sufficient, by (88.3), that $\mathsf{A}^{kpmq} = \mathsf{A}^{mqkp}$ when A is evaluated in the particular configuration on which the infinitesimal strain is superimposed. For this reciprocal relation to hold for *all* configurations, the condition (82.13) is necessary and sufficient. That is, *in order that Betti's theorem shall hold for infinitesimal deformations superimposed on any given configuration of an elastic material, it is necessary and sufficient that the elastic material be hyperelastic*[2]. The fact that BETTI's theorem does hold for hyperelastic materials makes it possible to extend immediately to the general theory of small deformations superimposed upon a given deformation a number of the results of the infinitesimal theory, e.g., formulae for mean strains, dimensional changes, etc.

As an example, consider two solutions $\boldsymbol{u}$ and $\bar{\boldsymbol{u}}$ corresponding to infinitesimal free motion. We then have $\boldsymbol{b}^* = \overline{\boldsymbol{b}^*} = \boldsymbol{t}_{\mathrm{R}}^* = \bar{\boldsymbol{t}}_{\mathrm{R}}^* = \boldsymbol{0}$. From (88.3) it follows that

$$\int_{\mathscr{B}} \varrho_0 (\boldsymbol{u}\cdot\ddot{\bar{\boldsymbol{u}}} - \ddot{\boldsymbol{u}}\cdot\bar{\boldsymbol{u}})\, dv_0 = 0. \tag{88.4}$$

In particular, if $\boldsymbol{u} = \boldsymbol{U}\cos\omega t$ and $\bar{\boldsymbol{u}} = \bar{\boldsymbol{U}}\cos\bar{\omega}t$, from (88.4) we conclude that

$$(\omega^2 - \bar{\omega}^2)\int_{\mathscr{B}} \varrho_0 \boldsymbol{U}\cdot\bar{\boldsymbol{U}}\, dv_0 = 0, \tag{88.5}$$

and hence that

$$\int_{\mathscr{B}} \varrho_0 \boldsymbol{U}\cdot\bar{\boldsymbol{U}}\, dv_0 = 0 \quad \text{if} \quad \omega^2 \neq \bar{\omega}^2: \tag{88.6}$$

That is[3], in the terms usual in the classical infinitesimal theory, *for infinitesimal free harmonic vibration about any configuration of a hyperelastic body, the normal functions corresponding to distinct proper frequencies are orthogonal.* A theorem showing that the proper frequencies are real was proved in Sect. 68bis on the assumption that the ground state is *infinitesimally stable*; as we have just shown, the orthogonality of the proper functions rests on the distinct assumption that there is a *stored-energy function.*

---

[1] LOVE [1927, *1*, § 121].

[2] TRUESDELL [1963, *74*]. That BETTI's theorem does in fact hold for infinitesimal deformations of a hyperelastic material from any state of strain was remarked by SIGNORINI [1936, *3*, p. 13] [1949, *21*, Cap. I, § 2, ¶ 7] and was proved by ZORSKI [1964, *98*, § 3].

[3] BEATTY [1964, *4*] [1965, *1*, § 2.2], generalizing a standard theorem of the infinitesimal theory.

For hyperelastic materials, boundary-value problems in statics are often equivalent to variational problems. If we assume that the body forces $\boldsymbol{b}$ are conservative and hence have a potential v such that $\boldsymbol{b} = -\operatorname{grad} v$, the differential equations (88.1), in the case of statics, become

$$\left.\begin{aligned} \operatorname{Div} \sigma_{\boldsymbol{F}}(\boldsymbol{F}) - \operatorname{grad} v = 0, \\ \left(\frac{\partial \sigma}{\partial x^k{}_{,\alpha}}\right)_{;\alpha} = \frac{\partial^2 \sigma}{\partial x^k{}_{,\alpha}\, \partial x^m{}_{,\beta}}\, x^m{}_{,\beta;\alpha} = v_{,k}\,. \end{aligned}\right\} \tag{88.7}$$

We consider first *boundary-value problems of place and traction*, in which the deformation $\boldsymbol{\chi}$ is prescribed upon a portion $\mathscr{S}_1$ of the boundary $\partial\mathscr{B}$ of $\mathscr{B}$ while the surface tractions $\boldsymbol{t}_{\mathrm{R}}$, per unit area in the reference configuration, are prescribed upon the remainder $\mathscr{S}_2$:

$$\boldsymbol{x} = \boldsymbol{\chi}(\boldsymbol{X}) = \boldsymbol{\chi}_0(\boldsymbol{X}) \quad \text{if} \quad \boldsymbol{X} \in \mathscr{S}_1, \tag{88.8}$$

$$\boldsymbol{T}_{\mathrm{R}} \boldsymbol{n}_{\mathrm{R}} = \varrho_{\mathrm{R}} \sigma_{\boldsymbol{F}}(\boldsymbol{F})\, \boldsymbol{n}_{\mathrm{R}} = \boldsymbol{t}_{\mathrm{R}}(\boldsymbol{X}) \quad \text{if} \quad \boldsymbol{X} \in \mathscr{S}_2. \tag{88.9}$$

The *place boundary-value problem* and the *traction boundary-value problem* correspond to the special cases $\mathscr{S}_1 = \partial\mathscr{B}$ and $\mathscr{S}_2 = \partial\mathscr{B}$, respectively.

Consider the integral

$$J(\boldsymbol{\chi}) = \int_{\mathscr{B}} \sigma(\boldsymbol{F})\, dm + \int_{\mathscr{B}} v(\boldsymbol{\chi})\, dm - \int_{\mathscr{S}_2} \boldsymbol{u} \cdot \boldsymbol{t}_{\mathrm{R}}\, ds_{\mathrm{R}}. \tag{88.10}$$

where $\boldsymbol{u}$ is the displacement (44.22). This integral represents a kind of total enthalpy for the body $\mathscr{B}$ fixed at the portion $\mathscr{S}_1$ of its boundary $\partial\mathscr{B}$ and subject to the assigned surface tractions $\boldsymbol{t}_{\mathrm{R}}$ on the portion $\mathscr{S}_2$ (cf. Sect. 81). We have the following ***first variational theorem***[1]: *The deformation $\boldsymbol{\chi}$ is a solution of the boundary-value problem of place and traction, specified by* (88.7), (88.8), (88.9), *if and only if the variation $\delta J(\boldsymbol{\chi})$ of* (88.10) *vanishes for all variations $\delta\boldsymbol{\chi}$ such that $\delta\boldsymbol{\chi}(\boldsymbol{X}) = \boldsymbol{0}$ when $\boldsymbol{X} \in \mathscr{S}_1$.* To prove this theorem, we compute the variation $\delta J(\boldsymbol{\chi})$ when $\delta\boldsymbol{\chi} = \boldsymbol{0}$ on $\mathscr{S}_1$. We note that $\delta\boldsymbol{u} = \delta\boldsymbol{\chi}$ and that $\delta\boldsymbol{F} = \nabla\delta\boldsymbol{\chi}$. Using the identity

$$\operatorname{Div}(\boldsymbol{S}\boldsymbol{v}) = (\operatorname{Div} \boldsymbol{S}^T) \cdot \boldsymbol{v} + \operatorname{tr}(\boldsymbol{S}\nabla\boldsymbol{v}), \tag{88.11}$$

applied to the case when $\boldsymbol{S} = \sigma_{\boldsymbol{F}}(\boldsymbol{F})^T$ and $\boldsymbol{v} = \delta\boldsymbol{\chi}$, by the help of the divergence theorem we obtain

$$\delta J(\boldsymbol{\chi}) = \int_{\mathscr{B}} (\operatorname{grad} v - \operatorname{Div} \sigma_{\boldsymbol{F}}) \cdot \delta\boldsymbol{\chi}\, dm + \int_{\mathscr{S}_2} (\varrho_{\mathrm{R}}\, \sigma_{\boldsymbol{F}}\, \boldsymbol{n}_{\mathrm{R}} - \boldsymbol{t}_{\mathrm{R}}) \cdot \delta\boldsymbol{\chi}\, ds_{\mathrm{R}}. \tag{88.12}$$

It is clear that (88.12) vanishes for all $\delta\boldsymbol{\chi}$ if and only if (88.7) and (88.9) hold, which proves the theorem.

A boundary-value problem of place and pressure[2] results if, retaining (88.8), we replace (88.9) by

$$\boldsymbol{T}\boldsymbol{n} = \varrho \boldsymbol{F} \sigma_{\boldsymbol{F}}(\boldsymbol{F})^T \boldsymbol{n} = -p\boldsymbol{n} \quad \text{if} \quad \boldsymbol{X} \in \mathscr{S}_2, \tag{88.13}$$

---

[1] It is difficult to trace the history of this principle, since in the older work the nature of the allowed variations was often left to be inferred from the result. For the case when $\mathscr{S}_2 = \partial\mathscr{B}$ the theorem is due in principle to GREEN [1839, *1*, pp. 253–256], whose work was corrected and extended by HAUGHTON [1849, *1*, p. 152] and KIRCHHOFF [1850, *2*, § 1]. Cf. the discussion of a somewhat more general variational principle by TRUESDELL [1952, *20*, § 33]. For the case when $\mathscr{S}_1 = \partial\mathscr{B}$, the theorem was proved by HADAMARD [1903, *4*, ¶ 264]. Both cases are presented, partly in words, by HELLINGER [1914, *1*, § 7a]. For the case of small deformations but possibly non-linear stress relation, the general principle has been attributed to GREENBERG [1949, *5*]; cf. also BUDIANSKY and PEARSON [1956, *4*].

[2] DUHEM [1905, *1*, Ch. I, § I].

where p is a prescribed constant pressure and where $\boldsymbol{n}$ is the unit normal at $\boldsymbol{x}=\boldsymbol{\chi}(\boldsymbol{X})$ to $\mathscr{S}_2$ in the deformed configuration. For such a problem the total enthalpy is defined by

$$J(\boldsymbol{\chi}) = \int\limits_{\mathscr{B}} \sigma(\boldsymbol{F})\, dm + \int\limits_{\mathscr{B}} v(\boldsymbol{\chi})\, dm + p \int\limits_{\mathscr{B}} \frac{1}{\varrho}\, dm. \tag{88.14}$$

instead of (88.10). Note that the value of the integral in the last term in (88.14) is the volume of $\mathscr{B}$ in the deformed configuration $\boldsymbol{\chi}$. With the help of the continuity equation in the form $\delta\varrho + \varrho \operatorname{div} \delta\boldsymbol{\chi} = 0$ the variation of (88.14) is again easily computed:

$$\delta J(\boldsymbol{\chi}) = \int\limits_{\mathscr{B}} (\operatorname{grad} v - \operatorname{Div} \sigma_{\boldsymbol{F}}) \cdot \delta\boldsymbol{\chi}\, dm + \int\limits_{\mathscr{S}_2} \delta\boldsymbol{\chi} \cdot (\varrho \boldsymbol{F} \sigma_{\boldsymbol{F}}^T + p\boldsymbol{1})\boldsymbol{n}\, ds. \tag{88.15}$$

From (88.15) it is clear that the first variational theorem remains valid when the boundary condition (88.9) is replaced by (88.13) and the varied integral (88.10) by (88.14).

Assume now that the stress-relation $(43.10)_2$, i.e., $\tilde{\boldsymbol{T}} = \mathfrak{t}(\boldsymbol{C})$, is invertible, so that the second Piola-Kirchhoff tensor $\tilde{\boldsymbol{T}}$ determines the right Cauchy-Green tensor $\boldsymbol{C} = \overset{-1}{\mathfrak{t}}(\tilde{\boldsymbol{T}})$. By (84.14), the material is hyperelastic if and only if the response function $\mathfrak{t}$ is the gradient of a scalar function $2\varrho_{\mathrm{R}}\bar{\sigma}(\boldsymbol{C})$ on the 6-dimensional space $\mathscr{S}$ of all symmetric tensors $\boldsymbol{C}$. It follows from this result that the gradient $\frac{\partial \mathfrak{t}(\boldsymbol{C})}{\partial \boldsymbol{C}}$ of $\mathfrak{t}(\boldsymbol{C})$, viewed as a linear transformation of the space $\mathscr{S}$, is symmetric. But the gradient of the inverse function $\overset{-1}{\mathfrak{t}}$ is the inverse of the gradient of $\mathfrak{t}$:

$$\frac{\partial \overset{-1}{\mathfrak{t}}(\tilde{\boldsymbol{T}})}{\partial \tilde{\boldsymbol{T}}} = \left(\frac{\partial \mathfrak{t}(\boldsymbol{C})}{\partial \boldsymbol{C}}\right)^{-1}, \qquad \tilde{\boldsymbol{T}} = \mathfrak{t}(\boldsymbol{C}). \tag{88.16}$$

Since the inverse of a symmetric transformation is again symmetric, we conclude that $\frac{\partial \overset{-1}{\mathfrak{t}}(\tilde{\boldsymbol{T}})}{\partial \tilde{\boldsymbol{T}}}$ is symmetric. On the assumption that the domain of $\overset{-1}{\mathfrak{t}}$ is simply connected, it follows that $\overset{-1}{\mathfrak{t}}$ is the gradient of a scalar function γ:

$$\boldsymbol{C} = \gamma_{\tilde{\boldsymbol{T}}}(\tilde{\boldsymbol{T}}), \qquad C^{\alpha\beta} = \frac{\partial \gamma}{\partial \tilde{T}_{\alpha\beta}}. \tag{88.17}$$

The function γ is called the *complementary strain-energy*, and (88.17) the *inverted stress-relation*. It is easy to verify the explicit formula

$$\gamma(\tilde{\boldsymbol{T}}) = \operatorname{tr}\left[\overset{-1}{\mathfrak{t}}(\tilde{\boldsymbol{T}})\,\tilde{\boldsymbol{T}}\right] - 2\varrho_{\mathrm{R}}\bar{\sigma}\left(\overset{-1}{\mathfrak{t}}(\tilde{\boldsymbol{T}})\right). \tag{88.18}$$

Cauchy's first law (43 A.6) may be expressed in terms of $\tilde{\boldsymbol{T}}$ by using $(43\,\mathrm{A}.11)_3$. When $\ddot{\boldsymbol{x}} = \boldsymbol{0}$ and $\boldsymbol{b} = -\operatorname{grad} v$, we obtain

$$\operatorname{Div}(\boldsymbol{F}\tilde{\boldsymbol{T}}) - \operatorname{grad}(\varrho_{\mathrm{R}} v) = \boldsymbol{0}. \tag{88.19}$$

For any given deformation $\boldsymbol{x} = \boldsymbol{\chi}(\boldsymbol{X})$ and any given field $\tilde{\boldsymbol{T}} = \tilde{\boldsymbol{T}}(\boldsymbol{X})$ of second Piola-Kirchhoff tensors in $\mathscr{B}$, consider the integral

$$\left.\begin{aligned} K(\boldsymbol{\chi}, \tilde{\boldsymbol{T}}) = & \int\limits_{\mathscr{B}} \left\{\frac{1}{2\varrho_{\mathrm{R}}}\left[\operatorname{tr}(\tilde{\boldsymbol{T}}\boldsymbol{C}) - \gamma(\tilde{\boldsymbol{T}})\right] + v(\boldsymbol{\chi})\right\} dm \\ & - \int\limits_{\mathscr{S}_2} \boldsymbol{u} \cdot \boldsymbol{t}_{\mathrm{R}}\, ds_{\mathrm{R}} - \int\limits_{\mathscr{S}_1} (\boldsymbol{\chi} - \boldsymbol{\chi}_0) \cdot \boldsymbol{F}\tilde{\boldsymbol{T}}\boldsymbol{n}_{\mathrm{R}}\, ds_{\mathrm{R}}, \end{aligned}\right\} \tag{88.20}$$

where $\boldsymbol{u}$ is the displacement vector. The boundary condition (88.9), when expressed in terms of $\tilde{\boldsymbol{T}}$, assumes the form

$$\boldsymbol{F}\tilde{\boldsymbol{T}}\boldsymbol{n}_{\mathrm{R}} = \boldsymbol{t}_{\mathrm{R}}(\boldsymbol{X}) \quad \text{if} \quad \boldsymbol{X} \in \mathscr{S}_2. \tag{88.21}$$

The following ***second variational theorem*** is due to E. REISSNER[1]: *The deformation* χ *and the field* $\tilde{\boldsymbol{T}}$ *of second Piola-Kirchhoff tensors give a solution of the boundary-value problem of place and traction specified by* (88.19), (88.17), (88.8), *and* (88.21) *if and only if the variation* $\delta K(\chi, \tilde{\boldsymbol{T}})$ *of* (88.20) *vanishes for all variations* $\delta\chi$ *and* $\delta\tilde{\boldsymbol{T}}$. For the proof the variation $\delta K(\chi, \tilde{\boldsymbol{T}})$ must be computed. An easy calculation, making use of the identity $\operatorname{tr}(\tilde{\boldsymbol{T}}\delta\boldsymbol{C}) = 2\operatorname{tr}(\tilde{\boldsymbol{T}}\boldsymbol{F}^T\delta\boldsymbol{F})$, of the identity (88.11) with the choice $\boldsymbol{S} = \tilde{\boldsymbol{T}}\boldsymbol{F}^T$ and $\boldsymbol{v} = \delta\chi$, and of the divergence theorem, gives the result

$$\left.\begin{aligned}\delta K(\chi, \tilde{\boldsymbol{T}}) = \int_{\mathscr{B}} \left\{\frac{1}{2\varrho_{\mathrm{R}}} \operatorname{tr}[(\boldsymbol{C} - \gamma_{\tilde{\boldsymbol{T}}})\,\delta\tilde{\boldsymbol{T}}] + \left[\operatorname{grad} v - \frac{1}{\varrho_{\mathrm{R}}} \operatorname{Div}(\boldsymbol{F}\tilde{\boldsymbol{T}})\right] \cdot \delta\chi\right\} dm \\ + \int_{\mathscr{S}_2} (\boldsymbol{F}\tilde{\boldsymbol{T}}\boldsymbol{n}_{\mathrm{R}} - \boldsymbol{t}_{\mathrm{R}}) \cdot \delta\chi\, ds_{\mathrm{R}} - \int_{\mathscr{S}_1} (\chi - \chi_0) \cdot \delta(\boldsymbol{F}\tilde{\boldsymbol{T}})\boldsymbol{n}_{\mathrm{R}}\, ds_{\mathrm{R}}.\end{aligned}\right\} \tag{88.22}$$

This variation vanishes for arbitrary $\delta\tilde{\boldsymbol{T}}$ and $\delta\chi$ if and only if (88.17), (88.19), (88.21), and (88.8) are all valid, which proves the theorem.

The variational problems described here can be modified in various ways. For example, when the boundary integrals on the right-hand side of (88.20) are omitted, the variational problem $\delta K(\chi, \boldsymbol{T}) = 0$ is still equivalent to the boundary-value problem of place and traction provided the variations $\delta\chi$ and $\delta\tilde{\boldsymbol{T}}$ are such that $\delta\chi = \boldsymbol{0}$ on $\mathscr{S}_1$ and $\delta(\boldsymbol{F}\tilde{\boldsymbol{T}}) = \boldsymbol{0}$ on $\mathscr{S}_2$.

For the problem of infinitesimal strain superimposed upon a given strain (Sect. 68), the variational theorems may be replaced by strict minimal principles, as in the classical infinitesimal theory. The details have been worked out by SHIELD and FOSDICK[2]. As an application, they have estimated the torsional rigidity of a cylinder subjected to extension and then twisted with its ends forced to remain plane. When certain inequalities restricting the first and second derivatives of the stored-energy function are satisfied, they have found that extension has a stiffening effect on a prism of incompressible isotropic material, no matter what the form of its cross-section. Cf. the results at the end of Sect. 70.

89. Stability. Infinitesimal static stability of a configuration of an elastic body under dead loading has been defined and studied in Sect. 68bis. When a stored-energy function exists, it is easy to define a more general, finite static stability. A configuration χ of a hyperelastic body subject to conservative body force is called *statically stable for the boundary conditions* (88.8) and (88.9) [or (88.13)] if the enthalpy (88.10) [or (88.14)] has a relative minimum at χ within the class of all configurations that satisfy the boundary condition (88.8)[3]. Specifically, χ is stable if

$$J(\chi^*) \geqq J(\chi) \tag{89.1}$$

for all χ^* that satisfy (88.8) and belong to a certain neighborhood of χ.

[1] E. REISSNER [1953, *23*]. A variational principle of this kind but with boundary conditions and boundary integrals unspecifed had been given by BORN [1906, *0*], Anhang § IV in two dimensions and more generally by HELLINGER [1914, *1*, § 7e]. The variant developed by MANACORDA [1954, *13*], who was followed in Sect. 232A of CFT, is vacuous because it presumes the relation $\boldsymbol{T}_{\mathrm{R}} = \mathfrak{h}(\boldsymbol{F})$ to be invertible for $\boldsymbol{F}$, in contradiction to the requirement (43.9) imposed by the principle of indifference.

[2] SHIELD and FOSDICK [1963, *64*].

[3] In the special case of a fixed boundary ($\mathscr{S}_1 = \partial\mathscr{B}$), this kind of stability was investigated by HADAMARD [1903, *4*, ¶ 269]. Stability for the boundary conditions of place and pressure was considered by DUHEM [1905, *1*, Ch. I, § I], but he obtained no theorem regarding finite deformations. Stability for each of the special boundary conditions $\mathscr{S}_1 = \partial\mathscr{B}$ and $\mathscr{S}_2 = \partial\mathscr{B}$ was considered by COLEMAN and NOLL [1959, *3*, § 15]. Insofar as it concerns the mixed boundary-value problem, the result in the text seems to be new.

As has been remarked by COLEMAN and NOLL[1], *a configuration that is stable for the boundary condition of traction is stable also for the boundary condition of place.* To see this, we may write out (89.1) by substituting (88.10) on each side, first for the boundary condition (88.9) and then for the boundary condition (88.8). If the former inequality holds for all comparison fields, it holds in particular for those such that $\boldsymbol{u}^* = \boldsymbol{u}$ on $\mathscr{S}_2$, and for these it reduces to the latter inequality. The converse need not hold.

According to elementary rules of the variational calculus, a necessary condition for stability is that $\delta J(\boldsymbol{\chi}) = 0$ when $\delta\boldsymbol{\chi} = \mathbf{0}$ on $\mathscr{S}_1$. Hence, by the first variational theorem of Sect. 88, the differential equations (88.7) and the boundary conditions (88.9) or (88.13) are satisfied when $\boldsymbol{\chi}$ is stable. Another necessary condition is that the second variation $\delta^2 J(\boldsymbol{\chi})$ satisfy the inequality

$$\delta^2 J(\boldsymbol{\chi}) \geqq 0 \tag{89.2}$$

for all variations $\delta\boldsymbol{\chi}$ that vanish on $\mathscr{S}_1$. Now, for boundary values of place and traction, $\delta^2 J(\boldsymbol{\chi})$ can easily be computed from (88.12). Since both (88.7) and (88.9) are valid, for the stable $\boldsymbol{\chi}$ all terms involving $\delta^2\boldsymbol{\chi}$ cancel. The remaining expression can be reduced by application of the divergence theorem to[2]

$$\left.\begin{aligned} \delta^2 J(\boldsymbol{\chi}) &= \int_{\mathscr{B}} \{\delta(\operatorname{grad} v)\cdot\delta\boldsymbol{\chi} + \operatorname{tr}[(\delta\sigma_{\boldsymbol{F}}^T)(\delta\boldsymbol{F})]\}\, dm, \\ &= \int_{\mathscr{B}} \left\{ v_{,km}\,\delta\chi^k\,\delta\chi^m + \frac{\partial^2\sigma}{\partial x^k{}_{,\alpha}\,\partial x^m{}_{,\beta}}\,\delta\chi^k{}_{,\alpha}\,\delta\chi^m{}_{,\beta} \right\} dm. \end{aligned}\right\} \tag{89.3}$$

For stability under dead loading, the body force $\boldsymbol{b}$ is not varied, so that $\delta(\operatorname{grad} v) = \mathbf{0}$. Substituting (89.3) into (89.2) then yields a condition of the form (68b.2). Schematically, for a hyperelastic body

$$\text{stability} \Rightarrow \text{infinitesimal stability}, \tag{89.4}$$

as is only to be expected. The converse, of course, is false. The condition of infinitesimal stability, which does not require existence of a stored-energy function, has been explored in detail in Sect. 68bis.

Nearly all rearches on stability up to now have concerned the case of dead loading. A more general concept of stability has been defined by BEATTY[3]. Namely, let $\boldsymbol{x} = \boldsymbol{\chi}(\boldsymbol{X}, t)$ be the motion whose stability is to be investigated, and let $\boldsymbol{x} = \boldsymbol{\upsilon}(\boldsymbol{X}, \tau; t)$ be another motion in some specified class, where $\boldsymbol{\upsilon}(\boldsymbol{X}, t; t) = \boldsymbol{\chi}(\boldsymbol{X}, t)$. Let the body force $\boldsymbol{b}$ and surface traction $\boldsymbol{t}$ be varied also, say $\boldsymbol{b} = \bar{\boldsymbol{b}}(\boldsymbol{x}, \tau; t)$ and $\boldsymbol{t} = \bar{\boldsymbol{t}}(\boldsymbol{x}, \tau; t)$. BEATTYS' definition of stability is then

$$\int_{\mathscr{B}} \varrho\,[\sigma(\boldsymbol{F}^*) - \sigma(\boldsymbol{F})]\, dv \geqq \int_t^\tau d\tau \left[\int_{\mathscr{B}} \varrho\,\boldsymbol{b}\cdot\frac{\partial\boldsymbol{\upsilon}}{\partial\tau}\, dv + \int_{\partial\mathscr{B}} \boldsymbol{t}\cdot\frac{\partial\boldsymbol{\upsilon}}{\partial\tau}\, ds \right], \tag{89.5}$$

where $\boldsymbol{F}^* = \Delta\boldsymbol{\upsilon}$.

When thermal phenomena are taken into account, for stability one must consider *thermal constraints* in addition to mechanical boundary conditions. A *state* of a body $\mathscr{B}$ is defined to be a pair $\{\boldsymbol{\chi}, \eta\}$, where $\boldsymbol{\chi}$ is a configuration and $\eta = \eta(\boldsymbol{X})$ a distribution of entropy over $\mathscr{B}$. The temperature relation (82.2) assigns to each state a temperature field $\theta = \theta(\boldsymbol{X})$ over $\mathscr{B}$.

Suppose a uniform temperature θ be prescribed. Then, given any configuration $\boldsymbol{\chi}$, we may use (82.3) to define an entropy distribution η such that the state $\{\boldsymbol{\chi}, \eta\}$ corresponds to the given uniform temperature θ. We take the specific free energy $\tilde{\psi}$ of (82.5) as the strain-energy function in the definition of the variational

[1] COLEMAN and NOLL [1959, *3*, § 15].
[2] This result generalizes a formula of HADAMARD [1903, *4*, ¶ 269].
[3] BEATTY [1964, *4*] [1965, *1*, § 1.2.1].

integrals (88.10) or (88.14). *Isothermal stability* is then defined, as in the beginning of this section, by the inequality (89.1).

Suppose now that the total entropy

$$H = \int_{\mathscr{B}} \eta \, dm \tag{89.6}$$

is prescribed. We take the energy function $\hat{\varepsilon}$ of (80.1) as the strain-energy function σ in the definition of the variational integrals (88.10) or (88.14). These integrals then depend not only on $\boldsymbol{\chi}$ but also on η. For example, in place of (88.10) we have

$$\hat{J}(\boldsymbol{\chi}, \eta) = \int_{\mathscr{B}} \hat{\varepsilon}(\boldsymbol{F}, \eta) \, dm + \int_{\mathscr{B}} v(\boldsymbol{\chi}) \, dm - \int_{\mathscr{S}_2} \boldsymbol{u} \cdot \boldsymbol{t}_{\mathrm{R}} \, ds_{\mathrm{R}}. \tag{89.7}$$

A state $\{\boldsymbol{\chi}, \eta\}$ is called *adiabatically stable* for the boundary conditions (88.8) and (88.9) if the integral (89.7) has a relative minimum within the class of all states that satisfy the boundary condition (88.8) and the thermal constraint (89.6). The rules of the variational calculus, using the method of Lagrange multipliers, show that for adiabatic stability it is necessary that the temperature be uniform in the body[1]. The differential equations (88.7) and the boundary conditions (88.8) and (88.9) also result as necessary conditions of stability when σ in these equations is taken to be the free-energy function $\tilde{\psi}$.

Let $\{\boldsymbol{\chi}, \eta\}$ be a state of uniform temperature θ which is isothermally stable for certain boundary conditions. Let $\boldsymbol{\chi}^*$ be a configuration in the neighborhood of $\boldsymbol{\chi}$ and satisfying the boundary conditions. Let η^* be an entropy distribution such that

$$\int_{\mathscr{B}} \eta^* \, dm = \int_{\mathscr{B}} \eta \, dm = H. \tag{89.8}$$

We define another entropy distribution η_1 by putting

$$\eta_1 = \tilde{\eta}(\boldsymbol{F}^*, \theta), \tag{89.9}$$

where $\boldsymbol{F}^*$ is the deformation gradient computed from $\boldsymbol{\chi}^*$. It follows from (82.5) and (89.9) that

$$\tilde{\psi}(\boldsymbol{F}^*, \theta) - \tilde{\psi}(\boldsymbol{F}, \theta) = \hat{\varepsilon}(\boldsymbol{F}^*, \eta^*) - \hat{\varepsilon}(\boldsymbol{F}, \eta) - (\eta^* - \eta)\,\theta - \beta \tag{89.10}$$

where

$$\beta = \hat{\varepsilon}(\boldsymbol{F}^*, \eta^*) - \hat{\varepsilon}(\boldsymbol{F}^*, \eta_1) - (\eta^* - \eta_1)\,\theta. \tag{89.11}$$

Integration of (89.10) over $\mathscr{B}$, using (89.8), gives

$$\tilde{J}(\boldsymbol{\chi}^*) - \tilde{J}(\boldsymbol{\chi}) = \hat{J}(\boldsymbol{\chi}^*, \eta^*) - \hat{J}(\boldsymbol{\chi}, \eta) - \int_{\mathscr{B}} \beta \, dm, \tag{89.12}$$

where the integrals $\tilde{J}$ and $\hat{J}$ are computed from (88.10) or (88.14) with $\sigma(\boldsymbol{F}) = \tilde{\psi}(\boldsymbol{F}, \theta)$ and $\sigma(\boldsymbol{F}) = \hat{\varepsilon}(\boldsymbol{F}, \eta)$, respectively. Now, if $\beta \geqq 0$, we find from (89.12) that $\tilde{J}(\boldsymbol{\chi}^*) \geqq \tilde{J}(\boldsymbol{\chi})$ implies $\hat{J}(\boldsymbol{\chi}^*, \eta^*) \geqq \hat{J}(\boldsymbol{\chi}, \eta)$. The condition $\beta \geqq 0$ is satisfied when $\hat{\varepsilon}(\boldsymbol{F}, \eta)$ is a convex function of η for each fixed $\boldsymbol{F}$. We have proved the following ***stability theorem of Coleman and Noll***[2]: *If the energy is a convex function of the entropy, then isothermal stability implies adiabatic stability for the same boundary conditions.* We note that the condition of convexity used here follows if it is assumed that the heat capacity be positive. It is also a consequence of the postulate of COLEMAN and NOLL mentioned in Sect. 81.

---

[1] GIBBS [1875, *1*, pp. 184—192]. Expositions have been given by TRUESDELL [1952, *20*, § 33] and by COLEMAN and NOLL [1959, *3*, pp. 119—128].

[2] COLEMAN and NOLL [1959, *3*, § 15].

When considering stability with *boundary conditions of place*, i.e. stability with $\mathcal{S}_1$ in (88.8) equal to $\partial\mathcal{B}$, we have $\int_{\mathcal{S}_2} \boldsymbol{\chi} \cdot \boldsymbol{t}_R \, ds_R = 0$, and we can replace $J(\boldsymbol{\chi})$ in (88.10) by the simpler integral

$$J'(\boldsymbol{\chi}) = \int_{\mathcal{B}} \sigma(\boldsymbol{F}) \, dm + \int_{\mathcal{B}} v(\boldsymbol{\chi}) \, dm. \tag{89.13}$$

For example, we say that a state $\{\boldsymbol{\chi}, \eta\}$ is *adiabatically stable with boundary conditions of place* if

$$\hat{J}'(\boldsymbol{\chi}) = \int_{\mathcal{B}} \hat{\varepsilon}(\boldsymbol{F}, \eta) \, dm + \int_{\mathcal{B}} v(\boldsymbol{\chi}) \, dm \tag{89.14}$$

has a relative minimum within the class of all states $\{\boldsymbol{\chi}^*, \eta^*\}$ that satisfy the boundary condition

$$\boldsymbol{\chi}^*(\boldsymbol{X}) = \boldsymbol{\chi}(\boldsymbol{X}) \quad \text{if} \quad \boldsymbol{X} \in \partial\mathcal{B} \tag{89.15}$$

and the thermal constraint (89.6).

We conclude this section with another theorem of COLEMAN and NOLL[1], which makes mathematical the idea, due to GIBBS[2], that a state with adiabatic stability for boundary conditions of place should be one which maximizes the entropy under appropriate constraints. ***Theorem of Gibbs, Coleman, and Noll:*** *A state* $\{\boldsymbol{\chi}, \eta\}$ *is adiabatically stable with boundary conditions of place if and only if every state* $\{\boldsymbol{\chi}^*, \eta^*\}$ *that obeys the constraint* (89.15) *and the equation*

$$\hat{J}'(\boldsymbol{\chi}^*, \eta^*) = \hat{J}'(\boldsymbol{\chi}, \eta) \tag{89.16}$$

also obeys the inequality

$$H^* = \int_{\mathcal{B}} \eta^* \, dm \leq \int_{\mathcal{B}} \eta \, dm = H. \tag{89.17}$$

Proof: We prove necessity by showing that if the hypothesis of the theorem does not hold, then the state $\{\boldsymbol{\chi}, \eta\}$ is not adiabatically stable, i.e. if there exists a state $\{\boldsymbol{\chi}_1, \eta_1\}$ obeying (89.15) and (89.16) but for which $H_1 > H$, then there exists a state $\{\boldsymbol{\chi}_2, \eta_2\}$ such that $H_2 = H$ but $\hat{J}'(\boldsymbol{\chi}_2, \eta_2) < \hat{J}'(\boldsymbol{\chi}, \eta)$. We do this by putting

$$\boldsymbol{\chi}_2 = \boldsymbol{\chi}_1, \tag{89.18}$$

and

$$\eta_2(\boldsymbol{X}) = \eta_1(\boldsymbol{X}) + \frac{H - H_1}{\int_{\mathcal{B}} dm}, \tag{89.19}$$

Clearly, since $\{\boldsymbol{\chi}_1, \eta_1\}$ obeys (89.16), so does $\{\boldsymbol{\chi}_2, \eta_2\}$. Furthermore $\{\boldsymbol{\chi}_2, \eta_2\}$ has the property $H_2 = H$. Since we assume that $\hat{J}'(\boldsymbol{\chi}_1, \eta_1) = \hat{J}'(\boldsymbol{\chi}, \eta)$, we have, by (89.14) and (89.18),

$$\hat{J}'(\boldsymbol{\chi}_2, \eta_2) - \hat{J}'(\boldsymbol{\chi}, \eta) = \hat{J}'(\boldsymbol{\chi}_2, \eta_2) - \hat{J}'(\boldsymbol{\chi}_1, \eta_1) = \int_{\mathcal{B}} [\hat{\varepsilon}(\boldsymbol{F}_1, \eta_2) - \hat{\varepsilon}(\boldsymbol{F}_1, \eta_1)] \, dm, \tag{89.20}$$

where $\boldsymbol{F}_1$ is the deformation gradient corresponding to $\boldsymbol{\chi}_1$. Since $H_1 > H$, (89.19) yields $\eta_2 < \eta_1$ at all $\boldsymbol{X}$. Hence, by (82.2) and the fact that $\theta > 0$, we have

$$\hat{\varepsilon}(\boldsymbol{F}_1, \eta_2) < \hat{\varepsilon}(\boldsymbol{F}_2, \eta_1) \tag{89.21}$$

at all $\boldsymbol{X}$. Substitution of (89.21) into (89.20) completes the proof of necessity.

---

[1] COLEMAN and NOLL [1959, *3*, § 13, Theorem 13].

[2] GIBBS [1875, *1*, pp. 56—62] replaced the rather vague ideas of CLAUSIUS about a trend in time by concrete assertions regarding extrema in statics.

Sufficiency is proved assuming that $\{\chi, \eta\}$ is not adiabatically stable, i.e. by starting with a state $\{\chi_1, \eta_1\}$ obeying $H_1 = H$ but for which $\hat{J}'(\chi_1, \eta_1) < \hat{J}'(\chi, \eta)$. We define a state $\{\chi_2, \eta_2\}$ by

$$\chi_2 = \chi_1, \tag{89.22}$$

$$\eta_2 = \check{\eta}(\boldsymbol{F}_1, \varepsilon_2) \tag{89.23}$$

where

$$\varepsilon_2(\boldsymbol{X}) = \varepsilon_1(\boldsymbol{X}) + \frac{\hat{J}'(\chi, \eta) - \hat{J}'(\chi_1 \eta_1)}{\int_{\mathscr{B}} dm}. \tag{89.24}$$

Since $\{\chi_1, \eta_1\}$ obeys (89.15), so does $\{\chi_2, \eta_2\}$. Substituting (89.22) and (89.24) into (89.14), we find that $\hat{J}'(\chi_2, \eta_2) = \hat{J}'(\chi, \eta)$. Equation (89.24) and the assumption $\hat{J}'(\chi_1, \eta_1) < \hat{J}'(\chi, \eta)$ yield $\varepsilon_2 > \varepsilon_1$ at all $\boldsymbol{X}$ and, using the fact that $\frac{\partial \eta}{\partial \varepsilon} = \theta^{-1} > 0$, we have $H_2 > H_1 = H$. Thus, if $\{\chi, \eta\}$ is not adiabatically stable, we find a state $\{\chi_2, \eta_2\}$ obeying (89.15) and (89.16) but not (89.17). Q.E.D.

90. Wave propagation. The theory of waves in general elastic materials has been developed in Sects. 71—78. In view of the condition (82.13) for the existence of a strain-energy function, from $(71.4)_2$ if follows at once that

$$\boldsymbol{Q}(\boldsymbol{n}) = [\boldsymbol{Q}(\boldsymbol{n})]^T, \qquad Q_{km}(\boldsymbol{n}) = Q_{mk}(\boldsymbol{n}): \tag{90.1}$$

For each direction of propagation, the acoustical tensor of a hyperelastic material is symmetric. Therefore, for a hyperelastic material, $\boldsymbol{Q}(\boldsymbol{n})$ has real proper numbers and at least three mutually orthogonal real proper vectors. Thus follows the ***theorem of Hadamard***[1]: *For any direction of propagation in a hyperelastic material every acoustic axis is real, and for each direction of propagation there is at least one orthogonal triad of acoustic axes.* Since $\boldsymbol{Q}(\boldsymbol{n})$ is symmetric, it is determined to within an arbitrary factor by its quadric [CFT, Sect. 21]; that quadric, used in the earlier researches to describe the properties of an elastic material in respect to the propagation of waves, is called *Fresnel's ellipsoid* or *the ellipsoid of polarization* if all proper numbers are positive.

A partial converse to HADAMARD's theorem has been obtained by BERNSTEIN[2]: *If for every direction of propagation the acoustical tensor of an elastic material is symmetric, then the material is hyperelastic.* To prove this theorem, we observe from (71.5) that (90.1) holds if and only if

$$(\mathsf{B}^{kpmq} - \mathsf{B}^{mpkq})\, n_p n_q = 0. \tag{90.2}$$

For this condition to hold for every $\boldsymbol{n}$, it is necessary and sufficient that $\mathsf{B}^{kpmq} - \mathsf{B}^{mpkq}$ be skew-symmetric with respect to the indices p and q, i.e.,

$$\mathsf{B}^{kpmq} - \mathsf{B}^{mpkq} + \mathsf{B}^{kqmp} - \mathsf{B}^{mqkp} = 0. \tag{90.3}$$

Hence

$$\mathsf{B}^{pkqm} - \mathsf{B}^{qkpm} + \mathsf{B}^{pmqk} - \mathsf{B}^{qmpk} = 0. \tag{90.4}$$

Adding these two equations yields

$$\mathsf{B}^{kpmq} + \mathsf{B}^{pkqm} - \mathsf{B}^{mqkp} - \mathsf{B}^{qmpk} = \mathsf{B}^{mpkq} - \mathsf{B}^{pmqk} + \mathsf{B}^{qkpm} - \mathsf{B}^{kqmp}. \tag{90.5}$$

---

[1] HADAMARD [1901, *1*, § 7] [1903, *4*, ¶ 268], generalizing a long sequence of earlier results, as mentioned in footnote 2, p. 268.

We have not been able to support the criticism of HADAMARD's theorem by DUHEM [1903, *3*] [1906, *1*, Ch. I, §§ II, IV].

[2] According to TRUESDELL [1961, *63*, § 12].

By four uses of the identities $(45.7)_3$ we see that the right-hand side vanishes, while the left-hand side reduces to $2\,(B^{kpmq}-B^{mqkp})$. Hence (84.15) holds, showing that the material is hyperelastic.

Rephrasing a theorem of KELVIN and TAIT [CFT, App., Sect. 37], we see that if the squared speeds are non-negative and if an orthogonal triad of acoustical axes exists, then the acoustical tensor is symmetric. From this fact and from BERNSTEIN's theorem we conclude that *if, in an elastic material, there is for every direction of propagation an orthogonal triad of acoustical axes, and if the corresponding speeds of propagation are real, then the material is hyperelastic.*

It does not seem to be possible to express BERNSTEIN's theorem entirely in terms of acoustical properties. The condition stated is only sufficient, not necessary. For example, if the speeds are purely imaginary, the conclusion still holds, but a criterion of this kind is not physically effective, since purely imaginary speeds cannot be distinguished from more general complex ones, neither kind being possible.

For isotropic materials a stronger result may be obtained[1]. From the explicit formula $(74.5)_2$ we see that if $\boldsymbol{n}$ is not normal to a principal direction of stress, so that $\cos\theta_1\cos\theta_2\neq 0$, then $Q\langle 12\rangle(\boldsymbol{n}) = Q\langle 21\rangle(\boldsymbol{n})$ if and only if

$$\frac{\partial t_1}{\partial \log v_2} - \frac{\partial t_2}{\partial \log v_1} = t_2 - t_1. \tag{90.6}$$

This condition and the others obtained from it by permutation of indices are equivalent to (86.14), whence it follows that if the *acoustical tensor $\boldsymbol{Q}(\boldsymbol{n})$ of an isotropic elastic material is symmetric for one single direction of propagation that is not normal to a principal direction of stress and strain, then the material is hyperelastic.* From this result and BERNSTEIN's theorem we conclude that there are three mutually exclusive kinds of isotropic elastic materials as far as wave propagation is concerned:

1. Those for which $\boldsymbol{Q}(\boldsymbol{n})$ is symmetric for the principal waves only.

2. Those for which $\boldsymbol{Q}(\boldsymbol{n})$ is symmetric for $\boldsymbol{n}$ parallel or normal to one axis of stress and strain, but for no other $\boldsymbol{n}$.

3. Those for which $\boldsymbol{Q}(\boldsymbol{n})$ is symmetric for all $\boldsymbol{n}$, these last being the hyperelastic materials.

Since for hyperelastic materials $\boldsymbol{Q}(\boldsymbol{n})$ is symmetric, the S-E condition (71.15) is necessary and sufficient that $\boldsymbol{Q}(\boldsymbol{n})$ be positive-definite for every $\boldsymbol{n}$. That is, *in a hyperelastic material the S-E condition is satisfied if and only if the squared wave speeds for every direction are positive.* In particular, if the S-E condition is satisfied, $U\neq 0$ always, so there can be no material singular surfaces.

HADAMARD's ***fundamental theorem on waves*** in Sect. 71 now takes the following more explicit form[2]: *In an infinitesimally stable configuration of a hyperelastic body, the speed of propagation corresponding to any acoustical axis is real.* Thus the condition of infinitesimal stability guarantees the possibility of *at least three real waves or material discontinuities normal to each direction at each point.*

In regard to wave propagation, the C-N condition is weaker than the S-E condition. Indeed, *the C-N⁺ condition is sufficient that for every direction, one wave speed be positive.* To prove this result[3], note that since $\boldsymbol{Q}(\boldsymbol{n})$ is symmetric, its proper numbers are the extrema of the bounded continuous function

$$F(\boldsymbol{l}) = \boldsymbol{l}\cdot\boldsymbol{Q}(\boldsymbol{n})\,\boldsymbol{l}, \tag{90.7}$$

[1] TRUESDELL [1961, *63*, § 12].

[2] HADAMARD [1901, *1*, § 7] [1903, *4*, ¶ 271].

[3] Given by NOLL in 1961, according to TRUESDELL [1961, *63*, § 12].

where $\boldsymbol{l}$ is a unit vector. By (71.5) and (52.16), $F(\boldsymbol{n})>0$. Hence at least one extreme of $F(\boldsymbol{l})$ must be positive; that is, $\boldsymbol{Q}(\boldsymbol{n})$ has at least one positive proper number, so that the corresponding wave-speeds are real and non-vanishing. Q.E.D. (The reader will recall the stronger theorems of Sect. 74 for isotropic elastic materials, whether or not they have a stored-energy function.)

If we set

$$P^{pq}=B_k{}^{pkq}, \tag{90.8}$$

from (84.15) it follows that $\boldsymbol{P}$ is symmetric. Consequently, if for each unit vector $\boldsymbol{n}$ we set

$$m(\boldsymbol{n})=\boldsymbol{n}\cdot\boldsymbol{P}\boldsymbol{n}, \tag{90.9}$$

then there are at least three mutually orthogonal directions for which $m(\boldsymbol{n})$ is an extreme. But by (90.8), (71.5), and (71.6),

$$m(\boldsymbol{n})=\operatorname{tr}\boldsymbol{Q}(\boldsymbol{n})=\varrho\sum_{a=1}^{3}U_a^2, \tag{90.10}$$

where the U_a^2 are the squared wave speeds for the direction $\boldsymbol{n}$. Thus follows a theorem due in principle to PASTORI[1]: ***In a hyperelastic material there are at each point at least three mutually orthogonal directions rendering the sum of the squares of the corresponding wave speeds an extreme.*** Clearly one extreme must be a minimum and one a maximum. The possibility that some wave speeds are purely imaginary is not excluded in the above statement.

In Sect. 76 has been given a method for finding the response coefficients $\aleph_\Gamma$ of an isotropic elastic material from the wave speeds. Supposing these coefficients so determined, we may substitute the resulting formulae into the conditions (86.12) and so find necessary and sufficient conditions that the measured speeds of principal waves be compatible with the theory of hyperelastic materials[2]:

$$\left.\begin{aligned}
&\frac{1}{\varrho_R\sqrt{III}}B_2+\frac{\partial}{\partial III}\left[\frac{U_{12}^2-U_{13}^2}{v_1^2(v_2^2-v_3^2)}\right]=0,\\
&\frac{1}{\varrho_R\sqrt{III}}B_1-\frac{\partial}{\partial III}\left[\frac{v_2^2U_{12}^2-v_3^2U_{13}^2}{v_1^2(v_2^2-v_3^2)}\right]=0,\\
&\frac{\partial}{\partial I}\left[\frac{U_{12}^2-U_{13}^2}{v_1^2(v_2^2-v_3^2)}\right]+\frac{\partial}{\partial II}\left[\frac{v_2^2U_{12}^2-v_3^2U_{13}^2}{v_1^2(v_2^2-v_3^2)}\right]=0,
\end{aligned}\right\} \tag{90.11}$$

where B_1 and B_2, defined by (76.9), are explicitly given functions of the speeds of the principal waves.

Simpler forms follow[3] by use of principal components. Indeed, from (74.7) and (86.4) we see that

$$U_{11}^2=v_1^2\frac{\partial^2\sigma}{\partial v_1\partial v_1},\qquad \frac{U_{12}^2(v_1^2-v_2^2)}{v_1^2}=v_1\frac{\partial\sigma}{\partial v_1}-v_2\frac{\partial\sigma}{\partial v_2},\ \ldots \tag{90.12}$$

Conditions of compatibility for this system are easily found:

$$\left.\begin{aligned}
&\frac{(v_1^2-v_2^2)\,U_{12}^2}{v_1^2}+\frac{(v_2^2-v_3^2)\,U_{23}^2}{v_2^2}+\frac{(v_3^2-v_1^2)\,U_{31}^2}{v_3^2}=0,\\
&\left(v_1\frac{\partial}{\partial v_1}-v_2\frac{\partial}{\partial v_2}\right)U_{11}^2=v_1^2\frac{\partial^2}{\partial v_1\partial v_1}\left[\frac{(v_1^2-v_2^2)\,U_{12}^2}{v_1^2}\right],\ \ldots,\\
&\left(v_1\frac{\partial}{\partial v_1}+v_2\frac{\partial}{\partial v_2}\right)\left[\frac{(v_1^2-v_2^2)\,U_{12}^2}{v_1^2}\right]=U_{11}^2-U_{22}^2,\ \ldots,
\end{aligned}\right\} \tag{90.13}$$

[1] PASTORI [1949, *12*, § 6]. The statement and proof reproduced here were given by TRUESDELL [1961, *63*, § 12].

[2] TRUESDELL [1961, *63*, § 12].

[3] GREEN [1963, *26*, § 3].

where the dots stand for expressions obtained from the preceding by cyclic permutation of indices. That there are 7 equations in the set (90.13) and but 3 in the set (90.11) arises from the fact that in (90.13) the conditions for compatibility of the wave speeds with the theory of isotropic elasticity are included, while for (90.11) they are presumed already satisfied. In distinction to the conditions of compatibility for more general elastic waves, those for hyperelastic materials connect the longitudinal with the transverse-wave speeds.

At the end of Sect. 71 we noted that the acoustical tensor for homentropic wave propagation in a generalized elastic material differs from that for homothermal wave propagation. In the case of a perfect material, we have the relations (82.15) and (82.17) connecting the tensors $\hat{\mathsf{A}}_{k\ m}^{\ \alpha\ \beta}$ and $\tilde{\mathsf{A}}_{k\ m}^{\ \alpha\ \beta}$, so that by (71.4) we see at once that the corresponding acoustical tensors $\hat{\boldsymbol{Q}}(\boldsymbol{n})$ and $\tilde{\boldsymbol{Q}}(\boldsymbol{n})$ are related as follows[1]:

$$\left.\begin{aligned}\hat{\boldsymbol{Q}}(\boldsymbol{n}) &= \tilde{\boldsymbol{Q}}(\boldsymbol{n}) + \boldsymbol{P}(\boldsymbol{n}) \otimes \boldsymbol{P}(\boldsymbol{n}),\\ \hat{Q}_{km}(\boldsymbol{n}) &= \tilde{Q}_{km}(\boldsymbol{n}) + P_k(\boldsymbol{n})\, P_m(\boldsymbol{n}),\end{aligned}\right\} \tag{90.14}$$

where

$$\left.\begin{aligned}P_k(\boldsymbol{n}) &= \sqrt{\frac{\varrho\,\varkappa_{(\boldsymbol{F})}}{\varrho_{\mathrm{R}}^2\,\theta}}\,\frac{\partial \hat{\mathfrak{h}}_k{}^{\alpha}}{\partial \eta}\, x^p{}_{,\alpha}\, n_p = \sqrt{\frac{\varrho\,\varkappa_{(\boldsymbol{F})}}{\theta}}\,\frac{\partial \hat{\theta}}{\partial x^k{}_{,\alpha}}\, x^p{}_{,\alpha}\, n_p,\\ &= \sqrt{\frac{\varrho\,\theta}{\varrho_{\mathrm{R}}^2\,\varkappa_{(\boldsymbol{F})}}}\,\frac{\partial \tilde{\mathfrak{h}}_k{}^{\alpha}}{\partial \theta}\, x^p{}_{,\alpha}\, n_p = \sqrt{\frac{\varrho\,\theta}{\varkappa_{(\boldsymbol{F})}}}\,\frac{\partial \tilde{\eta}}{\partial x^k{}_{,\alpha}}\, x^p{}_{,\alpha}\, n_p.\end{aligned}\right\} \tag{90.15}$$

Eq. (90.14) is *Duhem's identity*[2] connecting the homentropic and homothermal acoustical tensors of a perfect material. Since $\varkappa_{(\boldsymbol{F})}/\theta$ may be assumed to be positive [CFT, Sect. 265], $\boldsymbol{P}$ is real; therefore $\boldsymbol{P} \otimes \boldsymbol{P}$ is positive semi-definite; therefore if $\tilde{\boldsymbol{Q}}(\boldsymbol{n})$ is positive-definite, so is $\hat{\boldsymbol{Q}}(\boldsymbol{n})$. Thus we have established ***Duhem's theorem on wave speeds:*** *If, for a given perfect material, all homothermal wave speeds are real, so are all homentropic wave speeds.* The principal axes of $\hat{\boldsymbol{Q}}(\boldsymbol{n})$ differ from those of $\tilde{\boldsymbol{Q}}(\boldsymbol{n})$ except in the case when one axis is parallel to $\boldsymbol{P}$.

General conditions of compatibility for waves according to the infinitesimal theory of elasticity were derived at the end of Sect. 73. Since the acoustical tensor for any direction of propagation is symmetric, from (73.12) we read off the additional symmetry relation

$$\mathsf{K}_{\alpha\lambda\beta\mu} = \mathsf{K}_{\beta\lambda\alpha\mu}. \tag{90.16}$$

This condition does not suffice to yield the identity $\mathsf{L}_{\lambda\alpha\mu\beta} = \mathsf{L}_{\mu\beta\lambda\alpha}$, which is necessary and sufficient for existence of a strain-energy function in the infinitesimal theory. Indeed, from (73.14) we see that this relation assumes the form

$$\mathsf{K}_{\alpha\lambda\beta\mu} + \mathsf{K}_{\lambda\alpha\beta\mu} - \mathsf{K}_{\mu\lambda\beta\alpha} = \mathsf{K}_{\beta\mu\alpha\lambda} + \mathsf{K}_{\mu\beta\alpha\lambda} - \mathsf{K}_{\lambda\mu\alpha\beta}. \tag{90.17}$$

Putting (90.16) into (73.16) yields

$$\mathsf{K}_{\alpha\lambda\beta\mu} + \mathsf{K}_{\lambda\alpha\beta\mu} = \mathsf{K}_{\alpha\lambda\mu\beta} + \mathsf{K}_{\lambda\alpha\mu\beta}, \tag{90.18}$$

while use of this result and (90.16) reduces (90.17) to the form

$$\mathsf{K}_{\lambda\alpha\mu\beta} - \mathsf{K}_{\mu\lambda\beta\alpha} = \mathsf{K}_{\beta\mu\alpha\lambda} - \mathsf{K}_{\lambda\mu\alpha\beta}. \tag{90.19}$$

The identities (90.18) and (90.19) are TOUPIN and BERNSTEIN's ***second and third conditions of compatibility***[3].

Some remarks on second-order effects on wave propagation are given below in Sect. 93.

---

[1] DUHEM [1903, *3*] [1906, *1*, Ch. I, § 2]. Here we have given the derivation of TRUESDELL [1961, ***63***, § 13]. The results of GREEN [1962, ***23***, § 9] seem to be included as a special case.

[2] DUHEM [1903, *3*] [1906, *1*, Ch. 1, § 2]. The proof given above is that of TRUESDELL [1961, ***63***, § 13].

[3] TOUPIN and BERNSTEIN [1961, ***62***, § 4].

c) Solutions of special problems.

91. ERICKSEN's analysis of the deformations possible in every isotropic hyperelastic body. In Sects. 54 and 55 we have seen that any homogeneous strain is a possible deformation, subject to the action of surface loads only, in any homogeneous, isotropic, elastic body. ERICKSEN[1] has shown that conversely, *in compressible, homogeneous, isotropic, hyperelastic bodies, no other kind of deformation is possible, subject to surface tractions alone,* for an arbitrary form of the strain-energy function[2].

To prove this theorem, ERICKSEN starts from the stress relation in the form $(85.16)_3$; writing $\tilde{\sigma}(\boldsymbol{B}^{-1}) = \hat{\sigma}(I_1, I_2, I_3)$, where I_a here stands for the a'th principal invariant of $\boldsymbol{B}^{-1}$, and using the fact that[3] $\varrho/\varrho_R = \sqrt{I_3}$, we see that

$$T^k_m = -2\varrho_R \sqrt{I_3} \sum_{a=1}^{3} \frac{\partial \hat{\sigma}}{\partial I_a} \frac{\partial I_a}{\partial (B^{-1})^m_p} (B^{-1})^k_p. \tag{91.1}$$

Substitution of this result into $(16.6)_2$ yields the following equations of equilibrium subject to no external body force:

$$\left.\begin{aligned} &\sum_{a=1}^{3} \frac{\partial \hat{\sigma}}{\partial I_a} \left(\sqrt{I_3} \frac{\partial I_a}{\partial (B^{-1})^m_p} (B^{-1})^k_p\right)^{,m} + \\ &\quad + \sum_{a,b=1}^{3} \sqrt{I_3} \frac{\partial^2 \hat{\sigma}}{\partial I_a \partial I_b} I_b{}^{,m} \frac{\partial I_a}{\partial (B^{-1})^m_p} (B^{-1})^k_p = 0. \end{aligned}\right\} \tag{91.2}$$

In order that these equations be satisfied for every strain-energy function $\hat{\sigma}$, it is necessary and sufficient that the coefficient of each derivative vanish separately:

$$\left.\begin{aligned} &\left(\sqrt{I_3} \frac{\partial I_a}{\partial (B^{-1})^m_p} (B^{-1})^k_p\right)^{,m} = 0, \quad a = 1, 2, 3, \\ &\left(I_b{}^{,m} \frac{\partial I_a}{\partial (B^{-1})^m_p} + I_a{}^{,m} \frac{\partial I_b}{\partial (B^{-1})^m_p}\right) (B^{-1})^k_p = 0, \quad a, b = 1, 2, 3. \end{aligned}\right\} \tag{91.3}$$

In addition to satisfying these 12 equations, the tensor $\boldsymbol{B}^{-1}$ must be a metric tensor in Euclidean space; thus the Riemann tensor based on $\boldsymbol{B}^{-1}$ must vanish. This latter condition may be expressed in the form [CFT, Eq. (34.6)]

$$\left.\begin{aligned} R^*_{kqpm} \equiv \tfrac{1}{2}\big((B^{-1})_{km,pq} + (B^{-1})_{qp,km} - (B^{-1})_{kp,mq} - (B^{-1})_{qm,kp}\big) \\ + B^{rs}(A_{qpr}A_{kms} - A_{qmr}A_{kps}) = 0, \end{aligned}\right\} \tag{91.4}$$

where

$$A_{kmp} = A_{mkp} = \tfrac{1}{2}\big((B^{-1})_{kp,m} + (B^{-1})_{mp,k} - (B^{-1})_{km,p}\big). \tag{91.5}$$

It is obvious that (91.3) and (91.4) are satisfied by any tensor $\boldsymbol{B}^{-1}$ such that $(B^{-1})_{km,p} = 0$; our object is to prove that there are no other symmetric and positive-definite tensors $\boldsymbol{B}^{-1}$ that are solutions.

First, put $a = 3$ in $(91.3)_1$ and use $(9.12)_3$:

$$0 = \left(\sqrt{I_3} \frac{\partial I_3}{\partial (B^{-1})^m_p} (B^{-1})^k_p\right)^{,m} = \left(I_3^{\frac{3}{2}} B^p_m (B^{-1})^k_p\right)^{,m} = (I_3^{\frac{3}{2}})^{,k}; \tag{91.6}$$

---

[1] ERICKSEN [1955, *11*].

[2] As remarked at the beginning of Sect. 54, it follows *a fortiori* that the only deformations possible, subject to surface loads only, in every homogeneous, isotropic, elastic material, are homogeneous strains.

[3] Eqs. $(30.6)_2$ and (156.1) of CFT.

thus $I_3 = \text{const}$. Now put $a=1$ in $(91.3)_1$; using (91.6) and $(9.12)_1$, we see that

$$0 = \sqrt{I_3}\left(\frac{\partial I_1}{\partial (B^{-1})^m_p}(B^{-1})^k_p\right)^{,m} = \sqrt{I_3}(B^{-1})^{k,m}_m, \tag{91.7}$$

while by putting $a=3$, $b=1$ in $(91.3)_2$ and then using (91.6) and $(9.12)_3$ we see that

$$\left.\begin{aligned} 0 &= \left(I_1^{;m}\frac{\partial I_3}{\partial (B^{-1})^m_p} + I_3^{;m}\frac{\partial I_1}{\partial (B^{-1})^m_p}\right)(B^{-1})^k_p, \\ &= I_1^{;m} I_3 B^p_m (B^{-1})^k_p = I_3 I_1^{;k}. \end{aligned}\right\} \tag{91.8}$$

Since $\boldsymbol{B}^{-1}$ must be positive-definite, $I_3 > 0$, so that (91.7) and (91.8) imply

$$(B^{-1})^{mk}{}_{,m} = 0, \qquad (B^{-1})^m_{m,k} = 0. \tag{91.9}$$

From this fact and (91.5) we see that

$$A^k{}_{kp} = \tfrac{1}{2}\left(2(B^{-1})^k_{p,k} - (B^{-1})^k_{k,p}\right) = 0. \tag{91.10}$$

From (91.4), then,

$$\left.\begin{aligned} 0 &= R^{*km}{}_{mk}, \\ &= (B^{-1})^k_{k,m}{}^{,m} - (B^{-1})^k_{m,k}{}^{,m} + B^{rs}(A^p{}_{pr}A^m{}_{ms} - A^{pk}{}_r A_{kps}), \\ &= -B^{rs}A^{pk}{}_r A_{kps}, \\ &= -\left((B^{\frac{1}{2}})^{rq}A^{pk}{}_r\right)\left((B^{\frac{1}{2}})^s_q A_{kps}\right). \end{aligned}\right\} \tag{91.11}$$

Since this last expression reduces to a sum of squares in a Cartesian system, Eq. (91.11) is equivalent to

$$(B^{\frac{1}{2}})^s_q A_{kps} = 0; \tag{91.12}$$

since $\boldsymbol{B}^{\frac{1}{2}}$ is non-singular, equivalent to (91.12) is the condition

$$A_{kps} = 0. \tag{91.13}$$

By (91.5), then,

$$0 = A_{kps} + A_{skp} = (B^{-1})_{ps,k}, \tag{91.14}$$

and this result establishes ***Ericksen's theorem.***

Next we consider *incompressible materials*, for which, as we have seen in Sect. 57, four families of solutions in addition to homogeneous strain are possible, subject to surface tractions alone. ERICKSEN[1] has gone far toward showing that beyond these four families, there exist no further non-homogeneous deformations possible in every homogeneous, isotropic, incompressible, hyperelastic material held in equilibrium by application of suitable surface tractions only. Beginning from the constitutive equation $(86.18)_2$, we see that the condition of equilibrium subject to no body force is

$$\frac{1}{2\varrho}p_{,k} = \left(\frac{\partial\sigma}{\partial I}B^m_k\right)_{,m} - \left(\frac{\partial\sigma}{\partial II}(B^{-1})^m_k\right)_{,m} \equiv A_k, \tag{91.15}$$

where $I = I_{\boldsymbol{B}}$, $II = II_{\boldsymbol{B}}$. For the existence of a function p satisfying this equation, it is necessary and sufficient that the conditions of integrability $A_{k,p} = A_{p,k}$ be satisfied. Now

$$\left.\begin{aligned} A_{k,p} = {} & \overset{1}{a}_{kp}\frac{\partial\sigma}{\partial I} + \overset{2}{a}_{kp}\frac{\partial\sigma}{\partial II} + \overset{3}{a}_{kp}\frac{\partial^2\sigma}{\partial I^2} + \overset{4}{a}_{kp}\frac{\partial^2\sigma}{\partial I\,\partial II} + \overset{5}{a}_{kp}\frac{\partial^2\sigma}{\partial II^2} + \\ & + \overset{6}{a}_{kp}\frac{\partial^3\sigma}{\partial I^3} + \overset{7}{a}_{kp}\frac{\partial^3\sigma}{\partial I^2\,\partial II} + \overset{8}{a}_{kp}\frac{\partial^3\sigma}{\partial I\,\partial II^2} + \overset{9}{a}_{kp}\frac{\partial^3\sigma}{\partial II^3}, \end{aligned}\right\} \tag{91.16}$$

[1] ERICKSEN [1954, 5].

where

$$\left.\begin{aligned}
&a^1_{kp} = B^m_{k,mp}, \quad a^2_{kp} = -(B^{-1})^m_{k,mp}, \quad a^3_{kp} = I_{,p} B^m_{k,m} + (I_{,m} B^m_k)_{,p},\\
&a^4_{kp} = II_{,p} B^m_{k,m} - I_{,p}(B^{-1})^m_{k,m} + (II_{,m} B^m_k)_{,p} - (I_{,m}(B^{-1})^m_k)_{,p},\\
&a^5_{kp} = -II_{,p}(B^{-1})^m_{k,m} - (II_{,m}(B^{-1})^m_k)_{,p}, \quad a^6_{kp} = I_{,p} I_{,m} B^m_k,\\
&a^7_{kp} = II_{,p} I_{,m} B^m_k + I_{,p} II_{,m} B^m_k - I_{,p} I_{,m}(B^{-1})^m_k,\\
&a^8_{kp} = II_{,p} II_{,m} B^m_k - II_{,p} I_{,m}(B^{-1})^m_k - I_{,p} II_{,m}(B^{-1})^m_k,\\
&a^9_{kp} = -II_{,p} II_{,m}(B^{-1})^m_k.
\end{aligned}\right\} \tag{91.17}$$

In order that, for a given $\boldsymbol{B}$, we shall have $A_{k,p} = A_{p,k}$ for every strain-energy function $\bar{\sigma}$, it is necessary and sufficient that all the tensors $\boldsymbol{a}^\Gamma$ be symmetric. The problem, then, is to find all positive-definite symmetric tensors $\boldsymbol{B}$ such that the nine tensors $\boldsymbol{a}^\Gamma$ are symmetric, such that $\det \boldsymbol{B} = 1$, and such that the Riemann tensor based upon $\boldsymbol{B}^{-1}$ vanishes.

The special case when $I = \text{const.}$, $II = \text{const.}$ includes not only homogeneous deformations but also some others, not all of which are yet discovered[1].

Consider now the general case, in which I and II are not both constant, and exclude isolated points where $I_{,k} = II_{,k} = 0$. The condition that $\boldsymbol{a}^9$ be symmetric is

$$II_{,p} II_{,m}(B^{-1})^m_k = II_{,k} II_{,m}(B^{-1})^m_p. \tag{91.18}$$

Hence there is a scalar A such that

$$II_{,m}(B^{-1})^m_k = A\, II_{,k}. \tag{91.19}$$

Thus either $II_{,k} = 0$ or $II_{,k}$ is a proper vector of $\boldsymbol{B}^{-1}$ corresponding to the proper number A; in the latter case it is also a proper vector of $\boldsymbol{B}$, corresponding to the proper number A^{-1}. Similar analysis applied to $\boldsymbol{a}^6$ shows that either $I_{,k} = 0$ or

$$I_{,m}(B^{-1})^m_k = \bar{A}\, I_{,k}. \tag{91.20}$$

From (91.19) and (91.20) we see that $\boldsymbol{a}^8$ is symmetric if and only if

$$(A - \bar{A})(II_{,k} I_{,p} - I_{,k} II_{,p}) = 0. \tag{91.21}$$

If neither $I_{,k} = 0$ nor $II_{,k} = 0$, and if $A \neq \bar{A}$, then (91.21) asserts that $I_{,k}$ and $II_{,k}$ are parallel, but since by (91.19) and (91.20) they are proper vectors of $\boldsymbol{B}^{-1}$ corresponding to distinct proper numbers, they must be perpendicular. Thus $A = \bar{A}$ if neither $I = \text{const.}$ nor $II = \text{const.}$ It follows that I and II are functionally dependent. First, the assertion is trivial if either I or II is constant; second, in a region where two principal stretches are equal, I and II, being then equal to functions of one variable, are functionally dependent; third, if the principal stretches are distinct, then $I_{,k}$ and $II_{,k}$, being proper vectors corresponding to the same proper number of the symmetric tensor $\boldsymbol{B}^{-1}$, are parallel. Thus in all cases there is a non-constant function $\beta(\boldsymbol{x})$ such that $I = I(\beta)$, $II = II(\beta)$, or, equivalently, $v_a^{-2} = f_a(\beta)$, $a = 1, 2, 3$, where v_a is the a'th principal stretch. From what was shown above, we may thus write (91.19) and (91.20) in the forms

$$\beta_{,m} B^m_k = v_1^2 \beta_{,k}. \tag{91.22}$$

---

[1] Since $III = 1$, the stretches are constant, but the local rotation may vary if no further requirement is imposed. However, the Riemann tensor based upon $\boldsymbol{B}^{-1}$ must vanish. While ERICKSEN conjectured that this condition is sufficient to force the local rotation to be constant, R. L. FOSDICK pointed out to us that the deformation given in cylindrical co-ordinates by $r = aR$, $\theta = b\Theta$, $z = cZ$, $a^2bc = 1$, has constant strain invariants and is not homogeneous if $b \neq 1$. It does not contradict the conjecture stated at the end of this section since it is included as a special case of Family 3 of Sect. 57.

Therefore

$$\left.\begin{aligned} a^3_{kp} &= I' \beta_{,p} B^m_{k,m} + (I' v_1^2 \beta_{,k})_{,p}, \\ &= I' \beta_{,p} B^m_{k,m} + (I' v_1^2)' \beta_{,p} \beta_{,k} + I' v_1^2 \beta_{,kp}, \end{aligned}\right\} \tag{91.23}$$

where the prime denotes differentiation with respect to β. In order that $\boldsymbol{a}^3$ be symmetric, then,

$$I'(\beta_{,p} B^m_{k,m} - \beta_{,k} B^m_{p,m}) = 0. \tag{91.24}$$

Thus either $I = \text{const.}$ or there exists a scalar E such that

$$B^m_{k,m} = E \beta_{,k}. \tag{91.25}$$

Parallel analysis applied to $\boldsymbol{a}^5$ shows that either $II = \text{const.}$ or there exists a scalar D such that

$$(B^{-1})^m_{k,m} = D \beta_{,k}. \tag{91.26}$$

In order that $\boldsymbol{a}^4$ be symmetric, (91.25) and (91.26) must hold also when $I = \text{const.}$ or when $II = \text{const.}$, so these conditions hold always. For $\boldsymbol{a}^1$ to be symmetric, it is necessary that $B^m_{k,m}$ be the gradient of a scalar. By (91.25), then, $E = E(\beta)$, and, likewise, the symmetry of $\boldsymbol{a}^2$ requires that $D = D(\beta)$.

If we let $\boldsymbol{b}$ and $\boldsymbol{c}$ be unit proper vectors of $\boldsymbol{B}$ corresponding to the principal stretches v_2 and v_3, then, using (91.22), we may write $\boldsymbol{B}$ and $\boldsymbol{B}^{-1}$ in the forms [CFT, App., Eq. (37.12)$_3$]

$$\left.\begin{aligned} B^m_k &= v_1^2 \frac{\beta_{,k} \beta^{,m}}{\beta_{,q} \beta^{,q}} + v_2^2 b_k b^m + v_3^2 c_k c^m, \\ &= (v_1^2 - v_3^2) \frac{\beta_{,k} \beta^{,m}}{\beta_{,q} \beta^{,q}} + (v_2^2 - v_3^2) b_k b^m + v_3^2 \delta^m_k, \\ (B^{-1})^m_k &= \left(\frac{1}{v_1^2} - \frac{1}{v_3^2}\right) \frac{\beta_{,k} \beta^{,m}}{\beta_{,q} \beta^{,q}} + \left(\frac{1}{v_2^2} - \frac{1}{v_3^2}\right) b_k b^m + \frac{1}{v_3^2} \delta^m_k, \end{aligned}\right\} \tag{91.27}$$

where (91.27)$_2$ follows from (91.27)$_1$ because $\beta_{,k} \beta_{,q} \beta^{,q}$, b_k, and c_k are mutually orthogonal unit vectors. Substituting (91.27)$_2$ into (91.25) yields

$$(v_1^2 - v_3^2) \left(\frac{\beta_{,k} \beta^{,m}}{\beta_{,q} \beta^{,q}}\right)_{,m} + (v_2^2 - v_3^2)(b_k b^m)_{,m} = [E - (v_1^2)'] \beta_{,k}, \tag{91.28}$$

where we have used the facts that v_a^2 is a function of β and that $\boldsymbol{b}$ is orthogonal to grad β. Similarly, it follows from (91.27)$_3$ and (91.26) that

$$\left(\frac{1}{v_1^2} - \frac{1}{v_3^2}\right) \left(\frac{\beta_{,k} \beta^{,m}}{\beta_{,q} \beta^{,q}}\right)_{,m} + \left(\frac{1}{v_2^2} - \frac{1}{v_3^2}\right)(b_k b^m)_{,m} = \left[D - \left(\frac{1}{v_1^2}\right)'\right] \beta_{,k}. \tag{91.29}$$

Now all the scalar coefficients in (91.28) and (91.29) have already been shown to be functions of β. Hence for (91.28) and (91.29) to be compatible it is necessary that

$$\left.\begin{aligned} \left(\frac{\beta_{,k} \beta^{,m}}{\beta_{,q} \beta^{,q}}\right)_{,m} &= F(\beta) \beta_{,k} \quad \text{if} \quad v_1 \neq v_3, \\ (b_k b^m)_{,m} &= G(\beta) \beta_{,k} \quad \text{if} \quad v_2 \neq v_3. \end{aligned}\right\} \tag{91.30}$$

First let us consider the case when $v_1 \neq v_3$, so that (91.30)$_1$ must hold. Setting $a_k = \beta_{,k} (\beta_{,q} \beta^{,q})^{-\frac{1}{2}}$, we write (91.30)$_1$ in the form

$$a^m{}_{,m} a_k + a^m a_{k,m} = F(\beta) \sqrt{\beta_{,q} \beta^{,q}}\, a_k. \tag{91.31}$$

But $\boldsymbol{a}$ is a unit vector, so that $a_{k,m} a^m a^k = 0$; therefore, taking the scalar product of (91.31) by $\boldsymbol{a}$ shows that

$$a^m{}_{,m} = F(\beta) \sqrt{\beta_{,q} \beta^{,q}}, \tag{91.32}$$

whence by (91.31) we see that

$$\left.\begin{aligned} 0 &= a_{k,m} a^m, \\ &= \frac{\frac{1}{2}(\beta_{,q}\beta^{,q})(\beta_{,m}\beta^{,m})_{,k} - (\beta_{,qm}\beta^{,q}\beta^{,m})\beta_{,k}}{(\beta_{,r}\beta^{,r})^2}. \end{aligned}\right\} \tag{91.33}$$

Thus $\nabla[(\nabla\beta)^2]$ is parallel to $\nabla\beta$, whence it follows that $|\nabla\beta|$ and β are functionally dependent:

$$\sqrt{\beta_{,q}\beta^{,q}} = H(\beta), \tag{91.34}$$

say. Equation (91.32) now shows that div $\boldsymbol{a}$ is a function of β; therefore the surfaces $\beta = \text{const.}$ are surfaces of constant mean curvature. The total curvature of the surfaces $\beta = \text{const.}$ is $\frac{1}{2}[(a^m{}_{,m})^2 - a^{k,m} a_{m,k}]$; on the basis of the foregoing equations it is easy to show that this expression, too, is a function of β. Since the surfaces $\beta = \text{const.}$ are thus surfaces of constant mean and total curvature, they are necessarily planes, right circular cylinders, or spheres.

Now

$$a_{k,m} = \left(\frac{\beta_{,k}}{H}\right)_{,m} = \frac{\beta_{,km}}{H} - \beta_{,k}\beta_{,m}\frac{H}{H^2} = a_{m,k}. \tag{91.35}$$

This condition is sufficient that the surfaces $\beta = \text{const.}$ possess a normal congruence, which must satisfy the differential system

$$\frac{dx^k}{ds} = a^k. \tag{91.36}$$

Letting D/ds denote the intrinsic derivative with respect to arc length s, from (91.36) we have

$$\frac{D}{ds}\left(\frac{dx^k}{ds}\right) = a^k{}_{,m}\frac{dx^m}{ds} = a^k{}_{,m} a^m = 0, \tag{91.37}$$

where $(91.33)_1$ has been used. Hence the normal congruence must consist of straight lines. Therefore, still on the assumption that $v_1 \neq v_3$, *the surfaces* $\beta = \text{const.}$ *must be parallel planes, coaxial right circular cylinders, or concentric spheres.*

We shall rest content with outlining the remainder of the argument, for details referring the reader to ERICKSEN's paper.

Case 1: All principal stretches are distinct. In this case we must satisfy $(91.30)_2$, so that

$$b_{k,m} b^m + b_k b^m{}_{,m} = G(\beta)\beta_{,k}. \tag{91.38}$$

Taking the scalar product of this equation by $\boldsymbol{b}$ and using the fact that $\boldsymbol{b}$ is a unit vector orthogonal to $\nabla\beta$ shows that

$$\left.\begin{aligned} &b^m{}_{,m} = 0, \\ &b_{k,m} b^m = (b_{k,m} - b_{m,k}) b^m = \left(\frac{\partial b_k}{\partial x^m} - \frac{\partial b_m}{\partial x^k}\right) b^m = \left(\int G\, d\beta\right)_{,k}. \end{aligned}\right\} \tag{91.39}$$

Consider first the case when the surfaces $\beta = \text{const.}$ are parallel planes. Then it is possible to introduce a rectangular Cartesian system x, y, z such that $\beta = \beta(x)$ Use of (91.39) and the fact that $\boldsymbol{b}$ is a unit vector orthogonal to $\nabla\beta$ suffices to show that

$$b_x = 0, \quad b_y = \cos\psi, \quad b_z = \sin\psi, \tag{91.40}$$

where $\psi=\psi(x)$. From (91.27) we see then that

$$\|B^{km}\|=\left\|\begin{matrix} HK-J^2 & 0 & 0 \\ \cdot & \dfrac{K}{HK-J^2} & \dfrac{-J}{HK-J^2} \\ \cdot & \cdot & \dfrac{H}{HK-J^2} \end{matrix}\right\|. \tag{91.41}$$

where H, J, and K are the following functions of x:

$$\left.\begin{aligned} H&=(v_2^{-2}-v_3^{-2})\cos^2\psi+v_3^{-2}, \\ J&=(v_2^{-2}-v_3^{-2})\cos\psi\sin\psi, \\ K&=(v_2^{-2}-v_3^{-2})\sin^2\psi+v_3^{-2}. \end{aligned}\right\} \tag{91.42}$$

It can be shown that the Riemann tensor based on $\boldsymbol{B}^{-1}$ vanishes if and only if these functions are linear. Excluding the case when they are all constant, which corresponds to homogeneous strain, we may express (91.41) in the form $(57.9)_1$. Thus this class of deformations is that already studied as Family 2 of Sect. 57.

If we start with the assumption that the surfaces $\beta=$ const. are coaxial right circular cylinders and proceed in the same way, we may show that $\boldsymbol{B}$ has either the form $(57.2)_1$ or the form $(57.14)_1$, so that this class of deformations is exhausted by those already studied as Families 1 and 3 of Sect. 57.

Proceding by parallel steps from the assumption that the surfaces $\beta=$ const. are concentric spheres quickly leads to a contradiction. Thus there are no deformations of this kind.

Case 2: $v_2=v_3$. In order that there be a non-trivial strain, it is necessary that $v_1\neq v_3$; therefore, the proof above that the surfaces $\beta=$ const. are parallel planes, coaxial right circular cylinders, or concentric spheres remains valid. Moreover, by $(91.27)_2$,

$$B_k^m=(v_1^2-v_3^2)\frac{\beta_{,k}\beta^{,m}}{\beta_{,q}\beta^{,q}}+v_3^2\delta_k^m, \tag{91.43}$$

where $v_1v_3^2=1$. It is a straightforward matter to show that when the surfaces $\beta=$ const. are parallel planes or coaxial right circular cylinders, special cases of the results already obtained for Case 1 follow. If, however, the surfaces $\beta=$ const. are concentric spheres, we may choose a spherical co-ordinate system r, θ, φ such that $\beta=\beta(r)$, and from (91.43) it follows that

$$\|B^{km}\|=\left\|\begin{matrix} \dfrac{1}{v_3^4} & 0 & 0 \\ \cdot & \dfrac{v_3^2}{r^2} & 0 \\ \cdot & \cdot & \dfrac{v_3^2}{r^2\sin^2\theta} \end{matrix}\right\|, \tag{91.44}$$

where $v_3=v_3(r)$. It turns out that the Riemann tensor based on $\boldsymbol{B}^{-1}$ vanishes if and only if $v_3=r(r^3-A)^{-\frac{1}{3}}$, where A is a constant. Substituting this result into (91.44) leads to $(57.36)_1$, showing that this class of deformations is that already studied as Family 4 of Sect. 57.

Case 3: $v_1=v_3$. For a non-trivial strain, it is necessary that $v_1\neq v_2$. None of the foregoing analysis past (91.29) is applicable here. From $(91.27)_3$ we see that

$$(B^{-1})_m^k=\left(\frac{1}{v_2^2}-v_2\right)b_k b^m+v_2\delta_k^m. \tag{91.45}$$

It is conjectured that the Riemann tensor based on $\boldsymbol{B}^{-1}$ cannot vanish unless $v_2=1$, but this result has been proved by ERICKSEN only subject to the hypothesis that the field $\boldsymbol{b}$ is complex-lamellar, and the proof is not simple. We may summarize ERICKSEN'S analysis as follows: *Beyond homogeneous strains and the four families of deformations discussed in Sect. 57, the only deformations that can be maintained by the action of surface tractions alone in every homogeneous, isotropic, incompressible hyperelastic material are:*

1. Deformations in which the principal stretches are constant and both $B^m_{k,mp}$ and $(B^{-1})^m_{k,mp}$ are symmetric in k and p but do not vanish.
2. Deformations of the form (91.45) with $\boldsymbol{b}\cdot\operatorname{curl}\boldsymbol{b}\neq 0$.

Until recently it was guessed that there are no deformations of either kind. One deformation[1] satisfying the former condition is now known, but it is one included in Case 1. Some suspicion has thus been raised against the conjecture, still widely held, that homogeneous strain and the four families studied in Sect. 57 exhaust the deformations possible, subject to surface tractions alone, in arbitrary isotropic, incompressible, elastic materials. [Note added in proof. A further deformation included in Case 1 has been adduced by KLINGBEIL and SHIELD and by SYNGH and PIPKIN to show that the conjecture is untrue.]

92. Special properties of some exact solutions for isotropic materials. There is, at present, no solution known to be valid for all *hyperelastic* materials having certain symmetries that is not known to be valid, with at most minor modifications, for all *elastic* materials having the same symmetries. For this reason we have given in Sects. 54—59 as complete a presentation as we can today for the exact solutions, avoiding any use of the strain-energy function even in cases when it played a prominent if unnecessary part in the original papers cited. When we survey the extensive body of results developed there, we find no case in which the additional assumption that the material is hyperelastic, not merely elastic, leads to a crucial difference. All the universal relations, for example, perforce retain their forms unchanged, nor have we been able to find any new universal relation that holds for all forms of the strain-energy function of a hyperelastic material, but fails to hold for an elastic material that is not also a hyperelastic material. Of course, various complicated differential relations among various stress components in any given solution result from applying the general conditions (86.12), (86.13), or (86.14), but relations of this kind are difficult to interpret or apply. In short, we may say that *at present, on the basis of solutions of special problems, no straightforward test for the existence of a strain-energy function is known.*

For isotropic incompressible materials, however, it is possible to draw some more definite conclusions about certain solutions when the material is hyperelastic. We refer to the general solutions developed in Sect. 57.

Family 1 (Bending, stretching, and shearing of a rectangular block)[2]. If we substitute (86.17) into (57.3), we see that

$$\left.\begin{aligned} T\langle rr\rangle &= \varrho\int\left(\frac{\partial\bar\sigma}{\partial I}\frac{dI}{dr}+\frac{\partial\bar\sigma}{\partial II}\frac{dII}{dr}\right)dr,\\ &=\varrho\bar\sigma(I, II)+\text{const.}\end{aligned}\right\}\tag{92.1}$$

[1] In the example given in footnote 1, p. 338, $\|B^m_{k,mp}\|$ and $\|(B^{-1})^m_{k,mp}\|$ are diagonal matrices proportional to $1-b^2$.

[2] RIVLIN [1949, *15*, § 4 (footnote)] [1949, *16*, § 16], in somewhat lesser generality.

Accordingly, the condition (57.5) that the cylindrical faces $r=r_1$ and $r=r_2$ be rendered free of traction is

$$\bar{\sigma}(I, II)|_{r=r_1}=\bar{\sigma}(I, II)|_{r=r_2}\equiv\sigma_0. \tag{92.2}$$

To satisfy this condition it is sufficient, and for a general form of $\bar{\sigma}$ also necessary, that

$$I|_{r=r_1}=I|_{r=r_2}, \quad II|_{r=r_1}=II|_{r=r_2}. \tag{92.3}$$

By $(57.2)_{4,7}$ these conditions are satisfied if and only if

$$C=0, \quad r_1^2 r_2^2=\frac{A^2}{B^2}. \tag{92.4}$$

Thus there can be no transverse shear, and if r_1 and r_2 are given, the ratio A^2/B^2 is determined, or, if we prefer, for given r_1, A, and B the outer radius r_2 is determined. In view of the theorem stated just before (57.7), we may infer that *for isotropic hyperelastic materials it is always possible to effect bending of a rectangular block into annular form by applying terminal couples only; the curved faces of the block are then altogether free of traction, and the resultant tractions on the plane ends also vanish.* This property, however, is not known to characterize hyperelastic materials; the result (92.1) enables the theorem to be proved, but it has not been shown that no other form for $T\langle rr\rangle$ will suffice.

When (92.4) is satisfied, $(56.10)_3$ yields the following expression for the bending moment:

$$M=\tfrac{1}{2}\varrho(r_2^2-r_1^2)\sigma_0-\varrho\int_{r_1}^{r_2} r\bar{\sigma}(I, II)\,dr. \tag{92.5}$$

From (57.2) we see that when there is no transverse shearing, so that $C=0$, the neutral fibre $r=r_0$ is determined from the condition $B\langle\theta\theta\rangle=1$, viz, $r_0^2=1/B^2$. Thus (92.4) yields $A^2 r_0^2=r_1^2 r_2^2$. In the case of pure bending, we have $A^2 B^2=1$, or, by (92.4), $A^2=r_1 r_2$, so that the neutral fibre is given by the relation

$$r_0=\sqrt{r_1 r_2}, \tag{92.6}$$

just as in the classical infinitesimal theory.

Family 2 (Straightening, stretching, and shearing a sector of a hollow cylinder). Only when $C=0$, so that there is no shearing of the axial planes, does the solution for the deformation (57.8) simplify much. In that case, if we write $\bar{\sigma}(I, II)\equiv\hat{\sigma}(x)\equiv\check{\sigma}(R)$, from $(57.11)_2$ and $(57.9)_{3,4}$ we see that

$$T\langle yy\rangle-T\langle xx\rangle=-2\varrho x\hat{\sigma}'(x)=-\varrho R\check{\sigma}'(R). \tag{92.7}$$

Thus the condition (57.12) that the planes $\Theta=$const., including the plane ends of the block, be free of traction takes the form

$$\int_{R_1}^{R_2} R^2\check{\sigma}'(R)\,dR=\int_{x_1}^{x_2} x\hat{\sigma}'(x)\,dx=0, \tag{92.8}$$

but conditions on $\bar{\sigma}$ making it possible to determine A and B such as to satisfy this equation are not known.

Family 3 (Inflation, bending, torsion, extension, and shearing of an annular wedge). In the solution (57.15) corresponding to the family of deformations (57.13) no noteworthy simplification results, in general, when a strain-energy function exists. Only in the special case of pure torsion is there a special formula worth recording; namely, if we write $\hat{\sigma}(r) \equiv \bar{\sigma}(I, II)$, then putting (86.17) into $(57.19)_2$ yields[1]

$$\left.\begin{aligned} T &= \frac{2\pi}{D} \int_{r_1}^{r_2} r^2 \hat{\sigma}'(r)\, dr, \\ &= \frac{2\pi}{D} \left[r^2 \hat{\sigma}(r) \Big|_{r_1}^{r_2} - \int_{r_1^2}^{r_2^2} \hat{\sigma}(r)\, d(r^2) \right]. \end{aligned}\right\} \tag{92.9}$$

Family 4 (Inflation or eversion of a spherical sector). The existence of a strain-energy function does not appear to simplify any of the results corresponding to the deformation (57.35).

In summary, the only property of the exact solutions that may distinguish hyperelastic materials from more general elastic ones is the fact that *a rectangular block of incompressible hyperelastic material may always be bent into the form of a circular cylindrical annulus by the application of terminal couples while its ends are left free of resultant force,* but even this property is not a crucial one, for, insofar as is presently known, it may belong equally to certain elastic materials that are not hyperelastic. On the basis of the exact solutions, *no relation apt for experimental test for the existence of a strain-energy function has yet been found.*

Here we remark also a special property of the shear modulus of a hyperelastic material[2]. Namely, from (54.9) and (86.10) we see that in simple shear

$$\hat{\mu}(K^2) = 2\varrho \left(\frac{\partial \bar{\sigma}}{\partial I_{\boldsymbol{B}}} + \frac{\partial \bar{\sigma}}{\partial II_{\boldsymbol{B}}} \right)_{I_{\boldsymbol{B}} = II_{\boldsymbol{B}} = 3 + K^2,\ III_{\boldsymbol{B}} = 1} = 2\varrho \frac{d\bar{\sigma}^*}{dK^2}, \tag{92.10}$$

where $\bar{\sigma}^*(K^2) \equiv \bar{\sigma}(3 + K^2, 3 + K^2, 1)$. In particular, at a shear such that $\hat{\mu}(K^2) = 0$, the stored-energy function $\bar{\sigma}^*$ experiences an extreme, which may be a maximum or a point of inflection.

93. Second-order effects in isotropic materials. Let us assume that the strain-energy function is given by a power series in the principal invariants of the classical strain tensor $\boldsymbol{E} \equiv \frac{1}{2}(\boldsymbol{C} - \boldsymbol{1})$, as is possible in view of (86.7):

$$\left.\begin{aligned} \frac{\varrho_R \bar{\sigma}(\boldsymbol{B})}{\mu} \equiv \frac{\varrho_R \hat{\sigma}(\boldsymbol{E})}{\mu} &= \left(\frac{1}{2}\alpha_1 + 1\right) I_{\boldsymbol{E}}^2 - 2 II_{\boldsymbol{E}} + \\ &\quad + \beta_1 I_{\boldsymbol{E}}^3 + \beta_2 I_{\boldsymbol{E}} II_{\boldsymbol{E}} + \beta_3 III_{\boldsymbol{E}} + \cdots, \end{aligned}\right\} \tag{93.1}$$

where all terms of order 3 or less in the principal extensions have been written out explicitly, and where the reference configuration is assumed to be a natural state. We shall now calculate the Cauchy stress tensor $\boldsymbol{T}$ to the same order in the extensions. First, by the remark after (84.14) we have the following expression for the second Piola-Kirchhoff stress tensor $\tilde{\boldsymbol{T}}$:

$$\left.\begin{aligned} \frac{\tilde{\boldsymbol{T}}}{\mu} &= \alpha_1 I_{\boldsymbol{E}} \boldsymbol{1} + 2\boldsymbol{E} + 3\beta_1 I_{\tilde{\boldsymbol{E}}}^2 \boldsymbol{1} + \\ &\quad + \beta_2 II_{\tilde{\boldsymbol{E}}} \boldsymbol{1} + \beta_2 I_{\tilde{\boldsymbol{E}}} (I_{\tilde{\boldsymbol{E}}} \boldsymbol{1} - \tilde{\boldsymbol{E}}) + \beta_3 (II_{\tilde{\boldsymbol{E}}} \boldsymbol{1} - I_{\tilde{\boldsymbol{E}}} \tilde{\boldsymbol{E}} + \tilde{\boldsymbol{E}}^2) + \cdots, \end{aligned}\right\} \tag{93.2}$$

where we have used (9.11) and the fact that $\boldsymbol{E} = \tilde{\boldsymbol{E}}$, the classical infinitesimal strain tensor, to within an error which is of second order with respect to the principal

[1] RIVLIN [1949, *18*].
[2] TRUESDELL [1965, *36*].

extensions. By (44.23) and (43 A.11)$_9$ we see that

$$\left.\begin{aligned}\frac{\boldsymbol{T}}{\mu} &= \frac{\varrho}{\varrho_R}(\mathbf{1}+\boldsymbol{H})\frac{\tilde{\boldsymbol{T}}}{\mu}(\mathbf{1}+\boldsymbol{H}^T),\\ &=(1-I_{\tilde{\boldsymbol{E}}})(\mathbf{1}+\boldsymbol{H})(\alpha_1 I_{\boldsymbol{E}}\mathbf{1}+2\boldsymbol{E})(\mathbf{1}+\boldsymbol{H}^T)+\\ &\quad+[(3\beta_1+\beta_2)I_{\tilde{\boldsymbol{E}}}^2+(\beta_2+\beta_3)II_{\tilde{\boldsymbol{E}}}]\mathbf{1}-(\beta_2+\beta_3)I_{\tilde{\boldsymbol{E}}}\tilde{\boldsymbol{E}}+\beta_3\tilde{\boldsymbol{E}}^2+\cdots,\\ &=\alpha_1 I_{\boldsymbol{E}}\mathbf{1}+2\tilde{\boldsymbol{E}}+[(-\alpha_1+3\beta_1+\beta_2)I_{\tilde{\boldsymbol{E}}}^2+(\beta_2+\beta_3)II_{\tilde{\boldsymbol{E}}}]\mathbf{1}+\\ &\quad+(2\alpha_1-2-\beta_2-\beta_3)I_{\tilde{\boldsymbol{E}}}\tilde{\boldsymbol{E}}+(4+\beta_3)\tilde{\boldsymbol{E}}^2+\boldsymbol{H}\boldsymbol{H}^T\cdots.\end{aligned}\right\}\tag{93.3}$$

Comparing this result with the second-order stress relation (66.3) for general elastic materials, we conclude that

$$\left.\begin{aligned}\alpha_3&=-\alpha_1+3\beta_1+\beta_2, & \alpha_4&=\beta_2+\beta_3,\\ \alpha_5&=2\alpha_1-2-\beta_2-\beta_3, & \alpha_6&=4+\beta_3.\end{aligned}\right\}\tag{93.4}$$

Hence[1]

$$\mu(\alpha_4+\alpha_5)=2(\lambda-\mu),\tag{93.5}$$

and, conversely, the second-order theory of elasticity defined by the stress relation (66.3) reduces to second-order hyperelasticity if (93.5) is satisfied.

From these results we confirm the well known fact that *in the infinitesimal theory, every isotropic elastic material is hyperelastic,* and we see also that in the second-order theory this result no longer holds, for *in order that an isotropic elastic material be hyperelastic to the second order, it is necessary and sufficient that the four second-order elasticities be related to the two first-order elasticities through Eq.* (93.5). In the second-order theory for isotropic hyperelastic materials there are at most 5 independent elasticities instead of the 6 of the more general theory[2].

If we now look back at the known second-order solutions given in Sects. 66—67, we are hard pressed to find a quality by which the special case of hyperelasticity may be distinguished. For example, in the solution for torsion and extension of a cylinder, the two coefficients α_4 and α_5 indeed appear in the formulae (66.15) and (66.21) for the stresses and displacements, respectively, but in the resulting expressions (66.28), (66.27), (66.29), (66.32), and (66.39) for the overall extension, normal force, and moment, the coefficient α_5 does not appear at all. Hence we must conclude that *measurements of the overall Kelvin and Poynting effects in the torsion and extension of an arbitrary cylinder do not offer any possibility of test for the existence of a strain-energy function, to within terms of the second order in the extensions.* Effects of this kind, however, suffice to determine α_4 and α_6. E.g., on the assumption that μ and σ are known, we need only measure the change of length and volume in torsion of a cylinder with free ends and then use Eqs. (66.28) and (66.32). If the change of torsional rigidity effected by small extension is measured, we can then determine α_5 by Eq. (66.26). We could express the criterion (93.5) by solving the three equations just mentioned, but the result is complicated and hard to interpret.

Moreover, since the difference between elastic and hyperelastic materials depends upon the value of α_5, and since α_5 multiplies $I_{\tilde{\boldsymbol{E}}}\tilde{\boldsymbol{E}}$ in the stress relation (66.3), we see that in a deformation for which $I_{\tilde{\boldsymbol{E}}}=0$, this term vanishes, so the nature of α_5 cannot manifest itself. That is, *in deformations in which the first-*

[1] Sheng [1955, *24,* § 7].

[2] The same result may be inferred in a somewhat different way from the work theorems in Sect. 83.

order change of volume is zero, it is impossible to distinguish isotropic hyperelastic materials from isotropic elastic ones by second-order effects[1].

The only simple formula so far discovered as an explicit criterion for the existence of a strain-energy function in the second-order theory of isotropic materials occurs in connection with the speeds of principal waves[2].

The general conditions of compatibility have been given in Sect. 77. Supposing these be satisfied, we can form the quantity $\alpha_4+\alpha_5$ from $(77.17)_2$ and $(77.19)_2$, so converting (93.5) into a condition on the wave speeds. It is easier, however, to proceed a little differently. From (77.2) and (77.14) we see that

$$\left.\begin{aligned} \frac{\varrho_R}{\mu}\frac{\partial U_{12}^2}{\partial \delta_1} &= 3+\frac{1}{2}\alpha_5+\frac{1}{2}\alpha_6, \\ \frac{\varrho_R}{\mu}\left(\frac{\partial}{\partial \delta_1}-\frac{\partial}{\partial \delta_2}\right)U_{11}^2 &= \frac{2\lambda}{\mu}+4-\alpha_4+\alpha_5+2\alpha_6. \end{aligned}\right\} \tag{93.6}$$

Using these formulae, we may express (93.5) in the following form

$$\left(\frac{\partial}{\partial \delta_1}-\frac{\partial}{\partial \delta_2}\right)U_{11}^2-4\frac{\partial U_{12}^2}{\partial \delta_1}+6U_{\perp}^2=0, \tag{93.7}$$

which is *a necessary and sufficient condition that the second-order speeds of principal waves, assumed compatible with the theory of isotropic elastic materials, be possible for a material with a strain-energy function.* In this formula it is assumed that δ_1, δ_2, and δ_3 vary independently.

For isotropic incompressible materials, the strain-energy function is given to within terms of third order in the principal extensions by the expression[3]

$$\frac{\varrho\bar{\sigma}(\boldsymbol{B})}{\mu}=\frac{1}{2}\left(\frac{1}{2}+\beta\right)(I-3)+\frac{1}{2}\left(\frac{1}{2}-\beta\right)(II-3)+\cdots. \tag{93.8}$$

By (86.18) we see that in the second-order theory the response coefficients $\beth_1$ and $\beth_{-1}$ are given by (67.1). Thus *every isotropic incompressible elastic material is also hyperelastic to the second order.* Therefore, in incompressible isotropic materials there are no second-order effects by which the existence of a strain-energy function could be detected.

[Added in proof. By (66.54) and (93.5), we see that an elastic material is hyperelastic in the second-order theory if and only if, for every deformation subject to no body force and no surface traction, the second-order change of volume is given by a formula first found by ZENER[4]:

$$\left.\begin{aligned} V-V_R=&-\left(1+\frac{\varkappa_{\mathrm{app}}^{*\prime}(1)}{\varkappa}\right)\int_{\mathscr{B}}\frac{1}{2}I_{\tilde{\mathbf{E}}}^2\,dv_R- \\ &-\frac{\mu+\mu_{\mathrm{app}}^{*\prime}(1)}{\varkappa}\int_{\mathscr{B}}\left[\frac{1}{2}\left(I_{\boldsymbol{H}\boldsymbol{H}^T}-I_{\boldsymbol{H}^2}\right)-\frac{1}{3}I_{\tilde{\mathbf{E}}}^2\right]dv_R. \end{aligned}\right\} \tag{93.9)]}$$

[1] This statement seems to us clearer than the corresponding work theorem of SHENG [1955, *24*, § 7].

[2] TRUESDELL [1961, *63*, § 12] [1964, *88*]. GREEN [1963, *26*, § 3] derived (93.7) directly from $(90.13)_2$.

[3] Cf. CFT, Sect. 40. The observation is due to RIVLIN and SAUNDERS [1951, *14*, § 21].

[4] ZENER [1942, *8*] inferred this result by arguments we do not follow. A correct but intricate proof of a generalization to solids of the cubic system was given by TOUPIN and RIVLIN [1960, *56*, Eq. (6.17)]. The simple proof above, showing that ZENER's formula is a *criterion* for the existence of a stored-energy function, is due to WANG [1965, *44*].

Putting $I_{\tilde{\mathbf{E}}}=0$ in (93.9) yields a formula which might seem to contradict the fourth of the italicized statements following (93.5), but the identity (66.56) reduces the result to (66.55) and renders it empty as a criterion.

d) Special or approximate theories of hyperelasticity.

94. The nature of special or approximate theories. A great deal of the older literature of finite elasticity is filled by derivation of special or approximate stress relations[1]. Many physicists and engineers seem to be drawn to approximate rather from a revulsion against precise treatment than from any real gain. Truly, there are limits of precision in measurement, and a theory that presumes infinite accuracy may labor complications which are in fact illusory. In practice, however, it may be mathematically simpler and more economical to avoid unnecessary approximations, as indeed is the case so far in elasticity, where, except for one particular special theory to be discussed in the next section, the "approximations" have yielded little or nothing that is not known more generally and simply, either from the exact solutions and general theorems, or from the results of systematic procedures such as SIGNORINI'S expansion or the theory of infinitesimal deformation from an arbitrary configuration. The typical contribution of a paper proposing a special theory has been to discover a special case of some general property of elastic materials, such as the existence of Kelvin and Poynting effects, or some universal relation, and by comparison with rough measurements to conclude that the special theory in question is established by experiment. From the exact results we have taken pains to prove and explain in Sects. 54—62, the reverse conclusion follows, for we have proved such effects *typical* of all non-linear theories of elasticity, so that they neither establish nor controvert any particular one.

Judged by their practical contribution to our understanding of the behavior of elastic materials, the special or approximate theories do not deserve to be described here. Moreover, a fairly complete survey, comparison, and criticism of such theories proposed up to 1953 has been given by TRUESDELL[2], and the subsequent literature seems to be largely repetitive. Therefore we limit ourselves here to some general remarks.

There have been three approaches. (I) The result of a perturbation procedure at a certain stage, perhaps in disguise, is taken as definitive and made the basis of further study. This approach has the advantage of being generally correct to a certain order and in addition yielding results that are physically possible to a higher order. (II) The stress relation is supposed to be exactly linear[3], or exactly

[1] Some of these theories concern materials elastic but not hyperelastic; the unimportance of the whole subject renders the distinction unimportant *a fortiori*.

[2] TRUESDELL [1952, *20*, §§ 48—54] [1953, *22*]. We add here the remark that the first proposal of a special three-dimensional theory of finite elastic strain was made by PIOLA [1836, *1*, Eq. (152)]: on the basis of a molecular hypothesis, $\tilde{\boldsymbol{T}} \propto \boldsymbol{C}^{-1}$.

[3] Various authors, seeing the futility of expecting nature to prefer one strain measure to another for all materials, have suggested that we should select the strain measure afresh for each material in order to get a simple form of constitutive equation. E.g., GROSSMAN [1961, *22*, § 4] has laid down the following axiom: "The relation of stress to strain is quasi-linear, provided suitable measures of stress and strain are chosen." Indeed, with no restriction at all, we can say still more: *Every* invertible stress relation

$$\boldsymbol{T} = \boldsymbol{f}(\boldsymbol{B})$$

for an isotropic elastic material is linear, trivially, in an appropriately defined, particular strain measure. Indeed, let μ be a constant bearing the dimension of stress (e.g., we can take $\mu \equiv \operatorname{tr}[\boldsymbol{f}(\alpha \boldsymbol{1})]$ for some convenient number α), and set

$$\boldsymbol{M} \equiv \frac{1}{\mu} \boldsymbol{f}(\boldsymbol{B}).$$

Then $\boldsymbol{M}$, being a dimensionless, uniquely invertible, symmetric tensor-valued, isotropic function of $\boldsymbol{B}$, is a spatial strain measure (cf. CFT, Sect. 32). The original stress relation may now be written in the form

$$\boldsymbol{T} = \mu \boldsymbol{M},$$

indeed linear, in the special measure $\boldsymbol{M}$.

quadratic, or of some other special functional form, in some particular measure of strain. (III) The stored-energy function is supposed to have a particularly simple form in terms of a particular measure of strain. Neither of these latter approaches has any support, either experimental or theoretical, in the facts of elastic response. There are infinitely many correct ways of measuring strain, and in the general theory any of these may be adopted, and the results of different choices are equivalent [cf. CFT, Sect. 32]. A relation which is specially simple in terms of one strain measure, however, loses its simplicity when expressed in terms of a different one. There is no basis in experiment or logic for supposing nature prefers one strain measure to another. However, such theories if correctly formulated show how a material *might* behave. Thus they may be useful in establishing the possible existence of certain phenomena or in serving as counter-examples. Also, for certain special theories the conditions, generally transcendental equations, that arise but remain somewhat indefinite in the general solutions (Sects. 54, 57, 62, 92) may sometimes be solved explicitly or at least proved to have a solution of a certain kind.

The two most frequently encountered particular theories of compressible materials will be defined here for reference. In the theory of St. Venant and Kirchhoff[1] the strain-energy function is assumed to have just the same form as in the classical infinitesimal theory, except that the tensor $\tilde{\boldsymbol{E}}$ of infinitesimal strain, linear in the displacement gradient, is replaced by the quadratic Green-St. Venant tensor $\boldsymbol{E} \equiv \frac{1}{2}(\boldsymbol{C}-\boldsymbol{1})$. Thus for isotropic materials

$$\varrho_R\,\hat{\sigma}(\boldsymbol{E}) = \tfrac{1}{2}(\lambda+2\mu)\,I_{\boldsymbol{E}}^2 - 2\mu\, II_{\boldsymbol{E}}. \tag{94.1}$$

By (84.14), then,

$$\tilde{\boldsymbol{T}} = \varrho_R\,\hat{\sigma}_{\boldsymbol{E}}(\boldsymbol{E}) = \lambda I_{\boldsymbol{E}}\boldsymbol{1} + 2\mu\,\boldsymbol{E}, \tag{94.2}$$

so that by $(43\text{A}.11)_9$

$$\left.\begin{aligned} \boldsymbol{T} &= \frac{\varrho}{\varrho_R}\boldsymbol{F}(\lambda I_{\boldsymbol{E}}\boldsymbol{1} + 2\mu\boldsymbol{E})\boldsymbol{F}^T, \\ &= \frac{1}{\sqrt{III_{\boldsymbol{B}}}}\left\{-\mu II_{\boldsymbol{B}}\boldsymbol{1} + \left[\frac{1}{2}\lambda(I_{\boldsymbol{B}}-3) + \mu(I_{\boldsymbol{B}}-1)\right]\boldsymbol{B} + \mu III_{\boldsymbol{B}}\boldsymbol{B}^{-1}\right\}. \end{aligned}\right\} \tag{94.3}$$

A number of analyses of special stability problems and certain theories of plates have been based upon this stress relation or approximations to it.

Signorini[2] argued for a theory in which $\boldsymbol{T}$ turns out to be exactly equal to a quadratic function of the Almansi-Hamel strain tensor $\boldsymbol{A}$, given by the definition $\boldsymbol{A} \equiv \frac{1}{2}(\boldsymbol{1}-\boldsymbol{B}^{-1})$. By use of (85.16), he showed that the most general strain-energy function leading to such a stress relation is

$$\varrho_R\,\breve{\sigma}(\boldsymbol{A}) = \frac{(\mu+\frac{1}{2}c)(1-I_{\boldsymbol{A}}) + \frac{1}{2}(\lambda+\mu-\frac{1}{2}c)I_{\boldsymbol{A}}^2 + c\,II_{\boldsymbol{A}}}{\sqrt{1-2I_{\boldsymbol{A}}+4II_{\boldsymbol{A}}-8III_{\boldsymbol{A}}}}, \tag{94.4}$$

so that

$$\left.\begin{aligned} \boldsymbol{T} = {} & [\lambda I_{\boldsymbol{A}} + \tfrac{1}{2}(\lambda+\mu-\tfrac{1}{2}c)I_{\boldsymbol{A}}^2 + c\,II_{\boldsymbol{A}}]\boldsymbol{1} + \\ & + 2[\mu - (\lambda+\mu+\tfrac{1}{2}c)I_{\boldsymbol{A}}]\boldsymbol{A} - 2c\boldsymbol{A}^2. \end{aligned}\right\} \tag{94.5}$$

The stress relation is much simplified when $c=0$, but the results of this special theory are then rather untypical of elastic response[3]; nevertheless, this particular stored-energy function has been given much attention by the Italian school, especially in regard to the dependence of various coefficients upon temperature and to inequalities to be imposed upon the temperature-dependent coefficients in order to obtain physically plausible conclusions.

A different stored-energy function was proposed by Bordoni[4]. While Signorini seemed to prefer algebraic simplicity, and Bordoni sought to accomodate certain facts of physical

[1] Kirchhoff [1852, *1*, p. 770]. The theory had been described earlier in words by St. Venant [1844, *1*] [1847, *1*, § 2], but neither his attempt [1863, *1*, § 2] to put it in mathematical form nor that of Boussinesq [1870, *1*], which he later endorsed [1871, *1*, ¶ 7], is entirely successful.

[2] Signorini [1942, *6*, pp. 67—68] [1945, *4*, pp. 164—167] [1949, *21*, Cap. II], and numerous other publications; Tolotti [1942, *7*].

[3] Cf. the specific criticism of Truesdell [1952, *20*, § 53].

[4] Bordoni [1953, *3*—*5*]. A survey of the Italian work has been written by Grioli [1962, *24*, Ch. III, §§ 4—7].

experience, JOHN[1] proposed still another stored-energy function because he was able to base upon it a fairly extensive analytical theory.

These three theories are distinguished by scrupulous mathematical treatment. The more abundant older literature, here left uncited, has been summarized by RIVLIN[2]: "The course that was adopted by many workers was to postulate some form or other for the strain-energy function on grounds of simplicity and to proceed on that basis. Unfortunately, the views of different workers on what constituted the simplest form for the strain energy did not coincide. I reflect here on the remark made by Mephistopheles in Goethe's *Faust* ..."

"Denn eben wo Begriffe fehlen,
Da stellt ein Wort zur rechten Zeit sich ein."

"The word in this case is *simplest.*" While much of this work, growing from laboratories of engineering or applied mechanics, is usually labelled "physical" and put into direct contact with experiments, the "physical" laws of elasticity appear to differ from one laboratory to another.

A new source of approximate theories of finite elasticity is furnished by solid-state physicists concerned with "dislocations". While intensely occupied with the geometrical structure to be assigned to the body manifold, workers in this area seem to lack a corresponding appreciation of the nature and use of constitutive equations. If the infinitesimal theory of elasticity, due account being taken of the fact that for non-homogeneous bodies the infinitesimal strain tensor $\tilde{\boldsymbol{E}}$ cannot be derived from a global displacement field from a natural state, may serve well enough for infinitesimal deformations, which are the main concern of most writers in this field, for finite deformations an arbitrarily selected stored-energy function is scarcely more appropriate to non-homogeneous bodies than to homogeneous ones. Nevertheless, the few papers[3] that concern finite deformation seem to rest on adopting (94.1), evidently from its formal similarity to the stored-energy function for infinitesimal deformations. We remark again that there are infinitely many strain tensors other than $\boldsymbol{E}$ which reduce to the classical infinitesimal strain tensor for small strains and rotations, and that use of any one of them will lead to a theory different from the one based on (94.1) but neither worse nor better motivated. The conceptual problem is in no way eased by use of stress functions [CFT, Sects. 224—230], popular among writers on dislocations. By means of stress functions, the equations of equilibrium are solved, for any continuum, in a manner independent of the geometry of the reference configuration, but it is the constitutive equations that determine the partial differential equations to be satisfied by those functions. As yet, nothing beyond the statement of the differential equations (44.7) seems to be known about the dislocations compatible with elasticity or hyperelasticity as distinguished from the very special elastic materials characterized by (94.1). In addition, it is hard to get a clear idea of what is a reasonable problem in the theory of dislocations.

95. The Mooney-Rivlin theory for rubber. We have seen at the beginning of Sect. 67 that the most general stress relation for isotropic incompressible elastic materials, to within terms of third or higher order in the principal extensions $v_a - 1$, is obtained when the response functions $\beth_1$ and $\beth_{-1}$ are taken as constants. A stress relation of this kind, viz

$$\boldsymbol{T} = -p\boldsymbol{1} + \mu\left(\tfrac{1}{2} + \beta\right)\boldsymbol{B} - \mu\left(\tfrac{1}{2} - \beta\right)\boldsymbol{B}^{-1}, \tag{95.1}$$

where μ and β are constants, is a possible one even if no approximations are made; it defines the Mooney-Rivlin theory for rubber[4]. From $(86.18)_2$ we see that this

---

[1] JOHN [1960, *30*]. JOHN's theory has been applied to problems of instability and to the theory of pre-stressed plates and rings by SENSENIG [1963, *63*] [1964, *75*].

[2] RIVLIN [1959, *25*].

[3] SEEGER and MANN [1959, *26*], KRÖNER and SEEGER [1959, *17*, § 3], KRÖNER [1960, *35*, §§ 10—14], PFLEIDERER, SEEGER, and KRÖNER [1960, *40*].

[4] The considerations of MOONEY [1940, *4*, Eq. (14) and pp. 586—587] were somewhat different from those in the text, due to RIVLIN and SAUNDERS [1951, *14*, § *21*]. However, it must not be thought that the argument given in favor of (95.2) as an approximation is compelling. If instead of the $v_a - 1$ we choose the $v_a^2 - 1$ as our three expansion parameters, we obtain as first approximation the neo-Hookean theory of RIVLIN, mentioned below, and as second approximation a three-constant theory proposed and studied by SIGNORINI [1955, *25*, Cap. III] and MANACORDA [1956, *17*] [1959, *19*, §§ 7—8]. As remarked by TRUESDELL [1952, *20*, § 54, footnote 10], the neo-Hookean theory results also by taking the lowest-order terms in an

elastic material is hyperelastic, its strain-energy function being

$$\varrho\bar{\sigma}(\boldsymbol{B})=\tfrac{1}{2}\mu\left[\left(\tfrac{1}{2}+\beta\right)\left(I_{\boldsymbol{B}}-3\right)+\left(\tfrac{1}{2}-\beta\right)\left(II_{\boldsymbol{B}}-3\right)\right]. \tag{95.2}$$

The experimental results cited at the end of Sect. 57 show that predictions from this theory stand in fair agreement with measurements made on deformation of various rubbers through very great strains, far greater than could be expected on the basis of the considerations of order of magnitude used to suggest (95.1). While no physical material seems to be described to within errors of experiment by this theory, it gives a good rough picture of the elastic response of most kinds of rubber. As we have seen, it is a theory which suggests itself as a natural possibility, and the mathematical predictions from it are extraordinarily simple and easy to derive. Accordingly, it has been studied very much in recent years; it is used as the first illustration for every general result concerning isotropic incompressible substances, and many special solutions have been found for it which would be hard to extend to more general materials.

In the special case when $\beta=\frac{1}{2}$, the theory is called *neo-Hookean*[1]. The corresponding form of strain energy has been derived from various theories of long-chain molecules[2]. In the neo-Hookean theory certain typical effects of non-linear elasticity are altogether wanting, and it is not adequate for rubber[3] except, perhaps, in extremely slow deformations, so it will not be considered further here.

In order that $\bar{\sigma}(\boldsymbol{B})$ as given by (95.2) be positive for all $\boldsymbol{B}$ it is necessary and sufficient that

$$\mu>0, \qquad -\tfrac{1}{2}\leqq\beta\leqq+\tfrac{1}{2}, \tag{95.3}$$

and these inequalities in turn are easily seen to be equivalent to the B-E inequalities (51.22) [cf. (53.1)] and to the E-inequalities (52.3). Consequently the

expansion of the stored-energy function in powers of $\boldsymbol{E}$; however, the second approximation is not the theory of SIGNORINI.

The foregoing remarks should remind the reader of the speciousness of arguments based upon perturbation methods. The reasons that the Mooney-Rivlin theory is given attention here are two: (1) its mathematical simplicity, illustrated by numerous cases in which a definite answer, capable of experimental confirmation or disproof, emerges for problems too difficult to solve for more general materials, and (2) the agreement, if crude and sometimes merely qualitative, with data from experiments.

[1] RIVLIN [1948, *11*, § 9], KUBO [1948, *6*].

[2] There is a prolix literature, some of which was cited by TRUESDELL [1952, *20*, § 54] [1953, *25*, § 54]. An exposition of the molecular theories was given by TRELOAR [1958, *49*]. ISHIHARA, HASHITSUME, and TATIBANA [1951, *4*] have put forward a more general kinetic theory. As remarked, in effect, by TRUESDELL [1953, *25*, p. 603] and ZAHORSKI [1959, *41*], their result is compatible with the special continuum theory of SIGNORINI [1955, *25*, Cap. III]. A more elaborate statistical theory has been proposed by WANG and GUTH [1952, *23*].

This point is our closest approach to contact with molecular theories of materials, except for remarks on polymer solutions in Sect. 119 and on the kinetic theory of gases in Sect. 125. Our purpose in this treatise is to construct theories that represent accurately and in generality observed behavior common to many materials. Molecular theories would fall within our scope only to the extent that they are successful toward this end. Instead, the progress of application of molecular theories to facts of experiment has been just backward from it: Experimenters have had to search high and low to find materials or circumstances in which the predictions from molecular models do not fail too grossly of the truth. If physicists and chemists really placed on the direct evidence of experiments the supreme value they claim to, or their "philosophers of science" claim on their behalf, they would long ago have discarded molecular theories except for monatomic gases and for phenomena where the individual particle, rather than the gross sample, plays an undisputed part. It is the obstinate *belief* in molecules, though ordinarily undiscerned, that makes physicists and chemists wastrels of effort, time, numbers, and money in their truly religious desire to "explain" everything in nature by statistical theories based on molecular hypotheses.

[3] Among many experiments demonstrating this fact may be cited those of RIVLIN [1947, *9*] and of GUMBRILL, MULLINS, and RIVLIN [1953, *10*].

speeds of all principal waves are real, as may be seen in several different ways from the results in Sect. 78. The formulae from the second-order theory of waves given at the end of that section are not exact according to the Mooney-Rivlin theory, however; in fact, they result from it only by linearization. An exact treatment of waves within the Mooney-Rivlin theory[1] yields results which are typical of the general theory, as follows from the appropriate specialization of (78.12). In this as in other problems the Mooney-Rivlin theory gives formulae which are fully general for isotropic incompressible elastic materials as far as terms of the second order in the extensions and are physically possible, though not general, for extensions of any magnitude.

The difference is made particularly clear by considering, instead of singular surfaces or infinitesimal oscillations as above, the response to disturbances of finite amplitude. For a Mooney-Rivlin material, (74.22) reduces *exactly* to the linear wave equation in which the speed, a constant, is just the same as that of transverse principal waves for the same state of strain. For any given V_1 and V_2, then, the speed is independent of the magnitude or nature of the function $u(x, t)$. That is, *in a Mooney-Rivlin material subject to homogeneous strain, all disturbances parallel to a given transverse principal axis are propagated at a common speed and unchanged in form.* Perhaps this is a characterizing property of the Mooney-Rivlin material; in any case, the possibility of waves of permanent form is certainly unexpected in a theory of finite deformation.

For given principal stresses t_a and given p, it is an easy consequence of (95.3) that the relations (95.1) determine three unique real principal stretches v_a satisfying the condition $v_a \geqq 0$; these stretches are monotone increasing functions of p. Since $III = v_1^2 v_2^2 v_3^2$, this invariant is also such a function. Consequently there is one and only one real value of p such that $III_{\mathbf{B}} = 1$. It follows that the stress tensor $\boldsymbol{T}$ determines a unique real positive-definite left Cauchy-Green tensor $\boldsymbol{B}$ such that $III_{\mathbf{B}} = 1$. That is, *the stress relation of the Mooney-Rivlin theory is uniquely invertible for strains of any magnitude* [2].

From $(54.9)_1$ we see that $\hat{\mu}(K^2) = \mu$; that is, the generalized shear modulus of the Mooney-Rivlin material is the constant denoted by μ in (95.1). Accordingly, the shear stress is proportional to the amount of shear for shears of all magnitudes[3]. The normal-stress moduli (54.15) are also constants, but the constant β is not easy to characterize except in terms of a special deformation. Consider, for example, the case of pure torsion with twist D. By $(57.19)_1$, the torque is strictly proportional to the twist, for twists of any amount[4]. From (57.18) we see that if

---

[1] ERICKSEN [1953, *6*, § 5] treated exhaustively both principal and non-principal waves. Infinitesimal oscillations of various kinds have been studied by HAYES and RIVLIN [1961, *25*] and by FLAVIN [1962, *19*] [1963, *23*].

[2] RIVLIN [1948, *13*], the neo-Hookean material having been treated in [1948, *11*, §§ 5—6]. RIVLIN [1948, *11*A] obtained further relations of invertibility and uniqueness for the neo-Hookean material.

[3] This fact, while it has played a large part in drawing toward the neo-Hookean and Mooney-Rivlin theories the attention of workers primarily interested in the allegedly physical side of the subject, is really of no meaning, since it refers to a particular measure of shear. In "HOOKE's law in shear", the phrase used in the literature of experiment and kinetic theory, it is immaterial what measure of shear be selected, so long as the shear is small, but for large shears linearity is obviously not invariant under change of strain measure. E.g., if we select the angle of shear, θ, rather than its amount, $K = \tan\theta$, then for large shears the Mooney-Rivlin theory does *not* confirm "HOOKE's law in shear".

[4] RIVLIN [1947, *9*, Eq. $(8)_1$]. More generally, from $(57.19)_1$ we may read off the following theorem: In an isotropic, incompressible, elastic material, the torque required to effect simple torsion of a circular cylinder is proportional to the twist if and only if the shear stress required to effect simple shear is proportional to the amount of shear. As suggested by the remarks in the preceding footnote, this observation does not have much physical content, since it merely connects, in regard to the response of isotropic, incompressible, elastic materials, a particular measure of finite torsion with a particular measure of finite shear.

the radial stress $T\langle rr\rangle$ vanishes on the outer cylinder $r=r_2$, then the normal traction on the plane ends is given by[1]

$$-\frac{T\langle zz\rangle}{\mu D^2}=\frac{1}{2}\left(\frac{1}{2}+\beta\right)(r_2^2-r^2)+\left(\frac{1}{2}-\beta\right)r^2. \tag{95.4}$$

While in the general theory we have seen in Sect. 57 that the resultant normal force is always a pressure if the E-inequalities are adopted, in the Mooney-Rivlin theory these same inequalities [in the form (95.3)] force the normal traction itself to be everywhere a pressure. The tension at the outer edge, $r=r_2$, is $-\mu D^2(\frac{1}{2}-\beta)r_2^2$; this value serves for interpretation of the coefficient β, since it is a monotone decreasing function of β, vanishing only in the case of a neo-Hookean material, when $\beta=\frac{1}{2}$.

All of the exact solutions in Sects. 55 and 57 are simplified considerably for the Mooney-Rivlin theory. Leaving it to the reader to notice the effect of setting $\beth_I=\mu(\frac{1}{2}+\beta)$, $\beth_{-I}=-\mu(\frac{1}{2}-\beta)$ in the formulae already derived for general values of $\beth_\Gamma$, we turn instead to a sequence of problems which cannot be solved with comparable explicitness for the general theory.

Problem 1. Extension resulting from twisting of a solid circular cylinder (Poynting's problem)[2]. To calculate the stretch F that results from a twist of amount D/F applied to a solid circular cylinder free of resultant normal force on the plane ends, we have only to set $N=0$ in $(57.21)_2$. Since $\beth_\Gamma=$ const, we obtain at once the following quartic equation for F:

$$\left(F-\frac{1}{F^2}\right)\frac{(\frac{1}{2}+\beta)F+\frac{1}{2}-\beta}{(\frac{1}{2}+\beta)F+1-2\beta}=\frac{1}{4}\left(\frac{D}{F}\right)^2 R_2^2. \tag{95.5}$$

When the inequality $(95.3)_2$ is satisfied, it is possible to show that (95.5) for each value of $(D/F)R_2$ has exactly one real root F_0, and that that root is greater than 1 if $D\neq 0$. That is, the twisted cylinder always elongates, but only for small twists is the elongation proportional to the square of the twist. Cf. (57.24), which yields a formula for determining β from the elongation $F-1$ resulting from a small twist of amount D/F:

$$\beta=\frac{3}{2}-\frac{12}{R_2^2}\lim_{D\to 0}\frac{F-1}{(D/F)^2}. \tag{95.6}$$

If we let $f(D^2R_2^2/F^2)$ be the positive root of (95.5), then by $(57.21)_1$ we get the following relation between the torque and the twist:

$$T=\frac{1}{2}\pi R_2^4\mu\left[\frac{1}{2}+\beta+\frac{\frac{1}{2}-\beta}{f(D^2R_2^2/F^2)}\right]\frac{D}{F}. \tag{95.7}$$

While for a fixed F, as already remarked, the torque is strictly proportional to the twist for all twists, from (95.7) we see that when F is adjusted so as to leave the transverse sections free of resultant force, such a simple proportion holds only in the case of the neo-Hookean material, for which $\beta=\frac{1}{2}$. In general, by $(95.3)_2$, since f is a strictly increasing function, the quantity in brackets is a decreasing function of D^2/F^2. That is, *a cylinder with free ends softens in torsion.*

Problem 2. Torsion or eversion of a cylindrical tube[3]. In (57.13), set $C=1$, $E=0$, $AF=1$. In the condition that the inner and outer surfaces $R=R_1$ and

[1] This formula was obtained by RIVLIN [1947, *9*] and verified by him, with the experimental value $\beta=0.38$, in his classical experiment on the torsion of a rubber cylinder. His later and more accurate work, as mentioned at the end of Sect. 57, has shown that (95.4) is only rather roughly verified. RIVLIN's earlien work on torsion [1948, *11* B], which concerned the neo-Hookean material, was published later.

[2] An exhaustive treatment of this problem, according to the slightly more general theory of SIGNORINI, was given by MANACORDA [1955, *18*].

[3] RIVLIN [1949, *16*, §§ 7—8]. Cf. also the discussion of a special case by KROUPA [1955, *17*, § 4].

$R = R_2$ be free of traction, which follows at once from $(57.15)_1$, we may evaluate the quadratures and so obtain the following equation relating B, D, and F:

$$\log \frac{1 + \frac{BF}{R_2^2}}{1 + \frac{BF}{R_1^2}} + \frac{1}{1 + \frac{R_1^2}{BF}} - \frac{1}{1 + \frac{R_2^2}{BF}} = \frac{\left(\frac{1}{2} + \beta\right)(R_2^2 - R_1^2) F^2}{\frac{1}{2} + \beta + F^2\left(\frac{1}{2} - \beta\right)} \left(\frac{D}{F}\right)^2. \quad (95.8)$$

Two particular cases are of interest. First, the tube is turned inside out but not twisted. Then $D = 0$. A solution such that $A < 0$, and hence $F < 0$, is desired. Eq. (95.8) can be shown to have exactly one negative root BF for every given pair R_1, R_2, that root being independent of the constants μ and β. The condition (57.26) for the vanishing of the total force on the transverse sections can likewise be simplified; after use of (95.8), it becomes

$$\left.\begin{aligned} \frac{\left[\left(\frac{1}{2} + \beta\right) F + \left(\frac{1}{2} - \beta\right)\right](F^3 - 1)}{\left[\frac{1}{2} + \beta - \left(\frac{1}{2} - \beta\right) F^2\right] F} &= -\frac{1}{2} \cdot \frac{1}{\left(1 + \frac{R_1^2}{BF}\right)\left(1 + \frac{R_2^2}{BF}\right)}, \\ &\equiv H, \end{aligned}\right\} \quad (95.9)$$

say. For given R_2 and R_1, with BF being the negative root of (95.8), H is a known and positive quantity, and (95.9) is a quartic to be solved for F, the stretch produced by the eversion. In the special case when $\beta = \frac{1}{2}$, (95.9) reduces to a cubic having as its single real root the number $\sqrt[3]{|1 + H|}$, which is greater than 1. Thus, when $\beta = \frac{1}{2}$, eversion results in lengthening. Perhaps lengthening is to be expected for all values of β satisfying $(95.3)_2$, but the nature of the roots of (95.9) seems not to have been studied.

Second, the lengthening resulting from an assigned twist is sought. The twist is D/F, and now a positive value of B is desired. The equation expressing the vanishing of the normal force on the cross-sections is complicated and is not written down here. RIVLIN has determined the character of the roots BF of (95.8) when the right-hand side is given a fixed value.

Problem 3. Extension, inflation, and shear of a cylindrical tube[1]. Since $\beth_\Gamma =$ const, (59.7) may be integrated at once to yield

$$\left.\begin{aligned} g &= \frac{K}{\frac{1}{2} + \beta + \frac{1}{A^2}\left(\frac{1}{2} - \beta\right)} \log \frac{R}{(AR^2 + B)^{\frac{1}{2}}} + \text{const}, \\ h &= \frac{L}{\frac{1}{2} + \beta + A\left(\frac{1}{2} - \beta\right)} \log \{[\tfrac{1}{2} + \beta + A(\tfrac{1}{2} - \beta)] R^2 + B(\tfrac{1}{2} - \beta)\} + \text{const} \end{aligned}\right\} \quad (95.10)$$

if $B \neq 0$, while if $B = 0$

$$\left.\begin{aligned} g &= \frac{K}{\frac{1}{2} + \beta + \frac{1}{A^2}\left(\frac{1}{2} - \beta\right)} \cdot \frac{1}{R^2} + \text{const}, \\ h &= \frac{L}{\frac{1}{2} + \beta + A\left(\frac{1}{2} - \beta\right)} \log R + \text{const}, \end{aligned}\right\} \quad (95.11)$$

where K and L are arbitrary constants. The various surface tractions that must be applied in order to maintain the deformation are gotten by easy calculations based on (56.8)—(56.10).

[1] RIVLIN [1949, *16*, §§ 9—13]. Cf. also GREEN and ZERNA [1954, *11*, §§ 3.7—3.8].

Problem 4. Eversion of a spherical shell[1]. The quadratures in (57.39) can be effected explicitly. If we set

$$\left.\begin{aligned} g(A,R) \equiv & \left(\tfrac{1}{2}+\beta\right)(A-R^3)^{-\frac{4}{3}}(5R^3-4A)R \\ & -2\left(\tfrac{1}{2}-\beta\right)(A-R^3)^{-\frac{2}{3}}(2A-R^3)/R, \end{aligned}\right\} \tag{95.12}$$

then (57.39) is equivalent to the following algebraic equation to determine A:

$$g(A,R_2)=g(A,R_1). \tag{95.13}$$

For each pair of values R_2, R_1, there exists exactly one root A such that $A>R_2^3$, where $R_2>R_1$. Thus exactly one everted configuration is possible for any given spherical shell.

Problem 5. Radial oscillations of a cylindrical tube[2]. From (62.6) we readily find that

$$F(x,\gamma)=\frac{\mu}{2\varrho R_1^2}(1-x^2)\log\frac{1+\gamma/x^2}{1+\gamma}, \tag{95.14}$$

while (62.11) yields

$$W(u)=\frac{\mu}{\varrho R^2}\left(u+\frac{1}{u}-2\right). \tag{95.15}$$

Thus it becomes possible to solve (62.13) explicitly for the dimensionless amplitudes a_0 and b_0 of the thin shell:

$$\left.\begin{aligned} a_0=\Bigg[&\frac{\dot{x}_0^2\varrho R^2}{2\mu}+\frac{x_0^2}{2}+\frac{1}{2x_0^2}, \\ &-\frac{1}{2}\left(\frac{\dot{x}_0^4\varrho^2R^4}{\mu^2}+x_0^4+\frac{1}{x_0^4}+\frac{2x_0^2\dot{x}_0^2\varrho R^2}{\mu}+\frac{2\dot{x}_0^2\varrho R^2}{\mu x_0^2}-2\right)^{\frac{1}{2}}\Bigg]^{\frac{1}{2}}, \end{aligned}\right\} \tag{95.16}$$

while $b_0=1/a_0$. Evaluating the integral (62.14) now yields

$$T=\pi R\sqrt{\frac{\varrho}{\mu}}. \tag{95.17}$$

Thus for the limiting case of the thin shell the period is independent of the amplitude of vibration.

The foregoing results refer to free oscillations. Turning to motion resulting from a pressure impulse, we put (95.14) into (62.17):

$$\frac{\mu}{\varrho R_1^2}\log\frac{1+\gamma}{1+\gamma/x^2}=p. \tag{95.18}$$

The left-hand side is a monotone increasing function of x when $\gamma>0$; it increases from the value $-\infty$ when $x=0+$ to the value $\frac{\mu}{\varrho R_1^2}\log(1+\gamma)$ when $x=+\infty$. Therefore there exists a single positive root x if $p<\frac{\mu}{\varrho R_1^2}\log(1+\gamma)$; that is, if

$$P_1-P_2<\mu\log\frac{R_2}{R_1}. \tag{95.19}$$

Accordingly, from the more general results in Sect. 62, periodic motion results if (95.19) is satisfied, and the period is easily calculated from (62.18). If (95.19) is violated, no periodic motion is possible. It may be verified also that if (95.19) is not satisfied, no static solution for the given pressure difference is possible. Further details of the dynamic solution have been given by KNOWLES.

---

[1] ERICKSEN [1955, *10*, § 3]. A detailed study of inflation of a spherical shell was given by GREEN and ZERNA [1954, *11*, § 3.10].

[2] KNOWLES [1960, *34*, § 6] [1962, *42*].

Likewise, more specific results concerning the radial oscillation of a spherical shell may be obtained for the Mooney-Rivlin material[1].

Various other special problems have been solved for the Mooney-Rivlin theory[2].

III. Various Generalizations of Elasticity and Hyperelasticity.

96. Thermo-elasticity. It has been mentioned above that the form of the function $\mathfrak{g}$ in the constitutive equation (43.1) may depend upon the temperature or the entropy. Since generally no account has been taken of such dependence, most of our specific results are to be thought appropriate to circumstances when either θ or η is kept constant, or to materials for which the response function is insensitive to changes in θ or η. A relation between isentropic and isothermal elasticities has been calculated, and some of its consequences have been determined (Sects. 45, 71, 82, 90). In Sects. 79–81 we have presented thermodynamic arguments leading to the equations of hyperelasticity in various conditions. The heat flux $\boldsymbol{h}$ has remained unspecified.

Thermo-elasticity seeks to describe the effects of non-uniform temperature fields upon deformation and the effects of deformation upon temperature. Necessarily it rests upon a constitutive equation for the heat flux as well as one for the stress. Only very recently have such equations been proposed.

Traditionally the heat flux $\boldsymbol{h}$ is associated with a temperature gradient, grad θ (cf. the discussion of Fourier's law in Sect. 296 of CFT). It is only to be expected that the energy transferred in response to a given temperature gradient will depend upon the specific entropy η and the deformation gradient $\boldsymbol{F}$:

$$\boldsymbol{h} = \hat{\boldsymbol{h}}(\boldsymbol{F}, \eta, \operatorname{grad} \theta). \tag{96.1}$$

Thermo-elasticity may be based on (96.1) and the constitutive equations (80.1)–(80.3) for the energy density, the temperature and the stress[3]. In many circumstances in many real materials the dissipative effects of viscosity and heat conduction seem to be of about equal importance[4]. It seems reasonable, therefore,

[1] Guo Zhong-Heng and Solecki [1963, *33*].

[2] Adkins [1954, *1*, §§ 4–8] integrated a number of the equations derived at the end of Sect. 59 and gave alternative formulations in terms of functions of a complex variable. Among his examples are certain deformations of the regions contained between confocal elliptic cylinders or between eccentric circular cylinders. For a Mooney-Rivlin material the quantity ה_1 in $(60.11)_6$ is a constant; hence it cancels from (60.21), which becomes a partial differential equation devoid of material constants. A number of solutions of this equation have been found by Adkins [1955, *3*], who later [1958, *1*, §§ 5–6] established the connection of these results with his reciprocal theorem, proved in Sect. 60. Kroupa [1955, *17*, § 3] remarked that in plane strain the Mooney-Rivlin theory is indistinguishable from the neo-Hookean theory, as follows at once by putting $\lambda = 1$, $\beth_1 = \mu(\frac{1}{2} + \beta)$ and $\beth_{-1} = -\mu(\frac{1}{2} - \beta)$ in $(60.11)_6$, since then $\text{ה}_1 = -\mu$; more specifically, *in problems of plane strain for the Mooney-Rivlin theory, the form of the constitutive equation for the plane portion of the stress is independent of* β.

[3] A constitutive equation of the form (96.1), but without rendering explicit the dependence on η, was proposed by Pipkin and Rivlin [1958, *35*, § 17]; they obtained the reduced form (96.3). We shall discuss their illustrations below. The thermo-energetic theory we present generalizes that proposed by Green and Adkins [1960, *26*, §§ 8.5–8.8]; they employed the temperature, θ, rather than the specific entropy, η, as an independent variable. Later Green [1962, *23*] started from slightly different assumptions; in particular, not assuming a constitutive equation for $\boldsymbol{h}$, he effected perturbation of the energy equation directly.

A treatment of thermo-elasticity of finite deformation along traditional lines, without using a dissipation inequality, has been given by Sedov [1962, *60*, §§ 14–15].

[4] In the special case of a fluid, the theory of Green and Adkins reduces to the theory of a piezotropic fluid that conducts heat according to Fourier's law, with density-dependent thermal conductivity, but without viscosity. This theory has long been known to be inadequate to describe the propagation of sound in gases, where the dissipative effects of viscosity and heat conduction are about equally great. Cf., e.g., Lamb [1932, *5*, §§ 359–360].

to generalize the constitutive equation (80.3) for the stress slightly by putting

$$\boldsymbol{T}=\mathfrak{g}(\boldsymbol{F},\eta)+\mathsf{V}(\boldsymbol{F},\eta)[\boldsymbol{L}], \tag{96.2}$$

where $\mathsf{V}(\boldsymbol{F},\eta)[\boldsymbol{L}]$, a linear tensor function of the velocity gradient $\boldsymbol{L}$, accounts for linear viscosity. When the dependence on η is ignored, (96.2) becomes the constitutive equation of a linearly viscous material as defined in Sect. 41.

The function $\hat{\boldsymbol{h}}$ in (96.1) cannot be entirely arbitrary. First, it must be such as to satisfy the Clausius-Duhem inequality in the form $(79.11)_2$: Heat never flows against a temperature gradient. Therefore, if we define a scalar function $f(\boldsymbol{u})$ as follows:

$$f(\boldsymbol{u})=\boldsymbol{u}\cdot\hat{\boldsymbol{h}}(\boldsymbol{F},\eta,\boldsymbol{u}), \tag{96.3}$$

then $f(\boldsymbol{u})\geqq 0$ for each fixed $\boldsymbol{F}$ and all $\boldsymbol{u}$. Since $f(\boldsymbol{0})=0$, f must have a minimum at $\boldsymbol{u}=\boldsymbol{0}$. Consequently, if $\hat{\boldsymbol{h}}$ is differentiable with respect to $\boldsymbol{u}$ at $\boldsymbol{u}=\boldsymbol{0}$, it is necessary that $f_{\boldsymbol{u}}=0$ there. But $f_{\boldsymbol{u}}=\boldsymbol{h}+(\boldsymbol{h}_{\boldsymbol{u}})^T\boldsymbol{u}$. Hence

$$\hat{\boldsymbol{h}}(\boldsymbol{F},\eta,\boldsymbol{0})=\boldsymbol{0}: \tag{96.4}$$

When the temperature gradient vanishes, so does the heat flux. PIPKIN and RIVLIN[1], who first obtained (96.4), described it as showing the "non-existence of a piezo-caloric effect". That is, deformation alone cannot give rise to a flow of heat.

Second, the principle of material frame-indifference, extended so as to include thermo-energetic variables, allows us to obtain reduced forms corresponding to (96.1):

$$\left.\begin{aligned}\boldsymbol{h}&=\boldsymbol{R}\hat{\boldsymbol{h}}(\boldsymbol{U},\eta,\boldsymbol{R}^T\operatorname{grad}\theta),\\ &=\boldsymbol{F}\tilde{\boldsymbol{h}}(\boldsymbol{C},\eta,\boldsymbol{F}^T\operatorname{grad}\theta),\end{aligned}\right\} \tag{96.5}$$

where $\boldsymbol{R}$ is the rotation tensor, $\boldsymbol{U}$ is the right stretch tensor, and $\boldsymbol{C}$ is the right Cauchy-Green tensor, while the functions $\hat{\boldsymbol{h}}$ and $\tilde{\boldsymbol{h}}$ are related by the identity

$$\hat{\boldsymbol{h}}(\boldsymbol{U},\eta,\boldsymbol{u})=\boldsymbol{U}\tilde{\boldsymbol{h}}(\boldsymbol{U}^2,\eta,\boldsymbol{U}\boldsymbol{u}). \tag{96.6}$$

Note that

$$\boldsymbol{F}^T\operatorname{grad}\theta=\nabla\theta \tag{96.7}$$

is the gradient of θ with respect to the place $\boldsymbol{X}$ in the reference configuration.

Reduced forms of the constitutive equations (80.1) and (80.2) are

$$\varepsilon=\hat{\varepsilon}(\boldsymbol{U},\eta)=\bar{\varepsilon}(\boldsymbol{C},\eta), \tag{96.8}$$

$$\theta=\hat{\theta}(\boldsymbol{U},\eta)=\bar{\theta}(\boldsymbol{C},\eta). \tag{96.9}$$

A reduced form for (96.2) may be read off from (41.2):

$$\boldsymbol{R}^T\boldsymbol{T}\boldsymbol{R}=\mathfrak{g}(\boldsymbol{U},\eta)+\mathsf{V}(\boldsymbol{U},\eta)[\boldsymbol{R}^T\boldsymbol{D}\boldsymbol{R}]. \tag{96.10}$$

Another reduced form of (96.2) is

$$\tilde{\boldsymbol{T}}=\mathfrak{t}(\boldsymbol{C},\eta)+\tilde{\mathsf{V}}(\boldsymbol{C},\eta)[\boldsymbol{F}^T\boldsymbol{D}\boldsymbol{F}], \tag{96.11}$$

where $\tilde{\boldsymbol{T}}$ is the second Piola-Kirchhoff tensor. In components,

$$T^{km}=\frac{\varrho}{\varrho_R}x^k{}_{,\alpha}x^m{}_{,\beta}[\mathfrak{t}^{\alpha\beta}(C_{\mu\nu},\eta)+\tilde{\mathsf{V}}^{\alpha\beta\gamma\delta}(C_{\mu\nu},\eta)\,x^p{}_{,\gamma}x^q{}_{,\delta}D_{pq}]. \tag{96.12}$$

---

[1] PIPKIN and RIVLIN [1958, *35*, § 18].

These furnish a full set of properly invariant reduced constitutive equations of *thermo-elasticity with linear viscosity.*

While our presentation has delivered the equations, it has done so by assuming each one separately, after some motivation. COLEMAN and NOLL[1] have given a unified theory based on the entropy inequality[2] $\Gamma \geqq 0$, where Γ is given by (79.6), generalizing the arguments leading to the temperature and stress relations given in Sect. 79 so as to apply to the present theory. They have proposed a precise logical position for the entropy inequality by adopting the following postulate[3]: *A set of constitutive equations is admissible only if the entropy production Γ is non-negative for every process compatible with these constitutive equations.* In this way, the entropy inequality appears as a requirement that constitutive equations must satisfy, and hence it has a status similar to that of the principle of material frame-indifference. In order that $\Gamma \geqq 0$ for all parts of a given body, it is necessary and sufficient that the specific entropy production γ given by (79.8) be non-negative. By (79.3), (79.9), and (79.10) we have

$$\varrho\theta\gamma = \varrho(\theta\dot{\eta} - \dot{\varepsilon}) + \operatorname{tr}(\boldsymbol{T}\boldsymbol{D}) + \frac{1}{\theta}\boldsymbol{h}\cdot\operatorname{grad}\theta. \tag{96.13}$$

We now assume constitutive equations of the forms (80.1), (80.2), (96.2), and (96.1). Consider a homogeneous body subject to time-dependent but homogeneous deformations $\boldsymbol{F}$ and entropy distributions η. For such a body, by (80.2), the temperature is time-dependent but uniform, so that $\operatorname{grad}\theta = 0$, and (96.13) becomes

$$\varrho\theta\gamma = \varrho(\theta\dot{\eta} - \dot{\varepsilon}) + \operatorname{tr}(\boldsymbol{T}\boldsymbol{D}) = \eta\theta\gamma_1. \tag{96.14}$$

An argument analogous to the one leading to (80.7) gives

$$\left.\begin{aligned}\varrho\theta\gamma = \operatorname{tr}\{[\mathfrak{g}(\boldsymbol{F},\eta) - \varrho\boldsymbol{F}\hat{\varepsilon}_{\boldsymbol{F}}(\boldsymbol{F},\eta)^T]\boldsymbol{L}\} +&\\ + (\hat{\theta}(\boldsymbol{F},\eta) - \hat{\varepsilon}_\eta(\boldsymbol{F},\eta))\dot{\eta} + \operatorname{tr}\{\boldsymbol{L}\mathsf{V}(\boldsymbol{F},\eta)[\boldsymbol{L}]\} \geqq 0.&\end{aligned}\right\} \tag{96.15}$$

This inequality must be valid for *all* functions $\boldsymbol{F} = \boldsymbol{F}(t)$ and $\eta = \eta(t)$. If the values of $\boldsymbol{F}$ and η at a particular time are given, we can still assign $\dot{\eta}$ and $\dot{\boldsymbol{F}}$, and hence also $\boldsymbol{L}$, arbitrarily. Letting $\boldsymbol{F}$ and η be arbitrary but taking $\boldsymbol{L} = \boldsymbol{0}$ shows that the temperature relation (80.11) must hold. Next, for any given $\boldsymbol{F}$ and $\boldsymbol{L}$ (96.15) must still hold when $\boldsymbol{L}$ is replaced by $\alpha\boldsymbol{L}$,

$$\left.\begin{aligned}\alpha\operatorname{tr}\{[\mathfrak{g}(\boldsymbol{F},\eta) - \varrho\boldsymbol{F}\hat{\varepsilon}_{\boldsymbol{F}}(\boldsymbol{F},\eta)^T]\boldsymbol{L}\} +&\\ + \alpha^2\operatorname{tr}\{\boldsymbol{L}\mathsf{V}(\boldsymbol{F},\eta)[\boldsymbol{L}]\} \geqq 0.&\end{aligned}\right\} \tag{96.16}$$

This polynomial inequality in α is satisfied for all α, when $\boldsymbol{F}$ and $\boldsymbol{L}$ are fixed, if and only if the coefficient of α vanishes, while the coefficient of α^2 is non-negative. The former requirement yields the stress relation (80.10), while the latter yields the inequality

$$\operatorname{tr}\{\boldsymbol{L}\mathsf{V}(\boldsymbol{F},\eta)[\boldsymbol{L}]\} \geqq 0. \tag{96.17}$$

For linearly viscous fluids the inequality (96.17) is equivalent to the statement that both the shear viscosity and the bulk viscosity must be non-negative.

A thermo-elastic material has two isotropy groups: one obtained from the constitutive equation (96.2) for the stress, as before, and one defined in an analogous

---

[1] COLEMAN and NOLL [1963, *18*].

[2] In the form used here, including the heat absorption term, the inequality itself was first given by TRUESDELL and TOUPIN, Ineq. (258.3) of CFT.

[3] The developments in this section and in Sects. 96bis and 96ter are all based on this postulate.

way from the constitutive equation (96.1) for the heat flux. A unimodular transformation $\boldsymbol{H}$ belongs to the isotropy group $\mathscr{g}$ for the heat flux if and only if

$$\hat{\boldsymbol{h}}(\boldsymbol{F}\boldsymbol{H}, \eta, \boldsymbol{u}) = \hat{\boldsymbol{h}}(\boldsymbol{F}, \eta, \boldsymbol{u}) \tag{96.18}$$

is an identity in $\boldsymbol{F}$, η, and $\boldsymbol{u}$. Replacing $\boldsymbol{F}$ by $\boldsymbol{F}\boldsymbol{Q}^T$ in (96.18) and combining the result with (96.5), we see that an orthogonal transformation $\boldsymbol{Q}$ belongs to $\mathscr{g}$ if and only if

$$\boldsymbol{Q}\hat{\boldsymbol{h}}(\boldsymbol{F}, \eta, \boldsymbol{u}) = \hat{\boldsymbol{h}}(\boldsymbol{Q}\boldsymbol{F}\boldsymbol{Q}^T, \eta, \boldsymbol{Q}\boldsymbol{u}) \tag{96.19}$$

is an identity in $\boldsymbol{F}$ and $\boldsymbol{u}$. A material is *isotropic* with respect to heat flux if the isotropy group contains the full orthogonal group. In this case, (96.19) holds for *all* orthogonal $\boldsymbol{Q}$, i.e., $\hat{\boldsymbol{h}}$ is an isotropic, vector-valued function of one scalar, one vector and one tensor. Using (96.19) with $\boldsymbol{Q} = \boldsymbol{R}$ and $\boldsymbol{F} = \boldsymbol{U}$, we find that (96.5), for isotropic materials, reduces to

$$\boldsymbol{h} = \hat{\boldsymbol{h}}(\boldsymbol{V}, \eta, \operatorname{grad}\theta), \tag{96.20}$$

or, equivalently,

$$\boldsymbol{h} = \tilde{\boldsymbol{h}}(\boldsymbol{B}, \eta, \operatorname{grad}\theta). \tag{96.21}$$

By use of the representation theorem (13.8) we may express $\tilde{\boldsymbol{h}}$ in the following reduced form[1]:

$$\left.\begin{aligned} \boldsymbol{h} &= (\daleth_0 \boldsymbol{1} + \daleth_1 \boldsymbol{B} + \daleth_2 \boldsymbol{B}^2)\operatorname{grad}\theta, \\ h_k &= (\daleth_0 \delta_k^m + \daleth_1 B_k^m + \daleth_2 B_k^r B_r^m)\,\theta_{,m}, \end{aligned}\right\} \tag{96.22}$$

where the $\daleth_\Gamma$ are scalar functions of $\boldsymbol{B}$, grad θ, and η. By the representation theorem (11.24), $\daleth_\Gamma$ is equal to a function of the following six scalar invariants in addition to η:

$$I, II, III, (\operatorname{grad}\theta)^2, \operatorname{grad}\theta\cdot\boldsymbol{B}\operatorname{grad}\theta, \operatorname{grad}\theta\cdot\boldsymbol{B}^2\operatorname{grad}\theta. \tag{96.23}$$

For isotropy two different definitions are current. According to the one, the isotropy group must contain the full orthogonal group; according to the other, only the proper orthogonal group. In the case of a simple material, there is no distinction, since if $\boldsymbol{H}\in\mathscr{g}$, then $-\boldsymbol{H}\in\mathscr{g}$. In any case where vectors or tensors of odd order are concerned, as in (96.18), this result holds no longer, and the two concepts are different. Sometimes "isotropic with a center of symmetry" or "holohedral" is used for what we have called "isotropic"; sometimes "hemitropic" or "hemihedral" is used for the broader concept. In any case, if only the proper orthogonal group is considered, $\hat{\boldsymbol{h}}$ is a hemitropic rather than an isotropic function. The representation (96.22) then has to be supplemented by additional terms[2].

---

[1] Proposed without derivation by B.-T. CHU [1957, *6*, Eq. (10.13b)], derived by PIPKIN and RIVLIN [1958, *35*, Eq. (23.9)] under the assumption that $\tilde{\boldsymbol{h}}$ is a polynomial.

[2] GREEN and ADKINS [1960, *26*, §§ 8.5—8.8], followed by KOH and ERINGEN [1963, *42*, § 5], found these terms by the device of replacing each vector $\boldsymbol{v}$ by its dual, the skew tensor with components $e^{kpq}v_q$ and then using various known representation theorems for isotropic tensor-valued functions of a set of not necessarily symmetric tensors.

ENGLAND and GREEN [1961, *12*] worked out the theory of small deformations superimposed upon large according to the theory of GREEN and ADKINS (cf. also GUO ZHONG-HENG [1962, *33*]); they considered the problems of the half space and of the penny-shaped crack in a infinite medium. FLAVIN and GREEN [1961, *16*] considered the propagation of plane infinitesimal waves in a state of homogeneous strain with two equal extension ratios; they found that one transverse mode is unaffected by the thermal terms, but the other transverse mode and the longitudinal one are modified. Further work on thermo-elastic oscillations has been done by FLAVIN [1962, *19*].

In the theory presented here, any static deformation $\boldsymbol{F}$ may be maintained by application of suitable body forces and surface tractions. The equation of energy, (79.3), reduces in static problems for which $q=0$ to

$$\operatorname{div} \boldsymbol{h}=0. \tag{96.24}$$

Substitution of (96.1) into (96.24), for a known field $\boldsymbol{F}$, thus leads to a partial differential equation for the temperature θ.

Two interesting cases for isotropic incompressible materials have been worked out by PIPKIN and RIVLIN. First[1] they considered a solid cylinder subject to torsion and extension. The deformation presumed is the special case of (57.13) when $B=0$, $C=1$, $E=0$, $A=1/F$. (We shall here denote the azimuth angle by ϑ, to avoid confusion with θ, the temperature.) As we have seen in Sect. 57, suitable forces, equipollent to a torsional couple, if applied to the plane sections $z=$ const. suffice to maintain the rod in equilibrium, and the cylinders $r=$ const. are free of traction. If a uniform temperature gradient along the axis is presumed,

$$\theta_{,r}=0, \quad \theta_{,\vartheta}=0, \quad \theta_{,z}=K, \tag{96.25}$$

substitution into (96.22) yields

$$\left.\begin{aligned} &h_r=0, \\ &h_\vartheta=K D F r^2 [\daleth_1+(F^2+F^{-1}+D^2 r^2)\daleth_2], \\ &h_z=K[\daleth_0+F^2\daleth_1+F^2(F^2+D^2 r^2)\daleth_2]. \end{aligned}\right\} \tag{96.26}$$

By (80.2), η is a function of r alone, and so are all invariants in the list (96.23). Hence (96.24) is satisfied. Since $h_r=0$, the solution is just for the case when the curved face of the rod is insulated. The mean conductivity is easily calculated from the total flux of heat through the planes $z=$ const. The lines of heat flow are helices in the cylinders $r=$ const. The pitch of these helices, for any given r, depends on both the twist, D, and the extension, F:

$$\frac{h_\vartheta}{r h_z}=D F r \frac{\daleth_1+(F^2+F^{-1}+D^2 r^2)\daleth_2}{\daleth_0+F^2[\daleth_1+(F^2+D^2 r^2)\daleth_2]}. \tag{96.27}$$

Only in the special case when the numerator vanishes (as it will if $D=0$, or for materials such that $\daleth_1\equiv\daleth_2\equiv 0$) will the heat flow be purely axial. If $\daleth_0=F\daleth_2$, the lines of heat flow are material lines under the deformation corresponding to a change in the twist, D/F, with respect to the deformed rod.

Second[2], consider a circular-cylindrical tube subject to shear. The deformation is given by (59.5) in the case when $A=1$, $B=0$; suitable surface tractions suffice to maintain it in any isotropic, incompressible elastic material, provided g and h be determined by integrating (59.7). If we assume that $\theta=\tilde{\theta}(r)$, all the invariants (96.23) are functions of r alone, and (96.22) yields

$$\left.\begin{aligned} &h_r/\tilde{\theta}'=\daleth_0+\daleth_1+(1+K^2)\daleth_2, \\ &h_\vartheta/\tilde{\theta}'=r^2 g'[\daleth_1+(2+K^2)\daleth_2], \\ &h_z/\tilde{\theta}'=h'[\daleth_1+(2+K^2)\daleth_2], \end{aligned}\right\} \tag{96.28}$$

where $K^2\equiv(r g')^2+h'^2$. The field equation (96.24) reduces to

$$[\daleth_0+\daleth_1+(1+K^2)\daleth_2]\,\tilde{\theta}'=\text{const.}; \tag{96.29}$$

since $\daleth_\Gamma$ reduces to a given function of $\tilde{\theta}'(r)$ and r. Eq. (96.29) is an implicit equation for $\tilde{\theta}'(r)$ as a function of r.

Some progress may also be made toward finding a temperature distribution $\tilde{\theta}(r,z)$ such as to yield assigned temperatures on two sections $z=$ const. when the inner and outer curved surfaces of the tube are insulated.

Cf. also the last paragraphs of Sect. 97.

The theory of thermo-elasticity presented in this section does not satisfy the ***principle of equipresence*** (CFT, Sect. 293η), according to which *a quantity*

---

[1] PIPKIN and RIVLIN [1958, *35*, § 24].

[2] PIPKIN and RIVLIN [1958, *35*, § 25].

present as an independent variable in one constitutive equation should be so present in all, unless, of course, its presence contradicts some law of physics or rule of invariance[1]. This principle forbids us to eliminate any of the "causes" present from interacting with any other as regards a particular "effect". It reflects on the scale of gross phenomena the fact that all observed effects result from a common structure, such as the motions of molecules. The separation of various "causes" and "effects" into related pairs or groups is to follow, to the extent that it holds at all, from the general laws of physics, requirements of invariance, and material symmetry. COLEMAN and MIZEL[2], in a work of unusual depth, have applied the principle to thermo-elasticity. Their starting point is the following set of four constitutive equations:

$$\left.\begin{aligned} \varepsilon &= \tilde{\varepsilon}(\theta, \operatorname{grad}\theta, \boldsymbol{F}, \dot{\boldsymbol{F}}), \\ \eta &= \tilde{\eta}(\theta, \operatorname{grad}\theta, \boldsymbol{F}, \dot{\boldsymbol{F}}), \\ \boldsymbol{T} &= \tilde{\mathfrak{g}}(\theta, \operatorname{grad}\theta, \boldsymbol{F}, \dot{\boldsymbol{F}}), \\ \boldsymbol{h} &= \tilde{\boldsymbol{h}}(\theta, \operatorname{grad}\theta, \boldsymbol{F}, \dot{\boldsymbol{F}}). \end{aligned}\right\} \tag{96.30}$$

They are able to show that the entropy inequality requires the variables grad θ and $\dot{\boldsymbol{F}}$ to drop out of $(96.30)_{1,2}$, that θ is given in terms of ε and η by the classical temperature relation, and that the equilibrium part of the stress is given by the stress relation of hyperelasticity.

---

[1] In a restricted form, the principle of equipresence was first proposed under the name "BRILLOUIN's Principle" by TRUESDELL [1949, *23*, § 19] [1951, *17*, § 19], who gave historical, molecular, and phenomenological motivations for it. In a lecture [1960, *58*] he described it as expressing "invariance of manifestation", and this term has appeared in the secondary literature, which began to take notice of the principle at about that time. The general formulation was given by TRUESDELL [1959, *37*, p. 79] and explained by TRUESDELL and TOUPIN (CFT, Sect. 293η). Among their motivating remarks are, "Let it not be thought that this principle would invalidate the classical separate theories in the cases for which they are intended, or that no separation of effects remains possible. Quite the reverse: The various principles of invariance, stated above, when brought to bear upon a general constitutive equation have the effect of *restricting* the manner in which a particular variable, such as the spin tensor or the temperature gradient, may occur. The classical separations may always be expected, in one form or another, for small changes — not as assumptions, but as *proven consequences* of invariance requirements. The principle of equipresence states, in effect, that no restrictions *beyond* those of invariance are to be imposed in constitutive equations. It may be regarded as a natural extension of OCKHAM's razor as restated by NEWTON: 'We are to admit no more causes of natural things than such as are both true and sufficient to explain their appearances, for nature is simple and affects not the pomp of superfluous causes.' This more general approach has the added value of showing in what way the classical separations fail to hold when interactions actually occur."

An account of TRUESDELL's first applications of the principle is given in Sect. 125.

[2] COLEMAN and MIZEL [1964, *18*]. Earlier, KOH and ERINGEN [1963, *42*] had proposed a theory of the same kind for hemitropic materials ("isotropic" in their terminology), but in their work thermodynamic relations were assumed separately, not demonstrated from a single inequality as in the analysis of COLEMAN and MIZEL. Using some of the representation theorems cited in Sect. 13, KOH and ERINGEN wrote down explicit forms of the constitutive equations, containing 84 response coefficients, each one of which is a function of 33 scalar invariants. Their worked-out examples concern only special cases in which no elastic phenomena occur. We are unable to follow their discussion of the Gough-Joule effect, which they claim to have explained by exhibiting a formula according to which $\boldsymbol{h}$ does not necessarily vanish when grad θ does. (Specifically, in their theory such an occurrence is possible only in a hemitropic material that is not isotropic.) The real Gough-Joule effect, namely, the heating or cooling of a body as a result of deformation, is covered by the temperature relation $\theta = \hat{\theta}(\boldsymbol{F}, \eta)$, according to which an isentropic deformation generally gives rise to a change of temperature. Of course it may occur in isotropic materials.

To outline their procedure, first write $\boldsymbol{g} \equiv \operatorname{grad} \theta$, solve $(96.30)_1$ for θ, and use ε in place of θ as independent variable:

$$\left.\begin{aligned} \theta &= \check{\theta}(\varepsilon, \boldsymbol{g}, \boldsymbol{F}, \dot{\boldsymbol{F}}), \\ \eta &= \check{\eta}(\varepsilon, \boldsymbol{g}, \boldsymbol{F}, \dot{\boldsymbol{F}}), \\ \boldsymbol{T} &= \check{\mathfrak{g}}(\varepsilon, \boldsymbol{g}, \boldsymbol{F}, \dot{\boldsymbol{F}}), \\ \boldsymbol{h} &= \check{\boldsymbol{h}}(\varepsilon, \boldsymbol{g}, \boldsymbol{F}, \dot{\boldsymbol{F}}). \end{aligned}\right\} \tag{96.31}$$

These functions are then substituted into (96.13), yielding

$$\left.\begin{aligned} \gamma = \check{\eta}_{\boldsymbol{g}} \cdot \dot{\boldsymbol{g}} + \operatorname{tr}[\check{\eta}_{\dot{\boldsymbol{F}}} \ddot{\boldsymbol{F}}^T] + \left(\check{\eta}_\varepsilon - \frac{1}{\theta}\right)\dot{\varepsilon} + \\ + \operatorname{tr}[\check{\eta}_{\boldsymbol{F}} \dot{\boldsymbol{F}}^T] + \operatorname{tr}\left[\frac{1}{\varrho\theta} \check{\mathfrak{g}} \dot{\boldsymbol{F}} \boldsymbol{F}^{-1}\right] + \frac{1}{\varrho\theta^2} \check{\boldsymbol{h}} \cdot \boldsymbol{g}. \end{aligned}\right\} \tag{96.32}$$

It can be verified that at any given place and time, it is possible to find a motion, stress, and heat flux satisfying the differential equations of momentum and energy and giving to the seven quantities ε, $\boldsymbol{g}$, $\boldsymbol{F}$, $\dot{\varepsilon}$, $\dot{\boldsymbol{g}}$, $\dot{\boldsymbol{F}}$, $\ddot{\boldsymbol{F}}$ arbitrarily assigned values. According to the Clausius-Duhem inequality, the right-hand side of (96.32) must be non-negative, identically in these seven quantities. Since $\dot{\boldsymbol{g}}$ occurs only in the first term, and since the remaining terms can be given an arbitrary value, the first term must vanish for all $\dot{\boldsymbol{g}}$:

$$\check{\eta}_{\boldsymbol{g}}(\varepsilon, \boldsymbol{g}, \boldsymbol{F}, \dot{\boldsymbol{F}}) = \boldsymbol{0}. \tag{96.33}$$

Since $\ddot{\boldsymbol{F}}$ occurs only in the second term, that term also must vanish, identically in $\ddot{\boldsymbol{F}}$:

$$\check{\eta}_{\dot{\boldsymbol{F}}}(\varepsilon, \boldsymbol{g}, \boldsymbol{F}, \dot{\boldsymbol{F}}) = \boldsymbol{0}. \tag{96.34}$$

Combining (96.33) and (96.34) shows that the relation $\eta = \check{\eta}(\varepsilon, \boldsymbol{g}, \boldsymbol{F}, \dot{\boldsymbol{F}})$ must have the form of a *caloric equation of state:*

$$\eta = \check{\eta}(\varepsilon, \boldsymbol{F}). \tag{96.35}$$

In just the same way as earlier in this section, from (96.35) and the requirement that $\gamma \geqq 0$ we derive the *temperature relation:*

$$\theta = \check{\theta}(\varepsilon, \boldsymbol{F}) = \frac{1}{\check{\eta}_\varepsilon(\varepsilon, \boldsymbol{F})}, \tag{96.36}$$

equivalent to (80.11).

Let $\boldsymbol{T}^0$ be the stress in thermal and mechanical equilibrium:

$$\boldsymbol{T}^0 = \check{\mathfrak{g}}^0(\varepsilon, \boldsymbol{F}) \equiv \check{\mathfrak{g}}(\varepsilon, \boldsymbol{0}, \boldsymbol{F}, \boldsymbol{0}), \tag{96.37}$$

and decompose the function $\check{\mathfrak{g}}$ into a sum of two functions, as follows:

$$\check{\mathfrak{g}} = \check{\mathfrak{g}}^0 + \check{\mathfrak{g}}^{\mathrm{D}}. \tag{96.38}$$

Substituting (96.38) into the entropy inequality $\gamma \geqq 0$, replacing $\dot{\boldsymbol{F}}$ by $\alpha \dot{\boldsymbol{F}}$, and letting $\alpha \to 0$, Coleman and Mizel prove that the classical stress relation must hold:

$$\check{\mathfrak{g}}^0(\varepsilon, \boldsymbol{F}) = -\varrho \check{\theta}(\varepsilon, \boldsymbol{F}) \boldsymbol{F} \check{\eta}_{\boldsymbol{F}}(\varepsilon, \boldsymbol{F})^T, \tag{96.39}$$

equivalent to (80.10) as far as $\boldsymbol{T}_0$ is concerned. The final dissipation inequality is

$$\operatorname{tr}[\boldsymbol{T}^{\mathrm{D}} \boldsymbol{D}] + \frac{1}{\theta} \boldsymbol{h} \cdot \boldsymbol{g} \geqq 0, \tag{96.40}$$

where $\boldsymbol{T}^{\mathrm{D}}=\check{\mathfrak{g}}^{\mathrm{D}}(\varepsilon,\boldsymbol{g},\boldsymbol{F},\dot{\boldsymbol{F}})$. Thus $\boldsymbol{T}^{\mathrm{D}}$ is the dissipative part of the stress. The inequality (96.40) cannot generally be split into two parts, one mechanical and one thermal, unless further assumptions are made. E.g., if $\operatorname{tr}[\boldsymbol{T}^{\mathrm{D}}\boldsymbol{D}]$ is independent of $\boldsymbol{g}$ and if $\boldsymbol{h}\cdot\boldsymbol{g}$ is independent of $\dot{\boldsymbol{F}}$, as was assumed from the outset in theory presented at the beginning of this section, then the two separate inequalities (79.11) result. In the more general theory considered here, the proof that $\boldsymbol{h}$ must vanish with grad θ no longer holds unless additional assumptions are made.

The conditions derived here are sufficient as well as necessary. That is, the assumed constitutive equations (96.31) are compatible with the principle of non-negative entropy production *if and only if* (96.35), (96.36), (96.39) and (96.40) hold.

In the usual way, it is easy to show that the principle of material indifference requires the constitutive equations (96.30), after the reductions already obtained, to be equivalent to the following set:

$$\left.\begin{aligned}\varepsilon&=\bar{\varepsilon}(\eta,\boldsymbol{C}),\\ \theta&=\bar{\theta}(\eta,\boldsymbol{C})=\bar{\varepsilon}_{\eta}(\eta,\boldsymbol{C}),\\ \boldsymbol{F}^{T}\boldsymbol{T}\boldsymbol{F}&=2\varrho\boldsymbol{C}\bar{\varepsilon}_{\boldsymbol{C}}(\eta,\boldsymbol{C})\boldsymbol{C}+\bar{\mathfrak{g}}^{\mathrm{D}}(\eta,\boldsymbol{F}^{T}\operatorname{grad}\theta,\boldsymbol{C},\boldsymbol{F}^{T}\boldsymbol{D}\boldsymbol{F}),\\ \boldsymbol{h}&=\boldsymbol{F}\bar{\boldsymbol{h}}(\eta,\boldsymbol{F}^{T}\operatorname{grad}\theta,\boldsymbol{C},\boldsymbol{F}^{T}\boldsymbol{D}\boldsymbol{F}),\end{aligned}\right\}\tag{96.41}$$

generalizing (96.5) and (96.11).

All general principles have now been applied. The classical separation of effects, with temperature gradient "causing" flow of heat and deformation "causing" stress, does *not* hold.

Such a separation can be found by imposing requirements of material symmetry and by considering motions and temperature gradients that are in some sense small. COLEMAN and MIZEL have obtained results of this kind for the case when all dependence on $\boldsymbol{C}$ in (96.41) reduces to dependence on the specific volume, v, as is appropriate for a fluid. In this case

$$\left.\begin{aligned}\varepsilon&=\bar{\varepsilon}(\eta,v),\\ \theta&=\bar{\varepsilon}_{\eta}(\eta,v),\\ \boldsymbol{T}&=-\bar{\varepsilon}_{v}(\eta,v)\,\mathbf{1}+\bar{\mathfrak{g}}^{\mathrm{D}}(\eta,\operatorname{grad}\theta,v,\boldsymbol{D}),\\ \boldsymbol{h}&=\bar{\boldsymbol{h}}(\eta,\operatorname{grad}\theta,v,\boldsymbol{D}),\end{aligned}\right\}\tag{96.42}$$

where $\bar{\mathfrak{g}}^{\mathrm{D}}$ and $\bar{\boldsymbol{h}}$ satisfy the identities

$$\left.\begin{aligned}\boldsymbol{Q}\bar{\mathfrak{g}}^{\mathrm{D}}(\eta,\operatorname{grad}\theta,v,\boldsymbol{D})\boldsymbol{Q}^{T}&=\bar{\mathfrak{g}}^{\mathrm{D}}(\eta,\boldsymbol{Q}\operatorname{grad}\theta,v,\boldsymbol{Q}\boldsymbol{D}\boldsymbol{Q}^{T}),\\ \boldsymbol{Q}\bar{\boldsymbol{h}}(\eta,\operatorname{grad}\theta,v,\boldsymbol{D})&=\bar{\boldsymbol{h}}(\eta,\boldsymbol{Q}\operatorname{grad}\theta,v,\boldsymbol{Q}\boldsymbol{D}\boldsymbol{Q}^{T}).\end{aligned}\right\}\tag{96.43}$$

Putting $\boldsymbol{Q}=-\mathbf{1}$ shows that $\bar{\mathfrak{g}}^{\mathrm{D}}$ is even and $\bar{\boldsymbol{h}}$ is odd in grad θ:

$$\left.\begin{aligned}\bar{\mathfrak{g}}^{\mathrm{D}}(\eta,-\operatorname{grad}\theta,v,\boldsymbol{D})&=\bar{\mathfrak{g}}^{\mathrm{D}}(\eta,\operatorname{grad}\theta,v,\boldsymbol{D}).\\ \bar{\boldsymbol{h}}(\eta,-\operatorname{grad}\theta,v,\boldsymbol{D})&=-\bar{\boldsymbol{h}}(\eta,\operatorname{grad}\theta,v,\boldsymbol{D}),\end{aligned}\right\}\tag{96.44}$$

In particular,

$$\bar{\boldsymbol{h}}(\eta,\mathbf{0},v,\boldsymbol{D})=\mathbf{0}:\tag{96.45}$$

The heat flux must vanish with grad θ. Notice that here the result follows from a requirement of material symmetry, specifically, the assumption that the central inversion $-\mathbf{1}$ belongs to the isotropy group of $\bar{\boldsymbol{h}}$; the thermodynamic argument used at the beginning of this section suffices here to prove only that $\bar{\boldsymbol{h}}(\eta,\mathbf{0},v,\mathbf{0})=\mathbf{0}$.

COLEMAN and MIZEL then set up two systems of expansion. The terms of first order obtained in the two cases are as follows:

$$\left.\begin{aligned} \boldsymbol{T}^{\mathrm{D}} &= \lambda(\operatorname{tr}\boldsymbol{D})\mathbf{1} + 2\mu\boldsymbol{D},\\ \boldsymbol{h} &= \begin{cases} \varkappa \operatorname{grad}\theta, \\ \varkappa \operatorname{grad}\theta + \beta_1 \boldsymbol{D} \operatorname{grad}\theta + \beta_2(\operatorname{tr}\boldsymbol{D}) \operatorname{grad}\theta. \end{cases} \end{aligned}\right\} \tag{96.46}$$

In the former case, the error vanishes with $\sqrt{\operatorname{tr}\boldsymbol{D}^2 + (\operatorname{grad}\theta)^2}$; in the latter, with $\sqrt{\operatorname{tr}\boldsymbol{D}^2}$ and $|\operatorname{grad}\theta|$ jointly. In either case, the separation of stretching as the sole "cause" of dissipative stress results, to the first order, from the assumption of isotropy[1], as is clearly to be expected in view of $(96.44)_1$. Separation of temperature gradient as the sole "cause" of heat flux results with one measure of smallness but not with another[2]. COLEMAN and MIZEL determined also the second approximations according to both these schemes.

Beginning with the treatment of elasticity itself in Sects. 79—81, we have outlined theories, each more general than its predecessor, for representing interaction of thermal and mechanical phenomena. These theories are based upon the following principles:

1. Balance of linear momentum, moment of momentum, and energy.
2. The Clausius-Duhem entropy inequality.
3. The principle of material frame-indifference.
4. The principle of equipresence.
5. The principle of fading memory.
6. Invariance of material symmetry, if any.

We have set up for consideration *constitutive equations* of increasing generality; the six basic principles have been *neither "approximated" nor augmented*[3] in their application to each case. That these same principles serve for proof of definite results concerning still more general constitutive equations, we shall see in the next section.

96bis. COLEMAN'S general thermodynamics of simple materials. (Added in proof.) While this treatise was in press, COLEMAN[4] constructed the thermodynamics of irreversible deformation processes at the level of generality of the simple material obeying the principle of fading memory. In logical order this work belongs in a chapter of its own, near the beginning, and it renders either obsolete or very special all previous work on thermo-elasticity and much on hyperelasticity, but at this time the best we can do is outline it as an appendix.

COLEMAN adopts as basic principles no more than the balance of linear momentum (16.6), the symmetry of the stress $\boldsymbol{T}$ as an expression of the balance of

---

[1] Results of this sort were first obtained by TRUESDELL [1949, *23*] [1951, *17*], working with a more general thermomechanical theory of fluids, which we outline in Sect. 125.

[2] Cf. the work on pure heat conduction as possibly associated with higher gradients of the temperature by COLEMAN and MIZEL [1963, *17*] and also the elimination of more general long-range effects by GURTIN [1965, *20*].

[3] In particular, we have made not the slightest use of any of the ideas commonly associated with the term "thermodynamics of irreversible processes", for which see the article by MEIXNER and REIK in Volume III/2 of this Encyclopedia. That theory is ruled by an all-transcending linearity, augmented by a formal rule of symmetry, which is alleged to follow from statistical mechanics. To speak of non-linear response within the framework there seems to be either vacuous or self-contradictory. The only attempt we have seen to carry over some of the vocabulary to modern continuum mechanics is that of WEHRLI and ZIEGLER [1962, *72*, §§ 3—5] (cf. also ZIEGLER [1963, *89*, § 6]), who proposed a "principle of least dissipation", which seems to result in constitutive equations of a rather special kind. Cf. the criticism by DILL [1965, *15*].

[4] COLEMAN [1964, *15* and *16*].

moment of momentum, the balance of energy (79.3), and the Clausius-Duhem inequality $\Gamma \geqq 0$. By (79.6), the Clausius-Duhem inequality has the specific form

$$\dot{H} \geqq \int_{\partial \mathscr{P}} \frac{\boldsymbol{h}}{\theta} \cdot \boldsymbol{n}\, ds + \int_{\mathscr{P}} \frac{q}{\theta}\, dm, \tag{96b.0}$$

with the local equivalent

$$\dot{\eta} \geqq \frac{1}{\varrho\theta} \operatorname{div} \boldsymbol{h} - \frac{1}{\varrho\theta^2} \boldsymbol{h} \cdot \operatorname{grad} \theta + \frac{q}{\theta} \tag{96b.00}$$

[cf. the special cases (79.11) and (96.15)]. These *fundamental laws of thermomechanics* suffice, when applied to constitutive equations of a very general kind, to yield *a definite and explicit thermodynamics*. It cannot be emphasized too strongly that this work has *nothing in common* with the "Onsagerist" approach, as explained, e.g., by MEIXNER and REIK in Volume III/2 of this Encyclopedia. The main differences are: (1) COLEMAN, making no mention of "forces" and "fluxes", deals with the *specific fields* of thermomechanics: deformation gradient $\boldsymbol{F}$, temperature θ, internal energy ε, specific entropy η, stress $\boldsymbol{T}$, heat flux $\boldsymbol{h}$, and others *defined* in terms of them; (2) *no new laws* are proposed, either as alleged consequences of statistical mechanics or as phenomenological assumptions; (3) *no special functional form* is demanded of the constitutive equations. COLEMAN'S approach, resting upon principles of determinism, local action, material frame-indifference, equipresence, and fading memory, is the same in kind as that we have followed in this treatise for purely mechanical effects (Sects. 17, 26, 28, 38), and it incorporates the earlier work as special cases. The bodies considered may be homogeneous or inhomogeneous and may enjoy arbitrary material symmetry.

α) *The constitutive equations and their general reduction.* In accord with the principle of equipresence (Sect. 96), COLEMAN as his defining constitutive equations assumes that the stress, heat flux, internal energy, and entropy are determined by the histories of the deformation and temperature fields and the present value of the temperature gradient $\boldsymbol{g} = \operatorname{grad} \theta$:

$$\left.\begin{aligned}
\boldsymbol{T} &= \underset{s=0}{\overset{\infty}{\mathfrak{T}}} \left(\boldsymbol{F}^{(t)}(s), \theta^{(t)}(s); \boldsymbol{g}(t)\right), \\
\boldsymbol{h} &= \underset{s=0}{\overset{\infty}{\mathfrak{H}}} \left(\boldsymbol{F}^{(t)}(s), \theta^{(t)}(s); \boldsymbol{g}(t)\right), \\
\varepsilon &= \underset{s=0}{\overset{\infty}{\mathfrak{e}}} \left(\boldsymbol{F}^{(t)}(s), \theta^{(t)}(s); \boldsymbol{g}(t)\right), \\
\eta &= \underset{s=0}{\overset{\infty}{\mathfrak{h}}} \left(\boldsymbol{F}^{(t)}(s), \theta^{(t)}(s); \boldsymbol{g}(t)\right).
\end{aligned}\right\} \tag{96b.1}$$

For the mathematical development it is more convenient to use the specific free energy:

$$\psi \equiv \varepsilon - \theta\eta, \tag{96b.2}$$

and to write Λ for the ordered pair $(\boldsymbol{F}, \theta)$, where $\boldsymbol{F}$ is a tensor and θ is a scalar. With the definitions

$$\left.\begin{aligned}
\alpha\Lambda &\equiv \Lambda\alpha \equiv (\alpha\boldsymbol{F}, \alpha\theta), \\
\Lambda_1 + \Lambda_2 &\equiv (\boldsymbol{F}_1 + \boldsymbol{F}_2, \theta_1 + \theta_2), \\
\Lambda_1 \cdot \Lambda_2 &\equiv \operatorname{tr}(\boldsymbol{F}_1 \boldsymbol{F}_2^T) + \theta_1\theta_2 = \Lambda_2 \cdot \Lambda_1, \\
|\Lambda| &\equiv \sqrt{\Lambda \cdot \Lambda} = \sqrt{\operatorname{tr} \boldsymbol{F}\boldsymbol{F}^T + \theta^2},
\end{aligned}\right\} \tag{96b.3}$$

where $\Lambda_1=(\boldsymbol{F}_1,\theta_1)$, $\Lambda_2=(\boldsymbol{F}_2,\theta_2)$, the set of all Λ becomes a 10-dimensional vector space $\mathscr{A}$. The *stress-entropy tensor* $\boldsymbol{\Sigma}$ is defined as

$$\boldsymbol{\Sigma}=\left(\frac{1}{\varrho}\boldsymbol{T}(\boldsymbol{F}^T)^{-1},\,-\eta\right)=\left(\frac{1}{\varrho}(\boldsymbol{F}^{-1}\boldsymbol{T})^T,\,-\eta\right), \tag{96b.4}$$

so that, identically,

$$\boldsymbol{\Sigma}\cdot\dot{\Lambda}=\frac{1}{\varrho}\operatorname{tr}(\boldsymbol{T}\boldsymbol{L})-\eta\dot{\theta}. \tag{96b.5}$$

The function $\Lambda^{(t)}(s)$ defined by

$$\Lambda^{(t)}(s)=\left(\boldsymbol{F}^{(t)}(s),\,\theta^{(t)}(s)\right) \tag{96b.6}$$

is called the *total history* up to time t. The domain of each total history is the entire real non-negative axis $[0,\infty)$. The restriction $\Lambda_+^{(t)}(s)$ of $\Lambda^{(t)}(s)$ to the open interval $(0,\infty)$ may be called the *past history* corresponding to $\Lambda^{(t)}(s)$. Finally, set

$$\varDelta^{(t)}(s)\equiv\Lambda_+^{(t)}(s)-\Lambda(t)=\left(\boldsymbol{F}(t-s)-\boldsymbol{F}(t),\,\theta(t-s)-\theta(t)\right), \tag{96b.7}$$

so that for a body which has always been at rest at a given temperature, $\varDelta^{(t)}(s)$ is the *zero history:* $\varDelta^{(t)}(s)\equiv\boldsymbol{0}$. Then COLEMAN's constitutive equations (96b.1) may be written in the compact forms

$$\left.\begin{aligned}
\psi&=\mathop{\mathfrak{p}}_{s=0}^{\infty}\left(\Lambda^{(t)}(s);\,\boldsymbol{g}(t)\right),\\
&=\mathop{\mathfrak{p}}_{s=0}^{\infty}\left(\varDelta^{(t)}(s);\,\Lambda(t),\,\boldsymbol{g}(t)\right),\\
\boldsymbol{\Sigma}&=\mathop{\mathfrak{S}}_{s=0}^{\infty}\left(\Lambda^{(t)}(s);\,\boldsymbol{g}(t)\right),\\
&=\mathop{\mathfrak{S}}_{s=0}^{\infty}\left(\varDelta^{(t)}(s);\,\Lambda(t),\,\boldsymbol{g}(t)\right),\\
\boldsymbol{h}&=\mathop{\mathfrak{H}}_{s=0}^{\infty}\left(\Lambda^{(t)}(s);\,\boldsymbol{g}(t)\right),\\
&=\mathop{\mathfrak{H}}_{s=0}^{\infty}\left(\varDelta^{(t)}(s);\,\Lambda(t);\,\boldsymbol{g}(t)\right),
\end{aligned}\right\} \tag{96b.8}$$

where the functional $\mathfrak{S}$ is determined as an ordered pair of functionals:

$$\mathfrak{S}=\left(\frac{1}{\varrho}(\boldsymbol{F}^{-1}\mathfrak{T})^T,\,-\mathfrak{h}\right). \tag{96b.9}$$

With $\boldsymbol{b}$ and q so adjusted as to satisfy the equations of balance of linear momentum and energy, an *admissible thermodynamic process* is a set of six fields $\boldsymbol{\chi}$, $\boldsymbol{T}$, ε, η, θ, $\boldsymbol{h}$ for all $X\in\mathscr{B}$ and for $-\infty<t<\infty$ such that $\boldsymbol{T}=\boldsymbol{T}^T$, $\theta>0$, and the constitutive equations (96b.8) are satisfied.

COLEMAN then generalizes the principle of fading memory. To define the recollection $\|\Lambda(s)\|_h$ of a history with values in $\mathscr{A}$, he uses (38.8) with $|\Lambda|$ defined by $(96\text{b}.3)_{6,7}$. Again the set of all histories with finite recollection forms a Hilbert space $\mathscr{H}$. With these extended definitions, the *principle of fading memory* is now stated as in Sect. 38:

($C^{\mathfrak{n}}$) *There is an obliviator of order greater than* $\mathfrak{n}+\frac{1}{2}$ *such that for each value of* Λ *and* $\boldsymbol{g}$ *the response functionals* $\mathfrak{p}$, $\mathfrak{S}$, *and* $\mathfrak{H}$ *are defined and* $\mathfrak{n}$ *times continuously Fréchet-differentiable in a neighborhood of the zero history in the function space* $\mathscr{H}$.

For example:

$$\left.\begin{aligned}\overset{\infty}{\underset{s=0}{\mathfrak{p}}}\left(\boldsymbol{\Lambda}^{(t)}(s)+\boldsymbol{\Gamma}(s);\,\boldsymbol{\Lambda},\boldsymbol{g}\right)=\overset{\infty}{\underset{s=0}{\mathfrak{p}}}&\left(\boldsymbol{\Lambda}^{(t)}(s);\,\boldsymbol{\Lambda},\boldsymbol{g}\right)+\delta\overset{\infty}{\underset{s=0}{\mathfrak{p}}}\left(\boldsymbol{\Lambda}^{(t)}(s);\,\boldsymbol{\Lambda},\boldsymbol{g}\middle|\boldsymbol{\Gamma}(s)\right)+\\&+\frac{1}{2}\delta^2\overset{\infty}{\underset{s=0}{\mathfrak{p}}}\left(\boldsymbol{\Lambda}^{(t)}(s);\,\boldsymbol{\Lambda},\boldsymbol{g}\middle|\boldsymbol{\Gamma}(s),\boldsymbol{\Gamma}(s)\right)+\\&+\cdots+\frac{1}{n!}\delta^n\overset{\infty}{\underset{s=0}{\mathfrak{p}}}\left(\boldsymbol{\Lambda}^{(t)}(s);\,\boldsymbol{\Lambda},\boldsymbol{g}\middle|\boldsymbol{\Gamma}(s),\ldots,\boldsymbol{\Gamma}(s)\right)+\\&+o\left(\|\boldsymbol{\Gamma}(s)\|_h^n\right),\end{aligned}\right\}\quad(96\text{b}.10)$$

where the functional $\delta^n\mathfrak{p}$, the n'th order Fréchet differential of $\mathfrak{p}$, is continuous in $\boldsymbol{\Lambda}^{(t)}(s)$, $\boldsymbol{\Lambda}$, and $\boldsymbol{g}$ and, most importantly, is a bounded homogeneous polynomial of degree n in $\boldsymbol{\Gamma}(s)$. In COLEMAN'S first memoir[1], (C^1) is assumed; in the second[2], (C^2). COLEMAN assumes further that, for each fixed function $\boldsymbol{\Lambda}^{(t)}(s)$ in the neighborhood mentioned, $\mathfrak{p}$, $\mathfrak{S}$, and $\mathfrak{H}$ are continuously differentiable with respect to $\boldsymbol{\Lambda}$ and $\boldsymbol{g}$ in the ordinary sense. For example,

$$\left.\begin{aligned}\overset{\infty}{\underset{s=0}{\mathfrak{p}}}\left(\boldsymbol{\Lambda}^{(t)}(s);\,\boldsymbol{\Lambda}+\boldsymbol{\Omega},\boldsymbol{g}\right)&=\overset{\infty}{\underset{s=0}{\mathfrak{p}}}\left(\boldsymbol{\Lambda}^{(t)}(s);\,\boldsymbol{\Lambda},\boldsymbol{g}\right)+\partial_{\boldsymbol{\Lambda}}\overset{\infty}{\underset{s=0}{\mathfrak{p}}}\left(\boldsymbol{\Lambda}^{(t)}(s);\,\boldsymbol{\Lambda},\boldsymbol{g}\right)\cdot\boldsymbol{\Omega}+o(|\boldsymbol{\Omega}|),\\\overset{\infty}{\underset{s=0}{\mathfrak{p}}}\left(\boldsymbol{\Lambda}^{(t)}(s);\,\boldsymbol{\Lambda},\boldsymbol{g}+\boldsymbol{v}\right)&=\overset{\infty}{\underset{s=0}{\mathfrak{p}}}\left(\boldsymbol{\Lambda}^{(t)}(s);\,\boldsymbol{\Lambda},\boldsymbol{g}\right)+\partial_{\boldsymbol{g}}\overset{\infty}{\underset{s=0}{\mathfrak{p}}}\left(\boldsymbol{\Lambda}^{(t)}(s);\,\boldsymbol{\Lambda},\boldsymbol{g}\right)\cdot\boldsymbol{v}+o(|\boldsymbol{v}|).\end{aligned}\right\}\quad(96\text{b}.11)$$

The limits implied by the o-symbols in (96b.10) and (96b.11) are assumed to be approached uniformly in $\boldsymbol{\Lambda}$ and the function $\boldsymbol{\Lambda}^{(t)}(s)$ when these are in appropriate neighborhoods. The gradients $\partial_{\boldsymbol{\Lambda}}\mathfrak{p}$, which has values in $\mathscr{A}$, and $\partial_{\boldsymbol{g}}$, which has values in the three-dimensional vector space $\mathscr{V}$, are both assumed continuous in $\boldsymbol{\Lambda}^{(t)}(s)$, $\boldsymbol{\Lambda}$ and $\boldsymbol{g}$.

Although the constitutive functionals in (96b.8) depend on the choice of reference configuration, it is easy to see that the principle of fading memory is intrinsic: For a given material, if (C^n) holds for one reference configuration, then it holds for all.

Let $1(s)$ be the function whose value is 1 for all positive s. Basic to the present theory is a linear differential operator D mapping the scalar-valued functionals $\mathfrak{p}$ into vector-valued functions $D\mathfrak{p}$ defined by the formula[3]

$$\left.\begin{aligned}D\overset{\infty}{\underset{s=0}{\mathfrak{p}}}\left(\boldsymbol{\Lambda}^{(t)}(s)\right)\cdot\boldsymbol{\Omega}&=\partial_{\boldsymbol{\Lambda}}\overset{\infty}{\underset{s=0}{\mathfrak{p}}}\left(\boldsymbol{\Lambda}^{(t)}(s);\,\boldsymbol{\Lambda}\right)\cdot\boldsymbol{\Omega}-\delta\overset{\infty}{\underset{s=0}{\mathfrak{p}}}\left(\boldsymbol{\Lambda}^{(t)}(s);\,\boldsymbol{\Lambda}\middle|\boldsymbol{\Omega}1(s)\right),\\&=\frac{d}{d\nu}\overset{\infty}{\underset{s=0}{\mathfrak{p}}}\left(\boldsymbol{\Lambda}^{(t)}(s)-\nu\boldsymbol{\Omega}1(s);\,\boldsymbol{\Lambda}+\nu\boldsymbol{\Omega}\right)\Big|_{\nu=0},\end{aligned}\right\}\quad(96\text{b}.12)$$

which must hold for all members $\boldsymbol{\Omega}$ of $\mathscr{A}$.

In writing (96b.12) we have omitted the symbol $\boldsymbol{g}$ from the list of arguments of the functional $\mathfrak{p}$; the reader should here think of (96b.12) as holding for any preassigned value of $\boldsymbol{g}$. Such omission of $\boldsymbol{g}$, to be repeated in subsequent definitions involving $\mathfrak{p}$, is motivated by a theorem which will be soon stated.

Writing $\dot{\boldsymbol{\Lambda}}^{(t)}(s)\equiv\frac{d}{dt}\boldsymbol{\Lambda}(t-s)=-\frac{d}{ds}\boldsymbol{\Lambda}^{(t)}(s)$, COLEMAN[4] introduces the quantity $\sigma(t)$ defined by

$$\left.\begin{aligned}\sigma(t)&=-\frac{1}{\theta(t)}\,\delta\overset{\infty}{\underset{s=0}{\mathfrak{p}}}\left(\boldsymbol{\Lambda}^{(t)}(s);\,\boldsymbol{\Lambda}\middle|\dot{\boldsymbol{\Lambda}}^{(t)}(s)\right),\\&=\frac{1}{\theta(t)}\,\delta\overset{\infty}{\underset{s=0}{\mathfrak{p}}}\left(\boldsymbol{\Lambda}^{(t)}(s);\,\boldsymbol{\Lambda}\middle|\frac{d}{ds}\boldsymbol{\Lambda}^{(t)}(s)\right),\end{aligned}\right\}\quad(96\text{b}.13)$$

[1] COLEMAN [1964, *15*].
[2] COLEMAN [1964, *16*].
[3] COLEMAN [1964, *16*, Eq. (3.9)].
[4] COLEMAN [1964, *15*, Eq. (6.28)].

and calls it the *internal dissipation* at time t corresponding to the history $\varLambda^{(t)}(s)$. We note that if $\psi=\hat{\psi}(\boldsymbol{F},\theta)$, then $\delta\mathfrak{p}=0$, which implies that $\sigma(t)\equiv 0$; that is, *if a material obeys a caloric equation of state in the sense of classical thermodynamics* (CFT, Sects. 246 and 251), *the internal dissipation is zero.* The following theorem shows, among other things, that the Clausius-Duhem inequality determines the sign of σ when $\sigma\neq 0$.

COLEMAN's ***main theorem***[1]: *Assume that at each particle of a body $\mathscr{B}$ constitutive equations of the form* (96b.8) *hold, where* $\mathfrak{p}$, $\mathfrak{S}$, *and* $\mathfrak{h}$ *obey the principle of fading memory* (C^1). *In order that the Clausius-Duhem inequality* (96b.00) *hold for all smooth admissible thermodynamic processes in $\mathscr{B}$, it is necessary and sufficient that:*

I. *The functionals* $\mathfrak{p}$ *and* $\mathfrak{S}$ *be independent of* $\boldsymbol{g}$; i.e.,

$$\partial_{\boldsymbol{g}}\mathfrak{p}\equiv\boldsymbol{0} \quad\text{and}\quad \partial_{\boldsymbol{g}}\mathfrak{S}\equiv\boldsymbol{0}. \tag{96b.14}$$

II. *The functional* $\mathfrak{S}$ *be determined by the functional* $\mathfrak{p}$ *through the relation*

$$\mathfrak{S}\equiv D\mathfrak{p}, \tag{96b.15}$$

called the stress-entropy relation.

III. σ, *given by* (96b.13), *obey the following* internal dissipation inequality:

$$\sigma\geqq 0. \tag{96b.16}$$

IV. *For each smooth history* $\varLambda^{(t)}(s)$, *the dependence of* $\boldsymbol{h}$ *on* $\boldsymbol{g}$ *be such that*

$$\boldsymbol{h}\cdot\boldsymbol{g}\geqq-\varrho\,\theta^2\sigma. \tag{96b.17}$$

For proof of this powerful theorem we must refer the reader to COLEMAN's memoir.

Eqs. (96b.14) and (96b.15) assert that we may replace $(96\text{b}.8)_1$ and $(96\text{b}.8)_3$ by the following simpler constitutive equations:

$$\left.\begin{aligned}\psi&=\mathop{\mathfrak{p}}_{s=0}^{\infty}\big(\varLambda^{(t)}(s)\big)=\mathop{\mathfrak{p}}_{s=0}^{\infty}\big(\boldsymbol{F}^{(t)}(s),\theta^{(t)}(s)\big),\\ \boldsymbol{\Sigma}&=\mathop{\mathfrak{S}}_{s=0}^{\infty}\big(\varLambda^{(t)}(s)\big)=D\mathop{\mathfrak{p}}_{s=0}^{\infty}\big(\varLambda^{(t)}(s)\big).\end{aligned}\right\} \tag{96b.18}$$

These equations hold not only for smooth histories, but for all histories in the common domain of $\mathfrak{p}$ and $\mathfrak{S}$.

Because of the central role played by the operator D in the present theory, it is important to have a simple interpretation for it. We can write $(96\text{b}.18)_1$ in the form

$$\psi=\mathop{\mathfrak{p}}_{s=0}^{\infty}\big(\varLambda^{(t)}(s);\varLambda\big)=\mathop{\tilde{\mathfrak{p}}}_{s=0}^{\infty}\big(\varLambda_+^{(t)}(s);\varLambda\big) \tag{96b.19}$$

where $\tilde{\mathfrak{p}}$ is a functional with the same values ψ as $\mathfrak{p}$. The domains of $\mathfrak{p}$ and $\tilde{\mathfrak{p}}$ are very simply related, and it is easy to see that $\tilde{\mathfrak{p}}$ must enjoy whatever smoothness the principle of fading memory gives to $\mathfrak{p}$; i.e., Eqs. (96b.10) and (96b.11) must hold when $\mathfrak{p}$ and $\varLambda^{(t)}(s)$ are replaced by $\tilde{\mathfrak{p}}$ and $\varLambda_+^{(t)}(s)$ (and, of course, $\boldsymbol{g}$ is omitted). The functional $\delta\tilde{\mathfrak{p}}$ so obtained is essentially the same as $\delta\mathfrak{p}$. For all $\boldsymbol{\Gamma}(s)$ in $\mathscr{H}$,

$$\delta\mathop{\tilde{\mathfrak{p}}}_{s=0}^{\infty}\big(\varLambda_+^{(t)}(s);\varLambda\,|\,\boldsymbol{\Gamma}(s)\big)=\delta\mathop{\mathfrak{p}}_{s=0}^{\infty}\big(\varLambda^{(t)}(s);\varLambda\,|\,\boldsymbol{\Gamma}(s)\big). \tag{96b.20}$$

---

[1] COLEMAN [1964, *15*, § 6, Theorem 1].

The functional $\partial_\Lambda \tilde{\mathfrak{p}}$ is not $\partial_\Lambda \mathfrak{p}$, but rather $D\mathfrak{p}$:

$$\partial_\Lambda \underset{s=0}{\overset{\infty}{\tilde{\mathfrak{p}}}}\left(\Lambda_+^{(t)}(s);\Lambda\right)=D\underset{s=0}{\overset{\infty}{\mathfrak{p}}}\left(\Lambda^{(t)}(s)\right). \tag{96b.21}$$

Equivalent to $(96\text{b}.12)_2$ is the following equation, which holds for all $\boldsymbol{\Omega}$ in $\mathscr{A}$:

$$D\underset{s=0}{\overset{\infty}{\mathfrak{p}}}\left(\Lambda^{(t)}(s)\right)\cdot\boldsymbol{\Omega}=\frac{d}{d\nu}\underset{s=0}{\overset{\infty}{\tilde{\mathfrak{p}}}}\left(\Lambda_+^{(t)}(s);\Lambda+\nu\boldsymbol{\Omega}\right)\Big|_{\nu=0}. \tag{96b.22}$$

This formula[1] renders completely transparent the meaning of the operator D. It states that $D\underset{s=0}{\overset{\infty}{\mathfrak{p}}}\left(\Lambda^{(t)}(s)\right)$ *is the derivative of* ψ *with respect to the present value* $\Lambda(t)$ *keeping the past history* $\Lambda_+^{(t)}(s)$ *fixed.*

Putting $\Lambda^{(t)}(s)=\left(\boldsymbol{F}^{(t)}(s),\,\theta^{(t)}(s)\right)$ and $\Lambda_+^{(t)}(s)=\left(\boldsymbol{F}_+^{(t)}(s),\,\theta_+^{(t)}(s)\right)$, we can write (96b.19) in the more extended forms

$$\psi(t)=\underset{s=0}{\overset{\infty}{\mathfrak{p}}}\left(\boldsymbol{F}^{(t)}(s),\,\theta^{(t)}(s)\right)=\underset{s=0}{\overset{\infty}{\tilde{\mathfrak{p}}}}\left(\boldsymbol{F}_+^{(t)}(s),\,\theta_+^{(t)}(s);\,\boldsymbol{F},\,\theta\right). \tag{96b.23}$$

The second of these equations makes apparent the dependence of $\psi(t)$ on the past histories $\boldsymbol{F}_+^{(t)}(s)$ and $\theta_+^{(t)}(s)$ and the present values $\boldsymbol{F}=\boldsymbol{F}^{(t)}(0)$, $\theta=\theta^{(t)}(0)$ of the deformation gradient and temperature. Corresponding to each of the "total derivatives" D of (96b.23) we define two "partial derivatives"[2] $D_{\boldsymbol{F}}\mathfrak{p}$, $D_\theta\mathfrak{p}$, by the equations

$$\operatorname{tr}\left[D_{\boldsymbol{F}}\underset{s=0}{\overset{\infty}{\mathfrak{p}}}\left(\boldsymbol{F}^{(t)}(s),\,\theta^{(t)}(s)\right)\boldsymbol{A}^T\right]=\frac{d}{d\nu}\underset{s=0}{\overset{\infty}{\tilde{\mathfrak{p}}}}\left(\boldsymbol{F}_+^{(t)}(s),\,\theta_+^{(t)}(s);\,\boldsymbol{F}+\nu\boldsymbol{A},\,\theta\right)\Big|_{\nu=0}, \tag{96b.24}$$

which must hold for all tensors $\boldsymbol{A}$, and the equation

$$D_\theta\underset{s=0}{\overset{\infty}{\mathfrak{p}}}\left(\boldsymbol{F}^{(t)}(s),\,\theta^{(t)}(s)\right)=\frac{d}{d\nu}\underset{s=0}{\overset{\infty}{\tilde{\mathfrak{p}}}}\left(\boldsymbol{F}_+^{(t)}(s),\,\theta_+^{(t)}(s);\,\boldsymbol{F},\,\theta+\nu\right)\Big|_{\nu=0}. \tag{96b.25}$$

Since the functional $D\mathfrak{p}$ is determined by its "component functionals" $D_{\boldsymbol{F}}\mathfrak{p}$ $D_\theta\mathfrak{p}$ as follows: $D\mathfrak{p}=(D_{\boldsymbol{F}}\mathfrak{p},\,D_\theta\mathfrak{p})$, by (96b.4) and (96b.9) we deduce the following

Corollary[3]: *The stress-entropy relation* (96b.14) *is equivalent to a stress relation*

$$\boldsymbol{T}=\varrho\boldsymbol{F}\left[D_{\boldsymbol{F}}\underset{s=0}{\overset{\infty}{\mathfrak{p}}}\left(\boldsymbol{F}^{(t)}(s),\,\theta^{(t)}(s)\right)\right]^T \tag{96b.26}$$

and an entropy relation

$$\eta=-D_\theta\underset{s=0}{\overset{\infty}{\mathfrak{p}}}\left(\boldsymbol{F}^{(t)}(s),\,\theta^{(t)}(s)\right). \tag{96b.27}$$

This capital result generalizes the classical stress relation and temperature relation of thermo-elasticity to simple materials with memory. The functional differential operators $D_{\boldsymbol{F}}$ and D_θ replace the partial derivatives occurring in the standard formulae presuming caloric equations of state [Eqs. (251.4) of CFT, or, with a different choice of variables, Eqs. (80.10) and (80.11) of this treatise]; in equilibrium, of course, (96b.26) and (96b.27) reduce to the classic equations of thermostatics, as is easily seen from (96b.24) and (96b.25).

---

[1] COLEMAN [1964, *16*, Eq. (9.5)].
[2] COLEMAN [1964, *16*, Eq. (9.12)], simplifying the definition of COLEMAN [1964, *15*, § 7].
[3] COLEMAN [1964, *15*, § 7, Theorem 2].

For histories $\Lambda^{(t)}(s)$ that are smooth for all s, a chain rule permits us to express the material time derivative of ψ as follows:

$$\left.\begin{aligned}\dot{\psi}(t) = \frac{d}{dt}\overset{\infty}{\underset{s=0}{\tilde{\mathfrak{p}}}}\left(\Lambda_+^{(t)}(s);\, \Lambda(t)\right) = \partial_\Lambda \overset{\infty}{\underset{s=0}{\tilde{\mathfrak{p}}}}\left(\Lambda_+^{(t)}(s),\, \Lambda(t)\right)\cdot\dot{\Lambda}(t) + \\ + \delta\,\overset{\infty}{\underset{s=0}{\tilde{\mathfrak{p}}}}\left(\Lambda_+^{(t)}(s),\, \Lambda(t)\,|\,\dot{\Lambda}_+^{(t)}(s)\right).\end{aligned}\right\} \tag{96b.28}$$

By (96b.20)and (96b.21) it follows that[1]

$$\dot{\psi}(t) = D\,\overset{\infty}{\underset{s=0}{\mathfrak{p}}}\left(\Lambda^{(t)}(s)\right)\cdot\dot{\Lambda}(t) + \delta\,\overset{\infty}{\underset{s=0}{\mathfrak{p}}}\left(\Lambda^{(t)}(s),\, \Lambda(t)\,\Big|\, -\frac{d}{ds}\Lambda^{(t)}(s)\right). \tag{96b.29}$$

Using the stress-entropy relation (96b.15) and the definition (96b.13), one obtains from (96b.29) the important equation[2]

$$\dot{\psi} = \boldsymbol{\Sigma}\cdot\dot{\boldsymbol{\Lambda}} - \theta\sigma \tag{96b.30}$$

or, by (96b.5),

$$\dot{\psi} = \frac{1}{\varrho}\operatorname{tr}(\boldsymbol{T}\boldsymbol{D}) - \eta\dot{\theta} - \theta\sigma. \tag{96b.31}$$

The definition of σ by Eq. (96b.13) is appropriate only to histories that are smooth for all s. Eq. (96b.30) however, may be used[3] to define $\sigma(t)$ for any history sufficiently regular that $\dot{\psi}$ and $\dot{\Lambda}$ exist at time t, and it can be shown that the inequalities (96b.16) and (96b.17) must hold for this extended class of histories. This observation is important for the application of the internal dissipation inequality to static continuations, which involve at least one point at which the history, though continuous, is not differentiable.

β) *Inequalities for stress relaxation and cyclical processes.* It follows from (96b.31) and the internal dissipation inequality (96b.16) that[4]

$$\dot{\psi} \leqq \frac{1}{\varrho}\operatorname{tr}(\boldsymbol{T}\boldsymbol{D}) - \eta\dot{\theta}. \tag{96b.32}$$

In particular,

$$\boldsymbol{D} = \boldsymbol{0},\ \dot{\theta} = 0 \Rightarrow \dot{\psi} = -\theta\sigma \leqq 0. \tag{96b.33}$$

Therefore, *if at a given moment the strain and temperature are held fixed, the free energy at that moment cannot increase*, regardless of the past history.

Given any total history $\boldsymbol{\Lambda}^{(t)}(s)$, we can define a new total history $\boldsymbol{\Lambda}^{(t+\delta)}(s)$, $\delta \geqq 0$, by the relation

$$\boldsymbol{\Lambda}^{(t+\delta)}(s) \equiv \begin{cases}\boldsymbol{\Lambda}^{(t)}(0), & 0 \leqq s < \delta, \\ \boldsymbol{\Lambda}^{(t)}(s-\delta), & \delta < s < \infty.\end{cases} \tag{96b.34}$$

The new function $\boldsymbol{\Lambda}^{(t+\delta)}(s)$ is called the *isothermal static* continuation of $\boldsymbol{\Lambda}^{(t)}(s)$ by the amount δ, generalizing the purely kinematic statical continuation (39.1). Hence

$$\dot{\boldsymbol{\Lambda}} = (\dot{\boldsymbol{F}}, \dot{\theta}) = (\boldsymbol{0}, 0) \quad \text{when} \quad t < \tau \leqq t + \delta. \tag{96b.35}$$

If we let $\Lambda^{(t)}(s)$ and $\Lambda^{(t+\delta)}(s)$ be the difference histories corresponding to $\boldsymbol{\Lambda}^{(t)}(s)$ and $\boldsymbol{\Lambda}^{(t+\delta)}(s)$, respectively, we have

$$\Lambda^{(t+\delta)}(s) = \begin{cases}\boldsymbol{0}, & 0 \leqq s < \delta, \\ \Lambda^{(t)}(s-\delta), & \delta < s < \infty.\end{cases} \tag{96b.36}$$

---

[1] COLEMAN [1964, *15*, Eq. (6.16)] [1964, *16*, Eq. (3.12)].
[2] COLEMAN [1964, *15*, Eq. (8.1)].
[3] COLEMAN [1964, *16*, Eq. (4.3)].
[4] COLEMAN [1964, *15*, Eq. (8.2)].

For the free energy and stress-entropy corresponding to the isothermal static continuation $\boldsymbol{\Lambda}^{(t+\delta)}(s)$ according to (96b.17) and (96b.18) we write

$$\left.\begin{aligned}\psi^\delta &= \underset{s=0}{\overset{\infty}{\mathfrak{p}}}\left(\boldsymbol{\Lambda}^{(t+\delta)}(s);\boldsymbol{\Lambda}\right),\\ \boldsymbol{\Sigma}^\delta &= \left(\frac{1}{\varrho}\boldsymbol{T}^\delta(\boldsymbol{F}^T)^{-1}, -\eta^\delta\right) = \underset{s=0}{\overset{\infty}{\mathfrak{S}}}\left(\boldsymbol{\Lambda}^{(t+\delta)}(s);\boldsymbol{\Lambda}\right).\end{aligned}\right\} \tag{96b.37}$$

Let ψ^∞ and $\boldsymbol{\Sigma}^\infty = \left(\frac{1}{\varrho}(\boldsymbol{F}^{-1}\boldsymbol{T}^\infty)^T, -\eta^\infty\right)$ be the "equilibrium" values of ψ and $\boldsymbol{\Sigma}$ for a given material, i.e., the values these quantities have when $\boldsymbol{\Lambda}(\tau)$ has been held at the constant value $\boldsymbol{\Lambda} = (\boldsymbol{F}, \theta)$ for all $\tau \leqq t$:

$$\left.\begin{aligned}\psi^\infty &= \underset{s=0}{\overset{\infty}{\mathfrak{p}}}\left(\boldsymbol{0}(s);\boldsymbol{\Lambda}\right) = \psi^\infty(\boldsymbol{F},\theta) = \psi^\infty(\boldsymbol{\Lambda}),\\ \boldsymbol{\Sigma}^\infty &= \underset{s=0}{\overset{\infty}{\mathfrak{S}}}\left(\boldsymbol{0}(s);\boldsymbol{\Lambda}\right) = \boldsymbol{\Sigma}^\infty(\boldsymbol{F},\theta).\end{aligned}\right\} \tag{96b.38}$$

In obvious analogy to (39.3) is the assertion that[1]

$$\lim_{\delta\to\infty} \left\|\boldsymbol{\Lambda}^{(t+\delta)}(s) - \boldsymbol{0}(s)\right\|_h = 0. \tag{96b.39}$$

From this assertion and the assumed continuity of the functionals $\mathfrak{p}$ and $\mathfrak{S}$ in the function variable $\boldsymbol{\Lambda}^{(t)}(s)$ follows a generalization of the stress-relaxation theorem:

$$\lim_{\delta\to\infty}\psi^\delta = \psi^\infty, \qquad \lim_{\delta\to\infty}\boldsymbol{T}^\delta = \boldsymbol{T}^\infty, \qquad \lim_{\delta\to\infty}\eta^\delta = \eta^\infty. \tag{96b.40}$$

Thus the *free energy, stress, and entropy corresponding to an isothermal static continuation by amount δ of any total history all approach their equilibrium values as δ increases without limit.* Moreover, the first of these limits is approached *monotonically*. For, by (96b.33), *during the "static part" of an isothermal static continuation of any history* (the part when $t < \tau \leqq t + \delta$), *the rate of change of the free energy, being equal to minus the temperature times the internal dissipation, cannot be positive*[2]. In other words, if $\delta_1 > \delta_2$, then

$$\underset{s=0}{\overset{\infty}{\mathfrak{p}}}\left(\boldsymbol{\Lambda}^{(t+\delta_1)}(s);\boldsymbol{\Lambda}\right) \leqq \underset{s=0}{\overset{\infty}{\mathfrak{p}}}\left(\boldsymbol{\Lambda}^{(t+\delta_2)}(s);\boldsymbol{\Lambda}\right). \tag{96b.41}$$

Setting (96b.41) alongside of $(96\text{b}.40)_1$ written in the form

$$\lim_{\delta\to\infty}\underset{s=0}{\overset{\infty}{\mathfrak{p}}}\left(\boldsymbol{\Lambda}^{(t+\delta)}(s);\boldsymbol{\Lambda}\right) = \underset{s=0}{\overset{\infty}{\mathfrak{p}}}\left(\boldsymbol{0}(s);\boldsymbol{\Lambda}\right), \tag{96b.42}$$

we see that for each value of $\boldsymbol{\Lambda}$ the inequality

$$\underset{s=0}{\overset{\infty}{\mathfrak{p}}}\left(\boldsymbol{\Lambda}^{(t)}(s);\boldsymbol{\Lambda}\right) \geqq \underset{s=0}{\overset{\infty}{\mathfrak{p}}}\left(\boldsymbol{0}(s);\boldsymbol{\Lambda}\right) \tag{96b.43}$$

holds for all difference histories $\boldsymbol{\Lambda}^{(t)}(s)$ in the domain of definition of $\mathfrak{p}$. This proves COLEMAN'S ***theorem on minimum free energy***[3]: *Of all total histories ending with given values of $\boldsymbol{F}$ and θ, that corresponding to constant values of $\boldsymbol{F}$ and θ for all times has the least free energy.*

---

[1] Eq. (96b.39) in the form (39.3) was first proved by COLEMAN and NOLL under the hypothesis that $|\boldsymbol{\Lambda}^{(t)}(s)|$ be bounded; later, COLEMAN [1964, *15*, § 8] showed it sufficient that $|\boldsymbol{\Lambda}^{(t)}(s)|\, h(s)$ be bounded; and, finally, WANG showed, in effect, that (96b.39) holds whenever $\boldsymbol{\Lambda}^{(t+\delta)}(s)$ is in $\mathscr{H}$. Cf. the insertion in Sect. 39.

[2] COLEMAN [1964, *15*, § 8, Remark 9].

[3] COLEMAN [1964, *15*, § 8, Theorem 3].

For a given $\boldsymbol{\Lambda}$, $\mathfrak{p}$ can be minimized at the function $\Delta^{(t)}(s)=\mathbf{0}(s)$ only if the functional derivative of $\mathfrak{p}$ with respect to $\Delta^{(t)}(s)$ is zero at $\mathbf{0}(s)$; hence COLEMAN's theorem has the following corollary: *For all $\boldsymbol{\Lambda}$, and for all functions $\boldsymbol{\Psi}(s)$ in the domain of definition of $\mathfrak{p}$,*

$$\delta \underset{s=0}{\overset{\infty}{\mathfrak{p}}}\big(\mathbf{0}(s);\boldsymbol{\Lambda}\,|\,\boldsymbol{\Psi}(s)\big)=0. \tag{96b.44}$$

Let us denote by $D\underset{s=0}{\overset{\infty}{\mathfrak{p}}}\big(\mathbf{0}(s);\boldsymbol{\Lambda}\big)$ the value of $D\underset{s=0}{\overset{\infty}{\mathfrak{p}}}\big(\boldsymbol{\Lambda}^{(t)}(s)\big)$ when $\boldsymbol{\Lambda}^{(t)}(s)\equiv\boldsymbol{\Lambda}$. It follows from (96b.38) and the stress-entropy relation $(96\text{b}.18)_2$ that

$$\boldsymbol{\Sigma}^{\infty}=D\underset{s=0}{\overset{\infty}{\mathfrak{p}}}\big(\mathbf{0}(s);\boldsymbol{\Lambda}\big). \tag{96b.44a}$$

By $(96\text{b}.12)_1$, for all $\boldsymbol{\Omega}$ in $\mathscr{A}$,

$$D\underset{s=0}{\overset{\infty}{\mathfrak{p}}}\big(\mathbf{0}(s);\boldsymbol{\Lambda}\big)\cdot\boldsymbol{\Omega}=\partial_{\boldsymbol{\Lambda}}\underset{s=0}{\overset{\infty}{\mathfrak{p}}}\big(\mathbf{0}(s);\boldsymbol{\Lambda}\big)\cdot\boldsymbol{\Omega}-\delta\underset{s=0}{\overset{\infty}{\mathfrak{p}}}\big(\mathbf{0}(s);\boldsymbol{\Lambda}\,|\,\boldsymbol{\Omega}1(s)\big). \tag{96b.45}$$

But, by (96b.44), the last term in this equation is zero; hence

$$D\underset{s=0}{\overset{\infty}{\mathfrak{p}}}\big(\mathbf{0}(s);\boldsymbol{\Lambda}\big)=\partial_{\boldsymbol{\Lambda}}\underset{s=0}{\overset{\infty}{\mathfrak{p}}}\big(\mathbf{0}(s);\boldsymbol{\Lambda}\big), \tag{96b.46}$$

and (96b.44a) becomes[1]

$$\boldsymbol{\Sigma}^{\infty}=\partial_{\boldsymbol{\Lambda}}\underset{s=0}{\overset{\infty}{\mathfrak{p}}}\big(\mathbf{0}(s);\boldsymbol{\Lambda}\big),\quad\text{or}\quad \boldsymbol{\Sigma}^{\infty}=\psi^{\infty}_{\boldsymbol{\Lambda}}(\boldsymbol{\Lambda}). \tag{96b.47}$$

The function $\psi^{\infty}=\psi^{\infty}(\boldsymbol{F},\theta)$ defined in $(96\text{b}.38)_3$ is said to determine a "fundamental equation" or a "caloric equation of state" (CFT, Sects. 246–251). Eq. (96b.47) is equivalent to the two equations

$$\boldsymbol{T}^{\infty}=\varrho\boldsymbol{F}[\psi^{\infty}_{\boldsymbol{F}}(\boldsymbol{F},\theta)]^{T},\qquad \eta^{\infty}=-\psi^{\infty}_{\theta}(\boldsymbol{F},\theta), \tag{96b.48}$$

namely, the standard formulae of thermostatics (82.7) and (82.8). We have seen above that in general thermodynamic processes in general simple materials these formulae are *replaced* by the relations (96b.26) and (96b.27). Earlier, in Sect. 82, we showed that for certain *particular materials* results having forms equivalent to (96b.48) hold in all processes. As we have seen above, COLEMAN's general equations (96b.26) and (96b.27) reduce in equilibrium to relations of the form (96b.48). Beyond this, here we have found a limiting case of a particular process, namely, an isothermal static continuation, in which the *classical formulae hold for all simple materials.* We may summarize the result in the following ***theorem of the thermostatic limit***: *In any simple material obeying the principle of fading memory as stated above, the equations of classical thermostatics hold not only in equilibrium but also in the limit of approach to equilibrium through isothermal static continuation.* In contrast to what the reader may expect from thermodynamic treatments, the result just stated is not a disguised postulate; rather, it is a proved theorem, following from the minimizing property (96b.43) of constant histories and hence, ultimately, from the Clausius-Duhem inequality.

If $\boldsymbol{\Lambda}(t)$ is assigned arbitrarily when $t\leqq t_1$, then by $(96\text{b}.18)_4$ $\boldsymbol{\Sigma}(t)$ is assigned when $t\leqq t_1$, and for each $t_0<t_1$, we can compute the following integral, which represents a kind of total work-heat:

$$\left.\begin{aligned}\Upsilon(t_1,t_0)&\equiv\int_{t_0}^{t_1}\boldsymbol{\Sigma}(t)\cdot\dot{\boldsymbol{\Lambda}}(t)\,dt=\int_{t_0}^{t_1}\left[\frac{1}{\varrho}\operatorname{tr}(\boldsymbol{T}\boldsymbol{D})-\eta\dot{\theta}\right]dt,\\&=\psi(t_1)-\psi(t_0)+\int_{t_0}^{t_1}\theta\sigma\,dt,\end{aligned}\right\} \tag{96b.49}$$

[1] COLEMAN [1964, *15*, Eq. (8.27)].

where the first form follows by (96b.5), the second, by (96b.30). By the internal dissipation inequality (96b.15),

$$\Upsilon(t_1, t_0) \geqq \psi(t_1) - \psi(t_0). \tag{96b.50}$$

This inequality states that in any smooth admissible thermodynamic process, the *total work-heat is never less than the increase of free energy.* From it we derive the following corollary[1]: *If*

$$\boldsymbol{\Lambda}(t) = \begin{cases} \boldsymbol{\Lambda}_0 = \text{const.} \quad \text{when} \quad t \leqq t_0 \\ \text{an arbitrary smooth admissible function when } t_0 < t < t_1 \\ \boldsymbol{\Lambda}_0 \quad \text{when} \quad t = t_1 \end{cases} \tag{96b.51}$$

then

$$\Upsilon(t_1, t_0) \geqq 0. \tag{96b.52}$$

That is, if at some time t_0 we take a particle which has been at rest at all previous times in the local state $\boldsymbol{\Lambda}_0$ and carry it along a deformation-temperature path $\boldsymbol{\Lambda}(t)$ which returns to $\boldsymbol{\Lambda}_0$ at some time t_1, then the total work-heat is never negative.

Proof: Since by hypothesis $\boldsymbol{\Lambda}(t_1) = \boldsymbol{\Lambda}(t_0) = \boldsymbol{\Lambda}_0$, and the difference history $\Delta(s)$ up to time t_0 is $\mathbf{0}(s)$,

$$\psi(t_0) = \mathop{\mathfrak{p}}_{s=0}^{\infty}\left(\mathbf{0}(s); \boldsymbol{\Lambda}_0\right), \qquad \psi(t_1) = \mathop{\mathfrak{p}}_{s=0}^{\infty}\left(\Delta^{(t_1)}(s); \boldsymbol{\Lambda}_0\right). \tag{96b.53}$$

By (96b.43),

$$\psi(t_0) \leqq \psi(t_1), \tag{96b.54}$$

so that the corollary follows from (96b.50).

Now, it follows immediately from this corollary that if $\dot{\theta}(t) = 0$ for all t, and if $\boldsymbol{F}^{(t_0)}(s) \equiv \mathbf{0}$, $\boldsymbol{F}(t_0) = \boldsymbol{F}(t_1)$, then

$$\int_{t_0}^{t_1} \frac{1}{\varrho} \operatorname{tr}(\boldsymbol{T}\boldsymbol{D})\, dt \geqq 0. \tag{96b.55}$$

This inequality expresses COLEMAN'S ***theorem on isothermal cyclic processes***[2]: *In every isothermal process starting from a state of equilibrium, the total stress work around a closed path is non-negative.*

γ) On the status of Gibbs relations. If the total history $\boldsymbol{\Lambda}^{(t)}(s)$ and the rate $\dot{\boldsymbol{\Lambda}}(t) = (\dot{\boldsymbol{F}}(t), \dot{\theta}(t))$ are such that $\theta(t)\,\sigma(t)$ is negligible when compared with $\dot{\psi}(t)$, then in place of (96b.30) we can write

$$\dot{\psi}(t) = \boldsymbol{\Sigma} \cdot \dot{\boldsymbol{\Lambda}} = \frac{1}{\varrho} \operatorname{tr}(\boldsymbol{T}\boldsymbol{D}) - \eta\,\dot{\theta}. \tag{96b.56}$$

An equation of this form is called a *Gibbs relation* (CFT, Sect. 247). In linear theories of non-equilibrium thermodynamics[3], which are intended to describe a situation "close to equilibrium" in some undefined sense, a Gibbs relation is laid down as an axiom. Within the present theory, COLEMAN[4] has *proved* that such an assumption has a status in the sense of retardation, which is defined as in Eq. (40.5): *Let* $\boldsymbol{\Lambda}^{(t)}(s)$

---

[1] COLEMAN [1964, *15*, § 9, Remark 12].

[2] COLEMAN [1964, *15*, § 9, Theorem 4].

[3] See, for example, the article of MEIXNER and REIK in this Encyclopedia, Volume III, Part 2.

[4] COLEMAN [1964, *15*, § 10, Theorem 5].

be any total history for which $\dot{\psi}(t)$ exists. Let ψ^α and Λ^α refer to the retarded process, so that

$$\left.\begin{aligned}\dot{\psi}^\alpha &= \frac{d}{dt}\underset{s=0}{\overset{\infty}{\mathfrak{p}}}\left(\Lambda^{(t)}(\alpha s)\right),\\ \dot{\Lambda}^\alpha &= -\frac{d}{ds}\Lambda^{(t)}(\alpha s)\big|_{s=0} = \alpha\dot{\Lambda}.\end{aligned}\right\} \tag{96b.57}$$

Then as $\alpha\to 0$,

$$\dot{\psi}^\alpha = \Sigma^\infty\cdot\dot{\Lambda}^\alpha + o(\alpha), \tag{96b.58}$$

where Σ^∞ is the equilibrium value of the stress-entropy corresponding to $\Lambda(t)$, given by Eq. (96b.38)$_4$. The proof of this major theorem rests upon Eqs. (96b.30), (96b.12), and (96b.44a).

Let Σ^α be the stress-entropy corresponding to the retarded history $\Lambda^{(t)}(\alpha s)$:

$$\Sigma^\alpha = \underset{s=0}{\overset{\infty}{\mathfrak{S}}}\left(\Lambda^{(t)}(\alpha s)\right). \tag{96b.59}$$

Applying COLEMAN and NOLL's retardation theorem (40.6) when $\mathfrak{n}=0$ to the functions $\Lambda^{(t)}(\alpha s)$ with values in $\mathscr{A}$, by the same argument as that which yields (40.19) we find that

$$\Sigma^\alpha = \Sigma^\infty + o(\alpha). \tag{96b.60}$$

Hence (96b.58) can be written in the alternative form[1]

$$\dot{\psi}^\alpha = \Sigma^\alpha\cdot\dot{\Lambda}^\alpha + o(\alpha). \tag{96b.61}$$

Since $\Sigma^\alpha\cdot\dot{\Lambda}^\alpha$ and $\Sigma^\infty\cdot\dot{\Lambda}^\alpha$ are $O(\alpha)$, for sufficiently small α the error term $o(\alpha)$ in (96b.58) and (96b.61) may be ignored when compared to $\Sigma^\alpha\cdot\dot{\Lambda}^\alpha$ or $\Sigma^\infty\cdot\dot{\Lambda}^\alpha$. Thus in the limit of ultimate retardation, *a Gibbs relation holds asymptotically for any simple material obeying the principle of fading memory.* Comparison with (96b.30) shows (96b.61) to be equivalent to

$$\theta^\alpha\sigma^\alpha = o(\alpha). \tag{96b.62}$$

It is not difficult to see that (96b.13) and COLEMAN and NOLL's theorem (40.6) on retardation yields only the uninteresting conclusion $\theta^\alpha\sigma^\alpha = O(\alpha)$, which does not justify neglect of $\theta^\alpha\sigma^\alpha$ in comparison with $\Sigma^\alpha\cdot\dot{\Lambda}^\alpha$ or $\Sigma^\infty\cdot\dot{\Lambda}^\alpha$. Eq. (96b.44) is essential to the proof of the useful result (96b.62).

In Sect. 80 we have found one way to apply traditional thermostatic formulae in problems of deformation, namely, by restricting attention to the special materials there called "perfect". COLEMAN's work establishes a second and more important *rational position for the conventional "thermodynamics of irreversible processes".* That position, being asymptotic in the sense of retardation, is just the same as that of linear viscosity as given by Eq. (40.15) with $\mathfrak{n}=1$. In the particular case of a fluid, the theory of linear viscosity reduces to the theory based on the Navier-Stokes equations [Eq. (40.18) with $\mathfrak{n}=1$, or, for incompressible fluids, Eq. (121.29) with $\mathfrak{n}=1$]. We have seen already that if one further term in the series expansion in α be retained, the Navier-Stokes theory must be replaced by that of the theory of the fluid of second grade. A similar result holds for the thermodynamic relations of COLEMAN's theory. That is, in general, *a caloric equation of state is not valid* to the *second order* in the sense of retardation. A corresponding result in the kinetic theory of gases, a theory, the results of which are partly more general and partly less general than those for simple materials, is well known.

---

[1] COLEMAN [1964, *16*, Eq. (4.22)].

There is a second or "Onsagerist" part of the conventional "thermodynamics of irreversible processes", according to which linear "phenomenological relations" between otherwise undefined "forces" and "fluxes" are laid down and then restricted by "reciprocal relations". We cannot too strongly emphasize that COLEMAN's results give no status whatever to this part. Certain tensors in thermomechanics are symmetric, others are skew, others are neither, and for good reason. In this treatise we give, in each case, specific definitions of the variables used, and we *prove* such symmetry relations as we can. E.g., subject to stated assumptions (which may be, and often are, relaxed), the stress tensor is symmetric, the linear elasticity tensor and the initial value of the stress-relaxation tensor are symmetric and positive-definite, the linear viscosity tensor is isotropic (and hence also symmetric), etc. Further relations of symmetry remain, no doubt, to be *proved* or *characterized*, but until they are, they have no place in a treatise on a mathematical science.

The foregoing theorem shows that a Gibbs relation holds in a very slow process. In his second memoir, COLEMAN shows that a Gibbs relation holds also in a very fast process.

To make precise the concept of a fast process, we may note that if $\boldsymbol{\Xi}$ is any member of $\mathscr{A}$ and $\boldsymbol{\Lambda}^{(t)}(s)$ any total history, we can define a new total history $\boldsymbol{\Lambda}^{(t)*}$ by the relation

$$\boldsymbol{\Lambda}^{(t)*}(s) = \begin{cases} \boldsymbol{\Lambda}^{(t)}(0) + \boldsymbol{\Xi}, & s = 0, \\ \boldsymbol{\Lambda}^{(t)}(s), & 0 < s < \infty. \end{cases} \tag{96b.63}$$

COLEMAN calls $\boldsymbol{\Lambda}^{(t)*}(s)$ the *jump continuation*[1] *of* $\boldsymbol{\Lambda}^{(t)}(s)$ *with jump* $\boldsymbol{\Xi}$. The history $\boldsymbol{\Lambda}^{(t)*}(s)$ differs from $\boldsymbol{\Lambda}^{(t)}(s)$ by only a superimposed strain-temperature impulse occurring at time t and having amount $\boldsymbol{\Xi}$. The past histories corresponding to $\boldsymbol{\Lambda}^{(t)}(s)$ and $\boldsymbol{\Lambda}^{(t)*}(s)$ are both equal to $\boldsymbol{\Lambda}^{(t)}_{+}(s)$, while the corresponding present values are $\boldsymbol{\Lambda}^{(t)}(0) = \boldsymbol{\Lambda}$ and $\boldsymbol{\Lambda}^{(t)*}(0) = \boldsymbol{\Lambda} + \boldsymbol{\Xi}$. Hence, the corresponding present values of the free energy are given by[2]

$$\psi = \underset{s=0}{\overset{\infty}{\tilde{\mathfrak{p}}}}\left(\boldsymbol{\Lambda}^{(t)}_{+}(s); \boldsymbol{\Lambda}\right), \qquad \psi^* = \underset{s=0}{\overset{\infty}{\tilde{\mathfrak{p}}}}\left(\boldsymbol{\Lambda}^{(t)}_{+}(s); \boldsymbol{\Lambda} + \boldsymbol{\Xi}\right). \tag{96b.64}$$

If the function $\boldsymbol{\Lambda}^{(t)}(s)$, and hence the vector $\boldsymbol{\Lambda}$ and the function $\boldsymbol{\Lambda}^{(t)}_{+}(s)$, are assigned in advance, then ψ^* may be regarded as a function $\overline{\psi}^*$ of $\boldsymbol{\Xi}$:

$$\psi^* = \overline{\psi}^*(\boldsymbol{\Xi}). \tag{96b.65}$$

Holding $\boldsymbol{\Lambda}^{(t)}(s)$ fixed, we may consider the ordinary gradient $\overline{\psi}^*_{\boldsymbol{\Xi}}$ of $\overline{\psi}^*$, defined as usual by Eq. (9.2). On comparing that definition and Eqs. (96b.64) and (96b.65) with (96b.22), we see that whenever $\boldsymbol{\Lambda}^{(t)*}(s)$ is in the domain of $\tilde{\mathfrak{p}}$, $\overline{\psi}^*_{\boldsymbol{\Xi}}(\boldsymbol{\Xi})$ exists and is given by

$$\overline{\psi}^*_{\boldsymbol{\Xi}}(\boldsymbol{\Xi}) = D \underset{s=0}{\overset{\infty}{\mathfrak{p}}}\left(\boldsymbol{\Lambda}^{(t)*}(s)\right). \tag{96b.66}$$

In particular, the value $\overline{\psi}^*_{\boldsymbol{\Xi}}(\boldsymbol{0})$ of $\overline{\psi}^*_{\boldsymbol{\Xi}}(\boldsymbol{\Xi})$ at $\boldsymbol{\Xi} = \boldsymbol{0}$ is just $D \underset{s=0}{\overset{\infty}{\mathfrak{p}}}\left(\boldsymbol{\Lambda}^{(t)}(s)\right)$, and the stress-entropy relation (96b.15) may be written in the form[3]

$$\boldsymbol{\Sigma} = \overline{\psi}^*_{\boldsymbol{\Xi}}(\boldsymbol{0}) \tag{96b.67}$$

where $\boldsymbol{\Sigma}$ is the stress-entropy corresponding to $\boldsymbol{\Lambda}^{(t)}(s)$.

Eq. (96b.67), which gives the true stress-entropy resulting from an arbitrary history, has a striking formal similarity to the classical equation (96b.47) giving the equilibrium stress-entropy $\boldsymbol{\Sigma}^{\infty}$ corresponding to a constant history $\boldsymbol{\Lambda}^{(t)}(s) \equiv \boldsymbol{\Lambda}$.

---

[1] COLEMAN [1964, *16*, Eq. (5.1)].

[2] Dr. COLEMAN has kindly provided this new and simpler method of presenting the material in his memoir [1964, *16*, § 5].

[3] COLEMAN [1964, *16*; Eq. (5.11)].

Equation (96b.67) says: To find the value of $\boldsymbol{\Sigma}$ resulting from a history, one should superimpose on that history a small strain-temperature impulse, express the free energy as a function $\overline{\psi}^*$ of the intensity of the impulse, and then calculate $\boldsymbol{\Sigma}$ by the rules one uses to calculate $\boldsymbol{\Sigma}^\infty$ from ψ^∞ in classical thermostatics.

To make the analogy between (96b.67) and (96b.47) more apparent, let $\boldsymbol{\Sigma}^*$ be the stress-entropy corresponding to the jump continuation $\boldsymbol{\Lambda}^{(t)*}(s)$:

$$\boldsymbol{\Sigma}^* = \underset{s=0}{\overset{\infty}{\mathfrak{S}}}\left(\boldsymbol{\Lambda}^{(t)*}(s)\right). \tag{96b.68}$$

It follows from (96b.15) and (96b.66) that[1]

$$\boldsymbol{\Sigma}^* = \overline{\psi}^*_{\boldsymbol{\Xi}}(\boldsymbol{\Xi}), \tag{96b.69}$$

a relation which does indeed have the form of Eq. (96b.47).

We now allow $\boldsymbol{\Xi}$ in (96b.63) to depend smoothly on a parameter ν: $\boldsymbol{\Xi} = \boldsymbol{\Xi}(\nu)$. Then $\boldsymbol{\Lambda}^* = \boldsymbol{\Lambda}^{(t)*}(0)$ depends smoothly on ν. Writing $\boldsymbol{\Lambda}^* = \boldsymbol{\Lambda}^*(\nu)$, we put

$$\boldsymbol{\Lambda}^{*\prime} \equiv \frac{d}{d\nu}\boldsymbol{\Lambda}^*(\nu) = \frac{d}{d\nu}\left(\boldsymbol{\Lambda} + \boldsymbol{\Xi}(\nu)\right) = \boldsymbol{\Xi}'(\nu). \tag{96b.70}$$

The free energy ψ^* corresponding to $\boldsymbol{\Lambda}^{(t)*}(s)$ also depends on ν: that is, $\psi^* = \psi^*(\nu) = \overline{\psi}^*\left(\boldsymbol{\Xi}(\nu)\right)$. It follows from the principle of fading memory that if $\boldsymbol{\Lambda}^{(t)*}(s)$ is in the domain of $\mathfrak{p}$ and if $\boldsymbol{\Xi}'(\nu)$ exists, then the derivative $\psi^{*\prime} = \psi^{*\prime}(\nu)$ also exists and is given by

$$\psi^{*\prime} = \frac{d}{d\nu}\psi^*\left(\boldsymbol{\Xi}(\nu)\right) = \overline{\psi}^*_{\boldsymbol{\Xi}}\left(\boldsymbol{\Xi}(\nu)\right)\cdot\boldsymbol{\Xi}'(\nu). \tag{96b.71}$$

On substituting (96b.69) and (96b.70) into (96b.71) we obtain[2]

$$\psi^{*\prime} = \boldsymbol{\Sigma}^*\cdot\boldsymbol{\Lambda}^{*\prime}, \tag{96b.72}$$

an equation which has the form of a Gibbs relation.

COLEMAN[3] has discussed the physical interpretation of the Gibbs relation for fast processes, Eqs. (96b.69) and (96b.72). Let us take a material with arbitrary total history $\boldsymbol{\Lambda}^{(t)}(s)$ and study it at time t by changing $\boldsymbol{\Lambda}(t)$ so fast that in comparison with the history $\boldsymbol{\Lambda}^{(t)}(s)$, the new changes in $\boldsymbol{\Lambda}$ appear to be discontinuous[4]. The values of $\boldsymbol{\Lambda}$, ψ, and $\boldsymbol{\Sigma}$ at each "instant" ν of this new fast process will be $\boldsymbol{\Lambda}^*(\nu)$, $\psi^*(\nu)$, and $\boldsymbol{\Sigma}^*(\nu)$. Eq. (96b.69) states that if ψ^* is regarded as a function $\overline{\psi}^*$ of $\boldsymbol{\Lambda}^*$, then $\boldsymbol{\Sigma}^*$ is determinable from the function $\overline{\psi}^*$ by precisely the rules that give $\boldsymbol{\Sigma}$ in a perfectly elastic material (i.e. the rules used in statical situations)[5]. That the function $\overline{\psi}^*$ depends on the previous history $\boldsymbol{\Lambda}^{(t)}(s)$ means that the nature of the apparently perfectly elastic material which serves to interpret results in the rapid deformation depends on $\boldsymbol{\Lambda}^{(t)}(s)$: Corresponding to each previous history we have, in general, a different elastic material.

The conclusion just reached is exemplified by considering the behavior of a glass in fast experiments. A glass is but a simple fluid which, when studied rapidly, appears to be a perfectly elastic solid with a stored energy function; the various moduli of the elastic solid, such as the shear-modulus and coefficient of thermal expansion, depend, however, on the previous history of the glass. In a fast process the glass obeys a Gibbs relation of the form (96b.72). COLEMAN's derivations of (96b.69) and (96b.72) lead to the conclusion that any simple material obeying the principle of fading memory must exhibit, like a glass, elastic behavior with a stored energy function $\overline{\psi}^*$, when studied sufficiently rapidly.

---

[1] COLEMAN [1964, *16*, Eq. (5.14)].

[2] COLEMAN [1964, *16*, Eq. (5.19)].

[3] COLEMAN [1964, *16*, § 5].

[4] COLEMAN and GURTIN [1965, *9*, § 8] give a method of approximating jump continuations by histories which are continuous.

[5] This conclusion is not to be confused with the relatively trivial observation expressed by the stress-relaxation formula (39.12).

δ) *Consequence of material indifference.* Eqs. (96b.23), (96b.26), (96b.27), and (96b.2) tell us that with the exception of $\mathfrak{H}$ all of the constitutive functionals listed in Eq. (96b.1) and (96b.8) are independent of the temperature gradient $\boldsymbol{g}$.

Let $\mathfrak{f}$ stand for any one of the scalar-valued functionals $\mathfrak{p}$, $\mathfrak{e}$, and $\mathfrak{h}$. It is not difficult to see that the present theory is compatible with the principle of material frame-indifference if and only if the functionals $\mathfrak{f}$, $\mathfrak{T}$, and $\mathfrak{H}$ obey the following identities[1] in $\boldsymbol{F}^{(t)}(s)$, $\theta^{(t)}(s)$, and $\boldsymbol{g}$ for each function $\boldsymbol{Q}(s)$ whose values are orthogonal tensors:

$$\left.\begin{aligned}\mathop{\mathfrak{f}}_{s=0}^{\infty}\left(\boldsymbol{Q}(s)\boldsymbol{F}^{(t)}(s),\,\theta^{(t)}(s)\right)&=\mathop{\mathfrak{f}}_{s=0}^{\infty}\left(\boldsymbol{F}^{(t)}(s),\,\theta^{(t)}(s)\right),\\ \mathop{\mathfrak{T}}_{s=0}^{\infty}\left(\boldsymbol{Q}(s)\boldsymbol{F}^{(t)}(s),\,\theta^{(t)}(s)\right)&=\boldsymbol{Q}(0)\mathop{\mathfrak{T}}_{s=0}^{\infty}\left(\boldsymbol{F}^{(t)}(s),\,\theta^{(t)}(s)\right)\boldsymbol{Q}(0)^{-1},\\ \mathop{\mathfrak{H}}_{s=0}^{\infty}\left(\boldsymbol{Q}(s)\boldsymbol{F}^{(t)}(s),\,\theta^{(t)}(s);\,\boldsymbol{Q}(0)\boldsymbol{g}\right)&=\boldsymbol{Q}(0)\mathop{\mathfrak{H}}_{s=0}^{\infty}\left(\boldsymbol{F}^{(t)}(s),\,\theta^{(t)}(s)\right).\end{aligned}\right\}\qquad(96\text{b}.73)$$

These identities are satisfield if and only if there be new functionals $\bar{\mathfrak{f}}$, $\bar{\mathfrak{T}}$, $\bar{\mathfrak{H}}$ such that[2]

$$\left.\begin{aligned}\varphi&=\mathop{\bar{\mathfrak{f}}}_{s=0}^{\infty}\left(\boldsymbol{C}^{(t)}(s),\,\theta^{(t)}(s)\right),\\ \tilde{\boldsymbol{T}}&=\varrho_{\mathrm{R}}\mathop{\bar{\mathfrak{T}}}_{s=0}^{\infty}\left(\boldsymbol{C}^{(t)}(s),\,\theta^{(t)}(s)\right),\\ \boldsymbol{F}^{T}\boldsymbol{h}&=\mathop{\bar{\mathfrak{H}}}_{s=0}^{\infty}\left(\boldsymbol{C}^{(t)}(s),\,\theta^{(t)}(s);\,\boldsymbol{F}^{T}\boldsymbol{g}\right);\end{aligned}\right\}\qquad(96\text{b}.74)$$

here $\tilde{\boldsymbol{T}}$ is the second Piola stress tensor, defined by Eq. $(43\text{A}.11)_5$, $\boldsymbol{C}^{(t)}(s)$ is defined, as usual, by Eq. (29.5), and φ, standing for ψ, ε, and η in turn, is the value of $\mathfrak{f}$ in each case.

When φ represents ψ, $(96\text{b}.74)_1$ becomes

$$\psi=\mathop{\bar{\mathfrak{p}}}_{s=0}^{\infty}\left(\boldsymbol{C}^{(t)}(s),\,\theta^{(t)}(s)\right),\qquad(96\text{b}.75)$$

which may be written alternatively in terms of the past histories $\boldsymbol{C}^{(t)}_{+}(s)$, $\theta^{(t)}_{+}(s)$:

$$\psi=\mathop{\bar{\mathfrak{p}}}_{s=0}^{\infty}\left(\boldsymbol{C}^{(t)}_{+}(s),\,\theta^{(t)}_{+}(s);\,\boldsymbol{C},\,\theta\right).\qquad(96\text{b}.76)$$

We define "the instantaneous partial derivative" $D_{\boldsymbol{C}}\bar{\mathfrak{p}}$ of $\bar{\mathfrak{p}}$ by the following equation[3], which holds for all symmetric tensors $\boldsymbol{G}$:

$$\operatorname{tr}\left[\mathop{D_{\boldsymbol{C}}\bar{\mathfrak{p}}}_{s=0}^{\infty}\left(\boldsymbol{C}^{(t)}(s),\,\theta^{(t)}(s)\right)\boldsymbol{G}\right]=\frac{d}{d\nu}\mathop{\bar{\mathfrak{p}}}_{s=0}^{\infty}\left(\boldsymbol{C}^{(t)}_{+}(s),\,\theta^{(t)}_{+}(s);\,\boldsymbol{C}+\nu\boldsymbol{G},\,\theta\right)\Big|_{\nu=0}.\qquad(96\text{b}.77)$$

In order to relate the functional $D_{\boldsymbol{C}}\bar{\mathfrak{p}}$ of (96b.77) to the functional $D_{\boldsymbol{F}}\mathfrak{p}$ defined in (96b.24), we consider the following *jump continuation* $\boldsymbol{F}^{(t)*}(s)$ of an arbitrary history $\boldsymbol{F}^{(t)}(s)$ in the domain of $\mathfrak{p}$:

$$\boldsymbol{F}^{(t)*}(s)\equiv\left\{\begin{array}{ll}\boldsymbol{F}^{(t)}(0)+\nu\boldsymbol{A}, & s=0,\\ \boldsymbol{F}^{(t)}(s), & 0<s<\infty;\end{array}\right\}\qquad(96\text{b}.78)$$

---

[1] COLEMAN [1964, *15*, Eq. (13)].

[2] Eqs. $(96\text{b}.74)_{1,3}$ correspond to Eqs. (13.3a) and (13.3c) of COLEMAN [1964, *15*]; Eq. $(96\text{b}.74)_2$ is Eq. (6.11) of COLEMAN [1964, *16*].

[3] COLEMAN [1964, *16*, Eq. (6.15)].

here ν is some small number, and $\boldsymbol{A}$ is an arbitrary tensor. Letting $\boldsymbol{C}^{(t)}(s)$ and $\boldsymbol{C}^{(t)*}(s)$ correspond, respectively, to $\boldsymbol{F}^{(t)}(s)$ and $\boldsymbol{F}^{(t)*}(s)$, we find by calculation that

$$\boldsymbol{C}^{(t)*}(s)=\left\{\begin{array}{ll}\boldsymbol{C}^{(t)}(0)+\nu\boldsymbol{G}+\nu^2\boldsymbol{H}, & s=0,\\ \boldsymbol{C}^{(t)}(s), & 0<s<\infty,\end{array}\right\} \tag{96b.79}$$

where

$$\boldsymbol{G}=\boldsymbol{F}^T\boldsymbol{A}+\boldsymbol{A}^T\boldsymbol{F},\qquad \boldsymbol{H}=\boldsymbol{A}^T\boldsymbol{A},\qquad \boldsymbol{F}=\boldsymbol{F}^{(t)}(0). \tag{96b.80}$$

Hence, if we put $\boldsymbol{C}=\boldsymbol{C}^{(t)}(0)$ and $\boldsymbol{C}^*=\boldsymbol{C}^{(t)*}(0)$, we have

$$\boldsymbol{C}^*=\boldsymbol{C}+\nu\boldsymbol{G}+O(\nu^2), \tag{96b.81}$$

while the past history $\boldsymbol{C}_+^{(t)*}(s)$ is the same as $\boldsymbol{C}_+^{(t)}(s)$. By $(96.7)_1$, (96b.18), (96b.75), and (96b.79),

$$\left.\begin{array}{l}\overset{\infty}{\underset{s=0}{\tilde{\mathfrak{p}}}}\left(\boldsymbol{F}_+^{(t)}(s),\,\theta_+^{(t)}(s);\,\boldsymbol{F}+\nu\boldsymbol{A},\,\theta\right)=\overset{\infty}{\underset{s=0}{\mathfrak{p}}}\left(\boldsymbol{F}^{(t)*}(s),\,\theta^{(t)}(s)\right)=\overset{\infty}{\underset{s=0}{\bar{\mathfrak{p}}}}\left(\boldsymbol{C}^{(t)*}(s),\,\theta^{(t)}(s)\right),\\ \quad=\overset{\infty}{\underset{s=0}{\bar{\mathfrak{p}}}}\left(\boldsymbol{C}_+^{(t)*}(s),\,\theta_+^{(t)}(s);\,\boldsymbol{C}^*,\,\theta\right)=\overset{\infty}{\underset{s=0}{\bar{\mathfrak{p}}}}\left(\boldsymbol{C}_+^{(t)}(s),\,\theta_+^{(t)}(s);\,\boldsymbol{C}+\nu\boldsymbol{G}+O(\nu^2),\,\theta\right),\end{array}\right\} \tag{96b.82}$$

and therefore,

$$\frac{d}{d\nu}\overset{\infty}{\underset{s=0}{\mathfrak{p}}}\left(\boldsymbol{F}_+^{(t)}(s),\theta_+^{(t)}(s);\,\boldsymbol{F}+\nu\boldsymbol{A},\,\theta\right)\Big|_{\nu=0}=\frac{d}{d\nu}\overset{\infty}{\underset{s=0}{\bar{\mathfrak{p}}}}\left(\boldsymbol{C}_+^{(t)}(s),\,\theta_+^{(t)}(s);\,\boldsymbol{C}+\nu\boldsymbol{G},\,\theta\right)\Big|_{\nu=0}, \tag{96b.83}$$

or, by (96b.24), (96b.77) and $(96\text{b}.80)_1$

$$\left.\begin{array}{l}\operatorname{tr}\left[D_{\boldsymbol{F}}\overset{\infty}{\underset{s=0}{\mathfrak{p}}}\left(\boldsymbol{F}^{(t)}(s),\,\theta^{(t)}(s)\right)\boldsymbol{A}^T\right]=\operatorname{tr}\left[D_{\boldsymbol{C}}\overset{\infty}{\underset{s=0}{\bar{\mathfrak{p}}}}\left(\boldsymbol{F}^{(t)}(s),\,\theta^{(t)}(s)\right)(\boldsymbol{F}^T\boldsymbol{A}+\boldsymbol{A}^T\boldsymbol{F})\right],\\ \qquad=2\operatorname{tr}\left[\boldsymbol{F}D_{\boldsymbol{C}}\overset{\infty}{\underset{s=0}{\bar{\mathfrak{p}}}}\left(\boldsymbol{F}^{(t)}(s),\,\theta^{(t)}(s)\right)\boldsymbol{A}^T\right].\end{array}\right\} \tag{96b.84}$$

In writing the last of the above equations we have used the fact that $D_{\boldsymbol{C}}\overset{\infty}{\underset{s=0}{\bar{\mathfrak{p}}}}\left(\boldsymbol{C}^{(t)}(s),\right.$ $\left.\theta^{(t)}(s)\right)$ is, by definition, a symmetric tensor. Since (96b.84) must hold for *all* tensors $\boldsymbol{A}$,

$$D_{\boldsymbol{F}}\overset{\infty}{\underset{s=0}{\mathfrak{p}}}\left(\boldsymbol{F}^{(t)}(s),\,\theta^{(t)}(s)\right)=2\boldsymbol{F}D_{\boldsymbol{C}}\overset{\infty}{\underset{s=0}{\bar{\mathfrak{p}}}}\left(\boldsymbol{C}^{(t)}(s),\,\theta^{(t)}(s)\right), \tag{96b.85}$$

which equation gives the desired relation[1] between $D_{\boldsymbol{C}}\bar{\mathfrak{p}}$ and $D_{\boldsymbol{F}}\mathfrak{p}$.

On substituting (96b.85) into the stress relation (96b.26) we find that[2]

$$\boldsymbol{T}=2\varrho\boldsymbol{F}D_{\boldsymbol{C}}\overset{\infty}{\underset{s=0}{\bar{\mathfrak{p}}}}\left(\boldsymbol{C}^{(t)}(s),\,\theta^{(t)}(s)\right)\boldsymbol{F}^T, \tag{96b.86}$$

or, by Eq. $(43\text{A}.11)_5$,

$$\boldsymbol{T}=2\varrho_{\text{R}}D_{\boldsymbol{C}}\overset{\infty}{\underset{s=0}{\bar{\mathfrak{p}}}}\left(\boldsymbol{C}^{(t)}(s),\,\theta^{(t)}(s)\right). \tag{96b.87}$$

Hence we have the following theorem[3]: *The functionals* $\bar{\mathfrak{p}}$ *and* $\bar{\mathfrak{T}}$ *of* Eq. $(96\text{b}.74)_2$ *are related by the simple equation*

$$\bar{\mathfrak{T}}=2D_{\boldsymbol{C}}\bar{\mathfrak{p}}. \tag{96b.88}$$

---

[1] Dr. COLEMAN has kindly advised us that the factor 2 which appears here should be supplied also in Eqs. (6.26)—(6.30) of his paper [1964, *16*].

[2] COLEMAN [1964, *16*, Eq. (6.27)].

[3] COLEMAN [1964, *16*, Eq. (6.29)].

Let $\boldsymbol{T}^{\infty}$ and $\psi^{\infty} = \psi^{\infty}(\boldsymbol{C})$ be the stress and free energy of a material which has been at rest at all times at one temperature and one value $\boldsymbol{C}$ of the Cauchy-Green tensor. The argument used to obtain (96b.47) from $(96\text{b}.18)_2$ can be used to show that (96b.86) yields the classical stress relation (84.11) of hyperelasticity in the specific form

$$\boldsymbol{T}^{\infty} = 2\varrho \boldsymbol{F} \psi^{\infty}_{\boldsymbol{C}}(\boldsymbol{C}) \boldsymbol{F}^{T}, \tag{96b.89}$$

with stored-energy function $\bar{\sigma} = \psi^{\infty}$, as to be expected from (82.7), since the temperature is held constant. Eq. (96b.86) is thus a generalization of the classic stress relation of isothermal equilibrium to general thermodynamic processes in a simple material with fading memory.

COLEMAN[1] has expressed a number of the thermodynamic relations in terms of the relative deformation gradient $\boldsymbol{F}^{(t)}_{(t)}(s)$. While use of this measure of deformation simplifies essentially the theory of simple fluids (Sect. 105ff.), in thermodynamics formulae more rather than less complicated seem to result.

ε) *Implications for linear visco-elasticity.* Let us assume here that the temperature is constant at all times and temporarily cease to indicate $\theta^{(t)}(s)$ as one of the arguments of our constitutive functionals; then we can use results and notations of Subchapter CV of this treatise. In particular, let us consider the "small strain" approximation formula (41.21) of COLEMAN and NOLL, where ε is given by (41.10). COLEMAN[2] has shown that the principle of fading memory and his theorem on minimum free energy severely restrict the initial value of the stress-relaxation function $\mathsf{M}(s)$. Namely, according to COLEMAN'S ***theorem on the initial value of the stress-relaxation function,*** *the linear transformation* $\mathsf{M}(0)[\cdot]$ *is both symmetric and positive semi-definite: i.e., for all pairs* $\boldsymbol{B}$, $\boldsymbol{D}$ *of symmetric tensors*

$$\operatorname{tr}[\boldsymbol{D}\mathsf{M}(0)[\boldsymbol{B}]] = \operatorname{tr}[\boldsymbol{B}\mathsf{M}(0)[\boldsymbol{D}]], \tag{96b.90}$$

$$\operatorname{tr}[\boldsymbol{D}\mathsf{M}(0)[\boldsymbol{D}]] \geqq 0. \tag{96b.91}$$

In the component notation of (41.22), these formulae may be written

$$\mathsf{M}^{kmpq}(0) = \mathsf{M}^{pqkm}(0), \tag{96b.92}$$

$$\mathsf{M}^{kmpq}(0)\, D_{km} D_{pq} \geqq 0, \tag{96b.93}$$

where the inequality holds for all symmetric tensors $\boldsymbol{D}$.

Eq. (96b.92) has the form often called a "reciprocity relation". It cannot be emphasized too much that COLEMAN'S theorem rests, ultimately, on a *single thermodynamic postulate,* namely the Clausius-Duhem inequality. In particular, no appeal is made to "microscopic reversibility" or to any supplementary extremal postulate.

Although it is a consequence of the equilibrium-stress relation $(96\text{b}.48)_1$, as in hyperelasticity, that the *linear elasticity tensor* L *is symmetric,*

$$\mathsf{L}^{kmpq} = \mathsf{L}^{pqkm} \tag{96b.94}$$

[cf. Eqs. (82.13), (84.15), (84.16)], the Clausius-Duhem inequality does not seem to require that $\mathsf{M}(s)$ be symmetric for positive s.

The assertion that the linear elasticity tensor is positive-semidefinite is *not* a consequence of the Clausius-Duhem inequality, as may be seen by considering the classical theory of elasticity, in which the Clausius-Duhem inequality is always satisfied, trivially, but an additional requirement such as the GCN condition is needed (cf. Sect. 52) in order to insure that $\mathsf{L}^{pqkm} D_{pq} D_{km} \geqq 0$ for all symmetric $\boldsymbol{D}$.

In a recent work done in the framework of the linear theory of visco-elasticity defined by Eq. (41.23), GURTIN and HERRERA[3] study the consequences of the assumption that

$$\int_0^t \operatorname{tr}[\boldsymbol{T}(\tau)\boldsymbol{D}(\tau)]\, d\tau \geqq 0 \tag{96b.95}$$

[1] COLEMAN [1964, *16*, Appendix].
[2] COLEMAN [1964, *16*, § 8].
[3] GURTIN and HERRERA [1964, *41*].

for *all* processes starting from a state of rest: $\tilde{\boldsymbol{E}}(\tau) = \text{const.}$ when $-\infty < \tau < 0$. The inequality (96b.95), which does not appear to be a consequence of the Clausius-Duhem inequality, should not be confused with the conclusion (96b.55) of COLEMAN's theorem on isothermal cyclic processes. In (96b.95) there is no restriction on the final value of the strain, whereas in (96b.55) there is.

Among the results of HERRERA and GURTIN is a proof that (96b.55) requires not only that the linear transformation $(\mathsf{L} + \mathsf{M}(0))[\cdot]$ be symmetric and positive semi-definite, but also that for all symmetric tensors $\boldsymbol{D}$

$$|\operatorname{tr}[\boldsymbol{D}\mathsf{M}(s)\boldsymbol{D}]| \leqq |\operatorname{tr}[\boldsymbol{D}\mathsf{M}(0)\boldsymbol{D}]| \quad \text{if} \quad s \geqq 0. \tag{96b.96}$$

ζ) *Other choices of independent variables.* In classical thermostatics (e.g. CFT, Sect. 251) it is customary to assume that all functional transformations are invertible and to interconvert between various pairs of independent variables, as occasion may render convenient. Similar assumptions of invertibility may be introduced into COLEMAN's general theory. Since the energy rate $\dot{\varepsilon}$ is given explicitly by the equation of energy balance (79.3), in the general theory the energy history $\varepsilon^{(t)}(s)$ may be preferable to the temperature history $\theta^{(t)}(s)$ as independent variable, and the special experience gathered in gas dynamics suggests that the entropy history $\eta^{(t)}(s)$ may be the fittest thermal variable for studies of waves. The corresponding counterparts of Eqs. (96b.1), if we take advantage of the fact, which is included in the results of COLEMAN's fundamental theorem (96b.18), that $\boldsymbol{g}$ must drop out of $(96\text{b}.1)_{1,3,4}$, are

$$\left.\begin{aligned} \boldsymbol{T} &= \mathop{\tilde{\mathfrak{T}}}_{s=0}^{\infty}\left(\boldsymbol{F}^{(t)}(s),\, \varepsilon^{(t)}(s)\right),\\ \boldsymbol{h} &= \mathop{\tilde{\mathfrak{H}}}_{s=0}^{\infty}\left(\boldsymbol{F}^{(t)}(s),\, \varepsilon^{(t)}(s),\, \boldsymbol{g}\right),\\ \theta &= \mathop{\tilde{\mathfrak{t}}}_{s=0}^{\infty}\left(\boldsymbol{F}^{(t)}(s),\, \varepsilon^{(t)}(s)\right),\\ \eta &= \mathop{\tilde{\mathfrak{h}}}_{s=0}^{\infty}\left(\boldsymbol{F}^{(t)}(s),\, \varepsilon^{(t)}(s)\right), \end{aligned}\right\} \tag{96b.97}$$

and

$$\left.\begin{aligned} \boldsymbol{T} &= \mathop{\hat{\mathfrak{T}}}_{s=0}^{\infty}\left(\boldsymbol{F}^{(t)}(s),\, \eta^{(t)}(s)\right),\\ \boldsymbol{h} &= \mathop{\hat{\mathfrak{H}}}_{s=0}^{\infty}\left(\boldsymbol{F}^{(t)}(s),\, \eta^{(t)}(s),\, \boldsymbol{g}\right),\\ \theta &= \mathop{\hat{\mathfrak{t}}}_{s=0}^{\infty}\left(\boldsymbol{F}^{(t)}(s),\, \eta^{(t)}(s)\right),\\ \varepsilon &= \mathop{\hat{\mathfrak{e}}}_{s=0}^{\infty}\left(\boldsymbol{F}^{(t)}(s),\, \eta^{(t)}(s)\right). \end{aligned}\right\} \tag{96b.98}$$

To derive (96b.97) from (96b.8), we assume that for each fixed $\boldsymbol{F}^{(t)}(s)$ the functional transformation $(96\text{b}.1)_3$ mapping $\theta(\tau)$ onto $\varepsilon(\tau)$ is uniquely invertible; to derive (96b.98), that $(96\text{b}.1)_4$ is uniquely invertible for $\theta(\tau)$. COLEMAN's results[1], generalizing formulae of classical thermostatics, are

$$\left.\begin{aligned} \boldsymbol{T} &= -\varrho\theta\boldsymbol{F}\left[D_{\boldsymbol{F}} \mathop{\tilde{\mathfrak{h}}}_{s=0}^{\infty}\left(\boldsymbol{F}^{(t)}(s),\, \varepsilon^{(t)}(s)\right)\right]^{T},\\ \frac{1}{\theta} &= D_{\varepsilon} \mathop{\tilde{\mathfrak{h}}}_{s=0}^{\infty}\left(\boldsymbol{F}^{(t)}(s),\, \varepsilon^{(t)}(s)\right), \end{aligned}\right\} \tag{96b.99}$$

[1] COLEMAN [1964, *15*, §§ 11—12].

and

$$\left.\begin{aligned} \boldsymbol{T} &= \varrho \boldsymbol{F}\left[D_{\boldsymbol{F}} \mathop{\hat{\mathfrak{e}}}_{s=0}^{\infty}\left(\boldsymbol{F}^{(t)}(s), \eta^{(t)}(s)\right)\right]^T, \\ \theta &= D_\eta \mathop{\hat{\mathfrak{e}}}_{s=0}^{\infty}\left(\boldsymbol{F}^{(t)}(s), \eta^{(t)}(s)\right). \end{aligned}\right\} \quad (96\text{b}.100)$$

The functional operators $D_{\boldsymbol{F}}$, D_ε, and D_η are defined in analogy to Eqs. (96b.24) and (96b.25).

Since $\psi \equiv \varepsilon - \eta\theta$, we may write Eq. (96b.31) in the form[1]

$$\dot{\varepsilon} = \frac{1}{\varrho} \operatorname{tr}(\boldsymbol{T}\boldsymbol{D}) + \theta(\dot{\eta} - \sigma). \quad (96\text{b}.101)$$

From the internal dissipation inequality (96b.16) and from Eqs. (96b.101) and (79.3) may be read off the following inequalities equivalent to (96b.32)[2]:

$$\theta\dot{\eta} \geqq \dot{\varepsilon} - \frac{1}{\varrho} \operatorname{tr}(\boldsymbol{T}\boldsymbol{D}), \quad (96\text{b}.102)$$

$$\theta\dot{\eta} \geqq q + \frac{1}{\varrho} \operatorname{div} \boldsymbol{h}. \quad (96\text{b}.103)$$

By (96b.102) and (96b.101), in particular[3],

$$\boldsymbol{D} = \boldsymbol{0},\ \dot{\varepsilon} = 0 \Rightarrow \dot{\eta} = \sigma \geqq 0, \quad (96\text{b}.104)$$

$$\boldsymbol{D} = \boldsymbol{0},\ \dot{\eta} = 0 \Rightarrow \dot{\varepsilon} = -\theta\sigma \leqq 0. \quad (96\text{b}.105)$$

According to the former implication, in a rigid motion at constant internal energy ε the rate of increase of entropy equals σ and is non-negative; in particular, *the entropy in a rigid iso-energetic motion cannot decrease.* This explains the name "internal dissipation" for σ. Similarly, according to (96b.105), *in a rigid isentropic motion, the internal energy cannot increase.* These inequalities are to be set alongside (96b.33).

Ineq. (96b.102) had been laid down as a postulate, *the principle of positive internal production of entropy*, by COLEMAN in earlier work[4] and applied to show that in an *iso-energetic, isochoric motion with constant stretch history, the stress power of a simple fluid must be non-negative.* It follows as a corollary, generalizing a classic argument in the theory of linear viscosity (CFT, Sect. 298), that the viscosity function of a fluid cannot have negative values, as we assume in the present treatise, Ineq. (108.9).

It is clear that in a development starting first with (96b.97) and then with (96b.98) the arguments used here to prove the theorem on minimum free energy [following Eq. (96b.43)] would prove the following two theorems. ***Theorem on maximum entropy***[5]: *Of all total histories ending with given values of* $\boldsymbol{F}$ *and* ε, *that corresponding to constant values of* $\boldsymbol{F}$ *and* ε *gives rise to the greatest entropy*; i.e.,

$$\mathop{\tilde{\mathfrak{h}}}_{s=0}^{\infty}\left(\boldsymbol{F}^{(t)}(s) - \boldsymbol{F}, \varepsilon^{(t)}(s) - \varepsilon; \boldsymbol{F}, \varepsilon\right) \leqq \mathop{\tilde{\mathfrak{h}}}_{s=0}^{\infty}\left(\boldsymbol{0}(s), \boldsymbol{0}(s); \boldsymbol{F}, \varepsilon\right). \quad (96\text{b}.106)$$

Theorem on minimum internal energy[6]: *Of all histories ending with given values of* $\boldsymbol{F}$ *and* η, *that corresponding to constant values of* $\boldsymbol{F}$ *and* η *has the least internal energy.*

[1] COLEMAN [1964, *15*, Eq. (11.14)].
[2] COLEMAN [1964, *15*, Ineqs. (12.9) and (12.10)]. Cf. the remarks in CFT concerning PLANCK's postulate (264.7).
[3] COLEMAN [1964, *15*, Eqs. (12.11) and (11.22), and Remarks 31 and 22].
[4] COLEMAN [1962, *8*, § 7]. See his Remarks 5, 6, and 7.
[5] COLEMAN [1964, *15*, § 12, Remark 31].
[6] COLEMAN [1964, *15*, § 11, Remark 22].

Similarly, the theorem on isothermal cyclic processes [following Eq. (96b.55)] can be extended to the following ***general theorem on cyclic processes***[1]: *In every isothermal* ($\dot{\theta}=0$), *iso-energetic* ($\dot{\varepsilon}=0$), *or isentropic* ($\dot{\eta}=0$) *cyclic process starting from a state of equilibrium, the total stress work done in traversing a closed path is non-negative.* This theorem gives precise meaning and rational status to a century of conjecture and assertion regarding "reversibility", "irreversibility", and the "second law of thermodynamics".

The following integral appears easier to calculate in practice than that defined in (96b.49):

$$A(t_1, t_0) = \int_{t_0}^{t_1} \left[\frac{1}{\varrho\theta} \operatorname{tr}(\boldsymbol{T}\boldsymbol{D}) - \frac{\dot{\varepsilon}}{\theta}\right] dt. \tag{96b.107}$$

From (96b.101) and (96b.16) we easily obtain, alongside (96b.50), the following integrated dissipation inequality[2]:

$$A(t_1, t_0) \geqq \eta(t_0) - \eta(t_1). \tag{96b.108}$$

By use of constitutive equations in the form (96b.97) the theorem on *Gibbs relations for retarded motions*, as contained in Eqs. (96b.57) through (96b.61), may be rewritten in the form commonly used by the cultivators of linear irreversible thermodynamics. If we retard a process, replacing $\boldsymbol{F}^{(t)}(s)$ and $\varepsilon^{(t)}(s)$ by $\boldsymbol{F}^{(t)}(\alpha s)$ and $\varepsilon^{(t)}(\alpha s)$, then[3], to within an error $o(\alpha)$,

$$\theta^\infty \dot{\eta}^\alpha = \dot{\varepsilon}^\alpha - \frac{1}{\varrho^\alpha} \operatorname{tr}(\boldsymbol{T}^\infty \boldsymbol{D}^\alpha), \tag{96b.109}$$

where the equilibrium values $\boldsymbol{T}^\infty$, θ^∞, η^∞ are obtained as follows:

$$\left.\begin{aligned} \tilde{\eta}^\infty(\boldsymbol{F}, \varepsilon) &= \mathop{\tilde{\mathfrak{h}}}_{s=0}^{\infty}\big(\boldsymbol{F}1(s), \varepsilon 1(s)\big), \\ \theta^\infty &= \mathop{\tilde{\mathfrak{t}}}_{s=0}^{\infty}\big(\boldsymbol{F}^\alpha 1(s), \varepsilon^\alpha 1(s)\big) = [\tilde{\eta}_\varepsilon^\infty(\boldsymbol{F}^\alpha, \varepsilon^\alpha)]^{-1}, \\ \boldsymbol{T}^\infty &= \mathop{\tilde{\mathfrak{T}}}_{s=0}^{\infty}\big(\boldsymbol{F}^\alpha 1(s), \varepsilon^\alpha 1(s)\big) = -\varrho^\alpha \theta^\infty \boldsymbol{F}^\alpha [\eta_{\boldsymbol{F}}^\infty(\boldsymbol{F}^\alpha, \varepsilon^\alpha)], \end{aligned}\right\} \tag{96b.110}$$

and the superscript α's refer to the retarded process. That is, $\eta^\infty(\boldsymbol{F}, \varepsilon)$ is the specific entropy of the material when it is held at rest at the deformation $\boldsymbol{F}$ with specific energy ε, so that $\boldsymbol{F}^{(t)}(s) = \boldsymbol{F}$, $\varepsilon^{(t)}(s) = \varepsilon$. According to $(96b.110)_{2,3}$, the temperature θ^∞ and stress $\boldsymbol{T}^\infty$ are calculated from the caloric equation of state $(96b.110)_1$ by the classical rules.

COLEMAN's theory has been shown to include a large part of the thermodynamic assumptions made here and there regarding the temperature and entropy of media in motion[4]. In particular, it provides a basis for all assumptions of this kind made in the present treatise. Of course it could *not* include the specifically *static* parts of thermostatics, such as the GIBBS minimal principles and COLEMAN and NOLL's extension of them to elastic materials subject to finite deformation (Sect. 87).

---

[1] COLEMAN [1964, *15*, §§ 11—12, Remarks 25 and 33]. ["Total stress work" for the first and third cases denotes the left-hand side of (96b.55) while for the second case it denotes the right-hand of (96b.107) when $\dot{\varepsilon} = 0$.]

[2] COLEMAN [1964, *15*, Eq. (12.18)] (cf. also his Eq. (11.28)).

[3] COLEMAN [1964, *15*, § 12, Remark 34].

[4] According to LIANIS [1965, *24*], COLEMAN's theory includes as a special case the proposal of BERNSTEIN, KEARSLEY, and ZAPAS [1964, *7*], and their thermodynamic postulate can be proved as a theorem. LIANIS claimed also that certain relations alleged to follow from "ONSAGER's principle" are implied by COLEMAN's theory.

96ter. COLEMAN and GURTIN'S general theory of wave propagation in simple materials. (Added in proof.) In 1901 the kinds of singular surfaces that may exist and propagate in a linearly viscous fluid were proved by DUHEM[1] to be extremely limited. Acceleration waves are altogether impossible (CFT, Sect. 298), and shock waves can occur only in peculiar combinations of circumstances[2] connected with the presence or absence of heat conduction of suitable kind and amount. In 1906 PRANDTL substituted for DUHEM'S careful assertions and proofs some qualitative remarks and calculations of the "physical" kind, which seem to be the origin of the widespread belief, reaffirmed in practically every book on gas dynamics[3], that surfaces of discontinuity are wiped out by any "dissipative mechanism". While DUHEM'S theorems lend this belief some qualified status for linearly viscous fluids, these are exceptional as examples of dissipative media, and the belief itself is unsound. Wave propagation is, in general, easily compatible with irreversible changes[4]. In sufficiently retarded motion, as we have seen in Sect. 41, the constitutive equation of any simple material with fading memory in COLEMAN and NOLL'S sense is sufficiently approximated by some particular constitutive equation of BOLTZMANN'S infinitesimal theory of visco-elasticity when only infinitesimal deformations are considered. In particular cases of that theory, examples of solutions with shock waves began to be found some years ago[5], and recently HERRERA and GURTIN[6] calculated the acoustical tensor in full generality and found it to be the same as for an associated elastic material.

In a monumental study now going to press, COLEMAN and GURTIN, partly in collaboration with HERRERA[7], have constructed *a general theory of wave propagation in materials with memory*. Not only mechanical but also thermal effects are included, in the full generality of COLEMAN'S thermodynamics of simple materials (Sect. 96bis). Here we can do no more than state a few of their major results.

[1] DUHEM [1901, *1* and *2*] [1902, *1*].

[2] DUHEM'S results on shocks were corrected by SERRIN [1959, *27*, footnote 1, p. 220].

[3] On p. 15 of Volume VIII/1 of this Encyclopedia is the somewhat less categorical statement of OSWATITSCH: „Ein Temperatursprung beispielsweise kann aber nicht auftreten; er würde durch eine, wenn auch noch so geringe Wärmeleitung ‚verschmiert' werden. Eine weitere Ursache für eine solche ‚Verschmierung' liegt in der inneren Reibung, die in Gebieten sprunghafter Geschwindigkeitsänderungen groß wird." While PRANDTL [1906, *3*] in his analysis of a plane shock layer in an inviscid but heat conducting gas made no reference to the earlier and more complete theory of "quasi-waves" created by DUHEM, and while he implied (though he did not affirm) a general opinion that discontinuities are unlikely in *nature*, he nowhere wrote anything at all about the general presence or absence of discontinuities in *theories* including dissipative effects.

[4] This observation, it seems, was first made by LAMPARIELLO [1931, *4*]. "One might think that the impossibility of wave propagation were a necessary consequence of viscosity. That is not so." As an example, he showed that wave propagation in a perfectly flexible string with viscous damping satisfies the same laws as for the frictionless string. This result of LAMPARIELLO, which affords the first special case of the general theorem of COLEMAN and GURTIN, seems to have passed unnoticed in the physical literature.

[5] Studying a one-dimensional theory of integral type, in 1949 SIPS [1951, *16*] discovered the "very remarkable fact" that the laws of propagation are "almost identical" with those in elasticity; he found the speed of propagation to be that of a perfectly elastic material, "a general result, independent of the form of the relaxation function". Cf. also the preceding footnote and the work of LEE and KANTER [1953, *15*], and CHU [1962, *8*]. BARENBLATT and CHERNYI [1963, *4*, § 3] remarked that in its inability to transmit waves the linearly viscous fluid is exceptional rather than typical among dissipative media.

[6] HERRERA and GURTIN [1964, *42*]. According to expectation in a theory of infinitesimal deformation, waves of all orders obey the same laws of propagation, as has been confirmed by FISHER and GURTIN [1964, *27*].

[7] COLEMAN, GURTIN, and HERRERA [1965, *8*], COLEMAN and GURTIN [1965, *9*, *10*, *11*].

In Sect. 71, where thermodynamic influences are ignored for the most part, an acceleration wave was defined as a singular surface across which the motion $\boldsymbol{x}=\boldsymbol{\chi}(\boldsymbol{X},t)$ and its first derivatives are continuous, while some of the second derivatives $x^k{}_{,\alpha\beta}$ suffer jump discontinuities. Here we extend the definition by requiring that the thermal fields θ and η be continuous, while their first derivatives $\dot\theta$, $\dot\eta$, grad θ, and grad η may exhibit jump discontinuities across the singular surface. When, in addition,

$$[\dot\theta]=0 \quad\text{and}\quad [\nabla\theta]=\mathbf{0}, \tag{96t.1}$$

the acceleration wave is called *homothermal*; if, instead,

$$[\dot\eta]=0 \quad\text{and}\quad [\nabla\eta]=\mathbf{0}, \tag{96t.2}$$

the wave is called *homentropic*.

For the theory of waves it is convenient to write COLEMAN's stress relations (96b.26) and (96b.100)$_1$ in the forms

$$\left.\begin{aligned}\boldsymbol{T}_{\mathrm{R}}&=\mathop{\tilde{\mathfrak{T}}_{\mathrm{R}}}_{s=0}^{\infty}\left(\boldsymbol{F}^{(t)}(s),\,\theta^{(t)}(s)\right),\\ &=\mathop{\hat{\mathfrak{T}}_{\mathrm{R}}}_{s=0}^{\infty}\left(\boldsymbol{F}^{(t)}(s),\,\eta^{(t)}(s)\right),\end{aligned}\right\} \tag{96t.3}$$

where $\boldsymbol{T}_{\mathrm{R}}$ is the first Piola-Kirchhoff stress tensor, defined by Eqs. (43A.3) [cf. the special cases (45.18) for elasticity]. Generalizing Eqs. (45.20), we define fourth-order tensors $\tilde{\mathsf{A}}$ and $\hat{\mathsf{A}}$ which measure the sensitivity of the stress to isothermal and isentropic changes of the deformation history:

$$\left.\begin{aligned}\tilde{\mathsf{A}}&\equiv D_{\boldsymbol{F}}\mathop{\tilde{\mathfrak{T}}_{\mathrm{R}}}_{s=0}^{\infty}\left(\boldsymbol{F}^{(t)}(s),\,\theta^{(t)}(s)\right),\\ \tilde{A}_{k\ m}^{\ \alpha\ \beta}&\equiv D_{F^m{}_\beta}\mathop{\tilde{\mathfrak{T}}_{\mathrm{R}k}{}^{\alpha}}_{s=0}^{\infty}\left(F^{(t)p}{}_\mu(s),\,\theta^{(t)}(s)\right),\end{aligned}\right\} \tag{96t.4}$$

and

$$\left.\begin{aligned}\hat{\mathsf{A}}&\equiv D_{\boldsymbol{F}}\mathop{\hat{\mathfrak{T}}_{\mathrm{R}}}_{s=0}^{\infty}\left(\boldsymbol{F}^{(t)}(s),\,\eta^{(t)}(s)\right),\\ \hat{A}_{k\ m}^{\ \alpha\ \beta}&\equiv D_{F^m{}_\beta}\mathop{\hat{\mathfrak{T}}_{\mathrm{R}k}{}^{\alpha}}_{s=0}^{\infty}\left(F^{(t)p}{}_\mu(s),\,\theta^{(t)}(s)\right),\end{aligned}\right\} \tag{96t.5}$$

where the operator D is defined by Eq. (96b.12).

COLEMAN and GURTIN's main results in the three-dimensional theory are stated as four great theorems[1]:

1. ***Theorem on homothermal and homentropic acceleration waves:*** *In any simple material obeying Coleman and Noll's principle of fading memory, the Fresnel-Hadamard propagation condition* (71.3) *holds for homothermal or homentropic acceleration waves. The acoustical tensor* $\boldsymbol{Q}(\boldsymbol{n})$ *is given by Eq.* (71.4) *with* A

[1] COLEMAN and GURTIN [1965, *11*]. In the proofs of the first and third theorems the Clausius-Duhem inequality is not used, but it is essential to the other two.

If no dependence on thermodynamic quantities is allowed, the first theorem yields a common formula for the acoustical tensor. For the case when the response functional can be expanded in a series of multiple integrals, results equivalent to those stated in the first two theorems, except that thermodynamic influences were not considered, were obtained at the same time and independently by VARLEY [1965, *38*]. Again without taking into account thermodynamic effects, analogous results within the theory of finite linear visco-elasticity were derived by CHEN [1965, *5A*].

replaced by $\tilde{A}$ *for the former, by* $\hat{A}$ *for the latter, where* $\tilde{A}$ *and* $\hat{A}$ *are defined by* (96t.4) *and* (96t.5).

2. **Theorem on the symmetry of acoustic tensors.** *The acoustic tensors for homothermal and homentropic waves are symmetric. If the instantaneous heat capacity* $\varkappa$ *is defined by*

$$\varkappa \equiv \theta D_\theta \underset{s=0}{\overset{\infty}{\mathfrak{h}}} \left(\boldsymbol{F}^{(t)}(s),\, \theta^{(t)}(s)\right), \tag{96t.6}$$

where $\mathfrak{h}$ *is the entropy functional occurring in* $(96\text{b}.1)_4$, *then the homothermal and homentropic acoustical tensors* $\tilde{\boldsymbol{Q}}$ *and* $\hat{\boldsymbol{Q}}$ *satisfy Duhem's relation* (90.14) *with* $\boldsymbol{P}(\boldsymbol{n})$ *given by the definition*

$$\boldsymbol{P}(\boldsymbol{n}) \equiv \frac{1}{\varrho_R} \sqrt{\frac{\varrho\theta}{\varkappa}}\, D_\theta \underset{s=0}{\overset{\infty}{\tilde{\mathfrak{T}}_{\mathrm{R}}}} \left(\boldsymbol{F}^{(t)}(s),\, \theta^{(t)}(s)\right) \boldsymbol{F}^T \boldsymbol{n}. \tag{96t.7}$$

Just as the definitions (96t.4) and (96t.5) generalize the elasticity of an elastic material, the conductivity of a thermal conductor is generalized by the definition

$$\boldsymbol{K} \equiv \partial_{\boldsymbol{g}} \underset{s=0}{\overset{\infty}{\mathfrak{H}}} \left(\boldsymbol{F}^{(t)}(s),\, \theta^{(t)}(s);\, \boldsymbol{g}\right), \tag{96t.8}$$

where $\mathfrak{H}$ is the functional giving the heat flux $\boldsymbol{h}$. If $(96\text{b}.1)_2$ reduces to FOURIER's law (71.18), then $\boldsymbol{K}$ is the ordinary conductivity tensor. COLEMAN and GURTIN call the simple material a *definite conductor of heat* if $\boldsymbol{K}$ is positive-definite for all total histories and temperature gradients, while if $\mathfrak{H} \equiv \boldsymbol{0}$, they call the material a *non-conductor of heat*. Their further main theorems concern these two types of materials.

3. **Theorem on definite conductors.** *Every acceleration wave in a definite conductor of heat is homothermal.*

4. **Theorem on non-conductors.** *Every acceleration wave*[1] *in a non-conductor of heat is homentropic.*

These four theorems include and unite the two theorems of DUHEM stated at the end of Sect. 71 and HADAMARD's fundamental theorem in Sect. 90. They show that in any simple material with fading memory, provided it be a definite conductor or a non-conductor, *the speeds and amplitudes of acceleration waves are those of an appropriate hyperelastic material.* For a given simple material, the stored-energy function of the "appropriate" hyperelastic material depends upon the deformation and temperature histories at the particle considered.

For one-dimensional motion the results assume a simpler form, and COLEMAN and GURTIN[2] have calculated the corresponding modes of growth and decay. The results, while entirely explicit, are too elaborate to list here, so we give only one example. The amplitude $a(t)$ of a plane acceleration wave propagating into an infinite non-conductor or definite conductor which has always been at rest at constant temperature in a homogeneous configuration satisfies the equation

$$a(t) = \frac{A}{\left(\dfrac{A}{a(0)} - 1\right) e^{Bt} + 1}, \tag{96t.9}$$

where A and B are constants determined by the material and the temperature and deformation ahead of the wave. Of course, A and B depend on different

[1] The term "wave" here means strictly "propagating singular surface", excluding material discontinuities.

[2] COLEMAN and GURTIN [1965, *10*].

material parameters in the two cases. One expects the damping coefficient B to be positive; for conductors, B is greater the smaller is the thermal conductivity (96t.8). The number A is a critical amplitude. According to the agreement in sign and relative magnitude between the initial amplitude $a(0)$ and A, the amplitude $a(t)$ decreases monotonically to zero as $t\to\infty$, remains constant, or increases in magnitude monotonically to ∞ in a finite time. This result contains as limit cases that of W. A. GREEN for elasticity (Sect. 74) and a counterpart for infinitesimal visco-elasticity; the latter corresponds to infinitesimally small $a(0)$, the former to setting $B=CA$ and letting A approach 0 while C is held fixed, so that (74.13) results.

In earlier studies COLEMAN, GURTIN, and HERRERA[1] considered further aspects of wave propagation when thermal effects are neglected: shear waves, shock waves, and the equivalence of singular surfaces to infinitesimal oscillations. For plane shock waves COLEMAN, GURTIN, and HERRERA applied Eq. (205.7) of CFT and obtained

$$U^2=\frac{E_{[\boldsymbol{F}]}}{\varrho_R}, \tag{96t.10}$$

where $E_{[\boldsymbol{F}]}$ is an "instantaneous secant modulus" as defined by them.

More generally, VARLEY[2] calculated a partial differential equation for the change of amplitude in an acceleration wave of arbitrary form as it progresses. He was able to integrate the equation explicitly for plane, cylindrical, or spherical waves propagating into an infinite body at rest. In VARLEY's work, thermal effects are not included. The later results of COLEMAN and GURTIN[3] show the temperature history to influence strongly the coefficient B in (96t.9) and hence to have a decisive effect in regard to growth or decay.

97. Electromechanical theories. A theory of the interaction between deformation and polarization has been constructed by TOUPIN[4]. It is a generalization of the theory of perfect materials in that (80.1) and (80.3) are replaced by

$$\left.\begin{aligned} \varepsilon&=\hat{\varepsilon}(\boldsymbol{F},\boldsymbol{P}),\\ \boldsymbol{T}&=\mathfrak{g}(\boldsymbol{F},\boldsymbol{P})=\boldsymbol{T}^T,\end{aligned}\right\} \tag{97.1}$$

where $\boldsymbol{P}$ is the polarization vector. Thermo-energetic effects are left out of account, so that, in particular, $\boldsymbol{h}=\boldsymbol{0}$. The electromagnetic influence upon the motion is represented by a body force:

$$\varrho\boldsymbol{b}=-\mathfrak{E}\operatorname{div}\boldsymbol{P}+\overset{*}{\boldsymbol{P}}\times\boldsymbol{B}, \tag{97.2}$$

where $\mathfrak{E}$ is the electromotive intensity[5], $\boldsymbol{B}$ is the magnetic intensity (not the Cauchy-Green tensor as elsewhere in this treatise), and $\overset{*}{\boldsymbol{P}}$ is the convected time flux of $\boldsymbol{P}$ (cf. CFT, Sect. 150):

$$\overset{*}{\boldsymbol{P}}=\frac{\partial\boldsymbol{P}}{\partial t}+\dot{\boldsymbol{x}}\operatorname{div}\boldsymbol{P}+\operatorname{curl}(\boldsymbol{P}\times\dot{\boldsymbol{x}}). \tag{97.3}$$

---

[1] COLEMAN, GURTIN and HERRERA [1965, *8*], COLEMAN and GURTIN [1965, *9*]. PIPKIN [1965, *30*] has constructed an important special exact solution showing the formation of a shock in a particularly simple kind of finitely visco-elastic fluid with long-range memory.

[2] VARLEY [1965, *38*, Eq. (4.13) and § 5], broadly generalizing results of W. A. GREEN mentioned in Sect. 74.

[3] COLEMAN and GURTIN [1965, *10*].

[4] TOUPIN [1956, *31*] [1963, *72*]. Alternative formulations of the static theory, involving different decompositions of the stress and including a Lagrangean principle, have been given by TOUPIN [1960, *55*]. Cf. also the expositions by ERINGEN [1961, *15*] [1962, *18*, Ch. 11] [1963, *22*] and SEDOV [1965, *31*]. A theory combining the ideas of Sects. 97 and 98 (α) has been proposed by DIXON and ERINGEN [1965, *16*].

[5] So as to maintain consistency with classical notations and with TOUPIN's, it seems preferable here to make exception to our own scheme of notation laid down in Sect. 6.

The assumption (97.2) amounts to an expression for Lorentz force, $\varrho\boldsymbol{b}=Q\boldsymbol{E}+\boldsymbol{J}\times\boldsymbol{B}$, where $\boldsymbol{E}$ is the electric field, Q is the charge density and $\boldsymbol{J}$ is the total current density. In LORENTZ's view of the effects of polarization, $-\operatorname{div}\boldsymbol{P}$ plays the role of a charge density, Q, while $\overset{*}{\boldsymbol{P}}-\dot{\boldsymbol{x}}\operatorname{div}\boldsymbol{P}$ is the corresponding total current density, $\boldsymbol{J}$. Thus

$$\mathfrak{E}=\boldsymbol{E}+\dot{\boldsymbol{x}}\times\boldsymbol{B}. \tag{97.4}$$

The corresponding supply of energy, ϱq, is $\boldsymbol{J}\cdot\boldsymbol{E}-\varrho\boldsymbol{b}\cdot\dot{\boldsymbol{x}}$; thus

$$\varrho q=\overset{*}{\boldsymbol{P}}\cdot\mathfrak{E}. \tag{97.5}$$

(For further motivation, see Sect. 286 of CFT.) With these constitutive assumptions Eqs. (16.6) and (79.3) assume the forms

$$\left.\begin{aligned}\operatorname{div}\boldsymbol{T}-\mathfrak{E}\operatorname{div}\boldsymbol{P}+\boldsymbol{P}^*\times\boldsymbol{B}&=\varrho\ddot{\boldsymbol{x}},\\ \varrho\dot{\varepsilon}&=\operatorname{tr}(\boldsymbol{T}\boldsymbol{D})+\mathfrak{E}\cdot\overset{*}{\boldsymbol{P}}.\end{aligned}\right\} \tag{97.6}$$

Guided by the principle of equipresence (CFT, Sect. 293 η), we lay down the following constitutive equation for $\mathfrak{E}$:

$$\mathfrak{E}=\hat{\mathfrak{E}}(\boldsymbol{F},\boldsymbol{P}). \tag{97.7}$$

While the procedure of Sect. 80, based on the Clausius-Duhem inequality, could doubtless be applied here, the further work of TOUPIN, which appeared long before the results presented in Sect. 80 were known, follows the older approach through the equation of energy. The forms of the functions $\mathfrak{g}$ and $\hat{\mathfrak{E}}$ in $(97.1)_2$ and (97.7) are to be so adjusted that $(97.6)_2$ is satisfied in every process. Since $\hat{\mathfrak{E}}$ enters $(97.6)_2$ only in the combination $\mathfrak{E}\cdot\overset{*}{\boldsymbol{P}}$, it cannot be determined by consideration of energy except to within the quantity $-\boldsymbol{G}\times\overset{*}{\boldsymbol{P}}$, where the arbitrary vector $\boldsymbol{G}$ may be called the *gyration vector*.

The principle of material frame-indifference requires that

$$\hat{\varepsilon}(\boldsymbol{Q}\boldsymbol{F},\boldsymbol{Q}\boldsymbol{P})=\hat{\varepsilon}(\boldsymbol{F},\boldsymbol{P}), \tag{97.8}$$

identically in the orthogonal tensor $\boldsymbol{Q}$, the tensor $\boldsymbol{F}$, and the vector $\boldsymbol{P}$. Since $\boldsymbol{F}=\boldsymbol{R}\boldsymbol{U}$, choosing $\boldsymbol{Q}=\boldsymbol{R}^T$ in (97.8) yields[1]

$$\left.\begin{aligned}\hat{\varepsilon}(\boldsymbol{F},\boldsymbol{P})&=\hat{\varepsilon}(\boldsymbol{U},\boldsymbol{R}^T\boldsymbol{P})\\ &=\bar{\varepsilon}(\boldsymbol{C},\boldsymbol{\Pi}),\end{aligned}\right\} \tag{97.9}$$

where $\boldsymbol{C}$ is the Cauchy-Green tensor (23.4) and

$$\boldsymbol{\Pi}\equiv\frac{\varrho_R}{\varrho}\boldsymbol{F}^{-1}\boldsymbol{P}=\frac{\varrho_R}{\varrho}\boldsymbol{U}^{-1}\boldsymbol{R}^T\boldsymbol{P}. \tag{97.10}$$

It is easily verified that

$$\dot{\boldsymbol{\Pi}}=\frac{\varrho_R}{\varrho}\boldsymbol{F}^{-1}\overset{*}{\boldsymbol{P}}. \tag{97.11}$$

Substitution of (97.9) and (97.11) into $(97.6)_2$ now yields (cf. Sect. 80)

$$\varrho\operatorname{tr}[2\boldsymbol{F}\bar{\varepsilon}_{\boldsymbol{C}}\boldsymbol{F}^T\boldsymbol{D}]+\varrho\bar{\varepsilon}_{\boldsymbol{\Pi}}\cdot\dot{\boldsymbol{\Pi}}=\operatorname{tr}(\hat{\boldsymbol{T}}\boldsymbol{D})+\frac{\varrho}{\varrho_R}\hat{\mathfrak{E}}\cdot(\boldsymbol{F}\dot{\boldsymbol{\Pi}}). \tag{97.12}$$

---

[1] TOUPIN [1956, *31*, § 10], by a somewhat different argument.

Since $\boldsymbol{F}\bar{\varepsilon}_{\boldsymbol{C}}\boldsymbol{F}^T$ is symmetric, it follows from (97.12), (97.9), and (97.7) that

$$\left.\begin{aligned} \boldsymbol{T} &= 2\varrho\boldsymbol{F}\,\bar{\varepsilon}_{\boldsymbol{C}}(\boldsymbol{C},\boldsymbol{\Pi})\boldsymbol{F}^T, \\ T^{km} &= 2\varrho\frac{\partial\bar{\varepsilon}}{\partial C_{\alpha\beta}}x^k{}_{,\alpha}x^m{}_{,\beta}, \end{aligned}\right\} \tag{97.13}$$

$$\left.\begin{aligned} \mathfrak{E} &= \varrho_{\mathrm{R}}(\boldsymbol{F}^{-1})^T\bar{\varepsilon}_{\boldsymbol{\Pi}}(\boldsymbol{C},\boldsymbol{\Pi}) - \boldsymbol{G}\times\overset{*}{\boldsymbol{P}}, \\ \mathfrak{E}_k &= \varrho_{\mathrm{R}}\frac{\partial\bar{\varepsilon}}{\partial\Pi^\alpha}X^\alpha{}_{,k} - e_{kpq}G^p\overset{*}{P}{}^q. \end{aligned}\right\} \tag{97.14}$$

In (97.13) we recognize the stress relation for a perfect material, namely, (80.10) or its equivalent form (84.11), generalized to allow for the effect of polarization but specialized to the case when the entropy is held constant, while (97.14) is a new relation, its counterpart for the electromotive force. In contrast to the theory of perfect materials, *the form of the energy function $\bar{\varepsilon}$ determines the remaining constitutive equations only in part*, since the gyration vector $\boldsymbol{G}$ has no energetic significance.

The full set of equations of TOUPIN's theory is as follows:

1. Eq. (15.14) for the density.
2. CAUCHY's first law, in the form $(97.6)_1$.
3. The usual field equations of electromagnetic theory, viz [CFT, Sect. 278 and Eqs. (283.35)]

$$\left.\begin{aligned} \frac{\partial\boldsymbol{B}}{\partial t} + \operatorname{curl}\boldsymbol{E} &= \boldsymbol{0}, \\ \operatorname{div}\boldsymbol{B} &= 0, \\ \operatorname{curl}\mathfrak{H} - \frac{\partial\mathfrak{D}}{\partial t} &= \boldsymbol{0}, \\ \operatorname{div}\mathfrak{D} &= 0, \\ \mathfrak{H} &= \frac{\boldsymbol{B}}{\mu_0} + \dot{\boldsymbol{x}}\times\boldsymbol{P}, \\ \mathfrak{D} &= \varepsilon_0\boldsymbol{E} + \boldsymbol{P}, \end{aligned}\right\} \tag{97.15}$$

where μ_0 and ε_0 are dimensional constants.

4. The constitutive equations $(97.1)_1$, supplemented by an equation of the form

$$\boldsymbol{G} = \hat{\boldsymbol{G}}(\boldsymbol{F},\boldsymbol{\Pi}). \tag{97.16}$$

From the energy function $\bar{\varepsilon}$ follow the further constitutive equations (97.13) and (97.14).

From the equations of balance in integral form, corresponding jump conditions at surfaces of discontinuity may be derived[1].

While the theory so obtained is neither Galilean-invariant nor Lorentz-invariant, TOUPIN considers it to be satisfactory for circumstances in which the velocity at all points in the medium, with respect to some inertial frame, is small in comparison with the speed of light in a vacuum.

An isotropy group may be defined on the basis of (97.8). When there is a reference configuration for which this group contains the orthogonal group, the material is isotropic. For an isotropic material, TOUPIN[2] showed that $\bar{\varepsilon}$ is equal to

---

[1] TOUPIN [1963, *72*, Eq. (5.21)].

[2] TOUPIN [1956, *31*, § 11] called upon the fundamental theorem of CAUCHY stated in Sect. 11.

a function of $I_{\boldsymbol{B}}, II_{\boldsymbol{B}}, III_{\boldsymbol{B}}, \boldsymbol{P}\cdot(\boldsymbol{B}\boldsymbol{P}), \boldsymbol{P}\cdot(\boldsymbol{B}^2\boldsymbol{P})$, and P^2 where now $\boldsymbol{B}$ is the Cauchy-Green tensor, as usual.

Within this theory, a number of problems have been solved by TOUPIN and others.

Considering first static problems[1], TOUPIN obtained the solution for simple shear of a block of homogeneous isotropic material subject to a uniform electric field and uniform polarization in the plane of shear. Electrostriction occurs as an effect of second or higher order in the amount of shear. ERINGEN[2] gave a solution for uniform extension of a hollow cylindrical dielectric subject to radial electric field and radial polarization. Symmetrical expansion of a hollow spherical dielectric has been considered by VERMA[3]. For materials of arbitrary symmetry, TOUPIN found a position for the linear theories of VOIGT and HELMHOLTZ as approximations.

In the dynamic theory[4], TOUPIN set up the equations for small deformations and weak fields superimposed upon a given finite strain and strong electromagnetic field. A further linearization yields VOIGT's theory of piezo-electricity. TOUPIN generalized this theory and applied it to a rotating rigid dielectric, so obtaining an explanation of WILSON's experiment. For a special form of $\hat{\varepsilon}$, he obtained an explanation of the Faraday effect and of a new effect which accompanies it, a dispersion of transverse sound waves by application of a strong magnetic field in the direction of propagation. The usual theory of photo-elasticity in isotropic media follows as a different special case. TOUPIN derived also FRESNEL's formula for the dragging of light by a moving dielectric. Finally, he compared his results with those of BORN and HUANG and of MASON on crystal lattices; his attempt to place these only partially explained proposals on a rational footing was not successful.

PIPKIN and RIVLIN[5] proposed a theory in which the current density $\boldsymbol{j}$ may be influenced by the deformation:

$$\boldsymbol{j}=\hat{\boldsymbol{j}}(\boldsymbol{F}, \boldsymbol{E}). \tag{97.17}$$

The theory is appropriate to circumstances in which forces such as to maintain a given deformation are supplied and then adjusted so as to keep that deformation fixed, no matter what electric field be applied. $\boldsymbol{F}$ is then a known function of position, constant in time, but its value influences the current that flows in response to a given electric field. The only electromagnetic field equations not trivially satisfied (cf. CFT, Sect. 278) are

$$\operatorname{div} \boldsymbol{j}=0, \qquad \operatorname{curl} \boldsymbol{E}=\boldsymbol{0}. \tag{97.18}$$

In view of the latter, $\boldsymbol{E}=-\operatorname{grad}\varphi$. In regions where the potential φ is single-valued, we may set up the one-to-one correspondence

$$\theta \leftrightarrow -\varphi, \quad \boldsymbol{j} \leftrightarrow \boldsymbol{h}, \tag{97.19}$$

by which the field equations $(97.18)_1$ and (96.24) are transformed into one another. Boundary conditions of prescribed temperature correspond to those of prescribed potential; of prescribed normal flow of heat, to prescribed normal flow of current; while the entropy inequality $(79.11)_2$ is transformed into

$$\boldsymbol{j}\cdot\boldsymbol{E} \geqq 0, \tag{97.20}$$

the usual requirement that the Joulean heating be dissipative. The constitutive equation (97.17) is transformed into (96.1). Since $\boldsymbol{E}$ is assumed to be a frame-indifferent vector, $\hat{\boldsymbol{j}}$ must satisy the same functional equation as does $\hat{\boldsymbol{h}}$ in (96.1). Thus the two theories are fully isomorphic, establishing PIPKIN and RIVLIN's ***analogy between steady flow of heat and steady flow of electricity:*** *Under the constitutive equations* (96.1) *and* (97.17), *the permutations* (97.19) *and*

$$\hat{\boldsymbol{j}} \leftrightarrow \hat{\boldsymbol{h}} \tag{97.21}$$

transform the solution of a problem of steady heat flow into one of steady electric current flow, and conversely.

---

[1] TOUPIN [1956, *31*, §§ 12—17]. Cf. also ERINGEN [1962, *18*, § 113].

[2] ERINGEN [1961, *15*, § 7] [1962, *18*, § 114] [1963, *22*, § 7].

[3] VERMA [1964, *91*].

[4] TOUPIN [1963, *72*, §§ 6—12].

[5] PIPKIN and RIVLIN [1958, *35*, § 26] [1960, *41*].

Therefore, a separate mathematical treatment of the electric problem is superfluous. Exact solutions of two particular problems[1] of current flow may be read off from the results summarized at the end of Sect. 96.

In a similar spirit, ADKINS and RIVLIN[2] proposed as governing the propagation of electromagnetic waves in deformed isotropic elastic materials the following electromagnetic constitutive equations, in place of $(97.15)_{5,6}$:

$$\left.\begin{aligned}\mathfrak{D} &= (\mathfrak{n}_0 1 + \mathfrak{n}_1 \overline{\boldsymbol{B}} + \mathfrak{n}_2 \overline{\boldsymbol{B}}^2)\boldsymbol{E},\\ \boldsymbol{B} &= (\mathfrak{u}_0 1 + \mathfrak{u}_1 \overline{\boldsymbol{B}} + \mathfrak{u}_2 \overline{\boldsymbol{B}}^2)\mathfrak{H},\end{aligned}\right\} \tag{97.22}$$

where $\overline{\boldsymbol{B}}$ is the Cauchy-Green tensor (denoted by $\boldsymbol{B}$ elsewhere in this treatise), and where the coefficients $\mathfrak{n}_\Gamma$ and $\mathfrak{u}_\Gamma$ are scalar invariants of $\overline{\boldsymbol{B}}$. They presumed that forces be supplied sufficient to maintain $\overline{\boldsymbol{B}}$ at a fixed value, so that the field equations $(97.15)_{1,3}$ become a system of linear partial differential equations for $\mathfrak{H}$ and $\boldsymbol{E}$ with coefficients that are prescribed functions of position. They considered in detail the propagation of electromagnetic waves in a twisted rod.

98. Polar elastic materials.

α) Origins[3]. While heretofore in this treatise we have taken CAUCHY's second law (16.7) as the expression of the principle of balance of moments, it has long been known that non-symmetric stress tensors may occur in mechanics. In reality, CAUCHY's second law is a constitutive assumption, postulating that all torques are the moments of forces, or, in other words, that *there are neither body couples nor couple stresses.* An account of the matter has been given in Sects. 200—205 and 241 of CFT (cf. also Sect. 232), but, as will be seen below, it contains one error, and the whole subject is now understood from a simpler and more general standpoint[4].

Materials in which there may be couple stresses or body couples are called *polar.* The first theory of a polar elastic material was proposed by E. and F. COSSERAT[5] in 1907. DUHEM had noticed that various phenomena which seemed incompatible with ordinary continuum mechanics could be described as effects of direction, and he suggested that materials be visualized as sets of points having vectors attached to them, that is, as *oriented media.* The COSSERATS formulated a theory of elasticity corresponding to a special case of this idea, which they called *the theory of Euclidean action.* The vectors specifying the orientation, or *directors,* they restricted to be a rigid triad. As basic principles they laid down an action principle of classical type but with density of action allowed to depend upon the directors and their gradient fields, and, second, the requirement that the action be *invariant under proper orthogonal transformations.* For continua of one, two, and three dimensions, they developed their theory with elegance, precision, and

[1] In a later paper PIPKIN and RIVLIN [1961, *45*] presented the former problem in the electric context, integrating the equation $\boldsymbol{j} = \operatorname{curl} \boldsymbol{H}$ so as to obtain the magnetic field, $\boldsymbol{H}$. VERMA [1963, *77*] considered electrical conduction in a spherical shell.

[2] ADKINS and RIVLIN [1963, *3*].

[3] The following survey disregards the old suggestion, recently the subject of a doddering but noisy international revival, of a non-polar elastic medium in which the stress is unsymmetric from linear dependence upon the local rotation. This proposal, on which dozens of papers have been scattered and weighty conferences have been held, as a theory of materials is simply wrong, since it violates the principle of material frame-indifference. Of course, MACCULLAGH's theory of the luminiferous aether (CFT, Sect. 302), equivalent to MAXWELL's electromagnetic theory, rests upon taking the stress to be a function of the infinitesimal rotation, violating the principle of indifference, but the aether is not thought to be a material in the sense of classical mechanics.

[4] While one reviewer of CFT, echoed by another, deplored its authors' "predilection for the greatest possible generality", no sooner was the ink dry on the proofsheets than the experts in mechanics found the framework insufficiently general for modern work and began to enlarge it.

[5] E. and F. COSSERAT [1907, *1*] [1909, *1*].

thoroughness. It seems to have attracted slight attention[1]. The COSSERATS emphasized rods and shells, in which the concept of director suggests itself naturally, but the engineers of the period may not have had enough interest in an exact theory of finite deformation to follow so many pages of long calculation in general terms. The final end the COSSERATS seem to have had in mind for their three-dimensional model was an aether and radiation theory, replacing and uniting the Fresnel-Cauchy theory of elasticity and the MacCullagh-Maxwell aether theory. Although the COSSERATS' book was originally published as an appendix to an influential treatise on physics as a whole, the physicists of the day may not have appreciated the careful statement of hypotheses and rigorous mathematics that extended the work past 200 pages. The COSSERATS' masterpiece stands as a tower in the field. Even the recent recreators of continuum mechanics, while they knew of it, did not know its contents in detail. Had they mastered it, not only would time and effort of rediscovery have been spared, but also a paragon of method would have lain in their hands.

Theories of elastic materials of grade *2* or higher (in the sense of Sect. 28) had been proposed by several authors[2], but under the presumption that the stress tensor is symmetric. In 1956 TIFFEN and STEVENSON[3] considered briefly a theory of infinitesimal motion in a material having possibly unsymmetric stress and couple stress, and the corresponding three-dimensional theory for finite deformations was laid out by TRUESDELL and TOUPIN (CFT, Sect. 256). These four authors, however, started from an equation of energy that is not sufficiently general: While it has the same form as the COSSERATS', the spin tensor is taken as that of the motion rather than that of the director frame. These authors also overlooked an indeterminacy in the stress corresponding to such an equation[4]. The first correct theories, linearized and general, respectively, corresponding to an energy equation of the type used by TRUESDELL and TOUPIN were obtained by AERO and KUVSHINSKII[5] and by GRIOLI[6], respectively. TOUPIN[7], using a frame of ideas more familiar to readers of this treatise and of CFT, proposed a theory which was found to be equivalent[8] to GRIOLI'S.

Meanwhile, the Cosserat theory had been revived by GÜNTHER[9], who remarked on its relevance to the theory of "dislocations" (above, Sect. 44), and by ERICKSEN and TRUESDELL[10], who, not considering any constitutive equation, generalized and simplified the director concept by allowing the directors to be arbitrary and to turn and stretch in any way. MINDLIN[11] suggested that a continuum

---

[1] HEUN [1913, *2*, § *21*] and HELLINGER [1914, *1*, § 76] in their classic articles (which also seem to have been little noticed until recently) drew attention to the COSSERATS' work. Expositions of it were written by SUDRIA [1926, *1*] [1935, *3*] and by KUTILIN [1947, *6*, Gl. X], and the COSSERATS' book was cited occasionally in the literature, but usually as if it concerned, like their earlier paper, only the classical theory of finite elastic strain.

[2] CAUCHY [1851, *1*], ST. VENANT [1869, *1*], JARAMILLO [1929, *2*].

[3] TIFFEN and STEVENSON [1956, *30*].

[4] The error was noticed by ERINGEN [1962, *18*, § 40] and by MINDLIN and TIERSTEN [1962, *46*, § 1 and footnote, p. 395].

[5] AERO and KUVSHINSKII [1960, *4*].

[6] GRIOLI [1960, *28*] [1962, *24*, Ch. X]. GRIOLI's theory was studied further by BRESSAN [1963, *15*]. Cf. also KALISKI [1963, *39*].

[7] TOUPIN [1962, *64*].

[8] Elaborate analysis of MINDLIN and TIERSTEN [1962, *46*, § 3] shows the finite theories of TOUPIN and GRIOLI to be equivalent, but while TOUPIN's linearization yields AERO and KUVSHINSKII's infinitesimal theory, GRIOLI's [1960, *28*] [1961, *21*] [1962, *24*, Ch. X, §§ 5—6] does not, and they assert that GRIOLI overlooked certain terms.

[9] GÜNTHER [1958, *21*]. Cf. also SCHAEFER [1962, *58*].

[10] ERICKSEN and TRUESDELL [1958, *15*].

[11] MINDLIN [1964, *62*].

model embodying most if not all essentials of an arbitrarily detailed molecular or other structural picture of matter, as far as gross behavior is concerned, is gotten by regarding each element of the material as being itself a deformable continuum. These continua are fitted together smoothly, so that all the simplicity of a field theory results. The *macromedium* thus is a collection of particles, with each of which is associated a *micromedium*. In the special case when each micromedium deforms homogeneously, it may be replaced, as far as the theory is concerned, by n deformable directors, where n is the dimension of the micromedium[1].

The two streams of thought have been clarified and connected in a penetrating and elegant memoir of TOUPIN[2], the contents of which we outline in two subsections below.

Just as the classical theory of hyperelasticity, by its simplicity and definiteness, has always furnished one of the main elements of intuition in what is now called the theory of simple materials, so also for broader concepts of "material" we may expect the special cases corresponding to "elasticity" to be especially easy to develop and illuminating as guides.

A general framework for a mechanics of moments of arbitrarily high order was sketched by TRUESDELL and TOUPIN (CFT, Sects. 205, 232). Their work has been extended and modified in a series of papers by GREEN and RIVLIN[3], on which we shall remark at the end of this section.

Characteristic of polar media is their greater subtlety of response to wave propagation. Unlike the classical elastic material (Sect. 73), these are all *dispersive*. The existence of dispersion is ordinary in each kind of polar medium, if for different reasons. In the oriented medium the director frame itself furnishes a scale of length. In the material of grade 2, such a scale is provided by the reciprocals of the second derivatives of the deformation. Thus the mere presence of dispersion does not say anything about one of these kinds of materials as distinct from another. For the details in each case, and for solutions of particular problems, the reader is referred to the various works which have been cited in the course of this subsection.

With the growth of understanding how and why the stress tensor may be unsymmetric came the need for correspondingly general concepts of momentum and energy, not only in pure continuum mechanics[4] but also in the motivation through structural or semi-structural models[5], and even in pure statistical mechanics[6].

β) Toupin's theory of oriented hyperelastic materials[7]. For the formal apparatus of the theory of oriented materials, see Sect. 61 of CFT. We suppose that

[1] This approach to the theory of oriented media is entirely natural in the case of rods and shells. The cross-section of the three-dimensional rod may experience any deformation; if, however, it is assumed to deform homogeneously, then nothing else about it need be taken into account than the deformation of a pair of distinct lines through its centroid. In this way, if tacitly, the principal torsion-flexure axes are introduced in the classical theory.

[2] TOUPIN [1964, *84*].

[3] GREEN and RIVLIN [1964, *38* and *39*].

[4] VOIGT [1887, *2*, Chap. I, § 2] [1895, *1*], E. and F. COSSERAT [1909, *1*, § 53], HEUN [1913, *1*, § 21a], HELLINGER [1914, *7*, § 46], TRUESDELL [1952, *20*, § 26, footnote 67], BODASZEWSKI [1953, *2*, § 3], TRUESDELL and TOUPIN in CFT, Sects. 200—205, DAHLER and SCRIVEN [1961, *9*], TOUPIN [1962, *64*, § 2], KALISKI, PŁOCHOCKI, and ROGULA [1962, *39*].

[5] For polyatomic gases, GRAD [1952, *10*, § 4]; for crystal lattices (in varying degrees of precision), GÜNTHER [1958, *20*], KRÖNER [1960, *34*, § 2] [1962, *44*] [1963, *44*] [1965, *23*], RAJAGOPAL [1960, *42*], KRISHNAN and RAJAGOPAL [1961, *33*]; for anisotropic fluids, ERICKSEN [1960, *18* and *20*] [1961, *14*] [1962, *14*].

[6] DAHLER and SCRIVEN [1963, *19*].

[7] TOUPIN [1964, *84*, §§ 2—9]. Essentially this same theory had been outlined earlier in an unpublished lecture by ERICKSEN [1963, *21*, § I)].

a body consists not only of particles X, which are carried in each configuration into places $\boldsymbol{x}$, but also of other elements which are mapped onto vectors $\boldsymbol{d}_{\mathfrak{a}}$, called *directors at* $\boldsymbol{x}$. With the place $\boldsymbol{X}$ in some reference configuration taken as independent variable, the motion of an oriented material may be written in the form

$$\left.\begin{aligned} &\boldsymbol{x}=\boldsymbol{\chi}(\boldsymbol{X},t), \quad \boldsymbol{d}_{\mathfrak{a}}=\hat{\boldsymbol{d}}_{\mathfrak{a}}(\boldsymbol{X},t),\\ &x^k=x^k(X^1,X^2,X^3,t), \quad d^k_{\mathfrak{a}}=\hat{d}^k_{\mathfrak{a}}(X^1,X^2,X^3,t).\end{aligned}\right\} \tag{98.1}$$

In simpler applications $\mathfrak{a}$ will have the range *1, 2, 3*, though the formal structure, in most cases, is valid for a greater number of directors.

As the axioms defining his theory of oriented hyperelastic materials, TOUPIN lays down an action principle and a requirement of Euclid invariance. We consider these two principles in turn. The *action* associated with a part $\mathscr{P}$ of a body in the time interval $T: t_1 \leqq t \leqq t_2$ is given in terms of a density[1] L:

$$A=\int_T\int_{\mathscr{P}} L(\boldsymbol{X},t)\, dv_{\mathrm{R}}\, dt, \tag{98.2}$$

and TOUPIN assumes that L can be expressed as an assigned function of the simpler kinematical variables associated with the gross motion and with the motion of the directors:

$$L=\hat{L}(\boldsymbol{x},t,\boldsymbol{d}_{\mathfrak{a}},\dot{\boldsymbol{x}},\dot{\boldsymbol{d}}_{\mathfrak{a}},\boldsymbol{F},\boldsymbol{F}_{\mathfrak{a}},\boldsymbol{X}), \tag{98.3}$$

where $\boldsymbol{F}_{\mathfrak{a}} \equiv \nabla \hat{\boldsymbol{d}}_{\mathfrak{a}}$, the material gradient of the director-function $\hat{\boldsymbol{d}}_{\mathfrak{a}}$, having the components $d^k_{\mathfrak{a},\alpha}$. TOUPIN's action principle is

$$\left.\begin{aligned} &\delta A+\int_T\int_{\mathscr{P}} \varrho_{\mathrm{R}}(\boldsymbol{b}\cdot\delta\boldsymbol{\chi}+\boldsymbol{b}^{\mathfrak{a}}\cdot\delta\hat{\boldsymbol{d}}_{\mathfrak{a}})\, dv_{\mathrm{R}}\, dt+\\ &+\int_T\int_{\partial\mathscr{P}} (\boldsymbol{t}_{\mathrm{R}}\cdot\delta\boldsymbol{\chi}+\boldsymbol{t}^{\mathfrak{a}}_{\mathrm{R}}\cdot\delta\hat{\boldsymbol{d}}_{\mathfrak{a}})\, ds_{\mathrm{R}}\, dt-\\ &-\int_{\mathscr{P}} \varrho_{\mathrm{R}}(\boldsymbol{m}\cdot\delta\boldsymbol{\chi}+\boldsymbol{m}^{\mathfrak{a}}\cdot\delta\hat{\boldsymbol{d}}_{\mathfrak{a}})\, dv_{\mathrm{R}}\Big|_{t_1}^{t_2}=0,\end{aligned}\right\} \tag{98.4}$$

where the functions $\boldsymbol{\chi}$ and $\hat{\boldsymbol{d}}_{\mathfrak{a}}$ are varied independently. In (98.4), as henceforth, summation over diagonally repeated German indices is understood. The multipliers occurring in (98.4) are named as follows:

Symbol	Name
$\boldsymbol{b}$	body macroforce density
$\boldsymbol{b}^{\mathfrak{a}}$	body microforce density
$\boldsymbol{t}_{\mathrm{R}}$	macrostress vector in the reference configuration
$\boldsymbol{t}^{\mathfrak{a}}_{\mathrm{R}}$	microstress vector in the reference configuration
$\boldsymbol{m}$	macromomentum per unit mass
$\boldsymbol{m}^{\mathfrak{a}}$	micromomentum per unit mass

If we set

$$\left.\begin{aligned} &\varrho_{\mathrm{R}}\boldsymbol{m}^*\equiv\hat{L}_{\dot{\boldsymbol{x}}}, \quad \boldsymbol{T}_{\mathrm{R}}\equiv-\hat{L}_{\boldsymbol{F}},\\ &\varrho_{\mathrm{R}}\boldsymbol{m}^{\mathfrak{a}*}\equiv\hat{L}_{\dot{\boldsymbol{d}}_{\mathfrak{a}}}, \quad \boldsymbol{T}^{\mathfrak{a}}_{\mathrm{R}}\equiv-\hat{L}_{\boldsymbol{F}_{\mathfrak{a}}},\\ &\varrho_{\mathrm{R}}\boldsymbol{b}^{\mathfrak{a}*}\equiv-\hat{L}_{\boldsymbol{d}_{\mathfrak{a}}},\end{aligned}\right\} \tag{98.5}$$

[1] This function generalizes the "density of Euclidean action" introduced by E. and F. COSSERAT [1907, 1] [1909, *1*, § 51].

where subscripts attached to $\hat{L}$ denote partial gradients, then (98.4) is formally equivalent to the following Euler equations:

$$\left.\begin{aligned} \varrho_R \dot{\boldsymbol{m}}^* - \operatorname{Div} \boldsymbol{T}_R - \varrho_R \boldsymbol{b} &= \hat{L}_{\boldsymbol{x}}, \\ \varrho_R \dot{\boldsymbol{m}}^{\mathfrak{a}*} - \operatorname{Div} \boldsymbol{T}_R^{\mathfrak{a}} + \varrho_R (\boldsymbol{b}^{\mathfrak{a}*} - \boldsymbol{b}^{\mathfrak{a}}) &= \boldsymbol{0}; \end{aligned}\right\} \tag{98.6}$$

the following boundary conditions for all t:

$$\left.\begin{aligned} \boldsymbol{T}_R \boldsymbol{n}_R &= \boldsymbol{t}_R, \\ \boldsymbol{T}_R^{\mathfrak{a}} \boldsymbol{n}_R &= \boldsymbol{t}_R^{\mathfrak{a}}; \end{aligned}\right\} \tag{98.7}$$

and the following initial and final conditions ($t=t_1$ and $t=t_2$) throughout the body:

$$\left.\begin{aligned} \boldsymbol{m}^* &= \boldsymbol{m}, \\ \boldsymbol{m}^{\mathfrak{a}*} &= \boldsymbol{m}^{\mathfrak{a}}. \end{aligned}\right\} \tag{98.8}$$

Since the action principle (98.4) is assumed to hold for all time intervals T, we must apply (98.8) at each instant as well as at all points.

In the special case when (98.3) has the simple form $\hat{L} = \varrho_R [\frac{1}{2}\dot{x}^2 - \sigma(\boldsymbol{F}, \boldsymbol{X})]$, the quantities $\boldsymbol{m}^{\mathfrak{a}*}$, $\boldsymbol{T}_R^{\mathfrak{a}}$, and $\boldsymbol{b}^{\mathfrak{a}*}$ vanish by definition, while $\boldsymbol{m}^* = \dot{\boldsymbol{x}}$, the ordinary momentum per unit mass. Eqs. (98.6) and (98.7), under this specializing assumption, reduce to Cauchy's laws in the classical forms (43 A.6), (43 A.5), and $(98.5)_2$ reduces to the stress relation (82.11) of hyperelasticity, with strain-energy function $\boldsymbol{\sigma}$. As is well known[1], in hyperelasticity developed in this way the Cauchy stress $\boldsymbol{T}$, which is related to $\boldsymbol{T}_R$ through (43 A.3), is symmetric if and only if $\sigma(\boldsymbol{F}, \boldsymbol{X})$ is a frame-indifferent function of $\boldsymbol{F}$.

In Toupin's theory, as we shall now see, all the foregoing results are replaced by greatly more general and flexible counterparts.

First, we apply the postulated variational condition (98.4) to the following three special kinds of variations:

$$\left.\begin{aligned} &1.\ \delta\boldsymbol{\chi} = \boldsymbol{d}, && \delta\hat{\boldsymbol{d}}_{\mathfrak{a}} = \boldsymbol{0}. \\ &2.\ \delta\boldsymbol{\chi} = \boldsymbol{W}\boldsymbol{p}, && \delta\hat{\boldsymbol{d}}_{\mathfrak{a}} = \boldsymbol{W}\boldsymbol{d}_{\mathfrak{a}}. \\ &3.\ \delta\boldsymbol{\chi} = \dot{\boldsymbol{x}}, && \delta\hat{\boldsymbol{d}}_{\mathfrak{a}} = \dot{\boldsymbol{d}}_{\mathfrak{a}}. \end{aligned}\right\} \tag{98.9}$$

Here $\boldsymbol{d}$ is an arbitrary vector, $\boldsymbol{W}$ is an arbitrary skew-symmetric tensor, and $\boldsymbol{p}$ is the position vector. Using the notation $\boldsymbol{a} \wedge \boldsymbol{b} \equiv \boldsymbol{a} \otimes \boldsymbol{b} - \boldsymbol{b} \otimes \boldsymbol{a}$ for the exterior product of the vectors $\boldsymbol{a}$ and $\boldsymbol{b}$, set

$$\left.\begin{aligned} 2\boldsymbol{A} &\equiv \boldsymbol{p} \wedge \boldsymbol{m} + \boldsymbol{d}_{\mathfrak{a}} \wedge \boldsymbol{m}^{\mathfrak{a}} = -2\boldsymbol{A}^T, \\ E &\equiv \boldsymbol{m} \cdot \dot{\boldsymbol{x}} + \boldsymbol{m}^{\mathfrak{a}} \cdot \dot{\boldsymbol{d}}_{\mathfrak{a}} - \hat{L}/\varrho_R, \\ \varrho_R \boldsymbol{K} &\equiv \boldsymbol{p} \otimes \hat{L}_{\boldsymbol{x}} + \boldsymbol{d}_{\mathfrak{a}} \otimes \hat{L}_{\boldsymbol{d}_{\mathfrak{a}}} + \dot{\boldsymbol{x}} \otimes \hat{L}_{\dot{\boldsymbol{x}}} + \dot{\boldsymbol{d}}_{\mathfrak{a}} \otimes \hat{L}_{\dot{\boldsymbol{d}}_{\mathfrak{a}}} + \boldsymbol{F} \hat{L}_{\boldsymbol{F}}^T + \boldsymbol{F}_{\mathfrak{a}} \hat{L}_{\boldsymbol{F}_{\mathfrak{a}}}^T, \end{aligned}\right\} \tag{98.10}$$

or, in components,

$$\left.\begin{aligned} A_{kr} &\equiv p_{[k} m_{r]} + d_{\mathfrak{a}[k} m^{\mathfrak{a}}{}_{r]} = -A_{rk}, \\ E &\equiv m_k \dot{x}^k + m_k{}^{\mathfrak{a}} \dot{d}_{\mathfrak{a}}^k - \hat{L}/\varrho_R, \\ \varrho_R K^r{}_q &\equiv p^r \frac{\partial \hat{L}}{\partial x^q} + d_{\mathfrak{a}}^r \frac{\partial \hat{L}}{\partial d_{\mathfrak{a}}^q} + \dot{x}^r \frac{\partial \hat{L}}{\partial \dot{x}^q} + \dot{d}_{\mathfrak{a}}^r \frac{\partial \hat{L}}{\partial \dot{d}_{\mathfrak{a}}^q} + x^r{}_{,\alpha} \frac{\partial \hat{L}}{\partial x^q{}_{,\alpha}} + d_{\mathfrak{a},\alpha}^r \frac{\partial \hat{L}}{\partial d_{\mathfrak{a},\alpha}^q}. \end{aligned}\right\} \tag{98.11}$$

---

[1] As we mentioned in Sect. 84, Noll [1955, *19*, Theorem I of § 16] proved that in hyperelasticity

$$\sigma \text{ is frame-indifferent} \Leftrightarrow \boldsymbol{T} \text{ is frame-indifferent} \Leftrightarrow \boldsymbol{T} \text{ is symmetric.}$$

In terms of these quantities, necessary and sufficient conditions that (98.4) hold for the special variations (98.9) are easily calculated:

$$\int_{\mathscr{P}} \varrho_R \boldsymbol{m}\, dv_R \Big|_{t_1}^{t_2} - \int_T \int_{\mathscr{P}} \varrho_R \boldsymbol{b}\, dv_R\, dt - \int_T \int_{\partial\mathscr{P}} \boldsymbol{t}_R\, ds_R\, dt = \int_T \int_{\mathscr{P}} \hat{L}_{\boldsymbol{p}}\, dv_R\, dt, \tag{98.12}$$

$$\left.\begin{aligned} &2\int_{\mathscr{P}} \varrho_R \boldsymbol{A}\, dv_R \Big|_{t_1}^{t_2} - \int_T \int_{\mathscr{P}} \varrho_R (\boldsymbol{p} \wedge \boldsymbol{b} + \boldsymbol{d}_{\mathfrak{a}} \wedge \boldsymbol{b}^{\mathfrak{a}})\, dv_R\, dt - \\ &\quad - \int_T \int_{\partial\mathscr{P}} (\boldsymbol{p} \wedge \boldsymbol{t}_R + \boldsymbol{d}_{\mathfrak{a}} \wedge \boldsymbol{t}_R^{\mathfrak{a}})\, ds_R\, dt = 2\int_T \int_{\mathscr{P}} \varrho_R \text{-}[\boldsymbol{K}]\text{-}\, dv_R\, dt, \end{aligned}\right\} \tag{98.13}$$

$$\left.\begin{aligned} &\int_{\mathscr{P}} \varrho_R E\, dv_R \Big|_{t_1}^{t_2} - \int_T \int_{\mathscr{P}} \varrho_R (\boldsymbol{b} \cdot \dot{\boldsymbol{x}} + \boldsymbol{b}^{\mathfrak{a}} \cdot \dot{\boldsymbol{d}}_{\mathfrak{a}})\, dv_R\, dt - \\ &\quad - \int_T \int_{\partial\mathscr{P}} (\boldsymbol{t}_R \cdot \dot{\boldsymbol{x}} + \boldsymbol{t}_R^{\mathfrak{a}} \cdot \dot{\boldsymbol{d}}_{\mathfrak{a}})\, ds_R\, dt = -\int_T \int_{\mathscr{P}} \hat{L}_t\, dv_R\, dt, \end{aligned}\right\} \tag{98.14}$$

where $\text{-}[\boldsymbol{S}]\text{-} \equiv \frac{1}{2}(\boldsymbol{S} - \boldsymbol{S}^T)$ denotes the skew-symmetric part of a tensor $\boldsymbol{S}$. In classical hyperelasticity, $\boldsymbol{A}$ becomes the density of moment of momentum per unit mass: $\boldsymbol{A} = \frac{1}{2}\boldsymbol{p} \wedge \dot{\boldsymbol{x}}$, while E reduces to the total energy per unit mass: $E = \dot{x}^2 - \hat{L}/\varrho_R = \frac{1}{2}\dot{x}^2 + \sigma$. In the present greater generality, the concepts represented by E and $\boldsymbol{A}$ are extended so as to correspond to the action density (98.3): the rates of working and the moments of the micromomenta are included, and the macromomentum is freed, in the spirit of Hamiltonian mechanics, from its classical identification with the product of velocity by mass density. The moment of momentum of a particle is no longer merely the moment of its linear momentum, and the torque acting on a body is no longer the moment of the various forces acting upon it but is greater by an amount equal to the total couples obtained from the *couple densities* $\boldsymbol{d}_{\mathfrak{a}} \wedge \boldsymbol{b}^{\mathfrak{a}}$ and $\boldsymbol{d}_{\mathfrak{a}} \wedge \boldsymbol{t}_R^{\mathfrak{a}}$ in the interior and on the bounding surface, respectively. These couples result from regarding the microforce $\boldsymbol{b}^{\mathfrak{a}}$ and the microstress vector $\boldsymbol{t}_R^{\mathfrak{a}}$ as being applied, not at the point $\boldsymbol{x}$, but at the terminus of the director $\boldsymbol{d}_{\mathfrak{a}}$ at $\boldsymbol{x}$. Thus they may be called *microcouples*. In terms of these extended notions of momentum and energy, the vanishing of the left-hand sides of (98.12)—(98.14) signifies the *balance of momentum, moment of momentum, and energy*. The right-hand sides vanish, for all manifolds $T \times \mathscr{P}$, if and only if the following conditions hold:

$$\hat{L}_{\boldsymbol{x}} = \boldsymbol{0}, \qquad \boldsymbol{K} = \boldsymbol{K}^T, \qquad \hat{L}_t = 0, \tag{98.15}$$

or, in co-ordinate form,

$$\frac{\partial \hat{L}}{\partial x^k} = 0, \qquad K_{pq} = K_{qp}, \qquad \frac{\partial \hat{L}}{\partial t} = 0. \tag{98.16}$$

Now let us consider the effect of a change of frame. For the motion alone, this effect is decsribed by (17.6). The directors $\boldsymbol{d}_{\mathfrak{a}}$, however, must also be regarded as affected. If we assume that the $\boldsymbol{d}_{\mathfrak{a}}$ are frame-indifferent in the sense of Sect. 17, they must obey transformation laws of the form (17.3), i.e.,

$$\boldsymbol{d}_{\mathfrak{a}}^*(\boldsymbol{X}, t^*) = \boldsymbol{Q}(t)\, d_{\mathfrak{a}}(\boldsymbol{X}, t). \tag{98.17}$$

TOUPIN does *not* require that the action function $\hat{L}$ be frame-indifferent, presumably because the action includes the frame-dependent effects of inertia. Instead, he imposes the weaker requirement that $\hat{L}$ be invariant under transformations of the form (17.3), (98.17) when $\boldsymbol{c}(t)$ and $\boldsymbol{Q}(t)$ are independent[1] of t. In other words,

[1] In fact, as TOUPIN observed, it suffices to consider only *proper* orthogonal $\boldsymbol{Q}$.

the function $\hat{L}$ must be invariant under transformations of the form

$$\left.\begin{aligned} \boldsymbol{x}^* &= \boldsymbol{\chi}^*(\boldsymbol{X}, t^*) = \boldsymbol{c} + \boldsymbol{Q}\boldsymbol{\chi}(\boldsymbol{X}, t), \\ \boldsymbol{d}^*_{\mathfrak{a}} &= \boldsymbol{d}^*_{\mathfrak{a}}(\boldsymbol{X}, t^*) = \boldsymbol{Q}\boldsymbol{d}_{\mathfrak{a}}(\boldsymbol{X}, t), \\ t^* &= t - a, \end{aligned}\right\} \tag{98.18}$$

where **c** *is an arbitrary constant vector,* **Q** *an arbitrary constant, orthogonal tensor, and a an arbitrary constant.* Toupin calls invariance under transformations (98.18) *Euclid invariance.* It is easily verified that $\hat{L}$ is Euclid invariant if and only if (98.15) holds. Combined with the interpretation already established for (98.15), this result may be phrased as the Cosserat-Toupin ***fundamental equivalence theorem***[1]. *In an oriented hyperelastic material subject to the action principle* (98.4), *the action density* (98.3) *is Euclid invariant if and only if linear momentum, moment of momentum, and energy are balanced.*

Thus far we have used quantities appropriate to the reference configuration. We now convert the results to forms having immediate meaning in the present configuration of the body. Alongside (43A.3), now regarded as a definition of Cauchy's stress tensor $\boldsymbol{T}$, which yields here

$$\left.\begin{aligned} \boldsymbol{T} &= \frac{\varrho}{\varrho_{\mathrm{R}}} \boldsymbol{T}_{\mathrm{R}} \boldsymbol{F}^T = -\frac{\varrho}{\varrho_{\mathrm{R}}} \hat{L}_{\boldsymbol{F}} \boldsymbol{F}^T, \\ T_k{}^m &= \frac{\varrho}{\varrho_{\mathrm{R}}} T_{\mathrm{R}k}{}^{\alpha} x^m{}_{,\alpha} = -\frac{\varrho}{\varrho_{\mathrm{R}}} \frac{\partial \hat{L}}{\partial x^k{}_{,\alpha}} x^m{}_{,\alpha}, \end{aligned}\right\} \tag{98.19}$$

we set a corresponding definition of the *hyperstress tensor* H:

$$\mathsf{H} \equiv -\frac{\varrho}{\varrho_{\mathrm{R}}} \boldsymbol{d}_{\mathfrak{a}} \otimes \hat{L}_{\boldsymbol{F}_{\mathfrak{a}}} \boldsymbol{F}^T, \qquad \mathsf{H}_{km}{}^p = -\frac{\varrho}{\varrho_{\mathrm{R}}} d_{\mathfrak{a}k} \frac{\partial \hat{L}}{\partial d^m_{\mathfrak{a},\alpha}} x^p{}_{,\alpha}. \tag{98.20}$$

If we define $\boldsymbol{t}$ and $\boldsymbol{t}^{\mathfrak{a}}$ by the relations

$$\boldsymbol{t}\, ds = \boldsymbol{t}_{\mathrm{R}}\, ds_{\mathrm{R}}, \qquad \boldsymbol{t}^{\mathfrak{a}}\, ds = \boldsymbol{t}^{\mathfrak{a}}_{\mathrm{R}}\, ds_{\mathrm{R}} \tag{98.21}$$

(cf. Sect. 43A and CFT, Sect. 210), the stress boundary conditions (98.7) assume the forms

$$\left.\begin{aligned} \boldsymbol{T}\boldsymbol{n} &= \boldsymbol{t}, & \mathsf{H}\boldsymbol{n} &= \boldsymbol{d}_{\mathfrak{a}} \otimes \boldsymbol{t}^{\mathfrak{a}}, \\ T^{kp} n_p &= t^k, & \mathsf{H}^{kmp} n_p &= d^k_{\mathfrak{a}} t^{\mathfrak{a}m}. \end{aligned}\right\} \tag{98.22}$$

If, finally, we set

$$\left.\begin{aligned} \boldsymbol{S} &\equiv \boldsymbol{d}_{\mathfrak{a}} \otimes \boldsymbol{m}^{\mathfrak{a}}, & S^k{}_p &= d^k_{\mathfrak{a}} m^{\mathfrak{a}}_p, \\ \boldsymbol{L} &\equiv \boldsymbol{d}_{\mathfrak{a}} \otimes \boldsymbol{b}^{\mathfrak{a}}, & L^k{}_p &= d^k_{\mathfrak{a}} b^{\mathfrak{a}}_p, \end{aligned}\right\} \tag{98.23}$$

from (98.6) it follows that

$$\left.\begin{aligned} \varrho \dot{\boldsymbol{m}} &= \operatorname{div} \boldsymbol{T} + \varrho \boldsymbol{b}, \\ \varrho \dot{m}^k &= T^{kp}{}_{,p} + \varrho b^k, \end{aligned}\right\} \tag{98.24}$$

and

$$\left.\begin{aligned} \varrho(\dot{\boldsymbol{S}} - \boldsymbol{K} + \dot{\boldsymbol{x}} \otimes \boldsymbol{m}) &= \operatorname{div} \mathsf{H} + \boldsymbol{T}^T + \varrho \boldsymbol{L}, \\ \varrho(\dot{S}^{pq} - K^{pq} + \dot{x}^p m^q) &= \mathsf{H}^{pqr}{}_{,r} + T^{qp} + \varrho L^{pq}. \end{aligned}\right\} \tag{98.25}$$

Note that -[**S**]-, in the terms introduced earlier, is the micromoment of the micromomentum, or, for short, the *spin momentum.* If the action density $\hat{L}$ is such that

---

[1] Cf. the very special case obtained by E. and F. Cosserat [1909, *1*, §§ 51—54, §§ 61—64].

$\hat{L}_{\dot{\boldsymbol{x}}} = \varrho_R \dot{\boldsymbol{x}}$, then by (98.5) the equation of linear momentum (98.24) reduces to the form of CAUCHY's first law (16.6). Under the same assumption, it follows from (98.15) that the skew-symmetric part of (98.25) assumes the form

$$\left.\begin{aligned} \varrho \,\text{-}[\dot{\boldsymbol{S}}]\text{-} &= \operatorname{div} \mathsf{M} + \text{-}[\boldsymbol{T}^T]\text{-} + \varrho \,\text{-}[\boldsymbol{L}]\text{-}, \\ \varrho \dot{S}^{[pq]} &= \mathsf{M}^{pqr}{}_{,r} + T^{[qp]} + \varrho L^{[pq]}, \end{aligned}\right\} \tag{98.26}$$

where the *couple-stress tensor* M is defined as follows:

$$\mathsf{M}^{kpq} = \mathsf{H}^{[kp]q}. \tag{98.27}$$

When $\text{-}[\dot{\boldsymbol{S}}]\text{-} = \boldsymbol{0}$, we recover from (98.26) the COSSERATS' equations for balance of moment of momentum [CFT, Eq. (205.10)]; if $\mathsf{M} = \boldsymbol{0}$ and $\text{-}[\boldsymbol{L}]\text{-} = \boldsymbol{0}$ also, (98.26) reduces to CAUCHY's second law (16.7). More generally, (98.26) is the equation of balance for the spin momentum.

The symmetric part of (98.25) is more difficult to interpret. In the classical theory of hyperelasticity, it reduces, as does (98.25) itself, to the stress relation.

In everything said so far, the number of directors $\boldsymbol{d}_\mathfrak{a}$ is arbitrary. If there are no more than three directors, and if these are linearly independent, there is a reciprocal set $\boldsymbol{d}^\mathfrak{b}$ such that $\boldsymbol{d}^\mathfrak{b} \cdot \boldsymbol{d}_\mathfrak{a} = \delta^\mathfrak{b}_\mathfrak{a}$. In this case we may solve $(98.22)_{2,4}$ for $\boldsymbol{t}^\mathfrak{a}$, obtaining on $\partial\mathscr{B}$

$$\boldsymbol{t}^\mathfrak{a} = (\mathsf{H}\boldsymbol{n})^T \boldsymbol{d}^\mathfrak{a}, \qquad t^{\mathfrak{a}m} = d^\mathfrak{a}_k \mathsf{H}^{kmp} n_p. \tag{98.28}$$

The equation of energy balance (98.14) can then be written in the form

$$\left.\begin{aligned} \int_{\mathscr{P}} \varrho E \, dv \Big|_{t_1}^{t_2} &= \int_T \int_{\mathscr{P}} \varrho \{\dot{\boldsymbol{x}} \cdot \boldsymbol{b} + \operatorname{tr}[(\dot{\boldsymbol{d}}_\mathfrak{a} \otimes \boldsymbol{d}^\mathfrak{a}) \boldsymbol{L}]\} \, dv \, dt + \\ &\quad + \int_T \int_{\partial\mathscr{P}} \{\dot{\boldsymbol{x}} \cdot \boldsymbol{T}\boldsymbol{n} + \operatorname{tr}[(\dot{\boldsymbol{d}}_\mathfrak{a} \otimes \boldsymbol{d}^\mathfrak{a}) \mathsf{H}\boldsymbol{n}]\} \, ds \, dt \\ &= \int_T \int_{\mathscr{P}} \varrho (\dot{x}_k b^k + \dot{d}_{\mathfrak{a}k} d^\mathfrak{a}_r L^{rk}) \, dv \, dt + \\ &\quad + \int_T \int_{\partial\mathscr{P}} (\dot{x}_k T^{km} + \dot{d}_{\mathfrak{a}k} d^\mathfrak{a}_r \mathsf{H}^{rkm}) \, ds_m \, dt. \end{aligned}\right\} \tag{98.29}$$

A field equation for the energy E results by substituting (98.14) and (98.25) into the differential equivalent of (98.29). The result of this routine calculation, in contrast to its classical special case (CFT, Sect. 241), is not simple and hence is not enlightening, so we do not write it out. In a theory of hyperelasticity, the equation of energy serves but a single end, namely, to deliver the form of the stress relation, and in TOUPIN's theory this form, written here as Eqs. (98.19) and (98.20), follows straight from the fundamental axiom (98.4).

Finally, the assertion that L is Euclid invariant imposes restrictions on the form of the function $\hat{L}$. TOUPIN considers only the special case when

$$\left.\begin{aligned} L &= T - W, \\ T &= \tfrac{1}{2} \varrho_R (\dot{x}^2 + \nu^{\mathfrak{a}\mathfrak{c}} \dot{\boldsymbol{d}}_\mathfrak{a} \cdot \dot{\boldsymbol{d}}_\mathfrak{c}), \\ W &= \varrho_R \sigma(\boldsymbol{d}_\mathfrak{a}, \boldsymbol{F}, \boldsymbol{F}_\mathfrak{a}, \boldsymbol{X}), \end{aligned}\right\} \tag{98.30}$$

where $\nu^{\mathfrak{a}\mathfrak{c}} = \nu^{\mathfrak{c}\mathfrak{a}}$ and $\dot{\nu}^{\mathfrak{a}\mathfrak{c}} = 0$. These functional forms hold if and only if $\boldsymbol{m} = \dot{\boldsymbol{x}}$ and $\varrho_R \boldsymbol{m}^\mathfrak{a} = \nu^{\mathfrak{a}\mathfrak{c}} \dot{\boldsymbol{d}}_\mathfrak{c}$. The Euclid invariance of L may then be stated as the requirement that

$$\sigma(\boldsymbol{Q}\boldsymbol{d}_\mathfrak{a}, \boldsymbol{Q}\boldsymbol{F}, \boldsymbol{Q}\boldsymbol{F}_\mathfrak{a}, \boldsymbol{X}) = \sigma(\boldsymbol{d}_\mathfrak{a}, \boldsymbol{F}, \boldsymbol{F}_\mathfrak{a}, \boldsymbol{X}) \tag{98.31}$$

for all orthogonal tensors $\boldsymbol{Q}$. Using the polar decomposition $\boldsymbol{F}=\boldsymbol{R}\boldsymbol{U}$ and choosing $\boldsymbol{Q}=\boldsymbol{R}^T$ yields[1]

$$\left.\begin{aligned}\sigma(\boldsymbol{d}_{\mathfrak{a}}, \boldsymbol{F}, \boldsymbol{F}_{\mathfrak{a}}, \boldsymbol{X}) &= \sigma(\boldsymbol{R}^T\boldsymbol{d}_{\mathfrak{a}}, \boldsymbol{U}, \boldsymbol{R}^T\boldsymbol{F}_{\mathfrak{a}}, \boldsymbol{X})\\ &= \bar{\sigma}(\boldsymbol{F}^T\boldsymbol{d}_{\mathfrak{a}}, \boldsymbol{C}, \boldsymbol{F}^T\boldsymbol{F}_{\mathfrak{a}}, \boldsymbol{X}),\end{aligned}\right\} \tag{98.32}$$

generalizing (84.4) and (84.5).

Toupin shows that when the strain and the microstrain are small, the theory just developed reduces to Mindlin's linear theory of microstructure.

Another case of interest results when the directors, three in number, are assumed to form a rigid non-coplanar triad:

$$\boldsymbol{d}_{\mathfrak{a}} \cdot \boldsymbol{d}_{\mathfrak{c}} = g_{\mathfrak{a}\mathfrak{c}}, \tag{98.33}$$

where the $g_{\mathfrak{a}\mathfrak{c}}$ are the coefficients in an assigned, constant, positive-definite form. Such a medium, in which the microstructure consists only in a capacity for rotation independent of the macrostrain, was first considered by E. and F. Cosserat and is called a *Cosserat material.* Either by allowing in (98.4) only variations consistent with (98.33) and hence satisfying the condition $\delta\hat{\boldsymbol{d}}_{\mathfrak{a}} \cdot \hat{\boldsymbol{d}}_{\mathfrak{b}} + \hat{\boldsymbol{d}}_{\mathfrak{a}} \cdot \delta\hat{\boldsymbol{d}}_{\mathfrak{b}} = 0$, or by specialization in (98.10) and (98.11), we see that the field equation (98.25) and boundary conditions $(98.22)_2$ reduce to their skew-symmetric parts, namely

$$\left.\begin{aligned}\varrho\,\text{-}\{\dot{\boldsymbol{S}} + \dot{\boldsymbol{x}} \otimes \boldsymbol{m}\}\text{-} &= \operatorname{div} \mathsf{M} - \text{-}\{\boldsymbol{T}\}\text{-} + \varrho\,\text{-}\{\boldsymbol{L}\}\text{-},\\ \mathsf{M}\boldsymbol{n} &= \tfrac{1}{2}\boldsymbol{d}_{\mathfrak{a}} \wedge \boldsymbol{t}^{\mathfrak{a}}.\end{aligned}\right\} \tag{98.34}$$

For a Cosserat material the equation of energy takes an especially simple form. The directors $\boldsymbol{d}_{\mathfrak{a}}$, being linearly independent, have a reciprocal set $\boldsymbol{d}^{\mathfrak{a}}$. Set

$$2\widetilde{\boldsymbol{W}} = \dot{\boldsymbol{d}}^{\mathfrak{a}} \otimes \boldsymbol{d}_{\mathfrak{a}}, \qquad 2\widetilde{W}_{km} = \dot{d}^{\mathfrak{a}}_{m}\, d_{\mathfrak{a}k}. \tag{98.35}$$

Because of the constraint (98.33), $\widetilde{\boldsymbol{W}}$ is skew-symmetric: $\widetilde{\boldsymbol{W}} = -\widetilde{\boldsymbol{W}}^T$; in fact, $\widetilde{\boldsymbol{W}}$, the *microspin*, is the angular velocity tensor of the director frame. The equation of energy balance (98.29) now assumes the form

$$\left.\begin{aligned}\int_{\mathscr{P}} \varrho E\, dv \Big|_{t_1}^{t_2} &= \int_T \int_{\mathscr{P}} \varrho \{\dot{\boldsymbol{x}} \cdot \boldsymbol{b} + 2\operatorname{tr}(\widetilde{\boldsymbol{W}}\,\text{-}\{\boldsymbol{L}\}\text{-})\}\, dv\, dt +\\ &\quad + \int_T \int_{\partial\mathscr{P}} (\dot{\boldsymbol{x}} \cdot \boldsymbol{T}\boldsymbol{n} + 2\operatorname{tr}(\widetilde{\boldsymbol{W}}\mathsf{M}\boldsymbol{n}))\, ds\, dt\\ &= \int_T \int_{\mathscr{P}} \varrho (\dot{x}_k b^k - 2\widetilde{W}_{km} L^{[km]})\, dv\, dt +\\ &\quad + \int_T \int_{\partial\mathscr{P}} (\dot{x}_k T^{km} - 2\widetilde{W}_{kp} M^{kpm})\, ds_m\, dt.\end{aligned}\right\} \tag{98.36}$$

Therefore, in the special case of a Cosserat material, the symmetric parts of $\boldsymbol{L}$ and H fail to appear in the equation of energy and thus have no energetic significance.

γ) Toupin's theory of hyperelastic materials of second grade[2]. The *elastic energy* of a part $\mathscr{P}$ of a body is presumed given in terms of a density or *strain-energy*

[1] For a simple theory of elastic materials with second grade with couple-stresses but not necessarily having a stored-energy function, the corresponding reduction had been given by Duvaut [1964, *24*].

[2] Toupin [1964, *84*, §§ 10—11].

function σ which depends upon the second as well as the first derivatives of the deformation:

$$\left.\begin{aligned} E(\mathscr{P}) &= \int_{\mathscr{P}} \varrho_R \sigma \, dv_R, \\ \sigma &= \sigma(\boldsymbol{F}, \nabla \boldsymbol{F}, \boldsymbol{X}) = \sigma(x^k{}_{,\alpha}, x^m{}_{,\beta\gamma}, X^\delta). \end{aligned}\right\} \tag{98.37}$$

The basic mechanical law is the principle of virtual work[1] for the case of equilibrium:

$$\left.\begin{aligned} \delta E(\mathscr{P}) &= \int_{\mathscr{P}} \varrho_R \boldsymbol{b} \cdot \delta\boldsymbol{\chi} \, dv_R + \int_{\partial\mathscr{P}} (\boldsymbol{t}_R \cdot \delta\boldsymbol{\chi} + \boldsymbol{h}_R \cdot D\,\delta\boldsymbol{\chi}) \, ds_R, \\ &= \int_{\mathscr{P}} \varrho_R \, b_k \delta\chi^k \, dv_R + \int_{\partial\mathscr{P}} (t_{Rk} \, \delta\chi^k + h_{Rk} D\,\delta\chi^k) \, ds_R, \end{aligned}\right\} \tag{98.38}$$

where $D\,\delta\boldsymbol{\chi}$ is the derivative of $\delta\boldsymbol{\chi}$ in the direction normal to $\partial\mathscr{P}$:

$$D\,\delta\boldsymbol{\chi} = (\nabla\,\delta\boldsymbol{\chi}) \cdot \boldsymbol{n}_R, \qquad (D\,\delta\chi)^k = (\delta\chi^k)_{,\alpha} \, n_R^\alpha. \tag{98.39}$$

$\boldsymbol{t}_R$ is the traction and $\boldsymbol{h}_R$ the hypertraction per unit area of the reference configuration. We introduce the following notations:

$$\left.\begin{aligned} H_{Rk}{}^{\alpha\beta} &\equiv \varrho_R \frac{\partial\sigma}{\partial x^k{}_{,\alpha;\beta}}, \\ T_{Rk}{}^{\alpha} &\equiv \varrho_R \frac{\partial\sigma}{\partial x^k{}_{,\alpha}} - H_{Rk}{}^{\alpha\beta}{}_{;\beta}. \end{aligned}\right\} \tag{98.40}$$

[Cf. Eq. (82.11) of the classical theory.] Defining the tensor DH as follows:

$$(DH)_k{}^\beta \equiv H_{Rk}{}^{\alpha\beta}{}_{;\alpha} - H_{Rk}{}^{\alpha\beta}{}_{;\delta} n_R{}^\delta n_{R\alpha}, \tag{98.41}$$

TOUPIN finds as local equivalents to the work principle (98.38) a field equation having exactly the same form as CAUCHY's first law (43 A.6) in the case of equilibrium, but generalized boundary conditions:

$$\left.\begin{aligned} &T_{Rk}{}^\alpha n_{R\alpha} - (DH)_k{}^\beta n_{R\beta} + H_k{}^{\alpha\beta}(\bar{B}_{\alpha\beta} - \bar{B}^\delta_\delta n_{R\alpha} n_{R\beta}) = t_{Rk}, \\ &H_{Rk}{}^{\alpha\beta} n_{R\alpha} n_{R\beta} = h_{Rk}, \end{aligned}\right\} \tag{98.42}$$

where $\bar{\boldsymbol{B}}$ is the three-dimensional extension of the second fundamental tensor of the surface $\partial\mathscr{P}$ in the reference configuration[2].

Thus far the work has a formal character, but the reader of the earlier parts of this treatise will perceive what is emerging: a theory of couple stress as associated with effects of curvature. To see this more clearly, we convert the results to spatial form. As definition of $\boldsymbol{T}$, we take the classical formula (43 A.3), and for H, the *hyperstress*, a formal analogue:

$$H^{kpq} = \frac{\varrho}{\varrho_R} x^k{}_{,\alpha} H_R{}^{p\alpha\beta} x^q{}_{,\beta}. \tag{98.43}$$

---

[1] Cf. CFT, Sect. 232, where a more general principle is laid down.

[2] More precisely, let $\boldsymbol{B}^*$ be the second fundamental tensor of the surface $\partial\mathscr{P}$ in the reference configuration. Resolve any vector $\boldsymbol{u}$ defined at points on that surface into normal and tangential components, $\boldsymbol{u} = \boldsymbol{u}_n + \boldsymbol{u}_t$. Then $\bar{\boldsymbol{B}}$ is defined at points on $\partial\mathscr{P}$ as being the following linear transformation:

$$\bar{\boldsymbol{B}}\boldsymbol{u} = \boldsymbol{B}^* \boldsymbol{u}_t.$$

In components, if $\bar{X}^\Gamma$, $\Gamma = 1, 2$, are co-ordinates in any system on $\partial\mathscr{P}$, from which we construct a three-dimensional co-ordinate system with the third co-ordinate being normal distance from $\partial\mathscr{P}$, then $\bar{\boldsymbol{B}}$ has the components

$$\|\bar{B}^{\gamma\delta}\| = \begin{Vmatrix} B^{\Gamma\Delta} & 0 \\ 0 & 0 \end{Vmatrix}.$$

The field equations, of course, take the classical form (16.6) of CAUCHY'S first law for the case of equilibrium. Writing $\bar{\boldsymbol{b}}$ for the second fundamental tensor of the surface $\partial\mathscr{P}$ in the present configuration, we convert (98.42) as follows:

$$\left.\begin{aligned} T^{kp} n_p - (D h)^{kp} n_p + H^{pkq}(\bar{b}_{pq} - \bar{b}^r_r n_p n_q) = t^k, \\ H^{pkq} n_p n_q = h^k, \end{aligned}\right\} \tag{98.44}$$

where $(Dh)^{kp} \equiv (DH)^{k\beta} x^p{}_{,\beta}$, and where the stress vector $\boldsymbol{t}$ is related to $\boldsymbol{t}_{\mathrm{R}}$ in the classical way (Sect. 43 A) and the hyperstress $\boldsymbol{h}$ to $\boldsymbol{h}_{\mathrm{R}}$ in just the same way:

$$\boldsymbol{h}\, ds = \boldsymbol{h}_{\mathrm{R}}\, ds_{\mathrm{R}}. \tag{98.45}$$

The major result, and the surprise, comes when the spatial form of the *stress relation* $(98.40)_2$ is calculated:

$$T^{kp} = \varrho \left[\frac{\partial \sigma}{\partial x^k{}_{,\alpha}}\, x^p{}_{,\beta} + \frac{\partial \sigma}{\partial x^k{}_{,\alpha;\beta}}\, x^p{}_{,\alpha;\beta}\right] - H^{pkq}{}_{,q}. \tag{98.46}$$

Now the expression in brackets is symmetric in the indices k and p if and only if the strain-energy function is frame-indifferent. Consequently, a necessary and sufficient condition for frame-indifference of σ is the differential equation

$$\left.\begin{aligned} T^{[kp]} + M^{pkq}{}_{,q} = 0, \\ \{\boldsymbol{T}^T\} + \operatorname{div} \boldsymbol{M} = \boldsymbol{0}, \end{aligned}\right\} \tag{98.47}$$

where the *couple-stress tensor* $\boldsymbol{M}$ is defined as the skew part of $\boldsymbol{H}$:

$$M^{kpq} = H^{[kp]q}. \tag{98.48}$$

Thus in the present theory of hyperelastic materials of second grade, *the Cosserats' equation of moments* [CFT, Eq. (205.10)] *results as a consequence of the energy's being dependent upon the second derivatives of the deformation.* Note that body couples have not been introduced, although they easily could have been.

Considering variations of the form (98.18) in (98.38) yields the following conditions of compatibility for the assigned body forces and surface tractions:

$$\left.\begin{aligned} \int_{\mathscr{P}} \varrho \boldsymbol{b}\, dv + \int_{\partial\mathscr{P}} \boldsymbol{t}\, ds = \boldsymbol{0}, \\ \int_{\mathscr{P}} \boldsymbol{p} \times \varrho \boldsymbol{b}\, dv + \int_{\partial\mathscr{P}} [\boldsymbol{p} \times \boldsymbol{t} + \boldsymbol{n} \times \boldsymbol{h}]\, ds = \boldsymbol{0}. \end{aligned}\right\} \tag{98.49}$$

These equations express the balance of forces and the balance of moments, in the case of equilibrium, provided $\boldsymbol{n} \times \boldsymbol{h}$ be interpreted as a couple per unit area. Notice that $\boldsymbol{n} \cdot \boldsymbol{h}$, the normal component of $\boldsymbol{h}$, does not enter (98.49). It represents a self-equilibrated field of force. By an argument parallel to that which led to (98.32), we see that the strain-energy function σ is frame-indifferent if and only if

$$\left.\begin{aligned} \sigma(\boldsymbol{F}, \nabla\boldsymbol{F}, \boldsymbol{X}) &= \sigma(\boldsymbol{U}, \boldsymbol{R}^T \nabla\boldsymbol{F}, \boldsymbol{X}), \\ &= \bar{\sigma}(\boldsymbol{C}, \boldsymbol{F}^T \nabla\boldsymbol{F}, \boldsymbol{X}), \\ &= \bar{\sigma}(C_{\alpha\beta}, g_{kp} x^k{}_{,\gamma} x^p{}_{,\delta;\varepsilon}, g_{\lambda\mu}, X^\nu), \\ &= \tilde{\sigma}(C_{\alpha\beta}, C_{\gamma\delta,\varepsilon}, s, g_{\lambda\mu}, X^\nu). \end{aligned}\right\} \tag{98.50}$$

The last form follows because

$$2 g_{kp} x^k{}_{,\gamma} x^p{}_{,\delta;\varepsilon} = C_{\gamma\delta,\varepsilon} + C_{\gamma\varepsilon,\delta} - C_{\varepsilon\delta,\gamma}, \tag{98.51}$$

as is easily verified. In view of (29.25), the reduction $(98.50)_4$ is only to be expected.

Consider now a Cosserat material in which the director frame is constrained to rotate with the same rotation as the macrodeformation:

$$\boldsymbol{d}_{\mathfrak{a}}(\boldsymbol{X}, t) = \boldsymbol{R}(\boldsymbol{X}, t)\, \boldsymbol{D}_{\mathfrak{a}}(\boldsymbol{X}), \tag{98.52}$$

where $\boldsymbol{R}$ is the local rotation. Then spin and microspin are the same: $\boldsymbol{W} = \widetilde{\boldsymbol{W}}$, where $\widetilde{\boldsymbol{W}}$ is given by (98.35). The equation of energy (98.36) for a Cosserat material may now be written in the form

$$\left.\begin{aligned} \frac{d}{dt}\int_{\mathscr{P}} \varrho\, E\, dv = \int_{\mathscr{P}} \varrho[\dot{\boldsymbol{x}}\cdot\boldsymbol{b} + 2\operatorname{tr}(\boldsymbol{W}\text{-}[\boldsymbol{L}]\text{-})]\, dv\, + \\ + \int_{\partial\mathscr{P}} [\dot{\boldsymbol{x}}\cdot\boldsymbol{T}\boldsymbol{n} + 2\operatorname{tr}(\boldsymbol{W}\mathsf{M}\boldsymbol{n})]\, ds, \end{aligned}\right\} \tag{98.53}$$

identical in form with the equation of energy that corresponds to (98.38). It was an equation of this form that furnished the starting point for the Grioli-Toupin theory[1] and for its predecessors (above, p. 390). What has been shown, then, is that *the Grioli-Toupin material may be considered, at will, as a special case of the oriented hyperelastic material or of the hyperelastic material of grade two.* In favor of the Grioli-Toupin theory, in which the microrotation and macrorotation coincide, we can find no experimental evidence or theoretical advantage. Its importance lies rather in showing that *different physical motivations toward a more general theory of elasticity, and different formal apparatus, can lead to identical results.* It is all too common to hear questions or pronouncements from the physically inclined about what *really* happens in materials. This example shows, as far as the present case is concerned, that such questions and pronouncements are meaningless. If experiment should come to verify in the most minute detail every single prediction from this special theory, it would be impossible to say whether the material tested "really" has at each point a rigid director frame that rotates with the macrodeformation, or whether it "really" has a strain-energy function depending upon the second as well as the first derivatives of the deformation.

For the classical theory of elasticity, we have given in Sect. 50 an exhaustive analysis of the states of initial stress compatible with a given kind of symmetry. In particular, it is customary to assume the existence of a natural state, in which the stress vanishes, and there are various simple conditions sufficient to ensure that a hyperelastic solid have such a state. TOUPIN[2] has sketched analysis which indicates that for a material of grade 2, only if the strain-energy function has certain restricted forms will there exist any configuration in which both the stress and the hyperstress vanish.

[δ) (added in proof). *Green and Rivlin's multipolar elasticity*[3]. Recently GREEN and RIVLIN have developed a generalized theory of elasticity based on the principle of dissipation as proposed by COLEMAN and NOLL (see Sect. 96) rather than a variational principle.

They consider generalized thermodynamic processes which are specified by adding to the usual list of fields χ, $\boldsymbol{T}$, $\boldsymbol{b}$, θ, ε, $\boldsymbol{h}$, and q (cf. Sect. 79) multipolar displacement fields ${}_{\mathfrak{j}}\mathsf{D}$ ($\mathfrak{j} = 1, 2, \ldots$) with components ${}_{\mathfrak{j}}\mathsf{D}^k{}_{\alpha_1\alpha_2\ldots\alpha_{\mathfrak{j}}}$, multipolar body force fields ${}_{\mathfrak{j}}\mathsf{b}$ with components ${}_{\mathfrak{j}}\mathsf{b}^k{}_{p_1p_2\ldots p_{\mathfrak{j}}}$, and multipolar stress fields ${}_{\mathfrak{j}}\mathsf{T}$ with components ${}_{\mathfrak{j}}\mathsf{T}^{km}{}_{p_1p_2\ldots p_{\mathfrak{j}}}$. While the fields ${}_{\mathfrak{j}}\mathsf{b}$ and ${}_{\mathfrak{j}}\mathsf{T}$ are assumed to be frame-indifferent in the sense of Sect. 17, the multipolar displacement fields ${}_{\mathfrak{j}}\mathsf{D}$ are assumed to transform under changes of frame (17.1), (17.2) into fields ${}_{\mathfrak{j}}\mathsf{D}^*$ given by

$${}_{\mathfrak{j}}\mathsf{D}^{*k}{}_{\alpha_1\alpha_2\ldots\alpha_{\mathfrak{j}}}(\boldsymbol{X}, t^*) = Q^k{}_m(t)\; {}_{\mathfrak{j}}\mathsf{D}^m{}_{\alpha_1\alpha_2\ldots\alpha_{\mathfrak{j}}}(\boldsymbol{X}, t). \tag{98.54}$$

GREEN and RIVLIN define a generalized kinetic energy and a generalized rate of work and then lay down an equation of energy balance appropriate to their definitions. They obtain generalized versions of the laws of balance of momentum

---

[1] As shown by TOUPIN [1962, *64*, § 5], and more briefly later [1964, *84*, § 11], for this material $M^{[kpq]}$ is not determined by the strain-energy function, and that function must be of the form

$$\sigma = \bar{\sigma}(C_{\alpha\beta}, C_{\gamma[\delta,\lambda]}, g_{\nu\pi}, X^\mu).$$

[2] TOUPIN [1964, *84*, § 12].

[3] GREEN and RIVLIN [1964, *38* and *39*]. In later work, done partly in collaboration with NAGHDI, they have developed the matter further [1965, *18A* and *18B*]. Similar theories have been proposed by ERINGEN [1964, *25*] and by ERINGEN and SUHUBI [1964, *26* and *27*].

and moment of momentum by requiring that this energy balance equation be frame-indifferent.

The theory of multipolar elasticity of GREEN and RIVLIN rests upon a set of constitutive equations in which, in addition to the deformation gradient $\boldsymbol{F}$ and the entropy η, the multipolar displacements ${}_{\mathfrak{j}}D$ enter as independent variables. The heat flux is permitted to depend also on the gradients of arbitrary order of the temperature θ.

GREEN and RIVLIN first investigate the restrictions imposed upon their constitutive equations by the principle of material frame-indifference, and then they derive reduced forms for some of these equations. Second, assuming that the entropy production (79.6) be non-negative for all processes compatible with the constitutive equations, they derive a temperature relation of the same form as (80.11) and a set of relations which generalize the stress relation (80.10).

A theory of hyperelasticity of grade $\mathfrak{n}$ results as a special case of GREEN and RIVLIN's general theory when

$$ {}_{\mathfrak{j}}D = {}_{\mathfrak{j}}F, \quad \mathfrak{j} = 1, 2, \ldots \mathfrak{n}, \tag{98.55} $$

where ${}_{\mathfrak{j}}F$ is the $\mathfrak{j}$'th deformation gradient, with components (28.11). The choice (98.55) is permissible because the ${}_{\mathfrak{j}}F$ do in fact obey the transformation law (98.54) under changes of frame (cf. 28.14).

We have not had time to study the work of GREEN and RIVLIN with great care and hence have not been able to give it the detailed exposition it undoubtedly deserves, nor have we compared the results of GREEN and RIVLIN with those of TOUPIN, presented earlier in this section, and with those of CFT, Sects. 166, 205, and 232.

IV. Hypo-elastic materials.

99. Definition of a hypo-elastic material[1]. Hypo-elastic materials are simple materials subject to the following two constitutive restrictions:

(a) The defining response functional $\mathfrak{G}$ of (28.3) satisfies the identity

$$ \boldsymbol{T}(t) = \mathop{\mathfrak{G}}_{s=0}^{\infty} \left(\boldsymbol{F}^{(t)}(s)\right) = \mathop{\mathfrak{G}}_{s=0}^{\infty} \left(\boldsymbol{F}^{(t)}[\sigma(s)]\right) \tag{99.1} $$

[1] As mentioned in fine print on p. 250, the basic concept of hypo-elasticity is virtually suggested by CAUCHY's theory of initially stressed elastic media [1829, *1*, Eqs. (36) (37)], explained in Sect. 68, but there is no reference to *time rates* in his work, which seems to be directed toward infinitesimal static deformation. Special theories of hypo-elastic type, employing invariant time rates, were proposed by JAUMANN [1911, *4*, § IX] [1918, *1*, §§ 96—100] and LOHR [1917, *2*, §§ 21—22]; the general theory was mentioned by FROMM [1933, *2*, Eq. (53a)] as appropriate "zur Darstellung gewisser Erscheinungen der Nachwirkung", but he did not develop its properties. A visco-elastic theory including a special case of hypo-elasticity had been proposed earlier by ZAREMBA [1903, *11* and *12*], but without recognition of its relevance for elastic response. In this connection we leave intentionally uncited numerous non-invariant theories of elastic-plastic transition, intended to describe deformations that are in some sense "small". In the course of an exposition of results obtained by MURNAGHAN [1944, *1* and *2*] [1945, *1*] [1949, *9* and *10*] under uncertain and questionable assumptions, TRUESDELL [1953, *25*, § 56], mentioning expressly the work of CAUCHY and ZAREMBA and acknowledging the help of ERICKSEN, proposed a theory somewhat less general than the one he later [1955, *31*, § 1] put forward under the name *hypo-elasticity*. Essentially the same theory was considered thereupon by THOMAS [1955, *27* and *29*]. The theory explained in our text is the same as TRUESDELL's, although we have chosen to motivate and develop it in what seems to us a better way than the original, and in terms adjusted to those used in the rest of this treatise..

Expositions of aspects of hypo-elasticity have been given by PRAGER [1961, *47*, Ch. 8], GRIOLI [1962, *24*, Ch. 9], ERINGEN [1962, *18*, Ch. 8], GUO-ZHONG HENG [1963, *30* and *32*], and FREDRICKSON [1964, *33*, § 5.5].

for every $\boldsymbol{F}^{(t)}(s)$ in the domain of $\mathfrak{G}$ and every increasing function $\sigma(s)$ such that $\sigma(0)=0$, $\lim_{s\to\infty}\sigma(s)=\infty$.

(b) There is a tensor function $\mathfrak{g}$ such that

$$\dot{\boldsymbol{T}}=\mathfrak{g}(\boldsymbol{T},\boldsymbol{L}) \tag{99.2}$$

for every process possible in the material, where $\boldsymbol{T}$ is the stress and $\boldsymbol{L}$ the velocity gradient. The function $\mathfrak{g}$ is assumed to be continuously differentiable near $\boldsymbol{L}=\boldsymbol{0}$.

Property (a) states, in physical terms, that the stress $\boldsymbol{T}(t)$ at time t depends only on the order in which the body has occupied its past configurations but not on the time-rate at which these past configurations were traversed. The function $\sigma(s)$ represents a change of time-scale, and (99.1) asserts the invariance of the present stress under such a change[1].

Since a deformation that occurred in the distant past can be brought into the recent past by a change of time-scale, it follows from (99.1) that hypo-elastic materials need not have a fading memory (cf. Sect. 38). More precisely, property (a) is in general inconsistent with any of the exact forms of the principle of fading memory stated in Sect. 38. Roughly speaking, a hypo-elastic material may have permanent memory of all the configurations it has ever occupied. It will be recalled that elastic materials have a perfect memory for a fixed reference configuration, no matter how long ago that configuration may have been occupied, if it ever was. Since that configuration may be chosen arbitrarily, we are justified in saying that elastic materials have a perfect memory of *every* configuration they ever occupy[2]. It is the quality of permanent memory that justifies considering hypo-elasticity as a possible model for the kind of behavior that in physical materials is called "elastic". As we shall see, hypo-elasticity and the theory we have heretofore called "elasticity", while indeed they have some attributes in common, are different theories of elastic response.

A theory based on the constitutive assumption (a) alone would have scant predictive value, because the entire kinematical history of a body can rarely be known. It is the assumption (b) that can make hypo-elasticity a useful theory. This assumption states that the instantaneous rate of stress, $\dot{\boldsymbol{T}}$, is determined by the present velocity gradient $\boldsymbol{L}$ [cf. (24.1)] and the present stress $\boldsymbol{T}$. In other words, the stressing of a hypo-elastic material subject to given stress is determined uniquely by the stretching and spin to which it is instantaneously subjected. Assume that the deformation gradient $\boldsymbol{F}(\tau)$ is known for all times τ between some initial time 0 and the present time t. If also the initial stress $\boldsymbol{T}(0)$ is known, then the differential equation (99.2) will determine the present stress $\boldsymbol{T}(t)$ uniquely.

---

[1] Requirement (a) generalizes the original postulate of TRUESDELL [1953, *25*, § 56]: "As in the classical elasticity theory, suppose that there are *only two dimensional moduli* (§ 47), *a natural elasticity* μ_{En} *of dimension* $\mathsf{M}\mathsf{L}^{-1}\mathsf{T}^{-2}$ *and a reference temperature* θ_0 *of dimension* Θ," and [1955, *31*, § 1] "just as in the classical theory, *no mechanical modulus of our ideal material shall carry a dimension independent of stress* ... In particular, the absence of a modulus of dimension T makes what are usually called 'relaxation effects' impossible in our present theory."

The development of the theory given above is to some extent foreshadowed by that of NOLL [1955, *19*, § 14], starting from a more special basis.

[2] It is customary to say that elastic materials are oblivious of all states intermediate between the reference configuration and the present one. Such a view of elasticity is consistent with the one asserted in the text, for it merely reflects the fact that if the response of an elastic material to deformation from any one reference configuration is known, its response to deformation from any other reference configuration is *uniquely determined* by (43.3). The elastic material itself indeed remembers perfectly *every* configuration it may have occupied, but the theorist need take account of only one configuration other than the present one. These remarks indicate that the intuitive concept of memory is not unambiguous.

Hence, to determine $\boldsymbol{T}(t)$, all that need be known about the kinematical history previous to the initial instant 0 is the effect of this previous history on the initial stress. Thus, the unknowable previous history can be replaced by more readily available information on the initial stress.

The variables entering the equations of hypo-elasticity are all quantities defined in the present configuration of the material. The concept of finite strain is not used, nor is there any obvious way in which the theory would simplify if we were to assume the existence of a natural state. Thus a hypo-elastic material "in general has neither preferred state nor preferred stress; in its response to deformation it is, analytically speaking, entirely smooth"[1].

Since $\boldsymbol{L}=\dot{\boldsymbol{F}}\boldsymbol{F}^{-1}$ [cf. (24.1)], it follows that (99.2) is a special constitutive equation of the rate type (36.2). In the present case, the reduced equation (36.10) reads

$$\boldsymbol{R}^T\overset{\circ}{\boldsymbol{T}}\boldsymbol{R}=\mathfrak{f}(\boldsymbol{R}^T\boldsymbol{T}\boldsymbol{R},\boldsymbol{R}^T\boldsymbol{D}\boldsymbol{R};\boldsymbol{U}),\tag{99.3}$$

where $\overset{\circ}{\boldsymbol{T}}$ is the co-rotational stress rate (36.13), $\boldsymbol{D}$ is the stretching tensor, $\boldsymbol{R}$ the rotation tensor, and $\boldsymbol{U}$ the right stretch tensor, all taken with respect to some fixed reference configuration. The original equation (99.2) shows that $\dot{\boldsymbol{T}}$, and hence $\overset{\circ}{\boldsymbol{T}}$, must be determined independently of the choice of reference configuration. Now, given any orthogonal $\boldsymbol{R}$ and any symmetric $\boldsymbol{U}$, it is always possible to find a reference configuration such that the present deformation gradient is given by $\boldsymbol{F}(t)=\boldsymbol{R}\boldsymbol{U}$. Therefore, (99.3) must be an identity in $\boldsymbol{R}$ and $\boldsymbol{U}$. Taking $\boldsymbol{U}=\boldsymbol{1}$ and $\boldsymbol{R}=\boldsymbol{1}$, we see that (99.3) becomes

$$\overset{\circ}{\boldsymbol{T}}=\dot{\boldsymbol{T}}-\boldsymbol{W}\boldsymbol{T}+\boldsymbol{T}\boldsymbol{W}=\mathfrak{h}(\boldsymbol{T},\boldsymbol{D}).\tag{99.4}$$

Taking $\boldsymbol{U}=\boldsymbol{1}$ and $\boldsymbol{R}$ an arbitrary orthogonal tensor, we infer from (99.3) and (99.4) that $\mathfrak{h}$ obeys the identity

$$\boldsymbol{R}^T\mathfrak{h}(\boldsymbol{T},\boldsymbol{D})\boldsymbol{R}=\mathfrak{h}(\boldsymbol{R}^T\boldsymbol{T}\boldsymbol{R},\boldsymbol{R}^T\boldsymbol{D}\boldsymbol{R}),\tag{99.5}$$

i.e. that $\mathfrak{h}$ is an isotropic tensor function of its two symmetric tensor arguments[2].

Consider two kinematical histories defined by $\boldsymbol{F}^{(t)}(s)=\boldsymbol{F}(t-s)$ and $\tilde{\boldsymbol{F}}^{(t)}(s)=\tilde{\boldsymbol{F}}(t-s)=\boldsymbol{F}(t-\sigma(s))$. The requirement (99.1) states that these two histories should determine the same present stress $\boldsymbol{T}(t)$. It is easily seen that the deformation gradient $\tilde{\boldsymbol{F}}$ corresponds to

$$\tilde{\boldsymbol{D}}(t)=\dot{\sigma}(0)\,\boldsymbol{D}(t),\qquad \tilde{\boldsymbol{W}}(t)=\dot{\sigma}(0)\,\boldsymbol{W}(t),\qquad \dot{\tilde{\boldsymbol{T}}}(t)=\dot{\sigma}(0)\,\dot{\boldsymbol{T}}(t).\tag{99.6}$$

Writing (99.4) for the process corresponding to $\tilde{\boldsymbol{F}}$ gives

$$\overset{\circ}{\tilde{\boldsymbol{T}}}=\alpha\overset{\circ}{\boldsymbol{T}}=\mathfrak{h}(\boldsymbol{T},\tilde{\boldsymbol{D}})=\mathfrak{h}(\boldsymbol{T},\alpha\boldsymbol{D}),\tag{99.7}$$

where $\alpha=\dot{\sigma}(0)$. Combining (99.4) with (99.7) shows that

$$\mathfrak{h}(\boldsymbol{T},\alpha\boldsymbol{D})=\alpha\,\mathfrak{h}(\boldsymbol{T},\boldsymbol{D}).\tag{99.8}$$

Given any positive constant α, we can choose $\sigma(s)=\alpha s$, $\dot{\sigma}(0)=\alpha$. Therefore, (99.8) holds for every positive α, which means that $\mathfrak{h}(\boldsymbol{T},\boldsymbol{D})$ is positively homogeneous in $\boldsymbol{D}$. Since $\mathfrak{h}$ was assumed to be continuously differentiable in a neighborhood of $\boldsymbol{D}=\boldsymbol{0}$, it follows that $\mathfrak{h}(\boldsymbol{T},\boldsymbol{D})$ must be linear in $\boldsymbol{D}$.

---

[1] TRUESDELL [1955, *31*, § 1].

[2] Much confusion and some errors in the expositions of hypo-elasticity have been caused, apparently, by TRUESDELL's original failure to see that (99.5) *must* hold, although one of his footnotes [1955, *31*, p. 85] foreshadows such a conclusion.

We summarize: *The constitutive equation of a hypo-elastic material may be written in the form*

$$\overset{\circ}{\boldsymbol{T}} = \mathsf{H}(\boldsymbol{T})[\boldsymbol{D}], \tag{99.9}$$

where $\overset{\circ}{\boldsymbol{T}} = \dot{\boldsymbol{T}} - \boldsymbol{W}\boldsymbol{T} + \boldsymbol{T}\boldsymbol{W}$ *is the co-rotational stress rate,* $\boldsymbol{W}$ *is the spin, and* $\boldsymbol{D}$ *is the stretching. The tensor function* $\mathsf{H}(\boldsymbol{T})[\boldsymbol{D}]$ *is linear in* $\boldsymbol{D}$ *and isotropic in* $\boldsymbol{T}$ *and* $\boldsymbol{D}$.

Instead of the co-rotational stress rate $\overset{\circ}{\boldsymbol{T}}$ one can use the convected stress rate:

$$\overset{\triangle}{\boldsymbol{T}} = \dot{\boldsymbol{T}} + \boldsymbol{L}^T\boldsymbol{T} + \boldsymbol{T}\boldsymbol{L} = \overset{\circ}{\boldsymbol{T}} + \boldsymbol{D}\boldsymbol{T} + \boldsymbol{T}\boldsymbol{D}, \tag{99.10}$$

obtaining as equivalent to (99.9)

$$\overset{\triangle}{\boldsymbol{T}} = \bar{\mathsf{H}}(\boldsymbol{T})[\boldsymbol{D}], \tag{99.11}$$

where $\bar{\mathsf{H}}$ differs from H by the term $\boldsymbol{D}\boldsymbol{T} + \boldsymbol{T}\boldsymbol{D}$. The isotropic functions H and $\bar{\mathsf{H}}$ of (99.9) and (99.11) will be called *hypo-elastic response functions*.

The tensor $\overset{\circ}{\boldsymbol{T}}$ in (99.9) may be replaced by any of the infinitely many invariant measures[1] of stressing included in the form $\overset{\circ}{\boldsymbol{T}} + \mathfrak{t}[\boldsymbol{T}, \boldsymbol{D}]$, where $\mathfrak{t}[\boldsymbol{T}, \boldsymbol{D}]$ is an arbitrary bilinear isotropic tensor function of $\boldsymbol{T}$ and $\boldsymbol{D}$.

Component forms of (99.9) and (99.11) are

$$\overset{\circ}{T}_{km} = \dot{T}_{km} - W_{kp}T^p{}_m + T_k{}^p W_{pm} = \mathsf{H}_{km}{}^{rs}(T_{uv})\,D_{rs}, \tag{99.12}$$

$$\overset{\triangle}{T}_{km} = \dot{T}_{km} + \dot{x}^p{}_{,k}T_{pm} + T_{kp}\dot{x}^p{}_{,m} = \bar{\mathsf{H}}_{km}{}^{rs}(T_{uv})\,D_{rs}, \tag{99.13}$$

where

$$\mathsf{H}_{km}{}^{rs} = \mathsf{H}_{mk}{}^{rs} = \mathsf{H}_{km}{}^{sr}, \quad \bar{\mathsf{H}}_{km}{}^{rs} = \bar{\mathsf{H}}_{mk}{}^{rs} = \bar{\mathsf{H}}_{km}{}^{sr}. \tag{99.14}$$

In any co-ordinate system, the covariant components of the convected stress rate appearing on the left-hand side of (99.13) may be expressed in terms of partial derivatives alone:

$$\overset{\triangle}{T}_{km} = \partial_t T_{km} + \dot{x}^p\,\partial_p T_{km} + T_{pm}\,\partial_k\dot{x}^p + T_{kp}\,\partial_m\dot{x}^p, \tag{99.15}$$

as follows from (15.13) and the fact that all terms in (36.21) involving the Christoffel symbols cancel, or directly from (36.17).

In the case when the response function H in (99.9) is a polynomial function of $\boldsymbol{T}$, we may apply the representations (13.7) and (11.22) to show that

$$\left.\begin{aligned} \mathsf{H}(\boldsymbol{T})[\boldsymbol{D}] = {} & [\beth_1\operatorname{tr}\boldsymbol{D} + \beth_2\operatorname{tr}(\boldsymbol{T}\boldsymbol{D}) + \beth_3\operatorname{tr}(\boldsymbol{T}^2\boldsymbol{D})]\,\boldsymbol{1} + \\ & + [\beth_4\operatorname{tr}\boldsymbol{D} + \beth_5\operatorname{tr}(\boldsymbol{T}\boldsymbol{D}) + \beth_6\operatorname{tr}(\boldsymbol{T}^2\boldsymbol{D})]\,\boldsymbol{T} + \\ & + [\beth_7\operatorname{tr}\boldsymbol{D} + \beth_8\operatorname{tr}(\boldsymbol{T}\boldsymbol{D}) + \beth_9\operatorname{tr}(\boldsymbol{T}^2\boldsymbol{D})]\,\boldsymbol{T}^2 + \\ & + \beth_{10}\boldsymbol{D} + \beth_{11}(\boldsymbol{D}\boldsymbol{T} + \boldsymbol{T}\boldsymbol{D}) + \beth_{12}(\boldsymbol{D}\boldsymbol{T}^2 + \boldsymbol{T}^2\boldsymbol{D}), \end{aligned}\right\} \tag{99.16}$$

where $\beth_1, \ldots, \beth_{12}$ are polynomials in the principal invariants $I_{\boldsymbol{T}}, II_{\boldsymbol{T}}, III_{\boldsymbol{T}}$. In components

$$\left.\begin{aligned} \mathsf{H}^{kpmq} = {} & g^{kp}(\beth_1 g^{mq} + \beth_2 T^{mq} + \beth_3 T^{mr}T_r{}^q) + \\ & + T^{kp}(\beth_4 g^{mq} + \beth_5 T^{mq} + \beth_6 T^{mr}T_r{}^q) + \\ & + T^{ks}T_s{}^p(\beth_7 g^{mq} + \beth_8 T^{mq} + \beth_9 T^{mr}T_r{}^q) + \\ & + \tfrac{1}{2}\beth_{10}(g^{km}g^{pq} + g^{pm}g^{kq}) + \\ & + \tfrac{1}{2}\beth_{11}(g^{kq}T^{mp} + g^{pq}T^{mk} + g^{mp}T^{kq} + g^{mk}T^{pq}) + \\ & + \tfrac{1}{2}\beth_{12}(g^{kq}T^{mr}T_r{}^p + g^{pq}T^{mr}T_r{}^k + g^{mp}T^{kr}T_r{}^q + g^{mk}T^{pr}T_r{}^q). \end{aligned}\right\} \tag{99.17}$$

[1] Various such stress rates have been used in the literature. Cf. CFT, Sects. 148—151. Despite claims and whole papers to the contrary, any advantage claimed for one such rate over another is pure illusion.

It is likely that a representation of the form (99.16) is possible even when the dependence on $\boldsymbol{T}$ is not polynomial. Of course, the response function $\bar{H}$ in (99.11) also has a representation of the same form as (99.16). The response coefficients $\beth_\Gamma$ are not uniquely determined by the material.

Assume that a hypo-elastic response function H is prescribed, and consider the corresponding constitutive equation (99.9). This equation must be expected to define a class of materials rather than a single material in the sense of Sect. 19. Cf. the remarks in Sect. 36. An example of such a class, for a special H, will be given in the following section. In general, the problem of finding all materials, i.e. all response functionals, compatible with a given constitutive equation (99.9) of hypo-elasticity is unsolved. It is even conceivable that some hypo-elastic response functions H do not correspond to any material at all, in the sense in which the term "material" is used in this treatise. It appears likely that all hypo-elastic materials are isotropic, but a proof of this conjecture is not known[1].

For hypo-elasticity, BERNSTEIN[2] has introduced a concept of "material" different from the one used here. He considered *stress-deformation pairs* $[\boldsymbol{T}, \boldsymbol{F}]$, i.e. possible pairs of values for the stress tensor $\boldsymbol{T}$ and the deformation gradient $\boldsymbol{F}$. He called two such pairs $[\boldsymbol{T}_1, \boldsymbol{F}_1]$ and $[\boldsymbol{T}_2, \boldsymbol{F}_2]$ *equivalent* with respect to a given hypo-elastic response function H if functions $\boldsymbol{T}(t)$ and $\boldsymbol{F}(t)$ such that $\boldsymbol{T}(t_i) = \boldsymbol{T}_i$, $\boldsymbol{F}(t_i) = \boldsymbol{F}_i$, $i = 1, 2$, and such that (99.9) holds when $\mathring{\boldsymbol{T}}$ and $\boldsymbol{D}$ are computed from $\boldsymbol{T}(t)$ and $\boldsymbol{F}(t)$ could be found. By a hypo-elastic material he understood an assignment of a hypo-elastic response function H and a corresponding equivalence class of stress-deformation pairs. The reference configuration is assigned *a priori* in the description given here. A change of reference configuration must be accompanied with an appropriate change in the stress-deformation class. To a material in the sense of BERNSTEIN may correspond one, several, or no simple materials according to the definition in Sect. 28.

In accord with the general principle laid down in Sect. 30, *incompressible hypo-elastic materials*[3] are defined by constitutive equations of the form (99.9), (99.11), (99.12), or (99.13), except that $\boldsymbol{T}$ is replaced by $\boldsymbol{T}+p\boldsymbol{1}$. Since $\overset{\circ}{\overline{p\boldsymbol{1}}} = \dot{p}\boldsymbol{1}$, the form corresponding to (99.9) is

$$\mathring{\boldsymbol{T}} = -\dot{p}\boldsymbol{1} + H(\boldsymbol{T}+p\boldsymbol{1})[\boldsymbol{D}], \tag{99.18}$$

where only motions such that $\operatorname{tr}\boldsymbol{D} = 0$ are to be considered.

Special hypo-elastic materials are defined by restricting the nature of the dependence of the response function $H(\boldsymbol{T})[\;]$ upon $\boldsymbol{T}$. A material is said[4] to be of grade n if $H(\boldsymbol{T})[\;]$ is a polynomial of degree n in the components of $\boldsymbol{T}$. If $n=0$, this definition is not invariant under change of invariant stress rate, since the various invariant stress rates differ from one another by terms linear in $\boldsymbol{T}$. Hence we do not consider the definition to be physically relevant[5] unless $n \geqq 1$. The constitutive equation for a hypo-elastic material of grade 1 has the form

$$\left.\begin{aligned}\dot{\boldsymbol{S}} = \boldsymbol{W}\boldsymbol{S} - \boldsymbol{S}\boldsymbol{W} + \left(\frac{\lambda}{2\mu} + \gamma_0 I_{\boldsymbol{S}}\right)(\operatorname{tr}\boldsymbol{D})\boldsymbol{1} + {}\\ + (1+\gamma_1 I_{\boldsymbol{S}})\boldsymbol{D} + \gamma_2(\operatorname{tr}\boldsymbol{D})\boldsymbol{S} + \gamma_3(\operatorname{tr}(\boldsymbol{S}\boldsymbol{D}))\boldsymbol{1} + \tfrac{1}{2}\gamma_4(\boldsymbol{S}\boldsymbol{D}+\boldsymbol{D}\boldsymbol{S}),\end{aligned}\right\} \tag{99.19}$$

where $\boldsymbol{S} = \boldsymbol{T}/(2\mu)$, where λ and μ are constants that may be interpreted in the same way as are the elasticities of an isotropic elastic material in infinitesimal

[1] While the definitions in most of this treatise, resting upon the concept of isotropy *group*, lead to results the same as those obtained formerly by use of isotropic *functions*, a difference appears here. The hypo-elastic *response functions* H and $\bar{H}$ must always be isotropic functions, but such materials as correspond to hypo-elastic constitutive equations have not been proved to have for some reference configurations isotropy groups containing the orthogonal group.

[2] BERNSTEIN [1960, *8*, § :] [1961, *3*].

[3] NOLL [1955, *19*, Eq. (14.5)], correcting TRUESDELL [1955, *31*, Eq. (2.5)].

[4] TRUESDELL [1955, *31*, § 4].

[5] Thus we give no attention to the hypo-elastic material of JAUMANN [1911, *4*, § IX], which is the material of grade 0 as defined here, of to the material of grade 0 of TRUESDELL [1955, *31*, § 4] [1955, *32*], which results from taking the response function corresponding to a different time flux as being of degree 0 in $\boldsymbol{T}$.

deformation from a natural state, and where $\gamma_0, \gamma_1, \gamma_2, \gamma_3$ and γ_4 are dimensionless material constants.

100. Relation to elasticity. We investigate the response of a hypo-elastic material which initially occupies a prescribed configuration and is subject to a known initial stress $\boldsymbol{T}_0$. We take the initial configuration as the reference configuration.

Let

$$\boldsymbol{F} = \hat{\boldsymbol{F}}(\alpha), \quad \hat{\boldsymbol{F}}(0) = \boldsymbol{1}, \quad 0 \leqq \alpha < \infty, \tag{100.1}$$

be a given one-parameter family of deformation gradients. When α is interpreted as being the time, then (100.1) describes a deformation process, counted from the initial instant $\alpha = 0$. The corresponding stress $\boldsymbol{T} = \hat{\boldsymbol{T}}(\alpha)$ is the obtained by solving the differential equation (99.9), i.e.

$$\hat{\boldsymbol{T}}' - \hat{\boldsymbol{W}}\hat{\boldsymbol{T}} + \hat{\boldsymbol{T}}\hat{\boldsymbol{W}} = \mathsf{H}(\hat{\boldsymbol{T}})[\hat{\boldsymbol{D}}], \tag{100.2}$$

subject to the initial condition $\hat{\boldsymbol{T}}(0) = \boldsymbol{T}_0$. In (100.2), $\hat{\boldsymbol{T}}'$ denotes the derivative of $\hat{\boldsymbol{T}}(\alpha)$ with respect to α; $\hat{\boldsymbol{D}}$ and $\hat{\boldsymbol{W}}$ denote, respectively, the stretching and spin corresponding to the deformation process (100.1).

Now consider a deformation process that is described by

$$\boldsymbol{F}(t) = \hat{\boldsymbol{F}}(\alpha(t)), \quad \alpha(t_0) = 0, \quad t \geqq t_0, \tag{100.3}$$

where $\alpha(t)$ is an arbitrary smooth function of the time t when $t \geqq t_0$. It is easily seen that the stretching and spin corresponding to (100.3) are given by

$$\boldsymbol{D}(t) = \dot{\alpha}(t)\,\hat{\boldsymbol{D}}(\alpha(t)), \quad \boldsymbol{W}(t) = \dot{\alpha}(t)\,\hat{\boldsymbol{W}}(\alpha(t)), \tag{100.4}$$

respectively. If we put

$$\boldsymbol{T}(t) = \hat{\boldsymbol{T}}(\alpha(t)), \tag{100.5}$$

we see that $\dot{\boldsymbol{T}}(t) = \dot{\alpha}(t)\,\hat{\boldsymbol{T}}'(\alpha(t))$. It then follows from (100.2) and (100.4) that (100.5) satisfies the differential equation (99.9). Clearly, (100.5) also satisfies the initial condition

$$\boldsymbol{T}(t_0) = \boldsymbol{T}_0. \tag{100.6}$$

Therefore, (100.5) gives the actual stress produced from the initial stress $\boldsymbol{T}_0$ in response to the deformation process (100.3). The result (100.5) means that *a hypo-elastic material behaves like an elastic material when only deformations belonging to the one-parameter family* (100.1) are considered. Indeed, $\hat{\boldsymbol{T}}(\alpha)$ may be regarded as a response function which determines the stress when the deformation parameter α is given. This response function is obtained by solving the differential equation (100.2). If we consider a second deformation process, $\boldsymbol{F} = \tilde{\boldsymbol{F}}(\alpha)$, the result may be applied again, but then $\boldsymbol{T} = \tilde{\boldsymbol{T}}(\alpha)$, where in general $\hat{\boldsymbol{T}} \neq \tilde{\boldsymbol{T}}$. Going back to the defining equations (99.1), we may express this result in another way. For the two deformation processes $\boldsymbol{F} = \hat{\boldsymbol{F}}(\alpha)$ and $\tilde{\boldsymbol{F}} = \boldsymbol{F}(\alpha)$, where $\alpha = \alpha(t)$, we shall have

$$\left.\begin{aligned} \overset{\infty}{\underset{s=0}{\mathfrak{G}}}\left(\hat{\boldsymbol{F}}(\alpha(t-s))\right) &= \hat{\mathfrak{g}}(\alpha(t), \boldsymbol{T}_0), \\ \overset{\infty}{\underset{s=0}{\mathfrak{G}}}\left(\tilde{\boldsymbol{F}}(\alpha(t-s))\right) &= \tilde{\mathfrak{g}}(\alpha(t), \boldsymbol{T}_0), \end{aligned}\right\} \tag{100.7}$$

where $\hat{\mathfrak{g}}$ and $\tilde{\mathfrak{g}}$ are, in general, *different* functions. That is, *while a hypo-elastic material always behaves like an elastic material, in general it will behave like one elastic material for deformations belonging to one one-parameter family and like a* different *elastic material for deformations belonging to another one*[1].

For small values of the deformation parameter α it is possible to obtain an explicit asymptotic expression for the stress $\boldsymbol{T}=\hat{\boldsymbol{T}}(\alpha)$, as follows. To the deformation gradient $\boldsymbol{F}=\hat{\boldsymbol{F}}(\alpha)$ correspond the displacement gradient $\boldsymbol{H}=\boldsymbol{F}-\mathbf{1}$, the infinitesimal strain tensor $\tilde{\boldsymbol{E}}=\frac{1}{2}(\boldsymbol{H}+\boldsymbol{H}^T)$, and the infinitesimal rotation tensor $\tilde{\boldsymbol{R}}=\frac{1}{2}(\boldsymbol{H}-\boldsymbol{H}^T)$. An easy analysis shows that

$$\left.\begin{aligned} \alpha\,\hat{\boldsymbol{W}}(0) &= \tilde{\boldsymbol{R}} + o(\alpha), & \alpha\,\hat{\boldsymbol{D}}(0) &= \tilde{\boldsymbol{E}} + o(\alpha), \\ \alpha\,\boldsymbol{T}'(0) &= \boldsymbol{T}-\boldsymbol{T}_0 + o(\alpha^2) & \boldsymbol{T} &= \boldsymbol{T}_0 + o(\alpha), \end{aligned}\right\} \tag{100.8}$$

where $\hat{\boldsymbol{D}}(\alpha)$ and $\hat{\boldsymbol{W}}(\alpha)$ denote, as before, the stretching and spin when the time coincides with the parameter α. If we write (100.2) for $\alpha=0$, then multiply by α, and then substitute (100.8) into the result, we find that

$$\boldsymbol{T}-\boldsymbol{T}_0=\tilde{\boldsymbol{R}}\boldsymbol{T}_0-\boldsymbol{T}_0\tilde{\boldsymbol{R}}+\mathsf{H}(\boldsymbol{T}_0)[\tilde{\boldsymbol{E}}]+o(\alpha). \tag{100.9}$$

Apart from the error term, (100.9) reduces to (41.25) [or (68.16)] when we put $\mathsf{L}=\mathsf{H}(\boldsymbol{T}_0)$. Thus, a given hypo-elastic material behaves in all infinitesimal deformation processes like a *single* elastic material. In other words, *for infinitesimal deformations from an arbitrary configuration, hypo-elasticity reduces to a special case of elasticity.*

It was this fact that motivated TRUESDELL's original proposal of the theory[2]. While the formula (68.16) may be derived by differentiating the constitutive equation of elasticity, TRUESDELL revived CAUCHY's idea of considering the differential form itself as the basic law; no natural state or other reference configuration is implied, and the stress relation corresponding to an arbitrary initial stress is to be found by integration. As TRUESDELL[3] remarked, "the equations of the theory of hypo-elasticity reduce to those of the classical linear theory under the assumptions usual in formulating that theory," but in large deformations the stress is built up by summation of linear increments from the stresses in the infinitesimally preceding configurations occupied in the deformation process.

By (99.5) $\mathsf{H}(\boldsymbol{T}_0)[\tilde{\boldsymbol{E}}]$ is an isotropic tensor function of $\boldsymbol{T}_0$ and $\tilde{\boldsymbol{E}}$. In particular, if we put $\boldsymbol{T}_0=\mathbf{0}$ and $\mathsf{H}(\mathbf{0})=\mathsf{L}$, we find that $\mathsf{L}[\tilde{\boldsymbol{E}}]$ is an isotropic function of $\tilde{\boldsymbol{E}}$ and hence has a representation of the form (12.5) . Thus, when the initial stress is zero and when the error term $o(\alpha)$ is omitted, (100.9) reduces to CAUCHY's constitutive equation (41.27) for an *isotropic* elastic material subject to infinitesimal deformation. Therefore, a hypo-elastic material cannot behave like an anisotropic material in infinitesimal deformations from an unstressed state. It follows a fortiori that *anisotropic elasticity is not included in hypo-elasticity as a special case*[4].

[1] This theorem of NOLL, here published for the first time, should clear away the confusion regarding the relation of hypo-elasticity to elasticity created by certain expositors.

[2] The more explicit treatment in our text, here published for the first time, was constructed by NOLL.

[3] TRUESDELL [1955, *31*, § 3].

[4] This argument was attributed to NOLL by TRUESDELL [1963, *75*]. The theorem corrects contrary claims by HILL [1959, *12*, § 5] and PRAGER [1961, *47*, Ch. X, § 1]; the latter even asserted that "hypo-elastic behavior ... corresponds to a minimal requirement that a material must fulfill, if it is in any sense to qualify as elastic." However, that hypo-elastic materials are generally not elastic had been proved by BERNSTEIN [1960, *8*, § 5]. In formulating the theory originally, TRUESDELL [1955, *31*, § 3] had sought "a new concept of elastic behavior, mutually exclusive with the theory of finite [elastic] strain except in the linearized case" and had been surprised by the theorem of NOLL stated and proved in the following text.

All elastic materials satisfy the differential equation (45.17), which seems similar to the hypo-elastic constitutive equation (99.9) but is in fact of more general form. Indeed, from its definition (45.2) we see that $\mathsf{C}=\mathsf{C}(\boldsymbol{F})=\mathsf{C}(\boldsymbol{R},\boldsymbol{C})$, where possible dependence on $\boldsymbol{X}$ is not indicated explicitly. Hence, (45.17) is of the form

$$\overset{\circ}{\boldsymbol{T}}=\mathsf{H}(\boldsymbol{T},\boldsymbol{R},\boldsymbol{C})[\boldsymbol{D}]. \tag{100.10}$$

In the special case when the dependence of H on $\boldsymbol{C}$ and $\boldsymbol{R}$ is only apparent, *and in this case alone*[1], can an elastic material be hypo-elastic.

Consider, for example, an isotropic elastic material. By substituting (48.24) into (45.6) and putting the result into (45.17), we obtain the differential equation

$$\left.\begin{aligned}\overset{\circ}{T}{}^{km}&=2F^{km}{}_{pq}B^{qr}D^{p}_{r},\\ \overset{\circ}{\boldsymbol{T}}&=\mathfrak{f}_{\boldsymbol{B}}(\boldsymbol{B})[\boldsymbol{DB}]+\{\mathfrak{f}_{\boldsymbol{B}}(\boldsymbol{B})[\boldsymbol{DB}]\}^{T}.\end{aligned}\right\} \tag{100.11}$$

This same result has been derived already, in the context of infinitesimal displacements rather than time rates, as Eq. (69.3). Alternatively, it may be derived directly by differentiating the constitutive equation (47.4) for isotropic elastic materials. If (47.4) is invertible, we can substitute its inverse,

$$\boldsymbol{B}=\overset{-1}{\mathfrak{f}}(\boldsymbol{T}), \tag{100.12}$$

into (100.11) and so obtain an equation of the form (99.9), with

$$\mathsf{H}(\boldsymbol{T})[\boldsymbol{D}]=\mathfrak{f}_{\boldsymbol{B}}\big(\overset{-1}{\mathfrak{f}}(\boldsymbol{T})\big)\big[\boldsymbol{D}\overset{-1}{\mathfrak{f}}(\boldsymbol{T})+\overset{-1}{\mathfrak{f}}(\boldsymbol{T})\boldsymbol{D}\big]. \tag{100.13}$$

Therefore, *every isotropic elastic material with invertible stress relation is hypo-elastic*[2], its hypo-elastic response function $\mathsf{H}(\boldsymbol{T})[\boldsymbol{D}]$ being given explicitly in terms of its elastic response function $\mathfrak{f}(\boldsymbol{B})$ by (100.13).

We consider now the special case of an elastic fluid[3]. Differentiating the constitutive equation (50.1) with respect to time and using the equation of continuity (15.14) in the form $\dot{\varrho}=-\varrho\,\mathrm{tr}\boldsymbol{D}$ yields

$$\dot{\boldsymbol{T}}=\varrho p'(\varrho)(\mathrm{tr}\boldsymbol{D})\boldsymbol{1}. \tag{100.14}$$

When the stress $\boldsymbol{T}$ is hydrostatic, we have $\boldsymbol{TW}-\boldsymbol{WT}=\boldsymbol{0}$, and hence $\dot{\boldsymbol{T}}=\overset{\circ}{\boldsymbol{T}}$. Consequently (100.14) is the specialization of (100.11) to the case of a fluid. It follows from (50.1) that

$$p(\varrho)=-\tfrac{1}{3}\mathrm{tr}\,\boldsymbol{T}. \tag{100.15}$$

Therefore, when the function $p=p(\varrho)$ has an inverse, equation (100.14) is equivalent to

$$\overset{\circ}{\boldsymbol{T}}=f(\mathrm{tr}\,\boldsymbol{T})(\mathrm{tr}\,\boldsymbol{D})\boldsymbol{1}, \tag{100.16}$$

where the function f is related to the function p as follows:

$$f(-3p(\varrho))=\varrho p'(\varrho). \tag{100.17}$$

Since (100.16) is of the form (99.9), *every elastic fluid with an invertible pressure function is a hypo-elastic material.*

[1] TRUESDELL [1963, *75*].
[2] NOLL [1955, *19*, § 15b].
[3] The following is a slight generalization of an investigation of BERNSTEIN [1960, *8*, § 3].

Consider the particular constitutive equation (100.16) of hypo-elasticity, in which f is assumed to be an arbitrary smooth scalar function. (100.17) may be regarded as a differential equation having infinitely many solutions $p(\varrho)$. Each of these solutions corresponds to an elastic fluid with a constitutive equation (50.1). Thus we have an example of a single equation of hypo-elasticity corresponding to infinitely many materials. If ϱ_0 is the density at some initial time $t=0$ and p_0 an arbitrarily prescribed initial pressure, then there will be a unique solution $p=p(\varrho)$ of (100.17) that satisfies $p_0=p(\varrho_0)$. If we prescribe $\boldsymbol{D}=\boldsymbol{D}(t)$ arbitrarily, then the solution of (100.16) that satisfies the initial condition $\boldsymbol{T}(0)=-p_0\mathbf{1}$ will be given by $\boldsymbol{T}(t)=-p(\varrho(t))\mathbf{1}$, where $\varrho(t)$ is the solution of the equation of continuity $\dot{\varrho}+\varrho\,\mathrm{tr}\,\boldsymbol{D}=0$ that satisfies the initial condition $\varrho(0)=\varrho_0$. Thus, for any solution of (100.16), if the stress is ever hydrostatic, it must be always hydrostatic, and if the stress is once not hydrostatic, it may never become hydrostatic. Each hydrostatic solution of (100.16) corresponds to the response of infinitely many fluids (one for each initial density). Non-hydrostatic solutions, obviously, cannot be consistent with the stress relation for an elastic fluid, but further apparatus is needed before we can determine whether or not they can correspond to more general elastic materials.

Indeed, we have nearly but not quite determined the range of agreement between hypo-elasticity and elasticity: Among elastic materials having a natural state, only those whose response to infinitesimal deformations from that state is that of an isotropic material can be hypo-elastic; every isotropic elastic material with *invertible* stress relation $\boldsymbol{T}=\mathfrak{f}(\boldsymbol{B})$ is hypo-elastic; and some such materials with non-invertible stress relations, namely, fluids with invertible pressure functions, are hypo-elastic. In the case of a given hypo-elastic response function $\mathsf{H}(\boldsymbol{T})[\;]$, however, we have not yet presented a test for determining whether any solutions $\boldsymbol{T}=\boldsymbol{T}(\boldsymbol{T}_0, \boldsymbol{x}, t)$ are compatible with elasticity. Such a test has been found by BERNSTEIN[1].

First, we write (45.17) in a different notation:

$$\left.\begin{aligned} \dot{\boldsymbol{T}} &= \mathsf{E}(\boldsymbol{F})[\boldsymbol{L}], \\ \dot{T}^{km} &= \mathsf{E}^{kmpq}\dot{x}_{p,q}, \end{aligned}\right\} \tag{100.18}$$

say, where

$$\mathsf{E}^{kmpq} = -T^{km}g^{pq} + T^{kq}g^{mp} + T^{mq}g^{kp} + C^{kmpq}, \tag{100.19}$$

$C^{kmpq}(\boldsymbol{F})$ being the elasticity defined by (45.2). We may write the general constitutive equation (99.9) of hypo-elasticity in the parallel form

$$\left.\begin{aligned} \dot{\boldsymbol{T}} &= \mathsf{K}(\boldsymbol{T})[\boldsymbol{L}], \\ \dot{T}^{km} &= \mathsf{K}^{kmpq}\dot{x}_{p,q}, \end{aligned}\right\} \tag{100.20}$$

where

$$\mathsf{K}^{kmpq} = \tfrac{1}{2}(T^{kq}g^{mp} + T^{mq}g^{kp} - T^{kp}g^{mq} - T^{mp}g^{kq}) + \mathsf{H}^{kmpq}. \tag{100.21}$$

In view of (99.14) and (45.3), the tensors $\mathsf{E}(\boldsymbol{F})[\;]$ and $\mathsf{K}(\boldsymbol{T})[\;]$ satisfy the same symmetry identities:

$$\left.\begin{aligned} \mathsf{E}^{kmpq} &= \mathsf{E}^{mkpq}, \qquad \mathsf{K}^{kmpq} = \mathsf{K}^{mkpq}, \\ \mathsf{E}^{kmpq} - \mathsf{E}^{kmqp} &= T^{kq}g^{mp} + T^{mq}g^{kp} - T^{kp}g^{mq} - T^{mp}g^{kq}, \\ &= \mathsf{K}^{kmpq} - \mathsf{K}^{kmqp}. \end{aligned}\right\} \tag{100.22}$$

Now if a particular solution $\boldsymbol{T}=\boldsymbol{T}(\boldsymbol{T}_0, \boldsymbol{x}, t)$ of (100.2) is compatible with a stress relation $\boldsymbol{T}=\mathfrak{g}(\boldsymbol{F})$, then $\mathsf{K}(\boldsymbol{T})[\;]=\mathsf{K}(\mathfrak{g}(\boldsymbol{F}))[\;]=\mathsf{E}(\boldsymbol{F})[\;]$. From (100.20) it

[1] BERNSTEIN [1960, *8*, § 4].

follows that

$$\frac{\partial \mathfrak{g}^{km}}{\partial x^p{}_{,\alpha}} \dot{x}^p{}_{,q} x^q{}_{,\alpha} = K^{km}{}_p{}^q(\mathfrak{g}(\boldsymbol{F}))\dot{x}^p{}_{,q}. \tag{100.23}$$

In order for this equation to hold for all deformation processes $\boldsymbol{F}=\boldsymbol{F}(t)$, it is necessary that

$$\frac{\partial \mathfrak{g}^{km}}{\partial x^p{}_{,\alpha}} = K^{km}{}_p{}^l X^\alpha{}_{,l}. \tag{100.24}$$

Conditions of integrability for this system are

$$\left.\begin{aligned} 0 &= \frac{\partial^2 \mathfrak{g}^{km}}{\partial x^p{}_{,\alpha}\partial x^q{}_{,\beta}} - \frac{\partial^2 \mathfrak{g}^{km}}{\partial x^q{}_{,\beta}\partial x^p{}_{,\alpha}}, \\ &= \frac{\partial K^{km}{}_p{}^l}{\partial T^{rs}} \frac{\partial \mathfrak{g}^{rs}}{\partial x^q{}_{,\beta}} X^\alpha{}_{,l} - \frac{\partial K^{km}{}_q{}^l}{\partial T^{rs}} \frac{\partial \mathfrak{g}^{rs}}{\partial x^p{}_{,\alpha}} X^\beta{}_{,l} + \\ &\quad + K^{km}{}_p{}^l \frac{\partial X^\alpha{}_{,l}}{\partial x^q{}_{,\beta}} - K^{km}{}_q{}^l \frac{\partial X^\beta{}_{,l}}{\partial x^p{}_{,\alpha}}. \end{aligned}\right\} \tag{100.25}$$

Equivalently, by (100.24) and Eq. (17.5) of CFT,

$$\frac{\partial K^{kmpq}}{\partial T^{rs}} K^{rsjl} - \frac{\partial K^{kmjl}}{\partial T^{rs}} K^{rspq} - K^{kmpl} g^{jq} + K^{kmjq} g^{pl} = 0. \tag{100.26}$$

Conversely, if $K(\boldsymbol{T})[\;]$ satisfies the conditions (100.26), we may work backward and find the function $\mathfrak{g}(\boldsymbol{F})$ by integrating (100.24). In other words, there is a fourth-order tensor $E(\boldsymbol{F})[\;]$ such that $E(\boldsymbol{F})[\;]=K(\boldsymbol{T})[\;]$; equivalently, by (100.19) and (100.21), if we set

$$C^{kmpq} \equiv T^{km} g^{pq} - T^{kq} g^{mp} - T^{mq} g^{kp} + K^{kmpq}, \tag{100.27}$$

not only will C^{kmpq} be a function of $\boldsymbol{F}$, but also there will exist a function $K(\boldsymbol{C})$ such that (45.2) is satisfied. From this result and (45.6) we see that the elasticity B^{kmpq} has the form

$$B^{kmpq} = T^{km} g^{pq} - T^{kq} g^{mp} + K^{kmpq}. \tag{100.28}$$

Thus, if a given constitutive equation of hypo-elasticity is compatible with the theory of elasticity at all, by (100.27) or (100.28) we can write down at once the elasticities of the corresponding material. Moreover, by (84.15) or (84.16) and (100.28), a hypo-elastic material is hyperelastic if its response function satisfies not only (100.26) but also the identity

$$H^{kmpq} + T^{km} g^{pq} = H^{pqkm} + T^{pq} g^{km}. \tag{100.29}$$

Since K^{kmpq}, for a given hypo-elastic material, is a known function of $\boldsymbol{T}$, (100.26) generally will yield a condition restricting $\boldsymbol{T}$. Now the values assumed by $\boldsymbol{T}$ are determined by the initial stress, $\boldsymbol{T}_0$. Therefore, in general, *solutions of a given equation of hypo-elasticity* (100.20) *may correspond to an elastic stress relation for some values of the initial stress* $\boldsymbol{T}_0$ *but in general will fail to do so for all possible initial stresses.*

We may illustrate this fact by applying (100.26) to the particular constitutive equation (100.16), for which $H^{kmpq}=f(\operatorname{tr}\boldsymbol{T})g^{km}g^{pq}$. Substitution into (100.21), followed by substitution of the resulting form for K^{kmpq} into (100.26), leads to an equation in which f does not appear. A simple necessary condition results from taking the contracted product of this equation by $g_{pk}g_{jl}$; the result is simply $\boldsymbol{T}=\frac{1}{3}(\operatorname{tr}\boldsymbol{T})\boldsymbol{1}$. Therefore, in order for a solution of (100.18) to correspond to an elastic material, (50.1) *must* be satisfied[1]. Thus the solutions of the particular

[1] This conclusion generalizes slightly a result of BERNSTEIN [1960, *8*, § 3].

constitutive equation (100.16) of hypo-elasticity that do not correspond to elastic fluids are not compatible with the constitutive equation of *any* elastic material.

We may now summarize the results of our analysis of (100.26): *Elastic fluids with invertible pressure function satisfy a particular class of constitutive equations of hypo-elasticity. They are the only solutions of those equations in which the stress is ever hydrostatic. The non-hydrostatic solutions are incompatible with the theory of elasticity.* It appears doubtful, in fact, that such solutions can correspond to the response of any material whatever.

The results just given supplement the theorems stated and proved in the first half of this section in confirming the fact that *in general, hypo-elasticity and the classical theory of finite elastic strain embody different concepts of elasticity; neither includes the other.*

THOMAS and GREEN[1] have discussed relations between hypo-elasticity and the common incremental theories of plasticity. The stress rates or increments, usually undefined, which occur in those theories are to be replaced by invariant time fluxes in order to get constitutive relations satisfying the principle of material frame-indifference. The (corrected) differential equations of the Prandtl-Reuss theory then become special cases of hypo-elastic ones, and those of the Lévy-Mises theory may be approached by a limiting process. Formal compatibility with a yield condition may then be imposed as an additional requirement.

101. Work theorems. Consider a homogeneous deformation process (cf. Sect. 81) described by a one-parameter family of deformation gradients:

$$\boldsymbol{F} = \hat{\boldsymbol{F}}(\alpha), \qquad \alpha_1 \leqq \alpha \leqq \alpha_2. \tag{101.1}$$

As we have seen in the previous section, for a hypo-elastic material there is a response function $\hat{\boldsymbol{T}}(\alpha)$ such that the stress is given by $\boldsymbol{T}(t) = \hat{\boldsymbol{T}}(\alpha(t))$ for every process in which $\boldsymbol{T}$ has a prescribed value $\boldsymbol{T}_0 = \hat{\boldsymbol{T}}(\alpha_0)$ for some value α_0 of α. We use the notation

$$\boldsymbol{T}_1 = \hat{\boldsymbol{T}}(\alpha_1), \quad \boldsymbol{T}_2 = \hat{\boldsymbol{T}}(\alpha_2). \tag{101.2}$$

For any such process the value

$$\hat{\boldsymbol{T}}_{\mathrm{R}} = \hat{\boldsymbol{T}}_{\mathrm{R}}(\alpha) = |\det \hat{\boldsymbol{F}}(\alpha)|\, \hat{\boldsymbol{T}}(\alpha)\left(\boldsymbol{F}(\alpha)^{-1}\right)^T \tag{101.3}$$

of the Piola-Kirchhoff tensor entering into (81.8) also depends only on α. Thus *the actual work*

$$W_{12} = \frac{1}{\varrho_{\mathrm{R}}} \int_{\alpha_1}^{\alpha_2} \operatorname{tr}[\hat{\boldsymbol{T}}_{\mathrm{R}}(\alpha)^T \hat{\boldsymbol{F}}'(\alpha)]\, d\alpha \tag{101.4}$$

done by the surface tractions on a hypo-elastic body during a homogeneous deformation process depends only on the value of the stress for a particular α and not on the rate at which the process is traversed.

The following work theorem is due to BERNSTEIN[2]: *If for a hypo-elastic material the work* (101.4) *is non-negative for every closed deformation process, then the material* (in the sense of BERNSTEIN) *is actually hyperelastic.*

Proof. Let (101.1) describe a closed process, so that

$$\hat{\boldsymbol{F}}(\alpha_1) = \hat{\boldsymbol{F}}(\alpha_2) = \boldsymbol{F}_0. \tag{101.5}$$

Let $\boldsymbol{T}_{\mathrm{R}} = \hat{\boldsymbol{T}}_{\mathrm{R}}(\alpha)$ be the response function (101.3) for this process and the prescribed *initial* Piola-Kirchhoff tensor $\boldsymbol{T}_{\mathrm{R}1} = \hat{\boldsymbol{T}}_{\mathrm{R}}(\alpha_1)$. The same response function $\hat{\boldsymbol{T}}_{\mathrm{R}}(\alpha)$

[1] THOMAS [1955, *28* and *30*], GREEN [1956, *9* and *10*]. Cf. also ERINGEN [1962, *18*, § 89].
[2] BERNSTEIN [1960, *8*, § 6].

must be used if the process is traversed in the opposite direction, in which case $\boldsymbol{T}_{R1}$ is the final Piola-Kirchhoff tensor. The corresponding work (101.4) is the opposite of the work for the original process. If both are assumed to be non-negative, they must actually be zero. Hence, if the hypothesis of BERNSTEIN'S theorem holds, the work of every closed deformation process must actually be zero.

Consider, now, a closed deformation process (101.1), (101.5), and extend it by putting

$$\hat{\boldsymbol{F}}(\alpha) = \begin{cases} \boldsymbol{F}_0 + (\alpha_1 - \alpha)\boldsymbol{A} & \text{if} \quad \alpha_1 - \varepsilon \leqq \alpha \leqq \alpha_1, \\ \boldsymbol{F}_0 + (\alpha - \alpha_2)\boldsymbol{A} & \text{if} \quad \alpha_2 \leqq \alpha \leqq \alpha_2 + \varepsilon, \end{cases} \tag{101.6}$$

where $\boldsymbol{A}$ is an arbitrary tensor and $\varepsilon > 0$. The extended process $\hat{\boldsymbol{F}}(\alpha)$, $\alpha_1 - \varepsilon \leqq \alpha \leqq \alpha_2 + \varepsilon$ is again closed. The work $\overline{W}_{12}$ for the extended process is easily computed:

$$\overline{W}_{12} = W_{12} + \frac{1}{\varrho_R} \int_0^{\varepsilon} \operatorname{tr}\{[\hat{\boldsymbol{T}}_R(\alpha_2 + \sigma) - \boldsymbol{T}_R(\alpha_1 - \sigma)]\boldsymbol{A}\}\, d\sigma, \tag{101.7}$$

where $\hat{\boldsymbol{T}}_R(\alpha)$ is an extension to $\alpha_1 - \varepsilon \leqq \alpha \leqq \alpha_2 + \varepsilon$ of the original response function. Since both $\overline{W}_{12}$ and W_{12} must vanish, the integral on the right-hand side of (101.7) must be zero. Since ε is arbitrary, taking the limits as $\varepsilon \to 0$ yields $\operatorname{tr}\{[\hat{\boldsymbol{T}}_R(\alpha_2) - \hat{\boldsymbol{T}}_R(\alpha_1)]\boldsymbol{A}\} = 0$. Since $\boldsymbol{A}$ was arbitrary, it follows that $\hat{\boldsymbol{T}}_R(\alpha_2) = \hat{\boldsymbol{T}}_R(\alpha_1)$, and hence $\hat{\boldsymbol{T}}(\alpha_1) = \hat{\boldsymbol{T}}(\alpha_2)$: The initial and final values of the stress must coincide. Therefore, if two different values of the stress are compatible with a given deformation gradient, there can be no process that connects one value with the other value of the stress. According to BERNSTEIN'S definition of material, the two different stresses would actually correspond to different materials. We conclude that the deformation gradient determines the stress uniquely, and hence that the material is elastic. It follows from the first work theorem of Sect. 83 that the material must actually be hyperelastic, which completes the proof of BERNSTEIN'S theorem.

The following three work theorems, due to BERNSTEIN and ERICKSEN[1], are valid under the restrictive condition that $H(\boldsymbol{T})$ in (99.9), when regarded as linear transformation of the 6-dimensional space $\mathscr{S}$ of all symmetric tensors, be always invertible.

(1) *If the work* (101.4) *is non-negative for each deformation process such that the initial and final stresses coincide, there exists a function* $\theta(\boldsymbol{T}, \boldsymbol{T}_1)$ *such that the work* (101.4) *for every deformation process, is given by*

$$W_{12} = \theta(\boldsymbol{T}_2, \boldsymbol{T}_1). \tag{101.8}$$

---

[1] BERNSTEIN and ERICKSEN [1958, *6*], ERICKSEN [1958, *14*].

THOMAS [1956, *29*] [1957, *29*] found conditions such that certain special theories of hypo-elasticity lead to a relation of the form

$$\operatorname{tr} \boldsymbol{T}\boldsymbol{D} = \varrho\, \dot{\hat{\varepsilon}}(\boldsymbol{T}, \boldsymbol{D}, \varrho).$$

If there are no dissipative mechanisms, the value of the function $\hat{\varepsilon}$ is the internal energy ε. Cf. also VERMA [1958, *50*].

HILL [1959, *12*, § 4] [1961, *26* and *27*] [1962, *35*] discussed questions of uniqueness and stability for hypo-elastic materials such that H in (99.9) is self-adjoint:

$$H^{kmrs} = H^{rskm},$$

or, equivalently,

$$H^{kmrs} = \frac{\partial^2 \varphi}{\partial G_{km}\, \partial G_{rs}},$$

where $\varphi(\boldsymbol{G})$ is a homogeneous polynomial degree 2 in $\boldsymbol{G}$, with coefficients depending upon $\boldsymbol{T}$.

The function θ has the property

$$\theta(\boldsymbol{T}_1, \boldsymbol{T}_1) = 0, \tag{101.9}$$

i.e. the work is actually zero when the initial and final stresses coincide.

(2) *If there is an initial stress* $\boldsymbol{T}_1$ *such that the work* (101.4) *is non-negative whenever* $\boldsymbol{T}_1 = \hat{\boldsymbol{T}}(\alpha_1)$, *then* $\boldsymbol{T}_1 = \boldsymbol{0}$, *and there is a function* $\gamma(\boldsymbol{T})$ *such that the work* (101.4) *is given by*

$$W_{12} = \gamma(\boldsymbol{T}_2) \tag{101.10}$$

whenever $\boldsymbol{T}_1 = \boldsymbol{0}$.

(3) *In order that arbitrarily close to each stress* $\boldsymbol{T}_1$ *there be stresses* $\boldsymbol{T}_2$ *such that* $W_{12} = 0$ *for every process such that* $\boldsymbol{T}_1 = \hat{\boldsymbol{T}}(\alpha_1)$, $\boldsymbol{T}_2 = \hat{\boldsymbol{T}}(\alpha_2)$, *it is necessary and sufficient that there be functions* φ $(\boldsymbol{T})$ *and* ψ $(\boldsymbol{T})$ *such that*[1]

$$\left.\begin{aligned} \boldsymbol{T} &= \varphi \mathsf{H}(\boldsymbol{T})^T [\psi_{\boldsymbol{T}}(\boldsymbol{T})], \\ T_{km} &= \varphi \mathsf{H}^{pq}{}_{km}(\boldsymbol{T}) \frac{\partial \psi}{\partial T_{pq}}, \end{aligned}\right\} \tag{101.11}$$

The function ψ, if it exists, is called a *hypo-elastic potential.*

The proofs of these three theorems are too elaborate to be included here.

102. Acceleration waves. Waves in hypo-elastic materials are easily studied with the aid of concepts and methods already found useful for setting up the theory of wave propagation in elastic materials. The reader should keep present the contents of Sects. 71—78 and 90.

In addition to the assumptions defining an acceleration wave in any medium we lay down the postulate[2]

$$[\boldsymbol{T}] = \boldsymbol{0}. \tag{102.1}$$

Application of the kinematical conditions of compatibility [CFT, Eq. (180.5)], yields[3]

$$[\dot{\boldsymbol{T}}] = -U\boldsymbol{A}, \qquad [\operatorname{div} \boldsymbol{T}] = \boldsymbol{A}\boldsymbol{n}, \tag{102.2}$$

where U is the local speed of propagation, $\boldsymbol{A}$ is a tensor specifying the jumps in the derivatives of $\boldsymbol{T}$, and $\boldsymbol{n}$ is the unit normal to the surface of discontinuity. Taking the jump of (100.20) across the singular surface yields the equation

$$-U\boldsymbol{A} = \mathsf{K}(\boldsymbol{T})[-U\boldsymbol{a} \otimes \boldsymbol{n}], \qquad -UA^{km} = \mathsf{K}^{kmpq}(-U a_p n_q), \tag{102.3}$$

where $(71.3)_2$ has been used, while application of $(102.2)_2$ and $(71.1)_3$ to the jump of CAUCHY's first law (16.6) yields

$$\boldsymbol{A}\boldsymbol{n} = \varrho U^2 \boldsymbol{a}, \qquad A^{km} n_m = \varrho U^2 a^k. \tag{102.4}$$

By cancelling U from (102.3) we cast aside material singular surfaces across which div $\boldsymbol{T}$ is discontinuous[4]. From (102.3) and (102.4) we then obtain an equation that

[1] The transpose $\mathsf{H}(\boldsymbol{T})^T$ in $(101.11)_1$ is to be understood in the 6-dimensional space $\mathscr{S}$.

[2] Dynamical principles require only that

$$[\boldsymbol{T}]\boldsymbol{n} = \boldsymbol{0}$$

[cf. CFT, Eq. (205.5)]. For an acceleration wave in an elastic material, as has already been remarked in Sect. 71, (102.1) holds in virtue of the stress relation $\boldsymbol{T} = \mathfrak{g}(\boldsymbol{F})$, since, by definition, $\boldsymbol{F}$ is continuous across an acceleration wave.

[3] The variables $\boldsymbol{x}$, t used in writing Eq. (180.5) of CFT are replaced by $\boldsymbol{X}$, t, and the present configuration is taken as the reference configuration.

[4] HILL [1962, *34*, § 5]. For the general theory of material singular surfaces, see Sect. 188 of CFT.

the wave speeds U and corresponding amplitudes $\boldsymbol{a}$ must satisfy[1]:

$$\boldsymbol{Q}(\boldsymbol{n})\,\boldsymbol{a} = \varrho\,U^2\boldsymbol{a}, \qquad Q^{km}(\boldsymbol{n})\,a_m = \varrho\,U^2 a^k, \tag{102.5}$$

where

$$Q^{km}(\boldsymbol{n}) = K^{kpmq}\,n_p\,n_q. \tag{102.6}$$

Comparison of the propagation condition (102.6) of hypo-elasticity with that of elasticity, Eq. (71.3), shows that both are of the same kind. Hypo-elastic waves, like elastic ones, progress without absorption or dispersion; only the direction of the amplitude $\boldsymbol{a}$, not its magnitude, is determined; if a wave can propagate in one sense, it can propagate also in the opposite sense; etc. [2]

Despite this range of agreement, hypo-elasticity and elasticity are generally not at all the same. Since, as we have seen in Sect. 100, anisotropic elasticity is not included in hypo-elasticity as a special case, we might expect hypo-elastic materials that are not elastic to show a response having the symmetries we have come to regard as appropriate to isotropic materials, yet not so restricted as the classical theory of finite strain of isotropic elastic materials would require. Such is the case in wave propagation, as we now proceed to verify[3].

Just as in elasticity, so also in hypo-elasticity we may define a *principal wave* as one propagating in the direction of a principal axis of stress. If $\boldsymbol{n}_1$ is a unit vector corresponding to the direction for which the principal stress is t_1, from (102.6) and (99.16) we readily calculate the corresponding acoustical tensor $\boldsymbol{Q}(\boldsymbol{n}_1)$:

$$\begin{aligned}
\boldsymbol{Q}(\boldsymbol{n}_1) = \;&[\boldsymbol{\Box}_1 + \tfrac{1}{2}\boldsymbol{\Box}_{10} + (\boldsymbol{\Box}_2 + \boldsymbol{\Box}_4 + \boldsymbol{\Box}_{11})\,t_1 + \\
&+ (\boldsymbol{\Box}_3 + \boldsymbol{\Box}_5 + \boldsymbol{\Box}_7 + \boldsymbol{\Box}_{12})\,t_1^2 + \\
&+ (\boldsymbol{\Box}_6 + \boldsymbol{\Box}_8)\,t_1^3 + \boldsymbol{\Box}_9\,t_1^4]\,\boldsymbol{n}_1 \otimes \boldsymbol{n}_1 + \\
&+ \tfrac{1}{2}[\boldsymbol{\Box}_{10} + (\boldsymbol{\Box}_{11}+1)\,t_1 + \boldsymbol{\Box}_{12}\,t_1^2]\,\boldsymbol{1} + \\
&+ \tfrac{1}{2}(\boldsymbol{\Box}_{11}-1)\,\boldsymbol{T} + \tfrac{1}{2}\boldsymbol{\Box}_{12}\,\boldsymbol{T}^2.
\end{aligned} \tag{102.7}$$

Hence for any hypo-elastic material, *the acoustic axes for principal waves are the principal axes of stress; consequently every principal wave is either transverse or longitudinal.* The corresponding squared speeds of propagation follow at once from (71.7):

$$\begin{aligned}
\varrho\,U_{11}^2 = \;&\boldsymbol{n}_1 \cdot \boldsymbol{Q}(\boldsymbol{n}_1)\,\boldsymbol{n}_1 \\
= \;&\boldsymbol{\Box}_1 + \boldsymbol{\Box}_{10} + (\boldsymbol{\Box}_2 + \boldsymbol{\Box}_4 + 2\boldsymbol{\Box}_{11})\,t_1 + \\
&+ (\boldsymbol{\Box}_3 + \boldsymbol{\Box}_5 + \boldsymbol{\Box}_7 + 2\boldsymbol{\Box}_{12})\,t_1^2 + (\boldsymbol{\Box}_6 + \boldsymbol{\Box}_8)\,t_1^3 + \boldsymbol{\Box}_9\,t_1^4, \\
= \;&\boldsymbol{\Box}_1 + III\,\boldsymbol{\Box}_6 + III\,\boldsymbol{\Box}_8 + I\,III\,\boldsymbol{\Box}_9 + \boldsymbol{\Box}_{10} + \\
&+ [\boldsymbol{\Box}_2 + \boldsymbol{\Box}_4 - II\,\boldsymbol{\Box}_6 - II\,\boldsymbol{\Box}_8 + (III - I\,II)\,\boldsymbol{\Box}_9 + 2\boldsymbol{\Box}_{11}]\,t_1 + \\
&+ [\boldsymbol{\Box}_3 + \boldsymbol{\Box}_5 + I\,\boldsymbol{\Box}_6 + \boldsymbol{\Box}_7 + I\,\boldsymbol{\Box}_8 + (I^2 - II)\,\boldsymbol{\Box}_9 + 2\boldsymbol{\Box}_{12}]\,t_1^2, \\
\varrho\,U_{12}^2 = \;&\boldsymbol{n}_2 \cdot \boldsymbol{Q}(\boldsymbol{n}_1)\,\boldsymbol{n}_2 \\
= \;&\tfrac{1}{2}\boldsymbol{\Box}_{10} + \tfrac{1}{2}\boldsymbol{\Box}_{11}(t_1 + t_2) + \tfrac{1}{2}(t_1 - t_2) + \tfrac{1}{2}\boldsymbol{\Box}_{12}(t_1^2 + t_2^2).
\end{aligned} \tag{102.8}$$

---

[1] Bernstein [1962, 3]. Subject to assumptions partly more general and partly more special, a result of this form was obtained also by Hill [1962, 34, § 3].

[2] When a hypo-elastic material is also elastic, we may put (100.28) into (102.6) and so recover (71.3), as is only right. Indeed, in this case both (102.6) and (71.3) must reduce to (74.1).

[3] Truesdell [1963, 75]. A special case has been considered also by Nariboli [1964, 63], who determined corresponding conditions for growth and decay.

To show that these formulae reduce to (74.3) and (74.4) in the case of an isotropic elastic material with invertible stress relation would require explicit inversion of (47.9) so as to express all quantities in terms of the t_a rather than the v_a. Our interest here is in hypo-elastic materials that are not elastic. From (102.8) it is apparent that a greater freedom of undular response is allowed in hypo-elasticity than in isotropic elasticity. If we suppose the squared wave speeds known as functions of the principal stresses, $\varrho U_{ab}^2 = f_{ab}(t_1, t_2, t_3)$, then, besides the relations of invariance under permutation of labels,

$$\left.\begin{aligned} f_{22}(t_1, t_2, t_3) &= f_{11}(t_2, t_1, t_3), \ldots, \\ f_{11}(t_1, t_2, t_3) &= f_{11}(t_1, t_3, t_2), \ldots, \end{aligned}\right\} \tag{102.9}$$

the only conditions of compatibility are (76.5). That is, *any three squared longitudinal wave speeds and any six squared transverse wave speeds satisfying the identities* (76.5) *are compatible with the theory of principal waves in hypo-elasticity.* While the nine squared wave speeds, if compatible with isotropic elasticity, determine the three elastic response coefficients $\beth_\Gamma$ uniquely, knowledge of the hypo-elastic squared wave speeds as functions of the principal stresses is far from sufficient to determine the twelve hypo-elastic response coefficients ם$_\Gamma$, which, indeed, are not uniquely determined in any case.

We have seen that the acoustical tensor $\boldsymbol{Q}(\boldsymbol{n})$, for any hypo-elastic material just as for any isotropic elastic material, is symmetric if $\boldsymbol{n}$ is parallel to a principal direction of stress. We know also that if in a given elastic material $\boldsymbol{Q}(\boldsymbol{n})$ is symmetric for every $\boldsymbol{n}$, the material is in fact hyperelastic (Sect. 90). Are there hypo-elastic materials other than isotropic hyperelastic materials in which $\boldsymbol{Q}(\boldsymbol{n})$ is symmetric for every $\boldsymbol{n}$? First[1], by (102.6) and (102.4), following steps parallel to those leading from (90.2) to (90.4), we find that K^{kmpq} must satisfy an equation of the form (90.4). Use of (100.22) and (100.23) reduces this condition to the form (100.29). This result is consistent with the theorem in Sect. 90: When the hypo-elastic material is also elastic, it has a symmetric acoustical tensor, for all $\boldsymbol{n}$, if and only if it is hyperelastic. However, (100.29) may be satisfied even for non-elastic materials. E.g., if $H^{kmpq} = g^{km} T^{pq}$, (100.29) is plainly satisfied, yet by use of (100.26) it can be shown that solutions of the corresponding equations of hypo-elasticity are compatible with elasticity if and only if the stress is hydrostatic. Thus there are non-elastic solutions of the equations of hypo-elasticity such as to have real, orthogonal acoustic axes for all directions of propagation.

The growth and decay of hypo-elastic acceleration waves have been analysed by Varley and Dunwoody[2].

In evaluation of the theory of hypo-elasticity as a whole, Truesdell[3] has written, "It is totally inappropriate for materials regarded physically as 'anisotropic'. It gives to the physical notion of 'elastic isotropy' a mathematical form that is less restrictive than the one embodied in the classical theory of finitely elastic, isotropic materials. In heuristic terms, hypo-elasticity retains the directional aspects of elastic isotropy while relaxing the relations among magnitudes carried along by the notion of isotropy in the theory of finite strain. It is possible that hypo-elasticity may be appropriate to physical materials that show no sign of preferred states, preferred directions, or fading memory."

103. Solutions of special problems. In a state of time-dependent homogeneous strain $\boldsymbol{F}(t)$, both $\boldsymbol{D}$ and $\boldsymbol{W}$ are known functions of t, so that (99.9) becomes an ordinary differential equation for $\boldsymbol{T}$. Under various assumptions of smoothness upon the function $H(\boldsymbol{T})$, there exists a unique homogeneous stress $\boldsymbol{T}(t)$ such that $\boldsymbol{T}(0) = \boldsymbol{T}_0$, where $\boldsymbol{T}_0$ is arbitrary. This is the case, for example, when $H(\boldsymbol{T})$ is analytic. In general, however, the homogeneous stress satisfying the constitutive

[1] Bernstein [1962, *3*].
[2] Varley and Dunwoody [1965, *39*].
[3] Truesdell [1963, *75*].

equation will not be dynamically possible for an assigned field of body force $\boldsymbol{b}$. If $\boldsymbol{b}=\boldsymbol{0}$, it is necessary, as shown in Sect. 28, that the motion be accelerationless, and this fact restricts $\boldsymbol{F}$ to the special form (28.9). Consequently, the differential equation (99.9) will become singular after a finite time, determined by the initial velocity gradient[1]. E.g., if $H(\boldsymbol{T})$ is analytic, the solution $\boldsymbol{T}(t)$ will be analytic so long as $|t|<t_0$, in the notation used in Sect. 28, and will fail to be analytic, in general, for previous and subsequent times. The singularity is independent of the initial stress $\boldsymbol{T}_0$.

For example, consider the accelerationless dilatation $\boldsymbol{x}=(1+kt)\boldsymbol{X}$, where k is a constant. In this motion

$$\varrho=\frac{\varrho_0}{(1+kt)^3}, \tag{103.1}$$

so that the density remains finite in every interval of time from which $t=-1/k$ is excluded. We consider only hydrostatic solutions: $\boldsymbol{T}=-p\boldsymbol{1}$. Since

$$\boldsymbol{W}=\boldsymbol{0}, \qquad \boldsymbol{D}=\frac{k}{1+kt}\boldsymbol{1}=-\frac{1}{3}\dot{\overline{\log\varrho}}\,\boldsymbol{1}, \tag{103.2}$$

the constitutive equation (99.9) of hypo-elasticity reduces to

$$-\dot{p}\boldsymbol{1}=f(p)\left(-\dot{\overline{\log\varrho}}\right)\boldsymbol{1}. \tag{103.3}$$

Hence[2]

$$\log\frac{\varrho}{\varrho_0}=\int\limits_{p_0}^{p}\frac{d\xi}{f(\xi)}. \tag{103.4}$$

In terms of the representation (99.16), f has the form

$$\left.\begin{aligned} f(p)=\tilde{\beth}_1+\tfrac{1}{3}\tilde{\beth}_{10}-(\tilde{\beth}_2+\tilde{\beth}_4+\tfrac{2}{3}\tilde{\beth}_{11})\,p\\ +(\tilde{\beth}_3+\tilde{\beth}_5+\tilde{\beth}_7+\tfrac{2}{3}\tilde{\beth}_{12})\,p^2-(\tilde{\beth}_6+\tilde{\beth}_8)\,p^3+\tilde{\beth}_9\,p^4, \end{aligned}\right\} \tag{103.5}$$

where

$$\tilde{\beth}_\Gamma(p)\equiv\beth_\Gamma(-3p,\,3p^2,\,-p^3). \tag{103.6}$$

For a material of grade 1, by (99.19) we find that

$$\left.\begin{aligned} f(p)&=\lambda+\tfrac{2}{3}\mu+\alpha_{\mathrm{p}}\,p,\\ -\alpha_{\mathrm{p}}&\equiv 3\gamma_0+\gamma_1+\gamma_2+\gamma_3+\tfrac{1}{3}\gamma_4, \end{aligned}\right\} \tag{103.7}$$

and (103.4) becomes

$$\frac{\varrho}{\varrho_0}=\left\{\begin{array}{ll} \left(\dfrac{\lambda+\frac{2}{3}\mu+\alpha_{\mathrm{p}}p}{\lambda+\frac{2}{3}\mu+\alpha_{\mathrm{p}}p_0}\right)^{1/\alpha_{\mathrm{p}}} & \alpha_{\mathrm{p}}\neq 0,\\ \exp\dfrac{p-p_0}{\lambda+\frac{2}{3}\mu}, & \alpha_{\mathrm{p}}=0, \end{array}\right\} \tag{103.8}$$

where p_0 is the value of p corresponding to $\varrho=\varrho_0$. The forms of the curves representing p as a function of ϱ according to (103.8) and the corresponding formula for materials of grade 2 have been classified into fourteen types and illustrated

---

[1] TRUESDELL [1955, *32*, § 3] [1955, *31*, § 6].

[2] The result, as a formula, is due to MURNAGHAN [1944, *1*] [1945, *1*] [1949, *10*, § 3] [1951, *8*, Ch. 4, § 3]. The derivation of it from hypo-elasticity was given by TRUESDELL [1953, *25*, § 57]; cf. also his remarks [1952, *21*] on MURNAGHAN's treatment, which does not reveal the condition (103.1), necessary if a homogeneous dilatation is to be possible at all in a simple material free of body force. According to MURNAGHAN, (103.8) gives a good fit to the data obtained from measurements at extreme pressures. Similar formulae were obtained by GLEYZAL from various special assumptions [1949, *4*, § 3].

by TRUESDELL[1]. There is great variety in the possible response: The material may harden or soften in compression or extension, in any combination; either the pressure or the density may approach asymptotic values; etc.

A motion such that

$$[\boldsymbol{D}]=k\left\|\begin{matrix}-\sigma & 0 & 0\\ \cdot & -\sigma & 0\\ \cdot & \cdot & 1\end{matrix}\right\|,\qquad \boldsymbol{W}=\boldsymbol{0}, \tag{103.9}$$

is accelerationless provided that

$$k=\frac{k_0}{1+k_0 t},\qquad \sigma=\sigma_0\frac{1+k_0 t}{1-k_0\sigma_0 t}, \tag{103.10}$$

where k_0 and σ_0 are constants. In this motion, a rectangular block with sides parallel to the co-ordinate planes is stretched at the rate k in the z-direction and contracted at the rate σk in the transverse directions. It remains always a rectangular block until $k_0 t=-1$ or $k_0\sigma_0 t=1$, whichever time occurs sooner. At these singularities, the volume is reduced to zero (cf. CFT, Sect. 142). We seek solutions in which the stress, also, is biaxial:

$$[\boldsymbol{T}]=\left\|\begin{matrix}-Q & 0 & 0\\ \cdot & -Q & 0\\ \cdot & \cdot & S\end{matrix}\right\|. \tag{103.11}$$

Substitution into (99.12) and (99.16) yields the following system to be solved for Q and S:

$$\left.\begin{aligned}\frac{1}{k}\frac{dQ}{dt}&=g(S,Q,\sigma),\\ \frac{1}{k}\frac{dS}{dt}&=f(S,Q,\sigma),\end{aligned}\right\} \tag{103.12}$$

where

$$\left.\begin{aligned}g(S,Q,\sigma)={}&(1-2\sigma)(-\beth_1+\beth_4 Q-\beth_7 Q^2)+\\ &+(S+2\sigma Q)(-\beth_2+\beth_5 Q-\beth_8 Q^2)+\\ &+(S^2-2\sigma Q^2)(-\beth_3+\beth_6 Q-\beth_9 Q^2)+\\ &+\sigma\beth_{10}-2\sigma Q\beth_{11}+2\sigma Q^2\beth_{12},\\ f(S,Q,\sigma)={}&(1-2\sigma)(\beth_1+\beth_4 S+\beth_7 S^2)+\\ &+(S+2\sigma Q)(\beth_2+\beth_5 S+\beth_8 S^2)+\\ &+(S^2-2\sigma Q^2)(\beth_3+\beth_6 S+\beth_9 S^2)+\\ &+\beth_{10}+2S\beth_{11}+2S^2\beth_{12},\end{aligned}\right\} \tag{103.13}$$

the arguments of $\beth_\Gamma$ $(I_{\boldsymbol{T}}, II_{\boldsymbol{T}}, III_{\boldsymbol{T}})$ being $S-2Q$, Q^2-2SQ, $Q^2 S$.

In the system (103.12), k and σ are assigned functions of t, in accord with (103.10); indeed, for any other dependence of k and σ on time, the assumed homogeneous deformation will not be accelerationless and hence will not be possible in any simple material unless suitable body forces be supplied. A "quasi-static" solution is easily obtained by neglecting this fact. Indeed, if we set $Q=0$, we must have $g(S,0,\sigma)=0$, or, by (103.13),

$$\sigma=\hat{\sigma}(S)=\frac{\beth_1+S\beth_2+S^2\beth_3}{2\beth_1+\beth_{10}}, \tag{103.14}$$

[1] TRUESDELL [1955, *31*, § 7].

where the arguments of $\mathfrak{D}_\Gamma$ are S, 0, 0. Substitution of (103.14) into (103.12)$_2$, followed by use of the relation $k=\dot{z}/z$, yields the general solution of (103.12)$_2$:

$$\log\frac{z}{Z}=\int_{S_0}^{S}\frac{d\xi}{f(\xi,0,\hat{\sigma}(\xi))}\,. \tag{103.15}$$

While special cases of this formula have appeared in the literature[1], it is not correct in general[2], since, as can easily be shown, the function $Q=0$ can *never* satisfy (103.12) when k and σ are given by (103.10). That is, *accelerationless extension cannot be produced by tensile force alone in any hypo-elastic material.* Integration of (103.12), even for a material of grade 1, is elaborate; the solution has been obtained only in some special cases[3]. For small strains $Q/(2\mu)$ turns out to be very small, and the quasi-static formula (103.15) is therefore sufficiently accurate, but for large strains it is not.

The foregoing remarks illustrate the way in which hypo-elasticity has been studied. The primary aim has been to determine, by integrating the constitutive equation (99.9), the stress as a function of the strain in particular cases. It must be remembered, however, that while a stress so obtained is independent of the time taken to effect a given strain, it is not at all independent of the sequence of intermediate strains. "A relation between stress and strain is thus the outcome, not the assumption, of our theory, and the form of this relation depends on the manner in which the stress is applied in time[4]", as we have shown true in fully general terms in Sect. 100. GREEN[5] has illustrated this fact in the case of extension. He considered more general extensile motions in which a linear acceleration field is allowed. For the particular material of grade 1 he considered, GREEN found a particular exact solution that approaches the quasi-static solution as $t\to\infty$, just the opposite behavior from that for the accelerationless extension. Therefore, according to hypo-elasticity it is generally meaningless to ask for the relation of stress to final strain unless a particular strain path has been specified. That the quasi-static solution gives a *unique* stress relation, independent of strain path, is one more reason for doubting its correctness.

In both the examples given so far, the convective terms in $\mathring{\boldsymbol{T}}$ are zero, so that $\mathring{\boldsymbol{T}}=\dot{\boldsymbol{T}}$. We now take up a simple case when those terms are essential in determining the nature of the stress relation. This case, that of shearing, illustrates also the success of hypo-elasticity in revealing, as a by-product of the integration, a behavior similar to that postulated in theories of plastic yielding.

Accordingly, we consider rectilinear shearing, given by Eq. (30.44) with x, y, X, Y replaced by y, x, Y, X, respectively (cf. CFT, Sect. 45). The amount of shear, K, is given by $K=\varkappa t$, where $\varkappa$ is a constant, and we shall use the dimensionless variable K as a measure of thetime elapsed. The constitutive equation (99.12) yields a complicated differential system to be satisfied by the stress components, but from the general theory we know that a solution exists for all time[6]. We shall consider only solutions such that $T\langle xz\rangle=T\langle yz\rangle=0$, and these only in two special cases.

---

[1] MURNAGHAN [1944, *2*] [1945, *1*] [1949, *10*, § 4]. The analysis in the text was given by TRUESDELL [1955, *31*, § 10].

[2] ERICKSEN [1954, *6*], TRUESDELL [1955, *31*, § 10] [1955, *32*, § 4].

[3] A mistake in quadrature by TRUESDELL [1955, *32*, § 4] was corrected by ERINGEN [1962, *18*, Eq. (80.10)], who gave curves comparing the exact and quasi-static solutions for particular materials of grade 1.

[4] TRUESDELL [1955, *31*, § 14].

[5] GREEN [1956, *11*].

[6] Simple shear affords one of the rare cases when no singularity develops, since $\boldsymbol{F}_1$ in (28.9) is then such that $I_{\boldsymbol{F}_1}=II_{\boldsymbol{F}_1}=III_{\boldsymbol{F}_1}=0$.

First, suppose the material is of grade 1. Then from (99.19) we find that the stress relation must have the form[1]

$$\left.\begin{aligned}
S\langle xy\rangle &= \frac{A}{\sqrt{|\alpha_s|}}\,\frac{\sinh}{\sin}\sqrt{|\alpha_s|}\,K + S\langle xy\rangle_0\,\frac{\cosh}{\cos}\sqrt{|\alpha_s|}\,K,\\
\frac{S\langle zz\rangle - S\langle zz\rangle_0}{\gamma_3} &= \frac{S\langle yy\rangle - S\langle yy\rangle_0}{-1+\gamma_3+\frac{1}{2}\gamma_4} = \frac{S\langle xx\rangle - S\langle xx\rangle_0}{1+\gamma_3+\frac{1}{2}\gamma_4},\\
&= \int_0^K S\langle xy\rangle(\xi)\,d\xi,\\
&= \frac{2A}{|\alpha_s|}\,\frac{\sinh^2}{\sin^2}\,\frac{1}{2}\sqrt{|\alpha_s|}\,K + \frac{S\langle xy\rangle_0}{\sqrt{|\alpha_s|}}\,\frac{\sinh}{\sin}\sqrt{|\alpha_s|}\,K,\\
\alpha_s &\equiv (1+\gamma_1+\gamma_4)(\gamma_3+\tfrac{1}{2}\gamma_4),\\
A &\equiv \tfrac{1}{2}+\tfrac{1}{2}(1+\gamma_1+\gamma_4)(S\langle xx\rangle_0 - S\langle yy\rangle_0).
\end{aligned}\right\}\qquad(103.16)$$

In these expressions $\mathbf{S}_0$ is the value of $\mathbf{S}$ when $K=0$, and the upper or lower forms are to be used according as $\alpha_s>0$ or $\alpha_s<0$, while if $\alpha_s=0$, then

$$\left.\begin{aligned}
S\langle xy\rangle - S\langle xy\rangle_0 &= A K,\\
\int_0^K S\langle xy\rangle(\xi)\,d\xi &= \tfrac{1}{2}A K^2 + S\langle xy\rangle_0 K.
\end{aligned}\right\}\qquad(103.17)$$

For detailed discussion of these results we refer the reader to the original memoir[2] of TRUESDELL. Here we remark only a few properties of the solution when the block is initially unstressed, so that $\mathbf{S}_0=\mathbf{0}$. Then $A=\frac{1}{2}$, and (103.16) and (103.17) reduce to

$$\begin{aligned}
S\langle xy\rangle &= \begin{cases}
\dfrac{1}{2\sqrt{\alpha_s}}\sinh\sqrt{\alpha_s}\,K & \text{if } \alpha_s>0,\\[2mm]
\dfrac{1}{2}K & \text{if } \alpha_s=0,\\[2mm]
\dfrac{1}{2\sqrt{-\alpha_s}}\sin\sqrt{-\alpha_s}\,K & \text{if } \alpha_s<0,
\end{cases}\\
\frac{S\langle zz\rangle}{\gamma_3} &= \frac{S\langle yy\rangle}{-1+\gamma_3+\frac{1}{2}\gamma_4} = \frac{S\langle xx\rangle}{1+\gamma_3+\frac{1}{2}\gamma_4},\\
&= \begin{cases}
\dfrac{1}{\alpha_s}\sinh^2\dfrac{1}{2}\sqrt{\alpha_s}\,K & \text{if } \alpha_s>0,\\[2mm]
\dfrac{1}{4}K^2 & \text{if } \alpha_s=0,\\[2mm]
-\dfrac{1}{\alpha_s}\sin^2\dfrac{1}{2}\sqrt{-\alpha_s}\,K & \text{if } \alpha_s<0.
\end{cases}
\end{aligned}\qquad(103.18)$$

Thus the Poynting effect in shear occurs in hypo-elasticity much as in the theory of isotropic elastic materials (Sects. 54—55). The normal stresses are all of the order of the square of the amount of shear for small shears, and there is no hypo-elastic material of grade 1 such that these normal stresses can vanish, or even such that $S\langle yy\rangle = S\langle xx\rangle$. The most interesting aspect of the result is the behavior of $\mathbf{S}$ as a function of shear. For materials such that $\alpha_s \geqq 0$, each component of stress increases steadily in absolute value as K increases. That is, these materials *stiffen* in shear. However, if $\alpha_s<0$, the stress increment corresponding to a strain

[1] TRUESDELL [1955, *31*, § 8].
[2] Note the corrections on pp. 1019—1020 of the volume in which it appears.

increment decreases as K increases. Thus these materials *soften* in shear. Moreover, according to (103.18), the shear stress $S\langle xy\rangle$ passes through a maximum when the shear K reaches the value

$$K_s = \frac{\pi}{2\sqrt{-\alpha_s}}, \tag{103.19}$$

and the corresponding shear stress attains its maximum value:

$$S\langle xy\rangle\Big|_{\max} = \frac{1}{2\sqrt{-\alpha_s}} = \frac{K_s}{\pi}. \tag{103.20}$$

This behavior is indicated in Fig. 7. In the case when $\alpha_s < 0$, we are not to expect that the portion of the curve corresponding to shears greater than (103.19) will actually be traversed. Rather, some type of failure or instability must occur at that shear, or at a lesser one. It is to be noted that the magnitude of K_s is determined by that of α_s, which in turn is determined by the response of the material to *small* strain increments. That is, failure or a yield-like phenomenon is in principle *predicted*, within the framework of hypo-elasticity, by quantities measureable in experiments on small strain. This feature is to be contrasted with the usual theories of plasticity, in which the yield stress is of a magnitude assumed *a priori* and altogether inaccessible to experiments at smaller stresses.

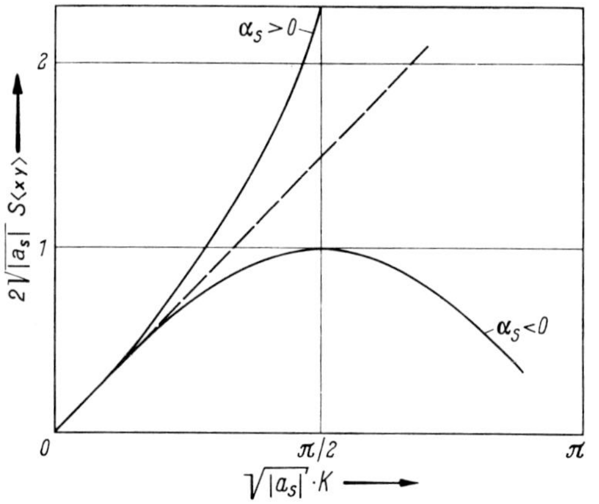

Fig. 7. Shearing of a hypo-elastic material of grade 1.

The theory of hypo-elasticity has never been advocated as a theory for describing any particular material. From the outset it has been regarded as a model of certain aspects of elastic response. In this light the result just obtained is to be interpreted. A theory based on smooth response to rates of change has been shown to imply, necessarily, an upper bound for the stress in a particular test. Again cautioning the reader that we neither claim to explain any particular experiment nor imply any connection with the usual theories of plasticity, we refer henceforth to the existence of the maximum as *hypo-elastic yield*.

According to the theory so far presented, at shears greater than K_s the shear stress decreases, ultimately falling to zero and even becoming negative. This behavior, which seems to differ from that of real materials, is not inherent to the idea of hypo-elastic yield, as we shall see now by study[1] of a special material of grade 2.

The particular constitutive equation of hypo-elasticity we shall consider is one proposed by THOMAS and GREEN[2].

$$\left.\begin{aligned} \operatorname{tr}\overset{\circ}{\boldsymbol{S}} &= \frac{1+\nu}{1-2\nu}\operatorname{tr}\boldsymbol{D}, \\ \overset{\circ}{\boldsymbol{S}}{}^* &= \boldsymbol{D}^* - \frac{\operatorname{tr}(\boldsymbol{D}^*\boldsymbol{S}^*)}{M^2}\,\boldsymbol{S}^*, \end{aligned}\right\} \tag{103.21}$$

[1] TRUESDELL [1956, *33*].
[2] THOMAS [1955, *28*], GREEN [1956, *9*].

where ν and M are dimensionless material constants and where the asterisk denotes a deviator (CFT, Sect. App. 38). These equations were first put forward as properly invariant corrections of the Prandtl-Reuss equations of plasticity theory, but of course there are infinitely many such forms, one for every choice of invariant stress rate. From $(103.21)_2$ it follows at once that

$$\overline{\operatorname{tr}(\boldsymbol{S}^*\boldsymbol{S}^*)}^{\bullet} = \overline{\operatorname{tr}(\boldsymbol{S}^*\boldsymbol{S}^*)}^{\circ} = 2\operatorname{tr}(\boldsymbol{D}^*\boldsymbol{S}^*)\left[1 - \frac{\operatorname{tr}(\boldsymbol{S}^*\boldsymbol{S}^*)}{M^2}\right]. \tag{103.22}$$

Hence if $\boldsymbol{S}^*$ is ever such that $\operatorname{tr}(\boldsymbol{S}^*\boldsymbol{S}^*) = M^2$, then the material derivative of $\operatorname{tr}(\boldsymbol{S}^*\boldsymbol{S}^*)$ vanishes; repeating the argument shows that the n'th material derivative of $\operatorname{tr}(\boldsymbol{S}^*\boldsymbol{S}^*)$ vanishes, so that in fact $\operatorname{tr}(\boldsymbol{S}^*\boldsymbol{S}^*)$ must remain constant for each particle. Since the Maxwell-v. Mises yield criterion (CFT, Sect. 300) is

$$\operatorname{tr}(\boldsymbol{S}^*\boldsymbol{S}^*) = M^2, \tag{103.23}$$

this fact suggested to THOMAS that constitutive equations of the form (103.21) can describe continuous transition from an "elastic" regime to a "plastic" one in which (103.23) holds. However, TRUESDELL[1] remarked that if $\boldsymbol{S}^*(t)$ and $\boldsymbol{D}^*(t)$ are continuous, the differential equation (103.22) has a unique solution, so that if (103.23) holds for one finite value of t, it must hold for all t. Thus it is impossible for a value of $\boldsymbol{S}^*$ satisfying (103.23) to be reached except, perhaps, after infinite time. This fact induced TRUESDELL to study (103.21) in the special case of accelerationless simple shearing, so as to illustrate the sense in which yield may be represented in hypo-elasticity.

In a simple shear of amount K, $(103.21)_1$ is satisfied identically if $\operatorname{tr}\boldsymbol{S} = 0$, and $(103.21)_2$ and the four non-trivial components of $(103.21)_2$ become

$$\left.\begin{aligned}
\frac{dS^{*\langle xy\rangle}}{dK} &= \frac{1}{2}\left(S^{*\langle yy\rangle} - S^{*\langle xx\rangle} + 1 - \frac{2S^{*\langle xy\rangle 2}}{M^2}\right),\\
\frac{dS^{*\langle xx\rangle}}{dK} &= S^{*\langle xy\rangle}\left(1 - \frac{S^{*\langle xx\rangle}}{M^2}\right),\\
\frac{dS^{*\langle yy\rangle}}{dK} &= -S^{*\langle xy\rangle}\left(1 + \frac{S^{*\langle yy\rangle}}{M^2}\right),\\
\frac{dS^{*\langle zz\rangle}}{dK} &= -S^{*\langle xy\rangle}\frac{S^{*\langle zz\rangle}}{M^2}.
\end{aligned}\right\} \tag{103.24}$$

This system is to be solved, subject to the requirement $\operatorname{tr}\boldsymbol{S}^* = 0$, with which it is plainly compatible. We consider only the case when $\boldsymbol{S}^* = \boldsymbol{0}$ if $K = 0$. Since $\operatorname{tr}\boldsymbol{S}^* = 0$, from (103.24) we see that

$$S^{*\langle yy\rangle} = S^{\langle yy\rangle} = -S^{*\langle xx\rangle} = -S^{\langle xx\rangle}, \qquad S^{*\langle zz\rangle} = S^{\langle zz\rangle} = 0, \tag{103.25}$$

and (103.24) reduces to the system

$$\frac{dS^{\langle xy\rangle}}{dK} = \frac{1}{2} + S^{\langle yy\rangle} - \frac{S^{\langle xy\rangle 2}}{M^2}, \qquad \frac{dS^{\langle yy\rangle}}{dK} = -S^{\langle xy\rangle}\left(1 + \frac{S^{\langle yy\rangle}}{M^2}\right). \tag{103.26}$$

An intermediate integral is

$$S^{\langle xy\rangle 2} = -S^{\langle yy\rangle}\left[1 + \left(\frac{1}{2} + M^2\right)\frac{S^{\langle yy\rangle}}{M^2}\right]. \tag{103.27}$$

Hence $S^{\langle yy\rangle} \leqq 0$ for all K: The normal stress on the shearing planes is always a tension. Therefore, from $(103.26)_1$ we see that $S^{\langle xy\rangle}$ is always less in magnitude than it would be according to infinitesimal elasticity theory, in which $S^{\langle xy\rangle} = \frac{1}{2}K$.

[1] TRUESDELL [1956, *33*].

For further interpretation of the results we recall the term *hypo-elastic yield* as already defined, and we shall say that *M-yield* occurs if (103.23) is satisfied. From $(103.26)_1$ it follows that $S_{\langle xy\rangle}(K)$ is a convex function with a single maximum, so that there will be hypo-elastic yield at some value of K, no matter what the value of M. At hypo-elastic yield, $S_{\langle xy\rangle}^2 = M^2(\frac{1}{2} + S_{\langle yy\rangle})$. Since

$$\operatorname{tr}(\boldsymbol{S}^* \boldsymbol{S}^*) = 2(S_{\langle xy\rangle}^2 + S_{\langle yy\rangle}^2), \qquad (103.28)$$

M-yield occurs (if at all) when $S_{\langle xy\rangle}^2 = \frac{1}{2}M^2 - S_{\langle yy\rangle}^2$. Comparison of these two results shows that *before M-yield, hypo-elastic yield must occur.* Therefore, a steady increase of shear stress up M-yield is impossible. The relation between hypo-elastic yield and M-yield is most easily seen from the graph of Eq. (103.27), shown in Fig. 8.

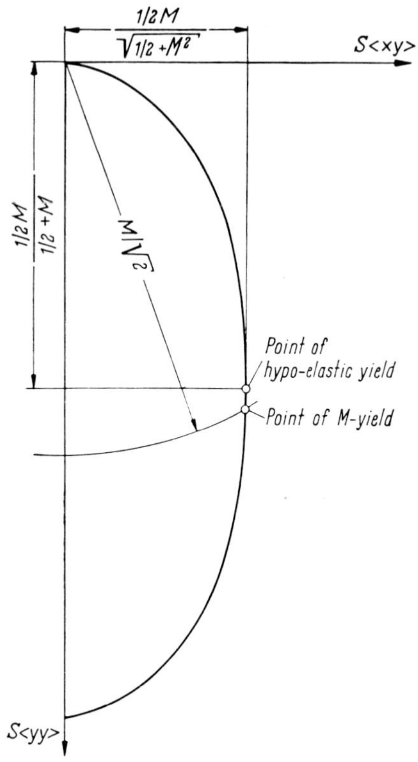

Fig. 8. Hypo-elastic yield and M-yield.

$S_{\langle xy\rangle}$ and $S_{\langle yy\rangle}$ are such as to lie on a semi-ellipse with center at $S_{\langle xy\rangle} = 0$, $S_{\langle yy\rangle} = \frac{1}{2}M^2/(\frac{1}{2} + M^2)$, and with semi-axes of length $\frac{1}{2}M^2/\frac{1}{2} + M^2$ and $\frac{1}{2}M/\sqrt{\frac{1}{2} + M^2}$. By (103.28) the locus $\operatorname{tr}(\boldsymbol{S}^* \boldsymbol{S}^*) = M^2$ is a circle with center at $S_{\langle xx\rangle} = S_{\langle yy\rangle} = 0$ and radius $M/\sqrt{2}$. M-yield occurs at the value of $S_{\langle xy\rangle}$ and $S_{\langle yy\rangle}$ corresponding to intersection of the circle and the ellipse. Since for all M the radius of the circle exceeds the distance from the origin to the nearer vertex of the semi-ellipse, we have fresh proof that M-yield cannot occur before hypo-elastic yield. Moreover, there is or is not a real point of intersection according as $M^2 \leqq \frac{1}{2}$ or $M^2 > \frac{1}{2}$. That is, *a necessary condition for M-yield is* $M^2 \leqq \frac{1}{2}$. Notice also from Fig. 8 that for small values of M, the stresses at M-yield and the stresses at hypo-elastic yield are virtually the same.

From (103.28) and (103.27),

$$\operatorname{tr}(\boldsymbol{S}^* \boldsymbol{S}^*) = -2 S_{\langle yy\rangle}\left(1 + \frac{S_{\langle yy\rangle}}{2M^2}\right); \qquad (103.29)$$

hence at M-yield

$$S_{\langle yy\rangle} = -M^2, \qquad S_{\langle xy\rangle} = M\sqrt{\tfrac{1}{2} - M^2} \leqq \tfrac{1}{4}, \qquad (103.30)$$

showing again the necessity that $M^2 \leqq \frac{1}{2}$ in order for M-yield to occur. At hypo-elastic yield, on the other hand,

$$S_{\langle yy\rangle} = -\frac{\frac{1}{2}M^2}{\frac{1}{2} + M^2} > -\frac{1}{2}, \qquad S_{\langle xy\rangle} = \frac{\frac{1}{2}M}{\sqrt{\frac{1}{2} + M^2}} < \frac{1}{2}. \qquad (103.31)$$

The smaller is M, the smaller are the magnitudes of the stresses at hypo-elastic yield, as well as those at M-yield, and by adjusting the value of M, any one of these stresses may be made as small as desired. Also, if $M^2 < \frac{1}{2}$,

$$\frac{S\langle xy\rangle|_{\text{hypo-el. yield}}}{S\langle xy\rangle|_{M\text{-yield}}} = \frac{\frac{1}{2}}{\sqrt{\frac{1}{4} - M^4}} > 1. \tag{103.32}$$

Thus, while hypo-elastic yield always occurs before M-yield, the corresponding shear stress is always greater, showing that M-yield cannot be attained by a steady increase of shear stress; rather, if M-yield occurs at all, the shear stress must first overshoot its M-yield value and then decline toward it as the shear increases to ∞. (M-yield does not occur at the point of overshoot, since $S\langle yy\rangle$ has not yet reached a large enough magnitude.) Moreover, the function of M in (103.32) approaches 1 as $M \to 0$: *For small values of M the shear stress at hypo-elastic yield is nearly the same as that at M-yield.* Thus for small values of M the shear stress rises quickly to its asymptotic value, very nearly $M/\sqrt{2}$, overshoots it imperceptibly, and levels off to it at an infinite amount of shear. If we let K_0 stand for the value of K at which $S\langle xy\rangle$ first reaches its asymptotic value, and K_s for the value of K at hypo-elastic yield, then, since $S\langle xy\rangle$ is a continuous function of K, we conclude from the foregoing statement that $K_0/K_s \to 1$ as $M \to 0$. Thus *the part of the stress-strain curve corresponding to values of K greater than K_0 is practically indistinguishable from a straight line,* if M is small.

To complete the integration, set

$$\tan\frac{1}{2}\varphi = \begin{cases} \sqrt{\dfrac{M^2+\frac{1}{2}}{M^2-\frac{1}{2}}}\tan\dfrac{\sqrt{M^2-\frac{1}{2}}}{2M}K & \text{if } M^2 > \dfrac{1}{2}, \\ K/\sqrt{2} & \text{if } M^2 = \dfrac{1}{2}, \\ \sqrt{\dfrac{\frac{1}{2}+M^2}{\frac{1}{2}-M^2}}\tanh\dfrac{\sqrt{\frac{1}{2}-M^2}}{2M}K & \text{if } M^2 < \dfrac{1}{2}. \end{cases} \tag{103.33}$$

Then the solution of (103.24) corresponding to vanishing initial stress is

$$\left.\begin{aligned} S\langle xy\rangle &= \frac{\frac{1}{2}M}{\sqrt{\frac{1}{2}+M^2}}\sin\varphi, \\ S\langle yy\rangle &= -\frac{M^2}{\frac{1}{2}+M^2}\sin^2\frac{1}{2}\varphi. \end{aligned}\right\} \tag{103.34}$$

Hypo-elastic yield occurs when $\varphi = \frac{1}{2}\pi$; calling the corresponding angle of shear ϑ_s, we see that

$$K_s = \tan\vartheta_s = \begin{cases} \dfrac{2M}{\sqrt{M^2-\frac{1}{2}}}\operatorname{Arctan}\sqrt{\dfrac{M^2-\frac{1}{2}}{M^2+\frac{1}{2}}} & \text{if } M^2 > \dfrac{1}{2}, \\ \sqrt{2} & \text{if } M^2 = \dfrac{1}{2}, \\ \dfrac{M}{\sqrt{\frac{1}{2}-M^2}}\log\dfrac{\frac{1}{2}+\sqrt{\frac{1}{4}-M^4}}{M^2} & \text{if } M^2 < \dfrac{1}{2}. \end{cases} \tag{103.35}$$

If $M^2 > \frac{1}{2}$, $S\langle xy\rangle$ becomes imaginary before M-yield can occur; if $M^2 \leq \frac{1}{2}$, M-yield occurs asymptotically as $K \to \infty$. The angle of shear ϑ_0 at which the shear stress first reaches its asymptotic value $(103.30)_2$ is given by

$$\tan\vartheta_0 = \frac{M}{\sqrt{\frac{1}{2}-M^2}}\log\frac{1}{2M^2}. \tag{103.36}$$

We have already mentioned that the portion of the stress-strain curve from ϑ_0 onward may be replaced, approximately, by a straight line when M is small.

The portion for small shear, however, does not approach a straight line. Indeed, if we let ϑ_1 be the angle of shear at which the stress-strain curve of the infinitesimal theory, $S\langle xy\rangle = \frac{1}{2}K$, crosses the asymptote, then

$$\frac{\vartheta_0}{\vartheta_1} \to \infty \quad \text{as} \quad M \to 0. \tag{103.37}$$

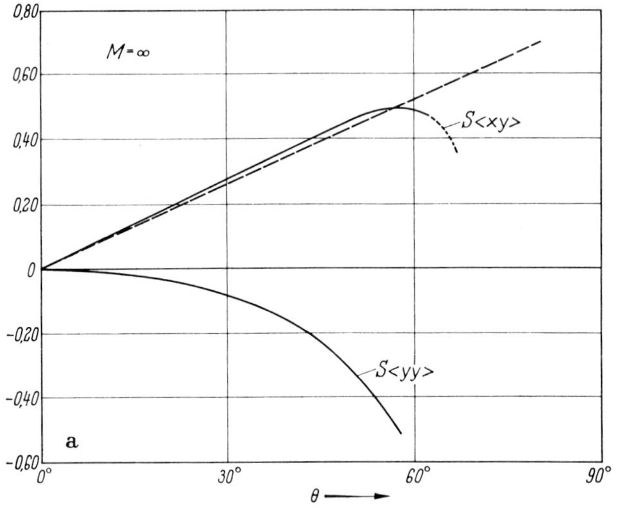

a

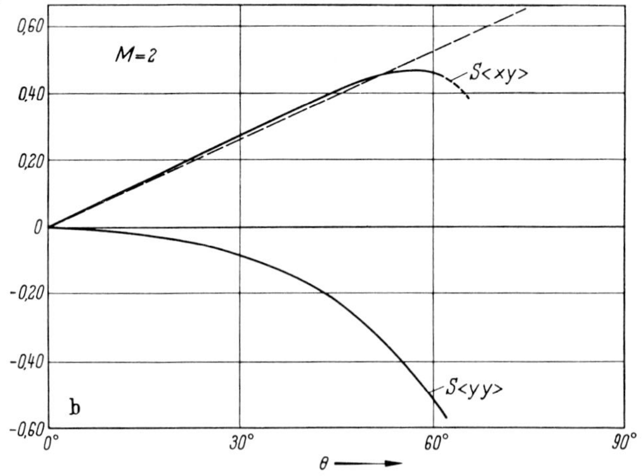

b

Fig. 9a, b. Hypo-elastic yield without M-yield for two special hypo-elastic materials of grade 2.

Therefore, the straight-line curve of the infinitesimal theory becomes a worse and worse approximation[1]. In other words, *the softening prior to M-yield is never negligible, no matter how small is the stress at M-yield.*

---

[1] The statement in the text refers to the interval $0 \leqq \vartheta < \vartheta_0$. Of course, the infinitesimal theory remains a good approximation for sufficiently small ϑ, but the *fraction* of the interval $0 \leqq \vartheta < \vartheta_0$ in which the percent error of the infinitesimal theory is less than an assigned amount becomes vanishingly small as $M \to 0$.

These results are illustrated in Fig. 9. The abscissa is the angle of shear, and the unit of stress is twice the shear modulus. If $2\mu M$ is to be identified with what is called a "yield stress" in theories of plasticity and in works on the strength of materials, values of M of about 1/100 or less are appropriate. For such values,

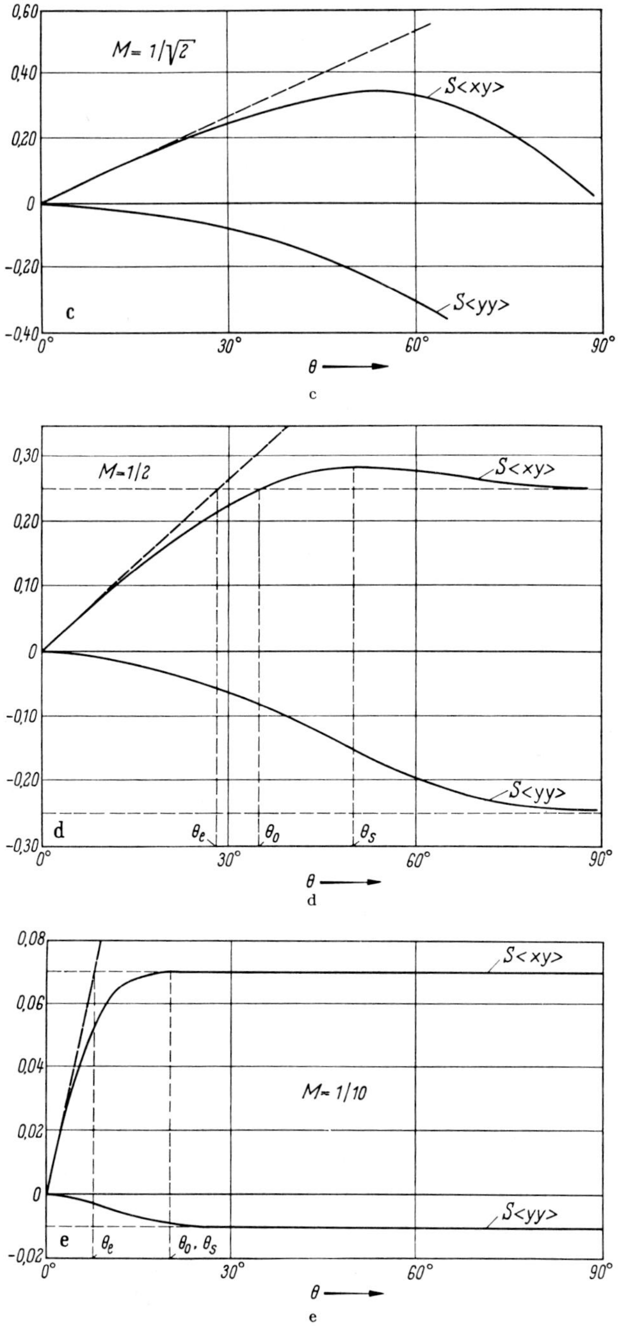

Fig. 9c–e. Hypo-elastic yield without or with M-yield for three special hypo-elastic materials of grade 2.

the stress-strain curves will be of the last type illustrated, but the angle of shear at which yield occurs will be much smaller.

The foregoing example illustrates the fact that for small strains, results predicted by properly invariant theories of large strains do not necessarily agree with formally similar theories set up expressly to describe small strains alone. A theory similar to a special case of hypo-elasticity, but linearized from the start and hence not frame-indifferent for large rotations, had been proposed by PRAGER[1]. In that theory, the stress $S\langle xy\rangle$ in simple shear approaches its asymptotic value steadily *from below*, so that there is no definite yield point. A similar result may be obtained in the present theory[2] by neglecting the normal stresses. While noticeable magnitudes of normal-stress differences are correctly associated with large strain, and while only small strains will be of interest in a case when M is very small, nevertheless it is *never* right to drop the normal-stress terms from $(103.24)_1$. These are the terms, *no matter how small they are*, that not only make the equations frame-indifferent but also give rise to the definite yield point. In this sense, in the example presented, yield in hypo-elasticity is *associated with effects of finite strain, no matter how small may be the strain itself at yield.* The dropping of small terms in equations is a custom so deeply ingrained as to have made, apparently, the foregoing distinction a hard one to grasp[3].

Various other problems[4] in hypo-elasticity have been studied, mostly for particular materials of grade 1.

E. Fluidity.

104. Scope and plan of the chapter. This chapter is a treatise on the theory of *fluid response*. The greater part, consisting in the first subchapter, concerns *simple fluids*, defined, in effect, as isotropic simple materials having the maximum possible material symmetries. We aim to present everything known about the general theory, as well as such approximate results as seem to be important for their physical interpretation.

In contrast to the chapter on elasticity, this one is brief. While the range of problems solved within the classical theory of finite elasticity has steadily broadened in the past decade, embracing an ever greater variety of cases, much of the work on fluids has tended to deepen a single narrow channel, namely, the theory of a small category of flows including and somewhat extending those of viscometric type. The growth of generality has been in the fluid rather than the flows. Roughly speaking, those problems solved exactly so far refer to peculiar bodies in which the deformation history, if rightly viewed, does not change along the path of the fluid particle, so that however great in variety the memory effects possible in the fluid, it is left little to remember in these cases, and the response becomes much simplified in consequence, allowing exact and explicit solution for solution for an arbitrary simple fluid. Methods of approximate solution, not yet

---

[1] PRAGER [1938, *4*] [1941, *2*] [1942, *4*].

[2] GREEN [1956, *9*, § 6].

[3] Cf., e.g., the remarks of FREUDENTHAL [1958, *17*, § 14], "These claims ... are ... flatly contradicted by the fact, confirmed in innumerable experiments ..." and of DRUCKER [1964, *23*, p. 416], "a flattening of the stress-strain curve predicted to occur at very large strains."

[4] TRUESDELL [1955, *31*, § 12] showed that torsional motion in a compressible material of grade 1 is impossible; his solution for incompressible materials [1955, *31*, § 13] is not free of initial hoop stress, as is remarked on p. 1020 of the volume.

Other problems have been considered by GREEN [1956, *9*, §§ 7—8], VERMA [1956, *36*] [1959, *37*] [1961, *67*] [1963, *79*], PARIA [1958, *33*], NARIBOLI [1961, *41*], SINHA [1961, *55*, *56*], VERMA and SINHA [1962, *69*] [1963, *80*], GUO ZHONG-HENG [1963, *31*], and ZAHORSKI [1963, *86*, § 6] [1963, *87* and *88*].

developed to nearly the extent they have been for elasticity, have not much widened the class of flows considered.

The second subchapter concerns other theories of fluid behavior. In some cases non-simple fluids are considered; in others, materials which may or may not be simple but which have been proposed for study within a different frame of concepts. Since these theories are still in an exploratory stage, we merely sketch the contents of the literature concerning them.

Again in contrast to elasticity, non-linear response of fluids was little cultivated until recently. While the classic memoir of STOKES (1845) begins with a principle defining the nature of viscosity in fairly general terms, he developed it only in the linear case. STOKES' principle was first further explored by REINER (1945), RIVLIN (1947—1949), and TRUESDELL (1947—1950). To RIVLIN is due the discovery of the exact solutions for viscometric flows of incompressible fluids. The concept of a simple fluid, after a major step had been taken by RIVLIN and ERICKSEN (1955), was formulated by NOLL (1957). The generality of RIVLIN'S special solutions was extended in two major stages, the first by RIVLIN himself (1956) and the second by COLEMAN and NOLL (1959). TRUESDELL (1948—1952) studied a theory of non-simple fluids designed to represent phenomena in rarefied gases. ERICKSEN (from 1960) has proposed and explored various theories of anisotropic fluids. The older theories of simple fluids are summarized in Sect. 119.

For parts of the subject presented in this chapter, there are several other expositions[1], but much of the contents is new.

I. Simple fluids.

a) General classes of flows.

105. The concept of a simple fluid. Simple fluids were defined in Sect. 32. We repeat the two physical assumptions on which this definition is based.

(i) *The present stress is determined by the history of the gradient of the deformation function.*

(ii) *A simple fluid has the maximum possible material symmetry.*

The most useful form of the constitutive equation of a compressible simple fluid is (32.8); of an incompressible simple fluid, (32.10). In both cases, the form

[1] TRUESDELL [1952, *20*, Ch. V], REINER [1956, *22*] [1958, *39*], OLDROYD [1956, *19*] [1959, *20*] [1961, *43*], GIESEKUS [1958, *19*] [1961, *17* and *18*], COLEMAN and NOLL [1961, *5*], RIVLIN [1961, *52*], SCHWARZL [1961, *53*], BHATNAGAR [1962, *5*], ERINGEN [1962, *18*, Chs. 7 and 10], KAPUR [1962, *40*], FREDRICKSON [1964, *33*], LODGE [1964, *57*], RIVLIN [1965, *30A*]. An early review was given by RIVLIN [1950, *14*]. While the scope, level, and quality of these works are various, one "fundamental" exposition is of such more than ordinary incorrectness as to deserve special mention: PHILIPPOFF [1957, *18*].

There are also numerous expositions of one-dimensional or linearized theories of "non-Newtonian flow" or "rheology".

A clear, short survey of the modern theory in relation to viscometric experiment has been given by MARKOVITZ [1965, *18*]. Cf. also the earlier review by LODGE [1960, *36*] of the experimental problem of determining normal stresses. Of especial interest also is the later discussion of the general problem of viscometric measurement and summary of the rheological properties of concentrated polymer solutions by LODGE [1964, *57*, Chs. 9 and 10].

By far the best exposition of theory and experiment on modern viscometry is the treatise of COLEMAN, MARKOVITZ, and NOLL [1965, *11*]. Their § 31 is an important history of experiment on materials now regarded as being non-linearly viscous fluids. They emphasize the originality, depth, and lasting importance of the pioneer work of MAXWELL. The early experiments now appear as floundering from want of clear objectives; the later ones, until comparatively recently, as misguided by faith in bad theory. In reference to much of the experiment done in this century but before the rise of the new continuum mechanics, COLEMAN, MARKOVITZ, and NOLL criticize the appeal to "plasticity", leading to "yield" that in many cases seems to be no more than a wishful extrapolation beyond the data measured.

of this constitutive equation is

$$\boldsymbol{T} = -p\boldsymbol{1} + \overset{\infty}{\underset{s=0}{\mathfrak{Q}}}\big(\boldsymbol{G}(s)\big), \tag{105.1}$$

where $\boldsymbol{T}$ is the stress at time t and where $\boldsymbol{G}(s)$, defined by Eq. $(31.10)_2$, differs from the history of the relative right Cauchy-Green tensor $\boldsymbol{C}_{(t)}(\tau)$ by only the unit tensor $\boldsymbol{1}$:

$$\boldsymbol{G}(s) = \boldsymbol{C}_{(t)}(t-s) - \boldsymbol{1} = \boldsymbol{C}_{(t)}^{(t)}(s) - \boldsymbol{1}. \tag{105.2}$$

We shall refer to the functions $\boldsymbol{G}(s)$ simply as *histories*. For compressible fluids the equilibrium pressure p is an assigned function of the density ϱ, and it is understood that the response functional $\mathfrak{Q}$ may also depend on ϱ as a parameter. For incompressible fluids the pressure p is indeterminate, and the functional $\mathfrak{Q}$ is determined only to within a scalar-valued functional. The indeterminacy of $\mathfrak{Q}$ may be removed by a normalization such as (32.12);

$$\operatorname{tr} \overset{\infty}{\underset{s=0}{\mathfrak{Q}}}\big(\boldsymbol{G}(s)\big) = 0. \tag{105.3}$$

When (105.3) is used, the indeterminate pressure coincides with the mean pressure (30.18).

As has been stated in Sect. 32, the value of the response functional $\mathfrak{Q}$ vanishes when the fluid has always been at rest, i.e. when $\boldsymbol{G}(s) \equiv \boldsymbol{0}$:

$$\overset{\infty}{\underset{s=0}{\mathfrak{Q}}}(\boldsymbol{0}) = \boldsymbol{0}. \tag{105.4}$$

Roughly speaking, the functional $\mathfrak{Q}$ describes the response of the fluid to disturbances from equilibrium.

The response functional $\mathfrak{Q}$ satisfies the isotropy relation (32.9):

$$\boldsymbol{Q}\overset{\infty}{\underset{s=0}{\mathfrak{Q}}}\big(\boldsymbol{G}(s)\big)\boldsymbol{Q}^T = \overset{\infty}{\underset{s=0}{\mathfrak{Q}}}\big(\boldsymbol{Q}\boldsymbol{G}(s)\boldsymbol{Q}^T\big) \tag{105.5}$$

for all orthogonal tensors $\boldsymbol{Q}$ and all histories $\boldsymbol{G}(s)$ in the domain of $\mathfrak{Q}$. The domain of $\mathfrak{Q}$ is assumed to be a suitable function space, for example the Hilbert space consisting of all functions $\boldsymbol{G}(s)$ such that $\boldsymbol{G}(0) = \boldsymbol{0}$ and such that the recollection (38.8), for some obliviator, is finite.

A component form of (105.1) is

$$T_{km}(t) = -p(t)\,g_{km}(t) + \overset{\infty}{\underset{s=0}{\mathfrak{Q}_{km}}}\big(C_{(t)pq}(t-s) - g_{pq}(t);\; g_{ru}(t)\big), \tag{105.6}$$

where

$$g_{ru}(t) = \boldsymbol{e}_r(\boldsymbol{x}) \cdot \boldsymbol{e}_u(\boldsymbol{x}). \tag{105.7}$$

Here $\boldsymbol{e}_r(\boldsymbol{x})$ may be any basis at the present position $\boldsymbol{x} = \boldsymbol{\chi}(X, t)$. When it is taken as the natural basis of a spatial co-ordinate system at $\boldsymbol{x}$, the quantities $g_{ru}(t)$ are components at $\boldsymbol{x}$ of the unit tensor $\boldsymbol{1}$.

Let $\boldsymbol{i}_r(\boldsymbol{x})$ be a basis which is orthonormal everywhere but may depend on $\boldsymbol{x}$. Replacing $\boldsymbol{e}_r$ by $\boldsymbol{i}_r$ in (105.6) and (105.7) gives $g_{ru} = \delta_{ru}$ and

$$T_{\langle km\rangle}(t) = -p(t)\,\delta_{km} + \overset{\infty}{\underset{s=0}{\mathfrak{Q}_{\langle km\rangle}}}\big(C_{(t)\langle pq\rangle}(t-s) - \delta_{pq}\big), \tag{105.8}$$

where the notation $\langle km\rangle$ indicates components with respect to the basis $\boldsymbol{i}_r(\boldsymbol{x})$ (cf. Sect. 6). The component functionals $\mathfrak{Q}_{\langle km\rangle}$ depend only on the material and not on the basis. Equation (105.8) coincides with (105.1) when the latter is reinterpreted in terms of matrices of components with respect to the orthonormal basis $\boldsymbol{i}_r(\boldsymbol{x})$. Reinterpretation of (105.4) in terms of such components yields a system of identities for the component functionals $\mathfrak{Q}_{\langle km\rangle}$. When an orthogonal curvilinear co-ordinate system is given, we may define $\boldsymbol{i}_r(\boldsymbol{x})$ in terms of the natural basis $\boldsymbol{e}_r(\boldsymbol{x})$ by

$$\boldsymbol{i}_k = \sqrt{g^{kk}}\,\boldsymbol{e}_k = \sqrt{g_{kk}}\,\boldsymbol{e}^k. \tag{105.9}$$

In this case the components of $\boldsymbol{T}$ and $\boldsymbol{C}_{(t)}$ in (105.8) coincide with the physical components (cf. Sect. 6).

In Sects. 106—109 we determine the *explicit stress systems* corresponding to certain special flows or families of flows, without taking up the question of whether these flows can occur in response to physically relevant systems of forces. In Sects. 111—113 we proceed to find circumstances in which certain of these flows are *dynamically possible*, subject to surface tractions alone or in conjunction with an assigned conservative body force field.

106. Lineal flows[1], viscometric flows[2]. Consider a motion of a body such that the velocity field $\dot{\boldsymbol{x}}(\boldsymbol{x}, t)$, in some Cartesian co-ordinate system x, y, z, has the components

$$\dot{x} = 0, \qquad \dot{y} = v(x, t), \qquad \dot{z} = 0. \tag{106.1}$$

We call such motions *lineal flows*[3]. For them the differential equations (21.15) reduce to

$$\frac{d\xi}{d\tau} = 0, \qquad \frac{d\eta}{d\tau} = v(\xi, \tau), \qquad \frac{d\zeta}{d\tau} = 0. \tag{106.2}$$

The solution that satisfies the initial conditions (21.16) is easily obtained:

$$\xi = x, \qquad \eta = y + \int_t^\tau v(x, \sigma)\, d\sigma, \qquad \zeta = z. \tag{106.3}$$

The matrix $[\boldsymbol{F}_{(t)}(\tau)]$ of the components of the relative deformation gradient $\boldsymbol{F}_{(t)}(\tau)$, computed according to (21.13), is

$$[\boldsymbol{F}_{(t)}(\tau)] = \left\| \begin{matrix} 1 & 0 & 0 \\ k(\tau) - k(t) & 1 & 0 \\ 0 & 0 & 1 \end{matrix} \right\|, \tag{106.4}$$

where

$$k(\tau) = \int_0^\tau v'(x, \sigma)\, d\sigma, \tag{106.5}$$

the prime denoting differentiation with respect to x. Eq. (106.4) shows that $\boldsymbol{F}_{(t)}(\tau)$ has the form

$$\boldsymbol{F}_{(t)}(\tau) = \boldsymbol{1} + \big(k(\tau) - k(t)\big)\,\boldsymbol{N}, \tag{106.6}$$

where $\boldsymbol{N}$ is a constant tensor with the component matrix

$$[\boldsymbol{N}] = \left\| \begin{matrix} 0 & 0 & 0 \\ 1 & 0 & 0 \\ 0 & 0 & 0 \end{matrix} \right\|. \tag{106.7}$$

---

[1] All the results presented here for simple fluids in lineal flows were first obtained by Coleman and Noll [1961, *5*, § 5], using a somewhat different formalism.

[2] Our definition of a viscometric flow is more general than that put forward by Coleman [1962, *10*, § 3] and used by Noll [1962, *50*], whose approach is followed here. Earlier results of this kind are included in Sect. 108.

[3] The terminology here differs from that of C F T, Sect. 68

Note that the square of $\boldsymbol{N}$ vanishes and that $\boldsymbol{N}$ has magnitude 1:

$$\boldsymbol{N}^2 = 0, \quad |\boldsymbol{N}| = 1. \tag{106.8}$$

Conversely, given any tensor satisfying (106.8), it follows from known results of linear algebra that there is an orthonormal basis such that the components of $\boldsymbol{N}$ with respect to that basis form the matrix (106.7).

The exponential of a tensor $\boldsymbol{A}$ is defined by the convergent power series expansion

$$e^{\boldsymbol{A}} = \sum_{n=0}^{\infty} \frac{1}{n!} \boldsymbol{A}^n. \tag{106.9}$$

Using (106.8) and (106.9), we see that (106.6) is equivalent to

$$\boldsymbol{F}_{(t)}(t-s) = \boldsymbol{F}^{(t)}_{(t)}(s) = e^{g(s)\boldsymbol{N}}, \tag{106.10}$$

where

$$g(s) = k(t-s) - k(t). \tag{106.11}$$

[The function $g(s)$ depends upon t as a parameter.] Substituting (106.10) into $\boldsymbol{C}^{(t)}_{(t)}(s) = \boldsymbol{F}^{(t)}_{(t)}(s)^T \boldsymbol{F}^{(t)}_{(t)}(s)$, from (105.2) we obtain

$$\boldsymbol{G}(s) = e^{g(s)\boldsymbol{N}^T} e^{g(s)\boldsymbol{N}} - \boldsymbol{1}. \tag{106.12}$$

Let $\mathfrak{Q}$ be the response functional of a simple fluid. We define

$$\underset{s=0}{\overset{\infty}{\mathfrak{K}}}\left(g(s);\, \boldsymbol{N}\right) = \underset{s=0}{\overset{\infty}{\mathfrak{Q}}}\left(e^{g(s)\boldsymbol{N}^T} e^{g(s)\boldsymbol{N}} - \boldsymbol{1}\right), \tag{106.13}$$

where $\mathfrak{K}$ is a functional of the scalar function $g(s)$ and a function of the tensor $\boldsymbol{N}$. It follows from (106.13), (105.5), and the properties of tensor exponentials[1] that $\mathfrak{K}$ satisfies the identity

$$\boldsymbol{Q} \underset{s=0}{\overset{\infty}{\mathfrak{K}}}(g(s);\boldsymbol{N})\, \boldsymbol{Q}^T = \underset{s=0}{\overset{\infty}{\mathfrak{K}}}\left(\alpha g(s);\, \frac{1}{\alpha} \boldsymbol{Q} \boldsymbol{N} \boldsymbol{Q}^T\right) \tag{106.14}$$

for all orthogonal tensors $\boldsymbol{Q}$ and all real and non-zero α. According to (106.13) and (105.1) the stress $\boldsymbol{T}$ is given by

$$\boldsymbol{T} = -p\boldsymbol{1} + \underset{s=0}{\overset{\infty}{\mathfrak{K}}}\left(g(s);\, \boldsymbol{N}\right). \tag{106.15}$$

In lineal flow, $\boldsymbol{N}$ has the components (106.7). Choosing the orthogonal tensor $\boldsymbol{Q}$ with the component matrix

$$[\boldsymbol{Q}] = \begin{Vmatrix} 1 & 0 & 0 \\ 0 & 1 & 0 \\ 0 & 0 & -1 \end{Vmatrix}, \tag{106.16}$$

we find that

$$\boldsymbol{Q} \boldsymbol{N} \boldsymbol{Q}^T = \boldsymbol{N}. \tag{106.17}$$

Using (106.14) with $\alpha = 1$ and the choice (106.16) for $\boldsymbol{Q}$, we see that (106.15) gives

$$\boldsymbol{Q} \boldsymbol{T} \boldsymbol{Q}^T = \boldsymbol{T}. \tag{106.18}$$

---

[1] Since $(\boldsymbol{Q}\boldsymbol{A}\boldsymbol{Q}^T)^n = \boldsymbol{Q}\boldsymbol{A}^n\boldsymbol{Q}^T$, from (106.9) we easily verify that if

$$\boldsymbol{f}(\boldsymbol{A}) = e^{\boldsymbol{A}^T} e^{\boldsymbol{A}},$$

then

$$\boldsymbol{Q}\boldsymbol{f}(\boldsymbol{A})\boldsymbol{Q}^T = \boldsymbol{f}(\boldsymbol{Q}\boldsymbol{A}\boldsymbol{Q}^T);$$

i.e., $\boldsymbol{f}$ is an isotropic function of $\boldsymbol{A}$.

This condition is satisfied if and only if the matrix of $\boldsymbol{T}$ has the form

$$[\boldsymbol{T}] = \begin{Vmatrix} T\langle 11\rangle & T\langle 12\rangle & 0 \\ T\langle 12\rangle & T\langle 22\rangle & 0 \\ 0 & 0 & T\langle 33\rangle \end{Vmatrix}. \tag{106.19}$$

Since the matrix of $\boldsymbol{N}$ has the specific form (106.7), it follows from (106.15) that the 6 components of $\boldsymbol{T}+p\boldsymbol{1}$ are functionals of $g(s)$ only. In view of (106.19), only 4 of these functionals can be non-zero. The shear stress $T\langle 12\rangle$ and the normal-stress differences $T\langle kk\rangle - T\langle mm\rangle$ are not affected by the hydrostatic term $p\boldsymbol{1}$. We denote the functional dependence of $T\langle 12\rangle$ and of two normal-stress differences on $g(s)$ by[1]

$$T\langle 12\rangle = \overset{\infty}{\underset{s=0}{\mathfrak{t}}}\big(g(s)\big), \tag{106.20}$$

$$T\langle 11\rangle - T\langle 33\rangle = \overset{\infty}{\underset{s=0}{\mathfrak{s}_1}}\big(g(s)\big), \tag{106.21}$$

$$T\langle 22\rangle - T\langle 33\rangle = \overset{\infty}{\underset{s=0}{\mathfrak{s}_2}}\big(g(s)\big). \tag{106.22}$$

The three scalar functionals $\mathfrak{t}$, $\mathfrak{s}_1$, $\mathfrak{s}_2$ are independent of the orientation of the Cartesian co-ordinate system used and depend only on the particular material under consideration. We call $\mathfrak{t}$, $\mathfrak{s}_1$, $\mathfrak{s}_2$ *material functionals*. They completely characterize the mechanical behavior of incompressible simple fluids in lineal flow. We may write Eqs. (106.20)—(106.22) in the following invariant form:

$$\left.\begin{aligned} \boldsymbol{T} &= T\langle 33\rangle\boldsymbol{1} + \overset{\infty}{\underset{s=0}{\mathfrak{t}}}\big(g(s)\big)\,(\boldsymbol{N}+\boldsymbol{N}^T) + \\ &\quad + \overset{\infty}{\underset{s=0}{\mathfrak{s}_1}}\big(g(s)\big)\,\boldsymbol{N}^T\boldsymbol{N} + \overset{\infty}{\underset{s=0}{\mathfrak{s}_2}}\big(g(s)\big)\,\boldsymbol{N}\boldsymbol{N}^T. \end{aligned}\right\} \tag{106.23}$$

As a second choice of $\boldsymbol{Q}$ in (106.14) we take

$$[\boldsymbol{Q}] = \begin{Vmatrix} 1 & 0 & 0 \\ 0 & -1 & 0 \\ 0 & 0 & 1 \end{Vmatrix} \tag{106.24}$$

and $\alpha = -1$. For this choice (106.7) yields

$$\frac{1}{\alpha}\,\boldsymbol{Q}\boldsymbol{N}\boldsymbol{Q}^T = -\boldsymbol{Q}\boldsymbol{N}\boldsymbol{Q}^T = \boldsymbol{N}. \tag{106.25}$$

It follows from (106.14) and (106.15) that the stress corresponding to $-g(s)$ is given by $\boldsymbol{Q}\boldsymbol{T}\boldsymbol{Q}^T$, where $\boldsymbol{T}$ is the stress corresponding to $g(s)$. Since the matrix of the components of $\boldsymbol{Q}\boldsymbol{T}\boldsymbol{Q}^T$ is

$$[\boldsymbol{Q}\boldsymbol{T}\boldsymbol{Q}^T] = \begin{Vmatrix} T\langle 11\rangle & -T\langle 12\rangle & 0 \\ -T\langle 12\rangle & T\langle 22\rangle & 0 \\ 0 & 0 & T\langle 33\rangle \end{Vmatrix}, \tag{106.26}$$

it follows from (106.20)—(106.22) that the material functionals $\mathfrak{t}$ and $\mathfrak{s}_\Gamma$, $\Gamma = 1, 2$, obey the identities

$$\overset{\infty}{\underset{s=0}{\mathfrak{t}}}\big(-g(s)\big) = -\overset{\infty}{\underset{s=0}{\mathfrak{t}}}\big(g(s)\big), \quad \overset{\infty}{\underset{s=0}{\mathfrak{s}_\Gamma}}\big(-g(s)\big) = \overset{\infty}{\underset{s=0}{\mathfrak{s}_\Gamma}}\big(g(s)\big). \tag{106.27}$$

---

[1] Coleman and Noll [1961, *5*, Eq. (5.22)].

In words, $\mathfrak{t}$ must be an *odd* functional and the $\mathfrak{s}_\Gamma$ must be *even* functionals[1]. It is not hard to see that the isotropy condition (106.14) imposes no further restrictions on the scalar material functionals $\mathfrak{t}$, $\mathfrak{s}_1$, and $\mathfrak{s}_2$. The condition (105.4) corresponds to

$$\overset{\infty}{\underset{s=0}{\mathfrak{t}}}(0)=0, \quad \overset{\infty}{\underset{s=0}{\mathfrak{s}_\Gamma}}(0)=0. \tag{106.28}$$

It is easily seen that lineal flows are isochoric and hence compatible with incompressibility. For incompressible simple fluids the three material functionals $\mathfrak{t}$, $\mathfrak{s}_1$ and $\mathfrak{s}_2$ determine the behavior of the material in lineal flow completely. For compressible simple fluids, it is necessary to introduce a fourth material functional, for example the functional whose value gives the mean pressure $\bar{p}=-\frac{1}{3}(T\langle 11\rangle+T\langle 22\rangle+T\langle 33\rangle)$. In addition, the four material functionals then depend parametrically on the density. We shall apply the results of this section only to problems involving incompressible fluids.

For lineal flows, the tensor $\boldsymbol{N}$ occurring in (106.12) is independent not only of the time lapse s but also of the present time t. The analysis leading to (106.19)—(106.23), (106.27), and (106.28) does not make use of this fact. All that is needed is that $\boldsymbol{N}$, for each t, satisfies (106.8) or, equivalently, that the matrix of $\boldsymbol{N}$ must have the form (106.7) with respect to some orthonormal basis $\boldsymbol{i}_r$, which need not be independent of t.

We say that a motion is *locally viscometric*[2] along the path-line of a particle if the history (105.2), for each present time t, has the form (106.12), where $\boldsymbol{N}$ satisfies (106.8). Both $\boldsymbol{N}$ and $g(s)$ may depend on t. The stress $\boldsymbol{T}$ in a locally viscometric motion of a simple fluid satisfies (106.19)—(106.23), in which $T\langle km\rangle$ are the components of $\boldsymbol{T}$ relative to a suitable orthonormal basis $\boldsymbol{i}_r(t)$.

We say that the motion of a body is a *viscometric flow* if it is locally viscometric along each path-line. As we shall see in Sects. 109—113, the viscometric flows include all flows that are commonly used to interpret viscometric experiments.

Suppose that, along some path-line, the relative deformation gradient $\boldsymbol{F}_{(0)}(\tau)$ has the form

$$\boldsymbol{F}_{(0)}(\tau)=\boldsymbol{Q}(\tau)\left(\mathbf{1}+k(\tau)\,\boldsymbol{N}_0\right)=\boldsymbol{Q}(\tau)\,e^{k(\tau)\boldsymbol{N}_0}, \tag{106.29}$$

where $\boldsymbol{Q}(\tau)$ is orthogonal and where $\boldsymbol{N}_0$ is a constant tensor that satisfies (106.8). Using (106.29) and (21.25) with $t'=0$, we obtain

$$\boldsymbol{F}_{(t)}(\tau)=\boldsymbol{Q}(\tau)\,e^{[k(\tau)-k(t)]\boldsymbol{N}_0}\boldsymbol{Q}(t)^T, \tag{106.30}$$

and hence

$$\boldsymbol{G}(s)+\mathbf{1}=\boldsymbol{C}_{(t)}^{(t)}(s)=\boldsymbol{F}_{(t)}(t-s)^T\boldsymbol{F}_{(t)}(t-s)=e^{g(s)\boldsymbol{N}^T}e^{g(s)\boldsymbol{N}}, \tag{106.31}$$

where $g(s)$ is given by (106.11) and where

$$\boldsymbol{N}=\boldsymbol{Q}(t)\,\boldsymbol{N}_0\boldsymbol{Q}(t)^T. \tag{106.32}$$

Therefore $\boldsymbol{N}$ satisfies (106.8). We have shown, then, that *a motion for which* (106.29) *holds is locally viscometric.*

107. Curvilineal flows[3]. We have seen in the previous section that the behavior of incompressible simple fluid in a locally viscometric flow is completely deter-

---

[1] COLEMAN and NOLL [1961, *5*, Eq. (5.26)].

[2] The first invariant definition of "viscometric flows" is due to COLEMAN [1962, *10*, § 3]; it corresponds to what we call in Sect. 108 "steady viscometric flows".

[3] The material in this section includes and in part generalizes results of NOLL [1962, *50*], which in turn generalize results of COLEMAN [1962, *10*] and of COLEMAN and NOLL [1959, *2* and *4*]. These authors considered only steady flows. The generalization to unsteady flows, not previously published, is by NOLL.

mined by the three scalar material functionals $\mathfrak{t}$, $\mathfrak{s}_1$, and $\mathfrak{s}_2$. We now describe a large class of viscometric flows, which includes the lineal flows of the previous section as a special case. We call *curvilineal flow* a flow whose velocity field $\dot{\boldsymbol{x}}(\boldsymbol{x}, t)$ satisfies the following three conditions:

(I) There is an orthogonal curvilinear co-ordinate system x^k such that the contravariant components $\dot{x}^k$ of $\dot{\boldsymbol{x}}$ are

$$\dot{x}^1 = 0, \qquad \dot{x}^2 = v(x^1, t), \qquad \dot{x}^3 = w(x^1, t). \tag{107.1}$$

(II) The ratio of the x^1-derivatives $v'(x^1, t)$ and $w'(x^1, t)$ does not depend on t, i.e. these derivatives have the form

$$v'(x^1, t) = f(x^1)\, q(x^1, t), \qquad w'(x^1, t) = h(x^1)\, q(x^1, t). \tag{107.2}$$

(III) The components g_{kk} of the unit tensor are constant along the path-lines of the particles. (The path-lines are the curves $\boldsymbol{\xi} = \boldsymbol{\xi}(\tau)$ defined by

$$\xi^1 = x^1, \quad \xi^2 = x^2 + \int_t^\tau v(x^1, \sigma)\, d\sigma, \quad \xi^3 = x^3 + \int_t^\tau w(x^1, \sigma)\, d\sigma.) \tag{107.3}$$

Condition (II) is satisfied, for example, when the flow is steady or when $w = 0$. Condition (III) is satisfied *a fortiori* when the g_{kk} depend only on x^1 or when $w = 0$ and the g_{kk} depend only on x^1 and x^3. L. Bragg has shown us analysis proving that the former case holds if and only if the co-ordinate system is cylindrical or Cartesian. The properties of some flows of these two kinds are developed in detail in Sects. 111, 112–113, 115, and 117.

When (107.1) holds, the formulae (107.3) give the solution of the differential equations (21.15) with the initial conditions (21.16). Hence (107.3) describes the relative deformation function for curvilineal flow. Let $\boldsymbol{e}_k = \boldsymbol{e}_k(\boldsymbol{x})$ be the natural basis of the co-ordinate system used, and let $\boldsymbol{e}^k$ be the basis dual to $\boldsymbol{e}_k$, so that $\boldsymbol{e}_k \cdot \boldsymbol{e}^m = \delta_k^m$. By (21.13), (107.3), and (107.2) we then have

$$\|\boldsymbol{e}^k(\boldsymbol{\xi}) \cdot \boldsymbol{F}_{(t)}(\tau)\, \boldsymbol{e}_m(\boldsymbol{x})\| = \|\xi^k{}_{,m}\| = \begin{Vmatrix} 1 & 0 & 0 \\ f\bar{q} & 1 & 0 \\ h\bar{q} & 0 & 1 \end{Vmatrix}, \tag{107.4}$$

where

$$\bar{q} = \int_t^\tau q(x^1, \sigma)\, d\sigma. \tag{107.5}$$

The unit vectors $\tilde{\boldsymbol{e}}_k$ in the direction of the co-ordinate lines are related to $\boldsymbol{e}_k$ and $\boldsymbol{e}^k$ by (cf. Sect. 6)

$$\boldsymbol{e}_k = \sqrt{g_{kk}}\, \tilde{\boldsymbol{e}}_k, \qquad \boldsymbol{e}^k = \sqrt{g_{kk}^{-1}}\, \tilde{\boldsymbol{e}}_k. \tag{107.6}$$

It follows from condition (III) that $g_{kk}(\boldsymbol{\xi}) = g_{kk}(\boldsymbol{x}) = g_{kk}$. Therefore, (107.4) is equivalent to

$$\|\tilde{\boldsymbol{e}}_k(\boldsymbol{\xi}) \cdot \boldsymbol{F}_{(t)}(\tau)\, \tilde{\boldsymbol{e}}_m(\boldsymbol{x})\| = [\mathbf{1}] + \bar{q} \begin{Vmatrix} 0 & 0 & 0 \\ \sqrt{g_{22}\, g_{11}^{-1}}\, f & 0 & 0 \\ \sqrt{g_{33}\, g_{11}^{-1}}\, h & 0 & 0 \end{Vmatrix}, \tag{107.7}$$

where $[\mathbf{1}]$ is the unit matrix. When $t = 0$, (107.7) reduces to

$$\|\tilde{\boldsymbol{e}}_k(\boldsymbol{\xi}(\tau)) \cdot \boldsymbol{F}_{(0)}(\tau)\, \tilde{\boldsymbol{e}}_m(\boldsymbol{x})\| = [\mathbf{1}] + k(\tau) \begin{Vmatrix} 0 & 0 & 0 \\ \alpha & 0 & 0 \\ \beta & 0 & 0 \end{Vmatrix}, \tag{107.8}$$

where

$$\left.\begin{aligned} k(\tau) &= \varkappa \int_0^\tau q(x^1, \sigma)\, d\sigma, \\ \varkappa &= \sqrt{g_{11}^{-1}(g_{22} f^2 + g_{33} h^2)}, \end{aligned}\right\} \tag{107.9}$$

and

$$\alpha = \frac{1}{\varkappa}\sqrt{g_{22} g_{11}^{-1}}\, f, \quad \beta = \frac{1}{\varkappa}\sqrt{g_{33} g_{11}^{-1}}\, h, \quad \alpha^2 + \beta^2 = 1. \tag{107.10}$$

Now, since both $\tilde{\boldsymbol{e}}_k(\boldsymbol{\xi})$ and $\tilde{\boldsymbol{e}}_k(\boldsymbol{x})$ are orthonormal bases, they are related by an orthogonal transformation $\boldsymbol{Q}(\tau)$:

$$\tilde{\boldsymbol{e}}_k(\boldsymbol{\xi}(\tau)) = \boldsymbol{Q}(\tau)\,\tilde{\boldsymbol{e}}_k(\boldsymbol{x}). \tag{107.11}$$

$\boldsymbol{\xi}(\tau)$ describes the path-line passing through $\boldsymbol{x}$ at $\tau = 0$, i.e. $\boldsymbol{\xi}(0) = \boldsymbol{x}$. If we substitute (107.11) into (107.8), we see that

$$\boldsymbol{Q}(\tau)^T \boldsymbol{F}_{(0)}(\tau) = \boldsymbol{1} + k(\tau)\,\boldsymbol{N}_0, \tag{107.12}$$

where $\boldsymbol{N}_0$ is a tensor whose matrix $[\boldsymbol{N}_0]$ with respect to the basis $\tilde{\boldsymbol{e}}_k(\boldsymbol{x})$ is given by

$$[\boldsymbol{N}_0] = \begin{Vmatrix} 0 & 0 & 0 \\ \alpha & 0 & 0 \\ \beta & 0 & 0 \end{Vmatrix}. \tag{107.13}$$

Using (107.13) and $(107.10)_3$, one can easily verify that $\boldsymbol{N}_0$ satisfies (106.8). Therefore, (107.12) shows that (106.29) holds and hence that the flow must be viscometric. We know already that for some orthonormal basis the matrix of $\boldsymbol{N}_0$ must have the form (106.7). An easy computation shows that such a basis is given by

$$\left.\begin{aligned} \boldsymbol{i}_1 &= \tilde{\boldsymbol{e}}_1 \\ \boldsymbol{i}_2 &= \alpha\,\tilde{\boldsymbol{e}}_2 + \beta\,\tilde{\boldsymbol{e}}_3 \\ \boldsymbol{i}_3 &= -\beta\,\tilde{\boldsymbol{e}}_2 + \alpha\,\tilde{\boldsymbol{e}}_3. \end{aligned}\right\} \tag{107.14}$$

When $\boldsymbol{x}$ is the position of the particle at an arbitrary time t, then the matrix of $\boldsymbol{N} = \boldsymbol{Q}(t)\,\boldsymbol{N}_0\,\boldsymbol{Q}(t)^T$ with respect to the basis $\boldsymbol{i}_k$ given by (107.14) always has the form (106.7). The tensor $\boldsymbol{N}$ itself need not be independent of t even though its matrix (106.7) is, since the basis (107.14) may rotate along the path-line of the particle.

In curvilineal flow of an incompressible simple fluid, the formulae (106.19)—(106.23) hold when the components $T\langle km\rangle$ are taken with respect to the rotating basis (107.14). These components are in general different from the physical components, i.e. from the components with respect to the basis $\tilde{\boldsymbol{e}}_k$. For these physical components, which we denote by $T^*\langle km\rangle$, we easily obtain from (107.14) and (106.19)—(106.22) the formulae

$$T^*\langle 12\rangle = \alpha\, \mathop{\mathfrak{t}}_{s=0}^{\infty}(g(s)), \qquad T^*\langle 13\rangle = \beta\, \mathop{\mathfrak{t}}_{s=0}^{\infty}(g(s)), \tag{107.15}$$

$$T^*\langle 23\rangle = \alpha\beta\, \mathop{\mathfrak{s}_2}_{s=0}^{\infty}(g(s)), \tag{107.16}$$

$$T^*\langle 11\rangle - T^*\langle 33\rangle = \mathop{\mathfrak{s}_1}_{s=0}^{\infty}(g(s)) - \beta^2\, \mathop{\mathfrak{s}_2}_{s=0}^{\infty}(g(s)), \tag{107.17}$$

$$T^*\langle 22\rangle - T^*\langle 33\rangle = (\alpha^2 - \beta^2)\, \mathop{\mathfrak{s}_2}_{s=0}^{\infty}(g(s)). \tag{107.18}$$

Here $g(s)$ is given by (106.11), in which (107.9) must be substituted for $k(\tau)$.

108. Steady viscometric flows. Consider the special case when the function $k(\tau)$ occurring in the formula (106.29) for the relative deformation gradient in a viscometric flow is linear in τ, so that

$$k(\tau)=\varkappa\tau. \tag{108.1}$$

If (108.1) holds, we say that the local viscometric flow along the path-line under consideration is *steady*[1]. For steady viscometric flows (106.11) becomes

$$g(s)=-\varkappa s. \tag{108.2}$$

If we define three functions τ, σ_1, and σ_2 by

$$\tau(\varkappa)\equiv \mathop{\mathfrak{t}}_{s=0}^{\infty}(-\varkappa s), \qquad \sigma_\Gamma(\varkappa)\equiv \mathop{\mathfrak{s}_\Gamma}_{s=0}^{\infty}(-\varkappa s), \qquad (\Gamma=1,2) \tag{108.3}$$

we find that the formulae (106.20)—(106.22) reduce to

$$T\langle 12\rangle=\tau(\varkappa), \tag{108.4}$$

$$T\langle 11\rangle-T\langle 33\rangle=\sigma_1(\varkappa), \tag{108.5}$$

$$T\langle 22\rangle-T\langle 33\rangle=\sigma_2(\varkappa). \tag{108.6}$$

It follows from (106.27) that τ is an odd function and that the σ_Γ are even functions:

$$\tau(-\varkappa)=-\tau(\varkappa), \qquad \sigma_\Gamma(-\varkappa)=\sigma_\Gamma(\varkappa). \tag{108.7}$$

The conditions (106.28) imply that

$$\tau(0)=0, \qquad \sigma_\Gamma(0)=0. \tag{108.8}$$

A further restriction on τ follows from the requirement that the dissipation of kinetic energy, measured by the stress power P [CFT, Eq. (217.4)] must be positive[2]. The stress power P is easily computed in the present case, and we are led to impose the inequality

$$P=\varkappa\tau(\varkappa)>0 \quad \text{if} \quad \varkappa\neq 0. \tag{108.9}$$

If it is assumed that τ is twice continuously differentiable, then (108.8) and (108.9) imply that $\tau(\varkappa)$ must be a strictly increasing function of $\varkappa$ in some interval around $\varkappa=0$. In this interval τ will then have a strictly increasing and odd inverse $\overset{-1}{\tau}$. In the remainder of this chapter, we set $\zeta\equiv\overset{-1}{\tau}$, and whenever ζ occurs, we assume that we are in the range in which it is defined. If ζ is not defined over the entire real axis, then the flows which we shall discuss cannot occur at high speeds. To obtain the basic equations (108.4)—(108.9), however, no restrictions on τ or σ_Γ are needed. In particular, these functions need not be invertible[3].

---

[1] The theory of steady viscometric flows in incompressible simple fluids was developed in several stages by COLEMAN and NOLL [1959, *2* and *4*] [1961, *5*] [1962, *10* and *50*]. References to earlier and more special studies are cited below for various detailed results.

[2] A thermodynamic argument showing that the stress power should be positive in *steady* viscometric flows was given by COLEMAN [1962, *10*, § 7]; he pointed out that P need not be positive in general flows. [Note added in proof: See the general thermodynamic theory presented in Sect. 96bis, especially the remarks at the end.]

[3] Cf. the remarks of TRUESDELL [1950, *16*, § 12] on multiple solutions and the discussion of singular viscometric functions by GIESEKUS [1962, *22*].

The three material functions τ, σ_1, and σ_2 completely determine the behavior of incompressible simple fluids in all steady viscometric flows[1] and hence are called the *viscometric functions*. The function τ is called the *shear-stress function*, while σ_1 and σ_2 are called the *normal-stress functions*. Instead of τ, the *viscosity function* $\tilde{\mu}$, defined by

$$\tilde{\mu}(\varkappa) \equiv \frac{1}{\varkappa}\,\tau(\varkappa) \tag{108.10}$$

is often more convenient. Like σ_1 and σ_2, it is an even function of $\varkappa$. The independent variable $\varkappa$ is called the *shearing*. The inverse ζ of the shear-stress function τ is called the *shearing function*.

Steady lineal flow corresponds to the case when the velocity component $v(x)$ in (106.1) does not depend on t. In this case, the shearing is given by

$$\varkappa = v'(x). \tag{108.11}$$

Steady curvilineal flow[2] corresponds to the case when the velocity components $v(x)$ and $w(x)$ in (107.1) do not depend on t. In this case, (107.2) is not a restriction, and we may put $q=1$, and

$$v'(x) = f(x), \qquad w'(x) = h(x). \tag{108.12}$$

The shearing, $\varkappa$, is then given by $(107.9)_2$. The formulae (107.15)—(107.18) for the physical components of the stress reduce to

$$T^{*\langle 12\rangle} = \alpha\,\tau(\varkappa), \qquad T^{*\langle 13\rangle} = \beta\,\tau(\varkappa), \tag{108.13}$$

$$T^{*\langle 23\rangle} = \alpha\beta\,\sigma_2(\varkappa), \tag{108.14}$$

$$\left.\begin{aligned} T^{*\langle 11\rangle} - T^{*\langle 33\rangle} &= \sigma_1(\varkappa) - \beta^2\sigma_2(\varkappa), \\ T^{*\langle 22\rangle} - T^{*\langle 33\rangle} &= (\alpha^2 - \beta^2)\,\sigma_2(\varkappa). \end{aligned}\right\} \tag{108.15}$$

It follows from (108.2), (106.31) and (106.32) that for steady viscometric *flows*[3]

$$\boldsymbol{C}_{(t)}^{(t)}(s) = \boldsymbol{Q}(t)\,\boldsymbol{C}_{(0)}^{(0)}(s)\,\boldsymbol{Q}(t)^T \tag{108.16}$$

and

$$\boldsymbol{C}_{(t)}^{(t)}(s) = e^{-\varkappa s \boldsymbol{N}^T}\, e^{-\varkappa s \boldsymbol{N}}, \tag{108.17}$$

where $\boldsymbol{N}$ satisfies (106.8) and hence has the matrix (106.7) with respect to some orthonormal basis which, in general, rotates along the path-line.

In Sect. 28 we have given as Eq. (28.21) a dimensionally invariant form of the general constitutive equation of a simple material. For simple fluids an equivalent and more useful form, corresponding to (105.1), is

$$\mathop{\mathfrak{Q}}_{s=0}^{\infty}(\boldsymbol{G}(s)) = \frac{\mu_0}{s_0}\mathop{\mathfrak{Q}^*}_{s^*=0}^{\infty}(\boldsymbol{G}(s_0 s^*)), \tag{108.18}$$

---

[1] MARKOVITZ [1957, *13*] seems to have been the first to recognize that in a variety of special theories the behavior of a fluid in various special steady viscometric flows is governed by a set of three material functions equivalent to the ones given here. CRIMINALE, ERICKSEN, and FILBEY [1958, *13*] and, independently, GIESEKUS (1958) [1961, *18*] [1962, *20*] showed that three material functions describe the behavior of fluids of the differential type (cf. Sect. 35) in steady viscometric flows, which the former authors called "laminar shear flows"; see also ERICKSEN [1959, *7*]. ERICKSEN [1960, *22*] recognized that the same is true for certain special flows in a theory of rate type described earlier by NOLL [1955, *19*, §§ 21—23]. The general theory is due to COLEMAN and NOLL [1959, *2*].

[2] NOLL [1962, *50*].

[3] Again the identity given in the footnote on p. 430 is used.

where μ_0 and s_0 are material constants such that

$$\dim \mu_0 = M L^{-1} T^{-1}, \qquad \dim s_0 = T, \tag{108.19}$$

where $\mathfrak{Q}^*$ is a dimensionless functional of its dimensionless tensor argument, and where s^* is a dimensionless dummy variable. The constants μ_0 and s_0 may be called the *natural viscosity* and *natural time-lapse*, respectively, of the simple fluid.

The general reduction (108.18) has its consequences upon the nature of the viscometric functions. If these are continuously differentiable at $\varkappa = 0$, they must have the forms[1]

$$\left.\begin{aligned} \tau(\varkappa) &= \varkappa\tilde{\mu}(\varkappa) = \mu_0 \varkappa V(K), \\ \sigma_1(\varkappa) &= \mu_0 s_0 \varkappa^2 N_1(K), \\ \sigma_2(\varkappa) &= \mu_0 s_0 \varkappa^2 N_2(K), \end{aligned}\right\} \tag{108.20}$$

where $K \equiv s_0 \varkappa$ and where V, N_1, and N_2, the *reduced viscometric functions*, are dimensionless, even functions of their dimensionless argument K. Thus the viscometric properties of a simple fluid are completely specified by the following quantities:

1. *The natural viscosity* μ_0, which is a constant having the dimensions $M L^{-1} T^{-1}$.
2. The *natural time-lapse* s_0, which is a constant having the dimension T.
3. The reduced viscometric functions V, N_1, and N_2, which are dimensionless even functions of a dimensionless argument.

(In case the constitutive functional $\mathfrak{Q}$ in (105.1) is allowed to depend upon the temperature as a parameter, the five quantities just listed are also temperature-dependent.)

Of course there are infinitely many legitimate choices for the constants μ_0 and s_0. TRUESDELL[2] has proposed the following definitions as being likely to serve in the interpretation of experiments. Let the *time-lapse function* $\tilde{s}(K)$ be so defined as to measure the relative importance of normal-stress and shear-stress effects and to vanish identically for such fluids, and only for those, as show no normal-stress effects in viscometric flows:

$$\left.\begin{aligned} \tilde{s}(K) &\equiv \frac{\sqrt{[\sigma_1(\varkappa)]^2 + [\sigma_2(\varkappa)]^2}}{\varkappa^2 \tilde{\mu}(\varkappa)}, \\ &= s_0 \frac{\sqrt{[N_1(K)]^2 + [N_2(K)]^2}}{V(K)}. \end{aligned}\right\} \tag{108.21}$$

Set[3]

$$\mu_0 = \tilde{\mu}(0), \qquad s_0 \equiv \tilde{s}(0). \tag{108.22}$$

The functions V, N_1, and N_2 are then subject to the restrictions

$$V(0) = 1, \qquad [N_1(0)]^2 + [N_2(0)]^2 = 1. \tag{108.23}$$

s_0, so defined, is a time which is large if the normal-stress effects are large in proportion to the shear-stress effects at low shearing.

The classical or "Navier-Stokes" theory of viscometry is gotten by supposing there are no normal-stress effects at all, and that V is constant:

$$\left.\begin{aligned} &\tilde{\mu}(\varkappa) \equiv \mu_0, \\ &V(K) \equiv 1, \quad N_1(K) \equiv N_2(K) \equiv 0. \end{aligned}\right\} \tag{108.24}$$

[1] TRUESDELL [1964, *87*, § 3] obtained these forms by direct application of the pi-theorem. His earlier work with special theories emphasized the importance of a natural time in phenomena of non-linear viscosity [1949, *22*, § 10] [1950, *16*, § 11] [1952, *20*, §§ 67, 72] [1952, *22*].

[2] TRUESDELL [1964, *87*, § 13] [1964, *85*].

[3] The special case for the Reiner-Rivlin theory was first proposed by TRUESDELL [1952, *20*, § 72].

The natural time-lapse, s_0, drops out of the classical theory, and the time-lapse function $\tilde{s}(K)$ vanishes identically. We shall have several occasions below to refer to various aspects of the classical theory.

Since calculation of the natural time-lapse by means of (108.22) requires the limiting values of $\sigma_1(\varkappa)/\varkappa^2$ and $\sigma_2(\varkappa)/\varkappa^2$ as $\varkappa \to 0$, while to measure those functions accurately large shearings are needed, experimental determination of s_0 is difficult. By extrapolation to 0 MARKOVITZ[1] has obtained the values of s_0 shown in the table alongside, in which appear also the corresponding values of the natural viscosity μ_0.

Table. *Natural constants of solutions of polyisobutylene in cetane.*

Concentration	μ_0 (poise)	s_0 (sec)
3.9%	4.8	0.026
5.4%	18.5	0.065
6.9%	60	0.14

109. Motions with constant stretch history. We now consider motions that satisfy the relation (108.16) for all $s \geqq 0$ but are not necessarily viscometric. Such motions are called *substantially stagnant motions* or *motions with constant stretch history*[2]. It is clear that (108.16) holds if and only if there is a time-dependent orthonormal basis $\boldsymbol{i}_k(t)$ such that the components with respect to $\boldsymbol{i}_k(t)$ of the history $\boldsymbol{C}_{(t)}^{(t)}(s)$ are independent of the present time and depend only on the time lapse s. In particular, when (108.16) holds, the principal stretches of the deformation carrying the configuration at time t into the configuration at time $t-s$ are independent of t. Roughly speaking, a motion with constant stretch history is one in which an observer moving with the particle can always orient himself in such a way as to see the stretch histories at that particle as independent of the present time.

The following ***fundamental theorem*** of NOLL gives a complete characterization of all motions with constant stretch history: *A motion has constant stretch history if and only if the deformation gradient $\boldsymbol{F}_{(0)}(\tau)$, relative to the configuration at some fixed time 0, is of the form*

$$\boldsymbol{F}_{(0)}(\tau) = \boldsymbol{Q}(\tau)\, e^{\tau\varkappa \boldsymbol{N}_0}, \qquad |\boldsymbol{N}_0| = 1, \qquad \boldsymbol{Q}(0) = \boldsymbol{1}, \tag{109.1}$$

where $\boldsymbol{N}_0$ is a constant tensor of unit magnitude and where $\boldsymbol{Q}(\tau)$ is an orthogonal tensor function which reduces to $\boldsymbol{1}$ *when* $\tau = 0$.

We first list a number of formulae which follow immediately from (109.1). We use the abbreviations

$$\boldsymbol{N} = \boldsymbol{Q}(t)\,\boldsymbol{N}_0\,\boldsymbol{Q}(t)^T, \qquad \boldsymbol{Z} = \dot{\boldsymbol{Q}}(t)\,\boldsymbol{Q}(t)^T. \tag{109.2}$$

Clearly, $\boldsymbol{Z}$ is a skew tensor. It is a measure of the rate of rotation of the basis $\boldsymbol{i}_k(t)$ relative to which $\boldsymbol{C}_{(t)}^{(t)}(s)$ has components independent of t. For the relative deformation gradient we obtain from (21.25)

$$\left.\begin{aligned} \boldsymbol{F}_{(t)}(\tau) &= \boldsymbol{Q}(\tau)\, e^{(\tau - t)\varkappa \boldsymbol{N}_0}\, \boldsymbol{Q}(t)^T, \\ &= \boldsymbol{Q}(\tau)\, \boldsymbol{Q}(t)^T e^{(\tau - t)\varkappa \boldsymbol{N}}. \end{aligned}\right\} \tag{109.3}$$

For the relative right Cauchy-Green tensor, (108.16) and (108.17) are valid when $\boldsymbol{N}$ is given by (109.2). For the velocity gradient $\boldsymbol{L}$, we obtain from (24.1), (109.3), and (109.2)

$$\boldsymbol{L} = \varkappa \boldsymbol{N} + \boldsymbol{Z}, \qquad \dot{\boldsymbol{L}} = \dot{\boldsymbol{Z}} + \varkappa(\boldsymbol{Z}\boldsymbol{L} - \boldsymbol{L}\boldsymbol{Z}). \tag{109.4}$$

---

[1] According to TRUESDELL [1964, *85*, § 6].

[2] These motions were introduced by COLEMAN [1962, *10* and *11*]. His results concerning them are generalized by those of NOLL [1962, *50*], which we present here.

The formulae of Sect. 25 yield the following results for the stretching $\boldsymbol{D}$, the spin $\boldsymbol{W}$, and the Rivlin-Ericksen tensors $\boldsymbol{A}_n$:

$$\boldsymbol{D} = \frac{\varkappa}{2}(\boldsymbol{N}+\boldsymbol{N}^T), \quad \boldsymbol{W} = \frac{\varkappa}{2}(\boldsymbol{N}-\boldsymbol{N}^T)+\boldsymbol{Z}, \tag{109.5}$$

$$\boldsymbol{A}_n = \varkappa^n \sum_{k=0}^{n} \binom{n}{k} (\boldsymbol{N}^k)^T \boldsymbol{N}^{n-k} = \boldsymbol{Q}(t)\,\boldsymbol{A}_n(0)\,\boldsymbol{Q}(t)^T. \tag{109.6}$$

Turning to the proof of NOLL's theorem, we first introduce the abbreviations

$$\boldsymbol{E}(t) = \boldsymbol{Q}(t)^T \boldsymbol{F}_{(0)}(t), \quad \boldsymbol{H}(\tau) = \boldsymbol{C}_{(0)}(-\tau) = \boldsymbol{C}_{(0)}^{(0)}(\tau). \tag{109.7}$$

It follows from $(23.12)_1$, (108.16), and (109.7) that

$$\boldsymbol{H}(s-t) = \boldsymbol{C}_{(0)}(t-s) = \boldsymbol{E}(t)^T \boldsymbol{H}(s)\,\boldsymbol{E}(t). \tag{109.8}$$

If we differentiate (109.8) with respect to t and then put $t=0$, we obtain

$$-\dot{\boldsymbol{H}}(s) = \boldsymbol{M}^T \boldsymbol{H}(s) + \boldsymbol{H}(s)\,\boldsymbol{M}, \quad \boldsymbol{M} = \dot{\boldsymbol{E}}(0). \tag{109.9}$$

Now, (109.9) is a differential equation for $\boldsymbol{H}(s)$. It must have a unique solution that satisfies the initial condition $\boldsymbol{H}(0)=\boldsymbol{1}$, which holds because of (109.7) and (23.9). We easily verify that this solution is given by

$$\boldsymbol{H}(s) = e^{-s\boldsymbol{M}^T} e^{-s\boldsymbol{M}} = \boldsymbol{C}_{(0)}^{(0)}(s). \tag{109.10}$$

Since the validity of (108.16) was assumed only for $s \geqq 0$, the conclusion (109.10) can first be asserted only for $s \geqq 0$. However, substitution of (109.10) back into (109.8) shows that $\boldsymbol{H}(\tau)$ is an analytic function of τ for *all* values of τ. Hence, by the principle of analytic continuation, (109.10) must also be valid for negative values of s. By (109.10) and $(109.7)_2$ we obtain

$$\boldsymbol{C}_{(0)}(\tau) = \boldsymbol{F}_{(0)}(\tau)^T \boldsymbol{F}_{(0)}(\tau) = e^{\tau \boldsymbol{M}^T} e^{\tau \boldsymbol{M}}, \tag{109.11}$$

which shows that $\boldsymbol{Q}(\tau) = \boldsymbol{F}_{(0)}(\tau)\, e^{-\tau \boldsymbol{M}}$ is orthogonal. Therefore, the assertion (109.1) of the theorem holds with $\varkappa = |\boldsymbol{M}|$, $\boldsymbol{N}_0 = \frac{1}{\varkappa}\boldsymbol{M}$. We note that the orthogonal tensor function $\boldsymbol{Q}(\tau)$ of the representation (109.1) need not be the same as the one in the assumed relation (108.16). However, (108.16) is always valid also for that $\boldsymbol{Q}(\tau)$ for which (109.1) holds, and hence the distinction is inessential. The tensor $\boldsymbol{N}_0$ of (109.1) is uniquely determined by the motion only if the proper numbers of $\boldsymbol{C}_{(0)}^{(0)}(\tau)$, for some value of τ, are distinct[1].

It is clear from the remarks made at the beginning of this section that steady local viscometric flows correspond to the special case when (106.8) holds, i.e., when $\boldsymbol{N}^2 = \boldsymbol{0}$. In this case, (108.17) and (109.6) reduce to

$$\boldsymbol{C}_{(t)}^{(t)}(s) = \boldsymbol{1} - s\varkappa(\boldsymbol{N}+\boldsymbol{N}^T) + s^2\varkappa^2 \boldsymbol{N}^T \boldsymbol{N}, \tag{109.12}$$

$$\boldsymbol{A}_1 = \varkappa(\boldsymbol{N}+\boldsymbol{N}^T), \quad \boldsymbol{A}_2 = 2\varkappa^2 \boldsymbol{N}^T \boldsymbol{N}. \tag{109.13}$$

The remaining Rivlin-Ericksen tensors $\boldsymbol{A}_n$, $n \geqq 3$, all vanish.

For motions with constant stretch history, by (108.17) $\boldsymbol{C}_{(t)}^{(t)}(s)$ is the product of the two entire functions $e^{-s\varkappa \boldsymbol{N}^T}$ and $e^{-s\varkappa \boldsymbol{N}}$ of s. Therefore, $\boldsymbol{C}_{(t)}^{(t)}(s)$ can be a polynomial in s only if $e^{-s\varkappa \boldsymbol{N}}$ is a polynomial in s. But $e^{-s\varkappa \boldsymbol{N}}$ is a polynomial in s only when $\boldsymbol{N}$ is nilpotent, i.e., if $\boldsymbol{N}^m = \boldsymbol{0}$ for some integer $m \geqq 0$. By a theorem of algebra, if $\boldsymbol{N}$ is nilpotent, then necessarily $\boldsymbol{N}^3 = \boldsymbol{0}$. Therefore, if $e^{-s\varkappa \boldsymbol{N}}$ is a

[1] The proof was given by NOLL [1962, *50*, § 2].

polynomial, its degree can be at most 2. Since the Rivlin-Ericksen tensors differ only by numerical factors from the coefficients of the power series expansion of $\boldsymbol{C}_{(t)}^{(t)}(s)$, we obtain the following results for motions with constant stretch history[1]: *If any one of the Rivlin-Ericksen tensors vanishes, then* $\boldsymbol{A}_{\mathfrak{n}}=\mathbf{0}$ *for* $\mathfrak{n}\geqq 5$, *and* $\boldsymbol{N}^3=\mathbf{0}$. *If* $\boldsymbol{A}_4=\mathbf{0}$, *then also* $\boldsymbol{A}_{\mathfrak{n}}=\mathbf{0}$ *for* $\mathfrak{n}\geqq 3$, *and* $\boldsymbol{N}^2=\mathbf{0}$. Thus, referring to the form $(109.1)_1$, we obtain the following classification of motions with constant stretch history:

1. Steady local viscometric flows: $\boldsymbol{N}_0^2=\mathbf{0}$. For these flows $\boldsymbol{A}_{\mathfrak{n}}=\mathbf{0}$ if $\mathfrak{n}\geqq 3$.
2. Flows in which $\boldsymbol{N}_0^3=\mathbf{0}$ but $\boldsymbol{N}_0^2\neq\mathbf{0}$. For these flows $\boldsymbol{A}_{\mathfrak{n}}=\mathbf{0}$ if $\mathfrak{n}\geqq 5$, but $\boldsymbol{A}_4\neq\mathbf{0}$.
3. Flows in which $\boldsymbol{N}_0$ is not nilpotent.

Flows with constant stretch history are of major importance because in them *the memory of a simple fluid, no matter how elaborate it may be, is left very little to remember*[2]; in particular, effects commonly associated with "relaxation" cannot occur, for, apart from an inessential rotation, the history of the motion does not appear to change in time for an observer situate upon a fluid particle. For any flow with constant stretch history, the stress is expressible as a function of the parameters defining the flow, namely, $\varkappa$ and $\boldsymbol{N}$. The analysis is parallel to that given in Sect. 106, leading to (106.15). If we put

$$\mathfrak{g}(\varkappa,\boldsymbol{N})=\mathop{\mathfrak{Q}}_{s=0}^{\infty}\left(e^{-\varkappa s\boldsymbol{N}^T}e^{-\varkappa s\boldsymbol{N}}\right), \tag{109.14}$$

we see that the stress is given by[3]

$$\boldsymbol{T}=-p\mathbf{1}+\mathfrak{g}(\varkappa,\boldsymbol{N}), \tag{109.15}$$

where $\mathfrak{g}(\varkappa,\boldsymbol{N})$ is an isotropic function of the (not necessarily symmetric) tensor $\boldsymbol{N}$ of unit magnitude. In the case of a compressible fluid, a dependence on the density ϱ as a parameter is understood. For incompressible fluids, the motion must be isochoric, i.e., $\operatorname{tr}\boldsymbol{D}=0$. It follows from $(109.5)_1$ that a motion with constant stretch history is isochoric if and only if

$$\operatorname{tr}\boldsymbol{N}=0. \tag{109.16}$$

The condition (109.16) is satisfied, in particular, when $\boldsymbol{N}$ is nilpotent.

The formula (109.15) is central, since it subsumes *every exact solution at present known to be valid for all simple fluids.* In particular, the results concerning the stresses in steady viscometric flows given in Sect. 108 could also have been obtained directly from (109.15). Some special flows with constant stretch history that are not viscometric will be mentioned in Sect. 118.

110. Dynamical preliminaries for the exact solutions of special flow problems. In Sects. 106—109 we have determined the form of the stress tensor in a simple fluid subjected to certain classes of flows, but we have not inquired whether these flows be dynamically possible, subject to physically reasonable fields of external body force $\boldsymbol{b}$. If we suppose, as henceforth we shall, that $\boldsymbol{b}$ has a single-valued potential $v(\boldsymbol{x},t)$, the analysis given in Sect. 30 (γ) becomes relevant. All results and formulae derived there for homogeneous incompressible simple materials in

---

[1] NOLL [1962, *50*, § 3], generalizing results of COLEMAN [1962, *10*].
[2] This capital observation, which goes far to explain *why* it was possible to discover the solutions for special flows of this kind by elementary means, is due to COLEMAN [1962, *10*].
[3] NOLL [1962, *50*, § 4].

general may be applied, *a fortiori*, to all homogeneous incompressible simple fluids. For a fluid, the stress system has certain symmetries, since

$$\mathop{\mathfrak{G}}_{s=0}^{\infty}\left(\boldsymbol{F}^{(t)}(s)\right)=\mathop{\mathfrak{Q}}_{s=0}^{\infty}\left(\boldsymbol{G}(s)\right), \tag{110.1}$$

where the functional $\mathfrak{Q}$ is isotropic, and the consequences of this fact furnish the subject of the preceding sections in this chapter. We now bring the two groups of results together.

The argument given at the beginning of Sect. 30 (γ) applies to fluids in particular, leading to the following somewhat stronger result: *For a given motion to be possible in every incompressible simple fluid subject to conservative body force, it must be circulation-preserving.* However, among the motions we have been discussing, few are universally possible in this way. While all of the steady fields and some of the unsteady ones are easily seen to satisfy (30.28) and hence to be indeed circulation-preserving, these fields contain arbitrary functions, and our task is now to determine those functions, if possible, in such a way that the necessary and sufficient condition (30.29) is satisfied for a given $\mathfrak{Q}$. To take a classic and familiar example, all lineal flows (106.1) are circulation-preserving if steady, since they are then accelerationless, but in order to be dynamically possible for the Navier-Stokes fluid the velocity profile $v(x)$ must be a quadratic function of x. As we shall see in the next section, a quadratic velocity profile is not dynamically possible in every simple fluid. The rectilinear *stream-line pattern* is common to all, but in order for this pattern to be possible for a particular fluid, the speeds assigned to the stream-lines must be severely restricted, in a way differing from one fluid to another. Our problem henceforth will be, for lineal flows, to find *those particular fields of speed* that are dynamically possible in the *particular fluid* defined by the response functional $\mathfrak{Q}$. We shall see that for some of the curvilineal flows (e.g. Couette flow and Poiseuille flow) the problem is of the same type. For others (e.g. general helical flow), the flow pattern itself will differ from one fluid to another.

For a flow pattern to be possible in all simple fluids is a rare exception. Even in the Navier-Stokes theory, the flow pattern is generally influenced by the viscosity, a material property, so that the flows mentioned above are in this sense degenerate also in classical hydrodynamics.

It is trivial to remark that any flow dynamically possible, subject to conservative body force, in *all* incompressible simple fluids, *must* satisfy (30.28), since the elastic fluid is a special simple fluid, and KELVIN's theorem yields (30.28) as a necessary condition in this example and hence a necessary condition in general. The condition (30.28), supplemented by boundary conditions and even by assignment of a stream-line pattern, does not determine a single flow, but rather an infinite family. The problems we consider in the following sections are more definite. In each case an infinite family of flows satisfying (30.28) is considered, and it is shown that a *particular member* of this class, determined by the form of the constitutive equation, is dynamically possible for a *particular* simple fluid. Which member it is, varies from one simple fluid to another. As in incompressible simple materials in general [Sect. 30 (γ)], only for homogeneous motions do the members possible turn out to be the same for every fluid. ~~We conjecture that *the only velocity fields that can be produced in every incompressible simple fluid by the action of surface tractions and a pressure field alone are homogeneous*.~~*

b) Special flow problems.

111. Problems for lineal flows. Simple shearing, channel flow, lineal oscillations. We consider now an incompressible simple fluid in lineal flow, described by (106.1). It follows from (106.15), (106.5), and (106.11) that the components of $\boldsymbol{T}_{\mathrm{E}}$, for lineal flow, depend on x and t only. Using this observation, (106.19), and (106.1), we find that the dynamical equations (30.26) reduce to

$$\partial_x T_{\mathrm{E}}\langle 12\rangle-\varrho\,\partial_y\varphi=\varrho\,\partial_t v,\qquad \partial_x T_{\mathrm{E}}\langle 11\rangle-\varrho\,\partial_x\varphi=0,\qquad \partial_z\varphi=0, \tag{111.1}$$

* [The conjecture is false.]

where $\partial_x, \partial_y, \partial_z$ denote the partial derivatives with respect to x, y, z and where ∂_t denotes the spatial time derivative. A simple analysis shows that the equations (111.1) are equivalent to

$$\partial_x T\langle 12\rangle + a(t) = \varrho\,\partial_t v, \qquad T\langle 11\rangle = a(t)\,y + b(t) + \varrho v, \tag{111.2}$$

where $a(t)$ and $b(t)$ are functions of t only.

The equations $(111.2)_1$, (106.20), (106.5), and (106.11) form a complete system of differential and functional equations for the velocity v and the shear stress $T\langle 12\rangle$. In a particular well set problem these equations and the boundary conditions can be expected to determine $T\langle 12\rangle$ and v uniquely. After v has thus been found, the equations (106.21), (106.22), and $(111.2)_2$ can be used to compute the normal stresses $T\langle kk\rangle$.

The normal-stress differences (106.21) and (106.22) depend only on x and t. This fact and $(111.2)_2$ imply that $T\langle 22\rangle$ and $T\langle 33\rangle$ are of the form

$$\left.\begin{aligned} T\langle 22\rangle &= a(t)\,y + f_2(x,t) + \varrho v,\\ T\langle 33\rangle &= a(t)\,y + f_3(x,t) + \varrho v. \end{aligned}\right\} \tag{111.3}$$

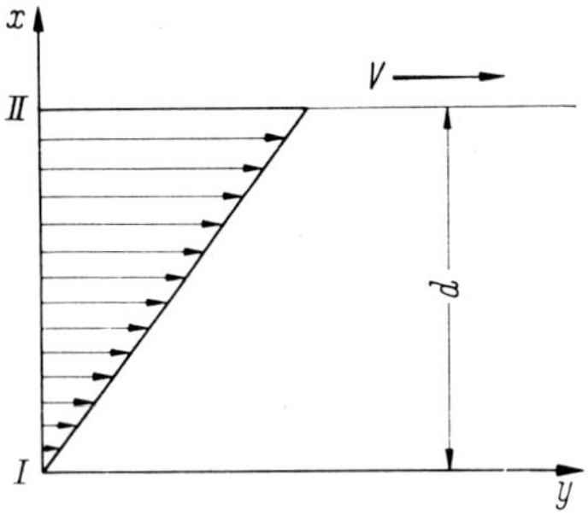

Fig. 10. Simple shearing.

In the case of steady lineal flow, which is a circulation-preserving motion, in fact accelerationless, the dynamical equations (111.2) are equivalent to

$$T\langle 12\rangle = -a x + c, \quad T\langle 11\rangle = a y + b(t) + \varrho v, \tag{111.4}$$

where a and c are constants.

We now investigate a number of particular flow problems that can be solved by use of (111.2), (111.3), and (111.4).

α) Simple shearing[1]. This is a steady lineal flow between an infinite plate I at rest and an infinite plate II moving with constant speed V parallel to plate I. The fluid is assumed to adhere to the plates. Also, it is assumed that there is no external body force and no driving force in the direction of the flow.

We choose the x-axis perpendicular to the plates with $x=0$ at plate I (see Fig. 10). The y-axis and the z-axis are chosen in the plane of plate I in such a way that plate II moves in the y direction. The distance between the plates is denoted by d. The boundary conditions are

$$v(0) = 0, \qquad v(d) = V. \tag{111.5}$$

The absence of driving force and body force means that $T\langle 22\rangle = T\langle yy\rangle$ is independent of y. By (111.3) this is possible only when $a=0$. It follows from (108.4) and $(111.4)_1$ that

$$T\langle 12\rangle = \tau(\varkappa) = c, \qquad \varkappa = \zeta(c) = \text{const.}, \tag{111.6}$$

and hence, by (108.11) and (111.5), that

$$v(x) = \frac{V}{d}\,x, \qquad \varkappa = \frac{V}{d}. \tag{111.7}$$

[1] Within the Reiner-Rivlin theory this problem was first solved by RIVLIN [1948, *14*, § 14] (cf. also the less specific remarks of REINER [1945, *3*, § 6] and an earlier solution by FROMM [1947, *3*, Eq. (5)] for a special fluid of the rate type). It was extended to fluids of the differential type by RIVLIN [1956, *25*, § 2], to simple fluids in general by COLEMAN and NOLL [1959, *2*, § 5]. Cf. also the discussion for certain fluids of the rate type by NOLL [1955, *19*, § 21].

Thus the velocity field has the form

$$\dot{x}=0, \quad \dot{y}=\varkappa x, \quad \dot{z}=0, \tag{111.8}$$

and hence it is a simple shearing as defined in CFT, Sect. 89.

According to (108.4)—(108.6) and $(111.4)_2$ the stress components are given by

$$\left.\begin{aligned} T\langle xy\rangle &= \tau(\varkappa), \qquad T\langle xx\rangle = b(t)+\varrho v, \\ T\langle yy\rangle &= \sigma_2(\varkappa)-\sigma_1(\varkappa)+b(t)+\varrho v, \\ T\langle zz\rangle &= -\sigma_1(\varkappa)+b(t)+\varrho v. \end{aligned}\right\} \tag{111.9}$$

Eq. $(111.9)_1$ gives the shearing traction per unit area that must be applied to the plates to produce the flow. The relation between this traction and the shearing $\varkappa=\frac{V}{d}$ is given by the shear-stress function, τ. Eqs. $(111.9)_{2,3,4}$ show that *shearing tractions alone are insufficient to produce a shearing flow between two plates; in addition, a stress normal to the plates must be applied.* The stress must exceed the normal stress in the direction of flow by the amount $\sigma_1(\varkappa)-\sigma_2(\varkappa)$ and the normal stress in the direction perpendicular to both plates and flow by the amount $\sigma_1(\varkappa)$. Cf. the more general and far less specific result (30.46) valid for arbitrary incompressible simple materials.

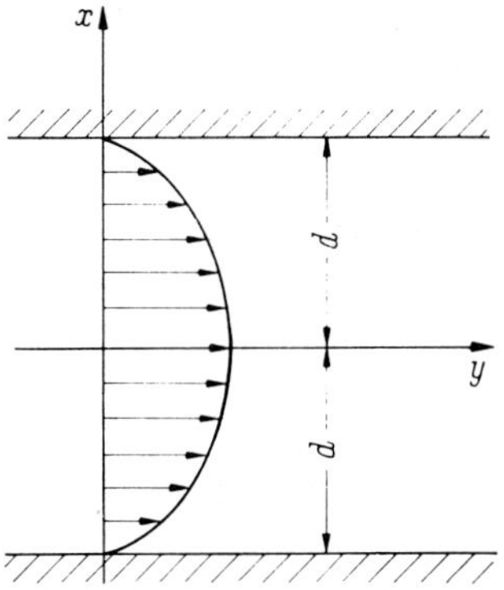

Fig. 11. Channel flow.

The necessity for the normal tractions is the simplest example of a normal-stress effect in non-linear viscosity. It shows some formal similarity but deep conceptual contrast to the Poynting effect in elasticity (Sects. 54—55). In particular, the presence or absence of the effect is not necessarily associated with rapid shearing: It is perfectly possible for a fluid to be such that $\sigma_1\equiv\sigma_2\equiv 0$, and then the effect is absent at all rates of shear. Such is the case in the classical theory of linearly viscous fluids but is not peculiar to it. The response functional $\mathfrak{Q}$ in (105.1) may be of very elaborate nature, including complicated memory effects, yet of such kind as to yield normal-stress functions that vanish identically. However, it is normal-stress effects that offer the simplest and most dramatic instances of demonstrably non-linear response, and most of the recent theoretical and experimental work on non-linear fluids concerns them, as we shall see in the rest of this subchapter.

β) Flow through a channel[1]. A steady lineal flow between two parallel infinite walls which are both at rest is called *channel flow*. The fluid is assumed to adhere to the walls. Setting again $x^1=x$, $x^2=y$, $x^3=z$, we place the x-axis perpendicular to the walls and the y and z axes in a plane halfway between the walls, with the y-axis parallel to the direction of flow (see Fig. 11). Let $2d$ be the distance between the walls. The boundary conditions are

$$v(+d)=v(-d)=0. \tag{111.10}$$

[1] Within the Reiner-Rivlin theory this problem may be solved by use of formulae given by RIVLIN [1948, *14*, § 14]. Later [1956, *25*, § 2] he extended the solution to fluids of the differential type. In the present degree of generality, it was given by COLEMAN and NOLL [1959, *2*, § 6]. Cf. also the discussion for certain fluids of the rate type by NOLL [1955, *19*, § 21].

From (108.4), (108.11), and $(111.4)_1$ we obtain $v'(x)=\zeta(-ax+c)$. Since ζ is an odd function, this conclusion is compatible with (111.10) only when $c=0$. Hence

$$\varkappa = v'(x) = -\zeta(ax). \tag{111.11}$$

Integration gives the *velocity profile* $v(x)$ *in terms of the shearing function*:

$$v(x) = \int_x^d \zeta(a\xi)\, d\xi. \tag{111.12}$$

The function $v=v(x)$ is even.

The total applied force f in the direction of the flow, exerted on a column of fluid with cross-section $\mathscr{A}$ and lying between two planes $y=y_1$ and $y=y_2$, is given by

$$f = \int_{\mathscr{A}} \{T\langle yy\rangle - \varrho v|_{y_2} - T\langle yy\rangle - \varrho v|_{y_1}\}\, dA. \tag{111.13}$$

By (111.3) we have

$$T\langle yy\rangle - \varrho v|_{y_2} - T\langle yy\rangle - \varrho v|_{y_1} = a(y_2 - y_1). \tag{111.14}$$

Hence (111.13) gives

$$f = a(y_2 - y_1)A, \tag{111.15}$$

where A is the area of $\mathscr{A}$. Since $A(y_2-y_1)$ is the volume of the column of fluid considered, it follows that *the constant a is the specific driving force,* i.e. the applied force per unit volume in the direction of the flow.

The volume discharge per unit time through a cross-section of the channel of unit depth is

$$Q = \int_{-d}^{+d} v(x)\, dx. \tag{111.16}$$

Substitution of (111.12) into (111.16) yields, after integration by parts and use of (111.10),

$$Q(a) = \frac{2}{a^2}\int_0^{ad} \xi\zeta(\xi)\, d\xi. \tag{111.17}$$

This formula gives *the discharge Q as a function of the specific driving force a if the shearing function ζ or, equivalently, the shear-viscosity function* (108.10), *is known.* Both Q and a are accessible to measurement. Equation (111.17) is easily solved for the function ζ in terms of the function $Q(a)$ when d is held fixed:

$$\zeta(ad) = \frac{1}{2ad^2}\frac{d}{da}[a^2 Q(a)]. \tag{111.18}$$

This formula may be used *to calculate the shearing function ζ and hence the shear-stress function τ and the viscosity function $\tilde{\mu}$ from an experimental determination of the discharge function $Q=Q(a)$.*

The normal stresses may be determined from (108.5), (108.6), $(111.4)_2$, and (111.11).

γ) Lineal oscillation[1]. An interesting unsteady lineal flow (106.1) is a *lineal oscillation.* It is defined by requiring that v satisfy the condition

$$v(x, t+\theta) = -v(x, t) \tag{111.19}$$

for some constant θ. The flow velocity is then periodic in time with period 2θ. The flow speed $|v|$ has period θ. It follows from (111.19) and (106.5) that

$$k(\tau+\theta) = -k(\tau) + \int_0^\theta v'(x, \sigma)\, d\sigma. \tag{111.20}$$

[1] COLEMAN and NOLL [1961, *5*, § 5].

Combining this result with (106.11), (106.20)–(106.22), and (106.27) we find that

$$T\langle xy\rangle|_{t+\theta} = -T\langle xy\rangle|_t, \tag{111.21}$$

$$\left.\begin{aligned} [T\langle xx\rangle - T\langle zz\rangle]|_{t+\theta} &= [T\langle xx\rangle - T\langle zz\rangle]|_t, \\ [T\langle yy\rangle - T\langle zz\rangle]|_{t+\theta} &= [T\langle yy\rangle - T\langle zz\rangle]|_t. \end{aligned}\right\} \tag{111.22}$$

Eq. (111.21) shows that *the shear stress is periodic with the same frequency as the velocity field*, while Eqs. (111.22) show that *the normal-stress differences oscillate with a frequency double that of the velocity field.* We note that harmonic oscillations cannot be expected to be possible at all for conservative body forces unless the functional $\mathfrak{t}$ of (106.20) is of very special form. The treatment of boundary-value problems for lineal oscillations appears to be difficult.

112. Helical flows, I. General equations.

α) General helical flows. We use cylindrical co-ordinates

$$x^1 = r, \quad x^2 = \theta, \quad x^3 = z, \tag{112.1}$$

and consider a flow having the contravariant velocity components

$$\dot r = 0, \quad \dot\theta = \omega(r, t), \quad \dot z = u(r, t), \tag{112.2}$$

where

$$\omega'(r, t) = f(r)\, q(r, t), \qquad u'(r, t) = h(r)\, q(r, t). \tag{112.3}$$

Such a flow is called *helical* because, in general, the streamlines are helices. Since $g_{11} = g_{rr} = 1$, $g_{22} = g_{\theta\theta} = r^2$, $g_{33} = g_{zz} = 1$, it is clear that the conditions (I), (II), (III) of Sect. 107 are satisfied. Hence a helical flow is a curvilineal flow.

Equations $(107.9)_2$ and (107.10) become

$$\varkappa = \sqrt{r^2 f(r)^2 + h(r)^2}, \tag{112.4}$$

$$\left.\begin{aligned} \alpha &= \frac{r}{\varkappa} f(r) = \frac{r f(r)}{\sqrt{r^2 f(r)^2 + h(r)^2}}, \\ \beta &= \frac{1}{\varkappa} h(r) = \frac{h(r)}{\sqrt{r^2 f(r)^2 + h(r)^2}}. \end{aligned}\right\} \tag{112.5}$$

It then follows from (107.15)–(107.18) that the physical components $T_E\langle km\rangle$ of the extra stress are functions of r and t only. Using this result and the formulae (6A.3) and (6A.4), we find that the dynamical equations (30.26) reduce to

$$\left.\begin{aligned} \partial_r T_E\langle rr\rangle + \frac{1}{r}(T_E\langle rr\rangle - T_E\langle\theta\theta\rangle) - \varrho\,\partial_r\varphi &= -\varrho r\omega^2, \\ r\,\partial_r T\langle r\theta\rangle + 2T\langle r\theta\rangle - \varrho\,\partial_\theta\varphi &= \varrho r^2\partial_t\omega, \\ \partial_r T\langle rz\rangle + \frac{1}{r} T\langle rz\rangle - \varrho\,\partial_z\varphi &= \varrho\,\partial_t u. \end{aligned}\right\} \tag{112.6}$$

A simple analysis shows that Eqs. (112.6) are satisfied if and only if

$$\partial_r(r^2 T\langle r\theta\rangle) = -r d(t) + \varrho r^3\partial_t\omega, \tag{112.7}$$

$$\partial_r(r T\langle rz\rangle) = -r a(t) + \varrho r\,\partial_t u, \tag{112.8}$$

$$T\langle rr\rangle = \varrho v + k(r, t) + a(t) z + d(t)\theta, \tag{112.9}$$

$$\partial_r k(r, t) + \frac{1}{r}(T\langle rr\rangle - T\langle\theta\theta\rangle) = -\varrho r\omega^2. \tag{112.10}$$

In a particular problem one would first use the equations (112.7), (112.8), (107.15), the boundary conditions, and the initial conditions to determine the velocity profile $\omega(r,t)$, $u(r,t)$ and the shearing stresses $T\langle r\theta\rangle$ and $T\langle rz\rangle$. The remaining stress components could then be computed from (107.16), (107.17) and (112.9), (112.10).

A *steady helical flow* is circulation-preserving, so we may find the stresses by the method of Sect. 30. Alternatively, Eqs. (112.7) and (112.8) can be integrated:

$$T\langle r\theta\rangle=\left(\frac{c}{r^2}-\frac{d}{2}\right),\qquad T\langle rz\rangle=\left(\frac{b}{r}-\frac{r\,a}{2}\right). \tag{112.11}$$

Here, a, b, c, and d are constants. After squaring and adding the two equations (108.13) we find from (112.11) and $(107.10)_3$ that

$$\tau(\varkappa)=\gamma,\qquad \varkappa=\zeta(\gamma), \tag{112.12}$$

where

$$\gamma=\gamma(r)=\sqrt{\left(\frac{c}{r^2}-\frac{d}{2}\right)^2+\left(\frac{b}{r}-\frac{r\,a}{2}\right)^2}. \tag{112.13}$$

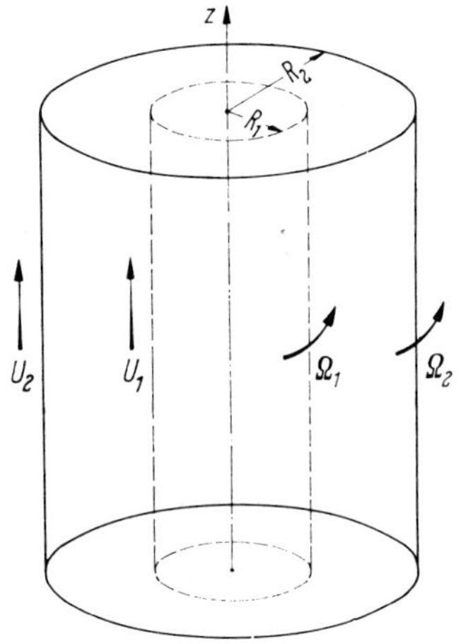

Fig. 12. Flow between concentric cylinders.

Using (108.13), (107.10), and (112.12), we obtain

$$\left.\begin{aligned}\alpha&=\alpha(r)=\left(\frac{c}{r^2}-\frac{d}{2}\right)\frac{1}{\gamma},\\ \beta&=\beta(r)=\left(\frac{b}{r}-\frac{r\,a}{2}\right)\frac{1}{\gamma},\end{aligned}\right\} \tag{112.14}$$

$$\left.\begin{aligned}f&=\omega'(r)=\frac{1}{r}\left(\frac{c}{r^2}-\frac{d}{2}\right)\frac{\zeta(\gamma)}{\gamma},\\ h&=u'(r)=\left(\frac{b}{r}-\frac{r\,a}{2}\right)\frac{\zeta(\gamma)}{\gamma}.\end{aligned}\right\} \tag{112.15}$$

For the normal stresses we find from (108.15), (112.9), (112.10), and (112.12):

$$T\langle rr\rangle-T\langle zz\rangle=\hat{\sigma}_1(\gamma)-\beta^2\hat{\sigma}_2(\gamma), \tag{112.16}$$

$$T\langle\theta\theta\rangle-T\langle zz\rangle=(\alpha^2-\beta^2)\hat{\sigma}_2(\gamma), \tag{112.17}$$

$$\left.\begin{aligned}T\langle rr\rangle=\varrho v+\int\left\{\frac{1}{r}\left[\alpha^2\hat{\sigma}_2(\gamma)-\hat{\sigma}_1(\gamma)\right]-\varrho r\omega(r)^2\right\}dr+{}\\ +az+d\theta+g(t),\end{aligned}\right\} \tag{112.18}$$

where the *modified normal stress functions* $\hat{\sigma}_1$ and $\hat{\sigma}_2$ are defined by

$$\hat{\sigma}_\Gamma(\gamma)=\sigma_\Gamma(\zeta(\gamma)),\qquad \Gamma=1,2. \tag{112.19}$$

β) Steady flows between coaxial cylinders[1]. We consider a steady helical flow between two infinite coaxial cylinders of radii R_1 and R_2, where $R_1<R_2$, which rotate about their axis with angular velocities Ω_1 and Ω_2 and slide in the direction of their axis with velocities U_1 and U_2, respectively (see Fig. 12). Assuming that

[1] A partial analysis for these flows, in the special case of a fluid of differential type, was given by RIVLIN [1956, *25*, § 4]. Cf. also FREDRICKSON [1960, *24*]. The results presented here slightly generalize those obtained by COLEMAN and NOLL [1959, *4*]. Graphical representations of some aspects of helical flow, with a view to experimental test, were described by TANNER [1963, *68*].

the fluid adheres to the cylinders, we have the boundary conditions

$$\omega(R_1)=\Omega_1, \quad u(R_1)=U_1, \quad \omega(R_2)=\Omega_2, \quad u(R_2)=U_2. \tag{112.20}$$

The constant d in (112.11)—(112.18) must vanish because the radial stress (112.18) is a single-valued function of position. By $(112.11)_1$ the moment M per unit height exerted on the fluid inside the cylinder $r=\text{const.}$ is $M=(2\pi r)\,r\,T\langle r\theta\rangle = 2\pi c$ and hence is independent of r. Thus the constant

$$c=\frac{M}{2\pi} \tag{112.21}$$

is determined by the torque M required to maintain the relative rotation of the bounding cylinders. An argument analogous to the one given in Sect. 111 (β) shows that the constant a in (112.11)—(112.18) is the specific driving force, i.e. the applied force per unit volume in the direction of the axis of the cylinders. Integrating $(112.15)_2$ and using $(112.20)_{2,4}$, we get

$$U_2-U_1=\int_{R_1}^{R_2}\left(\frac{b}{r}-\frac{r\,a}{2}\right)\frac{\zeta(\gamma(r))}{\gamma(r)}\,dr, \tag{112.22}$$

where

$$\gamma=\gamma(r)=\sqrt{\left(\frac{M}{2\pi r^2}\right)^2+\left(\frac{b}{r}-\frac{r\,a}{2}\right)^2}. \tag{112.23}$$

If we assume that (112.22) can be solved for the constant b:

$$b=F(U_2-U_1,\, M,\, a), \tag{112.24}$$

then the constants a, b, c, d are all expressed in terms of quantities which seem possible to measure.

The velocity profile can be found by integrating (112.15) and using $(112.20)_{1,2}$:

$$\omega(r)=\frac{M}{2\pi}\int_{R_1}^{r}\frac{\zeta(\gamma(\xi))}{\xi^3\gamma(\xi)}\,d\xi+\Omega_1, \tag{112.25}$$

$$u(r)=\int_{R_1}^{r}\left(\frac{b}{\xi}-\frac{\xi\,a}{2}\right)\frac{\zeta(\gamma(\xi))}{\gamma(\xi)}\,d\xi+U_1. \tag{112.26}$$

The boundary condition $(112.20)_3$ gives

$$\Delta\Omega=\Omega_2-\Omega_1=\frac{M}{2\pi}\int_{R_1}^{R_2}\frac{\zeta(\gamma(r))}{r^3\gamma(r)}\,dr, \tag{112.27}$$

which relates the angular velocity difference $\Delta\Omega$ of the bounding cylinders to the torque M and driving force a which produce the motion. The volume discharge Q per unit time through a cross section perpendicular to the cylinders is easily obtained from (112.26):

$$Q=\pi\left[R_2^2U_2-R_1^2U_1+\int_{R_1}^{R_2}\left(\frac{r^3a}{2}-r\,b\right)\frac{\zeta(\gamma(r))}{\gamma(r)}\,dr\right]. \tag{112.28}$$

The formula (112.18) with $(112.14)_1$ yields the following expression for the difference $\Delta T\langle rr\rangle = T\langle rr\rangle|_{R_2}-T\langle rr\rangle|_{R_1}$ of the normal stresses at the outer and inner cylinders:

$$\Delta T\langle rr\rangle=\varrho v\Big|_{R_1}^{R_2}+\int_{R_1}^{R_2}\left\{\left(\frac{M}{2\pi}\right)^2\frac{\hat\sigma_2(\gamma(r))}{\gamma(r)^2r^5}-\frac{\hat\sigma_1(\gamma(r))}{r}-\varrho r\omega(r)^2\right\}dr. \tag{112.29}$$

The formulae (112.27)—(112.29) give expressions for quantities that should be possible to measure in appropriate experiments.

113. Helical flows, II. Special cases: flows in circular pipes. In this section we consider a number of special helical flows, unsteady and steady, that are likely to be of use in viscometric and normal stress experiments.

α) Unsteady flow between coaxial cylinders. A helical flow between two infinite coaxial cylinders of radii R_1 and $R_2 (R_1 < R_2)$ is called "Couette flow" under the following circumstances: (i) The velocity component $u(r, t)$ in the direction of the cylinder axis is zero. (ii) The fluid adheres to the boundary cylinders, which rotate about their common axis with angular velocities $\Omega_1(t)$ and $\Omega_2(t)$.

The second condition means that

$$\omega(R_1, t) = \Omega_1(t), \qquad \omega(R_2, t) = \Omega_2(t). \tag{113.1}$$

The first condition implies that $h(r) = 0$ in (112.3) and that (112.4), (112.5) reduce to $\varkappa = r f(r)$, $\alpha = 1$, $\beta = 0$. We can adjust $q(r, t)$ in $(112.3)_1$ so that $f(r) = 1$, $q(r, t) = \omega'(r, t)$. Equation $(107.9)_1$ then becomes

$$k(\tau) = r \int_0^\tau \omega'(r, \sigma)\, d\sigma. \tag{113.2}$$

In the present case the $T\langle km\rangle$ in (106.20)—(106.22) coincide with the physical components, and we obtain

$$T\langle r\theta\rangle = \underset{s=0}{\overset{\infty}{\mathfrak{t}}}\big(g(s)\big), \qquad T\langle rz\rangle = T\langle \theta z\rangle = 0, \tag{113.3}$$

$$T\langle rr\rangle - T\langle zz\rangle = \underset{s=0}{\overset{\infty}{\mathfrak{s}_1}}\big(g(s)\big), \qquad T\langle \theta\theta\rangle - T\langle zz\rangle = \underset{s=0}{\overset{\infty}{\mathfrak{s}_2}}\big(g(s)\big), \tag{113.4}$$

where $g(s) = k(t-s) - k(t)$, by Eq. (106.11). In (112.7) we must put $d(t) = 0$ because the radial stress (112.9) is single-valued (we assume v to be single-valued). Substituting $(113.3)_1$ into (112.7), we obtain

$$-\,\partial_r\Big[r^2 \underset{s=0}{\overset{\infty}{\mathfrak{t}}}\Big(r\int_0^s \omega'(r, t-\sigma)\, d\sigma\Big)\Big] = \varrho r^3 \partial_t \omega(r, t). \tag{113.5}$$

Under suitable assumptions about the functional $\mathfrak{t}$, we expect that the functional equation (113.5), the boundary conditions (113.1), and suitable initial conditions will determine $\omega = \omega(r, t)$ uniquely.

The difference $\Delta T\langle rr\rangle = T\langle rr\rangle|_{R_2} - T\langle rr\rangle|_{R_1}$ of the normal stresses at the outer and inner cylinders can be obtained from (112.9), where $a(t) = d(t) = 0$, and from (112.10) and (113.4):

$$\Delta T\langle rr\rangle = \varrho v\Big|_{R_1}^{R_2} + \int_{R_1}^{R_2}\Big\{\frac{1}{r}\Big[\underset{s=0}{\overset{\infty}{\mathfrak{s}_2}}\big(g(s)\big) - \underset{s=0}{\overset{\infty}{\mathfrak{s}_1}}\big(g(s)\big)\Big] - \varrho r\omega(r, t)^2\Big\}\, dr. \tag{113.6}$$

Further analysis of the equations (113.3)—(113.6) does not seem possible unless special assumptions on the flow or the functionals $\mathfrak{t}$, $\mathfrak{s}_1$, $\mathfrak{s}_2$ be made[1].

[1] For certain special theories, and for the case of small oscillations, the problem has been discussed by OLDROYD [1951, *9*, § 3], MARKOVITZ [1952, *14*, § III A], JAIN [1960, *29*], and WALTERS [1961, *68*]. MARKOVITZ [1957, *13*] was the first to give the general results in terms of the viscometric functions, in the case when inertia is neglected. See also Sect. 123.

β) Steady flow between rotating cylinders ("Couette flow")[1]. This is a steady coaxial cylinder flow for which

$$U_1 = U_2 = a = 0 \tag{113.7}$$

in the equations of Sect. 112 (β). Equations (112.22) and (112.23) then give

$$b = 0, \quad \gamma = \frac{M}{2\pi r^2}. \tag{113.8}$$

According to (112.27) the angular velocity difference $\Delta\Omega = \Omega_2 - \Omega_1$ is related to torque M by[2]

$$\Delta\Omega = \int_{R_1}^{R_2} \frac{1}{r}\,\zeta\left(\frac{M}{2\pi r^2}\right) dr. \tag{113.9}$$

It is possible to invert (113.9) so as to calculate the shearing function ζ when $\Delta\Omega$ is known as a function of M. The result is[3]

$$\zeta\left(\frac{M}{2\pi R_1^2}\right) = \sum_{n=0}^{\infty} F\left(\left(\frac{R_1}{R_2}\right)^{2n} M\right), \tag{113.10}$$

where the function F is defined by

$$F(M) = 2M\,\frac{d\,\Delta\Omega}{d\,M}. \tag{113.11}$$

For steady Couette flow, the normal-stress difference (112.29) becomes

$$\Delta T_{\langle rr\rangle} = \varrho\, v\Big|_{R_1}^{R_2} + \int_{R_1}^{R_2} \left\{\frac{1}{r}\left[-\hat{\sigma}_1\left(\frac{M}{2\pi r^2}\right) + \hat{\sigma}_2\left(\frac{M}{2\pi r^2}\right)\right] - \varrho\, r\, \omega(r)^2\right\} dr. \tag{113.12}$$

By exactly the method used in deriving (113.10), it can be shown that[4]

$$\hat{\sigma}_2\left(\frac{M}{2\pi R_1^2}\right) - \hat{\sigma}_1\left(\frac{M}{2\pi R_1^2}\right) = \sum_{n=0}^{\infty} G\left(\left(\frac{R_1}{R_2}\right)^{2n} M\right), \tag{113.13}$$

where

$$G(M) = 2M\,\frac{d}{dM}\left(\Delta T_{\langle rr\rangle} + \int_{R_1}^{R_2} \varrho\, r\, \omega(r)^2\, dr\right). \tag{113.14}$$

When the gap between the two cylinders is very small, i.e., when

$$\frac{R_2 - R_1}{R_1} \ll 1,$$

---

[1] In the Reiner-Rivlin theory this problem was first solved by RIVLIN [1948, *14*, § 16]. The generalization to fluids of the differential type is included in his later results [1956, *25*, § 4]. For other special theories, steady Couette flow was considered by OLDROYD [1950, *11*, § 4] [1958, *32*, § 4], NOLL [1955, *19*, § 23], and DEWITT [1955, *9*]. In the generality presented here, the solution was first obtained by COLEMAN and NOLL [1959, *2*, § 8].

[2] This formula was derived by MOONEY [1931, *5*, § 3] from certain special assumptions. Earlier special work by REINER and RIWLIN [1927, *2*] and by FARROW, LOWE and NEALE [1928, *2*] suggests that an equivalent result must have been known to them.

[3] KRIEGER and ELROD [1953, *14*], PAWLOWSKI [1953, *22*]. See also COLEMAN and NOLL [1959, *2*, § *8*].

[4] MARKOVITZ, to be published.

the formulae (113.9) and (113.12) may be replaced by the following approximate formulae:

$$\Delta\Omega \approx \frac{R_2 - R_1}{R_1}\,\zeta\left(\frac{M}{2\pi R_1^2}\right), \tag{113.15}$$

$$\Delta T\langle rr\rangle \approx \frac{R_2 - R_1}{R_1}\left[\hat{\sigma}_2\left(\frac{M}{2\pi R_1^2}\right) - \hat{\sigma}_1\left(\frac{M}{2\pi R_1^2}\right) - \varrho R_1^2 \Omega_1 \Omega_2\right]. \tag{113.16}$$

For (113.16) it is assumed that v does not depend on r. The error in both (113.15) and (113.16) is of the order of $[(R_2 - R_1)/R_1]^2$. When the gap is small, the formula (113.15) may be used directly to find ζ after $\Delta\Omega$ has been determined experimentally as a function of M. Use of the formulae (113.10) and (113.13) is practicable only when the gap is large, because only in thi scase do the series converge rapidly. The formula (113.16) may be used to calculate the normal-stress function $\hat{\sigma}_2 - \hat{\sigma}_1$ from an experimental determination of $\Delta T\langle rr\rangle$ as a function of the torque M.

An apparatus for modern viscometric measurements on Couette flow has been described by Padden and DeWitt[1].

γ) Unsteady flow between sliding coaxial cylinders[2]. We consider a helical flow between two coaxial circular cylinders of radii R_1 and R_2 $(R_1 < R_2)$, and make the following assumptions:

(i) The angular velocity component $\omega(r, t)$ vanishes.

(ii) The specific driving force in the direction of the axis is zero.

(iii) The fluid adheres to the cylinders, which slide in the direction of their axis with velocities $U_1(t)$ and $U_2(t)$ respectively.

The third condition means that

$$u(R_1, t) = U_1(t), \qquad u(R_2, t) = U_2(t). \tag{113.17}$$

The first condition implies that $f(r) = 0$ in (112.3) and that (112.4), (112.5) reduce to $\varkappa = h(r)$, $\alpha = 0$, $\beta = 1$, so that

$$k(\tau) = \int_0^\tau u'(r, \sigma)\, d\sigma. \tag{113.18}$$

Equations (107.15)—(107.18) give

$$T\langle rz\rangle = \underset{s=0}{\overset{\infty}{\mathfrak{t}}}\big(g(s)\big), \qquad T\langle r\theta\rangle = T\langle \theta z\rangle = 0, \tag{113.19}$$

$$T\langle rr\rangle - T\langle \theta\theta\rangle = \underset{s=0}{\overset{\infty}{\mathfrak{s}_1}}\big(g(s)\big), \qquad T\langle zz\rangle - T\langle \theta\theta\rangle = \underset{s=0}{\overset{\infty}{\mathfrak{s}_2}}\big(g(s)\big), \tag{113.20}$$

where $g(s) = k(t - s) - k(t)$. Condition (ii) implies that $a(t) = 0$ in (112.8) and (112.9). Substitution of $(113.19)_1$ into (112.8) then yields the fundamental equation

$$-\partial_r\left[r \underset{s=0}{\overset{\infty}{\mathfrak{t}}}\left(\int_0^s u'(r, t - \sigma)\, d\sigma\right)\right] = \varrho\, r\, \partial_t u(r, t). \tag{113.21}$$

As in the case of Couette flow, under suitable assumptions on $\mathfrak{t}$ we expect that a solution $u = u(r, t)$ of (113.21) is uniquely determined by the boundary conditions (113.17) and suitable initial conditions.

---

[1] Padden and DeWitt [1954, *15*]. An improved apparatus of this kind has been constructed and used by Markovitz [1965, *18*]. In work done between 1942 and 1946, Garner, Nissan and Wood [1950, *1*, pp. 58—59] had studied normal-stress effects in Couette flow but had not reached any definite results.

[2] These results, due to Noll, are here published for the first time. The solution of Broer [1956, *3*, § 4] for flow due to an oscillating specific driving force is based on a special theory of the rate type that does not satisfy the principle of material indifference. This problem was approached within the Reiner-Rivlin theory by Narasimhan [1960, *38*, § 4]. See also Sect. 123.

The normal stress difference $\Delta T\langle rr\rangle = T\langle rr\rangle|_{R_2} - T\langle rr\rangle|_{R_1}$ can be obtained from $(113.20)_1$, (112.9) and (112.10), where now $\omega = a(t) = d(t) = 0$:

$$\Delta T\langle rr\rangle = \varrho v \Big|_{R_1}^{R_2} - \int_{R_1}^{R_2} \frac{1}{r} \mathop{\mathfrak{F}_1}_{s=0}^{\infty} \left(\int_0^s u'(r, t-\sigma)\, d\sigma \right) dr. \tag{113.22}$$

δ) Steady flow between coaxial pipes[1]. This is a flow of the type considered in Sect. 112 (β) for which

$$\Omega_1 = \Omega_2 = U_1 = U_2 = M = 0. \tag{113.23}$$

Equations (112.23) and (112.22) reduce to

$$\gamma(r) = \left| \frac{b}{r} - \frac{r a}{2} \right|, \tag{113.24}$$

$$0 = \int_{R_1}^{R_2} \zeta(\gamma(r))\, dr. \tag{113.25}$$

Equation (113.25) can be expected to determine the constant b in terms of the specific driving force a. It follows from (112.26) that the velocity profile is given by

$$u(r) = \int_{R_1}^{r} \zeta(\gamma(\xi))\, d\xi. \tag{113.26}$$

The volume discharge Q per unit time through a cross-section of the pipe is obtained by specialization of (112.28):

$$Q = -\pi \int_{R_1}^{R_2} r^2 \zeta(\gamma(r))\, dr. \tag{113.27}$$

The difference $\Delta T\langle rr\rangle$ of the normal stresses at the outer and inner pipe is obtained by specializing (112.29):

$$\Delta T\langle rr\rangle = \varrho v \Big|_{R_1}^{R_2} - \int_{R_1}^{R_2} \frac{1}{r} \hat{\sigma}_1(\gamma(r))\, dr. \tag{113.28}$$

ε) Flow down a circular pipe ("Poiseuille flow")[2]. This is a special steady helical flow through a fixed infinite circular pipe of radius R. Poiseuille flow may be regarded as the special flow of the type considered in Sect. 112 (β) for which

$$\Omega_1 = \Omega_2 = U_2 = M = 0, \tag{113.29}$$

$$R_1 = 0, \quad R_2 = R. \tag{113.30}$$

The velocity $U_1 = u(0)$ at the axis of the pipe is not prescribed, and the boundary condition $(112.20)_2$ must be replaced by the requirement that the stress components be continuous at the axis of the pipe. It then follows from (112.11) that we must put $b=0$ in (112.23), so that

$$\gamma(r) = \frac{r a}{2}, \tag{113.31}$$

where a is the specific driving force that produces the flow.

[1] Coleman and Noll [1961, *5*, § 4].

[2] In the Reiner-Rivlin theory this problem was first solved by Rivlin [1949, *17*]; the generalization to fluids of the differential type is included in his later work [1956, *25*, § 4]. Earlier, the solution had been obtained by Fromm [1947, *3*, Eq. (20)] for a special fluid of the rate type. For other special theories, the problem was treated by Noll [1955, *19*, § 22] and Oldroyd [1958, *32*, § 3]. For simple fluids, as presented here, the solution is due to Coleman and Noll [1959, *2*, § 7].

From (112.26) and (112.27) we get the velocity profile

$$u(r) = -\int_R^r \zeta\left(\frac{\xi a}{2}\right) d\xi. \tag{113.32}$$

The volume discharge Q per unit time[1] is obtained by specializing (112.28):

$$Q = Q(a) = \pi \int_0^R r^2 \zeta\left(\frac{r a}{2}\right) dr. \tag{113.33}$$

As in the case of channel flow [Sect. 111 (β)], the formula (113.31) can easily be inverted[2]:

$$\zeta\left(\frac{R a}{2}\right) = \frac{1}{\pi a^2 R^3} \frac{d}{d a}\left[a^3 Q(a)\right]. \tag{113.34}$$

Since $Q = Q(a)$ is not hard to measure experimentally, (113.34) should give a convenient way to calculate the shearing function ζ and hence the viscosity function $\eta = \eta(\varkappa) = \frac{\tau(\varkappa)}{\varkappa}$. To the best of our knowledge, adequate measurements of normal stresses in Poiseuille flows have yet to be made. Cf. Sect. 114 (β).

114. Normal-stress end effects. The flows considered in the previous section are bounded by *infinite* cylinders or pipes. Actual cylinders and pipes are of course finite, and we expect that actual flows are described by the equations of the previous section only approximately, with disturbances occurring near the ends of the bounding cylinders and pipes. In this section we consider two end effects which can occur only when the normal-stress functions are not zero. We assume that there is no external body force, so that $v = 0$.

α) *Climbing in a steady Couette flow*[3]. We consider a Couette flow apparatus in which the fluid is in contact at the upper end with an atmosphere of prescribed constant pressure p_0. The normal stress $T\langle zz\rangle$ in the axial direction is obtained from (112.16) and (112.18) where $\alpha = 1$, $\beta = 0$, $a = d = 0$, and $\gamma(r) = \frac{M}{2\pi r^2}$ [cf. (113.7), (113.8)]:

$$T\langle zz\rangle = \int_{R_1}^r \left\{\frac{1}{\xi}\left[\hat{\sigma}_2\left(\frac{M}{2\pi\xi^2}\right) - \hat{\sigma}_1\left(\frac{M}{2\pi\xi^2}\right)\right] - \varrho\xi\omega(\xi)^2\right\} d\xi - \hat{\sigma}_1\left(\frac{M}{2\pi r^2}\right) + g(t). \tag{114.1}$$

The value of $g(t)$ can be determined from the balance of the total force in the axial direction, i.e., from the relation

$$2\pi \int_{R_1}^{R_2} T\langle zz\rangle\, r\, dr = -p_0 \pi (R_2^2 - R_1^2). \tag{114.2}$$

---

[1] Equivalent relations derived from special assumptions occur in the rheological literature. E.g. HERMANS [1953, *12*, § 6.1].

[2] RABINOWITSCH [1928, *4*] attributed this formula to WEISSENBERG.

[3] Indication of the climbing effect in the Reiner-Rivlin theory was first mentioned by REINER [1945, *3*, § 6] in the context of simple shearing; he pronounced the phenomenon as being "against experience" and therefore, so as to prevent its occurrence, imposed a restriction implying that $\sigma_1 \equiv \sigma_2 \equiv 0$. In his work, however, it is not clear that the effect is independent of changes of volume. RIVLIN [1948, *14*, § 16] inferred the necessity for the climbing effect from his solution for Couette flow, and a detailed analysis, still within the Reiner-Rivlin theory, was given by SERRIN [1959, *29*]. Our treatment here generalizes that of ERICKSEN [1959, *7*].

We find that $g=g(t)$ does not depend on t and that it is given by

$$\left.\begin{aligned} g=-p_0+\frac{1}{R_2^2-R_1^2}\int_{R_1}^{R_2}\Big\{\varrho(R_2^2-r^2)\,r\,\omega(r)^2-\frac{R_2^2-r^2}{r}\,\hat{\sigma}_2\Big(\frac{M}{2\pi r^2}\Big)+\\ +\frac{R_2^2-r^2}{r}\,\hat{\sigma}_1\Big(\frac{M}{2\pi r^2}\Big)\Big\}\,dr. \end{aligned}\right\} \tag{114.3}$$

Denote the excess of the atmospheric pressure p_0 over the normal pressure $-T_{\langle zz\rangle}$ in the fluid by

$$N=p_0+T_{\langle zz\rangle}. \tag{114.4}$$

In general, N is not identically zero, and hence the free surface at the upper end of the fluid mass cannot be a plane $z=\text{const}$. The change of N in the radial direction is obtained from (114.1):

$$\left.\begin{aligned} \partial_r N=-\varrho\,r\,\omega(r)^2+\frac{1}{r}\Big[\hat{\sigma}_2\Big(\frac{M}{2\pi r^2}\Big)-\\ -\hat{\sigma}_1\Big(\frac{M}{2\pi r^2}\Big)\Big]+\frac{M}{\pi r^3}\,\hat{\sigma}_1'\Big(\frac{M}{2\pi r^2}\Big). \end{aligned}\right\} \tag{114.5}$$

If the normal stress functions are zero, we have

$$\partial_r N=-\varrho\,r\,\omega(r)^2<0. \tag{114.6}$$

We take this inequality as an indication[1] that the free surface will slope upwards from the inner to the outer cylinder when $\sigma_1\equiv\sigma_2\equiv 0$. The change of N can be positive,

$$\partial_r N>0, \tag{114.7}$$

only when the normal-stress functions are not all zero. If (114.7) holds, we expect that the free surface will slope upwards from the outer to the inner cylinder. In other words, the fluid will tend to climb up the inner cylinder. This phenomenon was observed in actual Couette flow by GARNER and NISSAN[2] in 1942. The climbing effect was publicized in a series of striking demonstrations by WEISSENBERG[3] and is often called "the Weissenberg effect". One of his diagrams is shown as Fig. 14, and Fig. 15 shows his photographs illustrating this effect and related phenomena in torsional flow, the theory of which will be represented in

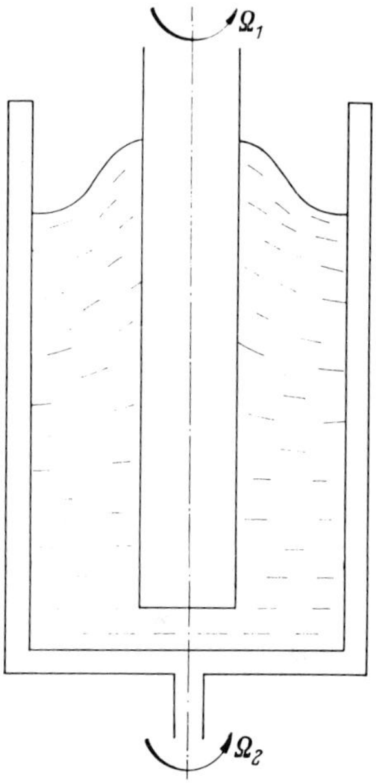

Fig. 13. Climbing in Couette flow.

[1] E.g., if also $\tau(\varkappa)\equiv 0$, the free surface has the classic parabolic form.

[2] GARNER and NISSAN [1946, *2*]. Cf. also WOOD, NISSAN and GARNER [1947, *18*] [1950, *1*, p. 47]; the introduction to the latter paper gives a history of experimental discovery of the climbing and swelling effects. The work of the Birmingham group, directed by GARNER, was done in 1942—1946 but not released for publication until 1949. The method of measuring variations of boundary pressure by inserting standpipes at various points, used in much subsequent apparatus, first appears in these early experiments. Actually, the climbing phenomenon itself was not new, since rotary stirrers had long ago been found inadequate for paints, but it was not understood that simple mechanical principles suffice to explain it. Likewise, the phenomenon of swelling on extrusion from a tube was known in the artificial fibre industry but thought to be of chemical or "plastic" origin or was subsumed under that turgid euphemism for ignorance, "thixotropy".

Other end effects have been discussed by WHITE [1964, *91*, § 5].

[3] WEISSENBERG [1947, *17*] [1949, *26* and *27*] [1950, *17*], FREEMAN and WEISSENBERG [1949, *2*].

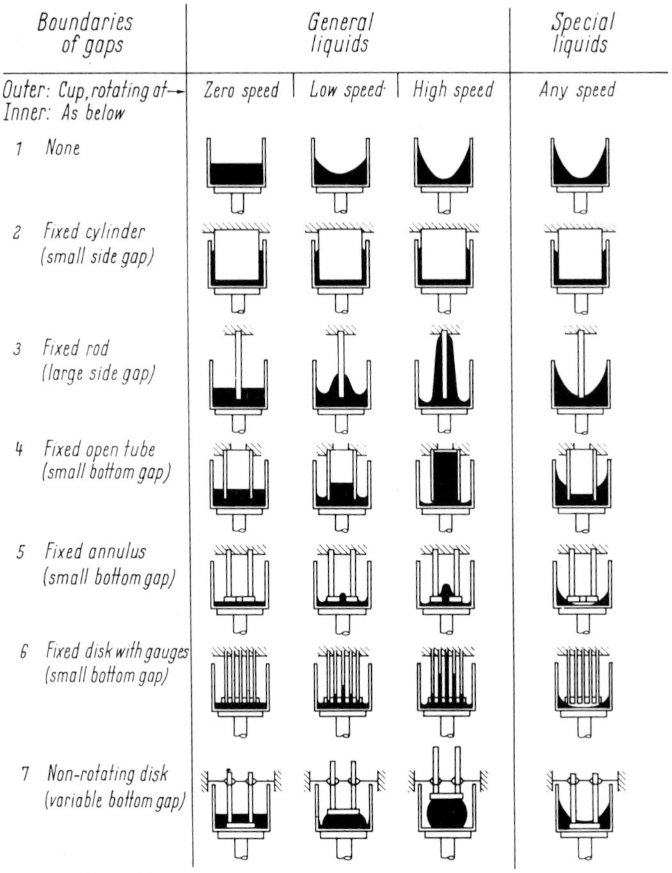

Fig. 14. WEISSENBERG's diagrams of normal-stress end effects (1947).

a Fixed rod. b Fixed open tube.

Fig. 15a–e. Demonstration experiments of WEISSENBERG (1948). (Photographs by courtesy of Dr. WEISSENBERG and the North-Holland Publishing Co.)

Sect. 115. Three later and clearer photographs from his circle are reproduced here as Fig. 16, while Fig. 17 gives some shots from a recent film by MARKOVITZ, in which some details of measurement are made plain.

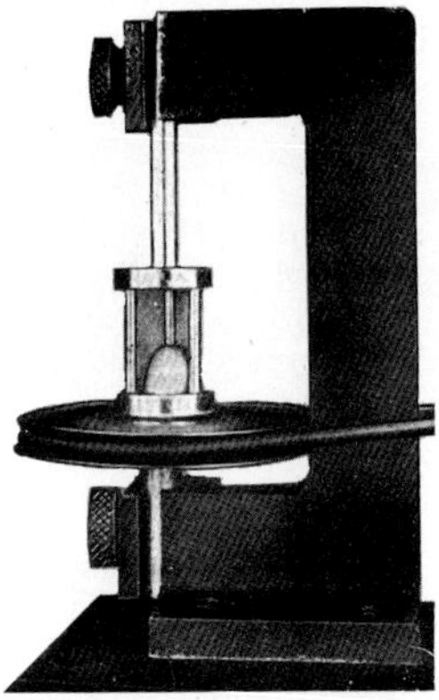

c Fixed annulus.

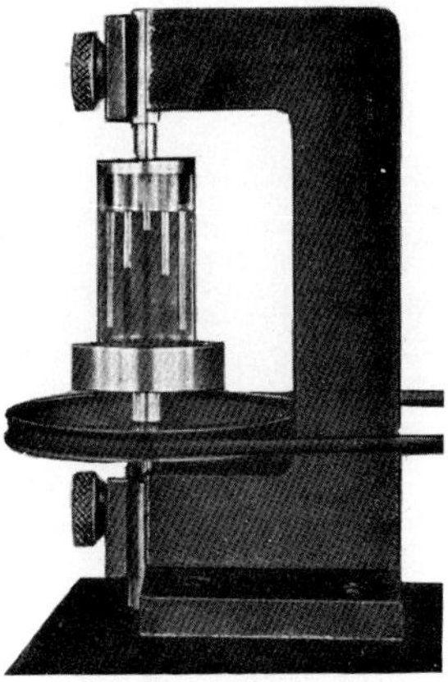

d Fixed disk with gauges.

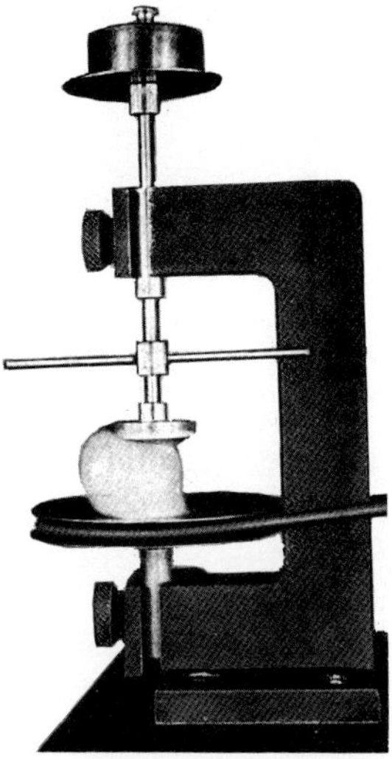

e Non-rotating disk.

Fig. 15c—e.

β) Swelling at exit from a pipe[1]. We consider a fluid which issues from a circular pipe into an atmosphere of prescribed constant pressure p_0. The normal stresses $T_{\langle zz\rangle}$ and $T_{\langle rr\rangle}$ in Poiseuille flow are obtained from (112.16) and (112.18),

[1] The existence of the swelling effect according to the Reiner-Rivlin theory was first mentioned by BRAUN and REINER [1952, *3*, § 4]. A detailed analysis, for this theory, was given by SERRIN [1959, *29*].

where $\alpha=0$, $\beta=1$, $d=0$, $\omega=0$, and $\gamma=\frac{ra}{2}$ [cf. (113.31)]:

$$T_{\langle rr\rangle}=-\int_0^r \frac{1}{\xi}\hat{\sigma}_1\left(\frac{a\xi}{2}\right)d\xi+az+g(t), \tag{114.8}$$

$$T_{\langle zz\rangle}=T_{\langle rr\rangle}+\hat{\sigma}_2\left(\frac{ar}{2}\right)-\hat{\sigma}_1\left(\frac{ar}{2}\right). \tag{114.9}$$

We assume that (114.8) and (114.9) give a good approximation to the real normal stresses even at the exit section, taken to be $z=0$. The value of $g(t)$ can then

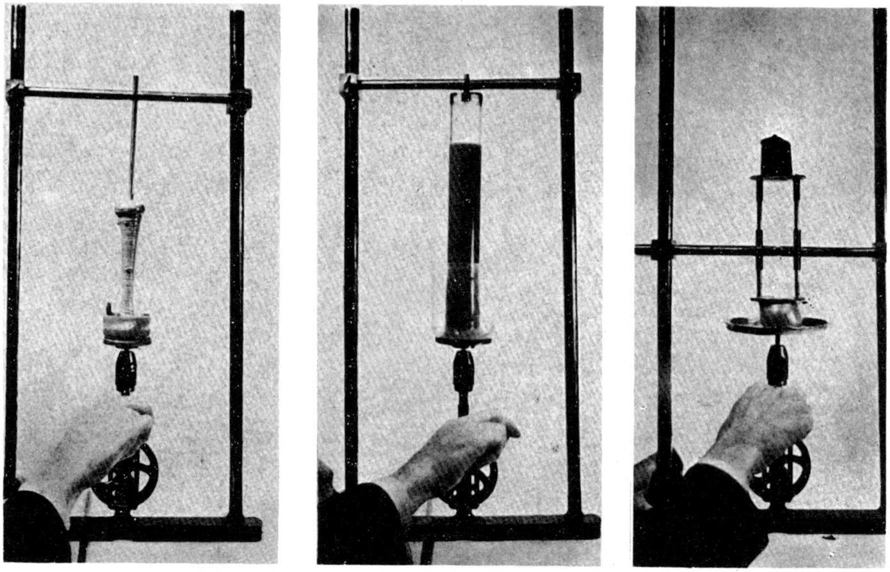

a b c

Fig. 16a—c. Fluid drawn upward by rotation. a Fixed rod. b Fixed open tube. c Non-rotating disk. [Photographs, published by JOBLING and ROBERTS (1958), by courtesy of Dr. WEISSENBERG and Academic Press.]

be computed from the balance of the total axial force at $z=0$, i.e. from

$$2\pi\int_0^R T_{\langle zz\rangle}\Big|_{z=0} r\,dr=-p_0\pi R^2. \tag{114.10}$$

The result is

$$g=-p_0+\frac{1}{R^2}\int_0^R\left[\frac{R^2+r^2}{r}\hat{\sigma}_1\left(\frac{ar}{2}\right)-2r\hat{\sigma}_2\left(\frac{ar}{2}\right)\right]dr. \tag{114.11}$$

Denote the excess of the atmospheric pressure p_0 over the radial pressure $-T_{\langle rr\rangle}$ at the wall of the pipe ($r=R$) by

$$P=p_0+T_{\langle rr\rangle}\big|_{r=R}. \tag{114.12}$$

From (114.8) and (114.11) we find

$$P=az+\frac{1}{R^2}\int_0^R r\left[\hat{\sigma}_1\left(\frac{ar}{2}\right)-2\hat{\sigma}_2\left(\frac{ar}{2}\right)\right]dr. \tag{114.13}$$

If the normal-stress function $\hat{\sigma}_1 - 2\hat{\sigma}_2$ is not zero, we can expect the excess pressure P at the exit section $z=0$ to be different from zero. We take this result as an indication that a stream issuing from the tube will swell or thin immediately after exit according as $P|_{z=0} < 0$ or $P|_{z=0} > 0$. Swelling will occur, in particular, if

$$2\hat{\sigma}_2(\gamma) - \hat{\sigma}_1(\gamma) > 0 \tag{114.14}$$

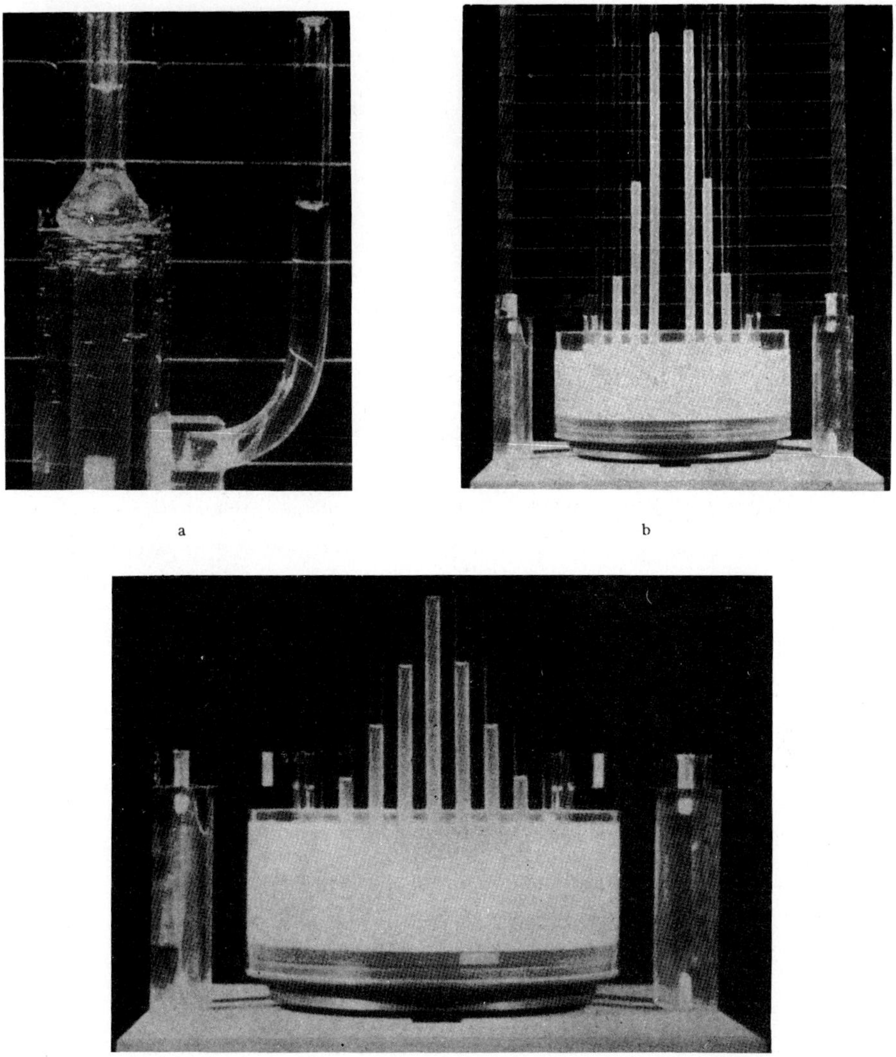

Fig. 17a—c. Photographs by courtesy of Dr. MARKOVITZ from the film "Rheological Behavior of Fluids", produced by Educational Services, Inc., Watertown, Mass., for the National Committee for Fluid Mechanics Films under grant from the U.S. National Science Foundation.

a The difference in normal thrusts on the bounding cylinder in the steady Couette flow of an 8.5% solution of polyisobutylene in decalin. The thrust on the inner (rotating) cylinder is indicated by the level of a surface of the fluid inside the inner tube. The thrust on the outer (stationary) cylinder is indicated by the level of the fluid in the side arm.

b The distribution of tractions in the cone-and-plate flow of a 2% solution of poly-(ethylene oxide) in water is indicated by the levels of the fluid in the vertical tubes.

c Same as (b), but torsional flow.

when $0<\gamma<\frac{1}{2}aR$. The swelling effect was first reported in actual flows by Merrington[1]; Fig. 18 includes his famous photograph of it. Since no thinning effect has been observed, the inequality (114.14) seems to hold in practice.

We note that it should be possible to check (114.13) quantitatively by measuring P at the wall of the pipe.

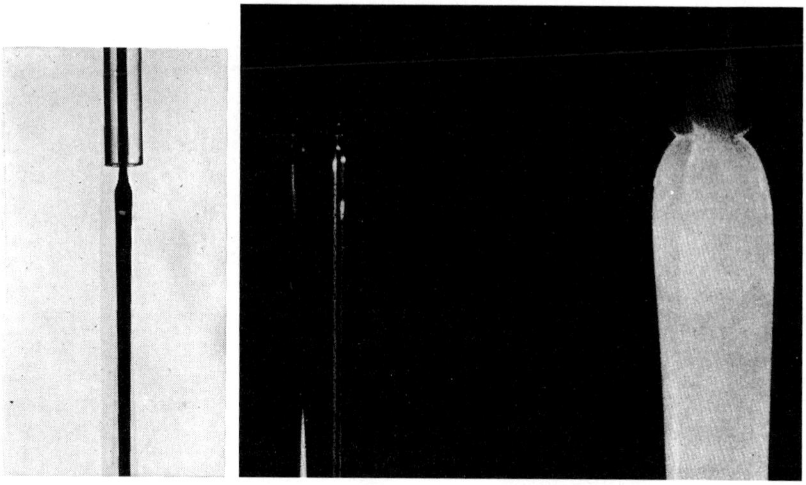

a b c

Fig. 18a—c. Swelling in steady flow from a pipe[2]. a Merrington's photograph (1943). b Glycerine emerging from a pipe. c 2% solution of poly-(ethylene oxide) in water, flowing from an orifice of the same diameter as in (b).

115. Steady torsional flows. For the flows to be considered in this section, the dynamical equations cannot be solved exactly when the body forces are conservative. Solutions can be obtained only after certain terms have been neglected. We expect that these solutions approximate the exact solutions, but a good theoretical justification is still lacking.

[1] Merrington [1943, *1*]. Garner, Nissan, and Wood [1950, *1*, pp. 38—40] stated that they had observed the swelling effect independently between 1942 and 1946. Earlier, Halton and Scott Blair [1936, *2*, p. 564] had noticed a similar effect in dough: "During extrusion, the dough cylinder swells ..." It is not clear, however, that they had established a steady flow. Measurements of the swelling have been reported by Metzner, Houghton, Sailor, and White [1961, *37*].

Merrington's phenomenon, occurring in steady flow, is not to be confused, as it sometimes has been, with one reported by Barus [1893, *1*, § 6]: "At the end of stated intervals of time (usually hours), the small cylinders of marine glue which had exuded were cut off with a sharp knife, and weighed. Now it was curious to note that these cylinders, left to themselves for about a day, showed a gradual and marked deformation, such that the originally plane bottom or surface of section eventually expanded into a symmetrical projecting conoid, with an acute apex angle of less than 45°. I take this to be an example of volume viscosity. The restitution of volume is greatest in the axis of the cylinder where the flow is a maximum, and where the matter has been crowded into the smallest space. As a whole, the experiment is somewhat puzzling, for it points to the occurrence of a notable amount of slowly reacting elasticity even in this truly viscous solid. Indeed, as time went on, a re-entrant conoid, corresponding to the projecting cone just described, gradually dimpled the second of the two surfaces of section. What is here indicated, therefore, is probably a surface of flow." A somewhat fuller description, along with a figure, was given by Barus [1893, *2*, § 6].

[2] Photograph (a) by courtesy of Dr. Merrington and Messrs. Arnold, London. Photographs (b) and (c) by courtesy of Dr. Markovitz from the film "Rheological behavior of Fluids", Educational Services, Inc., Watertown, Mass., for the National Committee for Fluid Mechanics Films under a grant from the U.S. National Science Foundation. A still more striking photograph has been published by Lodge [1964, *57*, Fig. 10.8], in which one of two fluids of about equal shear viscosity swells by 200% in diameter, while the other does not swell at all.

α) Flow between rotating parallel plates ("torsional flow")[1]. We consider a flow whose velocity field, in a cylindrical co-ordinate system, r, θ, z, has the contravariant components

$$\dot{r}=0, \quad \dot{\theta}=\omega(z), \quad \dot{z}=0. \tag{115.1}$$

After labeling the co-ordinates

$$x^1=z, \quad x^2=\theta, \quad x^3=r, \tag{115.2}$$

we see easily that the flow is curvilineal and hence that the equations of Sect. 107 and 108 apply. Equations $(107.9)_2$ and (107.10) reduce to

$$\varkappa=r\omega'(z), \quad \alpha=1, \quad \beta=0. \tag{115.3}$$

The components $T\langle km\rangle$ of (106.19) coincide with the physical components. Hence we have

$$T\langle rz\rangle=T\langle r\theta\rangle=0, \tag{115.4}$$

and, by $(115.3)_1$ and (108.4)–(108.6), the physical components $T_{\mathrm{E}}\langle km\rangle$ of the extra stress depend only on r and z. Using (6A.4), we see that the dynamical equations (30.26) reduce to

$$\left.\begin{aligned} &\partial_r T_{\mathrm{E}}\langle rr\rangle+\frac{1}{r}\left(T\langle rr\rangle-T\langle\theta\theta\rangle\right)-\\ &\qquad -\varrho\,\partial_r\varphi=-\varrho r\omega^2,\\ &\partial_z T\langle\theta z\rangle-\frac{\varrho}{r}\,\partial_\theta\varphi=0,\\ &\partial_z T_{\mathrm{E}}\langle zz\rangle-\varrho\,\partial_z\varphi=0. \end{aligned}\right\} \tag{115.5}$$

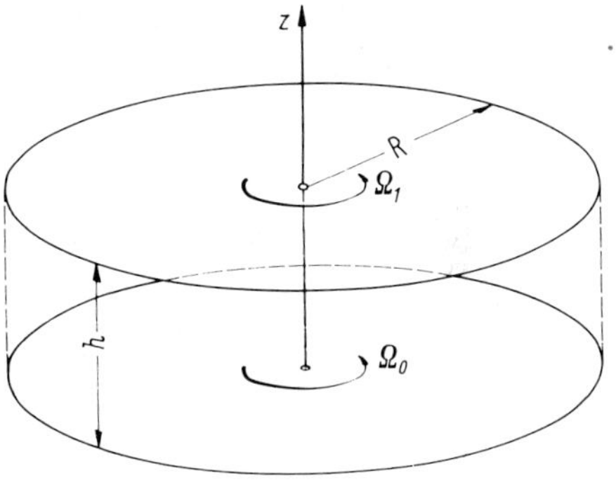

Fig. 19. Torsional flow.

We now assume that the flow takes place in a cylinder of radius R and height h and that it is bounded by two parallel rigid disks which rotate about a common axis with different angular velocities Ω_0 and Ω_1 as shown in Fig. 19. The boundary conditions expressing the assumption that the fluid adheres to the plates are

$$\omega(0)=\Omega_0, \quad \omega(h)=\Omega_1, \quad \Delta\Omega=\Omega_1-\Omega_0\neq 0. \tag{115.6}$$

We assume that the free surface $r=R$ of the fluid is in contact with an atmosphere of pressure p_0, which yields the condition

$$T\langle rr\rangle\big|_{r=R}=-p_0. \tag{115.7}$$

For simplicity, we assume also that the body force vanishes: $v=0$. If $v\neq 0$, the equations below must be modified by adding the term $-\varrho v$ at appropriate places. The condition (115.7), however, cannot be satisfied for general potentials v.

An analysis of the dynamical equations (115.5) shows that they are equivalent to the following conditions:

---

[1] Within the Reiner-Rivlin theory this problem was first solved by RIVLIN [1948, *14*, § 15] (cf. also TRUESDELL [1952, *20*, § 72D] for a slightly more general approach). Later RIVLIN [1956, *25*, § 3] generalized his solution to fluids of the differential type. For a special type of rate theory, the problem was solved by DEWITT [1955, *9*]. Presentation in terms of the viscometric functions is due to MARKOVITZ [1957, *13*]. COLEMAN and NOLL [1959, *2*, Introd.] remarked that the results can be extended to all simple fluids, as is seen here.

Unsteady torsional flow according to a special rate theory was considered by TORRE [1954, *26*, § 4].

(i) There are constants c and b such that

$$\omega(z) = c z + b, \qquad \varkappa = r c. \tag{115.8}$$

(ii) The stress components depend only on r and not on z.

(iii) The equation

$$\partial_r T\langle rr\rangle = \frac{1}{r}\,\sigma_2(c r) - \varrho r (c z + b)^2 \tag{115.9}$$

holds identically in r and z.

The values of the constants c and b are determined by (115.6):

$$c = \frac{\Delta\Omega}{h}, \qquad b = \Omega_0. \tag{115.10}$$

Since $\Delta\Omega \neq 0$ and hence $c \neq 0$, it is clear that the conditions (ii) and (iii) above are incompatible and hence that an exact solution does not exist. The difficulty is removed if we neglect inertia by putting $\varrho = 0$ in (115.9). It is then easy to obtain the following results: The torque M that must be applied to the plates is related to the difference $\Delta\Omega$ of the angular velocities by

$$M = 2\pi \int_0^R r^2 \tau\left(\frac{r}{h}\,\Delta\Omega\right) d r. \tag{115.11}$$

The normal stress distribution at the plates is given by

$$T\langle zz\rangle + p_0 = \sigma_1\left(r\,\frac{\Delta\Omega}{h}\right) - \int_r^R \frac{1}{\xi}\,\sigma_2\left(\xi\,\frac{\Delta\Omega}{h}\right) d\xi. \tag{115.12}$$

The total normal force F that must be applied to hold the plates in place is

$$F = -\pi R^2 p_0 + \pi \int_0^R r\left[2\sigma_1\left(r\,\frac{\Delta\Omega}{h}\right) - \sigma_2\left(r\,\frac{\Delta\Omega}{h}\right)\right] d r. \tag{115.13}$$

The sign of F is selected so that the force is directed away from the flow region when $F > 0$. In deriving this formula we have assumed, of course, that the velocity field (115.1) holds throughout the region of flow, that the region occupied by the fluid considered is strictly cylindrical, and that no special surface effects occur. In parallel-plate viscometers it seems doubtful that these requirements are strictly satisfied. However, even if "edge effects" disturb the flow pattern near the edge ($r = R$) of an apparatus, one can still expect (115.12) to be valid when r is not too close to R, provided that p_0 is taken to be an adjustable constant rather than the pressure of the outside atmosphere.

Equations (115.11)–(115.13) involve only the angular-velocity difference $\Delta\Omega$. The inertial term

$$\varrho r (c z + b)^2 = \varrho r \left(\frac{\Delta\Omega}{h}\, z + \Omega_0\right)^2, \tag{115.14}$$

however, depends also on Ω_0. Measurements in an actual flow may show that the values of M, $T\langle zz\rangle$, and F are not appreciably changed by a variation of Ω_0 when $\Delta\Omega$ is held constant. This may then be taken as an indication that neglect of inertia is justified. We note that for a given $\Delta\Omega$ the inertial term (115.14) can be minimized by choosing $\Omega_1 = -\Omega_0 = \frac{1}{2}\Delta\Omega$.

Apparatus for measuring normal stresses in a torsional flow was designed and constructed by GREENSMITH and RIVLIN[1] (Fig. 20).

β) Cone-and-plate flow[2]. We investigate a flow whose velocity field, in a spherical co-ordinate system r, θ, φ, has the contravariant components

$$\dot{r}=0, \qquad \dot{\theta}=0, \qquad \dot{\varphi}=\omega(\theta). \tag{115.15}$$

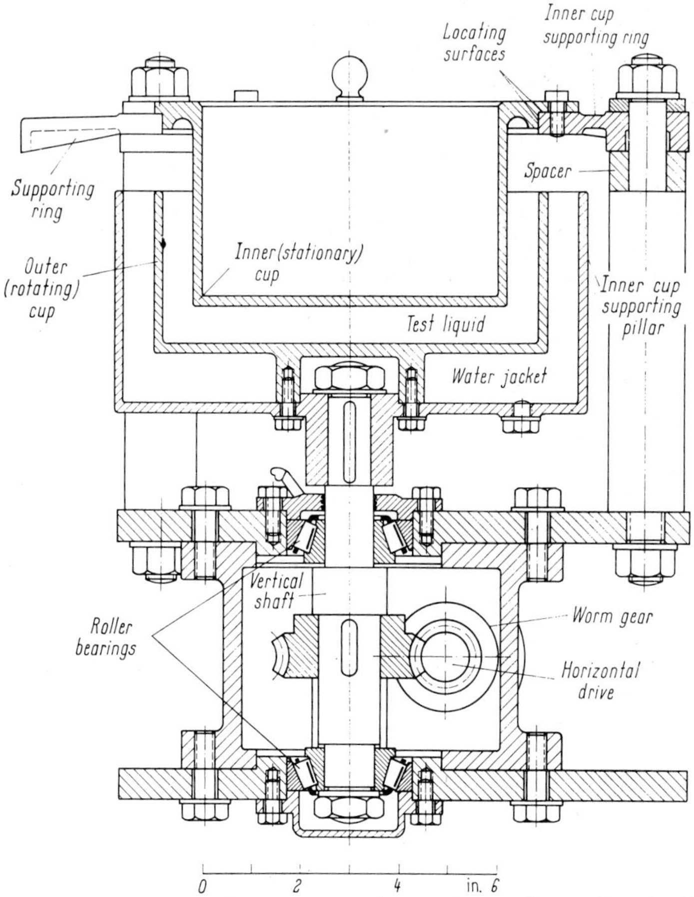

Fig. 20a. Diagram of GREENSMITH and RIVLIN's apparatus for measuring normal stresses in torsional flow (1951).

[1] Earlier (1942—1946) GARNER, NISSAN and WOOD [1950, *1*, pp. 60—61] had attempted to measure normal-stress effects in torsional flow but had failed to achieve sufficient accuracy. They observed that in a rotating vessel, fluid flows from the periphery of a stationary disc toward the center. Qualitatively, a secondary flow of this kind is suggested by Eq. (115.9) if $\sigma_2 > 0$. What seems to be the first satisfactory apparatus of any kind for normal stress measurements is that of GREENSMITH and RIVLIN [1951, *3*] [1953, *9*] (cf. also GREENSMITH [1952, *12*]), whose work is remarkable not only for its experimental accuracy but also for clear presentation and for analysis of the bearing of theory on experiment. Improved instruments for torsional flow have been constructed and used by ADAMS [1960, *1*], ADAMS and LODGE [1964, *1*], and MARKOVITZ [1965, *18*].

[2] Incomplete analyses of this problem, under special assumptions, were given by ROBERTS [1954, *20*], MARKOVITZ [1957, *13*], and ERICKSEN [1960, *22*, § 6]. Somewhat more general flows, for special types of fluids, were considered by OLDROYD [1958, *32*, § 6]. MARKOVITZ and WILLIAMSON [1957, *14*] obtained a solution within the Rivlin-Ericksen theory. Some of these authors considered flows between coaxial cones. The analysis presented here is restricted to the case when one of the cones reduces to a plane, because it is only in this case that the dynamical equations can be satisfied even approximately, as is explained in the text.

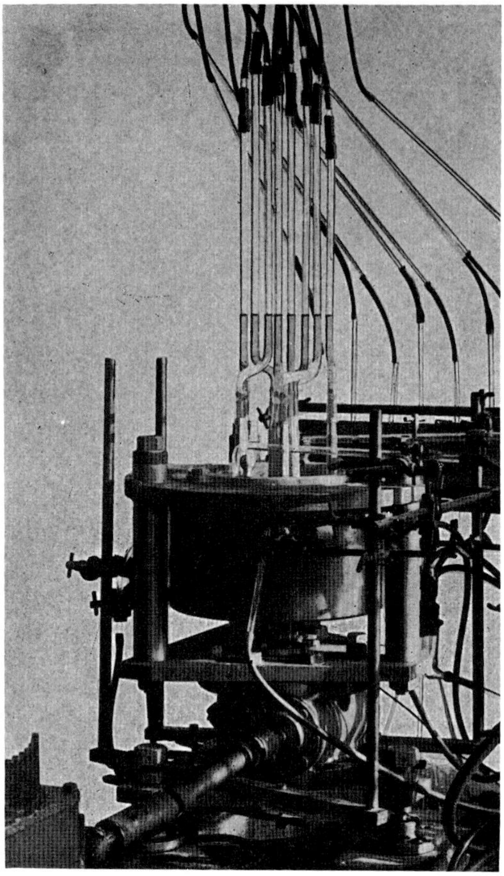

Fig. 20b. Photograph of GREENSMITH and RIVLIN's apparatus (by courtesy of Professor RIVLIN).

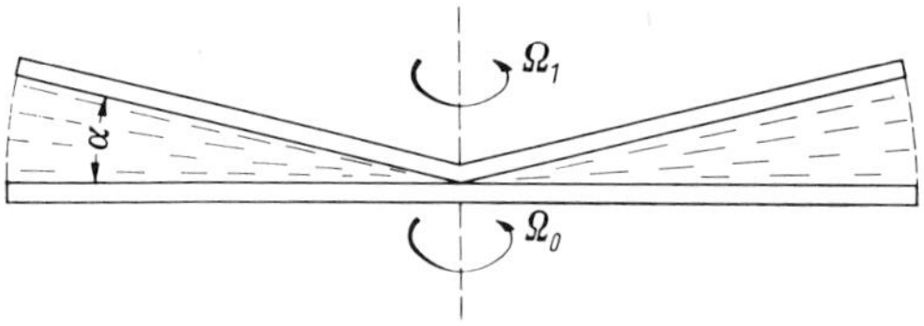

Fig. 21. Cone-and-plate flow.

Such a flow is curvilineal, and the equations of Sects. 107 and 108 apply when we use the labeling

$$\left.\begin{aligned} x^1 &= \theta, \\ x^2 &= \varphi, \\ x^3 &= r. \end{aligned}\right\} \tag{115.16}$$

Equations (112.4) and (112.5) reduce to

$$\left.\begin{aligned} \varkappa &= \sin\theta\,\omega'(\theta), \\ \alpha &= 1, \\ \beta &= 0. \end{aligned}\right\} \tag{115.17}$$

The components $T\langle km\rangle$ of (106.19) again coincide with the physical components. It follows from (108.4)—(108.6) that

$$\left.\begin{aligned} T\langle\theta\varphi\rangle &= \tau(\varkappa), \\ T\langle r\theta\rangle &= T\langle r\varphi\rangle = 0, \end{aligned}\right\} \tag{115.18}$$

and that the normal stress differences depend only on θ.

We now suppose that the fluid adheres to a rigid disk of radius R and a rigid cone as shown in Fig. 21. The angle between plate and cone is denoted by α. The plate and the cone are assumed to rotate about their common axis with different angular velocities Ω_0 and Ω_1. The boundary conditions then are

$$\left.\begin{aligned} \omega(\tfrac{1}{2}\pi) &= \Omega_0, \\ \omega(\tfrac{1}{2}\pi-\alpha) &= \Omega_1, \\ \Delta\Omega = \Omega_1 - \Omega_0 &\neq 0. \end{aligned}\right\} \tag{115.19}$$

We also assume that $r=R$ is a free surface in contact with an atmosphere of pressure p_0, which yields the condition

$$T\langle rr\rangle\big|_{r=R} = -p_0. \tag{115.20}$$

As in the case of torsional flow, we assume that the body force vanishes: $v=0$. By use of (6A.7) and (6A.8) it can be shown that the dynamical equations (30.26) are equivalent to

$$\varkappa = \zeta(\gamma(\theta)), \qquad \gamma(\theta) = \frac{c}{\sin^2\theta}, \tag{115.21}$$

$$\partial_r T\langle rr\rangle - \frac{1}{r}\left[\hat\sigma_1(\gamma(\theta)) + \hat\sigma_2(\gamma(\theta))\right] = -\varrho r \sin^2\theta\,\omega(\theta)^2, \tag{115.22}$$

$$\partial_\theta T\langle\theta\theta\rangle + \cot\theta\left[\hat\sigma_1(\gamma(\theta)) - \hat\sigma_2(\gamma(\theta))\right] = -\varrho r^2 \sin\theta\cos\theta\,\omega(\theta)^2. \tag{115.23}$$

The equations (115.22) and (115.23) are compatible only when the integrability condition corresponding to $\partial_{r\theta} T\langle\theta\theta\rangle = \partial_{\theta r} T\langle\theta\theta\rangle$ is satisfied. This condition is

$$c \cos\theta \left[\hat{\sigma}_1'(\gamma(\theta)) + \hat{\sigma}_2'(\gamma(\theta))\right] = -\varrho r^2 \sin^5\theta\, \omega(\theta)\, \omega'(\theta). \tag{115.24}$$

It follows from $(115.19)_3$ that $\omega'(\theta)$ cannot be identically zero, and hence (115.24) cannot be satisfied identically in r and θ. Therefore, an exact solution does not exist. Even when inertia is neglected ($\varrho = 0$) (115.24) will not hold unless the normal stress function $\hat{\sigma}_1 + \hat{\sigma}_2$ is a constant, so that $\hat{\sigma}_1'(\gamma) + \hat{\sigma}_2'(\gamma) = 0$. If $\hat{\sigma}_1 + \hat{\sigma}_2$ is not constant, the difficulties can be set aside by making the following two approximations:

(i) We neglect inertia, i.e. we put $\varrho = 0$ in (115.22) and (115.23).

(ii) We assume that the angle α between plate and cone is so small that $\cos\theta \approx 0$, $\sin\theta \approx 1$ when

$$\tfrac{1}{2}\pi - \alpha \leqq \theta \leqq \tfrac{1}{2}\pi. \tag{115.25}$$

With these approximations the compatibility condition (115.24) reduces to $0=0$. The equations (115.21)—(115.23) reduce to

$$\left.\begin{aligned} \varkappa = \omega'(\theta) = \zeta(c) = \text{const.}, \qquad & \partial_\theta T\langle\theta\theta\rangle = 0, \\ \partial_r T\langle rr\rangle = \frac{1}{r}\left[\sigma_1(\varkappa) + \sigma_2(\varkappa)\right]. & \end{aligned}\right\} \tag{115.26}$$

By use of (115.26) and the conditions (115.19) and (115.20) it is not difficult to obtain the following results: The torque M which must be applied to produce the flow and the angular velocity difference $\Delta\Omega$ are related by

$$M = \frac{2\pi}{3} R^3 \tau\left(\frac{\Delta\Omega}{\alpha}\right). \tag{115.27}$$

The normal stresses along the plate or the cone are

$$T\langle rr\rangle + p_0 = \sigma_1\left(\frac{\Delta\Omega}{\alpha}\right) - \log\left(\frac{R}{r}\right)\left[\sigma_1\left(\frac{\Delta\Omega}{\alpha}\right) + \sigma_2\left(\frac{\Delta\Omega}{\alpha}\right)\right]. \tag{115.28}$$

The total force F necessary to keep cone and plate in place is given by

$$F = -\pi R^2 p_0 + \frac{\pi}{2} R^2 \left[\sigma_1\left(\frac{\Delta\Omega}{\alpha}\right) - \sigma_2\left(\frac{\Delta\Omega}{\alpha}\right)\right]. \tag{115.29}$$

Remarks similar to the ones made at the end of subsection (α) apply here, too.

Instruments for measuring normal stresses in a cone-and-plate flow have been designed and constructed by WEISSENBERG[1] (Fig. 22) and others.

If a physical fluid were known to obey the equations of the theory of simple fluids perfectly, the velocity fields here described could not be maintained exactly in it. As was remarked by OLDROYD[2] in 1958, if a real fluid is confined within rigid conical boundaries in relative rotation, "some secondary flow is to be expected". In a lecture delivered in the same year, ERICKSEN[3] distinguished two

[1] The early experiments (1942—1946) of GARNER, NISSAN and WOOD [1950, *1*, pp. 58—59] on cone-and-plate flow were inconclusive. WEISSENBERG [1949, *28*] designed and demonstrated a "rheogoniometer" for effecting normal-stress measurements on cone-and-plate flow. Improved apparatus has been made and used by ROBERTS [1952, *19*], by ADAMS [1960, *1*] and ADAMS and LODGE [1964, *1*], and by MARKOVITZ [1965, *25*] and MARKOVITZ and BROWN [1964, *59*]. The instrument of ADAMS uses diaphragm-capacitance pressure gauges, which register more rapidly than the standpipes used in earlier work.

[2] OLDROYD [1958, *32*, § 6].

[3] ERICKSEN [1960, *22*, § VI].

kinds of secondary flows: that due to the inertia of the fluid ($\varrho\neq 0$) and that corresponding to failure to satisfy the condition (115.24) (or a more general condition arising when both bounding surfaces are cones), and he conjectured their general forms. Approximate calculation of such flows will be mentioned at the end of Sect. 123.

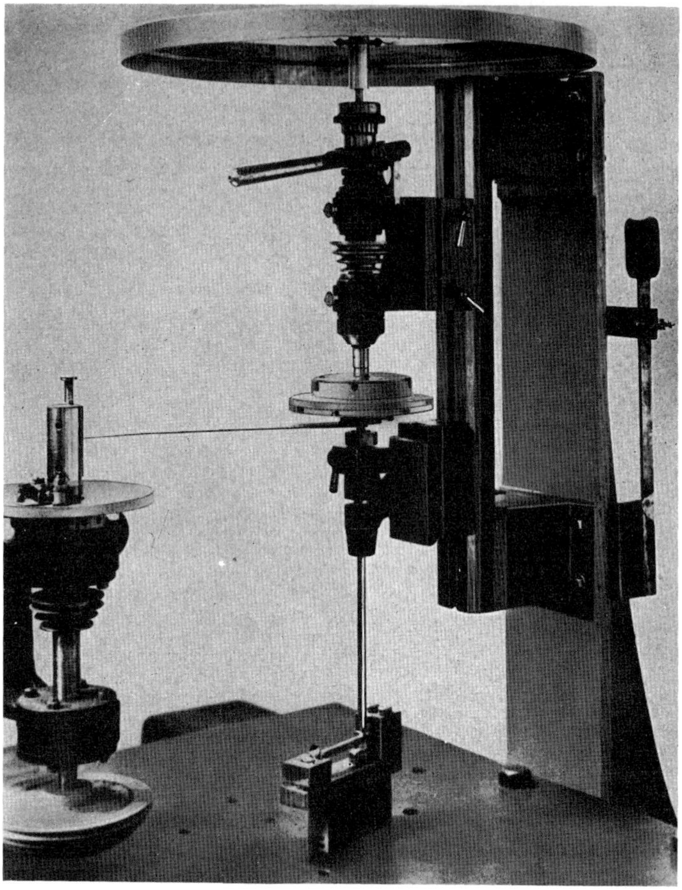

Fig. 22. WEISSENBERG's rheogoniometer (1948) (Photograph by courtesy of Dr. WEISSENBERG and the North-Holland Publishing Co.)

More generally, a flow pattern clearly different from that ordinarily observed in fluids or calculated according to the Navier-Stokes equations is sometimes said to exhibit a secondary flow[1]. A systematic theory for secondary flow in tubes is given in Sect. 122, below. Various perturbation schemes and approximations have been published for flows corresponding to other boundaries[2].

[1] E.g., at the International Congress of Rheology, Brown University, 1963, GIESEKUS showed a beautiful moving picture tracing the origin and growth of the flow produced by rotation of a sphere in a large vessel of polymer solution. This flow is entirely different from its counterpart in, say, water. HOPMANN and BARNETT [1964, *44A*] have reported secondary flows produced by rotation of a cone in a solution of polyisobutylene in cetane.

[2] LANGLOIS and RIVLIN [1959, *18*, §§ 2—3] (general), LANGLOIS [1963, *45*] (flow between concentric rotating spheres), GIESEKUS [1963, *25*] (single sphere). Analysis of secondary flows according to various special theories are included in some of the works cited in Sects. 119 and 123.

116. Position of the general and Navier-Stokes theories of viscometry. Steady Couette flow, steady flow between concentric pipes, Poiseuille flow, steady torsional flow, and cone-and-plate flow have two remarkable properties in common. First, some member of each class may be maintained in *any* simple fluid by applying suitable surface tractions only, at least approximately. This fact makes it possible to establish an experimental program with these classes of flows as a basis, since the form of the response functional does not affect the flow pattern, although of course it does influence the speed and the stress system. Second, the stresses and speeds in all these different flows, for any one fluid, are determined by the *same* three viscometric functions.

On the one hand, this is a loss, since no amount of experiment on flows of this class can determine any more about the response functional than the nature of these three functions. The vastly more detailed information about fluid response that may be embodied in the response functional can never be gotten by even the most accurate viscometric program[1]. Two fluids that behave just alike in one of the flows considered here may therefore behave entirely differently in some other situation, such as a stress relaxation experiment or a flow past an obstacle.

On the other hand, it is an economy, since it separates for observation a well-defined and important *aspect* of fluid behavior that is particularly simple. The experimenter concentrating his attention on these flows may forget altogether most of the physical properties of the fluids he is testing. In the modest program of viscometry, moreover, there is ample opportunity for cross check. In principle, the viscometric functions may be determined by sufficiently accurate measurements made on *any one* of the flows considered, and the results then suffice to predict in detail the speeds and stresses relevant to all the others. Further measurement then provides tests of consistency.

On this basis, a good deal of experiment has been and is being done. The fundamental work of Greensmith and Rivlin[2] was conceived in terms of the Reiner-Rivlin theory and hence is not sufficient to determine all three functions. A considerable literature[3], much of it interpreted in terms of special theories of at best questionable foundation, had meanwhile begun to grow. The first general appraisal of the experimental problem was made by Markovitz[4], who thereupon, with various collaborators, undertook a systematic program[5].

---

[1] Cf. the remarks of Truesdell [1965, *35*, pp. 25—27].

[2] Greensmith and Rivlin [1951, *3*] [1952, *12*] [1953, *9*].

[3] Roberts [1952, *19*] [1954, *20*] [1957, *21*], Padden and DeWitt [1954, *15*], Pilpel [1954, *17*], Philippoff [1956, *20*] [1961, *44*], Brodnyan, Gaskins, and Philippoff [1957, *5*], Markovitz and Williamson [1957, *14*], Jobling and Roberts [1958, *24*] [1959, *14* and *15*], Reiner [1960, *45*]. Greensmith and Rivlin [1953, *9*, § 1] described unpublished experiments of R. J. Russell from a thesis dated 1946.

While the works just cited concern mainly high-polymer solutions, Reiner and his collaborators claim to demonstrate the existence of normal-stress effects in air: Reiner [1957, *19*] [1958, *38*] [1959, *22*] [1961, *50*] [1962, *56*], Popper and Reiner [1958, *36*], Foux and Reiner [1964, *30*], Bousso [1964, *10*]. According to Taylor and Saffman [1957, *27*], the phenomena observed in Reiner's experiments are accounted for by the Navier-Stokes equations.

In the experimental literature the border between quantitative and qualitative is often blurred. In most of the papers cited, the data are not sufficiently accurate or detailed for real determination of viscometric functions. Some more obviously qualitative early experiments we cite in connection with the individual solutions. As late as 1962, in summarizing the whole field Markovitz and Brown [1964, *59*] wrote, "It appears that objections can be raised against every experiment that has been reported and that unequivocal conclusions must await consistent results obtained from a sufficient number of different experiments." For further discussion of experimental results, see the book of Coleman, Markovitz, and Noll [1965, *12*].

[4] Markovitz [1957, *13*].

[5] A description of the apparatus and the theoretical results used is given in Markovitz's lecture [1965, *25*].

According to the classical theory of viscometry[1], derived from the Navier-Stokes equations, as remarked at the end of Sect. 108 the three viscometric functions have the following forms:

$$\tilde{\mu}(\varkappa) = \mu_0 = \text{const.}, \qquad \sigma_1(\varkappa) \equiv \sigma_2(\varkappa) \equiv 0. \tag{116.1}$$

Thus in the classical theory, the three normal stresses are equal to one another, so that *there are no normal-stress effects in viscometric flows*. Moreover, the *shear stress is always proportional to the shearing*. These two statements, which together characterize the classical theory of viscometry, are in no way dependent upon one another, although only recently has the general theory become understood sufficiently to make this remark as obvious as now it is. The vanishing of the normal-stress functions σ_Γ renders the major part of our analysis in this sub-chapter vacuous and trivial. The constancy of the shear-viscosity function $\tilde{\mu}$ has the effect of simplifying drastically a number of the specific results. For example, (113.33) becomes

$$Q = \frac{\pi a R^4}{8 \mu_0}, \tag{116.2}$$

the famous *Hagen-Poiseuille efflux formula*, sometimes called "the law of the fourth power".

In all other results concerning discharges, torques, etc., similarly simple and explicit special cases follow for the classical theory. These formulae have been verified with precision by many experimenters for many fluids at various temperatures and pressures, in tests running from the present time back over more than a century.

The older authors, not possessing any more accurate theory with which to compare it, were hasty if not naive in pronouncing the Navier-Stokes theory to be fully confirmed by experiment. E.g., LAMB[2], adducing the authority of REYNOLDS, wrote that "this hypothesis [of linear viscosity] has been put to a very severe test by the experiments of POISEUILLE and others ... Considering the very wide range of values of the rates of distortion over which these experiments extend, we can hardly hesitate to accept the equations in question as a complete statement of the laws of viscosity." Apparently BATEMAN[3] was the first to see how precarious was the evidence. In regard to experiments on flow in tubes he wrote, "It should be remarked that a test of this kind only proves that a physical constant of dimension VL enters in some way into the equation of motion or the boundary conditions; it does not prove that these equations have the correct form." Developing this idea, TRUESDELL[4] remarked that "the 'experimental tests' which supposedly confirm the linear relation ... concern degenerate situations such as rectilinear shearing flow and serve merely to verify the interconnections of certain dimensionless parameters, thus failing entirely to discriminate between the classical linear viscosity and the myriad other possible laws which would give rise to the same relations between the same dimensionless parameters in the same degenerate situations."

The early work on the Reiner-Rivlin theory confirmed this suspicion. Indeed, RIVLIN[5] had written, "A fluid is generally recognized as non-Newtonian on the basis of viscosimetric experiments. For a Newtonian fluid in a steady state of laminar flow, a linear relation obtains between the shearing stress and rate of

---

[1] E.g., LAMB [1932, *5*, §§ 330—334].
[2] LAMB [1932, *5*, § 326]
[3] BATEMAN [1932, *1*, Part II, § 1.7].
[4] TRUESDELL [1953, *25*, p. 609].
[5] RIVLIN [1948, *14*, § 17]. Cf. the example given by ANDERSSON [1952, *2*].

shear. It is generally assumed that if such a relation is obtained experimentally, the fluid under investigation is Newtonian and that the whole of the classical hydrodynamic theory for viscous fluids is applicable. It is seen, from the results [given above], that this is by no means correct. Fluids, which appear Newtonian on the basis of experiments involving steady-state laminar flow, may, from the stand-point of the phenomenological theory, be non-Newtonian and distinguishable from Newtonian fluids by other steady-state experiments in which the flow is not laminar." In this passage, RIVLIN'S term "laminar" means, roughly, what we denote by "viscometric". His important observation may be made explicit in the full generality of the theory of simple fluids, presented in this treatise. E.g., if we invert $(108.20)_2$, we obtain

$$\varkappa = \zeta(T\langle xy\rangle) = \frac{T\langle xy\rangle}{\mu_0} W\left(\frac{s_0 T\langle xy\rangle}{\mu_0}\right), \tag{116.3}$$

where s_0 is the natural time-lapse, μ_0 is the natural viscosity, and W is a dimensionless material function related as follows to the dimensionless shear-viscosity function V:

$$W(\xi) = \frac{1}{\xi} \overset{-1}{Z}(\xi), \quad \text{where} \quad Z(\xi) \equiv \xi V(\xi). \tag{116.4}$$

Substitution into (113.33) yields[1]

$$Q = Q(a, R) = \frac{\pi a R^4}{2\mu_0} \int_0^1 u^3 W\left(\frac{a R s_0}{2\mu_0} u\right) du. \tag{116.5}$$

Therefore, if $Q \propto R^4$, it is necessary that $W(\xi) \equiv$ const., and since $W(0) = 1$, it follows that $W(\xi) \equiv 1$, and the Hagen-Poiseuille formula results. By (116.4) and $(108.20)_2$ we see that $W(\xi) \equiv 1$ if and only if $\tilde{\mu}(\varkappa) = \mu_0 =$ const. We have established the following ***theorem***: *In the theory of simple fluids, the Hagen-Poiseuille efflux formula* (116.2) *holds if and only if the shear-viscosity function is constant.* Not only does the shear-viscosity function fail to determine the response functional, it does not even have any influence on the normal-stress functions. *Far from establishing the validity of the Navier-Stokes equations for general flows, the classic "experimental tests" do not suffice even for the validity of the Navier-Stokes theory of viscometry.* Indeed, the common and just faith put in the Navier-Stokes equations for ordinary phenomena in ordinary fluids does not rest on "basic" experimental tests or "operational" definitions but rather upon a century of experience in the interrelation, with greater or lesser accuracy, of a vast variety of experimental data and theoretical formulae for a vast variety of natural fluids.

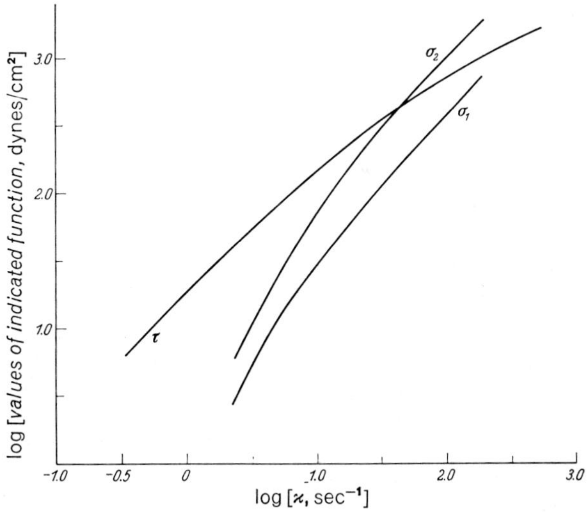

Fig. 23. Viscometric functions of 5.4% solution of polyisobutylene in cetane (by courtesy of Dr. MARKOVITZ).

[1] TRUESDELL [1964, 87, Eqs. 24—25].

Data sufficient for certain determination of all three viscometric functions, for any fluid, has only very recently been collected. Dr. MARKOVITZ has kindly released to us his unpublished results on a 5.4% solution of polyisobutylene in cetane (Fig. 23). The results for low shearing may be obtained by extrapolation. With an error which may be even greater than 10%, the natural time-lapse s_0 may be calculated from (108.22) and (108.21), yielding the value $s_0 = 0.068$ sec. The natural viscosity is $\mu_0 = 18.5$ poise. From these data the reduced viscometric functions may then be calculated by Eq. (108.20), with the results shown in Fig. 24. For the Navier-Stokes fluid, $V(K) \equiv 1$, while $N_1(K)$ and $N_2(K)$ have arbitrary constant values n_1 and n_2 such that $n_1^2 + n_2^2 = 1$. From Figs. 23 and 24 it is clear that the classical theory of viscometry has no discernible range of validity

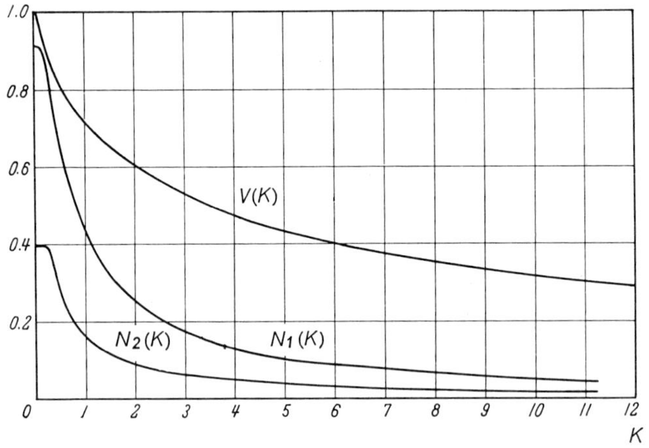

Fig. 24. Reduced viscometric functions for a 5.4% solution of polyisobutylene in cetane (by courtesy of Dr. MARKOVITZ).

for the particular fluid under study. The time-lapse function $\tilde{s}$, which is defined by (108.21) in such a way as to measure the relative magnitudes of the normal-stress and shear-stress effects, decreases very rapidly from its value s_0 when $K=0$ and seems to fall steadily to 0 as K grows large.

A more general position for the Navier-Stokes theory of viscometry with respect to the general theory will be established in Sect. 120.

In the foregoing sections we have proved that the three viscometric functions suffice for constructing the entire theory of simple fluids in viscometric flows. We have not shown the reduction to these three functions to be characteristic of simple materials. Therefore, consistency with the viscometric theory reported above is not known to be a sufficient condition that the material tested be a simple fluid.

117. Steady flow through tubes[1]. There is a one-to-one correspondence between points $\boldsymbol{x}$ in space and pairs $(z, \boldsymbol{p})$, where z is the co-ordinate of $\boldsymbol{x}$ with respect to a given z-axis and $\boldsymbol{p}$ is the vector giving the position of $\boldsymbol{x}$ with respect to the axis. Let $\mathscr{U}$ be the two-dimensional space of all vectors $\boldsymbol{p}$ perpendicular to the axis, and let $\boldsymbol{k}$ be the unit vector in the direction of the axis. A rectilinear shearing [CFT,

[1] For certain special fluids, such flows were analysed by OLDROYD [1949, *11*] [1951, *10*] [1958, *32*, § 5]; for the Reiner-Rivlin theory, by ERICKSEN [1956, *7*], GREEN and RIVLIN [1956, *13*], and BHATNAGAR and LAKSHMANA RAO [1960, *12*]. CRIMINALE, ERICKSEN and FILBEY [1958, *13*] extended the analysis to the case of fluids of the differential type.

Sect. 89] is a steady flow having a velocity field of the form

$$\dot{\boldsymbol{x}} = v(\boldsymbol{p})\,\boldsymbol{k}, \tag{117.1}$$

where v is a scalar field defined on some region $\mathscr{A}$ of the vector plane $\mathscr{U}$. If v satisfies the boundary condition

$$v(\boldsymbol{p}) = 0 \quad \text{when} \quad \boldsymbol{p} \in \partial\mathscr{A}, \tag{117.2}$$

then the rectilinear shearing (117.1) describes a flow through a tube of cross-section $\mathscr{A}$ in which the fluid adheres to the wall (Fig. 25). It is a particular kind of flow through the tube in that each particle moves always parallel to the wall, and with constant speed. Examples of such flows are channel flow [cf. Sect. 109 (β)], in which $\mathscr{A}$ is a strip bounded by two parallel straight lines, Poiseuille flow

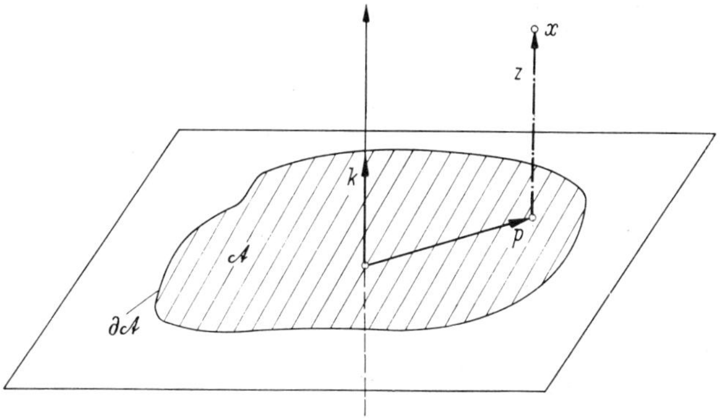

Fig. 25. Co-ordinates for flow in a tube.

[cf. Sect. 113 (ε)], in which $\mathscr{A}$ is a circular disc, and concentric pipe flow [cf. Sect. 113 (δ)], in which $\mathscr{A}$ is an annulus. Simple shearing is also a flow of the type (117.1), but it corresponds to the boundary condition (111.5), which is not a special case of (117.2).

Assuming a flow (117.1) given, we introduce an orthogonal co-ordinate system x^k by taking

$$x^1 = v(\boldsymbol{p}), \qquad x^2 = z, \qquad x^3 = \theta(\boldsymbol{p}), \tag{117.3}$$

where $\theta(\boldsymbol{p})$ is a solution of the partial differential equation

$$\nabla\theta \cdot \nabla v = 0. \tag{117.4}$$

The lines $\theta = \text{const.}$ are then the orthogonal trajectories of the lines $v = \text{const.}$ in the plane $\mathscr{U}$. The components g^{kk} of the unit tensor are given by

$$g^{11} = g_{11}^{-1} = (\nabla v)^2, \qquad g^{22} = g_{22} = 1, \qquad g^{33} = g_{33}^{-1} = (\nabla\theta)^2; \tag{117.5}$$

they do not depend on $x^2 = z$. It follows from the definitions of Sects. 107 and 108 that every rectilinear shearing (117.1) is a steady curvilineal flow. We have $\frac{dv}{dx^1} = f = 1$ and $h = 0$ in (108.12), and hence $(107.9)_2$ and (107.10) reduce to

$$\varkappa = \sqrt{(\nabla v)^2} = |\nabla v|, \qquad \alpha = 1, \qquad \beta = 0. \tag{117.6}$$

Let $\boldsymbol{i}$ be the unit vector in the direction of ∇v and $\boldsymbol{j}$ the unit vector in the direction of $\nabla\theta$. Then

$$\nabla v = \varkappa\,\boldsymbol{i}. \tag{117.7}$$

The physical components $T\langle km\rangle$ of the stress are components relative to the orthonormal basis $\boldsymbol{i}, \boldsymbol{k}, \boldsymbol{j}$. Using the notation

$$T\langle 11\rangle = T\langle vv\rangle, \qquad T\langle 12\rangle = T\langle vz\rangle, \qquad T\langle 22\rangle = T\langle zz\rangle, \qquad T\langle 33\rangle = T\langle \theta\theta\rangle, \tag{117.8}$$

we may write (106.19) in the equivalent form

$$\boldsymbol{T} = T\langle vz\rangle\,(\boldsymbol{i}\otimes\boldsymbol{k}+\boldsymbol{k}\otimes\boldsymbol{i}) + T\langle vv\rangle\,(\boldsymbol{i}\otimes\boldsymbol{i}) + T\langle zz\rangle\,(\boldsymbol{k}\otimes\boldsymbol{k}) + T\langle\theta\theta\rangle\,(\boldsymbol{j}\otimes\boldsymbol{j}). \tag{117.9}$$

By the results established in Sect. 108, *the stress deviator in an incompressible simple fluid undergoing steady rectilinear shearing is determined by the three viscometric functions.* The equations (108.4)—(108.6) become

$$T\langle vz\rangle = \tau(\varkappa), \qquad T\langle vv\rangle - T\langle\theta\theta\rangle = \sigma_1(\varkappa), \qquad T\langle zz\rangle - T\langle\theta\theta\rangle = \sigma_2(\varkappa). \tag{117.10}$$

Using (117.9), (117.10) and the relation $\boldsymbol{j}\otimes\boldsymbol{j} = \boldsymbol{1} - \boldsymbol{i}\otimes\boldsymbol{i} - \boldsymbol{k}\otimes\boldsymbol{k}$, we find that

$$\boldsymbol{T} = (\tau\boldsymbol{i} + \sigma_2\boldsymbol{k})\otimes\boldsymbol{k} + \boldsymbol{k}\otimes\tau\boldsymbol{i} + \varkappa\boldsymbol{i}\otimes\frac{\sigma_1}{\varkappa}\,\boldsymbol{i} + T\langle\theta\theta\rangle\boldsymbol{1}, \tag{117.11}$$

where the argument $\varkappa$ has been omitted. In order to compute div $\boldsymbol{T}$ from (117.11) we use the identity

$$\operatorname{div}(\boldsymbol{u}\otimes\boldsymbol{v}) = (\nabla\boldsymbol{u})\,\boldsymbol{v} + \boldsymbol{u}\operatorname{div}\boldsymbol{v} \tag{117.12}$$

three times and note that $\boldsymbol{i}$, τ, σ_1, σ_2 do not depend on z, so that their derivatives in the direction of $\boldsymbol{k}$ are zero. The result is

$$\operatorname{div}\boldsymbol{T} = \boldsymbol{k}\operatorname{div}(\tau\,\boldsymbol{i}) + \frac{\sigma_1}{\varkappa}\nabla(\varkappa\boldsymbol{i})\,\boldsymbol{i} + \varkappa\boldsymbol{i}\operatorname{div}\left(\frac{\sigma_1}{\varkappa}\,\boldsymbol{i}\right) + \nabla T\langle\theta\theta\rangle. \tag{117.13}$$

Taking the gradient of $(\nabla v)^2 = \varkappa^2$, and using the fact that the second gradient $\nabla\nabla v$ is a symmetric tensor, we obtain

$$\varkappa\nabla\varkappa = (\nabla\nabla v)(\nabla v) = \varkappa\nabla(\varkappa\boldsymbol{i})\,\boldsymbol{i}. \tag{117.14}$$

Substitution of (117.14) and (117.7) into (117.13) gives

$$\operatorname{div}\boldsymbol{T} = \boldsymbol{k}\operatorname{div}(\tilde{\mu}\nabla v) + \nabla v\operatorname{div}\left(\frac{\sigma_1}{\varkappa^2}\nabla v\right) + \frac{\sigma_1}{\varkappa}\nabla\varkappa + \nabla T\langle\theta\theta\rangle, \tag{117.15}$$

where $\tilde{\mu}$ is the viscosity function, defined by (108.10). If we put

$$h = T\langle\theta\theta\rangle + \int\frac{\sigma_1(\varkappa)}{\varkappa}\,d\varkappa - \varrho v, \tag{117.16}$$

we find from (117.15) that the dynamical equation (30.26) becomes

$$\boldsymbol{k}\operatorname{div}(\tilde{\mu}\nabla v) + \nabla v\operatorname{div}\left(\frac{\sigma_1}{\varkappa^2}\nabla v\right) + \nabla h = 0. \tag{117.17}$$

It follows from (117.17) that ∇h is independent of z and hence that h must have the form

$$h = z\,a + g(\boldsymbol{p}), \tag{117.18}$$

where a is a constant. Substituting (117.18) into (117.17), we find that the dynamical equations are equivalent to

$$\operatorname{div}(\tilde{\mu}(\varkappa)\nabla v) = -a \tag{117.19}$$

and

$$\nabla v\operatorname{div}\left(\frac{\sigma_1(\varkappa)}{\varkappa^2}\nabla v\right) + \nabla g = 0. \tag{117.20}$$

(117.20) states that g is constant along the lines $v=$ const. Hence, (117.20) is satisfied if and only if there is a function f such that

$$g(\boldsymbol{p})=f(v(\boldsymbol{p})), \quad \operatorname{div}\left(\frac{\sigma_1(\varkappa)}{\varkappa^2}\nabla v\right)=-f'(v). \tag{117.21}$$

If $\varkappa^{-2}\sigma_1(\varkappa)$ is proportional to $\tilde{\mu}(\varkappa)$, i.e. if

$$\sigma_1(\varkappa)=c\varkappa^2\tilde{\mu}(\varkappa)=c\varkappa\tau(\varkappa) \tag{117.22}$$

for some constant c, then every solution $v=v(\boldsymbol{p})$ of (117.19) also satisfies $(117.21)_2$, with the choice $f'(v)=ca=$ const. If (117.22) does not hold, not every solution of (117.19) can be expected to satisfy $(117.21)_2$. In fact, ERICKSEN[1] has shown that if $\tilde{\mu}$ is analytic and if (117.22) does not hold, then there are always solutions of (117.19) for which $(117.21)_2$ is not satisfied. He conjectured that if (117.22) does not hold, the only solutions of (117.19) that also satisfy $(117.21)_2$ are those for which the lines $v=$ const. are either concentric circles or parallel straight lines.

If the viscosity function $\tilde{\mu}(\varkappa)$ is a constant μ_0, then (117.19) reduces to the two-dimensional Poisson equation, $\mu_0\Delta v=-a$. It is a well known result of potential theory that, under suitable conditions on the region $\mathscr{A}$, (117.19) then has a unique solution $v=v(\boldsymbol{p})$ satisfying the boundary condition (117.2). The same can be expected to be true when $\tilde{\mu}(\varkappa)$ is not constant[2]. Unless (117.22) holds or the region $\mathscr{A}$ has one of the exceptional forms corresponding to channel flow, Poiseuille flow, or concentric pipe flow, the solution $v=v(\boldsymbol{p})$ cannot be expected to satisfy $(117.21)_2$. *For a general simple fluid, rectilinear flow through tubes is dynamically possible only for exceptional cross-sections; for general cross-sections, it is dynamically possible only in special simple fluids.*

The special simple fluids for which (117.22) holds include the linearly viscous fluids, for which $\tilde{\mu}(\varkappa)=\mu_0=$ const., $\sigma_1(\varkappa)=0$, and the fluids of the second grade, for which $\tilde{\mu}(\varkappa)=\mu_0=$ const., $\sigma_1(\varkappa)=\alpha\varkappa^2$ (cf. Sect. 123).

Notice that the conditions (117.19) and (117.20) do not involve the viscometric function $\sigma_2(\varkappa)$ at all. Thus it suffices, as far as concerns determination of compatible velocity fields, to consider any special class of simple fluids general enough to allow arbitrary forms to the functions $\tilde{\mu}(\varkappa)$ and $\sigma_1(\varkappa)$; in particular, it suffices[3] to consider only the Reiner-Rivlin theory (Sect. 119), although, of course, the stress system necessary to maintain a compatible flow will depend also upon $\sigma_2(\varkappa)$.

We assume now that (117.19) and $(117.21)_2$ are compatible and have a solution $v(\boldsymbol{p})$ for a given value of a. From (117.16), (117.18), $(117.21)_1$ and $(117.10)_3$ we derive

$$T\langle zz\rangle-\varrho v=za+\sigma_2(\varkappa)-\int\frac{\sigma_1(\varkappa)}{\varkappa}\,d\varkappa+f(v), \tag{117.23}$$

and hence

$$\partial_z[T\langle zz\rangle-\varrho v]=a. \tag{117.24}$$

The argument that led from (111.13) to (111.15) shows again that a is the *specific driving force*, i.e. the applied force per unit volume in the direction of the flow. The normal stresses can be computed from (117.23) and $(117.10)_{2,3}$.

When (117.19) and (117.21) are incompatible, rectilinear flow through the tube is not possible. An approximate solution of the problem for slow flow, yielding spiral streamlines, will be presented in Sect. 122.

---

[1] ERICKSEN [1956, *7*, § 3]. Cf. also STONE [1957, *24*].

[2] An extensive analysis of the differential equation (117.19) was made by OLDROYD [1947, *7* and *8*] [1948, *7* and *8*] [1949, *11*] [1951, *10*].

[3] This observation generalizes one of KAPUR and GOEL [1961, *32*], which elaborates a remark of CRIMINALE, ERICKSEN and FILBEY [1958, *13*, § III].

118. Steady extension[1]. Homogeneous motions with steady velocity gradient $\boldsymbol{L}$ have been defined in Sect. 30. The relative deformation is easily obtained from (21.15) and (21.16):

$$\boldsymbol{\xi} = e^{\boldsymbol{L}(\tau - t)}\boldsymbol{x}, \tag{118.1}$$

Hence

$$\boldsymbol{F}_{(t)}(\tau) = e^{\boldsymbol{L}(\tau - t)}. \tag{118.2}$$

It is clear that (109.1) holds with

$$\boldsymbol{Q}(\tau) = \boldsymbol{1}, \quad \varkappa = |\boldsymbol{L}|, \quad \boldsymbol{N}_0 = \boldsymbol{N} = \frac{1}{\varkappa}\boldsymbol{L}, \tag{118.3}$$

and hence that every steady flow with uniform velocity gradient is a motion with constant stretch history. Apart from the case when $\boldsymbol{L}^2 = \boldsymbol{0}$, which is a simple shearing (Sect. 111α), these motions are not viscometric flows, since they fall into the second or third of the categories listed at the end of Sect. 109.

In Sect. 30 it was shown that these motions are possible in every homogeneous incompressible material if $\boldsymbol{L}$ satisfies the conditions

$$\boldsymbol{L}^2 = (\boldsymbol{L}^2)^T, \qquad \operatorname{tr}\boldsymbol{L} = 0. \tag{118.4}$$

In the special case of a simple fluid in such a flow, use of (109.15) and $(118.3)_4$ reduces Eq. (30.35) for the stress to the form

$$\boldsymbol{T} = \varrho\left(\tfrac{1}{2}\boldsymbol{x}\cdot\boldsymbol{L}^2\boldsymbol{x} - v + h\right)\boldsymbol{1} + \mathfrak{q}(\boldsymbol{L}), \tag{118.5}$$

where $\mathfrak{q}$ is an isotropic function of $\boldsymbol{L}$ and where v is defined by (30.25).

For a steady extension (Sect. 30), $\boldsymbol{L}$ itself is symmetric and hence equals $\boldsymbol{D}$. In this case $\mathfrak{q}(\boldsymbol{D})$ has a representation of the form (12.2), so that (118.5) becomes

$$\boldsymbol{T} = \varrho\left(\tfrac{1}{2}\boldsymbol{x}\cdot\boldsymbol{D}^2\boldsymbol{x} + v - h\right)\boldsymbol{1} + \lambda_1\boldsymbol{D} + \lambda_2\boldsymbol{D}^2, \tag{118.6}$$

where λ_1 and λ_2 are scalar material functions of the two arguments

$$\left.\begin{aligned} \varkappa &\equiv |\boldsymbol{D}| = \sqrt{\operatorname{tr}\boldsymbol{D}^2} = \sqrt{d_1^2 + d_2^2 + d_3^2},\\ \delta &\equiv \det\boldsymbol{D} = d_1\, d_2\, d_3, \end{aligned}\right\} \tag{118.7}$$

d_1, d_2, d_3 being the proper numbers of $\boldsymbol{D}$ or *principal stretchings* (CFT, Sect. 83). The principal stresses are given by

$$t_k = \lambda_1(\varkappa, \delta)\, d_k + \lambda_2(\varkappa, \delta)\, d_k^2 + \tfrac{1}{2}\varrho\sum_{a=1}^{3}(d_a x_a)^2 + \varrho(v - h), \tag{118.8}$$

where x_a are the co-ordinates of $\boldsymbol{x}$ in a Cartesian system whose axes are parallel to the principal axes of stretching.

Consider now a fluid body that initially occupies a box whose edges are parallel to the co-ordinate axes. Since the principal axes of stretching are stationary, it follows from (118.1) that the body will occupy a box of the same type at all times t and that the lengths $b_k = b_k(t)$ of the edges of the box satisfy the equations

$$d_k = \dot{\overline{\log b_k}}. \tag{118.9}$$

Thus, the logarithmic rates of extension of the edges of the box are constant. The normal tractions that must be applied to the sides of the box to produce the extension can be computed from (118.8). The resulting formulae are rather

[1] COLEMAN and NOLL [1962, *12*]. A partial analysis for Reiner-Rivlin fluids had been given by REINER [1946, *3*]. SLATTERY [1964, *77*] presented a generalization to unsteady flows; he considered only a special class of D'ALEMBERT motions (CFT, Sect. 140).

complicated, but if inertia and body forces are neglected, we obtain

$$\left.\begin{aligned} t_1 - t_3 &= (d_1 - d_3)\,[\lambda_1(\varkappa,\delta) + (d_1 + d_3)\,\lambda_2(\varkappa,\delta)]\,,\\ t_2 - t_3 &= (d_2 - d_3)\,[\lambda_1(\varkappa,\delta) + (d_2 + d_3)\,\lambda_2(\varkappa,\delta)]\,,\end{aligned}\right\} \tag{118.10}$$

where t_k is the normal traction on the side perpendicular to the direction in which the stretching is d_k. The two equations (118.10) may be used to calculate the material functions λ_1 and λ_2 from an experimental determination of the dependence of $t_k - t_m$ on the logarithmic rates d_1, d_2, d_3.

We consider now the extension of a circular cylinder whose axis is parallel to the direction in which the stretching is d_1. We put

$$z = x_1, \qquad r = \sqrt{x_2^2 + x_3^2} \tag{118.11}$$

and choose

$$d_1 = d, \qquad d_2 = d_3 = -\tfrac{1}{2}d. \tag{118.12}$$

Then the body will retain the shape of a circular cylinder whose length $L = L(t)$ and radius $R = R(t)$ have the logarithmic extension rates

$$d = \dot{\overline{\log L}}, \qquad -\tfrac{1}{2}d = \dot{\overline{\log R}}. \tag{118.13}$$

From (118.8) we obtain for the normal traction t_R on the cylindrical boundary

$$t_R = \tfrac{1}{2}\varrho\,d^2(z^2 + \tfrac{1}{4}R^2) + \varrho\,v - \tfrac{1}{2}d\,\hat{\lambda}_1 - \tfrac{1}{4}d^2\hat{\lambda}_2 + \varrho\,h, \tag{118.14}$$

where $\hat{\lambda}_i = \hat{\lambda}_i(d) = \lambda_i\left(\frac{d}{2}\sqrt{6}, \frac{1}{4}\,d^3\right)$, $i = 1, 2$. We assume now that there are no body forces ($v = 0$) and that the bounding cross-sections of the cylinder are at

$$z = 0 \quad\text{and}\quad z = L = L(t). \tag{118.15}$$

We adjust h in (118.14) so that the mean value of t_R along the generators of the cylinder is zero. Then (118.14) reduces to

$$t_R = \tfrac{1}{2}\varrho\,d^2(z^2 - \tfrac{1}{3}L^2). \tag{118.16}$$

The normal tractions t_0 and t_L at the bounding cross-sections (118.15) turn out to be

$$t_0 = \tfrac{1}{2}\varrho\,d^2\left(\tfrac{1}{4}(r^2 - R^2) - \tfrac{1}{3}L^2\right) + \tfrac{3}{2}d\,\hat{\lambda}_1 + \tfrac{3}{4}d^2\hat{\lambda}_2, \tag{118.17}$$

$$t_L = \tfrac{1}{2}\varrho\,d^2\left(\tfrac{1}{4}(r^2 - R^2) + \tfrac{2}{3}L^2\right) + \tfrac{3}{2}d\,\hat{\lambda}_1 + \tfrac{3}{4}d^2\hat{\lambda}_2. \tag{118.18}$$

When inertia can be neglected ($\varrho = 0$), (118.16)—(118.18) reduce to

$$t_R = 0, \qquad t_0 = t_L = \tfrac{3}{2}d\,\hat{\lambda}_1 + \tfrac{3}{4}d^2\hat{\lambda}_2. \tag{118.19}$$

As remarked in Sect. 30, steady extension is only a special case of the dynamically possible homogeneous motions. It is degenerate in that the stress system (118.6) is possible in a Reiner-Rivlin fluid (Sect. 119). That is, whatever the forms of the material functions λ_1 and λ_2, we can find a Reiner-Rivlin fluid in which the stress system is exactly the same as (118.6). In other words, *in experiments on steady extension it is impossible to distinguish the class of simple fluids from the class of Reiner-Rivlin fluids.*

118 bis. Special flow problems for subfluids. (Added in proof.) We have seen in the previous sections the exact solutions of certain special flow problems for incompressible simple fluids. The corresponding problems for incompressible subfluids have been treated by WANG[1]. Here we consider the two representative cases: steady extension and simple shearing flow.

i) Steady extension. WANG has treated the case when the axes of extension coincide with the axes of a subfluid. Then in such a motion every local configuration is an undistorted state of the subfluid, and the stress matrix, taken relative to these axes, is diagonal. From (33b.2) it is easily verified that the stress tensor is again of the form (118.6), and the normal tensions are given by

$$\left.\begin{aligned} t_1 - t_3 &= (d_1 - d_3)\,[\lambda_1 + (d_1 + d_3)\,\lambda_2]\,, \\ t_2 - t_3 &= (d_2 - d_3)\,[\lambda_1 + (d_2 + d_3)\,\lambda_2]\,, \end{aligned}\right\} \qquad (118\text{b}.1)$$

except that λ_1 and λ_2 are in general functions of all three principal stretchings d_1, d_2, d_3. In particular, if the subfluid is a simple fluid, then the above equations reduce to (118.10). WANG has determined the restrictions on the functions λ_1 and λ_2 for the various types of subfluids. His results are listed in the following table.

Type No.	Restriction[2] on λ
14, 10, 9, 6, 4	No restriction
13, 8, 7, 3, 2	$\lambda(d_1, d_2, d_3) = \lambda(d_2, d_1, d_3)$
11, 1	λ is a symmetric function
12	λ is a rotationally symmetric function
5	$\lambda(d_1, d_2, d_3) = \lambda(d_1, d_3, d_2)$

ii) Simple shearing. From the velocity field (111.8), WANG notices that the material lines along the x and y axes and the material plane spanned by the y and z axes are invariant in the motion. Hence for subfluids of Types 1—10, there are suitable initial configurations for which the local configuration is invariant (hence also undistorted) in the motion. Then it is obvious that the stress components are given by (108.4)—(108.6), as for a simple fluid, except that the normal-stress and shear-stress functions, σ_1, σ_2 and τ depend, generally, also on the suitable initial configuration of the subfluid relative to the x, y, z axes. In this case, the stress components are also given by (111.9) with σ_1, σ_2 and τ corresponding to the subfluid considered.

For subfluids of Types 11—14, there are three preferred material lines in the material manifold. Hence the local configuration cannot remain invariant in simple shearing flow for all initial configurations. WANG has studied a problem for this case. He considers a subfluid of Type 14 for which the y and z axes coincide with two of the preferred axes of the subfluid, while its third preferred axis lies in the plane of x and y (i.e. the plane of shear).

For definiteness, let $\boldsymbol{e}_2$ and $\boldsymbol{e}_3$ be unit vectors in the directions of the y and z axes. Then the local configuration is characterized by

$$g_{12}(t) = \boldsymbol{e}_1(t) \cdot \boldsymbol{e}_2 . \qquad (118\text{b}.2)$$

From (33b.2) and (118b.2) WANG shows that the stress components are given by

$$T\langle 12\rangle = \tau\left(\varkappa, g_{12}(t)\right), \qquad (118\text{b}.3)$$

$$T\langle 11\rangle - T\langle 33\rangle = \sigma_1\left(\varkappa, g_{12}(t)\right), \qquad (118\text{b}.4)$$

$$T\langle 22\rangle - T\langle 33\rangle = \sigma_2\left(\varkappa, g_{12}(t)\right), \qquad (118\text{b}.5)$$

[1] WANG [1965, *42*].

[2] Here λ represents both λ_1 and λ_2.

where τ is an odd function while σ_1 and σ_2 are even functions:

$$\left.\begin{aligned}\tau(\lambda,\alpha)&=-\tau(-\lambda,-\alpha),\\ \sigma_\Gamma(\lambda,\alpha)&=\sigma_\Gamma(-\lambda,-\alpha),\quad \Gamma=1,2.\end{aligned}\right\}\tag{118b.6}$$

When $\varkappa=0$, the motion reduces to a static simple shear. Hence the static stress components are given by

$$T\langle 12\rangle=\varphi(g_{12})\equiv\tau(0,g_{12}),\tag{118b.7}$$

$$T\langle 11\rangle-T\langle 33\rangle=\psi_1(g_{12})\equiv\sigma_1(0,g_{12}),\tag{118b.8}$$

$$T\langle 22\rangle-T\langle 33\rangle=\psi_2(g_{12})\equiv\sigma_2(0,g_{12}).\tag{118b.9}$$

It is plausible that the shear stress function τ must satisfy the restriction

$$\varkappa[\tau(\varkappa,g_{12})-\varphi(g_{12})]\geqq 0.\tag{118b.10}$$

In this case, if the function φ is unbounded, then the shear stress (118b.3) in simple shearing is also unbounded since $g_{12}(t)$ attains all values in $(-1,1)$ while $t\in(-\infty,+\infty)$. Hence for every fixed $T\langle 12\rangle$ there are static solutions, found by inverting φ,

$$g_{12}=\overset{-1}{\varphi}(T\langle 12\rangle).\tag{118b.11}$$

Now suppose that φ is bounded,

$$K=\underset{\alpha\in(-1,1)}{\text{l.u.b.}}\ \varphi(\alpha).\tag{118b.12}$$

There the subfluids "yields" when $T\langle 12\rangle>K$. In this case simple shearing flow is a possible solution.

WANG has studied also certain inhomogeneous viscometric flows for incompressible subfluids. The solutions are local simple shearing flows, as we have seen for incompressible simple fluids.

c) Special simple fluids.

119. The older theories of non-linear viscosity. Since, as was stated in Sect. 104, the progress of the theory of non-linear viscosity has been mainly toward greater and greater generality of concept in a small class of flows, it is difficult to attribute particular results to particular persons. We mention here the three-dimensional non-linear constitutive equations proposed before 1959 and now seen to define, if sometimes imperfectly, special simple fluids[1].

In 1845 STOKES[2] laid down as basic hypothesis a "principle" which in mathematical form asserts that

$$\boldsymbol{T}=-p\mathbf{1}+\mathfrak{t}(\boldsymbol{D}),\tag{119.1}$$

where p is the pressure appropriate to equilibrium at the density and temperature at the place $\boldsymbol{x}$ at time t and where $\boldsymbol{D}$ is the stretching tensor. STOKES worked out

[1] Further literature on special theories has been cited and summarized by TRUESDELL [1952, *20*, §§ 59—72]. Cf. also the remarks of TRUESDELL [1956, *34*] and RIVLIN [1961, *52*] on the motivation of some of the researches since 1945.

There is no point in citing the vast rheological literature, since, attempting to force a one-dimensional framework upon circumstances now shown to be essentially three-dimensional, it neither reveals the typical physical phenomena nor points toward mathematical concepts sufficient for treating them.

Some early work on non-simple fluids was cited in footnote 2, p. 63. Cf. also Sects. 124 and 125.

[2] STOKES [1845, *1*, § 11].

the theory only for the case when $\mathfrak{r}$ is linear. Assuming the fluid to be isotropic, he found that

$$\boldsymbol{T} = -p\boldsymbol{1} + \lambda(\mathrm{tr}\,\boldsymbol{D})\boldsymbol{1} + 2\mu\boldsymbol{D}. \tag{119.2}$$

The resulting dynamical equations are called the *Navier-Stokes equations*. The "Stokes relation", viz $\lambda = -\frac{2}{3}\mu$, was inferred by STOKES from questionable hypotheses he later virtually withdrew[1].

Just a little earlier, ST. VENANT[2] had proposed a more general but also less definite theory. He postulated that the shear stress always points in the direction of the corresponding shearing. That is, $T_{km} = \varepsilon D_{km}$ if $k \neq m$, and for all systems of Cartesian co-ordinates. ST. VENANT proved that $T_{xx} - T_{yy} = \varepsilon(D_{xx} - D_{yy})$. Equivalently, $\boldsymbol{T} = \zeta\boldsymbol{1} + \varepsilon\boldsymbol{D}$. Hence, in particular, $\boldsymbol{T}$ and $\boldsymbol{D}$ are co-axial. While μ in the theory of STOKES is a constant or assigned function of the temperature, ST. VENANT allowed his coefficient ε to vary in space, although he did not specify the manner of its variation.

From hypotheses not altogether clear, KLEITZ[3] in 1866 proposed for turbulent flow of an incompressible fluid a special theory of the Stokes-St. Venant type, in which ε is a scalar function of $\boldsymbol{D}$. On the basis of a molecular hypothesis generalizing NAVIER'S, he concluded that ε is a power series in $\overline{II}_{\boldsymbol{D}}$, and he asserted that the empirical data available to him were accounted for by taking $\varepsilon = \varepsilon' + \varepsilon''(\overline{II}_{\boldsymbol{D}})^{\alpha}$, where ε', ε'', and α are constants. A result of this kind replaces the "power laws" of hydraulics by a properly invariant expression[4]. LÉVY[5] in 1867 proposed a theory of linear but non-simple fluids. BOUSSINESQ[6] replaced (119.1) by an apparently more general hypothesis:

$$\boldsymbol{T} = -p\boldsymbol{1} + \mathfrak{r}(\boldsymbol{D}, \boldsymbol{W}), \tag{119.3}$$

where $\boldsymbol{W}$ is the spin tensor and where $\mathfrak{r}(\boldsymbol{0}, \boldsymbol{W}) = \boldsymbol{0}$. The proposals of KLEITZ and LÉVY were criticized by ST. VENANT[7], whose authority seems to have sufficed to put an end to attempts to generalize the Navier-Stokes equations in France for half a century or more.

---

[1] A review of the literature concerned with the Stokes relation, which is now generally discredited except in textbooks by physicists, has been given by TRUESDELL [1952, *20*, § 61 A].

[2] ST. VENANT [1843, *1*].

[3] KLEITZ [1866, *1*] [1873, *1*, Ch. IV, §§ 34—35]. This same theory was revived by OLDROYD [1949, *11*, § 1], with "non-Newtonian" laminar flow as the intended application.

[4] KLEITZ [1873, *1*, Ch. IX, §§ 55—56, and Ch. II, § 22].

[5] According to ST. VENANT [1872, *3*]. Apparently LÉVY'S memoir was never published.

[6] BOUSSINESQ [1868, *1*, Note I].

[7] ST. VENANT'S reports [1869, *2*] [1872, *2*] were followed by a survey paper, from which it is worthwhile to quote [1872, *3*, § 8]:

"Tout se réduit ainsi ... à déterminer les valeurs diverses à ce coefficient ε du frottement fluide pour chaque point. Deux ingénieurs, bons analystes, l'ont tenté. L'Académie ... n'a point donné son approbation à leurs résultats principaux ...

"L'un et l'autre raisonnent en effet sur des mouvements fluides supposés continus et réguliers, ou exempts de cette *complication* que Navier signalait judicieusement comme ce qui empêchait ses formules de convenir aux applications ordinaires. Ils partent même tous deux de l'hypothèse de NAVIER ... Ils pensent seulement que NAVIER *n'a pas poussé l'approximation assez loin* ...

"Ni l'un ni l'autre de leurs résultats ... ne résout aucunement la question proposée."

However, ST. VENANT'S criticism [1872, *3*, § 9] is itself unsound. While he claimed that the results from the theories of KLEITZ and LÉVY do not conform to the experiments of POISEUILLE, neither he nor they obtained specific solutions on which a proper comparison could be based. In his argument toward rejecting KLEITZ'S theory, ST. VENANT was misled by a false formal analogy to elasticity theory that has tripped also some more recent authors (cf. the next footnote). LÉVY'S theory he considered to allow too much arbitrariness in the solutions.

Special theories within the definitions of STOKES and BOUSSINESQ were proposed by various authors in the second quarter of this century[1].

The modern period was opened by the revival of STOKES' definition by REINER[2], RIVLIN[3], and TRUESDELL[4]. The aims and methods of these three authors were somewhat different. REINER, assuming $\mathfrak{t}$ to be a matrix polynomial, obtained with the help of RACAH a representation of the form (12.6) for it, on the basis of which he discussed in some generality the normal stresses in shear flow[5].

[1] The theory proposed by GIRAULT [1931, *2*, Ch. III] and studied by VIGUIER [1947, *14* to *16*] [1948, *20*] [1949, *24* and *25*] [1951, *18*] is far from typical, as remarked by TRUESDELL [1952, *20*, § 59, footnote 3], since, guided by a false inference, GIRAULT demanded that $\mathfrak{t}(\boldsymbol{D})$ be an even function, with the consequence that all normal-stress effects are annulled. DE BACKER [1935, *1*] [1936, *1*] [1937, *1*, § 10], motivated by considerations in the kinetic theory of gases, proposed as constitutive equations certain spatial differential equations for the stress tensor. It seems that any one solution would correspond to a class of fluids satisfying (119.1). Likewise, the principle of minimum dissipation proposed by DE COLLE [1942, *1*] seems to presume STOKES's principle. Other special theories were proposed by SAKADI [1941, *3*] [1942, *5*], SETH [1944, *3*], and GARCIA [1947, *4*] [1948, *5*] [1949, *3*]. TRUESDELL [1949, *23*, § 18] [1951, *17*, § 18] [1952, *20*, § 60, footnote 4] remarked that the theories of SETH and GARCIA, based on false formal analogies to the theory of finite elastic strain, are not even dimensionally invariant.

[2] REINER [1945, *3*, § 4]. The dynamical equations were written out by BRAUN [1954, *3*A].

[3] RIVLIN [1947, *10*] [1948, *14*].

[4] TRUESDELL and SCHWARTZ [1947, *12*, § XI], TRUESDELL [1949, *22*, § 4] [1950, *16*, § 5].

[5] At the present writing, no exact solutions for flows of compressible fluids are known in any theory discussed in this treatise, except for the shearing mentioned in Sects. 119A and 124. Thus it is particularly interesting that REINER's aim was to explain dilatancy, which is a phenomenon of compressibility.

Dilatancy seems first to have been remarked by REYNOLDS, who defined it as "a definite change of bulk, consequent on a definite change of shape," and demonstrated it in two lectures [1885, *1*] [1887, *1*]. He considered only granular media: "sand, shingles, grain and piles of shot," and he found the phenomenon to be largely independent of whatever friction acts among the particles in contact. As an example, he remarked, "When the falling tide leaves the sand firm, as the foot falls on it the sand whitens or appears momentarily to dry around the foot. When this happens the sand is full of water ...; the pressure of the foot causing dilatation of the sand, more water is required ...; on raising the foot it is generally seen that the sand under the foot and around becomes momentarily wet; this is because, on the distorting forces being removed, the sand again contracts ..." In the demonstrations, REYNOLDS filled a rubber bag with wet sand, to which was attached a tube, and showed that when the bag was squeezed, the water rushed into it through the tube from the glass vessel. In various laboratories today may be found a simple apparatus for demonstrating the effect: A rod is allowed to stand freely in a tube of larger diameter, and the space between is filled with dry sand; to pull the rod out again, a very great resistance must be overcome.

REINER's concept of dilatancy amounts to the following: *a fluid is dilatant if application of non-zero shear stress necessarily produces a change of volume.* Set $\boldsymbol{V} \equiv \boldsymbol{T} + p\,\mathbf{1}$. Then the condition that non-vanishing pure shear stress be superimposed upon the equilibrium pressure is $\operatorname{tr} \boldsymbol{V} = \det \boldsymbol{V} = 0$, $II_{\boldsymbol{V}} \neq 0$. A fluid is dilatant if for any $\boldsymbol{D}$ compatible with such a state of stress, $\operatorname{tr} \boldsymbol{D} \neq 0$. While REINER in his treatment of dilatancy [1945, *3*, §§ 6–7] [1949, *13*, §§ 6 and 9] [1951, *12*, § 7] assumed that (119.4) may be inverted, a result stronger and more specific than his, although still within his theory, may be obtained directly. Namely, in order for a fluid to *fail* to be dilatant, it is necessary and sufficient that the equations $\operatorname{tr} \boldsymbol{V} = \det \boldsymbol{V} = 0$ have a solution $\boldsymbol{D}$ such that $\operatorname{tr} \boldsymbol{D} = 0$, while $II_{\boldsymbol{V}} \neq 0$. Now we easily see from (119.4) that the equations $\operatorname{tr} \boldsymbol{V} = 0$ and $\det \boldsymbol{V} = 0$ under these conditions take the forms

$$3\aleph_0 - 2\aleph_2 II_{\boldsymbol{D}} = 0, \tag{A}$$

$$\left.\begin{aligned} &\aleph_0^3 + (\aleph_0\aleph_1^2 - 2\aleph_0^2\aleph_2)\, II_{\boldsymbol{D}} + (\aleph_1^3 - 3\aleph_0\aleph_1\aleph_2)\, III_{\boldsymbol{D}} + \\ &\quad + \aleph_0\aleph_2^2\, II_{\boldsymbol{D}}^2 + \aleph_1\aleph_2^2\, II_{\boldsymbol{D}}\, III_{\boldsymbol{D}} + \aleph_2^3\, III_{\boldsymbol{D}}^2 = 0, \end{aligned}\right\} \tag{B}$$

while the inequality $II_{\boldsymbol{V}} \neq 0$ has the form

$$3\aleph_0^2 + (\aleph_1^2 - 4\aleph_0\aleph_2 + II_{\boldsymbol{D}}\aleph_2^2)\, II_{\boldsymbol{D}} - 3\aleph_1\aleph_2\, III_{\boldsymbol{D}} \neq 0, \tag{C}$$

where the arguments of $\aleph_\Gamma$ are 0, $II_{\boldsymbol{D}}$, $III_{\boldsymbol{D}}$, ϱ. Since $\aleph_0(0, 0, 0, \varrho) = 0$, $\boldsymbol{D} = \mathbf{0}$ is always a solution of (A) and (B), but it never satisfies (C). Intuitively we might expect the two conditions (A) and (B) to be incompatible or to determine certain values, possibly unique, for $II_{\boldsymbol{D}}$ and

RIVLIN exhibited exact solutions for certain special flows of incompressible fluids; these solutions, which had a deep and lasting effect on the development of the subject, have been presented in much generalized form in Sects. 111—113. RIVLIN's intended application was to high-polymer solutions. Incidentally, he gave a molecular motivation for normal-stress effects[1]. TRUESDELL, at that time interested mainly in rarefied gases, considered thermodynamic effects, dimensional invariance, scaling laws, and relations to results from the Maxwellian kinetic theory. His two theories of fluids are summarized in Sects. 119A and 125.

NOLL[2] showed that the principle of material frame-indifference reduces BOUSSINESQ's definition (119.3) to STOKES' definition (119.1) and also forces $\mathfrak{t}$ to be an isotropic function. That is, while REINER, RIVLIN, and TRUESDELL had introduced some special assumption of isotropy, in fact no anisotropic materials are included in the underlying definition. Therefore, with no loss in generality, we may regard the Reiner-Rivlin theory as based on the explicit constitutive equation

$$\boldsymbol{T} = -p\boldsymbol{1} + \aleph_0 \boldsymbol{1} + \aleph_1 \boldsymbol{D} + \aleph_2 \boldsymbol{D}^2, \tag{119.4}$$

where p is a function of the density and where the coefficients $\aleph_0$, $\aleph_1$, and $\aleph_2$ are functions of the density and of the three principal invariants of $\boldsymbol{D}$. For incompressible Reiner-Rivlin fluids, by a suitable choice of the indeterminate pressure p, (119.4) may be written

$$\boldsymbol{T} = -p\boldsymbol{1} + \aleph_1 (II_{\boldsymbol{D}}, III_{\boldsymbol{D}}) \boldsymbol{D} + \aleph_2 (II_{\boldsymbol{D}}, III_{\boldsymbol{D}}) \boldsymbol{D}^2, \tag{119.5}$$

only flows such that $I_{\boldsymbol{D}} = \operatorname{tr} \boldsymbol{D} = 0$ being admissible.

For viscometric flows, an easy computation using (106.10) and (106.7) shows that

$$[\boldsymbol{D}] = -\tfrac{1}{2} \dot{g}(0) \begin{Vmatrix} 0 & 1 & 0 \\ 1 & 0 & 0 \\ 0 & 0 & 0 \end{Vmatrix}, \quad II_{\boldsymbol{D}} = \tfrac{1}{4} [\dot{g}(0)]^2, \quad III_{\boldsymbol{D}} = 0, \tag{119.6}$$

where $[\boldsymbol{D}]$ is the matrix of $\boldsymbol{D}$ with respect to a suitable rotating orthonormal basis $\boldsymbol{i}_r(t)$. In the special case of curvilineal flow, the basis $\boldsymbol{i}_r(t)$ is given by (107.14), and $g(s) = k(t-s) - k(t)$, where $k(\tau)$ is given by $(107.9)_1$. Therefore

$$\dot{g}(0) = -\dot{k}(t). \tag{119.7}$$

$III_{\boldsymbol{D}}$. In the former case, the fluid is certainly dilatant. In the latter, there would be a fair chance for (C) to be satisfied by the solution, in which case the fluid would be non-dilatant.

Both possibilities may be illustrated by the special case when

$$\aleph_0 = \lambda \operatorname{tr} \boldsymbol{D}, \quad \aleph_1 = 2\mu, \quad \aleph_2 = \gamma, \tag{D}$$

where λ, μ, and γ are constants. If $\gamma = 0$, this fluid is the classical Navier-Stokes fluid. The conditions (A), (B), and (C) become

$$\gamma = 0, \tag{A'}$$

$$\mu^3 III_{\boldsymbol{D}} = 0, \tag{B'}$$

$$\mu^2 II_{\boldsymbol{D}} \neq 0. \tag{C'}$$

The system (A'), (B') is incompatible unless $\gamma = 0$. If $\gamma = 0$, any motion such that $II_{\boldsymbol{D}} \neq 0$ but $III_{\boldsymbol{D}} = 0$ corresponds to a state of pure shear stress. That is, the fluid defined by (D) is dilatant except in the special case when it reduces to the Navier-Stokes fluid, which is non-dilatant.

REINER did not discuss the compatibility of his assumptions with the dynamical equations, the equation of energy, and an assigned equation of state, as is necessary in any solution for compressible fluids.

[1] RIVLIN [1948, *16*] [1949, *19*].

[2] NOLL [1955, *19*, § 13].

Substituting (119.6) into (119.5), we see that the material functionals in (106.20)—(106.22), for Reiner-Rivlin fluids, reduce to

$$\left.\begin{aligned}\overset{\infty}{\underset{s=0}{\mathfrak{t}}}\big(g(s)\big)&=-\tfrac{1}{2}\dot{g}(0)\,\aleph_1\big(\tfrac{1}{4}\dot{g}(0)^2,0\big)\\ \overset{\infty}{\underset{s=0}{\mathfrak{s}_1}}\big(g(s)\big)&=\overset{\infty}{\underset{s=0}{\mathfrak{s}_2}}\big(g(s)\big)=\tfrac{1}{4}[\dot{g}(0)]^2\aleph_2\big(\tfrac{1}{4}\dot{g}(0)^2,0\big).\end{aligned}\right\}\qquad(119.8)$$

By (108.3) and (108.10), the viscosity function $\tilde{\mu}$ and the normal-stress functions σ_1, σ_2 become, for Reiner-Rivlin fluids,

$$\frac{1}{\varkappa}\tau(\varkappa)=\tilde{\mu}(\varkappa)=\frac{1}{2}\aleph_1\left(\frac{1}{4}\varkappa^2,0\right),\qquad \sigma_1(\varkappa)=\sigma_2(\varkappa)=\frac{1}{4}\varkappa^2\aleph_2\left(\frac{1}{4}\varkappa^2,0\right).\qquad(119.9)$$

Note that the two normal-stress functions are equal and hence that $T\langle 11\rangle = T\langle 22\rangle$ in every steady viscometric flow of any Reiner-Rivlin fluid. A universal relation of this kind suggests immediate check by experiment. Data was collected[1]; they seem to indicate that for polyisobutylene solutions, $T\langle 11\rangle$ and $T\langle 22\rangle$ are not equal when $\varkappa$ becomes appreciably large. This experimental evidence was the greatest motive for the rejection of the Reiner-Rivlin equation (119.4) as an adequate basis of a physical theory[2] and for the search for constitutive equations

[1] PADDEN and DEWITT [1954, *15*]. The incompatibility of their data with the Reiner-Rivlin theory was pointed out by MARKOVITZ [1957, *13*]. In a work dated 1958, GIESEKUS [1961, *18*, § V] was one of the first to conjecture that the experimental data might require the three viscometric functions to be independent.

[2] There are numerous papers on various aspects of the Reiner-Rivlin theory.

OLDROYD's analysis of rectilinear flow [1949, *11*] [1951, *10*] applies in particular to Reiner-Rivlin fluids. Cf. Sect. 117.

BRAUN and REINER [1952, *3*], to the extent that they considered incompressible fluids, derived anew results obtained earlier and more generally by RIVLIN; to the extent that compressible fluids are intended, their work is faulty from failure to take account of the static equation of state $p = p(\varrho)$ and other effects of compressibility.

BAKER and ERICKSEN [1954, *3*] found restrictions upon the response coefficients equivalent to the requirement that the proper numbers d_a of $\boldsymbol{D}$ and t_a of $\boldsymbol{T}$ satisfy the inequalities

$$(t_a - t_b)(d_a - d_b) > 0 \quad \text{if} \quad d_a \neq d_b.$$

ERICKSEN [1955, *12*] showed this condition to be sufficient but not necessary in order that work be expended in any motion in which the distortion tensor [CFT, Sect. 83] does not vanish. ERICKSEN [1957, *7*] found necessary conditions, for incompressible fluids, that there be no real characteristic directions. BERKER [1964, *6*] found the form of the traction on a fixed wall; for incompressible fluids a more general result of his is given in Sect. 123.

Conditions for superposability and self-superposability were obtained by GOLD and KRZYWOBLOCKI [1958, *20*, § 5] [1959, *9*, § 7] and by RATHNA and RAJESWARI [1961, *49*].

Variational principles equivalent to special theories of Reiner-Rivlin type were discussed by PAWLOWSKI [1954, *16*], BIRD [1960, *14*], JOHNSON [1960, *31*] [1961, *31*], SCHLECHTER [1962, *59*], and WEHRLI and ZIEGLER [1962, *72*, § 3]. LEIGH [1962, *45*] rediscovered a fact remarked by TRUESDELL [1949, *23*, § 29] [1951, *17*, § 23] [1952, *20*, § 61]: In a quadratic theory of Reiner-Rivlin type, it is impossible that the dissipation, $\operatorname{tr}(\boldsymbol{T}_E \boldsymbol{D})$, be positive for all non-vanishing $\boldsymbol{D}$ unless the theory reduces to the classical linear one. As shown by the theorem of ERICKSEN, stated just above, no such objection may be levelled at Reiner-Rivlin fluids in general.

The structure of plane shock layers according to the Reiner-Rivlin theory has been determined by GILBARG and PAOLUCCI [1953, *7*, § 5] and discussed by THEODORIDES [1956, *28*] [1957, *28*] [1958, *48*].

Turbulence in Reiner-Rivlin fluids has been discussed by LUMLEY [1964, *56*].

There is further literature, mainly Indian, concerning approximate solutions or solutions possible in the case when $\aleph_1$ and $\aleph_2$ are taken as constants: REINER [1949, *13*] [1951, *12*] [1952, *15*] [1954, *18*], SUBBA RAO and NIGAM [1954, *24*], JAIN [1955, *14* and *15*] [1957, *10*] [1958, *23*] [1961, *28* and *29*] [1962, *36*] [1964, *46*], NARASIMHAN [1956, *18*] [1960, *38*] [1961,

of greater generality, proposed in the last few years and explained in the earlier sections of this chapter. Results for the Reiner-Rivlin theory, sometimes attractive because of their simplicity, of course are included as special cases in all our work.

In a motion with constant stretch history, for Reiner-Rivlin fluids the relation (106.23) reduces to (119.5) with $\boldsymbol{D}$ having the form

$$\boldsymbol{D} = \tfrac{1}{2}\varkappa(\boldsymbol{N} + \boldsymbol{N}^T). \tag{119.10}$$

Several authors[1] have applied the results of the theory of finite elastic strain in an attempt to explain viscometric experiments. TRUESDELL[2] rejected these theories because the fluids under consideration show no evidence of having preferred reference configurations (e.g., undistorted states). The crux of the matter lies in the vagueness of the "elastic" theories for fluids. As remarked by GREENSMITH and RIVLIN[3], "none of these theories gives any indication of how the finite strain in the liquid is to be determined in terms of the state of flow." Nevertheless, in connection with his criticism TRUESDELL[4] had shown how the concept of "elasticity" of a fluid could be given a rational position. Instead of supposing as in real elasticity theory that the strain determining the stress be taken with respect to a fixed reference configuration, he suggested taking as reference the configuration occupied by the fluid at time $t - t^*$, where t^* is a fixed constant. That this idea is compatible with the theory of simple fluids is immediate from (105.1); one need only suppose that $\mathop{\mathfrak{Q}}\limits_{s=0}^{\infty}(\boldsymbol{G}(s)) = \mathfrak{f}(t^{*-1}\boldsymbol{G}(t^*))$, where $\mathfrak{f}$ is isotropic and t^* is a fixed time. This special case of the simple fluid may be called a *fluid of convected elasticity*[5]. This theory cannot account for general behavior in viscometric flow, since the universal relation (54.16) must hold[6]; in terms of the viscometric functions defined by Eqs. (108.4)—(108.6) and (108.10), this relation assumes the form[7]

$$\sigma_2(\varkappa) - \sigma_1(\varkappa) = t^* \varkappa^2 \tilde{\mu}(\varkappa). \tag{119.11}$$

The Reiner-Rivlin theory results in the limit as $t^* \to 0$. For finite values of t^*, the Reiner-Rivlin relation $(119.9)_3$ is avoided, and this was regarded as a virtue

---

39 and *40*], SRIVASTAVA [1956, *27*] [1958, *46* and *47*] [1959, *36*] [1960, *54*] [1961, *60*], MITHAL [1958, *29*] [1960, *37*] [1961, *38*], NIGAM [1958, *30*], SAKADI [1958, *41*], BHATNAGAR and LAKSHMANA RAO [1958, *7*] [1960, *12*], JONES [1960, *32*] [1964, *48*], KAPUR [1960, *33*] [1963, *40*], OBEROI and KAPUR [1960, *39*], RATHNA [1960, *43* and *44*] [1962, *54*], BHATNAGAR [1961, *4*], GRAEBEL [1961, *19*], JAIN and BALRAM [1961, *30*], RAJESWARI [1961, *48*] [1962, *52*] [1964, *69*], SLATTERY [1961, *57*] [1962, *61*] [1964, *78*], SLATTERY and BIRD [1961, *58*], ZIEGENHAUSEN, BIRD, and JOHNSON [1961, *69*], DATTA [1961, *11*] [1962, *13*], NANDA [1962, *49*] [1963, *56*], VARLEY [1962, *67*], IYENGAR [1964, *45*], LAKSHMANA RAO [1964, *52* and *53*], RINTEL [1964, *70*]. Results for the Reiner-Rivlin theory with constant response functions are included as special cases in those obtained or cited in Sect. 123, where the fluid of second grade is analysed.

Only after the leading theorists had abandoned the Reiner-Rivlin theory because the experiments had come to show its failure to describe *any* known physical fluid (except, of course, when it reduces to the Navier-Stokes theory) did literature about it to begin to gush from "applied" and "engineering" ambients.

[1] WEISSENBERG [1947, *17*] [1949, *26* and *27*] [1950, *17*], MOONEY [1951, *7*] [1953, *18*], besides various semi-chemical writers.

[2] TRUESDELL [1952, *20*, § 72].

[3] GREENSMITH and RIVLIN [1953, *9*, § 1].

[4] TRUESDELL [1952, *20*, § 72]. Cf. the related remarks of O.-E. MEYER [1874, *2*], BURGERS [1948, *1*, § 3], and ERICKSEN [1960, *22*, § III].

[5] TRUESDELL [1965, *37*].

[6] This was observed by TRUESDELL [1952, *20*, § 70].

[7] ERICKSEN [1960, *22*, § III].

by proponents of the theories of "elastic" type. However, there is no reason to suppose that (119.11) is any better[1].

The constitutive equation (119.1) defines a simple kind of a fluid of the differential type. General fluids of the differential type (cf. Sect. 35) were first considered by RIVLIN and ERICKSEN[2]. In the case of an incompressible fluid, a constitutive equation of the differential type is of the form

$$\boldsymbol{T} = -p\mathbf{1} + \mathfrak{r}(\boldsymbol{A}_1, \boldsymbol{A}_2, \ldots, \boldsymbol{A}_\mathfrak{n}), \tag{119.12}$$

where the $\boldsymbol{A}_\mathfrak{r}$ are the Rivlin-Ericksen tensors.

For viscometric flows, we obtain from (106.10) and $(106.8)_1$

$$\overset{(\mathfrak{n})}{\boldsymbol{F}}_{(t)}(t) = \boldsymbol{L}_\mathfrak{n} = (-1)^\mathfrak{n} \overset{(\mathfrak{n})}{g}(0)\,\boldsymbol{N} \tag{119.13}$$

and hence, by (25.2),

$$(-1)^\mathfrak{n}\boldsymbol{A}_\mathfrak{n} = \overset{(\mathfrak{n})}{g}(0)\,(\boldsymbol{N}+\boldsymbol{N}^T) + \left\{\sum_{\mathfrak{i}=1}^{\mathfrak{n}-1} \binom{\mathfrak{n}}{\mathfrak{i}} \overset{(\mathfrak{i})}{g}(0) \overset{(\mathfrak{n}-\mathfrak{i})}{g}(0)\right\} \boldsymbol{N}^T\boldsymbol{N}. \tag{119.14}$$

For any special constitutive equation of the form (119.12) the material functionals $\mathfrak{t}$, $\mathfrak{s}_1$, and $\mathfrak{s}_2$ of (106.20)—(106.22) may easily be obtained by substituting (119.14) into (119.12). The functionals $\mathfrak{t}$, $\mathfrak{s}_1$, and $\mathfrak{s}_2$ reduce then to functions of the derivatives $\overset{(\mathfrak{n})}{g}(0)$, $\mathfrak{n}=1, 2, \ldots$.

For steady viscometric flows, we have $\dot{g}(0) = -\varkappa$, $\overset{(\mathfrak{n})}{g}(0)=0$ when $\mathfrak{n}>1$, and (119.14) reduces to (109.13). For such flows, only the first two Rivlin-Ericksen tensors $\boldsymbol{A}_1$ and $\boldsymbol{A}_2$ are different from zero[3]. Thus if no other tests beyond those resting upon steady viscometric flows were available, we should be unable to distinguish between the general Rivlin-Ericksen fluid and the special one in which $\mathfrak{r} = \mathfrak{r}(\boldsymbol{A}_1, \boldsymbol{A}_2)$.

If we assume that $\mathfrak{r}$ is a polynomial function, we can use the representation (13.7) and write

$$\left.\begin{aligned}\mathfrak{r}(\boldsymbol{A}_1, \boldsymbol{A}_2, 0, \ldots, 0) &= \alpha_1\boldsymbol{A}_1 + \alpha_2\boldsymbol{A}_2 + \alpha_3\boldsymbol{A}_1^2 + \alpha_4\boldsymbol{A}_2^2 + \alpha_5(\boldsymbol{A}_1\boldsymbol{A}_2 + \boldsymbol{A}_2\boldsymbol{A}_1) + \\ &+ \alpha_6(\boldsymbol{A}_1^2\boldsymbol{A}_2 + \boldsymbol{A}_2\boldsymbol{A}_1^2) + \alpha_7(\boldsymbol{A}_1\boldsymbol{A}_2^2 + \boldsymbol{A}_2^2\boldsymbol{A}_1) + \alpha_8(\boldsymbol{A}_1^2\boldsymbol{A}_2^2 + \boldsymbol{A}_2^2\boldsymbol{A}_1^2),\end{aligned}\right\} \tag{119.15}$$

where $\alpha_1, \ldots, \alpha_8$ are simultaneous invariants of $\boldsymbol{A}_1$ and $\boldsymbol{A}_2$. Substituting (109.13) into (119.15) and observing (106.7), we find that, for fluids of the differential

[1] Here we may mention a futile controversy about "elastic" and "viscous" properties of fluids. RIVLIN [1948, *14*, § 17] wrote: "It should be remarked that any visco-elastic material, which is fluid-like in its ability to be continuously deformed, cannot be distinguished from a visco-inelastic fluid solely by means of steady-state experiments, in which the flow is laminar. For example, suppose the fundamental stress-strain-velocity relations for the material involve time derivatives of the stress components. For a steady state of flow, these vanish and the stress-strain-velocity relations are indistinguishable from those describing a true fluid." OLDROYD [1950, *11*, Added Note] justly remarked that this conclusion is false. Some confusion was added to the subject by qualitative remarks of TRUESDELL [1952, *20*, § 72] and of BRAUN and REINER [1952, *3*]. The whole controversy illustrates the empty wordplay coming from use of terms such as "elastic", "viscous", and "visco-elastic" in undefined, physical senses. It vanishes into thin air as soon as definite theories of mechanical response are laid down. In a steady flow the co-rotational stress rate $\overset{\circ}{\boldsymbol{T}}$ need not be zero, and invariant constitutive equations should involve $\overset{\circ}{\boldsymbol{T}}$ or one of its equivalents (cf. Sect. 36), not $\partial\boldsymbol{T}/\partial t$. Moreover, $\boldsymbol{A}_2$ generally is not zero in steady flows, and the terms in $\boldsymbol{A}_2$, as is clear from (119.16), suffice to yield response even in steady viscometric flows that is not at all the same as that following from the Reiner-Rivlin theory.

[2] RIVLIN and ERICKSEN [1955, *23*, Eq. (37.10)] showed, in effect, that a relation giving the stress as a function of the gradients of the accelerations of orders $1, 2, \ldots, \mathfrak{n}$ satisfies the principle of material indifference if and only if it is equivalent to Eq. (35.12) of this treatise.

[3] RIVLIN [1956, *25*].

type, the material functions $\tilde{\mu}$, σ_1, σ_2 are related to the coefficients α_Γ of the representation (119.15) by[1]

$$\left.\begin{aligned}\tilde{\mu}(\varkappa) &= \alpha_1 + 2\varkappa^2\alpha_5 + 4\varkappa^4\alpha_7,\\ \sigma_1(\varkappa) &= \varkappa^2(2\alpha_2 + \alpha_3 + 4\varkappa^2\alpha_4 + 4\varkappa^2\alpha_6 + 8\varkappa^4\alpha_8),\\ \sigma_2(\varkappa) &= \varkappa^2\alpha_3.\end{aligned}\right\} \tag{119.16}$$

For steady viscometric flows, the coefficients α_Γ depend, of course, only on $\varkappa$.

In order to obtain for fluids of the differential type the form of the relation (109.15), concerning general motions with constant stretch history, one need only substitute $(109.6)_1$ into (119.12).

The first to consider a properly invariant though special constitutive equation of a fluid of the rate type was ZAREMBA[2]. A variety of special fluids of the rate type were proposed and studied by later authors[3]. The most general constitutive equation for a compressible fluid of the rate type may be written in the form (36.27). If the fluid is incompressible, (36.27) must be replaced by

$$\overset{\triangle}{\boldsymbol{T}}_{\mathrm{E}\mathfrak{p}} = \mathfrak{h}(\boldsymbol{T}_\mathrm{E}, \overset{\triangle}{\boldsymbol{T}}_\mathrm{E}, \ldots, \overset{\triangle}{\boldsymbol{T}}_{\mathrm{E}\,\mathfrak{p}-1};\ \boldsymbol{A}_1, \boldsymbol{A}_2, \ldots, \boldsymbol{A}_\mathfrak{n}), \tag{119.17}$$

where $\boldsymbol{T}_\mathrm{E} = \boldsymbol{T} + p\boldsymbol{1}$ is an appropriate extra-stress. As we remarked in Sect. 36, the constitutive equation (119.17) must be regarded as defining a class of simple materials rather than a single one. Also, not for every function $\mathfrak{h}$ will (119.17) define a fluid. We shall see, however, that under certain conditions all fluids that are defined by (119.17) behave alike in motions with constant stretch history. For such motions, we have, by (109.3), $(109.2)_1$, and (109.15),

$$\boldsymbol{F}_{(t')}(t)^T\boldsymbol{T}_\mathrm{E}(t)\boldsymbol{F}_{(t')}(t) = \boldsymbol{Q}(t')\,e^{(t-t')\varkappa\boldsymbol{N}_0^T}\mathfrak{g}(\varkappa, \boldsymbol{N}_0)\,e^{(t-t')\varkappa\boldsymbol{N}_0}\boldsymbol{Q}(t')^T. \tag{119.18}$$

Applying the definition (36.17) to (119.18) and using $(109.2)_1$ and (109.15) again, we derive the formula

$$\overset{\triangle}{\boldsymbol{T}}_{\mathrm{E}\mathfrak{n}} = \varkappa^\mathfrak{n}\sum_{\mathfrak{k}=0}^{\mathfrak{n}}\binom{\mathfrak{n}}{\mathfrak{k}}\boldsymbol{N}^{\mathfrak{k}T}\boldsymbol{T}_\mathrm{E}\boldsymbol{N}^{\mathfrak{n}-\mathfrak{k}}. \tag{119.19}$$

Substitution of (119.19) and $(109.6)_1$ into (119.17) yields an equation involving only $\boldsymbol{T}_\mathrm{E}$, $\boldsymbol{N}$, and $\varkappa$. If this equation can be solved for $\boldsymbol{T}_\mathrm{E}$, we obtain a unique relation of the form (109.15), common to the whole class of simple fluids defined by (119.17) and governing their behavior in all motions with constant stretch history.

For steady viscometric flows[4], we have $\boldsymbol{N}^2 = \boldsymbol{0}$, and (119.19) reduces to

$$\left.\begin{aligned}&\overset{\triangle}{\boldsymbol{T}}_\mathrm{E} = \varkappa(\boldsymbol{T}_\mathrm{E}\boldsymbol{N} + \boldsymbol{N}^T\boldsymbol{T}_\mathrm{E}),\\ &\overset{\triangle}{\boldsymbol{T}}_{\mathrm{E}2} = 2\varkappa^2\boldsymbol{N}^T\boldsymbol{T}_\mathrm{E}\boldsymbol{N}, \qquad \overset{\triangle}{\boldsymbol{T}}_{\mathrm{E}\mathfrak{n}} = \boldsymbol{0} \quad \text{if} \quad \mathfrak{n} > 2.\end{aligned}\right\} \tag{119.20}$$

---

[1] RIVLIN [1956, *25*] first obtained several solutions to viscometric problems in terms of the coefficients $\alpha_1, \ldots, \alpha_8$. It was not realized until later that these solutions in fact depend on the α_Γ only through the three combinations yielding the material functions $\tilde{\mu}$, σ_1, and σ_2. Eqs. (119.16) were given by MARKOVITZ [1957, *13*], CRIMINALE, ERICKSEN, and FILBEY [1958, *13*, § III], GIESEKUS (1958) [1961, *18*], and COLEMAN and NOLL [1959, *4*, Appendix].

[2] ZAREMBA [1903, *11* and *12*] [1937, *4*]. We cite here and in the next footnote only properly invariant proposals, passing over in silence the great number of papers purporting to treat "small" motions by introducing time derivatives here and there.

[3] FROMM [1933, *2*] [1947, *3*] [1948, *4*], OLDROYD [1950, *11*] [1951, *9*] [1958, *32*], DEWITT [1955, *9*], NOLL [1955, *19*, §§ 6—12, 18—23], JUNG [1958, *26*].

[4] The above method of calculating the viscometric functions of a fluid of the rate type, here published for the first time, is due in principle to NOLL [1955, *19*, § 21]. OLDROYD [1965, *29*], using convected co-ordinates, has discussed viscometric flows in a general fluid of the mixed integral-rate type (defined at the end of Sect. 37).

If we substitute (119.20) and (109.13) into (119.17) and use matrices of the form (106.7) and (106.19) for $\boldsymbol{N}$ and $\boldsymbol{T}_{\mathrm{E}}$, respectively, we obtain a set of equations involving only $\varkappa$, $\tau = T_{\mathrm{E}\langle 12\rangle}$, $\nu_k = T_{\mathrm{E}\langle kk\rangle}$, $k=1, 2, 3$. If these equations can be solved for τ, ν_1, ν_2, and ν_3, then the viscometric functions τ, $\sigma_1 = \nu_1 - \nu_3$, and $\sigma_2 = \nu_2 - \nu_3$ can be uniquely determined from a knowledge of the response function $\mathfrak{h}$ in (119.17). As an illustration, we shall carry out this calculation explicitly in the special case when (119.17) reduces to[1]

$$\left.\begin{aligned}\mathring{\boldsymbol{T}}_{\mathrm{E}} = \beta_1 \boldsymbol{T}_{\mathrm{E}} + \beta_2 (\operatorname{tr} \boldsymbol{T}_{\mathrm{E}}) \mathbf{1} + \beta_3 \boldsymbol{A}_1 + \beta_4 (\boldsymbol{T}_{\mathrm{E}} \boldsymbol{A}_1 + \boldsymbol{A}_1 \boldsymbol{T}_{\mathrm{E}}) + \\ + \beta_5 (\operatorname{tr} \boldsymbol{T}_{\mathrm{E}}) \boldsymbol{A}_1 + \beta_6 \operatorname{tr} (\boldsymbol{T}_{\mathrm{E}} \boldsymbol{A}_1) \mathbf{1} + \beta_7 \boldsymbol{A}_2 + \beta_8 \boldsymbol{A}_1^2 + \beta_9 \operatorname{tr} (\boldsymbol{A}_1^2) \mathbf{1},\end{aligned}\right\} \tag{119.21}$$

where $\beta_1, \ldots, \beta_9$ are material constants. Substitution of (119.20) and (109.13) into (119.21) gives

$$\left.\begin{aligned}&\beta_1 \boldsymbol{T}_{\mathrm{E}} + \beta_2 (\operatorname{tr} \boldsymbol{T}_{\mathrm{E}}) \mathbf{1} + \beta_3 \varkappa (\boldsymbol{N} + \boldsymbol{N}^T) + \beta_4 \varkappa (\boldsymbol{T}_{\mathrm{E}} \boldsymbol{N}^T + \boldsymbol{N} \boldsymbol{T}_{\mathrm{E}}) + \\ &\quad + (\beta_4 - 1) \varkappa (\boldsymbol{T}_{\mathrm{E}} \boldsymbol{N} + \boldsymbol{N}^T \boldsymbol{T}_{\mathrm{E}}) + \beta_5 \varkappa (\operatorname{tr} \boldsymbol{T}_{\mathrm{E}}) (\boldsymbol{N} + \boldsymbol{N}^T) + 2\beta_6 \varkappa \operatorname{tr} (\boldsymbol{T}_{\mathrm{E}} \boldsymbol{N}) \mathbf{1} + \\ &\quad + 2\beta_7 \varkappa^2 \boldsymbol{N}^T \boldsymbol{N} + \beta_8 \varkappa^2 (\boldsymbol{N} + \boldsymbol{N}^T)^2 + \beta_9 \varkappa^2 \operatorname{tr} (\boldsymbol{N} + \boldsymbol{N}^T)^2 \mathbf{1} = \mathbf{0}.\end{aligned}\right\} \tag{119.22}$$

Using the matrices

$$[\boldsymbol{N}] = \begin{Vmatrix} 0 & 0 & 0 \\ 1 & 0 & 0 \\ 0 & 0 & 0 \end{Vmatrix}, \qquad [\boldsymbol{T}_{\mathrm{E}}] = \begin{Vmatrix} \nu_1 & \tau & 0 \\ \tau & \nu_2 & 0 \\ 0 & 0 & \nu_3 \end{Vmatrix}, \tag{119.23}$$

we find that (119.22) is equivalent to the following system of 4 scalar equations

$$\left.\begin{aligned}\beta_1 \tau + \beta_4 \varkappa \nu_1 + (\beta_4 - 1) \varkappa \nu_2 + \beta_5 \varkappa \nu + \beta_3 \varkappa = 0, \\ 2(\beta_4 + \beta_6 - 1) \varkappa \tau + \beta_1 \nu_1 + \beta_2 \nu + (2\beta_7 + \beta_8 + 2\beta_9) \varkappa^2 = 0, \\ 2(\beta_4 + \beta_6) \varkappa \tau + \beta_1 \nu_2 + \beta_2 \nu + (\beta_8 + 2\beta_9) \varkappa^2 = 0, \\ 2\beta_6 \varkappa \tau + \beta_1 \nu_3 + \beta_2 \nu + 2\beta_9 \varkappa^2 = 0,\end{aligned}\right\} \tag{119.24}$$

---

[1] When $\beta_2 = 0$, this equation becomes the constitutive equation proposed by OLDROYD [1958, *32*] [1961, *43*]; a special case of OLDROYD's theory was advocated by WILLIAMS and BIRD [1962, *75*]. In reporting on a program of experiment set up so as to test OLDROYD's theory, TANNER [1963, *69*] found that agreement is "fair when the ratio (apparent viscosity at zero shear rate/apparent viscosity at large shear rate) is less than about 6. For higher ratios the equation is inapplicable," and TANNER proposed a more general rate theory involving an arbitrary function.

When $\beta_7 = \beta_8 = \beta_9 = 0$, Eq. (119.21) corresponds to the "linear fluent body" of NOLL [1955, *19*, § 12]. NOLL's theory, in turn, generalized proposals by ZAREMBA [1903, *11* and *12*], FROMM [1933, *2*] [1947, *3*] [1948, *4*], and DEWITT [1955, *9*].

Viscometric flows of special fluids of the rate type were considered in several of the papers where the theories were first proposed; these papers are cited in the preceding sentence and in footnote 3, p. 482. Cf. also WILLIAMS and BIRD [1957, *31*], VERMA [1958, *51*] [1961, *66*], SHARMA [1959, *31*], DATTA [1961, *10*], MOHAN RAO [1962, *47* and *48*], WALTERS [1962, *71*] (the earlier work of BROER [1956, *3*], often cited, since it rests on a constitutive equation that is not frame-indifferent does not differ in kind from the vast literature on linearized visco-elasticity stretching back to MAXWELL's day and even earlier). Simple shearing of the general fluid of the rate type was considered by OLDROYD [1964, *65*]. Viscometric flows of certain fluids of the integral type were studied by PAO [1957, *16*] [1962, *51*], WALTERS [1960, *60*] [1962, *70*] [1964, *93*] and WALTERS and WATERS [1963, *81*]. Of course the results in all these cases are included, although not quite so simply as for the case when (119.21) holds, in the results given in Sects. 111 (α) 113 (β), 115 (β).

Some approximate solutions for OLDROYD's fluids were derived by SHARMA [1958, *42*] [1959, *30* and *32*], LESLIE [1961, *35*], RATHNA [1962, *54*], HERBERT [1963, *38*], KAPUR and GUPTA [1963, *41*], NANDA [1963, *57*], and FRATER [1964, *31* and *32*]; for certain fluids of the integral type, by WALTERS [1960, *59*] and THOMAS and WALTERS [1963, *70*] [1964, *82*].

Heat transfer in lineal flow of an Oldroyd fluid has been studied by JAIN [1962, *37*].

We do not cite most of the literature on approximate solutions for "Maxwell fluids", "visco-elastic fluids", etc., because the very terms that render the non-linear theories frame-indifferent are the ones neglected, knowingly or unknowingly.

where $\nu=\nu_1+\nu_2+\nu_3$. If $\beta_1+3\beta_2\neq 0$ and $\beta_1\neq 0$, these equations can be solved for τ, ν_1, ν_2, and ν_3. In this way one obtains for the viscosity function[1]

$$\tilde{\mu}(\varkappa)=\frac{\tau(\varkappa)}{\varkappa}=\frac{A+B\varkappa^2}{C+D\varkappa^2}, \tag{119.25}$$

where

$$\left.\begin{aligned} &A=-\beta_1\beta_3, \quad C=\beta_1^2, \\ &B=-(\beta_8+2\beta_9+2\beta_4\beta_9)+\frac{2(\beta_7+\beta_8+3\beta_9)}{\beta_1+3\beta_2}[\beta_1(\beta_4+\beta_5)+\beta_2+\beta_4\beta_2], \\ &D=2(\beta_4+\beta_6+\beta_4\beta_6)-\frac{2(2\beta_4+3\beta_6-1)}{\beta_1+3\beta_2}[\beta_1(\beta_4+\beta_5)+\beta_2+\beta_4\beta_2]. \end{aligned}\right\} \tag{119.26}$$

For the normal-stress functions we find

$$\left.\begin{aligned} \sigma_1(\varkappa)&=-\frac{1}{\beta_1}\{2(\beta_4-1)\varkappa\tau(\varkappa)+(2\beta_7+\beta_8)\varkappa^2\}, \\ \sigma_2(\varkappa)&=-\frac{1}{\beta_1}\{2\beta_4\varkappa\tau(\varkappa)+\beta_8\varkappa^2\}. \end{aligned}\right\} \tag{119.27}$$

It will be noted that the viscosity function passes from the value A/C at low shearing to the finite limit B/D for infinitely high shearing. Furthermore, despite the occurrence of 9 material constants, it is only the single constant β_7 that keeps the viscometric functions in any way independent. Indeed,

$$\tfrac{1}{2}\beta_1[\sigma_1(\varkappa)-\sigma_2(\varkappa)]=\varkappa^2\tilde{\mu}(\varkappa)-\beta_7\varkappa^2. \tag{119.28}$$

In theories for which $\beta_7=0$, any two viscometric functions determine the third, to within a factor of proportionality, just as in the "elastic" theories mentioned above [cf. Eq. (119.11)].

As being suitable for describing the climbing and swelling effects various special theories, some not even employing the concept of fluid, have been proposed. Most of these lead to a universal relation of the form[2]

$$\sigma_1(\varkappa)=k\sigma_2(\varkappa), \tag{119.29}$$

where k is some constant. As indicated by (119.9), according to the Reiner-Rivlin theory $k=1$. On the basis of arguments we do not understand, WEISSENBERG[3] inferred a result equivalent to

$$k=0, \tag{119.30}$$

a relation which, since a good deal of attention has been paid to it, may be given a name: *Weissenberg's assertion*. While at first conflicting opinions of the experimental data were current, the latest summaries[4] concerning polyisobutylene solutions present data contradicting (119.29) for every value of k. However, certain approximate calculations based on especially simple molecular models[5]

---

[1] Eqs. (119.25) and (119.27) contain as special cases results calculated by OLDROYD [1958, *32*, Eqs. (12) to (13)] for his particular rate theory.

[2] This common property of various special theories was noticed by MARKOVITZ [1957, *13*].

[3] WEISSENBERG [1947, *17*] [1949, *26* and *27*] [1950, *17*]. Cf. also GROSSMAN [1961, *22*].

[4] LODGE [1961, *36*], MARKOVITZ and BROWN [1963, *55*] [1964, *59*], ADAMS and LODGE [1964, *1*], MARKOVITZ [1965, *25*]. That earlier data contradicted the Reiner-Rivlin relation $k=1$ had been pointed out by MARKOVITZ [1957, *13*].

[5] RIVLIN's theory of long-chain molecules [1948, *16*] [1949, *19*] seems to yield the Reiner-Rivlin theory and hence to imply that $k=1$; however, LODGE's network theory [1956, *16*] is compatible with WEISSENBERG's assertion. TRUESDELL's exact solution [1956, *35*, Eq. $(34.4)_1$] for shear flow according to MAXWELL's kinetic theory of monatomic gases confirms WEISSENBERG's assertion in the limit as $t\to\infty$; in TRUESDELL's analysis, Maxwellian molecules are presumed, and there is no reason to expect that his results remain valid for more realistic intermolecular forces. Later work on molecular models seems to indicate that while the WEISSENBERG assertion is compatible with the crudest representations of structure, any sort of refinement suffices to render the three viscometric functions independent of one another. Cf. KOTAKA [1959, *16*], GIESEKUS [1962, *22*] [1964, *34*].

do lead to (119.29), and the matter is complicated by the fact that if

$$\lim_{\varkappa\to 0} \sigma_2(\varkappa)/\varkappa^2 \neq 0, \tag{119.31}$$

Eq. (119.29) will hold at low rates of shear for *any* simple fluid such that $\sigma_\Gamma(\varkappa)$ is sufficiently smooth at $\varkappa = 0$.

Virtually all the results for special flow problems for special simple fluids that have been obtained in the literature can be recovered by substituting (119.8), (119.9), (119.16), (119.25) or (119.27) into the formulae given in Sects. 105—117. However, we see no good reason for giving attention to any of these special constitutive equations. We have found no evidence that any of the special forms for the viscometric functions is appropriate to any one real fluid in general or to any class of particular circumstances for all real fluids. The special theories presented in this section seem artificial, distinguished only by the specious "simplicity" so well criticized by RIVLIN in the address from which we have quoted in Sect. 94. No such objection can be made against the theories of second-order and higher-order fluids, which will be treated in Sect. 121.

In all the special theories just discussed, the stress tensor is assumed symmetric. CONDIFF and DAHLER[1] have recently proposed an interesting linear theory of polar fluids, in which the skew part of the stress tensor is proportional to the difference between the spin and a microspin similar to that discussed for elastic materials in Sect. 98.

119A. Appendix. TRUESDELL'S theory of the "Stokesian" fluid. Up to now in discussion of simple fluids we have neglected possible dependence of the constitutive functional upon thermodynamic quantities such as the temperature. One of the early theories laid down particular assumptions such as to render this dependence both special and explicit. In an attempt to construct a model reflecting some of the properties of rarefied gases, and in particular their marked difference of response at low densities, TRUESDELL[2] proposed constitutive equations for the stress and heat flux equivalent to the following:

$$\left.\begin{aligned} \boldsymbol{T} &= \hat{\boldsymbol{T}}(\boldsymbol{D}, \varrho, \theta, R, \mu_0), \\ \boldsymbol{h} &= \hat{\boldsymbol{h}}(\operatorname{grad}\theta, \varrho, \theta, R, \varkappa_0), \end{aligned}\right\} \tag{119A.1}$$

where R, μ_0 and $\varkappa_0$ are constants having the physical dimension of the gas constant, viscosity, and thermal conductivity, respectively:

$$\left.\begin{aligned} \dim R &= L^2 T^{-2} \Theta^{-1}, \\ \dim \mu_0 &= M L^{-1} T^{-1}, \\ \dim \varkappa_0 &= M L T^{-3} \Theta^{-1}, \end{aligned}\right\} \tag{119A.2}$$

Θ being the dimension of temperature. The principle of material indifference requires that $\hat{\boldsymbol{T}}$ be isotropic in its dependence on $\boldsymbol{D}$, and $\hat{\boldsymbol{h}}$ in its dependence on grad θ. For consistency with

[1] CONDIFF and DAHLER [1964, *21*].

[2] TRUESDELL and SCHWARTZ [1947, *12*], TRUESDELL [1949, *22*] [1950, *16*, § 5] [1952, *20*, §§ 63—66] [1952, *22*, pp. 90—91]; cf. also CFT, Sect. 299. In previous presentations, p rather than ϱ was taken as an independent variable. Also listed as an argument was θ_0, a reference temperature, the presence of which has the effect of allowing f and f in (119A.4) and (119A.9) to depend upon the additional argument θ/θ_0. Further, so as to allow pressure-dependent response in a fluid not having a natural time-lapse, the mean pressure $\bar{p}$ was also admitted to the list of variables, with the effect that f and f may depend also upon the additional argument $\bar{p}/p$.

TRUESDELL began from an assumption about the functional dependence of the stress power:

$$\operatorname{tr}(\boldsymbol{T}\boldsymbol{D}) = \Phi(\boldsymbol{D}), \tag{A}$$

where scalar arguments of Φ are not written. He added an assumption of formal isotropy, which use of the principle of material indifference could have rendered redundant. It may be left to the reader to show that if in fact

$$\operatorname{tr}(\boldsymbol{T}\boldsymbol{D}) = \Phi(\dot{\boldsymbol{x}}, \boldsymbol{L}),$$

where Φ is an indifferent scalar, then

$$\boldsymbol{T} = \hat{\boldsymbol{T}}(\boldsymbol{D}),$$

where $\hat{\boldsymbol{T}}$ is an isotropic function.

classical hydrostatics, it is assumed also that

$$\hat{\boldsymbol{T}}(\boldsymbol{0}, \varrho, \theta, R, \mu_0) = -\hat{p}(\varrho, \theta)\,\boldsymbol{1}, \tag{119A.3}$$

where $\hat{p}$ is an assigned function. (In fact, $\hat{p}$ must depend also on dimensional constants such as R, but discussion of the form of $\hat{p}$ belongs rather to thermostatics than to the present theory.)

We write the dimensional matrix of the quantities related by $(119\text{A}.1)_1$:

	M	L	T	Θ
$\boldsymbol{T}$	1	−1	−2	0
$\boldsymbol{D}$	0	0	−1	0
ϱ	1	−3	0	0
θ	0	0	0	1
R	0	2	−2	−1
μ_0	1	−1	−1	0

The rank of this matrix is 4. Since the number of quantities related by $(119\text{A}.1)_1$ is 11, it must be equivalent to a dimensionless relation among 7 dimensionless ratios formed from the quantities occurring in it. Since such ratios are $\boldsymbol{T}/(R\varrho\theta)$ and $\mu_0\boldsymbol{D}/(R\varrho\theta)$, Eq. $(119\text{A}.1)_1$ is equivalent to a relation of the form

$$\frac{\boldsymbol{T}}{R\varrho\theta} = \mathfrak{f}\left(\frac{\mu_0\boldsymbol{D}}{R\varrho\theta}\right), \tag{119A.4}$$

where $\mathfrak{f}$ is a dimensionless isotropic function. By the representation theorem (12.6),

$$\boldsymbol{T} + \hat{p}(\varrho, \theta)\,\boldsymbol{1} = R\varrho\theta\,\beth_0\,\boldsymbol{1} + 2\mu_0\,\beth_1\,\boldsymbol{D} + \frac{\mu_0^2}{R\varrho\theta}\,\beth_2\,\boldsymbol{D}^2, \tag{119A.5}$$

where the $\beth_\Gamma$ are dimensionless functions of the following three dimensionless arguments:

$$\frac{\mu_0}{R\varrho\theta}\,I_{\boldsymbol{D}}, \qquad \frac{\mu_0{}^2}{(R\varrho\theta)^2}\,II_{\boldsymbol{D}}, \qquad \frac{\mu_0{}^3}{(R\varrho\theta)^3}\,III_{\boldsymbol{D}}. \tag{119A.6}$$

Moreover, by (119A.3),

$$\beth_0(0, 0, 0) = 0. \tag{119A.7}$$

If in the Reiner-Rivlin theory the response coefficients $\aleph_\Gamma$ in Eq. (119.4) are taken as strictly independent of θ, then the Stokesian and Reiner-Rivlin theory are *mutually exclusive* except when each reduces to the Navier-Stokes theory. Of course, by generalization each may be made to include the other. For example, if $\aleph_\Gamma$ is first allowed to depend on ϱ and θ, and then that dependence is suitably specialized, the Reiner-Rivlin theory yields the Stokesian theory as a special case. Such a derivation obscures the concept of the Stokesian fluid[1]. While in the Reiner-Rivlin theory the natural time-lapse generally serves as a measure of the importance of non-linear dependence on $\boldsymbol{D}$, in the Stokesian theory *there is no natural time-lapse*[2], just as in

[1] Confusion has been added also by the writers who refer to Reiner-Rivlin fluids as "Stokesian".

[2] This statement may have to be modified if the role of dimension-bearing constants in the thermal equation of state $p = \hat{p}(\varrho, \theta)$ is made explicit. In the special case of an ideal gas, $p = R\varrho\theta$, no dimensional constants beyond those in the list (119A.2) occur. A general thermal equation of state has the dimensionless form

$$\frac{p}{R\varrho_0\theta_0} = f\left(\frac{\varrho}{\varrho_0}, \frac{\theta}{\theta_0}\right),$$

where ϱ_0 and θ_0 are constants having the dimensions of density and temperature, respectively. (E.g., the equation of van der Waals may be written in the form

$$\frac{p}{R\varrho_0\theta_0} = \frac{\dfrac{\varrho}{\varrho_0}\cdot\dfrac{\theta}{\theta_0}}{1 - \dfrac{\varrho}{\varrho_0}} - \alpha\left(\frac{\varrho}{\varrho_0}\right)^2,$$

where α is dimensionless). The constant $\mu_0/(R\varrho_0\theta_0)$ has the dimensions of time. Therefore, if the list of independent variables in (119A.1) is enlarged to include all parameters in the equation of state, and if the equation of state is sufficiently general, the fully general Reiner-Rivlin theory results.

the classical Navier-Stokes theory. Indeed, it was the express aim of TRUESDELL to construct a theory of nonlinear viscosity that shares with the kinetic theory of monatomic gases the property of being devoid of a time constant. Dimensional invariance then forces *a particular dependence on density and temperature*, so that, necessarily, the theory is fully thermomechanical, coupling the effects of deformation and of change of temperature. Only when $\mathfrak{f}$ in (119A.4) is linear does the dependence on θ cancel out, the Stokesian theory then reducing to the Navier-Stokes theory. For more general dependence, by (119A.4), decreasing the product of density by temperature (in effect, decreasing the pressure), has the same effect on the ratio $\boldsymbol{T}/(R\varrho\theta)$ as does increase of $\boldsymbol{D}$.

The coupling of effects of temperature variation with those of stretching requires specification of the heat flux. From the representation theorem (13.8) we see that $\boldsymbol{h}$ must be parallel to grad θ; that is,

$$\boldsymbol{h} = k(\operatorname{grad}\theta, \varrho, \theta, R, \varkappa_0)\operatorname{grad}\theta, \tag{119A.8}$$

where k is a scalar function. The dimensional matrix for (119A.8) is as follows:

	M	L	T	Θ
$\boldsymbol{h}$	1	0	−3	0
grad θ	0	−1	0	1
ϱ	1	−3	0	0
θ	0	0	0	1
R	0	2	−2	−1
$\varkappa_0$	1	1	−3	−1

The rank of this matrix is 4. Since the number of quantities related by Eq. (119A.8) is 8, it is equivalent to a relation among 4 dimensionless ratios formed from the quantities occurring in it. Such ratios are $h\langle m\rangle/[\varkappa_0(\operatorname{grad}\theta)\langle m\rangle]$ and $\varkappa_0|\operatorname{grad}\theta|/[\varrho(R\theta)^{\frac{3}{2}}]$. Hence $(119\text{A}.1)_2$ is equivalent to a relation of the form

$$\boldsymbol{h} = \varkappa_0 f\left(\frac{\varkappa_0|\operatorname{grad}\theta|}{\varrho(R\theta)^{\frac{3}{2}}}\right)\operatorname{grad}\theta, \tag{119A.9}$$

where f is a dimensionless scalar function of its dimensionless argument[1]. If $\boldsymbol{h}$ is assumed to be an analytic function of grad θ, from (119A.9) we see that departures from FOURIER's law of heat conduction are at least of third order in the components of grad θ.

The viscometric functions for a Stokesian fluid have the forms

$$\left.\begin{aligned} \tilde{\mu}(\varkappa) &= \mu_0 \beth_1\left(-\left[\frac{1}{2}\,\frac{\mu_0\varkappa}{R\varrho\theta}\right]^2, 0\right), \\ \sigma_1(\varkappa) = \sigma_2(\varkappa) &= R\varrho\theta\left[\frac{1}{2}\,\frac{\mu_0\varkappa}{R\varrho\theta}\right]^2 \beth_2\left(-\left[\frac{1}{2}\,\frac{\mu_0\varkappa}{R\varrho\theta}\right]^2, 0\right). \end{aligned}\right\} \tag{119A.10}$$

In order to obtain a definite viscometric problem we must specify the thermal equation of state and must also delimit the density and temperature fields, either by boundary conditions or otherwise.

The Stokesian theory is easily generalized by allowing $\boldsymbol{T}$ to depend upon velocity gradients of all orders. As in the Rivlin-Ericksen theory, $\boldsymbol{T}$ turns out to be an isotropic function of the Rivlin-Ericksen tensors $\boldsymbol{A}_{\mathfrak{r}}$, and (119A.4) is replaced by

$$\frac{\boldsymbol{T} + \hat{p}(\varrho, \theta)\boldsymbol{1}}{R\varrho\theta} = \mathfrak{f}\left(\frac{\mu_0}{R\varrho\theta}\boldsymbol{A}_1, \left(\frac{\mu_0}{R\varrho\theta}\right)^2\boldsymbol{A}_2, \ldots, \left(\frac{\mu_0}{R\varrho\theta}\right)^{\mathfrak{n}}\boldsymbol{A}_{\mathfrak{n}}\right), \tag{119A.11}$$

where $\mathfrak{f}$ is a dimensionless isotropic function of its $\mathfrak{n}$ dimensionless arguments which vanishes when all of them vanish.

Since we are faced by a fully thermomechanical theory of a compressible fluid, even extremely simple solutions are hard to find. At present, but a single one is known[2]. We consider the case of steady lineal flow (Sect. 106) when ϱ is constant in space and time, while p, and hence θ, are functions of time only. Since $\boldsymbol{A}_{\mathfrak{r}}=$ const., from (119A.11) it follows that div $\boldsymbol{T} = \boldsymbol{0}$. Thus when $\boldsymbol{b} = \boldsymbol{0}$, the equations of motion are satisfied. Moreover, $\boldsymbol{A}_{\mathfrak{r}} = \boldsymbol{0}$ if $\mathfrak{n} \geqq 3$, so that the right-hand side of (119A.11), for this flow, reduces to an isotropic function

[1] TRUESDELL's conclusion [1950, *16*, § 14] that $f=$ const. is wrong.

[2] While not previously published, the results are easily suggested by TRUESDELL's solution for the corresponding problem in the Maxwellian kinetic theory [1956, *35*, Ch. V].

of $\boldsymbol{A}_1$ and $\boldsymbol{A}_2$ only, where $\boldsymbol{A}_1$ and $\boldsymbol{A}_2$ have the special forms (109.13). Since $\boldsymbol{h}=\boldsymbol{0}$ and $q=0$, the equation of energy (79.3) assumes the form of an ordinary differential equation for $\theta(t)$:

$$\frac{K}{\theta}\frac{d\theta}{dt}=\varkappa g\left(\frac{\mu_0\varkappa}{R\varrho\theta}\right), \tag{119A.12}$$

where the function g is determined by the function $\mathfrak{f}$ in (119A.11), and where K is a function of density and temperature, determined by the caloric equation of state $\varepsilon=\hat{\varepsilon}(\varrho,\theta)$. As to be expected, in general $\theta(t)\to\infty$ as $t\to\infty$. The arguments of (119A.11), therefore, all approach 0 as $t\to\infty$. If $\mathfrak{f}$ in (119A.11) is differentiable at the argument $\boldsymbol{0}, \boldsymbol{0}, \dots, \boldsymbol{0}$, we thus obtain the following asymptotic form for $\boldsymbol{T}$:

$$\boldsymbol{T}=-\hat{p}(\varrho,\theta)\,\mathbf{1}+\mu_0\varkappa(\boldsymbol{N}+\boldsymbol{N}^T)+o(1). \tag{119A.13}$$

Since θ becomes infinite, so does the value of $\hat{p}(\varrho,\theta)$ (under the usual assumptions regarding equations of state). The second term in (119A.13), which gives a linearly viscous shear stress, is constant. Thus, in the present example, the stress system ultimately settles into a state of ever larger hydrostatic pressure, upon which a constant shear stress is superimposed. The heating caused by the shearing eventually annuls all effects of "non-linear viscosity". No such result holds for the corresponding problem in the Reiner-Rivlin theory of incompressible fluids with response coefficients $\aleph_\Gamma$ that are independent of temperature.

120. Asymptotic approximations, I. Steady viscometric flows. In Sect. 108 we have seen that the Navier-Stokes theory of viscometry results as the special case when $V(s_0\varkappa)\equiv 1$, $N_1(s_0\varkappa)\equiv N_2(s_0\varkappa)\equiv 0$. It results also in much greater generality as an asymptotic approximation, in one of two ways[1]. First, hold the functions V, N_1, and N_2 and the constants μ_0 and s_0 fixed as $\varkappa\to 0$. Then, on the assumption that the reduced viscometric functions are continuous at $\varkappa=0$, we find from (108.20) and (108.23) that

$$\tilde{\mu}(\varkappa)\sim\mu_0,\qquad \sigma_1(\varkappa)\sim 0,\qquad \sigma_2(\varkappa)\sim 0: \tag{120.1}$$

For any given simple fluid, the Navier-Stokes theory of viscometry holds in the limit of slow shearing.

This way is not the only one to look at the matter. Again referring to (108.20), let us hold $\varkappa$, V, N_1, and N_2 fixed as $s_0\to 0$. Again (120.1) follows. That is, *if we consider a sequence of fluids having the same natural viscosity and reduced viscometric functions but different natural time-lapses, then for any fixed shearing the Navier-Stokes theory of viscometry holds in the limit of vanishing natural time-lapse.*

Both limit processes have familiar parallels in the comparison of linearly viscous fluids with perfect ones. The former process, referring only to the *flow*, recalls, "the less rapidly a fluid is sheared, the more closely it follows EULER's theory;" the latter, "the less viscous is the fluid, the more closely it follows EULER's theory." Both results are *asymptotic* in character, referring strictly to *limits* as $\varkappa\to 0$ or $s_0\to 0$. As such, they are exempt from objections grounded in the fact that neither $\varkappa$ nor s_0 is dimensionless.

Some of the older authors[2], misled by a specious formal analogy to the theory of elasticity, claimed that the Navier-Stokes theory holds for "small velocities", "small relative velocities", or "small rates of strain", just as the stress relation of the infinitesimal theory of elasticity is a sufficient approximation to a given (sufficiently smooth) finite stress relation if the strain and rotation are sufficiently small. Apparently the first person to see the error and a means of rectifying it was PRANDTL[3]:

---

[1] TRUESDELL [1964, *81*, § 3] considered these limit processes and also others in which both the flow and the fluid are varied. The idea had been sketched, though less clearly, by TRUESDELL [1949, *22*, § 8] [1950, *16*, § 11] [1952, *20*, §§ 66, 70] [1952, *22*, pp. 89—90]. Limit processes reducing the hygrosteric fluid of NOLL to the Stokesean fluid are considered by MARRIS [1964, *61* A].

[2] MEYER [1874, *3*, p. 109], BASSET [1888, *2*, § 466], LICHTENSTEIN [1929, *3*, Kap. 7, § 4], GIRAULT [1931, *2*, Ch. I], THOMPSON [1933, *5*], REINER [1945, *3*, p. 355].

[3] PRANDTL [1932, *6*, § 3].

„Was die Kleinigkeit (sic) der Ausdrücke

$$\frac{\partial u}{\partial x}, \ldots, \frac{\partial w}{\partial y} + \frac{\partial v}{\partial z}$$

betrifft, so muß man fragen, gegen Was diese klein sein sollen. Die entsprechenden Größen in der Elastizitätstheorie sind reine Zahlen und müssen deshalb klein gegen 1 sein. In der Hydrodynamik haben die obigen Größen jedoch die Dimension 1/Zeit und es ist die Frage, welche Zeit zum Vergleich herangezogen werden muß. Es ist ziemlich leicht einzusehen, daß dies die Maxwell'sche Relaxationszeit τ ist, die bei allen wenig zähen Flüssigkeiten unmeßbar klein ist. Die hydrodynamischen Deformationsgrößen müssen also sehr klein gegen den ungeheuer großen Wert $1/\tau$ sein. Hiesige Messungen mit Ölen von der Zähigkeit von rd. 5 cgs-Einheiten zeigten bei Deformationsgeschwindigkeiten 35000/sec noch keine Abweichungen von der Proportionalität (Drehmoment zwischen konzentrischen Zylindern bei sehr kleinem Spalt)."

While PRANDTL thought of Maxwellian visco-elasticity as the exact theory to which the Navier-Stokes theory is a first approximation, TRUESDELL[1] in independently retracing the criticism and the argument took the Reiner-Rivlin theory as exact and reached essentially the same conclusion. He introduced the *truncation number*,

$$\mathfrak{T} \equiv s_0 |\boldsymbol{D}|, \tag{120.2}$$

which in a steady viscometric flow has the value

$$\mathfrak{T} = s_0 \varkappa / \sqrt{2}, \tag{120.3}$$

and he remarked that for a given fluid $\mathfrak{T}$ furnishes a dimensionless scale for measuring the "smallness" of $\boldsymbol{D}$. From the general theory presented above, we see that the criterion of PRANDTL and TRUESDELL, with the truncation number given by (120.3), applies to *all simple fluids* in steady viscometric flows. For a given fluid having a shear-viscosity function differentiable at zero shearing, *the Navier-Stokes theory of viscometry*[2] *results if the shearing is sufficiently small in comparison with the reciprocal of the natural time-lapse*:

$$\mathfrak{T} \ll 1; \quad \text{equivalently, if} \quad \varkappa \ll \frac{1}{s_0}. \tag{120.4}$$

How small is "sufficiently small" depends, of course, on the nature of the viscometric functions for the particular fluid.

We have remarked in Sect. 108 that s_0, the natural time-lapse, drops out of the Navier-Stokes theory and need not be mentioned, as indeed it customarily is not, when that theory is formulated. However, when we come to make any

---

[1] TRUESDELL [1949, *22*, § 8] [1950, *16*, § 11] [1952, *20*, §§ 66, 70]. Earlier [1948, *19*] he had proposed use of a different truncation number, given essentially by the definition

$$\mathfrak{T} \equiv \frac{\mu_0 |\boldsymbol{D}|}{R \varrho \theta};$$

by (119A.4), this number is a scaling parameter for the theory of the Stokesian fluid. The criterion for validity of the Navier-Stokes theory of viscometry becomes

$$\varkappa \ll \frac{R \varrho \theta}{\mu_0}$$

in place of (120.4).

[2] To the reader of this treatise it should be obvious that the magnitude of the truncation number (120.2) gives little if any information regarding the validity of the Navier-Stokes theory as an approximation to the theory of the simple fluid in flows that are not viscometric.

dimensionally invariant statement about the *range of validity* of the Navier-Stokes theory as an approximation to the theory of simple fluids, the natural time-lapse of the more general fluid must be mentioned. Rather than saying that the natural time-lapse is not needed in the Navier-Stokes theory, we might say with less danger of confusion that *the natural time-lapse of the Navier-Stokes fluid is zero.*

In both the asymptotic processes presented at the beginning of this section, $\mathfrak{T}\to 0$. It would not be correct to infer that both are included as a special case of a single limit formula, since the mere statement that $\mathfrak{T}\to 0$ does not suffice to get unique asymptotic forms from (108.20) when μ_0, V, N_1 and N_2 are held fixed: Something has to be said about the behavior of $\varkappa$ as well.

However, use of the dimensionless variable $\mathfrak{T}$ enables us to calculate unique dimensionless *higher approximations* to certain dimensionless ratios formed from the viscometric functions. Indeed, if we assume V, N_1, and N_2 to be four times continuously differentiable at $s_0\varkappa=0$, from (108.20) and the fact that all three functions are even we find that

$$\left.\begin{aligned}\frac{\tilde{\mu}(\varkappa)}{\mu_0}&=1+V_1\mathfrak{T}^2+O(\mathfrak{T}^4),\\ \frac{\sigma_1(\varkappa)}{\mu_0\varkappa\mathfrak{T}}&=N_{10}+O(\mathfrak{T}^2),\\ \frac{\sigma_2(\varkappa)}{\mu_0\varkappa\mathfrak{T}}&=N_{20}+O(\mathfrak{T}^2),\end{aligned}\right\}\qquad(120.5)$$

where V_1, N_{10}, and N_{20} are dimensionless constants such that $N_{10}^2+N_{20}^2=1$. If we now consider once more a given fluid, so that μ_0, s_0, V_1, N_{10}, and N_{20} are given constants, we conclude that *while departure from the classical proportionality of shear stress to shearing is generally an effect of* third *order*[1] *in the shearing, the normal-stress effects are of* second *order*:

$$\left.\begin{aligned}\tau(\varkappa)=\varkappa\tilde{\mu}(\varkappa)&=\mu_0\varkappa\left[1+\tfrac{1}{2}V_1 s_0^2\varkappa^2+O(\varkappa^4)\right],\\ \sigma_1(\varkappa)&=\mu_0 s_0\varkappa^2\left[\frac{1}{\sqrt{2}}N_{10}+O(\varkappa^2)\right],\\ \sigma_2(\varkappa)&=\mu_0 s_0\varkappa^2\left[\frac{1}{\sqrt{2}}N_{20}+O(\varkappa^2)\right].\end{aligned}\right\}\qquad(120.6)$$

Thus, other things being equal, normal-stress effects can be expected to appear *at shearings low enough that the classical linearity of the shear-viscosity function, along with all the consequences of that linearity, continues to hold*[2].

The two limit processes and the corresponding higher approximations for the viscometric functions will now be raised to the generality of the simple fluid.

121. Asymptotic approximations, II. General flows. We now consider the special forms that various of the consequences of the principle of fading memory (Sect. 38) take when applied to simple fluids.

We have shown in Sect. 40 that the constitutive equation of a simple fluid may be approximated, to within terms of order $o(\|G(s)\|_h^n)$, by the constitutive

---

[1] For a special theory, this observation is due to BOUSSINESQ [1868, *1*].

[2] This point was emphasized by TRUESDELL [1952, *20*, § 72]. Notice that PRANDTL, in the passage quoted above, was entirely unaware of it, since he sought to find in the shear-viscosity function departures from the Navier-Stokes equations at high shearings. Indeed, none of the experts on the "physical" side of hydrodynamics published a word suggesting the slightest "intuition" of normal-stress effects until they were in fact discovered by experiment and explained by mathematics in the 1940's.

equations of a fluid of order $\mathfrak{n}$, i.e.,

$$\boldsymbol{T} = -p\mathbf{1} + \sum_{\mathfrak{k}=1}^{\mathfrak{n}} \underset{s=0}{\overset{\infty}{\mathfrak{P}_{\mathfrak{k}}}} (\boldsymbol{G}(s)), \tag{121.1}$$

where

$$\boldsymbol{G}(s) = \boldsymbol{C}_{(t)}(t-s) - \mathbf{1} \tag{121.2}$$

and where $\mathfrak{P}_{\mathfrak{k}}$ is an isotropic bounded homogeneous polynomial functional of degree $\mathfrak{k}$. As before, $\|\boldsymbol{G}(s)\|_h$ is the recollection of the history $\boldsymbol{G}$ corresponding to the obliviator h. For compressible fluids, a dependence of p and $\mathfrak{P}_{\mathfrak{k}}$ on the density, ϱ, is understood. For incompressible fluids, as usual, p is indeterminate.

When $\mathfrak{n} = 1$, (121.1) reduces to (37.8) in the case of a compressible fluid, and to

$$\boldsymbol{T} = -p\mathbf{1} + \int_0^\infty \zeta(s)\, \boldsymbol{G}(s)\, ds, \tag{121.3}$$

in the case of an incompressible fluid. In order to ascertain the physical meaning of the material function $\zeta(s)$ occurring in (121.3) consider stress-relaxation after a strain impulse as treated in Sect. 39. Substituting (39.6) into (121.2) and then putting the result into (121.3), we find that

$$\boldsymbol{T} = -p\mathbf{1} + (\boldsymbol{B}^{-1} - \mathbf{1}) \int_t^\infty \zeta(s)\, ds, \tag{121.4}$$

where $\boldsymbol{B}$ is the left Cauchy-Green tensor of the strain impulse. If the strain impulse is infinitesimal,

$$\boldsymbol{B}^{-1} - \mathbf{1} \approx -2\tilde{\boldsymbol{E}} \tag{121.5}$$

where $\tilde{\boldsymbol{E}}$ is the infinitesimal strain tensor. In this case, (121.4) reduces to

$$\boldsymbol{T} \approx -p\mathbf{1} + 2\bar{\mu}(t)\, \tilde{\boldsymbol{E}}, \tag{121.6}$$

where

$$\bar{\mu}(t) = -\int_t^\infty \zeta(s)\, ds, \qquad \zeta(t) = \frac{d}{dt}\bar{\mu}(t). \tag{121.7}$$

Comparison of (121.6) with (41.27) shows that $\bar{\mu}(t)$ is the time-dependent shear modulus of the fluid in stress-relaxation.

The formula (121.6) could also have been obtained from the constitutive equations (41.24) of isotropic linear visco-elasticity in the special case when $\lambda = 0$, $\bar{\lambda}(s) = 0$, $\mu = 0$ and when $\boldsymbol{T}$ is replaced by $\boldsymbol{T} + p\mathbf{1}$ to take account of incompressibility.

The function $\bar{\mu}(t)$ can be determined experimentally by suddenly imposing a small shear strain on a fluid body and then measuring the decaying shear stress necessary to maintain the strain. It is physically reasonable to assume that $\bar{\mu}(t)$ is positive and decreasing, whence it follows that

$$\zeta(t) \leqq 0. \tag{121.8}$$

For viscometric flows, by (106.31) and (106.8)

$$\boldsymbol{G}(s) = g(s)(\boldsymbol{N} + \boldsymbol{N}^T) + g(s)^2 \boldsymbol{N}^T \boldsymbol{N}. \tag{121.9}$$

Consider, now, the one-parameter family of flows obtained by replacing $g(s)$ by $\varkappa g(s)$, so that (121.9) is replaced by

$$\boldsymbol{G}(s) = \varkappa g(s)(\boldsymbol{N} + \boldsymbol{N}^T) + \varkappa^2 g(s)^2 \boldsymbol{N}^T \boldsymbol{N}. \tag{121.10}$$

The formula (121.10) would correspond, for example, to a family of curvilineal flows obtained by replacing $v(x^1, t)$ and $w(x^1, t)$ by $\varkappa v(x^1, t)$ and $\varkappa w(x^1, t)$ in (107.1). We investigate the behavior of the fluid in the limit as $\varkappa \to 0$. It is not hard to show that

$$o(\|G(s)\|_h^n) = o(\varkappa^n) \tag{121.11}$$

when $G(s)$ has the form (121.10). It follows that (121.3) can be expected to be valid to within an error of order $o(\varkappa)$. Substitution of (121.10) into (121.3) yields

$$T = -p\mathbf{1} + \varkappa(N + N^T)\int_0^\infty \zeta(s)\, g(s)\, ds + o(\varkappa). \tag{121.12}$$

If we use the basis for which (106.7) and (106.19) hold, we find that the material functionals in (106.20)—(106.22), to within terms of order $o(\varkappa)$, reduce to

$$\underset{s=0}{\overset{\infty}{\mathfrak{t}}}\big(\varkappa g(s)\big) = \varkappa \int_0^\infty \zeta(s)\, g(s)\, ds + o(\varkappa), \tag{121.13}$$

$$\underset{s=0}{\overset{\infty}{\mathfrak{s}_1}}\big(\varkappa g(s)\big) = o(\varkappa), \quad \underset{s=0}{\overset{\infty}{\mathfrak{s}_2}}\big(\varkappa g(s)\big) = o(\varkappa). \tag{121.14}$$

Thus, with error $o(\varkappa)$, *there are no normal-stress effects at low shearings* in the flows corresponding to (121.10).

Consider now the same problem for an incompressible fluid of second order. By (121.1), the constitutive equation is

$$T = -p\mathbf{1} + \int_0^\infty \zeta(s)\, G(s)\, ds + \underset{s=0}{\overset{\infty}{\mathfrak{P}_2}}\big(G(s)\big). \tag{121.15}$$

Substitution of (121.10) into (121.15) yields

$$\left.\begin{aligned} T = &-p\mathbf{1} + \varkappa(N + N^T)\int_0^\infty \zeta(s)\, g(s)\, ds + \\ &+ \varkappa^2\Big\{N^T N \int_0^\infty \zeta(s)\, [g(s)]^2\, ds + \underset{s=0}{\overset{\infty}{\mathfrak{P}_2}}\big(g(s)\,(N + N^T)\big)\Big\} + o(\varkappa^2). \end{aligned}\right\} \tag{121.16}$$

Since $\mathfrak{P}_2$ is an isotropic quadratic functional, it follows from the representation theorem (12.6) that

$$\left.\begin{aligned} \underset{s=0}{\overset{\infty}{\mathfrak{P}_2}}\big(g(s)\,A\big) = &\left[\underset{s=0}{\overset{\infty}{\mathfrak{q}_1}}\big(g(s)\big)\,(\operatorname{tr} A^2) + \underset{s=0}{\overset{\infty}{\mathfrak{q}_2}}\big(g(s)\big)\,(\operatorname{tr} A)^2\right]\mathbf{1} + \\ &+ \underset{s=0}{\overset{\infty}{\mathfrak{q}_3}}\big(g(s)\big)\,(\operatorname{tr} A)\, A + \underset{s=0}{\overset{\infty}{\mathfrak{q}}}\big(g(s)\big)\, A^2, \end{aligned}\right\} \tag{121.17}$$

where $\mathfrak{q}_1$, $\mathfrak{q}_2$, $\mathfrak{q}_3$, and $\mathfrak{q}$ are scalar quadratic functionals. Substituting $A = N + N^T$ into (121.17) and observing that $N^2 = \mathbf{0}$ and $\operatorname{tr} N = 0$, we find that (121.16) beomes

$$\left.\begin{aligned} T = &-p\mathbf{1} + \varkappa(N + N^T)\int_0^\infty \zeta(s)\, g(s)\, ds + \\ &+ \varkappa^2\Big\{N^T N \int_0^\infty \zeta(s)\, [g(s)]^2\, ds + (N + N^T)^2 \underset{s=0}{\overset{\infty}{\mathfrak{q}}}\big(g(s)\big)\Big\} + o(\varkappa^2), \end{aligned}\right\} \tag{121.18}$$

where all hydrostatic terms have been absorbed into $-p\mathbf{1}$. If we use again the basis for which (106.7) and (106.19) hold, we find the material functionals in

(106.20)—(106.22) have the forms[1]

$$\overset{\infty}{\underset{s=0}{\mathfrak{t}}}(\varkappa g(s)) = \varkappa \int_0^\infty \zeta(s) g(s)\, ds + o(\varkappa^2), \tag{121.19}$$

$$\overset{\infty}{\underset{s=0}{\mathfrak{s}_1}}(\varkappa g(s)) = \varkappa^2 \left\{ \int_0^\infty \zeta(s)\, [g(s)]^2\, ds + \overset{\infty}{\underset{s=0}{\mathfrak{q}}}(g(s)) \right\} + o(\varkappa^2), \tag{121.20}$$

$$\overset{\infty}{\underset{s=0}{\mathfrak{s}_2}}(\varkappa g(s)) = \varkappa^2 \overset{\infty}{\underset{s=0}{\mathfrak{q}}}(g(s)) + o(\varkappa^2). \tag{121.21}$$

We draw two important conclusions from (121.19)—(121.21). First, *to within an error $o(\varkappa^2)$, the shear-stress functional is the same as to within an error $o(\varkappa)$. Second, one normal-stress difference, namely*

$$T\langle 11\rangle - T\langle 22\rangle = \overset{\infty}{\underset{s=0}{\mathfrak{s}_1}}(\varkappa g(s)) - \overset{\infty}{\underset{s=0}{\mathfrak{s}_2}}(\varkappa g(s)) = \varkappa^2 \int_0^\infty \zeta(s) g(s)^2\, ds + o(\varkappa^2), \tag{121.22}$$

depends only on the material function $\zeta(s)$ and hence can be predicted from a knowledge of the shear modulus $\overline{\mu}(t)$ in stress relaxation. The first conclusion extends to unsteady viscometric flows the result obtained in the previous section for steady viscometric flows, namely, and in rough terms, that *the second-order correction to the classical theory is reflected in normal-stress effects but not in the relation between shear stress and shearing, or in its consequences such as formulae for velocity profiles and discharges.* Experimental trial of the second conclusion should be relatively easy. If (121.22) does not hold, the material tested is not described by the theory of the simple fluid.

For steady viscometric flows we identify the parameter $\varkappa$ with the shearing. Then, by (108.2), we must put $g(s) = -s$ in (121.10)—(121.22). For the viscometric functions (108.3), (108.10) we then obtain results of the form (120.6):

$$\tilde{\mu}(\varkappa) = \frac{\tau(\varkappa)}{\varkappa} = \mu_0 + o(\varkappa), \tag{121.23}$$

$$\sigma_1(\varkappa) = (2\alpha_1 + \alpha_2)\varkappa^2 + o(\varkappa^2), \tag{121.24}$$

$$\sigma_2(\varkappa) = \alpha_2 \varkappa^2 + o(\varkappa^2), \tag{121.25}$$

where the coefficients are related as follows[2]:

$$\mu_0 = -\int_0^\infty \zeta(s)\, s\, ds, \qquad 2\alpha_1 = \int_0^\infty \zeta(s)\, s^2\, ds, \qquad \alpha_2 = \overset{\infty}{\underset{s=0}{\mathfrak{q}}}(s). \tag{121.26}$$

The two material constants $-\mu_0$ and $2\alpha_1$ are the first two moments of the material function $\zeta(s)$. If the inequality (121.8) is assumed, and if $\zeta(s)$ is not identically zero, if follows that[2]

$$\mu_0 > 0, \qquad \alpha_1 < 0. \tag{121.27}$$

The first of these inequalities follows also from (108.9), which was imposed for thermodynamic reasons. The second does not seem to be physically obvious. It implies, for example, that for second-order fluids in steady Couette flow, the normal-stress difference (113.16) is positive, provided one of the cylinders be at rest.

---

[1] The generalization, not previously published, of a result of COLEMAN and NOLL [1961, *6*] is due to NOLL.

[2] COLEMAN and MARKOVITZ [1964, *17*, § 2].

We come now to the more specific results that follow as asymptotic approximations when a given flow with history $\boldsymbol{G}(s)$ is retarded (Sect. 40). We shall abbreviate the term "fluid of the differential type of grade $\mathfrak{n}$" to *fluid of grade* $\mathfrak{n}$, taking care not to confuse it with the (more general) fluid of *order* $\mathfrak{n}$, defined by (121.1).

The counterpart of (40.18) for incompressible fluids of grade $\mathfrak{n}$ is the constitutive equation[1]

$$\boldsymbol{T} = -p\boldsymbol{1} + \sum_{(\mathfrak{i}_1,\ldots,\mathfrak{i}_\mathfrak{k})} \mathfrak{f}_{\mathfrak{i}_1,\ldots,\mathfrak{i}_\mathfrak{k}}[\boldsymbol{A}_{\mathfrak{i}_1},\ldots,\boldsymbol{A}_{\mathfrak{i}_\mathfrak{k}}] \tag{121.28}$$

where the summation is to be extended over all indices satisfying (40.14). Using the representation theorem for multilinear tensor functions of Sect. 13 (α) and the relations (24.15), we find that for $\mathfrak{n}=4$ (121.28) reduces to[2]

$$\boldsymbol{T} = -p\boldsymbol{1} + \boldsymbol{S}_1 + \boldsymbol{S}_2 + \boldsymbol{S}_3 + \boldsymbol{S}_4 \tag{121.29}$$

where

$$\left.\begin{aligned} &\boldsymbol{S}_1 = \mu_0 \boldsymbol{A}_1, \qquad \boldsymbol{S}_2 = \alpha_1 \boldsymbol{A}_2 + \alpha_2 \boldsymbol{A}_1^2, \\ &\boldsymbol{S}_3 = \beta_1 \boldsymbol{A}_3 + \beta_2 (\boldsymbol{A}_2\boldsymbol{A}_1 + \boldsymbol{A}_1\boldsymbol{A}_2) + \beta_3 (\operatorname{tr}\boldsymbol{A}_2)\boldsymbol{A}_1, \\ &\boldsymbol{S}_4 = \gamma_1 \boldsymbol{A}_4 + \gamma_2(\boldsymbol{A}_3\boldsymbol{A}_1 + \boldsymbol{A}_1\boldsymbol{A}_3) + \gamma_3 \boldsymbol{A}_2^2 + \gamma_4(\boldsymbol{A}_2\boldsymbol{A}_1^2 + \boldsymbol{A}_1^2\boldsymbol{A}_2) + \\ &\qquad + \gamma_5(\operatorname{tr}\boldsymbol{A}_2)\boldsymbol{A}_2 + \gamma_6(\operatorname{tr}\boldsymbol{A}_2)\boldsymbol{A}_1^2 + [\gamma_7 \operatorname{tr}\boldsymbol{A}_3 + \gamma_8 \operatorname{tr}(\boldsymbol{A}_2\boldsymbol{A}_1)]\boldsymbol{A}_1. \end{aligned}\right\} \tag{121.30}$$

The constitutive equation (121.28) is valid for retarded motion and gives the stress to within terms of order $o(\alpha^\mathfrak{n})$ in the retardation factor α (cf. Sect. 40). The terms in (121.29) are arranged in such a way that

$$\boldsymbol{S}_\mathfrak{k} = O(\alpha^\mathfrak{k}), \qquad \mathfrak{k} = 1, 2, 3, 4. \tag{121.31}$$

Thus, e.g., $\boldsymbol{S}_3$ gives the terms to be added to $\boldsymbol{S}_1 + \boldsymbol{S}_2$ in order to yield the constitutive equation for the fluid of grade *3*.

For steady viscometric flow, by (109.12)

$$\boldsymbol{G}(s) = -s\varkappa(\boldsymbol{N} + \boldsymbol{N}^T) + s^2\varkappa^2 \boldsymbol{N}^T\boldsymbol{N}, \tag{121.32}$$

so that $\boldsymbol{G}(s)$, for fixed $\boldsymbol{N}$, depends only on $\varkappa s$. Hence, in view of (40.5), we can identify the shearing $\varkappa$ with the retardation factor α. Substitution of (109.13) into (121.30) yields, after repeated use of (106.7),

$$\left.\begin{aligned} &\boldsymbol{S}_1 = \mu_0\varkappa(\boldsymbol{N} + \boldsymbol{N}^T), \qquad \boldsymbol{S}_2 = \varkappa^2[2\alpha_1 \boldsymbol{N}^T\boldsymbol{N} + \alpha_2(\boldsymbol{N} + \boldsymbol{N}^T)^2], \\ &\boldsymbol{S}_3 = 2\varkappa^3(\beta_2 + \beta_3)(\boldsymbol{N} + \boldsymbol{N}^T), \\ &\boldsymbol{S}_4 = \varkappa^4\{4(\gamma_3 + \gamma_4 + \gamma_5)\boldsymbol{N}^T\boldsymbol{N} + 2\gamma_6(\boldsymbol{N} + \boldsymbol{N}^T)^2\}. \end{aligned}\right\} \tag{121.33}$$

---

[1] Like those given in Sect. 40, the present results follow from the general theory of COLEMAN and NOLL [1960, *17*, § 7] [1961, *5*, pp. 707—713]. Earlier LANGLOIS and RIVLIN [1959, *18*, § 2] had proposed for study the "slightly visco-elastic fluid" defined by the constitutive equation

$$\boldsymbol{T} = -p\boldsymbol{1} + \mu\boldsymbol{A}_1 + \varepsilon\mathfrak{p}(\boldsymbol{A}_1, \boldsymbol{A}_2, \ldots, \boldsymbol{A}_\mathfrak{n}),$$

where $\mathfrak{p}$ is an isotropic tensor polynomial and where results were contemplated in the limit as $\varepsilon \to 0$. They attempted to formulate, in analogy to the method in Sect. 65 for elastic materials, a systematic method of approximation in which the first step is the solution of a given problem for a linearly viscous fluid.

The formal approximation process proposed later by RIVLIN [1964, *71*] rests upon writing the velocity field in the form $\dot{\boldsymbol{x}} = \varepsilon\boldsymbol{u}$ and expanding all quantities in powers of ε. It leads to results of the same form as COLEMAN and NOLL's for steady flow but in general to different ones for unsteady flows, although both processes lead to (121.3) for infinitesimal motion. Cf. the discussion by RIVLIN [1965, *30A*, § 9].

[2] RIVLIN [1964, *71*, § 2] has an additional term proportional to $\operatorname{tr}(\boldsymbol{A}_1^3)$ in the formula for $\boldsymbol{S}_4$. This term can be eliminated by means of $(24.15)_3$.

By use of the basis for which (106.7) and (108.4)—(108.6) hold, we find the following expressions for the viscometric functions:

$$\left.\begin{aligned}\tilde{\mu}(\varkappa) &= \frac{1}{\varkappa}\,\tau(\varkappa) = \mu_0 + 2(\beta_2+\beta_3)\varkappa^2 + o(\varkappa^3),\\ \sigma_1(\varkappa) &= (2\alpha_1+\alpha_2)\varkappa^2 + [4(\gamma_3+\gamma_4+\gamma_5)+2\gamma_6]\varkappa^4 + o(\varkappa^4),\\ \sigma_2(\varkappa) &= \alpha_2\varkappa^2 + 2\gamma_6\varkappa^4 + o(\varkappa^4).\end{aligned}\right\} \quad (121.34)$$

Comparison with (121.23)—(121.25) shows that the constants μ_0, α_1, and α_2 defined by (121.26) coincide with those occurring in $(121.30)_{1,2}$. More generally, (121.34) furnishes an interpretation for the coefficients in power-series expansions of the viscometric functions, which can of course be written down straight off [cf. (120.6)], in terms of the coefficients occurring in the constitutive equations of fluids of grades *1*, *2*, *3*, and *4*.

Looking back at Sect. 120, we see that all the limit processes considered so far in the present section reduce in the case of viscometric flow to the former of the two simple limit processes presented there, i.e., to the limit of slow shearing. A generalization of the second limit process, corresponding to letting the natural time diminish, has been constructed by TRUESDELL[1]. As he remarked, his results could be obtained by adjusting the process of COLEMAN and NOLL. He preferred, instead, to start from the beginning and to present a heuristic[2] argument whereby, with only elementary mathematics, it is made clear how a functional may be replaced in an appropriate limit case by a function of the derivatives of its argument.

We begin from the dimensionless constitutive functional $\mathfrak{Q}^*$ as defined by (108.18). We shall consider a fixed deformation history $\boldsymbol{G}(s)$, and we shall consider the sequence of fluids obtained by holding fixed the dimensionless response functional $\mathfrak{Q}^*$ and the natural viscosity μ_0, while letting the natural time-lapse s_0 approach zero. First, if $\boldsymbol{G}(s)$ is $\mathfrak{n}$ times differentiable at $s=0$, then by (105.2) and (24.14) there is an interval $|s| = |s^*|\, s_0 < S$ such that, with arbitrarily small error,

$$\boldsymbol{G}(s) = \boldsymbol{G}(s^* s_0) \approx \sum_{\mathfrak{m}=1}^{\mathfrak{n}} \frac{(-1)^{\mathfrak{m}}}{\mathfrak{m}!}\, s^{*\mathfrak{m}} s_0^{\mathfrak{m}} \boldsymbol{A}_{\mathfrak{m}}. \quad (121.35)$$

The aim is now to choose s_0 so small that only the interval $|s| = |s^*|\, s_0 < S$ remains of any importance in determining the value of $\underset{s=0}{\overset{\infty}{\mathfrak{Q}^*}}\big(\boldsymbol{G}(s)\big)$, all the rest of the range being spread out into the distant past.

We shall say that a dimensionless functional $\mathfrak{Q}^*$ of a dimensionless tensor-valued function $\boldsymbol{H}(s^*)$ of a dimensionless scalar variable s^* is *obliviating* in the class $\mathscr{H}$ of functions $\boldsymbol{H}$ if we can find a constant L and another functional $\mathfrak{L}$ of functions defined only in the interval $[0, L]$ such that with arbitrarily small error

$$\underset{s^*=0}{\overset{\infty}{\mathfrak{Q}^*}}\big(\boldsymbol{H}(s^*)\big) \approx \underset{s^*=0}{\overset{L}{\mathfrak{L}}}\big(\boldsymbol{H}(s^*)\big). \quad (121.36)$$

---

[1] TRUESDELL (1962) [1964, *87*, § 13] [1964, *85*, § 5]. Approximating sequences in powers of a natural time first appear in the work of TRUESDELL [1950, *16*, § 11] [1952, *20*, § 69] on the Reiner-Rivlin fluid. COLEMAN and NOLL [1961, *5*, pp. 713—714] mentioned the possibility of a dimensional argument to get results of this kind for Rivlin-Ericksen fluids but did not carry it out, claiming that "such a formal procedure does not elucidate the physical significance behind our equations." Cf. also GIESEKUS [1958, *19*]. An expansion in powers of a time constant has been employed by PIPKIN [1964, *66*].

[2] Recall that proof of the basic approximation theorem of COLEMAN and NOLL (Sect. 40) has been omitted as being too difficult for the present treatise. The method of TRUESDELL has been generalized and made precise by the second theory of WANG, as explained at the end of Sect. 40.

Thus a functional is obliviating in $\mathscr{H}$ if it is insensitive to the behavior of $\boldsymbol{H}(s^*)$ when s^* is large, for all $\boldsymbol{H}\in\mathscr{H}$. Whether or not a given functional is obliviating depends upon the class of deformation histories considered. If, following GREEN and RIVLIN[1], we consider only histories $\boldsymbol{H}(s^*)$ that reduce to $\mathbf{0}$ when $s^*\geqq s_0^*$, then every functional is obliviating, with $L=s_0^*$ and with error equal to zero. In more general and typical deformation histories, only certain functionals will be obliviating, and both the value of L and the nature of $\mathfrak{L}$ will depend upon the magnitude of the error to be allowed in the approximation (121.36).

The assumption that the dimensionless constitutive functional $\mathfrak{Q}^*$ is obliviating is a strong one. It enables us, with scarcely any mathematical apparatus, to derive a sequence of approximating representations for $\mathfrak{Q}^*$. Take $\boldsymbol{H}(s^*)\equiv\boldsymbol{G}(s^* s_0)$. Then

$$\mathop{\mathfrak{Q}^*}_{s^*=0}^{\infty}\left(\boldsymbol{G}(s^* s_0)\right)\approx\mathop{\mathfrak{L}}_{s^*=0}^{L}\left(\boldsymbol{G}(s^* s_0)\right). \tag{121.37}$$

In all this, $\boldsymbol{G}$ and the allowed dimensionless errors are fixed, so that S and L are determined. Now take s_0 so small that $s_0 L<S$. Then $|s^*|\,s_0<S$ whenever $0\leqq s^*\leqq L$; hence we may use *both* (121.35) and (121.37). If we assume further that $\mathfrak{Q}^*$ is continuous at $\mathbf{0}$ in the topology defined by uniform convergence, then

$$\left.\begin{aligned}\mathop{\mathfrak{Q}^*}_{s^*=0}^{\infty}\left(\boldsymbol{G}(s^* s_0)\right)&\approx\mathop{\mathfrak{L}}_{s^*=0}^{L}\left(\sum_{\mathfrak{m}=1}^{\mathfrak{n}}\frac{(-1)^{\mathfrak{m}}}{\mathfrak{m}!}\,s^{*\mathfrak{m}}s_0^{\mathfrak{m}}\boldsymbol{A}_{\mathfrak{m}}\right),\\ &\approx\mathop{\mathfrak{Q}^*}_{s^*=0}^{\infty}\left(\sum_{\mathfrak{m}=1}^{\mathfrak{n}}\frac{(-1)^{\mathfrak{m}}}{\mathfrak{m}!}\,s^{*\mathfrak{m}}s_0^{\mathfrak{m}}\boldsymbol{A}_{\mathfrak{m}}\right),\end{aligned}\right\} \tag{121.38}$$

where at the second step the assumption (121.36) has been used again. But

$$\mathop{\mathfrak{Q}^*}_{s^*=0}^{\infty}\left(\sum_{\mathfrak{m}=1}^{\mathfrak{n}}\frac{(-1)^{\mathfrak{m}}}{\mathfrak{m}!}\,s^{*\mathfrak{m}}s_0^{\mathfrak{m}}\boldsymbol{A}_{\mathfrak{m}}\right)=\mathfrak{k}(s_0\boldsymbol{A}_1,\,s_0^2\boldsymbol{A}_2,\,\ldots,\,s_0^{\mathfrak{n}}\boldsymbol{A}_{\mathfrak{n}}), \tag{121.39}$$

where, since $\mathfrak{Q}^*$ is a dimensionless isotropic functional, $\mathfrak{k}$ is a dimensionless isotropic function of its $\mathfrak{n}$ arguments.

What has been indicated so far is that if in (108.18) we consider a fixed deformation history $\boldsymbol{G}$ in $\mathscr{H}$, a fixed dimensionless constitutive functional $\mathfrak{Q}^*$ that is obliviating in $\mathscr{H}$ and continuous at $\mathbf{0}$, and a fixed natural viscosity μ_0, then with arbitrarily small error

$$\boldsymbol{T}+p\mathbf{1}\approx\frac{\mu_0}{s_0}\,\mathfrak{k}(s_0\boldsymbol{A}_1,s_0^2\boldsymbol{A}_2,\ldots,s_0^{\mathfrak{n}}\boldsymbol{A}_{\mathfrak{n}}), \tag{121.40}$$

where $\mathfrak{k}$ is isotropic. If we add here the assumption that the function $\mathfrak{k}$ is at least $\mathfrak{n}$ times continuously differentiable[2] at the argument $\mathbf{0},\mathbf{0},\ldots,\mathbf{0}$, then we can express it, again with arbitrarily small error, as a polynomial of degree $\mathfrak{n}$ in its $\mathfrak{n}$ arguments. This polynomial can then be rearranged as *a polynomial in the natural time-lapse*, s_0:

$$\boldsymbol{T}+p\mathbf{1}\approx\mu_0\sum_{\mathfrak{q}=1}^{\mathfrak{n}}s_0^{\mathfrak{q}-1}\mathfrak{f}_{\mathfrak{q}}(\boldsymbol{A}_1,\boldsymbol{A}_2,\ldots,\boldsymbol{A}_{\mathfrak{n}})+\cdots, \tag{121.41}$$

where the dots stand for terms of degree greater than $\mathfrak{n}-1$ in s_0, and where $\mathfrak{f}_{\mathfrak{q}}$ is the most general isotropic polynomial of degrees $r_1, r_2, \ldots, r_{\mathfrak{n}}$ in the components of the tensors $\boldsymbol{A}_1, \boldsymbol{A}_2, \ldots, \boldsymbol{A}_{\mathfrak{n}}$ such that

$$r_1+2r_2+3r_3+\cdots+\mathfrak{n}r_{\mathfrak{n}}=\mathfrak{q}. \tag{121.42}$$

[1] GREEN and RIVLIN [1956, *12*, § 2] [1957, *8*, § 2].

[2] Note that in the rigorous treatment of COLEMAN and NOLL functional differentiability is assumed from the start.

The coefficients of the polynomials $\mathfrak{f}_{\mathfrak{q}}$ are dimensionless constants. In passing from (121.40) to (121.41) we incorporated in the unwritten error term all terms of degree higher than $\mathfrak{n}-1$ in s_0.

The results of this process of approximation will be seen to be formally the same as given by COLEMAN and NOLL'S formulae (121.28) and (40.14). The sense is different, since the ordering is in powers of the natural time-lapse. [Note added in proof. Since the foregoing was written, WANG has constructed a rigorous theory incorporating TRUESDELL'S as a special case. This theory has been sketched in Sects. 38 and 40.]

122. Secondary flows in tubes. If, subject to certain boundary conditions, the streamline pattern is different from that predicted by the Navier-Stokes equations, it is customary to speak of a *secondary flow.* In Sect. 114 we have discussed grossly some secondary flows that occur in experiments because of the finite size of laboratory apparatus. In these cases it is the nature of the end effects that differs, according as the normal-stress functions vanish or not. In Sect. 117 we have found evidence of a secondary flow that lies deeper than any matter of laboratory correction. Namely, a steady flow of an incompressible simple fluid through a tube of non-circular cross-section cannot be, in general, a rectilinear shearing (117.1). The pattern of straight streamlines, familiar for the linearly viscous fluid in an infinitely long tube, is not possible, apart from exceptional cases. We can therefore expect that the velocity field has a non-zero component perpendicular to the axis of the tube. We now present a method of approximation for calculating this secondary flow[1].

Assuming that the flow pattern is the same for all cross-sections, we have a velocity field of the form

$$\dot{\boldsymbol{x}} = v(\boldsymbol{p})\,\boldsymbol{k} + \boldsymbol{u}(\boldsymbol{p}), \tag{122.1}$$

where the notation of Sect. 117 is employed. The flow corresponding to the velocity component $\boldsymbol{u}(\boldsymbol{p})$ perpendicular to the axis is the *secondary flow.* In order that the flow be isochoric, it is necessary and sufficient that

$$\operatorname{div} \boldsymbol{u} = 0. \tag{122.2}$$

For simply connected cross-sections (122.2) is satisfied if and only if $\boldsymbol{u}$ can be derived[2] from a stream function[3] q:

$$\boldsymbol{u} = (\nabla q)^{\perp}; \tag{122.3}$$

---

[1] As mentioned in Sect. 117, ERICKSEN [1956, *7*] first indicated the existence of these secondary flows, and GREEN and RIVLIN [1956, *13*] first obtained an approximate solution for the flow in an elliptical pipe, within a very special theory. At the U.S. National Bureau of Standards a simple apparatus for observing the streamline pattern of flow in a tube was constructed by KEARSLEY, who confirmed the general form of the streamlines predicted by GREEN and RIVLIN. Operating within a special class of fluids, LANGLOIS and RIVLIN [1959, *18*, §§ 2—3] set up a general method of approximation and applied it to the elliptical pipe. The first systematic theory was obtained later by LANGLOIS and RIVLIN [1963, *46*] [1964, *71*]. The treatment in our text, due to NOLL and here published for the first time, especially from Eq. (122.16) onward, simplifies and shortens their work.

The existence of secondary mean flow in the turbulent motion of a fluid in a non-circular pipe, along with other phenomena, has been adduced by RIVLIN [1957, *20*] as evidence for an analogy between turbulent flow of a Newtonian fluid and smooth flow of a more general fluid. Similar departures from a straight streamline pattern in open-channel flow have been attributed to effects of non-linear viscosity by POSEY [1958, *37*].

Oscillating flow in non-circular tubes has been treated approximately by PIPKIN [1964, *66* and *68*], using an expansion in powers of a natural time-constant of the fluid.

[2] Since in this section no use is made of a reference configuration, we write ∇ for the spatial gradient, elsewhere denoted by "grad".

[3] Cf., e.g., CFT, Sect. 161.

here and in the remainder of this section we use the abbreviation

$$\boldsymbol{v}^{\perp} = \boldsymbol{k} \times \boldsymbol{v} \tag{122.4}$$

for any vector $\boldsymbol{v}$ in the plane $\mathscr{U}$. We have $(\boldsymbol{v}^{\perp})^{\perp} = -\boldsymbol{v}$.

Since the velocity (122.1) depends only on $\boldsymbol{p}$ and not on z, it is clear that the history (105.2) depends only on $\boldsymbol{p}$. Therefore, by (105.1), the extra-stress $\boldsymbol{T}_{\mathrm{E}} = \boldsymbol{T} + p\boldsymbol{1}$ depends only on $\boldsymbol{p}$. Also, the acceleration $\ddot{\boldsymbol{x}} = (\nabla \dot{\boldsymbol{x}})\dot{\boldsymbol{x}}$ [cf. (15.13)] depends only on $\boldsymbol{p}$. By the dynamical equation (30.26), grad φ is also a function of $\boldsymbol{p}$ only. Since then $\frac{\partial}{\partial z}$ grad $\varphi = \operatorname{grad} \frac{\partial \varphi}{\partial z} = 0$, it follows that

$$\varrho \frac{\partial \varphi}{\partial z} = -a = \text{const.}, \qquad \varrho \varphi = -a z + \zeta(\boldsymbol{p}). \tag{122.5}$$

It follows from (122.1) that the velocity gradient $\boldsymbol{L} = \nabla \dot{\boldsymbol{x}}$ is given by

$$\boldsymbol{L} = \boldsymbol{k} \otimes \nabla v + \nabla \boldsymbol{u}; \tag{122.6}$$

hence the acceleration $\ddot{\boldsymbol{x}} = \boldsymbol{L}\dot{\boldsymbol{x}}$ has the form

$$\ddot{\boldsymbol{x}} = (\boldsymbol{u} \cdot \nabla v)\boldsymbol{k} + (\nabla \boldsymbol{u})\boldsymbol{u}. \tag{122.7}$$

We employ the following decomposition of the extra-stress:

$$\boldsymbol{T}_{\mathrm{E}} = S(\boldsymbol{k} \otimes \boldsymbol{k}) + \boldsymbol{t} \otimes \boldsymbol{k} + \boldsymbol{k} \otimes \boldsymbol{t} + \boldsymbol{\Pi}, \tag{122.8}$$

where $\boldsymbol{t}$ is a vector in the two-dimensional space $\mathscr{U}$ and $\boldsymbol{\Pi}$ is a tensor on $\mathscr{U}$. S, $\boldsymbol{t}$, and $\boldsymbol{\Pi}$ are functions of $\boldsymbol{p}$ only. Using the identity (117.12), we find that

$$\operatorname{div} \boldsymbol{T}_{\mathrm{E}} = (\operatorname{div} \boldsymbol{t})\boldsymbol{k} + \operatorname{div} \boldsymbol{\Pi}. \tag{122.9}$$

It follows from (122.9), (122.7) and $(122.5)_2$ that the dynamical equation (30.26) is equivalent to the following system of two equations:

$$\left.\begin{aligned} \operatorname{div} \boldsymbol{t} + a &= \varrho \boldsymbol{u} \cdot \nabla v, \\ \operatorname{div} \boldsymbol{\Pi} - \nabla \zeta &= \varrho (\nabla \boldsymbol{u})\boldsymbol{u}. \end{aligned}\right\} \tag{122.10}$$

On the assumption that the body forces have no component in the z-direction, i.e., that the potential v depends only on $\boldsymbol{p}$, it follows from (122.5) and (30.27) that the derivative of the stress component

$$T_{\langle zz \rangle} = S - p = a z + S(\boldsymbol{p}) - \zeta(\boldsymbol{p}) + \varrho v(\boldsymbol{p}) \tag{122.11}$$

with respect to z is just the constant a. The argument that led from (111.13) to (111.15) shows a to be the *specific driving force* that produces the flow.

If the fluid adheres to the walls of the tube, we have the boundary conditions

$$v(\boldsymbol{p}) = 0, \qquad \boldsymbol{u}(\boldsymbol{p}) = \boldsymbol{0} \quad \text{when} \quad \boldsymbol{p} \in \partial \mathscr{A}, \tag{122.12}$$

which generalize (117.2). Since the stream function q is determined by $\boldsymbol{u}$ through (122.3) only to within an additive constant, we may replace $(122.12)_2$ by

$$q = 0, \qquad \frac{\partial q}{\partial n} = 0 \quad \text{when} \quad \boldsymbol{p} \in \partial \mathscr{A}, \tag{122.13}$$

where $\frac{\partial}{\partial n}$ denotes the derivative in the direction normal to the boundary curve $\partial \mathscr{A}$.

When the specific driving force a is zero, the dynamical equations (122.10) have the solution $v = 0$, $\boldsymbol{u} = \boldsymbol{0}$, $\varrho \varphi = p + \varrho v = 0$, which corresponds to the state of rest. We assume that, for values of a close to zero, solutions exist and that their dependence on a is smooth enough that they have expansions

$$\left.\begin{aligned} v &= \sum_{r=1}^{n} a^r v_r + O(a^{n+1}), \\ \boldsymbol{u} &= \sum_{r=1}^{n} a^r \boldsymbol{u}_r + O(a^{n+1}), \quad \boldsymbol{u}_r = (\nabla q_r)^{\perp}, \\ \zeta &= \sum_{r=1}^{n} a^r \zeta_r + O(a^{n+1}). \end{aligned}\right\} \tag{122.14}$$

We assume further that the derivatives with respect to $\boldsymbol{p}$ of the error terms $O(a^{n+1})$ are also of order $O(a^{n+1})$.

It is plausible that the flow will be slow when a is small. However, there is no reason to believe that the flow for a small specific driving force can be obtained from the flow for a larger specific driving force by a mere retardation. Therefore, the asymptotic approximations for slow flow discussed in Sect. 40 do not apply directly. Nevertheless, without being able to supply a rigorous justification, we shall assume that the extra-stress is given, to within an error of order $o(a^{\mathfrak{n}})$, by the constitutive equation (121.28) of an incompressible fluid[1] of grade $\mathfrak{n}$. We can evaluate from $(122.14)_{1,2}$ expansions for the Rivlin-Ericksen tensors $\boldsymbol{A}_{\mathfrak{k}}$ in terms of the v_r and q_r. Substituting these expansions into (121.28) and carrying out the decomposition (122.8), we can obtain expansions for $\boldsymbol{t}$ and $\boldsymbol{\Pi}$. Putting these expansions and the expansions (122.14) into the dynamical equations (122.10) and equating the coefficients of like powers of a, we can derive a successive system of partial differential equations for the v_r, q_r, and ζ_r. We expect to determine the v_r, q_r, and ζ_r from these differential equations and the boundary conditions

$$v_r = 0, \qquad q_r = 0, \qquad \frac{\partial q_r}{\partial n} = 0 \quad \text{when} \quad \boldsymbol{p} \in \partial \mathscr{A}, \tag{122.15}$$

which result from $(122.12)_1$ and (122.13). We shall carry out this procedure explicitly for the case when $\mathfrak{n} = 4$.

We obtain from (122.6) the expansion

$$\boldsymbol{L} = \sum_{r=1}^{n} a^r (\boldsymbol{k} \otimes \nabla v_r + \nabla \boldsymbol{u}_r) + O(a^{n+1}), \tag{122.16}$$

and hence from $(25.6)_2$:

$$\boldsymbol{A}_1 = \sum_{r=1}^{n} a^r [\boldsymbol{k} \otimes \nabla v_r + \nabla v_r \otimes \boldsymbol{k} + \nabla \boldsymbol{u}_r + (\nabla \boldsymbol{u}_r)^T] + O(a^{n+1}). \tag{122.17}$$

The Rivlin-Ericksen tensors $\boldsymbol{A}_{\mathfrak{k}}$ when $\mathfrak{k} > 1$ are best obtained from (122.17) and (122.16) by means of the recursion formula [CFT, Eq. (104.4)]

$$\boldsymbol{A}_{\mathfrak{k}+1} = (\nabla \boldsymbol{A}_{\mathfrak{k}}) \dot{\boldsymbol{x}} + \boldsymbol{A}_{\mathfrak{k}} \boldsymbol{L} + (\boldsymbol{A}_{\mathfrak{k}} \boldsymbol{L})^T, \tag{122.18}$$

where $(\nabla \boldsymbol{A}_{\mathfrak{k}}) \dot{\boldsymbol{x}}$ denotes the tensor with components $A_{\mathfrak{k}\, km,p} \dot{x}^p$. We note that, since the $\boldsymbol{A}_{\mathfrak{k}}$ depend only on $\boldsymbol{p}$, we have

$$(\nabla \boldsymbol{A}_{\mathfrak{k}}) \boldsymbol{k} = \frac{\partial \boldsymbol{A}_{\mathfrak{k}}}{\partial z} = \boldsymbol{0}. \tag{122.19}$$

It is evident from (122.16), (122.17), and (122.18) that $\boldsymbol{A}_{\mathfrak{k}} = O(a^{\mathfrak{k}})$. Therefore, by (121.29), (121.30), and (122.17) we have

$$\boldsymbol{T}_{\mathrm{E}} = \mu_0 \boldsymbol{A}_1 + O(a^2) = a \mu_0 [\boldsymbol{k} \otimes \nabla v_1 + \nabla v_1 \otimes \boldsymbol{k} + \nabla \boldsymbol{u}_1 + (\nabla \boldsymbol{u}_1)^T] + O(a^2). \tag{122.20}$$

---

[1] For fluids of the differential type it is easily seen that this conclusion is in fact valid.

Comparison of (122.20) with (122.8) shows that

$$t = a\mu_0 \nabla v_1 + O(a^2), \qquad \Pi = a\mu_0[\nabla \boldsymbol{u}_1 + (\nabla \boldsymbol{u}_1)^T] + O(a^2). \tag{122.21}$$

Since $(\nabla v)\cdot \boldsymbol{u} = O(a^2)$ and $(\nabla \boldsymbol{u})\boldsymbol{u} = O(a^2)$, we infer from (122.21) and $(122.14)_4$ that equating the coefficients of a in the dynamical equations (122.10) yields

$$\mu_0 \Delta v_1 = -1, \tag{122.22}$$

and

$$\mu_0 \operatorname{div}[(\nabla \boldsymbol{u}_1) + (\nabla \boldsymbol{u}_1)^T] - \nabla \zeta_1 = 0. \tag{122.23}$$

Equation (122.22) is a Poisson equation which has a unique solution v_1 that satisfies the boundary condition $v_1 = 0$ when $\boldsymbol{p} \in \partial \mathscr{A}$.

For any function $q = q(\boldsymbol{p})$ we have the easily verified identity

$$\operatorname{div}[(\nabla \boldsymbol{u}) + (\nabla \boldsymbol{u})^T] = (\nabla \Delta q)^{\perp}, \qquad \boldsymbol{u} = (\nabla q)^{\perp}. \tag{122.24}$$

Hence, since $\boldsymbol{u}_1 = (\nabla q_1)^{\perp}$, (122.23) becomes

$$\mu_0 (\nabla \Delta q_1)^{\perp} - \nabla \zeta_1 = 0. \tag{122.25}$$

Performing the $\perp$-operation (122.4) on (122.25) and then taking the divergence, we obtain

$$\mu_0 \Delta\Delta q_1 = 0. \tag{122.26}$$

This is a biharmonic equation for q_1. The only solution that satisfies the boundary conditions $(122.15)_{2,3}$ is $q_1 = 0$, and we have

$$q_1 = 0, \qquad \boldsymbol{u}_1 = \boldsymbol{0}. \tag{122.27}$$

For the Navier-Stokes theory, the entire velocity field is $v_1 \boldsymbol{k}$. This field has now been shown to serve as the leading term in the expansion (122.14) for an arbitrary simple fluid.

Specifically, it follows from (122.27) that $\dot{\boldsymbol{x}} = a\, v_1 \boldsymbol{k} + O(a^2)$. Hence, by (122.19), the first term on the right in the recursion formula (122.18) for $\mathfrak{k} = 1$ is of order $O(a^3)$. Later on we shall see that also $\boldsymbol{u}_2 = \boldsymbol{u}_3 = \boldsymbol{0}$ and hence that $\boldsymbol{u} = O(a^4)$. Thus we have $\dot{\boldsymbol{x}} = a\, v_1 \boldsymbol{k} + O(a^4)$, and hence, by (122.19), the first term on the right-hand side in the recursion formula (122.18) is actually always of an order higher than $O(a^5)$.

Using (122.27), we obtain from (122.16)—(122.18)

$$\left.\begin{aligned} \boldsymbol{A}_1^2 &= a^2[(\nabla v_1 \otimes \nabla v_1) + \varkappa_1^2(\boldsymbol{k}\otimes\boldsymbol{k})] + O(a^3),\\ \boldsymbol{A}_2 &= 2a^2(\nabla v_1 \otimes \nabla v_1) + O(a^3), \end{aligned}\right\} \tag{122.28}$$

where

$$\varkappa_1^2 = (\nabla v_1)^2. \tag{122.29}$$

Using (122.28) to determine the extra-stress $\boldsymbol{T}_{\mathrm{E}}$ by means of (121.29) to within terms of order $O(a^3)$, we can easily evaluate the coefficients of a^2 in the dynamical equations (122.10). The calculation, which is very similar to the one leading to (122.22) and (122.26), yields

$$\mu_0 \Delta v_2 = 0, \qquad \mu_0 \Delta\Delta q_2 = 0. \tag{122.30}$$

The only solutions of (122.30) satisfying the boundary conditions (122.15) are the zero solutions, and we conclude that

$$v_2 = 0, \qquad q_2 = 0, \qquad \boldsymbol{u}_2 = \boldsymbol{0}. \tag{122.31}$$

Therefore we have

$$\dot{\boldsymbol{x}} = a\, v_1 \boldsymbol{k} + O(a^3), \qquad \boldsymbol{u} = O(a^3). \tag{122.32}$$

In view of the results of Sect. 117, the conclusion (122.32) is not surprising. In the theory of fluids of the second grade the solution (122.32) is exact in the sense that the error terms $O(a^3)$ are zero. This is the case because for such fluids we have $\tilde{\mu}(\varkappa)=\mu_0$, $\sigma_1(\varkappa)=(2\alpha_1+\alpha_2)\varkappa^2$, and hence the condition (117.22) is satisfied. The differential equation (117.19) reduces to (122.22).

If we observe (122.31) and examine the recursion formula (122.18), we find that

$$\boldsymbol{A}_3=O(a^5) \tag{122.33}$$

and that $O(a^3)$ can be replaced by $O(a^4)$ in (122.28). The extra-stress (122.8) can then be determined to within terms of order $O(a^4)$ from (121.29) and $(121.30)_{1,2,3}$; we thus obtain

$$\left.\begin{aligned} t&=a\mu_0\nabla v_1+a^3[\mu_0\nabla v_3+2(\beta_2+\beta_3)\varkappa_1^2\nabla v_1]+O(a^4),\\ \boldsymbol{\Pi}&=a^2(2\alpha_1+\alpha_2)(\nabla v_1\otimes\nabla v_1)+a^3\mu_0[\nabla\boldsymbol{u}_3+(\nabla\boldsymbol{u}_3)^T]+O(a^4).\end{aligned}\right\} \tag{122.34}$$

From (122.32) it follows that $\boldsymbol{u}\cdot\nabla v=O(a^4)$ and $(\nabla\boldsymbol{u})\boldsymbol{u}=O(a^6)$. Equating the coefficients of a^3 in the dynamical equations (122.10) then yields

$$\mu_0\Delta v_3+2(\beta_2+\beta_3)\omega=0, \tag{122.35}$$

where

$$\omega=\operatorname{div}(\varkappa_1^2\nabla v_1)=\varkappa_1^2\Delta v_1+\nabla v_1\cdot\nabla(\varkappa_1^2), \tag{122.36}$$

and

$$\mu_0\operatorname{div}[\nabla\boldsymbol{u}_3+(\nabla\boldsymbol{u}_3)^T]-\nabla\zeta_3=0. \tag{122.37}$$

Assuming that v_1 has been determined as a solution of (122.22), one can evaluate ω from (122.36), and (122.35) becomes a Poisson equation, which has a unique solution v_3 that satisfies the boundary condition $(122.15)_1$. The argument that led from (122.23) to (122.27) shows that

$$q_3=0,\qquad \boldsymbol{u}_3=0. \tag{122.38}$$

Using this result and (122.16)—(122.18), we calculate

$$\left.\begin{aligned} \boldsymbol{A}_1^2&=a^2[(\nabla v_1\otimes\nabla v_1)+\varkappa_1^2(\boldsymbol{k}\otimes\boldsymbol{k})]+a^4[(\nabla v_1\otimes\nabla v_3)+(\nabla v_3\otimes\nabla v_1)+\\ &\quad+2(\nabla v_1\cdot\nabla v_3)(\boldsymbol{k}\otimes\boldsymbol{k})]+O(a^5),\\ \boldsymbol{A}_2&=2a^2(\nabla v_1\otimes\nabla v_1)+2a^4[(\nabla v_1\otimes\nabla v_3)+(\nabla v_3\otimes\nabla v_1)]+O(a^5),\\ \boldsymbol{A}_2^2&=4a^4\varkappa_1^2(\nabla v_1\otimes\nabla v_1)+O(a^6),\\ \operatorname{tr}\boldsymbol{A}_2&=2a^2\varkappa_1^2+4a^4(\nabla v_1\cdot\nabla v_3)+O(a^5),\\ \boldsymbol{A}_3&=O(a^5),\qquad \boldsymbol{A}_4=O(a^6).\end{aligned}\right\} \tag{122.39}$$

As before, we calculate the extra-stress (122.8) according to (121.29) and (121.30). The result is (we write only the coefficients of a^4)

$$\left.\begin{aligned} t&=a(\ldots)+a^3(\ldots)+a^4\mu_0\nabla v_4+O(a^5),\\ \boldsymbol{\Pi}&=a^2(\ldots)+a^4\{\mu_0[\nabla\boldsymbol{u}_4+(\nabla\boldsymbol{u}_4)^T]+(2\alpha_1+\alpha_2)[(\nabla v_3\otimes\nabla v_1)+\\ &\quad+(\nabla v_1\otimes\nabla v_3)]+\gamma\varkappa_1^2(\nabla v_1\otimes\nabla v_1)\}+O(a^5),\end{aligned}\right\} \tag{122.40}$$

where

$$\gamma=4(\gamma_3+\gamma_4+\gamma_5+\tfrac{1}{2}\gamma_6).$$

It follows from (122.32) and (122.38) that $(\boldsymbol{u}\cdot\nabla v)=O(a^5)$, $(\nabla\boldsymbol{u})\boldsymbol{u}=O(a^8)$. Equating the coefficients of a^4 in the dynamical equations (122.10) then gives

$$\mu_0\Delta v_4=0 \tag{122.41}$$

and

$$\left.\begin{aligned}\mu_0 \operatorname{div}[\nabla \boldsymbol{u}_4 + (\nabla \boldsymbol{u}_4)^T] + (2\alpha_1+\alpha_2) \operatorname{div}[(\nabla v_3 \otimes \nabla v_1) + (\nabla v_1 \otimes \nabla v_3)] + \\ + \gamma \operatorname{div}[\varkappa_1^2 (\nabla v_1 \otimes \nabla v_1)] - \nabla \zeta_4 = 0.\end{aligned}\right\} \tag{122.42}$$

From (122.41) and the boundary condition (122.15) we conclude that

$$v_4 = 0. \tag{122.43}$$

To simplify (122.42), we use the identity (122.24) and the easily verified identities

$$\left.\begin{aligned}\operatorname{div}[(\nabla v_3 \otimes \nabla v_1) + (\nabla v_1 \otimes \nabla v_3)] &= \nabla(\nabla v_1 \cdot \nabla v_3) + (\nabla v_1)\Delta v_3 + (\nabla v_3)\Delta v_1, \\ \operatorname{div}[(\nabla v_1) \otimes (\varkappa_1^2 \nabla v_1)] &= \tfrac{1}{4}\nabla(\varkappa_1^4) + \omega \nabla v_1,\end{aligned}\right\} \tag{122.44}$$

where ω is given by (122.36). From these identities and from (122.22) and (122.35) we infer that (122.42) is equivalent to

$$\left.\begin{aligned}\nabla\left\{\zeta_4 - (2\alpha_1+\alpha_2)\left[(\nabla v_1 \cdot \nabla v_3) + \frac{1}{\mu_0} v_3\right] - \frac{1}{4}\gamma \varkappa_1^4\right\} \\ = \mu_0 (\nabla \Delta q_4)^{\perp} + \delta \omega \nabla v_1,\end{aligned}\right\} \tag{122.45}$$

where

$$\delta = \gamma - \frac{2}{\mu_0}(2\alpha_1+\alpha_2)(\beta_2+\beta_3). \tag{122.46}$$

Performing the $\perp$-operation (122.4) on (122.45) and then taking the divergence, we find

$$\mu_0 \Delta\Delta q_4 = \delta (\nabla v_1)^{\perp} \cdot (\nabla \omega). \tag{122.47}$$

This is an inhomogeneous biharmonic equations for q_4, which has a unique solution satisfying the boundary conditions $(122.15)_{2,3}$.

We summarize: *The velocity field of a flow through a tube produced by the specific driving force a has the expansion*

$$\dot{\boldsymbol{x}} = (a v_1 + a^3 v_3)\boldsymbol{k} + a^4 (\nabla q_4)^{\perp} + O(a^5), \tag{122.48}$$

where first $v_1 = v_1(\boldsymbol{p})$ *is determined from* (122.22), *and then* $v_3 = v_3(\boldsymbol{p})$ *and* $q_4 = q_4(\boldsymbol{p})$ *are obtained from* (122.35) *and* (122.47), *respectively, subject to the boundary conditions*

$$v_1 = v_3 = q_4 = \frac{\partial q_4}{\partial n} = 0 \quad \text{when} \quad \boldsymbol{p} \in \partial \mathscr{A}. \tag{122.49}$$

It is possible, of course, to continue the procedure and to calculate the coefficients of a^5, a^6, etc. The computations become more and more complex. The coefficient of a^4 is the most interesting because it is the first that can be expected to give a non-zero contribution to the secondary flow.

For a given cross-section, the problem of determining the expansion (122.48) is solved for all fluids if it has been solved for just one, provided $\beta_2+\beta_3 \neq 0$ and $\delta \neq 0$ for this fluid. Indeed, if $\bar{v}_1, \bar{v}_3$, and $\bar{q}_4$ denote the solutions of (122.22), (122.35), and (122.47) when μ_0, $\beta_2+\beta_3$, and δ are all replaced by 1 in these equations, then

$$v_1 = \frac{1}{\mu_0}\bar{v}_1, \quad v_3 = \frac{\beta_2+\beta_3}{\mu_0^4}\bar{v}_3, \quad q_4 = \frac{\delta}{\mu_0^5}\bar{q}_4. \tag{122.50}$$

The streamline pattern of the secondary flow is given by the curves $q_4 = \text{const}$. By $(122.50)_3$, these are the same as the curves $\bar{q}_4 = \text{const}$. Therefore, the streamline pattern of the secondary flow depends only on the cross-section and is the same for all simple fluids, as long as terms of order $O(a^5)$ can be neglected[1].

[1] Thus, in particular, the streamline pattern corresponding to Eqs. (122.57) for the elliptical cross-section, obtained by GREEN and RIVLIN [1956, *13*] in their analysis of a very particular fluid, is here seen to be perfectly general.

For an elliptical cross-section, it is possible to evaluate explicitly the functions v_1, v_3, and q_4 entering into (122.48)[1]. We introduce Cartesian co-ordinates x, y in the plane $\mathcal{U}$ such that the elliptical boundary has the equation

$$\frac{x^2}{c^2}+\frac{y^2}{b^2}=1, \qquad c>b. \tag{122.51}$$

The solution v_1 of (122.22) that satisfies $v_1=0$ on $\partial\mathscr{A}$ is then given by

$$v_1=-\frac{c^2 b^2}{2\mu_0(c^2+b^2)}\left(\frac{x^2}{c^2}+\frac{y^2}{b^2}-1\right). \tag{122.52}$$

An easy calculation [cf. (122.29) and (122.36)] yields

$$\omega=\frac{-1}{\mu_0^3(c^2+b^2)^3}\left[b^4(3b^2+c^2)x^2+c^4(3c^2+b^2)y^2\right]. \tag{122.53}$$

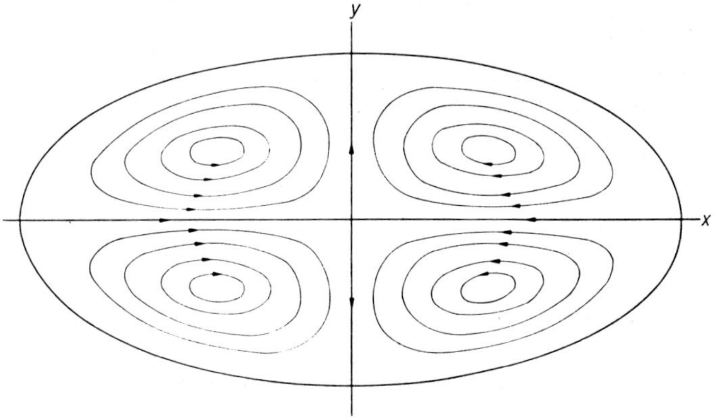

Fig. 26. Pattern of secondary flow in an elliptical tube.

Substituting (122.52) and (122.53) into (122.47), we find

$$\mu_0^5 \Delta\Delta q_4=\delta\frac{6c^2b^2(c^2-b^2)}{(c^2+b^2)^3}\,x\,y. \tag{122.54}$$

With a little calculation one can verify that

$$q_4=\frac{\delta}{\mu_0^5}A\left(\frac{x^2}{c^2}+\frac{y^2}{b^2}-1\right)^2 x\,y, \tag{122.55}$$

where

$$A=\frac{c^6b^6(c^2-b^2)}{4(c^2+b^2)^3(5c^4+6c^2b^2+5b^4)} \tag{122.56}$$

is the solution of (122.54) that satisfies the boundary condition $q_4=\frac{\partial q_4}{\partial n}=0$ on the ellipse (122.51). The components u_x and u_y of the secondary velocity field $\boldsymbol{u}=a^4(\nabla q_4)^{\perp}+O(a^5)$ are easily obtained from (122.55):

$$\left.\begin{aligned} u_x&=-\frac{\delta}{\mu_0^5}a^4A\left(\frac{x^2}{c^2}+\frac{y^2}{b^2}-1\right)\left(\frac{x^2}{c^2}+\frac{5y^2}{b^2}-1\right)x+O(a^5),\\ u_y&=\frac{\delta}{\mu_0^5}a^4A\left(\frac{x^2}{c^2}+\frac{y^2}{b^2}-1\right)\left(\frac{5x^2}{c^2}+\frac{y^2}{b^2}-1\right)y+O(a^5).\end{aligned}\right\} \tag{122.57}$$

The streamlines corresponding to this secondary flow are indicated in Fig. 26. The arrows show the direction of flow when $\delta>0$. If $\delta<0$, the arrows must be reversed.

[1] Langlois and Rivlin [1963, *46*, § 6].

The solution v_3 of (122.35) is also not hard to obtain:

$$v_3 = H(C x^2 + B y^2 + D)\left(\frac{x^2}{c^2} + \frac{y^2}{b^2} - 1\right), \tag{122.58}$$

where

$$\left.\begin{aligned} H &= -\frac{(\beta_2+\beta_3)}{6\mu_0^4}\,\frac{c^2 b^2}{(c^2+b^2)^3(c^4+b^4+6c^2b^2)},\\ C &= b^2[c^4(b^2+3c^2) - b^2(b^2+6c^2)(3b^2+c^2)],\\ B &= c^2[b^4(c^2+3b^2) - c^2(c^2+6b^2)(3c^2+b^2)],\\ D &= \frac{c^2 b^2}{c^2+b^2}(C+B). \end{aligned}\right\} \tag{122.59}$$

However, the contribution from v_3 only modifies the flow in the direction of the tube and has no bearing on the secondary flow; of course, it gives rise to normal thrusts on the wall[1].

123. Fluids of second grade. α) *Definition.* Fluids of grade n have been defined in Sect. 121. The constitutive equation of an *incompressible fluid of second grade* is obtained by omitting $\boldsymbol{S}_3$ and $\boldsymbol{S}_4$ from (121.29):

$$\boldsymbol{T} = -p\boldsymbol{1} + \mu_0 \boldsymbol{A}_1 + \alpha_1 \boldsymbol{A}_2 + \alpha_2 \boldsymbol{A}_1^2. \tag{123.1}$$

It was proved in Sect. 121 that in the sense of retardation, Eq. (123.1) approximates to the second order the constitutive equation of any incompressible simple fluid, while the Navier-Stokes formula, which may be obtained by omitting the terms proportional to $\boldsymbol{A}_2$ and $\boldsymbol{A}_1^2$, is the first-order approximation. We may expect, then, that the stress in any sufficiently slow flow of any incompressible simple fluid may be calculated from the flow with sufficient approximation by means of Eq. (123.1). The coefficients μ_0, α_1, and α_2 are determined by the response functional. However, works on hydrodynamics generally regard the Navier-Stokes formula, not as approximate, but rather as an exact definition of a particular fluid, flows of which are not restricted to any approximating sequence, and the coefficient μ_0 is interpreted, not as a number to be calculated from some more accurate theory, but rather as an empirical quantity, to be determined by comparing with experimental data results rigorously derived from the Navier-Stokes theory itself. The fluid of the second grade, being defined likewise by a frame-indifferent constitutive equation, may be approached in the same way, not as an approximation, but rather as a special case offering particular interest. In this section we shall consider the exact theory of fluids of second grade[2], but, as we shall see, the results in some cases are open to objections from which the Navier-Stokes theory is free.

The *viscometric functions* for fluids of second grade may be read off from (121.34):

$$\left.\begin{aligned} \tilde{\mu}(\varkappa) &= \mu_0,\\ \sigma_1(\varkappa) &= (2\alpha_1+\alpha_2)\varkappa^2,\\ \sigma_2(\varkappa) &= \alpha_2\varkappa^2. \end{aligned}\right\} \tag{123.2}$$

The viscosity remains a constant, as in the Navier-Stokes theory, but the normal stress functions do not generally vanish. The specializing features of various earlier theories, now known to contradict experiment, are avoided: The Reiner-

[1] These have been worked out by PIPKIN and RIVLIN [1963, *60*].

[2] Superposability of solutions in the theory of fluids of second grade has been considered by SHARMA [1964, *76*].

Rivlin relation $(119.9)_3$ and the Weissenberg assertion (119.30) would require that $\alpha_1=0$ and $2\alpha_1+\alpha_2=0$, respectively. A relation of the form (119.29) holds trivially, as is clear from the more general remark following (119.31). As far as viscometric flows are concerned, *the results predicted by the theory of fluids of second grade are the same as those from the theory of particular fluids with convected elasticity;* by Eq. (119.11), the response time t^* is given by[1]

$$t^* = -\frac{2\alpha_1}{\mu_0}. \tag{123.3}$$

Since $\mu_0 > 0$, this time is positive if and only if $\alpha_1 < 0$. This fact affords us a provisional interpretation for the coefficient α_1. The fluid of second grade may be regarded as responding to its past, present, or future experience according as $\alpha_1 < 0$, $\alpha_1 = 0$, or $\alpha_1 > 0$. Hence we see a different motivation for the inequality $(121.27)_2$.

The various results derived in the preceding sections for simple fluids in general take a more specific and definite form when applied to fluids of the second grade and afford a variety of interpretations for the coefficients α_1 and α_2 as well as methods of determining them, at least in principle, by experiment[2].

While modern work on viscometry often makes that subject seem an end in itself, it was not always so. Originally the aim of viscometry was *to determine the constitutive equation* for a particular material. The Navier-Stokes theory allows to an incompressible fluid but one material parameter beyond its density, namely, μ_0, the viscosity. A single viscometric experiment serves, in principle, to determine μ_0 once and for all, and thereby to determine the constitutive equation *for all flows* of the substance being tested. Further experiment serves, again in principle, only for cross-checks and confirmation. In Sects. 106—113 we have seen that for simple fluids in general, experiments on a single kind of steady viscometric flow, for the entire range of the shearing $\varkappa$, suffice to determine the three viscometric functions but nothing more. Experiments on other viscometric flows can serve, in principle, only as cross-checks on the nature of these three functions and cannot yield information anything like sufficient to determine the response functional of the fluid being tested. A single aspect of fluid behavior is selected by viscometric experiments, and the most painstaking and accurate program of measurement can deliver no information about such other physical properties as the fluid may have.

The special status of the Navier-Stokes fluid in regard to viscometry extends to the fluid of second grade, but no further[3]. In principle, from a single viscometric experiment the three constants (or functions of temperature) μ_0, α_1 and α_2 may be found, and these, in turn, fully specify the constitutive equation (123.1): *Viscometric measurements suffice to determine the form of the constitutive equation for all flows of a fluid of second grade.* This common property of the first and second approximations may be regarded as either an advantage or a disadvantage. On the one hand, it allows us to make definite predictions on the basis of scant measured data; on the other, it indicates that such predictions, since they are impossible in more general theories, can rarely be expected to square with more subtle experiment. Indeed, the constitutive equation for a fluid of third grade cannot be determined by viscometric measurements, since the three coefficients β_1, β_2, and β_3 in $(121.30)_3$ appear in Eqs. (121.34) in only the one combination $\beta_2+\beta_3$, so that β_1 has no viscometric significance at all.

[1] TRUESDELL [1965, *37*].

[2] Detailed discussion is presented by COLEMAN and MARKOVITZ, [1964, *17* and *60*].

[3] TRUESDELL (1962) [1964, *87*, § 13] [1964, *85*, § 6].

β) *Some particular unsteady flows.* The major interest in fluids of the second grade, granted their status as an approximation, lies in the possibility of solving for them various specific problems that presently seem beyond the reach of more general theories. To interpret these solutions, we assume that the fluid adheres to rigid bounding surfaces[1].

Since $\boldsymbol{A}_{\mathfrak{k}}=(-1)^{\mathfrak{k}}\boldsymbol{G}^{(\mathfrak{k})}(0)$, we infer from (121.9) that for viscometric flows, whether or not they be steady,

$$\left.\begin{aligned}\boldsymbol{A}_1&=-\dot{g}(0)(\boldsymbol{N}+\boldsymbol{N}^T),\\ \boldsymbol{A}_2&=\ddot{g}(0)(\boldsymbol{N}+\boldsymbol{N}^T)+2\dot{g}(0)^2\boldsymbol{N}^T\boldsymbol{N}.\end{aligned}\right\}\tag{123.4}$$

Substitution of (123.4) into (123.1) shows that for fluids of second grade the functionals $\mathfrak{t}$, $\hat{\mathfrak{s}}_1$, and $\hat{\mathfrak{s}}_2$ defined in Sect. 106 reduce to

$$\left.\begin{aligned}&\mathop{\mathfrak{t}}_{s=0}^{\infty}(g(s))=-\mu_0\dot{g}(0)+\alpha_1\ddot{g}(0),\\ &\mathop{\hat{\mathfrak{s}}_1}_{s=0}^{\infty}(g(s))=(2\alpha_1+\alpha_2)\dot{g}(0)^2,\qquad \mathop{\hat{\mathfrak{s}}_2}_{s=0}^{\infty}(g(s))=\alpha_2\dot{g}(0)^2.\end{aligned}\right\}\tag{123.5}$$

We consider first the case of *simple shearing*. By (106.11) and (106.5) we then have

$$\left.\begin{aligned}\dot{g}(0)&=-\dot{k}(t)=-\partial_x v(x,t),\\ \ddot{g}(0)&=\ddot{k}(t)=\partial_{tx}v(x,t).\end{aligned}\right\}\tag{123.6}$$

Combining (123.6), (123.5), (106.20), and $(111.2)_1$, we see that solving the dynamical equations reduces to finding solutions $v=v(x,t)$ of the following third-order linear partial differential equation[2]:

$$\mu_0\partial_{xx}v+\alpha_1\partial_{xtx}v+c(t)=\varrho\,\partial_t v.\tag{123.7}$$

The coefficient α_2 does not appear in this equation. Consequently, a simple shearing field if possible at all is shared by infinitely many fluids of second grade, though of course the surface tractions that must be applied in order to produce it will vary according to the value of α_2. In the case when $\alpha_1=0$, corresponding to a particular Reiner-Rivlin fluid, Eq. (123.7) is identical with the equation resulting from the Navier-Stokes theory, so that the simple shearings possible in a Reiner-Rivlin fluid of second grade are the same as those in the Navier-Stokes fluid, although, again, certain normal stresses, which depend upon the value of α_2, will have to be supplied in order to produce them. More generally, the presence of the term $\alpha_1\partial_{xtx}v$ in Eq. (123.7) gives rise to a distribution of speeds along the stream-lines that is wholly different from its classical counterpart, as we shall now see.

---

[1] BERKER [1964, *6*] has found a simple formula for the traction upon a surface to which the fluid adheres:

$$\boldsymbol{t}=[-p+(2\alpha_1+\alpha_2)w^2]\boldsymbol{n}+(\mu\boldsymbol{w}+\alpha_1\partial_t\boldsymbol{w})\times\boldsymbol{n},$$

where $\boldsymbol{w}$ is the vorticity: $\boldsymbol{w}=\operatorname{curl}\dot{\boldsymbol{x}}$. Not only is a resolution into normal and tangential components thus effected, but also, since $\boldsymbol{w}$ and $\partial_t\boldsymbol{w}$ are tangent to the boundary (CFT, Sect. 105), the tangential component is obtained by rotating $\mu\boldsymbol{w}+\alpha_1\partial_t\boldsymbol{w}$ clockwise through a right angle about the normal $\boldsymbol{n}$. The excess normal tension $(2\alpha_1+\alpha_2)w^2$ may be interpreted as an expression of the Poynting effect.

[2] TING [1963, *71*, Eq. $(2.5)_1$], MARKOVITZ and COLEMAN [1964, *60*, Eq. (7.6)] [1964, *61*, Eq. (2.3a)].

If there is no body force or driving force in the direction of flow, then $c(t)=0$, and (123.7) becomes[1]

$$\mu_0 \partial_{xx} v + \alpha_1 \partial_{xtx} v = \varrho \partial_t v. \tag{123.8}$$

A particular solution of (123.8) is given by[2]

$$v = V e^{-ax} \cos(\omega t - b x) \tag{123.9}$$

provided

$$\left.\begin{aligned} a &= \sqrt{\frac{\varrho\omega}{2\mu_0}\left(\frac{1}{\sqrt{1+\xi^2}} + \frac{\xi}{1+\xi^2}\right)}, \\ &= \sqrt{\frac{\varrho\xi}{2\alpha_1}\left(\frac{1}{\sqrt{1+\xi^2}} + \frac{\xi}{1+\xi^2}\right)}, \\ b &= \sqrt{\frac{\varrho\omega}{2\mu_0}\left(\frac{1}{\sqrt{1+\xi^2}} - \frac{\xi}{1+\xi^2}\right)}, \\ &= \sqrt{\frac{\varrho\xi}{2\alpha_1}\left(\frac{1}{\sqrt{1+\xi^2}} - \frac{\xi}{1+\xi^2}\right)}, \end{aligned}\right\} \tag{123.10}$$

where $\xi \equiv \alpha_1\omega/\mu_0$, and where the second forms, in each case, are valid only if $\alpha_1 \neq 0$ This solution may be regarded as representing a standing harmonic wave induced by an oscillating infinite plate at $x=0$ on a fluid body filling the half-space $x \geqq 0$. The normal stresses can be calculated with the help of $(111.2)_2$, (106.21), (106.22), and $(123.5)_{2,3}$. (Note that $b(t)$ in $(111.2)_2$ is not the b in $(123.10)_{3,4}$.)

In a lecture delivered in 1962, TRUESDELL[3] remarked that the term $\alpha_1 \partial_{xtx} v$ in (123.8) "has the effect of changing the whole picture of diffusion of velocity and vorticity from a boundary since the order of the governing differential equation is raised by one, and a new kind of instability becomes possible." To illustrate this prediction he later[4] discussed in detail the change in absorption and phase shift with driving frequency, according to the solution (123.10). In the classical theory $\alpha_1=0$, so that $a/\sqrt{\omega}$ and $b/\sqrt{\omega}$ reduce to equal constants:

$$a = b = \sqrt{\frac{\varrho\omega}{2\mu_0}}. \tag{123.11}$$

If $\alpha_1 \neq 0$, $a/\sqrt{\omega}$ and $b/\sqrt{\omega}$ become frequency-dependent, and their behavior as functions of the dimensionless variable ξ is entirely different according as $\alpha_1 > 0$ or $\alpha_1 < 0$. If $\alpha_1 > 0$, then the range of ξ is the closed half-line $[0, \infty)$, and a is a monotone function which approaches a limit as $\xi \to \infty$:

$$a \to \sqrt{\frac{\varrho}{\alpha_1}}. \tag{123.12}$$

The constant α_1 is thus determined by the absorption coefficient for very high frequency oscillation. If, on the other hand, $\alpha_1 < 0$, in agreement with the inequality $(121.27)_2$, then the range of ξ is $(-\infty, 0]$. In this case a first increases with frequency, but when $\xi = -1/\sqrt{3}$, the absorption coefficient experiences a maximum:

$$a^2_{\max} = -\frac{\varrho}{8\alpha_1}, \tag{123.13}$$

---

[1] COLEMAN and NOLL [1960, *17*, Eq. (7.12)]. This same equation was derived by BARENBLATT, ZHELTOV, and KOCHINA [1960, *5*, § 4] as governing flow through fissured rock.

[2] MARKOVITZ and COLEMAN [1964, *61*, § 2].

[3] TRUESDELL [1964, *87*, § 12].

[4] TRUESDELL [1964, *85*, § 6].

after which it falls off to zero. That is, a disturbance of very high frequency is propagated with scarcely any diminution. The circular frequency ω_{crit} at which the absorption peak occurs is related to the maximum absorption by the universal relation

$$\frac{a_{\max}^2}{\omega_{\text{crit}}} = \frac{\sqrt{3}}{8}\,\frac{\varrho}{\mu_0}, \tag{123.14}$$

no matter what is the value of α_1, so long as it be negative. If we denote the functions giving the absorption coefficient and phase shift in terms of α_1 and ξ by $\hat{a}(\alpha_1, \xi)$ and $\hat{b}(\alpha_1, \xi)$, from (123.10) we have $\hat{b}(\alpha_1, \xi) = \hat{a}(-\alpha_1, -\xi)$, so the nature of the phase shift in each case may be read off from the results already obtained for the absorption coefficient and depicted in Fig. 27. The maximum shearing (or vorticity) corresponding to a fixed amplitude V is given by

$$\left.\begin{aligned} \varkappa_{\max}(\omega) &= V\sqrt{a^2+b^2}, \\ &= V\sqrt{\frac{\varrho\xi}{\alpha_1\sqrt{1+\xi^2}}}, \\ &= V\sqrt{\frac{\varrho\omega}{\mu_0\sqrt{1+\alpha_1^2\omega^2/\mu_0^2}}} \end{aligned}\right\} \tag{123.15}$$

and thus, if $\alpha_1 \neq 0$, is always *less* than that given by the Navier-Stokes theory. A non-zero value of α_1, whichever its sign, *reduces the capacity of the fluid to be sheared but lets propagate more easily to great distances whatever shearing there may be.*

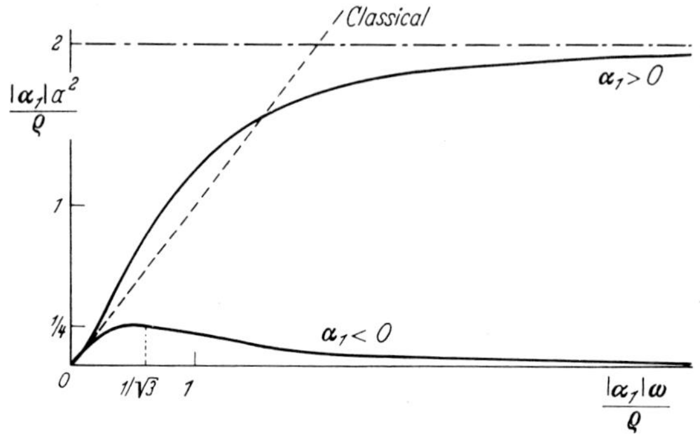

Fig. 27. Absorption and phase-shift as a function of frequency.

The phenomena just discussed may be interpreted as reflecting the existence of a natural time-lapse[1] (Sects. 108, 121), because the critical frequency $\omega_{\text{crit}}/2\pi$ is the reciprocal of such a time. Since the density ϱ and viscosity μ_0 also are material constants, it follows that the fluid has a critical length; for example, we may so regard the reciprocal of $a_{\max}$. The fluid has also a characteristic mass, but while $\mu_0/(a_{\max}\omega_{\text{crit}})$ and $\varrho/(a_{\max})^3$ and $(-\alpha_1)^{\frac{3}{2}}/\sqrt{\varrho}$ are material constants bearing the dimension M, no physical interpretation for them presents itself.

MARKOVITZ and COLEMAN[2] superimposed two solutions of the form (123.9) so as to obtain the flow of a fluid confined between two parallel plates, one of which is at rest while the other is driven sinusoidally. If the distance d between the plates is small enough that $ad \ll 1$ for the range of frequencies considered, their result for the shear stress agrees with that from the theory of infinitesimal viscoelasticity. Beyond this, they found that the normal-stress differences are in phase with the square of the velocity of the moving plate.

[1] Dimensional analysis of fluids of second grade made in general may be read off from the results at the ends of Sects. 108 and 121. Cf. TRUESDELL [1964, *87*, § 13] [1964, *85*, § 6], WHITE [1964, *96*, § 2].

[2] MARKOVITZ and COLEMAN [1964, *60*, § VII].

In the case of *unsteady helical flow* we obtain from (106.11) and (107.9)

$$\dot{g}(0) = -\varkappa q(r, t), \quad \ddot{g}(0) = \varkappa \partial_t q(r, t), \tag{123.16}$$

where $\varkappa$ and $q(r, t)$ are defined by (112.2)—(112.4). Substituting (123.16) into $(123.5)_1$ and combining the result with (107.15), (112.5), (112.3), (112.7), and (112.8), we find that solving the dynamical equations reduces to finding solutions $\omega = \omega(r, t)$ and $u = u(r, t)$ of the following third-order linear partial differential equations[1]:

$$\partial_r [r^3 \partial_r (\mu_0 \omega + \alpha_1 \partial_t \omega)] = -r d(t) + \varrho r^3 \partial_t \omega, \tag{123.17}$$

$$\partial_r [r \partial_r (\mu_0 u + \alpha_1 \partial_t u)] = -r a(t) + \varrho r \partial_t u. \tag{123.18}$$

Once a solution pair $\omega(r, t)$, $u(r, t)$ has been found, all stress components can be determined by using (107.15)—(107.18), (112.9), and (112.10).

A number of special solutions of (123.17)—(123.18) have been calculated by Markovitz and Coleman[2]. We give here a partial list of their results.

1. Axial oscillation. The fluid is contained between two coaxial infinite rigid cylinders of radii R_1 and R_2. Applied forces in the axial direction are assumed absent. The cylinder with radius R_2 is held at rest while the cylinder with radius R_1 oscillates sinusoidally in the axial direction:

$$u(R_1, t) = V \cos \nu t, \qquad u(R_2, t) = 0. \tag{123.19}$$

The following solution of (123.18) (with $a(t) = 0$) satisfies the boundary conditions (123.19) and corresponds to an axial oscillation of the fluid:

$$u(r, t) = V \operatorname{Re} \left\{ \frac{Y_0(\alpha R_2) J_0(\alpha r) - J_0(\alpha R_2) Y_0(\alpha r)}{Y_0(\alpha R_2) J_0(\alpha R_1) - J_0(\alpha R_2) Y_0(\alpha R_1)} e^{i\nu t} \right\}, \tag{123.20}$$

where α is given by

$$\alpha^2 = -\frac{i\nu\varrho}{\mu_0 + i\nu\alpha_1}, \tag{123.21}$$

and where J_0 and Y_0 denote Bessel functions in standard notation. Of course, when $d(t) = 0$, the angular velocity component ω is zero. When the gap between the two cylinders is small, i.e., when g and ξ as defined by

$$g \equiv (R_1 - R_2)/R_2, \qquad \xi = (r - R_2)/R_2, \tag{123.22}$$

are small in absolute value compared to 1, (123.20) reduces to the approximate formula

$$u = V \frac{\xi}{g} \operatorname{Re} \left\{ \left[1 - \frac{1}{2}\xi + \frac{1}{2}g + \frac{2 - \alpha^2 R_2^2}{6} \xi^2 - \frac{1}{4}\xi g - \frac{1 - 2\alpha^2 R_2^2}{12} g^2 + O(g^3) \right] e^{i\nu t} \right\}. \tag{123.23}$$

If the terms of order $O(g^2)$ are not written out, (123.23) reduces to

$$u(r, t) = V \frac{\xi}{g} \left(1 - \frac{1}{2}\xi + \frac{1}{2}g \right) \cos \nu t + O(g^2). \tag{123.24}$$

To within terms of order 1 in g the flow pattern is independent of the material constants and hence the same for all fluids.

---

[1] Markovitz and Coleman [1964, *61*, § 3]. A special case was mentioned by Coleman and Noll [1960, *17*, § 7].

[2] Markovitz and Coleman [1964, *61*, § 4, 5]. They interpret some of their results as affording new ways of determining experimentally the coefficients α_1 and α_2.

In the absence of body forces the difference between the normal tractions at the two cylinders is given by the approximating formula

$$\Delta T\langle rr\rangle = (2\alpha_1 + \alpha_2)\frac{V^2}{g R_2^2}\left[\left(1 - \frac{1}{2}g\right)\cos^2 \nu t + O(g^2)\right]. \tag{123.25}$$

2. *Angular oscillation.* The assumptions made here are the same as under (1) except that the cylinder with radius R_1 is assumed to oscillate, not in the axial direction, but about its axis, so that the boundary conditions (123.19) must be replaced by

$$\omega(R_1, t) = \Omega \cos \nu t, \qquad \omega(R_2, t) = 0. \tag{123.26}$$

A solution of (123.17) (with $d(t) = 0$) corresponding to an angular oscillation of the fluid is

$$\omega(r, t) = \frac{\Omega R_1}{r}\,\mathrm{Re}\left\{\frac{J_1(\alpha R_2)\,Y_1(\alpha r) - Y_1(\alpha R_2)\,J_1(\alpha r)}{J_1(\alpha R_2)\,Y_1(\alpha R_1) - Y_1(\alpha R_2)\,J_1(\alpha R_1)}\,e^{i\nu t}\right\} \tag{123.27}$$

where α is again given by (123.21) and where J_1 and Y_1 denote Bessel functions. An approximate formula useful when the gap is small is

$$\omega(r, t) = \Omega\frac{\xi}{g}\left[1 - \frac{3}{2}\xi + \frac{3}{2}g + O(g^2)\right]\cos \nu t. \tag{123.28}$$

In the absence of axial body forces the difference between the normal tractions at the cylinders obeys the formula

$$\Delta T\langle rr\rangle = -2\alpha_1\frac{\Omega^2}{g}\left[1 + O(g)\right]\cos^2 \nu t. \tag{123.29}$$

γ) *Instability, uniqueness, and nonexistence of lineal flow.* Although the special solutions presented above undoubtedly are useful for the interpretation of the behavior of real fluids under appropriate circumstances, COLEMAN, DUFFIN and MIZEL[1] have found grave defects in the theory of fluids of the second grade. Indeed, these authors showed that the differential equation (123.8) has physically unacceptable solutions when the inequalities (121.27) are valid[2].

A solution $v = v(x, t)$ defined on the half-strip $\mathscr{S}$ of the (x, t)-plane given by

$$0 \leqq x \leqq d, \quad 0 \leqq t < \infty, \tag{123.30}$$

corresponds to a lineal flow, starting at $t = 0$, of a fluid confined between two infinite parallel plates at $x = 0$ and $x = d$. A special solution of this type is simple shearing, (111.7). The corresponding boundary data are (111.5). Another solution is[3]

$$v = V\,\mathrm{Re}\left\{\frac{\sin \alpha x}{\sin \alpha d}\,e^{i\nu t}\right\}, \tag{123.31}$$

with the boundary values

$$v(0, t) = 0, \qquad v(d, t) = V\cos \nu t. \tag{123.32}$$

[1] COLEMAN, DUFFIN, and MIZEL [1965, *7*] gave also corresponding theorems for the case when the shear stress, $\mu_0\,\partial_x v + \alpha_1\,\partial_{xt} v$, rather than the velocity is prescribed on the plate $x = d$.

[2] For the case when $\alpha_1 > 0$, Eq. (123.8) has been studied by TING [1963, *71*]. He found that solutions of (123.8) are smoother than their counterparts in the classical theory, when (123.8) reduces to the equation of heat conduction. Since the argument of COLEMAN and MIZEL, given above in Sect. 121, and the experiments of MARKOVITZ and BROWN [1964, *59*] on polyisobutylene solutions both indicate that $\alpha_1 < 0$, TING's results do not seem to be relevant in the present context. They apply, however, in the theory of flow through fissured rocks as developed by BARENBLATT, ZHELTOV, and KOCHINA [1960, *5*, §§ 4—5] and BARENBLATT and CHERNYI [1963, *4*, § 3].

[3] MARKOVITZ and COLEMAN [1964, *60*, Eqs. (7.8) and (7.10)].

Both solutions are *bounded* in the strip $\mathscr{S}$, and the boundary data are bounded by V in either case. There are, however, solutions with bounded boundary values that do not remain bounded for all times t. In fact, if $\mu_0 > 0$ and $(-\alpha_1) > 0$, the solutions

$$v = A\, e^{a_n t} \sin \frac{n\pi}{d} x, \tag{123.33}$$

where n is an integer and

$$a_n = \frac{\mu_0 n^2 \pi^2}{(-\alpha_1)(n^2\pi^2 - \sigma^2 d^2)}, \qquad \sigma = \sqrt{\frac{\varrho}{(-\alpha_1)}}, \tag{123.34}$$

satisfy the boundary conditions

$$v(0, t) = v(d, t) = 0, \tag{123.35}$$

and are unbounded whenever $a_n > 0$, i.e., whenever

$$n > \frac{\sigma d}{\pi}. \tag{123.36}$$

It appears unlikely that such unbounded solutions are physically meaningful. Moreover, since Eq. (123.8) is linear, the solutions (123.33) may be used to show that *every bounded solution* $v(x, t)$ *of* (123.8) *is unstable*[1] in the following sense. *There is an unbounded solution* $v^*(x, t)$ *which has the same boundary values as* $v(x, t)$ *and has initial values* $v^*(x, 0)$ *differing arbitrarily little from the initial values* $v(x, 0)$. Indeed, for v^* we have merely to take

$$v^* = v + \varepsilon\, e^{a_n t} \sin \frac{n\pi}{d} x, \tag{123.37}$$

with n chosen large enough that (123.36) holds. We then have

$$|v^*(x, 0) - v(x, 0)| \leqq \varepsilon, \tag{123.38}$$

v and v^* have the same boundary values, and v^* is unbounded.

Coleman, Duffin, and Mizel obtained the following further results, proofs of which are too elaborate to be included here:

[1] Stability of steady lineal flows had been investigated along standard hydrodynamical lines by earlier authors, but they failed to find this capital result. Genensky [1960, *25*] considered the case when $\alpha_2 = 0$; his results were generalized by Kapur and Goel [1962, *41*]. They found that a necessary condition for stability of the profile $\dot{y} = v(x)$ is the existence of a value of x such that

$$\frac{\alpha_1}{\varrho} v''''(x) - v''(x) = 0.$$

Datta [1964, *22*] considered stability in Couette flow; he found the results to be very sensitive to the value of α_1.

Whether the instability found by Coleman, Duffin, and Mizel can be explained from thermodynamics is not known. As they mention, Truesdell, generalizing his observation of 1949 on Reiner-Rivlin fluids (stated in footnote 2, p. 479, above), had remarked in 1960 that there are no non-zero values of α_1 and α_2 such as to render the stress power non-negative in all flows of an incompressible fluid of second grade. Coleman [1962, *10*, § 7] had then shown that such a requirement would be unnecessarily strong. According to his results, the Clausius-Duhem inequality of thermodynamics implies that the stress power be positive in substantially stagnant flows of incompressible simple fluids, but not necessarily in all flows. The condition $\mu_0 \geqq 0$ is necessary and sufficient that all *steady viscometric flows* of an incompressible fluid of second grade be dissipative. Unpublished analysis of Noll shows that in order for all substantially stagnant flows of such a fluid to be dissipative, it is necessary and sufficient that $\mu_0 \geqq 0$ and $\alpha_1 + \alpha_2 = 0$. The experimental data on polyisobutylene solutions seems to contradict the latter condition. None of these results has evident bearing on the compatibility of unsteady lineal flows with the Clausius-Duhem inequality. Coleman, Duffin, and Mizel [1965, *7*, footnote in § 6] discuss this complicated matter.

(I) There is at most one solution of (123.8) *having given boundary values,*

$$v(0,t)=V_0(t), \quad v(d,t)=V_d(t), \tag{123.39}$$

and given initial values:

$$v(x,0)=\bar{v}(x). \tag{123.40}$$

(II) The only bounded solutions of (123.8) *compatible with the zero boundary data* (123.35) *are those given by* (123.33) *with*

$$n<\frac{\sigma d}{\pi}=\sqrt{\frac{\varrho}{(-\alpha_1)}}\,\frac{d}{\pi} \tag{123.41}$$

and linear combinations thereof. All other solutions are unbounded and of exponential order in the sense that

$$C_1 e^{\gamma_1 t}>\int_0^d |v(x,t)|^2\,dx>C_2 e^{\gamma_2 t}, \tag{123.42}$$

where C_1, C_2, γ_1, *and* γ_2 *are suitable positive constants. In particular, when*

$$d<\pi\sqrt{\frac{-\alpha_1}{\varrho}}, \tag{123.43}$$

there are no non-zero bounded solutions having zero boundary values.

(III) If

$$d=n\pi\sqrt{\frac{-\alpha_1}{\varrho}}, \tag{123.44}$$

then there are no solutions, bounded or unbounded, compatible with zero boundary data (123.35) *unless the initial data* (123.40) *satisfy the condition*

$$\int_0^d \bar{v}(x)\sin x\,dx=0. \tag{123.45}$$

The result (II) has the following consequence: Suppose that (123.43) holds and that at time $t=0$ the flow is lineal and non-zero. Suppose further that at time $t=0$ the bounding plates are brought to rest. Then, if the flow is to remain lineal, the velocity $v(x,t)$ must become arbitrarily large as $t\to\infty$. If the theory of fluids of the second grade is taken seriously, the only way of avoiding this physically unacceptable conclusion is to suppose that the flow ceases to be lineal.

The result (III) exhibits another characteristic length for a fluid of second grade, this time a critical width of channel, in which lineal flow cannot persist unless the distribution of speed is adjusted so as to satisfy (123.45). In such a channel, if a lineal flow is set up, and then the plates are brought to rest at $t=0$, the flow must cease to be lineal unless the initial field of speed happens to conform to (123.45).

COLEMAN and MIZEL[1] have confirmed this last expectation by analysis of the possible flows, lineal or not, of fluids of second grade. Assuming that the flow is lineal at $t=0$, that (123.35) holds at $t=0$, and (123.45) does not hold at $t=0$, they have proved that the following three conditions cannot all hold at all points in the flow:

$$\partial_t \dot{x}|_{t=0}=0, \quad \partial_{tz}\dot{y}|_{t=0}=0, \quad \partial_t \dot{z}|_{t=0}=0. \tag{123.46}$$

That is, if a non-zero lineal flow exists at some instant, in a stationary channel of critical width, then at that same instant must exist also some acceleration tending

[1] COLEMAN and MIZEL [1965, *13*].

to destroy the state of lineal flow. In fact, it is not known whether solutions exist at all.

At the beginning of this section we remarked that the theory of fluids of second grade, like the Navier-Stokes theory, could be regarded either as an approximation to the general theory, valid in sufficiently slow flows, or as a special theory which is of interest in its own right. The results presented in this section cast grave doubt upon the value of the latter alternative. The behavior of the fluid of second grade for unrestricted flow is simply not plausible for a physical theory.

It is not certain that these results render the fluid of second grade useless also as an approximation. For example, it could be argued that since disturbances of the form (123.33), however, small they may be initially, grow unbounded in time, they are excluded from the class of "slow" flows resulting by retardation of a given flow and hence are not relevant in the initial argument for the approximation procedure.

At present experiment is of no help, for in the only fluid for which all three viscometric functions are known, the range in which they are well approximated by those of a fluid of second grade is indiscernible, as may be seen by a glance at Figs. 23 and 24, pp. 467 and 468.

The mere fact that a second approximation is in some ways worse than the first is not without precedent. Consider, for example, approximation of a real function by a power series. If the function itself is monotone increasing from $-\infty$ to $+\infty$, the first approximation at a typical point, being a straight line, preserves this property, while every second approximation, being a parabola, fails to share it. E.g., infinitesimal elasticity predicts behavior that is not *a priori* ridiculous, however incorrect it may be, in large pure strains, but the second-order theory if interpreted outside the range of small strains leads to outright impossibility: either that large extension requires pressure, or that large compression requires tension. The price of closer approximation in a narrow range may be poorer approximation in a great one.

In any case the approximation procedure to which we have been referring concerns *the constitutive equation of the material, not the flows obtained by solving the differential equations of motion resulting from it.* The differential equations, of course, result from differentiating the constitutive equation. It remains possible that a properly approximate constitutive equation can yield differential equations insufficiently accurate, especially as $t\to\infty$. Nothing is known about the matter as yet[1].

Using the theory of the fluid of second grade, BHATNAGAR[2] and his collaborators have calculated and classified the secondary flows corresponding to various geometrical conditions. Cellular patterns are found in most cases. The number and nature of these depend strongly upon the coefficient α_1, and results according to the Reiner-Rivlin theory (Sect. 119) are not typical.

II. Other fluids.

124. KORTEWEG's theory of capillarity. So as to replace the classical theory of capillarity (this Encyclopedia, Volume X), which consists essentially in prescription of a jump condition at a surface separating homogeneous fluids of different densities, KORTEWEG[3] proposed smooth constitutive equations for stresses arising in response to density gradients. In the notations used in the present treatise, his

---

[1] A similar point arises in the kinetic theory of gases, where the Chapman-Enskog process yields higher approximations that are in some senses less satisfactory than the Navier-Stokes equations. Cf. the discussion at length by TRUESDELL [1956, *35*].

[2] BHATNAGAR and RAJESWARI [1962, *6*] (torsional oscillations between parallel plates) (cf. also SRIVASTAVA [1963, *67*]), BHATNAGAR and RATHNA [1963, *8*] (cone and plate, co-axial cones), BHATNAGAR and RAJESWARI [1963, *6*] (concentric rotating spheres). RAJESWARI and RATHNA [1962, *53*] have given an approximate calculation of the flow near a stagnation point and have derived boundary-layer equations. RATHNA [1962, *55*] has calculated approximately the flow induced by a rotating disk. Again the value of the coefficient α_1 has a decisive influence.

[3] KORTEWEG [1901, *4*].

starting point was the assumption

$$\boldsymbol{T}+p\boldsymbol{1}=\mathfrak{f}(\boldsymbol{L},\operatorname{grad}\theta,\operatorname{grad}\varrho,(\operatorname{grad})^2\varrho), \tag{124.1}$$

where p is given by a thermal equation of state: $p=\hat{p}(\varrho,\theta)$ and where $(\operatorname{grad})^2\equiv$ grad grad. The principle of material indifference[1] requires that $\mathfrak{f}$ depend on $\boldsymbol{L}$ only through $\boldsymbol{D}$, and that

$$\left.\begin{aligned}\boldsymbol{Q}\mathfrak{f}(\boldsymbol{D},\operatorname{grad}\theta,\operatorname{grad}\varrho,(\operatorname{grad})^2\varrho)\boldsymbol{Q}^T=\mathfrak{f}(\boldsymbol{Q}\boldsymbol{D}\boldsymbol{Q}^T,\boldsymbol{Q}\operatorname{grad}\theta,\\ \boldsymbol{Q}\operatorname{grad}\varrho,\boldsymbol{Q}(\operatorname{grad})^2\varrho\boldsymbol{Q}^T),\end{aligned}\right\} \tag{124.2}$$

identically in the orthogonal tensor $\boldsymbol{Q}$. That is, $\mathfrak{f}$ is an isotropic function of all its arguments. Taking $\boldsymbol{Q}=-\boldsymbol{1}$ shows also that $\mathfrak{f}$ is even in grad θ and grad ϱ. Without giving any reason, KORTEWEG dropped grad θ from the equations and secured even dependence on grad ϱ by assuming dependence on grad $\varrho\otimes$grad ϱ instead. Thus he arrived at an equation of the form

$$\boldsymbol{T}+p\boldsymbol{1}=\mathfrak{f}(\boldsymbol{D},\operatorname{grad}\varrho\otimes\operatorname{grad}\varrho,(\operatorname{grad})^2\varrho), \tag{124.3}$$

where $\mathfrak{f}$ is an isotropic function of its three arguments, which are symmetric tensors. KORTEWEG then assumed that $\mathfrak{f}$ is the sum of three linear functions, each of which depends on only one of the three arguments in (124.3). To each of these functions we may apply the representation theorem (12.5), so obtaining

$$\left.\begin{aligned}\boldsymbol{T}+p\boldsymbol{1}=\lambda(\operatorname{tr}\boldsymbol{D})\boldsymbol{1}+2\mu\boldsymbol{D}-\alpha(\operatorname{grad}\varrho)^2\boldsymbol{1}-\beta\operatorname{grad}\varrho\otimes\operatorname{grad}\varrho+\\ +\gamma(\Delta\varrho)\boldsymbol{1}+\delta(\operatorname{grad})^2\varrho,\end{aligned}\right\} \tag{124.4}$$

where $\lambda,\mu,\alpha,\beta,\gamma,\delta$ are functions of ϱ and θ alone. Eq. (124.4) is KORTEWEG's final reduced constitutive equation[2]. Equations of similar form with ϱ replaced by θ had been obtained by MAXWELL[3] in his celebrated work on rarefied gases and furnish one of the earliest indications that the theory based on the classical linear formula for viscous stresses is only, in some sense, an approximation. Had KORTEWEG chosen to retain the terms in grad θ and to insert dependence on $(\operatorname{grad})^2\theta$ in (124.1), he would have come out with four additional terms on the right-hand side of (124.4), terms which any reader of this treatise can write down by inspection. A formula of this kind, however, expresses only a special case of the (extended) starting assumption (124.1). Any modern reader, again, may write out a general solution to (124.2) (at least in the case of polynomial dependence) by using the representation theorems cited in Sect. 13.

The most interesting aspect of KORTEWEG's theory, the one that distinguishes it from most other theories of fluids, lies in its providing *a correction to classical hydrostatics* for figures of equilibrium of compressible fluids. As an illustration, KORTEWEG considered a spherical mass of fluid in equilibrium with purely radial variation of density: $\varrho=\hat{\varrho}(r)$, where r is the distance from a fixed center. Choosing

---

[1] In KORTEWEG's analysis the principle was not stated, and the reasons given did not distinguish it explicitly from mere co-ordinate invariance and from ideas of material symmetry. In fact, the principle of material indifference forces the isotropy group of (124.1) to contain the orthogonal group, as we see in the text. Thus, as in the theory based on the simpler assumption (119.1) of STOKES, only isotropic materials are included in the present theory.

[2] KORTEWEG suggested that the same constitutive equation, with ϱ replaced by the concentration of one of the constituents of a mixture of two incompressible fluids, should describe slow diffusion processes.

[3] MAXWELL [1876, 2, Eqs. (53), (54)].

spherical co-ordinates, we find from (124.4) that all shearing stresses vanish and that

$$\left.\begin{aligned} T_{\langle rr\rangle} &= -p-(\alpha+\beta)\varrho'^2+\gamma\left(\varrho''+\frac{2}{r}\varrho'\right)+\delta\varrho'', \\ T_{\langle\theta\theta\rangle} = T_{\langle\varphi\varphi\rangle} &= -p-\alpha\varrho'^2+\gamma\left(\varrho''+\frac{2}{r}\varrho'\right)+\delta\frac{\varrho'}{r}, \end{aligned}\right\} \tag{124.5}$$

where (6A.5) and (6A.6) have been used. The explicit equations of equilibrium, obtained by setting $\ddot{\boldsymbol{x}}=\mathbf{0}$ in (16.6) and using (6A.8), are

$$\left.\begin{aligned} b_{\langle\theta\rangle} = b_{\langle\varphi\rangle} &= 0, \\ \partial_r T_{\langle rr\rangle}+\frac{2}{r}\left(T_{\langle rr\rangle}-T_{\langle\theta\theta\rangle}\right)+\varrho b_{\langle r\rangle} &= 0. \end{aligned}\right\} \tag{124.6}$$

Thus equilibrium is not possible unless the external body force is radial. We assume that the temperature is constant or is an assigned function of r, and that $b_{\langle r\rangle}$ is an assigned function of r. Then substitution of (124.5) into (124.6) yields a simple quadrature for the difference of normal tensions on the surfaces $r=r_1$ and $r=r_2$:

$$T_{\langle rr\rangle}\Big|_{r_1}^{r_2} = -\int_{r_1}^{r_2}\varrho b_{\langle r\rangle}\,dr-\frac{2\delta\varrho'}{r}\Big|_{r_1}^{r_2}+2\int_{r_1}^{r_2}\frac{\beta+\dfrac{d\delta}{d\varrho}}{r}\varrho'^2\,dr. \tag{124.7}$$

If we suppose that the shell $r_1<r<r_2$ is a layer of transition between two homogeneous fluids, so that $\varrho'=0$ when $r\leqq r_1$ and when $r\geqq r_2$, the second term on the right-hand side vanishes. To obtain results appropriate to a thin shell of transition, we calculate the limit as $r_1\to r_0$ and $r_2\to r_0$. The first integral on the right-hand side vanishes. Under suitable assumptions of smoothness, the remaining integral yields a term proportional to $\mathfrak{t}/r_0$, as expected from the classical theory of capillarity.

More generally, KORTEWEG[1] allowed the surfaces of equal density to have any smooth form and calculated the difference of normal tensions on any two surfaces $\varrho=\varrho_1$ and $\varrho=\varrho_2$. In the limiting case of a thin shell, he found that this difference is proportional to the mean curvature, as is assumed in the classical theory of capillarity.

KORTEWEG, like other nineteenth-century experts on general mechanics, was content to demonstrate that his equations yield results equivalent to the classical theory in appropriate limit cases, but apparently he did not feel the second requirement of modern work, namely, to get different and new results as well. His theory seems not to have been taken up by any later writer.

125. TRUESDELL's theory of the "Maxwellian" fluid. So as to set up a phenomenological framework upon which effects said to occur in rarefied gases may be discussed and estimated, TRUESDELL[2] proposed, in effect, to take as independent variables the fields of velocity, density and temperature, and their material time derivatives and spatial gradients of all orders. These variables are

$$(\mathrm{grad})^p\overset{(m)}{\varrho},\quad (\mathrm{grad})^q\overset{(n)}{\theta},\quad (\mathrm{grad})^s\overset{(r+1)}{\boldsymbol{x}}, \tag{125.1}$$

[1] KORTEWEG [1901, *4*, Eq. (57)].

[2] TRUESDELL [1948, *18* and *19*] [1949, *23*] [1951, *17*] [1952, *20*, §§ 73—80]. Following the lines laid out in CFT, Sect. 307, the work of TRUESDELL is here corrected, condensed, and generalized.

where p, q, s, m, n, r run from 0 to any finite numbers. Because of the continuity equation in the form (15.14), we see that the time derivatives $\dot{\varrho}, \ddot{\varrho}, \dots, \overset{(m)}{\varrho}$ are functions of the spatial gradients of density and velocity, so there is no loss in generality in setting $m=0$ from the outset. While generalizing the theory of materials of grade $\mathfrak{n}$, defined in Sect. 28, in that variations of temperature are allowed to affect the stress, and generalizing also the proposal of KORTEWEG (Sect. 124), the present or "Maxwellian[1]" theory retains an important feature of the Stokesian (Sect. 119A) and kinetic theories of gases in excluding material constants having the dimensions of time. The list of dimension-bearing scalars is the same as for the "Stokesian" theory, namely, no more than a natural viscosity μ_0 and a natural thermal conductivity $\varkappa_0$:

$$\mu_0, \varkappa_0. \tag{125.2}$$

In accord with the principle of equipresence[2], *both the stress and the heat flux are assumed to be functions of the variables written in the two lists* (125.1) *and* (125.2).

The constitutive equations for stress and heat flux are then to be reduced by applying the principle of material indifference and the requirement of dimensional invariance. A separation of effects as due to kinematical and thermodynamic gradients, which might seem to be denied by the principle of equipresence, then follows in consequence of the requirements of invariance, supplemented by an ordering process, as we shall see now.

α) *Material frame-indifference.* The method of reduction introduced in Sect. 29 shows that the dependence of $\boldsymbol{T}$ and $\boldsymbol{h}$ upon the kinematical fields $(\operatorname{grad})^s \overset{(r+1)}{\boldsymbol{x}}$ must be equivalent to dependence upon the *generalized Rivlin-Ericksen tensors*, defined as follows:

$${}_{\mathfrak{j}}\boldsymbol{A}_{\mathfrak{n}} \equiv {}_{\mathfrak{j}}\overset{(\mathfrak{n})}{\boldsymbol{C}}_{(t)}(t), \tag{125.3}$$

where ${}_{\mathfrak{j}}\boldsymbol{C}$ is defined by (29.26), and where $\mathfrak{j} \geqq 1, \mathfrak{n} \geqq 1$. By (24.14), $\boldsymbol{A}_{\mathfrak{n}} = {}_1\boldsymbol{A}_{\mathfrak{n}}$. Furthermore, the functions giving $\boldsymbol{T}$ and $\boldsymbol{h}$ in terms of

$$(\operatorname{grad})^p \varrho, \quad (\operatorname{grad})^q \overset{(n)}{\theta}, \quad {}_{\mathfrak{j}}\boldsymbol{A}_{\mathfrak{r}} \tag{125.4}$$

are *isotropic*[3].

β) *Dimensional invariance.* If we set $R \equiv \varkappa_0/\mu_0$, we may replace the list (125.2) by

$$\mu_0, R, \tag{125.5}$$

which is more convenient for calculation. (R, which has the dimensions of the gas constant, $\mathsf{L}^2\mathsf{T}^{-2}\Theta^{-1}$, has been used already in connection with the Stokesian fluid in Sect. 119A.) By applying a dimensional argument parallel to that presented in detail in Sect. 119A, we show that the quantities

$$\frac{\boldsymbol{T}}{\varrho R\theta} \quad \text{and} \quad \frac{\boldsymbol{h}}{\varrho (R\theta)^{\frac{3}{2}}}, \tag{125.6}$$

[1] The name "Maxwellian" refers to MAXWELL's having been the first to see that stress in a gas at rest may be created by a temperature gradient. Cf. the terms in Eqs. (125.8) and (125.9) whose coefficients are $\beta_4, \beta_7, \beta_{14}$, and β_{17}.

[2] As was remarked in Sect. 96, it was in connection with the "Maxwellian" fluid that the principle of equipresence was first proposed.

[3] The remarks on "anisotropic" fluids in the original memoirs of TRUESDELL (e.g. [1951, *17*, § 22]) are vacuous; he did not at that time see that the principle of material indifference forces the Maxwellian fluid to be isotropic.

are expressible as dimensionless isotropic functions of the following dimensionless tensors:

$$\left.\begin{array}{l}\dfrac{\mu_0^p}{\varrho^{p+1}(R\theta)^{\frac{1}{2}p}}(\operatorname{grad})^p\varrho\,, \qquad \dfrac{\mu_0^{q+n}}{\theta\,\varrho^{n+q}(R\theta)^{\frac{1}{2}q+n}}(\operatorname{grad})^q\overset{(n)}{\theta}\,,\\[2ex] \dfrac{\mu_0^{\mathfrak{i}-1+\mathfrak{r}}}{\varrho^{\mathfrak{i}-1+\mathfrak{r}}(R\theta)^{\mathfrak{r}+\frac{1}{2}(\mathfrak{i}-1)}}\,{}_{\mathfrak{i}}\boldsymbol{A}_{\mathfrak{r}}\,.\end{array}\right\} \tag{125.7}$$

γ) *Separation of effects according to order and invariance.* If we assume that the response functions giving $\boldsymbol{T}$ and $\boldsymbol{h}$ are analytic functions of the arguments $(125.1)_1$ but not necessarily of the scalars ϱ, θ, μ_0, and R, it follows from the results just given that the quantities (125.6) are expressible as *analytic functions of the natural viscosity*, μ_0. Thus μ_0 emerges as a possible *ordering parameter.* (Cf. the special case presented in Sect. 119A and the parallel results concerning expansions in powers of the natural time-lapse in Sect. 120.) The quantities of orders 1 and 2 in μ_0 in the list (125.7) are as follows:

order 1:

Scalars: $\dfrac{\mu_0\dot\theta}{R\varrho\theta^2}$.

Vectors: $\dfrac{\mu_0}{\varrho\theta(R\theta)^{\frac{1}{2}}}\operatorname{grad}\theta$, $\quad\dfrac{\mu_0}{\varrho^2(R\theta)^{\frac{1}{2}}}\operatorname{grad}\varrho$.

Tensors: $\dfrac{\mu_0}{\varrho R\theta}\boldsymbol{A}_1$.

order 2:

Scalars: $\dfrac{\mu_0^2\ddot\theta}{\varrho^2R^2\theta^3}$.

Vectors: $\dfrac{\mu_0^2}{\varrho^2\theta(R\theta)^{\frac{3}{2}}}\operatorname{grad}\dot\theta$.

Tensors: $\dfrac{\mu_0^2}{\varrho^3R\theta}(\operatorname{grad})^2\varrho$, $\quad\dfrac{\mu_0^2}{\varrho^2R\theta^2}(\operatorname{grad})^2\theta$,

$\dfrac{\mu_0^2}{(R\varrho\theta)^2}\boldsymbol{A}_2$, $\quad\dfrac{\mu_0^2}{\varrho^2(R\theta)^{\frac{3}{2}}}\,{}_2\boldsymbol{A}_1$.

According to the results so far demonstrated, the stress tensor is given, to within terms multiplied by μ_0^3, by an isotropic quadratic function of the terms of order 1 in the above list, plus an isotropic linear function of the terms of order 2. Again adding a postulate generalizing (119A.3) so as to retain consistency with hydrostatics, by using representation theorems in Sect. 13 we thus obtain the following expression for the stress tensor:

$$\left.\begin{aligned}\frac{\boldsymbol{T}+\hat p(\varrho,\theta)\boldsymbol{1}}{R\varrho\theta} &= \varpi\boldsymbol{1}+\nu\frac{\mu_0}{R\varrho\theta}\boldsymbol{A}_1+\\ &+\beta_{13}\frac{\mu_0^2}{R\varrho^4\theta}\operatorname{grad}\varrho\otimes\operatorname{grad}\varrho+\beta_{14}\frac{\mu_0^2}{R\varrho^2\theta^3}\operatorname{grad}\theta\otimes\operatorname{grad}\theta+\\ &+\beta_{15}\frac{\mu_0^2}{R\varrho^3\theta^2}(\operatorname{grad}\varrho\otimes\operatorname{grad}\theta+\operatorname{grad}\theta\otimes\operatorname{grad}\varrho)+\\ &+\beta_{16}\frac{\mu_0^2}{R\varrho^3\theta}(\operatorname{grad})^2\varrho+\beta_{17}\frac{\mu_0^2}{R\varrho^2\theta^2}(\operatorname{grad})^2\theta+\\ &+\frac{\mu_0^2}{R^2\varrho^2\theta^2}(\beta_{18}\boldsymbol{A}_1^2+\beta_{19}\boldsymbol{A}_2)\,,\end{aligned}\right\} \tag{125.8}$$

where the scalars ϖ and ν have the forms

$$\left.\begin{aligned}\varpi = {} & \alpha_1 \frac{\mu_0}{R\varrho\theta} \operatorname{tr} \boldsymbol{A}_1 + \alpha_2 \frac{\mu_0 \dot{\theta}}{R\varrho\theta^2} + \\ & + \beta_1 \frac{\mu_0^2 \dot{\theta}^2}{R^2\varrho^2\theta^4} + \beta_2 \frac{\mu_0^2 \ddot{\theta}}{R^2\varrho^2\theta^3} + \beta_3 \frac{\mu_0^2}{R\varrho^4\theta} (\operatorname{grad} \varrho)^2 + \\ & + \beta_4 \frac{\mu_0^2}{R\varrho^2\theta^3} (\operatorname{grad} \theta)^2 + \beta_5 \frac{\mu_0^2}{R\varrho^3\theta^2} \operatorname{grad} \varrho \cdot \operatorname{grad} \theta + \\ & + \beta_6 \frac{\mu_0^2}{R\varrho^3\theta} \Delta\varrho + \beta_7 \frac{\mu_0^2}{R\varrho^2\theta^2} \Delta\theta + \\ & + \frac{\mu_0^2}{R^2\varrho^2\theta^2} [\beta_8 (\operatorname{tr} \boldsymbol{A}_1)^2 + \beta_9 \operatorname{tr} \boldsymbol{A}_1^2 + \beta_{10} \operatorname{tr} \boldsymbol{A}_2], \\ \nu = {} & \alpha_3 + \beta_{11} \frac{\mu_0 \dot{\theta}}{R\varrho\theta^2} + \beta_{12} \frac{\mu_0}{R\varrho\theta} \operatorname{tr} \boldsymbol{A}_1 .\end{aligned}\right\} \tag{125.9}$$

The material parameters $\alpha_1, \alpha_2, \alpha_3, \beta_1, \beta_2, \ldots, \beta_{19}$ are pure numbers.

When we come to write an expression for the heat flux to the same order in μ_0, we find that a term in the third-order tensor ${}_2\boldsymbol{A}_1$ must appear. A representation theorem for an isotropic vector-valued function of a third-order tensor does not seem to be known at present, but we conjecture that a linear function of this kind equals a linear combination of the three vectors formed by contracting the third-order tensor on two indices at a time. If we define the vectors $\boldsymbol{a}_\Gamma$ as these contractions, viz

$$a_{1k} \equiv {}_2A_{1\,k}{}^{m}{}_{m}, \qquad a_{2k} \equiv {}_2A_{1}{}^{m}{}_{km}, \qquad a_{3k} \equiv {}_2A_{1}{}^{m}{}_{mk}, \tag{125.10}$$

then the above-stated conjecture leads to the following second-order approximation for the heat flux:

$$\left.\begin{aligned}\frac{\boldsymbol{h}}{\varrho (R\theta)^{\frac{3}{2}}} = {} & \sigma \frac{\mu_0}{\varrho\theta (R\theta)^{\frac{1}{2}}} \operatorname{grad} \theta + \chi \frac{\mu_0}{\varrho^2 (R\theta)^{\frac{1}{2}}} \operatorname{grad} \varrho + \\ & + \delta_5 \frac{\mu_0^2}{\varrho^2\theta (R\theta)^{\frac{3}{2}}} \operatorname{grad} \dot{\theta} + \frac{\mu_0^2}{\varrho^2 (R\theta)^{\frac{3}{2}}} (\delta_6 \boldsymbol{a}_1 + \delta_7 \boldsymbol{a}_2 + \delta_8 \boldsymbol{a}_3) + \\ & + \delta_9 \frac{\mu_0^2}{\varrho^2\theta (R\theta)^{\frac{3}{2}}} \boldsymbol{A}_1 \operatorname{grad} \theta + \delta_{10} \frac{\mu_0^2}{\varrho^3 (R\theta)^{\frac{3}{2}}} \boldsymbol{A}_1 \operatorname{grad} \varrho ,\end{aligned}\right\} \tag{125.11}$$

where

$$\left.\begin{aligned}\sigma &= \gamma_1 + \delta_1 \frac{\mu_0 \dot{\theta}}{R\varrho\theta^2} + \delta_2 \frac{\mu_0}{R\varrho\theta} \operatorname{tr} \boldsymbol{A}_1 , \\ \chi &= \gamma_2 + \delta_3 \frac{\mu_0 \dot{\theta}}{R\varrho\theta^2} + \delta_4 \frac{\mu_0}{R\varrho\theta} \operatorname{tr} \boldsymbol{A}_1 .\end{aligned}\right\} \tag{125.12}$$

The material parameters $\gamma_1, \gamma_2, \delta_1, \delta_2, \ldots, \delta_{10}$ are pure numbers.

To generalize the theory so as to let the moduli depend on temperature, we may add a reference temperature θ_0 to the list (125.5). The effect is to replace the dimensionless constants $\alpha_\Gamma, \beta_\Gamma, \gamma_\Gamma, \delta_\Gamma$ by dimensionless functions of the dimensionless ratio θ/θ_0.

The formulae (125.8) and (125.11) include as special cases the results obtained at the second stage by the Chapman-Enskog process in the kinetic theory of gases[1]. In the kinetic theory, of course, the parameters α_Γ, β_Γ, γ_Γ, δ_Γ are specifically determined from the hypothecated law of intermolecular force. For example, it turns out that $\gamma_2=\delta_3=\delta_4=0$ for all models, disallowing a heat flux to arise in response to a density gradient in a fluid at uniform temperature[2]. The theory of the "Maxwellian" fluid formalizes a phenomenological basis for the type of results obtained in the kinetic theory, without the restriction to monatomic, moderately rarefied, perfect gases, a restriction from which the kinetic theory, to the extent it yields definite, mathematical results, seems as yet unable to free itself. There is no reason to regard the "Maxwellian" theory as restricted to gases, although it is set up in such a way as to emphasize effects of compressibility.

As far as gross phenomena are concerned, what is most important here is the emergence of a separation between the effects of deformation and temperature as a *theorem*, and not an obvious one, proved from the assumptions, while in older theories such separation was always assumed at the outset. To the first order in μ_0, temperature gradients give rise to heat flux alone and deformation to stress alone. When terms of second order in μ_0 are considered, this separation holds no longer, but the interaction terms in (125.8) and (125.11) are of a definite, restricted kind. For example, the second-order heat flux arising from deformation is associated with two distinct phenomena: expansion ($\operatorname{tr} \boldsymbol{A}_1 \neq 0$) and inhomogeneity of the velocity gradient (${}_2\boldsymbol{A}_1 \neq \boldsymbol{0}$). In a rectilinear shearing flow, for example, the heat flux is unaffected by the shearing. Likewise, from (125.8), a uniform temperature gradient cannot give rise to shearing stress, although it generally alters the normal stresses. These are but examples of the prohibitions imposed by material indifference and dimensional invariance on the interaction effects that the principle of equipresence might seem to permit. While the terms of third and higher order in μ_0 are far more elaborate[3], *they still allow interactions only of certain definite kinds.*

Like any theorem, the specific forms (125.8) and (125.11) have been derived from specific hypotheses. These are, first, that the original response functions are *analytic* functions of the arguments (125.1), and, second, that only terms proportional to the first three powers of μ_0 need be considered. If these assumptions are not made, the above-emphasized separation of effects will not generally follow. For example, if the assumption of analyticity is dropped, the tensor

$$\frac{\mu_0 \sqrt{R}}{\sqrt{\theta}} \; \frac{\operatorname{grad}\theta \otimes \operatorname{grad}\theta}{|\operatorname{grad}\theta|} \tag{125.13}$$

[1] The forms of the results in the kinetic theory are somewhat different because there, as an added (and unnecessary) approximation, the time derivatives $\dot{\theta}, \ldots, \ddot{\boldsymbol{x}}, \ldots$ are eliminated by means of the equations of energy and linear momentum with the stress and heat flux replaced by previously obtained approximations. In particular, gradients of the body force field $\boldsymbol{b}$ appear in the formula for $\boldsymbol{T}$ for this reason. TRUESDELL's original presentation of the Maxwellian fluid, too heavily influenced by the kinetic theory, included these gradients among the independent variables (but cf. his remark [1951, *17*, § 19, footnote 50]).

IKENBERRY and TRUESDELL [1956, *14*, §§ 15—16] observed that it is the formal approximation of time derivatives by space derivatives that brings into the Chapman-Enskog iterates certain terms that do not satisfy the principle of material indifference. The objections raised to some of these terms now seem like hints toward formulation of that principle: BOUSSINESQ [1868, *1*, Note I], BRILLOUIN [1900, *1*, § 14 (footnote), §§ 23, 27, 34, and 39], TRUESDELL [1949, *23*, § 30] [1951, *17*, § 30].

[2] However, M. BRILLOUIN [1900, *1*, § 36] included such a term; TRUESDELL [1949, *23*, § 31] [1951, *17*, § 31] [1952, *20*, § 76] named the corresponding phenomenon the "Brillouin effect" and suggested that experiments be performed to see if it exists in polyatomic gases.

[3] They are written out in the original papers of TRUESDELL [1948, *18* and *19*] [1949, *23*] [1951, *17*].

becomes a possible first-order stress arising from the temperature gradient alone, and the vector

$$\mu_0 R \theta \frac{\boldsymbol{a}_1}{|\boldsymbol{A}_1|} \tag{125.14}$$

becomes a possible first-order heat flux arising from inhomogeneity of the velocity alone. From the present point of view, then, the classical separation of effects appears to be, in part, an expression of *smoothness* of natural phenomena.

In Sect. 96, in connection with thermo-visco-elasticity, we have reviewed a later use of the principle of equipresence. There the principle of non-negative entropy production was brought to bear, and by its aid a separation of effects was shown to follow. The corresponding possibility for the theory of the Maxwellian fluid has not yet been explored. The entropy does not enter the theory as presented here. To include it, further constitutive assumptions, generalizing $(96.31)_{1,2}$ would be needed.

No specific problems have been solved in the theory of the Maxwellian fluid. The solution for rectilinear shearing, set up at the end of Sect. 119A for the Stokesian fluid, can easily be extended to the Maxwellian fluid, since most of the terms making the latter fluid more general than the former vanish for this particular flow.

If we now look back at the starting assumptions of the "Maxwellian" theory, we see that the dependence on space and time derivatives of all orders suggests an approximation to dependence on the history of the motion and temperature in a neighborhood of the particle in question. The "Maxwellian" theory thus appears as an early and unnecessarily elaborate attempt to express the general idea of material behavior we have formalized in Sect. 26. The Maxwellian theory is less general in that only fluids are considered but more general in that dependence on the temperature history $\theta(Z, t)$ at particles Z in a neighborhood of X is allowed. Thus (26.11) is to be replaced by

$$\left.\begin{aligned} \boldsymbol{T}(t) &= \mathfrak{F}(\chi_X^{(t)}, \theta^{(t)}), \\ \boldsymbol{h}(t) &= \mathfrak{h}(\chi_X^{(t)}, \theta^{(t)}). \end{aligned}\right\} \tag{125.15}$$

An isotropy group may be defined for each of these functionals as in Sect. 31. To define a fluid within this generalized theory, we require that these groups be the full unimodular group. From this generalized formulation, the theory of the "Maxwellian" fluid will emerge by an approximation process generalizing one of those given for simple fluids in Sect. 121.

[Added in proof. While TRUESDELL's work of 1949 now appears unconvincing as well as awkward in mathematics, none of the more recent theories has attempted anything approaching its generality. COLEMAN's thermodynamics of 1964, summarized in Sect. 96bis, corresponds to the special case of (125.15) when $\chi_X^{(t)}$ and $\theta^{(t)}$ over a neighborhood of X are specialized to $\boldsymbol{F}^{(t)}(s)$, $\theta^{(t)}(s)$, and grad $\theta(t)$ at X. The vast generality of Eqs. (125.15) may baffle our mathematical capacity in solving problems or may overreach the sane requirements of practice, but it is necessary if the constitutive equations of phenomenological theories are to be matched with those obtained from even so special a molecular model as that of a monatomic, moderately rarefied gas whose molecules repel each other in proportion to the inverse fifth power of the distance. This example illustrates the difference in kind, described in Sect. 3, between the continuum and the assembly of molecules as a model for a material.]

126. Anisotropic solids capable of flow. So as to allow for effects of anisotropy in a simple material capable of flow, NOLL[1] proposed the constitutive equation

$$\boldsymbol{T} + p\boldsymbol{1} = \mathfrak{f}(\boldsymbol{L}, \boldsymbol{R}, \varrho), \tag{126.1}$$

[1] NOLL [1955, *19*, § 13].

where $\boldsymbol{R}$ is the tensor of finite rotation from a fixed reference configuration. This constitutive equation defines a special material of the differential type. Since the stress is independent of the strain, this material is not "elastic" in the sense of Chap. D except when $\mathfrak{f}$ degenerates to a function of ϱ alone, yet the possible dependence on $\boldsymbol{R}$ allows response differing in direction. As a special case of (35.5) we have the reduced form

$$\boldsymbol{T}+p\mathbf{1}=\boldsymbol{R}\mathfrak{k}(\boldsymbol{R}^T\boldsymbol{D}\boldsymbol{R},\varrho)\boldsymbol{R}^T. \tag{126.2}$$

By (31.2), $\boldsymbol{H}$ belongs to the isotropy group g of this material if and only if

$$\mathfrak{f}(\boldsymbol{L}^*,\boldsymbol{R}^*,\varrho^*)=\mathfrak{f}(\boldsymbol{L},\boldsymbol{R},\varrho) \tag{126.3}$$

for all $\boldsymbol{F}$, where the asterisks indicates that $\boldsymbol{F}$ is to be replaced by $\boldsymbol{F}\boldsymbol{H}$. Since $\boldsymbol{L}^*=\boldsymbol{L}$, $\varrho^*=\varrho$, (126.3) reduces to

$$\mathfrak{f}(\boldsymbol{L},\boldsymbol{R}^*,\varrho)=\mathfrak{f}(\boldsymbol{L},\boldsymbol{R},\varrho), \tag{126.4}$$

where $\boldsymbol{R}^*$ is the rotational part of $\boldsymbol{F}\boldsymbol{H}$. In terms of the reduced response function $\mathfrak{k}$, (126.4) is equivalent to the requirement that

$$\boldsymbol{Q}\mathfrak{k}(\boldsymbol{D}',\varrho)\boldsymbol{Q}^T=\mathfrak{k}(\boldsymbol{Q}\boldsymbol{D}'\boldsymbol{Q}^T,\varrho) \tag{126.5}$$

for all symmetric tensors $\boldsymbol{D}'$ and all orthogonal tensors $\boldsymbol{Q}$ obtainable from an arbitrary tensor $\boldsymbol{F}$ by the formula $\boldsymbol{Q}=\boldsymbol{R}^T\boldsymbol{R}^*$. Equation (126.5) is of the form (8.7). Therefore, if $\mathfrak{k}$ is a function isotropic only with respect to the group $g'=\{+\mathbf{1},-\mathbf{1}\}$ and not with respect to any larger group, then we must have $\boldsymbol{R}^*=\pm\boldsymbol{R}$ for all $\boldsymbol{F}$. This is easily seen to be the case only when $\boldsymbol{H}=\pm\mathbf{1}$. Under these circumstances, by the results of Sect. 33, *the material defined by* (126.1) *is a triclinic simple solid.* If $\mathfrak{k}$ is isotropic with respect to any subgroup g' of the orthogonal group o, then by (31.7) g' is contained in the isotropy group g. In general, however, g is larger than g' because it can have non-orthogonal members. In particular, if $g'=o$, the full orthogonal group, the function $\mathfrak{k}$ is isotropic, and (126.2) reduces to (119.4). Therefore, *the material defined by* (126.1) *is a (special) simple fluid if and only if it is isotropic.*

In other words, NOLL's definition (126.1) delimits a class of simple materials, some of which are simple solids, that includes as a special and degenerate case the Reiner-Rivlin fluid.

Despite their being, generally, solids, these materials can flow. GREEN[1] has taken up NOLL's theory for the case of a transversely isotropic material. In such a material the isotropy group contains all $\boldsymbol{Q}$ such that $\boldsymbol{Q}\boldsymbol{e}_R=\boldsymbol{e}_R$ where $\boldsymbol{e}_R$ is a certain constant vector (Sect. 33). By invoking a representation theorem[2], GREEN inferred that $\boldsymbol{T}+p\mathbf{1}$ for such a material is given by an isotropic function of $\boldsymbol{D}$ and $\boldsymbol{n}\otimes\boldsymbol{n}$, where $\boldsymbol{n}=\boldsymbol{R}\boldsymbol{e}_R$. On the assumption that that function is a polynomial, from (13.7) we have then, putting $\boldsymbol{M}\equiv\boldsymbol{n}\otimes\boldsymbol{n}$,

$$\boldsymbol{T}+p\mathbf{1}=\alpha_1\boldsymbol{M}+\alpha_2\boldsymbol{D}+\alpha_3\boldsymbol{D}^2+\alpha_4(\boldsymbol{M}\boldsymbol{D}+\boldsymbol{D}\boldsymbol{M})+\alpha_5(\boldsymbol{M}\boldsymbol{D}^2+\boldsymbol{D}^2\boldsymbol{M}), \tag{126.6}$$

---

[1] GREEN [1964, *35*]. In a later study GREEN [1964, *36*] defined an "anisotropic simple fluid" as a simple material having a constitutive equation of the form

$$\boldsymbol{T}=\underset{s=0}{\overset{\infty}{\mathfrak{G}}}\left(\boldsymbol{F}_{(t)}^{(t)}(s),\boldsymbol{R}(t);\varrho(t)\right).$$

Of course these materials, too, may be solids.

[2] ADKINS [1958, *2*, § 6] [1960, *3*, § 5].

where it has been assumed that the material is incompressible, and where α_1, $\alpha_2, \ldots, \alpha_5$ are functions of the following invariants:

$$\operatorname{tr} \boldsymbol{D}^2, \quad \operatorname{tr} \boldsymbol{D}^3, \quad \operatorname{tr} \boldsymbol{M}, \quad \operatorname{tr}(\boldsymbol{M}\boldsymbol{D}), \quad \operatorname{tr} \boldsymbol{M}\boldsymbol{D}^2. \tag{126.7}$$

(Note that $\operatorname{tr}\boldsymbol{M}=n^2$, $\boldsymbol{M}^2=n^2\boldsymbol{M}$, $\operatorname{tr}(\boldsymbol{M}\boldsymbol{D})=\boldsymbol{n}\cdot\boldsymbol{D}\boldsymbol{n}$, $\operatorname{tr}\boldsymbol{M}\boldsymbol{D}^2=\boldsymbol{n}\cdot\boldsymbol{D}^2\boldsymbol{n}$.) This constitutive equation is of just the same form as one occurring in ERICKSEN'S theory of transversely isotropic fluids, which we shall discuss in Sect. 128. In particular, the stress in the fluid at rest, being given by $\boldsymbol{T}=-p\boldsymbol{1}+\alpha_1\boldsymbol{M}$, is not a hydrostatic pressure unless the additional requirement $\alpha_1=0$ is imposed. If $\alpha_1=0$, then $\boldsymbol{D}$ cannot vanish if any shear stress is present. In other words, so long as shear stress is applied, the material will continue to deform. In this sense, the material corresponds to an intuitive idea of a "fluid". While ERICKSEN'S theory lays down a differential equation to govern the change of $\boldsymbol{n}$ in time, in the Noll-Green theory $\boldsymbol{n}$ is determined at each time by the rotation $\boldsymbol{R}$ from the reference configuration in which the constant vector $\boldsymbol{e}_{\mathrm{R}}$ denotes the axis of transverse isotropy.

GREEN proceeded to solve the problem of steady simple shearing. His results, which are similar to, but not identical with ERICKSEN'S (Sect. 128), we now present in somewhat greater generality. At the end of Sect. 30 we have shown that simple shearing can be maintained in any incompressible simple material by applying suitable surface tractions. In the usual formulae for simple shearing,

$$x=X, \qquad y=Y+\varkappa t X, \qquad z=Z, \tag{126.8}$$

the reference configuration is taken as the configuration at time 0. The rotation $\boldsymbol{R}_0$ with respect to this configuration is given by [CFT, Eq. (45.11)]

$$[\boldsymbol{R}_0(t)]=\begin{Vmatrix} \dfrac{1}{\sqrt{1+\frac{1}{4}K^2}} & \dfrac{-\frac{1}{2}K}{\sqrt{1+\frac{1}{4}K^2}} & 0 \\ \dfrac{\frac{1}{2}K}{\sqrt{1+\frac{1}{4}K^2}} & \dfrac{1}{\sqrt{1+\frac{1}{4}K^2}} & 0 \\ 0 & 0 & 1 \end{Vmatrix}, \tag{126.9}$$

where $K=\varkappa t$. As t varies from $-\infty$ to $+\infty$, $[\boldsymbol{R}_0(t)]$ varies from

$$[\boldsymbol{R}_0(-\infty)]=\begin{Vmatrix} 0 & 1 & 0 \\ -1 & 0 & 0 \\ 0 & 0 & 0 \end{Vmatrix} \quad \text{to} \quad [\boldsymbol{R}_0(+\infty)]=\begin{Vmatrix} 0 & -1 & 0 \\ 1 & 0 & 0 \\ 0 & 0 & 0 \end{Vmatrix}. \tag{126.10}$$

Let it be supposed that the reference configuration with respect to which (126.2) holds is obtained from the configuration at time 0 by a certain rotation $\boldsymbol{R}_{\mathrm{r}}^{-1}$. Then the rotation occurring in (126.2) is given by

$$\boldsymbol{R}(t)=\boldsymbol{R}_0(t)\,\boldsymbol{R}_{\mathrm{r}}, \tag{126.11}$$

so that

$$\boldsymbol{R}(-\infty)=\boldsymbol{R}_0(-\infty)\,\boldsymbol{R}_{\mathrm{r}}, \qquad \boldsymbol{R}(+\infty)=\boldsymbol{R}_0(+\infty)\,\boldsymbol{R}_{\mathrm{r}}. \tag{126.12}$$

For any particular rotation $\boldsymbol{R}_{\mathrm{r}}$, the behavior of quantities such as $\boldsymbol{R}_{\mathrm{r}}\boldsymbol{e}_{\mathrm{R}}$ is easily calculated. (GREEN'S example may be obtained by taking $\boldsymbol{R}_{\mathrm{r}}$ as a rotation about an axis normal to the flow plane, while $\boldsymbol{e}_{\mathrm{R}}$ is a vector parallel to the flow plane.) The analysis at the beginning of Sect. 106, in the special case when $k(\tau)=\varkappa\tau$, can be applied to (126.2), with the result that

$$\boldsymbol{T}=-p\boldsymbol{1}+\boldsymbol{R}\mathfrak{t}\left(\tfrac{1}{2}\varkappa\boldsymbol{R}^{\mathsf{T}}(\boldsymbol{N}+\boldsymbol{N}^{\mathsf{T}})\boldsymbol{R}\right)\boldsymbol{R}^{\mathsf{T}} \tag{126.13}$$

where $\boldsymbol{N}$ is given by (106.7). In view of (126.9) and (126.11), Eq. (126.13) gives the stress explicitly as a function of time and of $\boldsymbol{R}_r$.

GREEN stated that he has obtained corresponding results for Poiseuille flow, again similar to those from ERICKSEN'S theory, to which we now turn our attention.

127. The anisotropic fluids of ERICKSEN, I. General theory. So as to describe the behavior of liquid crystals and suspensions of large molecules, ERICKSEN has proposed and developed a class of theories of media oriented by a single director, $\boldsymbol{d}$, whose behavior in time is governed by a differential equation. (For the general theory of oriented media, see CFT, Sect. 61).

ERICKSEN'S starting point is[1] the fully general equations of balance for momentum, moment of momentum, and energy for every part $\mathscr{P}$ of a body $\mathscr{B}$ in its present configuration:

$$\left.\begin{aligned}\frac{d}{dt}\int_{\mathscr{P}}\varrho\boldsymbol{m}\,dv&=\int_{\partial\mathscr{P}}\boldsymbol{T}\boldsymbol{n}\,ds+\int_{\mathscr{P}}\varrho\boldsymbol{b}\,dv,\\ \frac{d}{dt}\int_{\mathscr{P}}\varrho\boldsymbol{A}\,dv&=\int_{\partial\mathscr{P}}\mathsf{M}\boldsymbol{n}\,ds+\int_{\mathscr{P}}\varrho\boldsymbol{L}\,dv,\\ \frac{d}{dt}\int_{\mathscr{P}}\varrho E\,dv&=\int_{\partial\mathscr{P}}\boldsymbol{q}\cdot\boldsymbol{n}\,ds+\int_{\mathscr{P}}\varrho r\,dv,\end{aligned}\right\}\tag{127.1}$$

where the symbols not previously occurring in this treatise have the following interpretations:

$\boldsymbol{m}$ = momentum per unit mass
$\boldsymbol{A}=-\boldsymbol{A}^T$ = moment of momentum per unit mass
$\mathsf{M}=-\mathsf{M}^T$ (in the sense that $\mathsf{M}^{kpq}=-\mathsf{M}^{pkq}$)
= influx of moment of momentum.
$\boldsymbol{L}=-\boldsymbol{L}^T$ = supply of moment of momentum per unit mass
E = total energy per unit mass
$\boldsymbol{q}$ = influx of total energy
r = supply of total energy per unit mass.

(For the field equations and jump conditions corresponding to an equation of balance, see CFT, Sects. 157 and 193.) The momentum, moment of momentum, and energy of the medium are taken as sums of two parts, one being the classical quantity for an ordinary point-medium and the other, the contribution of the director motion[2]:

$$\left.\begin{aligned}\boldsymbol{m}&=\dot{\boldsymbol{x}}+\dot{\boldsymbol{d}},\\ 2\boldsymbol{A}&=\boldsymbol{p}\wedge\dot{\boldsymbol{x}}+\boldsymbol{d}\wedge\dot{\boldsymbol{d}},\\ E&=\varepsilon+\tfrac{1}{2}(\dot{x}^2+\dot{d}^2),\end{aligned}\right\}\tag{127.2}$$

where $\boldsymbol{a}\wedge\boldsymbol{b}\equiv\boldsymbol{a}\otimes\boldsymbol{b}-\boldsymbol{b}\otimes\boldsymbol{a}$, the exterior product of $\boldsymbol{a}$ and $\boldsymbol{b}$, so that the axial vector corresponding to $\boldsymbol{a}\wedge\boldsymbol{b}$ is the cross product, $\boldsymbol{a}\times\boldsymbol{b}$. Thus Eq. (127.2) defines the density of moment of momentum as being composed of two parts: (1) the moment, about a fixed point, of the density of the linear momentum of the motion, and (2) the moment, about the point occupied by the particle, of the density of the director momentum, regarded as located at the terminus of the director.

[1] ERICKSEN [1960, *18*, § 2] [1960, *20*, § II] [1961, *14*, § II].

[2] ERICKSEN [1963, *21*, § I] prefers not to lay down $(127.2)_1$ but rather to regard the principle of linear momentum as the condition that the equation of energy be invariant under translations.

Component forms of Eq. (127.2) are

$$\left.\begin{aligned} m^k &= \dot{x}^k + \dot{d}^k, \\ A^{km} &= p^{[k}\dot{x}^{m]} + d^{[k}\dot{d}^{m]}, \\ E &= \varepsilon + \tfrac{1}{2}(\dot{x}_k\dot{x}^k + \dot{d}_k\dot{d}^k). \end{aligned}\right\} \tag{127.3}$$

For the rate of change of any additive set function, it is possible to define an influx and a supply. Thus there exist $\boldsymbol{T}_{\mathrm{m}}$, $\boldsymbol{T}_{\mathrm{M}}$, $\boldsymbol{b}_{\mathrm{m}}$, $\boldsymbol{b}_{\mathrm{M}}$ such that[1]

$$\left.\begin{aligned} \frac{d}{dt}\int_{\mathscr{P}} \varrho\,\dot{\boldsymbol{d}}\,dv &= \int_{\partial\mathscr{P}} \boldsymbol{T}_{\mathrm{m}}\boldsymbol{n}\,ds + \int_{\mathscr{P}} \varrho\,\boldsymbol{b}_{\mathrm{m}}\,dv, \\ \frac{d}{dt}\int_{\mathscr{P}} \varrho\,\dot{\boldsymbol{x}}\,dv &= \int_{\partial\mathscr{P}} \boldsymbol{T}_{\mathrm{M}}\boldsymbol{n}\,ds + \int_{\mathscr{P}} \varrho\,\boldsymbol{b}_{\mathrm{M}}\,dv. \end{aligned}\right\} \tag{127.4}$$

By $(127.1)_1$,

$$\boldsymbol{T} = \boldsymbol{T}_{\mathrm{M}} + \boldsymbol{T}_{\mathrm{m}}, \qquad \boldsymbol{b} = \boldsymbol{b}_{\mathrm{M}} + \boldsymbol{b}_{\mathrm{m}}. \tag{127.5}$$

At this level of generality, nothing is gained or lost by this splitting of the momentum equation into two parts. The aim is only to introduce variables that will later appear in constitutive equations. If, as will be done, separate constitutive equations are given for $\boldsymbol{T}_{\mathrm{m}}$ and $\boldsymbol{T}_{\mathrm{M}}$, Eqs. (127.4) represent *a division of the total momentum into two parts, each of which is balanced without influence by the other.*

As motivation for setting up these forms, ERICKSEN[2] considered a medium composed of packets of rod-like molecules. A packet is idealized as a line segment of fixed mass M but variable in length and direction. A stiff but possibly extensible dumbbell molecule is a special case. The points $\boldsymbol{y}$ in such a line segment are given by

$$\boldsymbol{y} = \boldsymbol{x} + \lambda\boldsymbol{d}, \qquad \lambda \in I, \tag{127.6}$$

where I is some fixed interval. If μ is the mass measure for the line,

$$M = \int_I d\mu(\lambda). \tag{127.7}$$

Without loss of generality we can suppose the parametrization so selected that

$$\int_I \lambda\,d\mu = 0, \qquad \int_I \lambda^2\,d\mu = M. \tag{127.8}$$

Then, identifying $\boldsymbol{y}$ and $\boldsymbol{x}$ with their position vectors, we have

$$\int_I \boldsymbol{y}\,d\mu = M\boldsymbol{x}, \qquad \int_I \lambda\,\boldsymbol{y}\,d\mu = M\boldsymbol{d}. \tag{127.9}$$

Hence

$$\left.\begin{aligned} \frac{1}{M}\int_I \dot{\boldsymbol{y}}\,d\mu &= \dot{\boldsymbol{x}}, \\ \frac{1}{M}\int_I \boldsymbol{y}\wedge\dot{\boldsymbol{y}}\,d\mu &= \boldsymbol{x}\wedge\dot{\boldsymbol{x}} + \boldsymbol{d}\wedge\dot{\boldsymbol{d}}, \\ \frac{1}{2M}\int_I \dot{y}^2\,d\mu &= \frac{1}{2}(\dot{x}^2 + \dot{d}^2). \end{aligned}\right\} \tag{127.10}$$

The quantities on the right-hand sides are, per unit mass, the momentum, the moment of momentum with respect to the center of mass, and the kinetic energy of the rod. Here $\boldsymbol{x}$ is the position vector of the center of mass and $\dot{\boldsymbol{x}}$ is its velocity. The spin of the rod, described by $\dot{\boldsymbol{d}}$, is determined by forces, here left unspecified, that do not influence the motion of the center of mass. The formal identity of (127.10) with Eqs. (127.2) suggests that ERICKSEN'S

---

[1] Cf. OSEEN [1933, *3*].

[2] ERICKSEN [1961, *14*, § III]. Cf. also the more general considerations of ERICKSEN [1962, *14*].

material may be regarded, heuristically, as a dense assembly of little massy, inflexible lines capable of spinning and stretching. The rod has been assumed homogeneous in the sense that λ and $\mu(\lambda)$ are independent of t.

No such separation holds for the remaining field variables[1]. The influx $\boldsymbol{M}$ of moment of momentum, is regarded as arising solely from the moments of the two parts of the stress:

$$\left.\begin{aligned} 2\boldsymbol{M}\boldsymbol{n} &= \boldsymbol{p}\wedge\boldsymbol{T}_{\mathrm{M}}\boldsymbol{n}+\boldsymbol{d}\wedge\boldsymbol{T}_{\mathrm{m}}\boldsymbol{n},\\ M^{krq} &= p^{[k}T_{\mathrm{M}}^{r]q}+d^{[k}T_{\mathrm{m}}^{r]q}. \end{aligned}\right\} \tag{127.11}$$

The part $d^{[k}T_{\mathrm{m}}^{r]q}$ is the *couple-stress tensor* (cf. Sects. 203—205 of CFT, and Sect. 98 of this treatise).

Likewise, the influx of energy $\boldsymbol{q}$ is composed of a non-mechanical part, $\boldsymbol{h}$, as well as the transport of rate of working:

$$\left.\begin{aligned} \boldsymbol{q} &= \boldsymbol{T}_{\mathrm{M}}^{T}\dot{\boldsymbol{x}}+\boldsymbol{T}_{\mathrm{m}}^{T}\dot{\boldsymbol{d}}+\boldsymbol{h},\\ q^{k} &= \dot{x}^{r}T_{\mathrm{M}\,r}{}^{k}+\dot{d}^{r}T_{\mathrm{m}\,r}{}^{k}+h^{k}. \end{aligned}\right\} \tag{127.12}$$

Similarly

$$\left.\begin{aligned} \boldsymbol{L} &= \tfrac{1}{2}(\boldsymbol{p}\wedge\boldsymbol{b}_{\mathrm{M}}+\boldsymbol{d}\wedge\boldsymbol{b}_{\mathrm{m}})+\boldsymbol{L}_{0},\\ L^{kq} &= p^{[k}b_{\mathrm{M}}^{q]}+d^{[k}b_{\mathrm{m}}^{q]}+L_{0}^{kq}, \end{aligned}\right\} \tag{127.13}$$

and

$$r=\boldsymbol{b}_{\mathrm{M}}\cdot\dot{\boldsymbol{x}}+\boldsymbol{b}_{\mathrm{m}}\cdot\dot{\boldsymbol{d}}+q. \tag{127.14}$$

With these definitions and assumptions, CAUCHY'S first law of motion (16.6) splits into two field equations of identical form:

$$\left.\begin{aligned} \varrho\ddot{\boldsymbol{x}} &= \operatorname{div}\boldsymbol{T}_{\mathrm{M}}+\varrho\boldsymbol{b}_{\mathrm{M}},\\ \varrho\ddot{\boldsymbol{d}} &= \operatorname{div}\boldsymbol{T}_{\mathrm{m}}+\varrho\boldsymbol{b}_{\mathrm{m}}, \end{aligned}\right\} \tag{127.15}$$

use of which enables us to simplify the field equations corresponding to $(127.1)_{2,3}$, specialized by aid of $(127.3)_{2,3}$:

$$\left.\begin{aligned} \{\boldsymbol{T}_{\mathrm{M}}\} &= \{(\operatorname{grad}\boldsymbol{d})\,\boldsymbol{T}_{\mathrm{m}}^{T}\}+\varrho\boldsymbol{L}_{0},\\ T_{\mathrm{M}}^{[kr]} &= d^{[k}{}_{,q}T_{\mathrm{m}}^{r]q}+\varrho L_{0}^{kr}, \end{aligned}\right\} \tag{127.16}$$

where $\{\boldsymbol{A}\}\equiv\tfrac{1}{2}(\boldsymbol{A}-\boldsymbol{A}^{T})$, and

$$\left.\begin{aligned} \varrho\dot{\varepsilon} &= \operatorname{tr}(\boldsymbol{T}_{\mathrm{M}}\boldsymbol{L}^{T})+\operatorname{tr}[\boldsymbol{T}_{\mathrm{m}}(\operatorname{grad}\dot{\boldsymbol{d}})^{T}]+\operatorname{div}\boldsymbol{h}+\varrho q\\ &= \dot{x}_{r,k}T_{\mathrm{M}}^{rk}+\dot{d}_{r,k}T_{\mathrm{m}}^{rk}+h^{k}{}_{,k}+\varrho q. \end{aligned}\right\} \tag{127.17}$$

The field equations forming the basis of ERICKSEN'S theory are (15.14), (127.15), (127.16), and (127.17). In addition, ERICKSEN split $\boldsymbol{b}_{\mathrm{m}}$ into two parts:

$$\boldsymbol{b}_{\mathrm{m}}=\boldsymbol{b}_{\mathrm{m}}'+\boldsymbol{b}_{\mathrm{m}}'', \tag{127.18}$$

only the former of which he regarded as having an influence upon the supplies of moment of momentum and the energy, as follows:

$$\left.\begin{aligned} \boldsymbol{L}_{0} &= \tfrac{1}{2}(\boldsymbol{b}_{\mathrm{m}}'\wedge\boldsymbol{d}),\\ q &= -\boldsymbol{b}_{\mathrm{m}}'\cdot\dot{\boldsymbol{d}}+q_{0}. \end{aligned}\right\} \tag{127.19}$$

It will be noted that the stress tensor $\boldsymbol{T}_{\mathrm{M}}$ is generally unsymmetric.

---

[1] To motivate definitions and assumptions equivalent to these, ERICKSEN [1961, *14*] adduced results from a purely static theory (Sect. 128), reasons of formal symmetry, and similarity to an earlier theory proposed by ANZELIUS [1931, *1*].

To set up constitutive equations, ERICKSEN selected as dependent variables

$$\boldsymbol{T}_{\mathrm{M}}, \boldsymbol{T}_{\mathrm{m}}, \boldsymbol{h}, \boldsymbol{b}''_{\mathrm{m}}, \varepsilon, \tag{127.20}$$

and as independent variables

$$\varrho, \theta, \boldsymbol{d}, \dot{\boldsymbol{d}}, \boldsymbol{L}, \operatorname{grad}\theta. \tag{127.21}$$

Thus the theory is an immediate generalization of the Reiner-Rivlin theory (Sect. 119) with allowance for heat conduction and for effects of orientation. Application of the principle of material indifference shows, in effect, that (127.21) may be replaced by

$$\varrho, \theta, \boldsymbol{d}, \mathring{\boldsymbol{d}}, \boldsymbol{D}, \operatorname{grad}\theta, \tag{127.22}$$

where $\mathring{\boldsymbol{d}}$ is the co-rotational time flux of $\boldsymbol{d}$ [CFT, Sect. 148]:

$$\mathring{\boldsymbol{d}} = \dot{\boldsymbol{d}} - \boldsymbol{W}\boldsymbol{d}. \tag{127.23}$$

According to the principle of equipresence (Sects. 96, 125), each of the variables in the list (127.20) should be allowed to depend upon all of the variables in the list (127.22), unless forbidden to do so by some rule of invariance or some law of general mechanics and energetics. ERICKSEN adopted this principle only in part. He supposed that

$$\boldsymbol{T}_{\mathrm{m}} = \boldsymbol{0}, \tag{127.24}$$

so that the couple stress vanishes and

$$\boldsymbol{T} = \boldsymbol{T}_{\mathrm{M}}. \tag{127.25}$$

When $\boldsymbol{b}''_{\mathrm{m}} = \boldsymbol{0}$, so that $\boldsymbol{b}_{\mathrm{m}} = \boldsymbol{b}'_{\mathrm{m}}$, by (127.16) and (127.19) follows the restrictive relation

$$\left.\begin{aligned} \boldsymbol{T} - \boldsymbol{T}^{T} &= \varrho\, \boldsymbol{b}_{\mathrm{m}} \wedge \boldsymbol{d}, \\ T^{[km]} &= \varrho\, b_{\mathrm{m}}^{[k} d^{m]}. \end{aligned}\right\} \tag{127.26}$$

It seems likely that consistency with the entropy inequality requires the constitutive functions to be related to one another, as in the theory of thermo-elasticity (Sect. 96), but such restrictions have not yet been determined[1].

ERICKSEN further restricted attention to the case when $\boldsymbol{T}$, $\boldsymbol{h}$, and $\boldsymbol{b}_{\mathrm{m}}$ are linear in the variables $\dot{\boldsymbol{d}}$, $\boldsymbol{D}$, and grad θ, and when

$$\varepsilon = \varepsilon(\varrho, \boldsymbol{d}, \eta), \tag{127.27}$$

η being the specific entropy. Thus he obtained the following constitutive equations

$$\left.\begin{aligned} \boldsymbol{T} &= \boldsymbol{A}^{0} + \mathsf{A}^{1}[\mathring{\boldsymbol{d}}] + \mathsf{A}^{2}[\boldsymbol{D}] + \mathsf{A}^{3}[\operatorname{grad}\theta], \\ \boldsymbol{b}_{\mathrm{m}} &= \boldsymbol{b}^{0} + \boldsymbol{B}^{1}\mathring{\boldsymbol{d}} + \mathsf{B}^{2}[\boldsymbol{D}] + \boldsymbol{B}^{3}\operatorname{grad}\theta, \\ -\boldsymbol{h} &= \boldsymbol{c}^{0} + \boldsymbol{C}^{1}\mathring{\boldsymbol{d}} + \mathsf{C}^{2}[\boldsymbol{D}] + \boldsymbol{C}^{3}\operatorname{grad}\theta, \end{aligned}\right\} \tag{127.28}$$

or, in components

$$\left.\begin{aligned} T_{km} &= A^{0}_{km} + A^{1}_{kmp}\mathring{d}^{p} + A^{2}_{kmpq}D^{pq} + A^{3}_{kmp}\theta^{,p}, \\ b_{\mathrm{m}k} &= b^{0}_{k} + B^{1}_{km}\mathring{d}^{m} + B^{2}_{kmp}D^{mp} + B^{3}_{km}\theta^{,m}, \\ -h_{k} &= c^{0}_{k} + C^{1}_{km}\mathring{d}^{m} + C^{2}_{kmp}D^{mp} + C^{3}_{km}\theta^{,m}. \end{aligned}\right\} \tag{127.29}$$

---

[1] ERICKSEN [1960, *18*, § 4] wrote out the entropy inequality for this theory and remarked that restrictions should follow from it.

The forms of the tensors $\boldsymbol{A}^0, \boldsymbol{A}^1, \ldots, \boldsymbol{C}^3$ are restricted by the principle of material indifference (Sect. 19). While ERICKSEN drew attention to the possibility of constitutive equations expressing symmetry with respect to proper but not inproper rotations, he considered mainly the case when $\boldsymbol{T}$, $-\boldsymbol{h}$, and $\boldsymbol{b}_{\mathrm{m}}$ are expressible as isotropic functions of $\mathring{\boldsymbol{d}}$, $\boldsymbol{D}$, and grad θ, linear in each. Using representation theorems for such functions[1], ERICKSEN obtained the following explicit forms:

$$\left.\begin{aligned}
\boldsymbol{T} &= (\alpha_0 + \alpha_1 \operatorname{tr}\boldsymbol{D} + \alpha_2 \boldsymbol{d}\cdot\boldsymbol{D}\boldsymbol{d} + \alpha_3 \mathring{\boldsymbol{d}}\cdot\boldsymbol{d})\,\mathbf{1} + \\
&\quad + (\alpha_4 + \alpha_5 \operatorname{tr}\boldsymbol{D} + \alpha_6 \boldsymbol{d}\cdot\boldsymbol{D}\boldsymbol{d} + \alpha_7 \mathring{\boldsymbol{d}}\cdot\boldsymbol{d})\,\boldsymbol{d}\otimes\boldsymbol{d} + \\
&\quad + \alpha_8 \boldsymbol{D} + \alpha_9 (\boldsymbol{D}\boldsymbol{d})\otimes\boldsymbol{d} + \alpha_{10}\boldsymbol{d}\otimes(\boldsymbol{D}\boldsymbol{d}) + \\
&\quad + \alpha_{11}\boldsymbol{d}\otimes\mathring{\boldsymbol{d}} + \alpha_{12}\mathring{\boldsymbol{d}}\otimes\boldsymbol{d}, \\
\boldsymbol{b}_{\mathrm{m}} &= (\gamma_0 + \gamma_1 \operatorname{tr}\boldsymbol{D} + \gamma_2 \boldsymbol{d}\cdot\boldsymbol{D}\boldsymbol{d} + \gamma_3 \boldsymbol{d}\cdot\mathring{\boldsymbol{d}})\,\boldsymbol{d} + \\
&\quad + \gamma_4 \boldsymbol{D}\boldsymbol{d} + \gamma_5 \mathring{\boldsymbol{d}}, \\
-\boldsymbol{h} &= \beta_0 \operatorname{grad}\theta + \beta_1 (\boldsymbol{d}\cdot\operatorname{grad}\theta)\,\boldsymbol{d},
\end{aligned}\right\} \tag{127.30}$$

where, for consistency with (127.26),

$$\varrho\gamma_4 = \alpha_9 - \alpha_{10}, \qquad \varrho\gamma_5 = \alpha_{12} - \alpha_{11}. \tag{127.31}$$

The coefficients $\alpha_\Gamma, \beta_\Gamma, \gamma_\Gamma$ are functions of the scalars

$$\varrho, \theta, d, \tag{127.32}$$

where, as usual, $d \equiv |\boldsymbol{d}|$. Notice that in this theory deformation affects the stress but not the heat flux, while a temperature gradient gives rise to heat flux but has no effect on the stress. This separation of effects is a consequence of the assumed transverse isotropy of the material.

When the director acceleration $\ddot{\boldsymbol{d}}$ is small enough to be neglected, $(127.15)_2$ reduces to $\boldsymbol{b}_{\mathrm{m}} = \mathbf{0}$, where $\boldsymbol{b}_{\mathrm{m}}$ is given by $(127.30)_3$. Equivalently,

$$\mathring{\boldsymbol{d}} = \boldsymbol{f}(\varrho, \theta, \boldsymbol{d}, \boldsymbol{D}). \tag{127.33}$$

By (127.26), the stress is now symmetric. Eq. (127.33) is then used to eliminate[2] $\mathring{\boldsymbol{d}}$ from $(127.28)_1$ or its special case (127.30).

Corresponding constitutive equations for incompressible materials are easy to write down at any stage. One simply omits terms involving $\operatorname{tr}\boldsymbol{D}$ and replaces the terms proportional to $\mathbf{1}$ by an arbitrary hydrostatic pressure. For example, the counterpart of the theory outlined in the last paragraph above is defined by the following constitutive equations, in which it is assumed also that grad $\theta = \mathbf{0}$:

$$\left.\begin{aligned}
\boldsymbol{T} &= -p\mathbf{1} + (\lambda_1 + \lambda_2 \boldsymbol{d}\cdot\boldsymbol{D}\boldsymbol{d})\,\boldsymbol{d}\otimes\boldsymbol{d} + \\
&\quad + \lambda_3 \boldsymbol{D} + \lambda_4 (\boldsymbol{d}\otimes\boldsymbol{D}\boldsymbol{d} + \boldsymbol{D}\boldsymbol{d}\otimes\boldsymbol{d}), \\
T^k{}_m &= -p\,\delta^k_m + (\lambda_1 + \lambda_2 D_{pq} d^p d^q)\, d^k d_m + \\
&\quad + \lambda_3 D^k_m + \lambda_4 (d^k D^q_m d_q + D^k_q d^q d_m),
\end{aligned}\right\} \tag{127.34}$$

$$\left.\begin{aligned}
\mathring{\boldsymbol{d}} &= (\mu_1 + \mu_2 \boldsymbol{d}\cdot\boldsymbol{D}\boldsymbol{d})\,\boldsymbol{d} + \mu_3 \boldsymbol{D}\boldsymbol{d}, \\
\mathring{d}_k &= (\mu_1 + \mu_2 D_{pq} d^p d^q)\, d_k + \mu_3 D^m_k d_m.
\end{aligned}\right\} \tag{127.35}$$

---

[1] SMITH and RIVLIN [1957, *23*].

[2] Equations similar to those resulting from this process were derived by ANZELIUS [1931, *1*, Eq. (14)].

The theory so defined allows a certain effect of elasticity of structure in that the length of the director $\boldsymbol{d}$ may vary in time. If we forbid this effect, requiring $\boldsymbol{d}$ to be a unit vector, we impose the condition $\boldsymbol{d}\cdot\overset{\circ}{\boldsymbol{d}}=0$ upon (127.35), so that this equation becomes

$$\left.\begin{aligned}\overset{\circ}{\boldsymbol{d}}&=\lambda\left(\boldsymbol{D}\boldsymbol{d}-(\boldsymbol{d}\cdot\boldsymbol{D}\boldsymbol{d})\,\boldsymbol{d}\right),\\ \overset{\circ}{d}_k&=\lambda\,(D^m_k\,d_m-D_{pq}\,d^p\,d^q\,d_k)\,.\end{aligned}\right\}\qquad(127.36)$$

The theory[1] based on the constitutive equations (127.34) and (127.36) seems to be the simplest possible theory of anisotropic fluids included in the class introduced by ERICKSEN. Some special solutions for it will be presented in Sect. 129.

If in the theory of oriented hyperelastic materials, summarized in our Sect. 98 (β), we limit the number of directors to one and set $\boldsymbol{m}=\dot{\boldsymbol{x}}$, $\boldsymbol{d}_1=\boldsymbol{d}$, $\boldsymbol{m}^1=\dot{\boldsymbol{d}}$, the tensor $\boldsymbol{A}$ given by $(98.10)_1$ becomes identical with that given by $(127.2)_2$, and for suitably specialized action L Eq. $(98.10)_2$ reduces to $(127.3)_3$. However, the present theory is not one of hyperelastic materials; designed especially to represent certain dissipative effects, it cannot be derived from an action principle. It would be possible, of course, to generalize the present theory so as to include TOUPIN's as a special case. First, leaving $\boldsymbol{m}$ arbitrary, we should replace $(127.2)_{2,3}$ by more general expressions in terms of the micromomenta:

$$\left.\begin{aligned}2\boldsymbol{A}&=\boldsymbol{p}\wedge\boldsymbol{m}+\boldsymbol{d}_{\mathfrak{a}}\wedge\boldsymbol{m}^{\mathfrak{a}},\\ E&=\varepsilon+\tfrac{1}{2}(\boldsymbol{m}\cdot\dot{\boldsymbol{x}}+\boldsymbol{m}^{\mathfrak{a}}\cdot\dot{\boldsymbol{d}}_{\mathfrak{a}})\,.\end{aligned}\right\}\qquad(127.37)$$

Second, the variational principle should be replaced by direct mechanical constitutive equations, following the method used throughout this treatise.

With rates as well as deformation gradients taken as independent variables, a theory of visco-elastic oriented materials will result. We leave the development of this more elaborate theory to the reader.

128. The anisotropic fluids of ERICKSEN, II. Statics. Even in the simplest of the theories considered in the previous section, the stress acting on the fluid at rest need not be hydrostatic. From (127.34) we have

$$\boldsymbol{T}|_{\boldsymbol{D}=\boldsymbol{0}}=-p\boldsymbol{1}+\lambda_1\boldsymbol{d}\otimes\boldsymbol{d}.\qquad(128.1)$$

If consistency with classical hydrostatics is desired, it may be secured by setting $\lambda_1=0$.

ERICKSEN[2] has considered the static theory from a more general standpoint. As basic law he lays down a work principle:

$$\left.\begin{aligned}\delta\int_{\mathscr{B}}\varrho\sigma\,dv&=\int_{\partial\mathscr{B}}(\boldsymbol{t}_{\mathrm{M}}\cdot\delta\boldsymbol{x}+\boldsymbol{t}_{\mathrm{m}}\cdot D\boldsymbol{d})\,ds+\\ &\quad+\int_{\mathscr{B}}\varrho\,(\boldsymbol{b}_{\mathrm{M}}\cdot\delta\boldsymbol{x}+\boldsymbol{b}''_{\mathrm{m}}\cdot D\boldsymbol{d})\,dv,\end{aligned}\right\}\qquad(128.2)$$

where the work function σ is assumed to have the following functional dependence:

$$\sigma=\sigma(\boldsymbol{d},\ \mathrm{grad}\ \boldsymbol{d},\ \varrho,\ \theta)\,;\qquad(128.3)$$

where $\boldsymbol{t}_{\mathrm{M}}$, $\boldsymbol{t}_{\mathrm{m}}$, $\boldsymbol{b}_{\mathrm{M}}$, $\boldsymbol{b}''_{\mathrm{m}}$ are undetermined multipliers (cf. CFT, Sect. 232); and where

$$D\boldsymbol{d}=\delta\boldsymbol{d}+(\mathrm{grad}\ \boldsymbol{d})\,\delta\boldsymbol{x},\qquad(128.4)$$

---

[1] This is the theory ERICKSEN first presented in 1958 and included in a theory of intermediate generality [1960, *19*, § 2]. HAND [1961, *23*] showed that it includes as a special case a theory of dilute suspensions based on results concerning the motion of an ellipsoid in a linearly viscous fluids. Equations similar to (127.34) and (127.36) had been proposed by OSEEN [1933, *3*].

[2] ERICKSEN [1961, *14*, § IV]. In a somewhat less general theory, FRANK [1958, *16*] had defined and discussed various material symmetries.

the variational counterpart of $\dot{\boldsymbol{d}}$ [cf. Eq. (15.13)]. The variations are supposed also to be consistent with conservation of mass, so that $\delta\varrho + \operatorname{div}(\varrho\,\delta\boldsymbol{x}) = 0$.

To calculate the variations, assumed isothermal, note first that for any quantity Ψ,

$$\delta(\varrho\Psi\,dv) = \varrho D\Psi\,dv, \tag{128.5}$$

where

$$D\Psi \equiv \delta\Psi + (\operatorname{grad}\Psi)\cdot\delta\boldsymbol{x}.$$

Hence

$$\left.\begin{aligned} D\varrho &= -\varrho\operatorname{div}\delta\boldsymbol{x},\\ D\operatorname{grad}\boldsymbol{d} &= \operatorname{grad} D\boldsymbol{d} - \operatorname{grad}\boldsymbol{d}\operatorname{grad}\delta\boldsymbol{x}. \end{aligned}\right\} \tag{128.6}$$

Then

$$\left.\begin{aligned} \delta\int_{\mathscr{B}}\varrho\sigma\,dv &= \int_{\mathscr{B}}\varrho\,D\sigma\,dv,\\ &= \int_{\mathscr{B}}\varrho\left[\sigma_\varrho D\varrho + \sigma_{\boldsymbol{d}}\cdot D\boldsymbol{d} + \operatorname{tr}\left((\sigma_{\operatorname{grad}\boldsymbol{d}})^T(D\operatorname{grad}\boldsymbol{d})\right)\right]dv,\\ &= \int_{\mathscr{B}}\left\{\operatorname{tr}(\boldsymbol{T}_{\mathrm{M}0}^T\operatorname{grad}\delta\boldsymbol{x}) - \varrho\,\boldsymbol{b}'_{\mathrm{m}}\cdot D\boldsymbol{d} + \right.\\ &\qquad \left. + \operatorname{tr}(\boldsymbol{T}_{\mathrm{m}0}^T\operatorname{grad} D\,\boldsymbol{d})\right\}dv,\\ &= \int_{\partial\mathscr{B}}(\boldsymbol{T}_{\mathrm{M}0}^T\,\delta\boldsymbol{x} + \boldsymbol{T}^T{}_{\mathrm{m}0}D\,\boldsymbol{d})\cdot\boldsymbol{n}\,ds -\\ &\qquad - \int_{\mathscr{B}}\left[(\operatorname{div}\boldsymbol{T}_{\mathrm{M}0})\cdot\delta\boldsymbol{x} + (\operatorname{div}\boldsymbol{T}_{\mathrm{m}0} + \varrho\,\boldsymbol{b}'_{\mathrm{m}})\cdot D\boldsymbol{d}\right]dv, \end{aligned}\right\} \tag{128.7}$$

where

$$\left.\begin{aligned} \boldsymbol{T}_{\mathrm{M}0} &= -\varrho^2\sigma_\varrho\boldsymbol{1} - \varrho(\operatorname{grad}\boldsymbol{d})^T\sigma_{\operatorname{grad}\boldsymbol{d}},\\ &= -\varrho^2\sigma_\varrho\boldsymbol{1} - (\operatorname{grad}\boldsymbol{d})^T\boldsymbol{T}_{\mathrm{m}0},\\ \boldsymbol{T}_{\mathrm{m}0} &= \varrho\,\sigma_{\operatorname{grad}\boldsymbol{d}},\\ \boldsymbol{b}'_{\mathrm{m}} &= -\sigma_{\boldsymbol{d}}; \end{aligned}\right\} \tag{128.8}$$

in components,

$$\left.\begin{aligned} T_{\mathrm{M}0\,m}{}^k &= -\varrho^2\frac{\partial\sigma}{\partial\varrho}\delta^k_m - \varrho\frac{\partial\sigma}{\partial d^q{}_{,k}}d^q{}_{,m},\\ T_{\mathrm{m}0\,k}{}^m &= \varrho\frac{\partial\sigma}{\partial d^k{}_{,m}},\\ b'_{\mathrm{m}\,k} &= -\frac{\partial\sigma}{\partial d^k}. \end{aligned}\right\} \tag{128.9}$$

If the work principle (128.2) is to hold for arbitrary variations $\delta\boldsymbol{x}$ and $D\boldsymbol{d}$, comparison with (128.7) yields the differential equations

$$\left.\begin{aligned} \operatorname{div}\boldsymbol{T}_{\mathrm{M}0} + \varrho\,\boldsymbol{b}_{\mathrm{M}} &= \boldsymbol{0},\\ \operatorname{div}\boldsymbol{T}_{\mathrm{m}0} + \varrho(\boldsymbol{b}'_{\mathrm{m}} + \boldsymbol{b}''_{\mathrm{m}}) &= \boldsymbol{0}, \end{aligned}\right\} \tag{128.10}$$

and the boundary conditions

$$\left.\begin{aligned} \boldsymbol{T}_{\mathrm{M}0}\boldsymbol{n} &= \boldsymbol{t}_{\mathrm{M}},\\ \boldsymbol{T}_{\mathrm{m}0}\boldsymbol{n} &= \boldsymbol{t}_{\mathrm{m}}. \end{aligned}\right\} \tag{128.11}$$

These equations may be compared with (127.15) and (127.18). To render them consistent, we need only set

$$\left.\begin{aligned} \boldsymbol{T}_{\mathrm{M}} &= \boldsymbol{T}_{\mathrm{M}0} + \boldsymbol{T}'_{\mathrm{M}},\\ \boldsymbol{T}_{\mathrm{m}} &= \boldsymbol{T}_{\mathrm{m}0} + \boldsymbol{T}''_{\mathrm{m}}, \end{aligned}\right\} \tag{128.12}$$

where $\boldsymbol{T}'_{\mathrm{M}}$ and $\boldsymbol{T}''_{\mathrm{m}}$, which represent effects of friction, vanish when $\dot{\boldsymbol{x}} = \boldsymbol{0}$ and $\dot{\boldsymbol{d}} = \boldsymbol{0}$. As in the theory of hyperelasticity and its generalizations (Sects. 79—81, 96, 98), the stresses $\boldsymbol{T}_{\mathrm{M}0}$ and $\boldsymbol{T}_{\mathrm{m}0}$ are derived from a potential σ by differentiation, according to (128.8).

If the energy function σ actually depends upon grad $\boldsymbol{d}$, according to (128.8) neither of the stress tensors $\boldsymbol{T}_{\mathrm{M}}$ and $\boldsymbol{T}_{\mathrm{m}}$ is a hydrostatic tensor, in general, or even a symmetric tensor. The static properties of the anisotropic fluids of ERICKSEN are thus quite different from those of ordinary fluids.

In a later paper, ERICKSEN[1] has found the form assumed by the equations of this theory when the director motions are subjected to the constraints

$$\psi_{\mathfrak{i}}(\varrho, \boldsymbol{d}, \operatorname{grad} \boldsymbol{d}) = 0, \qquad \mathfrak{i} = 1, 2, \ldots, \mathfrak{p}. \tag{128.13}$$

As examples he mentions incompressibility of the material and inextensibility of the director:

$$\psi_1 \equiv \varrho - \varrho_0, \qquad \psi_2 \equiv d^2 - d_0^2, \tag{128.14}$$

where ϱ_0 and d_0 are constants. His results, which may be obtained by a simple generalization of the method presented in Sect. 30, are

$$\left.\begin{aligned}
\boldsymbol{T}_{\mathrm{M}0} &= -\varrho^2 \sigma_\varrho \mathbf{1} - \varrho\,(\operatorname{grad} \boldsymbol{d})^T \frac{\partial \sigma}{\partial \operatorname{grad} \boldsymbol{d}} - \\
&\quad - \sum_{\mathfrak{i}=1}^{\mathfrak{p}} q_{\mathfrak{i}} \left[\varrho \frac{\partial \psi_{\mathfrak{i}}}{\partial \varrho} \mathbf{1} + (\operatorname{grad} \boldsymbol{d})^T \frac{\partial \psi_{\mathfrak{i}}}{\partial \operatorname{grad} \boldsymbol{d}}\right], \\
\boldsymbol{T}_{\mathrm{m}0} &= \varrho \frac{\partial \sigma}{\partial \operatorname{grad} \boldsymbol{d}} + \sum_{\mathfrak{i}=1}^{\mathfrak{p}} q_{\mathfrak{i}} \frac{\partial \psi_{\mathfrak{i}}}{\partial \operatorname{grad} \boldsymbol{d}}, \\
\boldsymbol{b}'_{\mathrm{m}} &= -\frac{\partial \sigma}{\partial \boldsymbol{d}} - \sum_{\mathfrak{i}=1}^{\mathfrak{p}} \frac{q_{\mathfrak{i}}}{\varrho} \frac{\partial \psi_{\mathfrak{i}}}{\partial \boldsymbol{d}},
\end{aligned}\right\} \tag{128.15}$$

where the quantities $q_{\mathfrak{i}}$ are multipliers. Substitution of $(128.15)_2$ into $(128.15)_1$ yields the alternative form

$$\boldsymbol{T}_{\mathrm{M}0} = -\varrho\left(\varrho \frac{\partial \sigma}{\partial \varrho} + \sum_{\mathfrak{i}=1}^{\mathfrak{p}} q_{\mathfrak{i}} \frac{\partial \psi_{\mathfrak{i}}}{\partial \varrho}\right) \mathbf{1} - (\operatorname{grad} \boldsymbol{d})^T \boldsymbol{T}_{\mathrm{m}0}. \tag{128.16}$$

In the forms given here, the constraints (128.13) have not been reduced so as to satisfy the principle of material indifference.

ERICKSEN[2] has derived and interpreted conditions of compatibility to be satisfied by the fields $\boldsymbol{b}_{\mathrm{M}}$, $\boldsymbol{b}''_{\mathrm{m}}$, $\boldsymbol{t}_{\mathrm{M}}$, and $\boldsymbol{t}_{\mathrm{m}}$ in (128.10) and (128.11).

In a later paper[3], ERICKSEN has determined all functions $\sigma(\boldsymbol{d}, \operatorname{grad} \boldsymbol{d})$ giving rise to the same field equations as a given one. The difference of any two such functions he calls a *nilpotent energy*. The difference between the predictions based on two functions differing by a nilpotent energy manifests itself only in the boundary conditions.

129. The anisotropic fluids of ERICKSEN, III. Special flows. ERICKSEN has chosen to explore the consequences of his theories of anisotropic fluids by studying special cases. Rather than seeking the maximum generality, he has often selected the simplest possible special theory that reveals the phenomenon. Several of his examples concern the simplest of all the theories included in the framework set up in Sect. 128, namely, the theory of incompressible transversely isotropic fluids satisfying (127.34) and (127.35).

α) *Homogeneous steady flows.* We may call a motion homogeneous if not only $\boldsymbol{D}$ but also $\boldsymbol{d}$ is a function of time only. To the extent that thermal effects are left out of account, for *all* kinds of incompressible anisotropic fluids described

[1] ERICKSEN [1962, *15*, § 2].
[2] ERICKSEN [1962, *15*, § 3].
[3] ERICKSEN [1962, *16*].

in Sect. 127 the extra stress $\boldsymbol{T}+p\boldsymbol{1}$ becomes a function of time only. The theorem on homogeneous motions in Sect. 30 is then easily generalized. The motion is dynamically possible, subject to surface tractions alone, if and only if it is circulation-preserving. The kinematical condition is again (30.33), and again $p=\varrho(\zeta-v+h)$, with ζ given by (30.34). Eq. (127.33), with $\boldsymbol{D}(t)$ and $\boldsymbol{W}(t)$ being given functions of time, becomes *an ordinary differential equation for determining the homogeneous director field* $\mathbf{d}(t)$. The orientation of the fluid, as represented by $\mathbf{d}$, thus is determined from its initial value and from the gross motion.

Consider a simple shearing[1], given by (30.45), or, equivalently, by (106.1) with $v(x,t)=\varkappa x$, where $\varkappa>0$. Substitution into (127.36), account being taken of (127.23), yields

$$\left.\begin{aligned}\dot{d}_{\langle y\rangle}&=\tfrac{1}{2}\varkappa d_{\langle x\rangle}(\lambda+1-2\lambda d_{\langle y\rangle}^2),\\ \dot{d}_{\langle x\rangle}&=\tfrac{1}{2}\varkappa d_{\langle y\rangle}(\lambda-1-2\lambda d_{\langle x\rangle}^2).\end{aligned}\right\}\qquad(129.1)$$

When $\lambda\neq 0$, a first integral of this system is

$$\left.\begin{aligned}a d_{\langle y\rangle}^2&+b d_{\langle x\rangle}^2\\ &=\frac{1}{2\lambda}\left[(a+b)\lambda+a-b\right],\end{aligned}\right\}\qquad(129.2)$$

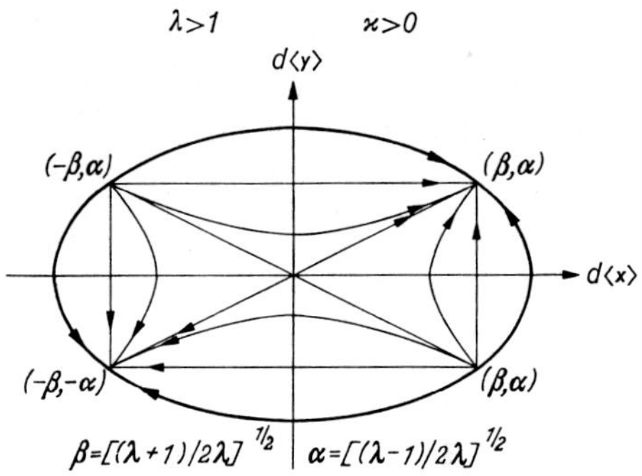

Fig. 28. ERICKSEN's stability diagram.

where a and b are arbitrary constants. For representative values of a/b, the integral curves are sketched in Fig. 28. The arrowheads indicate the direction of increasing time. Corresponding plots for the negatives of the indicated values of a and b may be obtained by interchanging $d_{\langle y\rangle}$ and $d_{\langle x\rangle}$ and reversing the arrowheads, as is plain from (129.1).

Consider the figure, appropriate to the case when $\lambda>1$. Set $\alpha\equiv[(\lambda-1)/(2\lambda)]^{\frac{1}{2}}$, $\beta\equiv[(\lambda+1)/(2\lambda)]^{\frac{1}{2}}$. There are unstable constant solutions, namely,

$$\left.\begin{aligned}&d_{\langle y\rangle}=d_{\langle x\rangle}=0,\quad d_{\langle z\rangle}=\pm 1\\ \text{and}\quad&\\ &d_{\langle y\rangle}=\pm\alpha,\quad d_{\langle x\rangle}=\mp\beta,\quad d_{\langle z\rangle}=0,\end{aligned}\right\}\qquad(129.3)$$

and stable constant solutions:

$$d_{\langle y\rangle}=\pm\alpha,\quad d_{\langle x\rangle}=\pm\beta,\quad d_{\langle z\rangle}=0.\qquad(129.4)$$

Other than the unstable solutions (129.3), every solution approaches (129.4) as $t\to\infty$. That is, the flow tends to turn the directors down into the flow plane and to align them at an angle φ with the direction of flow, where

$$\tan^2\varphi=\frac{\alpha^2}{\beta^2}=\frac{\lambda-1}{\lambda+1}<1.\qquad(129.5)$$

In contrast to the corresponding result obtained by GREEN for solids capable of flow (Sect. 126), this angle always lies between 0 and $\frac{1}{4}\pi$. Such behavior has been reported for the long chain molecules in polyisobutylene solutions.

By (127.34) we find that the stresses relax with time toward the values given by (108.4) with the following special forms for the viscometric functions:

$$\left.\begin{aligned}\tau(\varkappa)&=(\tfrac{1}{2}\lambda_3+\lambda_1\alpha^2\beta^2+\tfrac{1}{2}\lambda_4)\varkappa+(\operatorname{sign}\varkappa)\lambda_1\alpha\beta,\\ \sigma_1(\varkappa)&=\lambda_1\alpha^2+(\lambda_2\alpha^2+\lambda_4)\alpha\beta|\varkappa|,\\ \sigma_2(\varkappa)&=\lambda_1\beta^2+(\lambda_2\beta^2+\lambda_4)\alpha\beta|\varkappa|.\end{aligned}\right\}\qquad(129.6)$$

[1] ERICKSEN [1960, *19*, § 3].

If $\lambda_1 \neq 0$, the material experiences both shear-stress and unequal normal stresses when at rest. Thus there is a sort of yield in shear, such as is postulated in the theory of "Bingham materials". $\tau(\varkappa)$ is not continuous at $\varkappa=0$, and $\sigma_1(\varkappa)$ and $\sigma_2(\varkappa)$ are not differentiable there[1].

If $\lambda=1$, the stable and unstable constant solutions all reduce to the single solution

$$d_{\langle x\rangle}=\pm 1, \qquad d_{\langle y\rangle}=d_{\langle z\rangle}=0. \tag{129.7}$$

Thus it is possible for the directors to be and to remain aligned parallel to the flow.

If $0<\lambda<1$, Eqs. (129.1) can be shown to be the same as those governing the motion of a single rigid ellipsoid of revolution in a linearly viscous fluid. The axis of revolution is identified with the direction of $\boldsymbol{d}$, and $\lambda=(a^2-b^2)/(a^2+b^2)$, where a is the length of the major axis and b is the common length of the other two axes. As is clear from constructing a figure replacing Fig. 28, $\boldsymbol{d}$ varies periodically with time. Consequently, by (127.34), the stresses also vary periodically with time as well as depending upon the initial value $\boldsymbol{d}(0)$. ERICKSEN has calculated the average values of the stresses over a period and has found that if $\lambda_1=0$, these averages are indistinguishable from stresses obtained by the Navier-Stokes equations with a viscosity dependent upon $\boldsymbol{d}(0)$.

The foregoing considerations have been extended by ERICKSEN[2] to the more general class of homogeneous isochoric circulation-preserving flows for which

$$\dot{x}=2\varkappa' y, \qquad \dot{y}=2\varkappa'' x, \qquad \dot{z}=0. \tag{129.8}$$

If we set

$$\mathfrak{W}\equiv\frac{\varkappa'-\varkappa''}{\varkappa'+\varkappa''}, \tag{129.9}$$

then $|\mathfrak{W}|$ is the kinematical vorticity number of the flow (CFT, Sect. 91); any desired value of $\mathfrak{W}$ may be obtained by choice of $\varkappa'$ and $\varkappa''$. Rectilinear shearing results from the value $|\mathfrak{W}|=1$. Setting

$$\left.\begin{aligned} \tau&\equiv\lambda(\varkappa'+\varkappa'')t,\\ v&\equiv\mathfrak{W}/\lambda, \end{aligned}\right\} \tag{129.10}$$

we reduce (127.36) to the form

$$\left.\begin{aligned} d_{\langle y\rangle}'&=(-v+2d_{\langle y\rangle}^2-1)\,d_{\langle x\rangle},\\ d_{\langle x\rangle}'&=(v+2d_{\langle x\rangle}^2-1)\,d_{\langle y\rangle},\\ d_{\langle z\rangle}'&=-2d_{\langle x\rangle}\,d_{\langle y\rangle}\,d_{\langle z\rangle}, \end{aligned}\right\} \tag{129.11}$$

where a prime denotes differentiation with respect to τ. Here the cases when $\lambda(\varkappa'+\varkappa'')=0$ have been excluded as representing either rigid motion of the fluid or pure convection of the director. The system (129.11) has the following first integral:

$$A(2d_{\langle x\rangle}^2-1+v)+B(2d_{\langle y\rangle}^2-1-v)=0, \tag{129.12}$$

where A and B are arbitrary constants. The integral curves are ellipses, hyperbolas, or straight lines.

---

[1] The earlier special theories of simple fluids implied that $\tau(\varkappa)=\mu_0\varkappa+O(\varkappa^3)$ and $\sigma_\Gamma(\varkappa)=O(\varkappa^2)$, but only because of more or less hidden assumptions of regularity. As is clear from our development leading to the results in Sect. 108, in a simple fluid $\tau(\varkappa)$ may be any odd function, while the $\sigma_\Gamma(\varkappa)$ may be any even functions. Thus Eqs. (129.6) do not stand in contradiction with the theory of simple fluids. Whether the yield effect and the linear dependence of $\sigma_\Gamma(\varkappa)$ on $|\varkappa|$ are consistent with experimental data for any real fluid is uncertain.

[2] ERICKSEN [1960, *21*].

If $v^2<1$, essentially all solutions of (129.11) tend toward one of the constant solutions:

$$\left.\begin{aligned} d_{\langle x\rangle} &= \pm\sqrt{\tfrac{1}{2}(1-v)} \\ d_{\langle y\rangle} &= \mp\sqrt{\tfrac{1}{2}(1+v)} \end{aligned}\right\} \quad \text{if} \quad \lambda(\varkappa'+\varkappa'')>0, \tag{129.13}$$

$$\left.\begin{aligned} d_{\langle x\rangle} &= \pm\sqrt{\tfrac{1}{2}(1-v)} \\ d_{\langle y\rangle} &= \pm\sqrt{\tfrac{1}{2}(1+v)} \\ d_{\langle z\rangle} &= 0 \end{aligned}\right\} \quad \text{if} \quad \lambda(\varkappa'+\varkappa'')<0. \tag{129.14}$$

Thus $\boldsymbol{d}$ is turned, eventually, so as to lie in the flow plane and subtend a fixed angle with the co-ordinate axes.

If $v^2>1$, there are no stable constant solutions, and $\boldsymbol{d}$ varies periodically with time. While ERICKSEN described the motion as "tumbling" the director, he remarked that Eqs. (129.11) show that $\boldsymbol{d}$ rotates at varying angular velocity, spending most of the time nearly parallel to one of the co-ordinate axes.

Thus a slightly rotational flow ($\mathfrak{W}^2<\lambda^2$) tends to order the director as time goes on, while in a highly rotational flow ($\mathfrak{W}^2>\lambda^2$) the directors continue to rotate indefinitely.

ERICKSEN[1] has considered rectilinear shearing according to the more general theory based on (127.30) [or its special case, (127.35)], allowing the director to stretch as well as rotate. In this theory, $\boldsymbol{d}=\boldsymbol{0}$ is always a solution, and for it the constitutive equation (127.30) for the stress reduces to the Navier-Stokes formula (119.2). Corresponding to a given velocity field $\dot{y}(x,t)$, various different solutions $\boldsymbol{d}(x,t)$ are possible. For the case of rectilinear shearing, ERICKSEN has determined necessary and sufficient conditions that the solution $\boldsymbol{d}=\boldsymbol{0}$ be stable. These conditions take the form of inequalities restricting the coefficients in (127.30) and limiting the magnitude of $\varkappa$. Thus there is a critical shearing, beyond which the solution $\boldsymbol{d}=\boldsymbol{0}$ will fail to be stable. ERICKSEN has determined also a non-vanishing steady solution for values of $\varkappa$ exceeding the critical value. Again the normal-stress differences for small values of $\varkappa$ turn out to be approximately linear functions of $\varkappa$ for this solution.

In a later paper, ERICKSEN[2] has studied in detail the problem of the existence and stability of steady homogeneous director fields in a given steady homogeneous velocity field. He found the zero solution stable for small enough velocity gradients if it is assumed stable in a state of rest. Thus these fluids tend to behave like linearly viscous fluids, showing no effects of orientation, if the stretching and spin are small enough. For larger values of the stretching but not too great spin, flow tends to produce orientation, as we have verified above for a fluid with a director of fixed length. The corresponding stress system again exhibits normal-stress differences that are approximately linear functions of $\varkappa$ for small values of $\varkappa$.

β) *Non-homogeneous viscometric flows.* Returning to the simple theory based on (127.34) and (127.36), ERICKSEN[3] has solved the problem of flow in an infinitely long circular pipe. (For the corresponding solution for simple fluids, see Sect. 113 ε.) The assumed velocity field in cylindrical co-ordinates is of the form (112.2) with $\omega(r,t)\equiv 0$, $u(r,t)=f(r)$. The flow down the tube is thus rectilinear and accelerationless. ERICKSEN considered only solutions such that $\boldsymbol{d}=\boldsymbol{d}(r,t)$. Thus the extra stress, given by (127.34), is a function of r and t only, and moreover $T_{\langle r\theta\rangle}=T_{\langle\theta z\rangle}=0$. The dynamical equations for the case when $\boldsymbol{b}=\boldsymbol{0}$ become a special case of (56.6), from which we see that p is independent of θ and is a linear function of z. ERICKSEN assumed further that the stresses are steady and that the pressure gradient along the tube is independent of r and t; that is, $\partial_z p=-E=$ const. Then the dynamical equations (56.6) become equivalent to

[1] ERICKSEN [1960, *20*, § III].
[2] ERICKSEN [1962, *17*].
[3] ERICKSEN [1961, *13*].

the conditions

$$\left.\begin{aligned} T\langle rr\rangle &= -\int \frac{T\langle rr\rangle - T\langle\theta\theta\rangle}{r}\,dr + E z,\\ T\langle rz\rangle &= -\frac{Er}{2} + \frac{F}{r},\\ p = T_E\langle rr\rangle - T\langle rr\rangle &= T_E\langle rr\rangle + \int \frac{T\langle rr\rangle - T\langle\theta\theta\rangle}{r}\,dr - E z,\end{aligned}\right\} \tag{129.15}$$

where F is a constant.

For the assumed flow pattern, the r and z components of (127.36) become

$$\left.\begin{aligned} \dot{d}\langle r\rangle &= \tfrac{1}{2} f'\, d\langle z\rangle\,(\lambda - 1 - 2\lambda d\langle r\rangle^2),\\ \dot{d}\langle z\rangle &= \tfrac{1}{2} f'\, d\langle r\rangle\,(\lambda + 1 - 2\lambda d\langle z\rangle^2).\end{aligned}\right\} \tag{129.16}$$

These equations are of the form (129.1), so that the analysis of rectilinear shearing may be carried over to this flow, with $f'(r)$ replacing $\varkappa$, which now, of course, can no longer be restricted to positive values by a convention. ERICKSEN considered only the case when $|\lambda|>1$ and $\lambda_1 \neq 0$, since under these conditions the fluid was proved above to behave in rectilinear shearing much as a "Bingham material" is assumed to behave. Again it follows that as $t\to\infty$ the components of the director $\boldsymbol{d}$ tend to the limit values

$$d\langle r\rangle = \sin\varphi, \qquad d\langle\theta\rangle = 0, \qquad d\langle z\rangle = \cos\varphi, \tag{129.17}$$

where φ satisfies (129.5). On the convention that $0 \leqq \varphi \leqq \pi$, the angle φ lies in one of the following ranges:

$$\left.\begin{aligned} &0 < \varphi < \tfrac{1}{4}\pi && \text{if } f'>0,\ \lambda>1;\\ &\tfrac{3}{4}\pi < \varphi < \pi && \text{if } f'<0,\ \lambda>1;\\ &\tfrac{1}{4}\pi < \varphi < \tfrac{1}{2}\pi && \text{if } f'>0,\ \lambda<-1;\\ &\tfrac{1}{2}\pi < \varphi < \tfrac{3}{4}\pi && \text{if } f'<0,\ \lambda<-1.\end{aligned}\right\} \tag{129.18}$$

The problem is now to determine the effect of the director motion upon the form of the velocity profile f, which so far has been left unspecified.

ERICKSEN considered first the possibility of stable, steady flow. With $\boldsymbol{d}$ given by (129.17), Eqs. (127.34) yield the following stress system:

$$\left.\begin{aligned} T\langle rr\rangle &= -p + \lambda_1 \sin^2\varphi + A f',\\ T\langle\theta\theta\rangle &= -p,\\ T\langle zz\rangle &= -p + \lambda_1 \cos^2\varphi + B f',\\ T\langle rz\rangle &= C + D f',\\ T\langle r\theta\rangle &= T\langle\theta z\rangle = 0,\end{aligned}\right\} \tag{129.19}$$

where

$$\left.\begin{aligned} 2A &= (\lambda_2 \sin^2\varphi + \lambda_4)\sin 2\varphi \operatorname{sign} f',\\ 2B &= (\lambda_2 \cos^2\varphi + \lambda_4)\sin 2\varphi \operatorname{sign} f',\\ 2C &= \lambda_1 \sin 2\varphi \operatorname{sign} f',\\ 4D &= 2(\lambda_3 + \lambda_4) + \lambda_2 \sin^2 2\varphi.\end{aligned}\right\} \tag{129.20}$$

Thus C is an apparent yield stress in shear, and D is an apparent viscosity. From (129.15) and (129.19) we have

$$\left.\begin{aligned} T_{\langle rr\rangle} &= -\int \frac{1}{r}(\lambda_1 \sin^2\varphi + Af')\, dr + Ez + L, \\ C + Df' &= -\frac{Er}{2} + \frac{F}{r}, \\ p &= \lambda_1 \sin^2\varphi + Af' + \int \frac{\lambda_1 \sin^2\varphi + Af'}{r}\, dr - Ez. \end{aligned}\right\} \quad (129.21)$$

In order for the flow to be free of singularities at the center of the pipe, it would be necessary that $F=0$. ERICKSEN imposed the condition that the stress power be positive and found that the resulting inequalities are incompatible with the requirement that $F=0$, except, of course, in the case when $\lambda_1=\lambda_2=\lambda_4=0$, when the theory reduces to that of a linearly viscous fluid.

As the next possibility, ERICKSEN considered plug flow, in which the fluid moves as a rigid body: $f=f_0=\text{const}$. From (127.36), $\dot{\boldsymbol{d}}=\boldsymbol{0}$; hence $\boldsymbol{d}=\boldsymbol{d}(r, \theta, z-f_0 t)$. ERICKSEN restricted his attention to steady, axially symmetric solutions: $\boldsymbol{d}=\boldsymbol{d}(r)$. The dynamical equations (16.6), combined with (6A.4) and the constitutive equations (127.34), yield $r^2 T_{\langle r\theta\rangle}=\lambda_1 r^2 d_{\langle r\rangle} d_{\langle\theta\rangle}=\text{const}$. To avoid the singularity at $r=0$, ERICKSEN set $d_{\langle r\rangle} d_{\langle\theta\rangle}=0$ and gave reasons for considering only the solutions with $d_{\langle\theta\rangle}=0$. Thus

$$d_{\langle r\rangle}=\sin\psi, \quad d_{\langle\theta\rangle}=0, \quad d_{\langle z\rangle}=\cos\psi, \quad (129.22)$$

where $\psi(r)$ is an arbitrary angle, and Eqs. (127.34) now yield

$$\left.\begin{aligned} T_{\langle rr\rangle} &= -p + \lambda_1 \sin^2\psi, \\ T_{\langle\theta\theta\rangle} &= -p, \\ T_{\langle zz\rangle} &= -p + \lambda_1 \cos^2\psi, \\ T_{\langle rz\rangle} &= \tfrac{1}{2}\lambda_1 \sin 2\psi, \\ T_{\langle r\theta\rangle} &= T_{\langle\theta z\rangle} = 0, \end{aligned}\right\} \quad (129.23)$$

so that we may again use (129.15) and obtain

$$\left.\begin{aligned} T_{\langle rr\rangle} &= -\frac{1}{2}\int \frac{1}{r}\left[1-\sqrt{1-(Hr/\lambda_1)^2}\right] dr + Hz + K, \\ p &= \lambda_1 \sin^2\psi + \int \frac{\lambda_1 \sin^2\psi}{r}\, dr - Hz, \\ 2T_{\langle rz\rangle} &= -Hr, \end{aligned}\right\} \quad (129.24)$$

where H and K are constants of integration, and where only solutions such that $T_{\langle rz\rangle}$ remains finite at $r=0$ are allowed. Combining $(129.23)_4$ and $(129.24)_3$ yields a differential equation for $\sin\psi$; the solution finite at $r=0$ is

$$\sin^2\psi = \frac{1}{2}\left[1-\sqrt{1-\left(\frac{Hr}{\lambda_1}\right)^2}\right], \quad (129.25)$$

where

$$\left.\begin{aligned} 0 \leqq \psi \leqq \tfrac{1}{2}\pi \quad &\text{if} \quad H/\lambda_1 < 0, \\ \tfrac{1}{2}\pi \leqq \psi \leqq \pi \quad &\text{if} \quad H/\lambda_1 > 0. \end{aligned}\right\} \quad (129.26)$$

The solutions in this one-parameter family remain real only so long as

$$0 \leqq r \leqq r_1 \equiv |\lambda_1/H|. \quad (129.27)$$

ERICKSEN sought next to fit a central plug flow smoothly within a steady stable flow satisfying $(129.21)_1$, matching the velocity vectors and stress vectors on the common boundary cylinder $r=r_2$. Comparing $(129.21)_1$ and $(129.24)_1$ shows that

$$E=H. \tag{129.28}$$

ERICKSEN found also that in order for $T\langle rz\rangle$ to be continuous, it is necessary that $F=0$, so that

$$T\langle rz\rangle=-\tfrac{1}{2}Hr \tag{129.29}$$

throughout. Then

$$f=\begin{cases} -\frac{1}{4D}(Hr^2+4Cr)+M & \text{if } r\geqq r_2, \\ -\frac{1}{4D}(Hr_2^2+4Cr_2)+M & \text{if } r\leqq r_2, \end{cases} \tag{129.30}$$

where M is an arbitrary constant. In all this, r_2 is any number satisfying the inequality $r_2\leqq r_1$, where r_1 is given by (129.27). ERICKSEN then gave arguments in favor of choosing constants in such a way as to make

$$\psi(r_2)=\begin{cases} \varphi & \text{if } \lambda>1, \\ \frac{1}{2}\pi-\varphi & \text{if } \lambda<1. \end{cases} \tag{129.31}$$

Then

$$r_2=-\frac{2C}{H}, \tag{129.32}$$

and C is the value of $T\langle rz\rangle$ when $r=r_2$.

With this apparatus, ERICKSEN was able to exhibit a solution, conjectured to be the only stable solution, for flow of an anisotropic fluid in an infinitely long pipe of radius R, to whose walls the fluid adheres ("Poiseuille flow"). The specific driving force, H, is regarded as given, and C and D, as determined by $(129.20)_{3,4}$ to within sign, are material constants. If H is large enough that

$$r_2=-\frac{2C}{H}<R, \tag{129.33}$$

then choice of M in (129.30) yields

$$f=\begin{cases} -\frac{1}{4D}(r-R)[H(r+R)+4C] & \text{if } r\geqq r_2, \\ -\frac{1}{4D}(r_2-R)[H(r_2+R)+4C] & \text{if } r\leqq r_2. \end{cases} \tag{129.34}$$

It can be shown that φ, C, and D are uniquely determined. The velocity profile (129.34) is the same as that for a so-called Bingham material. If H is small enough that (129.33) is violated, the appropriate solution is a state of rest. The stress systems corresponding to the state of rest and to the plug flow are easily calculated from the results already given.

VERMA[1] has considered in the same way the problem of flow between rotating cylinders ("Couette flow"). He found that for sufficiently small applied torque, a solution not very different from that according to the Navier-Stokes equations exists, but for larger torques an inner annulus rotates as a rigid body, while the fluid outside it is sheared between this rigid core and the outer wall.

γ) *Unsteady homogeneous flows.* Supposing a steady simple shearing flow set up according to the theory described at the end of Part (α) of this section, ERICKSEN[2] has considered the

[1] VERMA [1962, *68*]. Flow in a convergent or divergent channel has been studied by LESLIE [1964, *53*]. VERMA has considered the flow between co-axial cones [1963, *78*] and helical flow [1964, *92*].

[2] ERICKSEN [1963, *20*].

recoil of the fluid when the shear stress is suddenly removed. The velocity field is assumed to be of the form (106.1), and the director $\boldsymbol{d}$ is assumed to lie in the plane of flow. Setting $f=\partial v(x,t)/\partial x$, from (127.34) and (127.35) we obtain the following forms of the constitutive equations for this flow:

$$\left.\begin{aligned}\partial_t d\langle x\rangle &= \tfrac{1}{2}(\mu_3-1)f d\langle y\rangle+(\mu_1+\mu_2 f d\langle x\rangle d\langle y\rangle)\, d\langle x\rangle,\\ \partial_t d\langle y\rangle &= \tfrac{1}{2}(\mu_3+1)f d\langle x\rangle+(\mu_1+\mu_2 f d\langle x\rangle d\langle y\rangle)\, d\langle y\rangle,\end{aligned}\right\} \tag{129.35}$$

and

$$\left.\begin{aligned}T\langle x x\rangle &= -p+(\lambda_1+\lambda_2 f d\langle x\rangle d\langle y\rangle)\, d\langle x\rangle^2+\lambda_4 f d\langle x\rangle d\langle y\rangle,\\ T\langle y y\rangle &= -p+(\lambda_1+\lambda_2 f d\langle x\rangle d\langle y\rangle)\, d\langle y\rangle^2+\lambda_4 f d\langle x\rangle d\langle y\rangle,\\ T\langle z z\rangle &= -p,\\ T\langle x y\rangle &= (\lambda_1+\lambda_2 f d\langle x\rangle d\langle y\rangle)\, d\langle x\rangle d\langle y\rangle+\tfrac{1}{2}(\lambda_3+\lambda_4 d^2) f,\\ T\langle y z\rangle &= T\langle z x\rangle = 0.\end{aligned}\right\} \tag{129.36}$$

CAUCHY's first law of motion (16.6) yields

$$\left.\begin{aligned}0 &= \partial_x T\langle x x\rangle,\\ \varrho\,\partial_t v &= -\partial_y p+\partial_x T\langle x y\rangle,\\ 0 &= -\partial_z p.\end{aligned}\right\} \tag{129.37}$$

By $(129.36)_{1,2}$, then,

$$\left.\begin{aligned}p &= (\lambda_1+\lambda_2 f d\langle x\rangle d\langle y\rangle)\, d\langle x\rangle^2+\lambda_4 f d\langle x\rangle d\langle y\rangle+A(t)\, y+B(t),\\ \varrho\,\partial_t v+A(t) &= \partial_x T\langle x y\rangle,\end{aligned}\right\} \tag{129.38}$$

where $T\langle x y\rangle$ is given by $(129.36)_4$. The initial conditions are

$$v(x,0)=2\varkappa X, \qquad d\langle x\rangle = N\cos\varphi, \qquad d\langle y\rangle = N\sin\varphi, \tag{129.39}$$

where φ is given by (129.5) and N is determined from $\varkappa$ according to the solution mentioned at the end of Part (α) of this section. The boundary conditions are

$$\left.\begin{aligned}T\langle x y\rangle(l,t) &= 0\\ v(0,t) &= 0\end{aligned}\right\}\quad t>0, \tag{129.40}$$

where l is the distance between the shearing planes.

While he was unable to solve the problem specified by these conditions, ERICKSEN calculated approximately certain features of the solution. The most important of these are:

$$\lim_{t\to\infty} f\, e^{-2\bar{\mu}_1 t} \text{ exists}, \tag{129.41}$$

where $\bar{\mu}_1$ is the value of μ_1, assumed negative, when $\boldsymbol{d}=\boldsymbol{0}$, and

$$\frac{1}{l}\int_0^\infty v(l,t)\,dt = -\frac{1}{2}\,\frac{T\langle x x\rangle - T\langle y y\rangle}{T\langle x y\rangle}, \tag{129.42}$$

where the ratio on the right is calculated from the original flow. The former result gives the ultimate rate of relaxation of the velocity profile, and the latter the mean total recoil.

130. Theories of diffusion. A general framework for a theory of mixtures of several interdiffusing materials was constructed by TRUESDELL[1] and has been presented in CFT, Sects. 158—159, 215, 243. Each place $\boldsymbol{x}$ is occupied simultaneously by $\mathfrak{k}$ particles $X_\mathfrak{a}$, one from each of the substances in motion:

$$\boldsymbol{x} = \boldsymbol{\chi}_\mathfrak{a}(X_\mathfrak{a}, t). \tag{130.1}$$

[1] TRUESDELL [1957, *30*]. KELLY [1964, *49*] has generalized the work in several regards: He gives integral equations of balance, jump conditions at singular surfaces, a moment of momentum equation including spin momentum and couple stresses, and corresponding equations for electromagnetism.

Each constituent has its own velocity and acceleration:

$$\dot{\boldsymbol{x}}_{\mathfrak{a}} \equiv \frac{\partial \chi_{\mathfrak{a}}}{\partial t}, \qquad \ddot{\boldsymbol{x}}_{\mathfrak{a}} \equiv \frac{\partial^2 \chi_{\mathfrak{a}}}{\partial t^2}, \tag{130.2}$$

and in general a backward prime with subscript $\mathfrak{a}$ denotes the material derivative following the motion of the constituent $\mathfrak{a}$. The density of the constituent $\mathfrak{a}$ is $\varrho_{\mathfrak{a}}$. The total density ϱ, the absolute concentration $c_{\mathfrak{a}}$, the mean velocity $\dot{\boldsymbol{x}}$, and the diffusion velocity $\boldsymbol{u}_{\mathfrak{a}}$ are defined as follows:

$$\varrho \equiv \sum_{\mathfrak{a}=1}^{\mathfrak{f}} \varrho_{\mathfrak{a}}, \qquad c_{\mathfrak{a}} \equiv \frac{\varrho_{\mathfrak{a}}}{\varrho}, \qquad \dot{\boldsymbol{x}} \equiv \sum_{\mathfrak{a}=1}^{\mathfrak{f}} c_{\mathfrak{a}} \dot{\boldsymbol{x}}_{\mathfrak{a}}, \qquad \boldsymbol{u}_{\mathfrak{a}} \equiv \dot{\boldsymbol{x}}_{\mathfrak{a}} - \dot{\boldsymbol{x}}. \tag{130.3}$$

The *supplies*[1] of mass, momentum, and energy for the constituent $\mathfrak{a}$, per unit mass of the mixture, are defined as follows:

$$\left.\begin{aligned} \varrho \hat{c}_{\mathfrak{a}} &\equiv \frac{\partial \varrho_{\mathfrak{a}}}{\partial t} + \operatorname{div}(\varrho_{\mathfrak{a}} \dot{\boldsymbol{x}}_{\mathfrak{a}}), \\ \varrho \hat{\boldsymbol{p}}_{\mathfrak{a}} &\equiv \varrho_{\mathfrak{a}} (\ddot{\boldsymbol{x}}_{\mathfrak{a}} - \boldsymbol{b}_{\mathfrak{a}}) - \operatorname{div} \boldsymbol{T}_{\mathfrak{a}}, \\ \varrho \hat{\varepsilon}_{\mathfrak{a}} &\equiv \varrho_{\mathfrak{a}} (\dot{\varepsilon}_{\mathfrak{a}} - q_{\mathfrak{a}}) - \operatorname{tr}(\boldsymbol{T}_{\mathfrak{a}} \operatorname{grad} \dot{\boldsymbol{x}}_{\mathfrak{a}}) - \operatorname{div} \boldsymbol{h}_{\mathfrak{a}}, \end{aligned}\right\} \tag{130.4}$$

where $\boldsymbol{T}_{\mathfrak{a}}$, $\varepsilon_{\mathfrak{a}}$, and $\boldsymbol{h}_{\mathfrak{a}}$ are the partial stress, partial specific internal energy, and partial heat flux, respectively, and where $\boldsymbol{b}_{\mathfrak{a}}$ is the body force density and $q_{\mathfrak{a}}$ the the heat absorption density for the constituent $\mathfrak{a}$. TRUESDELL showed that if the total stress and total heat flux for the mixture are suitably defined, the classical equations (15.14), (16.6), and (79.3) hold for the mixture if and only if

$$\left.\begin{aligned} &\sum_{\mathfrak{a}=1}^{\mathfrak{f}} \hat{c}_{\mathfrak{a}} = 0, \qquad \sum_{\mathfrak{a}=1}^{\mathfrak{f}} (\hat{\boldsymbol{p}}_{\mathfrak{a}} + \hat{c}_{\mathfrak{a}} \boldsymbol{u}_{\mathfrak{a}}) = \boldsymbol{0}, \\ &\sum_{\mathfrak{a}=1}^{\mathfrak{f}} [\hat{\varepsilon}_{\mathfrak{a}} + \hat{\boldsymbol{p}}_{\mathfrak{a}} \boldsymbol{u}_{\mathfrak{a}} + \hat{c}_{\mathfrak{a}} (\varepsilon_{\mathfrak{a}} + \tfrac{1}{2} u_{\mathfrak{a}}^2)] = 0. \end{aligned}\right\} \tag{130.5}$$

These equations assert the balance of total mass, total momentum, and total energy, respectively. TRUESDELL remarked that there are two kinds of constitutive equations for a mixture: those defining the individual constituents, and those defining the manner in which they exchange mass, momentum and energy among one another. The former, like those for simple materials, specify $\boldsymbol{T}_{\mathfrak{a}}$, $\boldsymbol{h}_{\mathfrak{a}}$, and $\varepsilon_{\mathfrak{a}}$ in terms of kinematic data; the latter specify the supplies $\hat{c}_{\mathfrak{a}}$, $\hat{\boldsymbol{p}}_{\mathfrak{a}}$, and $\hat{\varepsilon}_{\mathfrak{a}}$.

While the further study of TRUESDELL[2] was directed only toward the case when the supplies of momentum are linear functions of the diffusion velocities and did not consider at all the possible effects of diffusion on the constitutive equations for the constituents themselves, some aspects of the general problem have been approached, if tentatively, in a series of papers by ADKINS and GREEN. They consider only the case when no chemical reactions occur, so that $\hat{c}_{\mathfrak{a}} = 0$ and hence Eq. $(130.5)_1$ is satisfied identically, while Eqs. $(130.5)_{2,3}$ are simplified;

[1] As can be shown from Eqs. (130.4), the rates of increase, per unit mass of the mixture, of mass, momentum, and energy of the $\mathfrak{a}$'th constituent due to interactions with the others are $\hat{c}_{\mathfrak{a}}$, $\hat{\boldsymbol{p}}_{\mathfrak{a}} + \hat{c}_{\mathfrak{a}} \dot{\boldsymbol{x}}_{\mathfrak{a}}$ and $\hat{\varepsilon}_{\mathfrak{a}} + \hat{\boldsymbol{p}}_{\mathfrak{a}} \cdot \dot{\boldsymbol{x}}_{\mathfrak{a}} + \hat{c}_{\mathfrak{a}} (\varepsilon_{\mathfrak{a}} + \frac{1}{2} \dot{x}_{\mathfrak{a}}^2)$, respectively.

[2] TRUESDELL [1961, *64*] [1962, *66*]. Cf. also a more general suggestion in CFT, Sect. 295.

they do not mention the equation of energy[1]. For the most part, they consider only a binary mixture, so that $(130.5)_2$ reduces to

$$\hat{\boldsymbol{p}}_2 = -\hat{\boldsymbol{p}}_1. \tag{130.6}$$

Thus it suffices to set up only three constitutive equations: one for $\boldsymbol{T}_1$, one for $\boldsymbol{T}_2$, and one for $\hat{\boldsymbol{p}}_1$.

In his first work, ADKINS[2] considered a mixture of fluids and adopted the constitutive equation for $\hat{\boldsymbol{p}}_\mathfrak{a}$ proposed in CFT, Eq. $(295.11)_2$. Namely, the diffusive forces are regarded as arising solely from the relative motion of the constituents:

$$\hat{\boldsymbol{p}}_\mathfrak{a} = \boldsymbol{f}_\mathfrak{a}(\varrho_\mathfrak{b}, \dot{\boldsymbol{x}}_\mathfrak{b} - \dot{\boldsymbol{x}}_\mathfrak{a}), \qquad \mathfrak{a}, \mathfrak{b} = 1, 2, \ldots, \mathfrak{k}. \tag{130.7}$$

It is natural to extend the *principle of material indifference* (Sect. 19) so as to require that the supplies of momentum $\hat{\boldsymbol{p}}_\mathfrak{a}$ be frame-indifferent. By (17.8) a difference of two velocities is always indifferent. Hence the functions $\boldsymbol{f}_\mathfrak{a}$ in (130.7) must be isotropic:

$$\boldsymbol{f}_\mathfrak{a}(\varrho_\mathfrak{b}, \boldsymbol{Q}\boldsymbol{v}_\mathfrak{b}) = \boldsymbol{Q}\boldsymbol{f}_\mathfrak{a}(\varrho_\mathfrak{b}, \boldsymbol{v}_\mathfrak{b}), \tag{130.8}$$

for arbitrary orthogonal tensors $\boldsymbol{Q}$ and arbitrary values of the arguments. The representation theorem (13.8), applied to $\mathfrak{v} \equiv \boldsymbol{f}_\mathfrak{a}$, yields

$$\varrho\hat{\boldsymbol{p}}_\mathfrak{a} = \sum_{\mathfrak{b}=1}^{\mathfrak{k}} \mathcal{F}_{\mathfrak{a}\mathfrak{b}}(\boldsymbol{u}_\mathfrak{a} - \boldsymbol{u}_\mathfrak{b}), \tag{130.9}$$

where $\mathcal{F}_{\mathfrak{a}\mathfrak{b}}$ is a scalar function of the $\varrho_\mathfrak{c}$ and the scalar products $(\boldsymbol{u}_\mathfrak{c} - \boldsymbol{u}_\mathfrak{b}) \cdot (\boldsymbol{u}_\mathfrak{d} - \boldsymbol{u}_\mathfrak{b})$, and where, by $(130.5)_2$,

$$\sum_{\mathfrak{a}=1}^{\mathfrak{k}} (\mathcal{F}_{\mathfrak{a}\mathfrak{b}} - \mathcal{F}_{\mathfrak{b}\mathfrak{a}}) = 0. \tag{130.10}$$

ADKINS has remarked[3] that if one constituent of the mixture, say the $\mathfrak{k}+1$'st, is a rigid solid, then, since any quantity of momentum may be absorbed or given out by it in interaction with the other constituents, we may in effect drop the constraints (130.5) and their consequence, (130.10). As a result, special cases become more tractable. ADKINS has determined representation formulae for the supplies of momentum when (130.7) is generalized to include dependence on three vectors specifying preferred directions in the solid.

Among the examples considered by ADKINS[4] is that of two Reiner-Rivlin fluids (Sect. 119), one in steady rectilinear and the other in a steady lineal flow possibly having a component transverse to the first:

$$\dot{\boldsymbol{x}}_1 = U(x)\boldsymbol{j}, \qquad \dot{\boldsymbol{x}}_2 = v(x)\boldsymbol{i} + u(x)\boldsymbol{j}, \tag{130.11}$$

---

[1] They preferred as definitions of quantities associated with the mean motion what TRUESDELL called "inner" rather than "total" quantities. E.g., they regarded as the stress of the mixture

$$\boldsymbol{T}_\mathrm{I} \equiv \sum_{\mathfrak{a}=1}^{\mathfrak{k}} \boldsymbol{T}_\mathfrak{a},$$

while TRUESDELL set

$$\boldsymbol{T} \equiv \sum_{\mathfrak{a}=1}^{\mathfrak{k}} (\boldsymbol{T}_\mathfrak{a} - \varrho_\mathfrak{a}\boldsymbol{u}_\mathfrak{a} \otimes \boldsymbol{u}_\mathfrak{a}),$$

as is always done in the kinetic theory of gas mixtures. Unless boundary conditions on a boundary across which diffusion occurs are specified in terms of the total stress, the distinction remains a matter of preference. In particular, if a problem is set up in terms of the constituents alone, without mention of properties of the mixture, the difference is one of interpretation only. The field equations satisfied by the inner quantities are more complicated in form than the classical ones.

[2] ADKINS [1963, *1*, § 4].

[3] ADKINS [1963, *1*, § 5].

[4] ADKINS [1963, *1*, § 9]. In § 10 of this paper, flow between rotating cylinders is considered in much the same way.

where $\boldsymbol{i}$ and $\boldsymbol{j}$ are unit vectors in the directions of x and y, respectively. By (130.9),

$$\hat{\boldsymbol{p}}_1 = -\hat{\boldsymbol{p}}_2 = F[-v(x)\,\boldsymbol{i} - (u(x) - U(x))\,\boldsymbol{j}], \tag{130.12}$$

where F is a function of ϱ_1, ϱ_2, and $v^2 + (u-U)^2$. In flow with steady density, Eq. $(130.4)_1$ requires that

$$\varrho_2 v(x) = \text{const.} \tag{130.13}$$

If we suppose that each constituent obeys the constitutive equation (119.4) for a Reiner-Rivlin fluid, or, more generally, Eq. (119.12) for a Rivlin-Ericksen fluid, in a flow given by (130.11) we obtain explicit formulae for $\boldsymbol{T}_1 + p_1\boldsymbol{1}$ and $\boldsymbol{T}_2 + p_2\boldsymbol{1}$ in terms of the three functions $U(x)$, $v(x)$, and $u(x)$. On the assumption that the pressures p_1 and p_2 as well as the densities ϱ_1 and ϱ_2 are functions of x alone, substitution of (130.12) and the formulae for $\boldsymbol{T}_1$ and $\boldsymbol{T}_2$ into $(130.4)_2$ yields a fairly simple system of four ordinary differential equations. When (130.13) is subjoined, the five unknowns U, u, v, ϱ_1, and ϱ_2 are to be found by integration of this system. ADKINS has considered various relations among p_1, ϱ_1, p_2, and ϱ_2 under various specializing assumptions, to which the results seem to be rather sensitive.

In his second essay, ADKINS[1] allowed not only the diffusive forces but also the partial stresses to depend upon the motions of all constituents. Specifically, he proposed the linked constitutive equations

$$\left.\begin{aligned} \boldsymbol{T}_{\mathfrak{a}} &= \mathfrak{g}_{\mathfrak{a}}\left(\varrho_{\mathfrak{b}}, \operatorname{grad}\dot{\boldsymbol{x}}_{\mathfrak{b}}, \operatorname{grad}(\dot{\boldsymbol{x}}_{\mathfrak{b}})_{\mathfrak{c}}^{\cdot}\right), \\ \hat{\boldsymbol{p}}_{\mathfrak{a}} &= \boldsymbol{f}_{\mathfrak{a}}\left(\varrho_{\mathfrak{b}}, \dot{\boldsymbol{x}}_{\mathfrak{b}}, (\dot{\boldsymbol{x}}_{\mathfrak{b}})_{\mathfrak{c}}^{\cdot}, \operatorname{grad}\dot{\boldsymbol{x}}_{\mathfrak{b}}\right), \end{aligned}\right\} \tag{130.14}$$

where $\mathfrak{b}, \mathfrak{c} = 1, 2, \ldots, \mathfrak{k}$. ADKINS found forms to which these equations must reduce in order to satisfy the principle of material indifference.

In a third study, ADKINS[2] considered a Reiner-Rivlin fluid diffusing through an elastic material. He supposed the constitutive equations of the materials to be unaffected by diffusion, and he proposed the following constitutive equation for the supplies of momentum:

$$\hat{\boldsymbol{p}}_1 = -\hat{\boldsymbol{p}}_2 = \boldsymbol{f}(\varrho_1, \varrho_2, \boldsymbol{F}_1, \operatorname{grad}\dot{\boldsymbol{x}}_1, \operatorname{grad}\dot{\boldsymbol{x}}_2, \boldsymbol{u}_1 - \boldsymbol{u}_2), \tag{130.15}$$

where *1* refers to the elastic material and *2* to the fluid.

Suppose first[3] that an isotropic elastic solid is penetrated by an elastic fluid, both in equilibrium. The supplies of momentum are zero, and the constitutive equations for the two materials have the forms (47.9) and (50.3), where p_2, the pressure in the fluid, must be constant in order to satisfy $(130.4)_2$ in this case. If the elastic solid is assumed to be in a state of homogeneous strain, Eqs. (48.2) give its principal stresses. The total tensions on the principal planes of stress and strain for the solid are then

$$\tilde{t}_a = \bar{t}_a(v_1, v_2, v_3) - p_2(\varrho_2). \tag{130.16}$$

Since the present density ϱ_1 of the solid and its density ϱ_R in the reference configuration satisfy the usual relation

$$v_1 v_2 v_3 \varrho_1 = \varrho_R, \tag{130.17}$$

in (130.16) and (130.17) we have a system of four equations relating the four quantities v_a and ϱ_1 when ϱ_2 and the $\tilde{t}_a$ are prescribed. This problem does not illustrate the theory of diffusion but falls rather within the general theory of finite elastic deformation.

As an example of a truly diffusive problem, ADKINS considered an isotropic elastic slab subject to stretches $v_1 = v_1(X)$, $v_2 = v_3 = \text{const}$, across which a fluid is diffusing with velocity $\dot{\boldsymbol{x}}_2 = u(x)\,\boldsymbol{i}$. By $(130.4)_1$, $\varrho_2 u = k = \text{const}$. Since $x = \int v_1(X)\,dX$, we may take either x or X

[1] ADKINS [1963, *2*].

[2] ADKINS [1964, *2*].

[3] These examples and a further one, in which a fluid diffuses steadily through a slab of isotropic elastic material in simple shear, are given by ADKINS [1964, *2*, §§ 7—9].

as the single variable upon which the problem depends. By (48.2), the principal stresses in the elastic material are given explicitly in terms of v_1, and v_2, while by (119.4), or, more generally, by (119.12), the principal stresses in the fluid are given explicitly in terms of u'. The supply of momentum (130.15) reduces in this case to an isotropic function of $\boldsymbol{B}_1$ and $\boldsymbol{u}_2$ alone. By (13.8) then,

$$\left.\begin{aligned}\hat{\boldsymbol{p}}_1 &= (\varphi_0 \mathbf{1} + \varphi_1 \boldsymbol{B}_1 + \varphi_2 \boldsymbol{B}_1^2)\,\dot{\boldsymbol{x}}_2\\ &= (\varphi_0 + \varphi_1 v_1^2 + \varphi_2 v_1^4)\,\dot{\boldsymbol{x}}_2,\end{aligned}\right\} \tag{130.18}$$

where φ_Γ is a scalar function of the joint invariants of $\boldsymbol{B}_1$ and $\dot{\boldsymbol{x}}_2$ and is an assigned function of v_1, v_2, and u'; of course, φ_Γ depends also on ϱ_2 in general. The equations of motion (130.4) become now

$$\left.\begin{aligned}&\frac{d\,T_1\langle 11\rangle}{dx} + \varrho\,\hat{p}_1\langle 1\rangle = 0,\\ &\frac{d\,T_2\langle 11\rangle}{dx} - \varrho\,\hat{p}_1\langle 1\rangle = \varrho_2 u \frac{du}{dx}.\end{aligned}\right\} \tag{130.19}$$

Adding and integrating yields

$$T_1\langle 11\rangle + T_2\langle 11\rangle = k u + \text{const.} \tag{130.20}$$

Thus (130.19) may be replaced by the following system of two differential equations for the functions v_1 and ϱ_2:

$$\left.\begin{aligned}&\frac{d\,T_1\langle 11\rangle}{dX} + \frac{\varrho}{\varrho_2}(\varphi_0 + \varphi_1 v_1^2 + \varphi_2 v_1^4)\,k v_1 = 0,\\ &T_1\langle 11\rangle + T_2\langle 11\rangle = \frac{k^2}{\varrho_2} + \text{const.}\end{aligned}\right\} \tag{130.21}$$

When these are solved, u and ϱ_1 are to be obtained from the equations $\varrho_2 u = k$ and $\varrho_1 v_1 v_2^2 = \varrho_R$.

Next ADKINS[1] proposed far more general constitutive equations. Considering a mixture of *1* solid and $\mathfrak{n}-1$ fluids, so as to satisfy the principle of equipresence (Sects. 96, 125) he allowed all the $\boldsymbol{T}_\mathfrak{a}$ and all the $\hat{\boldsymbol{p}}_\mathfrak{a}$ to depend upon the gradients of velocity of every constituent, on the deformation gradient of the solid, and on the diffusion velocities of the fluids relative to the solid. The distinct material properties of the various constituents were accounted for by specifying the forms to which the equations for the several $\boldsymbol{T}_\mathfrak{a}$ must reduce in the case when no diffusion occurs. Again, reduced forms expressing the principle of material indifference and various material symmetries are found.

GREEN and ADKINS[2] in a later work retreated from this position of generality and required a greater differentiation between the constituents. For the case of a mixture of two Reiner-Rivlin fluids, they gave some reasons for supposing that the stress in fluid *1* is unaffected by the stretching of fluid *2*, though it may depend on the spin of that fluid as well as its own. While the spin is not frame-indifferent (CFT, Sect. 144), the difference of two spins is, and GREEN and ADKINS inferred that the partial stresses are in fact functions of this difference.

The foregoing summary makes it clear that the rational theory of diffusion is in its infancy, with many possibilities awaiting search. The uncertainty as to even the overall pattern for the right constitutive equations, reflected by the diversities among the five papers on the subject issued by the same author within a few months, suggest that some basic principle of the subject remains to be discovered.

[Added in proof. The "missing principle", surely, is a proper generalization of the Clausius-Duhem inequality. The problem is currently under study by GREEN and NAGHDI, BOWEN[3], and COLEMAN.]

[1] ADKINS [1964, *3*].

[2] GREEN and ADKINS [1964, *37*]. Their objection to the principle of equipresence as rendering it impossible to distinguish one constituent from another is unsound, since the mere fact that two response functions depend on the same variables does not mean that they have the same isotropy groups, and it is the isotropy groups that specify material symmetries.

[3] BOWEN [1965, *2*].

General references.

Of the books and reviews which lay any claim to covering a large part of the subject treated here, none can be recommended as trustworthy. Reference to works developing further matters of detail has been made at appropriate points in the treatise.

List of Works Cited.

This list, not attempting to be a complete bibliography of the subject, is limited to works consulted by the authors in writing the treatise. Italic numbers in parentheses following the reference indicate the sections in which it has been cited.

Addenda of 1965 and 1992 are listed on pp. 578—580.

1678 [*1*] HOOKE, R.: Lectures de Potentia Restitutiva, or of Spring Explaining the Power of Springing Bodies. London = R. T. GUNTHER: Early Science in Oxford **8**, 331—356 (1931). (*19A*)

1769 [*1*] EULER, L.: De aequilibrio et motu corporum flexuris elasticis iunctorum. Novi comm. acad. sci. Petrop. **13** (1768), 259—304 = Opera omnia II **11**, 3—16. (*66*)

1780 [*1*] COULOMB, C. A.: Recherches sur la meilleure manière de fabriquer les aiguilles aimantées ... Mém. math. phys. divers savans **9**, 165—264. (*66*)

1787 [*1*] COULOMB, C. A.: Recherches théoriques et expérimentales sur la force de torsion, et sur l'élasticité de fils de métal: Application de cette théorie à l'emploi des métaux dans les arts et dans différentes expériences de physique: Construction de différentes balances de torsion, pour mesurer les plus petits degrés de force. Observations sur les loix de l'élasticité et de la cohérence. Mém. acad. sci. Paris **1784**, 229—272. (*66*)

1807 [*1*] YOUNG, T.: Mathematical Elements of Natural Philosophy. A Course of Lectures on Natural Philosophy and the Mechanical Arts **2**, 1—86. London. (*66*)

1823 [*1*] CAUCHY, A.-L.: Recherches sur l'équilibre et le mouvement intérieur des corps solides ou fluides, élastiques ou non élastiques. Bull. Soc. Philomath. Paris 9—13 = Oeuvres (2) **2**, 300—304. (*43*)

1828 [*1*] CAUCHY, A.-L.: Sur les équations qui expriment les conditions d'équilibre ou les lois du mouvement intérieur d'un corps solide, élastique ou non élastique. Ex. de Math. **3**, 160—187 = Oeuvres (2) **8**, 195—226. (*26, 43*)

1829 [*1*] CAUCHY, A.-L.: Sur l'équilibre et le mouvement intérieur des corps considérés comme des masses continues. Ex. de Math. **4**, 293—319 = Oeuvres 9, 342—369. (*19A, 68, 99*)

1831 [*1*] POISSON, S.-D.: Mémoire sur les équations générales de l'équilibre et du mouvement des corps élastiques et des fluides (1829). J. École Poly. **13**, Cahier 20, 1—174. (*19A*)

1836 [*1*] PIOLA, G.: Nuova analisi per tutte le questioni della meccanica molecolare. Mem. Mat. Fis. Soc. Ital. Modena **21** (1835), 155—231. (*94*)

1839 [*1*] GREEN, G.: On the laws of reflection and refraction of light at the common surface of two non-crystallised media (1837). Trans. Cambridge Phil. Soc. **7** (1838—1842) 1—24 = Papers, 245—269. (*82A, 88*)

1841 [*1*] GREEN, G.: On the propagation of light in crystallised media (1839). Trans. Cambridge Phil. Soc. **7** (1838—1842), 121—140 = Papers, 293—311. (*82A, 84*)

1843 [*1*] ST. VENANT, A.-J.-C. B. DE: Note à joindre au mémoire sur la dynamique des fluides, présenté le 14 avril 1834. C. R. Acad. Sci. Paris **17**, 1240—1243. (*54, 119*)

1844 [*1*] ST. VENANT, A.-J.-C. B. DE: Sur les pressions qui se développent à l'intérieur des corps solides lorsque les déplacements de leurs points, sans altérer l'élasticité, ne peuvent cependant pas être considérés comme très petits. Bull. Soc. Philomath. **5**, 26—28. (*94*)

1845 [*1*] STOKES, G. G.: On the theories of the internal friction of fluids in motion, and of the equilibrium and motion of elastic solids. Trans. Cambridge Phil. Soc. **8** (1844—1849), 287—319 = Papers **1**, 75—129. (*26, 119*)

1847 [*1*] ST. VENANT, A.-J.-C. B. DE: Mémoire sur l'équilibre des corps solides, dans les limites de leur élasticité, et sur les conditions de leur résistance, quand les déplacements éprouvés par leurs points ne sont pas très-petits. C. R. Acad. Sci. Paris **24**, 260—263. (*94*)

1849 [*1*] HAUGHTON, S.: On the equilibrium and motion of solid and fluid bodies (1846). Trans. Roy. Irish Acad. **21**, 151—198. (*88*)

1850 [*1*] CAUCHY, A.-L.: Mémoire sur les systèmes isotropes de points matériels. Mém. Acad. Sci. Paris **22**, 615—654 = Oeuvres (1) **2**, 351—386. (*11*)

[*2*] KIRCHHOFF, G.: Über das Gleichgewicht und die Bewegung einer elastischen Scheibe. J. reine angew. Math. **40**, 51—88 = Ges. Abh. 237—279. (*82A, 88*)

1851 [*1*] CAUCHY, A.-L.: Note sur l'équilibre et les mouvements vibratoires des corps solides. C. R. Acad. Sci. Paris **32**, 323—326 = Oeuvres (1) **11**, 341—346. (*28, 98*)

[*2*] DA SILVA, D. A.: Memoria sobre a rotação das forças em torno dos pontos d'applicação. Mem. Ac. Sc. Lisboa (2a) $\mathbf{3}_1$, 61—231. (*44*)

1852 [*1*] KIRCHHOFF, G.: Über die Gleichungen des Gleichgewichts eines elastischen Körpers bei nicht unendlich kleinen Verschiebungen seiner Theile. Sitzgsber. Akad. Wiss. Wien **9**, 762—773. (Not repr. in Abh.) (*82, 82A, 94*)

1853 [*1*] MAXWELL, J. C.: On the equilibrium of elastic solids (1850). Trans. Roy. Soc. Edinb. **20** (1848/1853), 87—120 = Papers **1**, 30—73. (*66*)

1855 [*1*] ST. VENANT, A.-J.-C. B. DE: Mémoire sur la torsion des prismes ... (1853). Mém. Divers Savants Acad. Sci. Paris **14**, 233—560. (*66*)

[*2*] THOMSON, W. (Lord KELVIN): On the thermo-elastic and thermo-magnetic properties of matter. Quart. J. Math. **1** (1855—1857), 55—77 = (with notes and additions) Phil. Mag. (5) **5** (1878), 4—27 = Pt. VII of On the dynamical theory of heat. Papers **1**, 291—316. (*82A, 83*)

1856 [*1*] THOMSON, W. (Lord KELVIN): Elements of a mathematical theory of elasticity. Phil. Trans. Roy. Soc. Lond. **146**, 481—498. (*82A*)

1860 [*1*] NEUMANN, C.: Zur Theorie der Elasticität. J. reine angew. Math. **57**, 281—318. (*82, 82A*)

1863 [*1*] ST. VENANT, A.-J.-C. B. DE: Mémoire sur la distribution des élasticités autour de chaque point d'un solide ou d'un milieu de contexture quelconque, particulièrement lorsqu'il est amorphe sans être isotrope. J. Math. Pures Appl. (2) **8**, 257—295, 353—430. (*84, 94*)

[*2*] THOMSON, W. (Lord KELVIN): Dynamical problems regarding elastic spheroidal shells and spheroids of incompressible liquid. Phil. Trans. Roy. Soc. Lond. A **153**, 583—616 = Papers **3**, 351—394. (*82A, 83, 84, 85*)

1866 [*1*] KLEITZ, C.: Sur les forces moléculaires dans les liquides en mouvement avec application à l'hydrodynamique. C. R. Acad. Sci. Paris **63**, 988—991. (Partial abstract of [1873, *1*]). (*119*)

1867 [*1*] MAXWELL, J. C.: On the dynamical of theory gases (1866). Phil. Trans. Roy. Soc. Lond. A **157**, 49—88 = Phil. Mag. (4) **35** (1868), 129—145, 185—217 = Papers **2**, 26—78. (*35*)

[*2*] THOMSON, W. (Lord KELVIN), and P. G. TAIT: Treatise on Natural Philosophy, Part I. Cambridge. [The second edition, which appeared in 1879, has been reprinted by Dover Publications, N.Y., under the redundant as well as gratuitous title, "Principles of Mechanics and Dynamics".] (*54, 82A*)

1868 [*1*] BOUSSINESQ, J.: Sur l'influence des frottements dans les mouvements reguliers des fluides. J. Math. Pures Appl. (2) **13**, 377—438. (*119, 120, 125*)

[*2*] ST. VENANT, A.-J.-C. B. DE: Formules de l'élasticité des corps amorphes que des compressions permanentes et inégales ont rendus hétérotropes. J. Math. Pures Appl. (2) **13**, 242—254. (*68*)

1869 [*1*] ST. VENANT, A.-J.-C. B. DE: Note sur les valeurs que prennent les pressions dans un solide élastique isotrope lorsque l'on tient compte des dérivées d'ordre supérieur des déplacements très-petits que leurs points ont éprouvés. C. R. Acad. Sci. Paris **68**, 569—571. (*28, 98*)

[*2*] ST. VENANT, A.-J.-C. B. DE: Rapport sur un mémoire de M. MAURICE LEVY, relatif à l'hydrodynamique des liquides homogènes, particulièrement à leur écoulement rectiligne et permanent. C. R. Acad. Sci. Paris **68**, 582—592. (*28, 119*)

1870 [*1*] BOUSSINESQ, J.: Note complémentaire au mémoire sur les ondes liquides periodiques, presenté le 29 novembre 1869, et approuvé par l'Académie le 21 fevrier 1870. — Établissement de relations générales et nouvelles entre l'énergie interne d'un corps fluide ou solide, et ses pressions ou forces élastiques. C. R. Acad. Sci. Paris **71**, 400—402. (*84, 94*)

1871 [*1*] ST. VENANT, A.-J.-C. B. DE: Formules des augmentations que de petites déformations d'un solide apportent aux pressions ou forces élastiques, supposées considérables, qui déjà étaient en jeu dans son intérieur. J. Math. Pures Appl. (2) **16**, 275—307. (*94*)

1872 [*1*] BOUSSINESQ, J.: Théorie des ondes liquides periodiques (1869). Mém. Divers Savants **20**, 509—615. Abstracts in C. R. Acad. Sci. Paris **68**, 905—906 (1869); **70**, 360—367 (1870), supplemented by [1870, *1*]. (*82, 84*)

[*2*] ST. VENANT, A.-J.-C. B. DE: Rapport sur un mémoire de M. KLEITZ intitulé: Études sur les forces moléculaires dans les liquides en mouvement, et application à l'hydrodynamique. C. R. Acad. Sci. Paris **74**, 426—438. (*119*)

[*3*] ST. VENANT, A.-J.-C. B. DE: Sur l'hydrodynamique des cours d'eau. C. R. Acad. Sci. Paris **74**, 570—577, 649—657, 693—701, 770—774. (*119*)

1873 [*1*] KLEITZ, C.: Études sur les forces moléculaires dans les liquides en mouvement et application à l'hydrodynamique. Paris: Dunod. (*119*)

1874 [*1*] BOLTZMANN, L.: Zur Theorie der elastischen Nachwirkung. Sitzgsber. Akad. Wiss. Wien **70**², 275—306 = Ann. Physik, Ergänz. **7**, 624—654 (1876) = Wiss. Abh. **1**, 616—639. (*35, 41*)

[*2*] MEYER, O.-E.: Zur Theorie der inneren Reibung. J. reine angew. Math. **78**, 130—135. (*35, 41, 119*)

[*3*] MEYER, O.-E.: Theorie der elastischen Nachwirkung. Ann. Physik **151** = (6) **1**, 108—119. (*35, 41, 120*)

1875 [*1*] GIBBS, J. W.: On the equilibrium of heterogeneous substances. Trans. Conn. Acad. **3** (1875—1878), 108—248, 343—524 = Works **1**, 55—353. (*82, 82A, 89*)

[*2*] MEYER, O. E.: Zusatz zu der Abhandlung zur Theorie der inneren Reibung. J. reine angew. Math. **80**, 315—316. (*35, 41*)

1876 [*1*] BUTCHER, J. G.: On viscous fluids in motion. Proc. Lond. Math. Soc. **8** (1876/77), 103—135. (*35*)

[*2*] MAXWELL, J. C.: On stresses in rarified gases arising from inequalities of temperature. Phil. Trans. Roy Soc. Lond. **170**, 231—256 = Papers **2**, 680—712. (*124*)

1883 [*1*] MACH, E.: Die Mechanik in ihrer Entwicklung, historisch-kritisch dargestellt. Leipzig: Brockhaus. [There are many later editions and translations.] (*18*)

1885 [*1*] REYNOLDS, O.: On the dilatancy of media composed of rigid particles in contact. With experimental illustrations. Phil. Mag. (2) **20**, 469—481 = Papers **2**, 203—216. (*119*)

1886 [*1*] TODHUNTER, I., and K. PEARSON: A History of the Theory of Elasticity and of the Strength of Materials from Galilei to Lord Kelvin, **1** [This is mainly the work of PEARSON, as indicated in the enumeration of sections.] Cambridge: Cambridge Univ. Press. Reprinted, Dover Publications, New York, 1960. (*50*)

1887 [*1*] REYNOLDS, O.: Experiments showing dilatancy, a property of granular material, possibly connected with gravitation. Proc. Roy. Inst. Gt. Britain **11**, 354—363 = Papers **2**, 217—227. (*119*)

[*2*] VOIGT, W.: Theoretische Studien über die Elasticitätsverhältnisse der Krystalle. Abh. Ges. Wiss. Göttingen **34**, 100 pp. (*98*)

1888 [*1*] BARUS, C.: MAXWELL's theory of the viscosity of solids and its physical verification. Phil. Mag. (5) **26**, 183—217. (*2*)

[*2*] BASSET, A. B.: A Treatise on Hydrodynamics. 2 vols. Cambridge: Cambridge Univ. Press. (*120*)

[*3*] THOMSON, W. (Lord KELVIN): On the reflection and refraction of light. Phil. Mag. (5) **26**, 414—425. Reprinted in part as §§ 107—111 of Baltimore Lectures on Molecular Dynamics and the Wave Theory of Light. London: Clay & Sons 1904. (*51*)

1889 [*1*] PEARSON, K.: On the generalized equations of elasticity, and their application to the wave theory of light. Proc. Lond. Math. Soc. **20**, 297—350. (*35*)

[*2*] POINCARÉ, H.: Leçons sur la Théorie Mathématique de la Lumière. Paris. (*30, 82A*)

[*3*] VOIGT, W.: Über adiabatische Elasticitätsconstanten. Ann. Physik **36**, 743—759. (*82A*)

[*4*] VOIGT, W.: Über die innere Reibung der festen Körper, insbesondere der Krystalle. Göttinger Abh. **36**, No. 1. (*35*)

1891 [*1*] SCHOENFLIESS, A.: Kristallsysteme und Kristallstruktur. Leipzig. (*33*)

1892 [*1*] POINCARÉ, H.: Leçons sur la Théorie de l'Elasticité. Paris. (*30, 68, 82A*)

[*2*] VOIGT, W.: Über innere Reibung fester Körper, insbesondere der Metalle. Ann. Physik (2) **47**, 671—693. (*35*)

[*3*] VOIGT, W.: Bestimmung der Constanten der Elasticität und Untersuchung der inneren Reibung für einige Metalle. Göttinger Abh. **38**, No. 2. (*35*)

1893 [*1*] BARUS, C.: Note on the dependence of viscosity on pressure and temperature. Proc. Amer. Acad. Arts Sci. (2) **19** = **27**, 13—18. (*114*)

[*2*] BARUS, C.: Isothermals, isopiestics, and isometrics relative to viscosity. Amer. J. Sci. (3) **45** = **145**, 87—96. (*114*)

[*3*] CELLERIER, G.: Sur les principes généraux de la thermodynamique et leur application aux corps élastiques. Bull. Soc. Math. France **21**, 26—43. (*43*)

[*4*] TODHUNTER, I., and K. PEARSON: A History of the Theory of Elasticity and of the Strength of Materials from Galilei to Lord Kelvin, **2** [almost entirely the work of PEARSON] Cambridge: Cambridge Univ. Press. Reprinted Dover Publications, New York, 1960. (*66*)

[*5*] VOIGT, W.: Über eine anscheinend nothwendige Erweiterung der Theorie der Elasticität. Göttinger Nachr. 534—552 = Ann. Physik (2) **52**, 536—555. (*66*)

1894 [*1*] FINGER, J.: Das Potential der inneren Kräfte und die Beziehungen zwischen den Deformationen und den Spannungen in elastisch isotropen Körpern bei Berücksichtigung von Gliedern, die bezüglich der Deformationselemente von dritter, beziehungsweise zweiter Ordnung sind. Sitzgsber. Akad. Wiss. Wien (IIa) **103**, 163—200, 231—250. (*66, 86*)

[*2*] FINGER, J.: Über das Kriterion der Coaxialität zweier Mittelpunktsflächen zweiter Ordnung. Sitzgsber. Akad. Wiss. Wien (IIa) **103**, 1061—1065. (*48*)

[*3*] FINGER, J.: Über die allgemeinsten Beziehungen zwischen Deformationen und den zugehörigen Spannungen in aeolotropen und isotropen Substanzen. Sitzgsber. Akad. Wiss. Wien (IIa) **103**, 1073—1100. (*12, 48, 85, 86*)

1895 [*1*] VOIGT, W.: Über Medien ohne innere Kräfte und eine durch sie gelieferte mechanische Deutung der Maxwell-Hertzschen Gleichungen. Abh. Ges. Wiss. Göttingen **1894**, 72—79. (*98*)

1896 [*1*] COSSERAT, E., and F.: Sur la théorie de l'élasticité. Ann. Toulouse **10**, 1—116. (*4, 42, 43, 83, 84, 85*)

1900 [*1*] BRILLOUIN, M.: Théorie moléculaire des gaz. Diffusion du mouvement et de l'énergie. Ann. Chim. (7) **20**, 440—485. (*125*)

1901 [*1*] DUHEM, P.: Sur les théorèmes D'HUGONIOT, les lemmes de M. HADAMARD et la propagation des ondes dans les fluides visqueux. C. R. Acad. Sci. Paris **132**, 117—120. (*96t*)

[*2*] DUHEM, P.: Des ondes qui peuvent persister en un fluide visqueux. C. R. Acad. Sci. Paris **133**, 579—580. (*96t*)

[*3*] HADAMARD, J.: Sur la propagation des ondes. Bull. Soc. Math. France **29**, 50—60. (*71, 74, 90*)

[*4*] KORTEWEG, D. J.: Sur la forme que prennent les équations du mouvement des fluides si l'on tient compte des forces capillaires causées par des variations de densité considérables mais continues et sur la théorie de la capillarité dans l'hypothèse d'une variation continue de la densité. Arch. Néerl. Sci. Ex. Nat. (2) **6**, 1—24. (*124*)

[*5*] NATANSON, L.: On the laws of viscosity. Phil. Mag. (6) **2**, 342—356. (*19A, 35*)

[*6*] NATANSON, L.: Sur les lois de la viscosité. Bull. Int. Acad. Sci. Cracovie 95—111. (*19A, 35*)

[*7*] NATANSON, L.: Sur la double refraction accidentelle dans les liquides. Bull. Int. Acad. Sci. Cracovie 161—171. (*19A, 35*)

[*8*] NATANSON, L.: Über die Gesetze der inneren Reibung. Z. physik. Chem. **38**, 690—704. (*19A, 35*)

[*9*] REYNOLDS, O.: On the equations of motion and the boundary conditions for viscous fluids (1883). Papers **2**, 132—137. (*35*)

1902 [*1*] DUHEM, P.: Recherches sur l'hydrodynamique. Seconde Partie. Ann. Toulouse (2) **4**, 101—169. Also included in reprint, Paris: Gauthier Villars 1903. (*96t*)

[*2*] NATANSON, L.: Sur la propagation d'un petit mouvement dans un fluide visqueux. Bull. Int. Acad. Sci. Cracovie 19—35. (*19A*)

[*3*] NATANSON, L.: Sur la fonction dissipative d'un fluide visqueux. Bull. Int. Acad. Sci. Cracovie 448—494. (*19A*)

[*4*] NATANSON, L.: Sur la déformation d'un disque plastico-visqueux. Bull. Int. Acad. Sci. Cracovie 494—512. (*19A*)

[*5*] NATANSON, L.: Sur la conductibilité calorifique d'un gaz en mouvement. Bull. Int. Acad. Sci. Cracovie 137—146. (*19A*)

1903 [*1*] DUHEM, P.: Sur la viscosité en un milieu vitreux. C. R. Acad. Sci. Paris **136**, 281—283. (*41*)

[*2*] DUHEM, P.: Sur les équations du mouvement et la relation supplémentaire au sein d'un milieu vitreux. C. R. Acad. Sci. Paris **136**, 343—345. (*41, 71*)

[*3*] DUHEM, P.: Sur la propagation des ondes dans un milieu parfaitement élastique affecté de déformations finies. C. R. Acad. Sci. Paris **136**, 1379—1381. (*90*)

[*4*] HADAMARD, J.: Leçons sur la Propagation des Ondes et les Équations de l'Hydrodynamique. Paris: Hermann. (*31, 68b, 71, 74, 88, 89, 90*)

[*5*] NATANSON, L.: Über einige von Herrn B. WEINSTEIN zu meiner Theorie der inneren Reibung gemachten Bemerkungen. Physik. Z. **4**, 541—543. (*19A*)

[*6*] NATANSON, L.: Sur l'application des équations de Lagrange dans la théorie de la viscosité. Bull. Int. Acad. Sci. Cracovie 268—283. (*19A*)

[*7*] NATANSON, L.: Sur l'approximation de certaines équations de la théorie de la viscosité. Bull. Int. Acad. Sci. Cracovie 283—311. (*19A*)

[*8*] ZAREMBA, S.: Remarques sur les travaux de M. NATANSON relatifs à la théorie de la viscosité. Bull. Int. Acad. Sci. Cracovie 85—93. (*19A, 35*)

[9] Zaremba, S.: Sur une généralisation de la théorie classique de la viscosité. Bull. Int. Acad. Sci. Cracovie 380—403. (*19A, 35*)

[10] Zaremba, S.: Sur un problème d'hydrodynamique lié à un cas double refraction accidentale dans les liquides et sur les considérations théoriques de M. Natanson relatives à ce phénomène. Bull. Int. Acad. Sci. Cracovie 403—423. (*19A, 35*)

[11] Zaremba, S.: Sur une forme perfectionnée de la théorie de la relaxation. Bull. Int. Acad. Sci. Cracovie 594—614. (*19A, 35, 36, 99, 119*)

[12] Zaremba, S.: Le principe des mouvements relatifs et les équations de la mécanique physique. Bull. Int. Acad. Sci. Cracovie 614—621. (*19A, 35, 36, 99, 119*)

1904 [1] Duhem, P.: Recherches sur l'élasticité, Première Partie. De l'équilibre et du mouvement des milieux vitreux. Ann. École Normale (3) **21**, 99—141. Repr. Paris: Gauthier-Villars 1906. (*41*)

1905 [1] Duhem, P.: Recherches sur l'élasticité, Troisième Partie. La stabilité des milieux élastiques. Ann. École Norm. (3) **22**, 143—217. Repr. Paris: Gauthier-Villars 1906. (*52, 68b, 88, 89*)

[2] Poincaré, H.: Science and Hypothesis. Transl. W. J. G. London. (*3*)

[3] Poynting, J. H.: Radiation-pressure. Phil. Mag. **9**, 393—406 = Papers **2**, 335—346. (*54, 66*)

1906 [1] Duhem, P.: Recherches sur l'élasticité. Quatrième Partie. Propriétés générales des ondes dans les milieux visqueux et non-visqueux. Ann. École Normale (3) **23**, 169—225. Repr. Paris: Gauthier-Villars 1906. (*71, 90*)

[2] Jaumann, G.: Elektromagnetische Vorgänge in bewegten Medien. Sitzgsber. Akad. Wiss. Wien (IIa) **15**, 337—390. (*19A*)

[3] Prandtl, L.: Zur Theorie des Verdichtungsstoßes. Z. ges. Turbinenwesen **3**, 241—245 = Ges. Abh. **2**, 935—942. (*96t*)

1907 [1] Cosserat, E. and F.: Sur la mécanique générale. C. R. Acad. Sci. Paris **145**, 1139—1142. (*19A, 98*)

1909 [1] Cosserat, E., and F.: Théorie des Corps Déformables. Paris: Hermann, vi + 226 pp. Publ. also as pp. 953—1173 of O. D. Chwolson, Traité de Physique, transl. E. Davaux, 2nd ed., **2**, Paris 1909. (*19A, 98*)

[2] Poynting, J. H.: On pressure perpendicular to the shear-planes in finite pure shears, and on the lengthening of loaded wires when twisted. Proc. Roy. Soc. Lond. A **82**, 546—559 = Papers **2**, 358—371. (*54, 66*)

[3] Volterra, V.: Sulle equazioni integro-differenziali della teoria dell'elasticità. Rend. Lincei (5) $\mathbf{18}_1$, 295—301 = Opere **3**, 288—293. (*41*)

[4] Volterra, V.: Equazioni integro-differenziali della elasticità nel caso della isotropia. Rend. Lincei (5) $\mathbf{18}_1$, 577—586 = Opere **3**, 294—303. (*41*)

1910 [1] Kötter, F.: Über die Spannungen in einem ursprünglich geraden, durch Einzelkräfte in stark gekrümmter Gleichgewichtslage gehaltenen Stab. Sitzgsber. Preuss. Akad. Wiss. Teil **2**, 895—922. (*60, 66, 86*)

[2] Voigt, W.: Lehrbuch der Kristallphysik. Leipzig u. Berlin: Teubner. (*33*)

1911 [1] Almansi, E.: Sulle deformazioni finite dei solidi elastici isotropi, I. Rend. Accad. Lincei (5A) $\mathbf{20}_1$, 705—714. (*86*)

[2] Almansi, E.: Sulle deformazioni finite dei solidi elastici isotropi, III. Rend. Accad. Lincei (5A) $\mathbf{20}_2$, 289—296. (*86*)

[3] Houston, R. A.: A relation between tension and torsion. Phil. Mag. (6) **22**, 740—741. (*86*)

[4] Jaumann, G.: Geschlossenes System physikalischer und chemischer Differentialgesetze. Sitzgsber. Akad. Wiss. Wien (IIa) **120**, 385—530. (*19A, 99*)

1912 [1] Hamel, G.: Elementare Mechanik. Leipzig u. Berlin: Teubner. (*42*)

1913 [1] Heun, K.: Ansätze und allgemeine Methoden der Systemmechanik. Enz. math. Wiss. $\mathbf{4}^2$ (1904—1935), art. 11. (*98*)

[2] Poynting, J. H.: The changes in length and volume of an India-rubber cord when twisted. India-Rubber J., Oct., 4, p. 6 = Papers **2**, 424—425. (*54, 66*)

1914 [1] Hellinger, E.: Die allgemeinen Ansätze der Mechanik der Kontinua. Enz. Math. Wiss. $\mathbf{4}^4$, 602—694. (*42, 68b, 88, 98*)

1915 [1] Armanni, G.: Sulle deformazioni finite dei solidi elastici isotropi. Nouvo Cimento (6) **10**, 427—447. (*44*)

1916 [1] Almansi, E.: La teoria delle distorsioni e le deformazioni finite dei solidi elastici. Rend. Accad. Lincei (5) $\mathbf{25}_2$, 191—192. (*44*)

1917 [1] Lohr, E.: Entropieprinzip und geschlossenes Gleichungssystem. Denkschr. Akad. Wiss. Wien **93**, 339—421. (*99*)

1918 [1] Jaumann, G.: Physik der kontinuierlichen Medien. Denkschr. Akad. Wiss. Wien **95**, 461—562. (*99*)

1920 [1] JOUGUET, E.: Sur les ondes de choc dans les corps solides. C. R. Acad. Sci. Paris **171**, 461—464. (*71, 73*)

[2] JOUGUET, E.: Sur la célérité des ondes dans les corps solides. C. R. Acad. Sci. Paris **171**, 512—515. (*71, 73, 77*)

[3] JOUGUET, E.: Sur la variation d'entropie dans les ondes de choc des solides élastiques. C. R. Acad. Sci. Paris **171**, 789—791. (*71, 73*)

[4] JOUGUET, E.: Application du principe de Carnot-Clausius aux ondes de choc des solides élastiques. C. R. Acad. Sci. Paris **171**, 904—907. (*71, 73*)

1921 [1] JOUGUET, E.: Notes sur la théorie de l'élasticité. Ann. Toulouse (3) **12**, (1920), 47—92. (*71, 73*)

[2] WEITZENBÖCK, R.: Zur Tensoralgebra. Math. Z. **10**, 80—87. (*12*)

1924 [1] ARIANO, R.: Deformazioni finite dei sistemi continui. Rend. Palermo **48**, 97—120. (*48*)

1925 [1] ARIANO, R.: Deformazioni finite di sistemi continui. Memoria 1ª. Ann. di Mat. (4) **2**, 217—261. (*42*)

[2] BRILLOUIN, L.: Sur les tensions de radiation. Ann. Physique (10) **4**, 528—586. (*42, 66, 68, 77*)

1926 [1] SUDRIA, J.: Contribution à la théorie de l'action euclidienne. Ann. Fac. Sci. Toulouse (3) **17** (1925), 63—152. (*98*)

1927 [1] LOVE, A. E. H.: A Treatise on the Mathematical Theory of Elasticity. 4th ed. Cambridge. Reprinted by Dover Publications, New York, 1944. (*66, 88*)

[2] REINER, M., and M. RIWLIN: Die Theorie der Strömung einer elastischen Flüssigkeit im Couette-Apparat. Kolloid-Z. **43**, 1—5. (*113*)

1928 [1] ARIANO, R.: Deformazioni finite di sistemi continui. Memoria 2ª. Ann. di Mat. Pura Appl. (4) **5**, 55—71. (*42*)

[2] FARROW, F. D., G. M. LOWE, and S. M. NEALE: The flow of starch pastes. Flow at high and low rates of shear. J. Textile Inst. **19**, T 18—T 31. (*113*)

[3] MURNAGHAN, F. D.: On the energy of deformation of an elastic solid. Proc. Nat. Acad. Sci. U.S.A. **14**, 889—891. (*47*)

[4] RABINOWITSCH, B.: Über die Viskosität und Elastizität von Sohlen. Z. phys. Chem., Abt. A **145**, 1—26. (*113*)

1929 [1] ARIANO, R.: Deformazioni finite di sistemi continui. Memoria 3ª. Ann. di Mat. Pura Appl. (4) **6**, 265—282. (*42*)

[2] JARAMILLO, T. J.: A generalization of the energy function of elasticity theory. Diss. Univ. Chicago. (*98*)

[3] LICHTENSTEIN, L.: Grundlagen der Hydromechanik. Berlin: Springer. (*120*)

1930 [1] ARIANO, R.: Deformazioni finite di sistemi continui, isotropi. Rend. Ist. Lombardo (2) **63**, 740—754. (*42, 48, 86*)

[2] MISES, R. v.: Über die bisherigen Ansätze in der klassischen Mechanik der Kontinua. Proc. 3rd Int. Congr. Appl. Mech. Stockholm **2**, 1—9. (*3, 35*)

[3] SIGNORINI, A.: Sulle deformazioni termoelastiche finite. Proc. 3rd Int. Congr. Appl. Mech. **2**, 80—89. (*44, 47, 63, 64, 86*)

1931 [1] ANZELIUS, A.: Die Viskositätsanomalien der anisotropen Flüssigkeiten. Grundlegende Tatsachen. Uppsala Univ. Årsskr. Mat. Och Naturvet., 1—84. (*127*)

[2] GIRAULT, M.: Essai sur la Viscosité en Mécanique des Fluides. Publ. Sci. Tech. Min. de l'Air, No. 4, Paris. (*119, 120*)

[3] KAPLAN, C.: On the strain-energy function for isotropic bodies. Phys. Rev. (2) **38**, 1020—1029. (*66*)

[4] LAMPARIELLO, G.: Sull'impossibilità di propagazione ondosa nei fluidi viscosi. Rend. Accad. Lincei (6) **13**, 688—691. (*96 t*)

[5] MOONEY, M.: Explicit formulae for slip and fluidity. J. Rheology **2**, 210—222. (*113*)

[6] MURNAGHAN, F. D.: On finite deformations and the energy of deformation of a non-isotropic medium. Atti Congr. Internaz. Matem. Bologna (1928) **5**, 151—153. (*47*)

[7] WEISSENBERG, K.: Die Mechanik deformierbarer Körper. Abh. Akad. Wiss. Berlin No. 2. (*35*)

1932 [1] BATEMAN, H.: Part I, Ch. 3: General physical properties of a viscous fluid. Part II: Motion of an incompressible viscous fluid. Part IV: Compressible fluids. Report of the Committee on Hydrodynamics. Washington: Bull. Nat. Research Council No. 84. (*116*)

[2] CERUTI, G.: Sopra un'estensione della teoria elastica alla seconda approssimazione. Rend. Ist. Lombardo (2) **65**, 997—1012. (*66*)

[3] HOHENEMSER, K., and W. PRAGER: Fundamental equations and definitions concerning the mechanics of isotropic continua. J. Rheol. **3**, 245—256. (*35*)

[4] HOHENEMSER, K., and W. PRAGER: Über die Ansätze der Mechanik isotroper Kontinua. Z. angew. Math. Mech. **12**, 216—226. (*35*)
[5] LAMB, H.: Hydrodynamics. 6th ed. Cambridge Univ. Press. Reprinted, Dover Publications, New York, 1945. (*21, 74, 96, 116*)
[6] PRANDTL, L.: Extract from a letter to D. A. GRAVE, published in D. A. GRAVE: Physical foundations of hydrodynamics and aerodynamics [in Russian]. Isv. Akad. Nauk SSSR Otdel mat. est. nauk (Ser. VII) **1932**, 763—782. (*120*)
[7] SIGNORINI, A.: Sollecitazioni iperastatiche. Rend. Ist. Lombardo (2) **65**, 1—7. (*44*)

1933 [1] ARIANO, R.: L'isotropia nelle deformazione finite. Rend. Ist. Lombardo **66**, 1—13, 207—220. (*42, 48*)
[2] FROMM, E.: Stoffgesetze des isotropen Kontinuums, insbesondere bei zähplastischen Verhalten. Ing.-Arch. **4**, 432—466. (*36, 99, 119*)
[3] OSEEN, C. W.: The theory of liquid crystals. Trans. Faraday Soc. **29**, 833—899. (*127*)
[4] SIGNORINI, A.: Sulle deformazione finite dei sistemi a trasformazioni reversibili. Rend. Accad. Lincei (6) **18**, 388—394. (*86*)
[5] THOMPSON, J. H. C.: On the theory of visco-elasticity, and some problems of the vibrations of visco-elastic solids. Phil. Trans. Roy. Soc. Lond. A **231**, 339—407. (*120*)

1935 [1] DE BACKER, S.: Les fluides visqueux et les ondes propageables. C. R. Acad. Sci. Paris **200**, 899—901. (*119*)
[2] SETH, B. R.: Finite strain in elastic problems. Phil. Trans. Roy. Soc. Lond. (A) **234**, 231—264. (*48*)
[3] SUDRIA, J.: L'action Euclidienne de Deformation et de Mouvement. Mém. Sci. Phys. No. **29**. Paris: Gauther-Villars. 56 pp. (*98*)

1936 [1] DE BACKER, S.: Les fluides visqueux et les ondes propageables. Évolution d'un gaz monoatomique et polyatomique. Bull. Acad. Roy. Belg., Cl. Sci. (5) **22**, 1284—1295. (*119*)
[2] HALTON, P., and G. W. SCOTT BLAIR: A study of some physical properties of flour doughs in relation to their bread-making qualities. J. Phys. Chem. **40**, 561—580. (*114*)
[3] SIGNORINI, A.: Trasformazioni termoelastiche finite, caratteristiche dei sistemi differenziali, onde di discontinuità in particolare, onde d'urto e teoria degli esplosivi. Atti XXIV Riun. Soc. Ital. Progr. Sci. **3**, 6—25. (*42, 63, 64, 88*)

1937 [1] DE BACKER, S.: Les fluides visqueux et les ondes propageables. Bull. Acad. Roy. Belg. Cl. Sci. (5) **23**, 59—72, 262—273. (*119*)
[2] MURNAGHAN, F. D.: Finite deformation of an elastic solid. Amer. J. Math. **59**, 235—260. (*4, 42, 43, 48, 66, 85*)
[3] VAN DER WAERDEN, B. L.: Moderne Algebra I, 2. Aufl. Berlin: Springer. (*10*)
[4] ZAREMBA, S.: Sur une Conception Nouvelle des Forces Intérieures dans un Fluide en Mouvement. Mém. Sci. Math. No. 82. Paris: Gauthier-Villars. (*19A, 36, 119*)

1938 [1] BIRCH, F.: The effect of pressure upon the elastic parameters of isotropic solids, according to MURNAGHAN's theory of finite strain. J. Appl. Phys. **9**, 279—288. (*75*)
[2] BRILLOUIN, L.: Les Tenseurs en Mécanique et en Élasticité. 2nd ed. Paris: Masson 1960. (*42, 45, 66, 68, 77*)
[3] KILCHEVSKI, N.: A new theory of the mechanics of continuous media [in Ukrainian]. Zbirnik Inst. Mat. Akad. Nauk URSR, No. 1, 17—114. (*35*)
[4] PRAGER, W.: On isotropic materials with continuous transition from elastic to plastic state. Proc. 5th Internat. Congr. Appl. Mech. (Cambridge), pp. 234—237. (*103*)

1939 [1] KAPPUS, R.: Zur Elastititätstheorie endlicher Verschiebungen. Z. angew. Math. Mech. **19**, 271—285, 344—361. (*42*)
[2] WEYL, H.: The Classical Groups, their Invariants and Representations. Princeton Univ. Press. Reprinted 1953. (*10, 11*)

1940 [1] BIOT, M.: The influence of initial stress on elastic waves. J. Appl. Phys. **11**, 522—530. (*75*)
[2] BORN, M., and R. D. MISRA: On the stability of crystal lattices, IV. Proc. Cambridge Phil. Soc. **36**, 466—478. (*66*)
[3] BRILLOUIN, M.: Influence de la Température sur l'Élasticité d'un Solide. Mém. Sci. Math. 99. Paris: Gauthier-Villars. (*68*)
[4] MOONEY, M.: A theory of large elastic deformation. J. Appl. Phys. **11**, 582—592. (*54, 95*)

1941 [1] MURNAGHAN, F. D.: The compressibility of solids under extreme pressures. Kármán Anniv. Vol. 121—136. (*42*)

[2] PRAGER, W.: A new mathematical theory of plasticity. Rev. Fac. Sci. Univ. Istanbul A 5, 215—226. (*103*)

[3] SAKADI, Z.: On the extension of the differential equations of incompressible viscous fluid. Proc. Physico-Math. Soc. Japan (3) **23**, 27—33. (*119*)

[4] TIMOSHENKO, S. P.: Strength of Materials, II. 2nd ed. New York: Van Nostrand. (*66*)

1942 [1] DE COLLE, L.: Teorema di minimo relativo a fluidi viscosi generali. Rend. Ist. Lombardo Sci. Lett. Rend., Cl. Sci. Mat. Nat. **75** = (6) **3**, 343—352. (*119*)

[2] FINZI, B.: Propagazione ondosa nei continui anisotropi. Rend. Ist. Lombardo, Cl. Sci. Mat. Nat. **75** = (3) **6** (1941/42), 630—640. (*71*)

[3] HAY, G. E.: The finite displacement of thin rods. Trans. Amer. Math. Soc. **51**, 65—102. (*60*)

[4] PRAGER, W.: Fundamental theorems of a new mathematical theory of plasticity. Duke Math. J. **9**, 228—233. (*103*)

[5] SAKADI, Z.: On the extension of the differential equations of incompressible fluid, II. Proc. Physico-Math. Soc. Japan (3) **24**, 719—722. (*119*)

[6] SIGNORINI, A.: Deformazioni elastiche finite: elasticità di 2° grado. Atti 2° Congr. Mat. Ital. 1940, pp. 56—71. (*64, 94*)

[7] TOLOTTI, C.: Sulla più generale elasticità di 2° grado. Rend. Sem. Mat. Univ. Roma (5) **3**, 1—20. (*94*)

[8] ZENER, C.: Theory of lattice expansion introduced by cold-work. Trans. Amer. Inst. Mining Met. Engrs. **147**, 361—364. (*93*)

1943 [1] MERRINGTON, A. C.: Flow of visco-elastic materials in capillaries. Nature, Lond. **152**, 663. (*114*)

[2] REINER, M.: Ten Lectures on Theoretical Rheology. Jerusalem. (There are later, expanded editions.) (*35*)

[3] SIGNORINI, A.: Trasformazioni termoelastiche finite. Mem. 1ª, Ann. di Mat. (4) **22**, 33—143. (*42, 43, 47, 50*)

[4] TOLOTTI, C.: Orientamenti principali di un corpo elastico rispetto alla sua sollecitazione totale. Mem. Accad. Italia, Cl. sci. mat. nat. (7) **13**, 1139—1162. (*64*)

[5] TOLOTTI, C.: Deformazione elastiche finite: onde ordinarie di discontinuità e caso tipico di solidi elastici isotropi. Rend. Mat. e Applic. (5) **4**, 33—59. (*71, 74*)

[6] TOLOTTI, C.: Sul potenziale termodinamico dei solidi elastici omogenei ed isotropi per trasformazioni finite. Atti R. Accad. Italia **14**, 529—541. (*84*)

[7] UDESCHINI, P.: Sull'energia di deformazione. Rend. Ist. Lombardo (3) **7** = **76**, 25—34. (*83*)

1944 [1] MURNAGHAN, F. D.: The compressibility of media under extreme pressures. Proc. Nat. Acad. Sci. U.S.A. **30**, 244—247. (*99, 103*)

[2] MURNAGHAN, F. D.: On the theory of the tension of the elastic cylinder. Proc. Nat. Acad. Sci. U.S.A. **30**, 382—384. (*99, 103*)

[3] SETH, B. R.: On the stress-strain velocity relations in equations of viscous flow. Proc. Indian Acad. Sci. (A) **20**, 336—339. (*119*)

1945 [1] MURNAGHAN, F. D.: A revision of the theory of elasticity. Bol. Soc. Mat. Mexicana **2**, 81—89. (*99, 103*)

[2] PRAGER, W.: Strain hardening under combined stresses. J. App. Phys. **16**, 837—840. (*12, 43*)

[3] REINER, M.: A mathematical theory of dilatancy. Amer. J. Math. **67**, 350—362. Reprinted in Rational Mechanics of Materials, Intl. Sci. Rev. Ser. New York: Gordon & Breach 1965. (*4, 12, 111, 114, 119, 120*)

[4] SIGNORINI, A.: Recenti progressi della teoria delle trasformazioni termoelastiche finite. Atti Conv. Mat. Roma 1942, pp. 153—168. (*43, 63, 64, 71, 94*)

1946 [1] CATTANEO, C.: Su un teorema fondamentale nella teoria delle onde di discontinuità. Atti Accad. Sci. Lincei Rend., Cl. Sic. Fis. Mat. Nat. (8) **1**, 66—72, 728—734. (*68b, 71, 73*)

[2] GARNER, F. H., and A. H. NISSAN: Rheological properties of high-viscosity solutions of long molecules. Nature, Lond. **158**, 634—635. (*114*)

[3] REINER, M.: The coefficient of viscous traction. Amer. J. Math. **68**, 672—680. (*118*)

1947 [1] BHAGAVANTAM, S., and R. SURYANARAYANA: Third-order elastic coefficients of crystals. Nature, Lond. **160**, 750—751. (*66*)

[2] BIRCH, F.: Finite elastic strain of cubic crystals. Phys. Rev. (2) **71**, 809—824. (*66*)

[3] FROMM, H.: Laminare Strömung Newtonscher und Maxwellscher Flüssigkeiten. Z. angew. Math. Mech. **25/27**, 146—150. (*111, 113, 119*)

[4] GARCIÀ, G.: Ecuaciones exactas y soluciones exactas del movimiento y de las tensiones en los fluidos viscosas. Actas Acad. Ciencias Lima **10**, 117—170. (*119*)

[5] HANDELMAN, G. H., C. C. LIN, and W. PRAGER: On the mechanical behavior of metals in the strain-hardening range. Quart. Appl. Math. **4**, 397—407. (*43*)
[6] KUTILIN, D. I.: Theory of Finite Deformations [in Russian]. Moscow and Leningrad: OGIZ. 275 pp. (*42, 98*)
[7] OLDROYD, J. G.: Rectilinear plastic flow of a Bingham solid, I. Flow between eccentric circular cylinders in relative motion. Proc. Cambridge Phil. Soc. **43**, 396—405. (*117*)
[8] OLDROYD, J. G.: Rectilinear plastic flow of a Bingham solid, II. Flow between confocal elliptic cylinders in relative motion. Proc. Cambridge Phil. Soc. **43**, 521—532. (*117*)
[9] RIVLIN, R. S.: Torsion of a rubber cylinder. J. Appl. Phys. **18**, 444—449, 837. (*57, 95*)
[10] RIVLIN, R. S.: Hydrodynamics of non-Newtonian fluids. Nature, Lond. **160**, 611—613. (*119*)
[11] SWIFT, H. W.: Length changes in metals under torsional overstrain. Engineering **163**, 253—257. (*66*)
[12] TRUESDELL, C., and R. N. SCHWARTZ: The Newtonian mechanics of continua. U.S. Naval Ordnance Lab. Memo. 9223. (*28, 119, 119A*)
[13] VAN HOVE, L.: Sur l'extension de la condition de Legendre du calcul des variations aux intégrales multiples à plusieurs fonctions inconnues. Proc. Kon. Ned. Akad. Wet. **50**, 18—23. (*68b*)
[14] VIGUIER, G.: Les équations de la couche limite dans le cas de gradients de vitesse élevés. C. R. Acad. Sci. Paris **224**, 713—714. (*119*)
[15] VIGUIER, G.: L'écoulement d'un fluide visqueux avec gradients de vitesse élevés. C. R. Acad. Sci. Paris **224**, 1048—1050. (*119*)
[16] VIGUIER, G.: La couche limite de Prandtl avec importants gradients de vitesses. C. R. Acad. Sci. Paris **225**, 45—46. (*119*)
[17] WEISSENBERG, K.: A continuum theory of rheological phenomena. Nature, Lond. **159**, 310—311. (*114, 119*)
[18] WOOD, G. F., A. H. NISSAN, and F. H. GARNER: Viscometry of soap-in-hydrocarbon systems. J. Inst. Petrol. **33**, 71—94. (*114*)
1948 [1] BURGERS, J. M.: Non-linear relations between viscous stresses and instantaneous rate of deformation as a consequence of slow relaxation. Proc. Kon. Ned. Akad. Wet. **51**, 787—792. (*119*)
[2] COURANT, R., and K. O. FRIEDRICHS: Supersonic Flow and Shock Waves. New York: Interscience Publ. (*74*)
[3] ECKART, C.: The thermodynamics of irreversible processes, IV. The theory of elasticity and anelasticity. Phys. Rev. (2) **73**, 373—382. (*28*)
[4] FROMM, H.: Laminare Strömung Newtonscher und Maxwellscher Flüssigkeiten. Z. angew. Math. Mech. **28**, 43—54. (*119*)
[5] GARCIÁ, G.: Sur une formule, cardinale et canonique des tensions internes et sur l'équation cardinale, canonique de mouvement des fluides visqueux. Ann. Soc. Polonaise Math. **21**, 107—113. (*119*)
[6] KUBO, R.: Large elastic deformation of rubber. J. Phys. Soc. Japan **3**, 312—317. (*95*)
[7] OLDROYD, J. G.: Rectilinear plastic flow of a Bingham solid, III. A more general discussion of steady flow. Proc. Cambridge Phil. Soc. **44**, 200—213. (*117*)
[8] OLDROYD, J. G.: Rectilinear plastic flow of a Bingham solid, IV. Non-steady motion. Proc. Cambridge Phil. Soc. **44**, 214—228. (*117*)
[9] REINER, M.: Elasticity beyond the elastic limit. Amer. J. Math. **70**, 433—446. Reprinted in Foundations of Elasticity Theory. Intl. Sci. Rev. Ser. New York: Gordon & Breach 1965. (*4, 43, 47, 54*)
[10] RICHTER, H.: Das isotrope Elastizitätsgesetz. Z. angew. Math. Mech. **28**, 205—209. (*12, 42, 43, 47*)
[11] RIVLIN, R. S.: Large elastic deformations of isotropic materials, I. Fundamental concepts. Phil. Trans. Roy. Soc. Lond. A **240**, 459—490. (*95*)
[12] RIVLIN, R. S.: Large elastic deformations of isotropic materials, IV. Further developments of the general theory. Phil. Trans. Roy. Soc. Lond. A **241**, 379—397. Reprinted in Problems of Non-linear Elasticity. Intl. Sci. Rev. Ser. New York: Gordon & Breach 1965. (*4, 54, 55, 57, 86*)
[13] RIVLIN, R. S.: A uniqueness theorem in the theory of highly-elastic materials. Proc. Cambridge Phil. Soc. **44**, 595—597. (*95*)
[14] RIVLIN, R. S.: The hydrodynamics of non-Newtonian fluids, I. Proc. Roy. Soc. Lond. **193**, 260—281. Reprinted in Rational Mechanics of Materials. Intl. Sci. Rev. Ser. New York: Gordon & Breach 1965. (*4, 111, 113, 114, 115, 116, 119*)

[15] Rivlin, R. S.: Some applications of elasticity theory to rubber engineering. Proc. 2nd Tech. Conf. (London, June 23—25, 1948). Cambridge: Heffer. (*54*)

[16] Rivlin, R. S.: Normal stress coefficient in solutions of macromolecules. Nature, Lond. **161**, 567—569. (*119*)

[17] Stone, M. H.: Generalized Weierstrass approximation theorem. Math. Mag. **21**, 167—184, 237—254. (*37*)

[18] Truesdell, C.: A new definition of a fluid. U.S. Naval Ord. Lab. Mem. 9487. (*125*)

[19] Truesdell, C.: On the differential equations of slip flow. Proc. Nat. Acad. Sci. U.S.A. **34**, 342—347. (*120, 125*)

[20] Viguier, G.: Quelques remarques sur la couche limite de Prandtl. Son équation dans le cas de gradients de vitesse élevés. Recherche Aeron. No. 1, pp. 7—9. (*119*)

1949 [1] Bhagavantam, S., and D. Suryanarayana: Crystal symmetry and physical properties: Application of group theory. Acta Cryst. **2**, 21—26. (*66*)

[2] Freeman, S. M., and K. Weissenberg: Rheology and the constitution of matter. Proc. Intl. Congr. Rheology 1948. Amsterdam: North Holland Publ. Co., pp. II 12—II 14. (*114*)

[3] Garciá, G.: Ecuaciones cardinales canonicas exactas para los movimientos finitos y las tensiones en los fluidos viscosas. Actas Acad. Ciencias Lima **12**, 3—30. (*119*)

[4] Gleyzal, A.: A mathematical formulation of the general continuous deformation problem. Quart. Appl. Math. **6**, 429—437. (*43, 103*)

[5] Greenberg, H. J.: On the variational principles of plasticity. Grad. Div. Applied Math. Brown Univ. Rep. A 11—54, March. (*88*)

[6] Jahn, H. A.: Note on the Bhagavantam-Suryanarayana method of enumerating the physical constants of crystals. Acta Cryst. **2**, 30—33. (*66*)

[7] Kondo, K.: A proposal of a new theory concerning the yielding of materials based on Riemannian geometry. J. Japan Soc. Appl. Mech. **2**, 123—128, 146—151. (*34*)

[8] Milne-Thomson, L. M.: Finite elastic deformations. Proc. 7th Internat. Congr. Appl. Mech. (1948) **1**, 33—40. (*42*)

[9] Murnaghan, F. D.: A revision of the theory of elasticity. Anais Acad. Brasil Ci. **21**, 329—336. (*45, 99*)

[10] Murnaghan, F. D.: The foundations of the theory of elasticity (1947). Non-linear Problems in the Mechanics of Continua, pp. 158—174. New York. (*99, 103*)

[11] Oldroyd, J. G.: Rectilinear flow of non-Bingham plastic solids and non-Newtonian viscous liquids, I. Proc. Cambridge Phil. Soc. **45**, 595—611. (*117, 119*)

[12] Pastori, M.: Propagazione ondosa nei continui anisotropi e corrispondenti direzioni principali. Nuovo Cimento (9) **6**, 187—193. (*90*)

[13] Reiner, M.: Relations between stress and strain in complicated systems. Proc. Int. Congr. Rheology 1948. IV-44-IV-63. (*119*)

[14] Richter, H.: Verzerrungstensor, Verzerrungsdeviator, und Spannungstensor bei endlichen Formänderungen. Z. angew. Math. Mech. **29**, 65—75. (*42*)

[15] Rivlin, R. S.: Large elastic deformations of isotropic materials, V. The problem of flexure. Proc. Roy. Soc. Lond. A **195**, 463—473. Reprinted in Problems of Non-linear Elasticity. Intl. Sci. Rev. Ser. New York: Gordon & Breach 1965. (*4, 57, 92*)

[16] Rivlin, R. S.: Large elastic deformations of isotropic materials, VI. Further results in the theory of torsion, shear, and flexure. Phil. Trans. Roy. Soc. Lond. A **242**, 173—195. Reprinted in Problems of Non-linear Elasticity. Intl. Sci. Rev. Ser. New York: Gordon & Breach 1965. (*4, 55, 57, 59, 87, 92, 95*)

[17] Rivlin, R. S.: The hydrodynamics of non-Newtonian fluids, II. Proc. Cambridge Phil. Soc. **45**, 88—91. Reprinted in Rational Mechanics of Materials. Intl. Sci. Rev. Ser. New York: Gordon & Breach 1965. (*4, 113*)

[18] Rivlin, R. S.: A note on the torsion of an incompressible, highly-elastic cylinder. Proc. Cambridge Phil. Soc. **45**, 485—487. Reprinted in Problems of Non-linear Elasticity. Intl. Sci. Rev. Ser. New York: Gordon & Breach 1965. (*4, 57, 92*)

[19] Rivlin, R. S.: The normal-stress coefficient in solutions of macro-molecules. Trans. Faraday Soc. **45**, 739—748. (*119*)

[20] Sakadi, Z.: On elasticity problems when the second order terms of the strain are taken into account, II. Mem. Fac. Eng. Nagoya **1**, 95—107. (*66*)

[21] Signorini, A.: Trasformazioni termoelastiche finite. Memoria 2^a. Ann. di Mat. Pur. Appl. (4) **30**, 1—72. (*42, 63, 64, 87, 88, 94*)

[22] Truesdell, C.: A new definition of a fluid, I. The Stokesian fluid. Proc. 7th Internat. Congr. Appl. Mech. (1948) **2**, 351—364. (*108, 119, 119A, 120*)

[23] Truesdell, C.: A new definition of a fluid, II. The Maxwellian fluid. U.S. Naval Res. Lab. Rep. No. P-3553. (*96, 119, 125*)

[24] Viguier, G.: Les forces tangentielles de viscosité avec gradients de vitesse élevés. Experientia **5**, 397—398. (*119*)

[*25*] VIGUIER, G.: Nouvelles équations de la mécanique des fluides visqueux. Hrvatsko Prirodoslovno Drustvo. Glasnik Mat.-Fiz. Astr. (II) **4**, 193—200. (*119*)
[*26*] WEISSENBERG, K.: Geometry of rheological phenomena (1946—1947). The Principles of Rheological Measurement, pp. 36—65. London. (*114, 119*)
[*27*] WEISSENBERG, K.: Abnormal substances and abnormal phenomena of flow. Proc. Intl. Congr. Rheology 1948. Amsterdam: North Holland Publ. Co., pp. I-29—I-46. (*114, 119*)
[*28*] WEISSENBERG, K.: Specification of rheological phenomena by means of a rheogoniometer. Proc. Intl. Congr. Rheology 1948. Amsterdam: North Holland Publ. Co., pp. II 114—II 118. (*115*)
1950 [*1*] GARNER, F. H., A. H. NISSAN, and G. F. WOOD: Thermodynamics and rheological behavior of elastico-viscous systems under stress. Phil. Trans. Roy. Soc. Lond. A **243**, 37—66. (*113, 114, 115*)
[*2*] GREEN, A. E., and W. ZERNA: Theory of elasticity in general co-ordinates. Phil. Mag. (7) **41**, 313—336. (*42*)
[*3*] GREEN, A. E., and R. T. SHIELD: Finite elastic deformation of incompressible isotropic bodies. Proc. Roy. Soc. Lond. A **202**, 407—419. (*57*)
[*4*] GOLDENBLAT, I. I.: On a problem in the mechanics of finite deformation of continuous media [in Russian], C. R. Dokl. Acad. Sci. SSR **70**, 973—976. (*86*)
[*5*] HUANG, K.: On the atomic theory of elasticity. Proc. Roy. Soc. Lond. A **203**, 178—194. (*45*)
[*6*] KONDO, K.: On the dislocation, the group of holonomy and the theory of yielding. J. Japan. Soc. Appl. Mech. **3**, 107—110. (*34*)
[*7*] KONDO, K.: On the fundamental equations of the theory of yielding. J. Japan Soc. Appl. Mech. **3**, 184—188. (*34*)
[*8*] KONDO, K.: The mathematical analyses of the yield point, I. Uniform stress. J. Japan Soc. Appl. Mech. **3**, 188—195. (*34*)
[*9*] KONDO, K.: Mathematical analyses of the yield point, II. J. Japan Soc. Appl. Mech. **4**, 4—8. (*34*)
[*10*] LICHNÉROWICZ, A.: Eléments du Calcul Tensoriel. Paris: Armand Colin. Engl. transl., Elements of Tensor Calculus. New York: John Wiley & Sons. (*6*)
[*11*] OLDROYD, J. G.: On the formulation of rheological equations of state. Proc. Roy. Soc. Lond. A **200**, 523—541. Reprinted in Rational Mechanics of Materials. Intl. Sci. Rev. Ser. New York: Gordon & Breach 1965. (*19A, 36, 37, 113, 119*)
[*12*] OLDROYD, J. G.: Finite strains in an anisotropic elastic continuum. Proc. Roy. Soc. Lond. A **202**, 407—419. (*43, 69, 84*)
[*13*] RIVLIN, R. S.: On the definition of strain. Some Recent Developments in Rheology, pp. 125—129. London: United Trade Press. (*43*)
[*14*] RIVLIN, R. S.: Some flow properties of concentrated high-polymer solutions. Proc. Roy. Soc. Lond. A **200**, 168—176. (*104*)
[*15*] SIGNORINI, A.: Un semplice esempio di 'incompatibilità' tra la elastostatica classica e la teoria delle deformazioni elastiche finite. Acad. Naz. Lincei Rend. Cl. fis. mat. nat. (8) **8**, 276—281. (*63, 64*)
[*16*] TRUESDELL, C.: A new definition of a fluid, I. The Stokesian fluid. J. Math. Pures Appl. (9) **29**, 215—244. (*3, 13, 108, 119, 119A, 120, 121*)
[*17*] WEISSENBERG, K.: Rheology of hydrocarbon gels. Proc. Roy. Soc. Lond. A **200**, 183—188. (*114, 119*)
1951 [*1*] FUMI, F. G.: Third-order elastic coefficients of crystals. Phys. Rev. (2) **83**, 1274—1275. (*66*)
[*2*] GREEN, A. E., and R. T. SHIELD: Finite extension and torsion of cylinders. Phil. Trans. Roy. Soc. Lond. A **224**, 47—86. Reprinted in Problems of Non-linear Elasticity. Intl. Sci. Rev. Ser. New York: Gordon & Breach 1965. (*67, 70*)
[*3*] GREENSMITH, H. W., and R. S. RIVLIN: Measurements of the normal stress effect in solutions of polyisobutylene. Nature, Lond. **168**, 664—667. (*115, 116*)
[*4*] ISHIHARA, A., N. HASHITSUME, and M. TATIBANA: Statistical theory of rubber-like elasticity. IV. Two-dimensional stretching. J. Chem. Phys. **19**, 1508—1512. (*95*)
[*5*] KONDO, K.: Mathematical analyses of the yield point, III. Isotropic stress. J. Japan Soc. Appl. Mech. **4**, 35—38. (*34*)
[*6*] LODGE, A. S.: On the use of convected coordinate systems in the mechanics of continuous media. Proc. Cambridge Phil. Soc. **47**, 575—584. (*15*)
[*7*] MOONEY, M.: Secondary stresses in viscoelastic flow. J. Colloid Sci. **6**, 96—107. (*119*)
[*8*] MURNAGHAN, F. D.: Finite Deformation of an Elastic Solid. New York: John Wiley & Sons. (*42, 66, 103*)
[*9*] OLDROYD, J. G.: The motion of an elastico-viscous liquid contained between coaxial cylinders, I. Quart. J. Mech. Appl. Math. **4**, 271—282. (*36, 113, 119*)

[10] OLDROYD, J. G.: Rectilinear flow of non-Bingham plastic solids and non-Newtonian viscous liquids, II. Proc. Cambridge Phil. Soc. **47**, 410—418. (*117, 119*)

[11] REINER, M.: The theory of cross-elasticity [in Hebrew]. Hebrew Inst. Tech. Sci. Publ. **4**, 15—30. (*43*)

[12] REINER, M.: The rheological aspect of hydrodynamics. Quart. Appl. Math. **8**, 341—349. (*119*)

[13] RIVLIN, R. S.: Mechanics of large elastic deformations with special reference to rubber. Nature, Lond. **167**, 590—595. (*42*)

[14] RIVLIN, R. S., and D. W. SAUNDERS: Large elastic deformations of isotropic materials, VII. Experiments on the deformation of rubber. Phil. Trans. Roy. Soc. Lond. A **243**, 251—288. (*53, 55, 57, 67, 93, 95*)

[15] RIVLIN, R. S., and A. G. THOMAS: Large elastic deformations of isotropic materials, VIII. Strain distribution around a hole in a sheet. Phil. Trans. Roy. Soc. Lond. A **243**, 289—298. (*60*)

[16] SIPS, R.: Propagation phenomena in elastic-viscous media. J. Polymer Sci. **6**, 285—293 (*96t*)

[17] TRUESDELL, C.: A new definition of a fluid, II. The Maxwellian fluid. J. Math. Pures Appl. **30**, 111—155. (*13, 96, 119, 125*)

[18] VIGUIER, G.: Circulation d'un fluide visqueux incompressible. Bull. Acad. Roy. Belg. Cl. Sci. (5) **37**, 397—405. (*119*)

1952 [1] ADKINS, J. E., and R. S. RIVLIN: Large elastic deformations of isotropic materials, IX. The deformation of thin shells. Phil. Trans. Roy. Soc. Lond. A **244**, 505—531. (*60*)

[2] ANDERSSON, B.: On the stress-tensor of viscous isotropic fluids. Ark. Fysik **4**, 501—503. (*116*)

[3] BRAUN, I., and M. REINER: Problems of cross-viscosity. Quart. J. Mech. Appl. Math. **5**, 42—53. (*114, 119*)

[4] FUMI, F. G.: Physical properties of crystals: The direct-inspection method. Acta Cryst. **5**, 44—48. (*66*)

[5] FUMI, F. G.: The direct-inspection method in systems with a principal axis of symmetry. Acta Cryst. **5**, 691—695. (*66*)

[6] FUMI, F. G.: Third-order elastic coefficients in trigonal and hexagonal crystals. Phys. Rev. (2) **86**, 561. (*66*)

[7] GENT, A. N., and R. S. RIVLIN: Experiments on the mechanics of rubber, I. Eversion of a tube. Proc. Phys. Soc. Lond. B **65**, 118—121. (*57*)

[8] GENT, A. N., and R. S. RIVLIN: Experiments on the mechanics of rubber, II. The torsion, inflation, and extension of a tube. Proc. Phys. Soc. Lond. B **65**, 487—501. (*57*)

[9] GENT, A. N., and R. S. RIVLIN: Experiments on the mechanics of rubber, III. Small torsion of stretched prisms. Proc. Phys. Soc. Lond. B **65**, 645—648. (*70*)

[10] GRAD, H.: Statistical mechanics, thermodynamics, and fluid dynamics of systems with an arbitrary number of integrals. Comm. Pure Appl. Math. **5**, 455—494. (*98*)

[11] GREEN, A. E., R. S. RIVLIN, and R. T. SHIELD: General theory of small elastic deformations superposed on finite elastic deformations. Proc. Roy. Soc. Lond. A **211**, 128—154. (*68, 69, 70*)

[12] GREENSMITH, H. W.: Flow Properties of High Polymers. Thesis, Univ. London. (*115, 116*)

[13] KONDO, K.: On the geometrical and physical foundation of the theory of yielding. Proc. 2nd Japan Congr. Appl. Mech. pp. 41—47. (*34*)

[14] MARKOVITZ, H.: A property of Bessel functions and its application to the theory of two rheometers. J. Appl. Phys. **23**, 1070—1077. (*113*)

[15] REINER, M.: A possible cross-viscosity effect in air. Bull. Res. Council Israel **2**, 65. (*119*)

[16] RICHTER, H.: Zur Elastizitätstheorie endlicher Verformungen. Math. Nachr. **8**, 65—73. English transl. in Foundations of Elasticity Theory. Intl. Sci. Rev. Ser. New York: Gordon & Breach 1965. (*42, 43, 47*)

[17] RIVLIN, R. S., and D. W. SAUNDERS: The free energy of deformation for vulcanised rubber. Trans. Farady Soc. **48**, 200—206. (*55*)

[18] RIVLIN, R. S., and A. G. THOMAS: Rupture of rubber, I. Characteristic energy for tearing. J. Polymer Sci. **10**, 291—318. (*83*)

[19] ROBERTS, J. E.: The pressure distribution in liquids in laminar shearing motion and comparison with predictions from various theories. British Ministry of Supply Report, August. (*115, 116*)

[20] TRUESDELL, C.: The mechanical foundations of elasticity and fluid dynamics. J. Rational Mech. Anal. **1**, 125—300. Corrected reprint, Intl. Sci. Rev. Ser.

New York: Gordon & Breach 1965. (*4, 5, 19A, 28, 42, 45, 53, 54, 56, 64, 66, 68, 82, 82A, 84, 85, 86, 87, 88, 89, 94, 95, 98, 104, 108, 115, 119, 119A, 120, 121, 125*)

[*21*] TRUESDELL, C.: Review of MURNAGHAN [1951, *8*]. Bull. Amer. Math. Soc. **58**, 577—579. (*103*)

[*22*] TRUESDELL, C.: A program of physical research in classical mechanics. Z. angew. Math. Phys. **11**, 79—95. Reprinted along with [1952, *20*], Intl. Sci. Rev. Ser. New York: Gordon & Breach. 1965. (*28, 108, 119A, 120*)

[*23*] WANG, M. C., and E. GUTH: Statistical theory of networks of non-Gaussian flexible chains. J. Chem. Phys. **20**, 1144—1157. (*95*)

1953 [*1*] ADKINS, J. E., A. E. GREEN, and R. T. SHIELD: Finite plane strain. Phil. Trans. Roy. Soc. Lond. A **246**, 181—213. (*57, 59, 60*)

[*2*] BODASZEWSKI: O niesymetrycznym stanie napieçia i o jego zastosowaniach w mechanice ośrodków ciągłych. Arch. Mech. Stosow. **5**,351—396. (This work is characterized by KALISKI, PŁOCHOCKI, and ROGULA [1962, *39*] as "absolutely incorrect".)

[*3*] BORDONI, P. G.: Sopra le trasformazioni termoelastiche finite di certi solidi omogenei ed isotropi. Rend. Mat. e Applic. (5) **12**, 237—266. (*94*)

[*4*] BORDONI, P. G.: Deduzione di un'equazione di stato dei solidi dalla teoria delle trasformazioni termoelastiche finite. Rend. Accad. Lincei (8) **14**, 784—790. (*94*)

[*5*] BORDONI, P. G.: Trasformazioni adiabatiche di ampiezza finita. Ricerca Sci. **23**, 1569—1578. (*94*)

[*6*] ERICKSEN, J. L.: On the propagation of waves in isotropic incompressible perfectly elastic materials. J. Rational Mech. Anal. **2**, 329—337. Reprinted in Problems of Non-linear Elasticity. Intl. Sci. Rev. Ser. New York: Gordon & Breach 1965. (*48, 49, 72, 78, 95*)

[*7*] GILBARG, D., and D. PAOLUCCI: The structure of shock waves in the continuum theory of fluids. J. Rational Mech. Anal. **2**, 617—642. (*119*)

[*8*] GREEN, A. E., and E. W. WILKES: A note on the finite extension and torsion of a circular cylinder of compressible elastic isotropic material. Quart. J. Mech. Appl. Math. **6**, 240—249. (*66*)

[*9*] GREENSMITH, H. W., and R. S. RIVLIN: The hydrodynamics of non-Newtonian fluids, III. The normal stress effect in high-polymer solutions. Phil. Trans. Roy. Soc. Lond. A **245**, 399—429. (*115, 116, 119*)

[*10*] GUMBRELL, S. M., L. MULLINS, and R. S. RIVLIN: Departures of the elastic behaviour of rubbers in simple extension from the kinetic theory. Trans. Faraday Soc. **49**, 1495—1505 .(*95*)

[*11*] HEARMON, R. F. S.: "Third-order" elastic constants. Acta Crystal. **6**, 331—339. (*66*)

[*12*] HERMANS, J. J.: Dilute solutions of flexible chain molecules. Flow Properties of Disperse Systems, pp. 199—265. Amsterdam: North Holland Publ. Co. (*113*)

[*13*] HUGHES, D. S., and J. R. KELLY: Second-order elastic deformation of solids. Phys. Rev. (2) **92**, 1145—1149. (*66, 76, 77*)

[*14*] KRIEGER, I. M., and H. ELROD: Direct determination of the flow curves of non-Newtonian fluids, II. Shearing rate in the concentric cylinder viscometer. J. Appl. Phys. **27**, 134—136. (*113*)

[*15*] LEE, E. H., and I. KANTER: Wave propagation in finite rods of viscoelastic materials. J. Appl. Phys. **24**, 1115—1122. (*96t*)

[*16*] MANACORDA, T.: Sul legame sforzi-deformazione nelle trasformazioni finite di un mezzo continuo isotropo. Riv. Mat. Univ. Parma **4**, 31—42. (*47*)

[*17*] MIŞICU, M.: Echilibrul mediilor continue cu deformari mari. Stud. Cercet. Mec. Metal. **4**, 31—53. (*42, 63, 68b*)

[*18*] MOONEY, M.: A test of the theory of secondary viscoelastic stress. J. Appl. Phys. **24**, 675—678. (*119*)

[*19*] NIORDSEN, F.: Ändliga deformationer inom elasticitetsterion. Inst. Hallfasthetslära Kungl. Tekn. Högskolen Stockholm, Publ. nr. 106. (*42*)

[*20*] NOVOZHILOV, V. V.: Foundations of the Nonlinear Theory of Elasticity. Translated by F. BAGEMIHL, H. KOMM, and W. SEIDEL from a Russian book published in 1948. Rochester: Graylock. (*42*)

[*21*] NYE, J. F.: Some geometrical relations in dislocated crystals. Acta Metallurg. **1**, 153—162. (*34*)

[*22*] PAWLOWSKI, J.: Bestimmung des Reibungsgesetzes der nicht-Newtonschen Flüssigkeiten aus den Viskositätsmessungen mit Hilfe eines Rotationsviskosimeters. Kolloid-Z. **130**, 129—131. (*113*)

[*23*] REISSNER, E.: On a variational theorem for finite elastic deformations. J. Math. Phys. **32**, 129—135. Reprinted in Problems of Non-linear Elasticity. Intl. Sci. Rev. Ser. New York: Gordon & Breach 1965. (*88*)

[*24*] Rivlin, R. S.: The solution of problems in second order elasticity theory. J. Rational Mech. Anal. **2**, 53—81. Reprinted in Problems of Non-linear Elasticity. Intl. Sci. Rev. Ser. New York: Gordon & Breach 1965. (*66, 67*)

[*25*] Truesdell, C.: Corrections and additions to "The Mechanical Foundations of Elasticity and Fluid Dynamics". J. Rational Mech. Anal. **2**, 593—616. See [1952, *20*]. (*4, 5, 19A, 42, 66, 94, 95, 99, 103, 116*)

1954 [*1*] Adkins, J. E.: Some generalizations of the shear problem for isotropic incompressible materials. Proc. Cambridge Phil. Soc. **50**, 334—345. (*59, 95*)

[*2*] Adkins, J. E., A. E. Green, and G. C. Nicholas: Two-dimensional theory of elasticity for finite deformations. Phil. Trans. Roy. Soc. Lond. A **247**, 279—306. (*60*)

[*3*] Baker, M., and J. L. Ericksen: Inequalities restricting the form of the stress-deformation relations for isotropic elastic solids and Reiner-Rivlin fluids. J. Wash. Acad. Sci. **44**, 33—35. Reprinted in Foundations of Elasticity Theory. Intl. Sci. Rev. Ser. New York: Gordon & Breach 1965. (*51, 87, 119*)

[*4*] Browder, F. E.: Strongly elliptic systems of differential equations. Contrib. Th. Partial Diff. Eqns. Annals of Math. Studies No. 33, 15—51. (*68*)

[*5*] Ericksen, J. L.: Deformations possible in every isotropic incompressible perfectly elastic body. Z. angew. Math. Phys. **5**, 466—486. Reprinted in Problems of Non-linear Elasticity. Intl. Sci. Rev. Ser. New York: Gordon & Breach 1965. (*57, 91*)

[*6*] Ericksen, J. L.: Review of Truesdell [1953, *25*]. Math. Rev. **15**, 178. (*103*)

[*7*] Ericksen, J. L., and R. S. Rivlin: Large elastic deformations of homogeneous anisotropic materials. J. Rational Mech. Anal. *3*, 281—301. Reprinted in Problems of Non-linear Elasticity. Intl. Sci. Rev. Ser. New York: Gordon & Breach 1965. (*30, 57, 58*)

[*8*] Green, A. E.: A note on second-order effects in the torsion of incompressible cylinders. Proc. Cambridge Phil. Soc. **50**, 488—490. (*67*)

[*9*] Green, A. E., and E. B. Spratt: Second-order effects in the deformation of elastic bodies. Proc. Roy. Soc. Lond. A **224**, 347—361. (*63, 65, 67*)

[*10*] Green, A. E., and E. W. Wilkes: Finite plane strain for orthotropic bodies. J. Rational Mech. Anal. **3**, 713—723. (*58, 59, 60*)

[*11*] Green, A. E., and W. Zerna: Theoretical Elasticity. Oxford: Clarendon Press. (*42, 68, 70, 83, 95*)

[*12*] Kondo, K.: On the theory of the mechanical behavior of microscopically non-uniform materials. Res. Assn. Appl. Geometry (Tokyo), Res. Note No. (2) 4, 36 pp. (*34*)

[*13*] Manacorda, T.: Sopra un principio variazionale di E. Reissner per la statica dei mezzi continui. Boll. Un. Mat. Ital. (3) **9**, 154—159. (*88*)

[*14*] Morrey, C. B. jr.: Second order elliptic systems of differential equations. Contrib. Th. Partial Diff. Eqns. Annals of Math. Studies No. 33, 101—159. (*68*)

[*15*] Padden, F. J., and T. W. DeWitt: Some rheological properties of concentrated polyisobutylene solutions. J. Appl. Phys. **25**, 1086—1091. (*113, 116, 119*)

[*16*] Pawlowski, J.: Über eine Erweiterung des Helmholtzschen Prinzips. Kolloid-Z. **138**, 6—11. (*119*)

[*17*] Pilpel, N.: The viscoelastic properties of aqueous soap gels. Trans. Faraday Soc. **50**, 1369—1378. (*116*)

[*18*] Reiner, M.: Second order effects in elasticity and hydrodynamics. Bull. Res. Council Israel **3**, 372—379. (*119*)

[*19*] Rivlin, R. S., and C. Topakoglu: A theorem in the theory of finite elastic deformations. J. Rational Mech. Anal. **3**, 581—589. (*65*)

[*20*] Roberts, J. E.: Pressure distribution in liquids in laminar shearing motion and comparison with predictions from various theories. Proc. 2nd Internat. Congr. Rheology 1953, pp. 91—98. New York: Academic Press. (*115, 116*)

[*21*] Shimazu, Y.: Equation of state of materials composing the earth's interior. J. Earth Sci. Nagoya Univ. **2**, 15—172. (*42, 66, 69, 77*)

[*22*] Stoppelli, F.: Una generalizzazione di un teorema di Da Silva. Rend. Acad. Sci. Napoli (4) **21**, 214—225. (*44*)

[*23*] Stoppelli, F.: Un teorema di esistenza e di unicità relativo alle equazioni dell'elastostatica isoterma per deformazioni finite. Ricerche mat. **3**, 247—267. (*46*)

[*24*] Subba Rao, R., and S. D. Nigam: The effect of cross-viscosity on the performance of full journal bearing without side leakage. Z. angew. Math. Phys. **5**, 426—429. (*119*)

[*25*] Torre, C.: Kritik und Ergänzung des Maxwellschen Ansatzes für elastisch-zähe Stoffe. Verdrehung von Stäben als Beispiel. Öst. Ing.-Arch. **8**, 55—76. (*36*)

[26] TORRE, C.: Ergänzung zum Maxwellschen Ansatz für elastisch-zähe Stoffe. Verdrehung mit instationärer Spannungsänderung als Beispiel. Kolloid-Z. **138**, 11—18. (*36, 115*)

[27] TRUESDELL, C.: A new chapter in the theory of the elastica. Proc. First Midwest Conf. Solid. Mech 1953, pp. 52—55. (*44*)

1955 [1] ADKINS, J. E.: Finite deformation of materials exhibiting curvilinear aeolotropy. Proc. Roy. Soc. Lond. A **229**, 119—134. (*34, 58, 59*)

[2] ADKINS, J. E.: Some general results in the theory of large elastic deformations. Proc. Roy. Soc. Lond. A **231**, 75—90. (*59*)

[3] ADKINS, J. E.: A note on the finite plane-strain equations for isotropic incompressible materials. Proc. Cambridge Phil. Soc. **51**, 363—367. (*60, 95*)

[4] ADKINS, J. E., and R. S. RIVLIN: Large elastic deformations of isotropic materials, X. Reinforcement by inextensible cords. Phil. Trans. Roy. Soc. Lond. A **248**, 201—223. (*30, 43*)

[5] BILBY, B. A., R. BULLOUGH, and E. SMITH: Continuous distributions of dislocations: a new application of the methods of non-Riemannian geometry. Proc. Roy. Soc. Lond. A **231**, 263—273. (*34*)

[6] BILBY, B. A.: Types of dislocation sources. Defects in crystalline solids. Report of conf. at Bristol, 1954, pp. 123—133. London: The Physical Society. (*34*)

[7] CAPRIOLI, L.: Su un criterio per l'esistenza dell'energia di deformazione. Boll. Un. Mat. Ital. (3) **10**, 481—483. English translation in Foundations of Elasticity Theory. Intl. Sci. Rev. Ser. New York: Gordon & Breach 1965. (*83*)

[8] COTTER, B., and R. S. RIVLIN: Tensors associated with time-dependent stress. Quart. Appl. Math. **13**, 177—182. (*19A, 36*)

[9] DEWITT, T. W.: A rheological equation of state which predicts non-Newtonian viscosity, normal stresses, and dynamic moduli. J. Appl. Phys. **26**, 889—894. (*113, 115, 119*)

[10] ERICKSEN, J. L.: Eversion of a perfectly elastic spherical shell. Z. angew. Math. Mech. **35**, 381—385. (*57, 64, 95*)

[11] ERICKSEN, J. L.: Deformations possible in every compressible, isotropic, perfectly elastic material. J. Math. Phys. **34**, 126—128. Reprinted in Problems of Nonlinear Elasticity. Intl. Sci. Rev. Ser. New York: Gordon & Breach 1965. (*91*)

[12] ERICKSEN, J. L.: A consequence of inequalities proposed by BAKER and ERICKSEN. J. Wash. Acad. Sci. **45**, 268. (*119*)

[13] GREEN, A. E.: Finite elastic deformation of compressible isotropic bodies. Proc. Roy. Soc. Lond. A **227**, 271—278. (*59*)

[14] JAIN, M. K.: Boundary-layer effects in non-Newtonian fluids. Z. angew. Math. Mech. **35**, 12—16. (*119*)

[15] JAIN, M. K.: The motion of an infinite cylinder in rotating non-Newtonian liquid Z. angew. Math. Mech. **35**, 379—381. (*119*)

[16] KONDO, K., and collaborators: Memoirs of the Unifying Study of the Basic Problems of Engineering Sciences by Means of Geometry, I. Tokyo: Gakujutsu Bunken Fukyu-Kai. (*34*)

[17] KROUPA, F.: Plane deformation in the non-linear theory of elasticity. Czechosl. J. Phys. **5**, 18—29. (*95*)

[18] MANACORDA, T.: Sulla torsione di un cilindro circolare omogeneo e isotropo nella teoria delle deformazioni finite di solidi elastici incomprimibili. Boll. Un. Mat Ital. (3) **10**, 177—189. (*95*)

[19] NOLL, W.: On the continuity of the solid and fluid states. J. Rational Mech. Anal **4**, 3—81. Reprinted in Rational Mechanics of Materials. Intl. Sci. Rev. Ser. New York Gordon & Breach 1965. (*19A, 28, 36, 39, 42, 43, 48, 84, 85, 98, 99, 100, 108, 111, 113, 119, 126*)

[20] REINER, M.: The complete elasticity law for some metals according to POYNTING's observations. Appl. Sci. Res. A **5**, 281—295. (*59*)

[21] RIVLIN, R. S.: Further remarks on the stress-deformation relations for isotropic materials. J. Rational Mech. Anal. **4**, 681—702. (*11, 13*)

[22] RIVLIN, R. S.: Plane strain of a net formed by inextensible cords. J. Rational Mech. Anal. **4**, 951—974. (*43*)

[23] RIVLIN, R. S., and J. L. ERICKSEN: Stress-deformation relations for isotropic materials. J. Rational Mech. Anal. **4**, 323—425. Reprinted in Rational Mechanics of Materials. Intl. Sci. Rev. Ser. New York: Gordon & Breach 1965. (*11, 12, 13, 19A, 35, 45, 69, 119*)

[24] SHENG, P.-L.: Secondary Elasticity. Chin. Assoc. Adv. Sci. (Taipei) Monograph Series 1, I, No. 1. (*42, 65, 66, 93*)

[25] SIGNORINI, A.: Trasformazioni termoelastiche finite. Memoria 3ª, Solidi incomprimibili. Ann. di Mat. Pur. Appl. (4) **39**, 147—201. (*42, 63, 64, 65, 95*)

[26] STOPPELLI, F.: Sulla svilluppabilità in serie di potenze di un parametro delle soluzioni delle equazioni dell'elastostatica isoterma. Ricerche Mat. **4**, 58—73. (*46*)

[27] THOMAS, T. Y.: On the structure of the stress-strain relations. Proc. Nat. Acad. Sci. U.S.A. **41**, 716—720. (*19A, 99*)

[28] THOMAS, T. Y.: Combined elastic and Prandtl-Reuss stress-strain relations. Proc. Nat. Acad. Sci. U.S.A. **41**, 720—726. (*100, 103*)

[29] THOMAS, T. Y.: Kinematically preferred co-ordinate systems. Proc. Nat. Acad. Sci. U.S.A. **41**, 762—770. (*19A, 99*).

[30] THOMAS, T. Y.: Combined elastic and von Mises stress-strain relations. Proc. Nat. Acad. Sci. U.S.A. **41**, 908—910. (*100*)

[31] TRUESDELL, C.: Hypo-elasticity. J. Rational Mech. Anal. **4**, 83—133, 1019—1020. Reprinted in Foundations of Elasticity Theory. Intl. Sci. Rev. Ser. New York: Gordon & Breach 1965. (*99, 100, 103*)

[32] TRUESDELL, C.: The simplest rate theory of pure elasticity. Comm. Pure Appl. Math. N.Y.U. **8**, 123—132. Reprinted in Foundations of Elasticity Theory. Intl. Sci. Rev. Ser. New York: Gordon & Breach 1965. (*28, 99, 103*)

[33] WILKES, E. W.: On the stability of a circular tube under end thrust. Quart. J. Mech. Appl. Math. **8**, 88—100. (*68b*)

1956 [1] ADKINS, J. E.: Finite plane deformation of thin elastic sheets reinforced with inextensible cords. Phil. Trans. Roy. Soc. Lond. A **249**, 125—150. (*43*)

[2] BILBY, B. A., and E. SMITH: Continuous distributions of dislocations, III. Proc. Roy. Soc. Lond. A **236**, 481—505. (*34*)

[3] BROER, L. F. J.: On the hydrodynamics of visco-elastic fluids. Appl. Sci. Res. A **6** (1956/57), 226—236. (*113, 119*)

[4] BUDIANSKY, B., and C. E. PEARSON: On variational principles and GALERKIN's procedure for non-linear elasticity. Quart. Appl. Math. **14** (1956/57), 328—331. (*88*)

[5] DOYLE, T. C., and J. L. ERICKSEN: Nonlinear elasticity. Adv. Appl. Mech. **4**, 53—115. (*42*)

[6] ERICKSEN, J. L.: Stress deformation relations for solids. Canad. J. Phys. **34**, 226—227. (*83*)

[7] ERICKSEN, J. L.: Overdetermination of the speed in rectilinear motion of non-Newtonian fluids. Quart. Appl. Math. **14**, 318—321. Reprinted in Rational Mechanics of Materials. Intl. Sci. Rev. Ser. New York: Gordon & Breach 1965. (*117, 122*)

[8] ERICKSEN, J. L., and R. A. TOUPIN: Implications of HADAMARD's condition for elastic stability with respect to uniqueness theorems. Canad. J. Math. **8**, 432—436. Reprinted in Foundations of Elasticity Theory. Intl. Sci. Rev. Ser. New York Gordon & Breach 1965. (*68b*)

[9] GREEN, A. E.: Hypo-elasticity and plasticity. Proc. Roy. Soc. Lond. A **234**, 46—59. (*100, 103*)

[10] GREEN, A. E.: Hypo-elasticity and plasticity, II. J. Rational Mech. Anal. **5**, 725 734. (*100*)

[11] GREEN, A. E.: Simple extension of a hypo-elastic body of grade zero. J. Rational Mech. Anal. **5**, 637—642. (*103*)

[12] GREEN, A. E., and R. S. RIVLIN: The mechanics of non-linear materials with memory. Brown Univ. Report C11—17. (*28, 29, 31, 37, 121*)

[13] GREEN, A. E., and R. S. RIVLIN: Steady flow of non Newtonian fluids through tubes. Quart Appl. Math. **14**, 299—308. (*117, 122*)

[14] IKENBERRY, E., and C. TRUESDELL: On the pressures and the flux of energy in a gas according to Maxwell's kinetic theory, I. J. Rational Mech. Anal. **5**, 3—54. (*19A, 125*)

[15] KOPPE, E.: Methoden der nichtlinearen Elastizitätstheorie mit Anwendung auf die dünne Platte endlicher Durchbiegung. Z. angew. Math. Mech. **36**, 455—462. (*42*)

[16] LODGE, A. S.: A network theory of flow and stress in concentrated polymer solutions. Trans. Faraday Soc. **52**, 120—130. (*119*)

[17] MANACORDA, T.: Sul potenziale isotermo nella più generale elasticità di secondo grado per solidi incomprimibili. Ann. di Mat. (4) **40**, 77—86. (*94*)

[18] NARASIMHAN, M. H. L.: On the steady laminar flow of certain non-Newtonian liquids through an elastic tube. Proc. Indian Acad. Sci. A **43**, 237—246. (*119*)

[19] OLDROYD, J. G.: Non-Newtonian flow of liquids and solids. Rheology, Theory, and Application **1**, Ch. 16. New York: Academic Press. (*104*)

[20] PHILIPPOFF, W.: Flow birefringence and stress. J. Appl. Phys. **27**, 984—989. (*116*)

[*21*] Pearson, C. E.: General theory of elastic stability. Quart. Appl. Math. **14** (1956/57), 133—144. (*68b*)
[*22*] Reiner, M.: Phenomenological Macrorheology. Rheology, Theory and Applications **1**, Ch. 2. New York: Academic Press. (*104*)
[*23*] Rivlin, R. S.: Stress relaxation in incompressible elastic materials at constant deformation. Quart. Appl. Math. **14**, 83—89. (*39, 55, 57, 70*)
[*24*] Rivlin, R. S.: Large elastic deformations. Rheology, Theory and Applications **1**, Ch. 10. New York: Academic Press. (*42*)
[*25*] Rivlin, R. S.: Solution of some problems in the exact theory of visco-elasticity. J. Rational Mech. Anal. **5**, 179—188. (*111, 112, 113, 115, 119*)
[*26*] Seeger, A.: Neuere mathematische Methoden und physikalische Ergebnisse zur Kristallplastizität. Verformung und Fließen des Festkörpers (Koll. Madrid 1955). Berlin-Göttingen-Heidelberg: Springer. (*34*)
[*27*] Srivastava, A. C.: Beltrami motions in non-Newtonian fluids. J. Assoc. Appl. Phys. **3**, 69—72. (*119*)
[*28*] Theodorides, P.: Mehrparametrige Zähigkeit als Grundlage einer Quasi-Kontinuumstheorie der Kompressionsfront für mehratomige Gase. Z. angew. Math. Mech. Sonderheft, 538—546. (*119*)
[*29*] Thomas, T. Y.: Isotropic materials whose deformation and distortion energies are expressible by scalar invariants. Proc. Nat. Acad. Sci. U.S.A. **42**, 603—608. (*101*)
[*30*] Tiffen, R., and A. C. Stevenson: Elastic isotropy with body force and couple. Quart. J. Mech. Appl. Math. **9**, 306—312. (*98*)
[*31*] Toupin, R. A.: The elastic dielectric. J. Rational Mech. Anal. **5**, 849—915. Reprinted in Foundations of Elasticity Theory. Intl. Sci. Rev. Ser. New York: Gordon & Breach 1965. (*97*)
[*32*] Truesdell, C.: Das ungelöste Hauptproblem der endlichen Elastizitätstheorie. Z. angew. Math. Mech. **36**, 97—103. Russian transl. in Mekhanika **1** (41), 67—74 (1957). English transl. in Foundations of Elasticity Theory. Intl. Sci. Rev. Ser. New York: Gordon & Breach 1965. (*51, 52, 53, 74, 87*)
[*33*] Truesdell, C.: Hypo-elastic shear. J. Appl. Phys. **27**, 441—447. Reprinted in Foundations of Elasticity Theory. Intl. Sci. Rev. Ser. New York: Gordon & Breach 1965. (*103*)
[*34*] Truesdell, C.: Experience, theory, and experiment. Proc. 6th Hydraulics Conf. Bull. **36**, State Univ. Iowa Studies Engr., 3—18. (*119*)
[*35*] Truesdell, C.: On the pressures and the flux of energy in a gas according to Maxwell's kinetic theory, II. J. Rational Mech. Anal. **5**, 55—128. (*119, 119A, 123*)
[*36*] Verma, P. D. S.: Hypo-elastic pure flexure. Proc. Indian Acad. Sci. A **44**, 185—192. (*103*)
[*37*] Ziegler, H.: On the concept of elastic stability. Adv. Appl. Mech. **4**, 351—403. (*68b*)

1957 [*1*] Adkins, J. E., and A. E. Green: Plane problems in second-order elasticity theory. Proc. Roy. Soc. Lond. A **239**, 557—576. (*60*)
[*2*] Adkins, J. E.: Cylindrically symmetrical deformations of incompressible elastic materials reinforced with inextensible cords. J. Rational Mech. Anal. **5**, 189—202. (*43*)
[*3*] Barta, J.: On the non-linear elasticity law. Acta Tech. Acad. Sci. Hung. **18**, 55—65. (*48, 51*)
[*4*] Blackburn, W. S., and A. E. Green: Second-order torsion and bending of isotropic elastic cylinders. Proc. Roy. Soc. Lond. A **240**, 408—422. (*66*)
[*5*] Brodnyan, J. G., F. H. Gaskins, and W. Philippoff: On normal stresses, flow curves, flow birefringence, and normal stresses of polyisobutylene solutions. Part II. Experimental. J. Soc. Rheol. **1**, 109—118. (*116*)
[*6*] Chu, Boa-Teh: Thermodynamics of elastic and of some visco-elastic solids and non-linear thermoelasticity. Brown Univ. Div. Eng. Report No. 1, July. (*96*)
[*7*] Ericksen, J. L.: Characteristic direction for equations of motion of non-Newtonian fluids. Pac. J. Math. **7**, 1557—1562. (*119*)
[*8*] Green, A. E., and R. S. Rivlin: The mechanics of non-linear materials with memory, Part. I. Arch. Rational Mech. Anal. **1** (1957/58), 1—21, 470. Reprinted in Rational Mechanics of Materials. Intl. Sci. Rev. Ser. New York: Gordon & Breach 1965. (*28, 29, 31, 37, 121*)
[*9*] Hill, R.: On uniqueness and stability in the theory of finite elastic strain. J. Mech. Phys. Solids **5**, 229—241. (*52, 68, 68b, 87*)
[*10*] Jain, M. K.: The stability of certain non-Newtonian liquids contained between two rotating cylinders. J. Sci. Engr. Res. **1**, 195—202. (*119*)
[*11*] Manacorda, T.: Sul comportamento meccanico di una classe di corpi naturali. Riv. Mat. Univ. Parma **8**, 15—25. (*41, 43*)

[12] MANFREDI, B.: Sopra la più generale equazione reologica di stato per una classe di solidi naturali. Boll. Un. Mat. Ital. (3) **12**, 422—435. (*28*)

[13] MARKOVITZ, H.: Normal stress effect in polyisobutylene solutions, II. Classification of rheological theories. Trans. Soc. Rheol. **1**, 37—52. Reprinted in Rational Mechanics of Materials. Intl. Sci. Rev. Ser. New York: Gordon & Breach 1965. (*108, 113, 115, 116, 119*)

[14] MARKOVITZ, H., and R. B. WILLIAMSON: Normal stress effect in polyisobutylene solutions, I. Measurements in a cone and plate instrument. Trans. Soc. Rheol **1**, 25—36. (*115, 116*)

[15] NOLL, W.: On the foundation of the mechanics of continuous media. Carnegie Inst. Tech. Dept. Math. Rep. No. 17, June. (*28, 29, 31*)

[16] PAO, Y.-H.: Hydrodynamic theory for the flow of a visco-elastic fluid. J. Appl. Phys. **28**, 591—598. (*37, 119*)

[17] REICHHARDT, H.: Vorlesungen über Vektor- und Tensorrechnung. Berlin: VEB Deutscher Verlag der Wissenschaften. (*6*)

[18] PHILIPPOFF, W.: On normal stresses, flow curves, flow birefringence, and normal stresses of polyisobutylene solutions. Part I. Fundamental Principles. Trans. Soc. Rheol. **1**, 95—107. (*104*)

[19] REINER, M.: A centripetal-pump effect in air. Proc. Roy. Soc. Lond. A **240**, 173—188. (*116*)

[20] RIVLIN, R. S.: The relation between the flow of non-Newtonian fluids and turbulent Newtonian fluids. Quart. Appl. Math. **15** (1957/58), 212—215. Correction, Q. Appl. Math. **17** (1959/60), 447 (1960). (*122*)

[21] ROBERTS, J. E.: Normal stress effects in tetralin solutions of polyisobutylene. Nature **179**, 487—488. (*116*)

[22] SMITH, G. F., and R. S. RIVLIN: The anisotropic tensors. Quart. Appl. Math. **15**, 308—314. (*33*)

[23] SMITH, G. F., and R. S. RIVLIN: Stress-deformation relations for anisotropic solids. Arch. Rational Mech. Anal. **1** (1957/58), 107—112. (*33, 50, 127*)

[24] STONE, D. E.: On non-existence of rectilinear motion in plastic solids and non-Newtonian fluids. Quart. Appl. Math. **15**, 257—262. (*117*)

[25] STOPPELLI, F.: Su un sistema di equazioni integrodifferenziali interessante l'elastostatica. Ricerche Mat. **6**, 11—26. (*46*)

[26] STOPPELLI, F.: Sull'esistenza di soluzioni delle equazioni dell'elastostatica isoterma nel caso di sollecitazioni dotate di assi di equilibrio, I. Ricerche Mat. **6**, 241—287. (*46*)

[27] TAYLOR, G. I., and P. G. SAFFMAN: Effects of compressibility at low Reynolds number. J. Aero. Sci. **24**, 553—562. (*116*)

[28] THEODORIDES, P. J.: A basic approach to shock front analysis. Univ. Maryland Inst. Fluid Dyn. Appl. Math. Note, January. (*119*)

[29] THOMAS, T. Y.: Deformation energy and the stress-strain relations for isotropic materials. J. Math. and Phys. **35**, 335—350. (*101*)

[30] TRUESDELL, C.: Sulle basi della termomeccanica. Rend. Accad. Lincei (8) **22**, 33—88, 158—166. English transl. in Rational Mechanics of Materials. Intl. Sci. Rev. Ser. New York: Gordon & Breach 1965. (*130*)

[31] WILLIAMS, M. C., and R. B. BIRD: Steady flow of an Oldroyd visco-elastic fluid in tubes, slits, and narrow annuli. A. I. Ch. E. Journal. **8**, 378—382. (*119*)

1958 [1] ADKINS, J. E.: A reciprocal property of the finite plane strain equations. J. Mech. Phys. Solids **6**, 267—275. (*60, 95*)

[2] ADKINS, J. E.: Dynamic properties of resilient materials: Constitutive equations. Phil. Trans. Roy. Soc. Lond. A **250**, 519—541. (*13, 30, 50, 126*)

[3] ADKINS, J. E.: A three-dimensional problem for highly elastic materials subject to constraints. Quart. J. Mech. Appl. Math. **11**, 88—97. (*43*)

[4] ANGLES D'AURIAC, P.: Contribution à l'étude de l'élasticité des corps très déformables. Thèse. Univ. Paris = Arch. Mech. Stosow. **13**, 775—824 (1961). (*42, 54*)

[5] BERG, B. A.: Deformational anisotropy [in Russian]. Prikl. mat. Mekh. **22**, 67—77. English transl., J. Appl. Math. Mech. **22**, 90—103. (*69*)

[6] BERNSTEIN, B., and J. L. ERICKSEN: Work functions in hypo-elasticity. Arch. Rational Mech. Anal. **1** (1957/58), 396—409. (*101*)

[7] BHATNAGAR, P. L., and S. K. LAKSHMANA RAO: Problems on the motion of non-Newtonian viscous liquids. Proc. Indian Acad. Sci. A **45**, (1957), 161—171. (*119*)

[8] BILBY, B. A., R. BULLOUGH, L. R. T. GARDNER, and E. SMITH: Continuous distributions of dislocations, IV. Single glide and plane strain. Proc. Roy. Soc. Lond. A **244**, 538—577. (*34*)

[*9*] BILBY, B. A., L. R. T. GARDNER, and E. SMITH: The relation between dislocation density and stress. Acta Metallurg. **6**, 29—33. (*34*)

[*10*] BILBY, B. A., and L. R. T. GARDNER: Continuous distributions of dislocations, V. Twisting under conditions of single glide. Proc. Roy. Soc. Lond. A **247**, 52—108. (*34*)

[*11*] BLACKBURN, W. S.: Second-order effects in the torsion and bending of transversely isotropic incompressible elastic beams. Quart. J. Mech. Appl. Math. **11**, 142—158. (*67*)

[*12*] CARICATO, G.: Sulle trasformazioni di un sistema elastico atte a conservare l'isotropia. Rend. Mat. e Applic. (5) **17**, 313—318. (*50*)

[*13*] CRIMINALE W. O., jr., J. L. ERICKSEN, and G. L. FILBY, jr.: Steady shear flow of non-Newtonian fluids. Arch. Rational. Mech. Anal. **1** (1957/58), 410—417. (*108, 117, 119*)

[*14*] ERICKSEN, J. L.: Hypo-elastic potentials. Quart. J. Mech. Appl. Math. **11**, 67—72. (*101*)

[*15*] ERICKSEN, J. L., and C. TRUESDELL: Exact theory of stress and strain in rods and shells. Arch. Rational Mech. Anal. **1**, 296—323. Reprinted in Rational Mechanics of Materials. Intl. Sci. Rev. Ser. New York: Gordon & Breach 1965. (*98*)

[*16*] FRANK, F. C.: On the theory of liquid crystals. Disc. Faraday Soc. **25**, 19—28. (*128*)

[*17*] FREUDENTHAL, A. M., and H. GEIRINGER: The mathematical theories of the inelastic continuum. This Encyclopedia, Vol. VI, pp. 229—433. (*103*)

[*18*] GENT, A. N., and A. G. THOMAS: Forms for the stored (strain) energy function for vulcanized rubber. J. Polymer Sci. **28**, 625—628. (*55*)

[*19*] GIESEKUS, H.: Die rheologische Zustandsgleichung. Rheol. Acta **1** (1958/61), 2—20. (*104, 121*)

[*20*] GOLD, R. R., and M. Z. v. KRZYWOBLOCKI: On superposability and self-superposability conditions for hydrodynamic equations based on continuum, I. J. reine angew. Math. **199**, 139—164. (*119*)

[*21*] GÜNTHER, W.: Zur Statik und Dynamik des Cosseratschen Kontinuums. Abh. Braunschw. Wiss. Ges. **10**, 195—213. (*34, 44, 98*)

[*22*] HALMOS, P. R.: Finite Dimensional Vector Spaces, 2nd ed. Princeton Univ. Press. (*6, 33*)

[*23*] JAIN, M. K.: The collapse or growth of a spherical bubble or cavity in certain non-Newtonian liquid. Proc. 1st. Congr. Theor. Appl. Mech. (Kharagpur, 1956), pp. 207—212. (*119*)

[*24*] JOBLING, A., and J. E. ROBERTS: Goniometry of flow and rupture. Rheology Theory and Applications. New York: Academic Press. **2**, 503—535. (*116*)

[*25*] JOHN, F.: On finite deformations of elastic isotropic material. Inst. Math. Sci. New York Univ. Report IMM-NYU 250. (*42, 46, 66*)

[*26*] JUNG, H.: Zur Theorie der Maxwellschen Flüssigkeiten. Rheol. Acta **1** (1958/61), 280—285. (*119*)

[*27*] KONDO, K., and collaborators: Memoirs of the Unifying Study of the Basic Problems in Engineering and Physical Sciences by Means of Geometry, II. Tokyo: Gakujutsu Bunken Fukyu-Kai. (*34*)

[*28*] KRÖNER, E.: Kontinuumstheorie der Versetzungen und Eigenspannungen. Berlin-Göttingen-Heidelberg: Springer. (*34*)

[*29*] MITHAL, K. G.: Motion of a non-Newtonian fluid produced by the uniform rotation of a plate. Ganita **9**, 95—117. (*119*)

[*30*] NIGAM, S. D.: Rotation of an infinite plane lamina in non-Newtonian liquid: Motion started impulsively from rest. Bull. Calcutta Math. Sci. **50**, 65—67. (*119*)

[*31*] NOLL, W.: A mathematical theory of the mechanical behavior of continuous media. Arch. Rational Mech. Anal. **2** (1958/59), 197—226. Reprinted in Rational Mechanics of Materials. Intl. Sci. Rev. Ser. New York: Gordon & Breach 1965. (*15, 19A, 26, 27, 28, 29, 31, 32, 33, 35, 36*)

[*32*] OLDROYD, J. G.: Non-Newtonian effects in steady motion of some idealized elastico-viscous liquids. Proc. Roy. Soc. Lond. A **245**, 278—297. (*113, 115, 117, 119*)

[*33*] PARIA, G.: Love waves in hypoelastic body of grade zero. Quart. J. Mech. Appl. Math. **11**, 509—512. (*103*)

[*34*] PIPKIN, A. C., and R. S. RIVLIN: Note on a paper "Further remarks on the stress-deformation relations for isotropic materials." Arch. Rational Mech. Anal. **1** (1957/58), 469. (*11*)

[*35*] PIPKIN, A. C., and R. S. RIVLIN: The formulation of constitutive equations in continuum physics. Div. Appl. Math. Brown Univ. Report. Sept. (*11, 13, 29, 33, 96, 97*)

[36] Popper, B., and M. Reiner: Cross-stresses in air. Boundary Layer Research. Berlin-Göttingen-Heidelberg: Springer. (*116*)
[37] Posey, C. J.: Discussion on open channel flow. Trans. Amer. Soc. Civil Engr. **123**, 712—713. (*122*)
[38] Reiner, M.: The centripetal-pump effect in a vacuum pump. Proc. Roy. Soc. Lond. A **247**, 152—167. (*116*)
[39] Reiner, M.: Rheology. This Encyclopedia, Vol. VI, pp. 434—550. (*104*)
[40] Rice, M. H., R. G. McQueen, and J. M. Walsh: Compression of solids by strong shock waves. Solid State Physics **6**, 1—63. (*71*)
[41] Sakadi, Z.: Stationary motion of viscous fluid around a rotating solid sphere. Mem. Fac. Eng. Nagoya **10**, 42—45. (*119*)
[42] Sharma, S. K.: Propagation of sound waves in visco-elastic compressible fluids. J. Sci. Engr. Res. **2**, 253—258. (*119*)
[43] Signorini, A.: Estensione delle formole di Almansi a sistemi elastici anisotropi. Rend. Accad. Lincei (8) **25**, 246—253. (*86*)
[44] Smith, G. F., and R. S. Rivlin: The strain-energy function for anisotropic elastic materials. Trans. Amer. Math. Soc. **88**, 175—193. (*85*)
[45] Stoppelli, F.: Sull'esistenza di soluzioni delle equazioni dell'elastostatica isoterma nel caso di sollecitazioni dotate di assi di equilibrio, II, III. Ricerche Mat. **7**, 71—101, 138—152. (*46*)
[46] Srivastava, A. C.: The flow of a non-Newtonian liquid near a stagnation point. Z. angew. Math. Phys. **9**, 80—84. (*119*)
[47] Srivastava, A. C.: Rotation of a plane lamina in non-Newtonian fluids. Bull. Calcutta Math. Soc. **50**, 57—64. (*119*)
[48] Theodorides, P.: Parallel effects of bulk viscosity and time lag in kinetics of non-monatomic fluids. Z. angew. Math. Phys. **96**, 668—686. (*119*)
[49] Treloar, L. R. G.: The Physics of Rubber Elasticity. 2nd ed. Oxford Univ. Press. (*95*)
[50] Verma, P. D. S.: Deformation energy for hypoelastic materials of grade zero. J. Sci. Engr. Res. **2**, 251—252. (*101*)
[51] Verma, P. D. S.: Steady flow of linear fluent material past a fixed sphere. J. Assoc. Appl. Physicists **5**, 6—9. (*119*)

1959 [1] Capriz, G.: Sui casi di "incompatibilità" tra l'elastostatica classica e la teoria delle deformazioni elastiche finite. Riv. Mat. Univ. Parma **10**, 119—129. (*64*)
[2] Coleman, B. D., and W. Noll: On certain steady flows of general fluids. Arch. Rational Mech. Anal. **3**, 289—303. Reprinted in Rational Mechanics of Materials. Intl. Sci. Rev. Ser. New York: Gordon & Breach 1965. (*107, 108, 111, 113, 115*)
[3] Coleman, B. D., and W. Noll: On the thermostatics of continuous media. Arch. Rational Mech. Anal. **4**, 97—128. Reprinted in Foundations of Elasticity Theory. Intl. Sci. Rev. Ser. New York: Gordon & Breach 1965. (*51, 52, 81, 87, 89*)
[4] Coleman, B. D., and W. Noll: Helical flow of general fluids. J. Appl. Phys. **30**, 1508—1512. Reprinted in Rational Mechanics of Materials. Intl. Sci. Rev. Ser. New York: Gordon & Breach 1965. (*107, 108, 112, 119*)
[5] Dana, J. S.: Dana's manual of mineralogy, 17th ed., revised by C. S. Hurlbut, jr. New York: John Wiley. (*33*)
[6] Dewey, J.: Strong shocks and stress-strain relations in solids. Aberdeen Proving Ground Ballistic Res. Lab. Rep. No. 1074. (*71*)
[7] Ericksen, J. L.: Secondary flow phenomena in non-linear fluids. Tappi **42**, 773—775. (*108, 114*)
[8] Genensky, S. M., and R. S. Rivlin: Infinitesimal plane strain in a network of elastic cords. Arch. Rational Mech. Anal. **4**, 30—44. (*43*)
[9] Gold, R. R., and M. Z. v. Krzywoblocki: On superposability and self-superposability conditions for hydrodynamic equations based on continuum, II. J. reine angew. Math. **200**, 140—169. (*119*)
[10] Green, A. E., R. S. Rivlin, and A. J. M. Spencer: The mechanics of non-linear materials with memory, Part II. Arch. Rational Mech. Anal. **3**, 82—90. (*28, 37*)
[11] Green, A. E., and A. J. M. Spencer: The stability of a circular cylinder under finite extension and torsion. J. Math. Phys. **37**, 316—338. (*68b*)
[12] Hill, R.: Some basic principles in the mechanics of solids without a natural time. J. Mech. Phys. Solids **7**, 209—225. (*100, 101*)
[13] Jain, M. K.: Problems of cross-elasticity Proc. 2nd. Congr. Theor. Appl. Mech. (New Delhi, 1956), pp. 81—86. (*74*)
[14] Jobling, A., and J. E. Roberts: Flow testing of viscoelastic materials. Design and calibration of the Roberts-Weissenberg model R8 rheogoniometer. J. Polymer Sci. **36**, 421—431. (*116*)

[*15*] JOBLING, A., and J. E. ROBERTS: An investigation, with the Weissenberg rheogoniometer, of the stress distribution in flowing polyisobutylene solutions at various concentrations and molecular weights. J. Polymer Sci. **36**, 433—441. (*116*)
[*16*] KOTAKA, T.: Note on the normal stress effect in the solution of rodlike macromolecules. J. Chem. Phys. **30**, 1566—1567. (*119*)
[*17*] KRÖNER, E., u. A. SEEGER: Nicht-lineare Elastizitätstheorie der Versetzungen und Eigenspannungen. Arch. Rational Mech. Anal. **3**, 97—119. (*34, 94*)
[*18*] LANGLOIS, W. E., and R. S. RIVLIN: Steady flow of slightly visco-elastic fluids. Brown Univ. D.A.M. Tech. Rep. No. 3,. December. (*115, 121, 122*)
[*19*] MANACORDA, T.: Sulla propagazione di onde ordinarie di discontinuità nella elasticità di secondo grado per solidi incompribili. Riv. Mat. Univ. Parma **10**, 19—33. (*74, 78, 95*)
[*20*] OLDROYD, J. G.: Complicated rheological properties. Rheology of Disperse Systems, pp. 1—15. New York, etc.: Pergamon. (*104*)
[*21*] PIPKIN, A. C., and R. S. RIVLIN: The formulation of constitutive equations in continuum physics, I. Arch. Rational Mech. Anal. **4** (1959/60), 129—144. (*11, 13, 29, 33*)
[*22*] REINER, M.: The physics of air viscosity as related to gas-bearing design. First Intl. Sympos. Gas-Lub. Bearings, Off. Naval Res., Washington. (*116*)
[*23*] RIVLIN, R. S.: The constitutive equations for certain classes of deformations. Arch. Rational. Mech. Anal. **3**, 304—311. (*39*)
[*24*] RIVLIN, R. S.: The deformation of a membrane formed by inextensible cords. Arch. Rational Mech. Anal. **2** (1958/59), 447—476. (*43*)
[*25*] RIVLIN, R. S.: Mathematics and rheology, the 1958 Bingham Medal Address. Physics Today **12**, 32—34, 36. (*94*)
[*26*] SEEGER, A., u. E. MANN: Anwendung der nicht-linearen Elastizitätstheorie auf Fehlstellen in Kristallen. Z. Naturforsch. **14**a, 154—164. (*94*)
[*27*] SERRIN, J.: Mathematical principles of classical fluid mechanics. This Encyclopedia, Vol. VIII, Part I, pp. 125—263. (*12, 28, 96t*)
[*28*] SERRIN, J.: The derivation of stress-deformation relations for a Stokesian fluid. J. Math. Mech. **8**, 459—470. (*12*)
[*29*] SERRIN, J.: Poiseuille and Couette flow of non-Newtonian fluids. Z. angew. Math. Mech. **39**, 295—299. (*114*)
[*30*] SHARMA, S. K.: Rotation of a plane lamina in a visco-elastic liquid. Appl. Sci. Res. A **9** (1959/60), 43—52. (*119*)
[*31*] SHARMA, S. K.: Visco-elastic steady flow. Z. angew. Math. Mech. **39**, 313—322. (*119*)
[*32*] SHARMA, S. K.: Flow of a visco-elastic liquid near a stagnation point. J. Phys. Soc. Japan **14**, 1421—1425. (*119*)
[*33*] SPENCER, A. J. M., and R. S. RIVLIN: The theory of matrix polynomials and its application to the mechanics of isotropic continua. Arch. Rational Mech. Anal. **2** (1958/59), 309—336. (*11, 13*)
[*34*] SPENCER, A. J. M., and R. S. RIVLIN: Finite integrity bases for five or fewer symmetric 3×3 matrices. Arch. Rational Mech. Anal. **2** (1958/59), 435—446. (*11*)
[*35*] SPENCER, A. J. M.: On finite elastic deformations with a perturbed strain-energy function. Quart. J. Mech. Appl. Math. **12**, 129—145. (*69*)
[*36*] SRIVASTAVA, A. C.: Superposition in non-Newtonian fluids. Proc. 2nd Congr. Theor. Appl. Mech. (New Delhi, 1956), pp. 187—194. (*119*)
[*37*] TRUESDELL, C.: The rational mechanics of materials — past, present, future. Applied Mech. Rev. **12**, 75—80. Corrected reprint, Applied Mechanics Surveys. Washington: Spartan Books. 1965. (*65, 96*)
[*38*] URBANOWSKI, W.: Small deformations superposed on finite deformations of a curvilinearly orthotropic body. Arch. Mech. Stosow. **11**, 223—241. (*58, 69*)
[*39*] VERMA, P. D. S.: Hypo-elastic strain in rotating shafts and spherical shells. Proc. 2nd Congr. Theor. Appl. Mech. (New Delhi, 1956), pp. 99—110. (*103*)
[*40*] YAMAMOTO, M.: Phenomenological theory of visco-elasticity of three dimensional bodies. J. Phys. Soc. Japan **14**, 313—330. (*35, 36*)
[*41*] ZAHORSKI, S.: A form of the elastic potential for rubber-like materials. Arch. Mech. Stosow. **11**, 613—618. (*95*)
1960 [*1*] ADAMS, N.: Measurements of pressure gradients in cone-and-plate and in parallel plate viscometers. Rheol. Abstr. **3**, No. 3, 28—30. (*115*)
[*2*] ADKINS, J. E.: Symmetry relations for orthotropic and transversely isotropic materials. Arch. Rational Mech. Anal. **4**, 193—213. (*11, 13, 50*)
[*3*] ADKINS, J. E.: Further symmetry relations for transversely isotropic materials. Arch. Rational Mech. Anal. **5**, 263—274. (*11, 13, 50, 126*)

[4] AERO, E. L., and E. V. KUVSHINSKII: Fundamental equations of the theory of elastic media with rotationally interacting particles [in Russian]. Fizika Tverdogo Tela 2, 1399—1409. English Transl. Soviet Physics Solid State 2, 1272—1821 (1961). (*98*)

[5] BARENBLATT, G. I., IU. P. ZHELTOV, and I. N. KOCHINA: Basic concepts in the theory of seepage of homogeneous liquids in fissured rocks (strata). PMM **24**, 1286—1303 (Transl. of Priklad. Mat. Mekh. **24**, 852—864). (*23*)

[6] BERGEN, J. T.: Stress relaxation of polymeric materials in combined torsion and tension. Visco-elasticity: Phenomenological Aspects, pp. 108—132. New York: Academic Press. (*39*)

[7] BERGEN, J. T., D. C. MESSERSMITH, and R. S. RIVLIN: Stress relaxation for biaxial deformation of filled high polymers. J. Appl. Polymer Sci. **3**, 153—167. (*39, 70*)

[8] BERNSTEIN, B.: Hypo-elasticity and elasticity. Arch. Rational Mech. Anal. **6**, 89—104. Reprinted in Foundations of Elasticity Theory. Intl. Sci. Rev. Ser. New York: Gordon & Breach 1965. (*99, 100, 101*)

[9] BHAGAVANTAM, S.: Third order elasticity. Proc. 3rd Congr. Theor. Appl. Mech. (Bangalore, 1957), pp. 25—30. (*66*)

[10] BHAGAVANTAM, S., and E. V. CHELAM: Elastic behavior of matter under very high pressures. Uniform compression. Proc. Indian Acad. Sci. **52**, 1—19. (*69*)

[11] BHAGAVANTAM, S., and E. V. CHELAM: Elastic behavior of matter under very high pressures. General deformation. J. Indian Inst. Sci. **42**, 29—40. (*69*)

[12] BHATNAGAR, P. L., and S. K. LAKSHMANA RAO: Steady motion of non-Newtonian fluids in tubes. Proc. 3rd. Congr. Theor. Appl. Mech. (Bangalore, 1957), pp. 225—234. (*117, 119*)

[13] BILBY, B. A.: Continuous distributions of dislocations. Progress in Solid Mechanics **1**, 329—398. (*34*)

[14] BIRD, R. B.: New variational principle for incompressible non-Newtonian flow. Phys. Fluids **3**, 539—541. Comment, ibid. **5**, 502 (1962). (*119*)

[15] CHELAM, E. V.: Elastic behavior of matter under high pressures. Simple shear. J. Indian Inst. Sci. **42**, 41—46. (*69*)

[16] CHELAM, E. V.: Elastic behavior of matter under very high pressures. Considerations of stability. J. Indian Inst. Sci. **42**, 101—107. (*68b*)

[17] COLEMAN, B. D., and W. NOLL: An approximation theorem for functionals, with applications in continuum mechanics. Arch. Rational Mech. Anal. **6**, 355—370. Reprinted in Rational Mechanics of Materials. Intl. Sci. Rev. Ser. New York: Gordon & Breach 1965. (*38, 40, 121, 123*)

[18] ERICKSEN, J. L.: Anisotropic fluids. Arch. Rational Mech. Anal. **4** (1959/60), 231—237. Reprinted in Rational Mechanics of Materials. Intl. Sci. Rev. Ser. New York: Gordon & Breach 1965. (*98, 127*)

[19] ERICKSEN, J. L.: Transversely isotropic fluids. Kolloid-Z. **173**, 117—122. (*127, 129*)

[20] ERICKSEN, J. L.: Theory of anisotropic fluids. Trans. Soc. Rheol. **4**, 29—39. (*98, 127, 129*)

[21] ERICKSEN, J. L.: A vorticity effect in anisotropic fluids. J. Polymer Sci. **47**, 327—331. (*129*)

[22] ERICKSEN, J. L.: The behavior of certain visco-elastic materials in laminar shearing motions. Visco-elasticity: Phenomenological Aspects, pp. 77—91. New York: Academic Press. (*108, 115, 119*)

[23] FOUX, A., and M. REINER: Extension of metal wires in simple torsion. Technical Rep. Technion Res. Devel. Found. (*54, 66*)

[24] FREDRICKSON, A. G.: Helical flow of an annular mass of visco-elastic fluid. Chem. Engr. Sci. **11**, 252—259. (*112*)

[25] GENENSKY, S. M.: A general theorem concerning the stability of a particular non-Newtonian fluid. Quart. Appl. Math. **18** (1960/61), 245—250. (*123*)

[26] GREEN, A. E., and J. E. ADKINS: Large Elastic Deformations and Non-linear Continuum Mechanics. Oxford: Clarendon Press. (*13, 30, 34, 42, 43, 54, 55, 57, 58, 59, 60, 66, 68b, 85, 96*)

[27] GREEN, A. E., and R. S. RIVLIN: The mechanics of non-linear materials with memory. Part III. Arch. Rational Mech. Anal. **4** (1959/60), 387—404. (*28, 37*)

[28] GRIOLI, G.: Elasticità asimmetrica. Annali di Mat. Pura Appl. (4) **50**, 389—417. Summary, Proc. 10th Intl. Congr. Appl. Mech. Stresa 1960, pp. 252—254 (1962). (*98*)

[29] JAIN, M. K.: On rotational instability in visco-elastic liquids. Proc. 3rd. Congr. Theor. Appl. Mech. (Bangalore, 1957), pp. 217—224. (*113*)

[30] JOHN, F.: Plane strain problems for a perfectly elastic material of harmonic type. Communs. Pure Appl. Math. **13**, 239—296. (*94*)

[31] JOHNSON, M. W.: Some variational theorems for non-Newtonian flow. Phys. Fluids 3, 871—878. (*119*)
[32] JONES, J. R.: Flow of a non-Newtonian fluid in a curved pipe. Quart. J. Mech. Appl. Math. 13, 428—443. (*119*)
[33] KAPUR, J. N.: Some problems in hydrodynamics of non-Newtonian viscous liquids with variable coefficient of cross-viscosity. Proc. Natl. Inst. Sci. India A 25, (1959), 231—235. (*119*)
[34] KNOWLES, J. A.: Large amplitude oscillations of a tube of incompressible elastic material. Quart. Appl. Math. 18, 71—77. Reprinted in Problems of Non-linear Elasticity. Intl. Sci. Rev. Ser. New York: Gordon & Breach 1965. (*62, 95*)
[35] KRÖNER, E.: Allgemeine Kontinuumstheorie der Versetzungen und Eigenspannungen. Arch. Rational Mech. Anal. 4, (1959/60), 273—334. Reprinted in Foundations of Elasticity Theory. Intl. Sci. Rev. Ser. New York: Gordon & Breach 1965. (*34, 94, 98*)
[36] LODGE, A. S.: On the methods of measuring normal stress differences in shear flow. Rheol. Abstr. 3, No. 3, 21—23. (*104*)
[37] MITHAL, K. G.: Problems related to non-Newtonian fluids. Thesis. Lucknow Univ. (*119*)
[38] NARASIMHAN, M. N. L.: Mechanics of flow of real fluids through flexible tubes. J. Sci. Engr. Res. 4, 91—104. (*113, 119*)
[39] OBEROI, M. M., and J. N. KAPUR: On axially-symmetric non-Newtonian flows. Bull. Calcutta Math. Soc. 52, 165—172. (*119*)
[40] PFLEIDERER, H., A. SEEGER, u. E. KRÖNER: Nichtlineare Elastizitätstheorie geradliniger Versetzungen. Z. Naturforsch. 15a, 758—772. (*94*)
[41] PIPKIN, A. C., and R. S. RIVLIN: Electrical conduction in deformed isotropic materials. J. Math. Phys. 1, 127—130. (*97*)
[42] RAJAGOPAL, E. S.: The existence of interfacial couples in infinitesimal elasticity. Ann. Physik (7) 6, 192—201. (*98*)
[43] RATHNA, S. L.: Superposability of steady axi-symmetrical flows in a non-Newtonian fluid. Proc. Indian Acad. Sci. A 51, 155—163. (*119*)
[44] RATHNA, S. L.: Couette and Poiseuille flow in non-Newtonian fluids. Proc. Nat. Inst. Sci. India A 26, 392—399. (*119*)
[45] REINER, M.: Cross stresses in the laminar flow of liquids. Phys. Fluids 3, 427—432. (*116*)
[46] RIVLIN, R. S.: The formulation of constitutive equations in continuum physics, II. Arch. Rational Mech. Anal. 4 (1959/60), 262—272. (*13, 29, 33*)
[47] RIVLIN, R. S.: Constitutive equations for classes of deformations. Viscoelasticity: Phenomenological Aspects, pp. 93—108. New York: Academic Press. (*39*)
[48] RIVLIN, R. S.: Some topics in finite elasticity. Structural Mechanics, pp. 169—198. Oxford etc: Pergamon. (*42*)
[49] SEEGER, A., u. O. BUCK: Die experimentelle Ermittlung der elastischen Konstanten höherer Ordnung. Z. Naturforsch. 15a, 1056—1067. (*66*)
[50] SIROTIN, YU. I.: Anisotropic tensors [in Russian]. Dokl. Akad. Nauk SSSR 133, 321—324. English Transl., Soviet Phys. Doklady 5, 774—777 (1961). (*33*)
[51] SIROTIN, YU. I.: The construction of tensors with specified symmetry [in Russian]. Kristallografia 5, 171—179. English Transl., Soviet Phys. Cryst. 5, 157—165. (*33*)
[52] SMITH, G. F.: On the minimality of integrity bases for symmetric 3×3 matrices. Arch. Rational Mech. Anal. 5, 382—389. (*11*)
[53] SPENCER, A. J. M., and R. S. RIVLIN: Further results on the theory of matrix polynomials. Arch. Rational Mech. Anal. 4 (1959/60), 214—230. (*11, 13, 37*)
[54] SRIVASTAVA, A. C.: Rotatory oscillation of an infinite plate in non-Newtonian fluids. Appl. Sci. Res. A 9 (1959/60), 369—373. (*119*)
[55] TOUPIN, R. A.: Stress tensors in elastic dielectrics. Arch. Rational Mech. Anal. 5, 440—452. (*97*)
[56] TOUPIN, R. A., and R. S. RIVLIN: Dimensional changes in crystals caused by dislocations. J. Mathematical Phys. 1, 8—15. Reprinted in Problems of Non-linear Elasticity. Intl. Sci. Rev. Ser. New York: Gordon & Breach 1965. (*63, 68, 93*)
[57] TRUESDELL, C.: The Rational Mechanics of Flexible or Elastic Bodies, 1638—1788. L. Euleri Opera Omnia (2) 11_2. Zürich: Füssli. (*66*)
[58] TRUESDELL, C.: Modern theories of materials. Trans. Soc. Rheol. 4, 9—22. (*4, 96*)
[59] WALTERS, K.: The motion of an elastico-viscous liquid contained between concentric spheres. Quart. J. Mech. Appl. Math. 13, 325—333. (*119*)
[60] WALTERS, K.: The motion of an elastico-viscous liquid contained between coaxial cylinders (II). Quart. J. Mech. Appl. Math. 13, 444—461. (*119*)

1961 [*1*] Adkins, J. E., and A. E. Green: The finite flexure of an aeolotropic elastic cuboid. Arch. Rational Mech. Anal. **8**, 9—14. (*58, 59*)

[*2*] Adkins, J. E.: Large elastic deformation. Progress in Solid Mech. **2**, 2—60. (*43*)

[*3*] Bernstein, B.: Remarks on the materials of the rate type and the principle of determinism. Trans. Soc. Rheol. **5**, 35—40. (*37, 99*)

[*4*] Bhatnagar, P. L.: On two-dimensional boundary layer in non-Newtonian fluids with constant coefficients of viscosity and cross-viscosity. Proc. Indian Acad. Sci. **53**, 95—97. (*119*)

[*5*] Coleman, B. D., and W. Noll: Recent results in the continuum theory of visco-elastic fluids. Ann. N.Y. Acad. Sci. **89**, 672—714. (*38, 104, 106, 108, 111, 113, 121*)

[*6*] Coleman, B. D., and W. Noll: Normal stresses in second-order viscoelasticity. Trans. Soc. Rheol. **5**, 41—46. (*121*)

[*7*] Coleman, B. D., and W. Noll: Foundations of linear viscoelasticity. Rev. Mod. Phys. **33**, 239—249. Reprinted in Foundations of Elasticity Theory. Intl. Sci. Rev. Ser. New York: Gordon & Breach 1965. (*3, 38, 41*)

[*8*] Corneliussen, A. H., and R. T. Shield: Finite deformation of elastic membranes with application to the stability of an inflated and extended tube. Arch. Rational Mech. Anal. **7**, 273—304. (*60*)

[*9*] Dahler, J. S., and L. E. Scriven: Angular momentum of continua. Nature, Lond. **192**, 36—37. (*98*)

[*10*] Datta, S. K.: On the steady motion of an idealized elastico-viscous liquid through channels with suction and injection. J. Phys. Soc. Japan **16**, 794—797. (*119*)

[*11*] Datta, S. K.: Laminar flow of non-Newtonian fluid in channels with porous walls. Bull. Calcutta Math. Soc. **53**, 111—116. (*119*)

[*12*] England, A. H., and A. E. Green: Steady-state thermoelasticity for initially stressed bodies. Phil. Trans. Roy. Soc. Lond. A **253**, 517—542. (*96*)

[*13*] Ericksen, J. L.: Poiseuille flow of certain anisotropic fluids. Arch. Rational. Mech. Anal. **8**, 1—8. Reprinted in Rational Mechanics of Materials. Intl. Sci. Rev. Ser. New York: Gordon & Breach 1965. (*129*)

[*14*] Ericksen, J. L.: Conservation laws for liquid crystals. Trans. Soc. Rheol. **5**, 23—24. Reprinted in Rational Mechanics of Materials. Intl. Sci. Rev. Ser. New York: Gordon & Breach 1965. (*98, 127, 128*)

[*15*] Eringen, A. C.: On the foundations of electro-elastostatics. Contract Nonr.-1100 (02) Tech. Rep. No. 19, Purdue Univ., November. 65 pp. (*97*)

[*16*] Flavin, J. N., and A. E. Green: Plane thermoelastic waves in an initially stressed medium. J. Mech. Phys. Solids **9**, 179—190. (*96*)

[*17*] Giesekus, H.: Der Spannungstensor des visko-elastischen Körpers. Rheol. Acta **1** (1958/61), 395—403. (*104*)

[*18*] Giesekus, H.: Einige Bemerkungen zum Fließverhalten elasto-viskoser Flüssigkeiten in stationären Schichtströmungen. Rheol. Acta **1** (1958/61), 404—413. (*104, 108, 119*)

[*19*] Graebel, W. P.: Stability of a Stokesian fluid in Couette flow. Phys. Fluids **4**, 362—368. (*119*)

[*20*] Green, A. E.: Torsional vibrations of an initially stressed circular cylinder. Problems of Contin. Mech. (Muskhelisvili Anniv. Vol.), pp. 148—154. Phila.: S.I.A.M. (Russian Ed., pp. 128—134.) (*70*)

[*21*] Grioli, G.: Onde di discontinuità ed elasticità asimmetrica. Rend. Accad. Lincei (8) **29**, (1960), 309—312. (*98*)

[*22*] Grossman, P. U. A.: Weissenberg's rheological equation of state. Kolloid-Z. **174**, 97—109. (*94, 119*)

[*23*] Hand, G. C.: A theory of dilute suspensions. Arch. Rational Mech. Anal. **7**, 81—86. (*127*)

[*24*] Hayes, M., and R. S. Rivlin: Propagation of a plane wave in an isotropic elastic material subject to pure homogeneous deformation. Arch. Rational Mech. Anal. **8**, 15—22. (*52, 73, 74*)

[*25*] Hayes, M., and R. S. Rivlin: Surface waves in deformed elastic materials. Arch. Rational Mech. Anal. **8**, 358—380. (*73, 95*)

[*26*] Hill, R.: Uniqueness in general boundary-value problems for elastic or inelastic solids. J. Mech. Phys. Solids **9**, 114—130. (*101*)

[*27*] Hill, R.: Bifurcation and uniqueness in non-linear mechanics of continua. Problems Contin. Mech. (Muskhelisvili Anniv. Vol.), pp. 155—164. Phila.: S.I.A.M. (*101*)

[*28*] Jain, M. K.: The flow of a non-Newtonian liquid near a rotating disk. Appl. Sci. Res. A **10**, 410—418. (*119*)

[*29*] Jain, M. K.: Flow of non-Newtonian liquid near a stagnation point with and without suction. J. Sci. Engr. Res. **5**, 81—90. (*119*)

[30] JAIN, M. K., and M. BALRAM: Problems of cross-viscosity with large suction. J. Sci. Engr. Res. **5**, 259—270. (*119*)

[31] JOHNSON, M. W., jr.: On variational principles for non-Newtonian fluids. Trans. Soc. Rheol. **5**, 9—21. (*119*)

[32] KAPUR, J. N., and S. GOEL: Flow of visco-elastic liquids in tubes. Bull. Calcutta Math. Sci. **53**, 1—6. (*117*)

[33] KRISHNAN, R. S., and E. S. RAJAGOPAL: The atomistic and the continuum theories of crystal elasticity. Ann. Physik (7) **8**, 121—136. (*98*)

[34] KRÖNER, E.: Bemerkung zum geometrischen Grundgesetz der allgemeinen Kontinuumstheorie der Versetzungen und Eigenspannungen. Arch. Rational Mech. Anal. **7**, 78—80. Reprinted in Foundations of Elasticity Theory. Intl. Sci. Rev. Ser. New York: Gordon & Breach 1965. (*34*)

[35] LESLIE, F. M.: The slow flow of a viscoelastic liquid past a sphere, with an appendix by R. I. TANNER. Quart. J. Mech. Appl. Math. **14**, 36—48. (*119*)

[36] LODGE, A. S.: A new cone-and-plate and parallel plate apparatus for the determination of normal stress differences in steady shear flow. Rheol. Abstr. **4**, No. 3, 29. (*119*)

[37] METZNER, A. B., W. T. HOUGHTON, R. A. SAILOR, and J. L. WHITE: A method for the measurement of normal stresses in simple shearing flow. Rheol. **5**, 133—147. (*114*)

[38] MITHAL, K. G.: On the effects of uniform high suction on the steady flow of a non-Newtonian liquid due to a rotating disk. Quart. J. Mech. Appl. Math. **14**, 403—410. (*119*)

[39] NARASIMHAN, M. N. L.: Laminar non-Newtonian flow in an annulus with porous walls. Z. angew. Math. Mech. **41**, 44—54. (*119*)

[40] NARASIMHAN, M. N. L.: Laminar non-Newtonian flow in a porous pipe. Appl. Sci. Res. A **10**, 393—409. (*119*)

[41] NARIBOLI, G. A.: A note on the fracture of a hypo-elastic bar. Proc. Fifth Congr. Theor. Appl. Mech. (Roorkee, 1959), c. 55—60. (*103*)

[42] NOVOZHILOV, V. V.: Theory of Elasticity (Transl. from a Russian work publ. in 1958). Jerusalem: Israel Progr. Sci. Transl. (*42*)

[43] OLDROYD, J. G.: The hydrodynamics of materials whose rheological properties are complicated. Rheol. Acta **1** (1958/61), 337—344. (*104, 119*)

[44] PHILIPPOFF, W.: Elastic stresses and birefringence in flow. Trans. Soc. Rheol. **5**, 163—191. (*116*)

[45] PIPKIN, A. C., and R. S. RIVLIN: Electrical conduction in a stretched and twisted tube. J. Math. and Phys. **2**, 636—638. (*97*)

[46] PIPKIN, A. C., and R. S. RIVLIN: Small deformations superposed on large deformations in materials with fading memory. Arch. Rational Mech. Anal. **8**, 297—308. (*41*)

[47] PRAGER, W.: Introduction to Mechanics of Continua. Boston etc.: Ginn. German Transl. Einführung in die Kontinuumsmechanik. Basel: Birkhäuser. (*42, 47, 99, 100*)

[48] RAJESWARI, G. K.: Flow of non-Newtonian fluid between torsionally oscillating disks. Proc. Indian Acad. Sci. A **54**, 188—204. (*119*)

[49] RATHNA, S. L., and G. K. RAJESWARI: Superposability in non-Newtonian fluids with variable viscosity and cross-viscosity coefficients. Proc. Nat. Acad. Sci. India **27**. (*119*)

[50] REINER, M.: Research on cross stresses in the flow of rarefied air. Technical report. The Technion. 1 March. (*116*)

[51] RIVLIN, R. S.: Constitutive equations involving functional dependence of one vector on another. Z. angew. Math. Phys. **12**, 447—452. (*13, 29, 33*)

[52] RIVLIN, R. S.: Some reflections on non-linear visco-elastic fluids. Phénomènes de relaxation et du fluage en rhéologie non-linéaire, pp. 83—93. Paris: Ed. C.N.R.S. (*104, 119*)

[53] SCHWARZL, F.: Einige Betrachtungen zum gegenwärtigen Stand der makroskopischen Theorien des rheologischen Verhaltens. Rheol. Acta **1** (1958/61), 345—355. (*104*)

[54] SEEGER, A.: Recent advances in the theory of defects in crystals. Physica Status Solidi **1**, 669—698. (*34*)

[55] SINHA, S. B.: Torsional vibrations of an hypo-elastic cylinder. Bull. Acad. Polon. Sci. Sér. Sci. Tech. **9**, 197—200. (*103*)

[56] SINHA, S. B.: Torsional vibrations of a hypo-elastic circular cylinder. Arch. Mech. Stosow. **13**, 389—392. (*103*)

[57] SLATTERY, J.: Flow of a simple non-Newtonian fluid past a sphere. Appl. Sci. Res. A **10**, 286—294. (*119*)

[58] SLATTERY, J. C., and R. B. BIRD: Non-Newtonian flow past a sphere. Chem. Engr. Sci. **16**, 231—241. (*119*)
[59] SPENCER, A. J. M.: The invariants of six symmetric 3×3 matrices. Arch. Rational Mech. Anal. **7**, 64—77. (*11*)
[60] SRIVASTAVA, A. C.: Flow of non-Newtonian fluids at small Reynolds number between two infinite disks: one rotating and the other at rest. Quart. J. Mech. Appl. Math. **14**, 353—358. (*119*)
[61] TOUPIN, R. A.: Some relations between waves, stability, uniqueness criteria, and restrictions on the form of the energy function in elasticity theory. Lecture to the British Appl. Math. Coll. Newcastle-on-Tyne (widely circulated in manuscript but not published). (*52, 74*)
[62] TOUPIN, R. A., and B. BERNSTEIN: Sound waves in deformed perfectly elastic materials. Acousto-elastic effect. J. Acoust. Soc. Amer. **33**, 216—225. Reprinted in Problems of Non-linear Elasticity. Intl. Sci. Rev. Ser. New York: Gordon & Breach 1965. (*45, 52, 66, 68, 71, 73, 76, 77, 90*)
[63] TRUESDELL, C.: General and exact theory of waves in finite elastic strain. Arch. Rational Mech. Anal. **8**, 263—296. Reprinted in Problems of Non-linear Elasticity. Intl. Sci. Rev. Ser. New York: Gordon & Breach 1965 and also in Wave Propagation in Dissipative Materials. Berlin-Heidelberg-New York: Springer 1965. (*45, 48, 52, 71, 72, 73, 74, 75, 76, 77, 82, 84, 90, 93*)
[64] TRUESDELL, C.: Una teoria meccanica della diffusione. Celebraz. Archimedee Sec. XX (1960), III Simposio di mecc. e mat. applic. 161—168. (*130*)
[65] URBANOWSKI, W.: Deformed body structure. Arch. Mech. Stosow. **13**, 277—294. (*69*)
[66] VERMA, P. D. S.: Steady flow formation of linear fluent bodies with suction. J. Sci. Engr. Res. **5**, 17—22. (*119*)
[67] VERMA, P. D. S.: Solid rotating shafts II. Bull. Inst. Poly. Jasi (2) **7**, 41—44. (*103*)
[68] WALTERS, K.: The motion of an elastico-viscous liquid contained between coaxial cylinders (III). Quart. J. Mech. Appl. Math. **14**, 431—436. (*113*)
[69] ZIEGENHAUSEN, A. J., R. B. BIRD, and M. W. JOHNSON, jr.: Non-Newtonian flow around a sphere. Trans. Soc. Rheol. **5**, 47—49. (*119*)
1962 [1] ADKINS, J. E.: Syzygies relating the invariants for transversely isotropic materials. Arch. Rational Mech. Anal. **11**, 357—367. (*11*)
[2] AMARI, S.: A geometrical theory of moving dislocations and anelasticity. Res. Assn. Appl. Geom. (Tokyo) Res. Note (3), No. 52. (*34*)
[3] BERNSTEIN, B.: Conditions for second-order waves in hypo-elasticity. Trans. Soc. Rheol. **6**, 263—273. (*102*)
[4] BERNSTEIN, B., and R. A. TOUPIN: Some properties of the Hessian matrix of a strictly convex function. J. reine angew. Math. **210**, 65—72. (*51*)
[5] BHATNAGAR, P. L.: Non-Newtonian fluids. Proc. 49th Indian Sci. Congr. (Cuttack, 1961/62), pp. 1—30. (*104, 119*)
[6] BHATNAGAR, P. L., and G. K. RAJESWARI: The secondary flows induced in a non-Newtonian fluid between two parallel infinite oscillating plates. J. Indian Inst. Sci. **44**, 219—238. (*123*)
[7] BLATZ, P. J., and W. L. KO: Application of finite elasticity theory to deformation of rubbery materials. Trans. Soc. Rheol. **6**, 223—251. (*59*)
[8] CHU, BOA-TEH: Stress waves in isotropic viscoelastic materials. I. Brown Univ. Div. Engr. Dept. Defense ARPA Report, March. (*96t*)
[9] COLEMAN, B. D.: Mechanical and thermodynamical admissibility of stress-strain functions. Arch. Rational Mech. Anal. **9**, 172—186. (*52, 83*)
[10] COLEMAN, B. D.: Kinematical concepts with applications in the mechanics and thermodynamics of incompressible viscoelastic fluids. Arch. Rational Mech. Anal. **9**, 273—300. (*106, 107, 108, 109, 123*)
[11] COLEMAN, B. D.: Substantially stagnant motions. Trans. Soc. Rheol. **6**, 293—300. (*109*)
[12] COLEMAN, B. D., and W. NOLL: Steady extension of incompressible simple fluids. Phys. Fluids **5**, 840—843. (*118*)
[13] DATTA, S. K.: Slow steady rotation of a sphere in a non-Newtonian inelastic viscous fluid. Appl. Sci. Res. A **11** (1962/63), 47—52. (*119*)
[14] ERICKSEN, J. L.: Kinematics of macromolecules. Arch. Rational Mech. Anal. **9**, 1—8. (*98, 127*)
[15] ERICKSEN, J. L.: Hydrostatic theory of liquid crystals. Arch. Rational Mech. Anal. **9**, 379—394. (*128*)
[16] ERICKSEN, J. L.: Nilpotent energies in liquid crystal theory. Arch. Rational Mech. Anal. **10**, 189—196. (*128*)
[17] ERICKSEN, J. L.: Orientation induced by flow. Trans. Soc. Rheol. **6**, 275—291. (*129*)

[*18*] Eringen, A. C.: Non-linear Theory of Continuous Media. New York etc.: McGraw-Hill. xii + 477 pp. (*5, 42, 43, 97, 98, 99, 100, 103, 104*) Cf. the review by Pipkin [1964, *67*].

[*19*] Flavin, J. N.: Thermo-elastic Rayleigh waves in a prestressed medium. Proc. Cambridge Phil. Soc. **58**, 532—538. (*95, 96*)

[*20*] Giesekus, H.: Elasto-viskose Flüssigkeiten, für die in stationären Schichtströmungen sämtliche Normalspannungskomponente verschieden groß sind. Rheologica Acta **2**, 50—62. (*108*)

[*21*] Giesekus, H.: Strömungen mit konstanten Geschwindigkeitsgradienten und die Bewegung von darin suspendierten Teilchen. I und II. Rheologica Acta **2**, 101—122. (*30*)

[*22*] Giesekus, H.: Flüssigkeiten mit im Ruhestand singulärem Fließverhalten (quasiplastische Flüssigkeiten). Rheologica Acta **2**, 122—130. (*108, 119*)

[*23*] Green, A. E.: Thermoelastic stress in initially stressed bodies. Proc. Roy. Soc. Lond. A **266**, 1—19. (*70, 90, 96*)

[*24*] Grioli, S.: Mathematical Theory of Elastic Equilibrium (Recent Results). Ergeb. angew. Math. No. 7. Berlin-Göttingen-Heidelberg: Springer. (*42, 44, 46, 63, 64, 65, 94, 98, 99*)

[*25*] Guo Zhong-Heng: The problem of stability and vibration of a circular plate subject to finite initial deformation. Arch. Mech. Stosow. **14**, 239—252. (*60*)

[*26*] Guo Zhong-Heng: Certain problems of initially deformed plate. Arch. Mech. Stosow. **14**, 779—788. (*60*)

[*27*] Guo Zhong-Heng: Vibration and stability of a cylinder subject to finite deformation. Arch. Mech. Stosow. **14**, 757—768. (*68b*)

[*28*] Guo Zhong-Heng: The equations of motion of a circular plate subject to initial strain. Bull. Acad. Sci. Polon. Sér. Sci. Tech. **10**, 63—70. (*60*)

[*29*] Guo Zhong-Heng: The equation of motion of a plate subject to initial homogeneous finite deformation. Bull. Acad. Sci. Polon. Sér. Sci. Tech. **10**, 107—113. (*60*)

[*30*] Guo Zhong-Heng: A contribution to the theory of variated states of finite strain. Bull. Acad. Sci. Polon. Sér. Sci. Tech. **10**, 129—133. (*66*)

[*31*] Guo Zhong-Heng: Equations of small motion of a cylinder subject to a large deformation. Its natural vibration and stability. Bull. Acad. Sci. Polon. Sér. Sci. Tech. **10**, 177—182. (*68b*)

[*32*] Guo Zhong-Heng: Variated states of membranes subject to finite deformation. Bull. Acad. Sci. Polon. Sér. Sci. Tech. **10**, 307—312. (*60*)

[*33*] Guo Zhong-Heng: Displacement equations of isentropic motion of a body subject to a finite isothermal initial strain. Bull. Acad. Sci. Polon. Sér. Sci. Tech. **10**, 479—483. (*96*)

[*34*] Hill, R.: Acceleration waves in solids. J. Mech. Phys. Solids **10**, 1—16. (*52, 68b, 78, 102*)

[*35*] Hill, R.: Uniqueness criteria and extremum principles in self-adjoint problems of continuum mechanics. J. Mech. Phys. Solids **10**, 185—194. (*101*)

[*36*] Jain, M. K.: Forced flow of a non-Newtonian liquid against a rotating disk. Bull. Inst. Polit. Iasi. (2) **8** (12), 83—92. (*119*)

[*37*] Jain, M. K.: Heat transfer by laminar natural-convection flow of visco-elastic fluids between parallel walls. Arch. Mech. Stosow. **14**, 747—756. (*119*)

[*38*] Jones, E. E., and J. E. Adkins: Perturbation problems in finite elasticity. Perturbation of constraint conditions. J. Math. Mech. **11**, 341—370. (*43*)

[*39*] Kaliski, S., Z. Płochocki, and D. Rogula: The asymmetry of the stress tensor and the angular momentum conservation for a combined mechanical and electromagnetic field in a continuous medium. Bull. Acad. Sci. Polon. Sér. Sci. Tech. **10**, 135—141. (*98*)

[*40*] Kapur, J. N.: Some aspects of non-Newtonian flow. Math. Seminar **2** (1961), 181—204. (*104*)

[*41*] Kapur, J. N., and S. Goel: A stability theorem for general non-Newtonian fluid. Appl. Sci. Res. A **11**, 304—310. (*123*)

[*42*] Knowles, J. A.: On a class of oscillations in the finite-deformation theory of elasticity. J. Appl. Mech. **29**, 283—286. (*62, 95*)

[*43*] Kondo, K., and collaborators: Memoirs of the Unifying Study of Basic Problems in Engineering and Physical Sciences by means of Geometry, III. Tokyo: Gakujutsu Bunken Fukyu-Kai. (*34*)

[*44*] Kröner, E.: Dislocations and continuum mechanics. Appl. Mech. Rev. **15**, 599—606. (*44, 98*)

[*45*] Leigh, D. C.: Non-Newtonian fluids and the second law of thermodynamics. Phys. Fluids **5**, 501—502. (*119*)

[*46*] MINDLIN, R. D., and H. F. TIERSTEN: Effects of couple-stresses in linear elasticity. Arch. Rational Mech. Anal. **11**, 415—448. (*98*)
[*47*] MOHAN RAO, D. K.: Rectilinear motion of a Maxwell fluid. J. Indian. Inst. Sci. **45**, 19—20. (*119*)
[*48*] MOHAN RAO, D. K.: Flow of a Maxwell liquid between two rotating coaxial cones having the same vertex. Proc. Indian Acad. Sci. **56**, 198—205. (*119*)
[*49*] NANDA, R. S.: On the three-dimensional flow of certain non-Newtonian fluids. Arch. Mech. Stosow. **14**, 137—145. (*119*)
[*50*] NOLL, W.: Motions with constant stretch history. Arch. Rational Mech. Anal. **11**, 97—105. (*106, 107, 108, 109*)
[*51*] PAO, Y.-H.: Theories for the flow of dilute solutions of polymers and of non-diluted liquid polymers. J. Polymer Sci. **61**, 413—448. (*119*)
[*52*] RAJESWARI, G. K.: Laminar boundary layer on rotating sphere and spheroids in non-Newtonian fluids. Z. angew. Math. Phys. **13**, 442—460. (*119*)
[*53*] RAJESWARI, G. K., and S. L. RATHNA: Flow of a particular class of non-Newtonian visco-elastic and visco-inelastic fluids near a stagnation point. Z. angew. Math. Phys. **13**, 43—57. (*123*)
[*54*] RATHNA, S. L.: Slow motion of a non-Newtonian liquid past a sphere. Quart. J. Mech. Appl. Math. **15**, 427—434. (*119*)
[*55*] RATHNA, S. L.: Flow of a particular class of non-Newtonian fluids near a rotating disk. Z. angew. Math. Mech. **42**, 231—237. (*123*)
[*56*] REINER, M.: Research on cross stresses in the flow of different gases. Technion Res. Devel. Found. Report, 1st April. (*116*)
[*57*] RIVLIN, R. S.: Constraints on flow invariants due to incompressibility. Z. angew. Math. Phys. **13**, 589—591. (*24*)
[*58*] SCHAEFER, H.: Versuch einer Elastizitätstheorie des zweidimensionalen ebenen Cosserat-Kontinuums. Misz. Angew. Math. Festschrift Tollmien. Berlin: Akademie-Verlag. (*98*)
[*59*] SCHLECHTER, R. S.: On a variational principle for the Reiner-Rivlin fluid. Chem. Engr. Sci. **17**, 803—806. (*119*)
[*60*] SEDOV, L. I.: Introduction to the Mechanics of Continuous Media [in Russian]. Moscow. Gosud. Izdat. Fiz-Mat. Lit. (*96*)
[*61*] SLATTERY, J. C.: Approximations to the drag force on a sphere maving slowly through either an Ostwald-DeWaele or a Sisko fluid. A.I.Ch.E. Jonral **8**, 663—667. (*119*)
[*62*] SMITH, G. F.: Further results on the strain-energy function for anisotropic elastic materials. Arch. Rational Mech. Anal. **10**, 108—118. (*85*)
[*63*] SPENCER, A. J. M., and R. S. RIVLIN: Isotropic integrity bases for vectors and second-order tensors. I. Arch. Rational. Mech. Anal. **9**, 45—63. (*11*)
[*64*] TOUPIN, R. A.: Elastic materials with couple-stresses. Arch. Rational Mech. Anal. **11**, 385—414. (*98*)
[*65*] TRUESDELL, C.: Solutio generalis et accurata problematum quamplurimorum de motu corporum elasticorum incomprimibilium in deformationibus valde magnis. Arch. Rational Mech. Anal. **11**, 106—113. Addendum **12**, 427—428 (1963). (*30, 61*)
[*66*] TRUESDELL, C.: Mechanical basis of diffusion. J. Chem. Phys. **37**, 2336—2344. (*130*)
[*67*] VARLEY, E.: Flows of dilatant fluids. Quart. Appl. Math. **19** (1961/62), 331—347. (*119*)
[*68*] VERMA, P. D. S.: Couette flow of certain anisotropic fluids. Arch. Rational Mech. Anal. **10**, 101—107. (*129*)
[*69*] VERMA, P. D. S., and S. B. SINHA: A note on the propagation of symmetrical disturbance from a transverse cylindrical hole in an infinite plate. Bull. Calcutta Math. Soc. **54**, 75—78. (*103*)
[*70*] WALTERS, K.: A note on the rectilinear flow of elastico-viscous liquids through straight pipes of circular cross-section. Arch. Rational Mech. Anal. **9**, 411—414. (*119*)
[*71*] WALTERS, K.: Non-Newtonian effects in some elastico-viscous liquids whose behavior at small rates of shear is characterized by a general linear equation of state. Quart. J. Mech. Appl. Math. **15**, 63—76. (*119*)
[*72*] WEHRLI, C., and H. ZIEGLER: Einige mit dem Prinzip von der größten Dissipationsleistung verträgliche Stoffgleichungen. Z. angew. Math. Phys. **13**, 372—393. (*96, 119*)
[*73*] WESOŁOWSKI, Z.: Stability in some cases of tension in the light of the theory of finite strain. Arch. Mech. Stosow. **14**, 875—900. (*68b*)
[*74*] WESOŁOWSKI, Z.: Some stability problems of tension in the light of the theory of finite strains. Bull. Acad. Sci. Polon. Sér. Sci. Tech. **10**, 123—128. (*68b*)

[75] WILLIAMS, M. C., and R. B. BIRD: Three-constant Oldroyd model for viscoelastic fluids. Phys. Fluids **5**, 1126—1128. (*119*)
[76] WOO, T. C., and R. T. SHIELD: Fundamental solutions for small deformations superposed on finite biaxial extension of an elastic body. Arch. Rational. Mech. Anal. **9**, 196—224. (*70*)
[77] ZAHORSKI, S. L.: Equations of the theory of large elastic deformations in terms of the geometry of the undeformed body. Arch. Mech. Stosow. **14**, 941—956. (*43*)
[78] ZAHORSKI, S.: On a certain form of the equations of the theory of finite elastic strain. Bull. Acad. Sci. Polon. Sér. Sci. Tech. **10**, 415—420. (*43*)
[79] ZAHORSKI, S.: Experimental investigation of certain mechanical properties of rubber. Bull. Acad. Sci. Polon. Ser. Sci. Tech. **10**, 421—427. (*57*)
1963 [1] ADKINS, J. E.: Non-Linear diffusion. I. Diffusion and flow of mixtures of fluids. Phil. Trans. Roy. Soc. Lond. A **255**, 607—633. (*130*)
[2] ADKINS, J. E.: Non-linear diffusion, II. Constitutive equations for mixtures of isotropic fluids. Phil. Trans. Roy. Soc. Lond. A **255**, 635—648. (*130*)
[3] ADKINS, J. E., and R. S. RIVLIN: Propagation of electromagnetic waves in circular rods in torsion. Phil. Trans. Roy. Soc. Lond. A **255**, 389—416. (*97*)
[4] BARENBLATT, G. I., and G. G. CHERNYI: On moment relations on surfaces of discontinuity in dissipative media. PMM **27**, 1205—1218 (transl. of Priklad. Mat. Mekh. **27**, 784—793). (*96t*, *123*)
[5] BERNSTEIN, B., E. A. KEARSLEY, and L. J. ZAPAS: A study of stress relaxation with finite strain. Trans. Soc. Rheol. **7**, 391—410. (*37*, *39*)
[6] BHATNAGAR, P. L., and G. K. RAJESWARI: Mouvement secondaire d'un fluide non newtonien compris entre deux sphères concentriques tournant autour d'un axe. C. R. Acad. Sci. Paris **256**, 3823—3826. (*123*)
[7] BHATNAGAR, P. L., and G. K. RAJESWARI: Secondary flow of non-Newtonian fluids between two concentric spheres rotating about an axis. Indian J. Math. **5**, 93—112. (*123*)
[8] BHATNAGAR, P. L., and S. L. RATHNA: Flow of a fluid between two rotating co-axial cones having the same vertex. Quart. J. Mech. Appl. Math. **16**, 329—346. (*123*)
[9] BONDAR, V. D.: On the possibility of considering the deformed and stressed states of a medium as the initial state. PMM **27**, 185—196 (transl. of Priklad. Mat. Mekh. **27**, 135—141). (*47*)
[10] BRAGG, L. E., and B. D. COLEMAN: On strain energy functions for isotropic elastic materials. J. Math. and Phys. **4**, 424—426. (*87*)
[11] BRAGG, L. E., and B. D. COLEMAN: A thermodynamical limitation on compressibility. J. Math. and Phys. **4**, 1074—1077. (*52*)
[12] BRESSAN, A.: Sulla propagazione delle onde ordinarie di discontinuità nei sistemi a trasformazioni reversibili. Rend. Sem. Mat. Padova **33**, 99—139. (*74*)
[13] BRESSAN, A.: Termodinamica e magneto-viscoelasticità con deformazioni finite in relatività generale. Rend. Sem. Mat. Padova **34** (1964), 1—73. (*5*)
[14] BRESSAN, A.: Una teoria di relatività generale includente, oltre all'elettromagnetismo e alla termodinamica, le equazioni costitutive dei materiali ereditari. Sistemazione assiomatica. Rend. Sem. Mat. Padova **34** (1964), 74—109. (*5*)
[15] BRESSAN, A.: Sui sistemi continui nel caso asimmetrico. Ann. di Mat. Pur. Appl. (4) **62**, 169—222. (*98*)
[16] BRESSAN, A.: Onde ordinarie di discontinuità nei mezzi elastici con deformazioni finite in relatività generale. Riv. Mat. Univ. Parma (2) **4**, 23—40. (*5*)
[17] COLEMAN, B. D., and V. MIZEL: Thermodynamics and departures from FOURIER'S law of heat conduction. Arch. Rational Mech. Anal. **13**, 245—261. (*96*)
[18] COLEMAN, B. D., and W. NOLL: The thermodynamics of elastic materials with heat conduction and viscosity. Arch. Rational Mech. Anal. **13**, 167—178. (*96*)
[19] DAHLER, H. S., and L. E. SCRIVEN: Theory of structured continua. I. General considerations of angular momentum and polarization. Proc. Roy Soc. Lond. A **275**, 505—527. (*3*, *98*)
[20] ERICKSEN, J. L.: Recoil of orientable fluids. Proc. 7th Congr. Theor. Appl. Mech. Bombay 1961, pp. 211—218. (*129*)
[21] ERICKSEN, J. L.: Oriented solids. Proc. Sympos. Structural Dynamics. Tech. Doc. Rept. ASD-TDR-63-140. Wright-Patterson Air Force Base, Ohio. May. (*98*, *127*)
[22] ERINGEN, A. C.: On the foundations of electro-elastostatics. Intl. J. Engr. Sci. **1**, 127—153. (*97*)
[23] FLAVIN, J. N.: Surface waves in pre-stressed Mooney material. Quart. J. Mech. Appl. Math. **16**, 441—449. (*95*)
[24] FOSDICK, R. L., and R. T. SHIELD: Small bending of a circular bar superposed on finite extension or compression. Arch. Rational Mech. Anal. **12**, 223—248. (*67*, *68b*)

[25] GIESEKUS, H.: Die simultane Translations- und Rotationsbewegung einer Kugel in einer elastoviskosen Flüssigkeit. Rheol. Acta **3**, 59—71. (*115*)

[26] GREEN, A. E.: A note on wave propagation in initially deformed bodies. J. Mech. Phys. Solids **11**, 119—126. (*74, 90, 93*)

[27] GREENSMITH, H. W.: Rupture of rubber, X. The change in stored energy on making a small cut in a test piece held in simple extension. J. Appl. Polymer Sci. **7**, 993—1002. (*83*)

[28] GREUB, W.: Linear Algebra, 2nd Ed. Berlin-Göttingen-Heidelberg: Springer. (*6*)

[29] GUO ZHONG-HENG: Homographic representation of the theory of finite thermoelastic deformations. Arch. Mech. Stosow. **15**, 475—505. (*42*)

[30] GUO ZHONG-HENG: Some notes on hypoelasticity. Arch. Mech. Stosow. **15**, 683—690. (*99*)

[31] GUO ZHONG-HENG: Certain dynamical problems of an incompressible hypoelastic sphere. Arch. Mech. Stosow. **15**, 871—878. (*103*)

[32] GUO ZHONG-HENG: On the constitutive equation of hypoelasticity. Bull. Acad. Polon. Sci. Sér. Sci. Tech. **11**, 301—304. (*99*)

[33] GUO ZHONG-HENG, and R. SOLECKI: Free and forced finite-amplitude oscillations of an elastic thick-walled hollow sphere made of incompressible material. Arch. Mech. Stosow. **15**, 427—433. (*62, 95*)

[34] GUO ZHONG-HENG, and R. SOLECKI: Free and forced finite-amplitude oscillations of a thick-walled sphere of incompressible material. Bull Acad. Polon. Sci. Sér. Sci. Tech. **11**, 47—52. (*62*)

[35] GUO ZHONG-HENG, and W. URBANOWSKI: Stability of non-conservative systems in the theory of elasticity of finite deformations. Arch. Mech. Stosow. **15**, 309—321. (*68b*)

[36] GUO ZHONG-HENG, and W. URBANOWSKI: Certain stationary conditions in variated states of finite strain. Bull. Acad. Polon. Sci. Sér. Sci. Tech. **11**, 27—32. (*69*)

[37] HAYES, M.: Wave propagation and uniqueness in pre-stressed elastic solids. Proc. Roy. Soc. Lond. A **274**, 500—506. (*68*)

[38] HERBERT, D. M.: On the stability of viscoelastic liquids in heated plane Couette flow. J. Fluid Mech. **17**, 353—359. (*119*)

[39] KALISKI, S.: On a model of the continuum with essentially non-symmetric tensor of mechanical stress. Arch. Mech. Stosow. **15**, 33—45. (*98*)

[40] KAPUR, J. N.: Flow of Reiner-Rivlin fluids in a magnetic field. Appl. Sci. Res. B **10**, 183—194. (*119*)

[41] KAPUR, J. N., and R. C. GUPTA: Two dimensional flow of visco-elastic fluid near a stagnation point with large suction. Arch. Mech. Stosow. **15**, 711—717. (*119*)

[42] KOH, S. L., and A. C. ERINGEN: On the foundations of non-linear thermo-viscoelasticity. Int. J. Engr. Sci. **1**, 199—229. (*96*)

[43] KRÖNER, E.: Zum statischen Grundgesetz der Versetzungstheorie. Ann. Physik (7) **11**, 13—21 (*44*).

[44] KRÖNER, E.: On the physical reality of torque stresses in continuum mechanics. Int. J. Engr. Sci. **1**, 261—278. (*44, 98*)

[45] LANGLOIS, W. E.: Steady flow of a slightly visco-elastic fluid between rotating spheres. Quart. Appl. Math. **21**, 61—71. (*115*)

[46] LANGLOIS, W. E., and R. S. RIVLIN: Slow steady-state flow of visco-elastic fluids through non-circular tubes. Rend. Mat. **22**, 169—185. (*122*)

[47] LIANIS, G.: Finite elastic analysis of an infinite plate with an elliptic hole in plane strain. Proc. 4th U.S. Nat. Congr. Appl. Mech. 1962, pp. 667—683. (*60, 67*)

[48] LIANIS, G.: Small deformations superposed on an initial large deformation in finite linear viscoelastic material. Univ. Washington Dept. Aero. Engr. Rep. 63—64, April. (*41*)

[49] LIANIS, G.: Equivalence of constitutive equations for non-linear materials with memory. Univ. Washington Dept. Aero. Engr. Rep. 63—5, April. (*29*)

[50] LIANIS, G.: Problems of small strains superposed on finite deformation in viscoelastic solids. Purdue Univ. Sch. Aero. Engr. Rep. 63—5, June. (*41*)

[51] LIANIS, G.: Constitutive equations of viscoelastic solids under finite deformation. Purdue Univ. Sch. Aero. Engr. Rep. 63—11, December. (*41*)

[52] LOKHIN, V. V.: A system of defining parameters characterizing the geometrical properties of an anisotropic medium [in Russian]. Dokl. Akad. Nauk SSSR **149**, 295—297. English Transl., Soviet Physics Doklady **8**, 260—261. (*13*)

[53] LOKHIN, V. V.: General forms of the relations between tensor fields in an anisotropic continuous medium, the properties of which are described by vectors, tensors of the second rank, and antisymmetric tensors of the third rank [in Russian]. Dokl. Akad. Nauk SSSR **149**, 1282—1285. English. Transl., Soviet Physics Doklady **8**, 345—348. (*13*)

[*54*] LOKHIN, V. V., and L. I. SEDOV: Non-linear tensor functions of several tensor arguments PMM 27, 597—629 (transl. of Priklad. Mat. Mekh. 27, 393—417). (*13*)
[*55*] MARKOVITZ, H., and D. R. BROWN: Parallel plate and cone-plate normal stress measurements on polyisobutylene-cetane solutions. Trans. Soc. Rheol. 7, 137—154. (*119*)
[*56*] NANDA, R. S.: On the three-dimensional flow of certain non-Newtonian liquids. Appl. Sci. Res. A **11** (1962/63) 376—386. (*119*)
[*57*] NANDA, R. S.: Visco-elastic flow due to a vibrating plane. Arch. Mech. Stosow. **15**, 599—606. (*119*)
[*58*] NOLL, W.: La mécanique classique, basée sur un axiome d'objectivité. La Méthode Axiomatique dans les Mécaniques Classiques et Nouvelles. Colloque International, Paris (1959). Paris: Gauthier-Villars. pp. 47—56. (*18, 19*)
[*59*] NOVOZHILOV, V. V.: On the forms of the stress-strain relation for initially isotropic nonelastic bodies (geometric aspect of the question). PMM **27**, 1219—1243 (transl. of Priklad. Mat. Mekh. **27**, 794—812). (*13*)
[*60*] PIPKIN, A. C., and R. S. RIVLIN: Normal stresses in flow through tubes of non-circular cross-section. Z. angew. Math. Phys. **14**, 738—742. (*122*)
[*61*] PIPKIN, A. C., and A. S. WINEMAN: Material symmetry restrictions on non-polynomial constitutive equations. Arch. Rational Mech. Anal. **12**, 420—426. (*28*)
[*62*] SEDOV, L. I., and V. V. LOKHIN: The specification of point symmetry groups by the use of tensors [in Russian]. Dokl. Akad. Nauk SSSR **149**, 796—797. (*13*)
[*63*] SENSENIG, C. B.: Instability of thick elastic solids. N.Y. Univ. Courant Inst. Math. Sci. Rep. 310, June. = (condensed) Comm. Pure Appl. Math. **17**, 451—491 (1964). (*94*)
[*64*] SHIELD, R. T., and R. L. FOSDICK: Extremum principles in the theory of small elastic deformations superposed on large elastic deformations. Progress in Applied Mechanics (Prager Anniv. Vol.). New York: Macmillan, pp. 107—125. (*88*)
[*65*] SHIELD, R. T., and A. E. GREEN: On certain methods in the stability theory of continuous systems. Arch. Rational Mech. Anal. **12**, 354—460. (*68b*)
[*66*] SMITH, G. F., M. M. SMITH, and R. S. RIVLIN: Integrity bases for a symmetric tensor and a vector — The crystal classes. Arch. Rational Mech. Anal. **12**, 93—133. (*11, 33*)
[*67*] SRIVASTAVA, A. C.: Torsional oscillations of an infinite plate in second-order fluids. J. Fluid Mech. **17**, 171—181. (*123*)
[*68*] TANNER, R. I.: Helical flow of elastico-viscous liquids. Part I, Theoretical. Rheol. Acta **3**, 21—26. (*112*)
[*69*] TANNER, R. I.: Helical flow of elastico-viscous liquids. Part II, Experimental. Rheol. Acta **3**, 26—34. (*119*)
[*70*] THOMAS, R. H., and K. WALTERS: On the flow of an elastico-viscous liquid in a curved pipe under a pressure gradient. J. Fluid Mech. **16**, 228—242. (*119*)
[*71*] TING, T. W.: Certain non-steady flows of second-order fluids. Arch. Rational Mech. Anal. **14**, 1—26. (*123*)
[*72*] TOUPIN, R. A.: A dynamical theory of elastic dielectrics. J. Engr. Sci. **1**, 101—126. (*97*)
[*73*] TRUESDELL, C.: Reactions of the history of mechanics upon modern research. Proc. 4th U.S. Nat. Congr. Appl. Mech. 1962, pp. 35—47. (*44*)
[*74*] TRUESDELL, C.: The meaning of BETTI's reciprocal theorem. J. Res. Nat. Bur. Stand. B **67**, 85—86. (*88*)
[*75*] TRUESDELL, C.: Remarks on hypo-elasticity. J. Res. Nat. Bur. Stand. B **67**, 141—143. (*100, 102*)
[*76*] TRUESDELL, C., and R. A. TOUPIN: Static grounds for inequalities in finite elastic strain. Arch. Rational. Mech. Anal. **12**, 1—33. (*48, 51, 52, 64, 68, 87*)
[*77*] VERMA, P. D. S.: Electrical conduction in finitely deformed isotropic materials. Arch. Mech. Stosow. **15**, 3—6. (*97*)
[*78*] VERMA, P. D. S.: Flow of anisotropic fluids between rotating coaxial cones. Arch. Mech. Stosow. **15**, 767—773. (*129*)
[*79*] VERMA, P. D. S.: A few examples showing use of hypo-elasticity in plasticity. Proc. Nat. Acad. Sci. India **33**, 101—106. (*103*)
[*80*] VERMA, P. D. S., and S. B. SINHA: Solutions in hypo-elasticity of grade one. J. Sci. Engr. Res. **7**, 223—234. (*103*)
[*81*] WALTERS, K., and N. D. WATERS: A note on the formulation of simple equations of state for elastico-viscous liquids. Z. angew. Math. Phys. **14**, 742—745. (*119*)
[*82*] WESOŁOWSKI, Z.: The axially symmetric problem of stability loss of an elastic bar subject to tension. Arch. Mech. Stosow. **15**, 383—395. (*68b*)
[*83*] WESOŁOWSKI, Z.: An inverse method in stresses for solving problems of large elastic strain. Arch. Mech. Stosow. **15**, 857—896. (*44*)

[84] Wesołowski, Z.: The axial-symmetric problem of instability in the case of tension of the elastic rod. Bull. Acad. Polon. Sci. Sér. Sci. Tech. **11**, 53—58. (*68b*)

[85] Wesołowski, Z.: Duality of inverse methods in the theory of elasticity of finite deformations. Bull. Acad. Polon. Sci. Sér. Sci. Tech. **11**, 305—309. (*44*)

[86] Zahorski, S.: Some problems of motions and stability for hygrosteric materials. Arch. Mech. Stosow. **15**, 915—940. (*41, 103*)

[87] Zahorski, S.: Small additional motion superposed on the fundamental motion of a hypoelastic medium. Bull. Acad. Polon. Sci. Sér. Sci. Tech. **11**, 449—454. (*103*)

[88] Zahorski, S.: Hypoelastic stability in the case of simple extension. Bull. Acad. Polon. Sci. Sér. Sci. Tech. **11**, 455—461. (*103*)

[89] Ziegler, H.: Some extremum principles in irreversible thermodynamics, with application to continuum mechanics. Progress in Solid. Mech. **4**, 91—193. (*19A, 96*)

1964 [1] Adams, N., and A. S. Lodge: Rheological properties of concentrated polymer solutions. II. A cone-and-plate and parallel plate pressure distribution apparatus for determining normal stress differences in steady shear flow. Phil. Trans. Roy. Soc. Lond. A **256**, 149—184. (*115, 119*)

[2] Adkins, J. E.: Non-linear diffusion through isotropic highly elastic solids. Phil. Trans. Roy. Soc. Lond. A **256**, 301—316. (*130*)

[3] Adkins, J. E.: Diffusion of fluids through aeolotropic highly elastic solids. Arch. Rational Mech. Anal. **15**, 222—234. (*130*)

[4] Beatty, M.: Some static and dynamic implications of the general theory of elastic stability. Thesis. Johns Hopkins Univ. (see also Beatty [1965, *1*]). (*68b, 88, 89*)

[5] Bell, J. F.: Experiments on large amplitude waves in finite elastic strain. Proc. Intl. Sympos. Second-order Effects. Haifa 1962, pp. 173—186. (*74*)

[6] Berker, R.: Contrainte sur une paroi en contacte avec un fluide visqueux classique, un fluide de Stokes, un fluide de Coleman-Noll. C. R. Acad. Sci. Paris **258**, 5144—5147. (*119, 123*)

[7] Bernstein, B., E. A. Kearsley, and L. Zapas: Thermodynamics of perfect elastic fluids. J. Res. Nat. Bur. Stand. B **68**, 103—113. (*37, 96b*)

[8] Bland, D. R.: On shock waves in hyperelastic media. Proc. Intl. Sympos. Second-order Effects, Haifa 1962, pp. 93—108. (*71, 73, 77*)

[9] Bland, D. R.: Dilatational waves and shocks in large displacement isentropic dynamic elasticity. J. Mech. Phys. Solids **12**, 245—267. (*74*)

[10] Bousso, E.: Observations on the self-acting thrust airbearing effect. Proc. Intl. Sympos. Second-order Effects, Haifa 1962, pp. 483—492. (*116*)

[11] Bragg, L.: Monotonicity on curves in lieu of the C-N inequalities for finite elasticity. Arch. Rational Mech. Anal. **17**, 327—338. (*52, 85*)

[12] Brugger, K.: Thermodynamic definition of higher order elastic coefficients. Phys. Rev. (2) **133**, A1611—A1612. (*63*)

[13] Chacon, R. V. S., and R. S. Rivlin: Representation theorems in the mechanics of materials with memory. Z. angew. Math. Phys. **15**, 444—447. (*28, 40*)

[14] Chu, Boa-Teh: Finite amplitude waves in incompressible perfectly elastic materials J. Mech. Phys. Solids **12**, 45—57. (*74*)

[15] Coleman, B. D.: Thermodynamics of materials with memory. Arch. Rational Mech. Anal. **17**, 1—46. (*96b*)

[16] Coleman, B. D.: On thermodynamics, strain impulses, and viscoelasticity. Arch. Rational Mech. Anal. **17**, 230—254. (*96b*)

[17] Coleman, B. D., and H. Markovitz: Normal stress effects in second-order fluids. J. Appl. Phys. **35**, 1—9. (*121, 123*)

[18] Coleman, B. D., and V. J. Mizel: Existence of caloric equations of state in thermodynamics. J. Chem. Phys. **40**, 1116—1125. (*96*)

[19] Coleman, B. D., and W. Noll: Material symmetry and thermostatic inequalities in finite elastic deformations. Arch. Rational Mech. Anal. **15**, 87—111. (*33, 40, 50, 52*)

[20] Coleman, B. D., and W. Noll: Simple fluids with fading memory. Proc. Intl. Sympos. Second-order Effects, Haifa 1962, pp. 530—552. (*38, 39*)

[21] Condiff, D. W., and J. S. Dahler: Fluid mechanical aspects of antisymmetric stress. Phys. Fluids **7**, 842—854. (*119*)

[22] Datta, S. K.: Note on the stability of an elastico-viscous liquid in Couette flow. Phys. Fluids **7**, 1915—1919. (*123*)

[23] Drucker, D. C.: Survey on second-order plasticity. Proc. Intl. Sympos. Second-order Effects, Haifa 1962, pp. 416—423. (*103*)

[24] Duvaut, G.: Application du principe de l'indifférence matérielle à un milieu élastique matériellement polarisé. C. R. Acad. Sci. Paris **258**, 3631—3634. (*98*)

[25] Eringen, A. C.: Simple microfluids. Intl. J. Engr. Sci. **2**, 205—217. (*98*)

[26] Eringen, A. C., and E. S. Suhubi: Non-linear theory of simple microelastic solids-I. Intl. J. Engr. Sci. **2**, 189—204. (*98*)
[27] Eringen, A. C., and E. S. Suhubi: Non-linear theory of micro-elastic solids-II. Intl. J. Engr. Sci. **2**, 389—404. (*98*)
[28] Fisher, G. M. C., and M. E. Gurtin: Wave propagation in the linear theory of viscoelasticity. Brown Univ. Div. Appl. Math. Contract Nonr. 562 (25) Tech. Rep., 28, August. (*96t*)
[29] Foux, A.: An experimental investigation of the Poynting effect. Proc. Intl. Sympos. Second-order Effects, Haifa 1962, pp. 228—251. (*66*)
[30] Foux, A., and M. Reiner: Cross-stresses in the flow of rarefied air. Proc. Intl. Sympos. Second-order Effects, Haifa 1962, pp. 450—466. (*116*)
[31] Frater, K. R.: Flow of an elastico-viscous fluid between torsionally oscillating disks. J. Fluid Mech. **19**, 175—186. (*119*)
[32] Frater, K. R.: Secondary flow in an elastico-viscous fluid caused by rotational oscillations of a sphere. Part. I. J. Fluid Mech. **20**, 369—381. (*119*)
[33] Fredrickson, A. G.: Principles and Applications of Rheology. Englewood Cliffs: Prentice-Hall. (*99, 104*)
[34] Giesekus, H.: Statistical rheology of suspensions and solutions with special reference to normal stress effects. Proc. Intl. Sympos. Second-order Effects, Haifa 1962, pp. 553—584. (*119*)
[35] Green, A. E.: A continuum theory of anisotropic fluids. Proc. Cambridge Phil. Soc. **60**, 123—128. (*126*)
[36] Green, A. E.: Anisotropic simple fluids. Proc. Roy. Soc. Lond. A **279**, 437—445. (*126*)
[37] Green, A. E., and J. E. Adkins: A contribution to the theory of non-linear diffusion. Arch. Rational Mech. Anal. **15**, 235—246. (*130*)
[37A] Green, A. E., and P. M. Naghdi: A dynamical theory of interacting continua. Univ. Calif. Berkeley Div. Appl. Mech. Rep. AM-64-13. August.
[38] Green, A. E., and R. S. Rivlin: Simple force and stress multipoles. Arch. Rational Mech. Anal. **16**, 325—353. (*98*)
[39] Green, A. E., and R. S. Rivlin: Multipolar continuum mechanics. Arch. Rational Mech. Anal. **17**, 113—147. (*98*)
[40] Green, W. A.: Growth of plane discontinuities propagating into a homogeneously deformed elastic material. Arch. Rational Mech. Anal. **16**, 79—88. (*74, 77*)
[41] Gurtin, M., and I. Herrera R.: On dissipation inequalities and linear viscoelasticity. Brown Univ. Div. Appl. Math. O.N.R. Rep. No. 27, June. (*96b*)
[42] Hayes, M.: Uniqueness for the mixed boundary-value problem in the theory of small deformations superimposed upon large. Arch. Rational Mech. Anal. **16**, 238—242. (*68*)
[43] Herrera R., I., and M. E. Gurtin: A correspondence principle for viscoelastic wave propagation. Quart. Appl. Math. **22**, 360—364. (*96t*)
[44] Holden, J. T.: Estimation of critical loads in elastic stability theory. Arch. Rational Mech. Anal. **17**, 171—183. (*68b*)
[44A] Hoppman, W. H., II, and C. N. Barnett: Flow generated by a cone rotating in a liquid. Nature **201**, 1205—1206. (*115*)
[45] Iyengar, S. R. K.: The stability of a non-Newtonian liquid between rotating concentric cylinders in the presence of a transverse pressure gradient. J. Sci. Engr. Res. **8**, 31—40. (*119*)
[46] Jain, M. K.: Collocation method to study problems of cross-viscosity. Proc. Intl. Sympos. Second-order Effects, Haifa 1962, pp. 623—635. (*119*)
[47] John, F.: Remarks on the non-linear theory of elasticity. Seminari Ist. Naz. Alta Matem. 1962/1963, 474—482. (*44*)
[48] Jones, J. R.: Secondary flow of a non-Newtonian liquid between eccentric cylinders in relative motion. Z. angew. Math. Phys. **15**, 329—341. (*119*)
[49] Kelly, P. D.: A reacting continuum. Intl. J. Engr. Sci. **2**, 129—153. (*130*)
[50] Klingbeil, W. W., and R. T. Shield: Some numerical investigations on empirical strain energy functions in the large axi-symmetric extensions of rubber membranes. Z. angew. Math. Phys. **15**, 608—629. (*55*)
[51] Ko, W. L., and P. J. Blatz: Application of finite visco-elastic theory to the deformation of rubberlike materials. I. Uniaxial stress relaxation data. Calif. Inst. Tech. GALCIT SM 64—4. January. (*41*)
[52] Lakshmana Rao, S. K.: Self-modelling flows of non-Newtonian viscous liquids. Z. angew. Math. Mech. **44**, 65—66. (*119*)
[53] Lakshmana Rao, S. K.: Stagnation point-line vortex flow of non-linear viscous liquids. Z. angew. Math. Mech. **44**, 67—69. (*119*)

[*54*] LESLIE, F. M.: Hamel flow of certain anisotropic fluids. J. Fluid Mech. **18**, 595—601. (*129*)
[*55*] LIANIS, G.: Application of irreversible thermodynamics in finite viscoelastic deformations. Purdue Univ. Sch. Aero. Engr. Sci. Rep. 64—1. January. (*41*)
[*56*] LIANIS, G., and P. H. DEHOFF, jr.: Studies on constitutive equations of first and second order viscoelasticity. Purdue Univ. Rep. A & ES 64—10. September. (*40*)
[*57*] LODGE, A. S.: Elastic Liquids. London and New York: Academic Press. (*104, 114*)
[*58*] LUMLEY, J. L.: Turbulence in non-Newtonian fluids. Phys. Fluids **7**, 335—337. (*119*)
[*59*] MARKOVITZ, H., and D. R. BROWN: Normal stress measurements on a polyisobutylene-cetane solution in parallel plate and cone plate instruments. Proc. Intl. Sympos. Second-order Effects, Haifa 1962, pp. 585—602. (*115, 116, 119, 123*)
[*60*] MARKOVITZ, H., and B. D. COLEMAN: Incompressible second-order fluids. Adv. Appl. Mech. **8**, 69—101. (*123*)
[*61*] MARKOVITZ, H., and B. D. COLEMAN: Nonsteady helical flows of second-order fluids. Phys. Fluids **7**, 833—841. (*123*)
[*62*] MINDLIN, R. D.: Microstructure in linear elasticity. Arch. Rational Mech. Anal. **16**, 51—78. (*98*)
[*63*] NARIBOLI, G. A.: The growth and propagation of waves in hypoelastic media. J. Math. Anal. Appl. **8**, 57—65. (*102*)
[*64*] NOLL, W.: Euclidean geometry and Minkowskian chronometry. Amer. Math. Monthly **71**, 129—144. (*17*)
[*65*] OLDROYD, J. G.: Non-linear stress, rate of strain relations at finite rates of shear in so-called "linear" elasticoviscous liquids. Proc. Intl. Sympos. Second-order effects, Haifa 1962, pp. 520—529. (*119*).
[*66*] PIPKIN, A. C.: Alternating flow of non-Newtonian fluids in tubes of arbitrary cross-section. Arch. Rational. Mech. Anal. **15**, 1—13. (*121, 122*)
[*67*] PIPKIN, A. C.: Review of ERINGEN [1962, *17*]. Quart. Appl. Math. **22**, 172—173. (*5*)
[*68*] PIPKIN, A. C.: Annular effect in viscoelastic fluids. Phys. Fluids **7**, 1143—1146. (*122*)
[*68A*] PIPKIN, A. C.: Small finite deformations of viscoelastic solids. Rev. Mod. Phys. **36**, 1034—1041. (*41*)
[*69*] RAJESWARI, G. K.: On the effects of variable suction on the steady laminar flow due to rotating bodies of revolution in non-Newtonian fluid. Z. angew. Math. Mech. **44**, 193—202. (*119*)
[*70*] RINTEL, L.: Flow of non-Newtonian fluids at small Reynolds number between two discs: one rotating and the other at rest. Proc. Intl. Sympos. Second-order Effects, Haifa 1962, pp. 467—472. (*119*)
[*71*] RIVLIN, R. S.: Second and higher-order theories for the flow of a viscoelastic fluid in a non-circular pipe. Proc. Intl. Sympos. Second-order Effects, Haifa 1962, pp. 668—677. (*121, 122*)
[*72*] RIVLIN, R. S.: A note on the mechanical constitutive equations for materials with memory. Z. angew. Math. Phys. **15**, 652—654. (*29*)
[*74*] SEEGER, A.: Discussion to paper by Foux. Proc. Intl. Sympos. Second-order Effects, Haifa 1962, p. 251. (*66*)
[*75*] SENSENIG, C. B.: Non-linear theory for the deformation of pre-stressed circular plates and rings. N. Y. Univ. Courant. Inst. Math. Sci. Rep. IMM-NYU 326, June. (*94*)
[*76*] SHARMA, G. C.: Superposability in second order fluid. J. Sci. Engr. Res. **8**, 185—190. (*123*)
[*77*] SLATTERY, J. C.: Unsteady relative extension of incompressible simple fluids. Phys. Fluids **7**, 1913—1914. (*118*)
[*78*] SLATTERY, J. C.: Time-reversed flows. J. Fluid mech. **19**, 625—630. (*119*)
[*79*] SMITH, G. F., and R. S. RIVLIN: Integrity bases for vectors. The crystal classes. Arch. Rational Mech. Anal. **15**, 169—221. (*11, 33*)
[*80*] SPENCER, C. B.: Finite deformations of an almost incompressible elastic solid. Proc. Intl. Sympos. Second-order Effects. Haifa 1962, pp. 200—216. (*43, 69*)
[*81*] TADJBAKHSH, I. G., and R. A. TOUPIN: On the equations of finite deformations in deformed coordinates. Intl. Business Machines Res. Lab. (Yorktown Heights) Rep. R. C. 1111. January 30. (*15, 62*)
[*82*] THOMAS, R. H., and K. WALTERS: The stability of elastico-viscous flow between rotating cylinders, I, II. J. Fluid Mech. **18**, 33—43 and **19**, 557—560. (*119*)
[*83*] THURSTON, R. N., and K. BRUGGER: Third order elastic constants and the velocity of small amplitude elastic waves in homogeneously stressed media. Phys. Rev. (2) **133**, A 1604—A 1610. (*71, 77*)
[*84*] TOUPIN, R. A.: Theories of elasticity with couple-stress. Arch. Rational Mech. Anal. **17**, 85—112. (*98*)

[85] TRUESDELL, C.: The natural time of a visco-elastic fluid: its significance and measurement. Phys. Fluids **7**, 1134—1142. (*28, 108, 120, 121, 123*)

[86] TRUESDELL, C.: A theorem on the isotropy groups of a hyperelastic material. Proc. Nat. Acad. Sci. U.S.A. **52**, 1081—1083. (*85, 87*)

[87] TRUESDELL, C.: Second-order effects in the mechanics of materials. Proc. Intl. Sympos. Second-order Effects, Haifa 1962, pp. 1—47. (*108, 116, 121, 123*)

[88] TRUESDELL, C.: Second-order theory of wave propagation in isotropic elastic materials. Proc. Intl. Sympos. Second-order Effects, Haifa 1962, pp. 187—199. (*77, 93*)

[89] VAINBERG, M. M.: Variational Methods for the Study of Nonlinear Operators [transl. of a Russian book published in 1955]. San Francisco, London, Amsterdam: Holden Day. (*38*)

[90] VERMA, P. D. S.: On waves in finite elastic strain. Indian J. Mech. Math. **2**, 17—18. (*71*)

[91] VERMA, P. D. S.: Symmetrical expansion of a hollow spherical dielectric. Intl. J. Engr. Sci. **2**, 21—26. (*97*)

[92] VERMA, P. D. S.: Helical flow of anisotropic fluids. J. Phys. Soc. Japan **19**, 2214—2218. (*129*)

[93] WALTERS, K.: Non-Newtonian effects in some general elastico-viscous liquids. Proc. Intl. Sympos. Second-order Effects, Haifa 1962, pp. 507—513. (*119*)

[94] WESOŁOWSKI, Z.: Problems of radial and axial oscilations of an elastic cylinder of infinitesimal length. Proc. Vibration Problems Warsaw **5**, 19—29. (*62*)

[95] WESOŁOWSKI, Z.: The stability of an elastic orthotropic parallelepiped subject to finite elongation. Bull. Acad. Polon. Sci. Sér. Sci. Tech. **12**, 155—160. (*68b*)

[96] WHITE, J. L.: Dynamics of viscoelastic fluids, melt fracture and the rheology of fibre spinning. J. Appl. Polymer Sci. (in press). (*114, 123*)

[97] WINEMAN, A. S., and A. C. PIPKIN: Material symmetry restrictions on constitutive equations. Arch. Rational Mech. Anal. **17**, 184—214. (*11, 28*)

[98] ZORSKI, H.: On the equations describing small deformations superposed on finite deformations. Proc. Intl. Sympos. Second-order Effects, Haifa 1962, pp. 109—128. (*52, 68, 69, 70, 88*)

1965 [1] BEATTY, M. L.: Some static and dynamic implications of the general theory of elastic stability. Arch. Rational Mech. Anal. **19**, 167—188. (*68b, 88, 89*)

[2] BOWEN, R.: The thermodynamics and mechanics of diffusion. Arch. Rational Mech. Anal. [This paper did not appear as such. For later papers by BOWEN on related subjects, see the Index in Volume 50 of Arch. Rational Mech. Anal.]. (*130*)

[3] BRAGG, L.: On relativistic worldlines and motions, and on non-sentient response. Arch. Rational Mech. Anal. **18**, 127—166. (*5*)

[4] BRAUER, R.: On the relation between the orthogonal group and the unimodular group. Arch. Rational Mech. Anal. **18**, 97—99. (*33*)

[5] CARLSON, D. E., and R. T. SHIELD: Second and higher order effects in a class of problems in plane finite elasticity. Arch. Rational Mech. Anal. **19**, 189—214. (*67*)

[5A] CHEN, P. J.: Acceleration waves in rheological materials. Thesis, Univ. Washington. (*96t*)

[6] COLEMAN, B. D.: Simple liquid crystals. Arch. Rational Mech. Anal. **20**, 40—58. (*33b*)

[7] COLEMAN, B. D., R. J. DUFFIN, and V. J. MIZEL: Instability, uniqueness, and nonexistence theorems for the equation $u_t = u_{xx} - u_{xtx}$ on a strip. Arch. Rational Mech. Anal. **19**, 100—116. (*123*)

[8] COLEMAN, B. D., M. E. GURTIN, and I. HERRERA: Waves in materials with memory. I. The velocity of one-dimensional shock and acceleration waves. Arch. Rational Mech. Anal. **19**, 1—19. Reprinted in Wave Propagation in Dissipative Materials. Berlin-Heidelberg-New York: Springer. (*96t*)

[9] COLEMAN, B. D., and M. E. GURTIN: Waves in materials with memory. II. On the growth and decay of one-dimensional acceleration waves. Arch. Rational. Mech. Anal. **19**, 239—265. Reprinted in Wave Propagation in Dissipative Materials. Berlin-Heidelberg-New York: Springer. (*96t*)

[10] COLEMAN, B. D., and M. E. GURTIN: Waves in materials with memory. III. Thermodynamic influences in the growth and decay of acceleration waves. Arch. Rational Mech. Anal. **19**, 266—298. Reprinted in Wave Propagation in Dissipative Materials. Berlin-Heidelberg-New York: Springer. (*96t*)

[11] COLEMAN, B. D., and M. E. GURTIN: Waves in materials with memory. IV. Thermodynamics and the velocity of three-dimensional acceleration waves. Arch. Rational Mech. Anal. **19**, 317—338. Reprinted in Wave Propagation in Dissipative Materials. Berlin-Heidelberg-New York: Springer. (*96t*)

[12] COLEMAN, B. D., H. MARKOVITZ, and W. NOLL: Viscometry of Simple Fluids. Springer Tracts in Natural Philosophy. (*104, 116*)

[*13*] COLEMAN, B. D., and V. J. MIZEL: Breakdown of laminar shearing flows for second-order fluids in channels of critical width. Z. Angew. Math. Mech. **46**, 445—448. (*123*)

[*14*] COLEMAN, B. D., and C. TRUESDELL: Homogeneous motions of incompressible materials. Z. angew. Math. Mech. **45**, 547—551. (*29, 30*)

[*15*] DILL, E. H.: Review of ZIEGLER [1963, *89*]. Physics Today **18**, 95—96. (*19A, 96*)

[*16*] DIXON, R. C., and A. C. ERINGEN: A dynamical theory of polar elastic dielectrics, I and II. Intl. J. Engr. Sci. **3**, 359—398. (*97*)

[*17*] ERICKSEN, J. L.: Non-existence theorems in linearized elastostatics. J. Diff. Eqns. **1**, 446—451. (*68*)

[*18*] GREEN, A. E., and P. M. NAGHDI: A general theory of an elastic-plastic continuum. Arch. Rational Mech. Anal. **18**, 251—281. (*5*)

[*18A*] GREEN, A. E., P. M. NAGHDI, and R. S. RIVLIN: Directors and multipolar displacements in continuum mechanics. Intl. J. Engr. Sci. **2**, 611—620. (*98*)

[*18B*] GREEN, A. E., and R. S. RIVLIN: Multipolar continuum mechanics: functional theory I. Proc. R. Soc. London A **284**, 303—324. (*98*)

[*19*] GREEN, W. A.: The growth of plane discontinuities propagating into a homogeneously deformed elastic material. Arch. Rational Mech. Anal. **19**, 20—23. (*74*)

[*20*] GURTIN, M.: Thermodynamics and the possibility of long-range interaction in rigid heat conductors. Arch. Rational Mech. Anal. **18**, 335—342. (*28, 96*)

[*21*] GURTIN, M.: Thermodynamics and the possibility of spatial interaction in elastic materials. Arch. Rational. Mech. Anal. **19**, 339—352. (*28*)

[*22*] KNOWLES, J. K.., and M. T.. JAKUB: Finite dynamic deformations of an incompressible elastic medium containing a spherical cavity. Arch. Rational Mech. Anal. **18**, 376—387. (*62*)

[*23*] KRÖNER, E.: Das physikalische Problem der antisymmetrischen Spannungen und der sogenannten Momentenspannungen. Proc. 11th Intl. Congr. Appl. Mech. (München, 1964). (*44, 98*)

[*24*] LIANIS, G.: Integral constitutive equations of nonlinear thermo-visco-elasticity. Purdue Univ. Rep. A. & E. S. 65-1. January. (*96b*)

[*25*] MARKOVITZ, H.: Normal stress measurements on polymer solutions. Proc. 4th Intl. Congr. Rheology 1963, **1**, 189—212. (*104, 113, 115, 116, 119*)

[*26*] NOLL, W.: Proof of the maximality of the orthogonal group in the unimodular group. Arch. Rational Mech. Anal. **18**, 100—102. (*33*)

[*27*] NOLL, W.: Materially uniform simple bodies with inhomogeneities. Arch. Rational Mech. Anal. **27**, 1—32 (1967); **31**, 401 (1968); **38**, 405—406 (1970). (*34, 44*)

[*28*] NOLL, W.: Representations of certain isotropic tensor functions. Archiv der Mathematik **21**. 87—90 (1970). (*11, 13*)

[*29*] OLDROYD, J. G.: Some steady flows of the general elastico-viscous liquid. Proc. Roy. Soc. Lond. A **283**, 115—133. (*119*)

[*30*] PIPKIN, A. C.: Shock structure in a visco-elastic fluid. Q. Appl. Math. (in press) (*96t*)

[*30A*] RIVLIN, R. S.: Viscoelastic fluids. Research Frontiers in Fluid Dynamics. New York: Interscience. (in press) (*104, 121*)

[*31*] SEDOV, L. I.: Some problems of designing new models of continuous media. Proc. 11th Intl. Congr. Applied Mech. (München 1964). (*97*)

[*32*] SMITH, G. E.: On isotropic integrity bases. Arch. Rational Mech. Anal. **18**, 282—292. (*12*)

[*33*] SPENCER, J. A. M.: Isotropic integrity bases for vectors and second-order tensors. Part II. Arch. Rational Mech. Anal. **18**, 51—82. (*12*)

[*34*] THURSTON, R. N.: Effective elastic coefficients for wave propagation in crystals under stress. J. Acoust. Soc. Am. **37**, 348—356, 1147. (*45, 71*)

[*35*] TRUESDELL, C.: Rational mechanics of deformation and flow (Bingham Medal Address). Proc. 4th Intl. Congr. Rheol. 1963, **2**, 3—50. (*116*)

[*36*] TRUESDELL, C.: Instabilities of perfectly elastic materials in simple shear. Proc. 11th Intl. Congr. Appl. Mech. (München 1964), (in press). (*54, 74, 92*)

[*37*] TRUESDELL, C.: Fluids of the second grade regarded as fluids of convected elasticity. Appendix by C.-C. WANG. Phys. of Fluids **8**, 1936—1938. (*119, 123*)

[*38*] VARLEY, E.: Acceleration waves in viscoelastic materials. Arch. Rational Mech. Anal. **19**, 215—225. (*96t*)

[*39*] VARLEY, E., and J. DUNWOODY: The effect of non-linearity at an acceleration wave. J. Mech. Phys. Solids **13**, 17—28. (*102*)

[*40*] WANG, C.-C.: Stress relaxation and the principle of fading memory. Arch. Rational Mech. Anal. **18**, 117—126. (*38, 40*)

[*41*] WANG, C.-C.: The principle of fading memory. Arch. Rational. Mech. Anal. **18**, 343—366. (*38, 40*)

[*42*] WANG, C.-C.: A general theory of subfluids. Arch. Rational Mech. Anal. **20**, 1—40. (*33b, 50, 85b, 118b*)

[*43*] WANG, C.-C.: On the radial oscillations of a spherical thin shell in the finite elasticity theory. Q. Appl. Math. (in press). (*62*)

[*44*] WANG, C.-C.: Second-order change of volume in isotropic materials free from applied loads. Z. angew. Math. Mech. (in press). (*66, 93*)

Addendum.

The following additional papers bear on the subject of the treatise. The italic numbers in parentheses indicate the sections in which reference to the paper would have been made, had we seen it in time.

1951 *A* NOVOZHILOV, V. V.: The connection between stress and deformation in non-linearly elastic media [in Russian]. Priklad. Mat. Mekh. **15**, 184—194. (*47, 85*)

1963 *A* RATHNA, S. L., and P. L. BHATNAGAR: Weissenberg and Merrington effects in non-Newtonian fluids. J. Indian Inst. Sci. Bangalore **45**, 57—82. (*123*)

1964 *A* DUVAUT, G.: Lois de comportements pour un milieu isotrope materiellement polarisé de degré 2. C. R. Acad. Sci. Paris **259**, 3178—3179. (*98*)

B FINZI, L.: Sulle equazioni costitutive nella meccanica dei continui. Ist. Lombardo Rend. Sci. A **97** (1963), 644—649. (*26*)

C GRIOLI, G.: Sulla meccanica dei continui a trasformazioni reversibili con caratteristiche di tensione asimmetriche. Sem. Ist. Naz. Alta Mat. 1962—3, 535—555. (*98*)

D MINDLIN, R. D.: On the equations of elastic materials with micro-structure. Columbia Univ. Dept. Civil Engr. Rep. No. 51, June. (*98*)

E MINDLIN, R. D.: Stress functions for a Cosserat continuum. Columbia Univ. Dept. Civil Engr. Rep. No. 53, September. (*98*)

F TANNER, R. I.: Observations on the use of Oldroyd-type equations of state for viscoelastic liquids. Chem. Engr. Sci. **19**, 349—355. (*119*)

1965 *A* BLAND, D. R.: On shock structure in a solid. J. Inst. Maths Applics **1**, 56—75. (*74*)

B BLEUSTEIN, J. L.: Effects of micro-structure on the stress concentration at a spherical cavity. Columbia Univ. Dept. Civil Engr. Rep. No. 54, February. (*98*)

C BOGARDUS, E. H.: Third-order elastic constants of Ge, MgO, and fused SiO_2. J. Applied Phys. **36**, 2504—2513. (*63*)

D BROWN, W. F.: Basis of a rigorous theory of magnetostriction. Proc. Intl. Conf. Magnetism (Nottingham, 1964). (*97*)

E BROWN, W. F.: Theory of magnetoelastic effects in ferromagnetism. J. Applied Phys. **36**, 994—1000. (*97*)

F COLEMAN, B. D., and M. E. GURTIN: Thermodynamics and one-dimensional shock waves in materials with memory. Proc. Roy. Soc. A **292**, 562—514. (*71, 96t*)

G COLEMAN, B. D., and V. J. MIZEL: On the existence of a caloric equation of state. Proc. 4*th* Intl. Congr. Rheol. 1963, **3**, 34—36. (*96*)

H DAVISON, L. W.: Propagation of finite amplitude waves in elastic solids, Thesis, Calif. Inst. Tech. (*71, 74*)

I ERINGEN, A. C., and J. D. INGRAM: A continuum theory of chemically reacting media. Intl. J. Engr. Sci. **3**, 197—212. (*130*)

J ESHEL, N. N.: Effects of strain-gradient on stress-concentration at a cylindrical hole in an elastic solid. Columbia Univ. Dept. Civil Engr. Rep. No. 2, June. (*98*)

K GIESEKUS, H.: Some secondary flow phenomena in general visco-elastic fluids. Proc. 4*th* Intl. Congr. Rheol. 1963, **1**, 249—266. (*123*)

L GIESEKUS, H.: Sekundärströmungen in viskoelastischen Flüssigkeiten bei stationärer und periodischer Bewegung. Rheol. Acta **4**, 85—101. (*122, 123*)

M GINN, R. F., and A. B. METZNER: Normal stresses in polymeric solutions. Proc. 4*th* Intl. Congr. Rheol. 1963, **2**, 583—602. (*116*)

N GREEN, A. E.: A note on linear transversely isotropic fluids. Mathematika **12**, 27—29. (*126*)

O GREEN, A. E., and P. M. NAGHDI: Plasticity theory and multipolar continuum mechanics. Mathematika **12**, 21—26. (*5, 98*)

P HAYES, J. W., and R. I. TANNER: Measurements of the second normal stress difference in polymer solutions. Proc. 4*th* Intl. Congr. Rheol. 1963, **3**, 389—399. (*112*)

Q JOHN, F.: Estimates for the derivatives of the stresses in a thin shell and interior shell equations. Comm. Pure Appl. Math. **18**, 235—267. (*94*)

R LIANIS, G.: Small deformation superposed on an initial large deformation in visco-elastic bodies. Proc. 4*th* Intl. Congr. Rheol. 1963, **2**, 109—119. (*41*)

S LOCKETT, F. J.: Creep and stress-relaxation experiments for non-linear materials. Intl. J. Engr. Sci. **3**, 59—75. (**41**)

T MARKOVITZ, H., and B. D. COLEMAN: Nonsteady helical flow of second-order fluids. Proc. *4th* Intl. Congr. Rheol. 1963, **2**, 143—145. (*123*)

U MILLER, C. E., and W. H. HOPPMAN II: Velocity field induced in a liquid by a rotating cone. Proc. *4th* Intl. Congr. Rheol. 1963, **2**, 619—636. (*123*)

V MINDLIN, R. D.: Second gradient of strain and surface-tension in linear elasticity. Columbia Univ. Dept. Civil Engr. Rep. No. 1, March. (*98*)

W MORGAN, A. J. A.: On the construction of constitutive equations for continuous media. Arch. Mech. Stosow. **17**, 145—174. (*26*)

X NARASIMHAN, M. N. L.: Stability of flow of a non-Newtonian liquid between two rotating cylinders in the presence of a circular magnetic field. Proc. *4th* Intl. Congr. Rheol. 1963, **1**, 345—363. (*119*)

Y PIPKIN, A. C.: Some non-Newtonian effects in flow through tubes. Proc. *4th* Intl. Congr. Rheol. 1963, **1**, 213—222. (*122*)

Z PIPKIN, A. C., and R. S. RIVLIN: Mechanics of rate-independent materials. Z. angew. Math. Phys. **16**, 313—326. (*5*)

AA SINGH, M. and A. C. PIPKIN: Controllable states of elastic dielectrics. Arch. Rational Mech. Anal., **21**, 169—210 (*91, 97*)

BB RAMAKANTH, J.: Some problems of propagation of waves in prestressed isotropic bodies. Proc. Vibration Probl. **6**, 161—172. (*73*)

CC REINER, M.: Second-order stresses in the flow of gases. Proc. *4th* Intl. Congr. Rheol. 1963, **1**, 267—279. (*116*)

DD REINER, M.: Research on second order effects in the elastic response of metals. Technion Found. Report. (*70*)

EE RIVLIN, R. S.: Nonlinear viscoelastic solids. SIAM Review **7**, 323—340. (*28, 37*)

FF SEWELL, M. J.: On the calculation of potential functions defined on curved boundaries. Proc. R. Soc. London A **286**, 402—411. (*88*)

GG SHERTZER, C. R., and A. R. METZNER: Measurement of normal stresses in visco-elastic materials at high shear rates. Proc. *4th* Intl. Congr. Rheol. 1963, **2**, 603—618. (*113*)

HH VARLEY, E.: Simple waves in general elastic materials. Arch. Rational Mech. Anal. **20**, 309—328. (*74*)

II WANG, C.-C.: A representation theorem for the constitutive equation of a simple material in motions with constant stretch history. Arch. Rational Mech. Anal. **20**, 329—340. (*109*)

Second addendum, 1991

1828 [*2*] CAUCHY, A.-L.: Sur l'équilibre et le mouvement d'un système de points matériels sollicité par des forces d'attraction ou de répulsion mutuelle. Ex. de Math. **3**, 188—212 = Oeuvres (2) **8**, 227—252. (*250*)

1906 [*0*] BORN, M.: Untersuchungen über die Stabilität der elastischen Linie in Ebene und Raum unter verschiedenen Grenzbedingungen. Göttingen: Dieterische Univ.-Buchh. (*88*)

1948 [*11*A] RIVLIN, R. S.: Large elastic deformations of isotropic materials, II. Some uniqueness theorems for pure, homogeneous deformation. Phil. Trans. Roy. Soc. Lond. A **240**, 491—508. (*95*)

[*11*B] RIVLIN, R. S.: Large elastic deformations of isotropic materials, III. Some simple problems in cylindrical co-ordinates. Phil. Trans. Roy. Soc. Lond. A **240**. 509—525. (*95*)

1954 [*3*A] BRAUN, I.: The momentum equation of the Reiner liquid. Rend. Circ. Mat. Palermo (2) **2** (1953), 258—265. (*119*)

1962 [*22*A] GIESEKUS, H.: Die rheologische Zustandsgleichung elasto-viskoser Flüssigkeiten—insbesondere von Weissenberg-Flüssigkeiten — für allgemeine und stationäre Fliessvorgänge. Z. angew. Math. Mech. **42**, 32—61. (*36*)

1964 [*61*a] MARRIS, A. W.: The reduction of the simple hygrosteric fluid to the Stokesian fluid. J. Appl. Mech. **31**, 170—174. (*119*A)

Subject Index.

(English-German.)

Where English and German spelling of a word is identical the German version is omitted.